Microlithography

Microlithography

Science and Technology

Third Edition

Edited by
Bruce W. Smith and Kazuaki Suzuki

CRC Press
Taylor & Francis Group
Boca Raton London New York

CRC Press is an imprint of the
Taylor & Francis Group, an **informa** business

First published in paperback 2024

Third edition published 2020
by CRC Press
2385 NW Executive Center Drive, Suite 320, Boca Raton FL 33431

and by CRC Press
4 Park Square, Milton Park, Abingdon, Oxon, OX14 4RN

© 2020, 2024 Taylor & Francis Group, LLC

First edition published by CRC Press 1998

Second edition published by CRC Press 2007

CRC Press is an imprint of Taylor & Francis Group, LLC

Library of Congress Cataloging-in-Publication Data

Names: Smith, Bruce W., 1959- editor. | Suzuki, Kazuaki, editor.
Title: Microlithography : science and technology / edited by Bruce W. Smith and Kazuaki Suzuki.
Description: Third edition. | Boca Raton : CRC Press, 2020. | Includes bibliographical references and index. | Summary: "Like the bestselling original, this third edition of Microlithography is a self-contained text detailing both elementary and advanced aspects of submicron microlithography, offering a balanced treatment of theoretical and operating practices"-- Provided by publisher.
Identifiers: LCCN 2019053732 | ISBN 9781439876756 (hbk) | ISBN 9781315117171 (ebk)
Subjects: LCSH: Microlithography--Industrial applications. | Integrated circuits--Masks. | Metal oxide semiconductors, Complementary--Design and construction. | Manufacturing processes.
Classification: LCC TK7836 .M525 2020 | DDC 621.3815/31--dc23
LC record available at https://lccn.loc.gov/2019053732

ISBN: 978-1-4398-7675-6 (hbk)
ISBN: 978-1-03-283673-7 (pbk)
ISBN: 978-1-315-11717-1 (ebk)

DOI: 10.1201/9781315117171

Typeset in Times
by Deanta Global Publishing Services, Chennai, India

Visit the Taylor & Francis Web site at
http://www.taylorandfrancis.com

and the CRC Press Web site at
http://www.crcpress.com

Contents

Preface to Third Edition

Semiconductor micro- and nanolithography continues to provide the necessary support to drive Moore's Law with future nanometer-scale device generations. With technological innovation in imaging systems, materials, processing, modeling, and optimization, what had once been envisioned as technical barriers to advancement are continually surpassed to allow increasingly more capable devices. The drumbeat of smaller, faster, cheaper, and lower power continues as the world moves into new technology applications demanding tremendously increased processing and storage capacity. Combined with new forefronts in electronics and photonics, the applications for nano-electronic devices have grown well beyond the needs of microlithography covered in the first and second editions of *Microlithography: Science and Technology*.

To address the new technology that has evolved over the last several years, this Third Edition of *Microlithography: Science and Technology* has been completely revised. While providing a balanced treatment of theoretical and operational considerations, from fundamental principles to advanced topics of nanoscale lithography, this edition details the technology necessary for current and future device generations. It provides the necessary reference for students and engineers to learn the fundamental as well as understand the future requirements of the challenging technology behind lithography at the nano scale. It also provides the basis for more experienced engineers to understand the interdisciplinary nature of microlithography, which involves aspects of many areas of science and engineering.

The book is divided into 13 chapters, starting with an overview of the lithography requirements of semiconductor processing (Lithography, Etch, and Silicon Process Technology, Chapter 1) and exploring the details of all the technologies involved. Chapters include Optical Nanolithography (Chapter 2), Multiple Patterning Lithography (Chapter 3), EUV Lithography (Chapter 4), Alignment and Overlay (Chapter 5), Design for Manufacturing and Design Process Technology Co-optimization (Chapter 6), Chemistry of Photoresist Materials (Chapter 7), Photoresist and Materials Processing (Chapter 8), Optical Lithography Modeling (Chapter 9), Maskless Lithography (Chapter 10), Imprint Lithography (Chapter 11), Metrology for Nanolithography (Chapter 12), and Directed Self-Assembly of Block Copolymers (Chapter 13). The Third Edition has involved contributions from 29 renowned experts from the world's leading academic and industrial organizations to provide in-depth coverage of these technologies. As a result, we are certain that the Third Edition of *Microlithography: Science and Technology* will remain a highly valuable resource for students, engineers, and researchers well into the future.

Bruce W. Smith
Kazuaki Suzuki

Editors

Bruce W. Smith is a Distinguished Professor of engineering at the Rochester Institute of Technology. He has been involved in teaching and research in microelectronic and microsystems engineering for over 35 years. His areas of research include semiconductor processing, deep ultraviolet (DUV), vacuum ultraviolet (VUV), immersion, and extreme ultraviolet (EUV) lithography, thin films, optics, and microelectronic materials. He has authored over 250 technical publications, given over 100 technical talks, and received over 25 patents, licensing his technology both nationally and internationally. He has worked extensively with individuals and organizations in the semiconductor industry, including industrial partners in the Semiconductor Research Corporation, SEMATECH, and the IMEC. He is the recipient of numerous teaching and research awards, including the Institute of Electrical and Electronics Engineers (IEEE) Technical Excellence Award, the American Vacuum Society (AVS) Excellence in Leadership Award, the Society for Photo-optical Instrumentation Engineers (SPIE) Research Mentoring Award, and the Rochester Institute of Technology Trustees Scholarship Award. He has also been inducted into the Rochester Institute of Technology Innovator Hall of Fame. Professor Smith is a Fellow of the Institute of Electrical and Electronics Engineers, the Optical Society of America, and the Society for Photo-optical Instrumentation Engineers.

Kazuaki Suzuki majored in plasma physics and X-ray astronomy in the University of Tokyo, Japan. He has been a project manager for developing new concept exposure tools at Nikon Corporation, such as the early-generation KrF excimer laser stepper, the first-generation KrF excimer laser scanner, the electron beam projection exposure system, and the full-field extreme ultraviolet scanner. He received his Ph. D. in Precision Engineering from the University of Tokyo about the system design of exposure tools for microlithography. He has authored and coauthored many papers in the field of exposure tool and related technologies, including advanced equipment control by using metrology data. He also holds numerous patents in the same field. In the first decade of this century, he was a member of the program committee of the Society for Photo-optical Instrumentation Engineers (SPIE) Microlithography and other international conferences such as Micro & Nano Engineering in Europe and the International Microprocesses and Nanotechnology Conference in Japan. He was one of the associate editors of *Journal of Micro/Nanolithography, MEMS, and MOEMS* (*JM3*) from 2002 to 2009. He moved to Tokyo Tech Academy for Convergence of Materials and Informatics at Tokyo Institute of Technology (Tokyo Tech) in March 2019.

Contributors

Robert D. Allen
IBM Research Almaden

Eran Amit
KLA Israel
(Current affiliation: PlayerMaker Inc.)

Chris Bencher
Applied Materials, Inc.

John J. Biafore
KLA

Luigi Capodieci
Motivo, Inc.

Matthew Colburn
Facebook Reality Labs

Udo Dinger
Carl Zeiss SMT GmbH

Derren N. Dunn
IBM Research

Carlos Fonseca
Tokyo Electron, Ltd.

Bruno La Fontaine
ASML

Gregg M. Gallatin
Applied Math Solutions, LLC

Michael A. Guillorn
IBM Research

Winfried Kaiser
Carl Zeiss SMT GmbH

David Laidler
IMEC

Chi-chun Liu
IBM Research

Chris A. Mack
Fractilia, LLC

Stephan Müllender
Carl Zeiss SMT GmbH

Paul Nealey
University of Chicago

Mark Neisser
Kempur Microelectronics Inc.

Juan de Pablo
University of Chicago

Doug Resnick
Canon Nanotechnologies

Helmut Schift
Paul Scherrer Institut

Bruce W. Smith
Rochester Institute of Technology

Mark D. Smith
KLA

John Sturtevant
Mentor, A Siemens Business

Kazuaki Suzuki
Nikon Corporation
(Current affiliation: Tokyo Institute of
   Technology)

James Thackeray
Dow Chemical Company

Takumi Ueno
Hitachi Chemical Co., Ltd.
(Current affiliation: Shinshu University)

Obert R. Wood II
GLOBALFOUNDRIES Inc. (Retired)

Stefan Wurm
ATICE LLC

Kenji Yoshimoto
Kyoto University

1 Lithography, Etch, and Silicon Process Technology

Matthew Colburn, Derren N. Dunn, and Michael A. Guillorn

CONTENTS

1.1 INTRODUCTION

Since Gordon Moore made his prophetic observation [1] that transistor density is doubling every 2 years (commonly referred to as "Moore's law"), the semiconductor industry has been steadily pushing the limits of lithography, etch, and Si process technology to deliver denser integrated circuits. For over 50 years, feature scaling in integrated circuits has enabled a reduction in circuit area. It stands to reason that with this reduction in area, a commensurate reduction in parasitic resistance and capacitance would follow, resulting in an improvement in device and circuit performance. The theory that describes the relationship between feature scaling and device performance was established by Dennard and coworkers in 1974 [2]. Simply stated, by scaling all dimensions by a given factor, κ, as shown in Figure 1.1, while simultaneously increasing doping by κ and decreasing operating voltage by κ to maintain constant field, power can be reduced by κ^2 or operating speed can be increased commensurately. While there are limits to this scaling approach [3], this theory provided

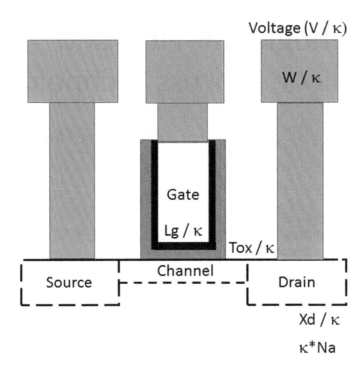

FIGURE 1.1 Schematic of scaled device in which key device dimensions are scaled by a factor of κ to reduce power or improve performance. (Adapted from Dennard, R.H. et al., *IEEE J. Solid-State Circuits*, 9, 256, 1974.)

the governing principles of technology development, ensuring that the value proposition behind Moore's law remains true. It is important to note that scaling can be leveraged to emphasize performance improvement or power reduction. This duality has enabled a singular base semiconductor process technology to be tailored for high-performance computational applications as well as for low-power mobile applications [4].

In spite of the practical framework that scaling theory provides, it is misleading to think of Moore's law as being driven solely by shrinking dimensions. Moore himself noted in a 1975 speech [1] that a "contribution of device and circuit cleverness" was required to account for the unabated progress in the performance of semiconductor products. As we enter the sixth decade of Moore's law scaling, the contribution of device and circuit cleverness coupled with continued advancements in lithography and patterning solutions have become equally important in maximizing the return on investment for developing new generations of semiconductor technologies.

Examples of "cleverness" in device architecture and process technology are found in abundance throughout the history of the semiconductor industry. The need to improve power dissipation drove the industry to transition from bipolar to complementary metal–oxide–semiconductor (CMOS) technology [5] in the 1980s. The pursuit of reduced parasitic junction capacitance led to the development of silicon-on-Insulator (SOI) device technology in the 1990s. The industry learned to manipulate channel mobility through stress engineering [6], and channel material engineering [4, 7] in the late 1990s and early 2000s, producing large gains in device performance through enhanced drive current. Degraded control of the fields in the channel region at scaled gate lengths, referred to as *short channel effects*, led to the pursuit of a reduction in effective gate oxide thickness (EOT) by pushing Si-based gate dielectrics beyond the point once thought physically possible [8]. This trend continued in the late 2000s with the implementation of high-permittivity, high-κ gate dielectrics and metal gate electrodes, producing further scaling of EOT [9] and reduced power due to a reduction in gate oxide leakage. In parallel with this revolution in gate processing, numerous advancements in junction engineering and silicide formation enabled significant improvements in junction depth control and contact resistance reduction [10].

Practical schemes featuring "double gate" fully depleted devices for improving short channel effect control have been studied since the early 1990s [11] and are illustrated in Figure 1.2. This work illustrated a path forward toward deeply scaled devices. Further conceptual work on such devices was carried out in the late 1990s [12], and experimental work was published in the early 2000s [13]. However, manufacturing concerns prevented widespread acceptance of this approach. In the late 1990s, work on transistors with three-dimensional (3-D) channel geometries featuring improved electrostatics began to transition from university labs [14] to industrial research facilities [15]. The operating principle of these devices was simple: raise the channel out of plane, allowing the gate electrode to wrap around the channel, and reduce the body thickness to form a "fin"-like structure, resulting in a superior geometry from an electrostatic point of view. This device, now referred to as the FinFET, continued to gain traction in the industry in the mid-2000s [16]. By the early 2010s, the FinFET became the standard device for all leading-edge CMOS technologies [17].

Innovation has not been limited to the front end of line (FEOL), that is, transistor device structure prior to contact formation. As feature density increased and critical dimensions dipped well below 250 nm, control of wafer topography became a formidable challenge. Planarization technology prior to interconnect metallization had been explored early in the semiconductor industry, relying on reflow of doped glass dielectric. However, reliability concerns and performance degradation proved to be shortcomings of these processes. In contrast, the use of chemical mechanical polishing (CMP) for planarization of high-quality dielectrics in the back end of line (BEOL) became a foundation on which all high-performance CMOS interconnect technologies were built. For a review, the author refers the reader to Krishnan et al. [18]. Improvements in BEOL topography through CMP enabled one of the most important materials innovations in the history of the semiconductor industry: the transition from Al interconnects to Cu interconnects.

The adoption of Cu interconnect metallurgy using the damascene [19] process provided a materials solution for significantly reducing interconnect resistance (R) with the added benefit of reduced reliability failures from electromigration [20], an example of which is shown in Figure 1.3. Additional performance gains were achieved by the introduction of several materials and structural innovations, a comparison of which can be seen by comparing the left and right sides of Figure 1.3. First, the adoption of reduced-permittivity, low-κ dielectrics led to reduced interconnect capacitance (C). Building off advances in low-κ dielectric materials, further reductions in permittivity were achieved through the introduction of airgap technology, achieving the lowest capacitance possible [22–24].

An important second innovation in BEOL architecture was the introduction of new ultrathin liner technology [20], which further reduced interconnect resistance at deeply scaled dimensions

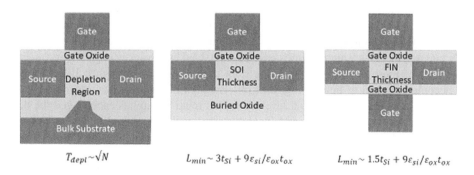

$$T_{depl} \sim \sqrt{N} \qquad L_{min} \sim 3t_{Si} + 9\varepsilon_{si}/\varepsilon_{ox}t_{ox} \qquad L_{min} \sim 1.5t_{Si} + 9\varepsilon_{si}/\varepsilon_{ox}t_{ox}$$

FIGURE 1.2 Comparison of depletion regions for Planar Bulk, Planar FDSOI, and FinFET. Note SOI Structures generally have thicker SOI Si thicknesses relative to FinFET Si thicknesses. (Adapted from Frank, D. J. et al., *Proceedings of the IEEE*, 89, 259, 2001.)

Thick Cap

Metal-past-via
Line-end extension

Tapered Via

Increasing
Resistance
Low-K dielectric

Robust dielectric

Thin Cap

No line-end
Extension

Perfect Profile
Aligned Via

Low Resistance
Airgap Dielectic

Robust dielectric

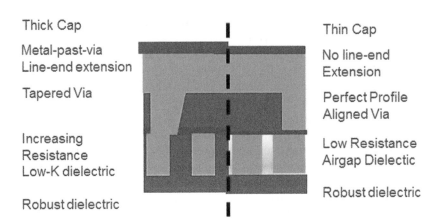

FIGURE 1.3   Representative schematic of a BEOL structure. (From Colburn, M., *SPIE Advanced Etch Conference Plenary*, 2015.)

by maximizing the cross-sectional area of the Cu portion of the interconnect while maintaining aggressive reliability specifications. Further innovations in microstructural engineering [25] demonstrated a path to maintain BEOL performance while continuing to achieve further dimensional scaling.

It is undeniable that there are physical limits to scaling within transistors as dimensions approach near-atomic dimensions. In the FEOL, short channel effects, quantum effects, and stochastic variation such as line-edge roughness, feature placement variation, and random dopant fluctuation [26] will ultimately limit device performance. Variants of planar silicon on insulator (SOI) devices to enhance low power performance have been proposed for the mobile/hand-held space [27]. More advanced 3-D channel architectures including gate all around (GAA) devices have been proposed to continue scaling in the high-performance space. In addition, a resurgent interest in vertical transistors [28], as well as higher-mobility transistor materials including III–V [29] materials, has also been proposed to extend device scaling improvements for both high-performance and low-power applications.

In the BEOL, challenges to continued device scaling are equally great. Nonlinear increases in resistivity at small dimensions [30] and dielectric breakdown due to increased electric field strength are reaching fundamental materials limits [31]. There are a limited number of variables that can be used to tune performance and address challenges in interconnect architecture: interconnect pitch, line width, aspect ratio, interconnect resistivity (metallization), interlayer dielectric constant, and dielectric thickness. Within these constraints, copper interconnect resistance increases exponentially as line width reduces below 100 nm primarily due to grain boundary scattering. The dielectric constant of a vacuum bounds the remaining improvements achievable by reducing dielectric constant. Within the confines of these increasing technologic challenges, fabrication of statistically yieldable and reliable device architectures have been enabled by advances in patterning technology.

In the face of these daunting obstacles, circuit area reduction through cell-level area scaling has become a subject of intense interest [32]. The area that a logic or memory cell occupies directly impacts density at the product level [33]. This relationship has encouraged the development of process technologies optimized for delivering smaller logic cells. To first order, logic cell area is dictated by the minimum gate pitch, the number of gates, the minimum wiring pitch transverse to the gates, and the number of wiring "tracks" required to complete the connections within the cell. By aggressively scaling these parameters, smaller logic cells can be synthesized, resulting in more compact circuit blocks, which occupy a smaller area and therefore have lower parasitic resistance and capacitance. However, a careful co-optimization of logic cell design, lithographic patterning techniques, and device process technology—referred to as design technology co-optimization (DTCO)—is required to achieve products that deliver the desired power, performance, and area scaling without compromising yield. DTCO has become a staple of the semiconductor industry since the 22 nm node [34]. Taking a holistic approach to patterning process development in view of device process technology and circuit design has become a new frontier in enabling device scaling [35].

Understanding the fundamental contributions of patterning process optimization to advanced device architectures is vital for anyone undertaking work in advanced CMOS technology development. This chapter reviews fundamental aspects of patterning process optimization by exploring essential elements of lithography, optical proximity correction (OPC), and reactive ion etch (RIE) technology. A discussion of modern patterning techniques used to push feature resolution beyond the limitations of optical systems is also provided. Finally, an overview of the current state of extreme ultraviolet (EUV) lithography is presented. This material is presented in the context of advanced integrated circuit manufacturing to provide a frame of reference for engineers and scientists looking for an introduction to this field.

1.2 LITHOGRAPHY FUNDAMENTALS

Semiconductor lithography has gone through a host of changes over the past 50 years. The industry transitioned from contact printing with broad-band wavelength arc lamps, to projection steppers capable of imaging an entire field, to the modern embodiment of 4× reduction projection scanners leveraging excimer laser sources. The maximum resolution of these systems is a function of the wavelength of light, λ, divided by numerical aperture (NA), NA, as shown in Figure 1.4.

Figure 1.4 chronicles the staggering improvement in resolution over the past 30 years as the industry progressed from g-line mercury lamps ($\lambda = 465$ nm) through i-line ($\lambda = 365$ nm) to ArF excimer lasers ($\lambda = 193$ nm) high–NA scanners. This plot includes the advent of 0.25 and 0.33 NA EUV scanners operating at $\lambda = 13.5$ nm. Modern scanners leverage 4× reduction projection optics with unprecedented precision and accuracy of both image formation and image placement. Modern scanners have aberrations lower than ~3.5 mλ RMS at center slit [37], demonstrated resolution near 0.25 k_1, and overlay demonstrated below 2 nm. As shown schematically, there are four principal components of an advanced projection system: illumination, objective lens, mask, and condenser lens.

Illumination systems used in modern optical lithography tools feature high-power, small-bandwidth [38] excimer lasers that project light into the optical system and onto the photo mask [39]. The illumination can be customized with programmable illuminators, comprised of an integrated

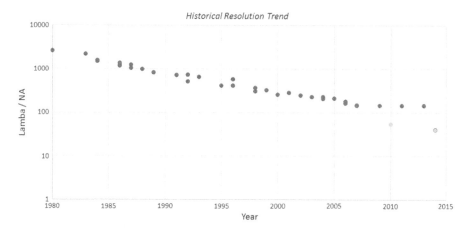

FIGURE 1.4 Historic trend of scale resolution power (λ/NA) including EUV. (From Colburn, M., *Lithography Solutions for the 22nm Node*, VLSI, Honolulu, HI, 2009. With permission.)

micromirror array, designed to optimize overall image fidelity through computational wave front engineering techniques [40]. Photomask images are reduced by a factor of four (4×) and projected onto the wafer. Modern lithography tools do not project the full image of the mask on the wafer at one time; rather, they move or scan the photomask four times (commensurate with the scale factor) faster than the wafer stage motion or scan. This scanning motion of the mask and wafer gives rise to the use of the term *scanner* for modern production lithography systems.

1.2.1 IMAGE FORMATION AND MODELING

Modern scanner optics are comprised of multiple optical elements required to accurately image features of a lithography mask and then project the resulting images to the wafer plane with a standard reduction of 4×, as shown in Figure 1.5. A fortunate simplifying feature of these systems is that they can be modeled with sufficient accuracy using Fraunhofer far-field approximations and Fourier optics [41]. In this approximation, any number of condenser, objective, projector, and transfer lens can be treated as a black box transfer function characterized by the optical state of waves entering the optical system from the object plane and exiting from the optical system to form an image in the object plane, as shown in Figure 1.6.

A common assumption in this approach is that the entrance and exit pupils of the optical system restrict the angular extent of waves passing through the lenses modeled by the optical system.

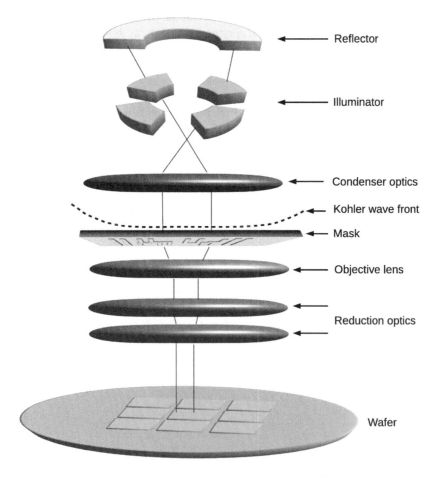

FIGURE 1.5 Schematic diagram of a modern stepper showing the path of light through an optical system using Kohler illumination from source to wafer.

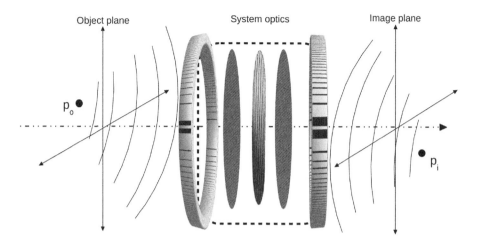

FIGURE 1.6 Schematic representation of a lithographic system as viewed from a Fourier optics perspective.

Additionally, it is also commonly assumed that geometric optics can be used to model the lens and apertures contained within the optical system; thus, the finite extent of the entrance and exit pupils is found by geometrically projecting the smallest aperture contained within the optical system to the object and image planes. Waves entering and exiting the optical system are then limited by the effective angular extent of the entrance aperture of the optical system and equivalently impacted by the effective angular extent of the exit aperture of the optical system.

In general, optical systems used in modern lithographic scanners are diffraction-limited systems. Diverging spherical waves from point source elements entering the optical system are transferred through the system and exit as converging spherical waves emanating from the exit pupil. Diffraction limits of the entrance aperture and the exit aperture of the optical system limit the spatial frequencies transferred by the system equivalently, because the entrance and exit pupils of the optical system are geometric projections of each other. Also, the effective diffraction limit of the optical system is dictated by the smallest aperture in the optical system projected onto the entrance or exit pupils.

In the following sections, essential elements of modern optical systems will be described in terms of a set of optical transfer functions, which in aggregate, allow computational lithographers to rapidly and accurately calculate the impact of lenses, aberrations, and diffraction limitations on the formation of images in the wafer plan from lithography mask images in the object plane shown in Figure 1.5.

1.2.2 Diffraction-limited Imaging: Abbe Imaging Theory

Lithographic projection systems share some essential common attributes with imaging processes observed in conventional light microscopes. Ernst Abbe developed an effective theory to predict observable images from diffraction-limited systems [41, 42]. Abbe assumed that imaging from an object close to an objective lens can be treated in two steps: first, light diffracted from the object and passed through the lens forms a diffraction pattern in the back focal plane of the objective lens; second, the objective lens of the system is of finite extent, so it limits the spatial frequencies that are transferred to the image plane of the system. In Abbe's approach, the objective lens can be treated as an optical system, as shown in Figure 1.7, in which the lens is both an aperture that limits diffracted waves passed by the system and a lens responsible for forming an image of the object.

In Figure 1.7, a representation of an optical system following Abbe imaging is shown. In this system, each point on the object diffracts light that is captured by the finite extent of the objective lens. Waves diffracted from the object then form a diffraction pattern in the back focal plane of the

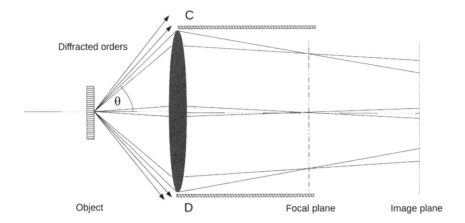

FIGURE 1.7 Optical system representing essential elements of a system following Abbe imaging theory.

objective lens, which, in a far-field approximation, propagates to form an image in the image plane shown in Figure 1.7. We can represent the amplitude distribution at the image plane, $U_i(x_i, y_i)$, from this imaging system as a superposition integral of wave amplitudes, $U_o(x_o, y_o)$, scattered by the object as follows [41, 42]:

$$U_i(x_i, y_i) = \int\limits_{-\infty}^{\infty} \int\limits_{-\infty}^{\infty} O(x_i, y_i; x_o, y_o) U_o(x_o, y_o) dx_o \, dy_o \qquad (1.1)$$

In this integral, a point in the image plane (x_i, y_i) is mapped from a point in the object plane (x_o, y_o), by a linear transformation that directly accounts for the magnification of the optical system, M, which can be either positive or negative depending upon whether or not the image is inverted [41].

$$(x_i, y_i) = (Mx_o, My_o) \qquad (1.2)$$

This integral can be understood as the convolution of the Fraunhofer diffraction pattern of the exit pupil of the system represented by the function O, with the idealized image of the object calculated from geometric optics, Uo. O can then be written as follows:

$$O(x_i, y_i; x_o, y_o) = C_o \int\limits_{-\infty}^{\infty} \int\limits_{-\infty}^{\infty} P(x, y) \exp\left[-i \frac{2\pi}{\lambda d_i} \big((x_i - Mx_o)x + (y_i - My_o)y\big)\right] dx \, dy \qquad (1.3)$$

In this equation [41], P(x,y) is a discrete function, referred to as the *pupil function*, that has a non-zero value for (x,y) points in the exit aperture of the optical system and is zero for all other points, d_i represents a spatial frequency in the object being imaged, and C_o is a normalization constant. Conventionally, this representation is rewritten in optical coordinates:

$$x' = \frac{x}{\lambda d_i}, \; y' = \frac{y}{\lambda d_i}, \; x'_o = Mx_o, \; y'_o = My_o$$

$$U_i(x_i, y_i) = \frac{1}{M^2} \int\limits_{-\infty}^{\infty} O(x_i - x'_o, y_i - y'_o) U_o\left(\frac{x'_o}{M}, \frac{y'_o}{M}\right) dx'_o \, dy'_o \qquad (1.4)$$

In general, this integral is thought of as the convolution of the object image predicted by geometric optics with the exit pupil function of the optical system [41]. This formulation allows computational lithographers to use methods developed for the analysis of linear systems to establish accurate, but computationally efficient, approaches to calculating aerial images for simulation OPC.

1.2.3 NUMERICAL APERTURE

One of the important defining characteristics of a diffraction-limited system that follows Abbe imaging is the concept of numerical aperture. Figure 1.8 shows a magnified area of the object illuminated in Figure 1.7.

If the two points shown in Figure 1.8 are treated as point sources, the minimum distance resolvable by the optical system can be calculated directly using the Rayleigh criteria [43]. In this approach, the minimum angular separation of two pinhole sources corresponds to the angle at which the maximum Airy intensity of one source overlaps the first minimum of the second.

Using Figures 1.7 and 1.8 in conjunction with the Rayleigh criteria, we can establish a relationship between the angle, θ, which represents the maximum acceptance half-angle of the optical system; the separation between diffracting points on the source d_i, λ; and the index of refraction, n, of the medium light is propagating through using the following relationship:

$$d_i = \frac{1.22\lambda}{2n\sin\theta} \tag{1.5}$$

The denominator in this equation, $2n\sin\theta$, is used as a defining characteristic of the resolving power of the optical systems that follow Abbe imaging [43]. The term $n\sin\theta$ is defined to be the NA of the optical system and describes the acceptance half-angle of the objective lens.

1.2.4 KOHLER ILLUMINATION

An important aspect of lithographic systems is the set of optical elements used to illuminate the mask or object of the system using partially coherent sources. In early optical studies, limitations to the ultimate performance of optical systems were observed when waves emanating from partially

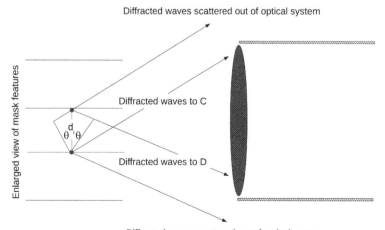

FIGURE 1.8 Enlarged section of the object shown in Figure 1.7. Two object points, separated by d_i and diffracting waves through the maximum acceptance angle of the lens, are used to define the numerical aperture of an optical system. (From Jenkins, F. A. and White, H. E., *Fundamentals of Optics*, McGraw-Hill, New York, 1957. With permission.)

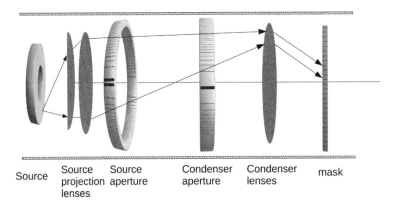

| Source | Source projection lenses | Source aperture | Condenser aperture | Condenser lenses | mask |

FIGURE 1.9 Diagram depicting the essential elements of Kohler illumination. Partially coherent source waves are defocused through a series of source and condenser optical elements to produce parallel illumination of a mask.

coherent points on the source were focused in the sample plane. To overcome these limitations, August Köhler developed a method of illuminating an object using a system of auxiliary and condenser lenses to illuminate the object with nearly parallel waves from the source.

In an idealized system, waves illuminating the mask will be spherical with radius of curvature selected to offset the impact of tilted illumination on mask features located at positions off the optical axis [44]. Also, this arrangement of source and condenser lenses has the added benefit that inhomogeneities in source intensity do not cause irregularities in the illumination field [42]. An important additional result is that condenser lens aberrations will have negligible impact on the degree to which waves from source points remain parallel [42].

Figure 1.9 shows a schematic diagram containing the essential optical elements of Köhler illumination. Köhler illumination provides uniform parallel illumination of the mask from partially coherent point sources and minimizes positional shifts in the image plane for features far away from the optical axis. Apertures in the system are used to optimize wave fronts from different source types and ensure optimal parallel illumination of the mask.

1.2.5 Partially Coherent Imaging

In the previous sections of this chapter, we have presented imaging results that implicitly assume that spatially coherent waves are propagated through an optical system. An implicit assumption behind these formulations is that the source is an idealized point source situated on the optic axis. Point sources are idealized constructs that are nearly impossible to realize in optical systems of interest to lithography. Sources typically used in modern lithographic systems are spatially extended sources that consist of multiple off-axis point sources that illuminate a mask object over a range of incident angles [44]. The earlier presentation of Fourier optics can be used to describe the propagation of waves emanating from each point in an extended source, but to truly understand the propagation from all points on an extended source interacting with a mask and scanner optics, one needs to utilize the machinery of partially coherent optics.

To understand partially coherent imaging, imagine that there are two points on a spatially extended source located at points r_1 and r_2. If the distance between points r_1 and r_2 is small relative to the wavelength λ of light emanating from each point, the intensity at a point in the image plane r_i due to light from source points r_1 and r_2 is written as follows [45]:

$$I(r_i) = I_{r_1}(r_i) + I_{r_2}(r_i) + 2\sqrt{I_{r_1}(r_i)}\sqrt{I_{r_2}(r_i)} \frac{\left\langle U^*(r_1) \middle| U(r_2)\right\rangle}{2\sqrt{I_{r_1}(r_i)}\sqrt{I_{r_2}(r_i)}} \qquad (1.6)$$

In this equation, $U*(r_1)$ and $U(r_2)$ are phasors emanating from points r_1 and r_2 and the term $\langle U^*(r_1)|U(r_2)\rangle$ is the cross-correlation of light emanating from points r_1 and r_2 normalized by the intensity of light from each point if it were an independent point source [42, 45]. In addition, the term $\dfrac{\langle U^*(r_1)|U(r_2)\rangle}{2\sqrt{I_{r_1}(r_i)}\sqrt{I_{r_2}(r_i)}}$ is called the *complex degree of coherence* and represents the intensity of light due to interference between waves from r_1 and r_2 at point r_i in the image plane.

There are two primary limiting cases for this equation. In the first, the distance between r_1 and r_2 approaches zero; then, light from source points 1 and 2 is completely correlated, and the intensity distribution in the image plane is the same as that for two coherent waves. In the second, the distance between points 1 and 2 approaches infinity, and light from points 1 and 2 is completely uncorrelated; therefore, the intensity distribution in the image plane at r_i is equivalent to a completely incoherent image [45]. Understanding partially coherent imaging concepts in lithography is particularly important, because illumination sources consisting of multiple points of light are routinely used. In addition, complex off-axis illumination sources are required to increase resolution and enable lithographic scaling.

1.2.6 Hopkins Diffraction Theory of Imaging and Transmission Cross Coefficients

Thus far, we have been discussing methods to model the propagation of light scattered by a mask through an optical system generating an image in the wafer plane. Each of these components can be tied together in a form usable for computation by using the Hopkins approach to the diffraction theory of imaging [46]. In this approach, the intensity in the image plane at a point (x,y) can be modeled by a four-dimensional integral as follows [46]:

$$I(x,y) = \int_{-\infty}^{\infty}\int_{-\infty}^{\infty}\int_{-\infty}^{\infty}\int_{-\infty}^{\infty} \text{TCC}(f',g':f'',g'')F(f',g')F^*(f'',g'')$$

$$\exp\left[-2\pi i\big((f'-f'')x+(g'-g'')y\big)\right]df'dg'df''dg'' \tag{1.7}$$

In this equation, the term $\text{TCC}(f',g':f'',g'')$ is referred to as the *transmission cross coefficient* (TCC) of the optical system, $F(f',g')$ is the Fourier transform of the mask, and $F^*(f'',g'')$ is the complex conjugate of the Fourier transform of the mask. This equation represents the partially coherent imaging of the mask denoted by F illuminated by two sets of partially coherent plane waves, (f',g') and (f'',g''). In addition, the TCC term in Equation 1.7 represents the propagation of source waves through the aberrated pupil function presented previously.

$$\text{TCC}(f',g':f'',g'') = \int_{-\infty}^{\infty}\int_{-\infty}^{\infty} P(f+f',g+g')\exp\left[\frac{-2\pi i}{\lambda}\phi(f+f',g+g')\right]$$

$$\exp\left[\frac{2\pi i}{\lambda}\phi(f+f'',g+g'')\right]df\,dg \tag{1.8}$$

In both the Hopkins intensity integrals and the TCC integrals, the integrals are taken over infinite domains, but these integrals become finite due to diffraction-limited optics. Also, in the case of the TCC, the intensity over the entrance aperture of the optical system is assumed to be constant, so the TCC integrals reduce to a constant multiplying the integral over a spatial frequency range, $f_2 + g_2 < r^2$, where r represents the pupil extent of the entrance aperture of the optical system.

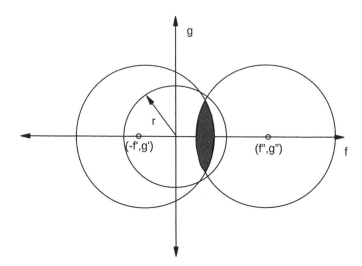

FIGURE 1.10 Illustration of the effective domain of integration for one pole of a dipole illuminator. The cross-hatched region illustrates the region over which TCC integration would be taken for source points centered at $(-f', g')$ and (f'', g''). (From Toh, K. K., *Tech. Rep. No. UCB/ERL M88/30*, University of California at Berkeley, Berkeley, CA, 1988. With permission.)

Figure 1.10 shows an example of the integration region used to calculate the TCC for the right-hand pole of a dipole source [47]. In this example, there are two source circles centered at $(-f', g')$ and (f'', g''), each of radius $NA/M\lambda_s$, where NA is taken at the entrance aperture of the objective system, M is the Gaussian magnification in the wafer plane, and λ_s is the mean wavelength of the source laser (Figure 1.11). These two source points are transmitted through the optical system of effective aperture of radius r that is capable of generating partially coherent images at the wafer plane. In a full TCC calculation for a dipole source like that shown in Figure 1.12, there would be a similar pole with source points similar to those at (f', g') and $(-f'', g'')$ that would generate the left-hand pole.

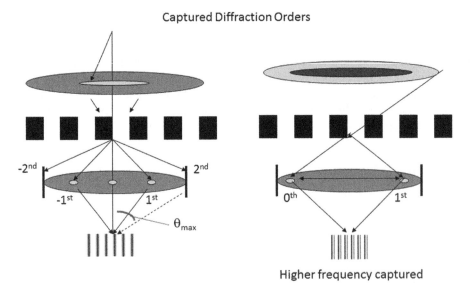

FIGURE 1.11 Comparison of diffracted orders captured for conventional and off-axis illumination [38]. Off-axis illumination captures higher frequencies that result in denser printable pattern (at a fixed λ/NA). (From Brunner, T. A. et al., *J. Micro. Nanolithogr. MEMS MOEMS*, 5, 5, 2006. With permission.)

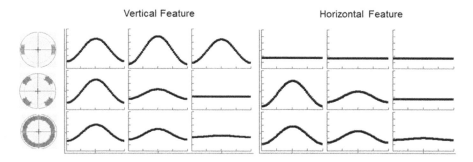

FIGURE 1.12 Comparison of aerial images for off-axis illumination. (a) Dipole Illumination 0.7/0.9; (b) quadrupole (0.75/0.97) Illumination; (c) annular (0.75/0.97) Illumination. One can see the horizontal and vertical contrast difference visually from the aerial images for three k_1 values: 0.45 (left column), 0.35 (middle column), and 0.28 (right column). (From Brunner, T. A. et al., *J. Micro. Nanolithogr. MEMS MOEMS*, 5, 5, 2006. With permission.)

1.2.7 IMPACT OF ILLUMINATOR

There are two types of illumination used in the industry: conventional illumination and off-axis illumination. As shown in Figure 1.11, conventional illumination affords three-beam interference (0th, +/− 1st), where the 0th order is a DC component. First-order diffraction allows for interference and transfer of pitch information from the mask to the wafer. For a coherent source, the maximum frequency (minimum pitch) captured by the lens is NA/λ. The highest-resolution image is $\lambda/2NA$, and the corresponding depth of focus is proportional to λ/NA^2 [48]. Traditionally, lithographers estimate the minimum critical dimension printable by a lithographic process as follows:

$$CD_{min} = k_1 \frac{\lambda}{2NA} \tag{1.9}$$

In Equation 1.9, the constant of proportionality, k_1, captures lithographic process components that contribute to the achievable critical dimension for a given lithographic process. This constant of proportionality is a measure of lithographic process performance that ranges from slightly larger than 0.25 to values of approximately 1. A k_1 value of 0.25 represents the ultimate resolution of an optical system and should be thought of as an asymptotic lower limit for a lithographic process. Larger k_1 values represent processes that are not delivering the maximum resolution for the scanner defined by λ and NA.

In its simplest form, with off-axis illumination (OAI), one can define classic annular, quadrupole, and dipole illuminators as shown in Figure 1.12. OAI leverages two-beam interference of zeroth and first order to provide contrast at high resolution at the expense of through-pitch performance. Annular illumination is defined by the inner and outer radius of a ring of intensity in the illuminator used to expose the mask. Historically, annular illumination is appropriate for designs that require a continuous range of supported feature sizes and pitches in orthogonal orientations requiring $k_1 < 0.5$. As k_1 is reduced toward a value of 0.25, annular illumination begins to yield smaller process windows, which require a change in illumination to overcome.

Quasar illumination is defined by four off-axis poles with a well-defined radial extent and fixed angular extent for each pole. Experience has demonstrated the benefit of this strategy for bidirectional design spaces at tighter pitches ($k_1 > 0.35$), but it suffers at very low k_1 ($k_1 \rightarrow 0.25$) due to lack of captured diffraction orders. A quadrupole with poles on axis, sometimes called a *c-quad configuration*, offers better resolution in both directions and can enhance tip-to-tip image formation but generally suffers at intermediate pitches.

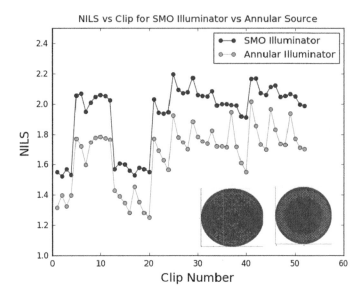

FIGURE 1.13 Comparison of NILS for an optimized illuminator and an annular illuminator. Through the entire clip range consisting of one-dimensional and two-dimensional patterns, the SMO illuminator provides superior NILS, which typically translates into larger process window.

Dipole illumination is defined by two on-axis poles with a well-defined radial extent and fixed angular extent for each pole. Dipoles are applied to design spaces with preferred or unidirectional features requiring tight across-chip critical dimension control and k_1 approaching 0.25, but they have degraded image formation for features aligned in the nonpreferred orientation. One drawback of dipole illumination is that forbidden pitch ranges develop as the radial extent or sigma range of each pole increases.

As we have shown in the previous examples, the use of OAI allows lithography engineers to tailor illumination to a particular design space. These examples represent an approach to illuminator design that focuses on simple optical elements customized to enable continued device scaling over a broad range of design styles.

Recent advancements in hardware and software have resulted in more sophisticated illuminations. Source mask optimization (SMO) technology leverages pixelated illumination and co-optimized masks to improve design space coverage and patterning process window on wafer. Utilizing sophisticated algorithms [49], flexible programmable illuminators [50, 51], and improvements in mask fabrication, it is now possible to increase design space coverage and process window simultaneously. An example of an SMO solution compared with a conventional illumination solution is shown in Figure 1.13.

Figure 1.13 shows an illuminator determined by SMO compared with a more traditional annular illuminator. In this plot, the SMO illuminator has yielded significantly higher normalized image log slope (NILS) response than the annular illuminator. Higher NILS translates into more robust imaging and process window for applicable resist systems. It is clear from Figure 1.13 that both illuminators are challenged for clips 1 to 5 and 10 to 20. These clips should be examined carefully by lithography engineers to ensure that there is sufficient process window for these constructs to enable targeted patterning yield specifications; otherwise, these constructs may need to be addressed through ground-rule constraints, retargeting, or new patterning process approaches.

1.2.8 SUB-RESOLUTION ASSIST FEATURES (SRAFs)

Another fundamental imaging technology, sub-resolution assist features, is critical for enabling through-pitch process window. The underlying principle behind SRAFs is quite simple. The image contrast of diffracted light in isolated spaces, approximately 2× the minimum pitch and larger, can be improved by adding additional spatial frequencies into the formation of an image. An example of the image with and without a sub-resolution feature is shown in Figure 1.14.

In the case of a 2× minimum pitch structure, a sub-resolution feature (too small to print) can add a higher spatial frequency of diffracted light into the capture angle of the lens. When combined with lower (greater than 2× minimum printable pitch) resolution, one can see improved contrast of the image. This improves the depth of focus and process window for the isolated structures. One consequence of SRAF usage is unintended printing of the higher-order pattern. It is necessary to balance the relative size of the SRAF (ease of mask fabrication) with the unintended (side-lobe) printing of the SRAF.

A natural question to ask at this point is how the introduction of SRAFs improves process window for isolated features. Shown in Figure 1.15 is a comparison of process window for an isolated trench feature with and without SRAFs using a typical annular illuminator. In Figure 1.15a and b, mask features from a baseline isolated trench and an isolated trench with two SRAFs per edge are shown. In these figures, the cross-hatched features are transmitting, and the white background does not transmit light.

The size of SRAFs in Figure 1.15b and their spacing from the main feature are chosen based on two criteria. First, the dimension of SRAFs features is chosen to boost the intensity of the main feature to a maximum value, such that the SRAF features do not print. Second, to first order, the spacing between the main feature and the SRAF features is chosen to approximate a fully nested grating. It is clear from Figure 1.15 that the introduction of SRAFs has resulted in a significant increase in depth of focus, from 195% to 272%, which is an increase of approximately 77%. The increase in dose latitude is more modest but still improved from 4.9% to 6.8%. The features shown in Figure 1.15 are an idealized case, but they do demonstrate the benefits of using SRAFs.

Aside from SRAF printing, there are some other challenges that computational pattering engineers face when using SRAFs, such as mask manufacturing rule constraints (MRCs), inability of mask processes to produce SRAFs of the prescribed size to produce maximum process window

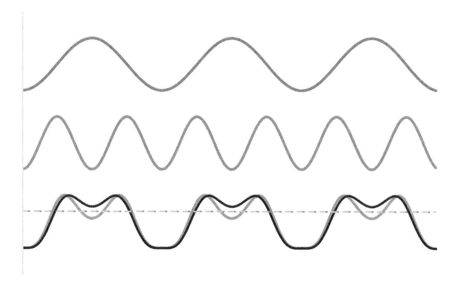

FIGURE 1.14 Schematic representation of the intensity for a 2× minimum pitch L/S without SRAF, 1× minimum pitch (higher frequency), and (bottom) a small trench at 2× minimum pitch with two different SRAFs. (Black) Mid-sized SRAF and (gray) large SRAF. One can see the improved sharpness of the SRAF isolated pitch pattern and that increasing the SRAF size too close to the anchor leads can lead to inadvertent printing. Note that the large SRAF image drops below the threshold for the resist in this thought experiment.

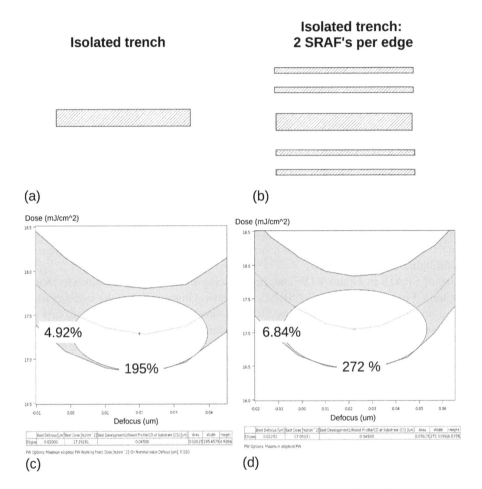

FIGURE 1.15 Comparison of isolated trench printing with and without SRAFs. (a) Schematic of horizontal trench feature without SRAFs. In this diagram, the cross-hatched feature transmits light and the white background does not. (b) Schematic diagram of a horizontal trench feature with two SRAFs per edge. (c) Ellipsoidal process window of isolated trench feature showing baseline dose-latitude and depth of focus. (d) Ellipsoidal process window with two SRAFs per edge showing a significant increase in depth of focus and a modest increase in dose latitude.

benefit, and challenges in ensuring that SRAF geometry and size distributions do not overwhelm shot count constraints introduced by mask writers.

1.2.9   OPTICAL PROXIMITY CORRECTION

Over time, lithographers had come to depend upon advances in optics and steppers to ensure that designed features could be printed on wafer with minimal modification. In the mid- to late 1980s, it was becoming clear that optical proximity effects were causing significant deviations from design targets transferred to mask versus what was actually printed on wafer. For example, tip-to-tip configurations were not printing on target if drawn in one-to-one dimensions in the mask plane. Additionally, critical dimensions on wafer for line-and-space patterns deviated from target depending upon the local pitch environment, and line-end configurations began to exhibit significant line-end pull-back.

To offset these effects, lithographers began to compensate by altering the size and shape of features on the mask to restore wafer printing to target specifications. To identify the correct changes to mask shapes, lithographers would use wafer measurements over a systematic array of mask shapes to identify the correct range to maintain wafer target dimensions. This practice became known as *rules-based optical proximity correction.* An early example of this approach is the addition of serifs to contact shapes by Starikov [52] to improve target printing of contact shapes. As device dimensions continued to decrease in size, the complexity of rules-based optical proximity correction schemes increased beyond what is feasible to correct through rules-based schemes. In response to this spiraling increase in complexity, model-based optical proximity correction (MBOPC) schemes were developed. These schemes leverage the full machinery of Fourier optics developed in the previous sections of this chapter. The main idea behind MBOPC is to treat mask shape creation as an optimization problem in which the differences between target critical dimensions and those predicted by compact models of a lithographic process -are minimized.

Figure 1.16 is a pictorial representation of a typical sequence of steps executed during MBOPC. In Figure 1.16a, a target layer is generated from drawn shapes by biasing shapes to compensate for known unit process biases such as RIE biases, changes due to wet chemical cleans, biases introduced by early stages of physical vapor line depositions, and so on. Next, as shown in Figure 1.16b, shapes on the target layer are copied to a correction layer and broken into line segments systematically based upon algorithms or rules-based strategies.

As shown in Figure 1.16c, predicted critical dimensions are then simulated using compact optical and resist models to generate an initial critical dimension (CD) between two segments. The difference between the target layer and the initial simulated contour is calculated and compared with a

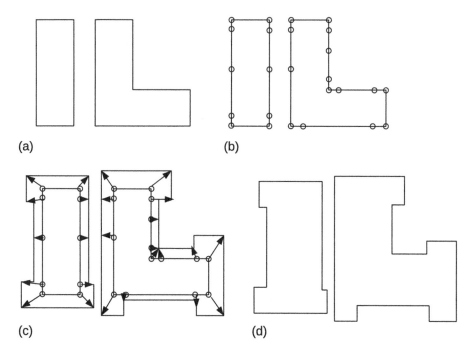

(a)　　　　　　　　　　　　　　(b)

(c)　　　　　　　　　　　　　　(d)

FIGURE 1.16 Shown is a typical sequence of steps used in optical proximity correction. (a) Target shapes are biased to create a target layer. Biases may be applied to offset known unit process biases due to reactive ion etch, wet processes, CD changes due to liners, etc. (b) Target layer shapes are broken into line segments identified by circles at each vertex. (c) Aerial images and resulting resist dimensions are calculated using compact models. Line segments are moved based upon calculated dimensions to minimize the difference between the target in (a) and the dimensions predicted by compact models. The process of moving segments, calculating resultant dimensions, and comparison with target is continued until a cost-function goal is reached or a set number of iterations has been completed. (d) Mask shapes resulting from model-based OPC that will be written to mask.

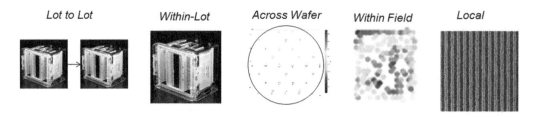

FIGURE 1.17 Elements of variation seen in lithography. (Adapted from Saulnier et al., 9422, 9422, 2015.)

target CD. Based upon the difference between simulation and target, line segments are moved as shown in Figure 1.16c. This process of calculating simulated contours and moving line segments continues until the difference between simulated contours and the target layer falls below a target tolerance or a prescribed number of iterations has been completed. If the target difference between simulated contours and the target layer is achieved, OPC is said to converge; otherwise, it has not converged.

Computationally, there are a number of different approaches to calculating contours used to compare with target layers. The majority of these methods utilize the benefits of Fourier imaging framework, described in earlier sections, to couple aerial images output by numerical approaches with compact resist models to enable full-chip OPC.

One of the most successful approaches used to efficiently generate aerial images using Hopkins imagining is the sum of coherent systems (SOCS) approach [53]. In this approach, the Hopkins imaging formulation is expanded by eigen functions into a set of coherent systems. Eigen function decomposition is accomplished by singular value decomposition (SVD) methods, and the resultant set of linear system outputs is then squared and summed to produce approximations to the aerial image intensity distribution due to a particular set of mask shapes [53].

Other methods to numerically calculate aerial image intensity distributions from mask shapes have been successfully demonstrated for use in OPC. For example, Stirnman and Rieger developed a technique that leverages zone sampling of layout density and treats optical proximity correction as a signal processing problem [54]. The main optical theory grounding this approach is the concept of "Structural Theory of Information" [55] to treat optical proximity correction as a series of nonlinear spatial filters. Both SOCS and the zonal approaches have been successfully applied to generating mask shapes through -MBOPC. One of the key computational problems to overcome irrespective of the approach used to numerically calculate aerial images, is how to sample target shapes efficiently to reduce runtime while maintaining accuracy. Two main approaches have been utilized, depending upon the size of critical features in a layout. These are a sparse [56] approach to aerial imaging sampling and pixel-based approaches [57].

In the sparse approach, aerial images are calculated at predefined evaluation points of line segments in Figure 1.16. Line segments are then moved based upon the output of compact resist models resulting from aerial image sampling at a discrete set of points normal to the active line segment. This approach has been used successfully for several technology nodes and is in use for implant, block, and far BEOL levels in modern technology nodes.

Pixel-based aerial imaging approaches [57] calculate a pixelated aerial image field for the active region of correction in an optical proximity correction run. Line segments in Figure 1.16 are then moved based upon sampling of the pixelated aerial image field and the results of contour evaluations from compact resist models (Figure 1.17).

1.2.10 ETCH MODELING

As OPC methodologies have been successfully applied to advanced technology nodes, computational patterning engineers have been asked to offset downstream process biases to wafer patterns during OPC. One of the most effective approaches to offsetting these biases is to use a model-based retargeting approach. In this approach, a compact model is calibrated to successfully predict a unit process bias based upon pattern size, local density, and other physical variables. In practice, among the most important downstream unit process biases to compensate are those due to RIE processes used to transfer patterns in resist to wafer. There are three primary steps in offsetting biases due to downstream biases: 1) generating a compact etch bias model that is capable of capturing biases from microloading and aspect ratio–dependent effects [58]; 2) generating OPC target layers that offset biases predicted by a compact etch bias model using typical design rule check (DRC) engines to generate a lithographic target layer [59, 60]; 3) generate final mask shapes using standard MBOPC.

This approach has been successfully applied to controlling the impact of etch biases on pattern transfer in 45 nm, 32 nm, and successive nodes [60]. One of the key challenges in applying model-based retargeting to correct for downstream unit process biases is to detect conditions that are unsupportable by lithographic processes. It is possible that an etch process bias or other unit process bias is so large that the resultant retargeted shapes are not supported by the targeted lithographic process. In these cases, the degree of retargeting applied is constrained through the application of augmented mask rule check code to ensure retargeted features do not endanger the printability of retargeted features. These approaches have been used successfully and continue to be areas of active research in OPC strategies based on model-based retargeting.

1.3 LITHOGRAPHY PROCESS AND TOLERANCE ASSESSMENT

Thus far, we have discussed the principles of image formation and OPC. While sufficient image quality is a requisite, the processes and materials used in patterning must be able to respond to the aerial image and tolerate the inherent variability that impacts imaging from other components of a patterning process. In particular, some of the most critical elements of designing a patterning process are the selection and formulation of a resist system along with the optimization of a patterning stack to enable effective printing and pattern transfer. Resist formulation and chemistry are covered in detail in another chapter of this book, so in the following, we will focus on other important aspects of patterning stack design.

The first part of the imaging solution is the bottom antireflective coating (BARC), which is intended to reduce the reflection of light at the resist–BARC interface. If light reflected from this interface is not minimized, critical dimensions may be reduced locally, and resist profiles may be roughened with an amplitude corresponding to the period of standing waves formed by the interference of the reflected wave at the resist–BARC interface. Figure 1.18 shows a typical lithographic film stack consisting of a resist film, BARC, and planarizing film. In Figure 1.18a, light propagates into the film stack, and a fraction of the light propagating through the stack is reflected at each interface in the stack, starting with the vacuum–resist interface, then the resist–BARC interface, and finally, a third reflection at the BARC–planarization interface. If the BARC film has been chosen to minimize reflected light at the resist–BARC interface, smooth, well-controlled resist profiles are obtained, as shown in Figure 1.18. If reflections at the resist–BARC interface are not minimized, standing waves may be produced that produce rough resist profiles, as shown in Figure 1.17c, or undercut profiles, as shown in Figure 1.18d.

To select an optimal BARC film for a patterning stack, lithography engineers need to minimize the coefficient of reflection, r_{rb}, at the resist–BARC interface. Optically, the reflectivity coefficient at an interface is determined by the following equation [42, 61, 62]:

$$r_{rb} = \frac{n_r - n_b}{n_r + n_b} \tag{1.10}$$

An optimal BARC film will minimize the total reflected light propagating back into the resist due to reflections at the resist–BARC interface and the BARC–planarization interface. Engineers can

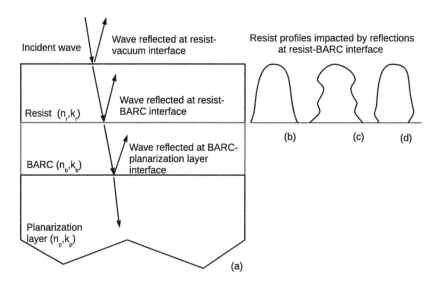

FIGURE 1.18 Shown is a typical patterning stack used in modern lithography processes that captures a range of resist profile outcomes depending upon the degree to which BARC films have been optimized to reduce reflections. (a) A typical lithography stack showing reflections at key interfaces in the stack: vacuum–resist, resist–BARC, and BARC–planarization interfaces. (b) An example of a smooth resist profile obtained by selecting a BARC film that minimizes light reflection at the resist–BARC interface. (c) Shown is a schematic representation of a resist profile affected by standing waves produced by a nonoptimal BARC film in which reflections at the resist–BARC interface are not minimized. (d) Shown is another example of the impact of nonoptimized BARC films. Reflected light at the resist–BARC interface can expose more resist at the resist–BARC interface than intended, leading to undercut profiles.

identify the properties of an optimal BARC film by finding the complex index of refraction, n_b, and film thickness, t, as shown by Mack et al. [62] using the following relationship:

$$R_{\text{stack}} = 0 = \left\| \frac{r_{rb} + r_{bp} \cdot e^{\frac{4\pi i n_b t}{\lambda}}}{1 + r_{rb} r_{bp} \cdot e^{\frac{4\pi i n_b t}{\lambda}}} \right\|^2 \tag{1.11}$$

Equation 1.11 is a deceptively simple equation cast in terms of the complex indices of refraction n_{ij} and a complex attenuation factor given by the exponential term in Equation 1.11. This equation will yield two independent equations when $R_{\text{stack}} = 0$, which can be used to solve for the three properties of an optimal BARC: the real part of the complex index of refraction, the imaginary part of the index of refraction, and the thickness of the BARC film [62].

In general, Equation 1.11 is solved numerically to identify a family of solutions identifying an optimal BARC for a given patterning stack. An instructive limiting case can be explored analytically to gain insight into optimal BARC selection. If we assume normal incidence light and that the resist and planarization film are nonabsorbing, we can find a BARC thickness at which $e^{\frac{4\pi i n_b t}{\lambda}} = -1$. In this case, the thickness is equivalent to $\lambda/4n_b$, and the BARC will be nonabsorbing with a real index of refraction as follows:

$$n_b = \sqrt{n_r n_p} \tag{1.12}$$

While single-layer BARC films have been used for several nodes, advanced process requirements drove the exploration of multilevel BARC solutions. Two primary issues drove the adoption of multilayer BARCs. First, as numerical aperture increases, depth of focus becomes increasingly challenged. Multilayer BARC solutions allow engineers greater leverage to tune out reflections from higher-NA illuminators as well as improving planarity. Additionally, multilayer coatings offer superior performance relative to single-layer BARC over larger ranges of NA [63]. As 3-D device structures with enhanced topography, such as FinFETs, became more prevalent, multilayer coatings that allow the resists' thickness to be amplified with anisotropic etch using inorganic/organic underlayer stacks become mandatory.

The resist systems used in modern process technologies are tailored for each application space. Each contains some common elements: i) adhesion promoters, ii) a switchable moiety, iii) an etch resistance moiety, iv) a quencher to mitigate undesired development, v) a photoacid generator (PAG), and vi) a casting solvent. For a review of a host of resist chemistries from literature, the reader is encouraged to consult references [64, 65]. Generally, high-resolution optical resists are chemically amplified. In chemically amplified resist (CAR) systems, photons are absorbed by the PAG. Acid is generated, and during the postexposure bake (PEB), the switchable moiety is deprotected, generating a polar group. In the case of positive tone resists, the deprotected region is developed in aqueous base developer. In a negative tone system, the deprotected region is insoluble in the developer. The unexposed region is dissolved during the develop process. When characterizing a lithographic process, one must consider the entire process flow. In addition to establishing the optimum antireflective coating stack, one must consider the effect of postapplication bake (affects resist density), the PEB (affects acid diffusion and developed dimension), the develop sequence (puddle versus dynamic, nozzle type, and sequence), and the rinse.

There are a multitude of contributors to variation. At the process level, when characterizing resist imaging performance, one can quantify the systematic hardware-induced variation one can expect based on the following simplified relationship:

$$\text{CDU} \propto \left(\frac{\partial \text{CD}}{\partial E} \right) \Delta E_{\text{Scanner}} + \left(\frac{\partial \text{CD}}{\partial F} \right) \Delta F_{\text{Scanner}}$$

$$+ \left(\frac{\partial \text{CD}}{\partial T} \right) \Delta T_{\text{Hotplate}} + \left(\frac{\partial \text{CD}}{\partial \text{CD}_{\text{mask}}} \right) \Delta \text{CD}_{\text{Mask}} + \text{LER} \tag{1.13}$$

There are certain specifications that one can apply to measure sensitivities in a resist response. Control of the dose delivered to the wafer plane is a primary factor in establishing sensitivity. Dose control can include scanner parameters and image flare. Focus tolerance is another concern; this can be impacted by errors in substrate surface mapping and variation in local topography. Resists can be sensitive to these parameters in differing degrees based on the diffusion that occurs during the PEB sequence. This can be impacted by the resist activation energy, PEB temperature, and hotplate control, including temperature ramping profile and uniformity. Another factor impacting the lithography is the mask error factor (MEF), which can be a result of errors in mask process, impacts from OPC convergence, and model accuracy. A final parameter to consider is line-edge roughness (LER). LER has been shown to be affected by a host of contributors such as diffusion, contrast, and formulation.

In addition to hardware contributions, there are systematic variations. OPC and on-mask print difference can lead to isolated feature-to-dense feature offsets.

$$\text{Bias}_{\text{dense-iso}} = \text{CD}_{\text{dense}} - \text{CD}_{\text{iso}} \tag{1.14}$$

Similar errors or convergence differences can affect tip-to-tip and tip-to-side variations. In addition to OPC and iso-dense bias errors, there are generally differences in dose-slope $\left(\frac{\partial \text{CD}}{\partial E} \right)$ between dense and isolated features. As control systems adjust dose to control targeting errors in monitor

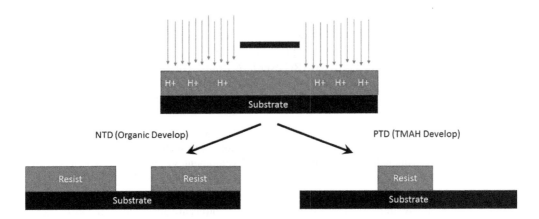

FIGURE 1.19 Comparison of NTD and PTD process flows.

structures, inevitably, dense and isolated features will respond differently. The error between dense and isolated structures will change by the following relation:

$$\text{Bias}_{\text{final}} = \text{Bias}_o + \left(\frac{\partial \text{CD}_{\text{final}}}{\partial E} - \frac{\partial \text{CD}_{\text{iso}}}{\partial E} \right) dE \qquad (1.15)$$

In addition to inherent process variations, temporal effects can also cause variation. Some of the variation shown earlier can manifest itself as a source of manufacturing variation. Here, variation is an analysis of spatial and temporal components. Local variation has many of the components of variation enumerated previously. Within-field variation reflects a convolution of the uniformity of scanner imaging, homogeneity in the mask features across a wide range of features, and accuracy of OPC and process models. The scanner can contribute through-slit and through-scan errors, which can result in systematic errors that may or may not be correctable. Across-wafer variation (AWV) incorporates natural variation in scanner, hotplate, and develop uniformity. Wafer-to-wafer variation such as AWV is dominated by tooling and process variation. Lot-to-lot variation can be impacted by tooling, process stability, process control strategies, and temporal variation on a longer time period. Today, the largest components of variation are driven by local sources such as LER, across-chip linewidth variation (ACLV), and AWV [66]. Sources of systematic variation, including tool and integrated process modules, have been identified through the use of advanced metrology and process monitoring techniques. Advanced manufacturing process control strategies have been developed to aggressively reduce these contributions [67] (Figure 1.19).

1.3.1 Transitions in Lithography

Historically, advances in lithography were driven by reductions in wavelength accompanied by improvements in imaging materials. For example, the transition from 365 nm steppers using Novolak resin resists to 248 scanners with coherent excimer sources ushered in the development of chemically amplified resists [68]. Furthermore, the development of top antireflective coatings (TARCs) and BARCs to address printing issues from standing waves became standard practice. As the industry transitioned to 193 nm, new materials solutions were required. Fluorinated aliphatic resins [69, 70] with improved transparency, new multilayer BARC systems [63, 71] to handle aggressive off-axis illumination, and the use of SRAF-based OPC strategies are some of the innovations required to deploy 193 nm lithography into production.

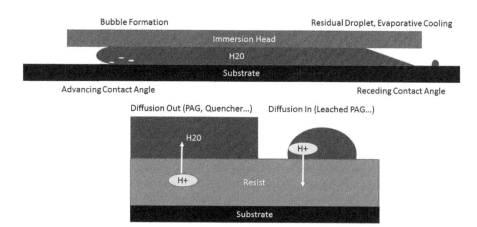

FIGURE 1.20 Challenges with immersion lithography: (a) bubble formation and droplet formation; (b) leaching showing out-diffusion of resist components (adversely affecting imaging) and in-diffusion of concentrated leachants from droplets (particles and imaging defects).

1.3.2 IMMERSION LITHOGRAPHY

Continued aggressive pitch scaling through the 32 and 28 nm nodes drove the development of immersion lithography [72] as a means to continue to leverage 193 nm scanners. Immersion lithography enables the numerical aperture of the optical system to increase beyond what is achievable with free space optics operating in air. This is achieved by coupling the objective lens of the optical system to the wafer through a layer of ultrapure deionized water. This technique results in an increase of the maximum frequency of the imaged orders and improves the resolution by $1/n_{H_2O}$ relative to a conventional scanner (Figure 1.20).

While resolution improved, the advent of immersion lithography [73] brought new challenges [74] in the way of scanner tooling and lithography materials. Significant engineering challenges needed to be addressed to place water in intimate contact with the wafer and the objective lens simultaneously. Serious concerns about material leaching [75, 76] from the resist into the immersion fluid, potentially contaminating the lens or the wafer, needed to be addressed. Eventually, innovations in immersion head design, scanner controls, and imaging material technology such as topcoats and water shedding agents [77] helped eliminate bubble formation and reduce component permeability.

1.3.3 NEGATIVE TONE IMAGING

Conventional chemically amplified resists developed with 0.26M TMAH respond as positive tone imaging materials. That is to say, the exposed material undergoes a chemical reaction and is removed. In negative tone development (NTD), the same reaction occurs, but the developer removes the unreacted portion of the imaging resist [78]. Negative tone imaging has specific benefits in terms of imaging small trenches and vias. To understand the underlying benefit of NTD, one must look at the image formation mechanism [79].

A simple thought experiment can be useful to elucidate the improved resolution attainable through a positive tone development (PTD) imaging process for minimum geometry features. In Figure 1.21, an aerial image is projected into a resist. Assuming that a single spatial frequency is captured at a dense pitch, one can see a simple sinusoidal image in the resist. In bright areas, acid is generated, and during PEB, it diffuses from high concentration to low concentration.

Naturally, the acid diffuses toward the unexposed area; thus, positive tone lithographic systems generally produce chrome-negative features such as trenches. If, however, the resist is developed in

FIGURE 1.21 In an NTD process, a bright-field image (black line) is formed. The acid diffuses down the aerial image (and subsequent concentration gradient), making the unexposed trench smaller than the associated aerial image. The resultant resist line profile, shown by the gray line, would produce a smaller trench when unexposed resist is developed.

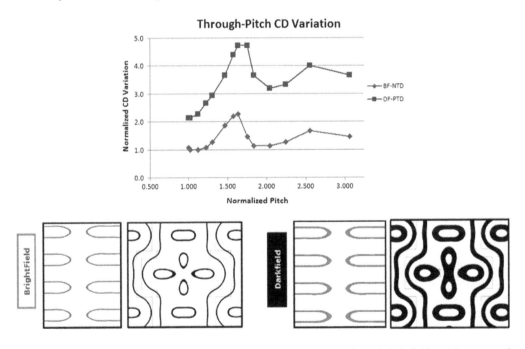

FIGURE 1.22 (a) On-wafer data comparing bright-field negative tone resist and dark-field positive tone resist validates the improvement process window. (b) Comparison of process variation simulations for a bright-field negative tone solution and dark-field positive tone resist solution. (From Landie, G., et al., 7972, 7972, 2011. With permission.)

a negative tone process, unexposed resist is dissolved, producing trench features that are generally smaller than the original exposed area. Thus, NTD represents a process that has inherent advantage for printing smaller trenches and vias. Likewise, PTD represents a process that has inherent advantage for printing smaller lines where acid diffuses into the line, resulting in smaller printed line CD. The fundamental improvement predicted by Brunner et al. [80] has been validated for narrow trench applications. In Figure 1.22, wafer data and simulated process variation for a negative tone process illustrate process improvements that are possible using NTD.

1.4 MULTIPLE PATTERNING AND OVERLAY OPTIMIZATION

As the required pitch for CMOS scaling continued to drive critical dimensions below the resolution limit of high-NA immersion scanners, the industry transitioned to multiple patterning schemes, as shown in Figure 1.23. These approaches, rather than printing features directly, developed schemes to use multiple exposures to decrease the effective pitch or used spacer techniques to decrease the pitch. *Double patterning* is the collective term used to describe multiple exposure processes used to generate a single design level. There are many techniques focused on achieving improvements in two-dimensional (2-D) imaging, tip-to-tip (T2T) control, frequency doubling, pitch splitting, and 2-D contact hole density enhancement. Double exposure can circumvent the k_1 diffraction limit and enhance 2-D printability. Initially, there were double-exposure-double-etch (DEDE) approaches for improved T2T and spatial frequency enhancement. Alternate techniques involving double expose, single develop, double-dipole lithography (DDL), and resist-freeze processes were also explored.

Of the aforementioned schemes, DDL was explored extensively as a way to push the limits of bidirectional imaging. This technique involved decomposing the illuminator-mask into two directionally specific orientations. As shown in Figure 1.24, a layout was decomposed by orientation and exposed sequentially (without removing the wafer from the scanner) to summate the two exposures, which are incoherently assembled in the resist film. This is similar to DEDE decomposition, shown in Figure 1.25, in which the design mandates a small T2T construct. The layout is thus decomposed by orientation and reassembled on wafer to generate the desired pattern.

Another approach to address aggressive minimum pitches below $0.5\lambda/NA$ is to use serial patterning technologies, including resist-freeze technology, resist-over-resist technology, and brute force litho-etch-litho-etch (LE2 or LELE) sequences.

Resist-freeze and resist-over-resist technology, as shown in Figure 1.26, was developed for line-space applications. LELE sequences were deployed for both line-space and via applications.

A unique embodiment of LELE sequence is the pack-unpack scheme shown in Figure 1.27, in which a dense via was printed on grid and then "selected" with subsequent exposure(s).

Overlay sensitivity was a defining characteristic of multiple patterning schemes. As shown by Hazelton et al. [83], critical dimension uniformity (CDU) in multiple patterning processes is

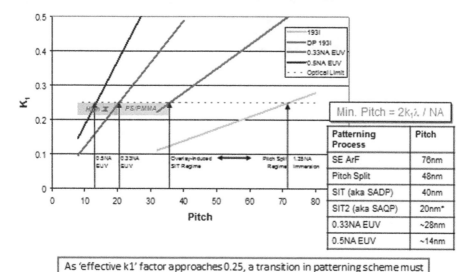

FIGURE 1.23   Natural transition points for patterning. (From Colburn, M., *SPIE Advanced Etch Conference Plenary*, 2015.)

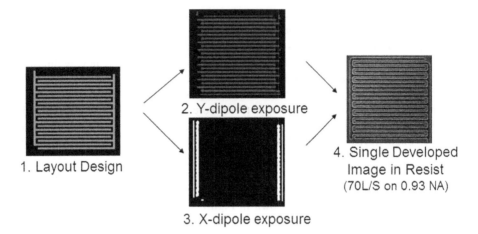

FIGURE 1.24 Dipole decomposition of a bidirectional layout showing a Y-dipole and an X-dipole component at the mask level, and a recombined image in resist after development. (Adapted from Colburn, M., *Lithography Solutions for the 22nm Node*, VLSI, Honolulu, HI, 2009. With permission.)

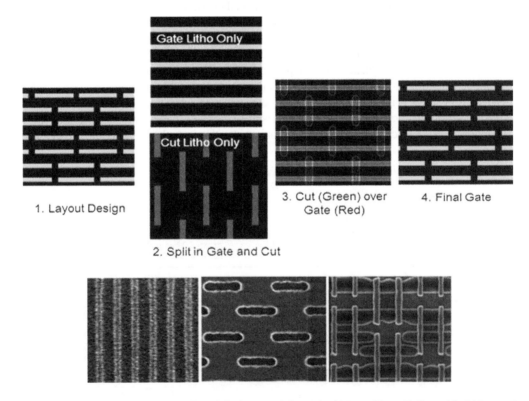

FIGURE 1.25 Double exposure double etch for improved tip-to-tip. (Adapted from Colburn, M., *Lithography Solutions for the 22nm Node*, VLSI, Honolulu, HI, 2009.)

defined in Figure 1.28. In practice, exposure 1 CDU and exposure 2 CDU are typically identical and matched where intralevel 3σ variation is smaller than expected (Figure 1.29). This analysis was extended to accommodate correlation of image placement error of exposure 1 and exposure 2 patterns that serves to tighten the nominal space variation relative to randomly distributed mask component, as shown in Figure 1.30.

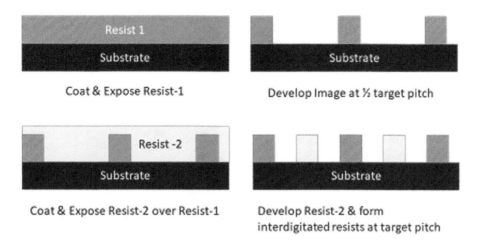

FIGURE 1.26   Pitch split with resist on resist technology. Example applications of pitch-split technology can be found in Hori et al., 2008 [82].

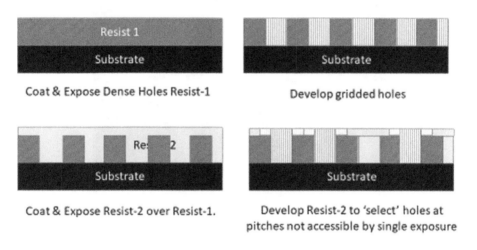

FIGURE 1.27   Pack-unpack (gridded contact holes).

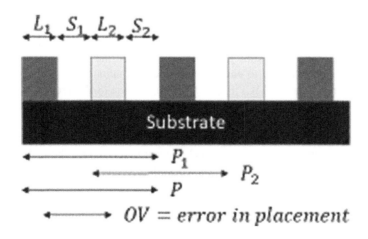

FIGURE 1.28   Space tolerance for pitch split.

$$\text{Space CDU} = \sqrt{\left(\frac{\text{CDU}_{L1}}{2}\right)^2 + \left(\frac{\text{CDU}_{L2}}{2}\right)^2 + \left(\text{OVL}_{3\sigma}\right)^2 + \left(3 \times \text{OVL}_{\text{MEAN}}\right)^2}$$

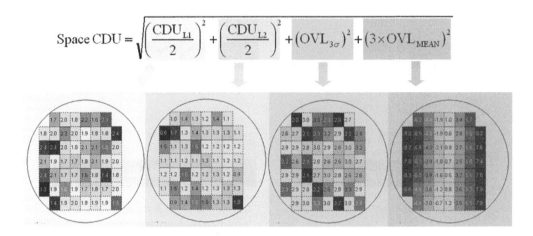

FIGURE 1.29   Analysis of independent quantities. (From Koay, C., et al., 2010.)

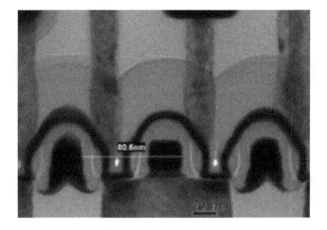

FIGURE 1.30   Demonstration of an SAC process using an Hf-based etch stop layer. The contact etch stops on the Hf layer, preventing shorting of the source or drain contact to the gate electrode.

1.4.1   SELF-ALIGNED INTEGRATED SOLUTIONS

As the demand for critical feature pitches decreased beyond the capability of 193 nm scanners and beyond the edge placement error tolerances achievable using pitch-splitting technology, the industry adopted "self-aligned" patterning strategies. There are two forms of self-aligned processes. One leverages the underlying structure to ensure the proper formation of a subsequent level. This is the case for self-aligned contact (SAC) [85] and self-aligned via (SAV) processes [86]. The second leverages an offset spacer feature to frequency double an initial lithographic pattern referred to as a *mandrel*. This is referred to as *self-aligned double patterning* (SADP) [87].

SAV processing makes use of the integration sequence practiced in most dual damascene interconnect integration schemes. In brief, the trench pattern is imaged on the wafer and etched into the underlying interlayer dielectric (ILD). The via pattern required to connect to the underlying structure is superimposed on the trench image penetrating through the ILD at the bottom of the preexisting trench. Following requisite cleans, a single metallization is performed. The basic dual damascene sequence requires exacting control of topography, CD, and overlay during the via imaging. SAV relieves some of the CD control and overlay requirements by introducing a hard mask on the wafer prior to the first trench patterning event. This hard mask remains in place during the via patterning. Using a pattern transfer process that is selective to the hard mask, via lithography patterns can overrun the edges of the original trench pattern without impacting CD in one axis while simultaneously achieving self-alignment in that same axis. SAV introduces some new challenges, including the deposition and removal of the hard mask film and the requirement to print asymmetric vias. In spite of this modest increase in complexity, SAV has found widespread adoption in all advanced CMOS process technologies.

SAC processing leverages the same general approach as SAV: preexisting features are used to define the location and dimension of the contact. Early demonstrations of SAC in a gate first 22 nm CMOS technology leveraged a Hf-based dielectric as an etch stop layer deposited over the FEOL structure prior to contact ILD deposition and CMP [88]. This approach allowed a conventional dielectric etch to be used to form the contact but enabled the CD control and overlay of the contact exposure to be relaxed, as shown in Figure 1.30.

Following the ILD etch, a selective etch was used to remove the Hf layer landing on the underlying silicide. SAC processing in process technologies using replacement metal gate (RMG) integration schemes has proved to be more difficult. In the RMG structure, an opportunity to fully encapsulate the gate in a hard mask layer never presents itself during the FEOL integration sequence. This is precluded by the fact that the dummy gate is removed from the FEOL structure late in the process flow, after the contact ILD is already deposited and planarized. Once the dummy gate is replaced by the gate stack, a scheme for recessing the metal gate and filling it with a suitable etch stop layer must be derived for SAC processing to work. This is challenging due to the heterogeneous nature of typical advanced RMG gate stacks, which include a gate dielectric, one or more work function setting metals, and a low-resistivity metal fill [89]. The recess of these films selective to the FEOL offset spacer, typically SiN based, and the contact ILD, typically SiO_2, is challenging. In spite of this difficulty, modern etch technology, advances in metrology, and control strategies have enabled this selective etch process to reach mainstream production, as demonstrated in Intel's 22 nm technology [90]. Once the recess is performed, a SiN capping layer is formed in a damascene-like process. The SAC etch is then executed using a process that is highly selective to SiN and Si, allowing the contact to land on the source and drain of the transistor without risk of shorting to the gate. An example of an RMG process featuring an SAC structure is shown in Figure 1.31.

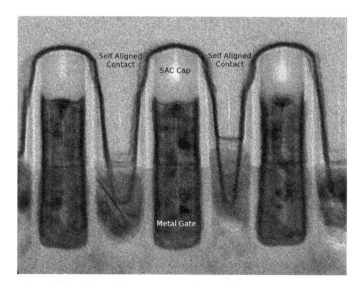

FIGURE 1.31 TEM image showing misaligned contacts avoiding the gate and contacting the S/D regions. The integration of the SAC cap can be seen directly above the gate electrode.

SADP was explored extensively for dynamic random-access memory (DRAM) production as a means to extend an incumbent lithography technology without requiring a reduction in λ. Based on a survey of the patent literature on this topic, one might conclude that interest in SADP spiked each time the industry struggled with the challenges of λ reduction. Examples of SADP or SADP-like processing can be found from the transition from i-line to 248 nm, 248 nm to 193 nm, 193 nm to 193 nm immersion, and so on. Widespread adoption of SADP for production applications finally arrived in the late 2000s when the demands of pitch resolution finally outstripped the pace of progress in wavelength reduction or the advent of further resolution enhancement techniques. In particular, two events coincided that moved this technology forward. First, the memory industry reached the limit of what was possible with 193 nm immersion and resolution enhancement technologies (RET) long before EUV scanners were ready for production [91]. Memory arrays feature regular grating-like patterns that naturally lend themselves to frequency doubling techniques. The transition from planar CMOS devices to the FinFET represents the second critical event that pushed SADP into mainstream production [89].

Traditionally, the lower-level metals in a CMOS circuit represented the densest and most challenging patterns to print. In the FinFET device, the pitch at which the fins are repeated is an important parameter that influences parasitic capacitance [92] and circuit density [93]. The first commercial application of FinFETs, Intel's 22 nm logic technology [90], illustrated a paradigm shift in patterning technology, as they found that a fin pitch denser than the metal pitch was required to deliver optimum density and performance. In this technology, a 60 nm fin pitch was leveraged in both the logic and memory portions of the chip. This density requirement exceeded the pitch limit of 193 nm immersion lithography (80 nm). While pitch split could theoretically deliver patterns of similar density, SADP was the logical choice due to its superior control of CD and pattern placement. Both these parameters are critical for producing a FinFET device with acceptable variation [94].

The concept of SADP has been extended to self-aligned multiple patterning, with demonstrations of quadruple (SAQP) and octuple (SAOP) demonstrating features' pitches down to 8 nm half-pitch [95]. Demonstrations of fin pitch below 30 nm using SAQP have been reported, illustrating the extendability as well as some of the challenges associated with this patterning approach [96].

Ultimately, SAQP and further extensions of self-aligned multiple patterning are limited by the accuracy of the lithography required to transform these patterns from grating-like structures into useful circuit patterns. While the ability to form denser gratings with these techniques is limited primarily by unit process technology and control methodology, customizing these patterns requires increasingly accurate lithography to form block or cut masks. Various schemes for enhancing the overlay margin have been proposed. For example, the use of varying-width mandrels was demonstrated early in FinFET technology development as a way to improve the overlay margin of the cut mask [97]. Ultimately, a method for self-aligning the cut or block mask is required to truly leverage the pitch resolution capabilities presented by SADP. While details of the process were not disclosed, a process for self-aligning a block mask to trenches created by a 24 nm-pitch SAQP process was demonstrated in 2013 [98]. Further examples of self-aligned customization schemes were later discussed in detail by researchers at Tokyo Electron Limited [99], clearly illustrating a potential path forward.

1.5 REACTIVE ION ETCH AND DEPOSITION PROCESSES

As is clearly evident in the previous section, RIE, plasma-enhanced chemical vapor deposition (PECVD), and atomic layer deposition (ALD) have become an integral part of modern patterning solutions. While SADP processes predate the invention of ALD, various forms of thermal and plasma-enhanced ALD (PEALD) have become the standard process technology used in SADP patterning to deposit the offset spacer. This stems from the superior conformality inherent in ALD processes. In addition, the ability to deposit high-quality films at low process temperatures is an advantage over CVD processes because it enables the use of a wide range of mandrel materials [100].

The importance of RIE in pattern transfer has been covered in great detail by many authors. The reader is encouraged to read the review paper by Donnelly and Kornblit [101] for an in-depth exploration of RIE applications in semiconductor processing. The demand for innovation in RIE has increased significantly, in part due to the reliance on self-aligned patterning schemes. In many of these patterning approaches, RIE and deposition processes determine the final dimension of the feature of interest. Advanced RIE processes can correct for systematic CD errors in the lithographically defined patterns. RIE can be used to reduce line width or narrow trench patterns. In addition, RIE can reduce LER and improve contact or via uniformity. Consequently, overall CD control has become increasingly complex and can no longer be treated as a strictly lithographic issue.

1.5.1 REACTIVE ION ETCH

RIE has played a central role in patterning for semiconductors for over 30 years. For many years, the most commonly used plasma reactor produced a capacitively coupled plasma (CCP) discharge. In the CCP reactor, two electrodes are arranged parallel to each other like a simple capacitor. A modest vacuum is created in the reactor, and a partial pressure of process gas is introduced through a nozzle or through a shower head pattern in one of the electrodes. A DC or, more commonly, an AC plasma discharge is created between the electrodes containing reactive neutrals and a sheath region with an induced electric field through which ions accelerate toward the surface of one of the electrodes. By placing the substrate on this electrode, the directionality of the electric field can be used to enhance anisotropy. For a detailed treatment of plasma fundamentals, the reader is encouraged to refer to Lieberman and Lichtenberg's textbook on plasma processing [102]. CCP etch systems have evolved significantly since their introduction, adding decoupled sources and one or more high-frequency excitations. This enhances control of the plasma generation and provides a means to modulate the electric field in the plasma sheath. Pulsed plasma technology has also been explored to enhance etch selectivity by achieving a finer degree of control of the ion energies and chemical species present in the plasma.

Another class of RIE reactors leverages high-density plasma (HDP) discharges to increase radical flux and produce enhanced control over the plasma dynamics. HDP sources feature advanced reactor designs that decouple plasma density from substrate bias. This technique improves the achievable selectivity of RIE by reducing the energy of the ions impinging on the wafer reducing the contribution of physical etching due to ion sputtering. Inductively coupled plasma (ICP) and evanescent wave plasma (EWP) afford methods of increasing plasma density, offering further selectivity enhancements with even lower substrate damage.

As dimensions continue to scale, atomic-level precision is mandatory. Atomic layer etch (ALE) is a term frequently used in an attempt to compare all chemical reaction–limited etch gas–phase etch processes to ALD. This comparison is not entirely accurate in that these etches are generally "very slow" but currently are not truly self-limiting. In contrast, ALD achieves atomic precision by alternating doses of precursor and catalyst, creating a true self-limiting process at the expense of throughput.

In addition to plasma-type gas flow, chuck configuration can play a critical role in the uniformity and selectivity of the overall RIE process. Gas distribution has become increasingly complex, evolving from a single gas mixture injection to various means of across-electrode gas distribution, which can produce custom gas mixtures as a function of position across the substrate [103]. Additionally, pulsed gas technology, sometimes referred to as quasi-atomic layer etch (q-ALE), offers a further method to improve selectivity [104].

As discussed previously in this chapter, SAV and SAC processes require highly selective RIE. There are several methods that can be employed to enhance the selectivity of this module beyond film choice in the integrated structure. As discussed earlier, reactor design, discharge type, and

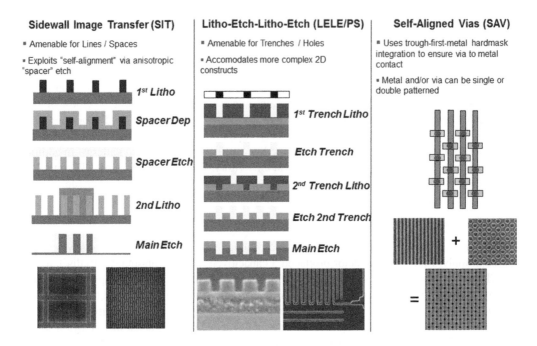

FIGURE 1.32 Self-aligned double patterning (SADP), pitch split litho-etch-litho-etch (LELE-PS), self-aligned via (SAV). (Adapted from Colburn, M., *Lithography Solutions for the 22nm Node*, VLSI, Honolulu, HI, 2009.)

plasma chemistry all play a role. These advanced etch processes also frequently leverage high degrees of polymerization to achieve a selectivity enhancement during the etch process. Such processes frequently require unique post-RIE cleans, which may also be plasma based.

RIE plays a pivotal role in the SADP process. Similarly to a standard patterning process, RIE may be used to transfer the mandrel lithography into one or more films present on the wafer, as shown in Figure 1.32. Following a film deposition using ALD or some similar technique, RIE is used to create the discrete offset spacer-like patterns. The lithographically defined mandrel sets the position of the spacer-defined patterns. However, RIE and deposition processes determine the final dimension of the feature of interest. An additional RIE, plasma etch, or wet-etch process is required to remove the original mandrel. A final RIE is used to transfer the spacer patterns in a highly anisotropic fashion, maintaining and in some cases adjusting the final CD of the SADP pattern. An RIE, plasma etch. or wet-etch process is then used to remove the mandrel prior to final pattern formation. A final RIE is used to transfer the spacer patterns in a highly anisotropic fashion maintaining and in some cases adjusting the final CD of the SADP pattern.

1.6   EXTREME ULTRAVIOLET (EUV) LITHOGRAPHY

As this book goes to press, EUV lithography is being introduced into foundries and internal device manufacturer lines. Consequently, one must discuss EUV technology, as it has entered mainstream lithography usage after decades of development and heavy investment from the industry. As discussed previously in this chapter, dimensional scaling and pitch reduction in semiconductor manufacturing have been driven in large part by reducing wavelength and increasing NA of lithography scanners. The improvement in resolution, R, is described by the Rayleigh equation:

$$R = \frac{k_1 \lambda}{\text{NA}} = \frac{0.25\lambda}{n\sin\theta} \tag{1.16}$$

For immersion scanners using 193 nm light, having an NA of 1.35, the practical pitch resolution is limited to 80 nm or a 40 nm line and space (L/S). For initial EUV scanners using 13.5 nm light and a 0.33 NA lens, the resolution is approximately 30 nm pitch or a 15 nm L/S, assuming $k_1 = 0.36$. There are several key differences that distinguish EUV systems from conventional 193 nm projection lithography, as shown in Figure 1.33.

In EUV, photons are generated using laser-pulsed plasma (LPP) technology [105], in which a continuous stream of tin (Sn) microdroplets is irradiated by focused pulses of a high-power CO_2 laser. These pulses expand and shape the droplet just prior to excitation and emission of 13.5 nm light. The resultant optical radiation is projected into the scanner's illumination system. Due to strong absorption at EUV wavelengths, the lens consists only of reflective elements contained within a high vacuum [106]. This makes the design and transmission of the light more difficult and less efficient than in 193 nm scanners.

Conventional masks used in lithography are operated in a transmission mode and illuminated telecentrically. In contrast, the EUV mask is a Si-Mo multilayer Bragg reflector containing a patterned absorber that is illuminated in a nontelecentric fashion [107]. This results in complex differences in horizontal and vertical imaging as well as differences in contrast between two edges. The objective lens collects the diffracted light within the 0.33 NA and projects it onto the wafer stage, forming the desired pattern on the wafer. At present, the total transmission of an EUV system is estimated as 4% of the generated EUV light [108].

There are three primary challenges in EUV lithography technology: source, mask, and resist [109]. The fundamental issue of the source stems from the inherently low conversion efficiency of CO_2 power to EUV radiation. As noted previously, the efficiency is relatively low, measuring 2.5–4% [110]. The portion of the energy that is not converted results in Sn debris. Sn is also collected within the collector optics in designed locations. Collectively, successful tin and debris mitigation strategies are critical to implementing EUV patterning [111], as this issue factors heavily into scanner stability and availability for production(i.e., uptime).

Mask fabrication remains a critical challenge for EUV. Unlike 193 nm masks, any pit or bump in the starting mask blank can generate a significant phase error. Consequently, well-characterized mask blanks with defined spatial maps are required, as zero-defect blanks are not practical economically. The Mo-Si multilayer used to form the Bragg reflector on the mask must be deposited in a defect-free and highly uniform manner. Upon this film, the absorber layer is deposited and ultimately patterned with the design information required for the image to be printed. The absorber

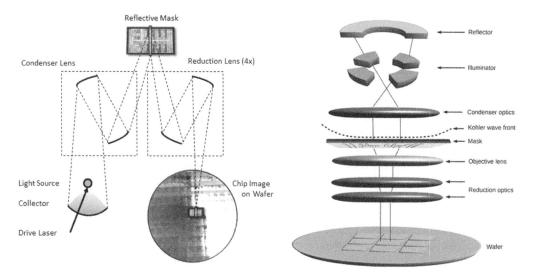

FIGURE 1.33   Schematic comparison of EUV and conventional 193 projection optics scanner.

itself can be patterned and repaired akin to conventional masks. However, a higher degree of precision is required, as necessitated by the higher resolving power of EUV. At the time of this writing, there are no viable approaches for repairing defects in the multilayer reflector. Once the pattern is etched into the mask, inspection of the resulting features represents an additional final challenge. An actinic inspection tool has been under development for some time. Inspection by scanning electron microscope (SEM)-based techniques is another strategy being pursued. Throughput with a single-SEM system is insufficient to meet the demands of large-scale mask production. SEM systems featuring various throughput improvements ranging from multiple columns to multiple beams operating within a single column are under development [112]. Ultimately, the challenges with the masks and blanks reduce to an economic challenge. The cost of entry into mask production is high, and market demand is limited to a few large customers. For the foreseeable future, we expect that optical inspection techniques will remain the upfront approach to inspecting EUV masks.

In the area of resist, EUV poses a unique challenge. The energy of the EUV photon is large relative to 193 nm, and consequently, for a given dose, shot noise becomes a larger contributor to overall variation [113]. Conventional resists have a photo-speed slower than desired for high-throughput production and have lacked the 2-D resolution that the scanners are fundamentally capable of imaging. Encouraging demonstrations of improved resolution at the expense of moderately degraded photospeed have been reported [114]. Even the most promising imaging materials are limited in terms of the aspect ratio they can produce with a reasonable pattern collapse margin. A host of inorganic and organic-inorganic hybrid resist platforms containing metal oxides are being proposed and developed [115]. These materials may help address some issues, including photospeed, LER, and pattern collapse, in part by providing a thinner but higher–etch resistance film.

In spite of these challenges, the results produced with EUV lithography have continued to justify the complexity of developing this technology. Numerous demonstrations of circuit patterning beyond the capabilities of 193 nm lithography have been reported [108]. More recently, electrical data from processes using EUV has been reported. Electrical data from interconnect structures produced using multiple patterning and EUV at a 10 nm technology ground rule demonstrated significantly lower variation in the EUV-based process. Demonstration of a 7 nm logic technology employing EUV extensively in many process modules has also been reported [96, 116]. This work featured single expose EUV in the contact patterning, replacing as many as five optical masks in an equivalent 193i multiple patterning approach.

1.7 SUMMARY

Semiconductor lithography and patterning process optimization have been the primary drivers for scaling over the past 50 years. The industry has adapted to challenges in device performance and power reduction by transitioning to new devices and implementing new material solutions. In many cases, this has driven the need for new innovations in patterning that go well beyond simple resolution enhancement. This has been particularly evident in the past. Integrated patterning solutions such as intralevel SADP and interlevel self-aligned integration strategies such as SAC and SAV are required to meet the process tolerance demands of ever more stringent technology definitions. Further, while pitch reduction remains an important component in the development of new patterning solutions, tolerance control and variability reduction are now the primary drivers in selecting the appropriate patterning solutions. With the advent of EUV, and the long-awaited reduction in wavelength, we will see the compression of multipatterning mask count to reduce the complexity of semiconductor chip integration. Commensurate with a second generation of EUV chips, the industry will incorporate novel device architectures such as nanowires, nanosheets, and potential vertical integration. Many of the techniques developed to extend 193 nm lithography are directly applicable to EUV. Beyond multiple exposure techniques and EUV-driven SADP, resolution enhancement technologies (RET) strategies for EUV employing SMO have already been demonstrated [107].

Cumulatively, these strategies offer a way to extend patterning to a point beyond where conventional semiconductor devices and interconnects will function.

Predictions of the end of Moore's law have been notoriously inaccurate, much like the predictions of the limits of lithography. As long as there is a value to scaling and advanced technology, the semiconductor industry will continue to develop novel patterning approaches to improve power and performance. A transition from dimensional scaling with a single level of CMOS devices may come to an end. However, the demand for lower-power products with higher storage capacity and improved processing efficiency will continue unabated. There is no apparent limit to the demand for data and advanced technology solutions to process increasingly complex data sets. This sobering fact will drive innovation in device and circuit technology and, in turn, continue to push the limits of lithography, etch, and Si process technology without end.

REFERENCES

1. Moore, G. (1965). Cramming more components onto integrated circuits. *Electronics, 38*, 114–117.
2. Dennard, R. H., Gaensslen, F. H., Rideout, V. L., Bassous, E., & LeBlanc, A. R. (1974). Design of ion-implanted MOSFET's with very small physical dimensions. *IEEE Journal of Solid-State Circuits, 9*(5), 256–268.
3. Frank, D. J., Dennard, R. H., Nowak, E., Solomon, P. M., Taur, Y., & Wong, H.-S. P. (2001, March). Device scaling limits of Si MOSFETs and their application dependencies. *Proceedings of the IEEE, 89*(3), 259–288.
4. Krishnan, S., Kwon, U., Moumen, N., Stoker, M. W., Harley, E. C. T., Bedell, S., & Nair, D. (2011). A manufacturable dual channel (Si and SiGe) high-k metal gate CMOS technology with multiple oxides for high performance and low power applications. In: *IEEE International Electron Devices Meeting* (pp. 1–28).
5. Mead, C., & Conway, L. (1980). *Introduction to VLSI systems*. Reading, MA: Addison-Wesley.
6. Ghani, T., Armstrong, M., & Auth, C. (2003). A 90nm high volume manufacturing logic technology featuring novel 45nm gate length strained silicon CMOS transistors. In: *IEEE International Electron Devices Meeting* (11.6.1–11.6.3).
7. Lee, M. L., Fitzgerald, E. A., Bulsara, M. T., Currie, M. T., & Lochtefeld, A. (2005). Strained Si, SiGe, and Ge channels for high-mobility metal-oxide-semiconductor field-effect transistors. *Journal of Applied Physics, 97*(1), 011101.
8. Hu, C. (1996, December). Gate oxide scaling limits and projection. In: *International Electron Devices Meeting. Technical Digest* (pp. 319–322).
9. Khare, M. (2007, September). High-k/metal gate technology: A new horizon. In: *2007 IEEE Custom Integrated Circuits Conference* (pp. 417–420).
10. Bai, P., Auth, C., Balakrishnan, S., Bost, M., Brain, R., Chikarmane, V., ... Bohr, M. (2004, December). A 65nm logic technology featuring 35nm gate lengths, enhanced channel strain, 8 Cu interconnect layers, low-k ILD and 0.57 μm^2 SRAM cell. In: *IEEE International Electron Devices Meeting, 2004. IEDM Technical Digest* (pp. 657–660).
11. Frank, D. J., Laux, S. E., & Fischetti, M. V. (1992, December). Monte Carlo simulation of a 30 nm dual-gate MOSFET: How short can Si go? In: *1992 International Technical Digest on Electron Devices Meeting* (pp. 553–556).
12. Wong, H. S. P., Frank, D. J., & Solomon, P. M. (1998, December). Device design considerations for double-gate, ground-plane, and single-gated ultra-thin SOI MOSFET's at the 25 nm channel length generation. In: *International Electron Devices Meeting 1998. Technical Digest (Cat. No. 98ch36217)* (pp. 407–410).
13. Guarini, K. W., Solomon, P. M., Zhang, Y., Chan, K. K., Jones, E. C., Cohen, G. M., ... Wong, H. S. (2001, December). Triple-self-aligned, planar double-gate MOSFETs: Devices and circuits. In: *International Electron Devices Meeting. Technical Digest (Cat. No. 01ch3722A)* (p. 19.2.1–19.2.4).
14. Huang, X., Lee, W.-C., Kuo, C., Hisamoto, D., Chang, L., Kedzierski, J., ... Hu, C. (1999, December). Sub 50-nm FinFET: PMOS. In: *International Electron Devices Meeting 1999. Technical Digest (Cat. No. 99ch36318)* (pp. 67–70).
15. Kedzierski, J., Fried, D. M., Nowak, E. J., Kanarsky, T., Rankin, J. H., Hanafi, H., ... Wong, H. S. P. (2001, December). High-performance symmetric-gate and CMOS-compatible v/sub t/ asymmetric-gate FinFET devices. In: *International Electron Devices Meeting. Technical Digest (Cat. No.01ch37224)* (pp. 19.5.1–19.5.4).

16. Kavalieros, J., Doyle, B., Datta, S., Dewey, G., Doczy, M., Jin, B., ... Chau, R. (2006). Tri-gate transistor architecture with high-k gate dielectrics, metal gates and strain engineering. In: *2006 Symposium on VLSI Technology, 2006. Digest of Technical Papers* (pp. 50–51).

17. Auth, C., Allen, C., Blattner, A., Bergstrom, D., Brazier, M., Bost, M., ... Mistry, K. (2012, June). A 22nm high performance and low-power CMOS technology featuring fully-depleted tri-gate transistors, self-aligned contacts and high density MIM capacitors. In: *2012 Symposium on VLSI Technology (VLSIT)* (pp. 131–132).

18. Krishnan, M., Nalaskowski, J. W., & Cook, L. M. (2010). Chemical mechanical planarization: Slurry chemistry, materials, and mechanisms. *Chemical Reviews, 110*(1), 178–204. Retrieved from https://doi.org/10.1021/cr900170z (PMID: 19928828).

19. Edelstein, D., Heidenreich, J., Goldblatt, R., Cote, W., Uzoh, C., Lustig, N., ... Slattery, J. (1997, December). Full copper wiring in a sub-0.25 /spl μm CMOS ULSI technology. In: *International Electron Devices Meeting. IEDM Technical Digest* (pp. 773–776).

20. Hu, C.-K., Gignac, L., Rosenberg, R., Liniger, E., Rubino, J., Sambucetti, C., Domenicucci, A., Chen, X., & Stamper, A. K. (2002). Reduced electromigration of Cu wires by surface coating. *Applied Physics Letters, 81*(10), 1782–1784.

21. Colburn, M. (2015). Surmounting industry inflection points. In: SPIE (Ed.), *Paper presented at SPIE Advanced Etch Conference Plenary*. Bellingham, WA.

22. Nitta, S., Ponoth, S., Breyta, G., Colburn, M., Clevenger, L., Horak, D., ... Edelstein, D. (2008). A multilevel copper/low-k/airgap BEOL technology. In: *Advanced Metallization Conference* (pp. 329–336).

23. Gosset, L., Farcy, A., de Pontcharra, J., Lyan, P., Daamen, R., Verheijden, G., ... Torres, J. (2005). Advanced Cu interconnects using air gaps. *Microelectronic Engineering, 82*(3), 321–332. Retrieved from http://www.sciencedirect.com/science/article/pii/S0167931705003618 (*Proceedings of the Ninth European Workshop on Materials for Advanced Metallization 2005*).

24. Natarajan, S., Agostinelli, M., Akbar, S., Bost, M., Bowonder, A., Chikarmane, V., ... Zhang, K. (2014, December). A 14nm logic technology featuring 2nd-generation FinFET, air-gapped interconnects, self-aligned double patterning and a 0.0588 μm² SRAM cell size. In: *2014 IEEE International Electron Devices Meeting* (pp. 3.7.1–3.7.3).

25. Pyzyna, A., Bruce, R., Lofaro, M., Tsai, H., Witt, C., Gignac, L., ... Park, D. G. (2015, June). Resistivity of copper interconnects beyond the 7 nm node. In: *2015 Symposium on VLSI Technology (VLSI Technology)* (pp. T120–T121).

26. Taur, Y., & Ning, T. H. (2004). *Fundamentals of modern VLSI devices*. New York: Cambridge University Press.

27. Planes, N., Weber, O., Barral, V., Haendler, S., Noblet, D., Croain, D., ... Haond, M. (2012, June). 28 nm FDSOI technology platform for high-speed low-voltage digital applications. In: *2012 Symposium on VLSI Technology (VLSIT)* (pp. 133–134).

28. Collaert, N., Alian, A., Arimura, H., Boccardi, G., Eneman, G., Franco, J., ... Thean, A. V.- Y. (2015). Ultimate nano-electronics: New materials and device concepts for scaling nano-electronics beyond the Si roadmap. *Microelectronic Engineering, 132*, 218–225.

29. Waldron, N., Merckling, C., Guo, W., Ong, P., Teugels, L., Ansar, S., ... Thean, A. V. Y. (2014, June). An InGaAs/InP quantum well finfet using the replacement fin process integrated in an RMG flow on 300mm Si substrates. In: *2014 Symposium on VLSI Technology (VLSI-Technology): Digest of Technical Papers* (pp. 1–2).

30. Rossnagel, S. M., & Kuan, T. S. (2002). Time development of microstructure and resistivity for very thin Cu films. *Journal of Vacuum Science and Technology, Part A, 20*(6), 1911–1915.

31. Lloyd, J. R., Liniger, E., & Shaw, T. M. (2005). Simple model for time-dependent dielectric breakdown in inter- and intralevel low-k dielectrics. *Journal of Applied Physics, 98*(8), 084109. Retrieved from https://doi.org/10.1063/1-2112171.

32. Yeric, G. (2011, December). Technology roadmaps and low power SoC design. In: *2011 International Electron Devices Meeting* (pp. 15.4.1–15.4.4).

33. Liebmann, L., Pileggi, L., Hibbeler, J., Rovner, V., Jhaveri, T., & Northrop, G. (2009). Simplify to survive: Prescriptive layouts ensure profitable scaling to 32nm and beyond. In: (Vol. 7275, pp. 7275–7275-9). Retrieved from https://doi.org/10.1117/12.814701.

34. Northrop, G. (2011, June). Design technology co-optimization in technology definition for 22nm and beyond. In: *2011 Symposium on VLSI Technology – Digest of Technical Papers* (pp. 112–113).

35. Liebmann, L., Chu, A., & Gutwin, P. (2015). The daunting complexity of scaling to 7nm without EUV: Pushing DTCO to the extreme. In: (Vol. 9427, pp. 9427–9427-12). Retrieved from https://doi.org/10.1117/12.2175509.

36. Colburn, M. (2009). *Lithography solutions for the 22nm node*. Honolulu, HI: VLSI Short Course.

37. Matsuyama, T., Williamson, D. M., & Ohmura, Y. (2006). The lithographic lens: Its history and evolution. In: (Vol. 6154, pp. 6154–6154-14). Retrieved from https://doi.org/10.1117/12.656163.

38. Brunner, T. A., Corliss, D. A., Butt, S. A., Wiltshire, T. J., Ausschnitt, C. P., & Smith, M. D. (2006). Laser bandwidth and other sources of focus blur in lithography. *Journal of Micro/Nanolithography, MEMS, and MOEMS, 5*(4), 5–5-7. Retrieved from https://doi.org/10.1117/1-2396926.

39. Rokitski, R., Rafac, R., Melchior, J., Dubi, R., Thornes, J., Cacouris, T., ... Brown, D. (2013). High power 120 W ArF immersion XLR laser system for high dose applications. In: (Vol. 8683, pp. 8683–8683-7). Retrieved from https://doi.org/10.1117/12.2012681.

40. Rosenbluth, A. E., Bukofsky, S. J., Fonseca, C. A., Hibbs, M. S., Lai, K., Molless, A. F., ... Wong, A. K. K. (2002). Optimum mask and source patterns to print a given shape. *Journal of Micro/Nanolithography, MEMS, and MOEMS, 1*, 1–1-18. Retrieved from https://doi.org/10.1117/1-1448500.

41. Goodman, J. (1968). *Introduction to Fourier optics.* San Francisco, CA: McGraw-Hill.

42. Born, M., & Wolf, E. (1980). *Principles of optics: Electromagnetic theory of propagation and interference of light,* 6th edition. San Francisco, CA: McGraw-Hill.

43. Jenkins, F. A., & White, H. E. (1957). *Fundamentals of optics.* New York: McGraw-Hill.

44. Mack, C. (2007). *Fundamental principles of optical lithography.* Chichester: John Wiley and Sons, Ltd.

45. Bahaa, E., Saleh, A., & Teich, M. (1991). *Fundamentals of photonics.* New York: John Wiley and Sons Inc.

46. Hopkins, H. (1953). On the diffraction theory of optical images. *Proceedings of the Royal Society of London Series A, 217*(1130), 408–432. Retrieved from http://rspa.royalsocietypublishing.org/content/217/1130/408.

47. Toh, K. K. (1988). Two dimensional images with effects of lens aberrations in optical lithography (Tech. Rep. No. UCB/ERL M88/30). Berkeley, CA: University of California at Berkeley.

48. Flagello, D. G., Mulkens, J., & Wagner, C. (2000). Optical lithography into the millennium: Sensitivity to aberrations, vibration and polarization. In: (Vol. 4000, pp. 4000–4000-12).

49. Rosenbluth, A. E., Melville, D. O., Tian, K., Bagheri, S., Tirapu-Azpiroz, J., Lai, K., ... Granik, Y. (2009). Intensive optimization of masks and sources for 22nm lithography. In: (Vol. 7274, pp. 7274–7274-15). Retrieved from https://doi.org/10.1117/12.814844.

50. Mulder, M., Engelen, A., Noordman, O., Streutker, G., van Drieenhuizen, B., van Nuenen, C., ... McIntyre, G. (2010). Performance of flexray: A fully programmable illumination system for generation of freeform sources on high NA immersion systems. In: (Vol. 7640, pp. 7640–7640-10). Retrieved from https://doi.org/10.1117/12.845984.

51. Himel, M. D., Hutchins, R. E., Colvin, J. C., Poutous, M. K., Kathman, A. D., & Fedor, A. S. (2001). Design and fabrication of customized illumination patterns for low-k1 lithography: A diffractive approach. In: (Vol. 4346, pp. 4346–4346-7). Retrieved from https://doi.org/10.1117/12.435682.

52. Starikov, A. (1989). Use of a single size square serif for variable print bias compensation in microlithography: Method, design, and practice. In: (Vol. 1088, pp. 1088–1088-14). Retrieved from https://doi.org/10.1117/12.953132.

53. Cobb, N. (1995). Sum of coherent systems decomposition by singular value decomposition (dissertation, Department of Electrical Engineering and Computer Science, University of California at Berkeley). Retrieved from http://www-video.eecs.berkeley.edu/papers/ncobb/socs.pdf.

54. Stirniman, J. P., & Rieger, M. L. (1994). Fast proximity correction with zone sampling. In: (Vol. 2197, pp. 2197–2197-8). Retrieved from https://doi.org/10.1117/12.175423.

55. Gabor, D. (1946). Theory of communication. *Journal of Institute of Electrical Engineers, 93*, 429–457.

56. Cobb, N., & Granik, Y. (2004). New concepts in OPC. In: (Vol. 5377, pp. 5377–5377-11). Retrieved from https://doi.org/10.1117/12.535605.

57. Cobb, N. (2005). Flexible sparse and dense OPC algorithms. In: (Vol. 5853, pp. 5853–5853-10). Retrieved from https://doi.org/10.1117/12.617198.

58. Granik, Y. (2003). Dry etch proximity modeling in mask fabrication. In: (Vol. 5130, pp. 5130–5130-6). Retrieved from https://doi.org/10.1117/12.504052.

59. Dunn, D. N., Mansfield, S., Stobert, I., Sarma, C., Lembach, G., Liu, J., & Herold, K. (2009). Etch aware optical proximity correction: A first step toward integrated pattern engineering. In: (Vol. 7274, pp. 7274–7274-9). Retrieved from https://doi.org/10.1117/12.814224.

60. Stobert, I., & Dunn, D. (2013). Etch correction and OPC: A look at the current state and future of etch correction. In: (Vol. 8685, pp. 8685–8685-11). Retrieved from https://doi.org/10.1117/12.2015000.

61. Hecht, E. (2017). *Optics,* 5th edition. Essex: Pearson Education Limited.

62. Mack, C. A., Harrison, D., Rivas, C., & Walsh, P. (2007). Impact of thin film metrology on the lithographic performance of 193-nm bottom antireflective coatings. In: (Vol. 6518, pp. 6518–6518-16). Retrieved from https://doi.org/10.1117/12.711488.

63. Burns, S., Pfeiffer, D., Mahorowala, A., Petrillo, K., Clancy, A., Babich, K., ... Larson, C. (2006). Silicon containing polymer in applications for 193 nm high NA lithography processes. In: (Vol. 6153, pp. 6153–6153-12). Retrieved from https://doi.org/10.1117/12.657197.

64. Levinson, H. J. (2005). *Principles of lithography*, 2nd edition. Bellingham, WA: SPIE Press.

65. Okoroanyanwu, U. (2015). *Molecular theory of lithography*. Bellingham, WA: International Society for Optics and Photonics.

66. Saulnier, N., Xu, Y., Wang, W., Sun, L., Cheong, L. L., Lallement, R., ... Colburn, M. (2015). EUV processing and characterization for BEOL. In: (Vol. 9422, pp. 9422–9422-12). Retrieved from https://doi.org/10.1117/12.2086126.

67. May, G. S., & Spanos, C. J. (2006). *Fundamentals of semiconductor manufacturing and process control*. Hoboken, NJ: John Wiley and Sons.

68. Ito, H., Willson, C. G., & Frechet, J. M. (1985). Positive-and negative-working resist compositions with acid generating photoinitiator and polymer with acid labile groups pendant from polymer backbone (US 4491628).

69. Allen, R., Wan, I. Y., Wallraff, G. M., Dipietro, R. A., Hofer, D. C., & Kunz, R. R. (1995). Resolution and etch resistance of a family of 193nm positive resists. *Journal of Photopolymer Science and Technology*, *8A*, 623–626.

70. Ito, H., Wallraff, G. M., Brock, P. J., Fender, N., Truong, H. D., Breyta, G., ... Allen, R. D. (2001). Polymer design for 157-nm chemically amplified resists. In: (Vol. 4345, pp. 4345–4345-12). Retrieved from https://doi.org/10.1117/12.436857.

71. Ogawa, T., Sekiguchi, A., & Yoshizawa, N. (1996). Advantages of a SiO_xN_y:H anti-reflective layer for ArF excimer laser lithography. *Japanese Journal of Applied Physics*, *35*(12S), 6360. Retrieved from http://stacks.iop.org/1347-4065/35/i=12S/a=6360.

72. Lin, B.-J. (2004). Immersion lithography and its impact on semiconductor manufacturing. In: (Vol. 5377, pp. 5377–5377-22). Retrieved from https://doi.org/10.1117/12.534507.

73. Gil, D., Bailey, T., Corliss, D., Brodsky, M. J., Lawson, P., Rutten, M., ... Robinson, C. (2004). First microprocessors with immersion lithography. In: (Vol. 5754, pp. 5754–5754-10). Retrieved from https://doi.org/10.1117/12.598855.

74. Kocsis, M., Heuvel, D. V. D., Gronheid, R., Maenhoudt, M., Vangoidsenhoven, D., Wells, G., ... Streefkerk, B. (2006). Immersion specific defect mechanisms: Findings and recommendations for their control. In: (Vol. 6154, pp. 6154–6154-12). Retrieved from https://doi.org/10.1117/12.660432.

75. Gronheid, R., Ercken, M., & Tenaglia, E. (2006). Dynamic leaching procedure on an immersion interference printer. In: (Vol. 6154, pp. 6154–6154-11). Retrieved from https://doi.org/10.1117/12.684420.

76. Rathsack, B. M., Scheer, S., Kuwahara, Y., Kitano, J., Gronheid, R., & Baerts, C. (2008). Finite element modeling of PAG leaching and water uptake in immersion lithography resist materials. In: (Vol. 6923, pp. 6923–6923-11). Retrieved from https://doi.org/10.1117/12.772850.

77. Sanders, D. P., Sundberg, L. K., Sooriyakumaran, R., Brock, P. J., DiPietro, R. A., Truong, H. D., ... Allen, R. D. (2007). Fluoro-alcohol materials with tailored interfacial properties for immersion lithography. In: (Vol. 6519, pP. 6519–6519-12). Retrieved from https://doi.org/10.1117/12.712768.

78. Tarutani, S., Kamimura, S., & Hideaki, T. (2009). Development of materials and processes for negative tone development toward 32-nm node 193-nm immersion double-patterning process. In: (Vol. 7273, pP. 7273–7273-8). Retrieved from https://doi.org/10.1117/12.814093.

79. Brunner, T. A., & Fonseca, C. A. (2001). Optimum tone for various feature types: Positive versus negative. In: (Vol. 4345, pP. 4345–4345-7). Retrieved from https://doi.org/10.1117/12.436866.

80. Brunner, T. A., Fonseca, C., Seong, N., & Burkhardt, M. (2004). Impact of resist blur on MEF, OPC, and CD control. In: (Vol. 5377, pp. 5377–5377-9). Retrieved from https://doi.org/10.1117/12.537472.

81. Landie, G., Xu, Y., Burns, S., Yoshimoto, K., Burkhardt, M., Zhuang, L., ... Vohra, V. (2011). Fundamental investigation of negative tone development (NTD) for the 22nm node (and beyond). In: (Vol. 7972, pp. 7972–7972-12). Retrieved from https://doi.org/10.1117/12.882843.

82. Hori, M., Nagai, T., Nakamura, A., Abe, T., Wakamatsu, G., Kakizawa, T., ... Shimokawa, T. (2008). Sub-40-nm half-pitch double patterning with resist freezing process*Proceedings, Advances in Resist Materials and Processing Technology*, pp. 6923–6928. Retrieved from https://doi.org/10.1117/12.772403.

83. Hazelton, A. J., Wakamoto, S., Hirukawa, S., McCallum, M., Magome, N., Ishikawa, J., ... Gaugiran, S. (2009). Double-patterning requirements for optical lithography and prospects for optical extension without double patterning. *Journal of Micro/Nanolithography, MEMS, and MOEMS*, *8*, 8–8.11. Retrieved from https://doi.org/10.1117/1-3023077.

84. seng Koay, C., Colburn, M. E., Izikson, P., Robinson, J. C., Kato, C., Kurita, H., & Nagaswami, V. (2010). Automated optimized overlay sampling for high-order processing in double patterning lithography. In: (Vol. 7638, pp. 7638–7638-10). Retrieved from https://doi.org/10.1117/12.846371.

85. Wei, M., Banerjee, R., Zhang, L., Masad, A., Reidy, S., Ahn, J., … Fazio, A. (2007, June). A scalable self-aligned contact NOR flash technology. In: *2007 IEEE Symposium on VLSI Technology* (pp. 226–227).

86. Narasimha, S., Chang, P., Ortolland, C., Fried, D., Engbrecht, E., Nummy, K., … Agnello, P. (2012, December). 22nm high-performance SOI technology featuring dual-embedded stressors, Epi-Plate High-K deep-trench embedded DRAM and self-aligned Via 15LM BEOL. In: *2012 International Electron Devices Meeting* (pp. 3.3.1–3.3.4).

87. Bencher, C., Chen, Y., Dai, H., Montgomery, W., & Huli, L. (2008). 22nm half-pitch patterning by CVD spacer self alignment double patterning (SADP). In: (Vol. 6924, pp. 6924–6924-7). Retrieved from https://doi.org/10.1117/12.772953.

88. Seo, S. C., Edge, L. F., Kanakasabapathy, S., Frank, M., Inada, A., Adam, L., … Paruchuri, V. K. (2011, June). Full metal gate with borderless contact for 14 nm and beyond. In: *2011 Symposium on VLSI Technology – Digest of Technical Papers* (pp. 36–37).

89. Auth, C. (2008, September). 45nm high-k+ metal gate strain-enhanced CMOS transistors. In: *2008 IEEE Custom Integrated Circuits Conference* (pp. 379–386).

90. Auth, C., Allen, C., Blattner, A., Bergstrom, D., Brazier, M., Bost, M., … Mistry, K. (2012, June). A 22nm high performance and low-power CMOS technology featuring fully-depleted tri-gate transistors, self-aligned contacts and high density MIM capacitors. In: *2012 Symposium on VLSI Technology (VLSIT)* (pp. 131–132).

91. Kwak, D., Park, J., Kim, K., Yim, Y., Ahn, S., Park, Y., … Kim, K. (2007, June). Integration technology of 30nm generation multi-level NAND flash for 64Gb NAND flash memory. In: *2007 IEEE Symposium on VLSI Technology* (p. 12–13).

92. Guillorn, M., Chang, J., Bryant, A., Fuller, N., Dokumaci, O., Wang, X., … Haensch, W. (2008, June). FinFET performance advantage at 22nm: An AC perspective. In: *2008 Symposium on VLSI Technology* (pp. 12–13).

93. Kawasaki, H., Basker, V. S., Yamashita, T., Lin, C. H., Zhu, Y., Faltermeier, J., … Ishimaru, K. (2009, December). Challenges and solutions of FinFET integration in an SRAM cell and a logic circuit for 22 nm node and beyond. In: *2009 IEEE International Electron Devices Meeting (IEDM)* (pp. 1–4).

94. Yamashita, T., Basker, V. S., Standaert, T., Yeh, C. C., Yamamoto, T., Maitra, K., … Leobandung, E. (2011, June). Sub-25nm FinFET with advanced fin formation and short channel effect engineering. In: *2011 Symposium on VLSI Technology – Digest of Technical Papers* (pp. 14–15).

95. Yaegashi, H. (2015). Toward 5nm node: Untoward scaling with multiple patterning. In: *First International DSA Symposium*.

96. Xie, R., Montanini, P., Akarvardar, K., Tripathi, N., Haran, B., Johnson, S., … Khare, M. (2016, December). A 7nm FinFET technology featuring EUV patterning and dual strained high mobility channels. In: *2016 IEEE International Electron Devices Meeting (IEDM)* (pp. 2.7.1–2.7.4).

97. Basker, V. S., Standaert, T., Kawasaki, H., Yeh, C. C., Maitra, K., Yamashita, T., … O'Neill, J. (2010, June). A 0.063 μm² FinFET SRAM cell demonstration with conventional lithography using a novel integration scheme with aggressively scaled fin and gate pitch. In: *2010 Symposium on VLSI Technology* (pp. 19–20).

98. Chawla, J. S., Chebiam, R., Akolkar, R., Allen, G., Carver, C. T., Clarke, J. S., … Yoo, H. J. (2013, June). Demonstration of a 12 nm-half-pitch copper ultralow-k interconnect process. In: *2013 IEEE International Interconnect Technology Conference – IITC* (pp. 1–3).

99. Mohanty, N., Smith, J. T., Huli, L., Pereira, C., Raley, A., Kal, S., … DeVillers, A. (2017). EPE improvement thru self-alignment via multi-color material integration. In: (Vol. 10147, pp. 10147–10147-13). Retrieved from https://doi.org/10.1117/12.2258108.

100. Beynet, J., Wong, P., Miller, A., Locorotondo, S., Vangoidsenhoven, D., Yoon, T.-H., … Maenhoudt, M. (2009). Low temperature plasma-enhanced ALD enables cost-effective spacer defined double patterning (SDDP). In: (Vol. 7520, pp. 7520–7520-7). Retrieved from https://doi.org/10.1117/12.836979.

101. Donnelly, V. M., & Kornblit, A. (2013). Plasma etching: Yesterday, today, and tomorrow. *Journal of Vacuum Science and Technology, Part A, 31*(5), 050825. Retrieved from https://doi.org/10.1116/1-4819316.

102. Lieberman, M. A., & Lichtenberg, A. J. (2005). *Principles of plasma discharges and materials processing.* Hoboken, NJ: John Wiley-Interscience.

103. Wise, R. (2013). Advanced plasma etch technologies for nanopatterning. *Journal of Micro/Nanolithography, MEMS, and MOEMS, 12*, 12–12-6. Retrieved from https://doi.org/10.1117/1.JMM.12.4.041311.

104. Tan, S., Yang, W., Kanarik, K. J., Lill, T., Vahedi, V., Marks, J., & Gottscho, R. A. (2015). Highly selective directional atomic layer etching of silicon. *ECS Journal of Solid State Science and Technology, 4*(6), N5010–N5012. Retrieved from http://jss.ecsdl.org/content/4/6/N5010.abstract.

105. Schafgans, A. A., Brown, D. J., Fomenkov, I. V., Sandstrom, R., Ershov, A., Vaschenko, G., ... Kool, R. (2015). Performance optimization of MOPA pre-pulse LPP light source. In: (Vol. 9422, pp. 9422–9422-11). Retrieved from https://doi.org/10.1117/12.2087421.
106. Pirati, A., Peeters, R., Smith, D., Lok, S., Minnaert, A. W. E., van Noordenburg, M., ... Kool, R. (2015). Performance overview and outlook of EUV lithography systems. In: (Vol. 9422, pp. 9422–9422-18). Retrieved from https://doi.org/10.1117/12.2085912.
107. Kim, R.-H., Wood, O., Crouse, M., Chen, Y., Plachecki, V., Hsu, S., & Gronlund, K. (2016). Application of EUV resolution enhancement techniques (RET) to optimize and extend single exposure bi-directional patterning for 7nm and beyond logic designs. In: (Vol. 9776, pp. 9776–9776-10). Retrieved from https://doi.org/10.1117/12.2219177.
108. Wagner, C., Kaiser, W. M., Mulkens, J., & Flagello, D. G. (2000). Advanced technology for extending optical lithography. In: (Vol. 4000, pp. 4000–4000-14). Retrieved from https://doi.org/10.1117/12.389046.
109. Yen, A. (2016). EUV lithography: From the very beginning to the eve of manufacturing. In: (Vol. 9776, pp. 9776–9776-59). Retrieved from https://doi.org/10.1117/12.2236044.
110. Mizoguchi, H., Nakarai, H., Abe, T., Nowak, K. M., Kawasuji, Y., Tanaka, H., ... Saitou, T. (2015). Performance of one hundred watt HVM LPP-EUV source. In: (Vol. 9422, pp. 9422–9422-13). Retrieved from https://doi.org/10.1117/12.2086347.
111. Fujimoto, J., Hori, T., Yanagida, T., & Mizoguchi, H. (2012). Development of laser-produced tin plasma-based EUV light source technology for HVM EUV lithography. *Physics Research International*, *2012*, 249495.
112. Kemen, T., Malloy, M., Thiel, B., Mikula, S., Denk, W., Dellemann, G., & Zeidler, D. (2015). Further advancing the throughput of a multi-beam sem. In: (Vol. 9424, pp. 9424–9424-6). Retrieved from https://doi.org/10.1117/12.2188560.
113. Gallatin, G. M., & Patrick Naulleau, R. B. (2007). Fundamental limits to EUV photoresist. In: (Vol. 6519, pp. 6519–6519-10). Retrieved from https://doi.org/10.1117/12.712346.
114. Nagahara, S., Carcasi, M., Nakagawa, H., Buitrago, E., Yildirim, O., Shiraishi, G., ... Tagawa, S. (2016). Challenge toward breakage of RLS trade-off for EUV lithography by photosensitized chemically amplified resist (PSCAR) with flood exposure. In: (Vol. 9776, pp. 9776–9776-18). Retrieved from https://doi.org/10.1117/12.2219433.
115. Grenville, A., Anderson, J. T., Clark, B. L., Schepper, P. D., Edson, J., Greer, M., ... Vandenberghe, G. (2015). Integrated fab process for metal oxide EUV photoresist. In: (Vol. 9425, pp. 9425–9425-8). Retrieved from https://doi.org/10.1117/12.2086006.
116. Loubet, N., Hook, T., Montanini, P., Yeung, C. W., Kanakasabapathy, S., Guillom, M., ... Khare, M. (2017, June). Stacked nanosheet gate-all-around transistor to enable scaling beyond FinFET. In: *2017 Symposium on VLSI Technology* (pp. T230–T231).

2 Optical Nanolithography

Bruce W. Smith

CONTENTS

2.1 INTRODUCTION

Optical lithography involves the creation of relief image patterns through the projection of radiation within or near the ultraviolet (UV) visible portion of the electromagnetic spectrum. Techniques of optical lithography, or photolithography, have been used to create patterns for engravings, photographs, and printing plates. In the 1960s, techniques developed for the production of lithographic printing plates were utilized in the making of microcircuit patterns for semiconductor devices. These early techniques of contact or proximity photolithography were refined to allow circuit resolution on the order of 3–5 micrometer (μm). Problems encountered with proximity lithography, such as mask and wafer damage, alignment difficulty, and field size, have limited its application for most photolithographic needs. In the mid-1970s, projection techniques minimized some of the problems encountered with proximity lithography and have led to the development of tools that currently allow resolution below 45 nm.

    Diagrammed in Figure 2.1 is a generic projection imaging system for lithography. The illumination system consists of a source and a condenser lens assembly that provides uniform illumination to

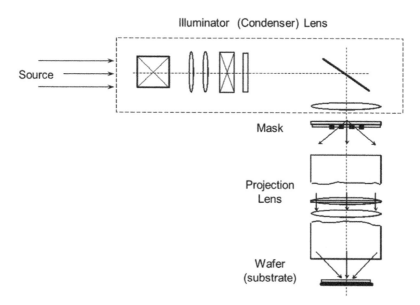

FIGURE 2.1 Schematic of a projection optical lithography system.

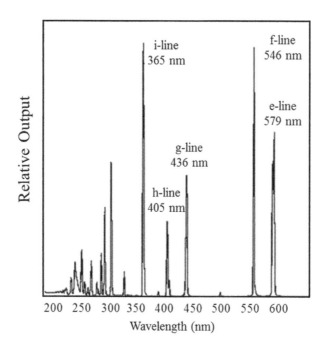

FIGURE 2.2 The spectral distribution of a mercury vapor lamp showing dominant wavelengths used in lithography (the g and i lines).

the mask. The illumination source outputs radiation in the blue ultraviolet portion of the electromagnetic spectrum. The mercury-rare gas discharge lamp is a source well suited for photolithography, and it is almost entirely relied on for production of radiation in the 350–450 nm range. The spectral distribution of a typical mercury lamp is shown in Figure 2.2. Because output below 365 mm is weak from a mercury or mercury-rare gas lamp, other sources have been utilized for shorter-wavelength exposure. The ultraviolet region from 150 to 300 nm is referred to as the *deep ultraviolet* (DUV).

Although a small number of lithographic techniques operating at these wavelengths have made use of gas discharge lamps, the use of a laser source is an attractive alternative. Several laser sources have potential for delivering high-power deep ultraviolet radiation for photoresist exposure. A class of lasers that has been shown to be well suited for photolithography is the excimer lasers. Excimer lasers using argon fluoride (ArF) and krypton fluoride (KrF) gas mixtures are most prominent, producing radiation at 193 and 248 nm, respectively. The optical configuration for projection microlithography tools most closely resembles a microscope system. Early microlithographic objective lenses were modifications of microscope lens designs that have now evolved to allow diffraction-limited resolution over large fields at high numerical apertures. Like a proximity system, a projection tool includes an illumination system and a mask, but it utilizes an objective lens to project images toward a substrate. The illumination system focuses an image of the source into the entrance pupil of the objective lens to provide maximum uniformity at the mask plane.

Both irradiance and coherence properties are influenced by the illumination system. The temporal coherence of a source is a measure of the correlation of the source wavelength to the source spectral bandwidth. As a source spectral bandwidth decreases, its temporal coherence increases. Coherence length, l_c, is related to source bandwidth as

$$l_c = \lambda^2 / \Delta\lambda$$

Interference effects become sufficiently large when an optical path distance is less than the coherence length of a source. Optical imaging effects such as interference (standing wave) patterns in photoresist become considerable as source coherent length increases.

The spatial coherence of a source is a measure of the phase relationships between photons or wavefronts emitted. A true point source, by definition, is spatially coherent, because all wavefronts originate from a single point. Real sources, however, are less than spatially coherent. A conventional laser that utilizes oscillation for amplification of radiation can produce nearly spatially coherent radiation. Lamp sources such as gas discharge lamps exhibit low spatial coherence, as do excimer lasers that require few oscillations within the laser cavity. Both temporal and spatial coherence properties can be controlled by an illumination system. Source bandwidth and temporal coherence are controlled through wavelength selection. Spatial coherence is controlled through manipulation of the effective source size imaged in the objective lens. In image formation, the control of spatial coherence is of primary importance because of its relationship to diffraction phenomena.

Current designs of projection lithography systems include (1) reduction or unit magnification, (2) refractive and/or reflective optics, and (3) array stepping or field scanning. Reduction tools allow a relaxation of mask requirements, including minimum feature size specification and defect criteria. This, in turn, reduces the contribution to the total process tolerance budget. The drawbacks for reduction levels greater than 4:1 include the need for increasingly larger masks and the associated difficulties in their processing. Both unit magnification (1:1) and reduction (M:1) systems have been utilized in lithographic imaging system design, each well suited for certain requirements. As feature size and control place high demands on 1:1 technology, reduction tools are generally utilized. A refractive projection system must generally utilize a narrow spectral band of a lamp-type source. Energy outside this range would be removed prior to the condenser lens system to avoid wavelength-dependent defocus effects or chromatic aberration. Some degree of chromatic aberration correction is possible in a refractive lens system by incorporating elements of various glass types. As wavelengths below 300 nm were pursued for refractive projection lithography, the control of spectral bandwidth became more critical. As few transparent optical materials exist at these wavelengths, chromatic aberration correction through glass material selection is difficult. Greater demands are therefore placed on the source, which may be required to deliver a spectral bandwidth on the order of a few picometers. Clearly, such a requirement would limit the application of lamp sources at these wavelengths, leading to laser-based sources as the only alternative for short-wavelength refractive systems. Reflective optical systems (catoptric) or combined refractive-reflective systems (catadioptric) can be used to reduce wavelength influence and reduce source requirements, especially at deep UV wavelengths.

By the mid-1990s, the transition from mercury lamp source technology to excimer laser deep UV technology was being driven by the need to resolve sub-250 nm geometry. The term *excimer* comes from "excited dimer," a class of molecules that exists only in the upper excited state but not in the ground state. The excimer molecule has a short upper-state lifetime, and it decays to the ground state through disassociation while emitting a photon. There are two types of excimer molecules: rare-gas excited dimers such as xenon (Xe2) and krypton (Kr2) and the rare-gas halogens such as XeF, XeCl, KrF, and ArF. The latter class of excimer molecules is of greater interest, because they emit deep-UV photons (351, 308, 248, and 193 nm, respectively). The F2 laser is not an excimer laser; it is a molecular laser. However, the principle of operation of the laser is similar to that of a KrF or ArF laser. The radiative lifetime of the upper laser state for KrF is about 9 ns, and the dissociative lifetime of the lower level is on the order of 1 ps. Population inversion in a KrF laser is therefore easily achieved. Typically, broadband (non-line-narrowed) KrF and ArF linewidths are about 300 and 500 pm, respectively, at full width at half maximum (FWHM), while the F2 laser linewidth is about 1 picometer (pm) at FWHM. As will be discussed in more detail, the refractive power in a dioptric or catadioptric lithography lens defines the allowable bandwidth, which is on the subpicometer level. To achieve this, laser "line-narrowing" is required. An effective commercial line-narrowing technique uses the combination of high magnification, large gratings, and narrow apertures to achieve bandwidths down to about 0.1 pm.

Since the early days of integrated circuit (IC) fabrication, photomasks have been used to project IC-level design images onto a photoresist-coated wafer. Although various maskless lithography approaches have been introduced over the years, most commercial IC applications have utilized a mask. Early photomask fabrication processes involved the hand drafting of chip patterns, transferring those patterns to acetate templates, and photographically reducing their images onto high-contrast emulsion-coated glass plates. As circuit dimensions decreased and chip density grew, computer-aided design (CAD) and optical flash pattern generators (PGs) evolved to meet the high-volume demands of the growing semiconductor industry. By the 1980s, electron beam (e-beam) photomask making allowed further improvements and higher throughput over optical approaches, while eliminating several of the intermediate steps necessary with the earlier methods. By the 1980s, emulsion masks were also replaced by photoresist-coated chromium-on-glass (COG) masks, meeting the transmission and durability needs of high-volume manufacturing. Early raster scanning e-beam platforms were eventually replaced by variable shaped beam (VSB) vector scanning systems in the late 1990s for IC device generations below about 130 nm.

To understand the underlying principles of optical lithography, fundamentals of both geometrical and physical optics need to be addressed. Because optical lithography using projection techniques is the dominant technology for current IC fabrication, the development of the physics behind projection lithography will be concentrated on in this chapter.

2.2 IMAGE FORMATION: GEOMETRICAL OPTICS

An understanding of optics where the wave nature of light is neglected can provide a foundation for further study into a more inclusive approach. Therefore, geometrical optics are introduced here, allowing investigation into valuable information about imaging [1]. This will lead to a more complete study of imaging through physical optics, where the wave nature of light is considered and interference and diffraction can be investigated.

Both refractive lenses and reflective mirrors play important roles in microlithography optical systems. The optical behavior of mirrors can be described by extending the behavior of refractive lenses. Although a practical lens will contain many optical elements, baffles, apertures, and mounting hardware, most optical properties of a lens can be understood through the extension of simple

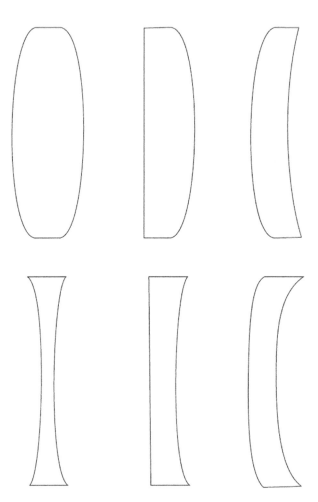

FIGURE 2.3 Single-element lens shapes. At the top are positive lenses—biconvex, planoconvex, and meniscus convex. At the bottom are negative lenses—biconcave, planoconcave, and meniscus concave.

single-element lens properties. The behavior of a simple lens will be investigated to gain an understanding of optical systems in general.

A perfect lens would be capable of an exact translation of an incident spherical wave through space. A positive lens would cause a spherical wave to converge faster, and a negative lens would cause a spherical wave to diverge faster. Lens surfaces are generally spherical or planar, and they may have forms including biconvex, plano-convex, biconcave, plano-concave, negative meniscus, and positive meniscus, as shown in Figure 2.3. In addition, aspheric surfaces are possible, which may be used in an optical system to improve its performance. These types of elements are generally difficult and expensive to fabricate and are not yet widely used. As design and manufacturing techniques improve, applications of aspherical elements will grow, including their use in microlithographic lens systems.

2.2.1 CARDINAL POINTS

Knowledge of the cardinal points of a simple lens is sufficient to understand its behavior. These points, the first and second focal points (F_1 and F_2), the principal points (P_1 and P_2), and the nodal points (N_1 and N_2), lie on the optical axis of a lens, as shown in Figure 2.4. The principal planes are also shown here, which contain respective principal points and can be thought of as the surfaces

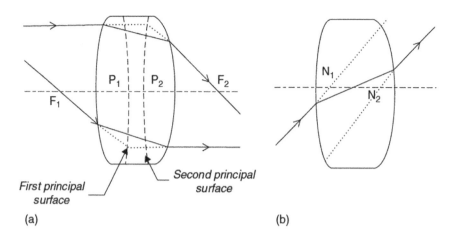

FIGURE 2.4   Cardinal points of a simple lens. (a) Focal points (F_1 and F_2) and principal points (P_1 and P_2). (b) Nodal points (N_1 and N_2).

where refraction effectively occurs. Although these surfaces are not truly planes, they are nearly so. Rays that pass through a lens act as if they refract only at the first and second principal planes and not at any individual glass surface. A ray passing through the first focal point (F_1) will emerge from the lens at the right parallel to the optical axis. For this ray, refraction effectively occurs at the first principal plane. A ray traveling parallel to the optical axis will emerge from the lens and pass through the second focal point (F_2). Here, refraction effectively occurs at the second principal plane. A ray passing through the optical center at the lens will emerge parallel to the incident ray and pass through the first and second nodal points (N_1, N_2). A lens or lens system can, therefore, be represented by its two principal planes and focal points.

2.2.2   Focal Length

The distance between a lens focal point and corresponding principal point is known as the effective focal length (EFL), as shown in Figure 2.5. The focal length can be either positive, when F_1 is to the left of P_1 and F_2 is to the right of P_2, or negative, when the opposite occurs. The reciprocal of the EFL ($1/f$) is known as the lens power. The front focal length (FFL) is the distance from the first focal point (F_1) to the leftmost surface of the lens along the optical axis. The back focal length (BFL) is the distance from the rightmost surface to the second focal point (F_2).

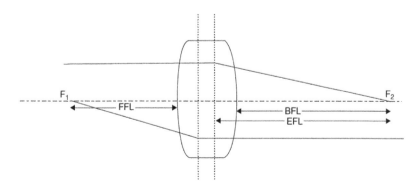

FIGURE 2.5   Determination of focal length for a simple lens, front focal length (FFL), and back focal length (EFL).

The lens maker's formula can be used to determine the EFL of a lens if the radii of curvature of surface (R_1 and R_2 for first and second surfaces), lens refractive index (n_i), and lens thickness (t) are known. Several sign conventions are possible. Distances measured toward the left will generally be considered as positive. R_1 will be considered positive if its center of curvature lies to the right of the surface, and R_2 will be considered negative if its center of curvature lies to the left of the surface. Focal length is determined by

$$\frac{1}{f} = (n_i - 1)\left[\frac{1}{R_1} - \frac{1}{R_2} + \frac{(n_i - 1)t}{n_i R_1 R_2}\right]$$

2.2.3 GEOMETRICAL IMAGING PROPERTIES

If the cardinal points of a lens are known, geometrical imaging properties can be determined. A simple biconvex is considered, such as the one shown in Figure 2.6, where an object is placed a positive distance s_1 from focal point F_1 at a positive object height y_1. This object can be thought of as consisting of many points that will emit spherical waves to be focused by the lens at the image plane. The object distance (d_1) is the distance from the principal plane to the object, which is positive for objects to the left of P_1.

The image distance to the principal plane (d_2), which is positive for an image to the right of P_2, can be calculated from the lens law:

$$\frac{1}{d_1} + \frac{1}{d_2} = \frac{1}{f}$$

For systems with a negative EFL, the lens law becomes

$$\frac{1}{d_1} + \frac{1}{d_2} = \frac{-1}{|f|}$$

The lateral magnification of an optical system is expressed as

$$m = \frac{y_2}{y_1} = \frac{-d_2}{d_1}$$

where y_2 is the image height, which is positive upward.

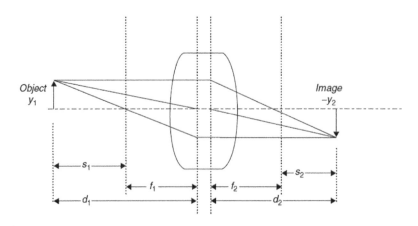

FIGURE 2.6 Ray tracing methods for finding image location and magnification.

The location of an image can be determined by tracing any two rays that will intersect in the image space. As shown in Figure 2.6, a ray emanating from an object point, passing through the first focal point F_1, will emerge parallel to the optical axis, being effectively refracted at the first principal plane. A ray from an object point traveling parallel to the optical axis will emerge after refracting at the second principal plane passing through F_2. A ray from an object point passing through the center of the lens will emerge parallel to the incident ray. All three rays intersect at the image location. If the resulting image lies to the right of the lens, the image is real (assuming light emanates from an object on the left). If the image lies to the right, it is virtual. If the image is larger than the object, magnification is greater than unity. If the image is erect, the magnification is positive.

2.2.4 APERTURE STOPS AND PUPILS

The light accepted by an optical system is physically limited by aperture stops within the lens. The simplest aperture stop may be the edge of a lens or a physical stop placed in the system. Figure 2.7 shows how an aperture stop can limit the acceptance angle of a lens. The numerical aperture (NA) is the maximum acceptance angle at the image plane that is determined by the aperture stop.

$$NA_{IMG} = n_i \sin\left(\theta_{max}\right)$$

Because the optical medium is generally air, $NA_{IMG} \sim \sin(\theta_{max})$. The field stop shown in Figure 2.8 limits the angular field of view, which is generally the angle subtended by the object or image from the first or second nodal point. The angular field of view for the image is generally the same as that for the object. The image of the aperture stop viewed from the object is called the *entrance pupil*, whereas the image viewed from the image is called the *exit pupil*, as seen in Figures 2.9 and 2.10. As will be viewed, the aberrations of an optical system can be described by the deviations in spherical waves at the exit pupil coming to focus at the image plane.

2.2.5 CHIEF AND MARGINAL RAY TRACING

We have seen that a ray emitted from an off-axis point, passing through the center of a lens, will emerge parallel to the incident ray. This is called the *chief ray*, and it is directed toward the

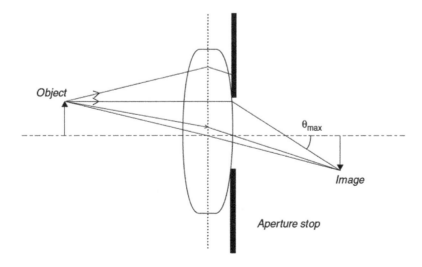

FIGURE 2.7 Limitation of lens maximum acceptance angle by an aperture stop.

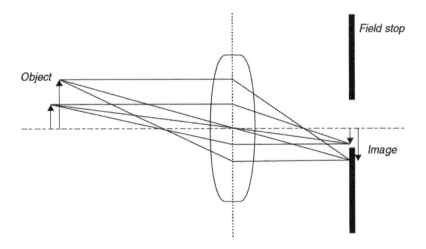

FIGURE 2.8   Limitation of angular field of view by a field stop.

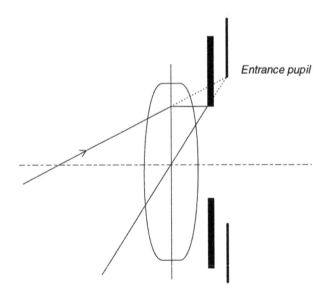

FIGURE 2.9   Location of the entrance pupil for a simple lens.

entrance pupil of the lens. A ray that is emitted from an on-axis point and directed toward the edge of the entrance pupil is called a *marginal ray*. The image plane can, therefore, be found where a marginal ray intersects the optical axis. The height of the image is determined by the height of the chief ray at the image plane, as seen in Figure 2.11. The marginal ray also determines the numerical aperture. The marginal and chief rays are related to each other by the Lagrange invariant, which states that the product of the image NA and image height is equal to the object NA and object height, or $NA_{OBJ}y_1 = NA_{IMG}y_2$. It is essentially an indicator of how much information can be processed by a lens. The implication is that as object or field size increases, NA decreases. To achieve an increase in both NA and field size, system complexity increases. Magnification can now be expressed as

$$m = \frac{NA_{OBJ}}{NA_{IMG}}$$

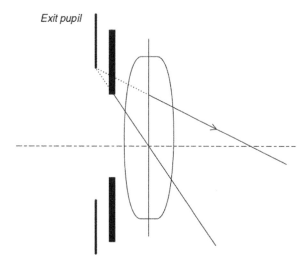

FIGURE 2.10 Location of the exit pupil for a simple lens.

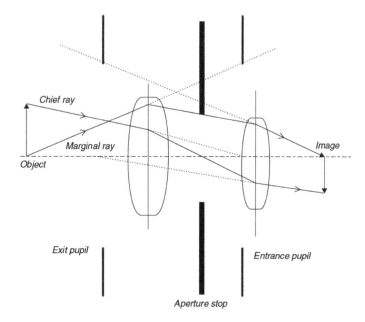

FIGURE 2.11 Chief and marginal ray tracing through a lens system.

2.2.6 Mirrors

A spherical mirror can form images in ways similar to refractive lenses. Using the reflective lens focal length, the lens equations can be applied to determine image position, height, and magnification. To use these equations, a sign convention for reflection needs to be established. Because refractive index is the ratio of the speed of light in vacuum to the speed of light in the material considered, it is logical that a change of sign would result if the direction of propagation was reversed. For reflective surfaces, therefore,

1. Refractive index values are multiplied by −1 upon reflection.
2. The signs of all distances upon reflection are multiplied by −1.

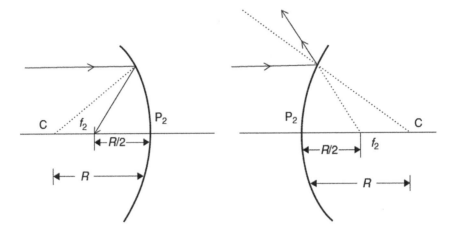

FIGURE 2.12   Location of principal and focal points for (a) concave and (b) convex mirrors.

Figure 2.12 shows the location of principal and focal points for two mirror types: concave and convex. The concave mirror is equivalent to a positive converging lens. A convex mirror is equivalent to a negative lens. The EFL is simplified because of the loss of the thickness term and sign changes to

$$ f = -\frac{1}{2} R $$

2.3   IMAGE FORMATION: WAVE OPTICS

Many of the limitations of geometrical optics can be explained by considering the wave nature of light. As it has been reasoned that a perfect lens translates spherical waves from an object point to an image point, such concepts can be used to describe deviations from non-geometrical propagation that would otherwise be difficult to predict.

An approach proposed by Huygens [2] allows an extension of optical geometric construction to wave propagation. Through use of this simplified wave model, many practical aspects of the wave nature of light can be understood. Huygens' principle provides a basis for determining the position of a wavefront at any instance based on knowledge of an earlier wavefront. A wavefront is assumed to be made up of an infinite number of point sources. Each of these sources produces a spherical secondary wave called a *wavelet*. These wavelets propagate with appropriate velocities that are determined by refractive index and wavelength. At any point in time, the position of the new wavefront can be determined as the surface tangent to these secondary waves. Using Huygens' concepts, electromagnetic fields can be thought of as sums of propagating spherical or plane waves. Although Huygens had no knowledge of the nature of the light wave or the electromagnetic character of light, this approach has allowed analysis without the need to fully solve Maxwell's equations.

The diffraction of light is responsible for image creation in all optical situations. When a beam of light encounters the edge of an opaque obstacle, propagation is not rectilinear, as might be assumed based on assumptions of geometrical shadowing. The resulting variation in intensity produced at some distance from the obstacle is dependent on the coherence of light, its wavelength, and the distance the light travels before being observed. The situation for coherent illumination is shown in Figure 2.13. Shown are a coherently illuminated mask and the resulting intensity pattern observed at increasing distances. Such an image in intensity is known as an *aerial image*. Typically, with coherent illumination, fringes are created in the diffuse shadowing between light and dark, a result of interference. Only when there is no separation between the obstacle and the recording plane does

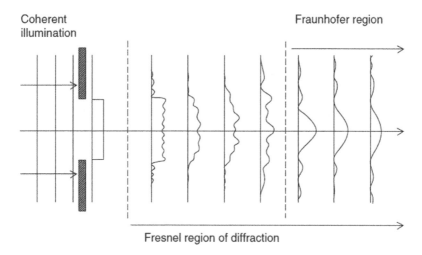

Coherent illumination

Fraunhofer region

Fresnel region of diffraction

FIGURE 2.13 Diffraction pattern of a coherently illuminated mask opening at near (Fresnel) and far (Fraunhofer) distances.

rectilinear propagation occur. As the recording plane is moved away from the obstacle, there is a region where the geometrical shadow is still discernible. Beyond this region, far from the obstacle, the intensity pattern at the recording plane no longer resembles the geometrical shadow; rather, it contains areas of light and dark fringes. At close distances, where geometric shadowing is still recognizable, near-field diffraction, or Fresnel diffraction, dominates. At greater distances, far-field diffraction, or Fraunhofer diffraction, dominates.

2.3.1 FRESNEL DIFFRACTION: PROXIMITY LITHOGRAPHY

The theory of Fresnel diffraction is based on the Fresnel approximation to the propagation of light, and it describes image formation for proximity printing, where separation distances between the mask and wafer are normally held to within a few microns [3]. The distribution of intensity resembles that of the geometric shadow. As the separation between the mask and wafer increases, the integrity of an intensity pattern resembling an ideal shadowing diminishes.

Theoretical analysis of Fresnel diffraction is difficult, and Fresnel approximations based on Kirchhoff diffraction theory are used to obtain a qualitative understanding [4]. Because our interest lies mainly with projection systems and diffraction beyond the near-field region, a rigorous analysis will not be attempted here. Instead, analysis of results will provide some insight into the capabilities of proximity lithography.

Fresnel diffraction can be described using a linear filtering approach that can be made valid over a small region of the observation or image plane. For this analogy, a mask function is effectively frequency filtered with a quadratically increasing phase function. This quadratic phase filter can be thought of as a slice of a spherical wave at some plane normal to the direction of propagation, as shown in Figure 2.14. The resulting image will exhibit "blurring" at the edges and oscillating "fringes" in bright and dark regions. Recognition of the geometrical shadow becomes more difficult as the illumination wavelength increases, the mask feature size decreases, or the mask separation distance increases. Figure 2.15 illustrates the situation where a space mask is illuminated with 365 nm radiation and the separation distance between mask and wafer is 1.8 μm. For relatively large features, on the order of 10–15 μm, rectilinear propagation dominates, and the resulting image intensity distribution resembles the mask. In order to determine the minimum resolvable feature width, some specification for maximum intensity loss and line width deviation must be made. These specifications are determined by the photoresist material and processes. If an intensity tolerance of

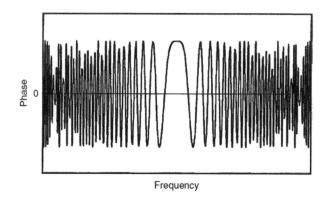

FIGURE 2.14 A quadratic phase function.

±5% and a mask space width to image width tolerance of ±20% are acceptable, a relationship for minimum resolution results:

$$w \approx 0.7\sqrt{\lambda s}$$

where

    w is space width
    λ is illumination wavelength
    s is separation distance

As can be shown, resolution below 1 µm should be achievable with separations of 5 µm or less. A practical limit for resolution using proximity methods is closer to 3–5 µm because of surface and mechanical separation control as well as alignment difficulties.

2.3.2 FRAUNHOFER DIFFRACTION: PROJECTION LITHOGRAPHY

For projection lithography, diffraction in the far-field or Fraunhofer region needs to be considered. No longer is geometric shadowing recognizable; rather, fringing takes over in the resulting intensity pattern. Analytically, this situation is easier to describe than Fresnel diffraction. When light encounters a mask, it is diffracted toward the object lens in the projection system. Its propagation will determine how an optical system will ultimately perform, depending on the coherence of the light that illuminates the mask.

Consider a coherently illuminated single-space mask opening as shown in Figure 2.16 The resulting Fraunhofer diffraction pattern can be evaluated by examining light coming from various portions of the space opening. Using Huygens' principle, the opening can be divided into an infinite number of individual sources, each acting as a separate source of spherical wavelets. Interference will occur between every portion of this opening, and the resulting diffraction pattern at some far distance will depend on the propagation direction θ. It is convenient for analysis to divide the opening into two halves ($d/2$). With coherent illumination, all wavelets emerging from the mask opening are in phase. If waves emitted from the center and bottom of the mask opening are considered (labeled W1 and W3), it can be seen that an optical path difference (OPD) exists as one wave travels a distance $d/2 \sin\theta$ farther than the other. If the resulting OPD is one half-wavelength or any multiple of one half-wavelength, waves will interfere destructively. Similarly, an OPD of $d \sin\theta$ exists between any two waves that originate from points separated by one-half of the space width. The waves from the top portion of the mask opening interfere destructively with waves from the bottom portion of the mask when

$$d \sin \theta = m\lambda \quad \left(m = \pm 1, \pm 2, \pm 3, \ldots\right)$$

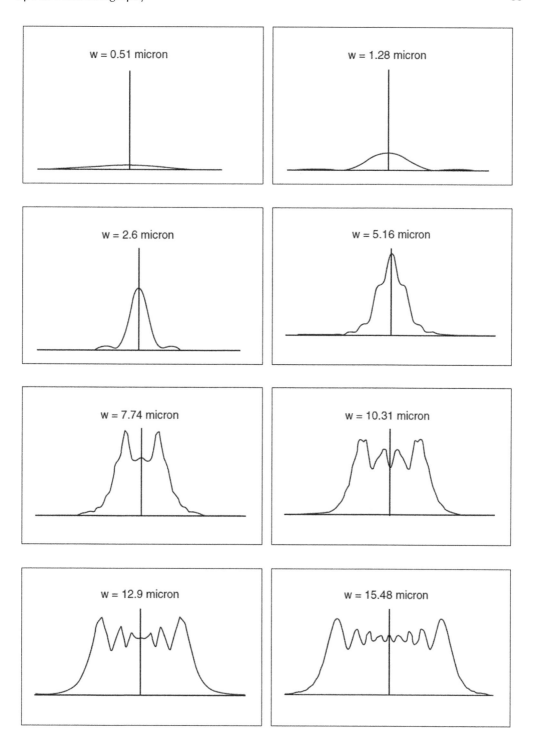

FIGURE 2.15 Aerial images resulting from frequency filtering of a slit opening with a quadratic phase function. The illumination wavelength is 365 nm and separation distance is 1.8 μm for mask opening sizes shown.

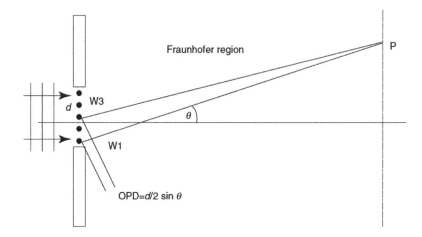

FIGURE 2.16 Determination of Fraunhofer diffraction effects for a coherently illuminated single mask opening.

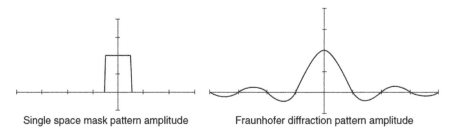

Single space mask pattern amplitude   Fraunhofer diffraction pattern amplitude

FIGURE 2.17 (a) A single-space mask pattern and (b) its corresponding Fraunhofer diffraction pattern. These are Fourier transform pairs.

where $|m| \leq = d/\lambda$. From this equation, the positions of dark fringes in the Fraunhofer diffraction pattern can be determined. Figure 2.17 is the resulting diffraction pattern from a single space, where a broad central bright fringe exists at positions corresponding to $\theta = 0$, and dark fringes occur where θ satisfies the destructive interference condition.

Although this geometric approach is satisfactory for a basic understanding of Fraunhofer diffraction principles, it cannot do an adequate job of describing the propagation of diffracted light. Fourier methods and scalar diffraction theory provide a description of the propagation of diffracted light through several approximations (previously identified as the Fresnel approximation), specifically [5, 6]:

1. The distance between the aperture and the observation plane is much greater than the aperture dimension.
2. Spherical waves can be approximated by quadratic surfaces.
3. Each plane wave component has the same polarization amplitude (with polarization vectors perpendicular to the optical axis).

These approximations are valid for optical systems with numerical apertures below 0.6 if illumination polarization can be neglected. Scalar theory has been extended beyond these approximations to numerical apertures of 0.7 [7], and full vector diffraction theory has been utilized for more rigorous analysis [8].

2.3.3 Fourier Methods in Diffraction Theory

Whereas geometrical methods allow determination of interference minimums for the Fraunhofer diffraction pattern of a single slit, the distribution of intensity across the pattern is most easily determined through Fourier methods. The coherent field distribution of a Fraunhofer diffraction pattern produced by a mask is essentially the Fourier transform of the mask function. If $m(x,y)$ is a two-dimensional mask function or electric field distribution across the $x-y$ mask plane, and $M(u,v)$ is the coherent field distribution across the $u-v$ Fraunhofer diffraction plane, then

$$M(u,v) = \mathcal{F}\{m(x,y)\}$$

will represent the Fourier transform operation. Both $m(x,y)$ and $M(u,v)$ have amplitude and phase components. From Figure 2.13, we could consider $M(u,v)$ the distribution (in amplitude) at the farthest distance from the mask.

The field distribution in the Fraunhofer diffraction plane represents the spatial frequency spectrum of the mask function. In the analysis of image detail, preservation of spatial structure is generally of most concern. For example, the lithographer is interested in optimizing an imaging process to maximize the reproduction integrity of fine feature detail. To separate out such spatial structure from an image, it is convenient to work in a domain of spatial frequency rather than of feature dimension. The concept of spatial frequency is analogous to temporal frequency in the analysis of electrical communication systems. Units of spatial frequency are reciprocal distance. As spatial frequency increases, pattern detail becomes finer. Commonly, units of cycles/mm or mm^{-1} are used, where 100 nm^{-1} is equivalent to 5 µm, 1000 mm^{-1} is equivalent to 0.5 µm, and so forth. The Fourier transform of a function, therefore, translates dimensional (x,y) information into spatial frequency (u,v) structure.

2.3.3.1 The Fourier Transform

The unique properties of the Fourier transform allow convenient analysis of spatial frequency structure [9]. The Fourier transform takes the general form

$$F(u) = \int_{-\infty}^{\infty} f(x) e^{-2\pi iux} dx$$

for one dimension. Uppercase and lowercase letters are used to denote Fourier transform pairs.

In words, the Fourier transform expresses a function $f(x)$ as the sum of weighted sinusoidal frequency components. If $f(x)$ is a real-valued, even function, the complex exponential ($e^{-2\pi iux}$) could be replaced by a cosine term, $\cos(2\pi ux)$, making the analogy more obvious. Such transforms are utilized but are of little interest for microlithographic applications, because masking functions, $m(x,y)$, will generally have odd as well as even components.

If the single slit pattern analyzed previously with Fraunhofer diffraction theory is revisited, it can be seen that the distribution of the amplitude of the interference pattern produced is simply the Fourier transform of an even, one-dimensional, nonperiodic, rectangular pulse, commonly referred to as a *rect function*, rect(x). The Fourier transform of rect(x) is a sinc(u), where

$$\text{sinc}(u) = \frac{\sin(\pi u)}{\pi u}$$

that is shown in Figure 2.17 The intensity of the pattern is proportional to the square of the amplitude, or a sinc$^2(u)$ function, which is equivalent to the power spectrum. The two functions, rect (x) and sinc(u), are Fourier transform pairs where the inverse Fourier transform of $F(u)$ is $f(x)$:

$$f(x) = \int\limits_{-\infty}^{\infty} F(u) e^{+2\pi iux} du$$

The Fourier transform is nearly its own inverse, differing only in sign.

The scaling property of the Fourier transform is of specific importance in imaging applications. Properties are such that

$$\mathcal{F}\left\{ f\frac{x}{b} \right\} = |b| F(bu)$$

and

$$\mathcal{F}\left\{ \mathrm{rect}\,\frac{x}{b} \right\} = |b| \mathrm{sinc}(bu)$$

where b is the effective width of the function. The implication of this is that as the width of a slit decreases, the field distribution of the diffraction pattern becomes more spread out with diminished amplitude values. Figure 2.18 illustrates the effects of scaling on a one-dimensional rect function.

A mask object is generally a function of both x and y coordinates in a two-dimensional space. The two-dimensional Fourier transform takes the form

$$F(u,v) = \int\limits_{-\infty}^{\infty}\int\limits_{-\infty}^{\infty} f(x,y) e^{-2\pi i(ux+vy)} dx\, dy$$

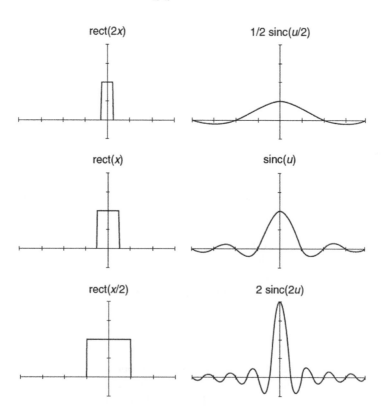

rect(2x)   1/2 sinc(u/2)

rect(x)   sinc(u)

rect(x/2)   2 sinc(2u)

FIGURE 2.18   Scaling effects on rect(x) and sinc(u) pairs.

The variables u and v represent spatial frequencies in the x and y directions, respectively. The inverse Fourier transform can be determined in a fashion similar to the one-dimensional case with a conventional change in sign.

In IC lithography, isolated as well as periodic lines and spaces are of interest. Diffraction for isolated features through Fourier transform of the rect function has been analyzed. Diffraction effects for periodic feature types can be analyzed in a similar manner.

2.3.3.2 Rectangular Wave

Where a single slit mask can be considered as a nonperiodic rectangular pulse, line/space patterns can be viewed as periodic rectangular waves. In Fraunhofer diffraction analysis, this rectangular wave is analogous to the diffraction grating. The rectangular wave function of Figure 2.19 has been chosen as an illustration where the maximum amplitude is A and the wave period is p, also known as the pitch. This periodic wave can be broken up into components of a rect function, with width half of the pitch, or half-pitch $p/2$, and a periodic function that will be called comb(x), where

$$\text{comb}\left(\frac{x}{p}\right) = \sum_{n=-\infty}^{\infty} \delta(x - np)$$

an infinite train of unit-area impulse functions spaced one pitch unit apart. (An impulse function is an idealized function with zero width and infinite height, having an area equal to 1.0.) To separate these functions, rect(x) and comb(x), from the rectangular wave, we need to realize that it is a convolution operation that relates them. Because convolution in the space (x) domain becomes multiplication in frequency,

$$m(x) = \text{rect}\left(\frac{x}{p/2}\right) * \text{comb}\left(\frac{x}{p}\right)$$

$$m(u) = \mathcal{F}\{m(x)\} = \mathcal{F}\left\{\text{rect}\left(\frac{x}{p/2}\right)\right\} \times \mathcal{F}\left\{\text{comb}\left(\frac{x}{p}\right)\right\}$$

By utilizing the transform properties of the comb function

$$\mathcal{F}\{\text{comb}(x/b)\} = |b|\,\text{comb}(bu)$$

the Fourier transform of the rectangular wave can be expressed as

$$\mathcal{F}\left\{\text{rect}\left(\frac{x}{p/2}\right) * \text{comb}\left(\frac{x}{p}\right)\right\} = M(u) = \frac{A}{2}\text{sinc}\left(\frac{u}{2u_0}\right)\sum_{n=-\infty}^{\infty}\delta(u - nu_0)$$

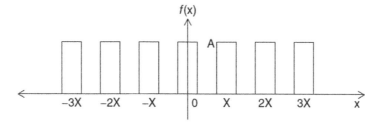

FIGURE 2.19   A periodic rectangular wave, representing dense mask features.

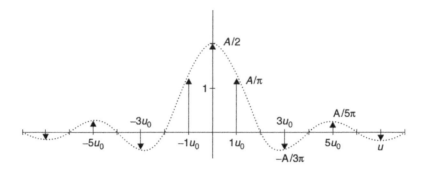

FIGURE 2.20   The amplitude spectrum of a rectangular wave, $A/2$ sinc$(u/2u_0)$. This is equivalent to the discrete orders of the coherent Fraunhofer diffraction pattern.

where $u_0 = 1/p$, the fundamental frequency of the mask grating. The amplitude spectrum of the rectangular wave is shown in Figure 2.20, where $A/2$ sinc$(u/2u_0)$ provides an envelope for the discrete Fraunhofer diffraction pattern. It can be shown that the discrete interference maxima correspond to $p \sin\theta = m\lambda$, where $m = 0, \pm 1, \pm 2, \pm 3$, and so on.

2.3.3.3   Harmonic Analysis

The amplitude spectrum of the rectangular wave can be utilized to decompose the function into a linear combination of complex exponentials by assigning proper weights to complex-valued coefficients. This allows harmonic analysis through the Fourier series expansion, utilizing complex exponentials as basis functions. These exponentials, or sine and cosine functions, allow us to represent the spatial frequency structure of periodic functions as well as non-periodic functions. Let us consider the periodic rectangular wave function $m(x)$ of Figure 2.19. Because the function is even and real-valued, the amplitude spectrum can be utilized to decompose $m(x)$ into the cosinusoidal frequency components:

$$m(x) = \frac{A}{2} + \frac{2A}{\pi}\left[\cos(2\pi u_0 x)\right] - \frac{2A}{3\pi}\left[\cos(2\pi(3u_0)x)\right]$$

$$+ \frac{2A}{5\pi}\left[\cos(2\pi(5u_0)x)\right] - \frac{2A}{7\pi}\left[\cos(2\pi(7u_0)x)\right] + \ldots$$

By graphing these components in Figure 2.21, it becomes clear that each additional term brings the sum closer to the function $m(x)$.

These discrete coefficients are the diffraction orders of the Fraunhofer diffraction pattern that are produced when a diffraction grating is illuminated by coherent illumination. These coefficients, represented as terms in the harmonic decomposition of $m(x)$ in Figure 2.21, correspond to the discrete orders seen in Figure 2.20. The zeroth order (centered at $u = 0$) corresponds to the constant DC term $A/2$. At either side are the $\pm$ first orders, where $u_1 = 1/p$. The $\pm$ second orders correspond to $u_2 = \pm 2/p$, and so on. It would follow that if an imaging system was not able to collect all diffracted orders propagating from a mask, complete reconstruction would not be possible. Furthermore, as higher-frequency information is lost, fine image detail is sacrificed. There is, therefore, a fundamental limitation to resolution for an imaging system determined by its inability to collect all possible diffraction information.

2.3.3.4   Finite Dense Features

The rectangular wave is very useful for understanding the fundamental concepts and Fourier analysis of diffraction. In reality, however, finite mask functions are dealt with rather than such infinite

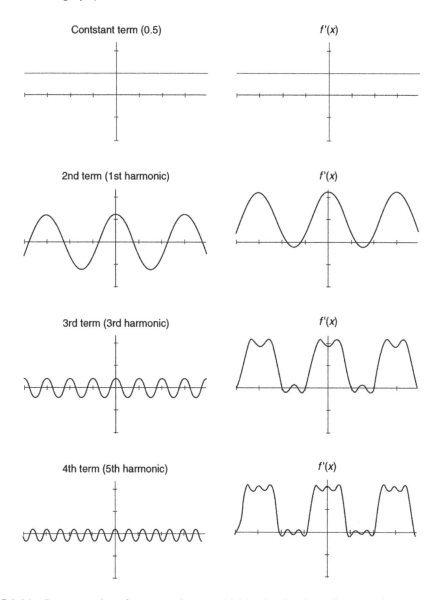

FIGURE 2.21 Reconstruction of a rectangular wave (right) using Fourier series expansion.

functions as the rectangular wave. The extent by which a finite number of mask features can be represented by an infinite function depends on the number of features present. Consider a mask consisting of five equal line/space pairs or a five-bar function, as shown in Figure 2.22. This mask function can be represented before as the convolution of a scaled rect (x) function and an impulse train comb(x). In order to limit the mask function to five features only, a windowing function must be introduced as follows:

$$m(x) = \left[\mathrm{rect}\left(\frac{x}{p/2}\right) * \mathrm{comb}\left(\frac{x}{p}\right) \right] \times \mathrm{rect}\left(\frac{x}{5p}\right)$$

As before, the spatial frequency distribution is a Fourier transform, but now each diffraction order is convolved with a sinc(u) function and scaled appropriately by the inverse width of the windowing function

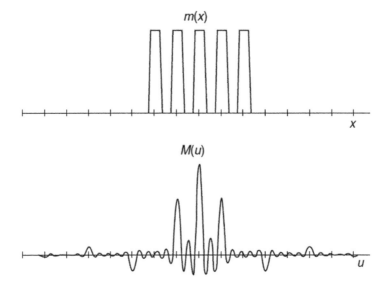

FIGURE 2.22 A five-bar mask function $m(x)$ and its corresponding coherent spatial frequency distribution $M(u)$.

$$M\left(u\right) = \left[\frac{A}{2}\operatorname{sinc}\left(\frac{u}{2u_0}\right)\sum_{n=-\infty}^{\infty}\delta\left(u-nu_0\right)\right] * 5\operatorname{sinc}\left(\frac{u}{u_0/5}\right)$$

As more features are added to the five-bar function, the width of the convolved sinc(u) is narrowed. At the limit where an infinite number of features is considered, the sinc(u) function becomes a $\delta(u)$, and the result is identical to the rectangular wave. At the other extreme, if a one-bar mask function is considered, the resulting spatial frequency distribution is the continuous function shown in Figure 2.17.

2.3.3.5 The Objective Lens

In a projection imaging system, the objective lens has the ability to collect a finite amount of diffracted information from a mask that is determined by its maximum acceptance angle or numerical aperture. A lens behaves as a linear filter for a diffraction pattern propagating from a mask. By limiting high-frequency diffraction components, it acts as a low-pass filter, blocking information propagating at angles beyond its capability. Information that is passed is acted on by the lens to produce a second inverse Fourier transform operation, directing a limited reconstruction of the mask object toward the image plane. It is limited not only by the loss of higher-frequency diffracted information but also by any lens aberrations that may act to introduce image degradation. In the absence of lens aberrations, imaging is referred to as *diffraction limited*. The influence of lens aberration on imaging will be addressed later. At this point, if an ideal diffraction-limited lens can be considered, the concept of a lens as a linear filter can provide insight into image formation.

2.3.3.6 The Lens as a Linear Filter

If an objective lens could produce an exact inverse Fourier transform of the Fraunhofer diffraction pattern emanating from an object, complete image reconstruction would be possible. A finite lens numerical aperture will prevent this. Consider a rectangular grating where $p\sin\theta = m\lambda$ describes the positions of the discrete coherent diffraction orders. If a lens can be described in terms of a two-dimensional pupil function $H(u,v)$, limited by its scaled numerical aperture, NA/λ, then

$$H(u,v) = 1 \quad \text{if } \sqrt{u^2 + v^2} < \frac{\text{NA}}{\lambda}$$

$$0 \quad \text{if } \sqrt{u^2 + v^2} > \frac{\text{NA}}{\lambda}$$

describes the behavior of the lens as a low-pass filter. The resulting image amplitude produced by the lens is the inverse Fourier transform of the mask's Fraunhofer diffraction pattern multiplied by this lens pupil function:

$$A(x,y) = \mathcal{F}\{M(u,v) \times H(u,v)\}$$

The image intensity distribution, known as the aerial image, is equal to the square of the image amplitude:

$$I(x,y) = |A(x,y)|^2$$

For the situation described, coherent illumination allows simplification of optical behavior. Diffraction at a mask is effectively a Fourier transform operation. Part of this diffracted field is collected by the objective lens, where diffraction is, in a sense, reversed through a second Fourier transform operation. Any losses incurred through limitations of a lens NA < 1.0 result in less than complete reconstruction of the original mask detail. To extend this analysis to real systems, an understanding of coherence theory is needed.

2.3.4   COHERENCE THEORY IN IMAGE FORMATION

Much has been written about coherence theory and the influence of spatial coherence on interference and imaging [10]. For projection imaging, three illumination situations are possible that allow the description of interference behavior. These are coherent illumination, where wavefronts are correlated and are able to interfere completely; incoherent illumination, where wavefronts are uncorrelated and unable to interfere; and partial coherent illumination, where partial interference is possible. Figure 2.23 shows the situation where spherical wavefronts are emitted from point sources that can be used to describe coherent, incoherent, and partial coherent illumination. With coherent illumination, spherical waves emitted by a single point source on axis result in plane waves normal to the optical axis when acted upon by a lens. At all positions on the mask, radiation arrives in phase. Strictly speaking, coherent illumination implies zero intensity. For incoherent illumination, an infinite collection of off-axis point sources results in plane waves at all angles ($\pm\pi$). The resulting illumination at the mask has essentially no phase-to-space relationship. For partially coherent illumination, a finite collection of off-axis point sources describes a source of finite extent, resulting in plane waves within a finite angle. The situation of partial coherence is of most interest for lithography, the degree of which will have a great influence on imaging results.

2.3.5   PARTIAL COHERENCE THEORY: DIFFRACTED-LIMITED RESOLUTION

The concept of degree of coherence is useful as a description of illumination condition. The Abbe theory of microscope imaging can be applied to microlithographic imaging with coherent or partially coherent illumination [11]. Abbe demonstrated that when a ruled grating is coherently illuminated and imaged through an objective lens, the resulting image depends on the lens numerical aperture. The minimum resolution that can be obtained is a function of both the illumination wavelength and the lens NA, as shown in Figure 2.24 for coherent illumination. Because no imaging is possible if no more than the undiffracted beam is accepted by the lens, it can be reasoned that a

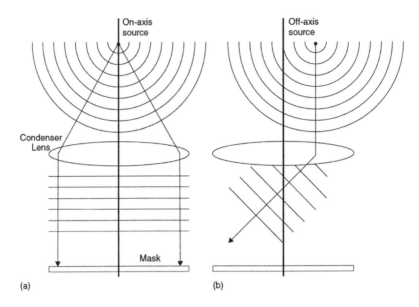

FIGURE 2.23 The impact of on-axis (a) and off-axis (b) point sources on illumination coherence. Plane waves result for each case and are normal to the optical axis only for an on-axis point.

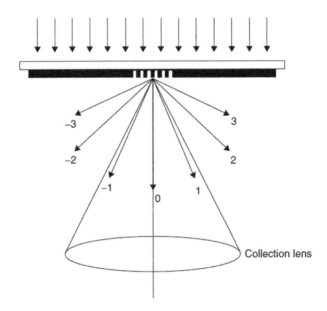

FIGURE 2.24 The condition for minimum diffraction-limited resolution for a coherently illuminated grating mask.

minimum of the first diffraction order is required for resolution. The position of this first order is determined as

$$\sin(\theta) = \frac{\lambda}{p}$$

Because a lens numerical aperture is defined as the sine of the half acceptance angle (θ), the minimum resolvable line width ($R = p/2$) becomes

$$R_{min} = \frac{p}{2} = 0.5\frac{\lambda}{NA}$$

Abbe's work made use of a smooth uniform flame source and a substage condenser to form its image in the object plane. To adapt to nonuniform lamp sources, Köhler devised a two-stage illuminating system to form an image of the source into the entrance pupil of the objective lens, as shown in Figure 2.25 [12]. A pupil at the condenser lens can control the NA of the illumination system. As the pupil is closed down, the source size (d_s) and the effective source size (d'_s) are decreased, resulting in an increase in the extent of coherency. Thus, Köhler illumination allows control of partial coherence. The degree of partial coherence (σ) is conventionally measured as the ratio of effective source size to full objective aperture size or the ratio of condenser lens NA to objective lens NA:

$$\text{Degree of coherence } (\sigma) = (d'_s/d_o) = (NA_C/NA_O)$$

In a Köhler illuminated system, the source is imaged into the pupil plane of the objective lens, and the object (the mask) is imaged at the image (the wafer) plane. This is shown in Figure 2.25, where rays are traced though the system to show these image locations. There are several NA locations in the configuration corresponding to object and image sides of the condenser lens and the projection lens. Specifically, θ_i and θ_i^* are the half collection angles associated with the condenser lens, where the associated NAs in a medium with refractive index n are $NA_i = \sin\theta_i$ and $NAi^* = \sin\theta_1^*$. The collection angles associated with the projection lens are θ_m (on the mask side of the lens) and θ_w (on the wafer side). The corresponding numerical apertures are $NA_m = n\sin\theta_m$ and $NA_w = n\sin\theta_w$. In a conventional projection system, air is the image medium, and NA is simply $\sin\theta_w$. For an immersion lithography system, NA is increased by the refractive index of the immersion fluid, which is about 1.44 at 193 nm with water. The reduction value of the projection lens is the ratio of NA_w/NA_m.

As σ approaches zero, a condition of coherent illumination exists. As σ approaches one, incoherent illumination exists. In lithographic projection systems, σ is generally in the range 0.3–0.9. Values below 0.3 can result in "ringing" in images; fringes that result from coherent interference effects similar to those shown as terms are added in Figure 2.21.

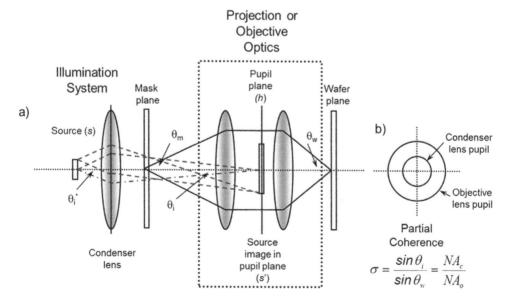

FIGURE 2.25 Schematic of Köhler illumination. The degree of coherence (σ) is determined as ds'/do or NA_c/NA_o.

Partial coherence can be thought of as taking an incoherent sum of coherent images. For every point within a source of finite extent, a coherent Fraunhofer diffraction pattern is produced that can be described by Fourier methods. For a point source on axis, diffracted information is distributed symmetrically and discretely about the axis. For off-axis points, diffraction patterns are shifted off axis, and, as all points are considered together, the resulting diffraction pattern becomes a summation of individual distributions. Figure 2.26 depicts the situation for a rectangular wave mask pattern illuminated with σ greater than zero. Here, the zeroth order is centered on axis but with a width >0, a result of the extent of partially coherent illumination angles. Similarly, each higher diffraction order also has width >0, an effective spreading of discrete orders. The impact of partial coherence is realized when the influence of an objective lens is considered. By spreading the diffraction orders about their discrete coherent frequencies, collection of higher–diffraction order energy is possible. If a situation exists where coherent illumination of a given mask pattern does not allow lens collection of diffraction orders beyond the zeroth order, partially coherent illumination would be preferred. Consider a coherently illuminated rectangular grating mask where ± first diffraction orders fall just outside a projection systems lens NA. With coherent illumination, imaging is not possible, as feature sizes fall below the $R=0.5\lambda$/NA limit. Through the use of partially coherent illumination, partial first–diffraction order information can be captured by the lens, and, together with zero-order energy, imaging is possible. Partial coherent illumination, therefore, is desirable, as mask features fall below $R=0.5\lambda$/NA in size. An optimum degree of coherence can be determined for a feature based on its size, the illumination wavelength, and the objective lens NA. Figure 2.27 shows the effect of partial coherence on imaging features of two sizes. The first case, Figure 2.27a, is one where aerial images for features are larger than the resolution possible for coherent illumination (here, 0.6λ/NA). As seen, any increase in partial coherence above $\sigma=0$ results in a degradation of the aerial image produced. At higher σ values, less first–diffraction order energy is collected, leading to demodulation by the zero order. In Figure 2.27b, features smaller than the resolution possible with coherent illumination (0.4λ/NA) are resolvable only as partial coherence levels increase above $\sigma=0$. It stands to reason that for every feature size and type, there exists a unique optimum partial coherence value that allows the greatest image improvement while also allowing the minimum degradation. Focus effects also need to be considered as partial coherence is optimized.

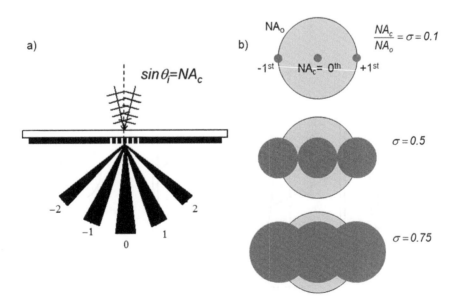

FIGURE 2.26 Spread of diffraction orders for partially coherent illumination. Resolution below 0.5λ/NA becomes possible.

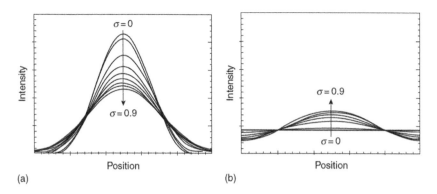

FIGURE 2.27 Intensity aerial images for features with various levels of partial coherence. Features corresponding to $0.6\lambda/$NA are relatively large and are shown on the left (a). Small features corresponding to $0.4\lambda/$NA are shown on the right (b).

2.4 IMAGE EVALUATION

The minimum resolution possible for equal lines and spaces with coherent illumination is that which satisfies the equation

$$R_{min} = hp_{min} = \frac{0.5\lambda}{NA}$$

which is commonly referred to as the *Rayleigh criterion* [13]. For nonequal lines and spaces, the equation is best defined in term of half-pitch (p/2), which is half of the sum of a line and space, as described earlier. Through incoherent or partially coherent illumination, resolution beyond this limit is made possible. Methods of image assessment are required to evaluate an image that is transferred through an optical system. As will be shown, such methods will also prove useful as an imaging system deviates from ideal and optical aberrations are considered.

2.4.1 OTF, MTF, AND PTF

The optical transfer function (OTF) is often used to evaluate the relationship between an image and the object that produced it [14]. In general, a transfer function is a description of an entire imaging process as a function of spatial frequency. It is a scaled Fourier transform of the point spread function (PSF) of the system. The PSF is the response of the optical system to a point object input, essentially the distribution of a point aerial image.

For a linear system, the transfer function is the ratio of the image modulation (or contrast) to object modulation (or contrast):

$$C_{image}(u)/C_{object}(u)$$

where contrast (C) is the normalized image modulation at frequency u

$$C(u) = (S_{max} - S_{min})/(S_{max} + S_{min}) \leq 1$$

Here, S is the image or object signal. To fulfill the requirements of a linear system, several conditions must be met. In order to be linear, the input of a system's response to the superposition of two inputs must equal the superposition of the individual responses. If $\Theta\{f(x)\} = g(x)$ represents the operation of a system on an input $f(x)$ to produce an output $g(x)$, then

$$\Theta\{f_1(x) + f_2(x)\} = g_1(x) + g_2(x)$$

represents a system linear with superposition. A second condition of a linear system is shift invariance, where a system operates identically at all input coordinates. Analytically, this can be expressed as

$$\Theta\{f(x-x_0)\} = g(x-x_0)$$

or a shift in input results in an identical shift in output. An optical system can be thought of as shift invariant in the absence of aberrations. Because the aberration of a system changes from point to point, the PSF can significantly vary from a center to an edge field point.

Intensities must add for an imaging process to the linear. In the coherent case of the harmonic analysis of a square wave in Figure 2.21, the amplitudes of individual components have been added rather than their intensities. Whereas an optical system is linear in amplitude for coherent illumination, it is linear in intensity only for incoherent illumination. The OTF, therefore, be can be used as a metric for analysis of image intensity transfer only for incoherent illumination. Modulation is expressed as

$$M = (I_{max} - I_{min})/(I_{max} + I_{min})$$

where I is image or object intensity as depicted in Figure 2.28. It is a transfer function for a system over a range of spatial frequencies. A typical OTF is shown in Figure 2.29, where modulation is plotted as a function of spatial frequency in cycles/mm. As seen, higher-frequency objects (corresponding to finer feature detail) are transferred through the system with lower modulation. The characteristics of the incoherent OTF can be understood by working backward through an optical system. We have seen that an amplitude image is the Fourier transform of the product of an object and the lens pupil function. Here, the object is a point, and the image is its PSF. The intensity PSF for an incoherent system is a squared amplitude PSF, also known as an Airy disk, shown in Figure 2.30. Because multiplication becomes a convolution via a Fourier transform, the transfer function of an imaging system with incoherent illumination is proportional to the self-convolution or the autocorrelation of the lens pupil function that is equivalent to the Fourier transform of its PSF. As seen in Figure 2.29, the OTF resembles a triangular function, the result of autocorrelation of a rectangular pupil function (which would be circular in two dimensions). For coherent

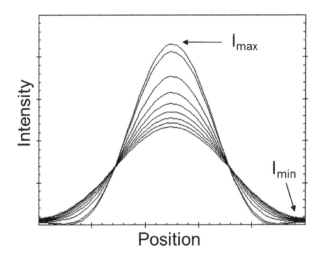

FIGURE 2.28 The decrease in modulation for a space feature aerial image, where I_{max} decreases and I_{min} increases with degradation.

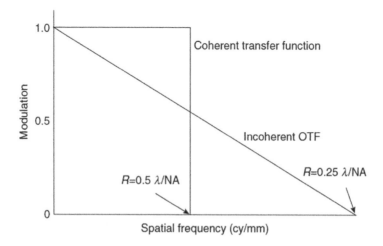

FIGURE 2.29 Typical incoherent optical transfer function (OTF) and coherent contrast transfer function (CTF).

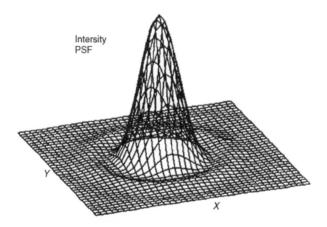

FIGURE 2.30 Intensity point spread function (PSF) for an incoherent system.

illumination, the coherent transfer function is proportional to the pupil function itself. The incoherent transfer function is twice as wide as the coherent transfer function, which is indicative that the cutoff frequency is twice that for coherent illumination. The limiting half-pitch resolution for incoherent illumination becomes

$$R_{min} = hp_{min} = \frac{0.25\lambda}{NA}$$

Although Rayleigh's criterion for incoherent illumination describes the point beyond which resolution is no longer possible, it does not give an indication of image quality at lower frequencies (corresponding to larger feature sizes). The OTF is a description of not only the limiting resolution but also the modulation at spatial frequencies up to that point.

The OTF is generally normalized to 1.0. The magnitude of the OTF is the modulation transfer function (MTF) that is commonly used. The MTF ignores phase information transferred by the system that can be described using the phase transfer function (PTF). Because of the linear properties of incoherent imaging, the OTF, MTF, and PTF are independent of the object. Knowledge of the pupil shape and the lens aberrations is sufficient to completely describe the OTF. For coherent and partially coherent systems, there are no such metrics that are object independent.

2.4.2   EVALUATION OF PARTIAL COHERENT IMAGING

For coherent or partially coherent imaging, the ratio of image modulation to object modulation is object dependent, making the situation more complex than for incoherent imaging. The concept of a transfer function can still be utilized, but limitations should be kept in mind. As previously shown, the transfer function of a coherent imaging system is proportional to the pupil function itself. The cutoff frequency corresponds exactly to the Rayleigh criterion for coherent illumination. For partially coherent systems, the transfer function is neither the pupil function nor its autocorrelation, resulting in a more complex situation. The evaluation of images requires a summation of coherent images correlated by the degree of coherence at the mask. A partially coherent transfer function must include a unique description of both the illumination system and the lens. Such a transfer function is commonly referred to as a *cross transfer function* or the *transmission cross coefficient* [15].

For a mask object with equal lines and spaces, the object amplitude distribution can be represented as

$$f(x) = a_0 + 2\sum_{n=1}^{\infty} a_n \cos(2\pi nux)$$

where
   x is image position
   u is spatial frequency

From partial coherence theory, the aerial image intensity distribution becomes

$$I(x) = A + B\cos(2\pi u_o x) + C\cos^2(2\pi u_o x)$$

which is valid for $u \geq (1+\sigma)/3$. The terms A, B, and C are given by

$$A = a_0^2 T(0,0) + 2a_1^2 \left[T(u_1,u_2) - T(-u_1,u_2) \right]$$

$$B = 4a_0 a_1 \mathrm{RE}\left[T(0,u_2) \right]$$

$$C = 4a_1^2 T(-u_1,u_2)$$

where $T(u_1,u_2)$ is the transmission cross coefficient, a measure of the phase correlation at two frequencies, u_1 and u_2. Image modulation can be calculated as $M = B/(A+C)$.

The concepts of an MTF can be extended to partially coherent imaging if generated for each object uniquely. Steel [16] developed approximations to an exact expression for the MTF for partially coherent illumination. Such normalized MTF curves (denoted as MTF_p curves) can be generated for various degrees of partial coherence [17], as shown in Figure 2.31. In systems with few aberrations, the impact of changes in the degree of partial coherence can be evaluated for any unique spatial frequency. By assuming a linear change in MTF between spatial frequencies u_1 and u_2, a correlation factor $G(\sigma,u)$ can be calculated that relates incoherent MTF_{INC} to partially coherent MTF_p:

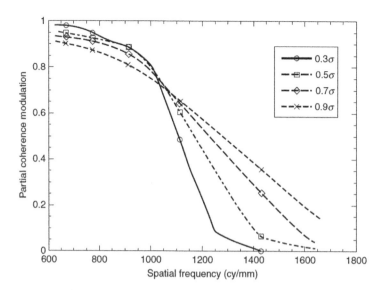

FIGURE 2.31 Partially coherent MTF$_p$ curves for σ values from 0.3 to 0.7 for a 365 nm, 0.37 NA diffraction-limited system.

$$u_1 \big| = (1-\sigma)\,\text{NA}/\lambda$$

$$u_2 \big| = (1+0.18\sigma)\,\text{NA}/\lambda$$

$$\big| = \frac{1}{1-(4/\pi)\sin(u\lambda/2\text{NA})} \quad u \le u_1$$

$$G(\sigma,u)\big| = \frac{1-(4/\pi)\sin(u_2\lambda/2\text{NA})(u-u_1)/(u_2-u_1)}{1-(4/\pi)\sin(u\lambda/2\text{NA})} \quad u_1 < u < u_2$$

$$< 1 \quad u_2 < u$$

The partially coherent MTF becomes

$$\text{MTF}_\text{P}(\sigma,u) = G(\sigma,u)\,\text{MTF}_\text{INC}(\sigma,u)$$

Using MTF curves such as those in Figure 2.31 for a 0.37 NA i-line system, partial coherence effects can be evaluated. With partial coherence of 0.3, the modulation at a spatial frequency of 1150 cycles mm^{-1} (corresponding to 0.43 µm lines) is near 0.35. Using a σ of 0.7, modulation increases by 71%. At 950 cycles mm^{-1}, however (corresponding to 0.53 µm lines), modulation decreases as partial coherence increases. The requirements of photoresist materials need to be addressed to determine the appropriate σ value for a given spatial frequency. The concept of critical modulation transfer function (CMTF) is a useful approximation for relating the minimum modulation required for a photoresist material. The minimum required modulation for resist with contrast γ can be determined as

$$\text{CMTF} = \frac{10^{1/\gamma}-1}{10^{1/\gamma}+1}$$

For a resist material with a γ of 2, a CMTF of 0.52 results. At this modulation, large σ values are best suited for the system depicted in Figure 2.31, and resolution is limited to somewhere near 0.38 µm. Optimization of partial coherence will be further addressed as additional image metrics are introduced.

As linearity is related to the coherence properties of mask illumination, stationarity is related to aberration properties across a specific image field. For a lens system to meet the requirements of stationarity, an isoplanatic patch needs to be defined in the image plane where the transfer function and PSF do not significantly change. Shannon [18] described this region as "much larger than the dimension of the significant detail to be examined on the image surface" but "small compared to the total area of the image." A real lens, therefore, requires a set of evaluation metrics; required for any lens will be a function of required performance and financial or technical capabilities. Although a large number of OTFs will better characterize a lens, more than a few may be impractical. Because an OTF will degrade with defocus, a position of best focus is normally chosen for lens characterization.

2.4.3 OTHER IMAGE EVALUATION METRICS

MTF or comparable metrics are limited to periodic features or gratings of equal lines and spaces. Other metrics may be used for the evaluation of image quality by measuring some aspect of an aerial image with less restriction on feature type. These may include measurements of image energy, image shape fidelity, critical image width, and image slope. Because feature width is a critical parameter for lithography, aerial image width is a useful metric for insight into the performance of resist images. A 30% intensity threshold is commonly chosen for image width measurement [19]. Few of these metrics, though, give an adequate representation of the impact of aerial image quality on resist process latitude.

Through measurement of the Aerial Image Log Slope (ILS), an indication of resist process performance can be obtained [20]. Exponential attenuation of radiation through an absorbing photoresist film leads to an exposure profile ($\delta e/\delta x$) related to aerial image intensity as

$$\text{Image Log Slope} = \text{ILS} = \frac{\delta}{\delta x} = \frac{\delta(\ln I)}{\delta x} \left[\text{at the mask edge}\right]$$

$$\text{Normalized Image Log Slope} = \text{NILS} = \text{ILS} \times \text{CD}$$

Because an exposure profile leads to a resist profile upon development, measurement of the slope of the log of an aerial image (at the mask edge) can be directly related to a resist image. Changes in this log aerial image gradient will, therefore, directly influence resist profile and process latitude. Using ILS as an image metric, aerial image plots such as those in Figure 2.27 can be more thoroughly evaluated. Shown in Figure 2.32 is a plot of image log slope versus partial coherence for features of size $R = 0.4$ to $0.6\lambda/NA$. As can be seen, for increasing levels of partial coherence, image log slope increases for features smaller than $0.5\lambda/NA$ and decreases for features smaller than $0.5\lambda/NA$. It is important to notice, however, that all cases converge to a similar image log slope value. Improvements for small features achieved by increasing partial coherence values cannot improve the aerial image in a way equivalent to decreasing wavelength or increasing NA.

To determine the minimum usable ILS value and optimize situations such as the one above for use with a photoresist process, resist requirements need to be considered. As minimum image modulation required for a resist (CMTF) has been related to resist contrast properties, there is also a relationship between resist performance and minimum ILS requirements. As bulk resist properties such as contrast may not be adequately related to process-specific responses such as feature size control, exposure latitude, or depth of focus, exposure matrices can provide usable depth-of-focus information for a resist-imaging system based on exposure and feature size specifications. Relating depth of field (DOF) to aerial image data for an imaging system can result in determination of a

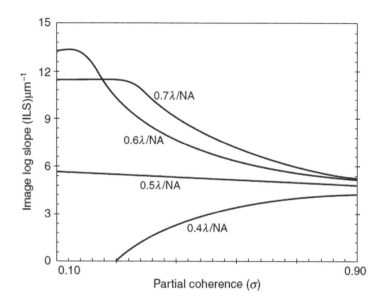

FIGURE 2.32 Image log slope (ILS) versus partial coherence for dense features from 0.4 to 0.7λ/NA in size.

minimum ILS specification. Although ILS is feature size dependent (in units of μm⁻¹), image log slope normalized by multiplying it by the feature width is not. A minimum Normalized Image Log Slope (NILS) can then be determined for a resist-imaging system with less dependence on feature size. A convenient rule-of-thumb value for minimum NILS is between 6 and 8 for a single-layer positive resist with good performance.

With image evaluation requirements established, Rayleigh's criterion can be revisited and modified for other situations of partial coherence. A more general form becomes

$$R = \frac{k_1 \lambda}{\text{NA}}$$

where k_1 is a process-dependent factor that incorporates everything in a lithography process that is not wavelength or NA. Its importance should not be minimized, as any process or system modification that allows improvements in resolution effectively reduces the k_1 factor. Diffraction-limited values are 0.25 for incoherent and 0.50 for coherent illumination, as previously shown. For partial coherence, k_1 can be expressed as

$$k_1 = \frac{1}{2(\sigma + 1)}$$

where the minimum resolution is that which places the ± first–diffraction order energy within the objective lens pupil, as shown in Figure 2.27.

2.4.4 DEPTH OF FOCUS

Depth of focus needs to be considered along with resolution criteria when imaging with a lens system. Depth of focus is defined as the distance along the optical axis that produces an image of some suitable quality. The paraxial (i.e. small angle) Rayleigh depth of focus generally takes the form

$$\text{DOF} = \pm \frac{k_2 \lambda}{n \sin^2 \theta}$$

where k_2 is also a process-dependent factor. In air, where the refractive index is 1.0, the denominator becomes NA^2. For immersion lithography, the index of the fluid needs to be considered. For a resist material of reasonably high contrast, k_2 may be on the order of 0.5. A process-specific value of k_2 can be defined by determining the resulting useful DOF after specifying exposure latitude and tolerances. DOF decreases linearly with wavelength and as the square of numerical aperture. As measures are taken to improve resolution, it is more desirable to decrease wavelength than to increase NA. Depth of focus is closely related to defocus, the distance along the optical axis from the location of best focus. The acceptable level of defocus for a lens system will determine the usable DOF. Tolerable levels of this aberration will ultimately be determined by the entire imaging system as well as the feature sizes of interest.

To understand the interdependence of image quality and focus can be thought of as deviations from a perfect spherical wave emerging from the exit pupil of a lens toward an image point. This is analogous to working backward through an optical system where a true point source in image space would correspond to a perfect spherical wave at the lens exit pupil. As shown in Figure 2.33, the deviation of an actual wavefront from an unaberrated wavefront can be measured in terms of an optical path difference (OPD). The OPD in a medium is the product of the geometrical path length and the refractive index. For a point object, an ideal spherical wavefront leaving the lens pupil is represented by a dashed line. This wavefront will come to focus as a point in the image plane. Compared to this reference wavefront, a defocused wavefront (one that would focus at a point some distance from the image plane) introduces error in the optical path distance to the image plane. This error increases with pupil radius. The resulting image will generally no longer resemble a point; instead, it will be blurred.

The acceptable DOF for a lithographic process can be determined by relating OPD to phase error. An optical path is best measured in terms of the number (or fraction) of corresponding waves. OPD is realized, therefore, as a phase-shifting effect or phase error (Φ_{err}) that can be expressed as

$$\Phi_{err} = \frac{2\pi}{\lambda} OPD$$

By determining the maximum allowable phase error for a process, an acceptable level of paraxial defocus can be determined. Consider again Figure 2.33. The optical path distance can be related to defocus (δ) as

$$OPD = \delta(1 - \cos\theta) = \frac{\delta}{2}\left(\sin^2\theta + \frac{\sin^4\theta}{4} + \frac{\sin^6\theta}{8} + \dots\right)$$

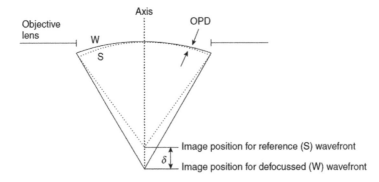

FIGURE 2.33 The optical path difference (OPD) for wavefronts corresponding to upper and lower most positions in the image plane. DOF decreases with larger OPD.

$$\text{OPD} \approx \frac{\delta}{2}\sin^2\theta = \frac{\Phi_{err}\lambda}{2\pi}$$

for small angles (the Fresnel approximation). Defocus can now be expressed as

$$\delta = \frac{\Phi_{err}\lambda}{\pi\sin^2\theta} = \left(\frac{\Phi_{err}}{\pi}\right)\frac{\lambda}{\text{NA}^2}$$

A maximum phase error term can be determined by defining the maximum allowable defocus that will maintain process specifications. DOF can, therefore, be expressed in terms of corresponding defocus (δ) and phase error (Φ_{en}/π) terms through use of the process factor k_2

$$\text{DOF} = \pm\frac{k_2\lambda}{n\sin^2\theta}$$

as previously seen.

If the distribution of mask frequency information in the lens pupil is considered, it is seen that the impact of defocus is realized as zero and first diffraction orders travel different optical path distances. For coherent illumination, the zero order experiences no OPD, while the $\pm$ first orders go through a pupil-dependent OPD. It follows that only features that have sufficiently important information (i.e., first diffraction orders) at the edge of the lens aperture will possess a DOF as calculated by the full lens NA. For larger features whose diffraction orders are distributed closer to the lens center, DOF will be substantially higher. For dense features of pitch p, an effective NA can be considered for each feature size that can subsequently be used for DOF calculation:

$$\text{NA}_{effective} \sim \frac{\lambda}{p}$$

As an example, consider dense 0.5 μm features imaged using coherent illumination with a 0.50 NA objective lens and 365 nm illumination. The first diffraction orders for these features are contained within the lens aperture at an effective NA of 0.365 rather than 0.50. The resulting DOF (for a k_2 of 0.5) is, therefore, closer to ±1.37 μm than to ±0.73 μm as determined for the full NA.

The distribution of diffraction orders needs to be considered in the case of partial coherence. By combining the wavefront description in Figure 2.33 with the frequency distribution description in Figure 2.26, DOF can be related to partial coherence, as shown in Figure 2.34. For coherent illumination, there is a discrete difference in optical path length traveled between diffraction orders. By using partial coherence, however, there is an averaging effect of OPD over the lens pupil. By distributing frequency information over a broad portion of the lens pupil, the difference in path lengths experienced between diffraction orders is reduced. In the limit of complete incoherence, the zero and first diffraction orders essentially share the same pupil area, effectively eliminating the effects of defocus (which is possible only in the absence of any higher-order diffraction terms). This can be seen in Figure 2.35, which is similar to Figure 2.32 except that a large defocus value has been incorporated. Here, it is seen that at low partial coherence values, ILS remains high, indicating that a greater DOF is possible.

2.5 IMAGING ABERRATIONS AND DEFOCUS

Discussion of geometrical image formation has so far been limited to the paraxial region that allows determination of the size and location of an image for a perfect lens. In reality, some degree of lens error or aberration exists in any lens, causing deviation from this first-order region. For microlithographic lenses, an understanding of the tolerable level of aberrations and interrelationships

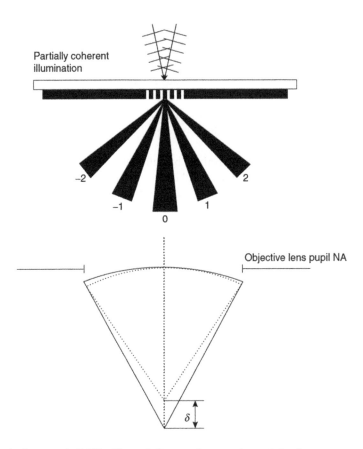

FIGURE 2.34 An increase in DOF will result from an increase in partial coherence as path-length differences are averaged across the lens pupil. In the limit for incoherent illumination, the zero and first diffraction orders fill the lens pupil, and DOF is theoretically infinite (in the absence of higher orders).

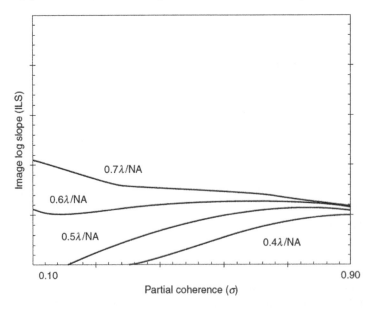

FIGURE 2.35 ILS versus partial coherence for dense features with λ/NA^2 of defocus. Degradation is minimal with higher σ values.

becomes more critical than for most other optical applications. To understand their impact on image formation, aberrations can be classified by their origin and effects. Traditionally referred to as the *Seidel aberrations*, these include monochromatic aberrations: spherical, coma, astigmatism, field curvature, and distortion as well as chromatic aberration. A brief description of each aberration will be given along with the effect of each on image formation. In addition, defocus is considered as an aberration and will be addressed. Although each aberration is discussed uniquely, all aberration types at some level will nearly always be present.

2.5.1 SPHERICAL ABERRATION

Spherical aberration is a variation in focus as a function of radial position in a lens. Spherical aberration exists for objects either on or off the optical axis. Figure 2.36 shows the situation for a distant on-axis point object where rays passing through the lens near the optical axis come into focus nearer the paraxial focus than rays passing through the edge of the lens. Spherical aberration can be measured as either a longitudinal (or axial) or a transverse (or lateral) error. Longitudinal spherical aberration is the distance from the paraxial focus to the axial intersection of a ray. Transverse spherical aberration is similar, but it is measured in the vertical direction. Spherical aberration is often represented graphically in terms of ray height, as in Figure 2.37, where longitudinal error (LA_r) is plotted against ray height at the lens (Y_r). The effect of spherical aberration on a point image is a blurring effect or the formation of a diffuse halo by peripheral rays. The best image of a point object is no longer located at the paraxial focus; instead, it is at the position of the circle of least confusion, or least blur. Longitudinal spherical aberration increases as the square of the aperture, and it is influenced by lens shape. In general, a positive lens will produce an undercorrection of spherical aberration (a negative value), whereas a negative lens will produce an overcorrection. As with most primary aberrations, there is also a dependence on object and image position. As an object changes position, for example, ray paths change, leading to potential increases in aberration levels. If a lens system is scaled up or down, aberrations are also scaled. This scaling would lead to a change in field size but not in numerical aperture. A simple system that is scaled up by 2× with a 1.5× increase in NA, for example, would lead to a 4.5× increase in longitudinal spherical aberration.

2.5.2 COMA

Coma is an aberration of object points that lie off axis. It is a variation in magnification with aperture that produces an image point with a diffuse comet-like tail. As shown in Figure 2.38, rays passing through the center and edges of a lens are focused at different heights. Tangential coma is measured as the distance between the height of the lens rim ray and the lens center ray. Unlike spherical aberration, comatic flare is not symmetric, and point image location is sometimes difficult.

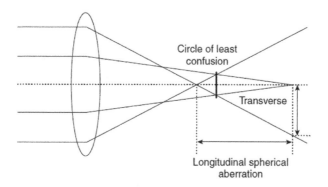

FIGURE 2.36 Spherical aberration for an on-axis point object.

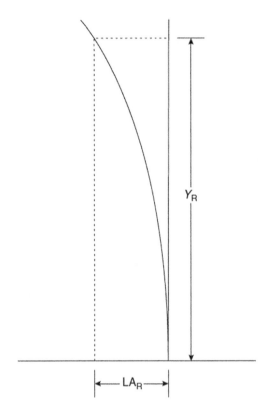

FIGURE 2.37 Longitudinal spherical aberration (LA_R) plotted against ray height (Y_R).

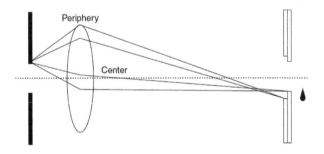

FIGURE 2.38 Coma for an off-axis object point. Rays passing through the center and edges of a lens are focused at different heights.

Coma increases with the square of the lens aperture and also with field size. Coma can be reduced, therefore, by stopping down the lens and limiting field size. It can also be reduced by shifting the aperture and optimizing field angle. Unlike spherical aberration, coma is linearly influenced by lens shape. Coma is positive for a negative meniscus lens and decreases to negative for a positive meniscus lens.

2.5.3 Astigmatism and Field Curvature

Astigmatism is also an off-axis aberration. With astigmatism present, rays that lie in different planes do not share a common focus. Consider, for instance, a plane that contains the chief ray and the optical axis, known as the *tangential plane*. The plane perpendicular to this is called the *sagittal plane*

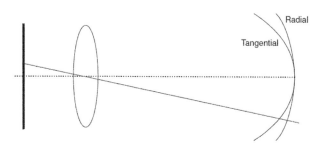

FIGURE 2.39 Astigmatism for an off-axis object point. Rays in different planes do not share a common focus.

that also contains the chief ray. Rays in the tangential plane will come to focus at the tangential focal surface, as shown in Figure 2.39. Rays in the sagittal plane will come to focus at the sagittal focal surface, and if these two do not coincide, the intermediate surface is called the *medial image surface*. If no astigmatism exists, all surfaces coincide, with the lens field curvature called the *Petzval curvature*. Astigmatism does not exist for on-axis points and increases with the square of field size. Undercorrected astigmatism exists when the tangential surface is to the left of the sagittal surface. Overcorrection exists when the situation is reversed. Point images in the presence of astigmatism generally exhibit circular or elliptical blur.

Field curvature results in a Petzval surface that is not a plane. This prevents imaging of point objects in focus on a planar surface. Field curvature and astigmatism are closely related and must be considered together if methods of field flattening are used for correction.

2.5.4 DISTORTION

Distortion is a radial displacement of off-axis image points, essentially a field variation in magnification. If an increase in magnification occurs as distance from field center increases, a pincushion or overcorrected distortion exists. For a decrease in magnification, barrel distortion results. Distortion is expressed either as a dimensional error or as a percentage. It varies as a third power of field size dimensionally or as the square of field size in terms of percentage. The location of the aperture stop will greatly influence distortion.

2.5.5 CHROMATIC ABERRATION

Chromatic aberration is a change in focus with wavelength. Because the refractive index of glass materials is not constant with wavelength, the refractive properties of a lens will vary. Generally, glass dispersion is negative, meaning that refractive index decreases with wavelength. This leads to an increase in refraction for shorter wavelengths and image blurring using multiple wavelengths for imaging. Figure 2.40 shows a longitudinal chromatic aberration for two wavelengths, a measure of

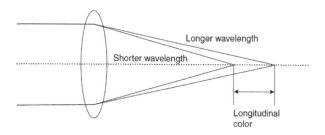

FIGURE 2.40 Chromatic aberration for an on-axis point using two wavelengths. For a positive lens, focal length is shortened with decreasing wavelength.

the separation of the two focal positions along the optical axis. For this positive lens, there is a shortening of focal length with decreasing wavelength or undercorrected longitudinal chromatic aberration. The effects of chromatic aberration are of great concern when light is not temporally coherent.

For most primary aberrations, some degree of control is possible by sacrificing aperture of field size. Generally, these methods are not sufficient to provide adequate reduction, and methods of lens element combination are utilized. Lens elements with opposite aberration sign can be combined to correct for specific aberration. Chromatic and spherical aberration can be reduced through use of an achromatic doublet, where a positive element (biconvex) is used in contact with a negative element (negative meniscus or planoconcave). On its own, the positive element possesses undercorrected spherical as well as undercorrected chromatic aberration. The negative element on its own has both overcorrected spherical and overcorrected chromatic aberration. If the positive element is chosen to have greater power as well as lower dispersion than the negative element, positive lens power can be maintained while chromatic aberration is reduced. To address the reduction of spherical aberration with the doublet, the glass refractive index is also considered. As shorter wavelengths are considered for lens systems, the choice of suitable optical materials becomes limited. At wavelengths below 300 nm, few glass types exist, and aberration correction, especially for chromatic aberration, becomes difficult.

Although aberration correction can be quite successful through the balancing of several elements of varying power, shape, and optical properties, it is difficult to correct a lens over the entire aperture. A lens is corrected for rays at the edge of the lens. This results in either overcorrection or undercorrection in different zones of the lens. Figure 2.41, for example, is a plot of longitudinal spherical aberration (LA) as a function of field height. At the center of the field, no spherical aberration exists. This lens has been corrected so that no spherical aberration exists at the edge of the field also. Other portions of the field exhibit undercorrection, and positions outside the field edge become overcorrected. The worst-case zone here is near 70%, which is common for many lens systems.

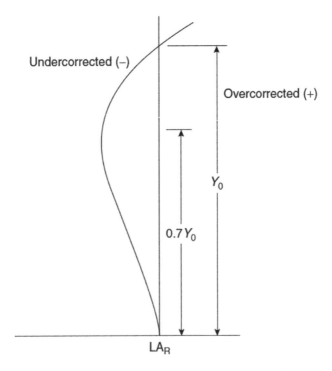

FIGURE 2.41   Spherical aberration corrected on axis and at the edge of the field. Largest aberration (—) is at a 70% zone position.

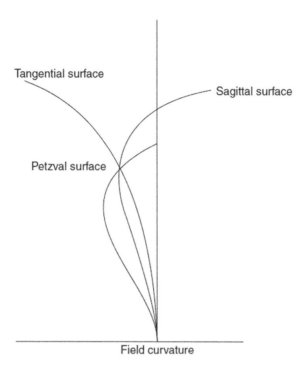

FIGURE 2.42 Astigmatism and field curvature plotted as a function of image height. One field position exists where surfaces coincide.

Figure 2.42 shows astigmatism or field curvature plotted as a function of image height. For this lens, there exists one position in the field where tangential and sagittal surfaces coincide or astigmatism is zero. Astigmatism is overcorrected closer to the axis (relative to the Petzval surface), and it is undercorrected farther out.

2.5.6 WAVEFRONT ABERRATION DESCRIPTIONS

For reasonably small levels of lens aberrations, analysis can be accomplished by considering the wave nature of light. As demonstrated for defocus, each primary aberration will produce unique deviations in the wavefront within the lens pupil. An aberrated pupil function can be described in terms of wavefront deformation as

$$P(r,\theta) = H(r,\theta)\exp\left[i\frac{2\pi}{\lambda}W(r,\theta)\right]$$

The pupil function is represented in polar coordinates, where $W(r,\theta)$ is the wavefront aberration function, and $H(r,\theta)$ is the pupil shape, generally circular. Each aberration can, therefore, be described in terms of the wavefront aberration function $W(r,\theta)$. Table 2.1 shows the mathematical description of $W(r,\theta)$ for primary aberrations: spherical, coma, astigmatism, and defocus. As an example, defocus aberration can be described in terms of wavefront deformation. Using Figure 2.33, the aberration of the wavelength w to the reference wavefront s is the OPD between the two. The defocus wave aberration $W(r)$ increases with aperture as [21]

$$W(r) = \frac{n}{2}\left(\frac{1}{R_s} - \frac{1}{R_w}\right)r^2$$

where R_s and R_w are radii of two spherical surfaces. Longitudinal defocus is defined as $(R_s - R_w)$. Defocus wave aberration is proportional to the square of the aperture distance, as previously seen.

TABLE 2.1

Mathematical Description for Primary Aberrations and Values of Peak-to-Valley Aberrations

Aberration	$W(r, \theta)$	W_{p-v}
Defocus	Ar^2	A[a]
Spherical	Ar^4	A
Balanced spherical	$A(r^4 - r^2)$	$A/4$
Coma	$Ar^3 \cos\theta$	$2A$
Balanced coma	$A(r^3 - 2r/3)\cos\theta$	$2A/3$
Astigmatism	$Ar^2 \cos^2\theta$	A
Balanced astigmatism	$(A/2)r^2 \cos^2\theta$	A

[a] The A coefficient represents the peak value of an aberration.

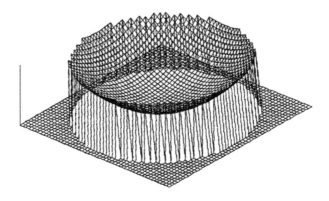

FIGURE 2.43 Defocus aberration (r^2) plotted as pupil wavefront deformation. Total OPD is 0.25λ.

Shown in Figures 2.43 through 2.46 are three-dimensional plots of defocus, spherical, coma, and astigmatism as wavefront OPD in the lens pupil. The plots represent differences between an ideal spherical wavefront and an aberrated wavefront. For each case, 0.25 waves of each aberration are present. Higher-order aberration terms also produce unique and related shapes in the lens pupil. Figure 2.47 shows a wavefront comprised of OPD from a combination of aberrations in one dimension and over the entire pupil.

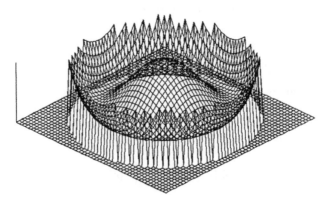

FIGURE 2.44 Primary spherical aberration (r^4).

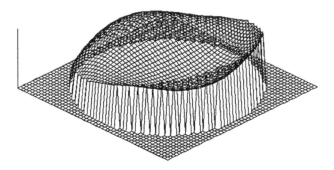

FIGURE 2.45 Primary coma aberration ($r^3 \cos\theta$).

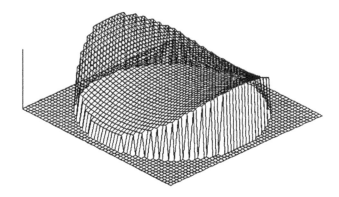

FIGURE 2.46 Primary astigmatism ($r^2 \cos^2\theta$).

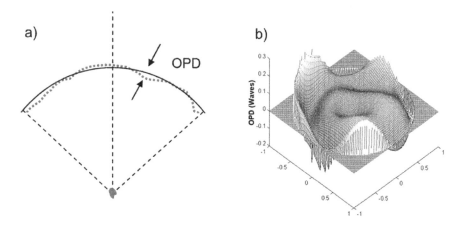

FIGURE 2.47 Aberration in a lens pupil represented as the optical path difference (OPD) at (a) a single rotational (azimuthal) location and (b) across the entire pupil. The full pupil OPD can be described as a composite of primary aberrations.

2.5.7 ZERNIKE POLYNOMIALS

Balanced aberrations are desired to minimize the variance within a wavefront. Zernike polynomials describe balanced aberration in terms of a set of coefficients that are orthogonal over a unit circle polynomial [22]. The polynomial can be expressed in Cartesian (x,y) or polar (r,θ) terms, and it can be applied to rotationally symmetrical nonsymmetrical systems. Because these polynomials are orthogonal, each term individually represents a best fit to the aberration data. Generally, fringe Zernike coefficient normalization to the pupil edge is used in lens design, testing, and simulation. Other normalizations do exist, including a renormalizing to the root mean square (RMS) wavefront aberration. The fringe Zernike coefficients are shown in Table 2.2 along with corresponding primary aberrations. The expansion of the polynomials are also shown in Figure 2.48.

TABLE 2.2

Fringe Zernike Polynomial Coefficients and Corresponding Aberrations

Term	Fringe Zernike Polynomial	Aberration
1	1	Piston
2	$r \cos(\alpha)$	X tilt
3	$r \sin(\alpha)$	Y tilt
4	$2r^2 - 1$	Defocus
5	$r^2\cos(2\alpha)$	Third-order astigmatism
6	$r^2 \sin(2\alpha)$	Third-order 45° astigmatism
7	$(3r^3 - 2r) \cos(\alpha)$	Third-order X coma
8	$(3r^3 - 2r) \sin(\alpha)$	Third-order Y coma
9	$(6r^4 - 6r^2)\,1$	Third-order spherical
10	$r^3 \cos(3\alpha)$	Third-order X three foil
11	$r^3 \sin(3\alpha)$	Third-order Y three foil
12	$(4r^4 - 3R^2) \cos(2\alpha)$	Fifth-order astigmatism
13	$(4R^4 - 3R^2) \sin(2\alpha)$	Fifth-order 45° astigmatism
14	$(10r^5 - 12r^3 + 3r) \cos(\alpha)$	Fifth-order X coma
15	$(10r^5 - 12r^3 + 3r) \sin(\alpha)$	Fifth-order Y coma
16	$20r^6 - 30r^4 + 12r^2 - 1$	Fifth-order spherical
17	$r^4 \cos(4\alpha)$	
18	$r^4 \sin(4\alpha)$	
19	$(5r^5 - 4r^3) \cos(3\alpha)$	
20	$(5r^5 - 4r^3) \sin(3\alpha)$	
21	$(15r^6 - 20r^4 + 6r^2) \cos(2\alpha)$	Seventh-order astigmatism
22	$(15r^6 - 20r^4 + 6r^2) \sin(2\alpha)$	Seventh-order 45° astigmatism
23	$(35r^7 - 60r^5 + 30r^3 - 4r) \cos(\alpha)$	Seventh-order X coma
24	$(35r^7 - 60r^5 + 30r^3 - 4r) \sin(\alpha)$	Seventh-order Y coma
25	$70r^8 - 140r^6 + 90r^4 - 20r^2 + 1$	Seventh-order spherical
26	$r^5 \cos(5\alpha)$	
27	$r^5 \sin(5\alpha)$	
28	$(6r^6 - 5r^4) \cos(4\alpha)$	
29	$(6r^6 - 5r^4) \sin(4\alpha)$	
30	$(21r^7 - 30r^5 + 10r^3) \cos(3\alpha)$	
31	$(21r^7 - 30r^5 + 10r^3) \sin(3\alpha)$	
32	$(56r^8 - 105r^6 + 60r^4 - 0r^2) \cos(2\alpha)$	Ninth-order astigmatism
33	$(56r^8 - 105r^6 + 60r^4 - 0r^2) \sin(2\alpha)$	Ninth-order 45° astigmatism
34	$(126r^9 - 280r^7 + 210r^5 - 60r^3 + 5r) \cos(\alpha)$	Ninth-order X coma
35	$(126r^9 - 280r^7 + 210r^5 - 60r^3 + 5r) \sin(\alpha)$	Ninth-order Y coma
36	$252r^{10} - 630r^8 + 560r^6 - 210r^4 + 30r^2 - 1$	Ninth-order spherical
37	$924r^{12} - 2772r^{10} + 3150r^8 - 1680r^6 + 420r^4 - 42r^2 + 1$	Eleventh-order spherical

Coefficients are normalized to the pupil edge.

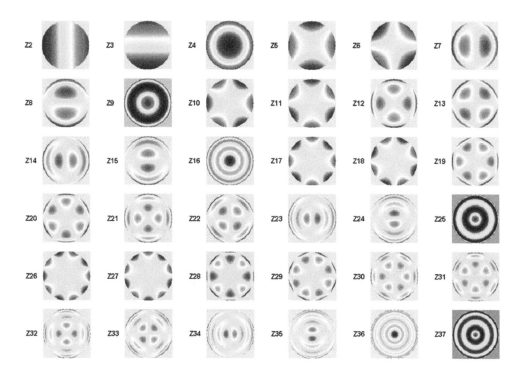

FIGURE 2.48 The Zernike polynomial expansion representations of individual aberrations including spherical terms (Z9, Z16, Z25, Z36, and Z37), astigmatism (Z5, Z6, Z12, Z13, Z21, Z22, Z32, and Z33), coma (Z7, Z8, Z14, Z15, Z23, Z24, 34, and Z35).

2.5.8 ABERRATION TOLERANCES

For OPD values less than a few wavelengths of light, aberration levels can be considered small. Because any amount of aberration results in image degradation, tolerance levels must be established for lens systems, dependent on application. This results in the need to consider not only specific object requirements and illumination but also resist requirements. For microlithographic application, resist and process capability will ultimately influence the allowable lens aberration level.

Conventionally, an acceptably diffraction-limited lens is one that produces no more than one quarter-wavelength ($\lambda/4$) wavefront OPD, as depicted in Figure 2.47. For many nonlithographic lens systems, the reduced performance resulting from this level of aberration is allowable. To measure image quality as a result of lens aberration, the distribution of energy in an intensity PSF (or Airy disk) can be evaluated. The ratio of energy at the center of an aberrated point image to the energy at the center of an unaberrated point image is known as the *Strehl ratio*, as shown in Figure 2.49. For an aberration-free lens, of course, the Strehl ratio is 1.0. For a lens with $\lambda/4$ OPD, the Strehl ratio is 0.80, nearly independent of the specific primary aberration types present. This is conventionally known as the *Rayleigh $\lambda/4$ rule* [23]. A general rule of thumb is that the effects on image quality are similar for identical levels of primary wavefront aberration. Table 2.3 shows the relationship between peak-to-valley (P-V) OPD, RMS OPD, and Strehl ratio. For low-order aberration, RMS OPD can be related to P-V OPD by

$$\text{RMS OPD} = \frac{(\text{P-V OPD})}{3.5}$$

The Strehl ratio can be used to understand a good deal about an imaging process. The PSF is fundamental to imaging theory and can be used to calculate the diffraction image of both coherent and

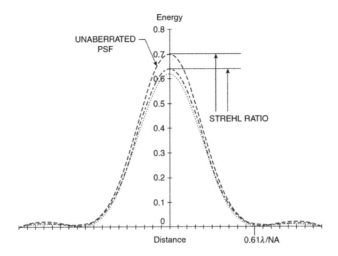

FIGURE 2.49   Strehl ratio for an aberrated point image.

TABLE 2.3
Relationship between Peak-to-Valley OPD, RMS OPD, and Strehl Ratio

P-V OPD	RMS OPD	Strehl Ratio[a]
0.0	0.0	1.00
$0.25RL = \lambda/16$	0.018λ	0.99
$0.5RL = \lambda/8$	0.036λ	0.95
$1.0RL = \lambda/4$	0.07λ	0.80
$2.0RL = \lambda/2$	0.14λ	0.4

[a] Strehl ratios below 0.8 do not provide for a good metric of image quality.

---

incoherent objects. By convolving a scaled object with the lens system PSF, the resulting incoherent image can be determined. In effect, this becomes the summation of the irradiance distribution of the image elements. Similarly, a coherent image can be determined by adding the complex amplitude distributions of the image elements. Figures 2.50 and 2.51 show the effects of various levels of aberration and defocus on the PSF for an otherwise ideal lens system. Figure 2.50a–c show PSFs for spherical, coma, and astigmatism aberration at 0.15λ OPD levels. It is seen that the aberrations produce similar levels of reduced peak intensities. Energy distribution, however, varies somewhat with aberration type. Figure 2.51 shows how PSFs are affected by these primary aberrations combined with defocus. For each aberration type, defocus is fixed at 0.25λ OPD.

To extend evaluation of aberrated images for partially coherent systems, the use of the PSF (or OTF) becomes difficult. Methods of aerial image simulation can be utilized for lens performance evaluation. By incorporating lens aberration parameters into a scalar or vector diffraction model, most appropriately through use of Zernike polynomial coefficients, aerial image metrics such as image modulation of ILS can be used. Figure 2.52 shows the results of a three-bar mask object imaged through an aberrated lens system at a partial value coherence of 0.5. Figure 2.52a shows aerial images produced in the presence of 0.15λ OPD spherical aberration with $\pm0.25\lambda$ OPD of defocus. Figure 2.52b and c show resulting images with coma and astigmatism, respectively.

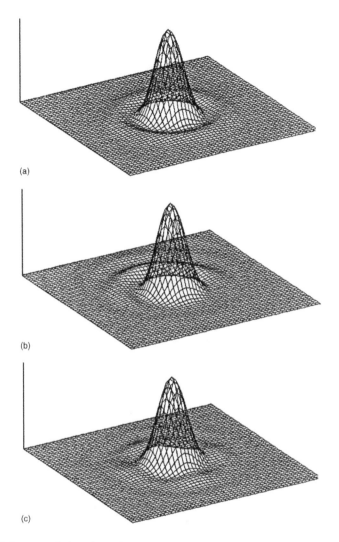

FIGURE 2.50 Point spread functions for 0.15λ of primary (a) spherical aberration, (b) coma, and (c) astigmatism.

Figure 2.52d shows the unaberrated aerial image through the same defocus range. These aerial image plots suggest that the allowable aberration level will be influenced by resist capability, as more capable resists and processes will tolerate larger levels of aberration.

2.5.9 MICROLITHOGRAPHIC REQUIREMENTS

It is evident from the preceding image plots that the Rayleigh $\lambda/4$ rule may not be suitable for microlithographic applications, where small changes in the aerial image can be translated into photoresist and result in substantial loss of process latitude. To establish allowable levels of aberration tolerances, photoresist requirements need to be considered along with process specifications. For a photoresist with reasonably high contrast and reasonably low NILS requirements, a balanced aberration level of 0.05λ OPD and a Strehl ratio of 0.91 would have been acceptable a short while ago [24]. As process requirements are tightened, demands on a photoresist process will be increased to maintain process latitude at this level of aberration. As shorter-wavelength technology is pursued, resist and process demands require that aberration tolerance levels be further reduced

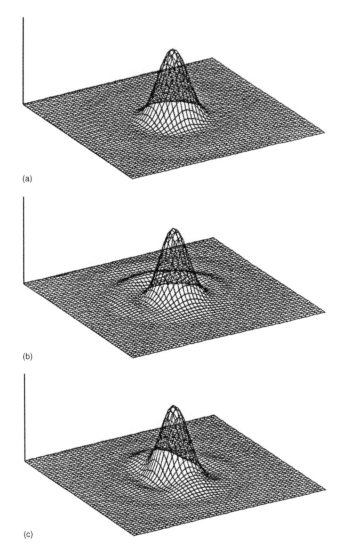

FIGURE 2.51 Point spread functions for 0.15λ of primary aberrations combined with 0.25λ of defocus: (a) spherical aberration, (b) coma, and (c) astigmatism.

to 10% of this level or better. For 193 nm lithography, this corresponds to an astonishing 1 nm wavefront deviation. Furthermore, the variation of any single aberration across the field is at the single nanometer level. It is also important to realize that aberrations cannot be strictly considered to be independent as they contribute to image degradation in a lens. In reality, aberrations are balanced with one another to minimize the size of an image point in the image plane. Although asymmetric aberrations (i.e., coma, astigmatism, and lateral chromatic aberration) should be minimized for microlithographic lens application, this may not necessarily be the case for spherical aberration. This occurs because imaging is carried out not through a uniform medium toward an imaging plane but instead, through several material media and within a photoresist layer. Figure 2.53 shows the effects of imaging in photoresist with an aberration-free lens using a scalar diffraction model and a positive resist model [25] for simulation. These are plots of resist feature width as a function of focal position for various levels of exposure. Focal position is chosen to represent the resist top surface (zero position) as well as a range below (negative) and above (positive) the top surface. This focus exposure matrix does not behave symmetrically throughout the entire focal range. Change

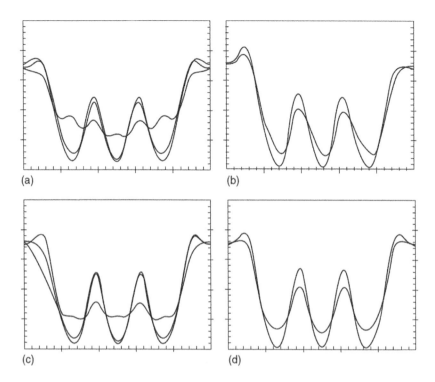

FIGURE 2.52 Aerial images for three-bar mask patterns imaged with a partial coherence of 0.5 (a) 0.15λ OPD of spherical aberration with ±0.25λ OPD of defocus. Note that optimal focus is shifted positively. (b) Coma with defocus. Defocus symmetry remains, but positional asymmetry is present. (c) Astigmatism with defocus. Optimal focal position is dependent on orientation. (d) Aerial images with no aberration present.

in feature size with exposure is not equivalent for positive and negative defocus amounts, as seen in Figure 2.53.

Figure 2.53d is a focus-exposure matrix plot for positive resist and an unaberrated objective lens. Figure 2.53a–e are plots for systems with various amounts of primary spherical aberration, showing how critical dimension (CD) slope and asymmetry are impacted through focus. For positive spherical aberration, an increase in through-focus CD slope is observed, whereas for small negative aberration, a decrease results. For this system, 0.03λ of negative spherical aberration produces better symmetry and process latitude than no aberration. The opposite would occur for a negative resist. It is questionable whether such techniques would be appropriate to improve imaging performance, because some degree of process dedication would be required.

Generally, a lithographic process is optimized for the smallest feature detail present; however, optimal focus and exposure may not coincide for larger features. Feature size linearity is also influenced by lens aberration. Figure 2.54 shows a plot of resist feature size versus mask feature size for various levels of spherical aberration. Linearity is also strongly influenced by photoresist response. These influences of photoresist processes and lens aberration on lithographic performance can be understood by considering the nonlinear response of photoresist to an aerial image. Consider a perfect aerial image with modulation of 1.0 and infinite ILS, such as that which would result from a collection of all diffraction orders. If this image is used to expose photoresist of any reasonable contrast, a resist image with near-perfect modulation could result. In reality, small-feature aerial images do not have unity modulation; instead, they have a distribution of intensity along the x–y plane. Photoresist does not behave linearly to intensity, nor is it a high-contrast-threshold detector. Imaging into a resist film is dependent on the distribution of the aerial image intensity and resist exposure properties. Resist image widths are not equal at the top and at the bottom of the resist.

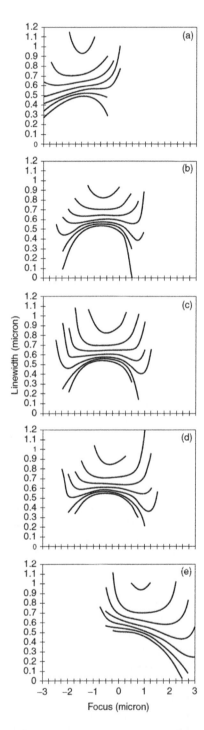

FIGURE 2.53   Focus-exposure matrix plots for imaging of 0.6 μm dense features, 365 nm, 0.5 NA, 0.3σ. Spherical aberration levels are (a) −0.2λ, (b) −0.05λ, (c) −0.03λ, (d) 0.00λ, (e) +0.20λ.

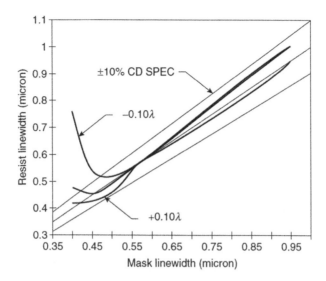

FIGURE 2.54 Resist linearity plots from 0.35 to 1.0 μm (with imaging system in Figure 2.53 for positive resist). Linearity is improved with the presence of positive spherical aberration.

Some unique optimum focus and exposure exist for every feature/resist process/imaging system combination, and any system or process changes will affect features differently.

2.6 OPTICAL MATERIALS AND COATINGS

Several properties of optical materials must be considered in order to effectively design, optimize, and fabricate optical components. These properties include transmittance, reflectance, refractive index, surface quality, chemical and mechanical stability, and purity. Transmittance, reflectance, and absorbance are fundamental material properties that are generally determined by the glass type and structure, and they can be described locally using optical constants.

2.6.1 OPTICAL PROPERTIES AND CONSTANTS

Transmittance through an optical element will be affected by the internal absorption of the material and external reflectances at its surfaces. Both these properties can be described for a given material thickness (t) through the complex refractive index

$$\hat{n} = n(1+ik)$$

where
  n is the real component of the refractive index
  k is the imaginary component, also known as the extinction coefficient

These constants can be related to a material's dielectric constant (ε), permeability (μ), and conductivity (σ) for real σ and $\varepsilon\alpha\sigma$ as

$$n^2(1-k^2) = \mu\varepsilon$$

$$n^2k = \frac{\mu\sigma}{\nu}$$

Internal transmittance for a homogeneous material is dependent on material absorbance (α) by Beer's law:

$$I(t) = I(0)\exp(-\alpha t)$$

where
  $I(0)$ is incident intensity
  $I(t)$ is transmitted intensity through the material thickness t

Transmittance becomes $I(t)/I(0)$. Transmittance cascades through an optical system through multiplication of individual element transmittance values. Absorbance as expressed by $-(1/t)$ ln(transmission) is additive through an entire system.

External reflection at optical surfaces occurs as light passes from a medium of one refractive index to a medium of another. For materials with nonzero absorption, surface reflection (from air) can be expressed as

$$R = \left[\frac{\left[n(1+ik)\cos\theta_i - n_1\cos\theta_t \right]}{\left[n(1+ik)\cos\theta_i + n_1\cos\theta_t \right]} \right]$$

where
  n and n_1 are the medium refractive indices
  θ_i is incident angle
  θ_t is transmitted angle

For normal incidence in air, this becomes

$$R = \frac{n^2(1+k^2) + 1 - 2n}{n^2(1+k^2) + 1 + 2n}$$

This simplifies for nonabsorbing materials in air to

$$R = \frac{(n-1)^2}{(n+1)^2}$$

Because refractive index is wavelength dependent, transmission, reflection, and refraction cannot be treated as constant over any appreciable wavelength range. The real refractive index for optical materials may behave as shown in Figure 2.55, where a large spectral range is plotted, and areas of index discontinuity occur. These transitions represent absorption bands in a glass material that generally occur in the UV and infrared (IR) regions. For optical systems operating in or near the visible region, refractive index is generally well behaved and can be described through the use of dispersion equations such as a Cauchy equation [26]:

$$n = a + \frac{b}{\lambda^2} + \frac{c}{\lambda^4} + \ldots$$

where the constants a, b, and c are determined by substituting known index and wavelength values between absorption bands. For optical systems operating in the UV or IR, absorption bands may limit the application of many otherwise suitable optical materials.

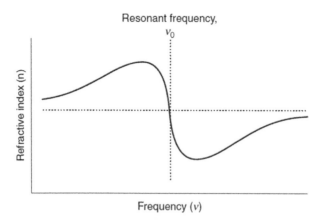

FIGURE 2.55 Frequency dependence of refractive index (n). Values approach 1.0 at high and low frequency extremes.

2.6.2 OPTICAL MATERIALS BELOW 300 nm

Optical lithography below 300 nm is made difficult because of the increase in absorption in optical materials. Few transparent materials exist below 200 nm, limiting design and fabrication flexibility in optical systems. Refractive projection systems are possible at these short wavelengths, but they require consideration of issues concerning aberration effects and radiation damage.

The optical characteristics of glasses in the UV are important when considering photolithographic systems containing refractive elements. As wavelengths below 250 nm are utilized, issues of radiation damage and changes in glass molecular structure become additional concerns. Refraction in insulators is limited by interband absorption at the material's band gap energy, E_g. For 193 nm radiation, a photon energy of $E \sim 6.4$ eV limits optical materials to those with relatively large band gaps. Halide crystals, including CaF_2, LiF, BaF_2, MgF_2, and NaF, and amorphous SiO_2 (or fused silica) are the few materials that possess large enough band gaps and have suitable transmission below 200 nm. Table 2.4 shows experimentally determined band gaps and UV cutoff wavelengths of several halide crystals and fused silica [27]. UV cutoff wavelength is determined as hc/E_g.

The performance of fused silica, in terms of environmental stability, purity, and manufacturability, makes it a superior candidate in critical UV applications such as photolithographic lens components, beam delivery systems, and photomasks. Although limiting the number of available materials to fused silica does introduce optical design constraints (for correction of aberrations including

TABLE 2.4
Experimentally Determined Band Gaps and UV Cutoff Wavelengths for Selected Materials

Material	E_y (eV)	$\lambda_c = hc/E_g$ (nm)
BaF_2	8.6	144
CaF_2	9.9	126
MgF_2	12.2	102
LiF	12.2	102
NaF	11.9	104
SiO_2	9.6	130

chromatic), the additional use of materials such as CaF_2 and LiF does not provide a large increase in design flexibility because of the limited additional refractive index range (n_i at 193 nm for CaF_2 is 1.492, for LiF is 1.521, and for fused silica is 1.561 [28]). Energetic particles (such as electrons and X-rays) and short-wavelength photons have been shown to alter the optical properties of fused silica [29]. Furthermore, because of the high peak power of pulsed lasers, optical damage through rearrangement is possible with excimer lasers operating at wavelengths of 248 and 193 nm [30]. Optical absorption and luminescence can be caused by a lack of stoichiometry in the fused silica molecular matrix. Changes in structure can come about through absorption of radiation and energy transfer processes. E' color centers in type III fused silica (wet fused silica synthesized directly by flame hydrolysis of silicon tetrachloride in a hydrogen-oxygen flame [31]) have been shown to exist at 2.7 eV (458 nm), 4.8 eV (260 nm), and 5.8 eV (210 nm) [32].

2.7 PHOTOMASKS AND PHOTOMASK MATERIALS

Whether for applications in optical DUV or soft x-ray EUV lithography, the requirements for photomask materials are generally wavelength dependent from near-UV, to DUV, and to EUV. A photomask begins as a mask blank, the foundation upon which circuit patterns are written. The mask blank is comprised of three basic components: a substrate, attenuating layers, and a radiation sensitive layer. Early emulsion masks combined an attenuating and photosensitive layer in a single high contrast film while later approaches have often further subdivided these basic components.

For use in an optical DUV lithography, the substrate is most commonly fused silica, a low-expansion synthetic quartz with high transmission into the UV to about 180 nm. The attenuating layer is usually referred to as "chrome" but is compositionally a graded or multilayer chromium oxi-nitride (CrON) film that provides high opacity and low reflectivity at exposure and inspection wavelengths. Used with no other attenuation, these masks are referred to as binary, where projected light is either transmitted or not, allowing for high resolution patterning of integrated circuit devices. For pattern transfer onto such a mask blank, a photoresist layer is coated over the chrome-on-glass (COG) substrate. The photoresists used in mask patterning are similar to those used for wafer patterning and tuned to the exposing wavelength (or radiation) used to pattern them. Photomask blanks are exposed in a mask pattern generator system, developed, etched, stripped of photoresist, cleaned, measured, inspected, and repaired if necessary.

For EUV lithography, the mask blank is also fused silica, generally doped with titanium dioxide and referred to as a low thermal expansion material (LTEM) or an ultra-low thermal expansion (ULE) substrate. EUV masks attenuate radiation through control of reflectivity rather than transmission so utilize a stack of multiple (forty or so) molybdenum (Mo) and silicon (Si) bilayers to reflect radiation. The multilayer stack is covered by a capping layer (like ruthenium, Ru), and an absorbing layer (such as tantalum boron nitride, TaBN), and then coated with photoresist. Mask fabrication follows similar steps to result in a reduction mask for projection in a reflective EUV lithography system. In the sections that follow, the steps for DUV and EUV mask fabrication will be described, together with descriptions of various mask types, variations, and applications. The choice of the a photomask substrate, thin films, and sensitizing layers are based on fundamental requirements, i.e. that materials are chosen so that a blank can be patterned to meet high resolution, high volume mask requirements and that radiation be sufficiently attenuated by the patterned mask to allow for the same needs of IC device lithography.

2.7.1 MASK SUBSTRATES

The choice of a mask substrate is a fundamental one, where materials chosen must meet both optical and mechanical requirements. It is obvious that a mask through which light is projected should be highly transmitting. But it may not be so obvious the degree to which any losses could be tolerated. For example, if a photomask substrate were to transmit 90% of the light directed to it, a pattern on

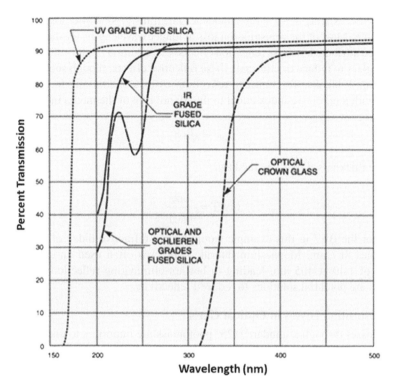

FIGURE 2.56 Transmission properties (per cm) of typical optical glass types from the ultraviolet to the visible.

that mask could theoretically still achieve a maximum contrast with an ideal opaque absorber. The same analysis could be done for a substrate that only passed 10% of the light. A first order analysis might simply suggest that the only loss in such a scenario is in throughput, or the amount of time needed to expose through such a mask. But by looking into the mechanisms for loss leads to larger concerns. Figure 2.56 shows the transmission characteristics of several types of optical glasses. Crown glasses like borosilicates (such as Scott BK7) are common materials used in many optical applications, including longer wavelength UV lithography lenses designed around g, h, and i-lines of the mercury (Hg) emission lamp. The transmission of these glasses below about 350 nm begins to fall and transmission below about 250 is negligible. But at wavelengths within the transmission window (the region where the material has its highest transmission), a maximum value of just over 90% is reached. Mechanisms behind this can stem from absorption or reflective losses. Fused silica including UV grades, as discussed earlier, provide higher transmission at shorter wavelengths.

2.7.1.1 Reflective Losses in Optical Glass

When illuminated at a normal incidence angle (i.e. perpendicular to the surface), reflection at an interface can be calculated using the *Fresnel* relationship for the two media:

$$R = \left| \frac{n_1 - n_2}{n_1 + n_2} \right|^2$$

Where n_1 and n_2 are the two material refractive indices. If the media are non-absorbing (or nearly so), the imaginary component of the refractive index drops out, allowing for evaluation using real values only. Take for example a borosilicate BK7 glass plate, with a refractive index (n) of 1.530 at the Hg h-line of 405 nm. The reflection at a surface of BK7 in air (n = 1.0) is 4.4%, as is the reflection in the glass existing to air at the other side. The transmission after loss from both surfaces is $(0.956)^2$, or 91.4%. Most of the transmission loss at longer wavelengths for the materials shown in Figure 2.56 can be attributed to these reflection effects. The addition of anti-reflective layers (ARLs) is often used on glass to reduce or eliminate these reflective effects, though such films are not common with photomask blanks. To eliminate reflection losses at media interfaces, a thin film non-absorbing layer with a refractive index equal to the square root of the media indices can be coated:

$$n_{ARL} = \sqrt{n_1 n_2}$$

at a quarter wave thickness (QWT) defined as:

$$QWT = \frac{m\lambda}{4n_{ARL}}$$

An ideal ARL for the BK7 in this example would have a refractive index of 1.24, coated at odd thickness multiples 81.7 nm. Magnesium fluoride (MgF_2) is often used an optical ARL, with a refractive index of 1.40 at 405 nm, leading to but not eliminating reflection at this wavelength. Multilayer ARLs are potential solutions to reduced reflectivity.

2.7.1.2 UV Absorption Losses in Optical Glass

The absorption losses through a standard 0.25″ photomask are important to understand with regard to transmission losses and material damage. The spectral absorbance bands of glass are related to regions of refractive index change, or dispersion. In the spectral transmission region of a typical glass material, the real and complex refractive indices increase toward shorter wavelength, as seen for fuse silica in Figure 2.57. The plot of extinction coefficient (or absorption index k, the imaginary refractive index) shows the region of strong dispersion and absorption just below 200 nm, correlating to the high index, low transmission properties of the material seen in Figure 2.56. This inherent material absorption presents challenges in using fused silica as an optical material below these wavelengths. The absorption of radiation through an optical material will lead to effects that are wavelength dependent. In the infrared, transmittance is influenced mainly by the hydroxyl (OH⁻ group) content during synthesis. In the visible region, at wavelengths below about 250 nm, the radiation absorbed by glass materials can lead to changes in molecular structure through rearrangement of molecular fragments. This can be seen for example with short UV exposure of flame hydrolyzed fused silica at wavelengths sufficient to cause molecular rearrangement, specifically at 2.7eV (458 nm), 4.8 eV (260 nm), and 5.8 eV (210 nm). The mechanisms involved with these

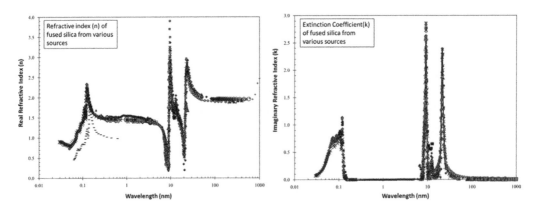

FIGURE 2.57 Refractive index (n) and extinction coefficient (or imaginary refractive index, k) of fused silica glass from UV to IR wavelengths, showing absorption edges of high dispersion and absorption.

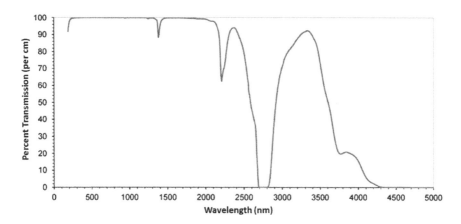

FIGURE 2.58   The internal transmission of UV grade fused silica used in photomask substrates.

rearrangements are dependent on the OH⁻ concentration, where hydroxyl group levels between about 300 and 1200 ppb lead to improved radiation stability. UV absorption of doped glasses (such as borosilicate or soda lime) is also influenced by the heavier elements used to increase refractive index such as lead, barium, titanium, niobium, or lanthanum. While doped glasses or common synthetic quartz were used for early generation photomasks, UV grade fused silica with impurity levels below 1000 ppb has been employed for decades with excimer grade fused silica having 500 ppb levels or lower are required today. Figure 2.58 shows the internal transmittance of a UV standard grade fused silica with >99.90% transmission at 248 nm and >99.50% transmission at 193 nm. An important secondary consideration regarding mask absorption is the requirement of high transmission at inspection wavelengths. Mask inspection is often not carried out at the lithography wavelength but instead at nearby or longer wavelengths. Radiation sources for the metrology systems used in defect, linewidth, repair, or overlay inspection must be compatible with the mask substrate.

Vacuum UV (VUV) lithography was at one point considered the next generation beyond DUV, making use of the 157 nm wavelength of F_2 excimer lasers. A host of problems prevented the technology at this wavelength from maturing fast enough for insertion for lithography generations below 65 nm. But the advances as a result of research into VUV optical materials did prove to benefit 193 nm lithography, especially as it was utilized in an immersion mode for 45 nm generations and well beyond. Fused silica begins to absorb at wavelengths below about 160 nm, limiting is usefulness at wavelength near or below this. Fluoride materials like magnesium fluoride, calcium fluoride, barium fluorine, and lanthanum fluoride retain high transmission below 160 nm and into the VUV. The hygroscopic nature of these materials would however limit their use as mask substrates, where the various process steps involved with mask fabrication would make them incompatible. There were some efforts made to use of magnesium fluoride for VUV mask blanks but greater success was achieved through doping of fused silica with fluorine. Fluorine doped fused silica, with Si-F bonds replacing hydroxyl bonds (Si-OH), allowed for an increase in the transmission of fused silica at VUV wavelengths [33]. Although the internal absorption of this fluorine-doped fused silica was sufficiently low at 157 nm for use in mask blanks (about 3–4% per cm), it would not be low enough to be used for optical lens elements. But this was a viable solution to avoid the hygroscopic deficiencies of magnesium fluoride mask blanks. Alas, the development of these materials for use at 157 nm never advanced as this wavelength generation ceased to exist once 193 nm water immersion lithography was proven viable.

2.7.1.3   Thermal Effects

Thermal expansion is a property common across all materials. In the semiconductor field, it is a concern especially during lithography steps as any dimensional change during exposure is captured at that particular instance, which will be measured against all previous and subsequent steps. During mask fabrication, any dimensional changes during the mask exposure step, whether e-beam or optical, will lead to pattern dimensional and placement errors. After exposure, pattern transfer to underlying layers can result in substrate temperature changes, such as with plasma etch processes. If a mask undergoes a multiple patterning process, such as with phase shift masks, any temperature differences between the first and subsequent steps will lead to errors. And during the lithographic exposure through a mask, temperature differences between exposures, during exposure, or across a mask will lead to dimensional change. The thermal expansion coefficient (TEC) is a measure of the linear change in size with temperature, generally measure in units of reciprocal degrees. Soda lime and similar glass types can have a TEC near about 9 ppm $°C^{-1}$ for example. Doped borosilicate glass is lower, at about 3.5 ppm $°C^{-1}$, and is often referred to as a low-expansion (LE) glass for low-end photomask application. To appreciate the impact that this can have on a mask substrate, consider a $6'' \times 6''$ borosilicate mask with a patterned area $5''$ across the plate. If the temperature of the mask is raised by just 0.5°C, the error across the full $5''$ pattern area is 0.22 μm. This amount of "run-out" across the plate would be unacceptable for any recent device generation. A fused silica mask, with a TEC of 0.55 ppm $°C^{-1}$ would lead to an improvement of over 6X, or about 35 nm over the same distance for a 0.5°C temperature change. This demonstrates the importance of temperature control during the steps of mask processing as well as mask use and also shows that the choice of fused silica as a mask substrate material may be driven as much by thermal expansion as by absorption. EUV masks are based on the attenuation of reflectivity. The requirements of the substrate are thus based mostly on mechanical characteristics rather than optical ones. Relaxed constraints on transmission allow for EUV substrates with very low thermal expansion characteristics. By doping fused silica with 7-9% titanium dioxide (TiO_2), ultra-low thermal expansion material (LTEM) substrates can be employed for EUV mask use [34]. Performance of these materials is on the order of 5 ppb $°C^{-1}$, more than 100X lower than fused silica. Following the same exercise of increasing the mask substrate temperature by 0.5°C, the dimensional change for such a LTEM substrate would be less than 0.1 nm per inch.

2.7.1.4   Glass Defects and Plate Flatness

The quality requirements of optical glass will be driven by application. Specifications for things like defects, damage threshold, and irregularity are taken into account when considering glass substrate materials. The same holds true for mask substrates, where defects in a mask blank are mostly intolerable for modern lithography applications. Inclusions, pinholes, scratches, bubbles, pits, and fractures can be present in glass plates at sizes upwards of several microns. Such anomalies can cause image shape errors, scatter, and undesired diffractive effects. Defects in a mask blank can also lead to increased susceptibility to radiation damage, limiting the useful life of a photomask especially in DUV applications. Photomasks must therefore meet the highest standards of optical materials such as those met with high purity fused silica (HPFS) manufactured by several suppliers. The highest grade of these materials may have inclusion cross sections below 0.03mm, homogeneity (index variation) better than 0.5 ppm, metal impurities below 10 ppb, and birefringence below 1 nm/cm. These levels of defects lead to blank manufacturing and inspection challenges. Inspection of glass blanks must detect both clear and opaque solid inclusions (minute foreign materials used during glass formation) and suspended void inclusions (bubbles), each which can present unique problems. Current mask blank inspection is carried out for the detection of defects down to about 2 μm with inspection times on the order of a few minutes per plate [35].

The flatness of a photomask substrate is measured in terms of surface irregularity, a specification shared with optical elements and surfaces. The tolerable flatness error for a photomask substrate is tied mainly to focus budget, which has decreased significantly over recent generations. As phase shift masks have become commonplace for high end IC manufacture, the limitation of short-range wavefront error can also be a consideration. This concern though is very short range and therefore less critical relative to substrate properties. The influence of optical surface

TABLE 2.5

Critical EUV Mask Blank Parameters for 22 nm Device Generation Lithography

EUV Blank Property	22 nm node requirement
Flatness (front and back)	<23 nm peak-to-valley (PV)
Front-side roughness	<0.15 nm RMS
Back-side roughness	<0.50 nm RMS
Defect Density	0.03/cm^2 @ 25 nm

irregularity defined by wavefront error is also reduced because of the lack of optical power of a flat mask. Instead, mask flatness requirements relative to focal depth are most important [36]. As an example, a variation in flatness of 2 μm across the active area of a mask results in an error of 125 nm at the wafer plane, where defocus scales as the square of magnification in a 4X reduction system. This is a significant portion of the focus budget for current lithography generations and thus not the acceptable flatness specification that it was a decade ago. Instead, flatness levels of 1 mm or less have become necessary.

Because the most demanding device dimensions are targeted, EUV mask flatness, roughness, and defect specifications are the tightest. Table 2.5 shows the requirements for each for 22 nm device technology [33]. There are flatness and roughness requirements are for both the front and backside. The roughness specification for the front-side is tighter than the backside because of the propagation of surface behavior through the film stack above, resulting in adverse optical effects.

2.7.2 MASK ABSORBERS

Historically, the attenuation through UV and DUV optical photomasks has been accomplished with chromium-based films. Generally referred to simply as "chrome," the nature of these materials is somewhat complex in nature, consisting of a graded or multilayer structure of chromium (Cr), chromium oxide (Cr_2O_3), and chromium nitride (CrN), not necessarily in stoichiometric form. Chrome masking films have exceptional attributes with respect to optical, thermomechanical, and adhesion properties and can be tailored through composition for low reflectivity at one or both surfaces. Early use of chromium metal films, or "shiny chrome," was soon replaced with low reflectivity "black chrome," so named because of the darkening effect of a think chromium oxide top layer. Early films were deposited to about 1000Å, sufficiently thick to form an optically dense layer. With decreasing lithography wavelength and tighter demands on resolution, it was realized that these materials could continue to support photomask technology for many generations.

2.7.2.1 Graded Chromium-Based Mask Films

The graded structure of a typical UV chrome film is seen in Figure 2.59, measured using x-ray photoelectron spectroscopy (XPS) [37]. The plot shows the concentration of the elements present through the films thickness from the top throughout the full depth. The composition of the film is mostly chromium metal with oxygen and nitrogen graded as chromium oxide (generally Cr_2O_3) and chromium nitride (CrN) near the top (air) surface. At a depth about halfway into the film, chromium metal dominates until the base substrate is reached somewhere around 1000Å. The carbon (C) and sulfur (S) at the top of the film is likely residual photoresist that was stripped from the surface prior to analysis. Such films are usually reactively sputter deposited from chromium targets to a target thickness ranging from about 50 to 100 nm, with thickness uniformity across a plate of better than 5 nm.

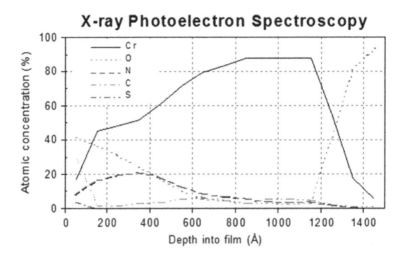

FIGURE 2.59 Compositional analysis of a typical graded chromium mask film, showing the presence of various elements through the film thickness.

2.7.2.2 Optical Density

An optical density (OD) of a masking layer above 2.0 is desirable where target of 3.0 is common, which can be achieved with film thicknesses above about 80 nm. Optical density is a logarithmic measure of opacity, allowing for the scaling of transmitted light with film thickness:

$$OD = -\log\left(\frac{1}{T}\right)$$

Opacity is simply reciprocal transmission, 1/T. An OD of 2.0 equates to a transmission is 1%, an OD or 3 is 0.1% and so on. The optical density of a commercial UV chrome film shown in Figure 2.60 is below, exhibiting an optical density near 3.0 for DUV wavelengths for a film thickness of 900 Å [37].

2.7.2.3 Reflectivity

The reflectivity from a chromium metal film is high, as seen in Figure 2.61 and on the order of 35% at UV wavelengths [37]. This is unacceptably high for use in projection lithography, where values below 10% are desirable.

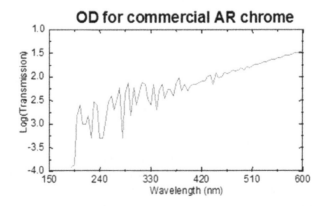

FIGURE 2.60 The optical density as a function of wavelength for a typical commercial chrome masking film.

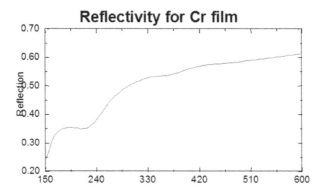

FIGURE 2.61 Reflectivity for pure chromium metal film, showing high reflectivity at UV lithography wavelengths.

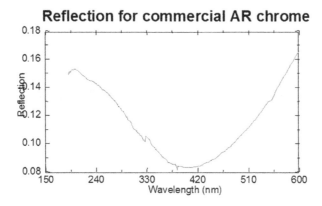

FIGURE 2.62 Reflectivity of a commercial low-reflective chrome film. (From Smith et al., 1999.)

By grading the chromium oxi-nitride (CrON) film structure, reflectivity below 10% can be achieved, as shown in Figure 2.62 for a commercial chromium based absorber optimized for use at 365 nm [37]. At this wavelength, about 8% reflectivity is possible.

Although such low reflective masking films are often referred to as AR, or anti-reflective, strictly speaking they are not. To be a true AR film, a separate and distinct layer must exist which when coated to a specific thickness based on optical constants (real and imaginary refractive indices, n and k), destructive interference results in the cancellation of reflective components. These conditions are not met with so-called AR chrome, instead a material of lower reflectivity (CrON) is graded with chromium metal so as to produce a top surface with sufficiently low reflectivity.

2.8 OPTICAL IMAGE ENHANCEMENT TECHNIQUES

2.8.1 OFF-AXIS ILLUMINATION

Optimization of the partial coherence of an imaging system has been introduced for circular illuminator apertures. By controlled distribution of diffraction information in the objective lens, maximum image modulation can be obtained. An illumination system can be further refined by considering illumination apertures that are not necessarily circular. Shown in Figure 2.63 is a coherently illuminated mask grating imaged through an objective lens. Here, the ±1 diffraction orders are distributed symmetrically around the zeroth order. As previously seen in Figure 2.34, when defocus is introduced, an OPD between the zeroth and the ± first order results. The acceptable depth of focus

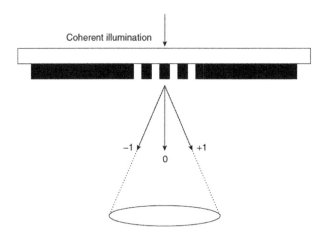

FIGURE 2.63 Coherently illuminated mask grating and objective lens. Only 0 and ± 1st diffraction orders are collected.

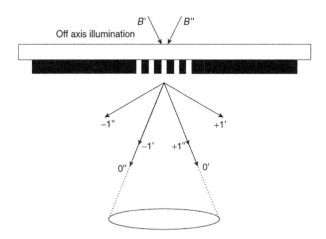

FIGURE 2.64 Oblique or off-axis illumination of a mask grating where 0 and 1st diffraction orders coincide in lens pupil.

is dependent on the extent of the OPD and the resulting phase error introduced. Figure 2.64 shows a system where illumination is obliquely incident on the mask at an angle, so that the zeroth and first diffraction orders are distributed on alternate sides of the optical axis. Using reasoning similar to that used for incoherent illumination, it can be shown that the minimum k factor for this oblique condition of partially coherent illumination is 0.25. The illumination angle is chosen uniquely for a given wavelength, NA, and feature size, and it can be calculated for dense features an $\sin^{-1}(0.5\lambda/p)$ for NA $=0.5\lambda/p$ where d is the feature pitch. The most significant impact of off-axis illumination is realized when considering focal depth. In this case, the zeroth and first diffraction orders now travel an identical path length regardless of the defocus amount. The consequence is a depth of focus that is effectively infinite.

Using off-axis illumination, source shaping such as dipole, quadrupole, annular, and more customized variations can lead to improvements in the printing of photoresist features. Although incoherent illumination leads to the limits of resolution for an optical system (and a $k_1 = 0.25$), the image quality at these limits will generally be poor. This is especially true as defocus and other image degradation mechanisms (e.g. aberrations) are considered. For projection optical lithography, incoherent illumination should be thought of as a starting point, where improvements in image quality can be achieved through the removal of select portions of a full $\sigma = 1$ illuminator. This approach leads to an understanding of how off-axis illumination and more customized illumination are an integral part of leading edge optical lithography.

Though a common treatment of off-axis illumination (OAI) is to start with a coherent on-axis point source, it can more intuitive to begin instead with a full circular incoherent source, as shown in Figure 2.65. For X-oriented features with $0.25 < k_1 < 0.5$, there is a central portion of a full sigma illuminator that will production diffraction energy that cannot be collected by the objective lens pupil. There are portions toward the edge of the illuminator however that will result in first diffraction order energy that can be collected by the objective lens. Because both 0th and 1st order energy is collected, it can be used in image production. The left edge of the illuminator results in collected $+1^{st}$ diffraction order energy and the right edge results in -1st order collected energy. If regions without more than 0th order are discarded, what remains is a two-component source often referred to as a dipole illuminator. If this exercise is carried out in the Y direction as well, a four-pole or *quadrupole* illuminator arises. If all mask orientations are allowed, the illuminator becomes *annular* in shape. Figure 2.66 shows a variety of off axis source shapes that may could be utilized if projection

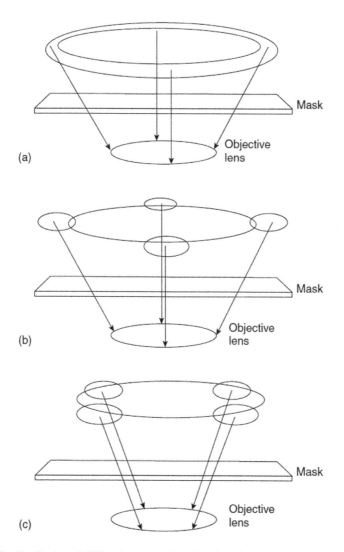

FIGURE 2.65 The distribution of diffraction energy based on locations of the illumination pupil.

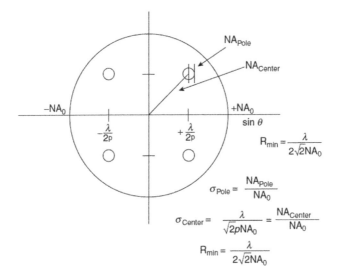

FIGURE 2.66   Several off-axis source shapes used in projection lithography.

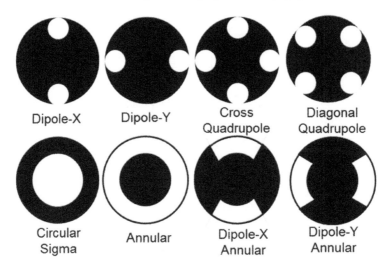

FIGURE 2.67   The diffracted energy for off-axis illumination (right) and on-axis illumination (left) and the corresponding collection regions in the objective lens pupil.

lithography. Variations in the pole shape result from the particular method of implementation and they may be circular, arcs, wedges, and other shapes.

The advantage to using only a select portion of the lens pupil is appreciated when considering that those portions removed only add background energy to an image. The consequence is a reduction in image modulation. An additional advantage of OAI is a large depth of focus that can be gained by distribution diffraction energy only at radial locations of the projection lens pupil. This is shown in Figure 2.67, where the defocus OPD is shown together with the diffraction order energy for on-axis low sigma illumination and off-axis dipole illumination. In the on-axis case, a large defocus aberration effect will result in image blur because of the phase difference between the diffraction energy travelling through the center of the pupil and that travelling through the edge of the pupil. For dipole illumination, each pole distributes 0th and 1st diffraction order energy at radial pupil locations only. Because of this, there is little difference in the phase of all orders as they are collected by the lens pupil. The end result is imaging with significant improvement in DOF.

Indeed, for poles that are single points, the DOF is infinite. More practical source sizes like those in Figure 2.66 may be necessary for throughput as well as the reduction in coherent image artifacts and the improvement will be lessened.

2.8.2 Isolated Line Performance

When considering grating features, optical analysis of off-axis and conventional illumination can be quite straightforward. When considering more isolated features, however, diffraction orders are less discrete. Convolving such a frequency representation with either illumination poles or an annular ring will result in diffraction information distributed over a range of angles. An optical angle of illumination that will place low-frequency information out at the full numerical apertures of the objective lens will distribute most energy at nonoptimal angles. Isolated line performance is, therefore, minimally enhanced by off-axis illumination. Any improvement is significantly reduced also as the pole or ring width is increased. When both dense and isolated features are considered together in a field, it follows that the dense to isolated feature size bias or proximity effect will be affected by off-axis illumination [38]. Figure 2.68 shows, for instance, the decrease in image CD bias between dense and isolated 0.35 μm features for increasing levels of annular illumination using a 0.55 NA i-line exposure system. As obscuration in the condenser lens pupil is increased (resulting in annular illumination of decreasing ring width), dense to isolated feature size bias decreases. As features approach $0.25\lambda/NA$, however, larger amounts of energy go uncollected by the lens, which may lead to an increase in this bias, as seen in Figure 2.69.

Off-axis illumination schemes have been proposed by which the modulation of nonperiodic features could be improved [39]. Resolution improvement for off-axis illumination requires multiple mask pattern openings for interference, leading to discrete diffraction orders. Small auxiliary patterns can be added close to an isolated feature to allow the required interference effects. By adding features below the resolution cutoff of an imaging system ($0.2\lambda/NA$, for example) and placing them at optimal distances so that their side lobes coincide with main feature main lobes ($0.7\lambda/NA$, for instance), it is possible to improve peak amplitude and ILS [40]. Higher-order lobes of isolated feature diffraction patterns can be further enhanced by adding additional $0.2\lambda/NA$ spaces at corresponding distances [41]. Various arrangements are possible, as shown in Figure 2.70. This figure shows arrangements for opaque line space patterns, an isolated opaque line, clear line space patterns, and an isolated clear space. The image enhancement offered by using these techniques is

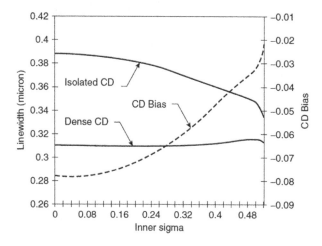

FIGURE 2.68 Image CD bias versus annular illumination. Inner sigma values correspond to the amount of obscuration in the condenser lens pupil. Partial coherence σ (outer) is 0.52 for 0.35 μm features using 365 nm illumination and 0.55 NA. Defocus is 0.5 μm. As central obscuration is increased, image CD bias increases.

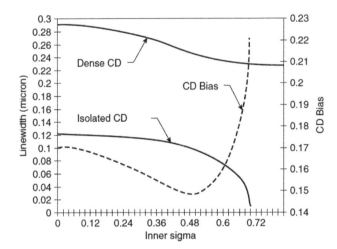

FIGURE 2.69 Similar conditions as Figure 2.57 for 0.22 μm features. Image CD bias is now reversed with increasing inner sigma.

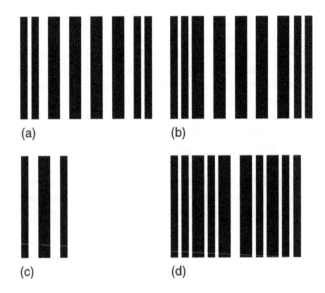

FIGURE 2.70 Arrangement for additional auxiliary patterns to improve isolated line and CD bias performance using OAL: (a) opaque dense features, (b) clear dense features, (c) opaque isolated features, and (d) clear isolated features.

realized as focal depth is considered. Figures 2.71 through 2.73 show the DOF improvement for a five-bar space pattern through focus. Figure 2.71 shows aerial images through λ/NA^2 (± 0.74 μm) of defocus for $0.5\lambda/NA$ features using conventional illumination with $\sigma = 0.5$. Figure 2.72 gives results using quadrupole illumination. Figure 2.73 shows aerial images through focus for the same feature width with auxiliary patterns smaller than $0.2\lambda/NA$ and off-axis illumination. An improvement in DOF is apparent with minimal intensity in the dark field. Additional patterns would be required to increase peak intensity, which may be improved by as much as 20%.

Another modification of off-axis illumination has been introduced that modifies the illumination beam profile [42]. This modified beam illumination technique fills the condenser lens pupil with weak quadrupoles where energy is distributed within and between poles, as seen in Figure 2.74. This has been demonstrated to allow better control of DOF of proximity effects for a variety of feature types.

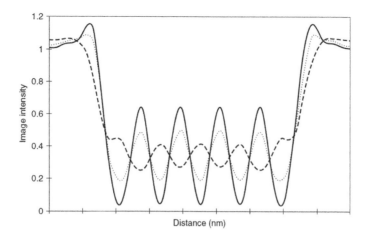

FIGURE 2.71   Aerial image intensity for 0.37 μm features through ±0.5λ/NA² of defocus using σ=0.5, 365 nm, and 0.5 NA.

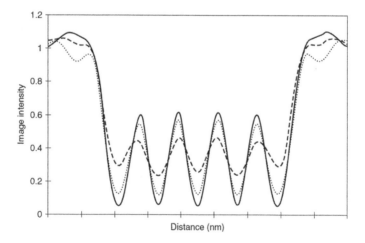

FIGURE 2.72   Aerial images as in Figure 2.71 using off-axis illumination, OAI.

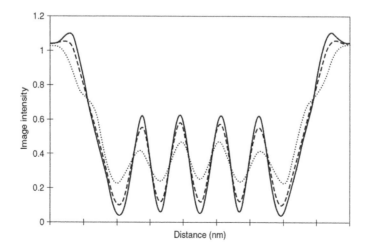

FIGURE 2.73   Aerial images for features with 0.08λ/NA auxiliary patterns and OAI. Note the improvement in minimum intensity of outermost features at greatest defocus.

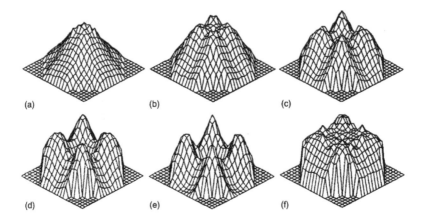

(a)   (b)   (c)

(d)   (e)   (f)

FIGURE 2.74 Modified illumination profiles for conventional and OAI. (Reproduced from Ogawa, T., Uematsu, M., Ishimaru, T., Kimura, M., and Tsumori, T., *SPIE*, 2197,19,1994. With permission.)

2.8.3 PHASE SHIFT MASKING

Up to this point, control of the amplitude of a mask function has been considered, and phase information has been assumed to be nonvarying. It has already been shown that the spatial coherence or phase relation of light is responsible for interference and diffraction effects. It would follow, therefore, that control of phase information at the mask may allow additional manipulation of imaging performance. Consider the situation in Figure 2.75, where two rectangular grating masks are illuminated with coherent illumination. The conventional "binary" mask in Figure 2.75a produces an electric field that varies from 0 to 1 as a transition is made from opaque to transparent regions. The minimum numerical aperture that can be utilized for this situation is one that captures the zero and $\pm$ first diffraction orders or $NA \geq \lambda/p$. The lens acts on this information to produce a cosinusoidal amplitude image appropriately biased by the zeroth diffraction orders. The aerial image is proportional to the square of the amplitude image. Now consider Figure 2.75b, where a π "phase shifter" is

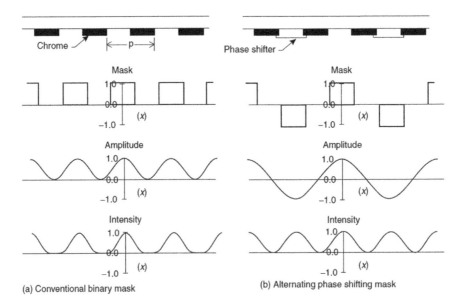

FIGURE 2.75 Schematic of (a) a conventional binary mask and (b) an alternating phase shift mask. The mask electric field, image amplitude, and image intensity are shown for each.

subtracted (or added) at alternating mask openings that create an electric field at the mask varying from −1 to +1, where a negative amplitude represents a π phase shift (a $\pi/2$ phase shift would be 90° out of the paper, $3\pi/2$ would be 90° into the paper, and so forth).

Analysis of this situation can be simplified if the phase-shift mask function is divided into separate functions, one for each phase component where $m(x)=m_1(x)+m_2(x)$.

The first function, $m_1(x)$, can be described as a rectangular wave with a pitch equal to four times the space width:

$$m_1(x) = \text{rect}\left(\frac{x}{p/2}\right) * \text{comb}\left(\frac{x}{2p}\right)$$

The second mask function, $m_2(x)$, can be described as

$$m_2(x) = \left[\text{rect}\left(\frac{x}{3p/2}\right) * \text{comb}\left(\frac{x}{2p}\right)\right] - 1$$

The spatial frequency distribution becomes

$$M(u) = \mathcal{F}\{m(x)\} = \mathcal{F}\{m_1(x)\} + \mathcal{F}\{m_2(x)\}$$

which is shown in Figure 2.76. It is immediately noticed that the zero term is removed through the subtraction of the centered impulse function, $\delta(x)$. Also, the distribution of the diffraction orders has been defined by a comb(u) function with one-half the frequency required for a conventional binary mask. The minimum lens NA required is that which captures the ± first diffraction orders, or $\lambda/2p$. The resulting image amplitude pupil filtered and distributed to the wafer is an unbiased cosine with a frequency of one-half the mask pitch. When the image intensity is considered ($I(x)=|A(x)|^2$), the result is a squared cosine with the original mask pitch. Intensity minimum points are ensured as the

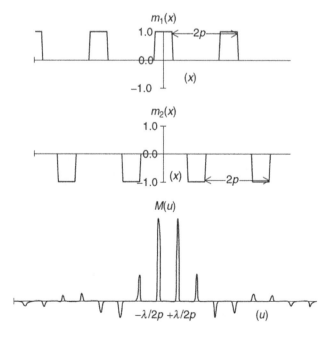

FIGURE 2.76 Spatial frequency distribution $m(u)$ resulting from coherent illumination of an alternating phase-shift mask as decomposed into $m_1(x)$ and $m_2(x)$.

amplitude function passes through zero. This "forced zero" results in minimum intensity transfer into photoresist, a situation that will not occur for the binary case as shown.

For coherent illumination, a lens acting on this diffracted information has a 50% decrease in the numerical aperture required to capture these primary orders. Alternatively, for a given lens numerical aperture, a mask that utilizes such alternating aperture phase shifters can produce a resolution twice that possible using a conventional binary mask. Next, consider image degradation through defocus or other aberrations. For the conventional case, the resulting intensity image becomes an average of cosines with decreased modulation. The ability to maintain a minimum intensity becomes more difficult as the aberration level is increased. For the phase-shifted mask case, the minimum intensity remains exactly zero, increasing the likelihood that photoresist can reproduce a usable image.

For the phase-shifted mask, because features one-half the size can be resolved, the minimum resolution can be expressed as

$$R_{min} = hp_{min} = 0.25\lambda/\text{NA}$$

As the partial coherence factor is increased from zero, the impact of this phase-shift technique is diminished to a point at which for incoherent illumination, no improvement is realized for phase shifting over the binary mask. To evaluate the improvement of phase-shift masking over conventional binary masking, the electric field at the wafer, neglecting higher-order terms, can be considered:

$$E(x) = \cos\left(\frac{\pi x}{\lambda}\right)$$

The intensity in the aerial image is approximated by

$$I(x) = \frac{1}{2}\left[1 + \cos\frac{2\pi x}{\lambda}\right]$$

which is comparable to that for off-axis illumination. In reality, higher-order terms will affect DOF. Phase-shift masking may, therefore, result in a lower DOF than for fully optimized off-axis illumination.

The technique of phase shifting alternating features on a mask is appropriately called *alternating phase-shift masking*. Phase information is modified by either adding or subtracting optional material from the mask substrate at a thickness that corresponds to a π phase shift [43, 44]. Figure 2.77 shows two wave trains traveling through a transparent refracting medium (a glass plate), both in phase on entering the material. The wavelength of light as it enters the medium from air is compressed by a factor proportional to the refractive index at that wavelength. Upon exiting the glass plate into air, the initial wavelength of the wavefronts is restored. If one wave train travels a greater path length than the other, a shift in phase between the two will result. By controlling the relationship between the respective optical path distances traveled over the area of some refracting medium with refractive index n_i, a phase shift can be produced as follows:

$$\Delta\phi = \frac{2\pi}{\lambda}(n_i - 1)t$$

where t is the shifter thickness. The required shifter thicknesses for a π phase shift at 365, 248, and 193 nm wavelengths in fused silica are 3720, 2470, and 1850 A, respectively. At shorter wavelengths, less phase-shift material thickness is required. Depending on the mask fabrication technique, this may limit the manufacturability of these types of phase-shift masks for short UV

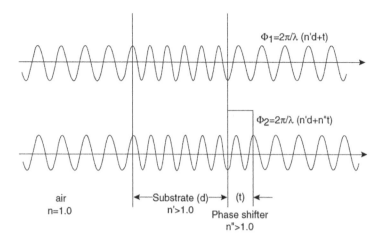

$\Phi_1 = 2\pi/\lambda \ (n'd+t)$

$\Phi_2 = 2\pi/\lambda \ (n'd+n''t)$

air
n=1.0

←—Substrate (d)—→| (t) |←—
n'>1.0
Phase shifter
n">1.0

FIGURE 2.77   Diagram of wave train propagation through phase-shifted and unshifted positions of a mask.

wavelength exposures. Generally, a phase shift can be produced by using either thin-film deposition and delineation or direct glass etch methods. Both techniques can introduce process control problems. In order to control phase shifting to within ±5°, a reasonable requirement for low–k factor lithography, i-line phase shifter thickness must be held to within 100 Å in fused silica. For 193 nm lithography, this becomes 50 Å. If etching techniques cannot operate within this tolerance level over large mask substrates (in a situation where an etch stop layer is not present), the application of etched glass phase-shift masks for IC production may be limited to longer wavelengths. There also exists a trade-off between phase errors allowed through fabrication techniques and those allowed through increasing partial coherence. As partial coherence is increased above zero, higher demands are placed on phase shifter etch control. If etch control ultimately places a limitation on the maximum partial coherence allowed, the issue of exposure throughput becomes a concern.

Variations in the alternating phase-shift mask have been developed to allow application to non-repetitive structures [45]. Figure 2.78 shows several approaches whereby phase-shifting structures are applied at or near the edge of isolated features. These rim phase-shifting techniques do not offer the doubling resolution improvement of the alternating approach, but they do produce a similar forced zero in intensity at the wafer because of a phase transition at feature edges. The advantage of these types of schemes is their ability to be applied to arbitrary feature types. As with the alternating phase-shift mask, these rim masks require film deposition and patterning or glass etch processing and may be difficult to fabricate for short UV wavelength applications. In addition, pattern placement accuracy of those features that are sub-0.25 k factor in size is increasingly challenging as wavelength decreases.

Other phase-shift mask techniques make use of a phase-only transition and destructive interference at edges [46]. A "chromeless" phase edge technique, as shown in Figure 2.78, requires a single mask patterning step and produces intensity minimums at the wafer mask plane at each mask phase transition. When used with a sufficiently optimized resist process, this can result in resolution well beyond the Rayleigh limit. Resist features as small as $k=0.20$ have been demonstrated with this technique, which introduces opportunities for application, especially for critical isolated feature levels. An anomaly of using such structures is the addition of phase transitions at every shifter edge. To eliminate resulting intensity dips produced at these edges, multiple-level masks have been used [47]. After exposure with the chromeless phase edge mask, a binary chrome mask can be utilized to eliminate undesired field artifacts. An alternative way to reduce these unwanted phase edge effects is to engineer into the mask additional phase levels, such as 60° and 120° [48]. To achieve such a phase combination, two-phase etch process steps are required during mask fabrication. This may

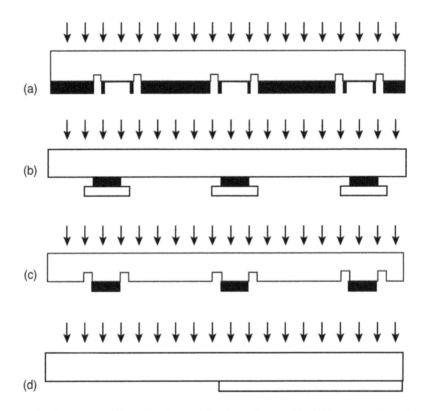

FIGURE 2.78 Various phase-shift mask schemes (a) etch outriggers, (b) additive rim shifters, (c) etched rim shifters, and (d) chromeless phase-shift mask.

ultimately limit application. Variations on these phase-shifting schemes include a shifter-shutter structure that allows control over feature width and reduces field artifacts, and a clear field approach using sub–Rayleigh limit gating or checkerboard structures [41].

Achieving the necessary 180° (or π) phase shift is carried out through multiple mask process steps, etching regions in a mask so that the light propagating through a phase shifted opening is out of phase with a neighboring opening by a half of a wavelength. The alternating phase shift mask (APSM) example shown in Figure 2.75 is for a simple line/space mask. More realistic IC geometry requires consideration of additional diversity in both X and Y directions. When phase shift masks are used for more complicated patterns, a second binary mask is often used in a dual exposure sequence to remove unwanted pattern regions or artifacts. This second mask is referred to as a "trim mask". An example of two such masks (a PSM and a trim mask) are depicted in the high-density gate designs shown in Figure 2.79 [49].

Each of these phase-shift masking approaches requires some level of added mask and process complexity. In addition, none of these techniques can be used universally for all feature sizes, shapes, or parity. An approach that can minimize mask design and fabrication complexity may gain the greatest acceptance for application to manufacturing. An attenuated phase-shift mask (APSM) may be such an approach, whereby conventional opaque areas on a binary mask are replaced with partially transmitting regions (5%–15%) that produce a π phase shift with respect to clear regions. This is a phase-shift mask approach that has evolved out of X-ray masking where attenuators inherently possess some degree of transparency [50]. As shown in Figure 2.80, such a mask will produce a mask electric field that varies from 1.0 to −0.1 in amplitude (for a 10% transmitting attenuator) with a shift in phase, represented by a transition from a positive electric field component to a negative. The electric field at the wafer possesses a loss of modulation, but it retains the phase change and transition through zero. Squaring the electric field results in an intensity with a zero minimum.

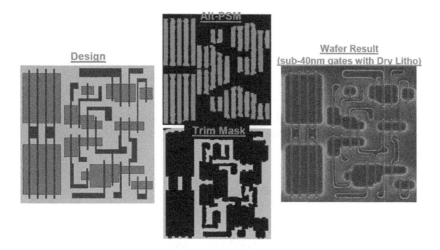

FIGURE 2.79 An APSM plus trim approach used for gate patterning showing the targeted design and the two masks used to create the desired pattern. The top mask (center) is the alternating phase shift mask, where color differences represent a p phase shift. Unwanted artifacts in x-y directions are removed with a second trim exposure, carried out immediately after the first.

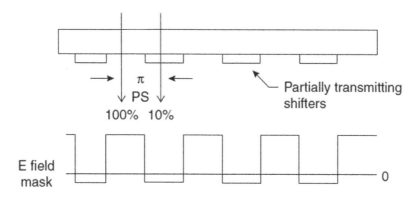

FIGURE 2.80 A 10% attenuated phase-shift mask. A π phase shift and 10% transmission are achieved in attenuated regions. The zero mask electric field ensures minimum aerial image intensity.

Work in areas of attenuated phase-shift masking has demonstrated both resolution and focal depth improvement for a variety of feature types. Attenuated phase shift mask efforts at 365, 248, and 193 nm have shown a near doubling of focal depth for features on the order of $k = 0.5$ [51, 52]. As such technologies are considered for IC mask fabrication, practical materials that can satisfy both the 180° phase shift and the required transmittance at wavelengths to 193 nm need to be investigated. A single-layer APSM material is most attractive from the standpoint of process complexity, uniformity, and control. The optimum degree of transmission of the attenuator can be determined through experimental or simulation techniques. A maximum image modulation or image log slope is desired while maintaining a minimum printability level of side lobes formed from intensity within shadowed regions. Depending on feature type and size and resist processes, APSM transmission values between 4% and 15% may be appropriate. In addition to meeting optical requirements to allow appropriate phase-shift and transmission properties, an APSM material must be able to be patterned using plasma etch techniques, have high etch selectivity to fused silica, be chemically stable, have high absorbance at alignment wavelengths, and not degrade with exposure. These requirements may ultimately limit the number of possible candidates for practical mask application.

Phase shifting in a transparent material is dependent on a film's thickness, real refractive index, and the wavelength of radiation, as seen earlier. To achieve a phase shift of 180°, the requirement film thickness becomes

$$t = \frac{\lambda}{2(n-1)}$$

The requirements of an APSM material demand that films are absorbing, that is, that they possess a nonzero extinction coefficient (k). This introduces additional phase-shifting contributions from film interfaces that can be determined by

$$\Phi = \arg\left(\frac{2n_2^*}{n_1^* + n_2^*}\right)$$

where

n_1^* is the complex refractive index ($n+k$) of the first medium
n_2^* is the complex refractive index of the second medium [53]

These additional phase terms are nonnegligible as k increases, as shown in Figure 2.81. In order to determine the total phase shift resulting from an absorbing thin film, materials and interface contributions need to be accounted for.

To deliver both phase-shift and transmission requirements, film absorption (α) or extinction coefficient (k) is considered:

$$\alpha = \frac{4\pi k}{\lambda}$$

where α is related to transmission as $T=e^{-\alpha t}$. In addition, mask reflectivity below 15% is desirable and can be related to n and k through the Fresnel equation for normal incidence:

$$R = \frac{(n-1)^2 k^2}{(n+1)^2 k^2}$$

In order to meet all optical requirements, a narrow range of material optical constants is suitable at a given exposing wavelength. Both chromium oxydinitride–based and molybdenum silicon

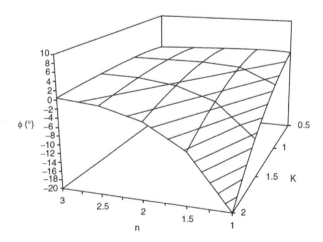

FIGURE 2.81 Additional phase terms resulting at interfaces as a function of n and k.

oxynitride–based materials have been used as APSM materials at 365 nm. For shorter-wavelength applications, these materials become too opaque. Alternative materials have been introduced that, through modification of material composition or structure, can be tailored for optical performance at wavelengths from 190 to 250 nm [54]. These materials include understoichiometric silicon nitride, aluminum-rich aluminum nitride, and other metal oxides, nitrides, and silicides. The usefulness of these materials in production may ultimately be determined by their ability to withstand short-wavelength exposure radiation. In general, understoichiometric films possess some degree of instability that may result in optical changes during exposure.

2.8.4 MASK BIASING AND FEATURE SHAPE CORRECTION

When considering one-dimensional imaging, features can often be described through the use of fundamental diffraction orders. Higher-order information lost through pupil filtering leads to less square-wave image reconstruction and a loss of aerial image integrity. With a high-contrast resist, such a degraded aerial image can be used to reconstruct a near-square-wave relief image. When considering two-dimensional imaging, the situation becomes more complex. Whereas mask to image width bias for a simplified one-dimensional case can be controlled via exposure/process or physical mask feature size manipulation, for two-dimensional imaging, there are high-frequency interactions that need to be considered. Loss or redistribution of high-frequency information results in such things as corner or contact rounding, which may influence device performance.

Other problems encountered when considering complex mask patterns are the fundamental differences between imaging isolated lines, isolated spaces, contacts, and dense features. The reasons for these differences are manyfold. First, a partially coherent system is not linear in either amplitude or intensity. As we have seen, only an incoherent system is linear in intensity, and only a coherent system is linear in amplitude. Therefore, it should not be expected that an isolated line and an isolated space feature will be complementary. In addition, photoresist is a nonlinear detector, responding differently to the thresholds introduced by these two feature types. This reasoning can be extended to the concept of mask biasing. At first guess, it may be reasoned that a small change in the size of a mask feature would result in a near-equivalent change in resist feature width or at least, aerial image width. Neither is possible, because in addition to the nonlinearity of the imaging system, biasing is not a linear operation.

Differences in image features of various types are also attributed to the fundamental frequency representation of dense versus isolated features. Dense features can be suitably represented by discrete diffraction orders using coherent illumination. Orders are distributed with some width for incoherent and partially incoherent illumination. Isolated features, on the other hand, can be represented as some fraction of a sinc function for coherent illumination, distributed across the frequency plane for incoherent and partially coherent illumination. In terms of frequency information, these functions are very different. Figure 2.82 shows the impact of partial coherence on dense to isolated feature bias for $0.6\lambda/NA$ features. Dense lines (equal lines and spaces) print smaller than isolated lines for low values of partial coherence. At high partial coherence values, the situation is reversed. There also exists some optimum where the dense to isolated feature bias is near zero. Variations in exposure, focus, aberrations, and resist process will also have effects.

Through characterization of the optical and chemical processes involved in resist patterning, image degradation can be predicted. If the degradation process is understood, small feature biases can be introduced to account for losses. This predistortion technique is often referred to as optical proximity correction (OPC), which is not a true correction in that lost diffraction detail is not accounted for. Mask biasing for simple shapes can be accomplished with an iterative approach, but complex geometry or large fields require rule-based computation schemes [55]. Generally, several adequate solutions are possible. Those that introduce the least process complexity are chosen for implementation. Figure 2.83a shows a simple two-dimensional mask pattern and the resulting simulated resist image for a $k_1 = 0.5$ process. Feature rounding is evident at both inside and outside

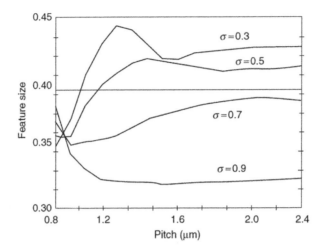

FIGURE 2.82 The variation in dense to isolated feature bias with partial coherence. For low σ values, dense features (2× pitch) print smaller than isolated features (6× pitch). For high σ values, the situation is reversed (365 nm, 0.5 NA, 0.4 µm, 0.5 NA, 0.4 µm lines, positive resist).

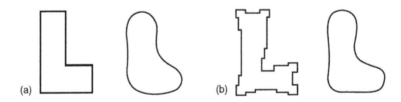

FIGURE 2.83 Simple two-dimensional mask patterns: (a) without OPC and the resulting resist image; (b) with OPD and the resulting resist image.

corners. The image degradation can be quantified by several means. Possible approaches may be to measure linear deviation, area deviation, or radius deviation. Figure 2.83b shows a biased version of the same simple pattern and the resulting simulated aerial image. Comparisons of the two images show the improvement realized with such correction schemes. The advantage of these techniques is the relatively low cost of implementation.

2.8.5 Mask Data Preparation (MDP) for Optical Proximity Correction (OPC)

The CAD data used for each lithography level of an IC needs to be translated into a format and a set of instructions that can be understood by the mask writer. Mask data preparation (MDP) is the process of translating completed design data which has undergone *design rule checking* (DRC), through the steps of adding optical proximity correction (OPC), separating phase shift mask (PSM) layers, and *fracturing* complex shapes into simple polygons understandable by the mask writer. The fracturing operation also includes the necessary sizing, biasing, scaling, and correction of data to produce necessary mask dimensions based on both mask and wafer process requirements.

At the mask level, added OPC patterns can include serifs, jogs, and sub-resolution assist featured at edges and corners. Phase mask features can require both added geometry and separation of phase and transmission layers, which needs to be addressed at both the pattern design and MDP steps. When multiple patterning exposure is employed at wafer level, the separation of patterns needs to be accounted for during the MDP as well. Examples of fundamental mask OPC are shown in Figure 2.84, where corner-rounding OPC features are added to a mask design [56].

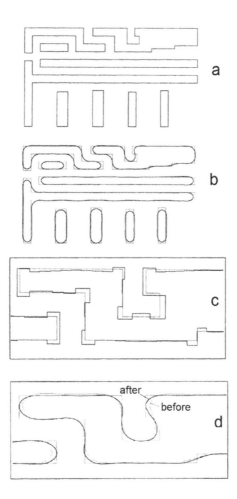

FIGURE 2.84 Mask OPC to correct for corner rounding with a) the initial design pattern, b) the pattern imaged using the initial mask showing corner rounding, c) the design pattern corrected for lithography effects, and d) the image from the mask before and after correction.

Additional corrections that will be made to mask data include those needed to account for the electron beam scatter and optical diffraction effects of the mask writer, as well as the process biases encountered in the exposure, develop, and pattern transfer steps of mask making. It is most desirable, and only practical, to automate each of these translation steps through the development of models and rules that describe each step of the processes. The result of the operation involved with MDP is a series of mask data formatting that progresses through the entire operation. As data is selected for backward comparison to the design or forward comparison to mask inspection, aerial image measurement, or wafer lithography, it may exist in a variety of forms – each quite unique in appearance.

2.8.6 Dummy Diffraction Mask

A novel technique of illumination control at the mask level was introduced that offered resolution improvement similar to that for off-axis illumination [57]. Though difficult to implement in a commercial lithography system, two separate masks were used to direct mask diffraction energy toward the objective lens pupil. In addition to a conventional binary mask, a second diffraction mask composed of line space or checkerboard phase patterns is created with 180° phase shifting between

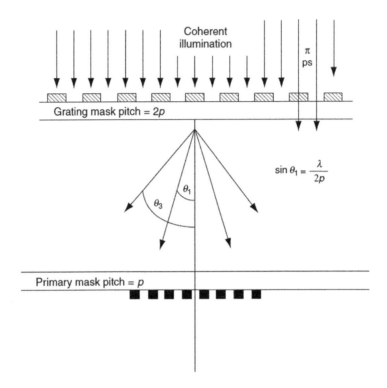

FIGURE 2.85 Schematic of a grating diffraction mask used to produce off-axis illumination for primary mask imaging.

patterns. Coherent light incident on the diffraction mask is diffracted by the phase grating as shown in Figure 2.85. When the phase grating period is chosen so that the angle of diffraction is $\sin^{-1}(\lambda/p)$, the first diffraction orders from the phase mask will deliver illumination at an optimum off-axis angle to the binary mask. There is no energy in the phase diffraction pattern on axis (no DC term), and higher orders have less energy than the first. For a line/space phase grating mask, illumination is delivered to the binary mask as with off-axis two-pole illumination. For a checkerboard phase grating mask, a situation similar to quadrupole illumination results.

A basic requirement for such an approach is that the phase mask and the binary mask are sufficiently far apart to allow far-field diffraction effects from the phase mask to dominate. This distance is maximized for coherent illumination on the order of $2p/\lambda$, where $2p$ is the phase mask grating period. As partial coherence is increased, a collection of illumination angles exists. This will decrease image contrast, as well as maximum intensity, and the required mask separation distance. The tolerance to phase error has been shown to be greater than $\pm 10\%$. Angular mis-registration of $10°$ may also be tolerable. The resolution capability for coherent illumination is identical to that for alternating phase shift-masking and off-axis illumination.

2.8.7 POLARIZATION EFFECTS ON MASK IMAGING

As discussed so far, the amplitude and phase components of light has been used for the design of a photomask. Light also possesses polarization characteristics that can be utilized to influence imaging performance [58]. Consider Figures 2.20 and 2.24, where zero and $\pm$ *first diffraction* orders are collected for an equal lines/space object. Here, the zeroth-order amplitude is $A/2$, and the $\pm$ first-order amplitudes are A/π. For a transverse electric (TE) state of linear polarization, these orders can be represented in terms of complex exponentials as

$$\text{zeroth order: } A/2 \begin{pmatrix} 0 \\ 1 \\ 0 \end{pmatrix} \exp\left[i2\pi/\lambda \left(0x + 0y + 1z \right) \right]$$

$$+\text{first order: } A/\pi \begin{pmatrix} 0 \\ 1 \\ 0 \end{pmatrix} \exp\left[i2\pi/\lambda \left(ax + 0y + cz \right) \right]$$

$$-\text{first order: } A/\pi \begin{pmatrix} 0 \\ 1 \\ 0 \end{pmatrix} \exp\left[i2\pi/\lambda \left(-ax + 0y + cz \right) \right]$$

For transverse magnetic (TM) polarization, these orders become

$$\text{zeroth order: } A/2 \begin{pmatrix} 0 \\ 1 \\ 0 \end{pmatrix} \exp\left[i2\pi/\lambda \left(0x + 0y + 1z \right) \right]$$

$$+\text{first order: } A/\pi \begin{pmatrix} c \\ 1 \\ -a \end{pmatrix} \exp\left[i2\pi/\lambda \left(ax + 0y + cz \right) \right]$$

$$-\text{first order: } A/\pi \begin{pmatrix} c \\ 0 \\ a \end{pmatrix} \exp\left[i2\pi/\lambda \left(-ax + 0y + cz \right) \right]$$

As previously shown, the sum of these terms produces the electric field at the wafer plane. The aerial images at the wafer plane for TE and TM polarization become

$$I_{\text{TE}}(x) = \frac{A}{\pi} + \frac{4A}{\pi} \cos^2\left(\frac{2\pi ax}{\lambda} \right) + \frac{2A}{\pi} \cos\left(\frac{2\pi ax}{\lambda} \right)$$

$$I_{\text{TM}}(x) = \frac{A}{\pi} + \frac{4A}{\pi^2} \left[a^2 + \left(c^2 - a^2 \right) \cos^2\left(\frac{2\pi ax}{\lambda} \right) \right] + \frac{2c}{\pi} \cos\left(\frac{2\pi ax}{\lambda} \right)$$

The normalized image log slope (NILS = ILS × line width) for each aerial image becomes

$$\text{NILS}_{\text{TE}} = 8$$

$$\text{NILS}_{\text{TE}} = 8 \frac{\sqrt{1 - a^2}}{1 + \left(16^2 / \pi^2 \right)}$$

The second term in the NILS$_{\text{TM}}$ equation is less than one, resulting in a lower resolution value for TM polarization as compared with TE polarization. Therefore, there can be some benefit to using TE polarization over TM polarization or nonpolarized light. This is discussed in more detail in Section 2.10.

The concept of polarization modulation built into a mask itself had been introduced as a potential step for mask modification. This would likely require the use of single crystalline mask materials and processes. Though difficult to realize, a polarized mask has been proposed as a means of accomplishing optimization of various feature orientations [59, 60].

2.9 OPTICAL SYSTEM DESIGN

In an ideal lens, the image formed is a result of all rays at all wavelengths from all object points, forming image plane points. Lens aberrations create deviations from this ideal, and a lens designer must make corrections or compensation. The degrees of freedom available to a designer include material refractive index and dispersion, lens surface curvatures, element thickness, and lens stops. Other application-specific requirements generally lead lens designers toward only a few practical solutions.

For a microlithographic optical system, Köhler illumination is generally used. A requirement for a projection lens is that two images are simultaneously relayed: the image of the reticle and the image of the source (or the illumination exit pupil). The projection lens cannot be separated from the entire optical system; consideration of the illumination optics needs to be included. In designing a lens system for microlithographic work, image quality is generally the primary consideration. Limits must often be placed on lens complexity and size to allow workable systems. The push toward minimum aberration, maximum numerical aperture, maximum field size, maximum mechanical flexibility, and minimum environmental sensitivity has led to designs that incorporate features somewhat unique to microlithography.

2.9.1 STRATEGIES FOR REDUCTION OF ABERRATIONS: ESTABLISHING TOLERANCES

Several classical strategies can be used to achieve maximum lens performance with minimum aberration. These might include modifying material indices and dispersion, splitting the power of elements, compounding elements, using symmetric designs, reducing the effective field size, balancing existing aberrations, or using elements with aspheric surfaces. Incorporating these techniques is often a delicate balancing operation.

2.9.1.1 Material Characteristics

When available, the use of several glass types of various refractive index values and dispersions allows significant control over design performance. Generally, for positive elements, high-index materials will allow reduction of most aberrations because of the reduction of ray angles at element surfaces. This is especially useful for the reduction of Petzval curvature. For negative elements, lower-index materials are generally favored, which effectively increase the extent to which correcting is effective. Also, a high value of dispersion is often used for the positive element of an achromatic doublet, whereas a low dispersion is desirable for the negative element. For microlithographic applications, the choice of materials that allows these freedoms is limited to those that are transparent at design wavelengths. For g-line and i-line wavelengths, several glass types transmit well, but below 300 nm, only fused silica and fluoride crystalline materials can be used. Without the freedom to control refractive index and dispersion, a designer is forced to look for other ways to reduce aberrations. In the case of chromatic aberration, reduction may not be possible, and restrictions must be placed on source bandwidth if refractive components are used.

2.9.1.2 Element Splitting

Aberrations can be minimized or balanced by splitting the power of single elements into two or more components. This allows a reduction in ray angles, resulting in a lowering of aberration. This technique is often employed to reduce spherical aberration: negative aberration can be reduced by splitting a positive element, and positive aberration can be reduced by splitting a negative element.

The selection of the element to split can often be determined through consideration of higher-order aberration contributions. Using this technique for microlithographic lenses has resulted in lens designs with a large number of elements.

2.9.1.3 Element Compounding

Compounding single elements into a doublet is accomplished by cementing the two and forming an interface. This technique enables control of ray paths and allows element properties not possible with one glass type. In many cases, a doublet will have a positive element with a high index combined with a negative element of lower index and dispersion. This produces an achromatized lens component that performs similarly to a lens with a high index and very high dispersion. This accomplishes both a reduction in chromatic aberration and a flattening of the Petzval field. Coma aberration can also be modified by taking advantage of the refraction angles at the cemented interface, where upper and lower rays may be bent differently. The problem with utilizing a cemented doublet approach with microlithographic lenses is again in suitable glass materials. Most available UV and deep UV glass materials have a low refractive index (~1.5), limiting the corrective power of a doublet. This results in a narrow wavelength band over which an achromatized lens can be corrected in the UV.

2.9.1.4 Symmetrical Design

An optical design that has mirror symmetry about the aperture stop is free of distortion, coma, and chromatic aberration. This is due to an exact canceling of aberrations on each side of the pupil. In order to have complete symmetry, unit magnification is required. Optical systems that are nearly symmetrical can result in substantial reduction of higher-order residuals of distortion, coma, and chromatic aberration. These systems, however, operate with unit magnification, a requirement for object-to-image symmetry. Because 1× imaging limits mask and wafer geometry, these systems can be limiting for very high-resolution applications but are widely used for larger-feature lithography.

2.9.1.5 Aspheric Surfaces

Most lens designs restrict surfaces to being spherically refracting or reflecting. The freedom offered by allowing the incorporation of aspheric surfaces can lead to dramatic improvements in residual aberration reduction. Problems encountered with aspheric surfaces include difficulties in fabrication, centering, and testing. Several techniques have been utilized to produce parabolic as well as general aspheres [61]. Lithographic system designs have started to take advantage of aspheric elements on a limited basis. The success of these surfaces may allow lens designs to be realized that would otherwise be impossible.

2.9.1.6 Balancing Aberrations

For well-corrected lenses, individual aberrations are not necessarily minimized; instead, they are balanced with respect to wavefront deformation. The optimum balance of aberration is unique to the lens design and is generally targeted to achieve minimum OPD. Spherical aberration can be corrected for in several ways, depending largely on the lens application. When high-order residual aberrations are small, correction of spherical aberration to zero at the edge of the aperture is usually best, as shown in Figure 2.41. Here, the aberration is balanced for minimum OPD and is best for diffraction-limited systems such as projection lenses. If a lens is operated over a range of wavelengths, however, this correction may result in a shift in focus with aperture size. In this case, spherical aberration may be overcorrected. This situation would result in a minimum shift in best focus through the full aperture range, but a decrease in resolution would result at full aperture.

Chromatic aberration is generally corrected at a 0.7 zone position within the aperture. In this way, the inner portion of the aperture is undercorrected, and the outer portion of the lens is overcorrected. Astigmatism can be minimized over a full field by overcorrecting third-order astigmatism and undercorrecting fifth-order astigmatism. This will result in the sagittal focal surface being located inside the tangential surface in the center of the field and vice versa at the outside of the field. Petzval field curvature is adjusted so that the field is flat with both surfaces slightly inward.

Corrections such as these can be made through control of element glass, power, shape, and position. The impact of many elements of a full lens design makes minimization and optimization very difficult. Additionally, corrections such as those discussed operate primarily on third-order aberration. Corrections of higher order and interactions cannot be made with single-element or surface modifications. Lens design becomes a delicate process, best handled with optical design programs that utilize local and global optimization. Such computational tools allow interaction of lens parameters based on a starting design and an optical designer experience. By taking various paths to achieve low aberration, high numerical aperture, large flat fields, and robust lithographic systems, several lens designs have evolved.

2.9.2 Basic Lithographic Lens Design

2.9.2.1 The All-Reflective (Catoptric) Lens

Historically, the 1× ring-field reflective lens used in a scanning mode was one of the earliest projection systems used in integrated circuit manufacture [62]. The reflective aspect of such catoptric systems has several advantages over refractive lens designs. Because most or all of the lens power is in the reflective surfaces, the system is highly achromatized and can be used over a wide range of wavelengths. Chromatic variation of aberrations is also absent. In addition, aberrations of special mirrors are much smaller than those of a refractive element. A disadvantage of a conventional catoptric system, such as the configurations lens shown in Figure 2.86, is the obscuration required for imaging. This blocking of light rays close to the optical axis is, in effect, a low-pass filtered system that can affect image modulation and depth of focus. The 1× Offner design of the ring-field reflecting system gets around this obscuration by scanning through restricted off-axis annulus of a full circular field, as shown in Figure 2.87. This not only eliminates the obscuration problems but also substantially reduces redial aberration variation. Because the design is rotationally symmetrical, all aberrations are constant around the ring. By scanning the image field through this ring, astigmatism, field curvature, and distortion are averaged. It can also be seen that this design is symmetric on both image and object sides. This results in 1× magnification, but it allows further cancellation of aberration.

Vignetting of rays by the secondary mirror forces operation off axis and introduces an increase in aberration level. Mechanically, at larger numerical apertures, reticle and wafer planes may be accessible only by folding the design. Field size is also limited by lens size and high-order aberration. Moreover, unit magnification limits both resolution and wafer size.

2.9.2.2 The All-Refractive (Dioptric) Lens

Early refractive microlithographic lenses resembled microscope objectives, and projection lithography was often performed using off-the-shelf microscope designs and construction. As IC device

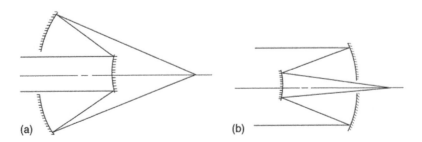

FIGURE 2.86 Two-mirror catoptric systems: (a) the Schwarzchild and (b) the Cassegrain configuration.

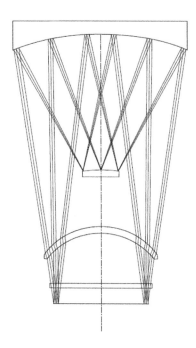

FIGURE 2.87   The 1× Offner ring-field reflective lens.

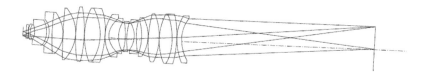

FIGURE 2.88   An all-refractive lens design for a 5× i-line reduction system.

areas grow, requirements for lens field sizes are increased. Field sizes greater than 25 mm are not uncommon for current IC technology, using lens numerical apertures above 0.50. Such requirements have led to the development of UV lenses that operate well beyond λ/4 requirements for diffraction-limited performance, delivering resolution approaching 0.15 μm.

Shown in Figure 2.88 is a refractive lens design for use in a 5× i-line reduction system [63]. The design utilizes a large number of low-power elements for minimization of aberration as well as aberration-canceling surfaces. The availability of several glass types at i-line and g-line wavelengths allows chromatic aberration correction of such designs over bandwidths approaching 10 nm. The maximum NA for these lens types is approaching 0.65 with field sizes larger than 30 nm.

Achromatic refractive lens design is not possible at wavelengths below 300 nm, and apart from chromatic differences of paraxial magnification, chromatic aberration cannot be corrected. Restrictions must be placed on exposure sources, generally limiting spectral bandwidth on the order of a few picometers. First-order approximations for source bandwidth based on paraxial defocus of the image by half of the Rayleigh focal depth also show the high dependence on lens NA and focal length. Chromatic aberration can be expressed as

$$\delta f = \frac{f(\delta n)}{(n-1)}$$

where

f is focal length
n is refractive index
δf is focus error or chromatic aberration

Combining this with the Rayleigh depth of focus condition

$$\text{DOF} = \pm 0.5 \frac{\lambda}{\text{NA}^2}$$

produces a relationship

$$\Delta \lambda \left(\text{FWHM} \right) = \frac{(n-1)\lambda}{2f\left(dn/d\lambda \right)\text{NA}^2}$$

where $dn/d\lambda$ is the dispersion of the lens material. Lens magnification, m, further affects required bandwidth as

$$\Delta \lambda \left(\text{FWHM} \right) = \frac{(n-1)\lambda}{2f\left(1+m \right)\left(dn/d\lambda \right)\text{NA}^2}$$

A desirable chromatic refractive lens from the standpoint of the laser requirements would, therefore, have a short focal length and a small magnification (high reduction factor) for a given numerical aperture. Requirements for IC manufacture, however, do not coincide. Shown in Figure 2.89 is an example of a chromatic refractive lens design [63]. This system utilizes an aspherical lens element that is close to the lens stop [64]. Because refractive index is also dependent on temperature and pressure, chromatic refractive lens designs are highly sensitive to barometric pressure and lens heating effects.

2.9.2.3   Catadioptric-Beam-Splitter Designs

Both the reflective (catoptric) and refractive (dioptric) systems have advantages that would be beneficial if a combined approach to lens design were utilized. Such a refractive-reflective approach is known as a *catadioptric* design. Several lens designs have been developed for microlithographic projection lens application. A catadioptric lens design that is similar to the reflective ring-field system is the 4× reduction Offner shown in Figure 2.90 [63]. The field for this lens is also an annulus or ring that must be scanned for full-field imaging. The design uses four spherical mirrors and two fold mirrors. The refractive elements are utilized for aberration correction, and their power is minimized, reducing chromatic effects and allowing the lens to be used with an Hg lamp at DUV wavelengths.

This also minimizes the sensitivity of the design to lens heating and barometric pressure. The drawbacks of this system are its numerical aperture, limited to sub-0.5 levels by vignetting, and the aberration contributions from the large number of reflective surfaces. The alignment of lens elements is also inherently difficult.

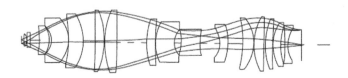

FIGURE 2.89   A chromatic all-refracting lens design for a 4×248 nm system.

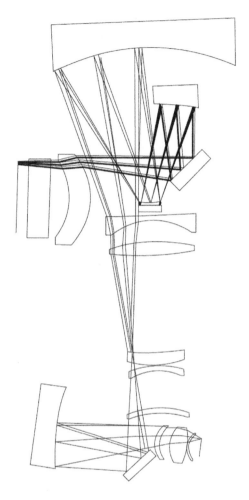

FIGURE 2.90   The 4× catadioptric MSI design.

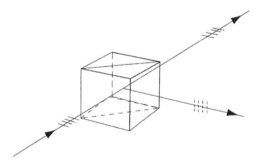

FIGURE 2.91   A polarizing beam-splitter. A linearly polarized beam is divided into TM and TE states at right angles.

   To avoid high amounts of obscuration of prohibitively low lens numerical apertures, many lens designs have made use of the incorporation of a beam-splitter. Several beam-splitter types are possible. The conventional cube beam-splitter consists of matched pairs of right-angle prisms, one with a partially reflecting film deposited on its face, optically cemented. A variation on the cube beam-splitter is a polarizing beam-splitter, as shown in Figure 2.91. An incident beam of linearly polarized light is divided, with TM and TE states emerging at right angles. Another possibility is a beam-splitter that is incorporated into a lens element, known as a *Mangin mirror*, as shown in Figure 2.92. Here, a partial reflector allows one element to act as both a reflector and a refractor. Although the use of a Mangin mirror does require central obscuration, if a design can achieve levels below 10% (radius), the impacts on imaging resolution and depth of focus are minimal [65].

The 4× reduction Dyson shown in Figure 2.93 is an example of a catadioptric lens design based on a polarizing beam-splitter [63]. The mask is illuminated with linearly polarized light that is directed through the lens toward the primary mirror [66]. Upon reflection, a waveplate changes the state of linear polarization, allowing light to be transmitted toward the wafer plane. Variations on this design use a partially reflecting beam-splitter, which may suffer from reduced throughput and a susceptibility coating damage at short wavelengths. Obscuration is eliminated, as is the low-NA requirement of the off-axis designs to prevent vignetting. The beam-splitter is well corrected for operation on axis, minimizing high-order aberrations and the requirement for an increasingly thin ring field for high NA as with the reduction Offner. The field is square, which can be used in a

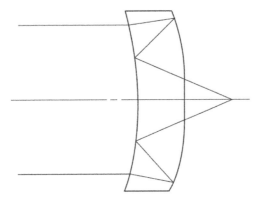

FIGURE 2.92 A Mangin mirror-based beam-splitter approach to a catadioptric system. A partially reflective surface allows one element to act as a reflector and a refractor.

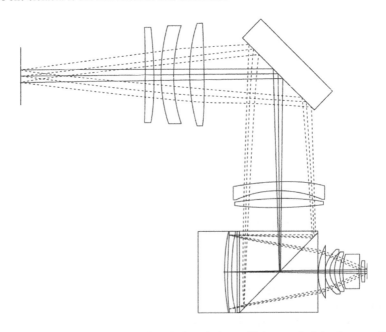

FIGURE 2.93 A 4× reduction Dyson catadioptric lens design utilizing a polarizing beam-splitter.

stepping mode, or rectangular for step and scanning. The simplified system, with only one mirror possessing most of the lens power, leads to lower aberration levels than for the reduction Offner design. This design allows a spectral bandwidth on the order of 5–10 nm, enabling operation with a lamp or laser source.

As previously seen, at high NA values (above 0.5) for high-resolution lithography, diffraction effects for TE and TM are different. When the vectorial nature of light is considered, a biasing between horizontally oriented and vertically oriented features results. Although propagation into a resist material will reduce this biasing effect [60], it cannot be neglected. Improvements on the reduction Dyson in Figure 2.93 have included elimination of the linear polarization effect by incorporating a second waveplate near the wafer plane. The resulting circular polarization removes the H-V biasing possible with linear polarization and also rejects light reflected from the wafer and lens surface, reducing lens flare. Improvements have also increased the NA of the Dyson design up to 0.7, using approaches that include larger-NA beam-splitter cubes, shorter image conjugates, increased mirror asphericity, and source bandwidths below 1 nm. This spectral requirement, along with increasingly small field widths to reduce aberration, requires that these designs be used only with excimer laser sources. Designs have been developed for both 248 and 193 nm wavelengths. Examples of these designs are shown in Figures 2.94 [61] and 2.95 [69]. Catadioptric 193 nm immersion lithography lens designs for numerical apertures above 1.0 are shown in Figure 2.96 [70].

2.10 POLARIZATION AND HIGH NA

As with any type of imaging, lithography is influenced by the polarization of the propagating radiation. In reality, the impact of polarization on imaging has been relatively low at NA values below 0.80 NA, as interfering rays are imaged into a photoresist with a refractive index greater than that for air. Because the refractive index of the resist (n_{PR}) is in the range of 1.60–1.80, the resulting angles

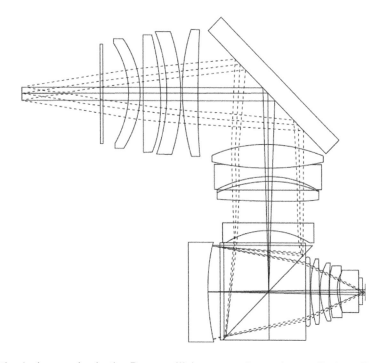

FIGURE 2.94 An improved reduction Dyson, utilizing a second waveplate to eliminate linear polarization effects at the wafer. (Reproduced from Williamson, McClay, D., Andresen, J., Gallatin, K., Himel, G., Ivaldi, M., Mason, C., McCullough, A., Otis, C., and Shamaly, J., *SPIE*, 2726, 780, 1996. With permission.)

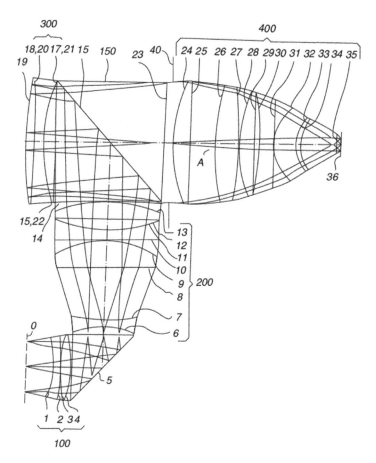

FIGURE 2.95 A reduction Dyson approach with the stop behind the beam-splitter. Numerical apertures to 0.7 can be achieved with a high degree of collumation. Spectral narrowing is likely to be needed.

of interference are reduced by Snell's law to NA/n_{PR}. Concerns with polarization have, therefore, been limited to the requirements of the optical coatings within lens systems and those lithography approaches making use of polarization for selection, such as the reduction Dyson lens designs seen in Figures 2.93 and 2.94. As immersion lithography has enabled numerical apertures above 1.0, the impact of polarization becomes more significant. For this reason, special attention needs to be paid to the influence of the polarization at most stages of the lithographic imaging process.

Polarized radiation results as the vibrations of a magnetic or electric field vector are restricted to a single plane. The direction of polarization refers to the electric field vector that is normal to the direction of propagation. Linear polarization exists when the direction of polarization is fixed. Any polarized electric field can be resolved into two orthogonally polarized components. Circular polarization occurs when the electric field vector has two equal orthogonal components, causing the resultant polarization direction to rotate about the direction of propagation. Circular polarization with a preferred linear component is termed *elliptical polarization*. Unpolarized radiation has no preferred direction of polarization.

2.10.1 IMAGING WITH OBLIQUE ANGLES

At oblique angles, radiation polarized in the plane of incidence exhibits reduced image contrast, as interference is reduced [71]. This is referred to as TM, p, or X polarization with respect to vertically oriented geometry. As angles approach $\pi/4$ [or $\sin^{-1}(1/\sqrt{2})$], no interference is possible, and image

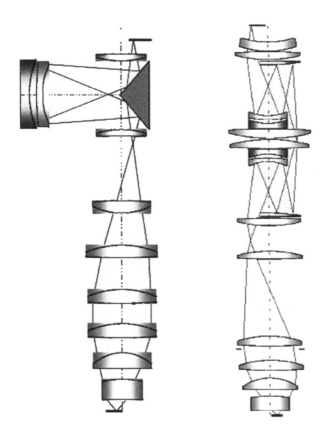

FIGURE 2.96 Examples of 193 nm high NA (>1.0) projection lens designs for immersion lithography.

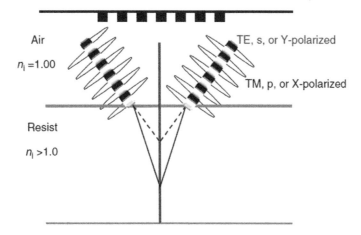

FIGURE 2.97 Interference in a photoresist film depicted in the two states of linear polarization. The mask features are oriented out of the paper.

contrast in air is reduced to zero. If the image is translated into a medium of higher index, the limiting angle is increased by the media index (n) as $\sin^{-1}(n/\sqrt{2})$.

For polarization perpendicular to the plane of incidence, complete interference exists, and no reduction in image contrast will result. Figure 2.97 shows the two states of linear polarization that contribute to a mask function oriented out of the plane of the page. TM polarization is in the plane of the page, and transverse electric (TE or Y) polarization is perpendicular. For nonlinear polarization, an image is formed as the sum of TE and TM image states.

2.10.2 POLARIZATION AND ILLUMINATION

At high NA values, methods can be used that avoid TM interference. Several approaches to remove this field cancellation from TM polarization have been proposed, including image decomposition for polarized dipole illumination [72]. Illumination that is consistently TE polarized in a circular pupil could achieve the optimum polarization for any object orientation. This is possible with an illumination field that is TE polarized over all angles in the pupil, known as *azimuthal polarization*, which is shown in Figure 2.98, along with TM or radial polarized illumination. Such an arrangement provides for homogeneous coupling of the propagating radiation regardless of angle or orientation. As an example, Figure 2.99 shows imaging results for four cases of illumination of line/space features as focus is varied. The conditions plotted are for TE polarized dipole, TE polarized cross-quadrupole, azimuthal annular, and unpolarized annular illumination. In this case, the TE polarized cross-quadrupole illumination results in superior image performance.

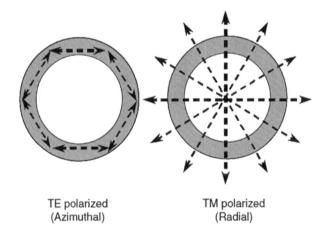

TE polarized   TM polarized
(Azimuthal)   (Radial)

FIGURE 2.98   TE or azimuthally polarized illumination (left) and TM or radially polarized illumination.

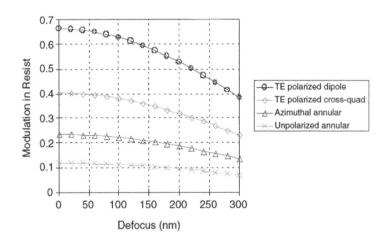

FIGURE 2.99   The modulation in a photoresist film for 45 nm features imaged at 1.20 NA (water immersion) under various conditions of illumination. The greatest modulation through focus is achieved using a TE polarized dipole illuminator.

2.10.3 Polarization Methods

Selection of a single linear state to polarize radiation requires a method that provides efficiency in the wavelength region of interest. Though many methods exist at longer wavelengths, the choices are more limited in the UV. Polarization can be achieved with crystalline materials that have a different index of refraction in different crystal planes. Such materials are said to be birefringent or doubly refracting. A number of polarizing prisms have been devised that make use of birefringence to separate two beams in a crystalline material. Often, they make use of total internal reflection to eliminate one of the planes. The Glan-Taylor, Glan-Thompson, Glan-Laser, beam-splitting Thompson, beam displacing, and Wollaston prisms are most widely used, and they are typically made of nonactive crystals such as calcite that transmit well from 350 to 2300 nm. Active crystals such as quartz can also be used in this manner if cut with the optic axis parallel to the surfaces of the plate.

Polarization can also be achieved through reflection. The reflection coefficient for light polarized in the plane of incidence is zero at the Brewster angle, leaving the reflected light at that angle linearly polarized. This method is utilized in polarizing beam-splitter cubes that are coated with many layers of quarter-wave dielectric thin films on the interior prism angle to achieve a high extinction ratio between the TE and TM components.

Wire grid polarization can also be employed as a selection method [73]. Wire grids, generally in the form of an array of thin parallel conductors supported by a transparent substrate, have been used as polarizers for the visible, infrared, and other portions of the electromagnetic spectrum. When the grid period is much shorter than the wavelength, the grid functions as a polarizer that reflects electromagnetic radiation polarized parallel to the grid elements, and it transmits radiation of the orthogonal polarization. These effects were first reported by Wood in 1902, and they are often referred to as *Wood's Anomalies* [74]. Subsequently, Rayleigh analyzed Wood's data and believed that the anomalies occur at combinations of wavelength and angle where a higher diffraction order emerges [75].

2.10.4 Polarization and Resist Thin Film Effects

To reduce the reflectivity at an interface between a resist layer and a substrate, a bottom antireflective coating (BARC) is coated beneath the resist, as discussed in detail in later chapters. Interference minima occur as reflectance from the BARC/substrate interface destructively interferes with the reflection at the resist/BARC interface. This destructive interference thickness repeats at quarter-wave thickness. Optimization of a single-layer BARC is possible for oblique illumination and also for specific cases of polarization, as seen in the plots of Figure 2.100. The issue with a single-layer AR film, however, is its inability to achieve low reflectivity across all angles and through both states of linear polarization. This can be achieved using a multilayer BARC design, as shown in Figure 2.101 [76]. By combining two films in a stack and optimizing their optical and thickness properties, reflectivity below 0.6% can be made possible for angles to 45° (or at 1.2 NA) for all polarization states, as shown in Figure 2.102.

2.11 IMMERSION LITHOGRAPHY

Ernst Abbe was the first to discover that the maximum ray slope entering a lens from an axial point on an object could be increased by a factor equal to the refractive index of the imaging medium. He first realized this in the late 1870s by observing an increase in the ray slope in the Canada balsam mounting compound used in microscope objectives at the time. To achieve a practical system employing this effect, he replaced the air layer between a microscope objective and a cover glass with oil having a refractive index in the visible near that of the glass on either side. This index matching prevents reflective effects at the interfaces (and total internal reflection at large angles),

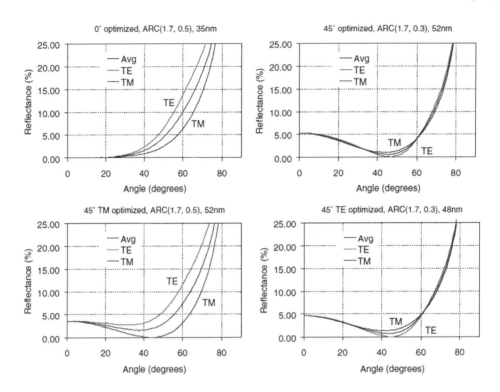

FIGURE 2.100 Reflectance plots for a BARC for various polarization states and conditions of optimization.

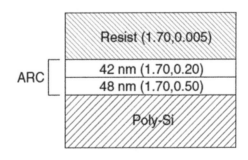

FIGURE 2.101 A two-layer BARC stack using matched refractive index (*n*) and dissimilar extinction coefficient (*k*) films.

leading to the term *homogenous immersion* for the system he developed. The most significant application of the immersion lens was in the field of medical research, where oil immersion objectives with high resolving power were introduced by Carl Zeiss in the 1880s. Abbe and Zeiss developed oil immersion systems by using oils that matched the refractive index of glass. This resulted in numerical aperture values up to a maximum of 1.4, allowing light microscopes to resolve two points distanced only 0.2 μm apart (corresponding to a k_1 factor value in lithography of 0.5).

The application of immersion imaging to lithography had not been employed until recently for several reasons. The immersion fluids used in microscopy are generally opaque in the UV and are not compatible with photoresist materials. Also, the outgassing of nitrogen during the exposure of DNQ/novolac (g-line and i-line) resists would prevent their application to a fluid-immersed environment. Most important, however, is the availability of alternative approaches to extend optical lithography. Operation at the 193 nm ArF wavelength leaves few choices other than immersion to continue the pursuit of optical lithography. Fortunately, polyacrylate

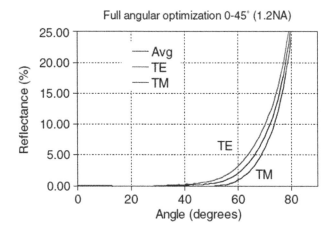

FIGURE 2.102 The optimized 193 nm reflection for the film stack of Figure 2.101 measured beneath the photoresist.

photoresists used at this wavelength do not outgas upon exposure (nor do 248 nm polyhydroxy styrene[PHOST] resists). The largest force behind the insurgence of immersion imaging into mainstream optical lithography has been the unique properties of water in the UV. Figure 2.103 shows the refractive index of water in the UV and visible region. Figure 2.104 shows the transmission for 1 mm and 1 cm water thickness values. As the wavelength decreases toward 193 nm, the refractive index increases to a value of 1.44, significantly larger than its value of 1.30 in the visible. Furthermore, the absorption remains low at 0.05 cm^{-1}. Combined with the natural compatibility of IC processing with water, the incentive to explore water immersion lithography at DUV wavelengths now exists.

The advantages of immersion lithography can be realized when the resolution is considered together with depth of focus. The minimum resolvable pitch for an optical imaging system is determined by wavelength and numerical aperture:

$$p = \frac{\lambda}{n \sin(\theta)}$$

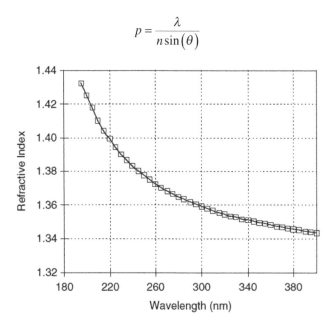

FIGURE 2.103 The refractive index of water in the ultraviolet.

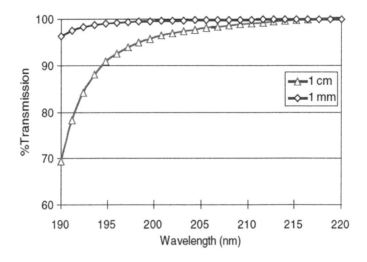

FIGURE 2.104   The transmission of water in the ultraviolet at 1 mm and 1 cm depths.

where n is the refractive index of the imaging media and θ is the half angle. As the refractive index of the medium is increased, the NA increases proportionately. At a $\sin\theta$ value of 0.93 (or 68°), which is near the maximum angle of any practical optical system, the largest NA allowed using water at 193 nm is 1.33. The impact on resolution is clear, but it is really just half of the story. The paraxial depth of focus for any media of index n takes the form

$$\text{DOF} = \pm \frac{k_2\lambda}{n\sin^2\theta}$$

Taken together, as resolution is driven to smaller dimensions with increasing NA values, the cost to DOF is significantly lower if refractive index is increased instead of the half angle. As an example, consider two lithography systems operating at NA values of 0.85, one being water immersion and the other imaging through air. Although the minimum resolution is the same, the paraxial DOF for the water immersion system is 45% larger than the air imaging system, a result of a half angle of 36° in water versus 58° in air.

2.11.1   CHALLENGES OF IMMERSION LITHOGRAPHY

Early concerns regarding immersion lithography included the possible adverse effects of small microbubbles that can form or be trapped during fluid filling, scanning, or exposure [77]. Though defects caused by trapped air bubbles remain a concern associated with fluid mechanics issues, the presence or creation of microbubbles has proved to be noncritical. In the process of forming a water fluid layer between the resist and lens surfaces, air bubbles are often created as a result of the high surface tension of water. The presence of air bubbles in the immersion layer could degrade the image quality because of the inhomogeneity-induced light scattering in the optical path. The scattering of air bubbles in water can be approximately described using a geometrical optics model [78]. As seen in Figure 2.105, an air bubble assumes a spherical shape in water when the hydraulic pressure because of gravity is ignored. The assumption is reasonable in lieu of immersion lithography, where a thin layer of water with thickness of about 0.5 mm is applied. The reflection/refraction at the spherical interface causes the light to scatter into various directions that can be approximated by flat surface Fresnel coefficients. However, the air bubble in water is a special case where the refractive index of the bubble is less than that of the surrounding medium, resulting in a contribution of total reflection to scattered irradiance at certain angles. The situation is described in Figure 2.106. For an arbitrary ray incident on a bubble, the angle of incidence is

$$i = \arcsin\left(s/a\right)$$

where

    a is the radius of the bubble

    s is the deviation from the center

The critical incident angle is

$$i_c = \arcsin\left(n_i/n_w\right)$$

where

    n_i is the refractive index of the air

    n_w is the refractive index of water

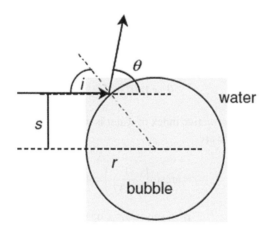

FIGURE 2.105 The geometric description of the reflection of an air bubble sphere in water.

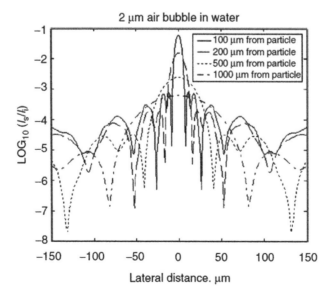

FIGURE 2.106 The ratio of the scattered intensity to the incident intensity for a 2 μm air particle in water at various separation values.

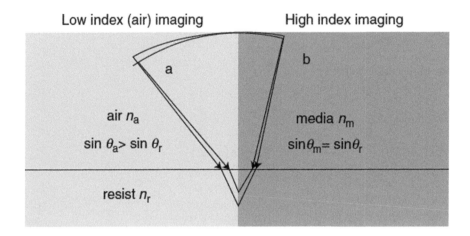

FIGURE 2.107   The effect of media index on the propagation angle for imaging equivalently sized geometry.

The corresponding critical scattering angle is

$$\theta_c = 180° - 2i_c$$

At a wavelength of 193 nm, the refractive index of water is $n_w = 1.437$. Therefore, the critical incident angle and critical scattering angle are

$$i_c = \arcsin\left(\frac{1}{1.437}\right) = 44°$$

$$\theta_c = 180° - 2i_c = 92°$$

The presence of total reflection greatly enhances adverse scattered light. In this case, the region covers all the forward directions. Hence, air bubbles in water cause strong scattering in all the forward directions. However, a complete understanding of scattering will require taking into account the effects of interference of the reflected light with other transmitted light. The rigorous solution of the scattering pattern can be numerically evaluated by partial wave (Mie) theory. In Mie scattering theory, the incident, scattered, and internal fields are expanded in a series of vector spherical harmonics [79]. At the wavelength of 193 nm, the scattering of an air bubble 2 μm in diameter was calculated according to Mie theory and plotted in Figure 2.107. At relatively short lateral distances from such a bubble and at distances beyond 100 μm, the scatter intensity becomes very low. The trapping of bubbles or the collection of several bubbles is a concern that needs to be addressed in the design of the liquid flow cell for an immersion lithography system [80].

2.11.2   HIGH-INDEX IMMERSION FLUIDS

Optical lithography is being pushed against fundamental physical and chemical limits, presenting the real challenges involved with resolution at dimensions of 32 nm and below. Hyper-NA optical lithography is generally considered to be imaging at angles close to 90° to achieve NAs above 1.0. Because of the small gains in NA at propagation angles in optics above 65°, values much above this are not likely in current or future lens design strategies. Hyper-NA is, therefore, forced upon material refractive index where the media with the lowest index create a weak link to system resolution. The situation is one where the photoresist possesses the highest refractive index and a photoresist top coat has the lowest refractive index:

$$n_{\text{photoresist}} > n_{\text{glass}} > n_{\text{fluid}} > n_{\text{top coat}}$$

The ultimate resolution of a lithography tool then becomes a function of the lowest refractive index. Resolution enhancement techniques (RET) methods that are already being employed in lithography can achieve k_1 process factors near 0.30, where 0.25 is the physical limit. It is, therefore, not likely that much ground will be achieved as a move is made into hyper-NA immersion lithography. The minimum half-pitch (hp) for 193 nm lithography following classical optical scaling, using a 68° propagation angle, becomes

$$hp_{\min} = \frac{k_1 \lambda}{n_i \sin \theta} = \frac{(0.25 \text{ to } 0.30)(193)}{n_i (0.93)} = \frac{52}{n_i} \text{ to } \frac{62}{n_i} \text{ nm}$$

for aggressive k_1 values between 0.25 and 0.30, where n_i is the lowest refractive index in the imaging path. Water as an immersion fluid is currently the weak link in a hyper-NA optical lithography scenario. Advances with second-generation fluid indices approaching 1.65 may direct this liability toward optical materials and photoresists. As resolution is pushed below 32 nm, it will be difficult for current photoresist refractive index values (~1.7) to accommodate. As photoresist refractive index is increased, the burden is once again placed on the fluid and the optical material. As suitable optical materials with refractive indices higher than that of fused silica are identified, the fluid is, once again, the weak link. This scenario will exist until a fluid is identified with a refractive index approaching 1.85 (limited by potential glass alternatives currently benchmarked by sapphire) and high-index polymer platforms.

To demonstrate the advantages of using higher–refractive index liquids, an imaging system using an immersion fluid is shown in Figure 2.107. The left portion of this figure depicts an optical wavefront created by a projection imaging system that is focused into a photoresist (resist) material with refractive index n_r. The refractive index of the imaging media is n_a (and, in this example, is air). The right portion of the figure depicts an optical wavefront focused through a medium of refractive index larger than the one on the left, specifically n_m. As the refractive index n_m increases, the effect of defocus, which is proportional to $\sin^2\theta$, is reduced. Furthermore, as shown in Figure 2.108, a refractive index approaching that of the photoresist is desirable to allow large angles into the photoresist film and also reduced reflection at interfaces between the medium and the resist. Ultimately, a small NA/n is desirable in all media, and the maximum NA of the system is limited to the smallest medium refractive index.

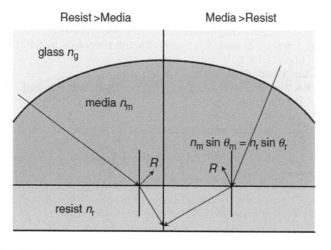

FIGURE 2.108 The effect of media index on reflection and path length.

TABLE 2.6
Absorption Peak (in eV and nm)
for Several Anions in Water

	eV	nm
I^-	5.48	227
Br^-	6.26	198
Cl^-	6.78	183
ClO_4^-	6.88	180
HPO_4^{2-1}	6.95	179
SO_4^{2-1}	7.09	175
$H_2PO_4^-$	7.31	170
HSO_4^-	7.44	167

In general, the UV absorption of a material involves the excitation of an electron from the ground state to an excited state. When solvents are associated, additional "charge-transfer-to-solvent" transitions (CTTS) are provided [81, 82]. The absorption wavelength resulting from CTTS properties and the absorption behavior of aqueous solutions of phosphate, sulfate, and halide ions follows the order

$$\text{Phosphates } PO_4^{3-} < \text{Sulfates } SO_4^{2-} < F^- < \text{Hydroxides } OH^- < Cl^- < Br^- < I^-$$

where phosphate anions absorb at shorter wavelengths than iodide. Table 2.6 shows the effect of these ions on the absorption peak of water, where anions resulting in a shift of this peak sufficiently below 193 nm are the most interesting. The presence of alkali metal cations can shift the maximum absorbance wavelength to lower values. Furthermore, the change in the absorption with temperature is positive and small (~500 ppm °C^{-1}), whereas the change with pressure is negative and small. Though challenging from an implementation standpoint, these anions represented an avenue for exploration into high refractive index fluids for 193 and 248 nm application [83].

REFERENCES

1. W. Smith. 1966. *Modern Optical Engineering*, New York: Mc-Graw-Hill.
2. C. Huygens. 1690. *Traitè de la lumierè, Leyden*. (English translation by S.P. Thompson, *Treatise on Light*, Macmillan, London, 1912).
3. A. Fresnel. 1816. *Ann. Chem. Phys.*, 1: 239.
4. J.W. Goodman. 1968. *Introduction to Fourier Optics*, New York: McGraw-Hill.
5. H. von Helmholtz. 1859. *J. Math.*, 57: 7.
6. G. Kirchhoff. 1883. *Ann. Phys.*, 18: 663.
7. D.C. Cole. 1992. "Extending scalar aerial image calculations to higher numerical apertures," *J. Vac. Sci. Technol. B*, 10(6): 3037.
8. D.G. Flagello and A.E. Rosenbluth. 1992. "Lithographic tolerances based on vector diffraction theory," *J. Vac. Sci. Technol. B*, 10(6): 2997.
9. J.D. Gaskil. 1978. *Linear Systems, Fourier Transforms and Optics*, New York: Wiley.
10. H.H. Hopkins. 1953. "The concept of partial coherence in optics," *Proc. R. Soc. A*, 208: 408.
11. R. Kingslake. 1983. *Optical System Design*, London: Academic Press.
12. D.C. O'Shea. 1985. *Elements of Modern Optical Design*, New York: Wiley.
13. Lord Rayleigh. 1879. *Philos. Mag.*, 8(5): 403.
14. H.H. Hopkins. 1981. "Introductory methods of image assessment," *SPIE*, 274: 2.

15. M. Born and E. Wolf. 1964. *Principles of Optics*, New York: Pergamon Press.
16. W.H. Steel. 1957. "Effects of small aberrations on the images of partially coherent objects," *J. Opt. Soc. Am.*, 47(5): 405.
17. A. Offner. 1979. "Wavelength and coherence effects on the performance of real optical projection systems," *Photogr. Sci. Eng.*, 23: 374.
18. R.R. Shannon. 1995. "How many transfer functions in a lens?" *Opt. Photon. News*, 1: 40.
19. A. Hill, J. Webb, A. Phillips, and J. Connors. 1993. "Design and analysis of a high NA projection system for 0.35 µm deep-UV lithography," *SPIE*, 1927: 608.
20. H.J. Levinson and W.H. Arnold. 1987. *J. Vac. Sci. Technol. B*, 5(1): 293.
21. V. Mahajan. 1991. *Aberration Theory Made Simple*, Bellingham, WA: SPIE Press.
22. F. Zernike. 1934. *Physica*, 1(7–12): 689.
23. Lord Rayleigh. 1964. *Scientific Papers*, Vol. 1, New York: Dover.
24. B.W. Smith. 1995. First International Symposium on 193 nm Lithography, Colorado Springs, CO, International SEMATECH.
25. PROLITH/2 KLA-Tencor FINLE Division, 2006.
26. M. Born and E. Wolf. 1980. *Principles of Optics*, Oxford: Pergamon Press.
27. F. Gan. 1992. *Optical and Spectroscopic Properties of Glass*, New York: Springer-Verlag.
28. *Refractive Index Information (Approximate)*. 1990. Acton, MA: Acton Research Corporation.
29. M. Rothschild, D.J. Ehrlich, and D.C. Shaver. 1989. *Appl. Phys. Lett.*, 55(13): 1276.
30. W.P. Leung, M. Kulkarni, D. Krajnovich, and A.C. Tam. 1991. *Appl. Phys. Lett.*, 58(6): 551.
31. J.F. Hyde. 1942. Method of making a transparent article of silica. U.S. Patent 2,272,342.
32. H. Imai, K. Arai, T. Saito, S. Ichimura, H. Nonaka, J.P. Vigouroux, H. Imagawa, H. Hosono, and Y. Abe. 1988. In *The Physics and Technology of Amorphous SiO₂*, R.A.B. Devine, ed., New York: Plenum Press, p. 153.
33. M. Rothschild, T.M. Bloomstein, T.H. Fedynyshyn, V. Liberman, W. Mowers, R. Sinta, M. Switkes, A. Grenville, and K. Orvek. 2003. "Fluorine—an enabler in advanced photolithography," *J. Fluorine Chem.*, 122(1): 3–10.
34. Arun J. Kadaksham, Ranganath Teki, Milton Godwin, Matt House, and Frank Goodwin. 2013. Low thermal expansion material (LTEM) cleaning and optimization for extreme ultraviolet (EUV) blank deposition. In *SPIE Advanced Lithography*. International Society for Optics and Photonics, pp. 86791R–86791R.
35. Harrie J. Stevens and C. Charles Yu. "Detecting inclusions in transparent sheets." U.S. Patent 6,388,745, issued May 14, 2002.
36. Kenneth Racette, Monica Barrett, Michael Hibbs, and Max Levy. 2005 "The effect of mask substrate and mask process steps on patterned photomask flatness." *Proc. SPIE*, 5752: 621–631.
37. Bruce W. Smith, Anatoly Bourov, Matthew Lassiter, and M. Cangemi. 1999. "Masking materials for 157 nm lithography." In 19th Annual Symposium on Photomask Technology Proceedings.
38. W. Partlow, P. Thampkins, P. Dewa, and P. Michaloski. 1993. *SPIE*, 1927: 137.
39. S. Asai, I. Hanyu, and K. Hikosaka. 1992. *J. Vac. Sci. Technol. B*, 10(6): 3023.
40. K. Toh, G. Dao, H. Gaw, A. Neureuther, and L. Fredrickson. 1991. *SPIE*, 1463: 402.
41. E. Tamechika, T. Horiuchi, and K. Harada. 1993. *Jpn. J. Appl. Phys.*, 32: 5856.
42. T. Ogawa, M. Uematsu, T. Ishimaru, M. Kimura, and T. Tsumori. 1994. *SPIE*, 2197: 19.
43. R. Kostelak, J. Garofalo, G. Smolinsky, and S. Vaidya. 1991. *J. Vac. Sci. Technol. B*, 9(6): 3150.
44. R. Kostelak, C. Pierat, J. Garafalo, and S. Vaidya. 1992. *J. Vac. Sci. Technol. B*, 10(6): 3055.
45. B. Lin. 1990. *SPIE*, 1496: 54.
46. H. Watanabe, H. Takenaka, Y. Todokoro, and M. Inoue. 1991. *J. Vac. Sci. Technol. B*, 9(6): 3172.
47. M. Levensen. 1993. *Phys. Today*, 46(7): 28.
48. H. Watanabe, Y. Todokoro, Y. Hirai, and M. Inoue. 1991. *SPIE*, 1463: 101.
49. Richard Schenker, Vivek Singh, and Yan Borodovsky. 2010. "The role of strong phase shift masks in Intel's DFM infrastructure development," *Proc. SPIE*, 7641: 76410S.
50. Y. Ku, E. Anderson, M.L. Shattenburg, and H. Smith. 1988. *J. Vac. Sci. Technol. B*, 6(1): 150.
51. R. Kostelak, K. Bolan, and T.S. Yang, 1993. In Proceedings of the OCG Interface Conference, p. 125.
52. B.W. Smith and S. Turget. 1994. *SPIE Optical/Laser Microlithography VII*, 2197: 201.
53. M. Born and E. Wolf. 1980. *Principles of Optics*, Oxford: Pergamon Press.
54. B.W. Smith, S. Butt, Z. Alam, S. Kurinec, and R. Lane. 1996. *J. Vac. Technol. B*, 14(6): 3719.
55. N. Cobb and Y. Granik. 2004. "New concepts in OPC," *Proc. SPIE*, 5388, Optical Microlithography XVII.
56. Michael L. Rieger and John P. Stirniman. 1996. Mask fabrication rules for proximity-corrected patterns. *Proc. SPIE*, 2884, 16th Annual BACUS Symposium on Photomask Technology and Management, 323.

57. H. Yoo, Y. Oh, B. Park, S. Choi, and Y. Jeon. 1993. *Jpn. J. Appl. Phys.*, 32: 5903.

58. B.W. Smith, D. Flagello, and J. Summa. 1993. *SPIE*, 1927: 847.

59. S. Asai, I. Hanyu, and M. Takikawa. 1993. *Jpn. J. Appl. Phys.*, 32: 5863.

60. K. Matsumoto and T. Tsuruta. 1992. *Opt. Eng.*, 31(12): 2656.

61. D. Golini, H. Pollicove, G. Platt, S. Jacobs, and W. Kordonsky. 1995. *Laser Focus World*, 31(9): 83.

62. A. Offner. 1975. *Opt. Eng.*, 14(2): 130.

63. D.M. Williamson, by permission.

64. J. Buckley and C. Karatzas. 1989. *SPIE*, 1088: 424.

65. J. Bruning, 1996. OSA Symposium on Design, Fabrucation, and Testing for Sub-0.25 Micron Lithographic Imaging.

66. H. Sewell. 1995. *SPIE*, 2440: 49.

67. D. Flagello and A. Rosenbluth. 1992. *J. Vac. Sci. Technol. B*, 10(6): 2997.

68. D. Williamson, J. McClay, K. Andresen, G. Gallatin, M. Himel, J. Ivaldi, C. Mason, A. McCullough, C. Otis, and J. Shamaly. 1996. *SPIE*, 2726: 780.

69. G. Fürter, Carl-Zeiss-Stiftung, by permission.

70. S. Owa, T. Fujiwara, Y. Ishii, K. Shiraishi, and S. Nagaoka. 2006. "Full field, ArF immersion projection tool," The 17th Annual SEMI/IEEE ASMC 2006 Conference, Boston, MA, pp. 63–70.

71. B.W. Smith, J. Cashmore, and M. Gower. 2002. "Challenges in high NA, polarization, and photoresists," *SPIE Opt. Microlith. XV*, 4691.

72. B.W. Smith, L. Zavyalova, and A. Estroff. 2004. "Benefiting from polarization-effects of high-NA on imaging," *Proc. SPIE Opt. Microlith. XVII*, 5377.

73. A. Estroff, Y. Fan, A. Bourov, B. Smith, P. Foubert, L.H. Leunissen, V. Philipsen, and Y. Aksenov. 2005. "Mask-induced polarization effects at high NA," *Proc. SPIE Opt. Microlith.*, 5754.

74. R.W. Wood. September 1902. "Uneven distribution of light in a diffraction grating spectrum," *Philos. Mag.*

75. Lord Rayleigh. July 1907. "On the remarkable case of diffraction spectra described by Prof. Wood," *Philos. Mag.*

76. B.W. Smith, L. Zavyalova, and A. Estroff. 2004. "Benefiting from polarization-effects of high-NA on imaging," *Proc. SPIE Opt. Microlit. XVII*, 5377.

77. B.W. Smith, Y. Fan, J. Zhou, A. Bourov, L. Zavyalova, N. Lafferty, F. Cropanese, and A. Estroff. 2004. "Hyper NA water immersion lithography at 193 nm and 248 nm," *J. Vac. Sci. Technol. B: Microelectron. Nanometer Struct.*, 22(6): 3439–3443.

78. P.L. Marston. 1989. "Light scattering from bubbles in water," *Ocean*, 89(Part 4), Acoust. Arct. Stud.: 1186–1193.

79. C.F. Bohren and D.R. Huffman. 1983. *Absorption and Scattering of Light by Small Particles*. Wiley.

80. Y. Fan, N. Lafferty, A. Bourov, L. Zavyalova, and B.W. Smith. 2005. "Air bubble-induced light scattering effect on image quality in 193 nm immersion lithography," *Appl. Opt.*, 44(19): 3904.

81. E. Rabinowitch. 1942. *Rev. Mod. Phys.*, 14: 112; G. Stein and A. Treinen. 1960. *Trans. Faraday Soc.*, 56: 1393.

82. M.J. Blandamer and M.F. Fox. 1968. *Theory and Applications of Charge-Transfer-To-Solvent Spectra*.

83. B.W. Smith, A. Bourov, H. Kang, F. Cropanese, Y. Fan, N. Lafferty, and L. Zavyalova. 2004. "Water immersion optical lithography at 193 nm," *J. Microlith. Microfab. Microsyst.*, 3(1): 44–45.

3 Multiple Patterning Lithography

Carlos Fonseca, Chris Bencher, and Bruce Smith

CONTENTS

3.1 INTRODUCTION

The need for continuous device scaling has significantly driven the development of new lithographic patterning techniques and imaging methods. While the general goal in patterning is to make circuit device dimensions smaller, there are key metrics that govern the evolution and adoption of advanced patterning technology. In general, integrated device manufacturers (IDMs) need to consider (1) device complexity, (2) device performance, and (3) implementation cost when developing and introducing advanced patterning techniques. In the absence of a cost metric, technical advances might yield admirable patterning approaches that could even outperform device performance targets for a given technology node. It is, however, important to consider cost for any new technique so as to strike a practical balance between process complexity and the achievement of performance targets. In light of this, this chapter addresses general trends and examples of the evolution in advanced lithography methods beyond traditional single patterning technology.

In prior technology nodes, lithography employed methods evolved from earlier generations with incremental improvements in wavelength, imaging optics, resolution enhancement technology (RET), and single patterning processes. RETs have included enhancements to the mask, the illumination conditions, and the target design constructs. Details regarding the design and implementation of these various technologies can be found in other chapters. To illustrate such lithographic evolution, historical patterning enhancement options that were employed over recent device generations (or nodes) are shown in Figure 3.1. Many of the enhancements and combinations depicted continue to be used, together with new and more complex pattern transfer approaches. This additional complexity often entails what is referred to today as *multiple patterning*.

As pattern feature dimensions shrink below the resolution limits of single exposure techniques, more complex approaches are considered, such as the splitting of feature geometries or groups of patterns prior to imaging. The splitting of these patterns into two (or more) mask levels gives rise to various multiple patterning processes. Instead of printing an entire intended circuit design level with a single mask, multiple exposure passes (using separate masks) can be a viable solution for

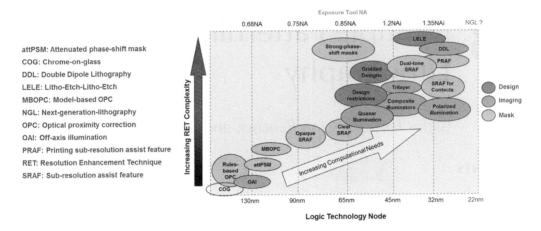

FIGURE 3.1   Early patterning enhancements options by technology node.

overcoming imaging limitations of single exposures. The ability to print multiple exposure passes to meet performance targets in a cost-effective manner has driven the search for the overall best techniques in multiple patterning.

Early forms of double patterning involved the concept of recording two separate exposure passes into a single layer of photoresist to produce stitched pattern features. The key benefit of splitting an IC design into two images is that the separated patterns could be exposed with uniquely optimized illumination conditions, yielding improved depth of focus (DOF), mask error enhancement factor (MEEF), and feature resolution. For example, highly dense patterns can benefit from off-axis illumination, while more isolated patterning can benefit from conventional on-axis illumination. Alternatively, dense circuit geometry with both horizontal and vertical nature can be separated for exposure with off-axis illumination conditions designed specifically for each feature orientation. Hsu et al. demonstrated this with double-dipole lithography (DDL) for a metal level by using two separate orthogonal dipole illumination conditions for two masks containing mostly one-dimensional (1-D) horizontal or vertical lines (i.e. one orientation per mask) [1]. As with most advanced patterning techniques in early development, the DDL method had limitations in printing complex two-dimensional patterns, which can require often severe circuit design restrictions for the effective decomposition of target layouts. For implementation into manufacturing, DDL needed to be cost competitive compared with more traditional single patterning approaches using improvments in scanner technology (i.e., higher numerical aperture), RET, and photoresist materials. Eventually, IC device design adjustments were made so that patterns could comply with the two-directional DDL restrictions, enabling it as a viable option for double patterning (Figure 3.2) [2].

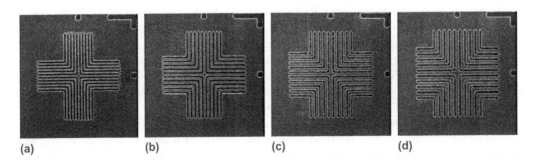

FIGURE 3.2   Multiple patterning example. Double-dipole lithography (DDL) example for 50 nm trenches at pitches of (a) 100 nm, (b) 125 nm, (c) 175 nm, and (d) 200 nm.

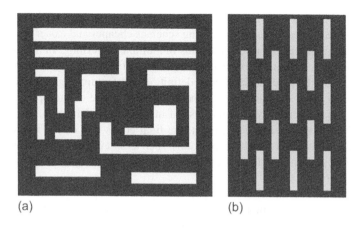

(a)   (b)

FIGURE 3.3   Design style options for (a) two-dimensional and (b) one-dimensional layout styles.

3.2   DESIGN INTERACTION IN MULTIPLE PATTERNING TECHNIQUES

The patterning technique chosen to print a given lithographic layer can have significant impact on the final design style, and vice versa. The nature of 1-D geometry generally makes it suited for multiple patterning techniques, because the layout can be split or decomposed more easily than random two-dimensional (2-D) layout styles. Such layout decomposition typically requires significant development using electronic design automation (EDA) tools, which can demand a significant amount of scripting and code development resources, together with time for robust decomposition algorithm implementation. Figure 3.3 shows representative design style options for 1-D and 2-D layout styles for a given lithography layer. Both Figure 3.3a and b represent the same circuit functionality. By inspection, one can easily determine that the 2-D style is likely to be more difficult to implement with multiple patterning techniques due to the varying pattern dimensions and spacing between features. A more repetitive and 1-D style option, such as the one shown in Figure 3.3b, offers a better fit for layout decomposition, since the pattern dimensions and spacing between neighboring features (i.e., pitch) are more regular throughout the layout.

As noted above, patterning schemes and design styles need to be developed simultaneously. This mutual interaction typically necessitates several cycles of learning to yield mutual changes to the patterning approach and the design layout. These cycles of learning are referred to as *design technology co-optimization* (DTCO). Additional details of DTCO and design for manufacturability (DFM) are addressed in Chapter 6, which is devoted to these topics. To fully understand the co-optimization of the design-patterning schemes and its implications, efforts typically require a detailed analysis of experimental data (lithographic metrics) across multiple design variations. It is important to realize that a significant amount of DTCO encompassing EDA and the integration scheme is almost always necessary before full chip–level demonstrations are realized with multiple patterning schemes.

3.3   FORMS OF MULTIPLE PATTERNING

Considering all aspects and options to imaging approaches, layout decomposition, materials selection, integration scheme, and target performance in multiple patterning schemes, one can conclude that a significantly large number of possible forms of multiple patterning can exist. As mentioned earlier, cost is a key determinant for identifying viable, manufacturing-worthy multiple patterning schemes for a given target layer. Therefore, final candidate multiple patterning schemes need to meet economic requirements. While these economic requirements act as a filtering process for choosing or developing a specific scheme, many options are typically explored, primarily for technical merits, in the initial phase of development. Once technical merits are established, the focus then shifts to making the technically viable patterning schemes cheaper and easier to implement.

Figure 3.4 depicts a general hierarchy of forms in multiple patterning. In this hierarchy, the patterning scheme is mostly driven by how feature edges are defined. Edge definition can be classified as direct or indirect. In the case of direct edge definition, the feature edges are defined directly by the exposure process. That is, mask edges, whether from a single or multiple reticles, define the final feature edges at the wafer level. In the case of indirect edge definition, the feature edges are not necessarily defined by mask edges. In fact, the final edges that define indirect edge-features (e.g., a slot contact) can come from film depositions, such as sidewall spacer processes. The natural evolution in multiple patterning has progressed from single patterning to what today is referred to as *integrated patterning*.

Integrated patterning essentially consists of forming feature edges by indirect methods. Most, if not all, of the final feature edges in a given layout are defined by a combination of multiple memorizations by means of multiple material stacks. As seen in Figure 3.4, once multiple material stacks and multiple memorizations are employed in a patterning scheme, it is probably targeting an integrated scheme. So, why use integrated patterning schemes? Higher patterning performance targets (smaller feature dimensions) drive the need for integrated schemes. Multiple memorization steps create an opportunity to improve on the patterning metrics such as critical dimension uniformity (CDU), line-edge roughness, local CDU, creating sharp corners, and so on. This opportunity arises from the fact that edges forming the final features can be decoupled, thus allowing extra degrees of freedom in the patterning process that defined the edges. One of the most important benefits in integrated patterning is the ability to create very small and sharp features by interacting edges from multiple memorization layers. Of course, the improved performance from more complex patterning schemes (such as integrated schemes) comes with higher manufacturing costs. It is important that

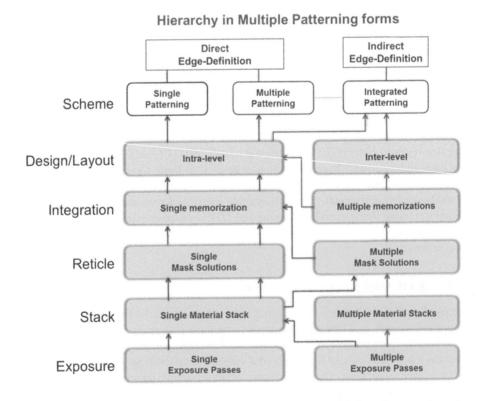

FIGURE 3.4   Hierarchy in multiple patterning. Edge definition drives and defines the patterning scheme.

the performance benefits are well balanced with manufacturing costs in order to implement high-volume manufacturing.

3.3.1 COMMERCIALIZATION EXAMPLES OF DOUBLE PATTERNING

Resolution enhancement techniques, such as off-axis illumination, phase shift masks, and optical proximity correction, could not keep up with the rate of device scaling, and from about 2005, most leading node device types transitioned to double patterning for critical layers. There are multiple forms of double patterning that can be exploited based on the layout challenges. These include (a) illumination splitting, (b) line and cut, (c1) scanner-based pitch reduction, and (c2) sidewall spacer–based pitch reduction.

3.3.1.1 Illumination Splitting

One of the first forms of double patterning adopted was *illumination splitting.* The benefit of splitting the design into two images is that one can expose different pattern types with different illumination modes. For example, dense patterns can benefit from off-axis illumination, while isolated patterning can benefit from conventional on-axis patterns. Likewise, the optimal illumination conditions for an array of small pitch parallel lines is different from the optimal illumination for random 2-D structures of varying pitch. Illumination splitting freed users from the compromises and trade-offs often made when trying to expose a wide variety of patterns in one single exposure. Below are some examples.

3.3.1.1.1 DRAM Example (circa 80 nm Nodes)

Bitline and wordline levels of a dynamic random-access memory (DRAM) have memory block layouts that contain critically dense parallel strings. They also contain larger isolated peripheral structures and thus, are optimally printed using different types of illumination. In an illumination split double patterning, the layout can be split into two exposures, placing the dense parallel lines on one mask to be exposed with an off-axis dipole illumination while the peripheral control circuits are placed on another mask to be exposed with a conventional on-axis illumination [3, 4]. The image splitting of this layout is generally straightforward and can be done without the need for EDA tools to optimize the pattern split. This technique was used in research and development (R&D). High-volume manufacturing implemented next-generation scanners and design rule restrictions to enable a low-cost single exposure solution (Figure 3.5).

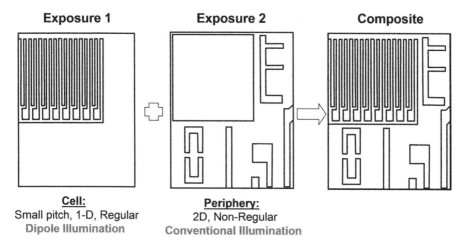

Exposure 1   **Exposure 2**   **Composite**

Cell:
Small pitch, 1-D, Regular
Dipole Illumination

Periphery:
2D, Non-Regular
Conventional Illumination

FIGURE 3.5 Example of illumination splitting double patterning [4].

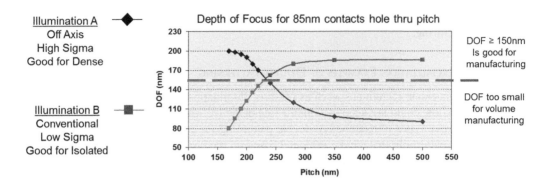

FIGURE 3.6 Example of illumination splitting double patterning [5].

3.3.1.1.2 Contact Pattern Example (circa 70 nm nodes)

DRAM contact layers frequently contained both critically dense contacts at the same as critically isolated contacts; however, these structures are optimized best under different illumination conditions. As an example, for a 70 nm node DRAM cell, dense contacts were shown to achieve better than 150 nm DOF with off-axis illumination, while isolated contacts achieved similar DOF using a conventional low-sigma on-axis illumination [5]. Splitting the layout into two masks, placing dense contacts onto one mask exposed with off-axis illumination while placing isolated contacts into a second mask exposed with on-axis illumination, enabled all structures to be printed with a manufacturable DOF better than 150 nm. The image splitting of this layout is generally straightforward and can be done without the need for EDA tools to optimize the pattern split (Figure 3.6).

3.3.1.1.3 ASIC/Microprocessor Metal Routing Layers (circa 32 nm nodes)

When 32 nm application-specific integrated circuit/microprocessors (ASIC/MPUs) came to production, the ultimate resolution of high-NA 193 nm immersion lithography was not enough for many of the more complex design layers. DDL was one popular double exposure technique, which saw very widespread use, with introduction around 32 and 28 nm nodes. As described earlier, DDL is a type of illumination splitting that takes a layout and decomposes the patterning into two images, with two masks, printed in succession into one photoresist. In general, DDL means that one mask is printed with dipole X and another mask printed with dipole Y. Unlike the earlier DRAM examples, where design splitting is done by separating dense and isolated polygons onto different masks, DDL co-optimizes the combined image, allowing polygons to get exposure from both masks. This necessitates an EDA tool to computationally co-optimize each mask and iteratively check the combined image for image contrast and DOF. An example of this is shown in Figure 3.7.

3.3.1.2 Line and Cut Double Patterning

Many RETs, such as off-axis illumination, phase masks, and optical proximity correction, help to improve the resolution scaling of arrays of lines, arrays of contacts, and edge placement for complex shapes but do little to help improve the scaling of tip-to-tip. The tip-to-tip distance between two neighboring lines can be one of the most challenging pattern challenges or "hotspots." While immersion scanners can resolve 40 nm half-pitch geometry in photoresist, it is challenging to print <90 nm tip-to-tip without risking a tip-to-tip bridge during defocus conditions through the resist layer. Figure 3.8 illustrates a tip-to-tip design layout, which is susceptible to line-end shortening or line-end extension with small amounts of defocus errors (as shown by the simulated developed photo-resist contour lines) and the risk of line-end bridging with larger defocus errors.

The inability to scale the tip-to-tip distance can challenge effective area scaling, especially in static random-access memory (SRAM blocks), which take up significant die area; some microprocessors' SRAM can be 25–50% of the total chip area. To address this, the industry turned to "line

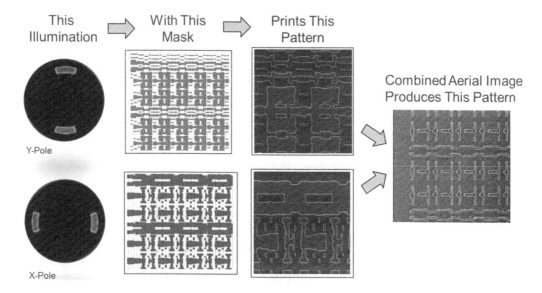

FIGURE 3.7 Example of EDA tool–enabled double-dipole lithography [6].

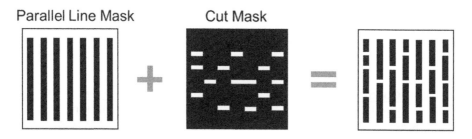

FIGURE 3.8 Illustration/simulation of resist contour lines compared with defocus conditions for a tip-to-tip layout.

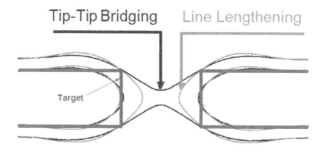

FIGURE 3.9 Example of line and cut double patterning, which typically consists of two complete cycles of litho and etch.

and cut" double patterning, which first performs a lithography and etch sequence to form continuous lines, and then performs a second lithography and etch sequence to "cut" the line ends. This double patterning technique enabled tip-to-tip design rules to be scaled down to about 30 nm. Line and cut double patterning has also been developed by several groups through the double exposure techniques, which add pattern assist features to a first exposure and then erase the pattern assist features in a second exposure (Figure 3.9) [7, 8].

3.3.1.2.1   ASIC/Microprocessor Gate Layer (circa 45 nm node)

Multiple device manufacturers developed transitional line and cut double patterning approaches for scaling their gate layer around the time of the 45 nm node. Figures 3.10 and 3.11 show industry examples of 45 nm SRAM block that scaled the tip-to-tip design rule to <55 nm using line and cut double patterning.

Reviewing these approaches, one can observe additional benefits when migrating from 65 nm single patterning to 45 nm line and cut double patterning. First, since the line ends are defined in a second patterning step, the exposure defining the lines can take advantage of a higher-contrast dipole illumination, which improves CD control distributions and line-edge roughness (as seen in Figure 3.12). Comparison studies have shown that switching to a dipole illumination for gate layer could improve CD distributions, and therefore device leakage variation, by ~3× compared with devices using 2-D resolving illumination modes for gate [11].

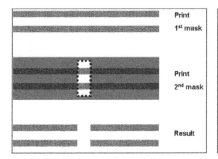
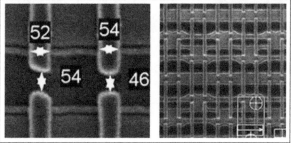

FIGURE 3.10   Line and cut double patterning example from IBM [9].

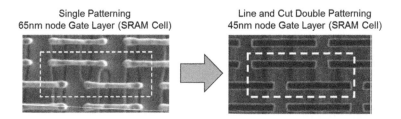

FIGURE 3.11   Line and cut double patterning example from Intel [10].

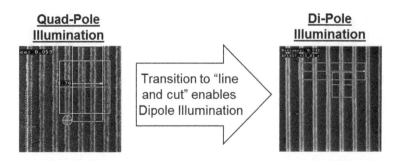

FIGURE 3.12   Comparison of two illumination modes: quadrupole (Quad-Pole), which can define line ends, versus dipole (Di-Pole), which cannot define line ends with good contrast.

3.3.1.3 Pitch Division Double Patterning

The previously discussed examples of double patterning can help to overcome critical challenges in a design, can enhance contrast, and can offer flexibility to optimize illumination modes for different shapes on a single layer. However, illumination splitting (such as DDL) and line and cut double patterning cannot overcome the diffracted limited resolution for pitch (as explained in Chapter 2). With 193 nm immersion lithography limited to a pitch of around 80 nm, it becomes necessary to implement pitch division double patterning for line pitches below 80 nm. There are two categories of pitch division double patterning: (1) sidewall spacer double patterning, also known as *self-aligned double patterning* (SADP), and (2) litho-etch-litho-etch double patterning (LELE).

3.3.1.3.1 Sidewall Spacer Double Patterning

Sidewall spacer DP or SADP is a technique that uses a sequence of deposition and etch steps to multiply the number of lines or equivalently, divide the pitch. A general process flow is outlined in the following and depicted in Figures 3.13a–c. First, a film stack is formed, normally comprising a chemical vapor deposited (CVD) sacrificial template material. This template material can be a silicon dielectric or amorphous carbon layer. In some cases, such as with an amorphous carbon template material, a bilayer template employing an additional silicon or metal dielectric layer can improve etch selectivity. The resulting template stack allows for the subsequent steps of patterning, etching, and deposition. As shown in Figure 3.13a, a photoresist layer has been coated on a template stack, patterned lithographically, and developed. For positive tone patterning, a trim step may be incorporated in the lithography operation to achieve a greater duty ratio (i.e., more line isolation). This might involve sizing during pattern transfer into an underlying bottom antireflective coating (BARC) layer or other approaches to shrink linewidth dimensions. Next, the resist pattern is etched into the template material, followed by a conformal CVD deposition of a dielectric spacer layer such as silicon nitride. The spacer layer is then anisotropically reactive ion etched to remove material from the topmost and bottommost spacer regions, leaving only the sidewall material remaining. The choice of the template and spacer materials must be chosen to allow for etch selectivity during this step and also the template strip step that follows. The resulting pattern is then transferred into the underlying layer; for example, a polysilicon gate layer.

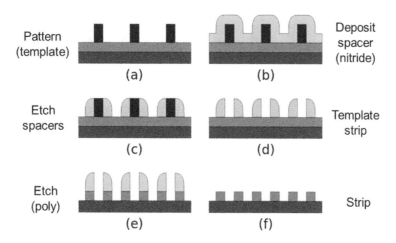

FIGURE 3.13A The process flow for positive tone self-aligned double patterning using a spacer material formed on a patterned template and selectively etching to leave sidewalls as for pattern transfer.

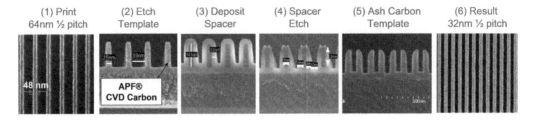

FIGURE 3.13B   SEM cross sections of SADP process flow starting with 64 nm half-pitch lithography and ending with 32 nm half-pitch array using CVD amorphous carbon template.

The sequence that would result in 32 nm patterning is summarized in the following and depicted in Figure 3.13b for an amorphous carbon template (advanced patterning film [APF]).

Step	Description/Function
0	Deposit "template" material, a sacrificial layer on which the spacers will be formed.
1	Perform lithography, such as 64 nm line/space (128 nm pitch).
2	Etch lines into template; bias the etch to obtain 32 nm lines on 128 nm pitch.
3	Deposit sidewall spacer over template; 32 nm in this case.
4	Spacer etch; an anisotropic etch that clears the top and bottom planes, leaving sidewall.
5	Strip template: result is 32 nm line/space (64 nm pitch); lines are made of spacer material.
6	Metrology: CD-scanning electron microscopy (SEM) image to verify that template CD and spacer CD are on target.

In the case of negative tone SADP processing, addition steps can be added to the process sequence. Prior to stripping the template material, a fill layer comprised of a similar material can be coated over the etched spacers, followed by a chemical mechanical planarization (CMP) step. Here, the template is not stripped; instead, the spacer material is removed using the appropriate reactive ion etching process. This is shown in Figure 3.13c, where patterned template material is used to transfer the negative tone pattern into the underlying substrate stack.

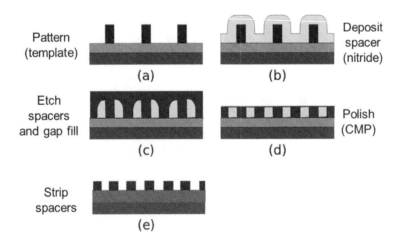

FIGURE 3.13C   Process flow for negative tone SADP using a template gap fill and CMP processes to allow for the transfer of patterned template material.

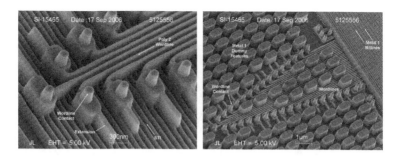

FIGURE 3.14 Example of SADP used in 50 nm node NAND Flash.

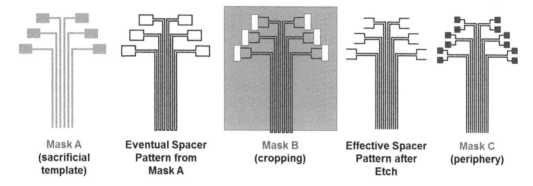

| Mask A (sacrificial template) | Eventual Spacer Pattern from Mask A | Mask B (cropping) | Effective Spacer Pattern after Etch | Mask C (periphery) |

FIGURE 3.15 Example of a three-mask SADP process flow, which can manufacture the patterns in Figure 3.14 [12].

3.3.1.3.1.1 NAND, Three Layers: Active, Bitline, and Wordline (circa 2007, sub-40 nm nodes) The NAND flash device was positioned well to exploit SADP pitch division, because the layout of three critical layers is predominantly long strings of parallel lines on all three critical levels: the active (aka STI), the wordline, and the bitline. Figure 3.14 shows evidence of SADP in 50 nm NAND flash and printing 200 nm pitch resist and using SADP to convert to 100 nm pitch device strings. The telltale sign for SADP is the tail sticking out of all the contact landing pads, which is evidence of a multiple patterning process.

The SADP process of pitch division cannot by itself produce working circuits, because the spacer pattern produces a continuous loop around the circumference of the template polygon, meaning that every line becomes a closed loop upon itself. A cut mask (also called a trim or crop mask) must be used to cut the spacer pattern down into discrete line traces. Additionally, the spacer pattern can only make features of a single linewidth. If large features such as contact landing pads are needed, one must apply a third mask to define the large features. An example of this three-mask process flow for producing a useful wordline array with contact landing pads is shown in Figure 3.15. Spacer double patterning has been used for several more technology nodes of NAND flash. This has enabled very rapid scaling down to 20 nm NAND flash, helping to increase device density and lower cost for the enablement of solid-state storage.

3.3.1.3.1.2 DRAM Active (circa 30 nm nodes) Many DRAM designs have an active (silicon) layer that consists of a densely packed array of staggered islands. Similarly to microprocessors with SRAM, this is a good candidate for the aforementioned line and cut double patterning to achieve small-island tip-to-tip. When DRAM moved to 30 nm nodes, the island pitch also required double patterning. Thus, the combination of two double patterning techniques can be used to drive DRAM

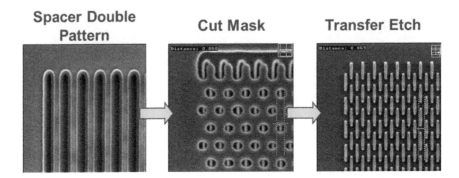

FIGURE 3.16 Example of a two-mask SADP process flow, which can manufacture DRAM active patterns [13].

active layer scaling. Figure 3.16 shows the combined use of (1) SADP, for generating 35 nm line and space line and (2) cut mask, for creating island pattern.

3.3.1.3.1.3 Spacer Mask Tones In the two aforementioned examples, the spacer double patterning is defining *lines*. It is also possible to use SADP to define trenches by changing the function of the trim mask from *cutting spacer* to a function of *blocking etch transfer*. Figure 3.17 shows how the masking step following spacer formation defines the final *tone* of the pattern formed. In a traditional dark-field patterning application, where one is defining a trench pattern, the second mask is used to *block the etch transfer* when doing a final etch transfer into a dielectric layer. This process flow can be used for defining metal trenches or slotted contacts/vias layers. In a traditional

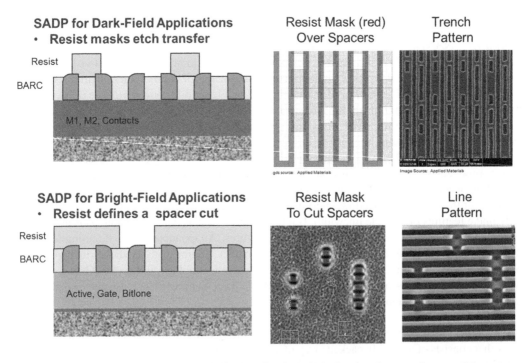

FIGURE 3.17 Examples of SADP in dark-field applications (top) showing the manufacture of damascene trenches or slotted contacts, and examples of SADP in bright-field applications (bottom) such as gate or fin patterning.

bright-field patterning application, where one is defining a line pattern, the second mask is used to *cut the spacers*, after which one would strip the resist before doing the final etch transfer into an active, gate layer, or tungsten bitline. It is therefore possible to use SADP for both bright-field and dark-field layers.

3.3.1.3.1.4 SADP for Complex Layers A great many SADP use cases have been applied to mostly parallel line structures such as NAND, DRAM active (with cut), DRAM bitline and word-lines, MPU fins, and some unidirectional metal layers. However, EDA tools have been developed to compute nonintuitive mandrel and trim mask combinations for achieving more complicated lay-outs. Figures 3.18 and 3.19 show R&D activities in both bright-field and dark-field tone patterning. However, for many reasons, the design of advanced nodes is trending toward more simple, unidirectional layout, with restricted design rules that can deliver good manufacturing process windows and yield. While more complicated patterns have the potential to deliver higher-performance circuits, the manufacturability related to all processing steps (etch, gap-fill, air-gap engineering, defect inspection, and litho) is moving designs more toward simple and restricted layout.

3.3.1.3.2 Litho-Etch-Litho-Etch Pitch Division

While these SADP examples can effectively enable pitch division, they can often require additional restricted design rules for layouts not available with existing IC libraries. An alternative form of pitch division that can alleviate this to some extent is LELE double patterning, which offers alternatives to some design rule restrictions. LELE double patterning involves, as the name implies,

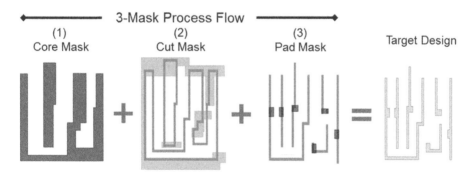

FIGURE 3.18   Example of three-mask SADP process flow for complex line pattern [13].

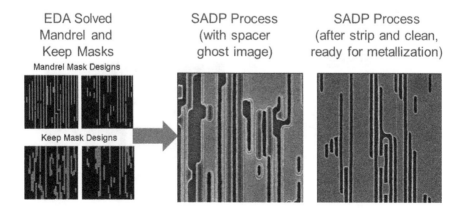

FIGURE 3.19   Example of two-mask SADP process flow for complex trench patterns [14].

two lithography steps and two etch steps [15, 16]. The LELE process generally uses the following operational sequence.

Step	Description/Function
0	Separate design into two masks (A and B) and perform optical proximity correction for each mask.
1	Perform lithography for first Mask A.
2	Etch Mask A pattern into substrate.
3	Strip Mask A photoresist.
4	Perform lithography for second Mask B.
5	Etch Mask B pattern into substrate.
6	Strip Mask B photoresist.

3.3.1.3.2.1 Double line LELE   For double patterning of lines (i.e., double line LELE), each pair of litho-etch sequences necessitates a sacrificial hard mask as a pattern transfer layer. A requirement of layer materials is that they must possess different etch properties to allow adequate etch selectivity. An example of a double line LELE process is seen in Figure 3.20, where two hard mask layers (Hard Mask 1 and Hard Mask 2) are later deposited onto an underlying substrate (i.e., gate material). The materials used as hard mask layers are most often silicon dielectrics, such as silicon dioxide, silicon nitride, or silicon oxynitride, deposited using CVD methods. The second hard mask must be selective against the etch process of the first hard mask so that it remains intact for second-layer patterning. The photoresist layers used in both sequences may be single resist films or combinations of resist films with BARCs. The use of BARC materials in photoresist processing is described in more detail in Chapter 8. After the first lithography and etch process "freezes" a first-layer pattern into the first hard mask material, a second layer of photoresist is coated and developed. A challenge

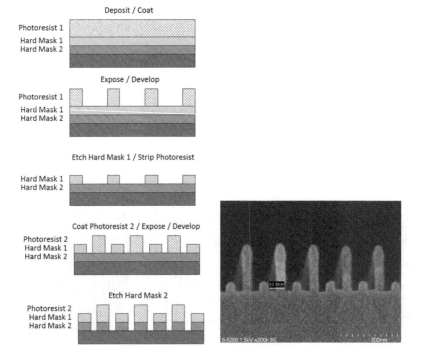

FIGURE 3.20   Process flow for double line LELE using two hard mask layers.

for second-layer lithography, which is more of a concern than with the first layer, is patterning over topography. While a thicker BARC film (or an additional organic planarizing layer) may be used to alleviate the problems associated with imaging over first-level patterning, the impact on process control during etch transfer may prevent this as an option. After the second-layer lithography step, the second photoresist pattern and the first-layer hard mask pattern are transferred into the second hard mask layer. Without optimization of the photoresist/BARC layers for topography during the second lithography step, pattern profiles can suffer, leading to unacceptable CD uniformity values after etch. Figure 3.20 also shows a SEM cross-sectional image of double line LELE after the second etch step but before stripping.

3.3.1.3.2.2 Double Space (Contact) LELE For double space (or contact hole) LELE, the multi-layer process can be somewhat simplified. As shown in Figure 3.21, a stack employing a single hard mask can suffice, since a first-layer litho-etch leaves hard mask material remaining for use with a second-layer litho-etch. In these cases, a first photoresist/BARC layer may be coated over a hard mask (a silicon dielectric, for example), followed by exposure, development, etch, and strip. The second-layer lithography then is carried out in a second photoresist/BARC layer that is optimized for the remaining topography, as in the case of double line LELE. The single hard mask is then etched with the second-layer pattern, and the photoresist is removed. Figure 3.21 also shows an SEM image of double contact LELE after the second etch and strip.

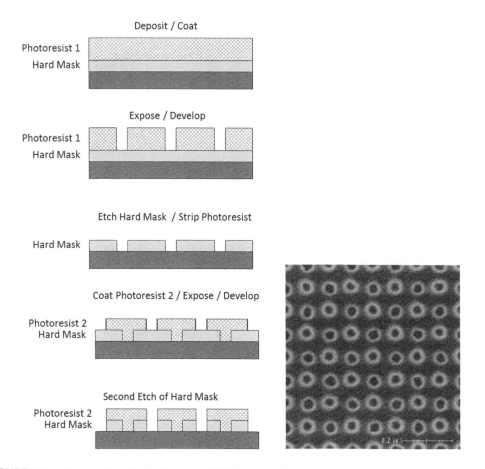

FIGURE 3.21	Process flow for double space LELE using two hard mask layers.

3.3.1.3.2.3 LELE Decomposition   The most challenging step of the LELE process flow is the design decomposition step, which must take the designer's IC layout and decide how to split up the content into two masks. For the example shown in the following, a pattern split can be done manually, but for chips with millions of cells, one must use an EDA tool to automate a pattern split. The EDA tool needs to be programmed to check that each individual mask does not violate the litho capabilities of a single mask (Figure 3.22) [17].

Design rule restrictions typically afflict LELE when there are "odd cycle conflicts." An odd cycle conflict occurs when there are three polygons at an intersection, which makes it impossible to split the pattern into two masks without having a pitch or resolution violation on one of the individual masks. The example in Figure 3.23 shows a simple example consisting of two vertical lines over one horizontal line; there is no way to split the pattern into two masks without one of the masks having a violation; therefore, one has two recourses: (1) adjust the design or (2) use three masks (LELELE).

Polygon stitching is another challenge related to LELE design splits. Polygon stitching is risky because it is difficult to control the CD at the overlapping regions, which receive double etch, and also because of the overlay between the two litho steps. Figure 3.24 shows a layout that has 12 polygon stitching locations; highlighted in boxed areas are one example of CD and placement errors typically associated with polygon stitching. The reason why this line in the red box needs stitching is because of the splitting solutions from nearby patterns, which then exert splitting on a simple straight line.

For line cuts, contact patterning, and via patterning, LELE is the industry process of record, and as pitch dimensions continue to shrink, an eventual migration to triple patterning or LE^3 (or LE³)

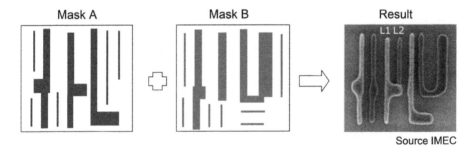

FIGURE 3.22   Example of LELE double patterning for pitch division of 2-D gate layouts.

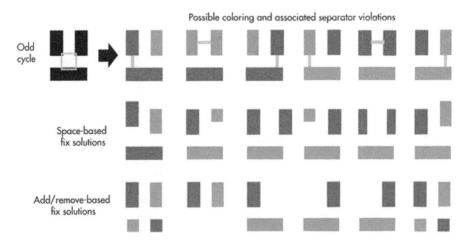

FIGURE 3.23   Example of odd-cycle conflict, which poses two-mask pattern splitting challenges for LELE double patterning [18].

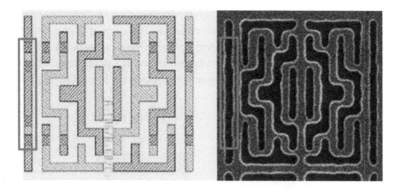

FIGURE 3.24 Example of LELE two-mask decomposition that requires polygon stitching [19].

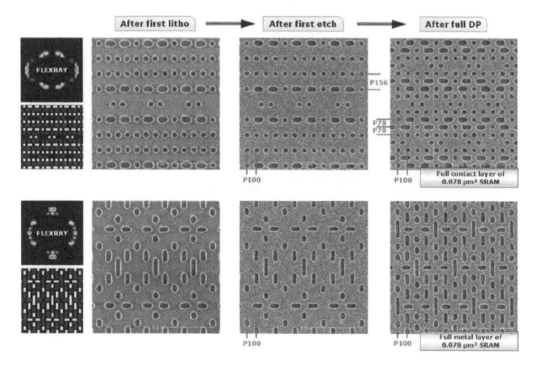

FIGURE 3.25 Example of LELE two-mask decomposition for 22 nm SRAM contacts [20].

will be forced; the industry is hoping that EUV will become commercially available before LE^4 (or LE⁴) is required. LELE process flow for cuts/contacts/vias is straightforward once the EDA tool has done the pattern split. The design goal is to avoid layout situations which generate pattern split-up violations, which forces an additional mask and cycle of litho-etch. Figure 3.25 shows two examples of complex contact patterns that were successfully split into only two masks.

3.3.1.4 Higher-Order Pitch Division

Multiple patterning (MP) beyond doubling can offer resolution improvement but at a cost of process complexity and control [21–23]. Theoretically, a final pattern is an assembly of carefully placed features using N patterning steps to result in pitch division of the same order. In some cases (such as multiple litho-etch), pitch splitting becomes additive, whereby each succession introduces a division as 2×, 3×, 4×, and so forth. In the case of sidewall spacer processes, division is exponential doubling, resulting in 2×, 4×, 8×, etc. Schematic examples of process flows for three sequences of litho-etch (LELELE) and double (2×) SADP (referred to here as an iterative spacer process) are shown in Figure 3.26. As with the examples described earlier for double patterning, the choice of materials is based on compatibility and selectivity between process steps. A triple line LELELE process utilizing three hard mask layers can make use of a third photoresist without a sacrificial hard mask to allow triple patterning into an underlying substrate layer. For a quadruple iterative

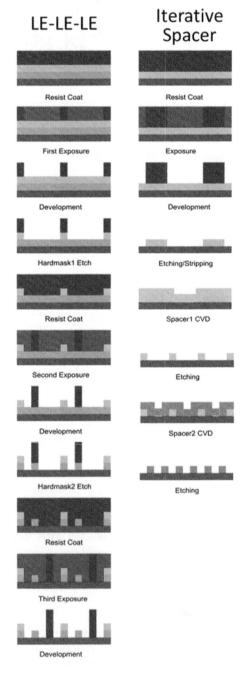

FIGURE 3.26 Examples of higher-order pitch division processes using three sequences of LELELE and a quadruple SADP process using iterative spacers.

spacer process, the first set of doubled patterns will go through a second sidewall spacer deposition and subsequent etch processes (one for removal of the spacer material and one for removal of the template material). In the case of self-aligned quadruple patterning (or SAQP), the pitch is divided twice. Additional materials may be needed for the film stack, such as secondary dielectric hard mask layers [23]. The extendibility beyond triple or quadrupole patterning will depend on cost and manufacturability.

3.3.1.4.1 CD Uniformity for LELE

CD uniformity analysis can be carried out by considering the multiple feature populations involved [21]. For a triple line LELELE process, the CDU becomes

$$\Delta CD_{\text{line}} = \left[\Delta CD^2 + \left(3\sigma_{CD_1,CD_2,CD_3} \right)^2 \right]^{1/2}$$

where CD_1, CD_2, and CD_3 are for each individual line population. This analysis is based on an assumption that CD uniformity and overlay standard deviation are the same for each patterning step.

For a triple space LELELE process, there are three overlay populations, $S1$–$S3$, which are likely to be the most significant source for CD nonuniformity:

$$\Delta CD_{\text{space}} = \left[2\Delta OL^2 + \frac{\Delta CD^2}{2} + \left(3\sigma_{\overline{S_1},\overline{S_2},\overline{S_3}} \right)^2 \right]^{1/2}$$

Although this is no worse by extension than the control of CDU for double space patterning, it will ultimately be the gating factor for implementation of multiple patterning LELE at ever decreasing dimensions.

3.3.1.4.2 CD Uniformity for Iterative Spacer

For iterative spacer CDU, overlay is removed from the analysis, since a single "master" exposure is used to form all subsequent subpatterns. Instead, for an example of quadruple patterning, there are four space CD populations ($S1$–$S4$), which are interdependent and largely influenced by deposition and etch process control. The space CDU for such an iterative spacer process becomes [21]

$$\Delta CD_{\text{space}} = \left[\frac{\Delta CD_{\text{litho}}^2}{2} + \frac{3\Delta CD_{sp1}^2}{2} + 2CD_{sp2}^2 + \left(3\sigma_{\overline{S_1},\overline{S_2},\overline{S_3},\overline{S_4}} \right)^2 \right]^{1/2}$$

where CD_{sp1} and CD_{sp2} are for each separate spacer process.

3.4 SUMMARY

As a result of numerous design rule restrictions imposed by double patterning, combined with the improved yields of regular layouts, the industry has seen a steady design migration from two-dimensional layouts from the past toward one-dimensional layouts dominated by line and cut double patterning. Over the past several generations, IC lithography has transitioned from the use of single patterning for 2-D active and 2-D gate layouts into a 1-D active and 1-D gate, manufactured principally by line and cut double patterning. Toward the 14/10/07 nm nodes and beyond, this trend continues, with SADP providing the pitch scaling for the lines, while LELE or LE^3 provides the pitch scaling for the cuts, contacts, and vias (Figure 3.27) [24, 25].

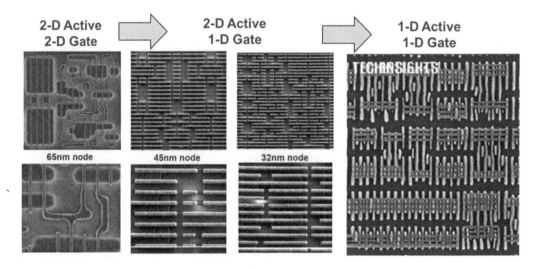

FIGURE 3.27 Multiple generations of semiconductor design evolution, showing single printing of 2-D active and gate patterns at 65 nm node, and evolution into SADP line and cut double patterning for 1-D active and gate patterns by 22 nm node.

REFERENCES

1. Stephen Hsu, Martin Burkhardt, Jungchul Park, Douglas Van Den Broeke, J. Fung Chen, "Dark field double dipole lithography (DDL) for 45 nm node and beyond", *Proc. SPIE* 6283, Photomask and Next-Generation Lithography, Mask Technology XIII, 62830U, 20 May 2006; doi: 10.1117/12.681854
2. Meng-Hsiu Wu et al., "Implementation of double dipole lithography for 45-nm node poly and diffusion layer manufacturing with 0.93NA", *Proc. SPIE* 6607, Photomask and Next-Generation Lithography Mask Technology XIV, 2007.
3. Won-Kwang Ma, "Double exposure to reduce overall line-width variation in 80 nm DRAM gate", *Proc. SPIE* 5377, Optical Microlithography XVII, 2004; doi: 10.1117/12.536358
4. Thorsten Winkler et al., "OPC for double exposure lithography", *Proc. SPIE* 5754, Optical Microlithography XVIII, 2005; doi: 10.1117/12.599441
5. Sang-Jin Kim, "Era of double exposure in 70 nm node DRAM cell", *SPIE* 5754, Optical Microlithography XVIII, 12 May 2004; doi: 10.1117/12.599458
6. Steven Hsu, "65-nm full-chip implementation using double dipole lithography", *SPIE* 5040, Optical Microlithography XVI, 26 June 2003; doi: 10.1117/12.485445
7. Shuji Nakao et al., "Innovative imaging of ultra-fine line without using any strong RET", *Proc. SPIE* 4346, Optical Microlithography XIV, 2001.
8. S. Manakli et al., "Complementary double exposure technique (CODE)", *J. Microlith. Microfab. Microsyst.*, 3: 305, 2004.
9. Henning Haffner et al., "Paving the way to full chip level double patterning", *Proc. SPIE* 6730, Photomask Technology 2007, 67302C, 5 October 2007; doi: 10.1117/12.746116
10. Clair Webb, "Intel design for manufacturing and evolution of design rules", *Proc. SPIE* 6925, Design for Manufacturability through Design-Process Integration II, 692503, 12 March 2008; doi: 10.1117/12.772052
11. Ewoud Vreugdenhil, "Yield Aware Design of gate layer for 45nm CMOS-ASIC using a high-NA dry KrF system", *Proc. SPIE* 6925, Design for Manufacturability through Design-Process Integration II, 69250D, 19 March 2008; doi: 10.1117/12.771998
12. Chris Bencher, "22nm half-pitch patterning by CVD spacer self-alignment double patterning (SADP)", *Proc. SPIE* 6924, Optical Microlithography XXI, 69244E, 7 March 2008; doi: 10.1117/12.772953
13. Huixiong Dai et al., "Implementing self-aligned double patterning on non-gridded design layouts", *Proc. SPIE* 7275, Design for Manufacturability through Design-Process Integration III, 72751E, 13 March 2009; doi: 10.1117/12.814423

14. Chris Bencher et al., "Density Multiplication Techniques for 15nm nodes", *Proc. SPIE* 7973, Optical Microlithography XXIV, 79730K, 23 March 2011; doi: 10.1117/12.881679

15. M. Maenhoudt, R. Gronheid, N. Stepanenko, T. Matsuda, D. Vangoidsenhoven, "Alternative process schemes for double patterning that eliminate the intermediate etch step", *Proc. SPIE* 6924, Optical Microlithography XXI, 69240P; 1 April 2008; doi: 10.1117/12.771884

16. Patrick Wong, Peter De Bisschop, Stewart Robertson, Nadia Vandenbroeck, John Biafore, Vincent Wiaux, Jeroen Van de Kerkhove, "Litho1-litho2 proximity differences for LELE and LPLE double patterning processes", *Proc. SPIE* 8326, Optical Microlithography XXV, 83260E, 13 March 2012; doi: 10.1117/12.916156

17. Geert Vandenberghe, "Lithography Options and Challenges towards the 32nm node", *IEEE Seminar*, 27 February 2008, https://www.academia.edu/21996196/Lithography_options_for_the_32nm_half_pitch_node_and_their_implications_on_resist_and_material_technology.

18. Michael White, "Make the leap from test chips to production designs at 20 nm", *Electron. Des.*, 4 January 2013, http://electronicdesign.com/digital-ics/make-leap-test-chips-production-designs-20-nm

19. Vincent Wiaux. "Double-patterning-compliant split and design". *SPIE Newsroom*, 2009. doi: 10.1117/2.1200902.1430.

20. Joost Bekaert et al., "Experimental verification of source-mask optimization and freeform illumination for 22-nm node static random access memory cells", *J. Micro Nanolithogr. MEMS MOEMS* 10(1), 013008, 1 January 2011; doi: 10.1117/1.3541778

21. Peng Xie, Bruce W. Smith, "Analysis of higher order pitch division for sub-32nm lithography", *Proc. SPIE* 7274, Optical Microlithography XXII, 72741Y, 16 March 2009; doi: 10.1117/12.816131

22. Andrew Carlson, Tsu-Jae King Liu, "Low-variability negative and iterative spacer processes for sub-30-nm lines and holes", *J. Micro. Nanolith. MEMS MOEMS* 8(1), 011009, 2009.

23. J. S. Chawla, K. J. Singh, A. Myers, D. J. Michalak, R. Schenker, C. Jezewski, B. Krist, F. Gstrein, T. K. Indukuri, H. J. Yoo, "Patterning challenges in the fabrication of 12 nm half-pitch dual damascene copper ultra low-k interconnects", *Proc. SPIE* 9054, Advanced Etch Technology for Nanopatterning III, 905404, 28 March 2014; doi: 10.1117/12.2048599

24. Kelin Kuhn, "Variation in 45nm and Implications for 32 nm and beyond", keynote presentation from 2009 2nd International CMOS Variability Conference, London, p. 82; http://download.intel.com/pressroom/pdf/kkuhn/Kuhn_ICCV_keynote_slides.pdf

25. UBM TechInsights, "Logic detailed structural analysis of the intel 22 nm ivy bridge processor", 2012, Figure 2.3.1.14, ubmtechinsights.com

4 EUV Lithography

Stefan Wurm, Winfried Kaiser, Udo Dinger,
Stephan Müllender, Bruno La Fontaine,
Obert R. Wood II, and Mark Neisser

CONTENTS

4.1 EUV DEVELOPMENT HISTORY AND OUTLOOK

4.1.1 INTRODUCTION

Extreme ultraviolet lithography (EUVL) is an optical lithography technology that has many simi-larities to conventional optical lithography. The optical imaging process follows Abbe and Rayleigh [1, 2], and EUVL systems make use of reduction optics, mask and wafer scanning, and alignment and focusing architectures that are similar to those found in current optical exposure tools. Two fundamental properties of an optical projection system are resolution (RES) and depth of focus (DOF) [1–3]:

$$RES = k_1 \cdot \frac{\lambda}{NA} \tag{4.1}$$

$$DOF = \pm k_2 \cdot \frac{\lambda}{NA^2} \tag{4.2}$$

where
λ is the optical wavelength
NA   is the numerical aperture of the projection system

The parameters k_1 and k_2 are empirically determined and correspond to those values that yield the desired critical dimension (CD) control within an acceptable integrated circuit (IC) manu-facturing process window. The classical Rayleigh limits for k_1 and k_2 are 0.5. In practice, the ultimate resolution and DOF depend on the characteristics of the lithographic process, such as the contrast of the resist system and the extent to which various optical enhancement techniques, such as phase shifting, optical proximity correction (OPC), and off-axis and polarized illumina-tion, are used [4, 5].

Historically, improvements in lithographic resolution have been achieved by incremental reduc-tions in wavelength and increases in NA. After conventional optical lithography at 193 nm wave-length increased NA close to the theoretical limit (1.0 in air), the industry introduced water-based immersion lithography, which helped push NA up to 1.35. As it became clear that EUVL would not be ready once 193 nm immersion lithography reached its single patterning resolution limit with the 28 nm node, self-aligned double patterning was introduced to continue scaling to smaller feature sizes. And recently, the industry had to move to self-aligned quadruple patterning for the 16/14 nm node as the manufacturing readiness of EUVL continued to be delayed. In addition, mul-tiple lithography and multiple etch patterning are also used, particularly for contact hole patterns. Now, IC manufacturers are targeting the 7 nm node for EUVL manufacturing introduction, as the complexity of multiple patterning may make it impossible to extend 193 nm multiple patterning as a manufacturing technology beyond quadruple patterning. With its 13.5 nm wavelength and the

NA of the first manufacturing tools at 0.33, even EUVL may have to be used in a double pattern-ing approach at the 7 nm node. Similarly to the development that optical lithography has seen over several decades, scaling to higher NAs will be the next step for EUVL, and optic designs with NAs of up to 0.5–0.6 are currently being studied [6, 7].

Although EUVL does have many similarities to conventional optical lithography, EUVL technology presents several unique challenges. This starts with the much shorter wavelength of ~13.5 nm that EUVL uses. Such short-wavelength radiation is readily absorbed in the atmosphere, which means that EUVL exposures must be done in a vacuum environment. The short wavelength also means that EUV is an all-reflective technology, using masks and projection optics mirrors that have to be coated with so-called EUV distributed quarter-wave Bragg reflectors. These reflectors are made up of alternating layers of high-Z and low-Z materials (Z being the atomic number); typi-cally, materials such as Mo and Si are used to form those special multilayer (ML) reflective coatings consisting of 40–60 Mo/Si bilayer pairs.

The highest reflectivity that can be obtained using distributed quarter-wave Bragg reflectors at EUV wavelength is around 70% [8]. This means that after reflection at two such mirror surfaces, the incident EUV photon flux is cut in half. This energy loss per reflecting surface limits the practi-cal number of optical elements that can be used and dictates the use of aspheric surfaces, making figure and finish specifications even more difficult to achieve. EUVL projection optics surfaces must meet atomic-level finish or roughness specifications on the order of 0.1 nm root mean square (rms) or less [9].

EUVL photomasks are also reflective, making the optical system nontelecentric, since the mask is illuminated slightly off the normal optical axis. Because of the high absorption of EUV light by all materials, pellicles (which are used in current optical technologies to protect the mask sur-face from defects) were not considered practical for EUV masks. However, recent progress in pel-licle membrane development has shown that very high EUV-wavelength transmission pellicle films (>90% one pass) are possible, and the industry is now targeting EUVL high-volume manufacturing (HVM) with a pellicle [10].

Producing defect-free EUV masks and keeping them defect free throughout their useful lifetime has been one of the most difficult challenges for EUVL. While this challenge cannot yet be con-sidered to be completely solved, the industry is very close to meeting the requirements for HVM introduction. This is due to the progress in reducing critical mask blank defects [11], defect mitiga-tion techniques developed by mask shops to render any remaining mask blank defects nonprintable [12], and the progress in developing a pellicle solution [10].

Because of the 30% energy loss per mirror surface, EUVL requires high-power EUV sources. However, it is significantly more difficult to generate EUV photons than visible light photons, and therefore, achieving the EUVL system throughput to meet the cost of ownership requirements of high-volume product IC manufacturing is more difficult. The poor efficiencies in converting electri-cal or laser energy into EUV light make EUV source technology much more complex. Improving source reliability/availability while continuing to increase the EUV power delivered has turned out to be the most difficult challenge for EUVL and the main obstacle to its introduction into HVM; however, steady progress has been made in addressing this critical challenge, and recent results indicate that the required productivity is now within reach [13].

Optics contamination has been an issue with almost all lithography technologies in the past, and this is not different for EUVL. Because of the much shorter wavelengths, many secondary electrons are generated by the absorbed EUV radiation; these can drive surface chemical reactions, leading to oxidation of the projection and illuminator mirror surfaces or to carbon deposition if water vapor and hydrocarbon partial pressures in the vacuum environment are not carefully controlled. And the harsh environment of the EUV source plasma can lead to erosion of the EUV source condenser mirror surface. None of the full-field EUV exposure tools in the field are operating yet at the high source power level needed to achieve productivity requirements (~80–100 W max. vs. a required ~200–250 W) for EUV introduction. Under current operating conditions, optics contamination in the projection optics box has not been an issue. EUV mask contamination/lifetime challenges will become clearer once they have to withstand sustained long-term high–source power operation. The lifetime of the condenser mirror in the EUV source has been one of the main concerns, but progress in developing in situ cleaning solutions has resulted in greatly increased collector lifetimes [14].

The industry has driven EUV resist resolution of chemically amplified resist (CAR) platforms and of new non-CAR platforms down to the 12–15 nm range [15]. However, simultaneously meeting the resolution, sensitivity, and linewidth roughness (LWR) requirements called for in the International Technology Roadmap for Semiconductors (ITRS) [16] remains a challenge. EUV resist outgassing, which has been seen as a potential critical issue for EUV resists that could result in damage to the projection optics, is well controlled and has become much less of a challenge. This is mostly due to the exposure tool capability to control the wafer stage and projection optics box environments, such that only very minute amounts of components outgassing from the resist will ever make it into the projection optics vacuum environment.

EUVL has demonstrated its outstanding imaging capability in abundance and is now in pilot line mode, with several chip makers preparing to use EUVL in manufacturing for the 7 nm node and beyond. If progress in productivity and availability of EUV exposure tools continues at the increased pace of recent years, it will be used in HVM.

4.1.2   EUVL Development History

EUVL development history, as illustrated in Figure 4.1, evolved through a sequence of phases that started in the late 1980s with the "Pioneer Works" phase. This phase was started by early concepts for EUVL emerging from research in Japan and the United States [17, 18] using so-called "soft X-rays" in the 10 to 30 nm wavelength range. In 1986, a research group from NTT reported the first images recorded with a soft X-ray reduction lithography setup [19]. In 1987, a group at the Lebedev Physical Institute [20] published a study on the properties of soft X-ray projection imaging, and in

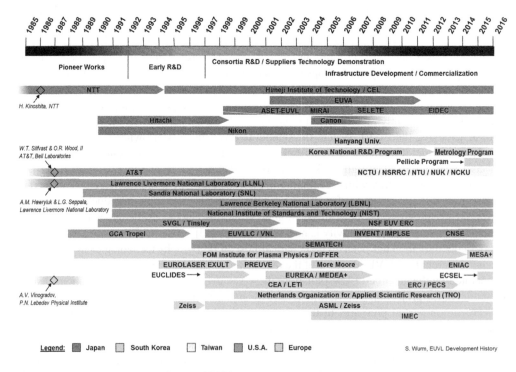

FIGURE 4.1   The development history of EUV.

1988, groups from Lawrence Livermore National Laboratories (LLNL) [21] and Bell Laboratories [22] made the first proposals to use EUV radiation for all-reflective projection lithography. Though Bragg quarter-wave reflectors were already used in those designs, the materials employed had to be different from the Mo/Si combination used for today's 13.5 nm EUV radiation. Interestingly, the first publications proposing soft X-ray exposure schemes used different mask technologies: one proposal was for a reflective mask, while the other suggested the use of a stencil mask type.

The first printing results demonstrated in 1986 resolved 0.5 μm features in polymethyl methacrylate (PMMA) resist using 12.4 nm synchrotron radiation, Schwarzschild optics, and a reflecting as well as a stencil mask [19, 23]. In 1990, Bell Laboratories accomplished the first nearly diffraction-limited printing of dense 50 nm features in PMMA resist using 14 nm synchrotron radiation, a Schwarzschild optic, and a reflective mask. Toward the end of the "Pioneer Works" phase, in 1991, a team from Sandia National Laboratories (SNL) produced the first soft X-ray imaging using a compact laser-produced plasma (LPP) source [24].

During the late "Pioneer Works" and the "early R&D" phases, a number of additional research organizations and companies in Japan, the United States, and more gradually also in Europe became active in soft X-ray research and development. Progress was characterized by the development of the first microsteppers with 10× demagnification, the so-called 10× microsteppers I and II, which were built at SNL in collaboration with Bell Laboratories. Also during this time, to avoid confusion with proximity X-ray lithography, a technology that was also under development in the early 1990s, the new lithography technology using soft X-ray radiation became known as EUV lithography. The culmination of the "early R&D" phase was the fabrication of the first functioning device patterned with EUVL by SNL in 1996 [25].

The third phase of EUV development, the "Consortia R&D/suppliers technology demonstration phase," started when the potential of EUV lithography to print smaller and smaller feature sizes became evident to semiconductor manufacturers, and several industry consortia started working on EUVL technology research and development. In 1997, a consortium of semiconductor manufacturers, called the Extreme Ultraviolet Limited Liability Company (EUV LLC), was formed in the United States to fund and guide the commercialization of EUVL [26]. The founding members of the EUV LLC were AMD, Intel, and Motorola, which were later joined by Micron, Infineon, and IBM. The EUV LLC developed key EUV component technologies through close cooperation among six semiconductor manufacturers, three national laboratories, and several other industry partners. The three national laboratories that partnered with the EUV LLC in the form of a combined virtual national laboratory (VNL) were SNL, LLNL, and Lawrence Berkeley National Laboratories (LBNL). The EUV LLC/VNL demonstrated all aspects of the technology and in the process built two full-field 0.1 NA four-mirror projection optics systems, which were used in the so-called engineering test stand (ETS) to produce the first full-field scanned images on a 200 mm wafer in 2001 [8, 27].

International SEMATECH started its EUVL program in 1998 to support mask modeling, microstepper optic development for two 0.3 NA microfield exposure tools (MET), and other topics [28]. Also in 1998, the European research program Extreme UV Concept Lithography Development System (EUCLIDES) was formed to evaluate EUVL. It focused on mirror substrates, high-reflectivity multilayer coatings, vacuum stages, and a comparison of plasma and synchrotron EUV sources. The French PREUVE program was also initiated in 1999 with a focus on developing a 0.3 NA MET, EUV sources, optics, multilayers, mask and resist modeling, and defect metrology. The PREUVE and EUCLIDES programs subsequently transitioned into the European MEDEA+ program [29].

In Japan, the Japanese Association of Super-Advanced Electronics Technologies (ASET) program was established in 1998 [30]. The EUV portion of the ASET program focused on multilayer, mask, resist, and process development but also developed a 0.3 NA MET. The ASET program was later joined by the Extreme Ultraviolet Lithography System Development Association (EUVA) program [31].

The EUV development "infrastructure development/commercialization" started in earnest in 2003, when after the successful technology demonstration by the EUV LLC/VNL, SEMATECH launched a focused infrastructure program targeted at resolving the final EUV infrastructure challenges to enable introduction of EUV lithography into manufacturing. In the process, SEMATECH's EUV program built up unique capabilities critical to EUV infrastructure development, including a Mask Blank Development Center (MBDC) and Resist Test Center (RTC) in Albany, New York [32, 33]. In addition, consortia in Japan and Europe started driving infrastructure readiness for manufacturing introduction, and suppliers in all regions made sufficient progress to enable EUV alpha tool sites through 2006–2007 at the College of Nanoscale Science and Technology (CNSE)/ SEMATECH [34] in Albany, NY, at the Interuniversity MicroElectronics Center (IMEC) in Leuven, Belgium [35], and at Semiconductor Leading Edge Technologies (SELETE) in Tsukuba, Japan [36]. The technology and process learning gained from those alpha tool operations accelerated resist and mask materials development and resulted in beta tools becoming available in the 2010/11 time frame [37].

In 2002, the first International EUVL Symposium was held in Dallas, TX. Since then, industry experts and researchers have been coming together at this annual confab to assess the status of EUVL and to identify the key challenges that need to be addressed to introduce EUVL into manufacturing. Starting with the 2003 meeting, the International Steering Committee of this conference has produced a consensus list of the most important critical issues/areas for EUV that needed increased efforts. Figure 4.2 shows a summary graph of those tables from 2003 to 2015. As can be seen, initially, mask, EUV source, and resist topics alternated among the top three spots until 2007. By 2008, EUV resist had made significant progress that it remained the third-ranked topic for 6 years, mostly due to the fact that imaging resolution dramatically improved during that time frame, breaking through the 20 nm barrier. However, as EUV introduction shifted out to future nodes, resist is back as the second most critical topic after source. This is probably to some extent a reflection of the improvements in reducing EUV mask defects and the potential availability of a pellicle solution. However, this is also an indication that requirements for EUV resists have become significantly tighter, given the resolution and linewidth requirements at smaller feature sizes and the increased need for faster resists to support throughput requirements. For the last 5 years, source has remained the number one area where EUV needs to improve.

4.1.3 EUV Manufacturing Outlook

As the following sections on EUV Optics and Multilayers (4.2), EUV Source (4.3), EUV Masks (4.4), EUV Resists (4.5), and EUV Extendibility (4.6) will show, EUV development has made outstanding progress since the first version of this EUV chapter was published almost a decade ago. Anybody following leading-edge manufacturing technology development over the past decade will have to admit that EUV represents some of the greatest achievements in advanced optics, new light sources, and materials developments in mask and imaging materials. However, these technological performance measures that define optics, source, mask, and resist performances must translate into productivity and yield numbers that meet or exceed manufacturing requirements.

Figure 4.3 shows an EUV exposure tool schematic that looks at the main system components only through the productivity and yield lenses. Clearly, achieving EUV source performance in terms of in-band EUV source power delivered and in terms of source uptime and reliability is critical for achieving productivity targets, as discussed in Section 4.3 (EUV Source). The number of optical surfaces required to achieve the desired imaging performance and the respective reflectivities of those surfaces are also critical parameters determining productivity; and, as discussed in Section 4.2 (EUV Optics/Multilayers) and in Section 4.6 (EUV Extendibility), when selecting a higher-NA version for the next generation of EUV exposure tools, the types and numbers of optics surfaces required for a given high-NA system design are key parameters in selecting optics designs.

Unlike in transmission lithography systems, the reflective mask used in EUV lithography needs to be as perfect a mirror surface as the imaging optic surfaces; and they need to achieve as high a

2003 / 32 hp	2004 / 32 hp	2005 / 32 hp	2006 / 32 hp	2007 / 22 hp
1. Source power and lifetime including condenser optics lifetime	1. Availability of defect free mask	1. Resist resolution, sensitivity & LER met simultaneously	1. Reliable high power source & collector module	1. Reliable high power source & collector module
2. Availability of defect free mask	2. Lifetime of source components & collector optics	2. Collector lifetime	2. Resist resolution, sensitivity & LER met simultaneously	2. Resist resolution, sensitivity & LER met simultaneously
3. Reticle protection during storage, handling and use	3. Resist resolution, sensitivity & LER met simultaneously	3. Availability of defect free mask	3. Availability of defect free mask	3. Availability of defect free mask
4. Projection and illuminator optics lifetime	• Reticle protection during storage, handling and use	4. Source power	4. Reticle protection during storage, handling and use	4. Reticle protection during storage, handling and use
5. Resist resolution, sensitivity and LER	• Source power	• Reticle protection during storage, handling and use	5. Projection and illuminator optics quality & lifetime	5. Projection and illuminator optics quality & lifetime
6. Optics quality for 32-nm half-pitch node	• Projection and illuminator optics lifetime	• Projection and illuminator optics quality & lifetime		

2008 / 22 hp	2009 / 22 hp	2010 / 22 hp	2011 / 22 hp	2012 / 22 hp
1. Long-term source operation with 100 W at IF and 5MJ/day	1. Mask yield & defect inspection/review infrastructure	1. Mask yield & defect inspection/review infrastructure	1. Long-term reliable source operation with 200 W at IF*	1. Long-term reliable source operation with a. 200 W at IF in 2014 b. 500 W–1,000 W in 2016
2. Defect free masks through lifecycle & inspection/review infrastructure	2. Long-term reliable source operation with 200 W at IF	1. Long-term reliable source operation with 200 W at IF	2. Mask yield & defect inspection/review infrastructure	2. Mask yield & defect inspection/review infrastructure
3. Resist resolution, sensitivity & LER met simultaneously	3. Resist resolution, sensitivity & LER met simultaneously	2. Resist resolution, sensitivity & LER met simultaneously	3. Resist resolution, sensitivity & LER met simultaneously	3. Resist resolution, sensitivity & LER met simultaneously
• Reticle protection during storage, handling and use	• EUVL manufacturing integration	• EUVL manufacturing integration	• EUVL manufacturing integration	• EUVL manufacturing integration
• Projection / illuminator optics and mask lifetime				

2013 / 22 hp	2014 / 16 hp	2015 / 16 hp
1. Long-term reliable source operation with a. 125 W at IF in 2014 b. 250 W in 2015	1. Reliable source operation with > 75% availability – 125 W at IF in 1H / 2015 (at customer)) – 250 W at IF in 1H / 2016 (HVM entry at customer)	1. Reliable source operation with > 85% availability – Expectation of 1500 average wafers per day in 2016
2. Mask yield & defect inspection/review infrastructure	2. Resist resolution, sensitivity & LER met simultaneously – Progress insufficient to meet 2015 introduction target	2. Resist resolution, sensitivity & LER met simultaneously – Increased focus needed on manufacturing performance (defectivity, pattern collapse,…)
4. Keeping mask defect free – Availability of pellicle mtg HVM req't – Minimize defect adders during use	3. Mask yield & defect inspection/review infrastructure – Enable high yield defect free mask blank supply chain	3. Mask yield & defect inspection/review infrastructure – Sustainability of mask tool supply chain remains critical
4. Resist resolution, sensitivity & LER met simultaneously	3. Keeping mask defect free – Availability of pellicle mtg HVM req't : need integrated industry strategy for solution – Minimize defect adders during use	4. Keeping mask defect free (by EUV pellicle) – Pellicle demonstration in the field (on NXE3300)

FIGURE 4.2 Critical issues/focus areas tables for EUV as identified by the ranked Steering Committees of the International EUVL Symposium from 2003 to 2015.

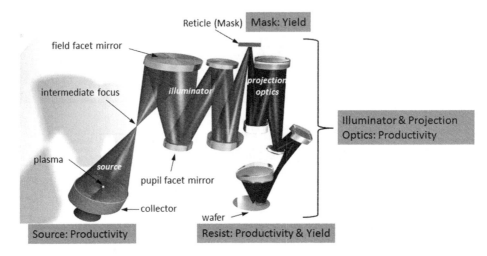

FIGURE 4.3 Main EUV system components seen through the lenses of productivity and yield. (Graphic of the optical train courtesy of Carl Zeiss.)

reflectivity at EUV wavelength as possible, so in that sense, they are no different from any of the projection optics mirrors. What makes the mask different from the projection optics mirrors is that the mask sits in the focal plane and can have no printable defect on its surface if it is to achieve manufacturing yield targets.

Imaging materials have a productivity and a yield component. As discussed in detail in Section 4.5, developing imaging materials that can meet resolution and LWR requirements while at the same time meeting productivity targets (photospeed) is very challenging. Historically, resist materials being used in manufacturing eventually required about 3× the dose used when they were still in development. For EUV, this will be more difficult, as any higher dose required to meet process variability requirements in manufacturing, for example, will have to be made up by increasing source power specifications. Much more so than for resist materials being used, for example in 193 imaging, resist pre and post processing will likely play a much larger role in meeting imaging process requirements for EUV manufacturing. This chapter will not discuss these aspects specifically, but Petrillo et al. [38] provides a good starting point for the interested reader. With mask defects looming large over the past decade as the main challenge to achieving yield in EUV manufacturing, resist defects have received far less attention. As EUV pilot lines started to ramp up a few years ago, the EUV process learning driven by those efforts has started to reduce EUV resist defects, and the expectation is certainly that EUV resists will meet defect requirements. So in summary, while EUV optics and EUV resists certainly do have an impact on EUV exposure tool productivity and yield, meeting those targets essentially boils down to meeting EUV source power and EUV mask blank defect requirements.

The diagram in Figure 4.4 illustrates the source power and mask blank defect requirements for introducing EUV into manufacturing for logic, foundry, and memory manufacturers. Memory manufacturers can live with a few mask blank defects and because of redundancy, still meet their yield targets. Therefore, memory manufacturer defect requirements are less stringent than those of logic and foundry manufacturers. However, memory manufacturers need a much higher source power than logic or foundry manufacturers to consider EUV for manufacturing insertion, as dynamic random-access memory and NAND chips to a large extent are commodity products; it will be interesting to see how this may change in the future, as logic and memory, for example in advanced computing, will have to move ever closer together to reduce interconnect bottlenecks.

For logic and foundry manufacturers to introduce EUV into manufacturing, mask blanks have essentially to be defect free, but they can consider introducing EUV at a lower source power than

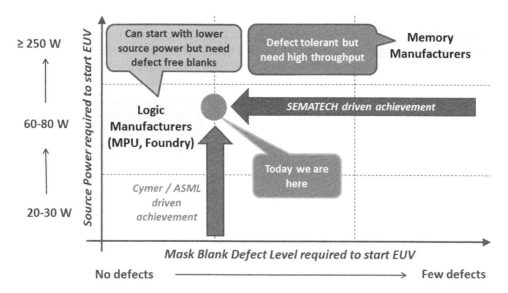

FIGURE 4.4 EUV source power and mask blank defect requirements for logic and memory to start EUV introduction into manufacturing. The industry consortium SEMATECH led the industry effort to reduce EUV mask blank defects between 2003 and 2013 [11].

memory manufacturers. With source productivity supporting close to 100 wafers per hour and mask defects in the low single digits, one can reasonably expect logic and foundry manufacturers to start phasing EUV into manufacturing. As Figure 4.4 shows, the industry is close to meeting the source power and mask blank defect targets for logic and foundry manufacturers.

The semiconductor industry is constantly evolving, and for a technology development, such as EUV, taking almost three decades in development, the assumptions for introducing such a technology will change. The background against which EUV technology initially has been developed was largely set by companies such as Intel and to a lesser extent by Motorola and AMD when founding the EUV LLC. At that time, the future manufacturing challenge for Intel that would need EUV lithography was seen as a microprocessor that could support real-time language translation using a ~10 GHz processor technology. However, long before any such development could start in earnest, the industry ran into a brick wall in scaling up clock frequencies. The answer to that was the development of multiple core processor technologies and a much reduced need for logic to use EUV.

Therefore, around 2005/2006, the leadership mantle for EUV manufacturing introduction passed from logic to memory manufacturers. Micron and Infineon were already part of the EUV LLC, but the real momentum change in favor of memory happened with Samsung taking a leadership role in helping to drive EUV technology development; memory companies such as SK Hynix and Toshiba also became more interested in EUV. Memory companies were seen as the potential first users of EUV technology in manufacturing for a few years, until it became clear that the two-dimensional (2-D) scalability of current memory technologies was being exhausted. In order to integrate more functionality in a given area to stay on the Moore's law curve [39], memory manufacturers had to go from 2-D scaling to three-dimensional (3-D) scaling; they have been doing this now for a few years, and memory products 64 layers deep are now reaching the market. Clearly, 3-D scaling cannot go on forever, as memory manufacturers face increasing complexity and cost to continue on this path. A paradigm change for memory makers could occur if new 2-D-scalable memory concepts become manufacturable; this would also make EUV much more attractive to them again, as they could fully benefit from the high resolution capability the technology can provide.

With the end of the first decade, the EUV manufacturing introduction leadership passed to foundry manufacturers. Foremost, this was and continues to be TSMC, by far the largest semiconductor foundry, but for some time GLOBALFOUNDRIES also contributed significantly to EUV technology development. The reason why EUV technology would be interesting to a leading-edge foundry is simple: EUV levels the playing field with the main competitors on the logic side, such as Intel or Samsung. It is much more difficult for a foundry to continue scaling using multiple patterning technology than it is for a vertically integrated device manufacturer controlling 100% of its design, product engineering, and manufacturing. Leading-edge fabless chip makers such as Qualcomm or Nvidia (to name just two) are among the main competitors of a company such as Intel. While it certainly is also challenging for Intel to continue shrinking devices while increasing device performance and reducing power consumption, its advantages as an integrated manufacturer will increase with each shrink, as long as EUV is not available to foundry manufacturers. EUV lithography would level the playing field for foundries to be able to help fabless companies to compete with Intel or Samsung. If EUV does not make it into manufacturing, and the multipatterning path continues, foundry companies will have to move from their current business model to a more dedicated customer model, to basically run captive manufacturing for fabless companies; however, the economics of this model may not be that attractive either. So, what could the EUV market look like in 10 years?

Let's first look at the best scenario, which is that there will continue to be three or four leading-edge manufacturers—including logic, foundry, and memory—which will have the resources to afford EUV; no competing technology to EUV will emerge as an attractive alternative; 2-D scaling will remain a key driver for future chip generations; and new 2-D scalable memory concepts will mature, so that EUV will be attractive to memory manufacturers. In this case, all the leading-edge companies can be expected to use EUV. But which EUV technology would they be able to use for 7 nm or 5 nm nodes, assuming that learning with some product (at low volume) would happen at the 7 nm node and the manufacturing ramp would then follow at the 5 nm node? The current 0.33 NA EUV tool generation is already struggling with single exposure resolution at ~ 13 nm half-pitch (HP). It will therefore be difficult to do 7 nm imaging with single exposure, and for the 5 nm node, EUV 0.33 NA would have to be used as a double patterning technology. While that has some impact on economics, the industry has lots of experience now with double patterning, and introducing EUV into manufacturing with double patterning at the 5 nm node should not be an issue; at the next node, EUV quadruple patterning could be used. The question is, at which node will a high-NA EUV exposure technology become attractive? This is hard to predict, as it requires correct answers for what technology/materials drivers will be relevant in 5–10 years, what the acceptable economics will be, and related to that, how long it will take to mature this high-NA EUV technology and how much it will cost. Most of those questions have no good answers yet, and the best one can do today is—as Section 4.6 does—describe how well we understand the challenges and what some of those time frames for development could look like.

The worst case for EUV manufacturing introduction would be that we have already exhausted much of the economically/technologically available improvement budget for EUV key technology components, such as EUV source power, mask blank defects, and imaging material resolution. If that were to be the case, it could result in further pushouts for EUV manufacturing introduction, such as we have seen for several nodes (32, 22, 16, and 11 nm nodes, which were all—in hindsight—unrealistic EUV manufacturing introduction targets). While it is hard to predict what would happen in this case, we could run into a "there is no alternative" scenario; EUV could be introduced into manufacturing even at reduced productivity/yield requirements, if the associated manufacturing cost increases could be absorbed by the whole electronic products supply chain. Reading through the content of the following sections, which provide details on the status of key EUV technologies, it should be clear that we are not yet in such a situation and that the boundaries of EUV technology can still be pushed quite a bit.

The second worst case would be if a competing manufacturing technology became available. One possible technology is nanoimprint lithography (NIL), which has quietly made much progress and is currently being driven by key industry layers to mature it for use in memory manufacturing [40].

If this effort is successful, it could take out the whole memory market as a potential EUV technology user, probably making it more expensive for the remaining logic and foundry manufacturers to use EUV. And finally, what if 2-D scaling at some point becomes technologically unfeasible or economically unaffordable? To some extent, we are already in this situation, as the number of leading-edge companies that can afford to stay in the game has continually shrunk over the past two decades, with only three or four leading-edge chip manufacturing companies left. Many of the companies dropping out of the race have developed strong businesses in less volatile markets, staying in manufacturing (as manufacturers themselves or through foundries) at more relaxed nodes—automotive would be a good example—or decided never to compete at the device level anyway but rather, to be a product company (Apple is probably the best example of this). Similarly, current leading-edge companies may also be driven by economics to migrate to different business models that are less focused on leadership through technology/feature size scaling but are instead more focused on products higher up the value chain.

In summary, given where we are on the ITRS, the competitive dynamic among leading-edge manufacturers (and fabless companies), and the maturity of EUV technology as outlined in the following sections, it is reasonable to expect the EUV manufacturing ramp-up at the 5 nm node, with some product opportunistically being shipped at the 7 nm learning node. So, hopefully, we are only 2–3 years away from seeing EUV lithography finally in HVM.

4.2 EUV OPTICS AND EUV MULTILAYERS

4.2.1 INTRODUCTION

As the next technology generation in lithography EUV is now in high volume manufacturing. At a wavelength of 13.5 nm, it enables ultimate resolution today down to 13 nm, and in the future, with high-NA EUV, to less than 8 nm.

The optical train in an EUV scanner comprises the collector (see Section 4.3), the illumination system, and the projection optics (see Figure 4.5). All three modules target optimal transmission to allow high productivity of the tool. The illuminator and projection optics determine the imaging performance. The illuminator (see Section 4.6) plays a key role in the image formation by shaping the light distribution in the pupil. In actual systems (ZEISS Starlith® 3400), it allows the imaging contrast for specific patterns to be optimized with high flexibility [41]. The projection optics catches and combines the diffracted waves leaving the mask and governs image quality and positioning (overlay) through aberration and flare.

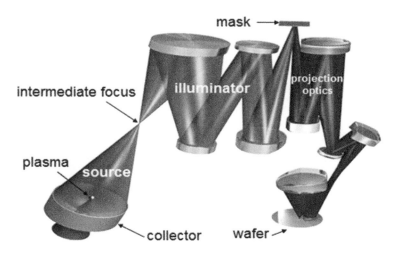

FIGURE 4.5 EUV optical train.

EUV optics has made significant progress over the years. Commercial full-field optical systems were developed first for alpha, beta, or preproduction tools with 0.25 NA (ZEISS ADT, Nikon EUV1, and ZEISS Starlith® 3100) and now for production tools with 0.33 NA (ZEISS Starlith® 3300/3400 family; see Figure 4.6), with the last group being produced now in significant numbers [42–46].

All EUV optical systems are mirror based due to the high absorption of any bulk material at this wavelength [47]. The design is constrained to use the fewest mirrors possible due to the limited reflectivity of each mirror. So, all these projection systems are folded six mirror designs. For 0.33 NA, this leads to large mirrors (up to ~0.5 m Ø) with strong aspheric surfaces, which are challenging to polish with extremely high surface quality. The next generation of EUV optics, the high-NA EUV optics, in the concept phase for a long time [48] and now in development, will feature ~0.55 NA, enabling a resolution of <8 nm HP, and show much larger dimensions with the need for even tighter requirements (see Figure 4.7 and Section 4.6.2). Because of mask shadowing effects [49], this system is designed to cover a half-field by using anamorphic imaging with a demagnification of 4× and 8× in the x and y direction, respectively [6, 50, 51]. By design tricks (e.g., central obscuration), higher transmission will be achieved, enabling a high-productivity tool [7].

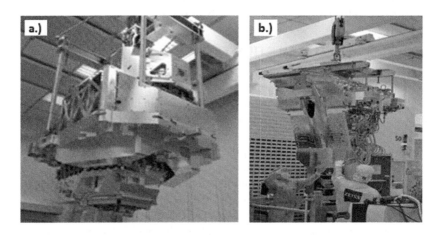

FIGURE 4.6   ZEISS Starlith® 3400 projection optics and illuminator.

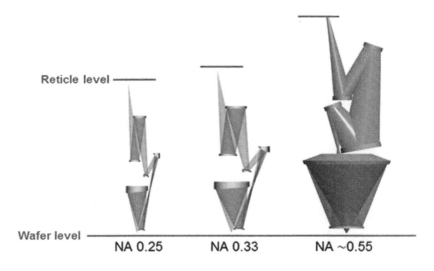

FIGURE 4.7   Design examples for NA 0.25, NA 0.33, and NA ~0.55 projection optics.

Optics	MET	ADT	3100	3300 / 3400
Photos show relative mirror size	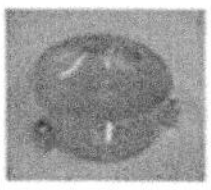			
Figure [pm rms] → aberrations	350	250	140	< 75
MSFR [pm rms] → flare	250	200	130	< 100
HSFR [pm rms] → light loss	300	250	150	< 100

FIGURE 4.8 Improvement of optical performance parameters for different product generations. MET = microexposure tool; ADT = alpha demo tool.

Strong technology advances were achieved in parallel to the product releases (0.33 NA), resulting in imaging performance down to 13 nm half-pitch resolution [9, 15]. As in all optical lithography systems, the key optical performance parameters of EUV projection optics fall into the categories of aberrations, flare, and transmission. Aberrations cause CD and overlay errors. The imaging performance needed requires a system wavefront deviation $<\lambda/50$. The primary origin of aberrations is surface figure, which is the long wave deviation of a mirror surface from the nominal (designed) form. The tight system specifications require figure numbers significantly below 100 pm rms. To achieve this target, highly sophisticated surface interferometry and polishing techniques were developed. All mirrors in the projection optics are made from low thermal expansion material (LTEM) to minimize thermal effects and to keep the optical performance constant during regular operations of the tool.

Flare was originally seen as the most critical performance specification of EUV optics, affecting CD and OPC [52]. To keep OPC efforts within reasonable limits, a flare level of less than 4% is needed. This requires the mid-spatial-frequency roughness (MSFR), a type of ripple on the mirror surface, to stay well below 100 pm rms.

High-spatial-frequency roughness (HSFR) of a mirror surface does not affect imaging but causes light loss of the reflecting surface through scattering. HSFR needs to be smaller than 100 pm rms to achieve the targeted system transmission. To reach these very low M/HSFR specifications—the so-called surface finish—specialized superpolishing techniques were developed.

The reflecting surface itself is a Bragg reflector film on the mirror substrate surface and is built by coating up to 100 quasi-quarter-wave layers of alternating materials (multilayer) with maximum difference in the (complex) refractive index (as Mo and Si for 13.5 nm).

These reflectors have a small spectral bandwidth and a theoretical maximum reflectivity of ~75% (in practice close to 70%); this means that after two mirror reflections, already more than half the light is lost. These reflector coatings have to preserve the surface figure of the mirror substrates and are spread over the curved surface in such a way as to match the local angular light distribution. They have to endure full-power production use over several years. For enabling a highly productive EUV exposure tool, mirrors with highest reflectivity and minimized light loss are essential (see Figure 4.8).

4.2.2 EUV Optics Figuring and Finishing

The increase in imaging performance comes at the price of extremely challenging requirements for the components in the optical column. In what follows, a concise review of the EUV-specific aspects from the viewpoint of ZEISS is given. A more detailed discussion can be found in

Dinger et al. [53, 54]and Murray et al. [55]. The early work within the EUV-LLC was reviewed in Taylor and Soufli [56], and some references to the progress by third parties are given in Marchetti [57], Miura et al. [58], Uzawa et al. [59], and Glatzel et al. [60].

4.2.2.1 Geometrical Challenges

4.2.2.1.1 Need for Aspheric Mirrors

Despite the small number of optical elements, the imaging performance of the wavefront errors of the optical system have to be controlled on an extraordinarily low level. Thus, in order to give the optics designer as many degrees of freedom as possible for balancing the aberrations, full aspheric designs are preferred.

4.2.2.1.2 Off-Axis Geometries and Freeboard

To avoid mutual obscurations of the mirrors, the all-reflective EUV projection systems are characterized by a folded beam path in off-axis geometry. For some of the mirrors, obscuration control puts tight constraints on the maximum distance from the optically used clear aperture to the physical edge of the mirrors, the so-called freeboard. Consequently, figuring and finishing processes have to be capable of provide high surface quality up to the very rim of the mirrors.

4.2.2.2 Surface Specification

Surface tolerances scale with certain powers (1 or 2) of the exposure light's wavelength. The small numbers, resulting from $\lambda = 13.5$ nm, are the major challenge in mirror manufacturing. In what follows, specifications will be discussed on a rule-of-thumb basis.

4.2.2.2.1 Specification Methodology

Surface error specification of EUV mirrors is usually done by means of a power spectral density (PSD); this is similar to other X-ray applications such as synchrotron optics or X-ray astronomy (see Figure 4.9). The PSD is obtained by a decomposition of the 2-D surface topography into its Fourier components and computing the power spectral density, i.e. the square of the Fourier amplitudes, normalized to the frequency sampling interval as a function of the spatial frequency.

The impact on optical performance can be roughly categorized by three major frequency bands: Figure, MSFR, and HSFR. The exact limits of the frequency bands depend on the optical layout and the position of the mirror in the optical system. In fact, since the spatial frequencies correspond to distinct scattering angles, in today's systems, roughness is specified in substantially more than

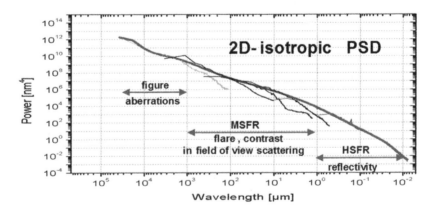

FIGURE 4.9 Schematic 2-D-isotropic PSD_2^{iso}. The rms roughness σ in a specific frequency band [f1,f2] is obtained from $\sigma^2 = 2\pi \int_{f1}^{f2} PSD_2^{iso} f df$; this representation of the PSD can be used with isotropic surfaces. (Reproduced from Dinger, U. et al., *Proc. SPIE*, 5193, 18, 2003. With permission.)

three frequency bands for optimal flare control. The rms error in a given frequency band is then obtained by integrating the PSD in this frequency band and taking the square root of the result to get amplitude numbers.

4.2.2.2.2 Figure

Figure errors typically cover the spatial wavelength band between the clear aperture of the mirrors down to approximately 1 mm. These errors are related to aberrations, which deteriorate the fidelity of the imaging (CD and overlay). As already stated, high-performance lithography systems require a *system* wavefront deviation of less than $\lambda/50$ rms—i.e. much less than the well-known $\lambda/14$ rms Maréchal criterion for diffraction limited imaging.

For a system of N mirrors, each having an rms figure error of $\sqrt{\langle \Delta h^2 \rangle}$, the total rms-wavefront error (WFE) that results is to a first approximation given by

$$\Delta WFE_{system} = \sqrt{N} * \Delta WFE_{single} = \sqrt{N} * 2 * \sqrt{\langle \Delta h^2 \rangle} \qquad (4.3)$$

Here, the errors of the mirrors are assumed to be statistically independent, allowing for some mutual compensation. The factor of 2 in the last term results from the reflection at the surface. Thus, to meet the $\lambda/50$ rms requirement in a six-mirror system, the rms figure error of a single mirror has to fulfill the condition $\sqrt{\langle \Delta h^2 \rangle} \leq 55$ pm. This number has to be met within the so-called subaperture, i.e. the footprint of the rays belonging to one field point.

4.2.2.2.3 Finish

MSFR covering roughly the spatial wavelength band between 1 mm and 1 µm results in in-field-of-view scattering of EUV light. This "flare" reduces the contrast in the structures of the aerial image at the wafer. To get uniform exposure conditions, not only the absolute value of the MSFR but also the variation of the MSFR over the mirrors and even the systems has to be controlled. HSFR covers the spatial wavelength band from approximately 1 µm down to the wavelength of the incident light. It results in out-of-field-of-view-scattering or, in other words, in a loss of light. Since EUV photons are very valuable, today's systems require tight control of HSFR. The total integrated scatter (TIS) resulting from reflection in a system of N mirrors, each having an equal rms roughness σ, is at normal incidence given by

$$TIS = N \left(\frac{4\pi}{\lambda} \sigma \right)^2 \qquad (4.4)$$

The TIS values add up linearly in the system, since their compensation by "backscattering" onto the correct ray path is very improbable. Thus, to keep TIS below 4%, resulting in 4% flare in the case of MSFR or 4% reflection loss (neglecting smoothening or roughening effects of the multilayers) for HSFR, the rms-roughness in the corresponding frequency bands has to fulfill the condition $\sigma \leq 90$ pm. Surfaces exhibiting such low roughness values—below 100 pm rms—are usually referred to as *superpolished*. In summary, the surface features of EUV mirrors have to be controlled in the complete spatial wavelength band extending from 10 nm to approximately 1 m within roughly $\sqrt{3} \cdot 100$ pm rms = 170 pm rms—i.e. within a few atomic layers. The factor $\sqrt{3}$ results from the fact that the powers in the different frequency bands add up linearly (cf. the capture of Figure 4.9).

4.2.2.3 Substrate Materials

In the projection optics, the extreme figure tolerances on atomic scales have to be maintained under the EUV-induced heat loads. This makes LTEM mandatory. This requirement leaves essentially two material classes: glass ceramics (e.g., ZERODUR® or CLEARCERAM®) and amorphous titania-doped fused silica (e.g., ULE®) [61–63].

Certain figuring and finishing processes, which have different removal rates for the ceramic and glassy components of glass ceramics, may "etch out" the microcrystallites, thereby increasing the HSFR. On the other hand, ULE® mirror blanks may exhibit so-called *striae*; that is, a vertically stratified structure of slightly different material composition. Again, this can translate into locally varying removal rates when the aspheric surface cuts through this structure. The striae tend to contribute to the MSFR regime, if not properly controlling the figuring and finishing processes. Depending on where a specific mirror is positioned in the system, it might be advantageous to choose one or the other material class [64].

4.2.2.4 Fabrication Schemes

Depending on the material, the desired geometry and surface shape are generated, for example, on precision diamond grinding or single diamond milling machines. Finishing, i.e. roughness reduction, is done by polishing with comparatively large tools using dedicated pads and slurries [55]. Figure is corrected by computer-controlled figuring technologies, shaping the surface locally based on the error maps from full-aperture interferometers. Various technologies, such as computer-controlled polishing (CCP), ion beam figuring (IBF), or magnetorheological figuring (MRF), can be used here [53].

The real challenge is to control the surface quality in all frequency bands simultaneously. Local figuring tends to degrade the roughness, for example by residual tool tracks (mainly MSFR) or etching effects (mainly HSFR), whereas large tool finishing degrades the aspheric figure due to the locally nonisotropic surface curvature. This PSD crosstalk calls for an iterative procedure, eventually converging to a sufficiently smooth PSD meeting all the system requirements.

4.2.2.5 Surface Metrology

The accuracy of any precision fabrication is constrained by the achievable accuracy of the accompanying metrology. EUV metrology has to cover the complete spatial frequency band reaching from roughly 1 m down to 10 nm with accuracy and precision on atomic scales, without leaving any spatial frequency gaps.

4.2.2.5.1 Figure

Figure is assessed by full-aperture interferometers (Figure 4.10), equipped with dedicated compensation optics to deal with the aspheric shape of the surfaces. These instruments yield 2-D surface

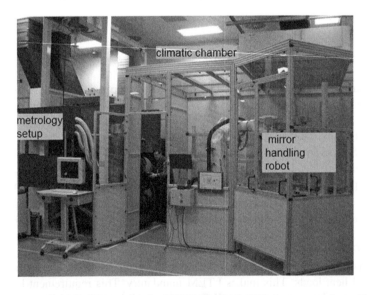

FIGURE 4.10 Full aperture interferometer for ZEISS Starlith® 3300 figure metrology. The climate-controlled instrument must be capable of characterizing 500 mm diameter mirrors. The weight of the large mirrors requires robotic loading.

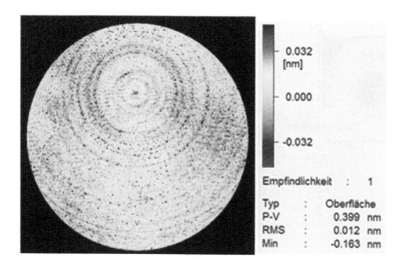

FIGURE 4.11 Production line validation test of interferometer performance demonstrating 12 pm rms reproducibility on a 200 mm diameter circular fraction of an aspheric off-axis mirror. The pixel size is 0.3 mm. "Empfindlichkeit" = sensitivity, "Oberfläche" = surface.

topography maps, which are fed forward to a locally computer-controlled figuring process. The instruments must handle meter-class optics.

Repeatability significantly below 100 pm has to be ensured to enable a converging figuring process on the 100 pm scale. Today, 20 pm rms is achieved in line production (Figure 4.11). Additionally, in order to make the fabrication process converge toward the correct surface shape, absolute accuracy on the 30 pm level is met in production for all error types that cannot be compensated during system alignment. This calls for highly dedicated calibration procedures. Besides intrinsic calibration errors of the interferometer, mount or gravitationally induced deformations can also contribute to inaccuracy if the optics are not tested in the mounting hardware and in their operational geometry.

4.2.2.5.2 Finish

MSFR is measured by microinterferometers. Here, the proper removal of the aspheric shape, the proper calibration of systematic interferometer errors, and the mitigation of vibrations are key to a converging finishing process. The HSFR domain is usually covered by atomic force microscopy (AFM). Since this is a scanning method, special care has to be taken regarding artifacts introduced by the scanning procedure and vibrations.

4.2.2.5.3 Absolute Positioning

The proper positioning of the off-axis aspheres on the physical substrates is controlled by 3-D coordinate measuring machines. Eventually, all the data are merged and digested into a continuous PSD covering the complete relevant frequency space without any gaps (Figure 4.12)

4.2.2.6 Results

The development of EUV manufacturing technologies at ZEISS started in the late 1990s, based on two decades of experience in the manufacturing of EUV and X-ray optics for spaceborne astronomy and synchrotrons [65]. The first EUV optics tackled at ZEISS as part of a round-robin test in mirror fabrication initiated by Sandia NL was ELT2—a component from a three-mirror microstepper optics design. In the early 2000s, the basic EUV fabrication and metrology technologies were developed within the two-mirror microexposure tool (MET) program, funded by SEMATECH and executed in close cooperation with the EUV LLC, LLNL, and LBNL. All fabrication and metrology processes were already capable of off-axis geometries. The fabrication of the subsequent off-axis full-field Alpha Demo Tool (ADT) projection optics was essentially based on these processes.

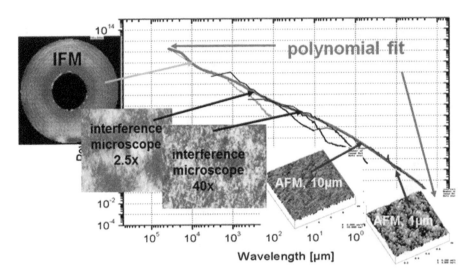

FIGURE 4.12 Gapless coverage of the PSD by a dedicated set of metrology tools using full aperture interferometers, interference microscopes, and atomic force microscopes. (Reproduced from Dinger, U. et al., *Proc. SPIE*, 5193, 18, 2003. With permission.)

During the ZEISS Starlith® 3100 and Starlith® 33x0 programs, the processes were continuously improved and adapted in view of quality and productivity.

Figure errors as well as MSFR were continuously reduced and achieved consistently over the last 15 years, enabling unprecedented, yet still improving, system resolution, overlay, and contrast. This progress is summarized in Figure 4.13, where the three different parts show:

- *Top graph*: System wavefront error after alignment (mainly driven by figure errors). Please note that 0.27 nm corresponds to $\lambda/50$. Each bar in the graph represents a shipped system for the tool generations shown.
- *Middle graph*: Shows progress in reducing MSFR. Each symbol represents a single mirror (all MSFR values are based on the same definition as for ZEISS Starlith® 3100/3300).
- *Bottom graph*: Shows progress in system flare reduction. Each bar represents a shipped system. The ZEISS Starlith® 33x0 systems consistently meet the 4% flare specification.

The progress in the HSFR will be discussed in the following section in the context of multilayer performance.

4.2.3 EUV Coatings

At the EUV wavelength of 13.5 nm, the refractive indices of all materials are close to unity. A single surface of an optical mirror used at normal incidence therefore reflects hardly any (<1%) light. For practical reasons, most of the optics need to be operated at near-normal incidence; therefore, it is necessary to coat these mirrors with a so-called artificial Bragg reflector, consisting of alternating layers of high and low index of refraction. This stack of bilayers forms the EUV multilayer, the number of contributing bilayers being limited by the absorption of the light in the coating.

The material combination of molybdenum (Mo) and silicon (Si) with typically 50 bilayers was found to be most suitable, with a theoretical peak reflectance of ~75% at normal incidence. In practice, the reflectance is lower, due to imperfections such as thickness errors, substrate and/or interface roughness, layer intermixing, and so on. The best values reported over the last 15 years on super-polished test substrates are in the 70% range [66–68], an example of which is shown in Figure 4.14.

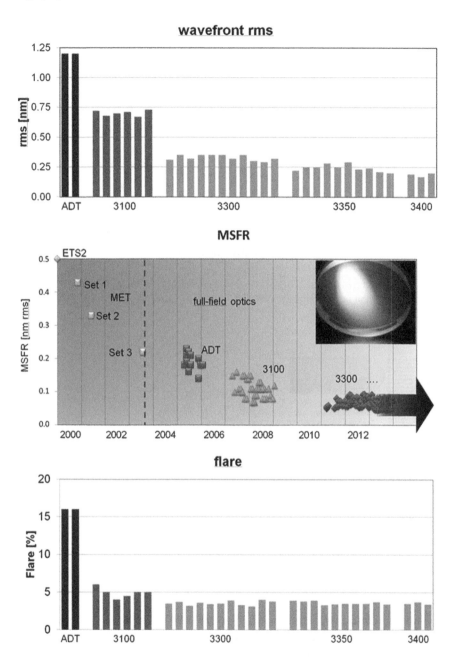

FIGURE 4.13 Evolution of optics performance with the EUV-tool generations. For a detailed explanation, see text.

High reflectance of the multilayer is only one of many requirements the coating has to fulfill, and other optical and materials demands have to be taken into the overall coating optimization.

4.2.3.1 Optimization of the Coating Properties for EUV

Whereas high reflectance of the coating is necessary for high transmission of the optical system and therefore, high throughput of exposed wafers, there are many more properties that need to be considered and optimized at the same time. In addition, these have to be realized on real optics, that is, large, heavy, and mostly curved substrates, instead of small, flat test samples.

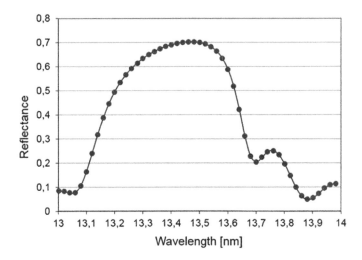

FIGURE 4.14   Best reflectivity value (70.3 ± 0.1% @ $\lambda = 13.5$ nm) for a 50-layer Mo/Si stack with B_4C and Y barrier layers measured at near normal incidence. (Courtesy of University of Twente. From Bosgra, J et al., *Appl. Opt.*, 51:36, 8541–8548, 2012.)

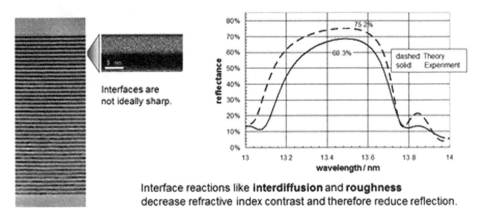

FIGURE 4.15   Reduction in reflectance due to intermixing of molybdenum and silicon: comparison of experiment and theory.

4.2.3.1.1   Reflectance

When examining a transmission electron microscopy (TEM) cross section of a Mo/Si multilayer (Figure 4.15), one finds that the interfaces between the materials are not well defined. Due to the energy of the deposited material and the cross sections of relevant chemical reactions, the formation of molybdenum silicides normally takes place. This reduces the optical contrast and therefore also the peak reflectance as well as the bandwidth (full width at half maximum [FWHM]).

By introducing thin, so-called barrier layers in between the molybdenum and silicon layers, it is possible to enhance the reflectance. Those extra layers consist preferably of materials with low EUV absorption, such as boron, carbon, or compounds of these with molybdenum or silicon, with typical thicknesses in the tenths of a nanometer range. The optimal thickness and material selection need to be determined experimentally (Figure 4.16).

4.2.3.1.2   Resistance to Elevated Operating Temperatures

The spectrum of most EUV sources contains only a fraction of EUV light within the acceptance band of the multilayer optical system. The part outside this band is denoted as *out-of-band* radiation

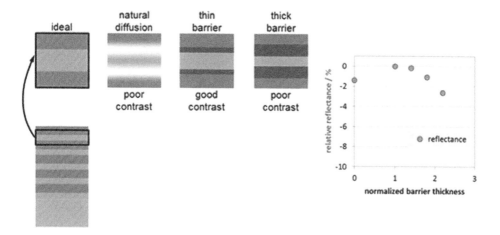

FIGURE 4.16   Experimental determination of the optimal barrier layer thickness.

and contains visible, infrared, and vacuum UV light. This light is partially absorbed by the coating and substrates, leading to an elevated operating temperature of the optic, especially in the illuminator. At these temperatures, the interdiffusion at the interfaces of the layers is enhanced and will result in a (fast) degradation of the coating; that is, a shift of the peak reflectance wavelength due to a change of the bilayer thickness (Mo_xSi_y has a larger density than the average density of separated Mo and Si). This layer intermixing can be avoided by using diffusion barrier layers (see earlier), although at increased thickness. These thicker barriers may cause slightly reduced maximum reflectance due to the reduced optical contrast.

4.2.3.1.3 Minimal Changes to Figure of the Mirror

As stated earlier, the figure of the mirror is polished within 100 pm and should not be changed significantly by the coating. The coating could cause such changes either due to the multilayer-induced substrate stress or due to lateral thickness errors. When depositing thin layers of different materials on each other, the difference in the atomic distances and in the interatomic forces leads to stress in the multilayer. Typical values for Mo/Si multilayers are in the −400 to −200 MPa range [69–71] and cause the deformation of typical substrates by up to several tens of nanometers.

This deformation can partially be compensated by the alignment of the mirror, but the remaining, noncorrectable residues are still in an unacceptable range of several hundreds of picometers. To reduce such deformation the Mo/Si coating can be modified to have low stress values by changing the Mo to Si ratio (with unacceptable losses in reflectance) and/or the barrier thicknesses; another possibility is the addition of an antistress layer underneath the multilayer with opposite stress values. This was first proposed by Mirkarimi and Montcalm [70, 72] and is preferably made from the same materials as used in the reflective coating [71].

The second possible contribution to the figure error is the deviation of the lateral layer thickness from the designed value. Considering a 0.1% thickness error, a deviation of just 7 pm per bilayer from the design value would add up to a figure change of 350 pm for the total multilayer with 50 bilayers. Again, long-range errors might be compensated by alignment, but this is difficult, if not impossible, for short-range errors.

Therefore, the precision at which the coating has to be applied needs to be in the very low picometer range on average per bilayer. When these requirements, originating from optical design considerations, are fulfilled, high reflectance of the multilayer is in general directly achieved. In practice, values down to ±1 pm per bilayer on a substrate radius of more than 225 mm have indeed been achieved (Figure 4.17).

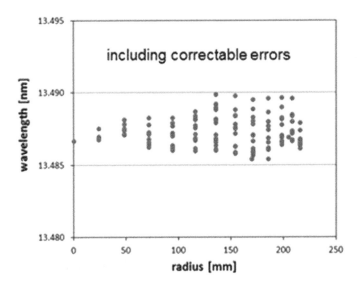

FIGURE 4.17 Typical peak wavelength spread of ±2.5 pm per bilayer on a 450 mm diameter substrate, measured at different azimuths. Using the Bragg equation, this translates into an averaged spread of approximately ±1 pm per bilayer.

4.2.3.1.4 Stability

The coating properties need to be stable for several years of operation. This means, for example, that the reflectance is not allowed to decrease by more than ~1% despite high-power EUV irradiation, oxidation, and carbon contamination due to limited vacuum conditions. As little as 1 nm of carbon on top of the coating results in a 1% reduction of the reflectance. Fortunately, this can be reversed by in situ cleaning with hydrogen radicals. A silicon-terminated multilayer would not survive these conditions, and a protective top coating needs to be applied. Not only should it be resistive to oxidation and to hydrogen radicals, but it should also not reduce the EUV reflectance significantly. Many materials, such as C, TiO_2, Rh, Ru, and RuO_2, have been investigated [73], including the interface to the underlying molybdenum.

4.2.3.2 Deposition Systems for EUV Multilayers

There are many ways to deposit thin layers on a substrate. So far, only three methods have proved to be viable for the demanding case of EUV multilayers: electron beam evaporation (EBE) [74], magnetron sputtering (MS) [75], and ion beam sputtering [10]. Each method has its advantages and disadvantages; for example, the last one is commonly used to coat the EUV mask, as it produces the fewest defects, although at the price of a slightly reduced peak reflectance [76]; more on EUV multilayers for masks can be found in Section 4.4.

Due to the geometry of EBE, i.e. the use of an electron beam to evaporate the coating material from a small crucible, the footprint of an EBE coater can be much smaller than that of an MS machine. In the latter case, large targets with a size (in one dimension) comparable to the substrate are used for each of the multilayer materials. This implies much higher initial costs for the coating materials. The low operational pressure of an EBE coater enables the use of an in situ X-ray reflection monitoring system, which allows continuous control of the layer thickness [73]. EBE was used to coat the mirrors for the ASML Alpha-Demo Tool [77] as well as the ZEISS Starlith® 3100 Systems [78].

However, a drawback of EBE is essentially the impossibility of using load locks for the mounting of the substrate into the machine. The necessary time to pump down the whole vessel not only

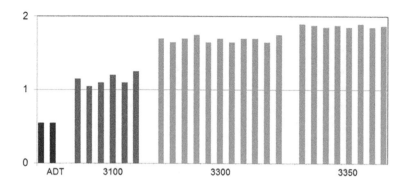

FIGURE 4.18 Evolution of the transmission levels for the consecutive EUV-tool generations, with the specified transmission for Starlith® 3100 set to 1.

makes the process slow and more expensive, but also may cause instabilities during the coating. In addition, the control of this semi-automated process, especially the in situ monitoring, is labor consuming. Given the advances in magnetron technology and its larger process stability, Carl Zeiss switched from EBE to MS using a fully automated process. The machine layout was developed for the special needs of Carl Zeiss and has the capability to coat large substrates up to a diameter of 680 mm with a maximum of six different materials. Currently, the mirrors for ZEISS Starlith® 3300 and 3400 are coated on improved machines [79].

By continuous improvements in polishing (HSFR) and coating technologies and advances in design over the last 15 years, the overall transmission of the consecutive EUV tool generations (illumination and projection optics) was constantly increased, as shown in Figure 4.18. This was a critical contribution to achieve the overall required exposure tool productivity and helped to keep EUV source power requirements within acceptable levels. Without this contribution, EUV sources would have to deliver even higher EUV power. The next section will discuss in detail the EUV source challenges and the industry achievements in this area.

4.3 EUV SOURCE

4.3.1 Introduction

It is clear why EUV light is desirable as a source for lithography. The shorter wavelength, compared with that used for immersion lithography tools, can provide improvement in resolution by a factor of up to 4. The challenge is to develop an EUV source with the right characteristics, which can be integrated into a lithography scanner. To understand what kind of conditions are needed for generating EUV light, we can observe naturally occurring light sources in this spectral region. Figure 4.19 shows a picture of our sun at a wavelength of 13.1 nm [80].

This emission comes from the sun's corona, where the temperature is about a million degrees. At this temperature, many of the electrons from the emitting ions have been removed. For instance, Fe ionized more than 10 times is responsible for the sun's emission in the EUV band between 10 and 20 nm. For incoherent sources of EUV light such as the sun, the emitters are typically excited through collisions and require electron temperatures of the order of the photon energy. For a lithography source operating at 13.5 nm, the photon energy is 92 eV, and, depending on the selected source emitter, the electrons should have a temperature greater than about $kT_e = 20$ eV (or 22,000 °C) for efficient production of EUV light. Other types of EUV sources can be created, where the energy is provided through high electric fields accelerating free electrons instead of using collisions or thermal processes. Such sources include synchrotrons, free-electron lasers, and high-harmonic generation by ultrashort pulse lasers [81, 82] (see Section 4.6).

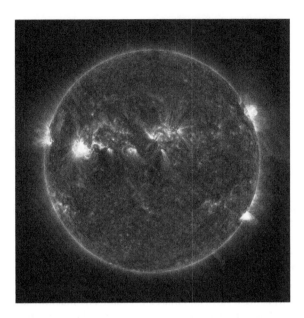

FIGURE 4.19   Image of the sun at a wavelength of 13.1 nm, taken by the Solar Dynamics Observatory (SDO).

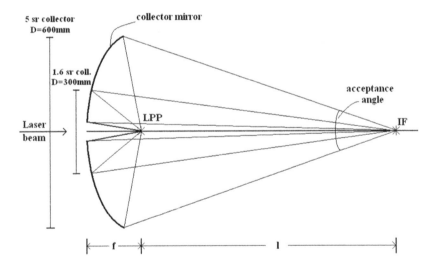

FIGURE 4.20   Schematic of normal-incidence collector for LPP sources.

4.3.1.1   Lithography Sources

For use in a lithography system, a source of EUV is not simply an emitting plasma. The EUV light needs to be collected and directed toward the lithography exposure tool using reflective collector optics as efficiently as possible. Two types of collector used with incoherent sources (discharge-produced plasma [DPP] and laser-produced plasma [LPP]) are shown in Figure 4.20 [83] and Figure 4.21 [84].

The LPP collectors can typically collect the EUV light from the plasma over a solid angle of approximately 5 sr, whereas the DPP collectors may collect only half as much light. This is due to the fact that the plasma in DPP sources is usually created close to solid electrodes, which tend to block some of the plasma emission. The light entering the EUV scanner also needs to have certain characteristics, which are specified at the image point of the plasma source, called the *intermediate focus* (IF).

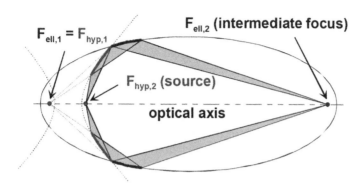

FIGURE 4.21   Schematic of grazing-incidence collector for DPP sources.

4.3.1.2   Requirements

To be useful for EUV lithography, the main source parameters that need to meet specific requirements are (i) power, (ii) spectral content, (iii) étendue, (iv) dose stability, and (v) lifetime. These parameters are closely coupled and should be carefully considered when engineering such a source.

4.3.1.2.1   Power

Source power is arguably the most important requirement. The power specification has evolved over the years and is driven by the need for lithography systems to print more pixels, at higher resolution, at ever-increasing speed, in order to satisfy Moore's law. In the late 1990s, some believed that EUV lithography would be necessary at the 70 nm node and beyond [26]. To meet the resolution necessary at that point, a four-mirror EUV system with a numerical aperture (NA) of 0.1 would have been sufficient, and EUV power of the order of 40 W at the intermediate focus would have met the throughput requirements [85]. During the following decade, the extension of deep ultraviolet (DUV) lithography to 193 nm with higher numerical apertures, the introduction of immersion lithography, and a number of resolution enhancement techniques (RETs) helped to push out the introduction of EUVL to the 32 nm node or beyond. It was also becoming clear that the EUV technology would take more time to be ready.

The EUV optical system would need to have a larger NA to meet the demand for higher resolution, which meant the addition of two mirrors to the imaging optics, bringing along a reduction of overall transmission of about 2. In 2004, the three scanner manufacturers jointly specified that power of 115 W at IF would be required to introduce EUV lithography into HVM [86]. The leading-edge manufacturers of integrated electronic devices have continued to make tremendous progress in packing more and more devices onto their chips using immersion lithography, introducing double and quadruple patterning techniques. We now project that EUV lithography will be used in HVM at the 7 nm node and beyond. In the meantime, the throughput required from lithography exposure tools has continued to climb, leading to EUV source power requirements of 350 W at introduction in HVM [13].

Another important factor to note in defining the power requirement is the resist sensitivity. As the critical dimensions (CDs) of features to print in resist become smaller, the dose required to print these features tends to increase. This is due to two main reasons. The aerial image of smaller features produced by the projection optics usually has a lower contrast, leading in many cases to a larger dose to print. With smaller CDs, stochastic effects also tend to lead to higher line-edge roughness unless a higher-dose resist is used. During the 1990s, the assumption for the resist dose was 5 mJ/cm^2 [26]; nowadays, 25 mJ/cm^2 is a more commonly used number [87].

4.3.1.2.2   Spectral Content

The EUV lithography exposure tool is composed of many reflective surfaces utilizing Mo/Si multilayer films. For this reason, the source spectrum needs to be centered at 13.5 nm within a bandwidth of 2%. Assuming that 11 reflections are needed from IF to the wafer, this specification ensures a transmission of approximately 1% of the light, assuming a peak mirror reflectivity of 67% [88].

It is interesting to note here that, unlike previous generations of lithography, the wavelength used for EUV lithography was not defined by the availability of high-power sources suitable for this application. Rather, it was determined by the availability of mirrors with adequate reflectivity in this spectral domain. As such, the spectral requirement is the most stringent requirement for EUV sources. Research is ongoing to develop multilayer mirror reflectors with high reflectivity at a shorter wavelength (in the range between 6.5 nm and 6.8 nm), along with sources emitting in that band, to provide a path to even higher-resolution lithography [89, 90].

The other important aspect of the source spectrum is its purity, meaning that a limited amount of light (<7%) should be emitted in the DUV/UV region (130 to 400 nm) because of the sensitivity of resists in this band [86] In addition, the amount of radiation produced in the infrared (IR) range should be minimized to avoid excessive heating of the wafer and projection optics.

4.3.1.2.3   Étendue

The source étendue is important for matching to the EUV lithography imaging system. It is defined as the product of the EUV-emitting plasma area and the solid angle subtended by the collector optics. The specification is generally accepted to be less than 1 to 3.3 mm^{2}sr and depends on the exact optical design of the illuminator and projection optics. If the étendue of the source is too large, not all of the light can be coupled to the optical system of the scanner.

4.3.1.2.4   Dose Stability and Control

Process control in lithography relies on the ability to control the exact energy delivered to the wafer to expose the resist. For critical features, the exposure latitude is of the order of only a few percent. To achieve sufficient control, the source itself should contribute a fraction of a percent of the dose error. In a pulsed source operation, this stability can be achieved with a high repetition rate and good pulse-to-pulse energy control. For instance, current LPP sources operate at a repetition rate of 50 kHz and are capable of providing dose stability <0.1% [91].

4.3.1.2.5   Lifetime

Another very important characteristic of the EUV source is its lifetime. This is critical in terms of cost of ownership and for the economic viability of EUV lithography. Early on, a target of 30,000 hours was set for the source [86]. For LPP sources, the collector is arguably the most critical component in terms of lifetime due to the harsh environment in which it resides. Debris, high-energy ions, and photons from the plasma can interact with the collector surface and have the potential to degrade its reflectivity over time.

4.3.1.3   Main Types of EUV Sources

There are two main types of EUV sources that have been developed for lithography: DPP and LPP. These are incoherent sources of EUV light pumped through collisional processes and emit over a large solid angle. Although both DPP and LPP sources have been developed for EUV lithography during the past 30 years [92], only LPP sources are still pursued commercially at this time. Free-electron-laser (FEL) sources are being investigated as candidates for providing power levels of the order of 1 kW per scanner; such sources would most likely be designed to serve multiple scanners in parallel [93, 94].

For metrology, the source requirements are quite different. In general, the power can be much lower, while brightness (power radiated per unit area and unit solid angle at the source) needs to be high. For this type of application, synchrotron light and high-harmonic generation become attractive. Such sources emit much more spatially coherent light. Their emission is produced through the acceleration of free electrons by high electromagnetic fields (e.g., undulator magnets or the electric field of ultrashort pulse lasers) [81]. In this chapter, we restrict our attention to sources for lithography.

4.3.2 DISCHARGE-PRODUCED PLASMA (DPP) SOURCES

DPP sources were the first to be developed for use in commercial lithography systems. They have several attractive features, such as size, relative simplicity, and lower cost compared with LPP sources.

4.3.2.1 Basic Concept

At the heart of all DPP sources is the principle of a plasma pinch. This is used to reach the very high temperatures needed for efficient production of EUV light. By passing a current through a localized plasma filament, an azimuthal magnetic field is induced according to Ampère's law, which leads to a compression of the plasma to a smaller diameter (through the $\bar{v} \times \bar{B}$ term of the Lorentz force), larger current density, and higher magnetic fields. The end result of this type of runaway effect is a hot and dense plasma capable of emitting short-wavelength light. The characteristics of the EUV emission are determined by the temperature achieved when the magnetic pressure is balanced by the plasma thermal pressure. Inside the pinch, the electron temperature (T_e), the linear electron density along the plasma pinch (N), the effective ionization stage (Z_{eff}), and the current (I) are connected through the Bennett relation [95]:

$$\frac{\mu_0 I^2}{4\pi} = \left(Z_{\text{eff}} + 1\right) N_i k T_e \tag{4.5}$$

Again, we can turn to nature for spectacular examples of plasma pinches. During the return stroke of a lightning strike, the negative charges accumulated in the clouds neutralize the excess positive charges accumulated on the ground through a current of the order of 30 kA travelling along the lightning arc channel. This current drives a plasma pinch effect whereby the ionized air can reach temperatures (kT) in the tens of eV. Once the charges have been neutralized, the current and its associated azimuthal magnetic field stop instantly, and along with them the short-wavelength flash of light. At that point, the hot channel of the lightning arc also expands abruptly, leading to a strong shock wave (thunder) [96]. Solar flares appearing in Figure 4.19 are other examples of naturally occurring hot plasma pinches.

4.3.2.2 General Design and Main Challenges of DPP Sources

One of the principal difficulties for engineering DPP EUV sources for lithography is managing the heat load and erosion of the electrodes. In many ways, this is what is driving the design choices for these types of sources. If the intense current flowing through the plasma pinch leads to electrode temperatures in excess of the melting temperatures of the materials used, significant erosion will occur. This results in discharge instability, reduced lifetime, and introduction of debris into the source chamber. The use of refractory materials and cooling of the electrodes provides much improvement, but the increasing need for more source power makes this challenge ever more daunting. In the end, thermal management, erosion, and debris management issues in the face of ever-increasing EUV power requirements were the most likely reasons leading to the demise of DPP sources for EUV lithography.

4.3.2.3 Types of DPP Sources

There are a few types of DPP sources that have been studied and used as EUV sources for lithography. Depending mainly on the emitter material, the geometry of the electrodes, and the characteristics of the high-power pulsed electrical circuit, different mechanisms drive the plasma pinch and determine the EUV emission of these sources.

4.3.2.3.1 Z-Pinch

A z-pinch, pulsed-plasma discharge developed by Xtreme Technologies (a joint venture of Lambda Physik and Jenoptik) was successfully used in the early EUV lithography MS-13 microexposure tools manufactured by Exitech [97]. The Xe-based source generated 35 W into 2 sr at a repetition rate of 1 kHz. The achieved conversion efficiency was 0.7% with a plasma radius of ≤0.9 mm and ~ mm length, satisfying the étendue requirements of 3.3 mm²sr [98, 99]. A later version of this type of source was also integrated in the first EUV exposure tool developed by Nikon: the EUV1 α-tool [100].

A schematic of the discharge chamber is depicted in Figure 4.22. The emitter gas, Xe, flows from a center channel at the back of the chamber through the main area between circular electrodes. At first, a cold plasma is generated at the surface of the insulator material surrounding the electrode at the back of the chamber. UV light and charged particles from this cold plasma propagate to the main part of the chamber and preionize the Xe gas. At that point, the main discharge is ignited, and a strong current pulse flows between the electrodes. Immediately, the pinch effect radially compresses and heats the plasma on the central axis of the device, resulting in EUV emission. Once the current pulse subsides, the plasma rapidly cools down and expands, ending the EUV emission. During this last phase, plasma debris (e.g., high- and low-energy ions and neutrals from the emitting gas or the electrodes) also travels outwards toward the intermediate focus.

To mitigate the effect of debris reaching the collector and leading to erosion and reduced lifetime, Xtreme Technologies developed a debris filter using a foil trap design (see Figure 4.23) [99]. With the implementation of debris mitigation techniques, which also include increasing the distance between the plasma source and the collector and the use of sputter-resistant materials for the collector coating, they were able to demonstrate source lifetime of 5–10 billion pulses. Power scaling of the z-pinch source was carried out through different means. Collection of the power from the

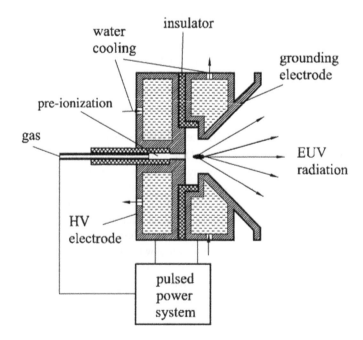

FIGURE 4.22 Schematic of the Xtreme Technologies gas discharge–produced plasma (GDPP) z-pinch source.

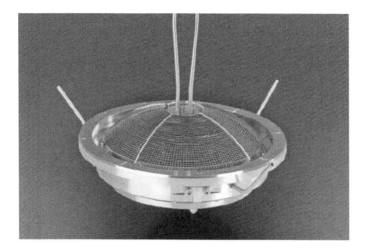

FIGURE 4.23   Foil trap to mitigate debris from the plasma reaching the collector.

FIGURE 4.24   DPP grazing-incidence collector with nine reflective shells.

plasma itself was improved by increasing the number of grazing-incidence shell reflectors used in the collector design. Figure 4.24 shows such a collector with nine shells [101].

Increasing the repetition rate and the energy per pulse and tailoring the plasma-emitting region all contributed to improvements of in-band EUV power production at IF. Finally, changing the plasma fuel from Xe to Sn led to a significant gain in output power. The graph in Figure 4.25 provides a clear view of the incremental progress made in terms of EUV power emitted by z-pinch gas discharge–produced plasmas (GDPP) during the years 2000 to 2005 [99]. The faster rise in power in 2004–2005 was due to a change of fuel from Xe to Sn, accompanied by an increase in conversion efficiency from 0.7% to 2%.

4.3.2.3.2 Dense Plasma Focus (DPF)

The dense plasma focus has been studied quite extensively for fusion applications [102, 103]. For this type of the source, the electrodes have a coaxial geometry, and the discharge has two main phases. Initially, the current between the anode and the cathode is radially symmetric, and the Lorentz force rapidly drives the current away from the insulator and toward the end of the coaxial electrodes. Once the current lines extend beyond the end of the electrodes, they develop a strong

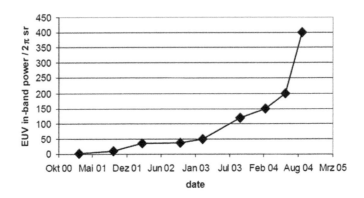

FIGURE 4.25 Progress of in-band EUV power (in 2 sr) for GDPP z-pinch sources during the period from 2000 to 2005.

axial component. At that point, a z-pinch develops in front of the electrodes with strong EUV emission [104].

Cymer has been the principal developer of DPF as EUV sources. Their Xe DPF source could produce 10 kHz EUV pulses of 60 mJ, using 12 J of electrical energy, corresponding to a conversion efficiency of 0.5%. They also demonstrated good performance with Sn-based plasmas [105]. The size of this source (0.4 × 2.5 mm) also met the étendue requirements considering the solid angle captured by DPP collectors (2 to 3 sr).

4.3.2.3.3 Capillary Discharge

The capillary discharge source is another example of a plasma z-pinch, produced with a simple design. As illustrated in Figure 4.26, a cylindrical insulator "capillary" is inserted between two cylindrical electrodes [106]. A low-inductance high-power circuit provides a rapid high-voltage pulse that ionizes the emitting Xe gas in the center of the capillary. The ensuing large current flowing through the ionized gas leads to the pinch effect and EUV plasma emission. Such sources were shown to be able to produce 10 W of EUV light into 2 sr at a repetition rate of 1 kHz.

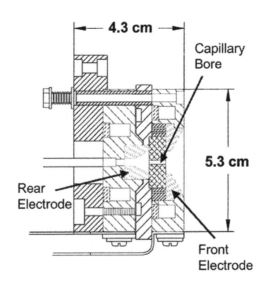

FIGURE 4.26 Schematic of the Sandia National Lab capillary discharge source.

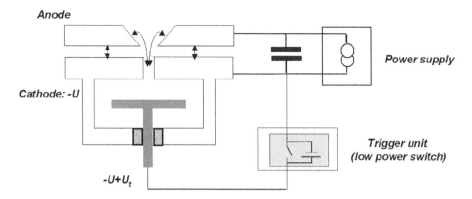

FIGURE 4.27   Schematic of a hollow cathode–triggered discharge.

4.3.2.3.4 Hollow Cathode–Triggered (HCT) Gas Discharge

The HCT gas discharge concept as an EUV light source was born out of the low-pressure high-current pseudo-spark switches for pulsed-power applications [107]. They differ from the other types of pinch plasmas in the fact that they don't require a separate pulsed-power circuit to drive them. The hollow cathode provides a way to trigger the high-current pinch plasma through the initial build-up of a virtual anode between two parallel plates, each with a borehole, as depicted in Figure 4.27 [108].

In the initial phase, a high-voltage pulse is applied to the two electrodes, and electrons from the cathodes are collected by the anode without breaking down the gas because the system is operated at the low-pressure side of the Paschen curve (10–100 Pa). This leads to an accumulation of ions (the virtual anode) gradually moving toward the cathode. Once this virtual anode gets close enough to the cathode, such that the electric field overcomes the breakdown threshold of the gas, a high current (10 kA) flows, leading to the pinch-plasma emission.

Philips Extreme UV, a joint venture of Philips and Fraunhofer ILT, developed such sources for EUV, using both Xe and Sn as emitting elements. Delivering a few joules to the discharge at a repetition rate of 7 kHz, they were able to achieve conversion efficiency of 0.8% for Xe and 2.5% for Sn.

As for the z-pinch and DPF source, the EUV-emitting, hot and dense plasma is not connected directly to the anode, which is of considerable advantage in terms of electrode erosion. Still, the temperature levels reached at the surface of the electrodes of the HCT discharge were such that any further increase in power would have led to serious erosion and the generation of source debris toward the collector and IF.

4.3.2.3.5 Laser-Induced Discharge Plasma

To avoid the issues associated with the HCT discharge plasma source, Philips Extreme UV decided to move to a new discharge design based on a pair of rotating electrodes coated with molten tin. Because the melting temperature of Sn is relatively low (231.9 °C), this helps to conduct the heat away from the plasma discharge. This design also provides a fresh Sn surface after each pulsed discharge [109]. Such sources were used to power the ASML ADT and provided up to 4 W of EUV at IF. Again, the in-band EUV power emitted by the plasma can be adjusted linearly with the repetition rate, up to about 8 kHz, as recorded in the graph in Figure 4.28 [110]. Unfortunately, the same scaling could not be easily realized in terms of IF power because of challenges with debris mitigation.

The development of this source concept based on the two Sn-coated rotating electrodes continued, with the integration of a low-power laser to trigger the discharge, which helped improve spatial stability and dose control (see Figure 4.29) [111]. It was named an LDP (for laser-induced discharge plasma) source.

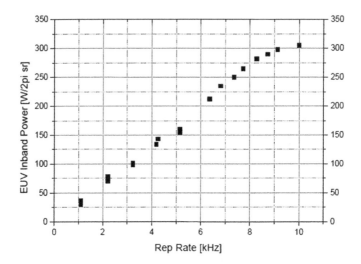

FIGURE 4.28 In-band EUV power (in 2 sr) of a Philips Extreme UV Sn DPP source as a function of the repetition rate.

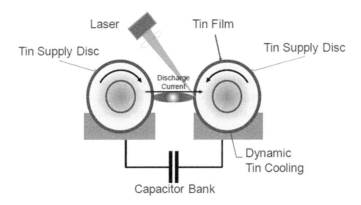

FIGURE 4.29 The laser-discharge plasma (LDP) source concept. Rotating disks coated with molten tin act as the electrodes for this source. A low-power laser is used to trigger the discharge.

Between 2005 and 2008, further consolidation of the EUV DPP source industry took place. Ushio acquired the Lambda Physik part of Xtreme Technologies and initiated a collaboration with Philips Extreme UV, and then assumed completed ownership of Xtreme Technologies. Ushio demonstrated operation of the Sn LDP source at power levels of up to 15 W, measured at IF (see Figure 4.30), and this source was integrated in one of the ASML NXE:3100 scanners [111]. In parallel, Cymer had been making progress with its laser-produced plasma source development and was reporting stable EUV power at IF of 10 W and announcing upgrades to 20 and 50 W [112] To this date, the Ushio LDP source has been the last discharge-based EUV source integrated into a commercial EUV lithography exposure tool.

4.3.3 LASER-PRODUCED PLASMA (LPP) SOURCES

LPP EUV sources are considerably more complex than DPP EUV sources and are very different in nature. Some of the notable differences of central importance are as follows. (i) The emitting plasma is smaller and created much further away from any surface compared with most DPP designs. This allows collection of EUV light over a larger solid angle while still meeting the étendue requirement. (ii) Although the overall wall-plug to EUV power efficiency may be lower in LPP, the conversion

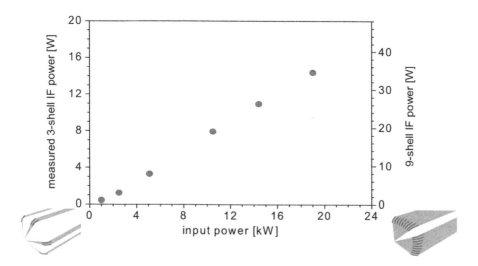

FIGURE 4.30 LDP EUV power measured at IF with a three-shell collector as a function of the input electrical pulse power. Power extrapolated for the full nine-shell collector is shown on the right y-axis.

efficiency from laser to EUV light is higher (up to ~5%). This means that less power is deposited in the source chamber, making thermal management less of an issue in LPP sources.

4.3.3.1 Basics

At the core of the LPP source is a target material that is irradiated with a high-irradiance laser pulse. During the early part of the pulse, as the electromagnetic field of the laser increases, multiphoton ionization of the target material first occurs. As the free charges are further accelerated by the laser field and transfer their energy through collisions, a dense and highly ionized plasma is created. For a large portion of the laser pulse, the plasma is efficiently heated through the inverse Bremsstrahlung mechanism (electron absorbing energy from the laser beam during a collision with an ion), as the plasma itself expands. This occurs in the corona region of the LPP, where the electron density is lower than the critical density (i.e., the electron density at which the plasma frequency, ω_p, is equal to the laser light frequency, ω).

$$\omega_p = \sqrt{\frac{n \cdot e^2}{\varepsilon_0 \cdot m_e}} \tag{4.6}$$

As it reaches this density, the laser light cannot propagate further, as the dielectric function $\varepsilon(\omega)=1 - (\omega_p/\omega)^2$ becomes negative, and is reflected by the critical surface. For CO_2 laser irradiance values of the order of 10^{10} to 10^{11} W/cm^2, Sn can be ionized multiple times (Sn^{+7} to Sn^{+10}) [88], and the electron temperature of the plasma can reach the desired range of kT$_e \geq 20$ eV, leading to efficient EUV emission around 13.5 nm.

4.3.3.2 Concept of Mass-Limited Target

Controlling source debris is also an extremely important aspect of the LPP source design. Early in the development of LPP sources for lithography, we realized that the total mass of target material used during the laser–plasma interaction would need to be limited to its minimum to avoid unnecessary generation of debris in the form of target fragments, clusters of neutrals and ions that don't contribute to the EUV emission.

Several designs aiming to limit the mass of the target exposed to the high-irradiance laser pulse have been conceived and tested through the years. From a solid surface (rotating cylindrical target to provide a fresh surface for every laser pulse) [113], to thin target tapes [114], spray jets [115], liquid filaments [116], and finally small droplets [117], the amount of fuel material that could be ablated by the laser pulse and end up as debris was reduced dramatically.

The main breakthrough came with the advent of droplet generation for the target material. Even now, considerable effort is going into the precise spatial and temporal control of smaller Sn droplet delivery to the laser–plasma interaction volume.

4.3.3.3 Main Components of LPP Sources

The LPP source comprises a few main components: a high-power CO_2 laser, a vacuum vessel to house the laser–plasma interaction that generates EUV light, a droplet generator (providing the target to be irradiated by the laser beam), and a normal-incidence collector. Figure 4.31 provides an illustration of the Cymer LPP EUV source vessel designed for the ASML NXE:3100 scanner [91]. A focusing lens concentrates the CO_2 laser light onto a Sn droplet that is delivered by the droplet generator (DG). The laser–plasma interaction takes place at the primary focus of the ellipsoidal collector mirror. The laser–droplet alignment and synchronization are carefully adjusted using the laser turning mirror and a DG steering mechanism, with multiple metrology modules and sensors providing feedback to maintain optimal performance. The ellipsoidal mirror collects EUV light from the plasma and redirects/images it onto the intermediate focus.

4.3.3.3.1 Fuel/Target System

Once the source is turned on, Sn droplets are generated at a repetition rate of 40 to 50 kHz without interruption (the same repetition rate as the laser). However, not all the droplets are used, and some are simply allowed to continue their path without being hit by the laser beam. A droplet catcher is located on the opposite side of the chamber from the DG to capture any unused Sn and avoid splatter inside the main chamber.

The droplets themselves are generated inside a reservoir where tin is heated above its melting temperature (231.9 °C) and then pushed through a filter and a fine nozzle by a pressurized gas, as depicted in Figure 4.32 [13] The pressure applied and the size of the nozzle determine the size and

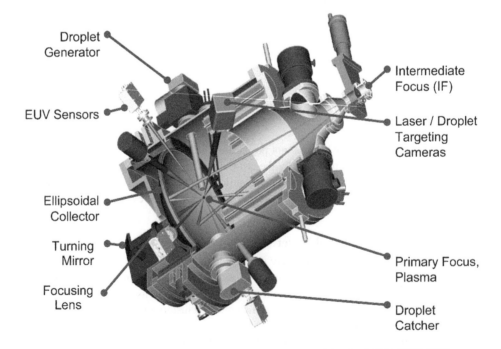

FIGURE 4.31   Schematic of the Cymer LPP EUV source vessel used for the ASML NXE:3100 scanners.

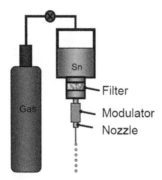

FIGURE 4.32 Schematic describing the droplet generator concept. Molten Sn kept in a heated reservoir is pushed through a filter and nozzle by gas pressure.

velocity of the droplets. A modulator on the nozzle of the DG is used to tune the delivery rate of the droplets. The typical size of a Sn droplet used in current EUV sources is of the order of 30 μm. This is significantly smaller than the laser beam diameter at its focal position (primary focus of the collector). For this reason, further "formatting" of the target needs to take place once the droplets have left the DG. This is achieved by the laser system.

4.3.3.3.2 High-Power Laser System

The laser is what provides the power to the EUV source. Both Cymer and Gigaphoton use CO_2 lasers as the main drivers of the EUV-emitting plasma. Considering that the best conversion efficiency values achieved from CO_2 light (10 μm) to EUV (13.5 nm) are of the order of 5%, and taking into account the overall transmission from plasma to IF, the laser power should be 25 kW for introduction of EUV sources in high-volume lithography manufacturing. Such high average pulsed CO_2 power is achieved using a master oscillator power amplifier (MOPA) architecture. The master oscillator is used to define the temporal and spectral characteristics of the pulse and also acts as the heartbeat for the source: all events are synchronized to this oscillator. A few power amplification stages, pumped using solid-state RF generators, are used to achieve the pulse energy levels required for the high target irradiance [118].

To achieve maximum conversion efficiency, the CO_2 light needs to be optimally coupled to the target. Since the droplets are typically smaller than the laser beam size at focus, an early laser pulse (a "pre-pulse") is used to first expand the target to a size commensurate with the main laser focal spot size before the main laser pulse interacts with the Sn fuel.

The exact value of the conversion efficiency achieved in the LPP EUV source depends on the intimate details of the dynamics of the laser power delivery to the target. This determines the hydrodynamics of the plasma and, in turn, how well the laser light can be absorbed by the plasma to create the right conditions for in-band emission at 13.5 nm.

The angular distribution of the EUV emission from the plasma is also a function of the exact characteristics of the laser–target interaction. The laser–target interaction and collector mirror configuration should be optimized to maximize the amount of clean EUV power delivered to IF.

4.3.3.3.3 Collector

One of the advantages of the LPP source is the isolated nature of the plasma, with open access to the EUV emission pattern over the full 4π sr. A normal incidence design with a solid angle of up to 5.5 sr provides approximately 2× more collection than the DPP collectors. A sample of such a collector is presented in Figure 4.33 [119]. The challenge is to achieve high reflectance over the full surface of the collector, accepting the fact that when the angle of incidence approaches 45°, only the s-polarized light is reflected. Of course, a multilayer reflector film is used to achieve high EUV

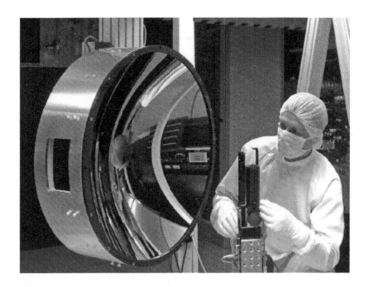

FIGURE 4.33 Normal-incidence (5.5 sr) EUV collector used in the ASML NXE:3100 source.

reflectance, but tradeoffs need be considered, as the maximum reflectivity of such multilayer films depends on the quality of the mirror surface in terms of nanoroughness [120]. The smoother the surface, the higher the EUV reflectivity, which comes at a price.

4.3.3.4 Source Lifetime

As for the DPP sources, achieving long collector lifetime is one of most (if not the most) difficult challenges for the introduction of EUV lithography into high-volume manufacturing. The source environment within the vacuum vessel needs to be taken into account: high-energy ions and neutrals from the plasma could strike the collector surface and cause irreversible damage. To protect the multilayer reflective film, hydrogen buffer gas within the vacuum chamber is introduced to slow down the high-energy ions, along with special collector capping layers. The attenuation factor as a function of Sn ion energy, for different hydrogen gas pressures, is plotted in Figure 4.34 [91].

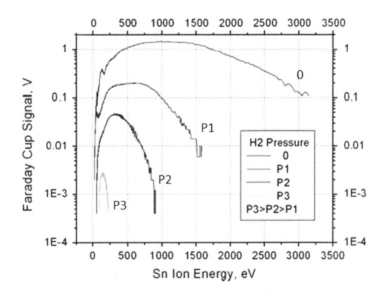

FIGURE 4.34 Sn ion mitigation. With increasing H_2 pressure, high-energy ions are stopped more effectively.

Considering also that kilograms of Sn are introduced every week inside the source vessel and that only 1 nm leads to >10% reflectivity drop, only a few parts per billion (ppb) of the tin accumulating on the surface of the collector could lead to serious reflectivity loss. The main enabler for overall collector lifetime is the design of a hydrogen gas flow within the source vacuum chamber. A strong flow of H_2 from the central opening in the collector mirror toward the plasma exerts a pressure on the Sn ions, stopping most of them from reaching the collector surface (Péclet protection). The hydrogen gas serves a second purpose: in the presence of high-energy photons, the H_2 molecules readily dissociate, creating hydrogen radicals that can scavenge any Sn that has reached the surface of the collector. The hydrogen atoms combine with Sn to form stannane, which is a gas that can be pumped out of the source vessel. The overall Sn-cleaning mechanism is based on the following two reactions:

$$H_2(g) \leftrightarrow 2H(g) \tag{4.7}$$

$$4H(g) + Sn(s) \leftrightarrow SnH_4(g) \tag{4.8}$$

Gigaphoton has developed a different debris mitigation technique based on a strong magnetic field (~1 T) designed to guide the Sn ions away from the collector surface. The company reports an efficiency of >98% reduction in the number of Sn ions reaching the collector [121].

Improvements in the efficacy of the debris mitigation techniques implemented in the ASML-Cymer source resulted in longer lifetime achieved in actual sources operating in the field, as can be seen, for example, in Figure 4.35, showing up to 120 billion pulses for the best-performing NXE:3100 collector [14] The relative reflectivity of one of the collectors used for the better part of a year in one of the NXE:3100 sources, operating at a power level of 10 W, stayed above 80% for more than 10^{11} pulses.

A modification of the ASML-Cymer source vessel configuration with a more vertical orientation, to accommodate a new illuminator design in the NXE:3300, introduced new challenges in terms of tin management. In parallel, the source power increased and the collector protection was adapted to improve lifetime, as shown in Figure 4.36 [122]. At 125 W, a factor of more than 10 compared with the NXE:3100 source power, a lifetime within a factor of 3 of that of the best NXE:3100 collector was achieved.

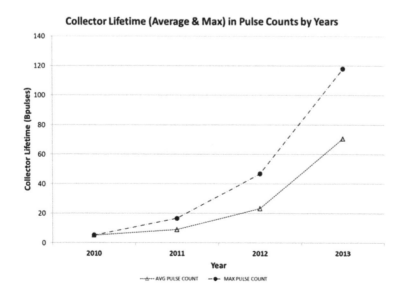

FIGURE 4.35 Improvements of NXE:3100 collector lifetime in terms of total number of pulses.

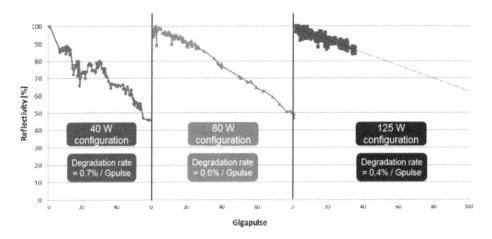

FIGURE 4.36 Improvements in collector lifetime as the NXE:3300 source power is increased.

To further improve the lifetime of its EUV collectors, ASML-Cymer has developed a separate in situ collector cleaning technique. It is based on the delivery of hydrogen radicals to the surface of the collector when the source is not in operation. A demonstration of the technique is given as an example in Figures 4.37 and 4.38 [14]. A picture of a used NXE:3100 collector covered with a significant amount of Sn is shown in the first figure. After exposure to the hydrogen radicals, the collector has been cleaned of the Sn that was covering its surface (second figure). The combination of continued improvements in Sn management inside the EUV source vessel and in situ cleaning provides a path toward the lifetime requirement of 33,000 hours mentioned earlier.

4.3.3.5 Power Scaling

We have seen that the collector lifetime has been improving and is capable of keeping up with the higher power levels required for HVM. In the end, the EUV power available for printing wafers is what matters. As mentioned earlier, 250 W of clean EUV power at the IF is the current requirement for EUV lithography to be used in production. The development of higher-power CO_2 laser drivers, together with higher conversion efficiency factors achieved through improved laser–plasma energy coupling, has resulted in an order-of-magnitude increase in EUV power at IF over the past few years. The actual EUV power achieved for different CO_2 laser architectures is plotted in Figure 4.39 [13]. Before 2012, no master oscillator (NOMO) was used, and the system relied on the

FIGURE 4.37 Sample NXE:3100 collector after use, showing Sn accumulation (light gray).

FIGURE 4.38 Same NXE:3100 collector as shown in Figure 4.37 after hydrogen radical cleaning.

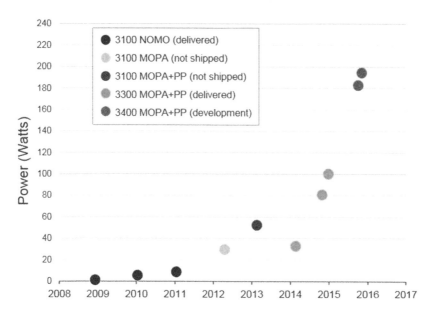

FIGURE 4.39 Progress in LPP EUV source power at IF since 2009. The introduction of the CO2 MOPA (master oscillator power amplifier) laser architecture with a pre-pulse (MOPA+PP) has led to major increases in usable EUV power at IF: >100 W at customer sites and 200 W on engineering sources.

presence of the Sn droplet in the optical path of the laser system to trigger the laser amplification. Such systems appeared limited to a few tens of watts at IF, with 10 W actually implemented in the field [118].

The MOPA and pre-pulse (MOPA+PP) architecture was first introduced on a NXE:3100 test system and then transferred to the NXE:3300 source, with over 100 W achieved in the field. EUV power of approximately 200 W has now been achieved with the new NXE:3400 MOPA+PP source architecture on an engineering system. Stable operation of such a source at 197 W for 1 hour is shown in Figure 4.40 [122].

Finally, dose control should be maintained as the EUV power is increased. A test-bench metric for the stability and control of the EUV dose is the fraction of simulated dies that meet a specific dose error. This is plotted in Figure 4.41, where 99.3% of the dies have a maximum dose error of <1% [122].

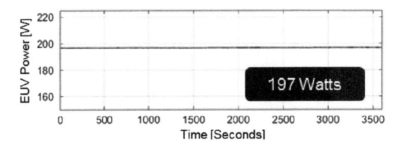

FIGURE 4.40 Stable operation of the ASML-Cymer LPP source at 197 W for 1 hour.

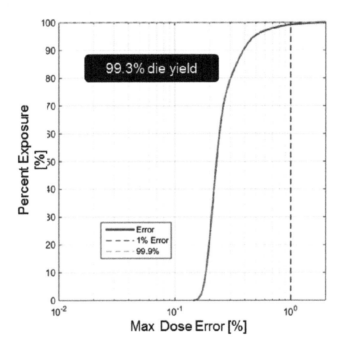

FIGURE 4.41 Fraction of simulated die exposures that meet a maximum dose error on a test-bench system. The results show that 99.3% of the dies have dose errors lower than 1%.

4.3.4 CONCLUSION

After many years of development, where multiple source designs have been considered, investigated, and tested on actual lithography exposure tools, the EUV source technology has converged to the LPP concept. At the 2016 SPIE Advanced Lithography Conference, several manufacturers of integrated electronic devices recognized significant progress of the source performance in terms of power and availability. It is now becoming clearer that insertion into production will occur at the 7 nm or 5 nm node.

4.4 EUV MASKS

4.4.1 EUV MASK TECHNOLOGY

EUV and 193 nm masks look quite similar at first glance; both employ a patterned metal film on glass substrates, and both are exposed in step-and-scan exposure tools that print 4× demagnified images of the mask on resist-coated silicon wafers. A schematic layout of an EUV mask substrate is shown in Figure 4.42. Because of the short EUV operating wavelength (13.5 nm), an EUV exposure tool must be operated in vacuum, an EUV imaging system must utilize reflective optics instead of transmitting optics (the EUV mask must also be used in reflection), and until recently, it was not thought possible to protect an EUV mask from particles added during the exposure process, for example with a thin pellicle film. The operation of a mask in an EUV exposure tool leads to contamination and defectivity problems because of its vacuum environment. Unless the hydrocarbon levels in the EUV exposure tool vacuum chamber are strictly controlled, EUV photons will crack the hydrocarbon molecules and produce a buildup of carbon on the mask surface (a 1 nm thickness of carbon will lead to ~1% loss in EUV reflectivity) [123]. The shipping, handling, and use of EUV masks are also highly problematic, because particles on the surface of a pellicle-less EUV mask will be exactly in focus and, if sufficiently large, will result in a printable defect on each exposed wafer. Because printable defects cannot be tolerated in semiconductor manufacturing, EUV masks must be frequently cleaned to remove carbon contamination and particle adders. The need for frequent particle removal can significantly shorten the lifetime of an EUV mask due to the changes in feature CD, EUV reflectivity, and rms surface roughness caused by the cleaning chemistry.

The necessity of operating an EUV mask in reflection also leads to several problems. An EUV mask must be illuminated at an angle to its normal, typically ~6°, to separate the incident and reflected light cones. If an EUV mask substrate is not perfectly flat, any out-of-plane mask flatness will lead to image placement errors at the wafer (1 nm of out-of-plane flatness of the mask can produce ~0.025 nm of image displacement at the wafer) [124]. The problems caused by EUV mask nonflatness were recognized early on in EUV lithography development, and the mask clamps in EUV exposure tools were arranged to be exceedingly flat and to hold EUV masks tightly to the clamp surface using electrostatic forces [125, 126]. The operation of an EUV mask in reflection means that the mask surface becomes part of the reflective imaging system. The EUV mask surface must be exceedingly smooth to minimize loss in EUV reflectivity due to scattering, and it must be coated with a complex and imperfectly reflective multilayer (ML) coating. The EUV reflective coating must be free of printable defects, and there is a risk that EUV radiation not reflected or scattered away by the EUV coating can lead to substrate heating; even if the mask has been fabricated on a low–thermal expansion material (LTEM), there is still some concern that mask heating could lead to thermally induced image placement errors.

A schematic cross section of a patterned EUV mask is shown in Figure 4.43. The 6025 form-factor substrate for an EUV mask is required to be comprised of a low thermal expansion material and its surface must be exceedingly smooth and free of pits, bumps, and scratches, because surface

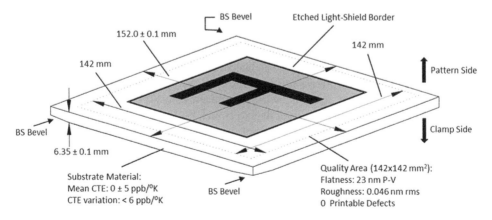

FIGURE 4.42 Schematic layout of an EUV mask substrate (pattern side up) showing the location of the three corner-bevels on the backside of the mask, the dimensions and a few key specs of the frontside quality area, and the overall substrate dimensions as specified in the SEMI standard P37-0613 [206].

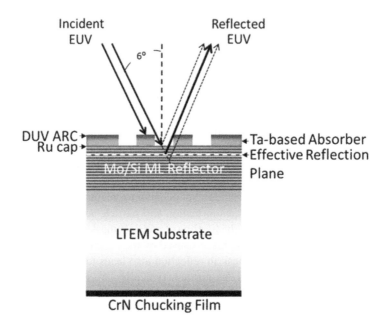

FIGURE 4.43 Schematic cross section of an EUV mask showing its LTEM substrate and thin conducting backside film (for electrostatic chucking), Ru-capped Mo/Si ML reflector with 40–50 bilayers, and patterned Ta-based absorber with DUV ARC capping layer.

imperfections on the surface of the substrate when coated with the reflective ML can results in printable phase defects. The substrate is provided with a thin conducting CrN coating on its backside to facilitate its electrostatic clamping to the mask stage in a EUV exposure tool. The substrate is transformed into a EUV mask blank by depositing a Mo/Si ML reflective coating with 40–50 bilayers of Mo and Si overcoated with a thin capping layer of Ru to protect the ML reflector from environmental degradation, a Ta-based absorbing layer, and a DUV antireflection coating (ARC). The Ta-based absorber/DUV ARC is patterned by e-beam writing and etch processing similar to that used to pattern current optical and DUV masks. In most cases, the Ru capping layer is used as an etch stop to protect the reflective ML during absorber etching and subsequent absorber defect repair. After etching, the Ta-based absorber performs the same function as the chrome layer on a DUV mask; that is, it provides high image contrast between areas coated with the Ta-based absorber and areas of bare Ru-capped Mo/Si ML. Because the Ta-based absorber has high DUV reflectivity it can reduce sensitivity in commonly used defect inspection tools. A DUV ARC is applied to the top of the absorber layer to improve inspection tool contrast. Because electron backscattering from the high–atomic number coatings on an EUV mask blank is larger than from an optical or DUV mask, an extra set of proximity effect corrections (using MPC software) must be applied to the patterns to ensure that they will be properly sized following the e-beam write [127].

A list of the key requirements for each of the layers on a EUV reticle, with up-to-date values for the parameter specs from the SEMI P37-0613 Specifications for Extreme Ultraviolet Lithography Substrates and Blanks [206], the ITRS Table LITH5c EUVL Mask Requirements for the 2016 calendar year [128], Naulleau, P., et al. (2011) [129], and JohnKadaksham, A., et al. (2013) [130] is shown in Table 4.1.

In the EUV spectral region, the index of refraction of all materials is close to unity and their transmission and reflection is almost zero. Fortunately, ML films composed of two materials with widely different values of EUV absorption can lead to enhanced reflection in a narrow wavelength band peaked at the Bragg wavelength, $\lambda = 2d\sin(\theta)$, where d is the ML period and θ is the angle of incidence. The combination of molybdenum (Mo) and silicon (Si) has an exceptionally high

TABLE 4.1

Compilation of the Key Requirements of Each of the Layers on an EUV Mask, with the Latest Values for the Parameter Specifications.

Layer	Material	Parameter	Specification
Backside	CrN, TaB	Sheet resistance	$<100\ \Omega/cm^2$
		BS Roughness $\lambda_{spatial} < 10\ \mu m$	<0.50 nm rms
Substrate	LTEM 6025	Mean CTE	0 ± 5 ppb/°C
		CTE variation	<6 ppb/°C
		FS flatness (PV)	<23 nm
		BS flatness (PV)	<23 nm
		FS roughness	0.046 nm rms
		FS local slope	$0.8\ \mu rad$
		Defect density @ 30 nm SiO_2 equivalent	≤ 0.003 defects/cm^2
Multilayer mirror	Mo/Si 40–50 bilayers	ML film stress	± 200 MPa
		Peak reflectance	$>65\%$
		Reflectance uniformity (PV)	$\leq 0.23\%$
		Mean center λ	13.53 nm
		Median center λ offset	± 0.06 nm
		Reflectivity FWHM	>0.50 nm
		Centroid λ uniformity	≤ 0.04 nm
		ML roughness	≤ 0.15 nm rms
		Defect density @ 50 nm SiO_2 equivalent	≤ 0.008 defects/cm^2
Capping layer	Si, Ru, TiO_2, etc.	Thickness loss over >100 cleaning cycles	<1 nm
Absorber layer	TaN, TaBN, etc.	Average reflectivity 13.465–13.535 nm	$\leq 0.50\%$
		Absolute contrast	$>97\%$
Antireflective coating	TaON, TaO, TaBO, etc.	DUV inspection contrast	$>60\%$

Source: values from the International Technology Roadmap for Semiconductors, Table LITH5c EUVL Mask Requirements, Semiconductor Industry Association, Washington, DC, 2016; Naulleau, P.P. et al., *Proc. SPIE*, 8166, 81660F-1, 2001; JohnKadaksham, A., et al. [130].

normal-incidence reflectivity at wavelengths near 13.5 nm and is employed on most of today's commercial EULV mask blanks.

A comparison of simulated and measured EUV reflectivity of a Mo/Si ML reflector with 40 bilayers of Mo and Si at 6° angle of incidence as a function of EUV wavelength is shown in Figure 4.44 (on the left). The simulations used to generate the data for Figure 4.44 were carried out using values for the optical constants of Mo and Si from the CXRO X-ray Database [131], and the EUV reflectivity measurements plotted in Figure 4.44 were made using the CXRO reflectometer on Advanced Light Source Beamline 6.3.2. Figure 4.44 shows that some additional improvement in peak EUV reflectivity and in the reflective bandwidth of Mo/Si ML coatings may someday be possible; that is, from ~65% today to ~75% in the ideal case. The primary cause for the lower measured reflectivity is the broadening of the Mo–Si interfaces due to interdiffusion, silicide formation, and increased interface roughness. Interface engineered Mo/Si MLs with intermixing barrier layers at the Mo–Si interfaces can produce coatings with peak reflectivity exceeding 70% [132].

A comparison of simulated and measured EUV reflectivity of a Mo/Si ML reflector with 40 bilayers of Mo and Si at 13.5 nm wavelength as a function of incidence angle is also shown in Figure 4.44 (on the right). The dot-dashed and dashed vertical lines in Figure 4.44 indicate the minimum and maximum angles that are utilized in a 0.33 numerical aperture (NA) EUV scanner when imaging at its diffraction limit. The figure shows that the peak reflectivity at larger angles of

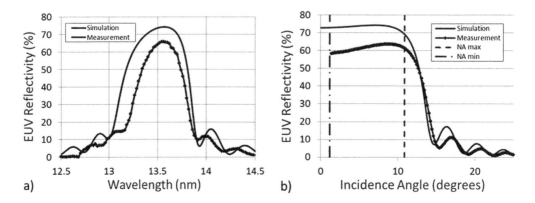

a)   Wavelength (nm)   b)   Incidence Angle (degrees)

FIGURE 4.44 Simulated and measured EUV reflectivity of a Mo/Si multilayer reflector with 40 bilayers versus wavelength at 6° angle of incidence (left) and versus incident angle at 13.5 nm wavelength (right). The dashed vertical lines show the minimum and maximum angles that will be utilized by a 0.33 NA EUV exposure tool when imaging at its diffraction limit.

incidence is beginning to fall off, producing reflective apodization, an effect that will become more pronounced in higher-NA EUV exposure tools in the future.

The calculated EUV mask reflectivity at 13.5 nm wavelength and 6° angle of incidence of a Ru-capped Mo/Si ML coating with 40 bilayers overcoated with a TaN absorber as a function of absorber thickness is shown in Figure 4.45. This figure shows that to meet the <0.5% average reflectivity loss spec shown in Table 4.1, a Ta-based absorber must be 60 to 80 nm thick, or light leakage through the absorber will result in lower image contrast. Since the height of the Ta absorber film leads to mask shadowing and results in a measurable horizontal/vertical print difference at the wafer, the shadowing error must be corrected by mask biasing. This and other 3-D mask effects are discussed in more detail in Section 4.4.6.

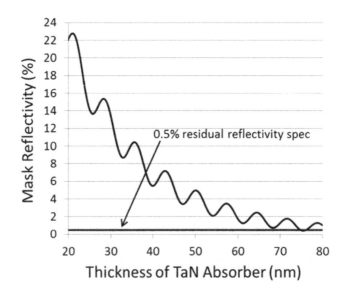

Thickness of TaN Absorber (nm)

FIGURE 4.45 Mask reflectivity at 13.5 nm wavelength and 6° incidence angle of a Ru-capped Mo/Si ML coating with 40 bilayers overcoated with a TaN absorber as a function of absorber thickness showing that Ta-based absorber films must be greater than 60 nm thick to meet the <0.5% average reflectivity specification for an EUV mask.

4.4.2 Mask Defects, Mitigation, and Repair

One of the biggest challenges to EUV lithography insertion continues to be the low yield of defect-free mask blanks. Imperfections on the surface of the blank substrate before ML deposition or defects added inside and on top of the ML during the deposition process can result in printable defects, depending on exactly where in the ML stack they are located and on their size. A printable defect caused by an imperfection in the surface of a blank substrate was found in EUV printing experiments as early as 1992, where a long line-feature was discovered to have been printed on the wafer with a mask that had no obvious corresponding absorber feature [133]. Subsequent inspection of the mask revealed a 2.7 nm deep, 8 μm wide scratch on the surface of its substrate. Since the scratch was close to the $\lambda/4$ depth needed for perfect cancellation of the reflected waves, this phase defect resulted in an unexpected high-contrast line in the resist image. The fact that this feature was not noticed prior to resist printing increased the urgency for at-wavelength inspection of EUV mask blanks. The first systematic study of EUV mask substrate defects was carried out via resist printing by Nguyen et al. [134, 135] in 1993 using an early two-mirror EUV imaging system at the Photon Factory in Tsukuba, Japan.

The defect density spec on EUV mask blanks needed for HVM is 0.003 d/cm^2 at 25 nm SiO$_2$-equivalent size (see Table 4.1). EUV mask blanks meeting the reflectivity and uniformity specs listed in Table 4.1 are available today, but the deposition equipment and processes used to produce Mo/Si ML-coated EUV mask blanks still have a disappointingly low yield. The low yield of defect-free mask blanks has been recognized by the industry as a critical issue for EUV lithography for more than a decade. As early as 2003, in an attempt to accelerate the development of a zero–defect adder Mo/Si deposition process, SEMATECH established the Mask Blank Development Center (MBDC) in Albany, New York [136]. During the MBDC's startup, SEMATECH, in partnership with Veeco Instruments, successfully transferred technology developed at Lawrence Livermore National Laboratory for depositing low-defect ML coatings on EUV blank substrates [137]. The objectives of the MBDC were to 1) accelerate equipment and process development for low-defectivity EUV mask blank manufacturing, 2) establish a user facility for metrology and evaluation of EUV mask blanks, and 3) encourage the development of a commercial supply infrastructure for EUV mask blanks. The user facility in Albany provided equipment for ion beam sputter deposition (IBD) of Mo/Si ML coatings, advanced mask cleaning R&D, and in-line mask blank metrology. The MBDC's metrology tools were used by SEMATECH to create Pareto charts of defect shape, size, and elemental composition, which, in turn, were used to determine the root cause of the defect generation and to identify all potential sources of contamination.

A record of the MBDC's progress in reducing defectivity on EUV mask blanks is shown in Figure 4.46. The champion plates from the MBDC's final deposition run in 2014 yielded two blanks with zero defects at 54 nm inspection sensitivity (SiO$_2$ equivalent) in a 135 mm × 135 quality area. And perhaps more importantly, single-digit defect numbers were achieved on over 50% of the mask blanks produced [11]. While exact numbers for the yield of low-defect EUV mask blanks at commercial blank suppliers are not available, it is widely believed that current yield numbers at commercial blank suppliers will not be able to meet industry needs once EUV lithography technology is introduced into HVM.

Given the time and effort that have been spent on developing the processes needed to manufacture totally defect-free EUV mask blanks, it is safe to say that totally defect-free EUV mask blanks will remain in short supply for the foreseeable future. While the repair of defects that occur during the absorber patterning process is relatively straightforward, repair of defects in the reflective ML-coating is highly problematic. Because of this, a variety of techniques have been developed to create an EUV mask with zero printable defects starting with an EUV mask blank that has no large defects and only a handful of small defects. The most important of these techniques are blank sorting, defect avoidance (pattern shift), defect compensation, and absorber defect repair followed by defect verification by wafer printing or with an actinic imaging microscope.

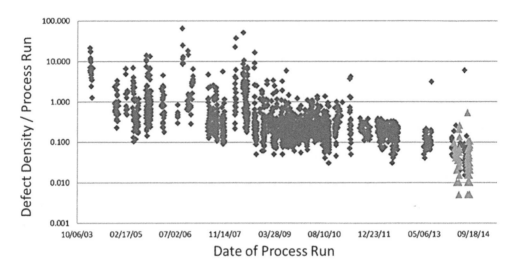

FIGURE 4.46 Record of EUV mask blank defect reduction progress at the SEMATECH Mask Blank Development Center in Albany, New York. (Reprinted from Antohe, A.O. et al., Proc. SPIE, 9422, 94221B-1, 2015. With permission.)

The technique of blank disposition or sorting assumes the existence of an inventory of well-characterized, high-quality mask blanks; that is, with precise substrate flatness data and with detailed maps of blank defect types, sizes, and accurate locations with respect to a set of fiducial marks. Then, depending on the type of circuit level to be printed, for example a contact or via level with minimal bright area or a first-metal level with a larger percentage of bright area, each blank in inventory can be compared with the requirements of the level to be printed, and the number of remaining mask defects that would need to be avoided, compensated, or repaired can be used to determine which blank should be assigned to each level. The technique of defect avoidance, illustrated in Figure 4.47, assumes that there exist one or more combinations of pattern shifts and rotations that will result in all known blank defects being completely covered with an absorber pattern [138] and hence, not printable. The technique of defect compensation, illustrated in Figure 4.48, can be used to minimize the effect of a blank defect by modification (local removal) of nearby absorber patterns [12].

If accurate information on the defect type, size, and especially location is known, in some cases, it is possible to modify nearby absorber patterns in such a way that the combination of the blank defect and the modified absorber pattern will minimize any through-focus printing effect of the blank defect. Finally, just as is the case with absorber defects on DUV and 193 nm masks, defects

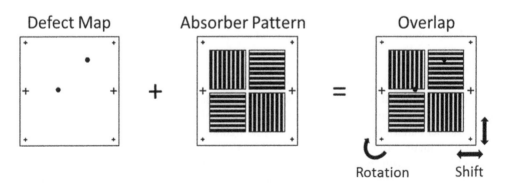

FIGURE 4.47 Sketch illustrating a defect avoidance technique that makes use of pattern shift and rotation to cover (hide) defects with absorber features that are sufficiently wide.

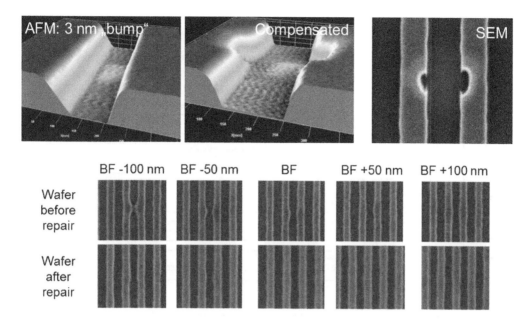

FIGURE 4.48 Example of a somewhat rare instance in which the effect of a blank defect can be compensated by local removal of nearby absorber patterns [207]. Top row: AFM image of a 3 nm bump defect on an EUV blank and SEM image of the mask after compensation. Middle and bottom rows: Series of SEM images of a resist-coated wafer (before and after defect compensation) showing the significantly reduced through-focus CD errors created by the compensated blank defect.

in the absorber pattern in EUV masks can be repaired using an in situ AFM to remove excess material in the case of an extrusion defect or an e-beam-assisted deposition tool to fill in missing areas of absorber in the case of an intrusion defect [139]. As is the case with DUV and 193 nm masks, after all defects have been mitigated by pattern shift, compensation, or repair, the repairs should be validated by wafer printing or with an actinic imaging microscope.

4.4.3 Inspection Tools and Techniques

From the start, the rate of progress in blank defect reduction was limited by the performance of the defect inspection tools that were available at that time. A chart showing the calendar year when different inspection techniques were first used to look for defects on blanks and patterned masks and to capture high-resolution aerial images of individual defects for defect repair and verification is shown in Figure 4.49. As mentioned in Section 4.4.2, the first observation of EUV blank substrate defects took place in EUV resist printing experiments using commercial wafer defect inspection tools. Defect detection via wafer printing is still widely used today, particularly when benchmarking the performance of a new EUV defect inspection technique. While the EUV resist printing/automated wafer inspection continues to be extremely important for inspection tool development and has actually improved in performance over time as the resolution of EUV exposure tools and the sensitivity of DUV and 193-nm wafer inspection tools have improved, the technique is still widely believed to be too costly and too time consuming to use in high volume manufacturing.

Very early in EUV blank development, it became clear that defects within the multilayer coating were important, but even more important were defects nucleated by particles or pits on the mask substrate. The process by which substrate surface defects become printable blank defects was first made clear in work by Gullikson et al. [140] at US National Laboratories. In that work, the authors predicted that pits and scratches in the mask substrate as small as 60 nm wide FWHM and with depths as small as 1.5 nm could result in a printable defect. Defects in the substrate surface were

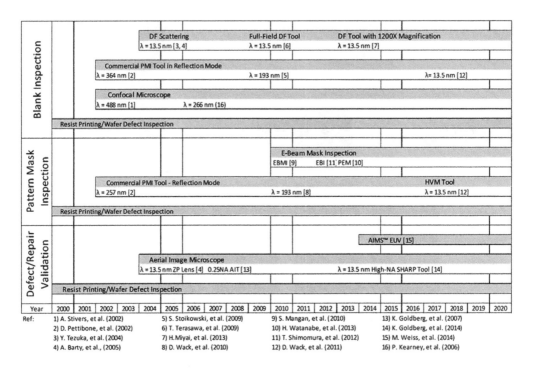

FIGURE 4.49 Chart showing the calendar year when different inspection techniques were first used to look for defects on EUV mask blanks and pattern masks and to capture high-resolution aerial images of individual defects for use in repair verification.

first studied using an inspection tool that utilized confocal microscopy at 488 nm wavelength [141]. The first commercial confocal microscope–based substrate inspection tool, a Lasertec M1350, was installed at the SEMATECH MBDC in 2003. After all upgrades were completed, this tool could capture 54 nm polystyrene latex (PSL) spheres on bare quartz substrates and 70 nm PSL spheres on Mo/Si ML-coated substrates [142]. A second-generation confocal microscope–based substrate inspection tool was installed in the SEMATECH MBDC in 2006. This Lasertec M7360 could capture 34 nm PSL on quartz and 41 nm PSL on ML blanks [143]. These substrate inspection tools were used by SEMATECH to evaluate blank substrates fabricated with different polishing techniques, to develop increasingly effective substrate cleaning/particle removal techniques, and to study different ML deposition conditions that could be used to either smooth or decorate substrate pits and bumps. These confocal microscope–based tools were particularly useful because they were equipped with diamond indenters that would allow "punch mark" fiducials to be created around individual defects so that they could be cross sectioned with a focused ion beam/scanning electron microscopy-energy-dispersive X-ray (FIB/SEM-EDX) tool and categorized in terms of shape, size, and elemental composition.

At about the same time, defects in substrates and ML-coated blanks were studied using a KLA-Tencor UVHR mask inspection system operating at a wavelength of 364 nm with a minimum pixel size of 150 nm by a partnership between KLA-Tencor, the EUV-LLC, Photronics, and Dupont Photomasks in 2002 [144]. Initially, the commercial mask inspection tools were not as sensitive as the inspection tools based on confocal microscopy, but by 2009, the sensitivity of a KLA-Tencor Teron 6xx mask inspection system operating in blank reflection mode at 193 nm wavelength was shown to be comparable to the sensitivity of the Lasertec M7360 [145]. Commercial EUV mask blanks available today can be supplied with a set of substrate fiducial marks and a high-resolution map of all substrate defects captured with a KLA-Tencor Teron 6xx inspection system operating in Phasur mode. In 2011, KLA-Tencor disclosed plans for the development of a commercial blank inspection system operating at 13.5 nm wavelength that would be available in the 2017–2020 timeframe [146].

Actinic inspection tools for ML-coated mask blanks based on dark-field scattering of EUV radiation at 13.5 nm wavelength were reported by MIRAI/AIST in Japan in 2004 [147] and by SEMATECH/LBNL/LLNL in the United States in 2005 [148]. In the tool developed in Japan, DF scattering of EUV radiation produced by a YAG LPP source was collected with a 26X Schwarzschild optic. In the dual-mode actinic tool developed in the United States, DF scattering of EUV radiation produced at the ALS synchrotron in Berkeley was collected with a channel plate, and BF scattering was imaged with a zone plate lens to capture an accurate image of individual defects. The MIRAI tool was further upgraded by Selete into a full-field inspection tool with a commercial EUV source in 2009 [149] and is currently being made into a commercial blank inspection tool by Lasertec/ EIDEC that when operated with 1200× high-magnification optics, can provide the location of individual defects with the accuracy needed for successful use of the defect inspection techniques described in Section 4.4.3 [150].

A record of the application of various inspection techniques to the inspection of patterned masks is also illustrated in the chart shown in Figure 4.49. The partnership of KLA-Tencor, the EUV-LLC, Photronics, and Dupont Photomasks demonstrated inspection of an EUV patterned mask with a KLA-Tencor DULV mask inspection system operating in reflection mode at a wavelength of 257 nm with a minimum pixel size of 125 nm in 2002 and reported the detection of defects as small as 80 nm in a patterned EUV mask using die-to-die inspection [144]. The capabilities of commercial patterned mask inspection tools operating at 193 nm wavelength with patterned masks were reported in 2010 [151]. In 2011, KLA-Tencor described plans for a commercial patterned mask inspection system operating at 13.5 nm that would be available in the 2017–2020 timeframe [146]. The use of an electron beam to inspect patterned EUV mask blanks has been reported by a number of groups. An electron beam mask inspection (EBMI) tool developed by Applied Materials has been used to inspect small areas of patterned masks in 10 hours at 10 nm pixel size [152]. A PEM tool that can detect 20 nm PSLs in 2 hours/100 mm square area has been demonstrated [153], and one that should be able to inspect patterned masks at 32 nm pixel size is being developed by EIDEC/ EBARA [154]. An e-beam tool developed by Hermes Microvision, an EBI eXplore 5400, has been used to inspect a 1250 µm × 1250 µm area at 16 nm pixel size in 6 minutes and 45 seconds and should be able to inspect a 100 cm^2 area in 747 hours at 15 nm pixel size [155]. All these e-beam inspection tools have been shown to be extremely sensitive and appear to be able to locate even the smallest defects in an EUV patterned mask, but they are all too slow for use in HVM—requiring many days to inspect a full-field EUV mask.

A record of the application of various inspection techniques to capture aerial images of individual defects is also illustrated in the chart shown in Figure 4.49. The SEMATECH/Berkeley actinic inspection tool (AIT), an EUV Fresnel zone plate microscope illuminated with a synchrotron EUV source, which was used to capture through-focus images of defects on EUV masks but was decommissioned at the end of 2012 [156], grew out of the actinic inspection tool developed by Barty et al. [148]. Its successor, the SEMATECH high NA actinic mask review project (SHARP) tool, which began operation in June 2013, can capture actinic images with zone plate lenses with numerical apertures from 0.25 to 0.625, has a fully programmable illuminator that can provide almost any desired pupil fill, and can capture through-focus images at a rate of ~8 defects per hour. The SHARP tool has been used to capture high-resolution images of individual defects in a EUV mask both pre and post repair and can capture images approximately five times faster than the AIT tool and with more than three times higher signal-to-noise ratio [157]. Carl Zeiss and the SEMATECH EUVL Mask Infrastructure Consortium are currently developing an EUV aerial image metrology system (AIMS™ EUV) [158]. The AIMS™ EUV will navigate to defect sites identified by blank inspection or patterned mask inspection tools and can capture aerial images under the same imaging conditions as will be used in a commercial EULV scanner. Aerial images can be captured for various focus levels. Based on their aerial images, it can be decided whether or not a potential defect needs to be repaired, and after repair, the success of the repair can be verified. The tool specs are as follows: Target Node: 16 nm HP, Scanner Emulation: up to 0.33 NA, CD Repro: $\leq$1.5 nm (3σ, mask level), Run Rate (seven focus planes per site): >27.5/hour @ 38.5% pupil fill or $\geq$51/hour @ 77% pupil fill. The tool has been fabricated with a reflective imaging system using scanner-quality optics, utilizes a commercial EUV DPP source, supports dual pod or standard mechanical interface (SMIF) pod mask loading, and has been designed to minimize add-on particles. The prototype AIMS™ EUV tool has recently achieved first light, and the first tool is expected to be available for customer use early in 2016 [159].

4.4.4 IN-FAB MASK HANDLING

The shipping, handling, and use of an EUV mask are more difficult than the same activities with a DUV or 193 nm mask because of the higher risk that a particle will be added to the frontside, due to the lack of a pellicle, or to the backside, because a particle between the backside of an EUV mask and the surface of the mask clamp in the scanner can produce enough out-of-plane distortion to result in an overlay error. Because of these risks, EUV masks now have their own custom mask carriers called *dual pods*. An exploded view of an EUV dual pod carrier is shown in Figure 4.50 [160]. An EUV mask is loaded pattern side down on the baseplate of the small metallic inner pod (metallic so that it will result in minimum outgassing). The baseplate of the inner pod is designed so that a small gap (~2.5 mm) between the surface of the inner pod baseplate and the patterned side of the mask is maintained. The trapped volume created by this gap is purposely kept small to limit any convection currents that might cause a particle to be carried to the mask surface. The inner pod cover is provided with filters that restrict the flow of air in and out of the pod during pump-down and vent as the mask is moved into and out of an exposure tool vacuum load lock. Once the inner pod cover is in place, the inner pod can be used to carry a mask through any subsequent steps (shipping, transport into and out of the exposure tool vacuum, etc.) without any need to touch the mask itself unless it is ready for EUV exposure. Because of this, a dual pod is often referred to as a *removable pellicle*. The outer pod of the EUV dual pod carrier is an RSP200 SMIF-compliant carrier that has been modified to hold an EUV inner pod [161].

When a clean, particle-free EUV mask, loaded in an EUV dual pod, is brought into a semi-conductor fab, it will require more complicated and costly handling than a normal (DUV or 193 nm) mask. A typical flow for in-fab mask handling is illustrated in Figure 4.51. Because of the fall-on particle risk, before being loaded into an EUV exposure tool, an EUV mask must be cleaned on both backside and frontside (backside first to avoid contaminating in-fab mask inspection tools) and then inspected on backside and frontside before being transferred to vacuum. Once the inner pod has been brought into vacuum, a robot handler is used to move the

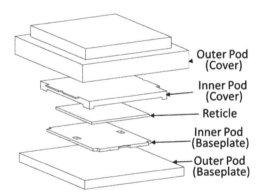

FIGURE 4.50   Schematic of EUV dual pod mask carrier with inner and outer pods exploded.

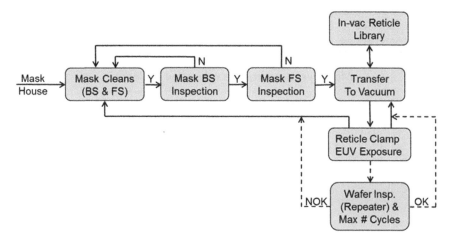

FIGURE 4.51 In-use mask handing flow chart required to maintain an EUV mask free of fall-on particles in the absence of a pellicle.

inner pod into an exposure tool in-vacuum mask library, or if the mask is needed immediately it can be removed from the inner pod, again by a robot to avoid adding particles, and loaded onto the exposure tool mask stage for EUV exposure. When an EUV mask has been clamped to the mask stage in an EUV exposure tool, its frontside is now at risk from fall-on particles and carbon contamination during the exposure process. Great care has been taken by the exposure tool manufacturer to minimize those two contamination risks [125, 126], but the consequences of a frontside particle adder on a critical feature are so great (a drop-off in yield starting the moment the particle arrives) that additional steps must be added to the in-fab mask handing flow to manage this eventuality. Monitor wafers must be printed at regular intervals, and those wafers must be inspected for defect repeaters (which would indicate the presence of a fall-on particle). Upon the appearance of a repeater on a monitor wafer or when the number of mask load, clamp, expose, unclamp, unload cycles exceeds the number of cycles known from experience to result in a significant buildup of carbon, a repeat of backside and frontside cleans and backside and frontside inspection will be needed to ensure that the particle adder is no longer present before the mask can be returned to the exposure tool.

The particle adder rate in ASML's current NXE EUV exposure tools is reported to be <0.002 particles per mask pass @ 92 nm [162]. Fortunately, this number has been found to gradually improve with tool use. Nevertheless, current EUV exposure tools tend to add approximately one particle every 24 hours or so [163]. The in-fab mask flow illustrated in Figure 4.51 indicates that extra tools for mask handling, backside and frontside cleaning, and backside and frontside mask inspection will need to be present in a fab with an EUV exposure tool.

A schematic of an EUV mask cleaning process is shown in Figure 4.52. Because EUV masks will need to be repeatedly cleaned to remove fall-on particles, the cleaning process must be able to remove hydrocarbon contamination and have high removal efficiency for fall-on particles without significantly shortening the life of the mask. A plot of multilayer properties as a function of the number of cleaning cycles for a particle removal cleaning process at the IBM Mask House in 2011 is shown in Figure 4.52 [138].

The plotted data shows that the mask surface has not been significantly roughened after 100 cleaning cycles. Recent journal articles describing the observation of damage to (peeling of) the Ru capping layer on Ru-capped Mo/Si-coated mask blanks [164] that had been repeatedly cleaned suggest that to ensure a long mask lifetime, the cleaning process must strike a balance between particle removal efficiency and avoiding changes to feature CD, mask CDU, mask EUV reflectivity, and mask rms surface roughness.

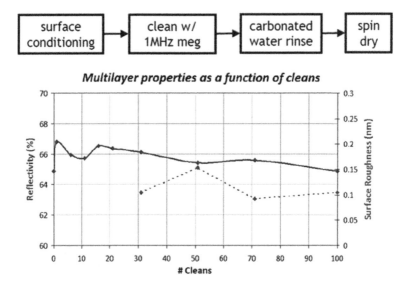

FIGURE 4.52 Schematic of typical EUV mask blank particle removal process (top) and plot of Ru-capped ML blanks' reflectivity and rms surface roughness as a function of the number of cleaning cycles (bottom). (From Gallagher, E., et al., *10th International Extreme Ultra-Violet Lithography (EUVL) Symposium*, 2011. With permission.)

4.4.5 PELLICLE DEVELOPMENT

Because EUV radiation is highly absorbed by all materials, a membrane pellicle for an EUV mask had not been thought to be practical, because the absorption in a membrane thick enough to have high mechanical stability would absorb much of the EUV radiation and significantly reduce the throughput of an EUV exposure tool. In 2006, Schroff et al. [165] attempted to fabricate an EUV pellicle comprised of a thin silicon film mounted on a hexagonal wire mesh and assumed that it would be possible to mount the pellicle sufficiently far from the mask surface that the shadow created by the wire mesh would have only a marginal impact on imaging. The early experimental work on this type of pellicle yielded highly wrinkled membranes and disappointingly low EUV transmission [165]. In 2009, Akiyama and Kubota of Shin Etsu Chemical Co. [166] described the fabrication of an EUV pellicle comprised of a 100 nm thick membrane of single-crystal silicon on a silicon honeycomb; however, the honeycomb that supported the pellicle was found to reduce the net transmission because of its high aspect ratio and was shown by Ko et al. [167] to result in nonuniform illumination of the mask. In 2013, ASML started the development of freestanding polysilicon pellicle films, and at the 2014 SPIE Advanced Lithography Symposium in San Jose, provided the preliminary set of EUV pellicle requirements for HVM insertion [10] shown in Table 4.2.

ASML has recently made enough progress on polysilicon membrane development that a full-size pellicle membrane has been fabricated with 85% EUV transmission (single pass) [168]. ASML arranged to have the impact of this pellicle on EUV exposures of 27 μm lines and spaces in a 0.25 NA NXE:3100 scanner evaluated and were able to confirm the 85% single-pass transmission number and show that the pellicle had little impact on image CDU (a slight increase of 0.16 nm), little or no impact on mask alignment, and no noticeable impact on resist LWR [168].

In early 2015, van den Brink introduced the mountable/demountable pellicle concept [169], illustrated in Figure 4.53, to the industry for the first time and made the claim that ASML's NXE pellicle was ready for industrialization [168].

At the present time, ASML's polysilicon pellicles, while promising, still have the following issues: 1) the transmission of current polysilicon membranes, ~85% in a single pass, is too low for

TABLE 4.2
Preliminary EUV Pellicle Requirements for HVM Insertion

	Item	Requirement
Pellicle material requirements	Pellicle film EUV transmission	90% single pass (81% double pass)
	EUV transmission spatial nonuniformity	<0.2%
	EUV transmission angular nonuniformity	<300 mrad max. local pellicle angle
	EUV intensity in scanning slit @ pellicle	5 W/cm² (250 W EUV source equivalent)
	Lifetime	~315 hours (production hours in an EUV+H$_2$ environment)
Pellicle + frame requirements	Standoff distance during exposure	2 ± 0.5 mm
	Maximum acceleration	100 m/s² during scanning
	Maximum ambient pressure rate of change	<3.5 mbar/s (peak during pump down/vent in the load lock)
	Mask reserved area for pellicle assembly (centered on substrate)	110.7 mm × 144.1 mm: inner 118.0 mm × 150.7 mm: outer
Pellicle impact on imaging performance: <0.1 nm CDU impact on wafer		

Source: Zoldesi, C., et al., *Proc. SPIE*, 9048, 90481N-1, 2014. With permission.

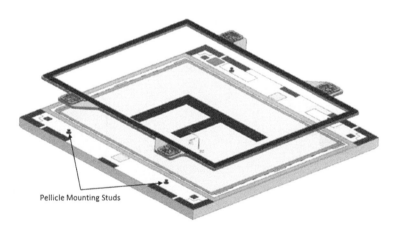

Pellicle Mounting Studs

FIGURE 4.53 Artist concept of ASML's mountable/demountable pellicle concept, which would protect mask frontside from fall-on particles and can be demounted, enabling any type of pattern mask inspection: optical, e-beam, or actinic [168].

economic scanner operation. In a recent presentation entitled "EUV mask particle adders during scanner exposure," Hyun et al. [163] of SK Hynix posed the question "What is more economical, transmission loss of 70% versus inspection and rework cost?"—suggesting that the time saved with a more productive (pellicle-less) scanner could be used for particle inspection and wafer rework and still have time left over; 2) the current polysilicon membranes are likely to become overheated and break when EUV source power exceeds 125 W; and 3) the consequences of a pellicle breaking while in a scanner are too dire to contemplate. Because polysilicon membranes are so brittle, when they break, they have been found to shatter so badly that it is unlikely that all the silicon fragments could ever be removed from the surface of a mask.

4.4.6 Alternative Mask Stacks and Materials

All current EUV masks are comprised of a multilayer film stack, which ideally provides high reflectivity for all occurring angles of incidence, and a patterned absorber or shifter layer, which defines the features to be printed. Because EUV reflective masks are illuminated at an oblique angle to separate incident and reflected light, their coating stack structure has an inordinately large impact on image quality and gives rise to several so-called 3-D mask effects, such as a horizontal–vertical print difference due to mask shadowing; reflective apodization, which results in a diffraction imbalance in the pupil; and through-focus pattern placement errors (telecentricity errors) that vary dramatically with pattern pitch. This section describes a variety of alternative EUV mask stacks and materials that could lead to the smaller 3-D effects that will be needed for use with 0.33 NA projection optics at the 7 nm and smaller technology nodes, or with EUV projection optics with higher than 0.33 NA, or with more extreme off-axis illumination schemes. Diagrams illustrating the three most well-known 3-D mask effects—mask shadowing, reflectivity apodization, and pitch-dependent through-focus pattern shifts (telecentricity errors)—are shown in Figure 4.54.

Mask shadowing refers to the fact that horizontal features, oriented perpendicular to the plane of illumination, are illuminated very differently than vertical features, oriented parallel to the plane of illumination. Mask shadowing leads to a horizontal–vertical print difference that must be corrected by applying an HV bias to the mask pattern. Reflective apodization results in an imbalance in diffraction orders and is created by the variation in EUV reflectivity with angle of incidence on the mask due to the Mo/Si ML reflector. Telecentricity errors lead to pitch-dependent shifts in best focus and in pattern placement at the wafer.

The search for an alternative architecture for the standard binary EUV reflective mask has been reinvigorated recently by the desire to extend EUV printing to sub-13 nm features and by the possibility that EUV projection optics at higher than the current 0.33 NA will be available in the not

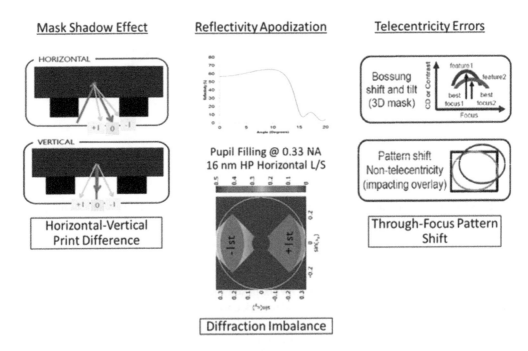

FIGURE 4.54 Sketches illustrating the three most important 3-D mask effects: mask shadowing, reflectivity apodization, and through-focus pattern shift with pitch. (Left: private communication Dr. Vicky Philipsen (IMEC). Middle: reprinted from Raghunathan, S., et al., *J. Vac. Sci. Technol. B*, 32, 06F805-1, 2014. Right: reprinted from Takai, K., et al., *Proc. SPIE*, 9235, 923515-1, 2014. All with permission.)

too distant future. Some of the alternative mask stacks that are being considered are illustrated in Figure 4.55. The dark features on the standard binary EUV mask are defined by a Ta-based absorber, which must be thick to absorb most of the incident and reflected light.

As shown in Figures 4.43 and 4.45, the total thickness of a Ta-based absorber and the antireflection coatings needed by mask inspection tools is typically between 60 and 80 nm, leading to orientation/pitch-dependent and slit-dependent (azimuthal) CD variations, which are currently mitigated by applying HV bias and other corrections to the mask features. The alternative mask stack illustrated in Figure 4.55 (a), the etched multilayer mask, which eliminates the absorber entirely, has been studied off and on by a number of groups since the beginning of EUV lithography development, starting with Tennant et al. at AT&T Bell Labs in 1991 [133], followed by Deng et al. at AMD in 2003 [170] and most recently by Takai et al. at Toshiba in 2013 [171]. Deng et al. [170] showed that the etched multilayer mask will require much less HV bias and will have smaller image placement errors than the standard bilayer mask even when the etch depth is the full 280 nm of the ML; because the effective reflection plane in the ML is only ~35 nm below its top surface, it will lead to less mask shadowing than the 60 to 80 nm thick absorber on a standard binary EUV mask. Some of the challenges when fabricating an etched ML mask are mask CD/profile control; ML pattern collapse, particularly during mask cleaning; defect inspection; and repair of ML pattern intrusion defects.

The alternating phase-shift mask (PSM) illustrated in Figure 4.55 (b) eliminates the absorber entirely and hence, should lead to less mask shadowing. This structure has been studied by Yan et al. at Intel in 2002 [172], where the phase shift was created by fabricating a step in the substrate prior to ML deposition, by Han et al. at Motorola in 2004 [173], by La Fontaine et al. at AMD in 2006 [174], and by Sun et al., at GLOBALFOUNDRIES in 2013 [175] using an etch stop layer embedded in the ML stack at the correct depth to produce a phase shift of 180 degrees. The alternating PSM fabrication technique described by Sun et al. [175] was used to create a phase-shift focus monitor mask, but one remaining concern with this type of mask is

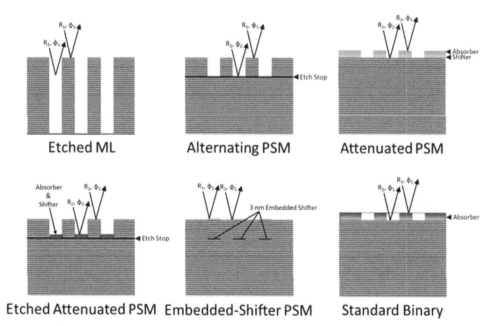

FIGURE 4.55 Illustrations for several alternative EUV mask structures: (a) etched multilayer mask, (b) alternative phase-shift mask, (c) attenuated phase-shift mask, (d) etched attenuated phase-shift mask, (e) embedded-shifter phase-shift mask, and (f) standard binary mask with alternative materials for ML reflector and patterned absorber.

that the edges of the bare ML that were created during the etch processing step when exposed to the ambient are likely to deteriorate (oxidize) over time and thus, limit the lifetime of this type of mask.

The attenuated PSM illustrated in Figure 4.55 (c) has also been studied by a number of groups, starting with a mask fabricated by Wood et al. at AT&T Bell Labs in 1997 [176]; by Deng et al. at AMD in 2003 [170], where the ML was etched leaving only four bilayers, in this case the etched region providing about 6% attenuated field intensity and 180 degree phase shift; and by Yan et al. at Intel in 2011 [177], where two semitransparent absorber materials were added to the stack to obtain the desired phase shift and absorption. Erdmann et al. at Fraunhofer in 2014 [178] reported that the standard attenuated PSM structure illustrated in Figure 4.55 (c) will not show any improvement in mask shadowing over the standard binary mask, because the absorber heights in the two types of masks are similar. Erdmann et al. [178] also claimed that the resolution of EUV lithography with 4× demagnification and standard binary and standard attenuated phase shift masks is limited to 13 nm and that further shrink with these masks will require increased demagnification.

The etched attenuated PSM is illustrated in Figure 4.55 (d), in which a given reflectivity and phase shift of the nominally dark features can be obtained by a partial etching of the multilayer combined with the addition of an absorber layer in the etched areas of the multilayer. This mask structure has been studied by Deng et al. at AMD in 2004 [179], by LaFontaine et al. at AMD in 2006 [174], by van Look et al. at IMEC in 2014 [180], and by Erdmann et al. at Fraunhofer in 2014 [178]. Erdmann et al. concluded that both the etched attenuated PSM illustrated in Figure 4.55 (d) and the embedded-shifter PSM illustrated in Figure 4.55 (e) provide a large design space and have the potential for sub-13 nm resolution at 4× demagnification [178]. Embedded-shifter PSMs and etched PSMs can provide smaller telecentricity errors and smaller mask topography–induced aberration effects and will have a reduced impact of the feature orientation of the images. The basic idea of the embedded-shifter PSM, shown in Figure 4.55 (e), is to produce dark features in the mask by a buried $\lambda/4$ phase-shift layer so that reflected light from both above and below this layer interfere destructively. Erdmann et al. at Fraunhofer also pointed out that since the total height of the embedded shifter layer is only ~3 nm, which is small compared with present binary mask absorber heights of ~60 to 80 nm, the embedded-shifter PSM should have significantly reduced shadow effect and other mask-induced imaging artifacts [178].

The alternative mask stack illustrated in Figure 4.55 (f) is similar to that of the standard binary mask but uses alternative materials for the multilayer reflector and for the patterned absorber. This structure has been studied recently by Wood et al. at GLOBALFOUNDRIES in 2015 [181], who suggested using a Ru/Si ML with 20 bilayers, wider spectral bandwidth and shallower effective reflectance plane, and a 25 nm thick Ni-based absorber, because this would result in significantly less, mask shadowing and fewer telecentricity errors at 0.33 and higher NA.

4.5 EUV RESISTS

EUV photoresist has the same functions as photoresist designed for other imaging wavelengths. It has to change solubility when exposed to light and processed, it has to be a "resist" when exposed to further processing such as etching, and it has to provide high-contrast patterning. In addition to this, it has to provide excellent photospeed; that is, it has to change solubility in response to very low-intensity patterns of light. In terms of typical photoresist dose to print units, production broadband and i-line resists have typical doses to print in the range of hundreds of mJ/cm^2, KrF and ArF resists have typical doses to print in the range of 30 to 70 mJ/cm^2 and EUV scanners are specified assuming photoresist with a dose to print of 15 to 20 mJ/cm^2 [37] So EUV photoresists will have to be the fastest production semiconductor photoresists ever used in semiconductor manufacturing.

4.5.1 EUV Resist Development Considerations

Because of the need for photospeed, almost all EUV resists in use today rely on chemically amplified technology of the sort used in typical KrF and ArF resists. These resists contain photoacid generators (PAGs). In KrF and ArF, these PAGs absorb light and then decompose, generating a strong acid. The photogenerated acid then catalyzes chemical reactions that change the solubility of the polymer that makes up the bulk of the photoresist. Because this sort of reaction is catalytic, one acid molecule can catalyze many solubility-changing reactions in the resist polymer, giving potentially very high photospeed. This mechanism has proved to give excellent imaging. Figure 4.56 shows an aerial image that is cleanly resolvable into good-quality 43 nm lines and spaces using today's chemically amplified resist technology to give a developed photoresist pattern with steep sidewalls, and a clear-cut boundary between exposed and unexposed areas. This shows the power of chemically amplified technology.

In order to get high photospeed in a resist, the resist should be designed so that most or all of the light absorbed by the film can be used productively. In KrF and ArF resists, any light absorbed by the polymer would not result in a photochemical reaction of the PAG, so part of the development process for these resists involved finding polymers that are transparent at the wavelength of interest but still have enough etch resistance. In these resists, the light absorption of PAGs is generally more than sufficient to provide good photospeed.

The situation in EUV is very different. Because EUV photons have much higher energy than KrF or ArF photons, their absorption mechanism is different. Instead of being absorbed by functional groups such as PAG chromophores, they are absorbed by atoms. Such absorption does not put a molecule into an excited state. Instead, the absorbing atom emits an electron. That electron has the energy of the photon minus the binding energy of the electron, or about 80 eV for typical EUV-generated electrons. This is many times the energy of several electronvolts needed to generate an acid from a PAG. The electron travels through the resist, losing energy and interacting with other atoms and functional groups as it moves, and sometimes generates secondary electrons. A simulated trajectory of such an electron is shown in Figure 4.57 [182].

The interaction of primary and secondary electrons with PAGs is what causes a PAG to decompose and generate a molecule of acid [183]. Primary and secondary electrons can be generated by any molecule in the resist, not just from the PAG. This is good, because PAG molecules in a resist

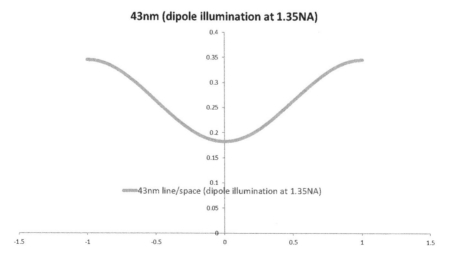

FIGURE 4.56 Aerial image of 43 nm lines and spaces using dipole illumination at 1.35 NA with immersion ArF lithography.

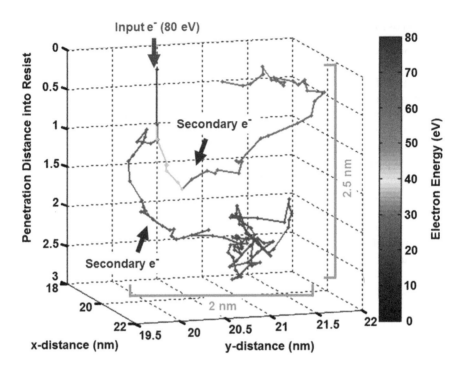

FIGURE 4.57 Simulated electron trajectory in EUV resist. The vertical line is the photon, and the other lines show the trajectory of the primary electron emitted and of one secondary electron. (Simulated trajectory provided by R. Brainard. Used with permission.)

absorb only a small fraction of incident EUV light, not enough to be of use in making an effective resist. So, the entire resist film has to be used as the EUV light absorber. Even using all the molecules in the film to absorb incident EUV light, it is not easy to get enough EUV light absorption for good photospeed [184, 185], which requires the absorbed energy being turned into primary and secondary electron energy that is used to catalyze the decomposition of PAGs into acids.

The EUV absorbance of a film can be calculated if the atomic composition of the film and its density are known.* The absorption per gram of individual atoms is shown in Figure 4.58. Generally, heavier elements absorb more than lighter ones. Typical elements in chemically amplified resists are carbon, hydrogen, and oxygen along with some heteroatoms such as nitrogen, sulfur, and fluorine. Calculating the absorption of some typical polymers related to those commonly found in resist produces the results shown in Table 4.3, where absorbance is expressed as the Dill parameter value. Since resist films are mostly polymer based, these are representative of what a resist film with a similar polymer would absorb. The absorption of a commercial EUV resist reported in 2010 is shown in Table 4.3 for reference. Also shown is the calculated absorption of zirconium and hafnium oxides, which should be representative of materials containing these two metals that have been used as the basis for novel EUV resists.

The maximum photospeed for a resist film has been calculated to occur when the film absorbs 43% of the impinging light [186–188]. In KrF and ArF resists, this much absorbance is not necessary to get good photospeed. Actual resist absorbances for chemically amplified resists are usually much smaller than this in order to get good sidewall angles. However, in EUV, resist absorbance values should be a significant fraction of 43% for the film to get reasonable photospeed. A fair absorbance target is 25% of incident EUV light being absorbed by a resist film. Values of absorbance needed

---

* Atomic scattering factors are available by element and photon energy in eV at http://henke.lbl.gov/optical_constants/asf. html. A wavelength of 13.4nm corresponds to 92.4.

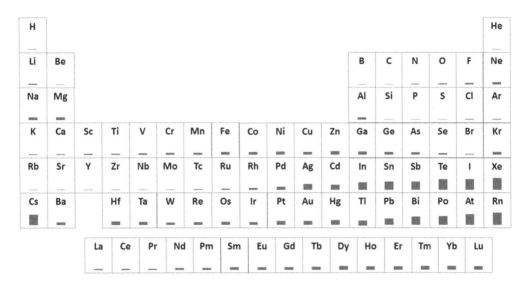

FIGURE 4.58 EUV absorption of individual elements. The bars show relative absorption per gram. If using a heavier element with higher EUV absorption increases the density of the film, the density will further increase the EUV absorption.

TABLE 4.3
Sample Calculated EUV Absorbances [184]

Material	Dill Absorbance on Unexposed Film (μm^{-1})
Poly(styrene)	2.8
Poly(methyl methacrylate)	5.2
FEVS P1101 (a commercial EUV resist) [212]	3.9
Pure zirconium oxide	13.8
Pure hafnium oxide	31.3

to absorb 43% and 25% of the EUV light are also shown in Table 4.4 for various film thicknesses. Thinner films need more absorbance than thicker films to absorb this much.

Line collapse will be an issue for EUV films if the aspect ratio is roughly more than two. Also, thinner films are needed to accommodate the lower depth of focus inherent in higher-resolution imaging. Since printed EUV lines and spaces are expected to be between 10 and 20 nm for initial implementations, EUV resists are expected to be used in very thin films, say 30 nm thick or less. It is clear from comparing the values in Table 4.3 with those in Table 4.4 that the challenge with EUV resists is getting enough light absorbance [184, 212] This is the opposite of historical experience in formulating resists for new wavelengths.

Progress has been made in adapting chemically amplified materials to the needs of EUV. Figure 4.59 shows some sample patterns of resist from different resist vendors. Resist vendors have met or come close to the required resolutions for the first EUV applications. However, the photospeed of these resists is slower than desired, and the linewidth roughness (LWR), line-edge roughness (LER), and contact hole critical dimension uniformity (CDU) are also higher than desired. As shown in Figure 4.60, there is a tradeoff between photospeed and line roughness. Slower resists give better results. No resists meet the LWR requirement in the ITRS roadmap at any photospeed [16]. This is a substantial challenge for the semiconductor industry. Some sort of post

TABLE 4.4

Values of Dill Absorbance Needed to Absorb 43% and 25% of Incident EUV Light as a Function of Film Thickness

Film Thickness (mm)	Dill Absorbance to Absorb 43% of Incident Radiation in the Film (μm^{-1})	Dill Absorbance to Absorb 25% of Incident Radiation in the Film (μm^{-1})
500	1.1	0.6
200	2.8	1.6
100	5.7	3.3
50	11.4	6.6
30	19.0	11.0
20	28.5	16.6

processing, pattern doubling, or roughness-tolerant device design will be needed to meet device requirements using EUV patterning.

EUV resists could benefit from higher absorbance, especially as operating resist thicknesses get reduced. From Figure 4.58, it is clear that heavier atoms generally provide better absorbance. It is also clear that as EUV imaging films get thinner, being able to use a metal-based resist will be a significant advantage. There is work underway on metal-based resists by several research groups. Notably, Ober and coworkers at Cornell have demonstrated EUV resists based on nanoparticles of hafnium and zirconium [189] Brainard and Freedman and coworkers at the State University of New York have demonstrated resists based on soluble organometallic compounds [190, 191], and the startup company Inpria has demonstrated EUV resists based on metal sol-gel chemistry [192].

A sol-gel material is one that starts with soluble compounds but crosslinks or "gels" through elimination of volatile moieties during a bake. In this case, the gelling process generates films with metal oxide content on a wafer during the post-apply bake. So, the Inpria results are basically using

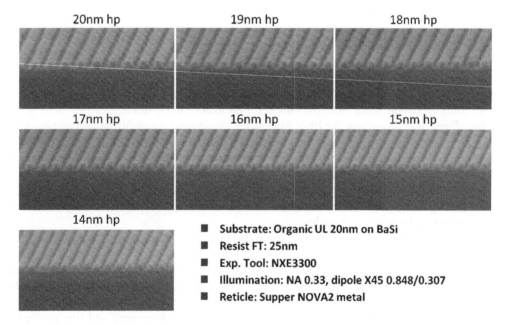

FIGURE 4.59 State of the art resist patterns. (Photo courtesy of Tokyo Ohka Kogyo Co., Ltd. Used with permission.)

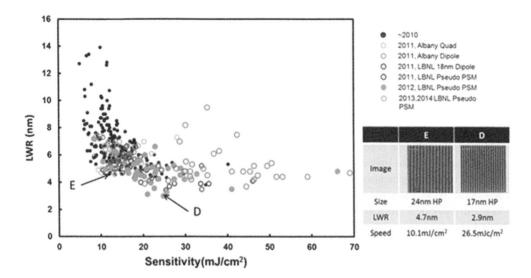

FIGURE 4.60 Experimental tradeoff between LWR and resist photospeed [211].

films that are modified metal oxides as resists. The work with metal-based resists has shown some very intriguing properties, with nanoparticle materials showing extremely fast photospeed, some sol-gel materials showing very high resolution, and some of the soluble organometallic materials showing excellent line smoothness. The tradeoff of photospeed with LWR and LER still is present, with the smooth materials being very slow and the fast materials having significant roughness. While showing a lot of promise, metal-based resists still face challenges in process integration. They also have to demonstrate parts per billion levels of other metals in their formation, so that metals that poison semiconductor devices are not present.

4.5.2 EUV-SPECIFIC RESIST CHALLENGES

EUV resists have the familiar challenge of more pattern collapse as films get thinner and compatibility with various substrates. But they face new challenges as well. One such challenge for EUV resists is out-of-band radiation. The radiation generation mechanism in an EUV light source produces many wavelengths of light, and since the optics are all reflective, potentially all the wavelengths can be focused on the resist. The non-EUV wavelengths that reach the resist film are called *out-of-band radiation*. These wavelengths will not have the resolution of the EUV wavelength and so will blur the aerial image if the resist is sensitive to those wavelengths. Tool manufacturers minimize the amount of this radiation using filters and design tricks, but some out-of-band radiation remains. EUV users can use special top coats that filter out some of this radiation, or preferably, EUV resists that are designed not to be sensitive to out-of-band radiation.

A second challenge for EUV resists is to get efficient transfer of energy from the EUV light into useful chemical reactions. In a chemically amplified resist, most of the photons will be absorbed by the polymer in the resist, not the photoacid generator. In order for the absorbed light to be photochemically useful, the primary and secondary electrons have to transfer their energy efficiently to photoacid generators. Research has been published describing possible mechanisms for this energy transfer [193], but nothing has been proved. Understanding and optimizing this energy transfer is important to enable further improvement of EUV resists. It requires understanding the mechanism of interaction of chemical compounds with electrons to provide excited states or electron-catalyzed reactions. It also requires understanding of other processes that cause nonproductive energy loss of traveling electrons. These interactions are undoubtedly related to chemical structure, but very little is known of the relationship between chemical structure and excitation by passing electrons.

Getting efficient transfer of energy to PAGs so that there are many chemical reactions per absorbed photon in a resist film means that one photon will change the solubility over some volume of resist. This volume is called *blur*. The size of this volume limits the possible resolution of resist. Higher-resolution resists require lower blur. EUV resists have the highest resolution of any resist type and so need the smallest blur, and this blur will need to get smaller as EUV resolution improves. But, lower blur means that there is a lower volume of reactions per photon, potentially making resist photospeed slower. So, as the feature sizes printed with EUV shrink, the required lower blur may lead to slower photospeed. This is a third challenge in developing better EUV resists.

A final EUV challenge is related to noise in imaging. Noise in imaging is already a significant issue. It is responsible for line-edge roughness and for contact hole nonuniformity for KrF and ArF resists. EUV resists may be noisier than ArF and KrF resists because the secondary electron process can add noise. Also, the noise does not shrink with feature size and will thus become larger relative to feature size as feature sizes shrink. Table 4.5 shows the simulation of an EUV resist using a calibrated photoresist simulation model for 26 nm dense contact holes. The predicted contact CDU matches the observed experimental value.

The photon shot noise comes from the fundamental physics of light. Shot noise 1 sigma is only 0.7%, which is basically insignificant compared with the other sources of the total CD variation of approximately 5% 1 sigma. The total variation comes mainly from random variations in the chemistry inside the resist film.

According to the ITRS roadmap [16], the projected hole sizes to be printed with EUV will be smaller than 26 nm. Shrinking printed dimensions will make the effect of shot noise larger and may also increase the chemical noise. Consider a 25 nm contact hole being imaged with no mask bias. The area of the hole is roughly 500 nm^2, so a 15 mJ/cm^2 EUV dose is $1.5 \times 500/100$ nm$^2 \times 680$ photons or 5100 photons. The standard deviation in the number of photons per contact hole will be the square root of 5100, or 71. So, the 3 sigma variation in dose just from quantum effects is 3×71 or 4% of the average dose. This is twice the 3 sigma shot noise of 2.1% calculated from the 1 sigma value shown in Table 4.5. This value is larger than that in Table 4.5 because the printed hole is smaller and the mask has no positive bias. This shows how shot noise can get substantially worse as holes get smaller. Shot noise will also be worse if imaging doses are smaller. It will get better if resist exposure latitude is better and if the resist absorbance of light is higher. In the past, achieving resolution with good profile has been the biggest challenge in photoresist development, but now reducing noise is the biggest issue facing photoresists in the future, and facing EUV photoresists in particular.

There are several methods that have been demonstrated to help in addressing noise. One is to use post processing to smooth features. Etching a feature can result in a smoother etched feature than the original printed feature. There are techniques to coat or flow printed resist features and make them smoother. Another option is to use multiple patterning. Not only can this double or quadruple the originally printed pitch, but it has been shown to give much smoother lines after patterning. One can also accommodate noise by using chip designs that offer immunity to certain kinds of noise.

TABLE 4.5

Simulation of Contact Hole Nonuniformity Using a Calibrated EUV Resist Model

	Photons	Acid	Polymer Protection	CD
Relative uncertainty (%)	0.70	0.85	1.28	4.98
Standard deviation	145	266	0.010	1.3 nm
Mean	20,817	31,296	0.740	26 nm

Relative uncertainty values are all one sigma [213].

For example, if a printed contact hole is self-aligned, then its CD after processing is determined by the features already on the wafer. This allows more printed CD variation than if the contact was not self-aligned. Of course, these methods can only accommodate a certain amount of printed CD variation, so there will always be pressure to improve printed LWR and contact hole CDU.

In order to reduce the noise inherent in resists, eventually, structured resist films may be necessary. In such a film, the PAGs, polymer molecules or whatever compounds are in the resist need to be in a regular array rather than being randomly scattered throughout the film. This is challenging to do but would provide good long-term benefits. Overall, EUV resists currently have reached the point where they are good enough to support pilot production to develop new generations of chips. There is no showstopper apparent that would prevent them from being used in some sort of volume production, assuming companies are willing to live with somewhat slower throughput for EUV tools than originally envisioned. In the long term, the use of EUV will start photoresist back on the path of continually shrinking printed feature sizes, a path that was on hold when resolution enhancement could only be achieved through multiple patterning. There will be a need in the future for innovative resist chemistries. We can expect much progress and new EUV resist results in the future.

4.6 EUV EXTENDIBILITY

4.6.1 INTRODUCTION

Of all the lithographic technologies that have been evaluated as possible replacements for optical and DUV projection lithography, EUV lithography has always stood out as having the greatest potential for extendibility. Just like optical and DUV projection lithography, EUV lithography can be extended by using a higher-NA imaging system, by further decreasing the EUV exposure wavelength (λ), or by employing one or more RETs that would enable patterning at a smaller value of k_1. The parameters NA, λ, and k_1 are related to CD by the well-known Rayleigh equation for resolution referenced in Equation 4.1. Reflective imaging systems with higher NA are possible at only a few specific NA/field-size combinations, and many of the highest-NA design options require a small central obscuration. Most higher-NA options will be accompanied by larger mask shadow effects, increased projection optics apodization, and larger telecentricity errors (see Section 4.4.6) unless the mask-to-wafer demagnification ratio is increased above 4. The status of current efforts to extend EUV lithography by increasing the NA of the imaging system will be discussed in Section 4.6.2. The throughput of an EUV exposure tool at shorter operating wavelengths depends on ML coating reflectivity and bandwidth, available in-band source power, and photoresist sensitivity. Viable alternatives at wavelengths shorter than 13.5 nm are available at only a few discrete wavelengths—the most likely candidates requiring LaN/B (B4C) ML coatings and a source at ~6.8 nm wavelength. The current status of efforts to extend EUV lithography by decreasing λ will be described in Section 4.6.3. Current efforts to extend EUV lithography by employing a variety of RETs, including mask optimization (MO), which involves aggressive mask pattern corrections and advanced coating stacks (OPC, subresolution assist features [SRAFS], and PSMs), source optimization (SO), which includes a variety of off-axis illumination (OAI) techniques and highly pixelated pupil fills, and source mask optimization (SMO), will be described in Section 4.6.4. Because extensive development work will be required for each of these extendibility options, insertion of EUV lithography into production with a full suite of RETs is unlikely before 2017–2018, EUV lithography with imaging system NAs higher than 0.33 is unlikely before 2020–2022, and EUV lithography at wavelengths shorter than 13.5 nm is unlikely before 2024–2026.

4.6.2 EXTENSION TO HIGHER NA

The 0.33 NA projection optics in current full-field EUV exposure tools will not support the printing of features much below 13 nm HP even when provided with the most aggressive OAI modes.

Multiple patterning EUV or an EUV exposure tool with larger than 0.33 NA projection optics will be required when smaller features and tighter pitches need to be printed to enable the current semiconductor industry roadmap. Table 4.6 shows that the printing of features smaller than 10 nm HP will require EUV projection optics with NAs above 0.4, and printing of features smaller than 7 nm HP will require projection optics with NAs at 0.55 and above. Even though designs for EUV reflective imaging systems with NAs as high as 0.7 can be found in the literature [45], the performance of these systems is limited by apodization caused by the reflective coating on the final folding mirror, the folding mirror that reflects light toward the projection optics exit mirror [45].

While some improvement in the performance of reflective EUV imaging systems can be obtained by adding two additional mirrors, their presence will lower the transmission of the system by more than a factor of 2 compared with the transmission of a six-mirror design [45]. In addition, the flare level of an eight-mirror imaging system will be about 30% higher than for a similar design with six mirrors. The only alternative is to reduce the angular spread on the final folding mirror by illuminating this mirror through a hole in the center of the final mirror and thus to introduce a central obscuration into the system [45]. Such a central obscuration would have to be kept as small as possible, that is, to no more than 20% of the pupil radius, if forbidden pitches are to be avoided [45].

Since no materials with adequate transparency are available in the EUV spectral region, EUV lithography must use a reflective mask. Because of this, an EUV mask must be illuminated at oblique incidence to allow a separation between incidence and reflected light. Without taking mechanical tolerances into account, incoming and reflected light cones at the mask will overlap at NAs slightly larger than 0.4. Resolving this problem would necessitate an increase in the chief ray angle at the mask (CRAO), which is currently fixed at 6° in 0.33 NA tools, or an increase in the magnification ratio of the projection optic system. Any significant increase in the current CRAO would require a new multilayer stack on the mask, since a more aggressive shadow correction would then be required, because the increased angles will begin to have a significant impact on image quality, telecentricity, and mask efficiency [194, 210].

To understand the effects of a larger CRAO on high-NA imaging, it is important to note that there is not only the chief ray angle itself but a whole range of angles of incidence on the mask. Large angles of incidence are attenuated by the current Mo/Si multilayer stack, because the reflectivity of the standard mask coating drops off at angles >11°. Consider the EUV diffraction pattern from a horizontal line-space grating illuminated with a y-oriented dipole. Due to the oblique angle of incidence, the bottom pole of the dipole will have a larger off-axis incidence angle, while the top pole will have a smaller incidence angle. Due to this asymmetry, the two poles of the dipole will experience very different imaging conditions. For a horizontal line-space grating pattern, the absorber will cast a shadow that depends on the thickness of the absorber, and even an absorber with zero thickness will still result in a shadow within the multilayer. Imaging simulations show that the 0th diffraction order of the bottom pole will be much more attenuated that the 0th order of the top pole. This imbalance in 0th orders will cause a through-focus pattern placement error (PPE) at the wafer, commonly referred to as a *telecentricity error*. Imaging simulations also show that the first diffraction orders from both the top and bottom poles of the dipole are much weaker than the corresponding 0th orders, which would lead to a contrast loss due to incomplete interference [49].

TABLE 4.6

Resolution Limits at 13.5 nm Wavelength versus the Numerical Aperture of the Imaging System

NA	0.33	0.40	0.45	0.50	0.55	0.60
Resolution @ $k_1 = 0.3$ single exposure (nm)	12.3	10.1	9.0	8.1	7.4	6.8

Three possible strategies are available for coping with the increasingly severe mask effects at higher NA: 1) adoption of a multilayer stack tuned for reflectivity over a broader angular bandwidth; 2) adoption of a multilayer stack optimized for a specific use case, for example, for one particular set of critical pitches; or 3) continued use of the standard multilayer stack but provision of projection optics with a higher magnification ratio. A discussion of potential alternative mask stacks, for example a multilayer stack that provides a wider reflective bandwidth, or a new absorber material that would allow use of a thinner patterned absorber, or a new embedded phase-shifting structure [178], like the one that was discussed in Section 4.4.6. Apparently, just increasing the CRAO alone does not lead to a simple design for a high-NA EUV optical system. Instead, to enable the required imaging performance at higher NA, the angular spread at the mask must be reduced.

The most straightforward way to reduce the angular spread at the mask is to increase the magnification ratio of the projection optics. With a larger magnification ratio, a wide range of high-NA design options become available [195]. A higher magnification ratio will allow the NA of the imaging system to be increased, but the change in magnification ratio will reduce the field size at the wafer if all other parameters are kept constant. For example, if the magnification ratio is allowed to double from 4× to 8×, a 0.5 NA six-mirror system that would provide 8.1 nm resolution in a 13×16.5 mm^2 image field (i.e., one-quarter of the current field size) will be possible, but its productivity would be only 55% that of a 4× system, or a 0.6 NA eight-mirror system that would provide 6.8 nm resolution in a 13×16.5 mm^2 image field will be possible, but its productivity would be only 22% that of a 4× system. In both these cases, printing a full-size 26×33 mm^2 image field on the wafer would require that four of the quarter-size fields be stitched together. Alternatively, a 0.6 NA eight-mirror system that could provide 6.8 nm resolution in a 26×16.5 mm^2 half-size image field would be possible with a 9″ mask but with a 30% drop in productivity, or a 26×33 mm^2 full-size image field would be possible with a 12″ mask but with a 40% drop in productivity. While the large drops in productivity and reductions in field size with the quarter-field systems are highly undesirable, based on the results of a survey of chip makers, mask makers, stepper suppliers, and mask tool and material suppliers at a SEMATECH High NA EUV Workshop at SEMICON West in San Francisco, California on July 8, 2013 [195], the larger mask size and need for the new mask infrastructure required for the half- and full-field systems appear to be completely unacceptable.

Because the angular spread at the mask needs to be reduced for horizontal lines and spaces (L&S) but not for vertical L&S, the magnification ratio of projection optics for use at higher NA only needs to be reduced in one direction, that is, in the y- or scanning direction. For example, an anamorphic 4×/8× magnification system could have 0.55 NA, a CRAO of 6°, and provide a 26×16.5 mm^2 half-size image field. The anamorphic magnification system [6] illustrated in Figure 4.61 would image a 6″ mask with a standard multilayer coating and absorber (no new mask infrastructure would be needed), and the image field would require only one stitch.

While not the only option for higher-NA imaging, an anamorphic design appears to be a highly attractive option, because it results in both a large field size and a high throughput. And, simulations have shown that the imaging performance of an anamorphic half-field system will be nearly identical to that of an 8× magnification isomorphic system.

Throughput estimates for an anamorphic 4×/8× magnification >0.5 NA six-mirror system, an 8× magnification quarter-field >0.5 NA six-mirror system, and current 0.33 NA projection optics from an NXE:3300 scanner as a function of source power/exposure dose are compared in Figure 4.62.

The maximum throughput of a high-NA quarter-field system will be limited to around 110 wafers/hour, even with a 2× faster wafer stage, no matter what source power is available [7]. If the acceleration of the wafer stage is increased by 2× and the mask stage by 4×, the throughput of an anamorphic 4×/8× high-NA system could reach 150 wafers/hour at 500 W of power when used with a 60 mJ/cm^2 resist and would saturate at ~180 wafers per hour at 1000 W of power [7].

A sketch of anamorphic magnification NA > 0.5 six-mirror projection system, illustrating the large increase in overall size compared with previous projection optics at NA = 0.25 and 0.33, is shown in Figure 4.63. The large size of the final projection optic mirror is dictated by the high NA.

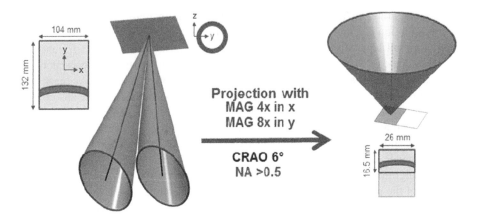

FIGURE 4.61 Illustration of 4×/8× anamorphic projection optic design with a CRAO=6° that utilizes a 6″ mask and provides a 26 mm × 16.5 mm image field at the wafer. (Reprinted from Kneer, B., et al., *Proc. SPIE*, 9422, 94221G-1, 2015. With permission.)

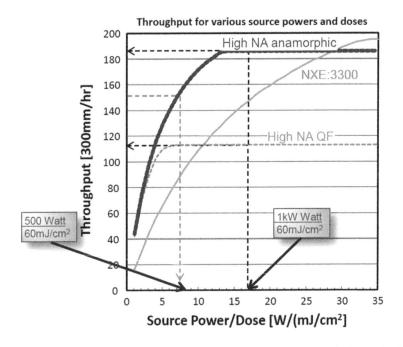

FIGURE 4.62 Throughput versus source power/dose for high-NA anamorphic 4×/8× projection optics at NA > 0.5, high-NA quarter-field projection optics at NA > 0.5, and current 0.33 NA projection optics in an NXE:3300 scanner. (Reprinted from van Schoot, J., et al., *Proc. SPIE*, 9422, 94221F-1, 2015. With permission from SPIE.)

The presence of a central obscuration in the final folding mirror and in the exit mirror reduces apodization due to the final folding mirror coating and improves transmission compared with that of previously unobscured projection optics at NA = 0.25 and 0.33. Further improvements in resolution and reductions in image flare will require significant improvements in mirror surface quality. While developments of an anamorphic magnification NA > 0.5 imaging system will provide a significant challenge to optics technology and manufacturing, no fundamental limitations have yet been discovered.

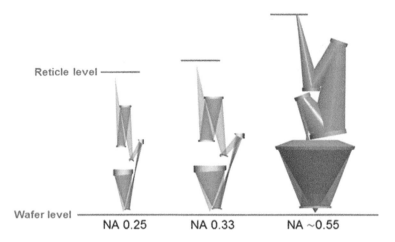

Reticle level

Wafer level

NA 0.25   NA 0.33   NA ~0.55

FIGURE 4.63 Design examples of an anamorphic high-NA projection optic at NA > 0.5 compared with previous projection optics at NA 0.25 and 0.33. The overall size of EUV projection optic designs has increased significantly over time. (Reprinted from Kneer, B., et al., *Proc. SPIE*, 9422, 94221G-1, 2015. With permission.)

4.6.3 EXTENSION TO SHORTER λ

In the same way as optical and DUV projection lithography was extended in the past by employing shorter exposure wavelengths, that is, using spectral lines of Hg at 436 nm (g-line), 405 nm (h-line), and 365 nm (i-line) in the early years and the output of excimer lasers at 248 nm (KrF) and 193 nm (ArF) more recently, EUV lithography could be extended by employing a wavelength shorter than the current $\lambda = 13.5$ nm. Obviously, a new operational wavelength for EUV lithography would only make sense if resolution were significantly improved, a useable DOF is maintained, and economically viable throughput is possible.

It turns out that viable candidates for a new EUV exposure wavelength are available only at a few discrete EUV wavelengths (near K-, L-, and M-shell absorption edges), which occur at 12.5 nm (Si), 11.4 nm (Be), 6.7 nm (B), 4.4 nm (C), and 3.1 nm (Sc). Plots of the maximum reflectivity of the most efficient ML coatings in the 3 to 14 nm wavelength range [196] are shown in Figure 4.64. Because

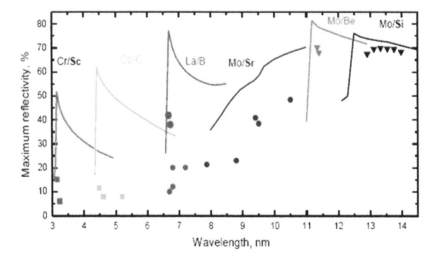

FIGURE 4.64 Maximum reflectivity of the most efficient multilayer coatings in the 3 to 14 nm wavelength range. (From Yulin, S., et al., *2012 EUV Litho Workshop*, 2012. With permission.)

TABLE 4.7

Calculations of the Throughput for an 11-Mirror ML Coated System with the Most Promising Performance in the 3 to 14 nm Wavelength Region

λ, Wavelength (nm)	13.5	9.5	6.63	4.4	3.1
Material pair	Mo/Si	Ru/Y	La/B$_4$C	Co/C	Cr/Sc
R, Reflectivity (%)	74	59	74	52.5	59
$\Delta\lambda$, Bandwidth (nm)	0.6307	0.1976	0.0623	0.0240	0.0123
$\Delta\lambda/\lambda$ (1 mirror) (%)	4.6721	2.0801	0.9397	0.5444	0.3941
R11 (11 mirrors)	3.64	0.30	3.64	0.08	0.30
$\Delta\lambda/\lambda$ (11 mirrors) (%)	2.311	0.9032	0.3403	0.2045	0.0129
FOM (11 mirrors)	1.000	0.032	0.147	0.002	<0.001

Source: Platonov, Y., et al., *2011 International Workshop on EUV and Soft X-ray Sources*, Dublin, 2011. With permission.

an economically viable throughput is critically important, only ML coatings with peak reflectivities >60% have much of a chance of being adopted; that is, Mo/Be at 11.4 nm and La/B at 6.67 nm. Even though the calculated reflectivities of Mo/Si at 13.5 nm and La/B at 6.67 nm are similar, the reflective bandwidth of La/B at 6.67 nm is much narrower (FWHM = 0.072) than the reflective bandwidth of Mo/Si at 13.5 nm (FWHM = 0.60). In addition, the difference between calculated and measured reflectivity due to interface quality (roughness and interdiffusion) is expected to be significantly larger in La/B than in Mo/Si, because the La-B bilayer period is 2× smaller and the La/B interfaces are more numerous; that is, 200 bilayers will be needed for maximum reflectivity in La/B compared with only ~40 bilayers in Mo/Si. Calculations of the throughput of 11-mirror ML-coated reflective imaging systems with the most promising ML coatings included in Figure 4.64 are shown in Table 4.7 [197]. The data in Table 4.7 shows that, from throughput considerations alone, the most viable new operational wavelength for EUV lithography would be ~6.8 nm. The data in Table 4.7 also shows that an 11-mirror La/B ML-coated system at 6.8 nm wavelength will have a throughput ~7× lower than an 11-mirror Mo/Si ML-coated system at 13.5 nm.

Since system throughput depends not only on ML reflectivity and bandwidth, but also on in-band source power and resist sensitivity, the laser-produced Sn plasma source used today at 13.5 nm wavelength would likely need to be replaced with a laser-produced Gd and/or Tb source at 6.8 nm [198]. The emission spectra of Gd and Tb in the 5–9 nm wavelength range generated with Q-switched Nd:YAG laser-produced rare-earth plasmas are shown in Figure 4.65.

Unfortunately, the conversion efficiency for a laser-produced Gd plasma source at 6.8 nm has already been shown to be approximately 5× smaller than the conversion efficiency of a laser-produced Sn plasma source at 13.5 nm. Furthermore, as shown in Figure 4.65, the dose required at ~6.8 nm wavelength will be 2.5–6× higher than the dose required at 13.5 nm because of the lower resist absorption at the shorter wavelength [199]. Given the sheer technical complexity required for a change in EUV wavelength, a new operational wavelength for EUV lithography is unlikely to be available before 2024–2026.

4.6.4 Extension to Smaller K_1

Current efforts to extend EUV lithography by employing a variety of RETs include mask optimization (MO), which involves aggressive mask pattern corrections and advanced coating stacks (OPC, SRAFs, and PSMs), source optimization (SO), which includes a variety of off-axis-illumination (OAI) techniques and highly pixelated pupil fills, and source-mask optimization (SMO).

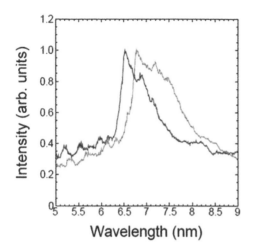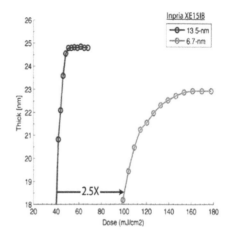

Wavelength (nm)

FIGURE 4.65 (left) Emission spectra of Gd and Tb in the 5 to 9 nm wavelength range generated with Q-switched Nd:YAG laser–produced rare-earth plasmas. (right) Performance of an inorganic photoresist at 13.5 and 6.7 nm recorded with the Berkeley Dose Calibration Tool. (Left: from Higashiguchi, T., *2012 International Workshop on EUV and Soft X-ray Sources*, 2012. With permission. Right: reprinted from Anderson, C., et al., Proc. SPIE, 8322, 9832212-1, 2012. With permission.)

The illumination system in an NXE:3300B scanner, which employs a Köhler-type design, allows changes to the illumination setting without efficiency loss by making use of the flexible fly's eye integration system illustrated in Figure 4.66 [200, 201].

The illuminator, called FlexPupil, is able to direct light from an intermediate focus (IF) following the EUV source to different pupil locations by superposing light from individual field facet mirrors located in a field plane after the source collector mirror. Since each field facet mirror can have its own orientation and tilt angle, the multiple mirror array can fill the pupil with complex arrangements of light spots and with a maximum system partial coherence of 0.8. This so-called FlexPupil illuminator is able to address the main OAI settings required for higher-resolution imaging and can support annular, dipole, and quadrupole pupil fills, as shown in Figure 4.66.

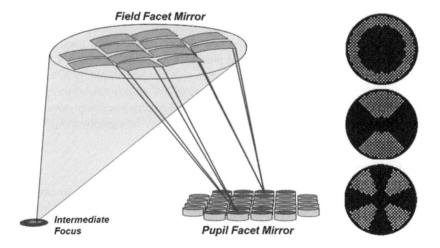

FIGURE 4.66 (left) Sketch of the flexible lossless illuminator system in an NXE scanner. (right) Illustrations of a few of the pupil fills that are available in an NXE scanner. (Left: reprinted from Migura, S., et al., Proc. SPIE, 9666, 96610T-1, 2015. With permission. Right: reprinted from Liu, X., et al., Proc. SPIE, 9048, 90480Q-1, 2014. With permission.)

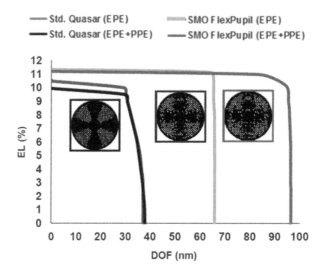

FIGURE 4.67 Comparison of overlapped exposure-latitude versus depth-of-focus curves for the printing of a 7 nm logic metal level using a 0.33 NA NXE scanner with a standard Quasar pupil and with two different free-form FlexPupils modeled using pattern placement–aware source mask optimization software. (Reprinted from Hsu, S., et al., Proc. SPIE, 9422, 94221L-1, 2015. With permission.)

EUV lithography faces many new imaging challenges, primarily because of the need to use reflective optics. The chief new challenges include 3-D mask topology effects inherent in non-telecentric optics, increased field- and pattern-dependent flare variations, and the need for new analysis metrics to quantify imaging performance beyond the typical CD-based process window metrics used in today's DUV and 193 nm applications. The oblique chief ray angle and 3-D mask topography introduces mask-side non-telecentricity, mask defocus detuning, and mask shadowing effects that lead to pattern- and slit-dependent shifts and biases. EUV's short 13.5 nm wavelength and residual mirror surface roughness increase the flare levels across the slit. An innovative SMO method has been developed recently that can significantly reduce edge placement errors (EPE) by exploiting the adjustability of the light distribution in the pupil provided by the NXE:3300 FlexPupil illuminator. The newly developed algorithms can mitigate the H-V bias, Bossung tilt, and pattern shifts due to shadowing and non-telecentricity and reduce the sensitivity to flare. Using the FlexPupil illuminator, it is possible to widen the overlapped process window, improve the image log slope (ILS), and reduce stochastic effects such as resist line-edge roughness.

Examples of the use of the newly developed placement-aware SMO with a 7 nm logic design, showing that the pattern placement error (PPE) distribution can be reduced by 37% and the DOF at 10% exposure latitude (EL) improved from 65 to 95 nm (a 46% improvement) using free-form illumination pupils are shown in Figure 4.67. Newly developed pattern placement–aware SMO and the FlexPupil illuminator can extend single-exposure 0.33 NA EUV lithography to $k_1 = 0.4$ and below.

4.6.5 Free-Electron-Laser EUV Source

A free-electron-laser (FEL) EUV source is a possible high-power alternative to current laser-produced plasma (LPP) EUV sources at both 13.5 nm (500 W) and 6.8 nm (1 kW) wavelengths [202, 203]. FEL sources are utilized today primarily by the scientific research community in applications where high brightness and coherence are more important than high average power. An FEL EUV source providing multiple tens of kilowatts of average power could be used in the future to power a semiconductor fab's entire fleet of EUV lithography scanners; its cost-of-ownership advantage over multiple LPP EUV sources could be as much as $60 million per year [204], and an FEL EUV source would avoid the difficult debris mitigation problems associated with LPP EUV sources. However, before an FEL EUV source with near 100% uptime can be guaranteed, both the accelerator design and the layout of the fab and scanners still need some additional development.

A sketch illustrating the key components of an FEL source for EUV lithography is shown in Figure 4.68. Typically, electrons are generated in an electron gun consisting of a laser-irradiated photocathode followed by an extraction field. Electron bunches exiting the electron gun/injector assembly would typically have ~ 10 MeV of energy, an electron beam emittance of the order of 0.1 mm-mrad, and a bunch charge of approximate 70–100 pC. The electron bunches from the gun/injector assembly would then enter a linear accelerator (Linac) consisting of a series of superconducting radio frequency (SRF) cavities fabricated from pure niobium and cooled with liquid helium to 1.8–2 °K, in which they would be accelerated to ~800 MeV before being delivered to a 2 cm period undulator, where they would generate self-amplified spontaneous emission at 13.5 nm (or 6.8 nm) wavelength. It is thought that an industrial FEL EUV source capable of supplying ~1 kW of power to a number of exposure tools simultaneously will need to operate in the 10–100 MHz repetition rate range and provide 20–50 kW of EUV output power to compensate for the inefficiencies in the optics used to modify the spatial coherence of the FEL output and distribute the EUV output power to individual EUV exposure tools. While a linear arrangement of the electron accelerator and undulator is the most common layout for a FEL today, a folded arrangement of electron accelerator and undulator, as illustrated in Figure 4.68, would not only reduce the facility's overall footprint but would also provide an opportunity to employ energy recovery. By recirculating the electron beam through the SRF cavities parallel to, but out of phase with, the accelerating bunches, the recirculated electrons could be decelerated, releasing much of their energy back into the SRF cavities. The advantages of this arrangement are twofold. First, depositing the electron beam energy back into the SRF cavities would greatly reduce the RF power requirements (by up to an order of magnitude). Second, the decelerated beam would be simpler to dispose of at a beam dump, generating less radiation (fewer neutrons) and making the facility safer and less costly.

Efficient distribution of the tens of kilowatts of EUV power from an FEL source to individual EUV scanners is likely to be a substantial challenge. A sketch illustrating a EUVL FEL source-compatible semiconductor fab complex is illustrated in Figure 4.69. A distance of at least 10 meters between the output of the FEL and the first EUV optic will be needed for beam expansion to avoid optical damage. Then, additional optics will be needed to reduce beam coherence to a manageable level. Because of this, if the optics between the FEL source and the input to each scanner (some of which can be grazing incidence) are only ~20% efficient, then a 50 kW FEL EUV source is likely to be required to power 10 1 kW scanners [205].

A comparison of capital and operational costs for 10 250 W LPP EUV sources versus one 10 KW FEL EUV source is presented in Table 4.8. The estimate for the capital cost (CapEx) of the 10 KW FEL EUV source shown in Table 4.8 assumed a redundant FEL configuration with a multipass, energy-recovery Linac and included an estimate for the cost of EUV beam distribution optics.

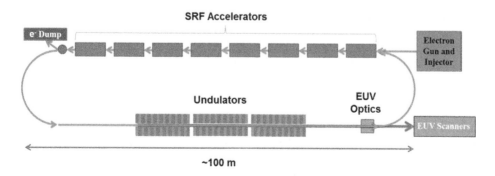

FIGURE 4.68 Schematic illustrating the key components of an FEL source for EUV lithography. (Reprinted from Hosler, E., et al., Proc. SPIE, 9422, 94220D-1, 2015. With permission.)

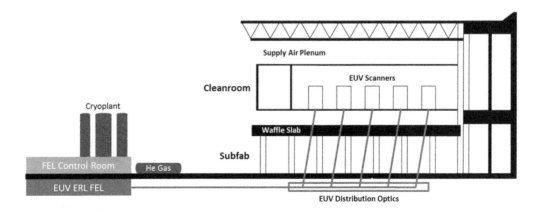

FIGURE 4.69   EUVL ERL FEL source-compatible semiconductor Fab complex and supporting equipment.

TABLE 4.8

Comparison of Costs between 10 250 W LPP EUV Sources and One 10 kW FEL EUV Source

Expense ($ Million)	10–250 W LPP Sources	10 kW FEL
OpEx	>85	23
CapEx	256	240
Total Cost First Year	>341	263
FEL CapEx Savings	–	<u>16</u>
FEL OpEx Savings	–	<u>>62</u>

Source: reproduced from Hosler, E., et al., Proc. SPIE, 9422, 94220D-1, 2015. With permission.

The primary CapEx items for a 10 kW FEL EUV source are the superconducting linear accelerator and associated RF systems, the two radiation vaults (so that the accelerators can be serviced independently), and the liquid He cryoplant. The CapEx for a 10 kW FEL EUV source turns out to be roughly the same as the CapEx for 10 discrete LPP EUV sources. The primary contributors to the operating cost (OpEx) of a 10 kW FEL EUV source are electrical power and labor.

The estimate for the utility costs for 10 LPP EUV sources shown in Table 4.8 was based on the conversion efficiency from CO_2 laser radiation at 10.6 μm to EUV radiation at 13.5 nm in a laser-produced Sn-droplet plasma, the cost of RF power for the CO_2 drive laser, and the cost for consumables. Estimates for the service costs (including collector replacement) for the 10 LPP sources were included in the number for OpEx. Table 4.8 shows that an FEL EUV source will result in potential cost savings of >$60 million per year and be able to supply substantially more EUV power. Many modern accelerator facilities operate with an uptime exceeding 95%, which is significantly higher than the uptime of today's LPP EUV sources, but is still unacceptably low, since when an FEL EUV source is offline, all EUV tools in the fab will be offline. One obvious solution to this problem would be to operate an FEL EUV source with as much redundancy as possible. Since the cryogenic plant is the primary contributor to the total accelerator expense, a redundant configuration would increase OpEx only slightly, would allow almost all accelerator service events to be carried out while source operation continues uninterrupted, and would allow operation at nearly 100% source uptime. The primary advantages of an FEL-based EUV lithography program are the large yearly OpEx savings, the ease of operation, and the unprecedented uptime.

REFERENCES

1. E. Abbe. 1873. Beiträge zur Theorie des Mikroskops und der mikroskopischen Wahrnehmung. *Arch. Mikrosk. Anat.* **Band 9**(Nr. 1): 413–468.
2. Lord Rayleigh. 1879. XXXI. Investigations in optics, with special reference to the spectroscope. *Philos. Mag. Ser. 5* **8**: 261–274.
3. B.J. Lin. 1986. Where is the lost resolution? Optical micorlithography V. *Proc. SPIE* **633**: 44–50.
4. H.J. Levinson. 2001. *Principles of Lithography*, Bellingham, WA: SPIE Press.
5. A.K.-K. Wong. 2001. *Resolution Enhancement Techniques*, Bellingham, WA: SPIE Press.
6. B. Kneer, S. Migura, W. Kaiser, J.T. Neumann, and J. van Schoot. 2015. EUV lithography optics for sub-9-nm resolution. *Proc. SPIE* **9422**: 94221G-1–94221G-10. doi: 10.1117/12.2175488
7. J. van Schoot, K. van Ingen Schenau, C. Valentin, and S. Migura. 2015. EUV lithography scanner for sub 8-nm resolution. *Proc. SPIE* **9422**: 94221F-1–94221F-12. doi: 10.1117/12.2087502
8. S. Wurm and C.W. Gwyn. 2007. EUV lithography. In *Microlithography*, 2nd Ed., ed. K. Suzuki and B.W. Smith, 383–464. Boca Raton, FL: CRC Press / Taylor & Francis Informa Group.
9. W. Kaiser. 2015. EUV Optics: achievements and future perspectives. In *14th International Extreme Ultra-Violet Lithography (EUVL) Symposium*, Maastricht, The Netherlands, October 5–7. http://euvl-symposium.lbl.gov/proceedings/2015
10. C. Zoldesi, K. Bal, B. Blum, G. Bock, D. Brouns, F. Dhalluin, N. Dziomkina, J.D. Arias Espinoza, J. de Hoogh, S. Houweling, M. Jansen, M. Kamali, A. Kempa, R. Kox, R. de Kruif, J. Lima, Y. Liu, H. Meijer, H. Meiling, I. van Mil, M. Reijnen, L. Scaccabarozzi, D. Smith, B. Verbrugge, L. de Winters, X. Xiong, and J. Zimmerman. 2014. Progress on EUV pellicle development. *Proc. SPIE* **9048**: 90481N-1–90481N-10. doi: 10.1117/12.2049276
11. A.O. Antohe, D. Balachandran, L. He, P. Kearney, A. Karumuri, F. Goodwin, and K. Cummings. 2015. SEMATECH produces defect-free EUV mask blanks: defect yield and immediate challenges. *Proc. SPIE* **9422**: 94221B-1–94221B-5. doi: 10.1117/12.2176126
12. L.L. Pang, C. Clifford, P. Hu, D. Peng, Y. Li, D. Chen, M. Satake, V. Tolani, and L. He. 2011. Compensation of EUV multilayer defects within arbitrary layouts by absorber pattern modification. *Proc. SPIE* **7969**: 79691E-1–79691E-14. doi: 10.1117/12.879556
13. A. Pirati, R. Peeters, D. Smith, S. Lok, M. van Noordenburg, R. van Es, E. Verhoeven, H. Meijer, A. Minnaert, J.-W. van der Horst, H. Meiling, J. Mallmann, C. Wagner, J. Stoeldraijer. G. Fisser, J. Finders, C. Zoldesi, U. Stamm, H. Boom, D. Brandt, D. Brown, I. Fomenkov, and M. Purvis. 2016. EUV lithography performance for manufacturing: status and outlook. *Proc. SPIE* **9776**: 97760A. doi: 10.1117/12.2220423
14. D.C. Brandt, I.V. Fomenkov, N.R. Farrar, B. La Fontaine, D.W. Myers, D.J. Brown, A.I. Ershov, N.R. Böwering, D.J. Riggs, R.J. Rafac, R. De Dea, R. Peeters, H. Meiling, N. Harned, D. Smith, and A. Pirati. 2014. LPP EUV source readiness for NXE 3300B. *Proc. SPIE* **9048**: 90480C-1–90480C-8. doi: 10.1117/12.2048184
15. R. Peeters, S. Lok, E. van Alphen, N. Harned, P. Kuerz, M. Lowisch, H. Meijer; D. Ockwell, E. van Setten, G. Schiffelers, J.-W. van der Horst, J. Stoeldraijer, R. Kazinczi, R. Droste, H. Meiling, and R. Kool. 2013. ASML's NXE platform performance and volume introduction. *Proc. SPIE* **8679**: 86791F-1–86791F-15. doi: 10.1117/12.2010932
16. International Technology Roadmap for Semiconductors, www.itrs2.net
17. H. Kinoshita and O. Wood. 2009. EUV lithography: an historical perspective. In *EUV Lithography*, ed. V. Bakshi, 1–54. Bellingham, WA: SPIE Press.
18. H. Kinoshita. 2016. *Extreme Ultraviolet Lithography – Principles and Basic Technologies*. Saarbruecken, Germany: Lambert Academic Publishing.
19. H. Kinoshita, R. Kaneko, K. Takei, N. Takeuchi, and S. Ishihara. 1986. Study on x-ray reduction projection lithography. In *Extended Abstracts of the 47th Autumn Meeting of the Japan Society of Applied Physics*, JSAP 28p-ZF-15 (in Japanese).
20. A.V. Vinogradow and N.N. Zorev. 1988. Possilble realization of projection x-ray lithography. *Sov. Phys. Dokl.* **33**: 682–684.
21. A.M. Hawryluk and L.G. Seppala. 1988. Soft x-ray projection lithography using an x-ray reduction camera. *J. Vac. Sci. Technol. B* **6**: 2162–2166. doi: 10.1116/1.584107
22. W.T. Silfvast and O.R. Wood II. 1988. Tenth micron lithography with a 10 Hz 37.2 nm sodium laser. *Microelectron. Eng.* **8**(3): 3–11. doi: 10.1016/0167-9317(88)90003-2
23. H. Kinoshita, K. Kurihara, Y. Ishii, and Y. Torii. 1989. Soft x-ray reduction lithography using multilayer mirrors. *J. Vac. Sci. Technol. B* **7**: 1648–1651. doi: 10.1116/1.584507

24. D.A. Tichenor, R.R. Freeman, J. Bokor, T.E. Jewell, W.M. Mansfield, G.D. Kubiak, M.E. Malinowski, O.R. Wood, D.L. White, D.L. Windt, D.M. Tennant, R.H. Stulen, S.J. Haney, J.E. Bjorkholm, W.K. Waskiewicz, A.A. MacDowell, K.W. Berger, and L.A. Brown. 1991. Diffraction-limited soft-x-ray projection imaging using a laser plasma source. *Opt. Lett.* **16**(20): 557–1559. doi: 10.1364/OL.16.001557

25. K.B. Nguyen, G.F. Cardinale, D.A. Tichenor, G.D. Kubiak, K. Berger, A.K. Ray-Chaudhuri, Y. Perras, S.J. Haney, R. Nissen, K. Krenz, R.H. Stulen, H. Fujioka, C. Hu, J. Bokor, D.M. Tennant, and L.A. Fetter. 1996. Fabrication of MOS devises with extreme ultraviolet lithography. In *OSA TOPS on Extreme Ultraviolet Lithography*, ed. G.D. Kubiak and D. Kania, Vol. 4, 208–211.

26. C.W. Gwyn and S. Wurm. 2009. EUV LLC – A historical perspective. In *EUV Lithography*, ed. V. Bakshi, 55–101. Bellingham, WA: SPIE Press.

27. D.A. Tichenor, W.C. Replogle, S.-H. Lee, W.P. Ballard, A.H. Leung, G.D. Kubiak, L.E. Klebanoff, S. Graham, J.E.M. Goldsmith, K.L. Jefferson, J.B. Wronosky, T.G. Smith, T.A. Johnson, H. Shields, L.C. Hale, H.N. Chapman, J.S. Taylor, D.W. Sweeney, J.A. Folta, G.E. Sommargren, K.A. Goldberg, P.P. Naulleau, D.T. Attwood, and E.M. Gullikson. 2002. Performance upgrades in the EUV engineering test stand. *Proc. SPIE* **4688**: 72–86. doi: 10.1117/12.472328

28. SEMATECH Inc., USA, http://www.sematech.org

29. Microelectronics Development for European Applications+ (MEDEA+), France, http://www.medeaplus.org

30. Association of Super-Advanced Electronics Technologies (ASET), Japan, http://www.aset.or.jp

31. Extreme UltraViolet Lithography System Development Association (EUVA), Japan, http://www.euva.or.jp

32. A. Ma, P. Kearney, D. Krick, R. Randive, I. Reiss, P. Mirkarimi, and E. Spiller. 2005. Progress towards the development of a commercial tool and process for EUVL mask blanks. *Proc. SPIE* **5751**: 168–177. doi: 10.1117/12.599936

33. K. Lowack, A. Rudack, K. Dean, M. Malloy, and M. Lercel. 2006. The EUV resist test center at SEMATECH-North. *Proc. SPIE* **6151**: 61512U-1–61512U-7. doi: 10.1117/12.657683

34. Colleges of Nanoscale Science and Technology (CNSE), Albany NY, http://www.sunycnse.com

35. Interuniversity MicroElectronics Center (IMEC), Belgium, http://www.imec.be

36. Semiconductor Leading Edge Technologies (SELETE), http://www.selete.co.jp

37. R. Peeters, S. Lok, J. Mallman, M. van Noordenburg, N. Harned, P. Kuerz, M. Lowisch, E. van Setten, G. Schiffelers, A. Pirati, J. Stoeldraijer, D. Brandt, N. Farrar, I. Fomenkov, H. Boom, H. Meiling, and R. Kool. 2014. EUV lithography: NXE platform performance overview. *Proc. SPIE* **9048**: 90481J-1–90481J-18. doi: 10.1117/12.2046909

38. K. Petrillo, K. Cho, A. Friz, C. Montgomery, D. Ashworth, M. Neisser, S. Wurm, T. Saito, L. Huli, A. Ko, and A. Metz. 2013. Resist process applications to improve EUV patterning. *Proc. SPIE* **8679**: 867911-1–867911-12. doi: 10.1117/12.2011566

39. G.E. Moore. 1965. Cramming more components onto integrated circuits. *Electronics* **28**(8): 114–117.

40. H. Hiura, Y. Takabayashi, T. Takashima, K. Emoto, J. Choi, and P. Schumaker. 2016. Nanoimprint system development and status for high volume semiconductor manufacturing. In *32nd European Mask and Lithography (EMLC) Conference*, Dresden, Germany, June 21–22.

41. J. Zimmermann, P. Graeupner, R. Gehrke, A. Winkler, C. Hennerkes, and S. Hsu. 2016. Flexible illumination for ultra-fine resolution with 0.33 NA EUV lithography. In *15th International Extreme UltraViolet Lithography (EUVL) Symposium*, Hiroshima, Japan, October 24–26. http://euvlsymposium.lbl.gov/proceedings/2016

42. H. Meiling, J.P. Benschop, R.A. Hartman, P. Kuerz, P. Hoghoj, R. Geyl, and N. Harned. 2002. EXTATIC, ASML's α-tool development for EUVL. *Proc. SPIE* **4688**: 52–63. doi: 10.1117/12.472308

43. K. Murakami, T. Oshino, H. Kondo, H. Chiba, K. Nomura, H. Kawai, Y. Kohama, K. Morita, K. Hada, Y. Ohkubo, and T. Miura. 2008. Development of EUV lithography tools in Nikon. *Proc. SPIE* **7140**: 71401C-1–71401C-8. doi: 10.1117/12.804706

44. H. Meiling, N. Buzing, K. Cummings, N. Harned, B. Hultermans, R. De Jonge, B. Kessels, P. Kürz, S. Lok, M. Lowisch, J. Mallman, B. Pierson, C. Wagner, A. Van Dijk, E. Van Setten, J. Zimmerman, P. Cheang, and A. Chen. 2009. EUVL: towards implementation in production. *Proc. SPIE* **7520**: 752008-1–752008-13. doi: 10.1117/12.845781

45. M. Lowisch, P. Kuerz, H.-J. Mann, O. Natt, and B. Thuering. 2010. Optics for EUV production. *Proc. SPIE* **7636**, pp. 763603-1–763603-11. doi: 10.1117/12.848624

46. M. Lowisch, P. Kuerz, O. Conradi, G. Wittich, W. Seitz, and W. Kaiser. 2013. Optics for ASML's NXE:3300B platform. *Proc. SPIE* **8679**: 86791H-1–86791H-9. doi: 10.1117/12.2012158

47. J.P.H. Benschop, W.M. Kaiser, and D.C. Ockwell. 1999. EUCLIDES, the EUROPEAN EUVL program. *Proc. SPIE* **3676**: 246–252.

48. W. Kaiser et al. 2008. The future of EUV. *SPIE*, February 2008 Handout and available on request from Carl Zeiss.
49. J.T. Neumann, P. Gräupner, W. Kaiser, R. Garreis, and B. Geh. 2013. Mask effects for high-NA EUV: impact of NA, chief-ray-angle, and reduction ratio. *Proc. SPIE* **8679**: 867915-1–867915-13. doi: 10.1117/12.2011455
50. Carl Zeiss SMT GmbH. 2010. International Publication Number WO 2012/034995 A2 (Priority Date September 9, 2010).
51. S. Migura, B. Kneer, J.T. Neumann, W. Kaiser, and J. van Schoot. 2014. EUV lithography optics for sub 9 nm resolution. In *13th International Extreme Ultra-Violet Lithography (EUVL) Symposium*, Washington, DC, October 27–29. http://euvlsymposium.lbl.gov/proceedings/2014
52. International SEMATECH. 1998. *2nd Next Generation Lithography Workshop*, Colorado Springs, CO, December 7–10.
53. U. Dinger, F. Eisert, H. Lasser, M. Mayer, A. Seifert, G. Seitz, S. Stacklies, F.-J. Stickel, and M. Weiser. 2000. Mirror substrates for EUV lithography: progress in metrology and optical fabrication technology. *Proc. SPIE* **4146**: 35–46. doi: 10.1117/12.406674
54. U. Dinger, G. Seitz, S. Schulte, F. Eisert, C. Muenster, S. Burkart, S. Stacklies, C. Bustaus, H. Hoefer, M. Mayer, B. Fellner, O. Hocky, M. Rupp, K. Riedelsheimer, P. Kuerz. 2003. Fabrication and metrology of diffraction limited soft x-ray optics for the EUV microlithography. *Proc. SPIE* **5193**: 18–28. doi: 10.1117/12.511489
55. P.G. Murray, T. Böhm, and H. Maltor. 2006. Use of nanocrystallite ceria in EUV lithography optics polishing. In *Technical Proceedings of the 2006 NSTI Nanotechnology Conference and Trade Show*, ed. M. Laudon and B.F. Romanowicz, Vol. 3, 218–220. Boca Raton, FL: CRC Press / Taylor & Francis Informa Group.
56. J.S. Taylor and R. Soufli. 2009. Specification, fabrication, testing and mounting of EUV optical substrates. In *EUV Lithography*, ed. V. Bakshi, 161–185. Bellingham, WA: SPIE Press.
57. L.A. Marchetti. 2003. Fabrication and metrology of 10X Schwarzschild optics for EUV imaging. *Proc. SPIE* **5193**: 1–10. doi: 10.1117/12.510318
58. T. Miura, K. Murakami, H. Kawai, Y. Kohama, K. Morita, K. Hada, and Y. Ohkubo. 2010. Nikon EUV development progress update. *Proc. SPIE* **7636**: 76361G-1–76361G-16. doi: 10.1117/12.846459
59. S. Uzawa, H. Kubo, Y. Miwa, T. Tsuji, H. Morishima, and K. Kajiyama. 2008. Canon's development status of EUV technologies. *Proc. SPIE* **6921**: 69210N-1–69210N-8. doi: 10.1117/12.769894
60. H. Glatzel, D. Ashworth, M. Bremer, R. Chin, K. Cummings, L. Girard, M. Goldstein, E. Gullikson, R. Hudyma, J. Kennon, B. Kestner, L. Marchetti, P. Naulleau, R. Soufli, and E. Spiller. 2013. Projection optics for extreme ultraviolet lithography (EUV) micro-field exposure tools (METs) with a numerical aperture of 0.5. *Proc. SPIE* **8679**: 867917-1–867917-16. doi: 10.1117/12.2012698
61. I. Mitra, J. Alkemper, U. Nolte, A. Engel, R. Mueller, S. Ritter, H. Hack, K. Megges, H. Kohlmann, W. Pannhorst, M.J. Davis, L. Aschke, and K. Knapp. 2003. Improved materials meeting the demands for EUV substrates. *Proc. SPIE* **5037**: 219–226. doi: 10.1117/12.484728
62. N. Kousuke. 2003. Low-thermal-expansion material for EUV applications. *SPIE* **5256**: 1271–1280. doi: 10.1117/12.518542
63. K.E. Hrdina, B.G. Ackerman, A.W. Fanning, C.E. Heckle, D.C. Jenne, and W.D. Navan. 2003. Measuring and tailoring CTE within ULE glass. *Proc. SPIE* **5037**: 227–235. doi: 10.1117/12.484925
64. U. Dinger, F. Eisert, S. Koehler, A. Ochse, J. Zellner, M. Lowisch, and T. Laufer. 2009. Projection objective. US Patent 7557902B2.
65. J.P.H. Benschop, U. Dinger, and D.C. Ockwell. 2000. EUCLIDES. First phase completed. *Proc. SPIE* **3997**: 34–47. doi: 10.1117/12.390073
66. S. Bajt, J.B. Alameda, T.W. Barbee Jr., W.M. Clift, J.A. Folta, B. Kaufmann, and E.A. Spiller. 2002 Improved reflectance and stability of Mo-Si multilayers. *Opt. Eng.* **41**(8): 1797–1804.
67. J. Bosgra, E. Zoethout, A.M.J. van der Eerden, J. Verhoeven, R.W.E. van de Kruijs, A.E. Yakshin, and F. Bijkerk. 2012. Structural properties of subnanometer thick Y layers in extreme ultraviolet multilayer mirrors. *Appl. Opt.* **51**(36): 8541–8548. doi: 10.1364/AO.51.008541
68. Fraunhofer-Institut für Werkstoff- und Strahltechnik. 2015. Weltrekord beim Reflexionsgrad von EUV-Lithografiespiegeln, Press Release Nr. XXXI – Fraunhofer IWS November 12. www.iws.fraunhofer.de/de/presseundmedien/presseinformationen/2015/presseinformation_2015-31.html (accessed March 19, 2017).
69. D.L. Windt, W.L. Brown, C.A. Volkert, and W.K. Waskiewicz. 1995. Variation in stress with background pressure in sputtered Mo/Si multilayer films. *J. Appl. Phys.* **78**(4): 2423–2430. doi: 10.1063/1.360164
70. P.B. Mirkarimi and C. Montcalm. 1998. Advances in the reduction and compensation of film stress in high reflectance multilayer coatings for extreme ultraviolet lithography. *Proc. SPIE* **3331**: 133–148. doi: 10.1117/12.309565

71. E. Zoethout, G. Sipos, R.W.E. van de Kruijs, A.E. Yakshin, E. Louis, S. Muellender, and F. Bijkerk. 2003. Stress mitigation in Mo/Si multilayers for EUV lithography. *Proc. SPIE* **5037**: 872–877. doi: 10.1117/12.490138

72. P.B. Mirkarimi. 1999. Stress, reflectance and temporal stability of sputter deposited Mo/Si and Mo/Be multilayer films for extreme ultraviolet lithography. *Opt. Eng.* **38**(7): 1246–1259. doi: 10.1117/1.602170

73. E. Louis, A.E. Yakshin, T. Tsarfati, and F. Bijkerk. 2011. Nanometer interface and materials control for multilayer EUV-optical applications. *Prog. Surf. Sci.* **86**(11–12): 255–294. doi: 10.1016/j.progsurf.2011.08.001

74. E. Spiller, A. Segmüller, J. Rife, and R.-P. Haelbich. 1980. Controlled fabrication of multilayer soft x-ray mirrors. *Appl. Phys. Lett.* **37**(11): 1048–1050. doi: 10.1063/1.91759

75. T.W. Barbee Jr. 1986. Multilayers for X-ray optics. *Opt. Eng.* **25**(8): 893–915. doi: 10.1117/12.7973929

76. E. Spiller, S.L. Baker, P.B. Mirkarimi, V. Sperry, E.M. Gullikson, and D.G. Stearns. 2003. High performance Mo-Si multilayer coatings for extreme ultraviolet lithography by ion-deposition. *Appl. Opt.* **42**(19): 4040–4058. doi: 10.1364/AO.42.004049

77. E. Louis, E. Zoethout, R.W.E. van de Kruijs, I. Nedelcu, A.E. Yakshin, S.A. van der Westen, T. Tsarfati, F. Bijkerk, H. Enkisch, and S. Muellender. 2005. Multilayer coatings for the EUV process development tool. *Proc. SPIE* **5751**: 1170–1177. doi: 10.1117/12.619856

78. E. Louis, E.D. van Hattum, S.A. van der Westen, P. Sallé, K.T. Grootkarzijn, E. Zoethout, F. Bijkerk, G. von Blanckenhagen, and S. Müllender. 2010. High reflectance multilayers for the EUV HVM projection optics. *Proc. SPIE* **7636**: 76362T-1–76362T-5. doi: 10.1117/12.846566

79. http://www.scia-systems.com/products/plasma-systems-pvd/multilayer-pvd-scia-multi-680.html

80. The active sun from the Solar Dynamics Observatory at 131 Å. 25 September 2011; https://svs.gsfc.nasa.gov

81. D. Attwood. 2007. *Soft X-Rays and Extreme Ultraviolet Radiation: Principles and Applications.* Cambridge: Cambridge University Press.

82. J.B. Murphy, D.L. White, A.A. MacDowell, and O.R. Wood. 1993. Synchrotron radiation sources and condensers for projection x-ray lithography. *Appl. Opt.* **32**(34): 6920–6929. doi: 10.1364/AO.32.006920

83. N.R. Böwering, A.I. Ershov, W.F. Marx, O.V. Khodykin, B.A.M. Hansson, E. Vargas L., J.A. Chavez, I.V. Fomenkov, D.W. Myers, and D.C. Brandt. 2006. EUV source collector. *Proc. SPIE* **6151**: 61513R-1–61513R-9. doi: 10.1117/12.656462

84. P. Marczuk and W. Egle. 2004. Source collection optics for EUV lithography. *Proc. SPIE* **5533**: 145–156. doi: 10.1117/12.549409

85. G.D. Kubiak, L.J. Bernardez II, and K.D. Krenz. 1998. High-power extreme-ultraviolet source based on gas jets. *Proc. SPIE* **3331**: 81–89. doi: 10.1117/12.309560

86. Canon, ASML and Nikon. 2004. Joint requirements. In *SEMATECH EUV Source Workshop*, Santa Clara, CA, February 22. www.sematech.org

87. P. Naulleau, C. Anderson, W. Chao, S. Bhattarai, A. Neureuther, K. Cummings, S.-H. Jen, M. Neisser, and B. Thomas. 2014. EUV resists: pushing to the extreme. *J. Photopolym. Sci. Technol.* **27**(6): 725–730. doi: 10.2494/photopolymer.27.725

88. B. Wu and A. Kumar. 2009. *Extreme Ultraviolet Lithography.* New York: McGraw-Hill.

89. I.A. Makhotkin, E. Zoethout, E. Louis, A.M. Yakunin, S. Müllender, and F. Bijkerk. 2012. Wavelength selection for multilayer coatings for lithography generation beyond extreme ultraviolet. *J. Micro/Nanolith. MEMS MOEMS* **11**(4): 040501. doi: 10.1117/1.JMM.11.4.040501

90. Y. Platonov, J. Rodriguez, M. Kriese, E. Gullikson, T. Harada, T. Watanabe, and H. Kinoshita. 2011. Multilayers for next generation EUVL at 6.x nm. *Proc. SPIE* **8076**: 80760N-1–80760N-9. doi: 10.1117/12.889519

91. I.V. Fomenkov, B. La Fontaine, D.J. Brown, I. Ahmad, P. Baumgart, N.R. Bowering, D.C. Brandt, A.N. Bykanov, S. De Dea, A.I. Ershov, N.R. Farrar, D.J. Golich, M.J. Lercel, D.W. Myers, C. Rajyaguru, S. Srivastava, Y. Tao, and G.O. Vaschenko. 2012. Development of stable extreme-ultraviolet sources for use in lithography exposure systems. *J. Micro/Nanolith. MEMS MOEMS*, **11**(2): 021110-1. doi: 10.1117/1.JMM.11.2.021110

92. V. Bakshi, ed. 2009. *EUV Lithography.* Bellingham, WA: SPIE Press.

93. A. Endo, K. Sakaue, M. Washio, and H. Mizoguchi. 2014. Optimization of high average power FEL beam for EUV lithography application. In *Proc. of the 36th International Free Electron Laser Conference*, 990–992. Basel, Switzerland, August 25–29; http://accelconf.web.cern.ch/AccelConf/FEL2014/papers/proceed.pdf (accessed March 19, 2017).

94. G. Stupakov and M.S. Zolotorev. 2014. FEL oscillator for EUV lithography. SLAC-PUB-15900. http://www.slac.stanford.edu/cgi-wrap/getdoc/slac-pub-15900.pdf (accessed March 19, 2017).

95. W.H. Bennett. 1955. Self-focusing streams. *Phys. Rev.* **98**(6): 1584–1593. doi: 10.1103/PhysRev.98.1584

96. E.M. Bazelyan and Y.P. Raizer. 2000. *Lightning Physics and Lightning Protection*. Boca Raton, FL: CRC Press.

97. M. Whitfield, A. Brunton, V. Truffert, P. Gruenewald, J. Cashmore, P. Richards, and M. Gower. 2003. MS-13 EUV Microstepper. In *2nd International Extreme Ultra-Violet Lithography (EUVL) Symposium*, Antwerp, Belgium, September 30–October 2. http://euvlsymposium.lbl.gov/proceedings/2003

98. U. Stamm. 2004. EUV Source Development at Xtreme Technologies – An update. In *3rd International Extreme Ultra-Violet Lithography (EUVL) Symposium*, Miyazaki, Japan, November 2–4. http://euvlsymposium.lbl.gov/proceedings/2004

99. U Stamm, J. Kleinschmidt, K. Gabel, G. Hergenhan, C. Ziener, G. Schriever, I. Ahmad, D. Bolshukhin, J. Brudermann, R. de Bruijn, T.D. Chin, A. Geier, S. Gotze, A. Keller, V. Korobotchko, B. Mader, J. Ringling, and T. Brauner. 2005. EUV sources for EUV lithography in alpha-, beta-, and high volume chip manufacturing: an update on GDPP and LPP technology. *Proc. SPIE* **5751**: 236–247. doi: 10.1117/12.599544

100. K. Tawarayama, H. Aoyama, S. Magoshi, Y. Tanaka, S. Shirai, and H. Tanaka. 2009. Recent progress of EUV full-field exposure tool in selete. *Proc. SPIE* **7271**: 727118. doi: 10.1117/12.813627

101. G. Bianucci, A. Bragheri, G.L. Cassol, R. Ghislanzoni, R. Mazzoleni, and F.E. Zocchi. 2011. Enabling the 22 nm node via grazing incidence collectors integrated into the DPP source for EUVL HVM. *Proc. SPIE* **7969**: 79690B-1–79690B-9. doi: 10.1117/12.879396

102. N.V. Filippov, T.I. Filippova, and V.P. Vinogradov. 1962. Dense high-temperature plasma in a non-cylindrical z-pinch compression. *Nucl. Fusion Suppl.* **2**: 577–587 (in Russian).

103. J.W. Mather. 1971. Dense plasma focus. In *Methods in Experimental Physics, Part B Plasma Physics*, ed. R.H. Lovberg and H.R. Griem, Vol. 9, 187–249. Cambridge, MA: Academic Press. doi: 10.1016/S0076-695X(08)60862-5

104. I.V. Fomenkov, N. Böwering, C.L. Rettig, S.T. Melnychuk, I.R. Oliver, J.R. Hoffman, O.V. Khodykin, R.M. Ness, and W.N. Partlo. 2004. EUV discharge light source based on a dense plasma focus operated with positive and negative polarity. *J. Phys. D: Appl. Phys.* **37**(23): 3266–3276. doi: 10.1088/0022-3727/37/23/007

105. I.V. Fomenkov, R.M. Ness, I.R. Oliver, S.T. Melnychuk, O.V. Khodykin, N.R. Boewering, C.L. Rettig, and J.R. Hoffman. 2004. Performance and scaling of a dense plasma focus light source for EUV lithography. *Proc. SPIE* **5374**: 168–182. doi: 10.1117/12.538690

106. N.R. Fornaciari, H. Bender, D. Buchenauer, M.P. Kanouff, S. Karim, G.D. Kubiak, C.D. Moen, G.M. Shimkaveg, W.T. Silfvast, and K.D. Stewart. 2001. Development of a high-average-power extreme-ultraviolet electric capillary discharge source. *Proc. SPIE* **4343**: 226–231. doi: 10.1117/12.436652

107. M.A. Gundersen and G. Schaefer. 2012. *Physics and Applications of Pseudosparks*, New York: Springer Science & Business Media.

108. J. Pankert, K. Bergmann, J. Klein, W. Neff, O. Rosier, S. Seiwert, C. Smith, S. Probst, D. Vaudrevange, G. Siemons, R. Apetz, J. Jonkers, M. Loeken, E. Bosch, G.H. Derra, T. Kruecken, and P. Zink. 2003. Physical properties of the HCT EUV source. *Proc. SPIE* **5037**: 112–118. doi: 10.1117/12.483611

109. J. Pankert, R. Apetz, K. Bergmann, M. Damen, G. Derra, O. Franken, M. Janssen, J. Jonkers, J. Klein, H. Kraus, T. Krücken, A. List, M. Loeken, A. Mader, C. Metzmacher, W. Neff, S. Probst, R. Prümmer, O. Rosier, S. Schwabe, S. Seiwert, G. Siemons, D. Vaudrevange, D. Wagemann, A. Weber, P. Zink, and O. Zitzen. 2006. EUV sources for the alpha-tools. *Proc. SPIE* **6151**: 61510Q-1–61510Q-9. doi: 10.1117/12.657066

110. N. Harned, M. Goethals, R. Groeneveld, P. Kuerz, M. Lowisch, H. Meijer, H. Meiling, K. Ronse, J. Ryan, M. Tittnich, H.-J. Voorma, J. Zimmerman, U. Mickan, and S. Lok. 2007. EUV lithography with the alpha demo tools: status and challenges. *Proc. SPIE* **6517**: 651706-1–651706-12. doi: 10.1117/12.712065

111. M. Yoshioka, Y. Teramoto, J. Jonkers, M.C. Schürmann, R. Apetz, V. Kilian, and M. Corthout. 2011. Tin DPP source collector module (SoCoMo) ready for integration into Beta scanner. *Proc. SPIE* **7969**: 79691G-1–79691G-9. doi: 10.1117/12.879386

112. D.C. Brandt, I.V. Fomenkov, A.I. Ershov, W.N. Partlo, D.W. Myers, R.L. Sandstrom, B.M. La Fontaine, M.J. Lercel, A.N. Bykanov, N.R. Böwering, G.O. Vaschenko, O.V. Khodykin, S.N. Srivastava, I. Ahmad, C. Rajyaguru, P. Das, V.B. Fleurov, K. Zhang, D.J. Golich, S. De Dea, R.R. Hou, W.J. Dunstan, C.J. Wittak, P. Baumgart, T. Ishihara, R.D. Simmons, R.N. Jacques, and R.A. Bergstedt. 2011. LPP source system development for HVM. *Proc. SPIE* **7969**: 79691H-1–79691H-8. doi: 10.1117/12.882208

113. G.D. Kubiak, D.A. Tichenor, M.E. Malinowski, R.H. Stulen, S.J. Haney, K.W. Berger, L.A. Brown, J.E. Bjorkholm, R.R. Freeman, W.M. Mansfield, D.M. Tennant, O.R. Wood II, J. Bokor, T.E. Jewell, D.L. White, D.L. Windt, and W.K. Waskiewicz. 1991. Diffraction-limited soft x-ray projection lithography with a laser plasma source. *J. Vac. Sci. Technol. B* **9**(6): 3184–3188. doi: 10.1116/1.585313

114. S.J. Haney, K.W. Berger, G.D. Kubiak, P.D. Rockett, and J. Hunter. 1993. Prototype high-speed tape target transport for a laser plasma soft-x-ray projection lithography source. *Appl. Opt.* **32**(34): 6934–6937. doi: 10.1364/AO.32.006934

115. D.A. Tichenor, G.D. Kubiak, W.C. Replogle, L.E. Klebanoff, J.B. Wronosky, L.C. Hale, H.N. Chapman, J.S. Taylor, J.A. Folta, C. Montcalm, R.M. Hudyma, K.A. Goldberg, and P.P. Naulleau. 2000. EUV engineering test stand. *Proc. SPIE* **3997**: 48–69. doi: 10.1117/12.390083

116. B.A.M. Hansson, L. Rymell, M. Berglund, O.E. Hemberg, E. Janin, J. Thoresen, S. Mosesson, J. Wallin, and H.M. Hertz. 2002. Status of the liquid-xenon-jet laser-plasma source for EUV lithography. *Proc. SPIE* **4688**: 102–109. doi: 10.1117/12.472274

117. M. Richardson, C.-S. Koay, K. Takenoshita, C. Keyser, and M. Al-Rabban. 2004. High conversion efficiency mass-limited Sn-based laser plasma source for extreme ultraviolet lithography. *J. Vac. Sci. Technol. B* **22**(2): 785–790. doi: 10.1116/1.1667511

118. D.C. Brandt, I.V. Fomenkov, N.R. Farrar, B. La Fontaine, D.W. Myers, D.J. Brown, A.I. Ershov, R.L. Sandstrom, G.O. Vaschenko, N.R. Böwering, P. Das, V.B. Fleurov, K. Zhang, S.N. Srivastava, I. Ahmad, C. Rajyaguru, S. De Dea, W.J. Dunstan, P. Baumgart, T. Ishihara, R.D. Simmons, R.N. Jacques, R.A. Bergstedt, P.I. Porshnev, C.J. Wittak, R.J. Rafac, J. Grava, A.A. Schafgans, Y. Tao, K. Hoffmann, T. Ishikawa, D.R. Evans, and S.D. Rich. 2013. CO_2/Sn LPP EUV sources for device development and HVM. *Proc. SPIE* **8679**: 86791G-1–86791G-8. doi: 10.1117/12.2011212

119. T. Feigl, M. Perske, H. Pauer, T. Fiedler, S. Yulin, M. Trost, S. Schröder, A. Duparré, N. Kaiser, A. Tünnermann, N.R. Böwering, A.I. Ershov, K. Hoffmann, B. La Fontaine, and K.D. Cummings. 2012. Optical performance of LPP multilayer collector mirrors. *Proc. SPIE* **8322**: 832217-1–832217-8. doi: 10.1117/12.919735

120. D.R. Kania, D.P. Gaines, D.S. Sweeney, G.E. Sommargren, B. La Fontaine, S.P. Vernon, D.A. Tichenor, J.E. Bjorkholm, F. Zernike, and R.N. Kestner. 1996. Precision optical aspheres for extreme ultraviolet lithography. *J. Vac. Sci. Technol. B* **14**(6): 3706–3708. doi: 10.1116/1.588652

121. J. Fujimoto, T. Hori, T. Yanagida, and H. Mizoguchi. 2012. Development of laser-produced tin plasma-based EUV light source technology for HVM EUV lithography. *Phys. Res. Int.* **2012**: Article ID 249495. doi: 10.1155/2012/249495

122. M.A. Purvis, A. Schafgans, D.J.W. Brown, I. Fomenkov, R. Rafac, J. Brown, Y. Tao, S. Rokitski, M. Abraham, M. Vargas, S. Rich, T. Taylor, D. Brandt, A. Pirati, A. Fisher, H. Scott, A. Koniges, D. Eder; S. Wilks, A. Link, and S. Langer. 2016. Advancements in predictive plasma formation modeling. *Proc. SPIE* **9776**: 97760K-1–97760K-12. doi: 10.1117/12.2221991

123. M.E. Malinowski, P.A. Grunow, C. Steinhaus, W.M. Clift, L.E. Klebanoff. 2001. Use of molecular oxygen to reduced EUV-induced carbon contamination of optics. *Proc. SPIE* **4343**: 347–356. doi: 10.1117/12.436677

124. P. Yan. 2005. Masks for extreme ultraviolet lithography. In *Handbook of Photomask Manufacturing Technology*, ed. Syed Rizvi, 245–246. Boca Raton, FL: CRC Press.

125. H. Meiling, H. Meijer, V. Banine, R. Moors, R. Groeneveld, H.-J. Voorma, U. Mickan, B. Wolschrijn, B. Mertens, P. van Baars, P. Kürz, and N. Harned. 2006. First performance results of the ASML alpha demo tool. *Proc. SPIE* **6151**: 615108-1–615108-12. doi: 10.1117/12.657348

126. T. Miura, K. Murakami, K. Suzuki, Y. Kohama, K. Morita, K. Hada, and Y. Ohkubo. 2007. Nikon EUVL development progress update. *Proc. SPIE* **6517**: 651707-1–651707-10. doi: 10.1117/12.711267

127. K. Standiford and C. Bürgel. 2013. A new mask linearity specification for EUV masks based on time dependent dielectric breakdown requirements. *Proc. SPIE* **8880**: 88801M-1–88901M-7. doi: 10.1117/12.2023109

128. International Technology Roadmap for Semiconductors (ITRS). 2016. Update Edition, Lithography, Table LITH5c EUVL Mask Requirements, Semiconductor Industry Association (SIA), 1101 K Street NW, Washington, DC, 20005, USA.

129. P.P. Naulleau, K.A. Goldberg, E. Gullikson, I. Mochi, B. McClinton, and A. Rastegar. 2011. Accelerating EUV learning with synchrotron light: mask roughness challenges ahead. *Proc. SPIE* **8166**: 81660F-1–81660F-7. doi: 10.1117/12.900488

130. A. JohnKadaksham, R. Teki, M. Godwin, M. House, and F. Goodwin. 2013. Low thermal expansion material (LTEM) cleaning and optimization for extreme ultraviolet (EUV) blank deposition. *Proc. SPIE* **8679**: 86791R-1–86791R-7. doi: 10.1117/12.2011718

131. B. Henke, E. Gullikson, and J. Davis. 1993. X-ray interactions: photoabsorption, scattering, transmission, and reflection at E = 50–30000 eV, Z = 1–92. *At. Data Nucl. Data Tables* **54**: 181–342. http://henke.lbl.gov/optical_constants/

132. M. Shriaishi, N. Kandaka, and K. Murakami. 2003. Mo/Si multilayers deposited by low-pressure rotary magnet cathode sputtering for extreme ultraviolet lithography. *Proc. SPIE* **5037**: 249–256. doi: 10.1117/12.484436

133. D.M. Tennant, L.A. Fetter, L.R. Harriott, A.A. MacDowell, P.P. Mulgrew, J.Z. Pastalan, W.K. Waskiewicz, D.L. Windt, and O.R. Wood. 1993. Mask technology for soft-x-ray projection lithography at 13 nm. *Appl. Opt.* **32**(34): 7007–7011. doi: 10.1364/AO.32.007007

134. K.B. Nguyen, T. Mizota, T. Haga, H. Kinoshita, and D.T. Attwood. 1994. Imaging of extreme ultraviolet lithographic masks with programmed substrate defects. *J. Vac. Sci. Technol. B* **12**: 3833–3840. doi: 10.1116/1.587450

135. K.B. Nguyen, A.K. Ray-Chaudhuri, R.H. Stulen, K. Krenz, L.A. Fetter, D.M. Tennant, and D.L. Windt. 1995. Printability of substrate and absorber defects on extreme ultraviolet lithography masks. *J. Vac. Sci. Technol. B* **13**: 3082–3088. doi: 10.1116/1.588327

136. P.A. Kearney, R.V. Randive, A. Ma, D. Krick, A. Weaver, I. Reiss, D. Abraham, P.B. Mirkarimi, and E. Spiller. 2004. Overcoming substrate defect decoration effects in EUVL mask blank development. *Proc. SPIE* **5567**: 800–806. doi: 10.1117/12.569271

137. P.A. Kearney, C.E. Moore, S.I. Tan, S.P. Vernon, and R.A. Levesque. 1997. Mask blanks for extreme ultraviolet lithography: ion beam sputter deposition of low defect density Mo/Si multilayers. *J. Vac. Sci. Technol. B* **15**: 2452–2454. doi: 10.1116/1.589665

138. E. Gallagher, K. Badger, L. Kindt, M. Lawliss, G. McIntyre, A. Wagner, and J. Whang. 2011. EUV masks: ready or not? In *10th International Extreme Ultra-Violet Lithography (EUVL) Symposium*, Miami, Florida, October 17–19. http://euvlsymposium.lbl.gov/proceedings/2011

139. J.H. Peters, S. Perlitz, U. Matejka, W. Harnisch, D. Hellweg, M. Weiss, M. Waiblinger, T. Bret, T. Hofmann, K. Edinger, and K. Kornilov. 2011. EUV defect repair strategy. In *10th International Extreme Ultra-Violet Lithography (EUVL) Symposium*, Miami, Florida, October 17–19. http://euvlsymposium.lbl.gov/proceedings/2011

140. E.M. Gullikson, C. Cerjan, D.G. Stearns, P.B. Mirkarimi, and D.W. Sweeney. 2002. Practical approach for modeling extreme ultraviolet mask defects. *J. Vac. Sci. Technol. B* **20**: 81–86. doi: 10.1116/1.1428269

141. A.R. Stivers, T. Liang, M.J. Penn, B. Lieberman, G.V. Shelden, J.A. Folta, C.C. Larson, P.B. Mirkarimi, C.C. Walton, E.M. Gullikson, and M. Yi. 2002. Evaluation of the capability of a multi-beam confocal inspection system for inspection of EUVL mask blanks. *Proc. SPIE* **4889**: 408–417. doi: 10.1117/12.468199

142. J.-P. Urbach, J.F.W. Cavelaars, H. Kusunose, T. Liang, and A.R. Stivers. 2003. EUV substrate and blank inspection with confocal microscope. *Proc. SPIE* **5256**: 556–565. doi: 10.1117/12.518388

143. P. Kearney, W.-I. Cho, C.-U. Jeon, E. Gullikson, A. Jia, T. Tamura, A. Tajimac, and H. Kusunose. 2006. State of the art EUV mask blank inspection with a Lasertec M7360 at the SEMATECH MBDC. In *5th International Extreme Ultra-Violet Lithography (EUVL) Symposium*, Barcelona, Spain, October 16-18. http://euvlsymposium.lbl.gov/proceedings/2006

144. D.W. Pettibone, A. Veldman, T. Liang, A.R. Stivers, P.J.S. Mangat, B. Lu, S.D. Hector, J.R. Wasson, K.L. Blaedel, E. Fisch, and D.M. Walker. 2002. Inspection of EUV reticles. *Proc. SPIE* **4688**: 363–374. doi: 10.1117/12.472311

145. S. Stokowski, J. Glasser, G. Inderhees, and P. Sankuratri. 2010. Inspecting EUV mask blanks with a 193-nm system. *Proc. SPIE* **7636**: 76360Z-1–76360Z-9. doi: 10.1117/12.850825

146. D. Wack, Y. Xiong, and G. Inderhees. 2011. Solutions for EUV mask and blank inspections. In *10th International Extreme Ultra-Violet Lithography (EUVL) Symposium*, Miami, Florida, October 17–19. http://euvlsymposium.lbl.gov/proceedings/2011

147. Y. Tezuka, M. Ito, T. Terasawa, and T. Tomie. 2004. Actinic detection and signal characterization of multilayer defects on EUV mask blanks. *Proc. SPIE* **5567**: 791–799. doi: 10.1117/12.568379

148. A. Barty, Y. Liu, E. Gullikson, J.S. Taylor, and O. Wood. 2005. Actinic inspection of multilayer defects on EUV masks. *Proc. SPIE* **5751**: 651–659. doi: 10.1117/12.598488

149. T. Terasawa, Y. Tezuka, M. Ito, and T. Tomie. 2004. High speed actinic EUV mask blank inspection with dark-field imaging. *Proc. SPIE* **5446**: 804–811. doi: 10.1117/12.557814

150. H. Miyai, T. Suzuki, K. Takehisa, H. Kusunose, T. Yamane, T. Terasawa, H. Watanabe, and I. Mori. 2013. The capability of high magnification review function for EUV actinic blank inspection tool. *Proc. SPIE* **8701**: 870118-1–870118-7. doi: 10.1117/12.2030712

151. D. Wack, Q.Q. Zhang, G. Inderhees, and D. Lopez. 2010. Mask inspection technologies for 22 nm HP and beyond. *Proc. SPIE* **7636**: 76360V-1–76360V-9. doi: 10.1117/12.850766

152. S. Mangan, C.C. Lin, G. Hughes, R. Brikman, A. Goldenshtein, V. Kudriashov, A. Litman, L. Shoval, and I. Englard. 2011. Study of EUV mask e-beam inspection conditions for HVM. *Proc. SPIE* **8166**: 816612-1–816612-8. doi: 10.1117/12.899067

153. S. Yamaguchi, M. Naka, T. Hirano, M. Itoh, M. Kadowaki, T. Koike, Y. Yamazaki, K. Terao, M. Hatakeyama, K. Watanabe, H. Sobukawa, T. Murakami, T. Karimata, K. Tsukamoto, T. Hayashi, R. Tajima, N. Kimura, and N. Hayashi. 2011. Performance of EBeyeM for EUV mask inspection. *Proc. SPIE* **8166**: 81662F-1–81662F-8. doi: 10.1117/12.898790

154. H. Watanabe, R. Hirano, S. Iida, T. Amano, T. Terasawa, M. Hatakeyama, T. Murakami, S. Yoshikawa, and K. Terao. 2013. EUV patterned mask inspection system using projection electron microscope techniques. *Proc. SPIE* **8880**: 88800U-1–88800U-9. doi: 10.1117/12.2027566

155. T. Shimomura, S. Narukawa, T. Abe, T. Takikawa, N. Hayashi, F. Wang, L. Ma, C.-W. Lin, Y. Zhao, C. Kuan, and J. Jau. 2012. Electron beam inspection of 16 nm HP node EUV masks. *Proc. SPIE* **8522**: 85220L-1–85220L-7. doi: 10.1117/12.976017

156. K.A. Goldberg and I. Mochi. 2010. Wavelength-specific reflections: a decade of EUV actinic mask inspection research. *J. Vac. Sci. Technol. B* **28**: C6E1-10. doi: 10.1116/1.3498757

157. K.A. Goldberg, M.P. Benk, A. Wojdyla, I. Mochi, S.B. Rekawa, A.P. Allezy, M.R. Dickinson, C.W. Cork, W. Chao, D.J. Zehm, J.B. Macdougall, P.P. Naulleau, and A. Rudack. 2014. Actinic mask imaging recent result and future directions from the SHARP EUV microscope. *Proc. SPIE* **9048**: 904804-1–904104-10. doi: 10.1117/12.2048364

158. M.R. Weiss, D. Hellweg, J.H. Peters, S. Perlitz, A. Garetto, and M. Goldstein. 2014. Actinic review of EUV masks: first results from the AIMS™ EUV system integration. *Proc. SPIE* **9048**: 90480X-1–90480X-9. doi: 10.1117/12.2046302

159. M.R. Weiss, D. Hellweg, M. Koch, J.H. Peters, S. Perlitz, A. Garetto, K. Magnusson, R. Capelli, and V. Jindal. 2015. Actinic review of EUV masks: status and recent results of the AIMS™ EUV system. *Proc. SPIE* **9422**: 942219-1–942219-9. doi: 10.1117/12.2086265

160. L. He, J. Lystad, S. Wurm, K. Orvek, J. Sohn, A. Ma, P. Kearney, S. Kolbow, and D. Halbmaier. 2009. Protection efficiency of a standard compliant EUV reticle handling solution. *Proc. SPIE* **7271**: 727101-1–727101-10. doi: 10.1117/12.814304

161. SEMI E152-0214 Mechanical Specifications of EUV pod for 150 nm EUVL reticles, Semiconductor Equipment and Materials International, SEMI® 3081 Zanker Road, San Jose, CA.

162. A. Pirati, R. Peeters, D. Smith, S. Lok, A.W.E. Minnaert, M. van Noordenburg, J. Mallmann, N. Harned, J. Stoeldraijer, C. Wagner, C. Zoldesi, E. van Setten, J. Finders, K. de Peuter, C. de Ruijter, M. Popadic, R. Huang, M. Lin, F. Chuang, R. van Es, M. Beckers, D. Brandt, N. Farrar, A. Schafgans, D. Brown, H. Boom, H. Meiling, and R. Kool. 2015. Performance overview and outlook of EUV lithography systems. *Proc. SPIE* **9422**: 94221P-1–94221P-18. doi: 10.1117/12.2085912

163. Y. Hyun, J. Kim, K. Kim, S. Koo, S. Kim, Y. Kim, C. Lim, and N. Kwak. 2015. EUV mask particles adders during scanner exposure. *Proc. SPIE* **9422**: 94221U-1–94221U-7. doi: 10.1117/12.2085626

164. S. Lee, J. Kim, S.-W. Koh, I. Jang, J. Choi, H. Ko, H.-S. Seo, S.-S. Kim, B.G. Kim, C.-U. Jeon. 2014. Durability of Ru-based EUV masks and the improvement. *Proc. SPIE* **9048**: 90480J-1–90480J-7. doi: 10.1117/12.2046556

165. Y.A. Shroff, M. Goldstein, B. Rice, S.H. Lee, K.V. Ravi, and D. Tanzil. 2006. EUV pellicle development for mask defect control. *Proc. SPIE* **6151**: 715104-1–615104-10. doi: 10.1117/12.656551

166. S. Akiyama, and Y. Kubota. 2009. Realization of EUV pellicle with single crystal silicon membrane. In *8th International Extreme Ultra-Violet Lithography (EUVL) Symposium*, Prague, Czech Republic, October 18–21. http://euvlsymposium.lbl.gov/proceedings/2009

167. K. Ko, E.-J. Kim, J.-W. Kim, J.-T. Park, C.-M. Lim, and H. Oh. 2012. Effect of extreme-ultraviolet pellicle support for patterned mask. *Proc. SPIE* **8322**: 832230-1–832230-8. doi: 10.1117/12.918019

168. C. Zoldesi. 2015. EUV pellicle is ready for the next step: industrialization. Oral presentation at Extreme Ultraviolet (EUV) Lithography VI, San Jose, California. Ms. Zoldesi failed to provide a manuscript for her paper, but did supply us with a copy of the PowerPoint slides for her presentation, which is where Figure 4.55 originated.

169. M. van den Brink. 2014. Many ways to shrink: the right moves to 10 nanometers and beyond. Keynote Presentation at *SPIE Photomask Technology*, Monterey, CA.

170. Y. Deng, B.M. La Fontaine, H.J. Levinson, and A.R. Neureuther. 2003. Rigorous EM simulation of the influence of the structure of mask patterns on EUVL imaging. *Proc. SPIE* **5037**: 302–313. doi: 10.1117/12.484986

171. K. Takai, K. Murano, T. Kamo, Y. Morikawa, and N. Hayashi. 2014. Capability of etched multilayer EUV mask fabrication. *Proc. SPIE* **9235**: 923515-1–923515-8. doi: 10.1117/12.2067892

172. P. Yan. 2002. EUVL alternating phase shift mask imaging evaluation. *Proc. SPIE* **4889**: 1099–1105. doi: 10.1117/12.468103

173. S. Han, E. Weisbrod, J.R. Wasson, R. Gregory, Q. Xie, P.J.S. Mangat, S.D. Hector, W.J. Dauksher, and K.M. Rosfjord. 2004. Development of phase shift masks for extreme ultraviolet lithography and optical evaluation of phase shift materials. *Proc. SPIE* **5374**: 261–270. doi: 10.1117/12.535503

174. B. La Fontaine, A.R. Pawloski, O. Wood, Y. Deng, H.J. Levinson, P. Naulleau, P.E. Denham, E. Gullikson, B. Hoef, C. Holfeld, C. Chovino, and F. Letzkus. 2006. Demonstration of phase-shift masks for extreme-ultraviolet lithography. *Proc. SPIE* **6151**: 61510A-1–61510A-8. doi: 10.1117/12.652212

175. L. Sun, S. Raghunathan, V. Jindal, E. Gullikson, P. Mangat, I. Mochi, K.A. Goldberg, M.P. Benk, O. Kritsun, T. Wallow, D. Civay, and O. Wood. 2014. Application of phase shift focus monitor in EUVL process control. *Proc. SPIE* **8679**: 86790T-1–86790T-12. doi: 10.1117/12.2011342

176. O.R. Wood II, D.L. White, J.E. Bjorkholm, L.E. Fetter, D.M. Tennant, A.A. MacDowell, B. LaFontaine, and G.D. Kubiak. 1997. Use of attenuated phase masks in extreme ultraviolet lithography. *J. Vac. Sci. Technol. B* **15**: 2448–2451. doi: 10.1116/1.589664

177. P. Yan, M. Leeson, S. Lee, G. Zhang, E. Gullikson, and F. Salmassi. 2011. Extreme ultraviolet-embedded phase-shift mask. *J. Micro/Nanolith. MEMS MOEMS* **10**(3): 033011-1–033011-10. doi: 10.1117/1.3616060

178. A. Erdmann, T. Fühner, P. Evanschitzky, J.T. Neumann, J. Ruoff, and P. Gräupner. 2013. Modeling studies on alternative EUV mask concepts for higher NA. *Proc. SPIE* **8679**: 86790Y-1–86790Y-12. doi: 10.1117/12.2011432

179. Y. Deng, B. La Fontaine, A.R. Pawloski, and A.R. Neureuther. 2004. Simulation of fine structures and defects in EUV etched multilayer masks. *Proc. SPIE* **5374**: 760–769. doi: 10.1117/12.537229

180. L. Van Look, V. Philipsen, E. Hendrickx, G. Vandenberghe, N. Davydova, F. Wittebrood, R. de Kruif, A. van Oosten, J. Miyazaki, T. Fliervoet, J. van Schoot, and J.T. Neumann. 2015. Alternative EUV mask technology to compensate for mask 3D effects. *Proc. SPIE* **9658**: 965801-1–965801-11. doi: 10.1117/12.2197213

181. O. Wood, S. Raghunathan, P. Mangat, V. Philipsen, V. Luong, P. Kearney, E. Verduijn, A. Kumar, S. Patil, C. Laubis, V. Soltwisch, and F. Scholze. 2015. Alternative materials for high numerical aperture extreme ultraviolet lithography masks stacks. *Proc. SPIE* **9422**: 942201-1–942201-12. doi: 10.1117/12.2085022

182. J. Torok, R. Del Re, H. Herbol, S. Das, I. Bocharova, A. Paolucci, L.E. Ocola, C. Ventrice Jr., E. Lifshin, G. Denbeaux, and R.L. Brainard. 2013. Secondary electrons in EUV lithography. *J. Photopolym. Sci. Technol.* **26**(5): 625–634. doi: 10.2494/photopolymer.26.625

183. Comment: It's also possible for an electron to interact with some other material than PAG and generate a reactive species, such as a free radical, that then reacts with a PAG molecule.

184. M. Neisser, K.Y. Cho, and. K. Petrillo. 2012. The physics of EUV photoresist and how it drives strategies for improvement. *J. Photopolym. Sci. Technol.* **25**(1): 87–94. doi: 10.2494/photopolymer.25.87

185. J.W. Thackeray. 2011. Materials challenges for sub-20-nm lithography. *J. Micro/Nanolith. MEMS MOEMS* **10**(3): 033009-1. doi: 10.1117/1.3616067

186. A.R. Gutiérrez and R.J. Cox. 1986. Maximizing light absorption at the bottom of a film. *Polym. Photochem.* **7**(6): 517–521. doi: 10.1016/0144-2880(86)90020-5

187. G. A. Thommes and V.J. Webers. 1985. Spectral response of photosensitive systems. 1. General effect of radiation attenuation throughout coating thickness. *J. Imaging Sci.* **29**: 112–116.

188. B.W. Smith. 1998. Resist processing. In *Microlithography Science and Technology*, ed. J.R. Sheats and B.W. Smith, 515–566. New York: Marcel Dekker.

189. J. Jiang et al., Nanoparticle euv photoresists: patterning via a ligand exchange mechanism? 2014 International Symposium on Extreme Ultraviolet Lithography, Washington DC.

190. J. Passarelli, M. Murphy, R. Del Re, M. Sortland, J. Hotalen, L. Dousharm, Y. Ekinci, M. Neisser, D. Freedman, R. Fallica, D.A. Freedman, and R.L. Brainard. 2015. Organometallic carboxylate resists for extreme ultraviolet with high sensitivity. *J. Micro/Nanolith. MEMS MOEMS* **14**(4): 043503. doi: 10.1117/1.JMM.14.4.043503

191. B. Cardineau, R. Del Re, M. Marnell, H. Al-Mashat, M. Vockenhuber, Y. Ekinci, C. Sarma, D.A. Freedman, and R.L. Brainard. 2014. Photolithographic properties of tin-oxo clusters using extreme ultraviolet light (13.5 nm). *Microelectron. Eng.* **127**: 44–50.

192. A. Grenville. 2013. Advances in directly patternable metal oxides for EUV resist. In *12th International Extreme Ultra-Violet Lithography (EUVL) Symposium*, Toyama, Japan, October 6–10. http://euvlsymposium.lbl.gov/proceedings/2013

193. F. Ogletee. 2015. Combined experimental and theoretical investigation in EUVL radiation chemistry fundamentals. In *14th International Extreme Ultra-Violet Lithography (EUVL) Symposium*, Maastricht, The Netherlands, October 5–7. http://euvlsymposium.lbl.gov/proceedings/2015

194. J.T. Neumann, P. Gräupner, J. Ruoff, W. Kaiser, R. Garreis, and B. Geh. 2012. 3D reticle effects for high-NA EUV lithography. In *11th International Extreme Ultra-Violet Lithography (EUVL) Symposium*, Brussels, Belgium, September 30–October 4. http://euvlsymposium.lbl.gov/proceedings/2012

195. P.A. Kearney, O. Wood, E. Hendrickx, G. McIntyre, S. Inoue, F. Goodwin, S. Wurm, J. van Schoot, and W. Kaiser. 2014. Driving the industry towards a consensus on high-numerical aperture (high-NA) extreme ultraviolet (EUV). *Proc. SPIE* **9048**: 90481Q-1–90481Q-9. doi: 10.1117/12.2048397

196. S. Yulin, T. Feigl, V. Nesterenko, M. Schürmann, M. Perske, H. Pauer, T. Fiedler, and N. Kaiser. 2012. EUV multilayer coatings: potential and limits. In *2012 International Workshop on EUV Lithography*, Maui, HI, June 4–8. https://www.euvlitho.com

197. Y. Platonov, J. Rodriguez, M. Kriese, E. Gullikson, T. Harada, T. Watanabe, and H. Kinoshita. 2011. Multilayers for next generation of EUVL at 6.X nm. *Proc. SPIE* **8076**: 80760N-1–80760N-9. doi: 10.1117/12.889519

198. T. Higashiguchi, Y. Suzuki, M. Kawasaki, H. Ohashi, N. Nakamura, R. Hirose, T. Ejima, W. Jiang, T. Miura, A. Endo, C. Suzuki, K. Tomita, M. Nishikino, S. Fujioka, H. Nishimura, A. Sinahara, D. Nakamura, A. Takahashi, T. Okada, S. Torii, T. Makimura, B. Li, P. Dunne, and G. O'Sullivan. 2013. Efficient light sources at BEUV & water window soft x-ray wavelengths. In *2013 International Workshop on EUV and Soft X-Ray Sources*, Dublin, Ireland, November 7–10. https://www.euvlitho.com

199. C. Anderson, D. Ashworth, L.M. Baclea-An, S. Bhattari, R. Chao, R. Claus, P. Denham, K. Goldberg, A. Grenville, G. Jones, R. Miyakawa, K. Murayama, H. Nakagawa, S. Rekawa, J. Stowers, and P. Naulleau. 2012. The SEMATECH Berkeley MET: demonstration of 15-nm half pitch in chemically amplified EUV resist and sensitivity of EUV resists at 6.x nm. *Proc. SPIE* **8322**: 832212-1–832212-13. doi: 10.1117/12.917386

200. S. Migura, B. Kneer, J.T. Neumann, W. Kaiser, and J. van Schoot. 2015. Anamorphic high-NA EUV lithography optics. *Proc. SPIE* **9661**: 96610T-1–96610T-7. doi: 10.1117/12.2196393

201. X. Liu, R. Howell, S. Hsu, K. Yang, K. Gronlund, F. Driessen, H.-Y. Liu, S. Hansen, K. van Ingen Schenau, T. Hollink, P. van Adrichem, K. Troost, J. Zimmermann, O. Schumann, C. Hennerkes, and P. Gräupner. 2014. EUV source-mask optimization for 7 nm node and beyond. *Proc. SPIE* **9048**: 90480Q-1–90480Q-11. doi: 10.1117/12.2047584

202. E.A. Schneidmiller, V.F. Vogel, H. Weise, and M.V. Yurkov. 2012. Potential of the FLASH free electron laser technology for the construction of a kW-scale light source for next-generation lithography. *J. Micro/Nanolith. MEMS MOEMS* **11**(2): 021122-1. doi: 10.1117/1.JMM.11.2.021122

203. U. Dinger, D. Tuerke, A. Meseck, M. Patra, E. Sohmen, and A. Jankowiak. 2012. Concept study on an accelerator based source for 6.x nm lithography. In *2012 International Workshop on EUV and Soft X-ray Sources*, Dublin, Ireland, October 8–10. https://www.euvlitho.com

204. E.R. Hosler, O.R. Wood, W.A. Barletta, P.J.S. Mangat, and M.E. Preil. 2015. Considerations for a free-electron laser-based extreme-ultraviolet lithography program. *Proc. SPIE* **9422**: 94220D-1–94220D-15. doi: 10.1117/12.2085538

205. E.R. Hosler, O.R. Wood II, and M.E. Preil. 2016. Extending extreme-UV lithography technology. In *SPIE Newsroom Lasers & Sources*. doi: 10.1117/2.1201601.006323.

206. SEMI P37-0613 Specification for Extreme Ultraviolet Lithography Substrates and Blanks, Semiconductor Equipment and Materials International, SEMI® 3081 Zanker Road, San Jose, CA.

207. T. Bret, R. Jonckheere, D. Van den Heuvel, C. Baur, M. Waiblinger, and G. Baralia. 2002. Closing the gap for EUV mask repair. *Proc. SPIE* **8322**: 83220C-1–83220C-9. doi: 10.1117/12.918322

208. S. Raghnuathan, O. Wood, P. Mangat, E. Verduijn, V. Philipsen, E. Hendrickx, R. Jonckheere, K. Goldberg, M. Benk, P. Kearney, Z. Levinson, and B. Smith. 2014. Experimental measurement of telecentricity errors for high-numerical aperture extreme ultraviolet mask images. *J. Vac. Sci. Technol. B* **32**: 06F801-1–06F801-8.

209. T. Kamo, T. Takai, K. Murano, Y. Morikawa, and N. Hayashi. 2014. Status and outlook for etched multilayer EUV mask. In *13th International Extreme Ultra-Violet Lithography (EUVL) Symposium*, Washington, DC, October 27–29. http://euvlsymposium.lbl.gov/proceedings/2014

210. V. Philipsen, E. Hendricks, R. Jonckheere, G. Vandenberghe, N. Davydova, T. Fliervoet, and J.T. Neumann. 2012. Impact of mask stack on high-NA EUV imaging. In *11th International Extreme Ultra-Violet Lithography (EUVL) Symposium*, Brussels, Belgium, September 30 – October 4. http://euvlsymposium.lbl.gov/proceedings/2012

211. M. Neisser, K. Cummings, S. Valente, C. Montgomery, Y.-J. Fan, K. Matthews, J. Chun, and P.D. Ashby. 2015. Novel resist approaches to enable EUV lithography in high volume manufacturing and extensions to future nodes. *Proc. SPIE* **9422**: 94220L-1–94220L-10. doi: 10.1117/12.2086307

212. R. Gronheid, C. Fonseca, M.J. Leeson, J.R. Adams, J.R. Strahan, C.G. Willson, and B.W. Smith. 2009. EUV resist requirements: absorbance and acid yield. *Proc. SPIE* **7273**: 727332-1–727332-8. doi: 10.1117/12.814716

213. M. Neisser. 2013. Resist technology – Recent progress and future prognosis. In International Symposium on Extreme Ultraviolet Lithography, Toyama, Japan.

5 Alignment and Overlay

David Laidler and Gregg M. Gallatin

CONTENTS

5.1   INTRODUCTION

This chapter discusses alignment in an exposure tool and its net result, overlay. Relevant concepts are described, and standard industry terminology is defined. The discussion has purposely been kept broad and tool non-specific. The content should be sufficient to make understanding the details and issues related to alignment and overlay in particular tools relatively straightforward.

5.2   OVERVIEW

As discussed in other chapters, integrated circuits are constructed by successively depositing and patterning layers of different materials on a silicon wafer. The patterning process consists of a combination of exposure and development of photoresist followed by etching and doping of the underlying layers and deposition of another layer. This process results in a complex and, on the scale of microns, very non-homogeneous material structure on the wafer surface. It is critical that the patterns in each layer are positioned accurately with respect to each other, and this is referred to as *overlay.*

Generally, the overlay between layers is required to be less than about a third of the minimum feature size or critical dimension (CD) in the layers. For example, at the 65 nm technology node, the overlay tolerance is on the order of 20 nm, and at the 14 nm node on the order of 5 nm. The complexity of advanced patterning processes in use today has brought even more challenges to overlay. Pattern density and complexity, along with the number of masking layers, have increased significantly from node to node. The underlying implication is that as CD decreases, so does the layer-to-layer overlay tolerance, with the added burden of having to align more masking layers.

The 22 nm node was generally recognized as the one that required double patterning on selected layers in order to meet pattern density requirements. Various double patterning strategies are currently being employed to achieve the required pattern density, but at the expense of added process complexity. In a conventional single patterning process, one pattern is printed in resist, whereas in double or multiple patterning, two or more patterns are superimposed in a single layer through an appropriate integration scheme. Such multiple patterning schemes add intralayer placement requirements to the conventional layer-to-layer pattern placement requirements, as shown in Figure 5.1. It is important to keep in mind that although there is added complexity and difficulty in multiple patterning processing, all the mechanisms contributing to overlay error remain the same; only the complexity and the strategies to manage it have increased. An optimized alignment strategy is required to manage both the within-layer and the layer-to-layer overlay combinations.

We often hear talk of the incredible improvements in CD resolution that have occurred over the years pushed by device scaling, but we hear little talk of the corresponding overlay improvements that have taken place over the same time period, which are equally, if not more, impressive. If, for example, we look back to 1985, when the typical production resolution was 1 μm (1000 nm), the corresponding overlay requirement was of the order of 300 nm, one-third of the CD, and this is a rule of thumb that has stayed true through to today. In 2017, with device dimensions below 14 nm, the corresponding overlay requirements are of the order of 5 nm or less. Over this time frame, the basic mechanisms have remained largely the same, but we have seen incredible improvements in hardware performance. As an example, in present-day exposure tools, the intrinsic tool capability for single-tool overlay is well within 2 nm, and with multiple tools the number is roughly a factor of two larger. As well as this, what makes these improvements in overlay capability even more impressive is that at the same time, the wafer size has increased from 100 to 300 mm, and for immersion lithography, water has been added between the lens and the wafer, which introduced its own issues that needed to be fully understood, compensated, and controlled.

The topic of overlay has been extensively published throughout the last four decades, topics ranging from improvements to the exposure tool, in particular the alignment system, overlay modeling, overlay sources of error, process overlay, and overlay metrology. Previous authors have discussed

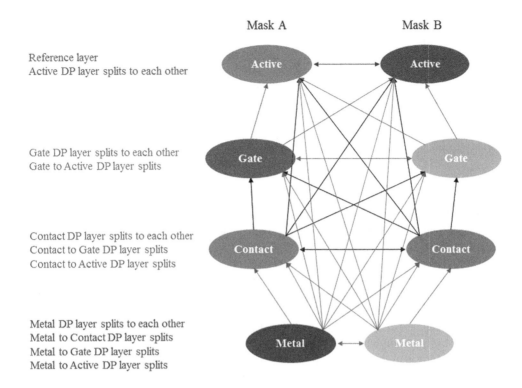

FIGURE 5.1 Possible critical overlay combinations in an integration flow using double patterning (DP).

extensively the theoretical aspects of overlay, ranging from alignment systems and overlay modeling, to overlay metrology [1–9]. In this chapter, we take a practical approach to overlay, focusing mostly on experimental results, with the aim that combining experimental results with the theoretical background presented later in the chapter will help to fill the gaps in understanding overlay. In Section 5.5 (Scanner Alignment Sequence of Events), we use the example of a lithography step, starting from the incoming wafer, through scanner alignment and exposure, to overlay metrology. In this sequence of lithography events, we examine in each step the mechanics of each step and how they impact and build up to the final overlay error. Throughout, use cases and experimental results will be used to demonstrate the effects and to give the reader a feel for how the overlay numbers arise and how to decompose overlay sources of error. A more detailed discussion of the alignment sensor, overlay modeling, and overlay metrology is also included in this chapter. The aim of this chapter is to give the reader a feel for how optimum overlay is achieved, the numbers that can be expected, and why achieving better than 5 nm overlay in a high-volume manufacturing (HVM) environment requires the tightest control in every step of wafer processing, not just lithography [10–18].

5.3 ALIGNMENT BASICS

It is important to accurately define what we actually mean by the terms *alignment* and *overlay* and to be consistent in their use. Often, the two words are used interchangeably and incorrectly, but each has its own specific meaning, and to aid understanding, it is important that we use them correctly. Alignment can be defined as *the method used to position two layers with respect to each other*, so here we are really talking about the alignment system of the exposure tool and the marks used for alignment. Overlay can be defined as *the resulting pattern placement error between two layers or layer splits*, the error that we really achieve on the wafer between layers and splits. To achieve good overlay in production, alignment is a very important part, but it is only a part of overlay control;

there are many other issues that contribute to the overlay observed on production wafers, and these relate not just to the scanner but also to the process integration flow. Control of overlay requires extensive knowledge of how the overlay error is built up through the various lithography process steps, from incoming wafer to overlay metrology, and the ability to relate overlay characteristics to the sources of overlay error. This section is not intended to be tool specific, although in certain cases this is hard to avoid, but the focus has been placed on state-of-the-art scanners. Overlay control is far more than just alignment; it is about matching of overlay fingerprints, stability of those fingerprints over time, and reduction or compensation of process signatures. Alignment is obviously an important part of overlay, but just a part of it.

5.3.1 Alignment

The purpose of alignment is to ensure that the layer currently being patterned is perfectly positioned with respect to any previously patterned layers, within the tolerances required by the technology. The alignment sensor measures the positions of a set of *alignment marks* patterned in a previous layer patterning step by reference to the *wafer stage* interferometers of the scanner, which define the *ideal* grid of the scanner. These alignment mark positions are programmed in the scanner recipe, and the alignment sensor determines the aligned position of each mark and from this, the deviation of each mark from its expected position. Alignment is accomplished by tool adjustments that minimize the difference between the detected and the expected stage positions over the specified alignment marks, based on built-in *scanner models*. The process of determining the position, orientation, and distortion of the patterns already on the wafer and then placing the projected image in the correct relation to these patterns is termed *alignment*. The result, or how accurately each successive patterned layer is matched to the previous layers, is termed *overlay*.

The alignment process requires, in general, both the translational and rotational positioning of the wafer and/or the projected image as well as some distortion of the image to match the actual shape of the patterns already present. The fact that the wafer and the image need to be positioned correctly to get one pattern on top of the other is obvious. The requirement that the image often needs to be distorted to match the previous patterns is not at first obvious but is a consequence of the following realities. No exposure tool or aligner projects an absolutely perfect image. All images produced by all exposure tools are slightly distorted with respect to their ideal shape. In addition, different exposure tools distort the image in different ways. Silicon wafers are not perfectly flat or perfectly stiff, and any tilt or distortion of the wafer during exposure, either fixed or induced by the wafer chuck, results in distortion of the as-printed patterns. Any vibration or motion of the wafer relative to the image that occurs during exposure and is unaccounted for or uncorrected by the exposure tool will "smear" the image in the photoresist. Thermal effects in the reticle, the projection optics, and/or the wafer will also produce distortions, as will the process integration flow.

The net consequence of all this is that the shape of the first-level pattern printed on the wafer is not ideal, and all subsequent patterns must, to the extent possible, be adjusted to fit the overall shape of the first-level printed pattern to meet overlay requirements. Different exposure tools have different capabilities to account for these effects, but in general, the distortions or shape variations that can be accounted for include X and Y magnification and skew. These distortions, when combined with translation and rotation, make up the complete set of linear transformations in the plane. They are defined and discussed in detail in Section 5.8.

Since the problem is to successively match the projected image to the patterns already on the wafer and not simply to position the wafer itself, the exposure tool must effectively be able to detect or infer the relative position, orientation, and distortion of both the wafer patterns themselves and the projected image. The position, orientation, and distortion of the wafer patterns are always measured directly, whereas the image position orientation and distortion are sometimes measured directly and sometimes inferred from the reticle position after a baseline reticle to image calibration has been performed.

5.3.2 ALIGNMENT MARKS

It is difficult to directly sense the circuit patterns themselves, and therefore, alignment is accomplished by adding specific marks, known as *alignment marks*, to the circuit patterns, typically in the scribeline. These alignment marks can be used to determine the reticle position, orientation, and distortion and/or the projected image position, orientation, and distortion. They can also be printed on the wafer along with the circuit pattern and hence can be used to determine the wafer pattern position, orientation, and distortion. Alignment marks generally consist of one or more clear or opaque lines on the reticle, which then become "trenches" or "mesas" when printed on the wafer. But, more complex structures such as gratings, which are simply periodic arrays of trenches and/or mesas, and checkerboard patterns are also used. Alignment marks are usually located along either the edges or the scribeline of each field, or a few "master marks" are distributed across the wafer. Although alignment marks are necessary, they are not part of the chip circuitry, and therefore, from the chip maker's point of view, they waste valuable wafer area or "real estate." This drives alignment marks to be as small as possible, and they are often less than a few hundred microns on a side. In principle, it would be ideal to align to the circuit patterns themselves, but this has so far proved to be very difficult to implement in practice. The circuit pattern printed in each layer is highly complex and varies from layer to layer. This approach therefore requires an adaptive pattern recognition algorithm. Although such algorithms exist, their speed and accuracy are not equal to those obtained with simple algorithms working on signals generated by dedicated alignment marks.

The alignment marks are specially designed targets provided by the scanner tool supplier and are normally specific to each supplier. They are designed for a specific alignment system and are placed in the scribeline between the product dies. Driven by overlay requirements of the advanced nodes, there has been an effort to aggressively drive down the size of both the alignment and overlay targets. Alignment marks can have a length ranging from approximately ~800 to 100 μm, and in recent years, their height has been reduced from fitting into an 80 μm scribe to fit in a 40 μm scribe to improve use of silicon real estate.

5.3.3 ALIGNMENT SYSTEMS

In order to "see" the alignment marks, alignment systems are incorporated into the exposure tool; generally, there are separate sensors for the wafer, the reticle, and/or the projected image itself. Depending on the overall alignment method, each of these sensors may be entirely separate systems, or they may be effectively combined into a single sensor. For example, a sensor that can "see" the projected image directly would nominally be "blind" with respect to wafer marks, and hence, a separate wafer sensor is required. But, a sensor that "looks" at the wafer through the reticle alignment marks themselves is essentially performing reticle and wafer alignment simultaneously, and hence, no separate reticle sensor is necessary. Note that in this case, the positions of the alignment marks in the projected image are being inferred from the position of the reticle alignment marks, and a careful calibration of reticle to image positions must have been performed previous to the alignment step.

Also, there are two generic system-level approaches for incorporating an alignment sensor into an exposure tool, termed *through-the-lens* and *not-through-the-lens* or *off-axis*. See Figure 5.2.

In the through-the-lens (TTL) approach, the alignment sensor looks through the same, or mostly the same, optics that are used to project the aerial image onto the wafer. In the off-axis (OA) approach, the alignment sensor uses its own optics, which are completely or mostly separate from the image projection optics. The major advantage of TTL is that, it can provide "common mode rejection" of opto-mechanical instabilities in the exposure tool. That is, if the projection optics move, then, to first order, the shift in the position of the projected image at the wafer plane matches the shift in the image of the wafer as seen by the alignment sensor. This cancellation helps desensitize the alignment process to opto-mechanical instabilities. The major disadvantage of TTL is that it requires the

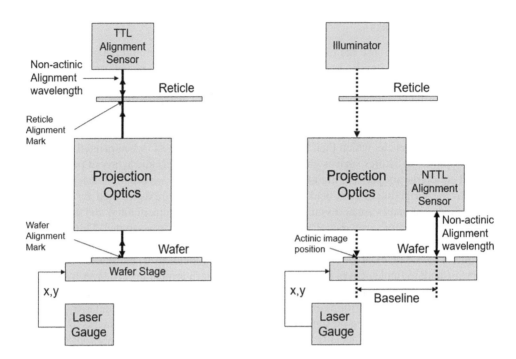

FIGURE 5.2 Through-the-lens (TTL) system on the left and not-through-the-lens (NTTL) or off-axis system on the right.

projection optics to be simultaneously good for exposure as well as alignment. Since alignment and exposure generally do not work at the same wavelength, the imaging capabilities of the projection optics for exposure must be compromised to allow sufficiently accurate performance of the alignment sensor. The net result is that neither the projection optics nor the alignment sensor is providing optimum performance. The major advantage of the OA approach is precisely that it decouples the projection optics and the alignment sensor, therefore allowing each to be independently optimized. Also, since an OA sensor is independent of the projection optics, it is compatible with different tool types such as i-line, deep ultraviolet (DUV), and extreme ultraviolet (EUV). Its main disadvantage is that opto-mechanical drift is not automatically compensated, and hence the "baseline" between the alignment sensor and the projected image must be recalibrated on a regular basis. The calibration procedure is illustrated in Figure 5.3. The TTL approach requires this same projected image to alignment sensor calibration to be made as well, but it does not need to be repeated as often.

Further, as implied earlier, essentially all exposure tools use sensors that detect the wafer alignment marks optically. That is, the sensors project light at one or more wavelengths onto the wafer and detect the scattering/diffraction from the alignment marks as a function of position in the wafer plane. Many types of alignment sensor are in common use, and their optical configurations cover the full spectrum from simple microscopes to heterodyne grating interferometers. Also, since different sensor configurations operate better or worse on given wafer types, most exposure tools "sport" more than one sensor configuration to allow good overlay on the widest possible range of wafer types. For detailed descriptions of various alignment sensor configurations, see [19–31].

Various configurations that an alignment sensor can take and the nominal signal shapes that it will produce are described below. Only the optical sensors are considered; that is, sensors that project light, either infrared, visible, or UV, onto the wafer and detect the scattered/diffracted light.

For each, the alignment sensor detects the positions of a set of alignment marks, and the signal it produces must depend in one way or another on the mark position. Thus, it follows that all alignment sensors, in a very general sense, produce a signal that can be considered to represent some

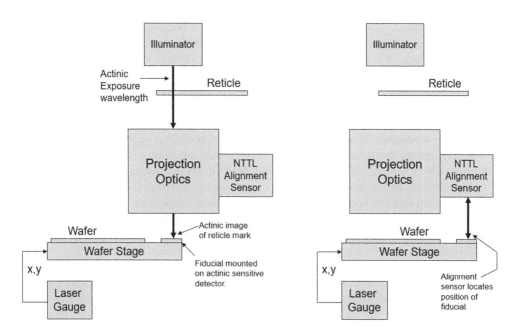

FIGURE 5.3 Illustration of the calibration procedure for an NTTL or off-axis alignment sensor. On the left, the actinic image sensor on the wafer stage senses the position of the projected image in terms of x,y laser gauge (or grid) coordinates. On the right, the NTTL alignment sensor senses the position of the actinic image sensor in terms of x,y laser gauge (or grid) coordinates. The difference between these two positions is the baseline offset.

type of image of the alignment mark. This image can be thoroughly conventional, as in a standard microscope, or it can be rather unconventional, as in a scanned grating interference sensor.

Simplified diagrams of each basic configuration are included in the figures for completeness. The simplicity of these diagrams is in stark contrast to the schematics of real alignment systems, whose complexity almost always belies the very simple concept they embody. For specific designs, see [19–31].

The following are common differentiators among basic alignment sensor types:

- **Scanning-vs.-staring**: A staring sensor simultaneously detects position-dependent information over a finite area on the wafer. A standard microscope represents an example of a staring sensor. The term *staring* comes directly from the idea that all necessary data for a single mark can be collected with the sensor simply staring at the wafer. A scanning sensor, on the other hand, can effectively "see" only a single point on the wafer and hence must be scanned either mechanically or optically to develop the full wafer position–dependent signal. Mechanical scanning may amount simply to moving the wafer in front of the sensor. Optical scanning can be accomplished by changing the illuminating light in such a way as to move the illuminating intensity pattern on the wafer. Optical scanning may involve the physical motion of some element in the sensor itself, such as a steering mirror, or it may not. For example, if the illumination spectrally contains only two closely spaced wavelengths, then for certain optical configurations, the intensity pattern of the illumination will automatically sweep across the wafer at a predictable rate.
- **Bright-field-vs.-dark-field**: All sensors illuminate the mark over some range of angles. This range can be large, as in an ordinary microscope, or it can be small, as in some grating sensors. If the range of angles over which the sensor detects the light scattered/ diffracted from the wafer is the same as the range of illumination angles, it is called

bright-field detection. The reason for this terminology is because for a flat wafer with no mark, the specularly reflected light will be collected, and hence the signal is bright where there is no mark. A mark scatters light out of the range of illumination angles, and hence marks send less light to the detectors and appear dark relative to the non-mark areas. In dark-field detection, the range of scatter/diffraction angles that are detected is distinctly different from the range of illumination angles. In this case, specularly reflected light is not collected. and so a non-mark area appears dark, and the mark itself appears bright.

- **Phase-vs.-amplitude**: Since light is a wave, it carries both phase and amplitude information, and sensors that detect only the amplitude or only the phase or some combination of both have been and are being used for alignment. A simple microscope uses both, since the image is essentially the Fourier transform of the scattered/diffracted light, and the Fourier transform is a result of both the amplitude and the phase information. A sensor that senses the position of the interference pattern generated by the light scattered at two distinct angles is detecting only the phase, whereas a sensor that looks only at the total intensity scattered into a specific angle or range of angles is detecting only the amplitude.

- **Broadband-vs.-laser**: Either sensors use broadband illumination, that is, a full spectrum of wavelengths spread over a few hundred nanometers generated by a lamp, or they use one or perhaps two distinct laser wavelengths. The advantage of the laser illumination is that it is a very bright coherent source and hence allows the detection of weakly scattering alignment marks and phase detection. The disadvantage is that because it is coherent, the signal strength and shape are very sensitive to the detailed thin film structure of the alignment mark. Hence, small changes in the thickness and/or shape of the alignment mark structure can lead to large changes in the alignment signal strength and shape. In certain cases, thin film effects lead to destructive interference and hence, no measurable alignment signal at all. To mitigate this, a second, distinctly different laser wavelength is often used, so that if the signal vanishes at one wavelength, it should not at the same time vanish at the other wavelength. This, of course, requires either user intervention to specify which wavelength should be used in particular cases or an algorithm that can automatically switch between the signals at the different wavelengths depending on signal strength and/or signal symmetry. The advantage of broadband illumination is that it automatically averages out all the thin film effects and hence is very insensitive to the details of the mark's thin film structure. Thus, the signal has a stable intensity and shape even as the details of the alignment mark structure vary. This is a good thing, since the sensor is only trying to find the mark position; it is not nominally trying to determine any details of the mark shape. Its disadvantage is that generally, broadband sources are not as bright as laser sources, and so it may be difficult to provide enough illumination to accurately sense weakly scattering alignment marks. Also, phase detection with a broadband source is difficult, since it requires equal path interference.

5.3.4 ALIGNMENT METHOD

The job of an alignment sensor is to determine the position of each of a given subset of the alignment marks in a coordinate system fixed with respect to the exposure tool. This position data is then used in either of two generic ways, termed *global* and *field-by-field*, to perform alignment. In global alignment, the marks in only a few fields are located by the alignment sensor(s), and all this data is combined in a best-fit sense to determine the optimum alignment of all the fields on the wafer. In field-by-field alignment, the data collected from a single field is used to align only that field. Field-by-field alignment is mentioned only for completeness but was used only in some early steppers. It is important to be aware that all today's state-of-the-art scanners utilize some form of global alignment. Global alignment is both faster, because not all the fields on the wafer are located, and less sensitive to noise, because it combines all the data together to find a best overall fit and utilizes the stage accuracy of the scanner. But since the results of the best fit are used in a feedforward or dead reckoning approach, it does rely on the overall opto-mechanical stability of the exposure tool. A detailed discussion of global alignment is presented in a separate section.

Alignment is normally implemented as a two-step process; that is, a fine alignment step with an accuracy of a few nanometers follows an initial coarse alignment step with an accuracy of a few microns. When a wafer is first loaded into the exposure tool, the uncertainty in its position in exposure tool coordinates is often on the order of tens of microns. The coarse alignment step uses a few large alignment targets and has a capture range equal to or greater than the initial wafer position uncertainty. The coarse alignment sensor is generally very similar to the fine alignment sensor in configuration, but in some cases, these two sensors can be combined into two modes of operation of a single sensor. The output of the coarse alignment step is the wafer position to within several microns, or less, which is within the capture range of the fine alignment system. Prior to coarse alignment, a step known as *pre-alignment* is performed, in which the edge and notch of the wafer are detected mechanically or optically so that it can be brought into the capture range of the coarse alignment sensor.

5.3.5 Alignment versus Leveling and Focusing

Along with image distortion, alignment requires positioning the wafer in all six degrees of freedom: three translational and three rotational. But, adjusting the wafer so that it lies in the projected image plane, that is, leveling and focusing the wafer, which involves one translational degree of freedom (motion along the optic axis) and two rotational degrees of freedom (orienting the plane of the wafer to be parallel to the projected image plane) are generally considered separate from "alignment" as used in the standard sense. Only in-plane translation (two degrees of freedom) and rotation about the projection optic axis (one degree of freedom) are commonly meant when referring to alignment. The reason for this separation in nomenclature is the difference in accuracy required. The accuracy required for in-plane translation and rotation generally needs to be on the order of about 20% to 30% of the minimum feature size or CD to be printed on the wafer. Current state-of-the-art CD values are on the order of 40 nanometers, and thus, the required alignment accuracy is on the order of a few tens of nanometers. On the other hand, the accuracy required for out-of-plane translation and rotation is related to the total usable depth of focus of the exposure tool which is generally only a few times the CD value. Thus, out-of-plane focusing and leveling of the wafer requires less accuracy than in-plane alignment. Also, the sensors for focusing and leveling are completely separate from the alignment sensors, and focusing and leveling does not usually rely on special fiducial patterns, that is, alignment marks on the wafer. Only the wafer surface needs to be sensed.

5.3.6 Other Scanner Hardware Issues That Impact Overlay

The factors that affect overlay are the standard ones of measurement and control. The position, orientation, and distortion of the patterns already on the wafer must be inferred from a limited number of measurements, and the position orientation and distortion of the pattern to be exposed must be controlled using a limited number of adjustments. For actual results and analysis from particular tools, see [24–32]. Here, we present simply a list of the basic sources of error relating to the exposure tool.

- **Alignment system**: Noise and inaccuracies in the ability of the alignment system to determine the positions of the alignment marks. This includes not only the alignment sensor itself but also the stages and laser gauges that serve as the coordinate system for the exposure tool, as well as the calibration and stability of the alignment system axis to the projected image, which is true for both OA and TTL. Also, the electronics and algorithm that are used to collect and reduce the alignment data to field and grid terms. Finally, it must be remembered that the alignment marks are not the circuit pattern, and the exposure tool is predicting the circuit pattern position, orientation, and distortion from the mark positions. Errors in this prediction due to the nonperfection of the initial calibration of the mark-to-pattern relationship or changes in the relationship due to thermal and/or mechanical effects and simplifications in algorithmic representation, such as the linear approximation to the nonlinear distortion, all contribute to overlay error.

- **Projection optics**: Variations and/or inaccuracies in the determination of the distortion induced in the projected pattern by the optical system. Thermo-mechanical effects change the distortion signature of the optics. At the nanometer level, this signature is also dependent on the actual aberrations of the projection optics, which causes different linewidth features to print at slightly different positions. In machine-to-itself overlay, the optical distortion is nominally the same for all exposed levels, so this effect tends to be minimal in this case. In machine-to-machine overlay, the difference in the optical distortion signatures of the two different projection optics is generally not trivial and thus, can be a significant contributor to overlay errors.
- **Illumination optics**: Nontelecentricity in the source pupil when coupled with focus errors and/or field nonflatness will produce image shifts and/or distortion. Variation in the source pupil intensity across the field also can shift the printed alignment mark position with respect to the circuit position.
- **Reticle**: Reticle metrology errors, i.e., errors in the mark-to-pattern position measurements. Variation in the mark-to-pattern position caused by reticle mounting and/or reticle heating. Particulate contamination of the reticle alignment marks can also shift the apparent mark position.
- **Wafer stage**: Errors in the position and rotation of the wafer stage, both in plane and out of plane, during exposure contribute to overlay errors. Also, wafer stage vibration contributes. These are rigid body effects; there are also nonrigid body contributors, such as wafer and wafer stage heating, which can distort the wafer with respect to the exposure pattern, and also chucking errors, which "stretch" the wafer in slightly different ways each time it is mounted.
- **Reticle stage**: Essentially all the same considerations as for wafer stage apply to the reticle stage but with some slight mediation due to the reduction nature of the projection optics.
- **Projection optics**: Errors in the magnification adjustment cause pattern mismatch. Heating effects can alter the distortion signature in uncontrollable ways.
- **Wafer temperature**: Thermal stability of the wafer must be maintained from pre-alignment through to alignment and exposure in order to avoid variability in correctable errors, primarily scaling.
- **Lens and reticle heating**: Lens and reticle heating can contribute to considerable overlay errors if not adequately calibrated and lead to drifts in overlay across the wafer and through the batch.
- **Immersion cooling**: Water evaporation at the edges of the immersion hood and at the edges of the wafer on immersion scanners can change the temperature of the wafer during exposure, leading to residual overlay errors. In state-of-the-art tools, this is compensated by having multiple temperature zones in the exposure chuck.

5.4 OVERLAY ERRORS AND CORRECTIONS

The measured "overlay" error is the position of a pattern at any point on the wafer with respect to a reference; it is a relative quantity. While overlay is a vector quantity, conventionally it is measured in x and y directions with respect to a reference and is expressed in dx and dy. Typically, dx and dy are numbers associated with overlay of the current layer of a process with respect to a previous layer. Using a physical meaningful mathematical model for overlay analysis has been a standard practice in lithography for decades. The models can be as simple as a 6- or 10-parameter linear model describing both stage and image field characteristics, or for more advanced exposure tools, higher-order terms have been developed for the compensation of non-linear grid and image field shape variation. All these aim to better describe the behavior of the exposure tool and process such that adjustment or corrections can be made to improve and obtain optimum overlay performance.

It is important to keep in mind that the measured dx and dy are in fact the combined effect of overlay, process, and metrology uncertainty. Therefore, the metrology technique used can have a significant impact on the total overlay budget. For measurements based on the interferometric laser stage, typically made during wafer alignment and feedforward corrections, the pattern position is relative to the position determined by the interferometer-controlled wafer stage. The wafer stage accuracy can directly impact the alignment position acquisition accuracy and also the positional accuracy during exposure, both of which in turn impact overlay accuracy. For advanced exposure tools, stage accuracy in fact drives the limits of overlay. For today's most advanced scanners, stage accuracy is in the order of 2 nm or less.

5.4.1 OVERLAY ERROR COMPONENTS

The overlay error components or correctable terms are the overlay errors that can be modeled and if removed, leave just the residual fingerprint. The correctable terms can be fed forward to the scanner in order to improve overlay performance. Figure 5.4 shows what these correctable overlay errors look like physically and how they are represented on a vector map.

5.4.2 OVERLAY MODELING

In order to understand the resulting overlay performance post-exposure and measurement, it is necessary to model the resulting overlay measurements, using the same model as used by the scanner for alignment and exposure of the wafer. From modeling the resulting overlay performance, it is possible to break down the performance into grid and intrafield terms, a very important part of overlay analysis, and to visualize the residual overlay fingerprint.

The combination of Equations 5.1 and 5.2 forms the 6-parameter linear wafer distortion sub-model that is obtained from the pure linear 10-parameter wafer deformation model in Equations 5.3 and 5.4 by equating equivalent grid and intrafield parameters:

$$\Delta x = T_x + \mu_x \cdot \left(X + x_{\text{field}} \right) - \theta \cdot \left(Y + y_{\text{field}} \right) \tag{5.1}$$

$$\Delta y = T_y + \mu_y \cdot \left(Y + y_{\text{field}} \right) + \left(\theta + \psi \right) \cdot \left(X + x_{\text{field}} \right) \tag{5.2}$$

$$\Delta x = T_x + \mu_x^{\text{grid}} \cdot X - \theta^{\text{grid}} \cdot Y + \mu_x^{\text{field}} \cdot x_{\text{field}} - \theta^{\text{field}} \cdot y_{\text{field}} \tag{5.3}$$

$$\Delta y = T_y + \mu_y^{\text{grid}} \cdot Y + \left(\theta^{\text{grid}} + \psi^{\text{grid}} \right) \cdot X + \mu_y^{\text{field}} \cdot y_{\text{field}} + \left(\theta^{\text{field}} + \psi^{\text{field}} \right) \cdot x_{\text{field}} \tag{5.4}$$

Here Δx and Δy are the differences between the measured and expected alignment mark positions in global stage coordinates, with X and Y the expected grid or field center positions and x_{field} and y_{field} the expected intrafield mark positions, for example, relative to the grid or field center positions. The parameters T_x, T_y, μ_x^{grid}, μ_y^{grid} and θ^{grid}, ψ^{grid} are the grid translation, scaling (magnification), and rotation errors in the x and y directions, and μ_x^{field}, μ_y^{field}, and θ^{field}, ψ^{field} are the intrafield scaling (magnification) and rotation errors, respectively. These equations are derived and discussed in Section 5.7. To get Equations 5.1 and 5.2 from Equations 5.3 and 5.4, the assumption is made that

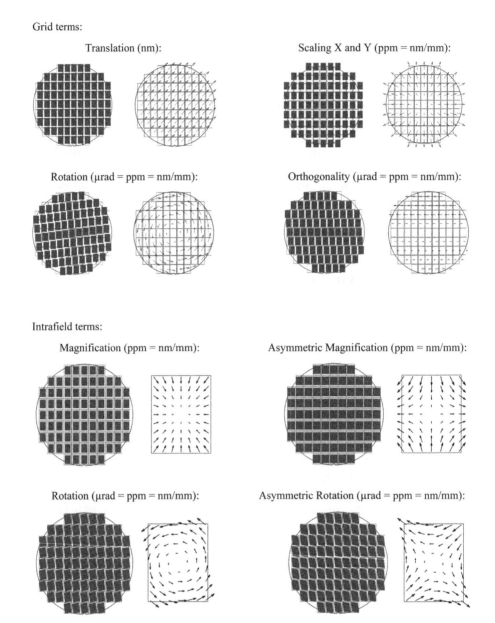

FIGURE 5.4 Correctable overlay errors; what they look like physically and how they are represented on a vector map.

the field and grid terms are the same; that is, $\mu_x^{\text{grid}} = \mu_x^{\text{field}} = \mu_x$, $\mu_y^{\text{grid}} = \mu_y^{\text{field}} = \mu_y$, $\theta^{\text{grid}} = \theta^{\text{field}} = \theta$ and $\psi^{\text{grid}} = \psi^{\text{field}} = \psi$.

Adjustments are applied internally to the relevant scanner parameters, based on this model, prior to wafer exposure, so that the overlay error of the layer currently being exposed is minimized with respect to the previously patterned layer. The model terms coming from the wafer alignment are translation x and y, scaling x and y, rotation, and orthogonality.

In addition to the standard linear model, there are additional corrections that can be fed back to the scanner to compensate for wafer process deformation. Figure 5.5 shows the standard linear grid model on the left, additional non-linear grid correction or corrections per exposure (CPE) in the

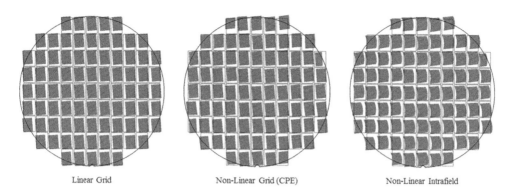

Linear Grid Non-Linear Grid (CPE) Non-Linear Intrafield

FIGURE 5.5 Linear grid, nonlinear grid, and nonlinear intrafield corrections.

center, and non-linear intrafield correction on the right. The linear grid correction has already been described and is the standard alignment model. The non-linear grid correction (CPE) is based on feedback to the scanner of additional corrections, based on extensive metrology data, and depends on a stable process deformation signature. For this correction, separate translation and intrafield corrections are calculated for each field, and these corrections are added on to the model after alignment and prior to exposure. Non-linear intrafield correction is similar and adds additional terms to each field.

5.4.3 CORRECTABLE ERRORS

When the scanner aligns the wafer, and we measure and model the resulting overlay, there will invariably be remaining what are termed *correctable errors*. The scanner needs to align a wafer with good repeatability; in fact, it can be argued that the accuracy is not important. This is not a problem as long as our alignment is stable and always produces the same correctable errors. If we were then to correct these remaining correctable terms by feeding them back to the scanner, we would be able to achieve overlay on our product wafers that comes close to the residuals, which are the overlay errors with all correctable terms perfectly removed. To achieve overlay that meets the requirements for each technology node, we need to have small and stable residual errors and stable systematic components across a batch and from batch to batch. Residual errors are minimized by optimizing stage and lens matching and in the case of dual stage systems, running with chuck dedication, which means that at each layer or split, a specific wafer is always exposed on the same wafer chuck. If the systematic components, often referred to as *correctable errors*, are small and, most importantly, stable, then production overlay can be optimized and controlled by statistical process control (SPC) and in today's advanced fabs advanced process control (APC). However, if the systematic components vary considerably from wafer to wafer and batch to batch, using process corrections to correct for them will offer little or no benefit.

The following example in Figure 5.6 (left) shows a situation where the systematic component, in this case translation x, has an offset but is relatively stable across the batch, so feeding back this value to the scanner as a correction would significantly improve the overlay.

However, the example in Figure 5.6 (right) shows the situation where there is significant variation from wafer to wafer, so feeding back the average value as a correction would have no impact on the overlay performance.

5.4.4 OVERLAY RESIDUAL SIGNATURES OR FINGERPRINTS

The overlay residual signature or fingerprint is what is visible on the wafer vector map after removal of the correctable terms. The residuals are what is left after the correctable errors, which are really

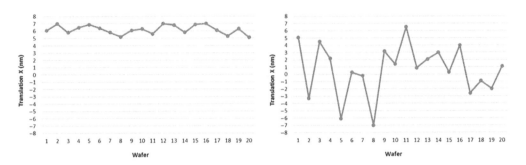

FIGURE 5.6   Translation variation through batch: stable (left), unstable (right).

the alignment error made by the scanner, are removed, or what is left behind and does not fit the model. The magnitude of the residuals should always be in line with the stage accuracy of the scanner and the metrology repeatability of the technique being used to measure the overlay error. If we see large random residuals, this is a sign of metrology noise. Any resulting signatures that can be observed are typically either from the scanner, due to immersion or other thermal effects, or more likely from process wafer deformation, caused by the process integration flow [32–41].

5.5   SCANNER ALIGNMENT SEQUENCE OF EVENTS

In this section, alignment and overlay are described by a sequence or series of events as a wafer passes through the scanner. At each scanner step, contributions to overlay are discussed using experimental results to illustrate these errors. The section discusses what scanner misalignment really means, what it can and cannot cause in terms of resulting overlay, and that it is only one component of overlay. At the same time, it highlights that resulting overlay performance is far from just a scanner issue, and for advanced nodes in particular, process effects are becoming more and more the dominant contributor. With regard to overlay, the sequence of events as a wafer passes through the scanner can be broken down into three distinct phases: pre-exposure, wafer exposure, and post-exposure overlay analysis, as illustrated in Figure 5.7. Each step in the sequence can introduce specific contributions to the final resulting overlay. A good grasp of this sequence of events is critical to understanding overlay errors and forms the basis for overlay analysis, characterization, and understanding how the best possible overlay performance is achieved.

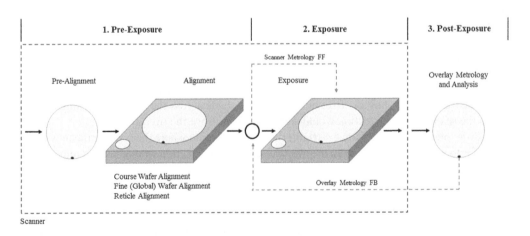

FIGURE 5.7   Scanner alignment sequence of events; stages that a wafer passes through.

5.5.1 Pre-Exposure

The pre-exposure stage can be broken down into three phases: pre-alignment, coarse alignment, and fine (global) alignment. In addition, the reticle needs to be aligned. All these impact the resulting overlay and make their own particular contribution, which if not optimum, can be observed in post-exposure analysis.

5.5.1.1 Wafer Pre-Alignment

The first phase a wafer passes through on arriving at the scanner from the coat and develop track is pre-alignment. The pre-aligner is a separate unit or module on which the wafer is placed by the robot when it enters the scanner prior to the wafer being placed onto the wafer exposure chuck. The pre-aligner, sometimes referred to as the *temperature stabilization unit*, serves two functions. The first is to allow the wafer to be placed on the wafer stage mechanically, within tight tolerances, and the second is to stabilize and match the temperature of the wafer to that of the exposure chuck.

After the wafer arrives on the pre-aligner, it is aligned mechanically by use of both the wafer notch and edge, which are used to determine the exact center and orientation of the wafer. For first layer exposures, which are performed without wafer alignment, the pre-aligner ensures that it is exposed within tight positional tolerances and that the relationship with the notch and wafer grid is maintained. Without this, it would not be possible to align subsequent layers to this first layer exposure. This makes it a very important scanner parameter that needs to be monitored in production, because a bad pre-aligner, either with an offset or with poor repeatability, could result in thousands of scrapped wafers. For subsequent layer exposures, which all have wafer alignment, accurate pre-alignment ensures that the alignment marks are within the capture range of the coarse alignment sensor. Today, for state-of-the-art scanners, wafer pre-alignment accuracy is of the order of 2 μm. This means that for a wafer with a first layer exposure, on which a second layer exposure is made without wafer alignment, the resulting overlay error will be in the region of 2 μm. This assumes exposure on the same tool. Typically, as for most parameters, this number increases if different scanners are used for exposure of the two layers.

The other function of the pre-aligner is to stabilize the temperature of the wafer and to match it to that of the exposure chuck so that there are no thermal changes when the wafer is loaded onto the exposure chuck, which would directly impact the resulting overlay. This is a very important and often neglected area of overlay control. Prior to the wafer arriving on the pre-aligner, the last step on the coat and develop track is a chill or temperature stabilization plate, which must be at the same temperature as both the pre-aligner and the exposure chuck so that the thermal stability of the wafer is maintained throughout the wafer's transit through the scanner. If the temperature of the wafer at the fine alignment step is different from when the wafer is actually exposed, the modeled expansion determined from fine alignment will not be correct at the time of exposure, leading to scaling errors. Also, if there are local thermal variations in wafer temperature during exposure, these can lead to localized overlay errors and large overlay residuals.

5.5.1.2 Coarse Wafer Alignment

Once the wafer has been pre-aligned and its temperature stabilized, the robot then transfers the wafer to the exposure chuck, where it is vacuum clamped ready for coarse alignment. An important aspect of overlay control is how well and how repeatably the wafer is clamped to the exposure chuck. All wafers must be clamped repeatably each time they come into the scanner. If they are not, this can change the shape of the wafer each time it is exposed, leading to local shifts in x and y and non-correctable residual signatures. The physical shape of the wafer, any warp and bow, and the backside surface can also play a part here. When a warped wafer is clamped to the exposure chuck, depending on whether it is convex or concave, this can cause local shifts in x and y, leading to residual overlay errors. The coarse alignment step brings the alignment marks within the capture range of the fine alignment sensor and is also critical for ensuring that the correct fine alignment marks are captured. Sometimes, scanners use the same alignment marks for both coarse and fine alignment, but they can in some cases use separate dedicated marks for each purpose. The objective, though, remains the same: to bring the fine alignment marks within the capture range of the fine alignment sensor, which we can refer to as the *real* alignment and results in the overlay that we later observe on the wafer. Coarse alignment typically uses only two pairs of scribeline marks, one on each side of the wafer, and normally has no impact on the final overlay performance that we observe on the exposed wafer. However, if it goes wrong, it can lead to wafers being rejected by the scanner and not exposed at all, or being grossly misaligned, by which we mean overlay errors of several microns and readily detectable.

5.5.1.3   Fine (Global) Wafer Alignment

Following coarse alignment, the wafer remains on the exposure chuck, and fine (global) alignment is performed; the fine alignment step can be thought of as the "real" alignment, as it is largely responsible for the overlay that we observe on the wafer following exposure. During the fine alignment step, the alignment sensor is used to measure the positions of the alignment marks on the wafer with respect to the stage grid of the scanner. The recorded stage position information comes from interferometers, which control the movement of the wafer stage. Alignment marks are typically placed in the scribeline of production reticles, so by definition, they are exposed together with every field on the wafer. However, it is important to note that only a subset of these alignment marks are actually used for wafer alignment. The alignment marks to be used, including location, layer, and number, are programmed into the scanner recipe that is used for exposure. The number of alignment marks used varies and is a tradeoff between alignment accuracy and throughput. A typical number used in production is between 16 and 32 x and y mark pairs, a number that can be measured on state-of-the-art scanners without impacting the throughput; an important consideration, as accuracy needs to be achieved while maintaining maximum throughput, which today exceeds 250 wph (wafers per hour). Some alignment mark designs require separate alignment marks for each orientation, x and y, whereas others provide information on both orientations from the same mark. The latter type can enable the number of alignment marks to be doubled and hence obtain more information about the shape of the wafer without any impact on throughput.

Once all the wafer alignment mark positions have been measured using the alignment sensor, the difference between the shape of the clamped wafer and the reference grid, known in the coordinates of the scanner stage, is modeled within the scanner, typically using the standard linear alignment model described in Section 5.4.2 on overlay modelling. For state-of-the-art scanners, more sophisticated modeling exists and can be used to compensate for wafer processing effects, such as wafer process deformation caused by rapid thermal processing (RPT) or high-temperature steps. However, using high-order models is not without tradeoffs; therefore, high-order models must be used with caution.

What has been described is effectively the entire wafer alignment process. This process relies entirely on a predefined subset of the alignment marks and a chosen model employed within the scanner. Once the adjustments are modeled and calculated, the alignment process is complete. If a standard linear model is used, any alignment contribution to total overlay can only be systematic, as exposure follows the model being used, and final overlay relies entirely on the accuracy of the stage. If the scanner alignment sensor incorrectly detects one or more of the alignment mark positions, the incorrect alignment mark positions will be modeled along with the positions of all the other marks and result in incorrect values for the model terms. In consequence, the scanner will expose the wafer with incorrect modeled terms, resulting in systematic correctable errors on the wafer. The following examples are given for the purpose of further understanding the overlay error associated with alignment. In Figure 5.8, the measured vector maps (top) and residual vector maps (bottom) are shown from a single batch of wafers: first, middle and last. With reference to the measured data, it can be seen that for the first two wafers there is a small amount of correctable error, but on the last wafer

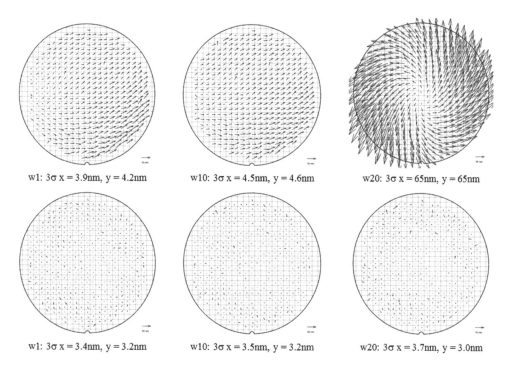

| w1: 3σ x = 3.9nm, y = 4.2nm | w10: 3σ x = 4.5nm, y = 4.6nm | w20: 3σ x = 65nm, y = 65nm |
| w1: 3σ x = 3.4nm, y = 3.2nm | w10: 3σ x = 3.5nm, y = 3.2nm | w20: 3σ x = 3.7nm, y = 3.0nm |

FIGURE 5.8 Measured (top) and residual (bottom) overlay errors, showing an example of an "alignment error."

the scanner has made an "alignment error", resulting in this case a very large rotation, observable in the resulting overlay.

However, with reference to the residual vector maps, it can be seen that they are all the same, clearly showing that the scanner introduced this correctable error during wafer alignment. Of course, this should never happen in production; the example was simply chosen to illustrate the point of what constitutes an "alignment error." What alignment cannot introduce is non-correctable residual signatures. Figure 5.9 shows what a scanner "misalignment" cannot cause, where here the error is coming from wafer process deformation caused by a high-temperature deposition step. Note that the measured and residual overlay vector maps are almost identical, showing that there is almost no correctable error. The only thing that scanner "misalignment" could have introduced when using the standard linear model is correctable errors, in which case the measured and residual vector maps would have been quite different.

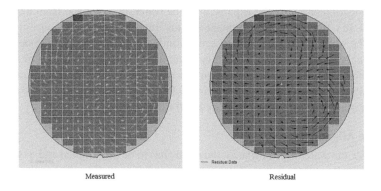

Measured   Residual

FIGURE 5.9 Measured and residual overlay error caused by wafer process deformation.

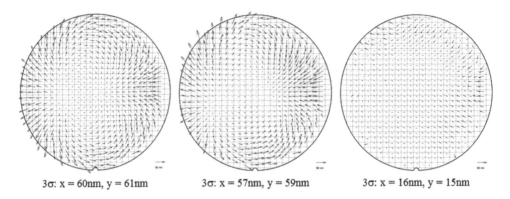

3σ: x = 60nm, y = 61nm   3σ: x = 57nm, y = 59nm   3σ: x = 16nm, y = 15nm

FIGURE 5.10 Scanner alignment mark measurement (left), measured overlay using scanner linear align-ment model (middle), and measured overlay using scanner third-order alignment model (right).

To emphasize this point, the following example discusses a wafer that has suffered from process deformation between the previous patterning layer and the current one. The wafer deformation, or shape change when clamped on the exposure chuck of the scanner, was determined by measur-ing the positions of all the alignment marks on the wafer. This allows us to map the shape of the wafer and the difference between the wafer grid and the "ideal" scanner reference grid as shown in Figure 5.10 left; the alignment residual vector map. As can be seen, the wafer has a non-correctable shape. If such a wafer is exposed after aligning using the standard 6-parameter linear alignment model, the same magnitude of overlay and signature is obtained, since the model is not able to correct for this, as shown in Figure 5.10 middle. In this example, a wafer was chosen that has been through an extreme wafer process deformation in order to demonstrate the issue, and for both the alignment residuals and the measured overlay, the magnitude is approximately 60 nm 3σ. Finally, in Figure 5.10 right is shown the measured overlay when using a third-order alignment model. Note that although significantly improved, the performance is still an order of magnitude above the base-line capability of the scanner.

After wafer alignment has been performed and before the wafer is exposed, the alignment sensor also measures the position of a stage fiducial, so that the wafer position is referenced to this fiducial. A fiducial is basically a mark, the same as on a wafer, that has been rigidly attached to the wafer stage and provides a stable reference point. At this point, we have all the information necessary to expose the wafer except for aligning the reticle and making the connection between the wafer and the reticle. This is done by aligning the reticle to the same fiducial on the wafer stage. In this way, the reticle is "indirectly" aligned to the wafer, and the four intrafield terms can be modeled: sym-metric and asymmetric magnification and rotation. Note that the intrafield terms, except for a cor-rection to magnification based on the expansion of the wafer, do not come from the wafer but from the stage fiducial. In a typical wafer alignment scheme, intrafield information from the wafer is not detected. Now, the wafer is ready to be exposed, "blind" stepped, following the determined model values. Any changes in the wafer shape at this stage, after wafer alignment, will result in overlay errors, as the wafer will be different from when it was aligned. The critical assumption here is that there are no changes to the wafer between the time when alignment was performed and the comple-tion of exposure; otherwise, the whole strategy breaks down.

5.5.1.4 Reticle Alignment
In the same way as the wafer is pre-aligned, the reticle has to be pre-aligned and its temperature sta-bilized. Reticle pre-alignment is normally performed using dedicated marks prior to the reticle being clamped to the reticle stage. Just as with wafer clamping, it is essential that the reticle is clamped uniformly and repeatably to the reticle stage; otherwise, varying intrafield residual signatures will

be introduced. As described in the previous section, once the wafer has been aligned to the fiducial on the wafer stage, which acts as a fixed reference, the reticle must be aligned to the same fiducial, so that reticle and wafer are indirectly aligned to each other. Typically, the four corners of the reticle are aligned to the wafer stage fiducial, so that the four intrafield terms can be modeled and compensated during exposure. If we consider differences between immersion and dry lithography, there is one important difference that occurs here. Because the alignment between the reticle and the stage fiducial is made through the lens and on an immersion scanner, there is water between the lens and the wafer/stage, so the alignment between them is through water. This contrasts with the wafer alignment, which is always performed off axis and is therefore always dry and not through water.

5.5.2 WAFER EXPOSURE

The wafer is now ready to be exposed according to the determined model values, and the scanner exposes the wafer under stage interferometer control, relying on the accuracy of the scanner wafer stage. Typical stage accuracy for state-of-the-art scanners is below 2 nm 3σ, and overlay performance is built around this incredibly accurate and repeatable stage performance. Scanner dynamic performance is part of this and relates to the synchronization of the scanning motion between the wafer and reticle stages.

Other issues that can impact overlay control and that occur during exposure are lens, reticle, and wafer heating. All these have to be calibrated and compensated for so that they do not degrade overlay performance either across the wafer or across the batch. Many alignment systems will compensate for some of the effects of lens and reticle heating on every wafer, because the reticle is aligned to the stage fiducial for every wafer, but if the effect is severe, this can still lead to changes across the wafer from the start of the exposure routing to the end. The magnitude of lens and reticle heating depends on the process being used. For example, if a traditional positive developer (PTD) process is being used for a contact hole layer that is dark field (DF), this can give rise to considerable reticle heating, as the light is partially absorbed by the quartz and chrome, but almost no lens heating. However, if a negative developer (NTD) process is being used for the same contact hole layer, which will then be light field (LF), this can give rise to considerable lens heating but almost no reticle heating. In both situations, as long as the correction mechanisms exist and can be calibrated, they should not impact overlay control, but any observed drifts in overlay across a wafer or batch could be attributable to heating that has not been adequately compensated. Wafer heating due to the exposure light had until recently been considered insignificant, but for today's advanced nodes and beyond, this is an area that could also become significant and need to be compensated.

In immersion lithography, another area that can impact overlay is cooling of the wafer due to evaporation of the water from the surface of the wafer as the wafer is being exposed, and in particular at the edges of the wafer, where residual water is extracted. This was a major issue when immersion lithography was introduced, but in today's state-of-the-art tools it is no longer an issue due to both the introduction of temperature-controllable zones within the exposure chuck itself, which compensate for this cooling, and additional tool calibrations. It is interesting to note that before immersion lithography, it was relatively straightforward to identify wafer deformation caused by process integration steps, but with the introduction of immersion lithography this became much less clear. This is because the effects are similar in being thermally related, and both cause increased overlay residuals. However, that being said, if there are any immersion cooling issues affecting the overlay performance, they tend to be much smaller in magnitude than process wafer deformation.

5.5.3 POST-EXPOSURE OVERLAY METROLOGY AND ANALYSIS

After exposure, the next step is to evaluate the resulting overlay performance. The analysis is performed using the same 10-parameter model described in the previous section and used internally by the scanner for wafer alignment and exposure. One of the most important aspects of understanding and being able to analyze overlay error is the ability to break down the error into "grid" or across-wafer error and "intrafield" or within-field error, and this is done by analysis using the 10-parameter model. From this are determined the "correctable" errors, which can be seen as the errors made by the scanner during the alignment phase, and the grid and intrafield "residuals". These correctable terms and residuals allow us to understand what is going on with both the scanner and the process. When looking at the results for a single wafer, we first want to see the magnitude of the correct-able terms and of the grid and intrafield residuals, but it is not just the magnitude of the residuals that we are interested in; we also need to look at what is called the *signature* or *fingerprint* of these residuals. With respect to grid residuals, if we ignore any process effects and talk about a perfect wafer, then the residuals should be in line with the baseline performance of the scanner and are a combination of the scanner stage accuracy and the overlay metrology precision. This means that for a state-of-the-art scanner, the 3σ residuals should be of the order of 2 nm and should not show any visible signature. Any increase above 2 nm and/or visible residual signature means that there is pro-cess wafer deformation and/or poor metrology repeatability, increasing the residuals. It is important to understand that for both grid and intrafield residuals, the best performance comes when both lay-ers are exposed on the same scanner and in the case of dual wafer stage systems, on the same wafer chuck. If they are not, there will always be a penalty, however small. If, for example, both layers are exposed on different chucks of a dual chuck scanner, the residuals will increase from around 2 to 3 nm and will show a small residual signature, and the same is true if the layers are exposed on different scanners, the effect probably being slightly larger. These differences come from small differences in wafer clamping, calibrations, wafer stage performance, and immersion cooling. With respect to intrafield residuals, these are due to relative distortion between the two exposed layers. They can come from the reticles themselves, reticle clamping repeatability of the scanner, dirt on the reticle table, lens distortion across the slit of the scanner, and use of different illumination modes for exposing each layer. As for grid residuals, if two layers are exposed on different scanners, there will be an increase in intrafield residuals due to a slightly different distortion signature across the lens and reticle clamping.

In production, it is not ideal to have to dedicate a single scanner to expose all the layers of a production batch, as this leads to production logistic problems. If a scanner goes down, the batches dedicated to that scanner cannot be processed. Therefore, it is important for scanners to be precisely matched to each other so that preferably all layers of a batch can be processed on any scanner in the fab. With today's improved calibration procedures, this situation is becoming closer, but at the same time the required tolerances are becoming tighter, so for certain critical layers and products, to get the tightest overlay possible, it may always be necessary to ensure that both layers are exposed on the same scanner. An example of this is double patterning (DP), where the layer is broken down into two "splits" exposed separately and for which very tight overlay is required. Typically, these two "splits" need to be exposed on the same scanner with chuck dedication, which means that both chucks are used, but each wafer in the batch is always exposed on the same chuck. This is another example of nothing having changed in the way that we align our wafers, just increased complexity.

Typically for a new product, or intermittently in production, a test wafer will be exposed and measured and the correctables and residuals analyzed as described above. The determined correct-ables will then be fed back as an adjustment to the scanner, so that the overlay performance comes closer to the residuals, which is the ultimate achievable performance, with no remaining correctable errors. When a wafer is aligned by the scanner and the results measured and analyzed post-expo-sure, there will always be some correctable error that needs to be fed back to the scanner. As long as these correctable terms are stable from wafer to wafer and from batch to batch, this is not a problem, as they can be corrected. So, the fact that the alignment sensor and applied model did not produce the best possible overlay is not actually a problem. What is important is the stability from wafer to wafer and from batch to batch, because without this it is not possible to actually apply these correc-tions and get any benefit. Although it is unusual for there to be no correctable error when a product wafer is initially exposed, it is important to be aware of their typical magnitude, because very large values are typically a sign that something is wrong, and it is dangerous to just blindly apply such corrections. For example, translation would be expected to be below 10 nm for a new product exposure without feedback; similarly, scaling, rotation, and orthogonality would be expected to be below approximately 0.03 ppm. Anything larger than these is often a sign of a problem. Take, for example, translation. If our first wafer showed a value of 50 nm in one direction, something is wrong; either the alignment mark coordinates in the scanner recipe are incorrect or the process has damaged the marks. Similarly, for scaling, if we see 1 ppm, this is probably a sign that the process has damaged the marks. Applying these large values without further analysis is dangerous and could lead to problems being hidden, which could later cause problems such as instability.

Another issue with feeding back correctable terms is the significance of the numbers; what is worth applying and what is not. Take, for example, the correctable errors shown below. All the terms except for translation, which is expressed in "nm" are expressed in ppm or μrad. Now, ppm is equivalent to nm/mm, as is μrad for the small numbers that we are referring to, so it is easier to visualize the significance of a correctable by thinking of it as nm/mm. If we have 0.01 ppm of grid scaling error, if we think of this as 0.01 nm/mm, then it is easy to see that for a 300 mm wafer, this would cause 1.5 nm of overlay error at the edge of the wafer, so although small, it is significant for advanced nodes, whereas if we had 0.001 nm/mm, this would equate to 0.15 nm at the edge of the wafer, which today is insignificant. Also, to judge the significance, we have to look at stability through the batch. Translation through the batch typically varies by ±0.5 nm, and so it is not possible to correct by better than 0.5 nm. It's a similar story for intrafield terms, but here we have to look at what this would cause at the edge of the field rather than the edge of the wafer. Today's scanner field is typically 26×33 mm, so if we have 0.1 nm/mm x magnification that would mean 1.3 nm at the edge of the field, which is significant, but 0.01 nm/mm would mean 0.13 nm, which is probably not.

In practice, in production, once the initial correctable terms are established for a layer, product, and scanner, they are constantly monitored and adjusted as and when necessary by an APC system to maintain optimum overlay for all batches. After reading this section, it is hoped that the reader will take away what happens to a wafer when it is processed through the scanner, what the potential issues are, and how the performance is analyzed after exposure. This is what alignment and overlay are all about, and it is hoped that the reader will have a grasp of what misalignment on the scanner means and what it can and cannot cause.

5.6 PROCESS EFFECTS ON OVERLAY

In addition to the overlay errors caused by the tool itself, and in particular the alignment system, overlay errors introduced by the process integration flow increasingly take up a large portion of the available overlay budget. As discussed previously, baseline scanner overlay performance for a full batch of wafers achievable today is around 2 nm 3σ; however, state-of-the-art overlay achieved in production is more of the order of 5 to 6 nm 3σ. In advanced processing, the process-related contribution can account for more than half of the total overlay budget.

In the course of wafer processing, wafers endure high-temperature, deposition, etch, and planarization steps, each of which can severely impact the integrity of the alignment marks and the shape of the wafers themselves, both effects resulting in process-related overlay error. As discussed previously, the function of the alignment system is to measure the position of a set of *alignment marks* patterned in a previously patterned layer; any deformed or damaged marks will directly impact the ability of the alignment system to correctly determine the alignment mark positions. Therefore, process-related alignment mark deformation is an area in overlay control that cannot be neglected. Similarly, process-related wafer deformation, wafer shape changes, can equally impact overlay, often significantly. Wafer process deformation potentially impacts the ability of the scanner to match the actual shape of the previously patterned wafer grid. This is due to the fact that the shape can no longer be described by the linear alignment model typically used, as the grid of the wafer no longer matches that of the scanner. Non-linear high-order scanner alignment models have been developed to compensate for wafer process deformation, a typical one being third-order.

5.6.1 Process Related Mark Damage and Mark Visibility

Alignment mark damage is generally related to process effects, such as chemical mechanical polishing (CMP), etch, and deposition. All of these steps have the potential to introduce asymmetry into the alignment mark shape, which can lead to misdetection of the correct alignment mark position and therefore degrade overlay accuracy. An important aspect of alignment on product is the visibility of the marks themselves to the alignment system. This is something that must be carefully considered when designing an integration flow. Alignment signal strength directly relates to the alignment system's ability to see the marks and to accurately detect them and can be impacted by the film transparency of the layers above the alignment marks and the depth of the marks themselves. Generally, signal strength can be very low and alignment accuracy still maintained if the shape of the alignment marks is close to ideal, but not if they have been damaged by the process. An important aspect of the design of an alignment system is its robustness to alignment mark damage or changes in shape. Although production processes will be optimized to minimize any mark damage, the marks will never be "perfect" or always identical and will have small variations, as the process is optimized primarily for the product. A capable alignment system must be robust to these changes and still be able to accurately detect the alignment mark position. At the same time, it is important that the alignment mark designs themselves are robust to the process steps involved in the integration flow, such as CMP, etch, and deposition, and this typically means sub-segmenting the marks to match the design rules and make them look more like the product from a pattern density point of view.

5.6.2 Process Related Wafer Deformation

Process wafer deformation is, as the name suggests, due to changes that occur to the shape of the wafer between one patterning step and another. This leads to errors that cannot be corrected by the standard scanner linear alignment model, as the grid of the wafer no longer matches that of the scanner, in a nonlinear way. Wafer deformation can be caused by many process steps. Deposition steps, for example, can introduce uniform and nonuniform stress in wafers, which when clamped no longer match the grid of the scanner and hence cause overlay errors. If a film deposition step such as nitride introduces uniform stress to the wafer, causing the wafers to expand or contract, this is readily corrected by the standard scanner linear alignment model, and it will not affect the resulting overlay. However, if the nitride deposition introduces non-linear stress, then when clamped, the changes in wafer shape are no longer correctable using the standard linear alignment model, resulting in non-correctable overlay errors, as seen in high residuals. Rapid thermal processing (RTP) steps are another major cause of wafer deformation, often in combination with layers containing stress, that have to be carefully characterized. In fact, in advanced overlay on product, it could be argued that all process steps need to be qualified for their impact on overlay; RTP, CMP, etch, deposition, and so on. Figure 5.11 shows an example of the non-correctable overlay error observed due to a badly calibrated RTP tool and when the RTP process was correctly calibrated. This is an extreme example but chosen to show and highlight the important part and impact that the process has on overlay.

There is much discussion today on how to mitigate wafer process deformation. One solution offered by the scanner suppliers is to use high-order alignment models in the scanner, which will attempt to correct for the non-linear wafer deformation caused by the process. Another offered by the scanner and metrology tool suppliers is to understand the wafer deformation caused by the process and to feed forward additional corrections to the scanner, often referred to as CPE. Both of these solutions are far from ideal, as process-related wafer deformation can vary from wafer to

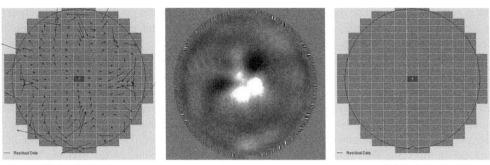

Residual Overlay 3σ ~ 150nm   Residual Overlay 3σ ~ 3nm

FIGURE 5.11 Residual overlay error resulting from miscalibrated RTP process (left), nanotopography map showing "sliplines" caused by the RTP process (center), and residual overlay error with well-calibrated RTP process (right).

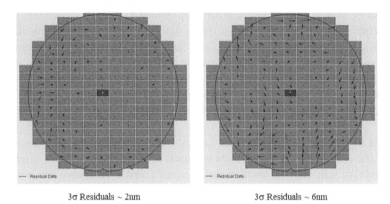

3σ Residuals ~ 2nm   3σ Residuals ~ 6nm

FIGURE 5.12 Residual fingerprints for a wafer aligned using the standard linear alignment model (left) and a high-order alignment model (right).

wafer and from batch to batch. In principle, when high-order alignment is used, each wafer is corrected independently, but if the type of signature is not consistent, it may not be able to adequately compensate all wafers. If a CPE strategy is being used, this necessitates that the signature from wafer to wafer is constant; it will not improve overlay, and in fact could degrade it, if this is not the case. A better solution is to minimize these process-induced wafer deformations as much as possible, or even eliminate them, so that the standard scanner linear alignment model can continue to be used. Although high-order models can be used, what is not obvious is that we lose many of the assumptions that we have always been able to rely on. Once they are introduced, misalignment can introduce signatures during exposure, and it becomes more difficult to separate wafer process deformation from alignment and model-induced signatures. An example of this is shown in Figure 5.12, where the residuals are shown on the left for a wafer exposed using the standard linear alignment model and on the right for a high-order alignment model. As can be seen in this example, where the wafer itself did not suffer from significant process deformation, the high-order alignment model has actually introduced more variation, and a clear signature is visible [15, 42–65].

5.7 ALIGNMENT SIGNAL REDUCTION ALGORITHMS

If the only degrading influence on the signal is zero mean Gaussian white noise, then the optimum algorithm for determining the signal centroid is to correlate the signal with itself and find the position of the peak of the output. Equivalently, since the derivative, i.e., the slope, of a function at its peak is zero, we can correlate the signal with its derivative and find the position where the output of the correlation crosses through zero. One proof that this is the optimal algorithm involves using the technique of maximum likelihood.

Wafer alignment signals are generated by scattering/diffracting light from an alignment mark on the wafer. For nongrating marks, the signal from a single alignment mark will generally appear as one or perhaps several localized "bumps" in the alignment sensor signal data. If sufficient symmetry is present in the alignment sensor as well as the alignment mark structure itself, this reduces to finding the centroid or "center of mass" of the signal. For grating marks, the signal is often essentially perfectly periodic and usually sinusoidal with perhaps an overall slowly varying amplitude envelope. In this case, the "centroid" can be associated with the phase of the periodic signal as measured relative to a predefined origin. In general, all the algorithms discussed later can be applied to periodic as well as isolated signal "bumps." But for periodic signals, the Fourier algorithm is perhaps the most appropriate.

As discussed earlier, real signals collected from real wafers will, of course, be corrupted by noise and degraded by the asymmetry present in real marks and real alignment sensors. Since the noise contribution can be treated statistically, it is straightforward to develop algorithms that minimize on average its contribution to the final result. If this were the only problem facing alignment, simply increasing the number of alignment marks scanned would allow overlay to become arbitrarily accurate. But, as discussed earlier, process variation both across a wafer and from wafer to wafer changes not only the overall signal amplitude but also the mark symmetry and hence, the signal symmetry. This effect is not statistical and is generally not predictable. Potential algorithms and general approaches for dealing with explicitly asymmetric marks and signals are given in Galatin et al. [66, 67].

It is simplest to proceed with the general analysis in continuum form. For completeness, the adjustments to the continuum form that must be made to use discrete, that is, sampled, data are briefly discussed. We will work in one dimension, since x and y values are computed separately. Finally, the portions of the signal representing alignment marks can be positive "bumps" in a nominally zero background, as would occur in dark-field imaging, or they can be negative "bumps" in a nominally nonzero background, as would occur in bright-field imaging. For simplicity, we will make the tacit assumption that we are dealing with dark-field-like signals. In the grating case, the signals are generally sinusoidal, which can be viewed either way, and we will treat this case separately.

Let $I(x)$ be a perfect, and hence symmetric, signal bump (or bumps) as a function of position, x. The noise, $n(x)$, will be taken to be additive, white, i.e, uncorrelated, and spatially and temporally stationary, i.e, constant in space and time.

When a new wafer is loaded, and a given alignment mark is scanned, the signal will be translated by an unknown amount s relative to some predetermined origin of the coordinate system. Thus, the actual detected signal, $D(x)$, is given by $I(x)$ shifted a distance s; that is, $D(x)=I(x-s)$. It is the purpose of the alignment sensor to determine the value of s from the detected signal. The position of the centroid or center of mass of the pure signal is defined as

$$C = \frac{\int x D(x)dx}{\int D(x)dx} = \frac{\int x I(x-s)dx}{\int I(x-s)dx} \tag{5.5}$$

To show that $s=C$, so that we can find s simply by computing the centroid, let $y=x-s$; then

$$C = \frac{\int (y+s)I(y)dy}{\int I(y)dy} = \underbrace{\frac{\int y I(y)dy}{\int I(y)dy}}_{=0} + s\left(\underbrace{\frac{\int I(y)dy}{\int I(y)dy}}_{=1}\right) \tag{5.6}$$

The first term vanishes because

$$\int y I(y)\,dy = \int \left[(\text{odd})\times(\text{even})\right] dy = 0$$

and we are assuming symmetric limits of integration, which for all practical purposes can be set to $\pm\infty$. Thus,

$$s = C \tag{5.7}$$

and the shift position can be found by computing the centroid.

In the presence of noise, the actual signal is the pure signal, shifted by the unknown amount s, with noise added:

$$D(x) = I(x-s) \rightarrow D(x) = I(x-s) + n(x) \tag{5.8}$$

Below we discuss standard algorithms for computing a value for C from the measured data $D(x)$, which, based on the above discussion, amounts to determining an estimated value of s, which we will label s_E. Along with using the measured data, some of the algorithms also make use of any a priori knowledge of the ideal signal shape $I(x)$.

The digitally sampled real data is not continuous, and we will use the convention $D_i = D(x_i)$ to label the signal values measured at the sample positions x_i, where $i = 1, 2, \ldots, N$, with N the total number of data values for a single signal.

5.7.1 THRESHOLD ALGORITHM

Consider an isolated single bump in the signal data that represents an "image" of an alignment mark. The threshold algorithm attempts to find the bump centroid by finding the midpoint between the two values of x at which the bump has a given value called the *threshold*. In the case where the signal contains multiple "bumps" representing multiple alignment marks, the algorithm can be applied to each bump separately, and the results can be combined to produce an estimate of the net bump centroid.

For now, let $D(x)$ consist of a single positive bump plus noise, and let D_T be the specified threshold value. Then, if the bump is reasonably symmetric and smoothly varying and D_T has been chosen appropriately, there will be two and only two values of x, call them x_L and x_R, which satisfy

$$D_T = D(x_L) = D(x_R) \tag{5.9}$$

The midpoint between x_L and x_R, which is the average of x_L and x_R, is taken as the estimate for s; that is,

$$s_E = \frac{1}{2}(x_L + x_R) \tag{5.10}$$

As shown later, this algorithm is very sensitive to noise, since it uses only two points out of the entire signal. A refinement that eliminates some of this noise dependence is to average the result from multiple threshold levels. Taking $D_{T_1}, D_{T_2}, \ldots, D_{T_N}$ to be N different threshold levels with $s_{E_1}, s_{E_2}, \ldots s_{E_N}$ being the corresponding centroid estimate for each, the net centroid is taken to be

$$s_E = \frac{1}{N}(s_{E_1} + s_{E_2} + \cdots + s_{E_N}) \tag{5.11}$$

This definition of s_E weights all the N threshold estimates equally. A further refinement of the multiple threshold approach is to weight the separate threshold results nonuniformly. This weighting can be based on intuition and/or modeling and/or experimental results that indicate that certain threshold levels tend to be more reliable than others. In this case,

$$s_E = w_1 s_{E_1} + w_2 s_{E_2} + \cdots + w_N s_{E_N} \tag{5.12}$$

where $w_1 + w_2 + \cdots + w_N = 1$

5.7.1.1 Noise Sensitivity of the Threshold Algorithm

Noise can lead to multiple threshold location results. Generally, the threshold level should be set to around 50% of the signal peak value.

Let x_{L0} and x_{R0} be the true noise free threshold positions; that is,

$$I(x_{L0}) = I(x_{R0}) = D_T \tag{5.13}$$

Now, let Δ_L and Δ_R be the deviations in threshold position caused by noise, so that $x_L = x_{L0} + \Delta_L$ and $x_R = x_{R0} + \Delta_R$. Substituting this into the threshold equation and assuming the Δs are small gives

$$D_T = D(x_L) = I(x_{L0} + \Delta_L) + n(x_{L0} + \Delta_L)$$

$$\simeq I(x_{L0}) + I'(x_{L0})\Delta_L + n(x_{L0}) + n'(x_{L0})\Delta_L$$

and $\tag{5.14}$

$$D_T = D(x_R) = I(x_{R0} + \Delta_R) + n(x_{R0} + \Delta_R)$$

$$\simeq I(x_{R0}) + I'(x_{R0})\Delta_R + n(x_R) + n'(x_{R0})\Delta_R$$

where the prime on $I(x)$ and $n(x)$ indicates differentiation with respect to x. Using $I(x_{L0}) = I(x_{R0}) = D_T$ and solving for the Δs yields

$$\Delta_L = \frac{n(x_{L0})}{I'(x_{L0}) + n'(x_{L0})}$$

$$\tag{5.15}$$

$$\Delta_R = \frac{n(x_{R0})}{I'(x_{R0}) + n'(x_{R0})}$$

The derivative of uncorrelated noise has an RMS slope of infinity. The discrete nature of real sampled data will mitigate this, but still, in order to obtain reasonable answers using this algorithm, the noise must be well behaved.

Assuming the $n(x)$ is smooth enough so that we can make the approximation $n' \ll I'$ gives

$$s_E = \frac{x_L + x_R}{2} + \frac{n(x_{L0})}{2I'(x_{L0})} + \frac{n(x_{R0})}{2I'(x_{R0})} \tag{5.16}$$

The RMS error σ_s in the single threshold algorithm as a function of the RMS noise σ_n, assuming the noise is spatially stationary and uncorrelated from the left to the right side of the bump and that $I'(x_{L0}) = I'(x_{R0}) \equiv I'$, is then

$$\sigma_s = \frac{\sigma_n}{\sqrt{2}I'} \tag{5.17}$$

This result shows explicitly that the error will be large in regions where the slope I' is small, and thus, it is best to choose the threshold to correspond to large slopes. If the results from N different threshold levels are averaged, and the slope I' is essentially the same at all the threshold levels, then

$$\sigma_s \cong \frac{\sigma_n}{\sqrt{2NI'}} \qquad (5.18)$$

5.7.1.2 Discrete Sampling and the Threshold Algorithm

For discretely sampled data, only rarely will any of the D_i values correspond exactly to the threshold value. Instead, there will be two positions on the left side of the signal and two positions on the right where the D_i values cross over the threshold level. Let the i values between which the crossover occurs on the left be i_L and i_L+1 and on the right i_R and i_R+1; then, the actual threshold positions can be determined by linear interpolation between corresponding sample positions.

5.7.2 CORRELATOR ALGORITHM

The correlator algorithm is somewhat similar to the variable weighting threshold algorithm in that it uses most or all of the signal data, but nominally, not uniformly. It can be derived using the method of least squares. The mean square difference between $D(x)$ and $I(x-s)$ is given by

$$\int \left(D(x) - I(x-s) \right)^2 dx \qquad (5.19)$$

To minimize this requires finding s_E such that

$$0 = \left[\frac{\partial}{\partial s} \int \left(D(x) - I(x-s) \right)^2 dx \right]_{s=s_E} \qquad (5.20)$$

Taking the derivative inside the integral and using the fact that $\int I(x) I'(x) dx = 0$ if I vanishes at the endpoints of the integration gives

$$0 = \int I'(x-s_E) D(x) dx \qquad (5.21)$$

Note that at the peak of the bump the derivative is zero, whereas at the edges the slope has the largest absolute value. Using the derivative of the bump as the "weighting" function in the correlation shows explicitly that essentially all the information about the bump centroid comes from its edges, with essentially no information coming from its peak. Simply put, if the signal is shifted a small amount, the largest change in signal value occurs in the regions with the largest slope, i.e., the "edges," and there is essentially no change in the value at the peak. The "edges" are therefore the most sensitive to the bump position and hence contain the most position information.

5.7.2.1 Noise Sensitivity of the Correlator Algorithm

In the presence of noise, the value of s is still determined by finding the value s_E for which

$$0 = \int I'(x-s_E) D(x) dx$$
$$= \int I'(x-s_E) \left(I(x) + n(x) \right) dx \qquad (5.22)$$

But now, because of the noise, s_E will differ by a small amount, δs, from the true value s, which we can take to be 0, so that now, with noise, we have $s_E = \delta s$ and hence

$$0 = \int I'(x - \delta s)\big(I(x) + n(x)\big)dx \tag{5.23}$$

Let $x \to x + \delta s$, use $I(x + \delta x) \cong I(x) + I'(x)\delta s$ and $\int I'(x)I(x)dx = 0$, and rearrange to get

$$\delta s = \frac{\int I'(x)n(x + \delta s)dx}{\int I'(x)^2 dx} \tag{5.24}$$

which is in general not equal to zero and amounts to the error in s_E for the particular noise function $n(x)$. Assuming spatially stationary noise—i.e., the noise at $x + \delta s$ is statistically equivalent to the noise at x—then yields

$$\sqrt{\langle \delta s^2 \rangle} = \sigma_s = \sqrt{\frac{\int\int I'(x_1)I'(x_2)\langle n(x_1)n(x_2)\rangle dx_1 dx_2}{\left[\int(I'(x))^2 dx\right]^2}} \tag{5.25}$$

For the case where the noise is uncorrelated, so that $\langle n(x_1)n(x_2)\rangle = \sigma_n^2 \delta(x_1 - x_2)$, this reduces to

$$\sigma_s = \frac{\sigma_n}{\sqrt{\int(I'(x))^2 dx}} \tag{5.26}$$

This result shows explicitly again that the error in the s_E is larger when the slope of the ideal bump shape is small. Note that $\langle n(x_1)n(x_2)\rangle = \sigma_n^2 \delta(x_1 - x_2)$ requires σ_n to have units of $I \times \sqrt{\text{length}}$, since the delta function has units of 1/length and $n(x)$ has units of I.

5.7.2.2 Discrete Sampling and the Correlator Algorithm

For discrete sampling, the integration is replaced by summation, i.e,

$$\int I'(x - x_0)D(x)dx \to \sum_i I'_{i-i_0} D_i \Delta x \tag{5.27}$$

As with the threshold algorithm, discrete sampling means that only rarely will an exact zero crossing in the output of the correlation occur exactly at a sample point, and generally, two consecutive values of i_0 will straddle the zero crossing, so that the true zero crossing occurs between them. With sufficiently dense sampling, linear interpolation between the two values of i_0 will then yield the optimum zero crossing position between the two sample positions.

5.7.3 FOURIER ALGORITHM

This algorithm is based on Fourier analyzing the signal. It is perhaps most straightforward to apply it to signals that closely approximate sinusoidal waveforms, but it can in fact be applied to any signal. We first discuss the algorithm for nonsinusoidal signals and then show the added benefit that accrues when it is applied to sinusoidal signals such as would be produced by a grating sensor, as discussed, for example, in Gatherer and Meng [68].

Assuming that $I(x)$ is real and symmetric, its Fourier transform

$$\tilde{I}(\beta) \equiv \frac{1}{\sqrt{2\pi}} \int_{-\infty}^{+\infty} I(x)e^{i\beta x}dx \tag{5.28}$$

is real and symmetric, i.e.,

$$\tilde{I}(\beta) = \tilde{I}^*(\beta): \text{ Real}$$

$$\tilde{I}(\beta) = \tilde{I}(-\beta): \text{ Symmetric}$$

(5.29)

The parameter β is the spatial frequency in radians/(unit length)$=2\pi\times$cycles/(unit length).

The Fourier transform of the measured signal, $\tilde{D}(\beta)$, is related to $\tilde{I}(\beta)$ by a phase factor,

$$\tilde{D}(\beta) \equiv \frac{1}{\sqrt{2\pi}} \int_{-\infty}^{+\infty} D(x) e^{i\beta x} dx$$

$$\equiv \frac{1}{\sqrt{2\pi}} \int_{-\infty}^{+\infty} I(x-s) e^{i\beta x} dx$$

(5.30)

$$\equiv \frac{1}{\sqrt{2\pi}} \int_{-\infty}^{+\infty} I(x') e^{i\beta(x'+s)} dx'$$

$$= e^{i\beta s} \tilde{I}(\beta)$$

where in the first step we have used $D(x) = I(x-s)$ as given above and in the second step we have changed variables, letting $x' = x + s$.

Remembering that $\tilde{I}(\beta)$ is real, we can then calculate s by scaling the arctangent of the ratio of the imaginary to the real component of the Fourier transform as follows:

$$s = \frac{1}{\beta} \arctan\left(\frac{\text{Im}\left(\tilde{D}(\beta)\right)}{\text{Re}\left(\tilde{D}(\beta)\right)}\right)$$

(5.31)

This can be proved by first noting that since $\tilde{I}(\beta)$ is real, we have $\text{Im}\left(\tilde{D}(\beta)\right) = \sin(\beta s)\tilde{I}(\beta)$ and $\text{Re}\left(\tilde{D}(\beta)\right) = \cos(\beta s)\tilde{I}(\beta)$. Therefore, $\tilde{I}(\beta)$ cancels in the ratio, leaving $\sin(\beta s)/\cos(\beta s) = \tan(\beta s)$. Then, taking the arctangent and dividing by the spatial frequency, β, leaves the shift, s, as desired.

Like the correlator algorithm, the Fourier algorithm can also be derived using least squares. Substituting the Fourier transform representations

$$I(x) = \frac{1}{\sqrt{2\pi}} \int \tilde{I}(\beta) e^{-i\beta x} d\beta$$

$$D(x) = I(x-s_0)$$

(5.32)

$$= \frac{1}{\sqrt{2\pi}} \int \tilde{I}(\beta) e^{-i\beta(x-s_0)} d\beta$$

into the least squares integral $\int \left(D(x) - I(x-s)\right)^2 dx$, taking a derivative with respect to s, and setting the result equal to zero for $s = s_E$ yields

$$0 = \int \beta \left|\tilde{I}(\beta)\right|^2 \sin\left(\beta(s_E - s_0)\right)$$

(5.33)

Using $\sin\left(\beta(s_E - s_0)\right) \cong \beta(s_E - s_0)$ for s_E close to s_0 and $\beta s_0 = \arctan\left(\text{Im}[\tilde{D}(\beta)] / \text{Re}[\tilde{D}(\beta)]\right)$, we obtain the same result.

There are several interesting aspects to the above result. Although the right-hand side can be evaluated for different values of β, for real symmetric signals, they all yield the same value of s. Thus, in the absence of any complicating factors such as noise or inherent signal asymmetry, any value of β can be used, and the result will be the same.

5.7.3.1 Noise Sensitivity of the Fourier Algorithm

In the presence of noise, the Fourier transform of the signal data takes the form

$$\tilde{D}(\beta) = e^{i\beta s}\tilde{I}(\beta) + \tilde{n}(\beta) \tag{5.34}$$

Substituting this for $\tilde{D}(\beta)$ into the result given earlier for s yields

$$s(\beta) = \frac{1}{\beta}\arctan\left(\frac{\tilde{I}(\beta)\sin(\beta s) + \mathrm{Im}\left[\tilde{n}(\beta)\right]}{\tilde{I}(\beta)\cos(\beta s) + \mathrm{Re}\left[\tilde{n}(\beta)\right]}\right) \tag{5.35}$$

where s is now a function of β. That is, different β values will yield different estimates for s. The best estimate will be obtained from a weighted average of the different s values. This weighted average can be written as

$$
\begin{aligned}
s_E &= \int f(\beta)s(\beta)d\beta \\
&= \int \frac{f(\beta)}{\beta}\arctan\left(\frac{\tilde{I}(\beta)\sin(\beta s) + \mathrm{Im}\left[\tilde{n}(\beta)\right]}{\tilde{I}(\beta)\cos(\beta s) + \mathrm{Re}\left[\tilde{n}(\beta)\right]}\right)d\beta \\
&\approx s\int f(\beta)d\beta + \int \frac{f(\beta)}{\beta}\frac{\left(\cos(\beta s)\mathrm{Im}\left[\tilde{n}(\beta)\right] - \sin(\beta s)\mathrm{Re}\left[\tilde{n}(\beta)\right]\right)}{\tilde{I}(\beta)}d\beta
\end{aligned}
\tag{5.36}
$$

where $f(\beta)$ is the weighting function. In the last step, we have assumed that $f(\beta)$ is large in regions where the signal to noise ratio is large, i.e., $\tilde{I} > \tilde{n}$, and it is essentially zero in regions where the signal to noise ratio is small, i.e., $\tilde{I} < \tilde{n}$. Assuming zero mean noise, so that $\langle \tilde{n}(\beta)\rangle = 0$, we must have $\int f(\beta)d\beta = 1$ for s_E to be equal to the true answer, s, on average, i.e., $\langle s_E\rangle = s$. The error in s_E is then given by the second term, and we have

$$\sigma_s^2 = \int \frac{f(\beta_1)}{\beta_1\tilde{I}(\beta_1)}\frac{f(\beta_2)}{\beta_2\tilde{I}(\beta_2)}\left\langle\begin{array}{l}\left(\cos(\beta_1 s)\mathrm{Im}\left[\tilde{n}(\beta_1)\right] - \sin(\beta_1 s)\mathrm{Re}\left[\tilde{n}(\beta_1)\right]\right) \\ \times\left(\cos(\beta_2 s)\mathrm{Im}\left[\tilde{n}(\beta_2)\right] - \sin(\beta_2 s)\mathrm{Re}\left[\tilde{n}(\beta_2)\right]\right)\end{array}\right\rangle d\beta_1 d\beta_2. \tag{5.37}$$

Using the fact that $n(x)$ is real gives $\mathrm{Re}\left[\tilde{n}(\beta)\right] = \frac{1}{2}\left[\tilde{n}(\beta) + \tilde{n}(-\beta)\right]$ and $\mathrm{Im}\left[\tilde{n}(\beta)\right] = \frac{1}{2i}\left[\tilde{n}(\beta) - \tilde{n}(-\beta)\right]$. Assuming $n(x)$ is uncorrelated, i.e. $\langle n(x)n(x')\rangle = \sigma_n^2\delta(x - x')$, it follows that

$$\left\langle\mathrm{Re}\left[\tilde{n}(\beta_1)\right]\mathrm{Re}\left[\tilde{n}(\beta_2)\right]\right\rangle = \frac{\sigma_n^2}{2}\left[\delta(\beta_1 - \beta_2) + \delta(\beta_1 + \beta_2)\right]$$

$$\left\langle\mathrm{Im}\left[\tilde{n}(\beta_1)\right]\mathrm{Im}\left[\tilde{n}(\beta_2)\right]\right\rangle = \frac{\sigma_n^2}{2}\left[\delta(\beta_1 - \beta_2) - \delta(\beta_1 + \beta_2)\right] \tag{5.38}$$

$$\left\langle\mathrm{Re}\left[\tilde{n}(\beta_1)\right]\mathrm{Im}\left[\tilde{n}(\beta_2)\right]\right\rangle = 0$$

Substituting this and letting $f(\beta) = f(-\beta)$, since $\tilde{I}(\beta) = \tilde{I}(-\beta)$, yields

$$\sigma_s^2 = \sigma_n^2 \int \left(\frac{f(\beta)}{\beta \tilde{I}(\beta)} \right)^2 d\beta \tag{5.39}$$

We can find the optimum form for $f(\beta)$ by letting $f(\beta) = a(\beta)/\int a(\beta) d\beta$, so that $\int f(\beta) d\beta = 1$ is automatically satisfied, then replacing a with $a + \Delta a$ and expanding in powers of Δa, and finally demanding that the first order in Δa terms vanish for all Δa. This yields the following relation:

$$\int \frac{a(\beta) \Delta a(\beta)}{\left(\beta \tilde{I}(\beta) \right)^2} d\beta = \left(\frac{\int \Delta a(\beta) d\beta}{\int a(\beta) d\beta} \right) \int \left(\frac{a(\beta)}{\beta \tilde{I}(\beta)} \right)^2 d\beta \tag{5.40}$$

which is satisfied by letting $a(\beta) = \left(\beta \tilde{I}(\beta) \right)^2$, which then gives

$$f(\beta) = \frac{\left(\beta \tilde{I}(\beta) \right)^2}{\int \left(\beta \tilde{I}(\beta) \right)^2 d\beta}. \tag{5.41}$$

Substituting this then gives

$$s_E = \frac{1}{\int \left(\beta \tilde{I}(\beta) \right)^2 d\beta} \int \beta \left(\tilde{I}(\beta) \right)^2 \arctan \left(\frac{\text{Im} \left[\tilde{D}(\beta) \right]}{\text{Re} \left[\tilde{D}(\beta) \right]} \right) d\beta \tag{5.42}$$

$$\sigma_s = \frac{\sigma_n}{\sqrt{\int \left(\beta \tilde{I}(\beta) \right)^2 d\beta}}$$

Thus, the optimum weighting is proportional to the power spectrum of the signal $\tilde{I}^2$, as one would expect when the noise is uncorrelated. Also, the β factor shows that there is no information about the position of the bump for $\beta \sim 0$. This is a consequence of the fact that $\beta = 0$ corresponds to a constant value in x that carries no centroid information. Finally, $I'(x)$ in Fourier or β space is given by $i\beta \tilde{I}(\beta)$, and thus, σ_s has the same basic form in both the correlator and the Fourier algorithm. Note that $\langle n(x) n(x') \rangle = \sigma_n^2 \delta(x - x')$ requires σ_n to have units of $I \times \sqrt{\text{length}}$, since the delta function has units of 1/length and $n(x)$ has units of I. This is exactly what is required for σ_s to have units of length, since $\tilde{I}$ has units of $I \times \text{length}$.

5.7.3.2 Discrete Sampling and the Fourier Algorithm

The main effect of having discretely sampled rather than continuous data is to replace all the integrals in the preceding analysis with sums, that is, to replace true Fourier transforms with discrete Fourier transforms (DFTs) or their fast algorithmic implementation, fast Fourier transforms (FFTs).

5.7.3.3 Application of the Fourier Algorithm to Grating Sensors

In many grating sensors and in some nongrating sensors the pure mark signal is not an isolated bump but a sinusoid of a specific known frequency, say β_0, multiplied possibly by a slowly varying envelope function. The information about the mark position in this case is encoded in the phase of the sinusoid. The total detected signal will, as usual, be corrupted by noise and other effects, which in general add sinusoids of all different frequencies, phases, and amplitudes to the pure β_0 sinusoid.

But, since we know that the mark position information is contained only in the β_0 frequency component of the signal, all the other frequency components can simply be ignored in a first approximation. They are useful only as a diagnostic for estimating the goodness of the signal. That is, if all the other frequency components are small enough that the signal is almost purely a β_0 sinusoid, then the expectation is that the mark is clean and uncorrupted and the noise level is low, in which case one can have high confidence in the mark position predicted by the signal. On the other hand, if all the other frequency components of the signal are as large as or larger than the β_0 frequency component, then it is likely that the β_0 frequency component is severely corrupted by noise and the resulting centroid prediction is suspect.

Using the preceding result for computing s from the Fourier transform of the signal but using only the β_0 frequency component in the calculation yields

$$s = \frac{1}{\beta_0} \arctan\left(\frac{\text{Im}\left(\tilde{D}(\beta_0) \right)}{\text{Re}\left(\tilde{D}(\beta_0) \right)} \right) \tag{5.43}$$

and in the presence of noise we have

$$s_E = \frac{1}{\beta_0} \arctan\left(\frac{\tilde{I}(\beta_0)\sin(\beta_0 s) + \text{Im}\left[\tilde{n}(\beta_0) \right]}{\tilde{I}(\beta_0)\cos(\beta_0 s) + \text{Re}\left[\tilde{n}(\beta_0) \right]} \right)$$

$$\simeq s + \frac{\cos(\beta_0 s)\,\text{Im}\left[\tilde{n}(\beta_0) \right] - \sin(\beta_0 s)\,\text{Re}\left[\tilde{n}(\beta_0) \right]}{\tilde{I}(\beta_0)} \tag{5.44}$$

for $\tilde{n}(\beta_0) \ll \tilde{I}(\beta_0)$. The effect of noise on the grating result is given by

$$\sigma_s^2 = \frac{\sigma_n^2}{\sqrt{\left(\beta_0 \tilde{I}(\beta_0) \right)^2 \Delta\beta}} \tag{5.45}$$

where $\Delta\beta$ is the frequency resolution of the sensor.

5.8   GLOBAL ALIGNMENT ALGORITHM

The purpose of the global alignment algorithm is to combine all the separate alignment mark position measurements into an optimum estimate of the correctable components of the field and grid distortions along with the overall grid and field positions. These "correctable" components generally consist of some or all the linear distortion terms described below plus, in some cases, higher-order nonlinear terms. As discussed in previous sections, each field will be printed with roughly the same rotation, magnification, skew, etc., with respect to the expected field. The linear components of the average field distortion are referred to collectively as *field terms*. The position of a given reference point in each field, such as the field center, defines the grid, and these points will also have some amount of rotation, magnification, skew, etc. with respect to the expected grid. The linear components of the grid distortion are referred to collectively as *grid terms*. In global fine alignment, where the alignment marks on only a few fields on the wafer are measured, both field and grid terms need to be determined from the alignment data to perform overlay. In field-by-field alignment, where each field is aligned based only on the data from that field, the grid terms are not directly relevant. Here, we consider only global fine alignment.

Let x and y be standard orthogonal Cartesian coordinates in two dimensions. Consider an arbitrary combination of translation, rotation, and distortion of the points in the plane. This will carry each original point (x,y) to a new position, (x',y'); that is,

$$x \to x' = f(x, y)$$

$$y \to y' = g(x, y) \tag{5.46}$$

The functions f and g can be expressed as power series in the x and y coordinates with the form

$$x' = f(x, y) = T_x + M_{xx}x + M_{xy}y + Cx^2 + Dxy + \cdots$$

$$y' = g(x, y) = T_y + M_{yx}x + M_{yy}y + Ey^2 + Fxy + \cdots \tag{5.47}$$

where the $T, M, C, D, E, F, \ldots$ coefficients are all constant, that is, independent of x and y.

The T terms represent a constant shift of all the points in the plane by the amount T_x in the x direction and T_y in the y direction. The M terms represent shifts in the coordinate values, which depend linearly on the original coordinate values. The remaining C, D, E, F, and higher-order terms all depend nonlinearly on the original coordinate values. Using matrix-vector notation, we can then write Equations 5.46 and 5.47 as a single equation of the form

$$\begin{pmatrix} x' \\ y' \end{pmatrix} = \underbrace{\begin{pmatrix} T_x \\ T_y \end{pmatrix}}_{\text{CONSTANT}} + \underbrace{\begin{pmatrix} M_{xx} & M_{xy} \\ M_{yx} & M_{yy} \end{pmatrix} \cdot \begin{pmatrix} x \\ y \end{pmatrix}}_{\text{LINEAR TERM}} + \begin{pmatrix} \text{Nonlinear} \\ \text{Terms} \end{pmatrix} \tag{5.48}$$

where the "+" and "·" indicate standard matrix addition and multiplication, respectively.

The constant term is a translation, which has separate x and y values. The linear term involves four independent constants: M_{xx}, M_{xy}, M_{yx}, and M_{yy}. These can be expressed as combinations of the more geometric concepts of rotation, skew, x-magnification (x-mag), and y-magnification (y-mag). Each of these "pure" transformations can be written as a single matrix:

$$\text{Rotation} = \begin{pmatrix} \cos(\theta_z) & -\sin(\theta_z) \\ \sin(\theta_z) & \cos(\theta_z) \end{pmatrix}$$

$$\text{Skew} = \begin{pmatrix} 1 & 0 \\ \sin(\psi) & 1 \end{pmatrix}$$

$$x\text{-Mag} = \begin{pmatrix} m_x & 0 \\ 0 & 1 \end{pmatrix} \tag{5.49}$$

$$y\text{-Mag} = \begin{pmatrix} 1 & 0 \\ 0 & m_y \end{pmatrix}$$

Here θ is the rotation angle and ψ is the skew angle both measured in radians, and m_x and m_y are the x and y magnifications, respectively, both of which are unitless. Both skew and rotation are area preserving, since their determinants are unity, whereas x-mag and y-mag change the area by factors of m_x and m_y, respectively.

Skew has been defined previously to correspond geometrically to a rotation of just the x axis by itself, i.e, x-skew. Instead of using rotation and x-skew, we could have used rotation and y-skew or the combination x-skew and y-skew. Similarly, instead of using x-mag and y-mag, the combinations isotropic magnification, i.e, "iso-mag" and x-mag or iso-mag and y-mag, could have been used. Which combinations are chosen is purely a matter of convention.

The net linear transformation matrix, M, can be written as the product of the mag, skew, and rotation matrices. Since matrix multiplication is not commutative, the exact form that M takes in this case depends on the order in which the separate matrices are multiplied. But since most distortions encountered in an exposure tool are small, we only need to consider the infinitesimal forms of the matrices, in which case the result is commutative.

Using the approximations

$$\cos(\varphi) \cong 1$$

$$\sin(\varphi) \cong \varphi$$

$$m_x = 1 + \mu_x$$

$$m_y = 1 + \mu_y$$

(5.50)

and then expanding to first order in all the small terms, θ, ψ, μ_x, and μ_y, we get

$$M = (x\text{-Mag}) \cdot (y\text{-Mag}) \cdot (\text{Skew}) \cdot (\text{Rotation})$$

$$= \begin{pmatrix} m_x & 0 \\ 0 & 1 \end{pmatrix} \cdot \begin{pmatrix} 1 & 0 \\ 0 & m_y \end{pmatrix} \cdot \begin{pmatrix} 1 & 0 \\ \sin(\psi) & 1 \end{pmatrix} \cdot \begin{pmatrix} \cos(\theta) & -\sin(\theta) \\ \sin(\theta) & \cos(\theta) \end{pmatrix}$$

$$= \begin{pmatrix} m_x \cos(\theta) & -m_x \sin(\theta) \\ m_y (\sin(\theta) + \sin(\psi)\cos(\theta)) & m_y (\cos(\theta) - \sin(\psi)\sin(\theta)) \end{pmatrix}$$

(5.51)

$$\cong \begin{pmatrix} 1 + \mu_x & -\theta \\ \theta + \psi & 1 + \mu_y \end{pmatrix}$$

$$\cong \underbrace{\begin{pmatrix} 1 & 0 \\ 0 & 1 \end{pmatrix}}_{\substack{\text{IDENTITY} \\ \text{MATRIX}}} + \begin{pmatrix} \mu_x & -\theta \\ \theta + \psi & \mu_y \end{pmatrix}$$

Thus, the transformation takes the infinitesimal form

$$\begin{pmatrix} x' \\ y' \end{pmatrix} \cong \begin{pmatrix} x \\ y \end{pmatrix} + \underbrace{\begin{pmatrix} T_x \\ T_y \end{pmatrix} + \begin{pmatrix} \mu_x & -\theta \\ \theta + \psi & \mu_y \end{pmatrix} \cdot \begin{pmatrix} x \\ y \end{pmatrix}}_{= \begin{pmatrix} \Delta x \\ \Delta y \end{pmatrix}}$$

(5.52)

To be the most general, we will solve for all six linear distortion terms discussed above: x and y translation, rotation, skew, x magnification, and y magnification. Note that not all exposure tools can correct for all these terms, and thus the algorithm must be adjusted accordingly. We will consider a generic alignment system that measures and returns the x and y position values of each of N_M alignment marks in each of N_F fields on a wafer. Let $m = 1,2,...,N_M$ label the marks in each field and $f = 1,2,...,N_F$ label the fields. We will use the following matrix-vector notation for position, as measured with respect to some predefined coordinate system fixed with respect to the exposure tool:

$$r_{mf} = \begin{pmatrix} x_{mf} \\ y_{mf} \end{pmatrix} = \text{Expected position of mark } m \text{ in field } f$$

$$r'_{mf} = \begin{pmatrix} x'_{mf} \\ y'_{mf} \end{pmatrix} = \text{Measured position of mark } m \text{ in field } f$$

$$R_f = \begin{pmatrix} X_f \\ Y_f \end{pmatrix} = \text{Expected position of field } f \text{ reference point} \qquad (5.53)$$

$$R'_f = \begin{pmatrix} X'_f \\ Y'_f \end{pmatrix} = \text{Measured position of field } f \text{ reference point}$$

To be explicit, we must now choose a reference point for each field. It is the difference between the measured and expected positions of this reference point that defines the translation of the field. A suitable choice would be the center of the field, but this is not necessary. Basically, any point within the field can be used, although this is not to say that all points are equal in this regard. Different choices will result in different noise propagation and round off error in any real implementation, and the reference point must be chosen to minimize these effects to the extent necessary. We will take the "center of mass" of the mark positions to be the reference point, i.e., we define the position of the reference point of field f by

$$R_f = \frac{1}{N_M} \sum_m r_{mf} \qquad (5.54)$$

If the alignment marks are symmetrically arrayed around a field, then R_f as defined in Equation 5.54 corresponds to the field center.

The analysis is simplified if we assume that the field terms are defined with respect to the field reference point; that is, field rotation, skew, and x and y magnification do not affect the position of the reference point. This can be done by writing

$$r_{mf} = R_f + d_m \qquad (5.55)$$

which effectively defines d_m as the position of mark m measured with respect to the reference point. The field terms are applied to d_m and the grid terms to R_f. Combining the previous two equations yields the following constraint:

$$\sum_m d_m = 0 \qquad (5.56)$$

Remember that the inherent assumption of the global fine alignment algorithm is that all the fields are identical, and so d_m does not require a field index, f. But, the measured d_m values will vary from field to field, and so we have for the measured data

$$r'_{mf} = R'_f + d'_{mf} \qquad (5.57)$$

The implicit assumption of global fine alignment is that, to the overlay accuracy required, we can write

$$r'_{mf} = T + G \cdot R_f + F \cdot d_m + n_{mf} \qquad (5.58)$$

where R_f and d_m are the expected grid and mark positions and

$$T = \begin{pmatrix} T_x \\ T_y \end{pmatrix} = \text{Grid Translation}$$

$$G = \begin{pmatrix} G_{xx} & G_{xy} \\ G_{yx} & G_{yy} \end{pmatrix} = \begin{pmatrix} 1 + \mu_x^{\text{grid}} & -\theta^{\text{grid}} \\ \theta^{\text{grid}} + \psi^{\text{grid}} & 1 + \mu_y^{\text{grid}} \end{pmatrix}$$

$$= \text{Grid rotation skew and mag matrix} \qquad (5.59)$$

$$F = \begin{pmatrix} F_{xx} & F_{xy} \\ F_{yx} & F_{yy} \end{pmatrix} = \begin{pmatrix} 1 + \mu_x^{\text{field}} & -\theta^{\text{field}} \\ \theta^{\text{field}} + \psi^{\text{field}} & 1 + \mu_y^{\text{field}} \end{pmatrix}$$

$$= \text{Field rotation skew and mag matrix}$$

The term n_{mf} is noise, which is nominally assumed to have a zero-mean Gaussian probability distribution and is uncorrelated from field to field and from mark to mark. The field translations are, by definition, just the shifts of the reference point of each field and so are given by

$$\left[\text{Translation of field } f \right] = T + G \cdot R_f$$

$$= \begin{pmatrix} T_x \\ T_y \end{pmatrix} + \begin{pmatrix} G_{xx} & G_{xy} \\ G_{yx} & G_{yy} \end{pmatrix} \cdot \begin{pmatrix} X_f \\ Y_f \end{pmatrix} \qquad (5.60)$$

Throughout this analysis, the "+" and "·" indicate standard matrix addition and multiplication, respectively.

In the equation for r'_{mf}, the unknowns are the field and grid terms. The expected positions and the measured positions are known. Thus, the equation must be inverted to solve for the combined field and grid terms, which amounts to 10 nominally independent numbers (two from the translation vector and four each from the grid and field matrices). The nominal independence of the 10 terms must be verified in each case, since some exposure tools and/or processes will, for example, have no skew (so that term is explicitly zero), or the grid and field isotropic magnification terms will automatically be equal, and so on. We will take all 10 terms to be independent for the remainder of this discussion. Appropriate adjustment of the results for dependent or known terms is straightforward.

Solving for the 10 terms from the expected and measured position values is generally done using some version of a least-squares fit. The least-squares approach, in a strict sense, applies only to Gaussian distributed uncorrelated noise. Since real alignment measurements are often corrupted by "flyers" or "outliers," that is, data values that are not part of a Gaussian probability distribution, some alteration of the basic least-squares approach must be made to eliminate or at least reduce their effect on the final result. Iterative least-squares uses weighting factors to progressively reduce the contribution from data values that deviate significantly from the fitted values. For example, if σ is the RMS deviation between the measured and fitted positions, one can simply eliminate all data values that fall outside some specified range measured in units of σ; for example, all points outside a $\pm 3\sigma$ range could be eliminated and the fit then recalculated without these points. This is an all or nothing approach; a data value is either used, that is, it has weight 1 in the algorithm, or not, that is, it has weight 0. A refinement of this approach allows the weight values to be chosen anywhere in the range from 0 to 1. Often, a single iteration of this procedure is not enough, and it must be repeated several times before the results stabilize. Procedures of this type, that is, ones that attempt, based on some criteria, to reduce or eliminate the effect of "flyers" on the final results, go by the general name of "robust statistics." Under this heading, there are also some basic variations on the least-squares approach itself, such as "least median of squares" or the so-called "L_1" approach, which minimizes the sum of absolute values rather than the sum of squares. An excellent and complete discussion of all these considerations is given by Branham [69]. Which, if any, of these approaches is used is exposure tool dependent. The optimum approach that needs to be applied in a particular case must be determined from the statistics of the measured data, including overlay results. Finally, it is not the straightforward software implementation of the least-squares solution derived below that is difficult; it is all the ancillary problems that must be accounted for that present the difficulty in any real application, such as the determination and elimination of flyers, allowing for missing data, determining when more fields are needed and which fields to add, etc. More sophisticated approaches to eliminating flyers are discussed by Nakajima et al. [70].

For the purposes of understanding the basic concept of global alignment, we will simply assume a single iteration of the standard least-squares algorithm in the derivation that follows.

Substituting the matrix-vector form for the field and grid terms into the equation for r'_{mf}, rearranging terms, and separating out the x and y components yields

$$n_{xmf} = x'_{mf} - T_x - X_f G_{xx} - Y_f G_{xy} - d_{xm} F_{xx} - d_{ym} F_{xy}$$

and   (5.61)

$$n_{ymf} = y'_{mf} - T_y - X_f G_{yx} - Y_f G_{yy} - d_{xm} F_{yx} - d_{ym} F_{yy}$$

The x and y terms can be treated separately, and with the equations written again in matrix-vector form, but clustered this time with respect to the grid and field terms, we have for the x equations

$$
\underbrace{\begin{pmatrix} n'_{x11} \\ n'_{x21} \\ n'_{x31} \\ \vdots \\ n'_{xN_M N_F} \end{pmatrix}}_{\text{error} \equiv \varepsilon_x}
= \underbrace{\begin{pmatrix} x'_{11} \\ x'_{21} \\ x'_{31} \\ \vdots \\ x'_{N_M N_F} \end{pmatrix}}_{\text{data} \equiv D_x}
- \underbrace{\begin{pmatrix} 1 & X_1 & Y_1 & d_{x1} & d_{y1} \\ 1 & X_1 & Y_1 & d_{x2} & d_{y2} \\ 1 & X_1 & Y_1 & d_{x3} & d_{y3} \\ \vdots & \vdots & \vdots & \vdots & \vdots \\ 1 & X_{N_F} & Y_{N_F} & d_{xN_M} & d_{yN_M} \end{pmatrix}}_{A}
\cdot \underbrace{\begin{pmatrix} T_x \\ G_{xx} \\ G_{xy} \\ F_{xx} \\ F_{xy} \end{pmatrix}}_{\text{unknowns} \equiv U_x}
\qquad (5.62)
$$

and for the y equations

$$
\underbrace{\begin{pmatrix} n'_{y11} \\ n'_{y21} \\ n'_{y31} \\ \vdots \\ n'_{yN_M N_F} \end{pmatrix}}_{\text{error} \equiv \varepsilon_y}
= \underbrace{\begin{pmatrix} y'_{11} \\ y'_{21} \\ y'_{31} \\ \vdots \\ y'_{N_M N_F} \end{pmatrix}}_{\text{data} \equiv D_y}
- \underbrace{\begin{pmatrix} 1 & X_1 & Y_1 & d_{x1} & d_{y1} \\ 1 & X_1 & Y_1 & d_{x2} & d_{y2} \\ 1 & X_1 & Y_1 & d_{x3} & d_{y3} \\ \vdots & \vdots & \vdots & \vdots & \vdots \\ 1 & X_{N_F} & Y_{N_F} & d_{xN_M} & d_{yN_M} \end{pmatrix}}_{A}
\cdot \underbrace{\begin{pmatrix} T_y \\ G_{yx} \\ G_{yy} \\ F_{yx} \\ F_{yy} \end{pmatrix}}_{\text{unknowns} \equiv U_y}
\qquad (5.63)
$$

Using the indicated notation, these equations reduce to

$$\varepsilon_x = D_x - A \cdot U_x$$

and   (5.64)

$$\varepsilon_y = D_y - A \cdot U_y$$

The standard least-squares solutions are found by minimizing the sum of the squares of the errors, which can be written as

$$\varepsilon_x^T \cdot \varepsilon_x = \left(D_x - A \cdot U_x\right)^T \cdot \left(D_x - A \cdot U_x\right)$$

and   (5.65)

$$\varepsilon_y^T \cdot \varepsilon_y = \left(D_y - A \cdot U_y\right)^T \cdot \left(D_y - A \cdot U_y\right)$$

The superscript "T" indicates the matrix transpose. Taking derivatives with respect to the elements of the unknown vectors, that is, taking derivatives one by one with respect to the field and grid terms, and setting the results to zero to find the minimum yields, after some algebra, the following result:

$$U_x = (A^T \cdot A)^{-1} \cdot A^T \cdot D_x$$

and   (5.66)

$$U_y = (A^T \cdot A)^{-1} \cdot A^T \cdot D_y$$

where the superscript "-1" indicates the matrix inverse.

Note that the A matrix is fixed for a given set of fields and marks. Thus, the combination $\left(A^T \cdot A\right)^{-1} \cdot A^T$ can be computed for a particular set of fields and marks, and the result can simply be matrix multiplied against the column vector of x and y data to produce the best-fit field and grid terms. Alignment is then performed by using this data in the $r_{mf} = T + G \cdot R_f + F \cdot d_m$ equation in a feed-forward sense to compute the position, orientation, and linear distortion of all the fields on the wafer.

5.9 ALIGNMENT MARK SIGNAL GENERATION MODELING: THE SCATTERING MATRIX

The widths, depths, and thicknesses of the various "blocks" of material that make up an alignment mark are usually from a few tenths to several times the sensing wavelength in size. The solution to Maxwell's equations for each particular case is necessary to make valid predictions of signal shape, intensity, and mark offset. Because of the overall complexity of wave propagation and boundary condition matching in an average alignment mark, it is essentially impossible to intuitively predict or understand how the light will scatter and diffract from a given mark structure. Thus, to really understand the details of why a particular mark structure produces a particular signal shape and a particular offset requires actually solving Maxwell's equations for that structure. Also, the amplitude and phase of the light scattered/diffracted in a particular direction depend sensitively on the details of the mark structure. Variations in the thickness or shape of a given layer by as little as a few nanometers or in its index by as little as a few percent can significantly alter the alignment signal shape and detected position. Thus, again, to truly understand what is happening requires detailed knowledge of the actual three-dimensional structure as well as its variation in real marks.

In general, all the codes and algorithms used for the purpose of alignment mark modeling are based on rigorous techniques for solving multilayer grating diffraction problems, and they essentially all couch the answer in the form of a "scattering matrix," which is nothing but the optical transfer function of the alignment mark. It is beyond the scope of this discussion to describe in detail the various forms that these algorithms take, and we refer the reader to the literature for details: see [71–81].

Although we will not discuss how a scattering matrix is computed, it is worthwhile understanding what a scattering matrix is and how it can be used to determine alignment signals for different sensor configurations.

The two key aspects of Maxwell's equations and the properties of the electromagnetic field that we will need are:

- The electromagnetic field is a vector field. That is, the electric and magnetic fields have a magnitude and a direction. The fact that light is a vector field, i.e, it has polarization states, should not be ignored when analyzing the properties of alignment marks, as this can, in many cases, lead to completely erroneous results.
- Light obeys the wave equation, i.e., it propagates. But, it must also obey Gauss's law. In other words, Maxwell's equations contain more physics than just the wave equation.

It is convenient to use the natural distinction between wafer in-plane directions, x and y, and the out-of-plane direction or normal to the wafer, z, to define the two basic polarization states of the electromagnetic field, which we will refer to as TE, for tangential electric, and TM, for tangential magnetic. For TE polarization, the electric field vector, $\vec{E}$, is tangent to the surface of the wafer; that is, it has only x and y components, $\vec{E} = \hat{e}_x E_x + \hat{e}_y E_y$. For TM polarization, the magnetic field vector, $\vec{B}$, is tangent to the surface of the wafer; that is, it has only x and y components, $\vec{B} = \hat{e}_x B_x + \hat{e}_y B_y$. See Figure 5.13.

We use the convention that $\hat{e}_x$, $\hat{e}_y$ and $\hat{e}_z$ are the unit vectors for the x, y, and z directions, respectively. We can work with just the electric field for both polarizations, since the corresponding magnetic field can be calculated unambiguously from it, and use the notation $\vec{E}_{TE}$ for the TE polarized waves and $\vec{E}_{TM}$ for the TM polarized waves.

For completeness, Maxwell's equations, in MKS units, for a homogeneous static isotropic non-dispersive nondissipative medium take the form

$$\vec{\partial} \cdot \vec{E} = 0$$

$$\vec{\partial} \cdot \vec{B} = 0$$

$$\vec{\partial} \times \vec{E} = -\partial_t \vec{B} \tag{5.67}$$

$$\vec{\partial} \times \vec{B} = \frac{n^2}{c^2} \partial_t \vec{E}$$

where
  n is the index of refraction as a function of position in three dimensions and
  c is the speed of light in vacuum.

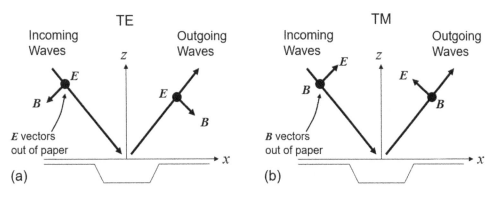

FIGURE 5.13 Definition of the two fundamental polarization states: (a) TE (or "s-polarization") and (b) TM (or "p-polarization").

Because Maxwell's equations as presented above are linear, the illumination or incoming amplitudes of the incoming electromagnetic waves illuminating the mark, B_{TE} and B_{TE}, and the amplitudes of the outgoing electromagnetic waves scattered and diffracted from the mark, A_{TE} and A_{TM}, are linearly related to one another. For each wavelength of the alignment light, λ, this relation can conveniently be written in the form

$$\underbrace{\begin{pmatrix} A_{TE}(\vec{\beta},k) \\ A_{TM}(\vec{\beta},k) \end{pmatrix}}_{\text{Outgoing Waves}} = \int \underbrace{\begin{pmatrix} S_{TE,TE}(\vec{\beta},\vec{\beta}') & S_{TE,TM}(\vec{\beta},\vec{\beta}') \\ S_{TM,TE}(\vec{\beta},\vec{\beta}') & S_{TM,TM}(\vec{\beta},\vec{\beta}') \end{pmatrix}}_{\text{Mark Scattering Matrix}\equiv S} \cdot \underbrace{\begin{pmatrix} B_{TE}(\vec{\beta}',k) \\ B_{TM}(\vec{\beta}',k) \end{pmatrix}}_{\text{Incoming Waves}} d^2\beta' \qquad (5.68)$$

where $k = 2\pi/\lambda$. The vector $\vec{\beta}$ has x and y components given by

$$\beta_x = k\sin(\theta)\cos(\phi)$$
$$\beta_y = k\sin(\theta)\sin(\phi) \qquad (5.69)$$

where θ and ϕ are standard spherical polar coordinates defined with respect to the z axis. The subscripts "TE" and "TM" refer to the two fundamental polarizations with respect to the wafer surface: transverse electric where the electric field is parallel to the wafer surface and transverse magnetic where the magnetic field is parallel to the wafer surface. Each element of the matrix S is a complex number, which depends on $\vec{\beta}$ and $\vec{\beta}'$, and which can be interpreted as the coupling from a particular incoming wave to a particular outgoing wave. For example, $S_{TE,TE}(\vec{\beta},\vec{\beta}')$ is the coupling from the incoming TE wave with tangential propagation vector $\vec{\beta}'$ to the outgoing TE wave with tangential propagation vector $\vec{\beta}$. In the same way, $S_{TE,TM}(\vec{\beta},\vec{\beta}')$ is the coupling from the incoming TM wave at $\vec{\beta}'$ to the outgoing TE wave at $\vec{\beta}$. Note that since elements of S are complex numbers, and complex numbers have an amplitude and a phase, the elements of S account for both the amplitude of the coupling, i.e., the magnitude of the outgoing wave given the magnitude of the incoming wave, and the phase shift that occurs when the incoming waves are coupled to outgoing waves. Note that for stationary and optically linear media, there is no cross-coupling of different temporal frequencies, $f_{in} = f_{out}$. The diagonal elements of S with respect to the tangential propagation vector are those for which $\vec{\beta} = \vec{\beta}'$, and these elements correspond to specular reflection from the wafer. The off-diagonal elements, i.e., those with $\vec{\beta} \neq \vec{\beta}'$, are nonspecular waves, i.e., the waves that have been scattered/diffracted by the alignment mark. See Figure 5.14 for an illustration of S (a) and how it separates into propagating and evanescent sectors (b).

The value of each element of S depends on the detailed structure of the mark, that is, on the thicknesses, shapes, and indices of refraction of all the material "layers" that make up the alignment mark as well as on the wavelength(s) and polarization of the light. The scattering matrix for a perfectly symmetric mark has an important property: it is centro-symmetric, as illustrated in Figure 5.15.

It follows from this that the scattering matrix must be computed for each particular mark structure and for each wavelength used by the alignment sensor, but once this matrix has been computed, the alignment signals that are generated by that particular mark for all possible sensor types and configurations that use those wavelengths are completely contained in S. In standard terminology, S is the optical transfer function or scattering matrix of the mark and completely describes its optical properties. Calculating S for a given mark and set of wavelengths generally requires using full-blown numerical Maxwell equation solver codes. In those cases where it is sufficient to calculate the scattering and diffraction from the mark in just two dimensions, say x and z, such codes can run

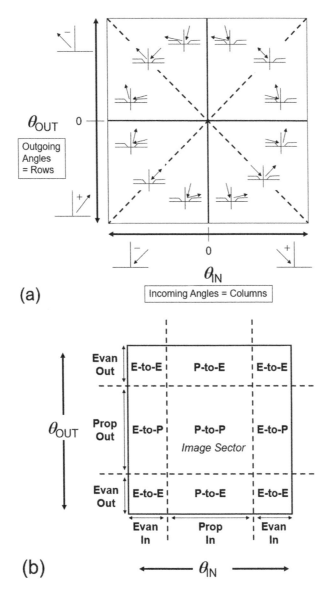

FIGURE 5.14 (a) Illustration of the meaning of the scattering matrix in terms of incoming and outgoing plane wave amplitudes. (b) Illustration of the propagating (P) and evanescent (E) sectors of the scattering matrix. "P-to-P" is the incoming propagating to outgoing propagating sector, "P-to-E" is incoming propagating to outgoing evanescent sector, etc. The P-to-P sector is the sector that contributes to the alignment signal, but in order to satisfy Maxwell's equations, all the sectors shown must be included in the calculation of the scattering matrix itself.

rather quickly, but for cases that require full three-dimensional scattering calculations, the relevant codes are still rather slow.

Computing the scattering matrix for the widest possible range of alignment mark structures is necessary to determine how robust a given alignment sensor design will be. For a sensor design using a given number or spread of wavelengths and polarization states, the requirement is that for all relevant or reasonable alignment mark structures, the sensor should produce a usable alignment signal.

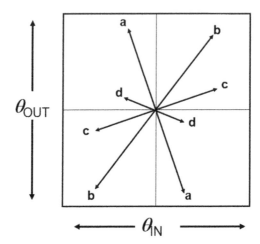

FIGURE 5.15 For a symmetric alignment mark, the scattering matrix must be centro-symmetric. That is, the values of the scattering matrix at the two positions marked "a" must be identical, the values at the two positions marked "b" must be identical, and so on.

REFERENCES

1. Alex Starikov et al., "Accuracy of overlay measurements: tool and asymmetry effects", *Opt. Eng.*, 31, 1298 (1992).
2. David J. Cronin and Gregg M. Gallatin, "Micrascan II overlay error analysis", *Proc. SPIE*, 2197, 932 (1994).
3. Nobutaka Magome and Hidemi Kawaii, "Total overlay analysis for designing future aligner", *Proc. SPIE*, 2440, 902 (1995).
4. A. C. Chen et al. "Overlay performannce of 180 nm ground rule generation x-ray lithography aligner", *J. Vac. Sci. Technol. B*, 15, 2476 (1997).
5. Frank Bornebroek et al., "Overlay performance in advanced processes", *Proc. SPIE*, 4000, 520 (2000).
6. Ramon Navarro et al., "Extended ATHENA™ alignment performance and application for the 100nm technology node", *Proc. SPIE*, 4344, 682 (2001).
7. Chen-Fu Chien et al., "Sampling strategy and model to measure and compensate overlay errors", *Proc. SPIE*, 4344, 245 (2001).
8. Jeroen Huijbregtse et al., "Overlay performance with advanced ATHENA™ alignment strategies", *Proc. SPIE*, 5038, 918 (2003).
9. Stephen J. DeMoor et al., "Scanner overlay mix and match matrix generation: capturing all sources of variation", *Proc. SPIE*, 5375, 66 (2004).
10. Leland Chang et al., "Moore's law lives on. *IEEE Circuits Devices Mag.*, 19(1), 35–42 (2003).
11. Kevin Lucas et al., "Interactions of double patterning technology with wafer processing, OPC and design flows", *Proc. SPIE*, 6924, 692403, Optical Microlithography XXI (March 12, 2008).
12. Rani S. Ghaida, Mukul Gupta, and Puneet Gupta, "Framework for exploring the interaction between design rules and overlay control", *J. Micro/Nanolith. MEMS MOEMS*, 12(3) (2013).
13. Mircea Dusa et al., "Pitch doubling through dual-patterning lithography challenges in integration and litho budgets", *Proc. SPIE*, 6520, 65200G, Optical Microlithography XX (March 27, 2007).
14. William H. Arnold, "Toward 3 nm overlay and critical dimension uniformity: an integrated error budget for double patterning lithography", *Proc. SPIE*, 6924, 692404, Optical Microlithography XXI (March 18, 2008).
15. Christopher P. Ausschnitt and Scott D. Halle, "Combinatorial overlay control for double patterning", *J. Micro/Nanolith. MEMS MOEMS*, 8(1), 011008-1–011008-8 (2009).
16. Huixiong Dai et al., "Alignment and overlay improvements for 3x nm and beyond process with CVD sidewall spacer double patterning", *Proc. SPIE*, 7274, 72743G, Optical Microlithography XXII (March 16, 2009).
17. M. A. van den Brink, C. G. de Mol, and R. A. George, "Matching performance for multiple wafer steppers using an advanced metrology procedure", *Proc. SPIE*, 0921, 180, Integrated Circuit Metrology, Inspection, and Process Control II (January 1, 1988).

18. M. Philipps, *Enabling Production at 14nm and Beyond*, Nikon LithoVision, 2014.
19. D. C. Flanders et al., "A new interferometric alignment technique", *Appl. Phys. Lett.*, 31, 426 (1977).
20. G. Bouwhuis and S. Wittekoek, "Automatic alignment system for optical projection printing", *IEEE Trans. Electr. Dev.*, ED-26, 723 (1979).
21. D. R. Bealieu and P. P. Hellebrekers, "Dark field technology: a practical approach to local alignment", *Proc. SPIE*, 772, 142 (1987).
22. M. Tabata and T. Tojo, "High-precision interferometric alignment using checker grating", *J. Vac. Sci. Technol. B*, 7, 1980 (1987).
23. Masanori Suzuki and Atsunobu Une, "An optical-heterodyne alignment technique for quarter-micron x-ray lithography", *J. Vac. Sci. Technol. B*, 7, 1971 (1989).
24. Norio Uchida et al., "A mask-to-wafer alignment and gap setting method for x-ray lithography using gratings", *J. Vac. Sci. Technol. B*, 9, 3202 (1991).
25. G. Chen et al., "Experimental evaluation of the two-state alignment system", *J. Vac. Sci. Technol. B*, 9, 3222 (1991).
26. Stefan Wittekoek et al., "Deep-UV wafer stepper with through-the-lens wafer to reticle alignment", *Proc. SPIE*, 1264, 534 (1990).
27. Kazuya Ota et al., "New alignment sensors for a wafer stepper", *Proc. SPIE*, 1463, 304 (1991).
28. Dohoon Kim et al., "Base-line error-free non-TTL alignment system using oblique illumination for wafer steppers", *Proc. SPIE*, 2440, 928 (1995).
29. Rina Sharma et al., "Photolithographic mask aligner based on modified moire technique", *Proc. SPIE*, 2440, 938 (1995).
30. Stan Drazkiewicz et al., "Micrascan adaptive x-cross correlative independent off-axis modular (AXIOM) alignment system", *Proc. SPIE*, 2726, 886 (1996).
31. Justin L. Kreuzer, "Self referencing mark independent alignment sensor", United States Patent #6628406 (September 30, 2003).
32. Don MacMillen and W. D. Ryden, "Analysis of image field placement deviations of a 5x microlithographic reduction lens", *Proc. SPIE*, 0334, 78, Optical Microlithography I: Technology for the Mid-1980s (September 13, 1982).
33. John D. Armitage Jr. and Joseph P. Kirk, "Analysis of overlay distortion patterns", *Proc. SPIE*, 0921, 207, Integrated Circuit Metrology, Inspection, and Process Control II (January 1, 1988).
34. Edward A. Mc Fadden and Christopher P. Ausschnitt, "A computer aided engineering workstation for registration control", *Proc. SPIE*, 1087, 255, Integrated Circuit Metrology, Inspection, and Process Control III (July 19, 1989).
35. Yan A. Borodovsky, "Overlay improvement through overlay modeling", *Proc. SPIE*, 2726, 311, Optical Microlithography IX (June 7, 1996).
36. Jan Mulkens et al., "High order field-to-field corrections for imaging and overlay to achieve sub 20-nm lithography requirements", *Proc. SPIE*, 8683, 86831J, Optical Microlithography XXVI (April 12, 2013).
37. Chun Yen Huang et al., "Using intrafield high-order correction to achieve overlay requirement beyond sub-40 nm node", *Proc. SPIE*, 7272, 72720I, Metrology, Inspection, and Process Control for Microlithography XXIII (March 23, 2009).
38. Martin A. van den Brink, Judon M. D. Stoeldraijer, and Henk F. Linders, "Overlay and field-by-field leveling in wafer steppers using an advanced metrology system", *Proc. SPIE*, 1673, 330, Integrated Circuit Metrology, Inspection, and Process Control VI (June 1, 1992).
39. Pary Baluswamy et al., "Sub-40nm high-volume manufacturing overlay uncorrectable error evaluation", *Proc. SPIE*, 8681, 868120, Metrology, Inspection, and Process Control for Microlithography XXVII (April 10, 2013).
40. M. Adel et al., "The challenges of transitioning from linear to high-order overlay control in advanced lithography", *Proc. SPIE*, 6827, 682722, Quantum Optics, Optical Data Storage, and Advanced Microlithography (November 21, 2007).
41. Hung Ming Lin et al., "Improve overlay control and scanner utilization through high order corrections", *Proc. SPIE*, 6922, 69222R, Metrology, Inspection, and Process Control for Microlithography XXII (March 24, 2008).
42. Jaap H. M. Neijzen et al., "Improved wafer stepper alignment performance using an enhanced phase grating alignment system", *Proc. SPIE*, 3677, 382, Metrology, Inspection, and Process Control for Microlithography XIII (June 14, 1999).
43. Peter Dirksen et al., "Effect of processing on the overlay performance of a wafer stepper", *Proc. SPIE*, 3050, 102, Metrology, Inspection, and Process Control for Microlithography XI (July 7, 1997).

44. Ramon Navarro et al., "Extended ATHENA alignment performance and application for the 100-nm technology node", *Proc. SPIE*, 4344, 682, Metrology, Inspection, and Process Control for Microlithography XV (August 22, 2001).

45. Frank Bornebroek et al., "Overlay performance in advanced processes", *Proc. SPIE*, 4000, 520, Optical Microlithography XIII (July 5, 2000).

46. Paul C. Hinnen et al., "Advances in process overlay", *Proc. SPIE*, 4344, 114, Metrology, Inspection, and Process Control for Microlithography XV (August 22, 2001).

47. Digh Hisamoto et al., "FinFET – a self-aligned double-gate MOSFET scalable to 20 nm", *IEEE Trans. Electr. Dev.*, 47(12) (December 2000).

48. David Laidler, "Identifying sources of overlay error in FinFET technology", *Proc. SPIE*, 5752, 80, Metrology, Inspection, and Process Control for Microlithography XIX (June 21, 2005).

49. David Laidler et al., "Sources of overlay error in double patterning integration schemes", *Proc. SPIE*, 6922, 69221E, Metrology, Inspection, and Process Control for Microlithography XXII (March 24, 2008).

50. Young-Sun Hwang et al., "Improvement of alignment and overlay accuracy on amorphous carbon layers", *Proc. SPIE*, 6152, 615222, Metrology, Inspection, and Process Control for Microlithography XX (March 24, 2006).

51. David Laidler et al., "Advances in process overlay: ATHENA alignment system performance on critical process layers", *Proc. SPIE*, 4689, 397, Metrology, Inspection, and Process Control for Microlithography XVI (July 1, 2002).

52. David Laidler et al., "Knowledge-based APC methodology for overlay control", *Proc. SPIE*, 5044, 32, Advanced Process Control and Automation (June 30, 2003).

53. Christopher P. Ausschnitt, Jaime D. Morillo, and Roger J. Yerdon, "Combined level-to-level and within-level overlay control", *Proc. SPIE*, 4689, 248, Metrology, Inspection, and Process Control for Microlithography XVI (July 1, 2002).

54. Nelson M. Felix et al., "Overlay improvement roadmap: strategies for scanner control and product disposition for 5-nm overlay", *Proc. SPIE*, 7971, 79711D, Metrology, Inspection, and Process Control for Microlithography XXV (April 20, 2011).

55. David Laidler et al., "A single metrology tool solution for complete exposure tool setup", *Proc. SPIE*, 7638, 763809, Metrology, Inspection, and Process Control for Microlithography XXIV (April 01, 2010).

56. Timothy A. Brunner et al., "Characterization of wafer geometry and overlay error on silicon wafers with nonuniform stress", *J. Micro/Nanolith. MEMS MOEMS*, 12(4), 043002.

57. K. T. Turner et al., "Monitoring process-induced overlay errors through high-resolution wafer geometry measurements", *Proc. SPIE*, 9050, 905013, Metrology, Inspection, and Process Control for Microlithography XXVIII (April 2, 2014).

58. Timothy A. Brunner et al., "Patterned wafer geometry (PWG) metrology for improving process-induced overlay and focus problems", *Proc. SPIE*, 9780, 97800W, Optical Microlithography XXIX (March 15, 2016).

59. Harry J. Levinson, *Principles of Lithography*, Third Edition, Society of Photo Optical, 2010. ISBN: 9780819483249.

60. David Laidler et al., "Impact of process decisions and alignment strategy on overlay for the 14 nm node", *Proc. SPIE*, 8683, 868306, Optical Microlithography XXVI (April 12, 2013).

61. David Laidler et al., "Mix and match overlay optimization strategy for advanced lithography tools (193i and EUV)", *Proc. SPIE*, 8326, 83260M, Optical Microlithography XXV (February 21, 2012).

62. Philippe Leray et al., "Overlay metrology for double patterning processes", *Proc. SPIE*, 7272, 72720G, Metrology, Inspection, and Process Control for Microlithography XXIII (March 23, 2009).

63. S. Wakamoto et al., "Improved overlay control through automated high-order compensation", *Proc. SPIE*, 6518, 65180J, Metrology, Inspection, and Process Control for Microlithography XXI (April 05, 2007).

64. C. P. Ausschnitt and P. Dasari, "Multi-patterning overlay control", *Proc. SPIE*, 6924, 692448, Optical Microlithography XXI (March 07, 2008).

65. Tomonori Dosho et al., "On-product overlay improvement with an enhanced alignment system", *Proc. SPIE*, 10147, 1014716, Optical Microlithography XXX (March 17, 2017).

66. Gregg M. Gallatin et al., "Modeling the images of alignment marks under photoresist", *Proc. SPIE*, 772, 193 (1987).

67. Gregg M. Gallatin et al., "Scattering matrices for imaging layered media", *J. Opt. Soc. Am. A*, 5, 220 (1988).

68. Alan Gatherer and Teresa H. Meng, "Frequency domain position estimation for lithographic alignment", *Proc. IEEE Int'l Conf. Acoustics, Speech and Signal Processing*, 3, 380 (1993).

69. Richard L. Branham, *Scientific Data Analysis*, Springer-Verlag, 1990.

70. Shinichi Nakajima et al., "Outlier rejection with mixture models in alignment", *Proc. SPIE*, 5040, 1729 (2003).

71. Norman Bobroff and Alan Rosenbluth, "Alignment errors from resist coating topography", *J. Vac. Sci. Technol. B*, 6, 403 (1988).

72. Chi-Min Yuan et al., "Modeling of optical alignment images for semiconductor structures", *Proc. SPIE*, 1088, 392 (1989).

73. J. Gamelin et al., "Exploration of scattering from topography with massively parallel computers", *J. Vac. Sci. Technol. B*, 7, 1984 (1989).

74. Gregory L. Wojcik et al., "Laser alignment modeling using rigorous numerical simulations", *Proc. SPIE*, 1463, 292 (1991).

75. Alfred K. Wong et al., "Experimental and simulation studies of alignment marks", *Proc. SPIE*, 1463, 315 (1991).

76. Chi-Min Yuan and Andrzej Strojwas, "Modeling optical microscope images of integrated-circuit structures", *J. Opt. Soc. Am. A*, 8, 778 (1992).

77. Xun Chen et al., "Accurate alignment on asymmetrical signals", *J. Vac. Sci. Technol. B*, 15, 2185 (1997).

78. Jaap H. Neijzen et al., "Improved wafer stepper alignment performance using an enhanced phase grating alignment system", *Proc. SPIE*, 3677, 382 (1999).

79. Takashi Sato et al., "Alignment mark signal simulation system for the optimum mark feature selection", *J. Micro/Nanolith. MEMS MOEMS*, 4(2), 023002 (April 26, 2005).

80. Chin B. Tan, Swee H. Yeo, and Andrew Khoh, "Modeling of wafer alignment marks using geometrical theory of diffraction (GTD)", *Proc. SPIE*, 5752, 977, Metrology, Inspection, and Process Control for Microlithography XIX (June 21, 2005).

81. Boris Menchtchikov et al., "Computational scanner wafer mark alignment", *Proc. SPIE*, 10147, 101471C, Optical Microlithography XXX (March 30, 2017).

6 Design for Manufacturing and Design Process Technology Co-Optimization

John Sturtevant and Luigi Capodieci

CONTENTS

6.1 DESIGN FOR MANUFACTURING (DFM)

The semiconductor industry is a multibillion dollar enterprise that has enjoyed decades of continuous improvement in critical metrics such as minimum feature size, cost per transistor, and circuit performance. The ecosystem has historically featured so-called integrated device manufacturers (IDMs) spanning both integrated circuit (IC) design and manufacturing functions, as well as semiconductor foundries that produce chips for fabless design companies. For many years initially, there existed a virtual wall of separation between the circuit design and manufacturing communities, even when they were both housed inside the same IDM. While performance scaling for integrated circuits continued down to 28 and 20 nm technology nodes, sustaining the geometric scaling of the actual circuit features is facing its ultimate physical limits with respect to manufacturability and yield. The unsung hero of the arduous challenge to keep IC performance on track has been and continues to be a heterogeneous set of computationally intensive computer-aided design (CAD) methodologies collectively known as design technology co-optimization or design tor manufacturing. The abstraction layer of the design rule manual (DRM) and its process design kit (PDK) collaterals has evolved into a two-way collaboration tool, whether within the few remaining IDMs or between fabless and foundry. Successful leading-edge designs require and benefit from in-depth knowledge of manufacturing processes, and conversely, high-yielding process technologies are "design-aware"; that is, they adapt to specific design styles and requirements.

6.1.1 ELECTRONIC DESIGN OVERVIEW

Most of this book focuses on semiconductor manufacturing, and in particular lithography technology, but it is important to understand how this technology is connected to the upstream integrated circuit design components. A detailed description of the entire digital and analog design flows is beyond the scope of this work, but an overview is helpful to set the context for design for manufacturing and design technology co-optimization. Electronic design automation (EDA) software tools are utilized to design, verify, and communicate all the information in the IC design flow.

Many of today's complex system-on-a-chip (SOC) designs are comprised of macroscopic blocks of analog circuits, digital logic circuits, memory arrays, microprocessors, signal processors, power management modules, and much more (Figure 6.1). A phenomenal amount of digital information processing capability can now be designed and manufactured on thumbnail-sized chips, and this is in very large part the result of the interplay between electronic design and lithography in manufacturing. Analog circuits have unique challenges, such as device matching and current gain requirements, but in general are designed at larger dimensions than are found in logic or memory circuits and as such, do not necessarily pose as formidable a challenge for DFM. We will focus on the digital design flow for illustration.

Macroscopic chip layout is determined by floor planning of major functional blocks accounting for efficient power distribution and input/output communication wiring between the blocks. This ultimately determines the overall size of the chip, subject to manufacturing constraints dictated by lithography scanner field size, as well as packaging requirements. Additionally, the end use application of the chip may constrain the chip size. Given the fact that most (except quite notably lithography) wafer fabrication processes are wafer or batch of wafer driven, the chip size obviously has a direct impact on manufacturing cost.

Digital design starts with functional requirements at a systems level, which are subsequently expressed abstractly in a register transfer level (RTL) human readable text file that describes the flow of digital signals between information storage nodes and the logical operations to be performed on those signals. This RTL behavioral description is fed into a physical design flow, which first logically synthesizes a "netlist" describing the necessary complementary metal–oxide–semiconductor (CMOS) circuit elements, such as resistors, capacitors, transistors, and logic gates that can perform useful logic signal operations such as AND, NAND, OR, and XOR. More complex functions such as adders, flip flops, latches, buffers, and many more are enabled by interconnecting the transistors in specific fashion, resulting in a library of physical layouts called *standard cells*. In the fab, these cells will be fabricated with gates and one or more metal interconnect layers in the so-called *front end of line* (FEOL) patterning steps. There are many different possible layouts that a standard cell designer can utilize to accomplish a given function, and a critical metric for comparing such designs is the cell height, often expressed in "tracks" or grid rows. Ideally, the fewer tracks required, the

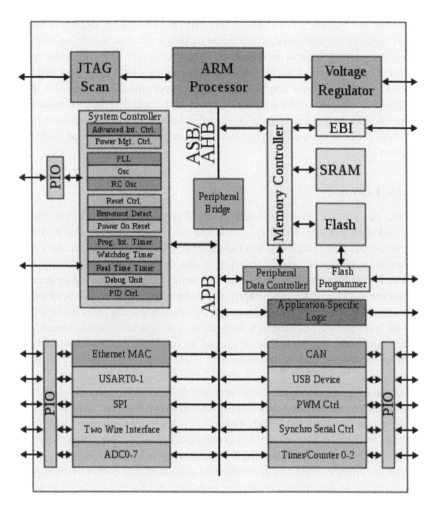

FIGURE 6.1   System-on-a-chip: schematic block diagram.

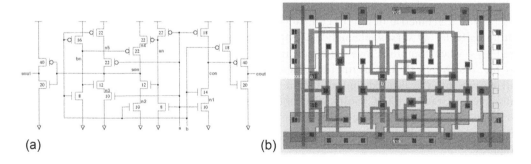

(a)   (b)

FIGURE 6.2   Example of standard cell: schematic (a), physical design layout (b).

better it is for cost, but standard cell design involves a complex set of tradeoffs between size, delay, and leakage current or power consumption. Figure 6.2 shows an example standard cell layout.

Standard cell libraries are constructed on regular horizontal or vertical grids so that they can efficiently be connected together into larger logic blocks. This function is performed by automated place and route software tools. These connections are accomplished in the fab by multiple layers of "routing" metal, the so-called *back end of line* (BEOL) patterning steps. For advanced logic technologies, as many as 10 total metal layers might be utilized between cells and routing interconnections. Each metal routing layer is predominantly unidirectional, with successive layers oriented orthogonal to their predecessors. Ideally, the tighter the routing metal can be packed in on a single layer, the fewer total levels of layer are required, which lowers manufacturing cost. But an important physical design consideration for the BEOL layers is the metal resistance and oxide insulator capacitance, which leads to a characteristic RC signal delay for the circuit that must be understood. Additionally, small metal lines with high current densities can potentially lead to electromigration reliability problems.

6.1.2 Physical Layout

The result of the design process is ultimately a set of physical polygon layers, each expressed by a standardized database file. Initially, this standard was graphic database system or GDS, later evolving to GDSII and eventually an OASIS format file. This file represents the designer's intent for wafer manufacturing, and each layer of the design is first physically realized on a photomask, which is constructed at 4× the desired wafer target dimension in accordance with the reduction factor used by wafer lithography scanning equipment. By the performance of successive lithography and implantation, etch, and deposition steps, the full circuit with billions of elements will be built up with conductors, insulators, and semiconducting materials to realize the desired chip function. It is typical in advanced 10 nm logic processes to utilize more than 50 different masking layers, and with double and triple patterning, this can imply nearly 100 photomasks.

So, the GDS data file expresses for each layer the design intent to manufacturing. How does manufacturing communicate its capability and limitations to the designer? Historically, the primary communication vehicle has been the DRM, which describes in text and pictorially the constraints on polygon layouts for a given layer as well as between multiple layers [1]. Single-layer rules are primarily related to polygon widths and spacing to neighboring polygons. The two-layer design rules include linewidth in common, edge-to-edge overlay or enclosure, and common overlap area. Design rules have always been predominantly restrictive in nature, but there has been a slow migration to more prescriptive recommendations, as we shall see.

Prior to sending the GDS file to the mask manufacturing facility, a step called *tape-out* since the data was initially housed on reels of magnetic tape, it is necessary to verify the design to be manufactured. Design rule checking (DRC) software enables the automatic inspection of a proposed geometric design to confirm compliance with the DRM rules (see Figure 6.3). Additionally, layout versus schematic (LVS) software performs checks that the physical layout structures, when combined together according to rules defined in a so called *process technology file*, match exactly the original electrical circuit as defined in the gate-level netlist. This will ensure that the desired logical functions will be accomplished by the physical layout. Subsequent extraction (XRC) software enables also the estimate of electrical performance of the manufactured layout. These software tools, as well as the other tools needed earlier in the design flow, are available from multiple EDA vendors, such as Cadence, Synopsys, and Mentor Graphics.

Design rule checking has been an indispensable operation for all technology nodes, and checks have become increasingly more complicated to properly express the growing number of complex manufacturing limitations. Thus, the design rule manuals have ballooned to several hundred pages or more in recent years. As we shall see, whereas DRC checks were sufficient at one point to guarantee manufacturing success, there have arisen a great number of additional required and optional operations necessary to ensure that a given design can rapidly reach maximum yield in fabrication. DFM can be thought of as the entire collection of these tools and methods.

M1 Design Rule (Metal 1)

M1.W.1	Width	A	0.120
M1.S.1	Space	B	0.120
M1.A1	Area	C	0.058
M1.A2	Enclosed Area	D	0.200

DRC Code: ERROR_M1_Space = SPACE M1 < 0.12

Design Manual Illustration:

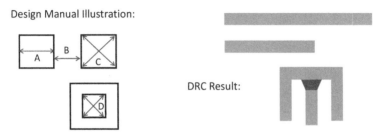

DRC Result:

FIGURE 6.3   Example of design rule specification.

6.1.3   DEVICE/DESIGN SCALING AND TECHNOLOGY "NODES"

Fundamental to the economics of semiconductor manufacturing and its relationship to electronic design is the concept of metal–oxide–semiconductor field-effect transistor (MOSFET) device scaling, sometimes referred to as *Dennard scaling* [2]. The observation for many generations of technology has been that as transistors get smaller, the gate power density remains constant, and operating voltage and current are reduced in proportion to gate length. In fact, the entire concept of a technology generation, or node, is born of this scaling law and its well-known cousin Moore's law, which in 1965 accurately predicted that transistors would steadily shrink in a cost-effective manner, leading to lower manufacturing cost per transistor and more transistors per chip [3].

The National Technology Roadmap for Semiconductors, later the International Technology Roadmap [4] for Semiconductors, was established in 1992 to provide the semiconductor design and manufacturing ecosystem with a projection for the timing and implications for the various steps in this scaling of device dimensions. By this time, the industry had already realized more than an order of magnitude of minimum polygon dimension reduction, but the creation of the roadmap firmly established and raised the visibility of the concept of technology "nodes."

In the earliest days of semiconductor manufacturing, it was possible to lower manufacturing cost by shrinking a design linearly in accordance with a scaling of all design rules. Such "dumb shrinks" were typically offered after a base node was yielding well and offered cost advantages if comparable yield could be maintained, since more die could be printed on a single wafer. Thus, the 130 nm node was shrunk by 15% to a 110 nm half-node, or a 65 nm node became a 55 nm half-node. Below 45 nm, however, the manufacturing process windows become so small that there is insufficient margin to apply simple shrinks with fixed lithography processes, and designs can no longer be easily ported to a smaller technology node. The result of the linear shrink nodes and the added half-nodes has been a prolonged period that has seen approximately a new node per year. That pace has recently slowed somewhat due to a variety of factors, principal among them the difficulty in realizing a decisive total cost advantage for shrinks due to multiple patterning, which adds design, mask, and processing cost for each layer. Initially, the name of the technology node equated to one-half the minimum pitch utilized for either the gate or the metal/contact layers. In reality, however, the node name has long ceased to be a quantitative descriptor of actual dimensions in the physical design;

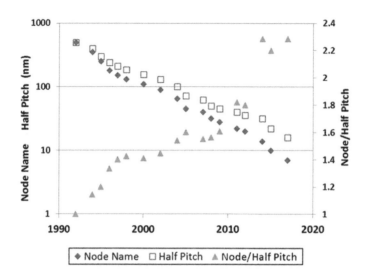

FIGURE 6.4   Discrepancy between node name and half-pitch dimensions.

it is really a simple way to designate the "next" shrink of design rules and manufacturing processes. Figure 6.4 illustrates the growing discrepancy between the node name and the minimum half-pitch dimensions. Thus, for instance, a 20 nm node chip features exactly zero features of 20 nm size.

6.1.4   Yield Loss Mechanisms in Manufacturing

The goal of the fab process, of course, is to manufacture a given integrated circuit chip with as high a yield as possible, and to do so immediately upon introduction into manufacturing. This has a direct impact on overall profitability. In practice, however, yield is never 100% from the start of manufacturing at a new node. Often, node level yield over time exhibits an S-shaped learning curve, characterized by low yields initially, then a sharp improvement, and finally, saturation at some terminal yield. Thus, enabling a more rapid ramp to as high a terminal yield as possible is an endeavor of particular interest to the entire design and manufacturing ecosystem.

Manufacturing yield merely refers to the percentage of die surviving through the full manufacturing and test cycle, which can exceed 2 months and thousands of individual process, metrology, and test operations. Design yield encompasses both functional and parametric aspects, and of course, manufacturing and design yield loss mechanisms are increasingly closely coupled. For many generations of technology, manufacturing yield was largely determined by random particulate defects resulting in an open or a short defect causing the chip to malfunction. At the 130 nm node, the primary yield loss mechanisms were determined to be more systematic in nature and principally tied to the limitations in the patterning process that gave rise to so-called *hotspots*. These systematic modes drive not only functional yield loss but also parametric yield loss, as fab process variability interacts with design limitations to result in electrical performance characteristics (leakage, speed bin, and voltage operating range) that do not meet the end use specifications [5]. Systematic defects can be predicted and therefore in principle mitigated through the use of model-based methods, as we shall describe in the following sections.

6.1.5   Critical Area Analysis and Critical Feature Analysis

There have been a large number of yield prediction models for semiconductor manufacturing, which enable the yield estimation for a given very-large-scale integration (VLSI) chip design and an assumed distribution of particulate defect sizes. Such approaches are based on critical area

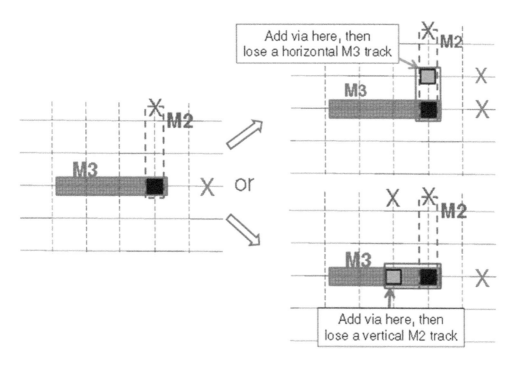

FIGURE 6.5   Example of insertion of redundant vias for interconnect layers.

analysis (CAA), which postulates that yield will suffer in accordance with the probability that a given defect of a certain size will "land" in a circuit region resulting in an open or a shorting of two polygons. So, different designs can have varying sensitivity to random defectivity, and it was realized that the same geometrical processing engines used to perform DRC were useful at identifying locations within the chip that would be most susceptible to random defects: the *critical areas*. This opened the door to methods that could be employed to alter the design in order to decrease the design defect sensitivity. One such manual approach was for the designer to use CAA to guide the tradeoffs between different layout styles to balance susceptibility to shorts and opens. More automated methods included fanning out of metal lines as space allows in order to help avoid metal–metal bridging. Additionally, so-called *redundant vias* can be automatically inserted at an interconnect node to lower the likelihood of a single via fail rendering the chip nonfunctional (Figure 6.5).

6.1.6   RECOMMENDED DESIGN RULES

Below 65 nm, the effectiveness of such critical area–based DFM layout optimization methods was somewhat diminished, as systematic yield detraction related to the patterning process dominated and therefore needed to be addressed first. It became successively more difficult to alter layouts through CAA without introducing new patterning weak spots, and geometrical constraints became more and more complex, as shown in Figure 6.6. Thus, the concept of binary design rules began to morph to more continuous functions expressed through recommended layout rules. These recommendations provide a pathway for the foundry to communicate to the designer that while a certain spacing, for instance, must always be greater than X, it would be ideal if it could be greater than Y, where Y > X. See Figure 6.7. Such an approach opens the opportunity to provide a quantitative assessment of layout compliance with recommendations such that various layout options could have an associated "DFM score" to enable comparisons, as illustrated in Figure 6.8. Equation-based DRC provides a framework for so-called critical feature analysis (CFA) tools to deliver such a quantitative assessment. The underlying assumption is that the fab can at least approximately model

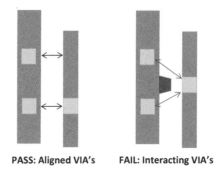

PASS: Aligned VIA's   FAIL: Interacting VIA's

FIGURE 6.6   Example of complex geometrical constraints for via/metal geometries.

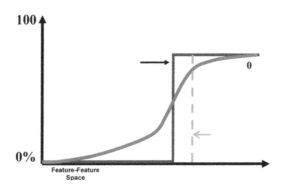

FIGURE 6.7   From discrete binary design rules to *continuous* recommended rules.

EXAMPLE: Enclosure Rule

611CR_GF	M(x+1) minimum overlap past Vx for two opposite sides with two other sides <0.004 um, where x=1-7	>=	0.04

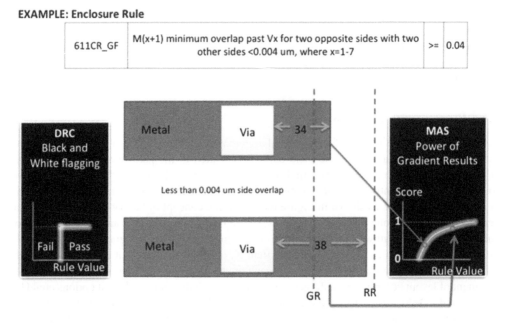

FIGURE 6.8   Extraction of a DFM score from equation-based design rule checking.

the impact of continuously variable physical layout parameters on yield. Since many recommended rules can effectively compete with one another, there are mechanisms that enable relative prioritization among the recommendations, again relying upon fab knowledge of this [6–8].

6.1.7 RESTRICTED DESIGN RULES

The transition from binary DRs to recommended rules was a useful path for the industry to provide incremental margin when possible and convenient for designers and as enabled by EDA software tools. But recommendations can be, and often are, ignored; thus, it was perhaps inevitable that the industry would experience the paradigm shift that began at the 32 nm node with the introduction of so-called restricted design rules (RDR). The introduction of RDRs also coincided with the deployment of various lithographic resolution enhancement techniques (RETs) that benefited or required such things as feature orientation or pitch limitations. Thus, concepts such as "forbidden pitch" (REF) and eventually gridded pitch, or unidirectional metal, or orientation dependent poly minimum dimensions, came into play at 32 and 22 nm nodes and have remained with us since. While there was in many cases a slight overall chip area penalty to pay, the benefits in reduced complexity and variability, as well as lithographic process window, have proved to be a worthwhile tradeoff.

The natural progression has continued from restricted design rules to prescriptive design rules, which switch the communication paradigm from describing what is forbidden to what is allowed [9]. And thus, today, design rules are characterized as a mixture of restricted and prescriptive rules, with restricted still dominating, but with layouts far more restricted today than in earlier technology generations. It has been speculated that the ultimate culmination of this evolution is a small library of allowed patterns available to the designer, which ensure correct-by-construction layouts preguaranteed to have sufficient manufacturability. This goal remains elusive for a variety of complex cost and flexibility tradeoff reasons.

6.1.8 ELECTRICAL TEST DIAGNOSTICS AND DFM

Electrical testing of chips is performed in line after initial metallization and at end of line before and after packaging. In-line tests with first-level metal include transistor characterization and basic defectivity through resistance tests on special in-scribe patterns. At end of line, the entire chip function and performance must be confirmed over a range of voltage inputs. Failures can often be distinguished between open or short or other, but the process of identifying precisely the three-dimensional coordinates of the fail for random logic designs is difficult and very time consuming. Additionally, physical failure analysis may not necessarily elucidate the likely mechanism.

It has become increasingly common, therefore, to utilize in-line electrical scan test diagnosis in conjunction with DFM design analysis to identify pathways to yield improvement through mitigation of systematic defects. Such an approach to yield learning starts with automatic test program generation (ATPG) tools to create electrical scan tests using in-chip circuits, then linking test failure data on specific nets to potential defect suspects through statistical analysis, and eventually drilling down to physical layout locations on specific layers, where it can be determined whether DFM violations are driving electrical failure.

Test data from failed devices is used to perform diagnosis. A diagnosis tool uses the design data, production test patterns, and tester data to identify the defects causing test failures. The diagnosis tool provides information such as defect classifications and suspected defect locations, showing, for example, that a particular device failed manufacturing test due to an open defect in a net segment that spans metal3, via3, and metal2, and giving the coordinates of each of these polygons. The flow shown in Figure 6.9 combining volume scan diagnosis, DFM analysis, and statistical analysis can reduce the time to determine the root cause of yield loss by 75–90% compared with traditional failure analysis methods. The ability to include and continuously add DFM analysis results enables the separation of design- and process-oriented yield loss [10, 11].

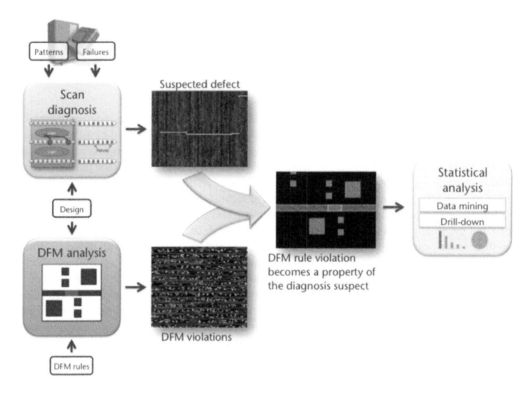

FIGURE 6.9   Combining test scan diagnostics with DFM analysis.

6.2   COMPUTATIONAL LITHOGRAPHY

6.2.1   LITHOGRAPHY RAYLEIGH CRITERIA

For initial technology generations, a design that passed DRC could be "taped out" to the mask shop for direct generation on a photomask, using typically an e-beam patterning process, and that mask in turn could be utilized in the wafer lithography process. The lithography process obeys the Rayleigh imaging criteria, relating the lithography equipment wavelength (λ) and numerical aperture (NA) to the minimum pattern resolution.

$$\text{Resolution} = k_1 \lambda / \text{NA}$$

Thus, the fundamental key to reduced design dimensions has been exposure wavelength reduction and NA increase. The k_1 factor can be thought of as a measure of the information loss associated with the lithography transfer function between mask and wafer. For k_1 greater than approximately 0.6, the design intent could be replicated on wafer with sufficient fidelity by using a photomask containing the target pattern directly. For lower k_1 values, proximity effects cause a loss of fidelity such that the mask pattern must be intentionally predistorted in order to reproduce the design intent on wafer. Figure 6.10 illustrates this loss of fidelity associated with the lithographic transfer function from mask to wafer as a function of k_1 [12]. Note that even at the lithographically trivial k_1 of 1.0, there is still a significant difference between predicted wafer contour and mask corners.

Industry cognizance of the existence of proximity effects in lithography came as early as the 500 nm technology node, and the curious effect was initially viewed as not likely to diminish yield. With shrinking dimensions, however, microprocessor performance binning was impacted, and researchers began to explore the origin of the effect, and offered materials and design data processing pathways to reduce its magnitude. For the latter, simple rules were applied with geometrical

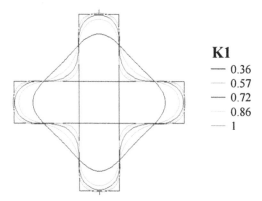

$K1$
— 0.36
--- 0.57
— 0.72
⋯ 0.86
— 1

FIGURE 6.10   Loss of image fidelity associated with lithography transfer function.

processing engines to bias line edges or to add serifs or hammerheads to line ends and corners. The tools to perform these post-tape-out operations are in fact typically the same tools used to perform the DRC checks. A new paradigm was formed, as it was now necessary to distinguish between original design intent layer and the transformed layer to be used in manufacturing to achieve the design intent.

As design geometries and hence, lithographic k_1 shrunk, driving process windows to become smaller, new RETs arose, such as phase-shift masks (PSM) and off-axis illumination (OAI). The latter was shown to be effective in enhancing the depth of focus (DOF) for dense patterns, and the geometrical processing engines also began to be used to place sub-resolution assist features (SRAFs) adjacent to isolated patterns. These SRAFs were initially added by the same geometric processing engines and put on a layer separate from the design intent. All these RET techniques, while enlarging the process window, led to further distortions in the printing, which needed to be corrected to preserve design intent on wafer. Thus, the 350 nm, 250 nm, and 180 nm technology nodes featured a proliferation of rule-based optical proximity correction (OPC).

Fueled largely by the increasing complexity of the rules as well as strict across-chip linewidth variation (ACLV) requirements for polysilicon gates on microprocessor chips, the 130 nm node was the first to utilize models in addition to rules for OPC. Later termed *computational lithography*, models have become an integral part of the mask data preparation and verification flow, and at today's production nodes are applied to nearly all masking levels, including implants. With new RETs such as model-assisted SRAF, double exposure, and source mask optimization (SMO), as well as geometrically increasing number of shapes within an approximately fixed chip size and finer critical dimension (CD) control tolerances, the computational complexity associated with model-based OPC has exploded. With 10 nm soon in production, ArF (193 nm) lithography will have been used for eight generations of high-volume manufacturing. A key enabler for improved resolution at fixed wavelength has been the exposure tool NA, which has more than doubled from 0.6 at 130 nm, to 1.2 with immersion lithography for 45 nm, to the maximum achievable1.35 for 32 nm and beyond. EUV lithography, with 13.5 nm wavelength, has been delayed for multiple nodes and may not be utilized until the 5 nm node. This long pause in wavelength or NA scaling has placed a premium on computational lithography to deliver the continued march to smaller technology nodes.

Against a backdrop of fixed wavelength and NA, k_1 has been lowered through a wide range of pathways, from RET-enabled process latitude expansion to OPC, to multipatterning, and these enablers have all impacted design practices and design-to-manufacturing data flows. Increasingly, it has been necessary for process and design evolution to occur in concert with one another, as enabled by DFM.

Today's typical data preparation flow for a single exposure layer is shown in Figure 6.11. Double- and triple-patterning layers, which are now commonly deployed at 14 nm and below, add a

DRC

Fill Patterns

Retargeting

SRAF Insertion

OPC

Post OPC Verification

Data Fracture

FIGURE 6.11   Mask data preparation flow (single layer).

decomposition step after DRC and feature multipattern layer–aware OPC prior to post-OPC verification on the overall synthesized layer contour.

6.2.2 LITHOGRAPHY SIMULATION

Model-based techniques have become an integral part of the design, mask data preparation, and verification flows. With RETs such as model-assisted SRAF, double exposure, and SMO, a geometrically increasing number of shapes within an approximately fixed chip size, and finer CD control tolerances, the computational complexity associated with model-based OPC has exploded. The result has been an evolution in model accuracy, simulation methodology, and computing platform in order to meet the often competing demands of accuracy and acceptable runtime. At the core of computational lithography reside the compact models that enable rapid but accurate prediction of the patterning process for full-chip layouts. Models for OPC and post-OPC verification describe the entire patterning process, including mask, optics, resist, and etch. This section explores in some detail the nature and evolution of these so-called computational lithography models.

It has been 40 years since Dill and coworkers introduced the mathematical framework for describing the exposure and development of conventional diazonaphthoquinone novolac positive tone photoresists [13]. This work laid the foundation for the evolution of lithography simulation software, which has become an indispensable part of the lithography development engineer tool suite for semiconductor DFM research and development. The University of California at Berkeley developed SAMPLE as the first lithography simulation package and enhanced it throughout the 1980s [14]. Several commercial technology computer-aided design (TCAD) software packages were introduced to the industry throughout the 1990s, all enabling increasingly accurate simulation of relatively small layout windows [15, 16].

As industry pioneer and PROLITH founder Chris Mack has described [17, 18], lithography simulation has allowed engineers to perform virtual experiments not easily realizable in the fab. It has enabled cost reduction through the comparison and narrowing of process options early in the technology development cycle and has been used to troubleshoot problems encountered in manufacturing.

A less tangible, but nonetheless invaluable, benefit has been the development of "lithographic intuition" and fundamental system understanding for photoresist chemists, lithography researchers, process development engineers, and as we shall see, even chip designers. Thus, patterning simulation is used for everything from design rule determination and validation to antireflection layer optimization to mask defect printing assessment, as well as countless other applications. At the heart of these successful uses of simulation have been the models that mathematically represent the distinct steps in the patterning sequence: aerial image formation, exposure of photoresist, postexposure bake (PEB), develop, and pattern transfer. Increasingly sophisticated "first principles" models have been developed to describe the physics and chemistry of these processes, with the result that critical dimensions and three-dimensional (3-D) profiles can be accurately predicted for a variety of processes through a broad range of process variations. (These models certainly do include multiple approximations and simplifications and are not truly atomic- or molecular-level representations, though interesting work has been conducted on such mesoscale models [19]). The quasi-rigorous mechanistic nature of TCAD models, applied in three dimensions, implies an extremely high compute load. This is especially true for chemically amplified systems, which involve complex coupled reaction–diffusion equations [20]. Despite the steady improvements in computing power, this complexity has continued to relegate these models to use for small-area simulations on the order of tens of microns or less of XY design layout space.

While TCAD simulation tools were evolving through the 1990s and accommodating the newly emerging chemically amplified deep ultraviolet (DUV) photoresist chemistries, a new approach to modeling patterning systems was being conceived at the University of California at Berkeley and at a small handful of startup companies. Rieger and Stirniman reported fast large area simulation using zone-sampling behavioral modeling in 1994 [21]. Convolution-based imaging was introduced shortly thereafter [22], and Cobb's seminal paper in 1996 introduced a mathematical framework for full-chip proximity correction. This work used a sum of coherent systems (SOCS) approximation to the Hopkins optical model and a simple physically based, empirically parameterized resist model [23]. Eventually, these two development paths resulted in commercial full-chip OPC offerings from EDA leaders Synopsys and Mentor Graphics. Later, both Cadence and Brion introduced OPC tools as well. It is interesting to note that while the term *optical* proximity correction was originally proposed, it was well known that proximity effects arise from a number of other process steps. The label "PPC" was offered to more appropriately describe the general problem, but OPC had by that time been established as the preferred moniker.

The onset of demand for full-chip model-based OPC corresponded with the increase in computing capability such that with the application of some simplifying assumptions and a limited scope of required prediction capability, several orders of magnitude of increase in layout area over TCAD tools could be reasonably accommodated. An important simplification was the reduction in problem dimensionality. TCAD models provide extremely valuable insight into 3-D imaging questions, such as how photoresist developed profiles may depend upon the Z-axis concentration gradients in residual casting solvent, PAG, and quencher, and how those gradients in turn may depend upon soft-bake temperature. TCAD models foster an understanding of the appearance of standing wave nodes or antinodes at the substrate–resist interface or the resist–air interface when designing an optimum antireflection solution. Unlike the problems above, Z-dimension information was not initially of primary importance in full-chip OPC simulations. In fact, the vast majority of the huge information content offered from a full-chip 3-D simulation, if such a thing were possible, would be underutilized in the design correction realm. Rather, the prediction of a single Z-plane contour across the XY layout is what is required. Indeed, the third dimension is becoming increasingly important to understand for post-OPC verification, in particular for different patterning failure modes. This consideration will be covered at the end of this section.

Optical models for a single plane did not require dramatic simplification, but the photoresist and etch [24] process models used in full chip are fundamentally different from those used in TCAD applications. Starting with the Cobb threshold approach, these models are variously referred to as "semi-empirical," "black box," "compact," "phenomenological," "lumped," or "behavioral" but are fundamentally characterized by a mathematical formulation that provides a transfer function between inputs and measured outputs of interest. An important facet of these models is that the user does not need access to sophisticated physiochemical characterization methods. Rather, the inputs required to calibrate the model, typically critical dimension measurements, are all readily available in the fab. The optical system has important setup information, such as numerical aperture and illumination profile, that is input to the model, but the photoresist system typically does not. Thus, it is the photoresist process model that differs most significantly between small-area TCAD and full-chip OPC applications. As will be shown, there are mathematical facets of these process models, which can be related to physical phenomena that are operative in the chemical system, but a detailed mechanistic chemical and kinetic understanding is not necessary to yield very useful simulation results.

The nature of the OPC application is such that some of the predictive power of TCAD simulation is not relevant, given that the patterning process by definition must be static in manufacturing as successive designs are built. There are relevant domains of variability where a model needs to dynamically predict, but these are largely limited to errors in mask dimension, dose, focus, and overlay. Dose can serve as a proxy for a variety of different manufacturing process excursions, such as PEB time and temperature.

For approximately 15 years now, process modeling has been used in full-chip OPC and has evolved from an incremental process window/yield expander at the 130 nm technology nodes to an absolutely vital element in the data preparation process for subsequent nodes. Figure 6.12 illustrates the evolution of Rayleigh criteria parameters from the 500 nm node through today's 10 nm and projects forward to the 5 nm "node." (As mentioned previously, the term technology "node" has become somewhat inappropriate since approximately the 32/28 nm designation and is merely now a marketing surrogate for year of technology introduction). It can be seen that the Rayleigh k_1 value at OPC introduction was approximately 0.55, well below the previously assumed "limit for practical manufacturing." k_1 has steadily dropped with each successive node, save for costly reprieves with the introduction of 193 nm lithography and double and triple patterning. Indeed zero yield in the fab would be virtually guaranteed below 90 nm without the application of model-based OPC. It is interesting to note that when extreme ultraviolet (EUV) is eventually adopted, it will correspond to a lower k_1 than the 130 nm node using 248 nm exposure wavelength; hence, model-based OPC will be a necessity from the outset. In fact, it is possible that if the 5 nm node is the first to utilize EUV, it will do so with double patterning from the start, since the single patterning k_1 value is likely to be <0.30.

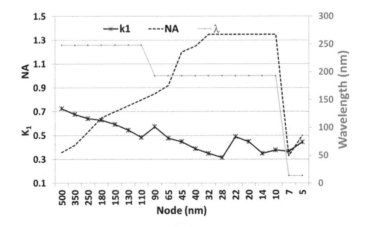

FIGURE 6.12 Evolution of Rayleigh $k1$ resolution from the 500 nm node down to 10 and 5 nm.

While difficult to quantify, there is certainly incremental yield entitlement if more accurate models can be used in the mask data preparation. Today, OPC and post-OPC verification simulation systems are a mission-critical element in multibillion-dollar design and fab operations. As such, they must be stable, highly reliable, and cost effective. Fab cycle time is critical to overall manufacturing cost and vital for enabling rapid time to market for new designs. The models that drive these solutions must faithfully represent the patterning of all designs that the factory will manufacture. And to the extent that such a representation is available, these models have been increasingly valuable for defining the design options and constraints needed for manufacturing robustness.

6.2.3   OPC and Post-OPC Verification

Prior to the introduction of accurate process models, OPC for 250 and 180 nm technology nodes utilized simple rules-based one-dimensional (1-D) biasing, line-end hammerheads, and corner serifs. The latter two approaches attempted to improve the corner fidelity loss shown in Figure 6.10. These rules could be efficiently applied to the full chip using the same geometry processing engines that are used in design rule checking. For the 130 and 90 nm technology nodes, the rules became too complex, and particularly for the gate layer, the CD control requirements dictated that models be engaged in the data preparation flow.

A considerable number of rule-based (typically table-drive) layout modifications are still implemented before the application of model-based OPC. These classes of operations are denoted as "retargeting" or "silicon targeting," and they are typically executed inside the foundry environment. The input of retargeting is a physical design (GDS) layout, and the output is a "true" (in the sense of physical dimensions) layout where potentially many, if not, all critical and near-critical features have been modified in size to represent what the actual silicon patterning process will generate. This target layout is also the primary input to the model-based OPC step (along with auxiliary geometrical information, as will be described in the following section).

The first step when applying model-based OPC is to break the target design polygon edges into smaller fragments or segments, which will be manipulated to render the final output on a mask than upon exposure will yield an acceptable physical manifestation of the target design. Next, the full chip is simulated using the optical and process models to provide a contour, which is compared to the design target shape. The edge placement error (EPE) for each fragment is calculated, and each fragment is in turn moved in or out; then the full chip is resimulated. This process is repeated until the simulated EPE drops below a user-specified tolerance. It is common for convergence to require between 6 and 10 full-chip simulation iterations; thus, it is critical for the simulation process to be highly efficient. Initial OPC tools utilized so-called "sparse" simulation, where the simulation sites were based on the target design polygons. In order to accurately simulate the EPE for a point of interest along the design, a sufficient number of simulation sites within the range of neighboring pattern influence, or "optical diameter," must be specified. As technology progressed and designs became smaller, the density of simulation sites greatly increased, driving up full-chip simulation time. Starting at the 45 nm technology node, several design layers, such as metal, became so dense relative to the optical diameter that it became more efficient to simulate on a uniform layout grid independently of the target shapes. This "dense" simulation methodology is similar to what TCAD simulators had used from their inception.

The number of layers requiring model-based OPC has steadily grown since 90 nm, and by the 32/28 nm node, essentially all layers utilized model-based OPC. With the introduction of multiple patterning at the 20 nm node, the number of OPC masks in the manufacturing sequence rose sharply, and today's 10 nm manufacturing features well over 70 OPC-corrected masks. Each layer may use different exposure tool settings (such as NA or illumination source shape), mask technology (binary, phase shift), photoresist materials and processes, and etch chemistry and process conditions. Each layer will therefore typically require a unique OPC model.

The initial OPC approach for most layers had been to simulate the nominal lithographic dose and focus condition so as to result in lowest EPE at this set point. Process window OPC (PWOPC) was introduced at 65 nm and utilizes multiple dose and/or defocus condition models during correction to mitigate manufacturing variation, which could result in pinch or bridge errors. It was shown that while nominal EPE would suffer slightly, catastrophic pinch or bridge errors could be avoided as the process deviated from the nominal condition, and such an approach has been broadly applied for metal layer patterning [25, 26]. Several other examples of PWOPC have been presented, which illustrate tradeoffs that can be made to account for manufacturing variation and maximize yield [27]. In addition, interlayer constraints, such as via enclosure by metal, can also be considered during PWOPC. PWOPC has been demonstrated for multipatterning as well, where the effect of overlay between layers needs to be considered [28, 29].

Since OPC execution is a rate-limiting step for the delivery of masks to the wafer fab, it represents an important component in the critical design-to-silicon time window. It is typical that the maximum allowed cycle time for OPC processing per layer is 24 hours. But, several factors have conspired to greatly increase computational load with decreasing design dimensions. The first is the increased design density, as referenced previously. Next is the expectation for improved accuracy, with subnanometer errors the common target today. This drives finer fragmentation and an increased number of OPC iterations. As indicated earlier, it has become common for metal layers to utilize PWOPC, and this drives additional simulations. Finally, the advent of multiple-exposure techniques such as double-dipole and then double and triple patterning requires essentially a 2–3$\times$ increase in compute load.

There are several strategies that have been employed in order to achieve what is typically a 24 hour turnaround time expectation for OPC data preparation. The migration to dense simulation has been effective in avoiding simulation explosion with design node, and some OPC tools efficiently utilize design hierarchy to eliminate redundant simulations. OPC tools also are designed to be highly scalable so that multiple inexpensive, high-performance central processing units (CPUs) can be utilized in parallel. It is common today for advanced technologies to employ over 1000 CPUs per layer job, and near-term projections exceed 5000 CPU. Several approaches to leverage high-performance computing hardware platforms, such as coprocessor acceleration and field-programmable gate arrays (FPGAs), have recently been deployed to address the growing compute challenge [30].

As model-based OPC deployment grew at 90 and 65 nm nodes, it became necessary to verify OPC corrections for full chip. Thus, dense simulation began to be used to confirm the performance of the corrected layout through process variations such as dose, focus, and mask manufacturing CD errors. These verification tools can use the same process model employed for OPC or a separate verification model. It is common to check for patterning failure hotspots, such as pinching or bridging, based upon quantitative analysis of the simulated contour shoreline. Additionally, interlayer edge placement analysis is possible, including checks such as metal-via area overlap, poly endcap past active, gate CD statistics over the active region, and undesired layer bridging. For interlayer checks, the additional process variable of overlay can be introduced. This is especially useful for multipatterning processes, where edge placement errors can lead to particular pathologies within chip. Recent reports on the use of full-chip two-layer simulation to study the intersection of CD and overlay variability introduced the concept of a relative edge placement error process window [31].

6.2.4 OPC PROCESS MODELING REQUIREMENTS

There are three principal requirements for compact process models used in OPC and post-OPC full-chip verification. These models must be accurate and predictable, they must be easy to calibrate, and finally, they must enable efficient full-chip simulation for rapid turnaround of fab mask data preparation operations.

6.2.4.1 Accuracy and Predictability

It is important to distinguish between these two related but distinct terms, which are linked to the typical flow utilized for generating OPC models: calibration and verification. Accuracy refers to the CD difference between the calibrated model prediction and the wafer result for the set of test patterns that is used to train the model. Thus, the accuracy of the model is dependent upon several factors. First is the intrinsic "correctness" in mathematically representing the patterning process steps; in other words, the ability to represent the patterning trends through target size, pitch, and pattern shape for 1-D and two-dimensional (2-D) structures at the process settings used to generate the data. So, the extent to which different varieties of test patterns are used to train the model will stress the globality of the model; thus, a model that has to predict three unique but similar structure CDs is likely to have better "accuracy" than a model that must predict 300 highly varied layout topologies. So, test pattern design coverage is important whenever model accuracy is in question.

There are many different metrics that can be utilized to quantitatively express the accuracy of the model, such as error range, chi-squared goodness of fit, and many others. One of the most useful metrics for characterizing OPC models, however, is the root mean square error value, or errRMS, associated with the test pattern ensemble. EPE corresponds to the simulated and measured CD error, and w is the user specified weighting of the ith measurement location.

$$\text{errRMS} = \sqrt{\frac{\sum w_i (CD_{\text{sim}} - CD_{\text{meas}})^2}{\sum w_i}} \qquad (6.1)$$

The duty of the OPC or post-OPC verification model is to correctly predict the patterning for every possible layout configuration that can appear per the design rules in the full chip. Thus, model predictability refers to the verification or validation on structures and or process conditions not explicitly utilized in the calibration exercise. The number of such unique design constructs for low-k_1 lithography is tremendous; perhaps seven or eight orders of magnitude more than could ever be reasonably used to train the model. In addition, the model must properly account for variability in the relevant manufacturing parameters, which will impact the final patterns in silicon. Thus, a robust and predictable model must not only be accurate on the training data set but must also be capable of extrapolating and interpolating to the myriad layouts and process conditions that will be encountered throughout the life cycle of a technology in manufacturing. The manufacturing variables best known to manifest critical random or quasi-systematic deviations, which in turn drive predictable CD changes, include focus, exposure, and mask CD.

An additional consideration that is closely related to predictability is the portability of the model across other parameter variations that may be required during the life cycle of the process. It is recognized, for instance, that if an entirely new photoresist material, PEB temperature, or etch recipe is implemented for manufacturing, a new model calibration will be required. But if some aspect of the exposure step is slightly altered, such as NA or illumination source intensity/polarization, it is desirable to "port" the same process model and change only the specific optical parameters that were changed. This is particularly helpful in early process development, when a current–technology node process model is used to simulate next-node printing with whatever new RET capabilities may become available. (In the past, this was most notably increased NA, but with the maximum NA now fixed at 1.35, it is more related to exploring changes in multiexposure masking and illumination schemes.) The degree to which the model can faithfully decouple optical exposure from resist processing is related not only to the details of the resist model but also to the nature of the approximations "upstream" in representing the mask and optical system. These will be addressed in the following sections.

6.2.4.2 Model Calibration Methodology

Full-chip patterning models must be easy to generate in a reproducible fashion using experimental measurement tooling that is conveniently available in the fab. This typically eliminates many of the spectroscopic measurement methods, such as Fourier transform infrared spectroscopy (FTIR), time-of-flight secondary ion mass spectrometry (TOF-SIMS), and UV-visible spectroscopy (UV-VIS), as well as dissolution rate monitoring (DRM) techniques, that are used to calibrate TCAD process models. These methods directly interrogate the patterning materials to determine their dynamic properties throughout the process flow. Full-chip models rely instead principally upon data from the final patterns themselves, and this data most often is derived from scanning electron microscopy (SEM) CD measurements. The CD SEM is uniquely suited to the task, since it is high throughput, high precision, and spatially specific, able to provide CD information with nm-level registration to the desired pattern location.

Proper selection of a representative sampling of allowed design variability is critical to calibrating a good model. Sophisticated techniques to generate test patterns that are representative of real design constructs have been developed by some users. Several workers have reported on the value of ensuring a full range of aerial image parameters in the test suite [32]. The number of distinct test patterns required to train a robust model has steadily increased with each technology node as greater accuracy at lower k_1 is demanded. Additionally, the number of process conditions (dose and focus) at which calibration data must be collected has increased. Finally, in order to improve the data signal to noise, multiple repeat measurements (across wafer, across field, and across tool) are typically made. The result has been an explosion in the number of CD measurements required and further motivates the use of contour images as described previously.

OPC models can describe the entire patterning process, including optics, resist, and etch, in a single lumped representation or can be discretized to characterize each module. The framework for simulating the aerial image formation in the photoresist film is well established and as mentioned, does not differ significantly between TCAD and full-chip tools. It is difficult, however, to definitively calibrate the optics, since the aerial image is not directly measurable. (Aerial image metrology systems (AIMS) tools are capable of emulating the aerial image for specific locations on a given reticle, but in practice these systems lack sufficient throughput, are not often conveniently available in the fab, and introduce their own response signature, which is not easily deconvolved.) Thus, one approach is to calibrate relevant optical parameters using a constant threshold of the aerial image compared with the developed photoresist CDs. With the optical model thus fixed, a subsequent tuning of resist model parameters is enabled. An alternative method is to simultaneously tune both optical and resist parameters in a single lumped calibration to the measurement data.

While details vary depending upon the exact software being used for OPC, there are several different classes of parameters associated with the calibration of the optical, resist, and etch processes. There are parameters that are directly measurable or known as designed values. These are primarily associated with the optical system and would include, for example, wavelength, NA, illumination profile, and film stack optical constants. While all these values may be input to the model as is, to the extent that their accuracy is not perfect, they can also be adjusted over a small range during the optimization. Care must be taken, however, in allowing these parameters to move too far from their design values, as this may result in a less physical model. Recent research into the application of Bayesian statistics is enabling more predictive models by more explicitly utilizing uncertainty in measured CD data and input parameters [33]. It is well known that the optical proximity effect is highly dependent upon the illumination profile. It is common today to input the in situ measured pupilgram instead of the as-designed version [34].

A second class of parameters are those associated with physical phenomena, where direct measurement is not done; rather, the model contains mathematical proxies for the parameter, but usually without a direct mapping correlation. These are the parameters that are most often associated with the complex photoresist PEB, develop, and etch chemical kinetics. A final class of calibration options includes software knobs for altering the approximations used in the model, such as number of optical kernels, or optical diameter, and resist or etch modelform.

The objective function most commonly used in the model fitting is minimization of the errRMS given in Equation 6.1. For each of the test pattern measurement locations, a comparison is made between the simulated and measured values. Different test patterns can be weighted differently (w_i) according to user preference; for instance, certain known critical design pitches can be weighted higher. This errRMS value is determined for the myriad different combinations of model fitting parameters, and the model with the lowest errRMS is chosen as optimum. With a large number of fitting parameters, the optimization process can become very time consuming; therefore, sophisticated optimization methods such as gradient descent, quasi-Newton, or genetic search allow efficient exploration of the RMS response surface to locate the global minimum [35].

In practice, a full calibration flow includes training patterns and distinct verification patterns. One approach is to start with a master set of patterns and use one half to train the model and the other half to verify. The goal in such a case would be for the errRMS fitness on the verification set to be as low as that of the calibration test suite. Another method is to include complex 2-D structures in the verification patterns and then compare the simulated contour with the experimental contour. If verification fitness is significantly worse than calibration fitness, it may indicate a need to expand the range of design topologies incorporated into the training set or to choose a model with additional fitting parameters.

6.2.4.2.1 Metrology Challenges Systematic and Random Errors

As mentioned above, full-chip models rely upon fab-measured CD data for a large variety of test patterns, and this data most often comes from SEM measurements. Multiple repeat measurements (across wafer and across field) are often made to provide a better statistical representation of the mean CD for each measurement gauge [36]. For an ensemble of different test patterns, there will be an experimental noise "floor," which will ultimately limit how accurate a model can be.

It is interesting to note that the standard error in the determination of the mean for typical OPC calibration structures is 0.5 nm for 1-D and 1.5 nm for 2-D. Thus, the 95% confidence interval associated with various measurement/simulation locations (gauges) is in the range of 1 nm for 1-D and 3 nm for 2-D. In order to achieve sub-1 nm RMS final model accuracy for state-of-the-art models, it is advisable to minimize the standard error in the determination of each data point mean, and five to nine data repeats are required to minimize the contribution of random errors. While a high CD variability across repeats may indicate poor process latitude, it may also be a sign of a poorly optimized SEM recipe setup, which always needs to be verified. Design-based metrology has enabled robust site and algorithm selection to be accomplished offline without the need for a setup wafer [37].

The degrees of freedom in the model will interact with the metrology noise such that it is possible to "overfit" the physical phenomena and start fitting the experimental noise. It has recently been shown, however, that various statistical information criteria can be used to differentiate competing models on the basis of predictive power, sample size, and model fitting parameters [38].

Whereas random CD errors can be minimized with sampling and robust setup, it is also important to consider systematic CD errors due to the proximity signature of the CD SEM method itself. Resist shrinkage under e-beam exposure can occur at different rates for dense and isolated photoresist lines [39], and there is an additional intrinsic difference in secondary electron scatter capture rate for isolated versus dense features. The result is that there are different physical to CD-SEM offsets for different feature types. These differences are in turn dependent upon the focus condition in the scanner, which leads to different resist sidewall angles. It has been shown that the metrology proximity effect can reach 1.5 nm for nominal exposure condition and upwards of 5 nm for defocus conditions [40]. Recently, researchers have reported on analytical models that explain the CD SEM bias effect [41].

6.2.4.2.2   Use of 2-D Contours

In addition to CD measurements, SEM output data can include contours for complex 2-D patterns. The latter are particularly useful, since it is not always possible to definitively associate a CD measurement with every layout feature, but the extracted contour allows the model to be trained on a significantly higher number of design points, including complex critical patterns such as memory cells. Additionally, contour-based metrology can inform the model regarding pattern failure modes such as pinching and bridging. SEM contour-based OPC model calibration has been demonstrated to yield very predictive results and represents an advantage in time to model due to greatly reduced need for discrete CD values from specific measurements. Contours have been successfully utilized to train optical, resist, etch, and even mask (using mask CD SEM contours) models [42–52]. The benefits of such usage include significant metrology turnaround time reduction and more predictive models.

6.2.5 Model Components

6.2.5.1 Optical Models

Full-chip simulations commonly employ the SOCS approximation to represent the intensity as a sum of convolutions of the mask with different basis functions or so-called *optical kernels*. There is a strong linear dependence of simulation time on the number of SOCS decomposition kernels used in the simulation. In addition, there is a quadratic dependence on the optical diameter (OD) associated with the model. Since the magnitude of the kernel eigenvalue coefficients decays quickly as kernel count increases, in practice it is often the case that 100 or fewer optical kernels are used with an OD <2.0 μm. For 28 nm, approximately 30 kernels were sufficient, but that number grows to 50 for 20 nm models and approaches100 for 10 nm models. The optical model accurately represents the wafer film stack optical behavior as well as the impact of the illumination and projection optics of the scanner and if desired, the impact of design and lens manufacturing aberrations.

6.2.5.1.1   Source Mask Optimization

Off-axis illumination was introduced in conjunction with SRAFs to enhance the depth of focus for critical features, and the types of available spatial illumination profiles grew from annular to quadrupole to dipole and variants of the same. For the 20 nm node, lithographic scanner manufacturers introduced a nearly infinite range of customized grayscale illumination profiles through sophisticated optics, thus enabling the ultimate level of control for fine tuning the source to best image the critical layout patterns. SMO software was introduced to take advantage of this customization ability on the scanners. SMO software can optimize the process window for specific layout features; the corresponding sourcemap file can be read directly by the computational lithography model and as such, poses no specific new challenges [53].

6.2.5.1.2   Interlayer Topography Proximity Effect Modeling

Historically, the masking layers that define ion implantation blocks have been regarded as "noncritical," since the CDs are typically much larger than the poly, contact, and metal layers. Simple proximity bias rules were initially used to control the implant layer CDs to target, and model-based OPC for some implant layers was adopted at 65 nm. But at 20 and 14 nm, the elimination of intralayer proximity bias is insufficient to deliver adequate CD control. This is due to the effect of the underlying active and poly layer topography, which in the absence of an antireflection control scheme, can lead to CD variation in excess of 200 nm, far larger than intralayer effects. The CD variation is caused by a complex mixture of thin film interference, bulk reflectivity, and reflections from underlying edges and tapered sidewalls, and as such, poses a difficult modeling challenge. Nevertheless, compact models have been developed that can account for both for intra- and interlayer proximity and can dramatically improve the CD control [54–56].

6.2.5.2   Photomask Models

For the first several generations of OPC deployment, it was typical for the wafer OPC models to be calibrated based upon an assumed exact match of the calibration test pattern layout and the actual test mask. However, it is known that there are three different types of errors related to the representation of the photomask that can express errors in OPC models: systematic 1-D mask CD errors, 2-D shape differences between as-designed and actual mask test patterns used in calibration, and approximations inside the model itself relative to 3-D electromagnetic field (EMF) effects. Since the mask manufacturing process is usually invariant for the life of the wafer technology, it had initially been acceptable to lump the systematic mask proximity effects into the resist model. This implies, however, that any substantial change to the mask process will require that the OPC model be recalibrated. Recent work on mask process proximity modeling, however, is enabling a departure from this paradigm [57–59].

These systematic mask considerations will be addressed in the sections below.

6.2.5.2.1   Mask CD Effects

Historically, OPC models were calibrated based upon an assumed exact match of the physical test mask and the test pattern layouts representing those patterns. However, it is known that systematic proximity effects such as corner rounding, linearity bias, and isolated-to-dense bias are manifested in the mask patterning process. Because the mask manufacturing process is usually invariant for the life of the wafer technology, it has been acceptable in the past to lump the systematic mask proximity effects into the resist process model. This implies, however, that any substantial change to the mask process will require the OPC model to be recalibrated. More significantly, the OPC model incorrectly ascribes mask behavior to the photoresist model, which will necessarily limit the predictive capability of the model to some extent. So, a mask process model (MPC) is calibrated based on mask CD or contour measurements; then, this MPC model is used to describe the mask input to the wafer OPC calibration flow. Figure 6.13 shows the improvement in systematic mask errors that can be accomplished through MPC. Ultimately, the usage model for MPC is to correct the mask proximity effects during mask manufacture, so that the signature never appears on wafer.

It is well known that due to the finite resolution of the mask writing process, the physical mask edges are not sharp corners but are rounded with a characteristic corner rounding that can be regarded as systematic for a given process. For 2-D features such as contact holes or line ends, this rounding can have a substantial impact on wafer patterning. Convex and concave corner rounding can be empirically tuned during model calibration and yields corner rounding values of approximately 8–10 nm (1×), which is consistent with direct mask SEM corner rounding measurements [60].

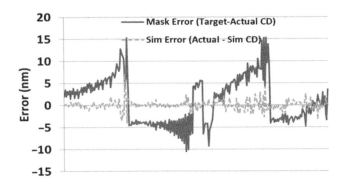

FIGURE 6.13   Improvement in systematic mask error reduction achieved by MPC.

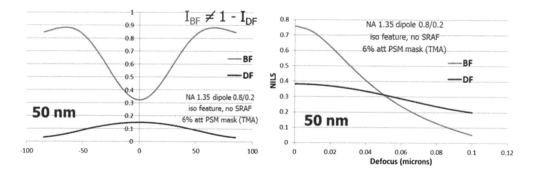

FIGURE 6.14   Advantage of attPSM mask over binary mask imaging.

6.2.5.2.2   Mask 3-D Effects

The Kirchhoff or flat mask approximation assumes that the mask absorber is sufficiently thin that the diffracted light can be computed by means of scalar or vector diffraction theory. This is in contrast to rigorous EMF simulation, which accounts explicitly for the topography and refractive indices of the mask materials and solves Maxwell's equations in 3-D, a highly computation-intensive operation not suitable for full-chip scale. Initial generations of full-chip simulation therefore utilized this flat mask approximation, but a variety of different simplification methods have been offered to enable a reduction of this 3-D EMF system to simpler 1-D or 2-D representations [61–66]. The most broadly deployed method is the domain decomposition method (DDM), which generates a library of 3-D EMF signals specific to incident angle and polarization for a given physical mask type. Chromium-based absorbing stacks were utilized for early-generation binary transmission masks, but for 32/28 nm, a thinner absorber, OMOG (opaque molybdenum silicide on glass), was introduced. This diminished aspect ratio reduced the contribution of mask 3-D effects, but for 20 nm with the expanded usage of 6% attenuated PSM, which is 40% thicker than OMOG, it led to more pronounced 3-D EMF effects, such as pattern-dependent best focus shifts. Thus, 20 nm was the first generation to broadly deploy mask 3-D models for OPC, which resulted in significantly improved prediction of critical dimensions through focus.

The return of negative tone develop has impacted the migration away from 6% attenuated PSM (attPSM) to thinner opaque OMOG. For imaging mask spaces to print wafer spaces with positive tone systems, OMOG was demonstrated in many cases to deliver improved process window over attPSM at 45–28 nm [67]. For imaging mask lines to print wafer spaces with negative tone develop, attPSM has an advantage over a binary intensity mask in the case of sub-50 nm features, as shown in Figure 6.14. So, the 20 and 14 nm nodes saw a return of 6% attPSM mask blank for contact and metal layers, where small photoresist spaces are the critical structures of interest.

In order to generate a DDM signal library, it is necessary to specify the physical mask topology and optical refractive indices as well as the scanner source map so that appropriate oblique incidence angles and polarizations can be utilized. It was shown that model accuracy is very sensitive to absorber sidewall angle, a parameter that is not always accurately known but can be tuned empirically to the wafer CD data due to the unique response of each feature to sidewall angle [68].

6.2.5.3   Resist Models

There are two basic types of resist process models used in full-chip simulations: those that apply a variable threshold to the aerial image in some manner and those that transform the aerial image shape [69]. Initial full-chip models used with sparse, site-based simulation were based upon the aerial image cutline of intensity versus position, with the simplest form, of course, a constant threshold. Increasing accuracy was realized by defining the threshold as a polynomial in various simulated image properties associated with the aerial intensity profile. Initially, I_{min} and I_{max} were utilized, then image slope was added, then image intensity at neighboring sites, and finally, a variety of functions used to calculate pattern density surrounding the site under consideration [70]. Thus, multiple different modelforms are possible.

The advent of dense grid full-chip simulation was accompanied by a new type of resist model, which applied a constant threshold to a 2-D resist surface [71]. The resist surface is generated by applying a variety of different fast mathematical operators to the aerial image surface. These operators are designed to represent important physical/chemical mechanisms operative in chemically amplified photoresists, including diffusion and acid–base neutralization. (See Equation 6.2.) The user can specify a modelform that selects which operators and k, n, and p values are desired; thus, as with the variable threshold model, a huge number of different forms are possible. The linear coefficients C_i and continuous parameters b and s are found by minimizing the objective function during calibration.

$$R(x, y) = T = \sum_{i=0}^{N-1} C_i F_i(x, y)$$

$$F(x, y) = \left[\left(\nabla^k I_{\pm b}(x, y) \right)^n \otimes G_{s,p}(x, y) \right]^{1/n}$$

(6.2)

A typical model fit result after applying a CM1 model is shown in Figure 6.15. It is interesting to note that the accuracy of OPC models has roughly scaled with the critical dimensions: an early paper by Conrad et al. [72] on a 250 nm process reported errors of 17 nm 3σ for nominal and 40 nm 3σ for defocus models. The model accuracy for today's 10 nm processes is on the order of 10× lower than these values. It can be seen that CD errRMS of 1 nm is achieved, and the errRMS value is maintained below 2.5 nm, throughout the defined focus and dose window.

In the late 1980s, IBM pioneered the use of the very first chemically amplified photoresist processes utilizing solvent develop of tert-butoxycarbonyloxystyrene (TBOC) photoresist to render a negative tone image. A fascinating review of this history is provided by Brock [73]. This process was quickly replaced by aqueous tetramethylammonium hydroxide (TMAH) develop positive tone imaging, but the intrinsic process window advantage of negative tone imaging was recognized and leveraged with the onset of crosslinking-based negative tone TMAH develop processes. It is fair to say, however, that the vast majority of all 248 and 193 nm manufacturing for 250 nm and below has been accomplished with positive tone aqueous TMAH develop photoresist systems. As the industry presses to 20 nm and below, however, every possible pathway toward improved process latitude

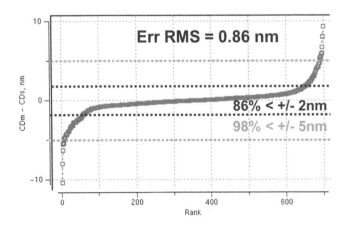

FIGURE 6.15 Typical resist model calibration using CM1.

is being explored or reexplored. Such is the desperate need for process window that the strong inertia behind 0.26 N TMAH fab plumbing is now being overcome, and solvent-based negative-acting imaging systems are being pursued to take advantage of the improved aerial image properties afforded by using a dark mask line to print the small wafer spaces required for metal and via layers. This is illustrated in Figure 6.14, where it can be seen that the NILS for a bright mask shape is superior to that for a dark mask shape.

Negative-acting photoresists have recently returned for the printing of hole layers such as contact and metal due to the significantly improved normalized image log-slope properties of the aerial image versus a positive-acting darkfield mask. It has been observed that these negative-acting photoresists can exhibit significantly different patterning behavior, and compact models have been introduced that account for their unique shrinkage and develop characteristics [74].

As process margins continue to narrow at lower k_1, models will need to more faithfully predict all failure modes which onset at the process window corners. In addition to pinching and bridging, models will need to accurately predict sub-resolution assist feature scumming or dimpling, side-lobe dimpling, and aspect ratio–induced mechanical pattern collapse.

6.2.5.3.1 Patterning Failure

We earlier introduced the concept of a "hotspot," and it is, of course, important that full-chip post-OPC verification models accurately predict such pattern weak points. The most commonly observed, and easiest to accurately simulate, are pinching or bridging, but additional failure modes include pattern toploss, which can lead to failure during etch, SRAF printing, and aspect ratio–induced pattern collapse. In order to accurately predict such 3-D phenomena, it is necessary to consider additional planes besides the one associated with the model used to predict critical dimensions for OPC.

6.2.5.3.1.1 Pinching/Bridging It has long been important for the OPC model to accurately predict the onset of pattern pinching or bridging at specific layout locations through the lithographic process window of dose, focus, and mask CD variation. Such imaging weaknesses originate in the aerial image, and it has become increasingly necessary to simulate patterning across the full reticle field, accounting for subtle differences in mask CD, local exposure dose, and local focus deviation. See Figure 6.16.

6.2.5.3.1.2 Resist Height: Toploss/Scumming OPC models dictate mask edge movement to achieve target by predicting photoresist and/or etched pattern contours at a single vertical plane, which is thought to correspond most closely to the plane indicated by CD SEM metrology. With

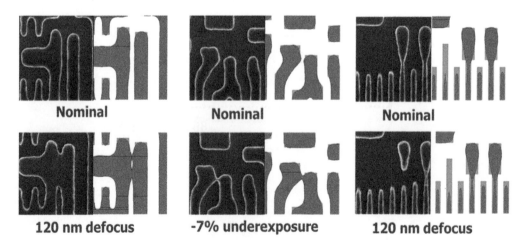

FIGURE 6.16 CD variations across focus and exposure.

diminished available process latitude at low k_1, it is increasingly difficult to maintain sufficient resist height after develop to act as an ample etch barrier. So, for instance, post-etch metal bridging can occur in locations where the OPC model predicts acceptable CD at one plane, but the final resist height is greatly diminished. Thus, it is increasingly important at 20 and 14 nm to have models that accurately predict the photoresist behavior through the entire vertical span of the imaging layer. Models have been developed that can rapidly predict such toploss, and these models can be referenced in OPC correction to better balance CD control and toploss-driven yield-limiting hotspots. Similarly, models can be calibrated to accurately predict the onset of resist scumming at the bottom of developed trenches or vias.

6.2.5.3.1.3 Sub-Resolution Assist Feature (SRAF) Printing SRAFs have been used extensively since 130 nm, historically placed by rules with relatively few SRAF widths and spacing to target features. In order to realize the maximum process window entitlement, however, model-based SRAF placement has been developed for 20 nm and below. Such an approach must be informed by an accurate simulation of the limiting conditions in which SRAFs will print on the wafer. Thus, similarly to toploss and scumming prediction, models must accurately predict the onset of SRAF printing, which occurs at the top of the photoresist in dark-field imaging positive or bright-field negative develop and at the bottom of the photoresist plane for the opposite case.

While full-chip OPC models based upon a 2-D resist contour simulation have to date been sufficient to meet the task of correction and verification, it may become necessary to have a level of 3-D awareness in these photoresist models [75]. One example is for an etch model to predict bias as a function of lithographic focus, which imparts resist profile changes.

6.2.5.4 Etch Models

Etch proximity effects are known to operate locally (i.e., on a scale similar to those of "optical" proximity effects) as well as over longer distances, approaching mm scale. Such long-distance loading effects are not easily accounted for in full-chip data preparation, but shorter-range effects can be compensated effectively. The two primary phenomena impacting CD control are aspect ratio–dependent etch rates (ARDE) and microloading [76]. With ARDE, the etch rate and therefore, the bias are seen to depend upon the space being etched, while microloading dictates that the etch bias is dependent upon the density of resist pattern within a region of interest. Different kernel types can accurately represent these phenomena and when used in combination, can yield a very accurate representation of the etch bias as a function of feature type [24]. Recently, it has been observed that etching of identical vias in differing trench environments can lead to different final dimensions, and this behavior can be represented by adding underlayer-aware kernels to the compact etch model [77]. Stobert et al. recently provided a nice overview of the application of etch models to full-chip OPC [78].

6.2.6 FUTURE LITHOGRAPHY PROCESS MODELS

6.2.6.1 Extreme Ultraviolet (EUV) Lithography

EUV lithography has been developed and anticipated for use in manufacturing for more than 15 years now, but implementation has been delayed by multiple technical challenges, which have thus far made multipatterning with the incumbent 193 nm immersion lithography a more cost-effective preference. But it is anticipated that for 7 nm and beyond, EUV may prove viable. The changes required to computational lithography for the wavelength change from 193 to 13.5 nm were relatively straightforward, related to 1) higher background flare, 2) nontelecentric lens design used with the all-reflective scanner optics, 3) 3-D EMF shadowing effects now resulting in differences between horizontal- and vertical-oriented features and varying across the scanner exposure slit (and hence, die location), and 4)the so-called "black-border" effect at the edge of the reticle exposure field, which results in unwanted partial exposure at the field edges and especially corners. All these solutions have been in place for several years, and in general, it is expected that there will not be any significant discontinuities in design practices or data preparation flows necessary explicitly for EUV [79–83]. However, given the seeming perpetual uncertainty about EUV deployment in manufacturing, the industry has had to plan to continue extensions of 193i, even for 7 nm, with the possibility of migrating to EUV. This leads to certain design rule and design style choices that might not necessarily be required for EUV but are needed for risk mitigation [84, 85].

6.2.6.2 Directed Self-Assembly (DSA) Patterning

Directed self-assembly patterning represents a partial departure from the "top-down" methods that have evolved to enable k_1 lowering for semiconductor manufacturing. DSA relies upon the unique property of a specific class of materials called block copolymers to self-aggregate when cast in a thin film and thermally equilibrated with a bake. This phase separation generally results in random patterns that are not useful, but in the presence of an underlying "guiding" pattern template, the material will organize itself into extremely useful regular repeating line/space or hole patterns, differentiated by significantly different plasma etch rates and chemical reactivity. Thus, it is possible to utilize DSA for the patterning of certain periodic structures. The CD of the pattern is determined by the block copolymer material, and the alignment is essentially perfect relative to the template. DSA processes are being targeted for possible use on via or FinFET layers at the 5 nm technology node [86].

6.3 MODEL-BASED DFM

6.3.1 LITHO-FRIENDLY DESIGN (LFD)

As discussed earlier, full-chip simulation to verify post-OPC layout correction has been a vital component of mask data preparation flows for several generations of technology manufacturing [87–91]. Such tools are used prior to committing mask manufacture in order to safeguard the considerable investment associated with mask and wafer fabrication. It was realized early on (between the 90 nm to 65 nm nodes), however, that these same simulations could be immensely useful upstream in the design for manufacturing flow to enable physical layout designers to assess and correct potential manufacturing hotspots existing for the gate and metal interconnect layers. Torres introduced the useful concept of the process variability (PV) band to quantify and conveniently illustrate to the designer the extrema of envisioned printing accompanying manufacturing variation of dose, focus, and mask CD [92]. Figure 6.17. PV bands have become a widely accepted process-variability visualization technique in our industry.

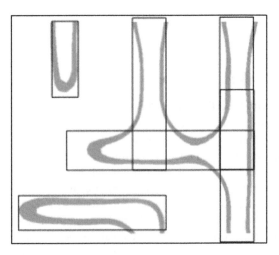

FIGURE 6.17 Process variability (PV) bands.

The field of model-based DFM, also called LFD, has thus greatly facilitated communication between the foundry manufacturing and layout design communities. An LFD "kit" is analogous to a DRC rule deck and includes the encrypted model and OPC information necessary to enable a designer to simulate patterning at a specific foundry and, more importantly, to make improvements early on to ensure more robust manufacturing. For the application of model-based DFM, it is useful to distinguish between front-end layers used in standard cells and back-end layers generated by automated place and route tools.

Front-end standard cells consume relatively small area and as such, can utilize full simulation to identify hotspots. The applied hotspot fix needs to be DRC compliant and must not in turn produce a new hotspot. Since it is not often obvious how to accomplish effective layout polygon manipulation consistent with these objectives, EDA tools can internally assess multiple options and provide semi-automated layout recommendations to the designer. Alternatively, for an integrated device manufacturer, changes to the OPC recipe might be considered based upon such analysis.

For routed layers, a full-chip solution is needed; therefore, faster methods have been developed for use during place and route (PnR) and final signoff. These "fast LFD" methods can involve pattern matching of known lithographic hotspots or other approaches to greatly reduce the amount of area to be simulated and can result in nearly 100% correction in far less than 1 CPU hour per square millimeter of layout area [93].

Historically, PnR hotspot fixes necessarily involved ripping up and rerouting the areas containing an identified hotspot, which had no guarantee of not producing new hotspots and could also require an additional cycle of timing closure. A new, more efficient approach is now available that uses local layout manipulation. It uses an integrated tool solution during the P&R process that accesses the full signoff DRC/PM/LFD deck for pattern recognition and analysis, and contains built-in intelligence for analyzing the encrypted problematic pattern definitions supplied by the foundries.

As designers must take on more and more responsibility for ensuring that designs can be manufactured with increasingly complex production processes, EDA software must evolve to fill the knowledge gap. LFD tools with model-based hints capability are one example of how EDA systems can be the bridge between design and manufacturing [94–98].

6.3.2 Pattern-Based DFM

Model-based methods represented a huge improvement on the increasingly insufficient and complex design rule abstractions, which can be mathematically interpreted as a set of linear constraints over the geometrical dimensions of the parametric space determined by adjacent polygons and edges (or portions of an edge) and which cannot therefore properly capture (but only approximate) intrinsic process nonlinearities.

While model-based methodologies put the power of fab process simulation into the hands of designers, the obvious drawback of such LFD approaches is the cycle time associated with such compute-intensive contour generation operations. Additionally, accurate process models are not always available so early in the technology development life cycle, meaning that large numbers of compute cycles are being spent using less than desired accuracy.

Against this backdrop, Capodieci et al. introduced the concept of pattern-based DRC [99], and this approach was later extended and generalized to topological patterns [100]. A geometric and/or topological pattern is capable of unambiguously defining (and classifying) specific 2-D geometrical configurations as either manufacturable or a yield detractor more comprehensively than conventional design rules. Named "DRC Plus" and illustrated in Figures 6.18 and 6.19, the approach first utilizes ultrafast image-based pattern matching algorithms to locate 2-D patterns leading to potential hotspots and then applies a conventional DRC rule specific to that pattern type. Of course, just like traditional DRC rules, DRC Plus relies on analysis from multiple sources, including lithography simulation, layout scoring methods, OPC practices, and fab hotspot observations to identify the

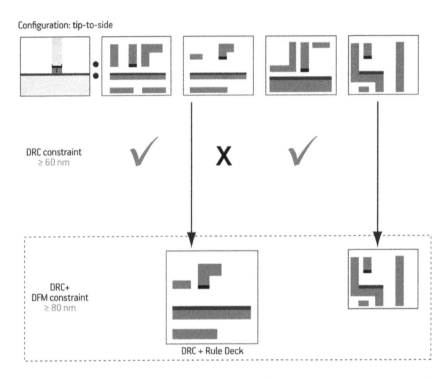

FIGURE 6.18   Example of DRC Plus rule augmentation with pattern variants and corresponding recommended rules.

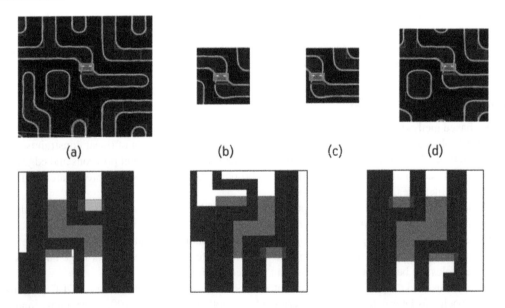

FIGURE 6.19   DRC Plus rules characterization through lithography simulation and pattern clustering.

problematic 2-D patterns. The output of DRC Plus is the same as traditional DRC, so the method has been relatively easy to integrate into existing design-to-silicon flows.

A number of applications of pattern-based DFM methods have been offered in recent years, including combinations of pattern- and model-based methods to provide quantitative differentiation of the relative manufacturability of different patterns, analogous to the previously described scoring

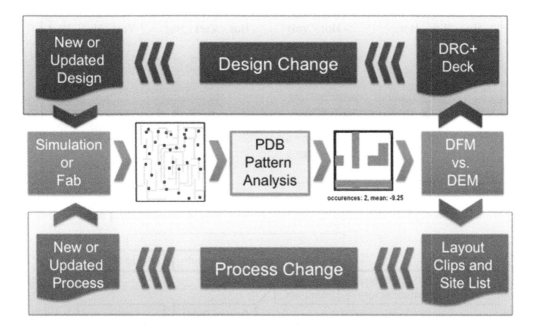

FIGURE 6.20 Pattern database integrates both physical design and process optimization.

methods. Ultrafast geometrical and topological pattern-matching capabilities have been integrated into virtually all physical verification tools and also extended to place-and-route flows in order to generate correct-by-construction (as much as possible) layout configurations during the physical design phase.

One additional key benefit of the adoption of pattern-based DFM for advanced nodes (from 32 nm on) has been the ability to "accumulate" (i.e., store and retrieve) a set of patterns corresponding to yield detractors into a persistent high-capacity database. Such a database becomes, in the period between technology development, yield ramp, and high-volume manufacturing, a versatile knowledge base that can act as a centralized information exchange among design, process, and manufacturing [101] (Figure 6.20). In a typical application, the database is queried to extract a set of patterns (or a set of representatives for families of patterns) given specific characteristics (based on geometric, OPC, and manufacturability features). These patterns can be simulated and/or used in the fab for process monitoring, SEM metrology, and failure analysis. Results can be reinserted into the knowledge base repository, thus accelerating yield cycles of learning.

6.4 DESIGN TECHNOLOGY CO-OPTIMIZATION (DTCO)

We have shown how design for manufacturing methods were born out of an opportunity to improve (minimize) layout susceptibility to random yield detractors but then migrated to address the short-falls of the traditional design-to-silicon flow, which unsuitably portrayed manufacturability as a purely binary proposition. The drive to lower and lower k_1 patterning led to more gray regions in the complex interplay between pattern layouts and in-fab process windows limitations, which led to systematic yield loss. The ability to simulate the lithography process accurately opened an entirely new communication channel between manufacturing and design; in effect, the ability to describe a continuous response surface as opposed to the earlier binary depiction.

A more apt descriptor, DTCO, has recently been introduced to highlight the need for both design and manufacturing technologies to be optimized with mutual awareness and as early as possible in the technology life cycle and subsequently, on a continued basis. DTCO does not refer to a specific software tool or step in an EDA flow but rather, to a series of interaction opportunities among the

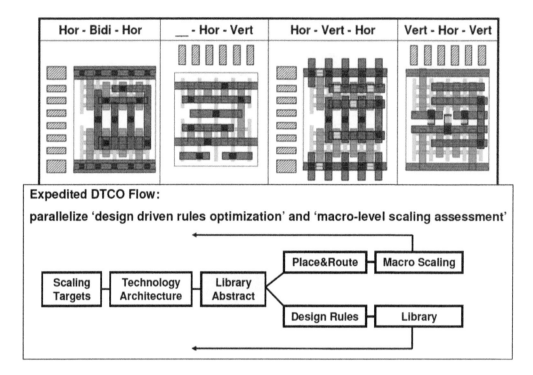

FIGURE 6.21 DTCO methodology iterates among selected variants of cell layout and parametric process configurations.

three key stakeholders in the design-to-fabrication flow: circuit and physical designers, device and process integration technologists, and manufacturing engineering.

Although already used in various forms for past technology nodes (see Figure 6.21), DTCO practices have been given a coherent methodological treatment by Liebmann et al. in the transition between 10 and 7 nm nodes [84]. The stated objective of the DTCO process is to "holistically" perform iterative optimizations over a heterogeneous parametric space, which includes circuit design, physical layout, patterning and fabrication processes, and IC performance targets. The key idea behind this approach is that a global model of the solution space can help avoid suboptimal solutions for the individual components of the problem. The proposed DTCO methodology, as outlined by Liebmann et al. [102], includes four phases:(a) identification of technology scaling targets, (b) cell architecture exploration and definition, (c) cell-level refinement and critical design rules identification, and (d) block-level refinement and P&R strategy identification.

Scaling targets in the first phase refer not only to linear critical dimensions (minimum line and spaces, unidimensional pitches, etc.) but also to 2-D constructs, which will be used in the second phase to build the fundamental cell architecture. The identification of these *special* (layout) constructs allows both a generalization of the design rules and at the same time, a restriction of the actual layout configurations that will be allowed in a given technology. During the second and third DTCO phases, cell architecture variants are explored and characterized using process simulations (and/or focused silicon testing) until almost all layout components necessary for a technology are identified. In the fourth DTCO phase, cells (and their variants) are assembled (P&R) into blocks in order to mimic a realistic IC product, which is again tested using process simulations and silicon test vehicle. This phase also allows a validation and/or optimization of router automation techniques and often leads to the discovery of new algorithmic requirements for P&R tools. The novelty element and the main strength of a rigorous DTCO methodology is the time granularity of joint design/ process validation and the incremental reduction in intrinsic systematic variability from special constructs to cell architecture to cell libraries and finally, to fully routed layout blocks. At the same time, process technology verification is not carried out solely on the basis of test structures based on an abstracted Design Rule Manual, but rather on more and more realistic (product-like) layout content.

In spite of successful deployment for virtually all nodes at and below 32 nm, DTCO has also nontrivial engineering costs, as it requires highly skilled interdisciplinary (design/process) teams to be available throughout the whole development cycle. While some automation has been deployed to enable and speed up DTCO cycles of learning, a general software solution is not available for next-generation nodes. Nevertheless, even with partial automation support, DTCO remains the de facto state-of-the-art technique for a sustained IC scaling roadmap.

REFERENCES

1. Starikov, A. 2003. "Design rules for real patterns," *Proc. Electronic Design Processes Workshop*, IEEE CS Press, pp. 46–73.
2. Dennard, Robert H., Gaensslen, Fritz, Yu, Hwa-Nien, et al., October 1974. "Design of ion-implanted MOSFET's with very small physical dimensions" (PDF), *IEEE J. Solid State Circuits*, SC-9 (5).
3. Moore, G. 1965. "Cramming more components onto integrated circuits", *Electronics*, 38 (8).
4. www.itrs2.net
5. Strojwas, A. 2006. "Conquering process variability: a key enabler for profitable manufacturing in advanced technology nodes". *Proc. IEEE International Symposium on Semiconductor Manufacturing*.
6. Paek, S.W., Jang, D.H., Park, J.H., et al., 2009. "Enhanced layout optimization of sub-45nm standard, memory cells and its effects", *Proc. SPIE*, 7275, 72751M.
7. Balasinski, A., 2013. *Design for Manufacturability: From 1D to 4D for 90-22 nm Technology Nodes*, Springer.
8. Pathak, P., Madhavan, S., Capodieci, L., 2012. "Framework for identifying recommended rules and DFM scoring model to improve manufacturability of sub-20nm layout design". *Proc. SPIE*, 8327, 83270U-13.
9. Liebmann, L., 2009. "Simplify to survive, prescriptive layouts provide profitable scaling to 32 nm and beyond", *Proc. SPIE*, 7275, 7275A.
10. Mekkoth, J., Krishna, M., Qian, J., et al., 2006. "Yield learning with layout-aware advanced scan diagnosis," *Proc. International Symposium of Testing and Failure Analysis (ISTFA)*, 206.
11. Malik, S., Mahavan, S., Schuyermyer, C., et al., 2013. "Deriving feature failure rate from silicon volume diagnostics data", *IEEE Des. Test*, July/August, 26.
12. Postnikov, S., Lucas, K., Wimmer, K., 2001. "Impact of optimized illumination upon simple lambda-based design rules for low k1 lithography", *Proc. SPIE*, 4344, 797.
13. Dill, F., Neureuther, A., Tuttle, J.A., et al., 1975. "Modeling projection printing of positive photoresists", *IEEE Trans. Electr. Dev.*, ED-22 (7), 456.
14. Oldham, W., Nandgaonkar, S., Neureuther, A. et al., 1979. "A general simulator for VLSI lithography and etching processes: part I-application to projection lithography", *IEEE Trans. Electr. Dev.*, ED-26 (4), 717.
15. Mack, C., 1985. "PROLITH: a comprehensive optical lithography model", *Proc. SPIE*, 538, 207.
16. Pistor, T., 1998. "A new photoresist simulator from panoramic technology", White Paper Berkeley, CA. Available at: www.panoramictech.com/Products/Resist/PanoramicResistSimulator.pdf.
17. Mack, C., 2005. "30 years of lithography simulation", *Proc. SPIE*, 5754, 1.
18. Mack, C., 2001. "Lithography simulation: a review", *Proc. SPIE*, 4440, 59.
19. Schmid, G., Burns, S., Stewart, M.D. et al., 2002. "Mesoscale simulation of positive tone chemically amplified photoresists", *Proc. SPIE*, 4690, 381.
20. Zuniga, M., Walraff, G., Neureuther, A., 1995. "Reaction diffusion kinetics in DUV positive-tone resist systems", *Proc. SPIE*, 2438, 113.
21. Stirniman, J., Rieger, M., 1994. "Fast proximity correction with zone sampling", *Proc. SPIE*, 2197, 294.
22. Bernard, D., Li, J., Rey, J., et al., 1996. "Efficient computational techniques for aerial imaging simulation", *Proc. SPIE*, 2726, 273.
23. Cobb., N., Zakhor, A., Miloslavsky, E., 1996. "Mathematical and CAD framework for proximity correction", *Proc. SPIE*, 2726, 208.
24. Granik, Y., 2001. "Correction for etch proximity: new methods and applications", *Proc. SPIE*, 4346, 98.

25. Sturtevant, J., Word, J., LaCour, P., et al., 2005. "Considerations for the use of defocus models in OPC", *Proc. SPIE*, 5756, 427.

26. Word, J., Sakajiri, K., 2006. "OPC to improve lithographic process window", *Proc. SPIE*, 6156, 61561I.

27. Azpiroz, J., Krasnoperova, A., Siddiqui, S., et al., 2009. "Improving yield through the application of process window OPC", *Proc. SPIE*, 7274, 27411.

28. Landié, G., Pena, J., Postnikov, S., et al., 2013. "Model-based stitching and inter-mask bridge prevention for double patterning lithography", *Proc. SPIE*, 8683, 868316.

29. Gheith, M., Hong, L., Word, J., 2009. "OPC for reduced process sensitivity in the double patterning flow", *Proc. SPIE*, 7274, 727419.

30. Kingsley, T., Sturtevant, J., McPherson, S., et al., 2008. "Advances in compute hardware platform for computational lithography", *Proc. SPIE*, 6520, 652018.

31. Gupta, R., Shang, S., Sturtevant, J., 2015. "Full-chip two-layer CD and overlay process window analysis", *Proc. SPIE*, 9427, 94270H.

32. Viswanathan, R., Abdo, A., 2010. "The feasibility of using image parameters for test pattern selection during model calibration", *Proc. SPIE*, 7640, 76401E.

33. Burbine, A., Sturtevant, J., Fryer, D., 2015. "Bayesian inference for OPC modeling", *Proc. SPIE*, 9780.

34. Sturtevant, J., Hong, L., Jayaram, S., et al., 2008. "Impact of illumination source symmetrization in OPC", *Proc. SPIE*, 7028, 70283M.

35. Snyman, J., 2005. *Practical Mathematical Optimization: An Introduction to Basic Optimization Theory and Classical and New Gradient-Based Algorithms*, Springer Publishing. ISBN 0-387-24348-8.

36. Han, G., Brendler, A., Mansfield, S., et al., 2007. "Statistical optimization of sampling plan and its relation to OPC model accuracy", *Proc. SPIE*, 6518, 651808.

37. Tabery, C., Capodieci, L., Haidinyak, C., et al. 2005. "Design-based metrology: advanced automation for CD-SEM recipe generation", *Proc. SPIE*, 5752, 527.

38. Burbine, A., Fryer, D., Sturtevant, J., 2015. "Akaike information criteria to select well-fit resist models", *Proc. SPIE*, 9427, 94270J.

39. Akerman, L., Eytan, G., Uchida, R., et al., 2006. "Investigation of possible ArF resist slimming mechanisms", *J. Microlith., Microfab., Microsyst.*, 5(4), 043006.

40. Rana, N., Archie C., Lu, W., et al., 2009. "The measurement uncertainty challenge of advanced patterning development", *Proc. SPIE*, 7272, 727203.

41. Mack, C., Raghunathan, A., Sturtevant, J., et al., 2016. "Modeling metrology for calibration of OPC models". *Proc. SPIE*, 9778, 97781Q.

42. Vasek, J., Menedeva, O. Sturtevant, J., et al., 2007. "SEM-contour based OPC model calibration through the process window", *Proc. SPIE*, 6518, 65180D.

43. Shindo, H., Sugiyama, A., Komura, H., et al., 2009. "High-precision contouring from SEM image in 32-nm lithography and beyond", *Proc. SPIE*, 7275, 72751F.

44. Filitchkin, P., Do, T., Sturtevant, J., et al., 2009. "Contour quality assessment for OPC model calibration", *Proc. SPIE*, 7272, 72722Q.

45. Granik, Y., 2005. "Calibration of compact OPC models using SEM contours", *Proc. SPIE*, 5992, 59921V1.

46. Hibino, D., Shindo, H., Sturtevant, J., et al., 2010. "High-accuracy OPC-modeling by using advanced CD-SEM based contours in the next-generation lithography", *Proc. SPIE*, 7638-67.

47. Fischer, D., Han, G., Oberschmidt, J., et al., 2008. "Challenges of implementing contour modeling in 32nm technology". *Proc. SPIE*, 6922, 69220A.

48. Weisbuch, F., Koh, K., Jantzen, K., 2014. "Bringing SEM-contour based OPC to production", *Proc. SPIE*, 9052-78.

49. Weisbuch F., Jantzen, K., 2015. "Enabling scanning electron microscope contour-based optical proximity correction models", *J. Micro/Nanolith. MEMS MOEMS*, 14(2), 021105.

50. Weisbuch, F., Omran, A., Jantzen, K., 2015. "Calibrating etch model with SEM contours", *Proc. SPIE*, 9426.

51. Hitomi, K., Miller, M., Halle, S., et al. 2015. "Application of SEM-based contours for OPC model weighting and sample plan reduction", *Proc. SPIE*, 9426, 94260Y, Optical Microlithography XXVIII.

52. Hibino, D., Sturtevant, J., De Bisschop, P., et al., 2011. "High-accuracy optical proximity correction modeling using advanced critical dimension scanning electron microscope–based contours in next-generation lithography", *J. Micro/Nanolith. MEMS MOEMS*, 10(1), 013012.

53. El-Sewefy, O., Lafferty, N., Sturtevant, J., et al., 2016. "Source mask optimization using 3D mask and compact resist models", *Proc. SPIE*, 9780, 978019.

54. Michel, J., Sturtevant, J., Granik, Y., et al., 2013. "Full chip implant correction with wafer topography OPC modeling in 2X bulk technologies", *Proc. SPIE*, 8880, 88801J.
55. Michel, J., Sturtevant, J., Granik, Y., et al., 2013. "Wafer topography modeling for ionic implantation mask correction dedicated to 2x nm FDSOI technologies", *Proc. SPIE*, 8683, 86830I, Optical Microlithography XXVI.
56. Oh, M., Chung, N., Sturtevant, J., 2012. "Full-chip correction of implant layer accounting for underlying topography", *Proc. SPIE*, 8326, 83262R, Optical Microlithography XXV.
57. Tejnil, E., Hu, Y., Sahouria, E., et al., 2008. "Advanced mask process modeling for 45-nm and 32-nm nodes", *Proc. SPIE*, 6924, 69243H.
58. Bork, I., Buck, P., Reddy, M., et al. 2015. "A fully model-based MPC solution including VSB shot dose assignment and shape correction", *Proc. SPIE*, 9635, 96350U, Photomask Technology 2015.
59. Zine, N., Farys, V., Armeanu, A., et al., 2015. "Accurate mask model implementation in OPC model for 14nm nodes and beyond", *Proc. SPIE*, 9635, 96350W, Photomask Technology 2015.
60. Chou, S., Shin, J., Shu, K., et al. 2004. "Study of mask corner rounding effects on lithographic patterning for 90-nm technology node and beyond", *Proc. SPIE*, 5446, 508.
61. Adam, K., Neureuther, A., 2002. "Domain decomposition methods for the rapid electromagnetic simulation of photomask scattering", *J. Microlithogr. Microfab., Microsyst.*, 1, 253.
62. Yang, M., 2009. "Analytical optimization of high-transmission attenuated phase-shifting reticles", *J. Micro/Nanolith. MEMS MOEMS*, 8, 013015.
63. Lam, M., Adam, K., 2007. "Understanding the impact of rigorous mask effects in the presence of empirical process models used in OPC", *Proc. SPIE*, 6520, 65203M.
64. McIntyre, G., Hibbs, M., Tirapu-Aspirov, J., et al., 2010. *J. Micro/Nanolith. MEMS MOEMS*, 9, 013010.
65. Cheng, J., Schramm, J., Sturtevant, J., et al., 2012. "OPC model prediction capability improvements by accounting for mask 3D-EMF effect", *Proc. SPIE*, 8326, 83261R, Optical Microlithography XXV.
66. Lam, M., Adam, K., Fryer, D., et al., 2013. "Accurate 3DEMF mask model for full-chip simulation", *Proc. SPIE*, 8683, 86831Dm.
67. Yuan, L., Zhou, L., Zhang, L., et al., 2010. "OMOG mask topography effect on lithography modeling of 32nm contact hole patterning", *Proc. SPIE*, 7640, 76402K1.
68. Sturtevant, J., Tejnil, E., Lin, T., et al., 2013. "Impact of 14-nm photomask uncertainties on computational lithography solutions", *J. Micro/Nanolith. MEMS MOEMS*, 13(1), 011004.
69. Sturtevant, J., 2012. "Challenges for patterning process models applied to large scale", *J. Vac. Sci. Technol. B*, 30, 030802.
70. Granik, Y., Cobb, N., 2003. "New process models for OPC at sub-90 nm nodes", *Proc. SPIE*, 5050, 11166.
71. Granik, Y., Medvedev, D., Cobb, N., 2007. "Towards standard process models for OPC", *Proc. SPIE*, 6520, 652043.
72. Conrad, E., Cole, D., Barouch, E., et al., 1999. "Model considerations, calibration issues, and metrology methods for resist-bias models", *Proc. SPIE*, 3677, 940.
73. Brock, D., 2007. "Patterning the world: the rise of chemically amplified photoresists", *Chem. Heritage Mag.*
74. Ao, C., Foong, Y., Zhang, D., et al., 2015. "Evaluation of compact models for negative-tone development layers at 20/14nm nodes", *Proc. SPIE*, 9426, 94261P, Optical Microlithography XXVIII.
75. Zuniga, C., Deng, Y., Granik, Y., 2015. "Resist profile modeling with compact resist model", *Proc. SPIE*, 9426, 94261R, Optical Microlithography XXVIII.
76. Bates, R., Goeckner, M., Overzet, L., 2014. "Correction of aspect ratio dependent etch disparities", *J. Vac. Sci. Technol. A*, 32(5). doi: 10.1116/1.4890004.
77. Jung, S., Do, T., Sturtevant, J., 2015. "Finding practical phenomenological models that include both photoresist behavior and etch process effects", *Proc. SPIE*, 9428, 94280I, Advanced Etch Technology for Nanopatterning IV.
78. Stobert, I., Dunn, D., 2013. "Etch correction, and OPC, a look at the current state and future of etch correction", *Proc. SPIE*, 8685, 868504, Advanced Etch Technology for Nanopatterning II.
79. Maloney, C., Word, J., Fenger, G., et al., 2014. "Feasiblity of compensating for EUV field edge effects through OPC", *Proc. SPIE*, 9048, 90480V.
80. Fenger, G., Lorusso, G., Hendrickx, E., et al., 2010. "Design correction in extreme ultraviolet lithography", *J. Micro/Nanolith. MEMS MOEMS*, 9 (4), 043001.
81. Mailfert, J., Zuniga, C., Philipsen, V., et al., 2012. "3D mask modeling for EUV lithography", *Proc. SPIE*, 8322, 83224.

82. Zuniga, C., Habib, M., Word, J., et al., 2011. "EUV flare and proximity modeling and model-based correction", *Proc. SPIE*, 7969, 79690T.

83. Lorusso, G., Van Roey, F., Hedrickx, E., et al., 2009. "Flare in EUV lithography: metrology, out-of-band radiation, fractal point-spread function, and flare map calibration", *J. Micro/Nanolith. MEMS MOEMS*, 8(4), 041505.

84. Liebmann, L., Chu, A., Gutwin, P., 2015. "The daunting complexity of scaling to 7nm without EUV: pushing DTCO to the extreme", *Proc. SPIE*, 9427, 942702.

85. Chava, B., Rio, D., Sherazi, Y., et al., 2015. "Standard cell design in N7: EUV vs immersion", *Proc. SPIE*, 9427, 94270E-1.

86. Ma, Y., Wang, Y., Word, J., et al., 2016. "Directed Self Assembly (DSDA) compliant flow with immersion lithograph: from material to design and patterning", *Proc. SPIE*, 9777, 97770N.

87. Arb, K., Reid, C., Li, Q., et al., 2012. "OPC and verification for LELE double patterning", *Proc. SPIE*, 8522, 85221B.

88. Sturtevant, J., Jayaram, S., Hong, L., 2009. "Process variability band analysis for quantitative optimization of exposure conditions", *Proc. SPIE*, 7275, 72751Q.

89. Sturtevant, J., Jayaram, S., Hong, L., et al., 2008. "Novel method for optimizing lithography exposure conditions using full-chip post-OPC simulation", *Proc. SPIE*, 6924, 69243P.

90. Sturtevant, J., Jayaram, S., Hong, L., 2008. "Exposure tool specific post-OPC verification", *Proc. SPIE*, 6925, 69250T.

91. Jiswal, O., Kuncha, R., Bharat, T., et al., 2010. "Electrical validation of through-process optical proximity correction verification limits", *J. Micro/Nanolith. MEMS MOEMS*, 9(4), 041303.

92. Torres, A., Berglund, N., 2005. "Integrated circuit DFM framework for deep sub-wavelength processes", *Proc. SPIE*, 5756, 39.

93. Beylier, C., Moyroud, C., Trouiller, Y., et al., 2012. "Fully integrated litho aware PnR design solution", *Proc. SPIE*, 8327, 83270A, Design for Manufacturability through Design-Process Integration VI.

94. Park, J., Kim, N., Kang, J., et al., 2015. "High coverage of litho hotspot detection by weak pattern scoring", *Proc. SPIE*, 9427, 942703, Design-Process-Technology Co-optimization for Manufacturability IX.

95. Deng, X., Du, C., Hong, L., et al., 2015. "An efficient lithographic hotspot severity analysis methodology using Calibre Pattern Matching and DRC application", *Proc. SPIE*, 9427, 94270Y, Design-Process-Technology Co-optimization for Manufacturability IX.

96. Rabie, A., Madkour, K., George, K., et al., 2014. "Model based multilayers fix for litho hotspots beyond 20 nm node", *Proc. SPIE*, 9053, 90530F, Design-Process-Technology Co-optimization for Manufacturability VIII.

97. Kang, J., Kim, B., Ha, N., et al., 2013. "Model based hint for litho hotspot fixing beyond 20nm node", *Proc. SPIE*, 8684, 86840N, Design for Manufacturability through Design-Process Integration VII.

98. Park, J., Kim, N., Kang, J., et al., 2015. "High coverage of litho hotspot detection by weak pattern scoring", *Proc. SPIE*, 9427, 942703.

99. Dai, V., Yang, J., Capodieci, L., et al., 2007. "DRC Plus: augmenting stardard DRC with pattern matching on 2D geometries", *Proc. SPIE*, 6521, 65210A.

100. Dai, V., Lai, Y., Capodieci, L., et al., 2014. "Systematic physical verification with topological patterns", *Proc. SPIE*, 9053, 905304.

101. Xu, J., Krishnamoorthy, K., Capodieci, L., et al., 2015. "Design layout analysis and DFM optimization using topological patterns", *Proc. SPIE*, 9427, 94270Q.

102. Liebmann, L., Vaidyanathan, K., Pileggi, L., 2016. *Design Technology Co-optimization in the Era of Sub-Resolution IC Scaling*, SPIE Press.

7 Chemistry of Photoresist Materials

Takumi Uemo, Robert D. Allen, and James Thackeray

CONTENTS

7.1 INTRODUCTION

The design requirements of successive generations of very-large-scale integrated (VLSI) circuits have led to a reduction in lithographic critical dimensions. The aim of this chapter is to discuss the progress of the resists for present and future lithography. It is worthwhile describing the history and the trend of lithography and resists. It is evident from Figure 7.1 that three turning points of resist materials have been reached [1]. The first turning point was the replacement of a negative resist composed of cyclized rubber and a bisazide by a positive photoresist composed of a diazonaphthoquinone (DNQ) and a novolac resin. This was induced by the change of exposure system from a contact printer to a g-line (436 nm) reduction projection step-and-repeat system—the so-called *stepper*. The cyclized rubber system has poor resolution due to swelling during the development and low sensitivity due to a lack of absorption at the g-line. The DNQ-based positive photoresist shows sensitivity at the g-line and high resolution using aqueous alkali development.

The performance of the g-line stepper was improved by increasing numerical aperture (NA). Then, shorter-wavelength i-line (365 nm) lithography was introduced. The DNQ-novolac resist can be used for i-line lithography; therefore, much effort has been made to improve the resolution capability of the DNQ-novolac resist as well as the depth-of-focus latitude. The effect of novolac resin and DNQ chemical structure on dissolution inhibition capability has been investigated mainly to get high dissolution contrast, which will be discussed in Section 7.2. The progress of this type of resist and an i-line stepper is remarkable, achieving resolution below the exposure wavelength of the i-line (0.365 µm). However, i-line lithography has difficulty accomplishing 0.3-µm processes (64 M dynamic random access memory [DRAM]), even using a high-NA i-line stepper in conjunction with a DNQ-novolac resist.

Several competing lithographic technologies have been proposed for next-generation engineering of the i-line: KrF lithography (deep UV lithography) and electron-beam lithography [2]. The wavefront engineering includes off-axis illumination (OAI), pupil filtering [3], and phase-shifting

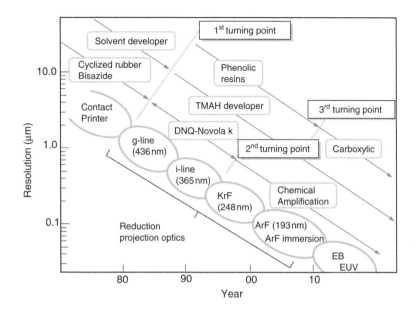

FIGURE 7.1 Development trend of lithography and resists.

lithography (discussed in Chapter 2). DNQ-novolac resists can be used for OAI and pupil filtering, although higher sensitivity is necessary. For phase-shifting lithography, bridging of patterns at the end of the line-and-space patterns occurs when a positive resist is used [4]. Therefore, negative resists with high sensitivity and high resolution are required for phase-shifting lithography. In KrF lithography, positive and negative resists with high sensitivity, high resolution, and high transmittance at the exposure wavelength are needed. Chemical amplification resists have been used in KrF lithography, which is the second turning point. ArF resists are currently used for large-volume production. In Section 7.7, Robert Allen has provided an update on the recent progress in resist materials.

7.2 DNQ-NOVOLAC POSITIVE PHOTORESISTS

The positive photoresist composed of DNQ and novolac resin was the workhorse for semiconductor fabrication [5]. It is surprising that this resist has a resolution capability below the exposure wavelength of i-line (365 nm). In this section, the photochemical reaction of DNQ and improvement of the resist performance by newly designed novolac resins and DNQ inhibitors are discussed.

7.2.1 PHOTOCHEMISTRY OF DNQ

The Wolff rearrangement reaction mechanism of DNQ was proposed by Suess [6] in 1942, as shown in Scheme 7.1. It is surprising that the basic reaction was already established half a century ago, although Suess [6] suggested that the chemical structure of the final product was

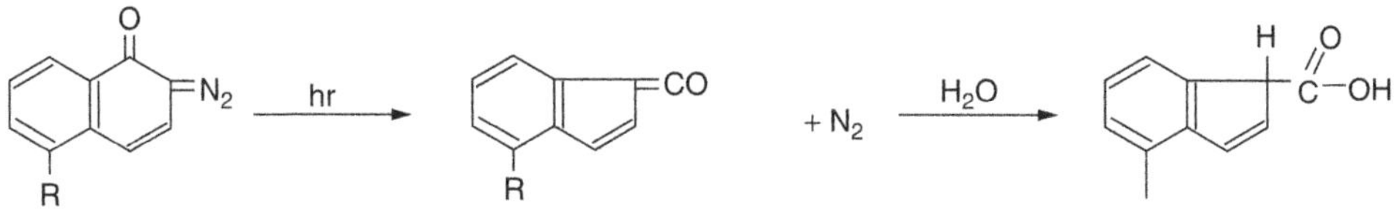

SCHEME 7.1 Photolysis mechanism for DNQ-PAC proposed by Suess.

3-indenecarboxylic acid (3-ICA), which was corrected to 1-indenecarboxylic acid (1-ICA) [7]. Packansky and Lyerla [7] showed direct spectroscopic evidence of 1-indenoketene intermediate formation at 77 K using infrared (IR) spectroscopy. Similar results were also reported by Hacker and Turro [8]. Packansky and Lyerla [7] also investigated the reaction of ketene intermediates using IR and ^{13}C nuclear magnetic resonance spectroscopy [7]. The reactivity of the ketene depends on the conditions, as shown in Scheme 7.2. Under ambient conditions, ketenes react with water trapped in the novolac resin to yield 3-indenecarboxylic acid. However, UV exposure in vacuo results in ester formation via a ketene–phenolic OH reaction.

Vollenbroek et al. [9, 10] also investigated the photochemistry of 2,1-diazonaphthoquinone (DNQ)-5-(4-cumylphenyl)-sulfonate and DNQ-4-(4-cumylphenyl)-sulfonate using photoproduct analysis. They confirmed that the photoproduct of DNQ is indenecarboxylic acid and its dissolution in aqueous base gives the formation of indenyl carboxylate dianion, which decarboxylates in several hours. They showed that films of mixtures of novolac and indenecarboxylic acid showed no difference in dissolution rate compared with that of exposed photoresist.

Many attempts to detect the intermediates in the photochemistry of DNQ by time-resolved spectroscopy have been reported. Nakamura et al. [11] detected a strong absorption intermediate at

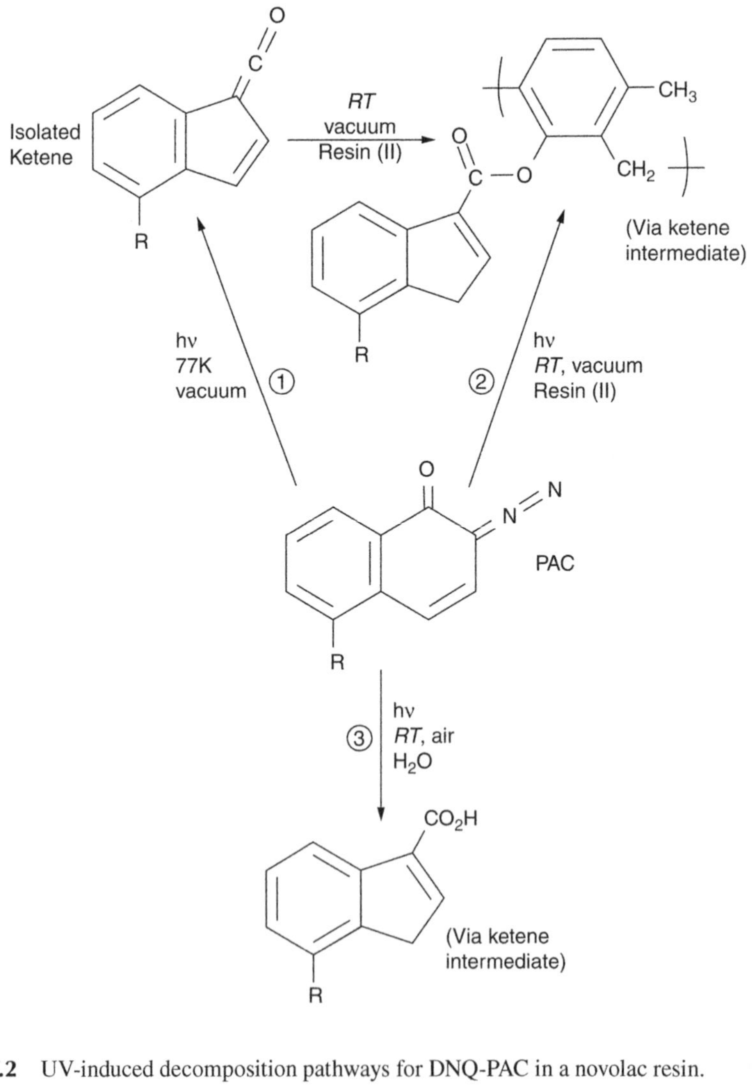

SCHEME 7.2   UV-induced decomposition pathways for DNQ-PAC in a novolac resin.

350 nm, which was assigned as a ketene intermediate formed by DNQ sulfonic acid in solution. Shibata et al. [12] observed the transient absorption at 350 nm of hydrated ketene as an intermediate of DNQ-5-sulfonic acid. Similar results were also reported by Barra et al. [13] and Andraos et al. [14]. Tanigaki and Ebbsen [15] observed the oxirene intermediate as a precursor of ketene, which was also confirmed by spectroscopic analysis in an Ar matrix at 22.3 K. It is still controversial whether ketocarbene is a reaction intermediate [9], whereas the existence of the ketene intermediate is confirmed. Because most of the attempts to detect the intermediates were performed in solution, further studies are needed to confirm the reaction intermediates in the resist film. Sheats [16] described reciprocal failure, intensity dependence on sensitivity, in DNQ-novolac resists with 364 nm exposure, which is postulated to involve the time-dependent absorbance of the intermediate ketene.

7.2.2 Improvement in Photoresist Performance

7.2.2.1 Novolac Resins

A group at Sumitomo Chemical has made a systematic study of novolac resin to improve the performance of the positive resists [17–22]. It is generally accepted that a resist with high sensitivity gives low film-thickness retention of the unexposed area after development and low heat resistance (Figure 7.2). Novolac resins have been designed with a molecular structure and a molecular weight different from existing materials, although the control of synthesis in novolac resin is considered to be difficult due to poor reproducibility. Sumitomo Chemical investigated the relation between lithographic performance and the characteristics of novolac resins, such as the isomeric structure of cresol, the position of the methylene bond, the molecular weight, and the molecular weight distribution (Figure 7.3). To clarify the lithographic performance, measurements of the dissolution rates of unexposed and exposed resists were attempted. It is not always easy to judge improvement in contrast by γ-value in exposure-characteristic curves where the remaining film thickness after development is plotted as a function of the logarithm of exposure dose. Because the difference in dissolution rate between exposed and unexposed resists is two to five orders of magnitude, it should be noted that the measurement of dissolution rate made the difference in resist performance clearer than that of the exposure characteristic curves used in their early work [17].

With increasing molecular weight, the dissolution rate of novolac resin ($\_R_n$), unexposed (R_0), and exposed ($\_R_p$) resist decreases as shown in Figure 7.4. Therefore, the contrast remains almost constant for changes in the molecular weight of the novolac resin [19].

When the para–meta ratio of novolac resin increases, the dissolution rates of novolac resin (R_n) and the resist films of unexposed (R_0) and exposed (R_p) resists decrease (Figure 7.5) [19]. However, the decrease in dissolution rate of an unexposed resist is larger than that of the exposed region for

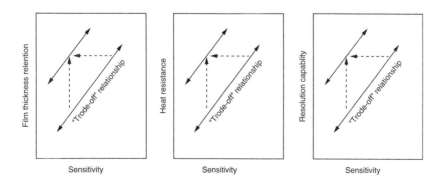

FIGURE 7.2 Tradeoff relationships for various performances of a photoresist. Dotted line indicates the improvement of the performance. (From Hanabata, M., et al., *J. Vac. Sci. Technol. B,* 7, 640, 1989.)

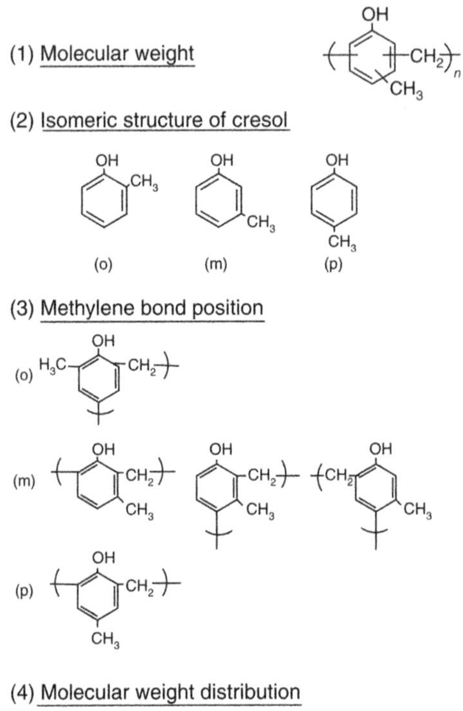

(1) Molecular weight

(2) Isomeric structure of cresol

(o)   (m)   (p)

(3) Methylene bond position

(o)

(m)

(p)

(4) Molecular weight distribution

FIGURE 7.3   Factors of novolac resins that influence resist characteristics. (From Hanabata, M., et al., *J. Vac. Sci. Technol. B*, 7, 640, 1989.)

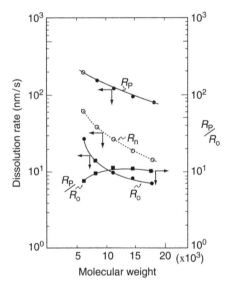

FIGURE 7.4   Effect of molecular weight of novolac resins on dissolution rates. The resists were composed of novolac resins synthesized from *m*-cresol (80%) and *p*-cresol (20%) and 3HBP-DNQ ester, where 3HBP is 2,3,4-trihydroxybenzophenone. R_n, dissolution rate of novolac resins; R_0, dissolution rate of unexposed film; Rp, dissolution rate of exposed (60 mJ/cm^2) film. (From Hanabata, M., et al., *J. Vac. Sci. Technol. B*, 7, 640, 1989.)

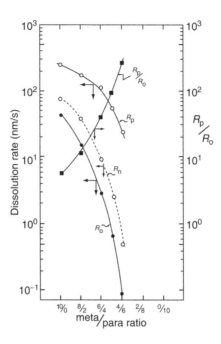

FIGURE 7.5 Effect of meta:para cresol ratio in novolac resins on dissolution rates. The molecular weight of these novolac resins is almost the same. R_n, dissolution rate of novolac resins; $R0$, dissolution rate of unexposed film; R_p, dissolution rate of exposed (60 mJ/cm²) film. (From Hanabata, M., et al., *J. Vac. Sci. Technol. B*, 7, 640, 1989.)

novolac resin with a high para–meta ratio. Therefore, the resist contrast was improved by using the novolac resin with high para–meta ratio at the expense of its sensitivity. The reason for the decrease in dissolution rate with high *p*-cresol content may be ascribed to high polymer regularity and rigidness, leading to slow diffusion of the developer.

Figure 7.6 shows the dependence of dissolution rate on the S4 value of *m*-cresol novolac resin, which represents the ratio of "unsubstituted carbon-4 in benzene ring of cresol to carbon-5," which indicates the fraction of ortho–ortho methylene bonding to high-ortho bonding [19]. With increasing S4 values, the content of type-B structure increases in novolac resin. The dissolution rate of an unexposed resist (R_0) shows a drastic decrease with increasing S4 value, as shown in Figure 7.6, whereas the dissolution rate of novolac (R_n) resin decreases slightly. It should be noted that the dissolution rate of the exposed area remains constant for various S4 values. Therefore, a high-contrast resist is obtained without sensitivity loss using a novolac resin with a high S4 value. Azo coupling of novolac resin with diazonaphthoquinone via base-catalyzed reaction during development, as shown in Scheme 7.3, can explain the difference in resist performance with different S4 values. High-ortho novolac has more vacant para positions compared with a normal novolac resin, and these vacant positions enhance the electrophilic azo-coupling reaction.

The effect of the molecular weight distribution of novolac resin on resist performance is shown in Figure 7.7 [18]. The dissolution rate of novolac resin increases with increasing molecular weight distribution (M_w/M_n). The discrimination between exposed and unexposed area is large at a certain M_w/M_n value, indicating that the optimum molecular weight distribution gives high-contrast resist.

On the basis of their systematic studies on novolac resins, Hanabata et al. [19] proposed the "stone wall" model for a positive photoresist with alkali development, as shown in Figure 7.8 [19]. In exposed parts, indenecarboxylic acid formed by exposure and low–molecular weight novolac resin dissolve first into the developer. This increases the surface contact area of high–molecular weight novolac with the developer, leading to dissolution promotion. In unexposed areas, an azo-coupling

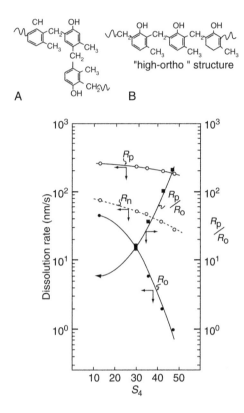

FIGURE 7.6 Effect of content of "unsubstituted carbon-4 in benzene ring of cresol," S4, in novolac resins synthesized from *m*-cresol on dissolution rates. The molecular weight of these novolac resins is almost the same. (From Hanabata, M., et al., *J. Vac. Sci. Technol. B,* 7, 640, 1989.)

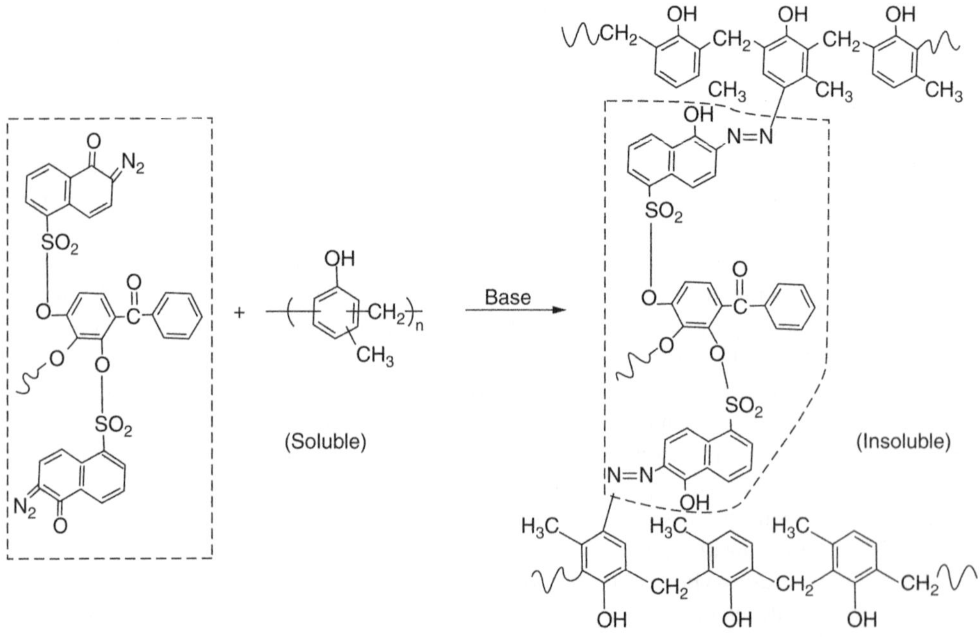

SCHEME 7.3 Dissolution-inhibition azo-coupling reaction of DNQ-PAC with novolac resin.

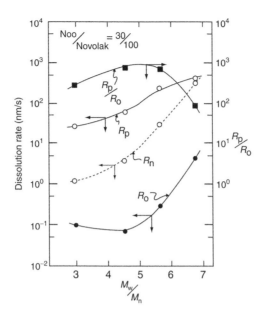

FIGURE 7.7 Effect of molecular weight distribution of novolac resins on dissolution rates. The molecular weight of these novolac resins is almost the same. (From Hanabata, M., et al., *Proc. SPIE, 920*, 349, 1988.)

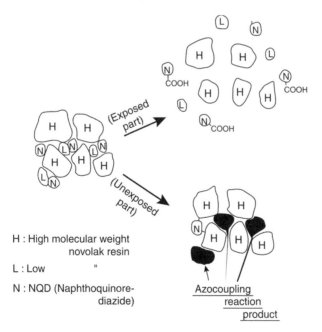

H : High molecular weight novolak resin

L : Low   "

N : NQD (Naphthoquinore-diazide)

Azocoupling reaction product

FIGURE 7.8 "Stone wall" model for development of positive photoresist. (From Hanabata, M., et al., *J. Vac. Sci. Technol. B*, 7, 640, 1989.)

reaction of low–molecular weight novolac resin with DNQ retards the dissolution of low–molecular weight resin. This stone wall model gave clues to the design of a high-performance positive photoresist in the authors' following works.

They extended the study on novolac resins by synthesizing from various alkyl (R)-substituted phenolic compounds, including phenol, cresol, ethylphenol, butylphenol, and copolymers of these, which were investigated to see the effect of their composition on the resist performance [20].

The requirements for resists were that the dissolution rate of exposed areas should be larger than 100 nm/s and the dissolution-rate ratio of exposed to unexposed area should be larger than 7. To meet the requirements, they proposed the selection principles of phenol compounds for novolac resin synthesis: (1) the average carbon number in substituent R per one phenol nucleus must be 0.5–1.5 (i.e., 0.5 < [C]/[OH] < 1.5); (2) the ratio of para-unsubstitution R with respect to OH group must be 50%.

The Sumitomo group tried to clarify the roles of individual molecular weight parts of novolac resin (low–molecular weight (150–500), middle–molecular weight (500–5000), and high–molecular weight novolac [greater than 5000]) in resist performance [21]. It was found that a novolac resin with a low content of middle–molecular weight component showed high performance, such as resolution, sensitivity, and heat resistance. Then, "tandem-type novolac resin," shown in Figure 7.9, was proposed.

The advantage of tandem-type novolac resins can be explained again by the stone wall model. The low–molecular weight novolacs and DNQ molecules are stacked between high–molecular weight novolacs. In exposed areas, dissolution of indenecarboxylic acid and low–molecular weight novolacs promotes dissolution of high–molecular weight novolacs due to an increase in surface contact with the developer. In unexposed areas, the azo-coupling reaction of DNQ compounds with low–molecular weight novolacs retards the dissolution.

Therefore, phenolic compounds can be used instead of low–molecular weight novolacs if the compounds have moderate hydrophobicity and azo-coupling capability with DNQ.

Hanabata et al. [22] showed high-performance characteristics of the resists composed of phenolic compounds, high–molecular weight novolacs, and a DNQ compound with high heat resistance.

Studies of the effects of novolac molecular structures on resist performance have also been reported by several groups. Kajita et al. [23] of JSR investigated the effect of novolac structure on dissolution inhibition by DNQ. They found that the use of meta-methyl-substituted phenols, especially 3,5-dimethylphenol, was effective to obtain a higher ratio of intra-/intermolecular hydrogen bonds, and the ratio could be controlled by selecting the phenolic monomer composition. Interaction between ortho–ortho units of novolac resin and the DNQ moiety of the photo-active compound (PAC) plays an important role in dissolution inhibition in alkali development. It was found that naphthalene sulfonic acid esters without the diazoquinone moiety also showed dissolution inhibition. This dissolution-inhibition effect also depended upon the structure of the novolac resin,

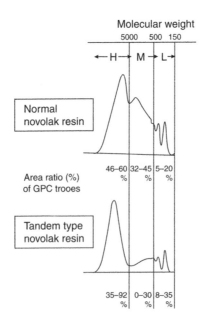

FIGURE 7.9 Gel-permeation chromatography traces of a normal novolac and a "tandem type" novolac resin. (From Hanabata, M., et al., *Proc. SPIE,* 1466, 132, 1991.)

indicating the importance of interaction between the naphthalene moiety and the novolac resin. They proposed a host–guest complex composed of a DNQ moiety and a cavity or channel formed with aggregation of several ortho–ortho linked units, as shown in Figure 7.10, where the complex is formed via electrostatic interaction.

Honda et al. [24] proposed the dissolution-inhibition mechanism of novolac–PAC interaction called *octopus-pot* and the relationship between the novolac microstructure and the DNQ inhibitor. The mechanism involves two steps. The first step is a static molecular interaction between novolac and DNQ via macromolecular complex formation during spin-coating. A secondary dynamic effect during the development process enhances the dissolution inhibition via the formation of cation complexes having lower solubility.

The addition of DNQ to novolac caused the OH band to shift to a higher frequency (blue shift) in the IR spectra, which suggests a disruption of the novolac hydrogen bonding by the inhibitor and concomitant hydrogen bonding with the inhibitor. The magnitude of the blue shift increases monotonically with the dissolution-inhibition capability, as shown in Figure 7.11. An increase in *p*-cresol content in *m*/*p*-cresol novolacs leads to an increase in ortho–ortho bonding because ortho positions only are available for reaction on the *p*-cresol nucleus. The magnitude of the blue shift for the *p*-cresol trimer was found to be dependent on the DNQ concentration and goes through a maximum at a mole ratio (*p*-cresol:DNQ) of 18. This suggests that a complex involving six units

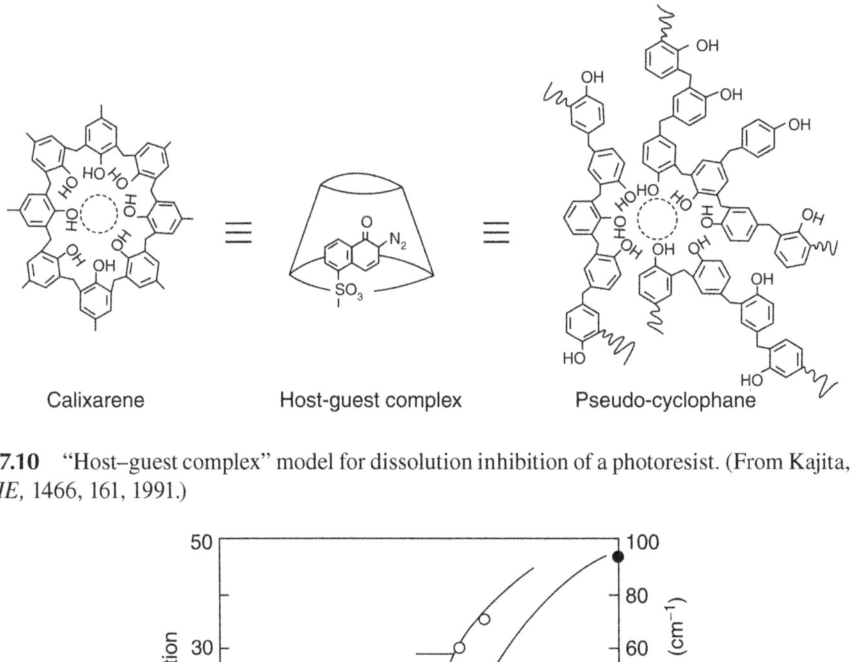

Calixarene   **Host-guest complex**   **Pseudo-cyclophane**

FIGURE 7.10   "Host–guest complex" model for dissolution inhibition of a photoresist. (From Kajita, T., et al., *Proc. SPIE,* 1466, 161, 1991.)

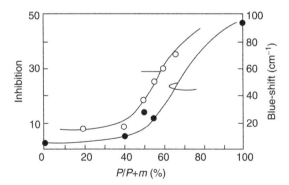

FIGURE 7.11   Correlation of dissolution inhibition and blue shift with *p*-cresol content in *m*/*p*-cresol novolacs. The x-axis indicates the *p*-cresol content in the feed stock for novolac synthesis. The inhibition was defined as the ratio of dissolution rate of the synthesized novolac to that of unexposed resist that was formulated with this novolac and 4HBP-DNQ ester, where 4HBP is 2,3,4,4-tetrahydroxybenzophenone (average esterification level = 2.75). The ester content in solid film is 20 wt.%. (From Honda, K., et al., *Proc. SPIE,* 1262, 493, 1990.)

of *p*-cresol trimer and one molecule of DNQ is formed, probably through intermolecular hydrogen bonding. The model of the macromolecular complex, the octopus-pot model, is schematically depicted in Figure 7.12.

To improve the dissolution-inhibition effect, Honda et al. [25] synthesized novolac resin with a *p*-cresol trimer sequence of novolac incorporated into a polymeric chain; they copolymerized *m*-cresol with a reactive precursor, which was prepared by attaching two units of *m*-cresol to the terminal ortho position of the *p*-cresol trimer. This kind of novolac can exhibit a higher degree of dissolution inhibition at a lower content of DNQ-PAC.

Secondary dynamic effects of dissolution chemistry during development were investigated for a series of novolac resins with different structures with various quaternary ammonium hydroxides [26, 27]. Phenolic resins used were (1) conventional *m*/*p* cresol (CON), (2) high-ortho, ortho *m*/*p* cresol novolac oligomer (hybrid pentamer: HP), (3) high-ortho, ortho *m*/*p* cresol made from polymerization of HP with *m*-cresol (HON), (4) novolac from xylenol feed stock (PAN), and (5) polyvinylphenol (PVP). UV spectral change as a function of dissolution time is shown in Figure 7.13.

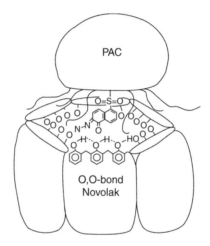

FIGURE 7.12   "Octopus-pot" model of macromolecular complex of ortho-ortho bonded novolac microstructure with DNQ-PAC. (From Honda, K., et al., *Proc. SPIE,* 1262, 493, 1990.)

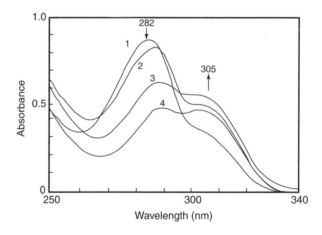

FIGURE 7.13   UV absorption spectral change of hybrid pentamer (HP; see Figure 7.15) film with development time with 0.262 N tetramethylammonium hydroxide solution. Development time: (1) 50 s; (2) 100 s; (3) 200 s; (4) 500 s. (From Honda, K., et al., *Proc. SPIE,* 1672, 305, 1992; Honda, K., et al., *Proc. SPIE,* 1925,197, 1993.)

The absorbance at 282 nm decreases with decreasing novolac film thickness. A new absorption band appeared at 305 nm, which was assigned to the cation complex between novolac and tetra-methylammonium cation (Figure 7.14). The relation between cation complex formation rate and dissolution rate of HP is shown in Figure 7.15 for three types of quaternary ammonium hydroxide. The dissolution rate can be monitored by absorption at 282 nm of the aromatic absorption band of novolac resin. The rate of the complex formation is approximately the same as the rate of dissolution when using a water rinse. This evidence supports cation diffusion as the rate-determining step for dissolution. The CON and HON novolacs relatively quickly build a high concentration of quaternary ammonium complex, while PAN shows slower formation of the complex despite lower molecu-lar weight than in CON novolac, as shown in Figure 7.16. Because the cation diffusion typically

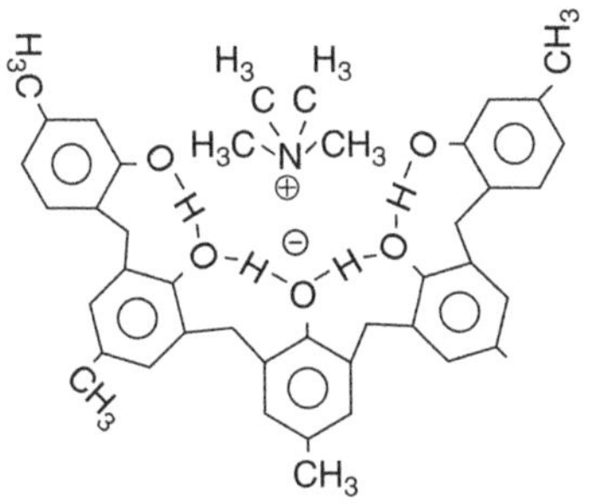

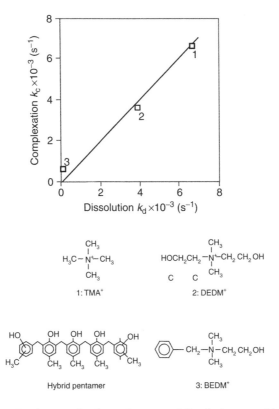

FIGURE 7.14 Schematic structure of tetramethylammonium ion complex of hybrid pentamer (HP; see Figure 7.15). (From Honda, K., et al., *Proc. SPIE,* 1925, 197, 1993.)

FIGURE 7.15 Correlation of cation complex formation rate and dissolution rate with hybrid pentamer. (From Honda, K., et al., *Proc. SPIE,* 1925, 197, 1993.)

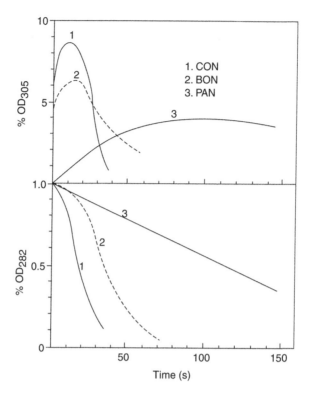

FIGURE 7.16 Spectroscopic dissolution rate monitoring (SDRM) curves of various types of novolac films in 0.262 N TMAH solution. (1) CON, a conventional *m/p*-cresol novolac; (2) HON, a high ortho-ortho *m/p*-cresol novolac made from polymerization of hybrid pentamer (HP; see Figure 7.15) with *m*-cresol; (3) PAN, a novolac made from xylenol feed stock. (From Honda, K., et al., *Proc. SPIE,* 1672, 305, 1992; Honda, K., et al., *Proc. SPIE,* 1925, 197, 1993.)

controls dissolution, polymer flexibility and microstructure exert a strong influence on cation diffusion rate. The effect of developer cations has also been studied by other workers [27, 28].

The studies described here suggest that the dissolution behavior of novolac resins is quite important in the development of positive photoresists. To understand the dissolution mechanism, theoretical and experimental studies of dissolution behavior of phenolic resins have been reported [28–34].

7.2.2.2 DNQ

The effects of chemical structures of DNQ inhibitors on resist performance are addressed next. The effects should be investigated in correlation with novolac structures. DNQ compounds are usually synthesized by the esterification reaction of phenol compounds with DNQ sulfonyl chloride, and many DNQ-PACs are reported. Kishimura and coworkers [35] have reported on a dissolution-inhibition effect of DNQ-PACs derived from polyhydroxybenzophenones and several *m*-cresol novolac resins. The number of DNQ moieties in the resist film and the average esterification value of DNQ-PACs were the same for each type of ballast molecule. The distance between DNQ moieties in the DNQ-PAC and the degree of dispersion of DNQ moieties in the resist film are important to enhance the dissolution-inhibition effect.

Tan et al. [36] of Fuji Photo Film described DNQ-sulfonyl esters of novel ballast compounds. The PAC structure was designed to minimize the background absorption at 365 nm and to enable a resist formulation to be optimized with low PAC loading. They proposed 3,3,3′,3′-tetramethyl-1,1′-spiroindane-5,6,7,5′,6′,7′-hexol (Figure 7.17) as a polyhydroxy ballast compound.

Nemoto et al. [37] investigated the effect of DNQ proximity and the hydrophobicity of a variety of trifunctional PACs on the dissolution characteristics of positive photoresists. They found that the

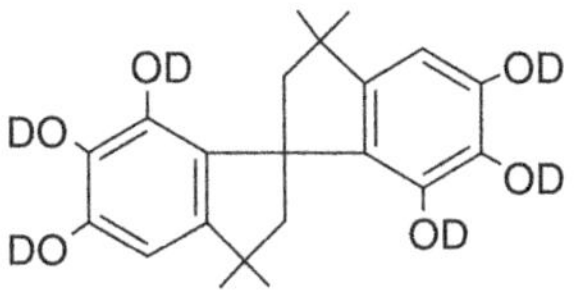

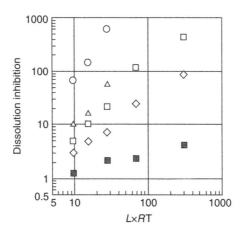

FIGURE 7.17 OD DNQ-PAC made from 3,3,3′3 ′-tetramethyl-1,1′-spiroindane-5,6,7,5′,6,7′-hexol polyhydroxy ballast compound, where D is 2,1-diazonaphthoquinone-5- sulfonyl.

new index $L \times RT$ is linearly related to dissolution inhibition for various novolac resins, as shown in Figure 7.18, where L is the average distance between DNQ groups in PAC molecules estimated by molecular orbital (MO) calculations, and RT is the retention time in reverse high-performance liquid chromatography (HPLC) measurements. RT can be a measure of hydrophobicity of the PAC.

Similar results were reported by Uenishi et al. [38] of Fuji Photo Film. They used model backbones without hydroxy groups and fully esterified DNQ-PACs. The inhibition capability was found to be correlated with the retention time on reverse-phase HPLC, a measure of hydrophobicity, and with the distance of the DNQ moiety of DNQ-PACs.

The Fuji Photo Film group extended their investigation of PAC structure effects on dissolution behavior to include the effects of the number of unesterified OH groups of DNQ-PAC on image performance [39]. PACs generally lose their inhibition capability with increasing number of unesterified OH groups compared with fully esterified PACs, whereas certain particular PACs still remained strongly inhibiting even when an OH was left unesterified. Such PACs lost inhibition when one more OH was left unesterified (Figure 7.19) and gave a large dissolution discrimination upon exposure, which resulted in a high-resolution resist. DNQ-PACs with steric crowding around OH groups seem to be the structural requirement in addition to hydrophobicity and remote DNQ configuration. The PAC provides good solubility in resist solvent. Hanawa et al. also proposed PACs obtained by selective esterification of OH groups of ballast molecules with sterically hindered OH groups (Figure 7.20) [40, 41]. These PACs give higher sensitivity, γ-value, and resolution than those of fully esterified PACs. Hindered OH groups are effective for scum-free development.

Workers of IBM and Hoechst-Celanese reported a DNQ-PAC derived from phenolphthalein that showed advantageous scumming behavior; even at high defocus, the patterns did not web or scum as

FIGURE 7.18 Influence of the resin structure on the relation between the dissolution inhibition and $L \times RT$ (see text). (O): 73MX35 novolac resin synthesized from 7/3 molar ratio of *m*-cresol to 3,5-xylenol; (Δ): 82MX35 novolac resin synthesized from 8/2 molar ratio of *m*-cresol to 3,5-xylenol; (□): 91MX35 novolac resin synthesized from 9/1 molar ratio of *m*-cresol to 3,5-xylenol; (O): 100M novolac resin synthesized from *m*-cresol; (■): polyhydroxystyrene. (From Kishimura, S., et al., *Polym. Eng. Sci.*, 32, 1550, 1992.)

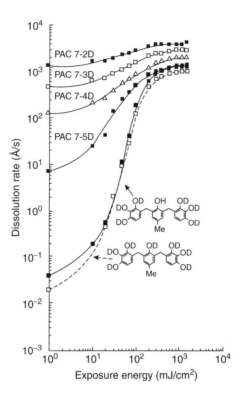

FIGURE 7.19 Dissolution rate change with exposure dose for DNQ-PAC with different esterification degree. DNQ-PAC with hindered OH shows a high performance of photoresist.

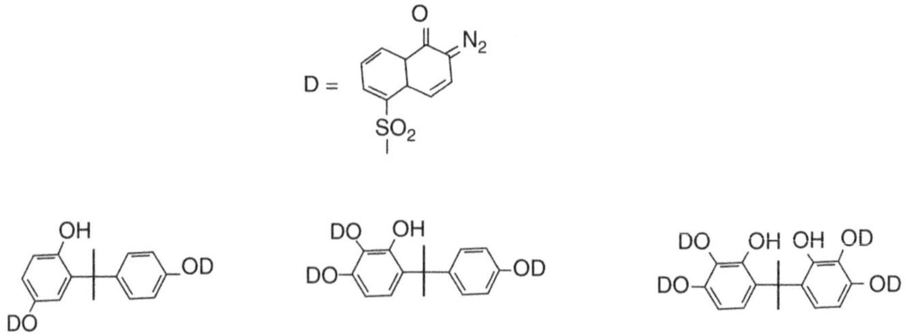

FIGURE 7.20 DNQ-PACs with steric hindrance OH groups obtained by selective esterification of OH groups with DNQ-sulfonylchlorides. (From Uenishi, K., et al., *Proc. SPIE,* 1672, 262, 1992.)

is usually observed [42]. This may be due to a base-catalyzed hydrolysis of a lactone ring that leads to a more soluble photoproduct, as shown in Scheme 7.4.

The polyphotolysis model [43] for DNQ resists stimulated study on PAC molecules. The model suggests that more DNQ groups in a single PAC molecule improve the resist contrast. Some results support the model, but some do not. Hanawa et al. [40] reported that two DNQ in a PAC showed a contrast similar to that of three DNQ in a PAC. Uenishi et al. [39] also investigated the effect of the number of DNQ in a single PAC and the number of OH groups unesterified. Although the number of DNQ in a PAC is important to improve the contrast in some cases, it is not simple, because the distance of DNQ groups and the hydrophobicity of the PAC also affect the dissolution-inhibition capability, as described.

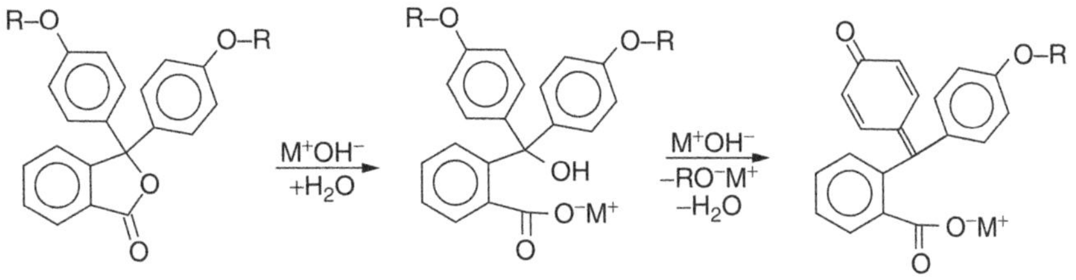

$$\text{R–O} \cdots \xrightarrow[+\mathrm{H_2O}]{\mathrm{M^+OH^-}} \cdots \xrightarrow[\substack{-\mathrm{RO^-M^+} \\ -\mathrm{H_2O}}]{\mathrm{M^+OH^-}} \cdots$$

SCHEME 7.4 Lactone-ring opening during alkali development.

7.2.3 PERSPECTIVE OF DNQ RESISTS

As described, enormous data on novolac resins and DNQ compounds has been accumulated. Several models, such as the stone wall model by the Sumitomo group, the host–guest model by JSR, and the octopus-pot model by OCG, were proposed to explain the dissolution behavior of high-performance photoresists. There are some agreements and contradictions between them. The naphthalene sulfonyl group plays a major role in novolac dissolution inhibition for both reports by Honda et al. (OCG) [24] and Kajita et al. (JSR) [23]. It is generally accepted that high-ortho novolac shows a high inhibition effect by DNQ-PAC. For base-induced reactions, Hanabata et al. (Sumitomo Chemical) [19] reported azo coupling, whereas Honda et al. [24] reported that azo coupling cannot explain the high dissolution-inhibition effect for high-ortho novolac. However, the models proposed above are instructive to understand resist performance and give clues for the development of high-performance resists.

Table 7.1 shows the summary of the Sumitomo group's work on novolac resins. Although these results are impressive, they were obtained under certain conditions. For example, the S4 effect was investigated for novolacs obtained from m-cresol. Therefore, the effects of m/p-cresol, molecular weight, and molecular weight distribution were not given. The effects of combinations of various

TABLE 7.1

Effect of Five Factors of Novolac Resins on Dissolution Rates and Resist Performance

Factors	Dissolution Rate		R_p/R_0 Ratio	Sensitivity
	Unexposed (R_p)	Exposed (R_0)		
Molecular weight, MW: ↑	↓	↓		(0)
	(0)	(x)		↓
				(x)
Isomeric structure of cresol para: ↑	↓	↓	(0)	↓
	(0)	(x)	↑	(x)
Methylene bond position, S_4: ↑	↓		(0)	
	(0)		↑	
Molecular weight distribution, M_w/M_n: ↑	(x)	(0)	(0)	(0)
	↓↑	↓	↑↓	↑
	(0)		(x)	
DNQ/novolac ratio: ↑	↓	(0)	(0)	(0)
	(0)	↑↓	↑	↑↓
		(x)		(x)

(0), Improvement; (x), deterioration; (↑), increase; (↓), decrease; (→), no change.

novolac resins and DNQ-PAC on resist performance were not fully understood. Further study of these effects may improve the resist performance.

7.3 CHEMICAL-AMPLIFICATION RESIST SYSTEMS

Deep-UV lithography is one of several competing candidates for future lithography to obtain resolution below 0.30 μm. DNQ-based positive photoresists are not suitable for deep-UV lithography due to absorption and sensitivity. Absorption of both novolac resins and DNQ-PACs is high and does not bleach at around 250 nm, resulting in resist profiles with severely sloping sidewalls. Much attention has been focused on chemical-amplification resist systems, especially for deep-UV lithography. These resists are advantageous because of high sensitivity, which is important because the light intensity of deep-UV exposure tools is lower than that of conventional i-line steppers.

In chemical-amplified resist systems, a single photoevent initiates a cascade of subsequent chemical reactions. The resists are generally composed of an acid generator that produces acid upon exposure to radiation and acid-labile compounds or polymers that change the solubility in the developer by acid-catalyzed reactions. As shown in Figure 7.21, the photogenerated acid catalyzes the chemical reactions that change the solubility in a developer. The change from insoluble to soluble is shown. The quantum yield for the acid-catalyzed reaction is the product of the quantum efficiency of acid generation multiplied by the catalytic chain length. In chemically amplified resists (CARs), acid generation is the only photochemical event. Therefore, it is possible to design the acid-labile base polymer with high transmittance in the deep-UV region. Because the small amount of acid can induce many chemical events, it is expected that yields of the acid-catalyzed reaction along the film thickness can be alleviated, even for concentration gradients of photogenerated acid along the film thickness. Another important point of CAR systems is the drastic polarity change resulting from an acid-catalyzed reaction, which can avoid swelling during the development and give high contrast.

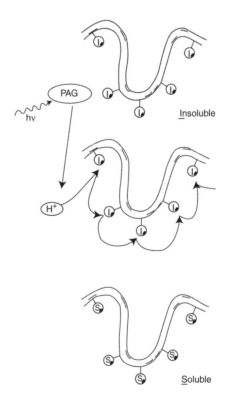

FIGURE 7.21   Schematic representation of a positive chemically amplified soluble resist.

Many chemical-amplification resist systems have been proposed since the report by the IBM group [44, 45]. Most of these are based on acid-catalyzed reactions [46–51], although some with base-catalyzed reactions have been reported [52, 53]. In the following sections, acid generators and resists are classified by acid-catalyzed reaction and discussed.

7.3.1 Acid Generators

7.3.1.1 Onium Salts

Most well-known acid generators are onium salts, such as iodonium and sulfonium salts, which were invented by Crivello [54]. The photochemistry of diaryliodonium and triarylsulfonium salts has been studied in detail by Dektar and Hacker [55–59]. The primary products formed upon irradiation of diphenyliodonium salts are iodobenzene, iodobiphenyl, acetanilide, benzene, and acid, as shown in Scheme 7.5 [55]. The reaction mechanism for product formation from diaryliodonium salts is shown in Scheme 7.6, where the bar indicates reaction in a cage. The mechanism for direct photolysis in solution can be described by three types of processes: in-cage reactions, cage-escape reactions, and termination reactions. The photolysis products are formed by heterolysis of diaryliodonium salts to a phenyl cation and iodobenzene, and also by homolysis to phenyl radical and iodobenzene radical cations. Direct photolysis favors product formation by a heterolytic cleavage pathway. In-cage recombination produces iodobiphenyl, whereas the cage-escaped reaction produces iodobenzene, acetanilide, and benzene.

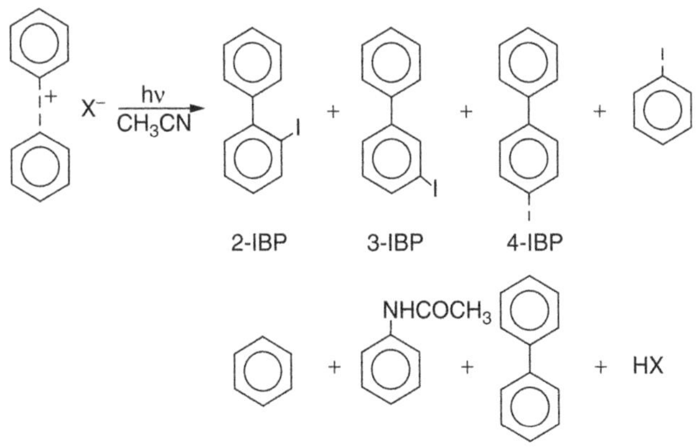

SCHEME 7.5 Photoproducts from diphenyliodonium salts.

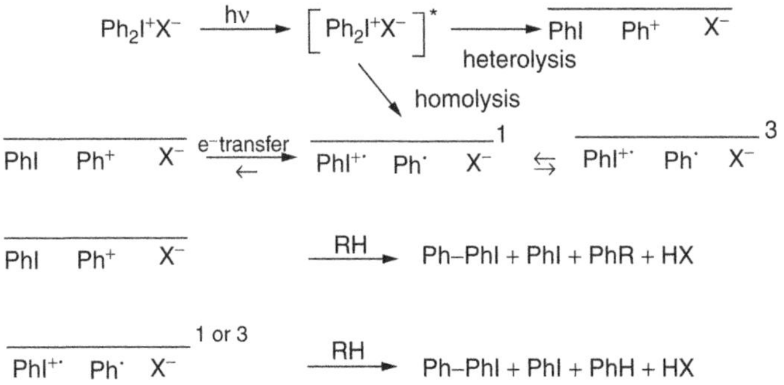

SCHEME 7.6 Mechanism of product formation from direct photolysis of diphenyliodonium salts. The bars indicate in-cage reactions.

Direct photolysis of triphenyl sulfonium salts produces new rearrangement products: phenyl-thiobiphenyl, along with diphenylsulfide, as shown in Scheme 7.7 [56]. The reaction mechanism is shown in Scheme 7.8. The heterolytic cleavage gives a phenyl cation and diphenyl sulfide, whereas homolytic cleavage gives the singlet phenyl radical and diphenylsulfinyl radical cation pair. These pairs of intermediates then produce the observed photoproducts by an in-cage recombination mechanism, leading to phenylthiobiphenyl. Diphenylsulfide is formed by direct photolysis either in cage or in a cage-escaped reaction. Other products are formed by cage-escaped or termination reactions.

The difference in photolysis of onium salts in the solid state and in solution has been studied from the viewpoint of cage and cage-escape reactions [58]. Photolysis of triphenylsulfonium salts in the solid state shows remarkable counter-ion dependence, and a cage:escape ratio as high as 5:1 is observed, whereas the ratio in solution is about 1:1. In the solid state, the cage-escape reaction with solvent and termination processes involving solvent cannot occur. Because the environment is rigid, in-cage recombination processes are favored. Thus, for the nonnucleophilic MF (PF$_6$, AsF$_6$, SbF$_6$, etc.) and triflate anions (CF$_3$SO$_3$), the in-cage recombination to give phenylthiobiphenyls predominates.

Photolysis of onium salts in a polymer matrix, poly(tert-butoxycarbonyloxystyrene) (t-BOC-PHS), was studied [59]. Environments with limited diffusion favor the recombination reaction to yield in-cage products; t-BOC-PHS films are such an environment, but there are fewer in-cage products than expected. This can be explained if sensitization of an onium salt by excited t-BOC-PHS occurs. McKean et al. [60, 61] measured the acid-generation efficiency in several polymer matrices using the dye-titration method. The quantum yield from triphenylsulfonium salts in poly(4-butoxycarbonyloxystyrene) (Pt-BOC-PHS) is lower than that in solution. The acid-generation efficiency from triphenylsulfonium salts in polystyrene, poly(4-vinylanisole), and poly(methylmethacrylate) was measured. They pointed out that the compatibility of sulfonium salts and the polymer matrix with respect to polarity affects the acid-generation efficiency.

SH = CH$_3$CN, Z = –NHCOCH$_3$
SH = CH$_3$OH, Z = –OCH$_3$
SH = C$_2$H$_5$OH, Z = –OC$_2$H$_5$

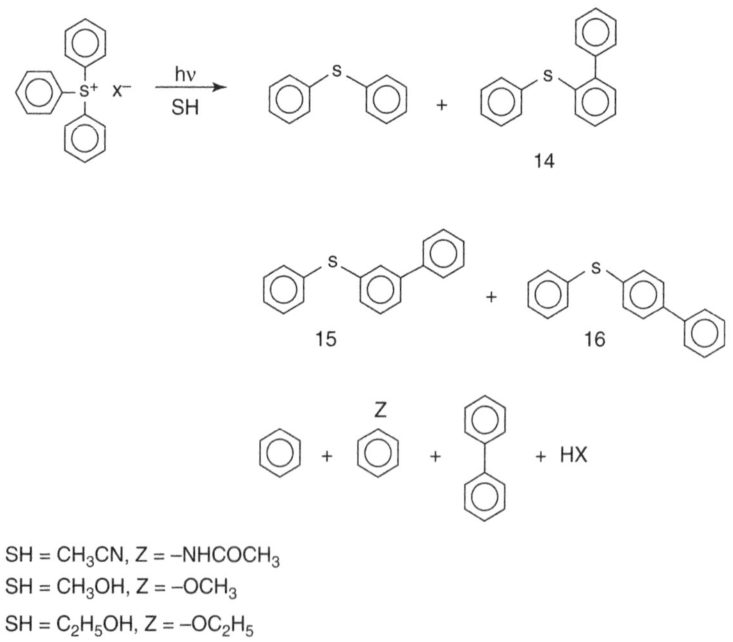

SCHEME 7.7   Photoproducts from direct irradiation of triphenylsulfonium salts.

$$Ph_3S^+ \; X^- \qquad \xrightarrow{\; h\nu \;} \qquad [Ph_3S^+ \; X^-]^* \qquad \underline{\hspace{4cm}} \qquad (1)$$

$$[Ph_3S^+ \; X^-]^* \qquad \xrightarrow{\hspace{2cm}} \qquad Ph_2S \qquad Ph^+ \qquad X^- \qquad (2)$$

$$\underline{Ph_2S \qquad Ph^+ \qquad X^-} \qquad \xrightarrow{\hspace{2cm}} \qquad Ph_2S^{+\cdot} \qquad Ph^\cdot \qquad X^- \qquad (3)$$

$$\underline{Ph_2S \qquad Ph^+ \qquad X^-} \qquad \xrightarrow{\hspace{2cm}} \qquad 14 \; + \; 15 \; + \; 16 \; + \; H^+ \qquad (4)$$

$$\underline{Ph_2S^{+\cdot} \qquad Ph^\cdot \qquad X^-} \qquad \xrightarrow{\hspace{2cm}} \qquad 14 \; + \; 16 \; + \; H^+ \qquad (5)$$

$$\underline{Ph_2S \qquad Ph^+ \qquad X^-} \qquad \xrightarrow{\hspace{2cm}} \qquad Ph_2S \; + \; Ph^+ \; + \; X^- \qquad (6)$$

$$\underline{Ph_2S^{+\cdot} \qquad Ph^\cdot \qquad X^-} \qquad \xrightarrow{\hspace{2cm}} \qquad Ph_2S^{+\cdot} \; + \; Ph^\cdot \; + \; X^- \qquad (7)$$

$$Ph^+ \; + \; RH \qquad \xrightarrow{\hspace{2cm}} \qquad Ph_R \; + \; H^+ \qquad (8)$$

$$Ph_2S^{+\cdot} \; + \; RH \qquad \xrightarrow{\hspace{2cm}} \qquad Ph_2S^+ \; -H \; + \; H^\cdot \qquad (9)$$

$$Ph_2S^+ \; -H \qquad \xrightarrow{\hspace{2cm}} \qquad Ph_2S \; + \; H^+ \qquad (10)$$

$$Ph^\cdot \; + \; RH \qquad \xrightarrow{\hspace{2cm}} \qquad PhH \; + \; R^\cdot \qquad (11)$$

$$Ph^\cdot \; + \; R^\cdot \qquad \xrightarrow{\hspace{2cm}} \qquad Ph - R \qquad (12)$$

$$Ph^\cdot \; + \; Ph^\cdot \qquad \xrightarrow{\hspace{2cm}} \qquad Ph - Ph \qquad (13)$$

$$R^\cdot \; + \; R^\cdot \qquad \xrightarrow{\hspace{2cm}} \qquad R - R \qquad (14)$$

SCHEME 7.8 Mechanism of direct photolysis of triphenylsulfonium salts. The bars indicate in-cage reactions.

7.3.1.2 Halogen Compounds

Halogen compounds such as trichloromethyl-s-triazene have been known as free-radical initiators for photopolymerization [46, 62]. These halogen compounds were also used as acid generators for an acid-catalyzed solubilization composition [63]. Homolytic cleavage of the carbon–halogen bond produces a halogen-atom radical followed by hydrogen abstraction, resulting in the formation of hydrogen-halide acid, as shown in Scheme 7.9 [64].

Calbrese et al. [65] found that the halogenated acid generator 1,3,5-tris(2,3-dibromopropyl)-1,3,5-triazine-(1,3,5)trione could be effectively sensitized to 365 and 436 nm chemistry of photoresist materials using electron-rich sensitizers. They proposed a mechanism for sensitization involving

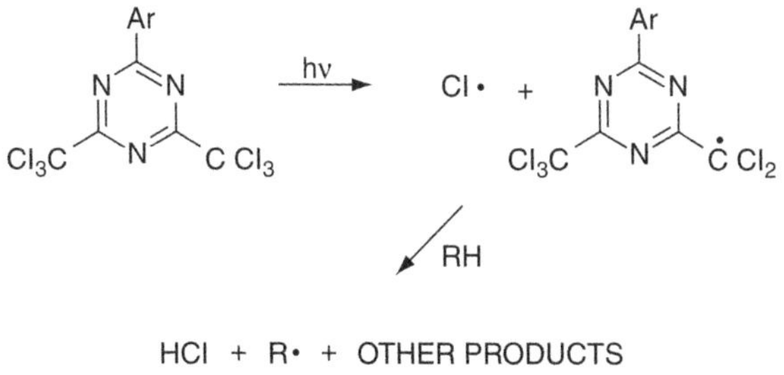

HCl + R• + OTHER PRODUCTS

SCHEME 7.9 Mechanism of acid formation from trichloromethyl triazene.

(1)   SH $\xrightarrow{\text{h}\nu}$ SH*

(2)   SH* + RX $\longrightarrow$ SH$^{+\cdot}$ + RX$^{-\cdot}$

(3) $\begin{cases} \text{RX}^{-\cdot} \longrightarrow \text{R}^{\cdot} + \text{X}^{-} \\ \text{SH}^{+\cdot} \longrightarrow \text{S}^{\cdot} + \text{H}^{+} \end{cases}$

SCHEME 7.10   Acid formation mechanism from halogen compounds via electron transfer reaction.

electron transfer from excited sensitizers to photoacid generators in these systems (Scheme 7.10). The energetics of electron transfer is described by

$$\Delta G = E\left(\frac{\text{SH}}{\text{SH}^{+\cdot}}\right) - E\left(\frac{\text{RX}^{-\cdot}}{\text{RX}}\right) - E_{00}\left(\text{SH} - \text{SH}^{*}\right) - C \tag{7.1}$$

where
  ΔG is the free enthalpy
  $E(\text{SH/SH}^{+\cdot})$ is the energy required to oxidize the sensitizer
  $E(\text{RX}^{-}/\text{RX})$ is the energy required to reduce the acid generator
  $E_{00}(\text{SH–SH*})$ is the electronic energy difference between the ground-state and the excited-state
        sensitizer
  C is the coulomb interaction of the ion pair produced

Electrochemical redox potentials and spectroscopic data support this mechanism [65]. Based on similar data for *p*-cresol as a model for phenolic resins in these resists, light absorbed by the resin when such resists are exposed to deep UV may contribute to the sensitivity via electron transfer from the resin to brominated isocyanate to produce an acid.

7.3.1.3   *o*-Nitrobenzyl Esters

Houlihan and coworkers [66] have described acid generators based on 2-nitrobenzyl-sulfonic acid esters. As shown in Scheme 7.11, the mechanism of the photoreaction of nitrobenzyl ester involves insertion of an excited nitro-group oxygen into a benzylic carbon–hydrogen bond. Subsequent

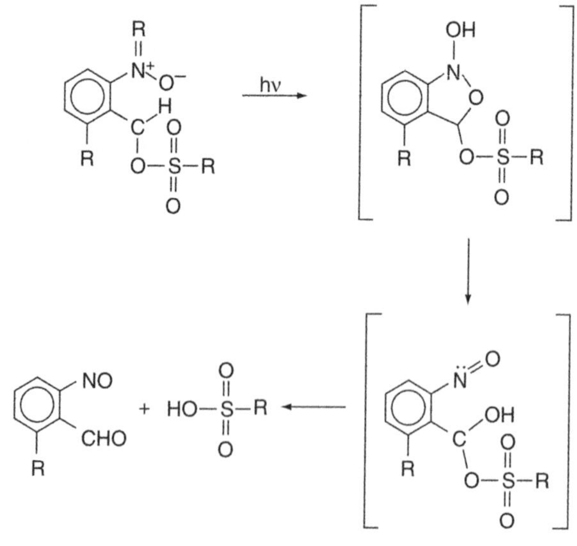

SCHEME 7.11   Mechanism of acid formation from *o*-nitrobenzyl esters.

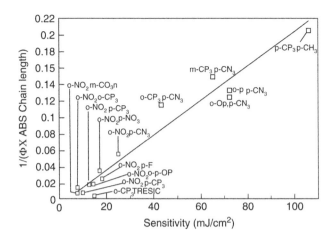

FIGURE 7.22 Plot of the lithographic sensitivity vs. 1/(FXcatalytic chain lengthXABS/mm). F is quantum yield of acid generation; ABS is the absorbance of the resist film. (From Houlihan, F. M., et al., *Chem. Mater.*, 3, 462, 1991.)

rearrangement and cleavage generates nitrosobenzaldehyde and sulfonic acid. They also made a study of thermal stability and acid-generation efficiency on varying the substituents on 2-nitrobenzylbenzenesulfonates [67, 68]. A plot of the reciprocal of quantum yield of acid generation, catalytic chain length, and absorbance per micron was made versus sensitivity (Figure 7.22). A reasonably linear plot over the whole range of esters is obtained, indicating that three basic parameters in Figure 7.22 determine the resist sensitivity [68].

7.3.1.4 *p*-Nitrobenzyl Esters

Yamaoka and coworkers [69] have described *p*-nitrobenzylsulfonic acid esters such as *p*-nitrobenzyl-9,10-diethoxyanthracene-2-sulfonate as bleachable acid precursors. Photodissociation of the *p*-nitrobenzyl ester proceeds via intramolecular electron transfer from the excited singlet state of the 9,10-diethoxyanthracene moiety to the *p*-nitrobenzyl moiety followed by heterolytic bond cleavage at the oxygen–carbon bond of the sulfonyl ester, as shown in Scheme 7.12, where dimethoxyanthracene-2-sulfonate is described. This mechanism was supported by the fact that the transient absorptions assigned to dimethoxyanthracene-2-sulfonate radical cation and nitrobenzyl radical anion are detected by laser spectroscopy [70].

7.3.1.5 Alkylsulfonates

Ueno et al. have shown that tris(alkylsulfonyloxy)benzene can act as a photoacid generator upon deep-UV [71] and electron-beam irradiation [72]. Schlegel et al. [73] found that 1,2,3-tris(methanesulfonyloxy)benzene (MeSB) gives a high quantum yield (number of acid moieties generated per photon absorbed in a resist film) when utilized in a novolac resin. The quantum yield would be more than 10 when calculated on the basis of the number of photons absorbed only by the sulfonate. This strikingly high quantum efficiency can be explained in terms of the sensitization mechanism from excited novolac resin to the sulfonates, presumably via an electron transfer reaction, as shown in Scheme 7.13 [74].

This mechanism was supported by the following model experiment. When a resist composed of bisphenol A protected with *t*-butoxycarbonyl (*t*-BOC-BA), cellulose acetate as a base polymer, and MeSB is deprotected with a photogenerated acid, absorbance at 282 nm increases due to bisphenol A formation, which can be used as a "detector" for the acid-catalyzed reaction. Because cellulose acetate shows no absorption at the exposure wavelength (248 nm), there is no possibility of sensitization from excited polymer to MeSB, leading to negligible change of

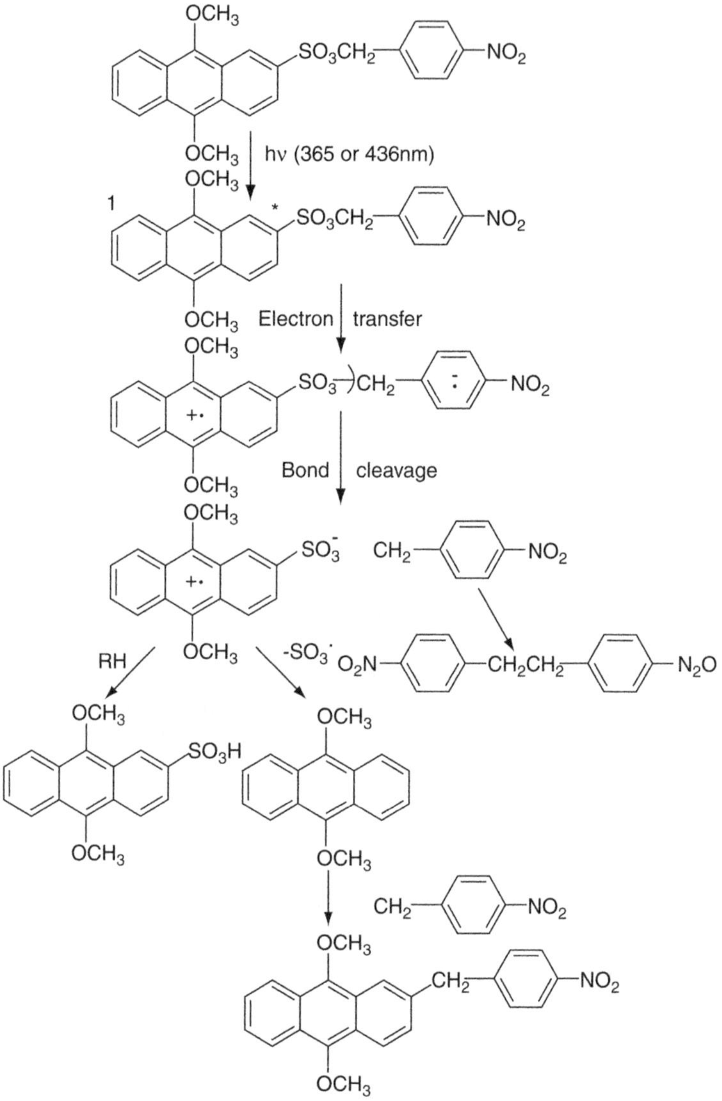

SCHEME 7.12 Mechanism of product formation from direct photolysis of *p*-nitrobenzyl-8,10-dimethoxyanthracene-2-sulfonate.

absorbance at 248 nm. On the contrary, when trimethylphenol (TMP) as a model compound of a novolac resin was added to the system, the deprotection reaction proceeded: the absorbance at 282 nm increased with exposure time as well as content of TMP, as shown in Figure 7.23. In addition, it was found that spectral sensitivity resembled the absorption spectra of novolac resin (Figure 7.24), indicating sensitization by novolac resin. The effect of chemical structure on acid-generation efficiency has been measured for various alkylsulfonates of pyrogallol backbone (Figure 7.25) and methanesulfonates of mono-, di-, and trihydroxybenzenes and their isomers [75]. The difference in the number of sulfonyl groups per benzene ring may affect the reduction potential or electron affinity of sulfonates, leading to a change in rates of the electron transfer reactions. The quantum yield for sulfonates is higher for smaller alkyl sizes. The acid-generation efficiency is higher for methanesulfonates derived from trihydroxybenzenes than for dihydroxy-benzene and monohydroxybenzene derivatives.

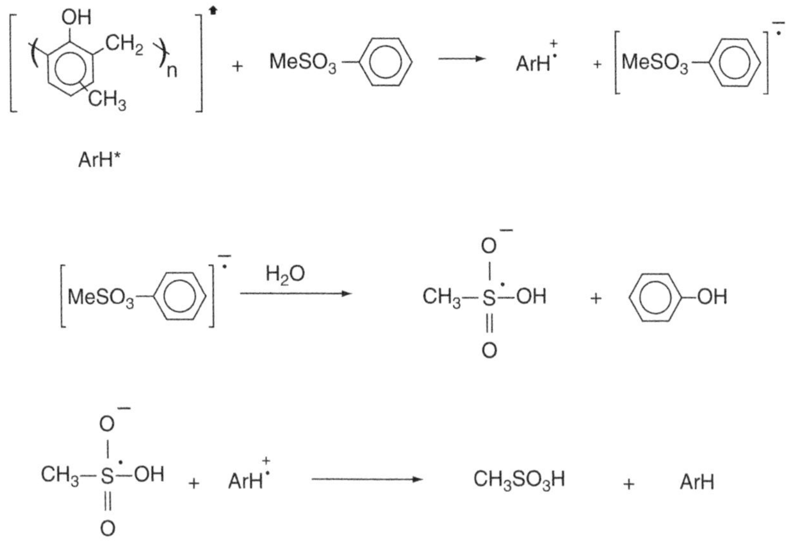

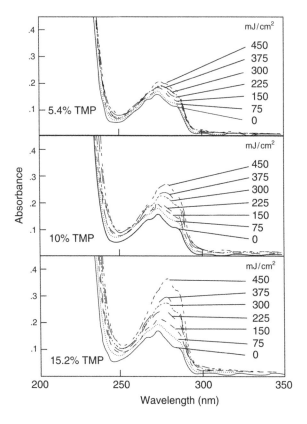

SCHEME 7.13 Mechanism of sulfonic acid generation from alkylsulfonates by electron transfer reaction.

FIGURE 7.23 Absorption spectra of resists MeSB/TMP/t-BOC-BA/CA=5/x/15/80-x (wt ratio) after a sequence coating-exposure-baking 80 °C/10 min with different exposure doses. MeSB: 1,2,3-tris (methane-sulfonyloxy)benzene; TMP: trimethyl- phenol; t-BOC-BA: bisphenol A protected with t-butoxycarbonyl; CA: cellulose. Film thickness: ~ 1.5 mm. (From Schlegel, L., et al., *Chem. Mater.*, 2, 299, 1990).

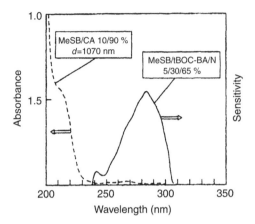

FIGURE 7.24 Solid line: spectral sensitivity curves of the resist MeSB/*t*-BOC-BA/novolac in the deep-UV region. The ordinate scale corresponds to logarithmic decrease of dose values. Dotted line: absorption spectrum of MeSB in cellulose film.

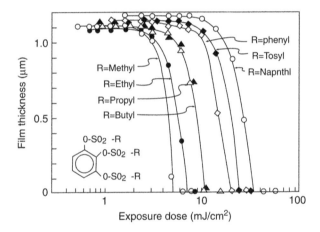

FIGURE 7.25 Exposure characteristic curves for resists with different sulfones. Novolac resin/*t*-BOC-BA/ sulfonate = 100/13.3/1.65 (mole ratio). (From Ueno, T., et al., *Polym. Eng. Sci.*, 32, 1511, 1992.)

7.3.1.6 α-Hydroxymethylbenzoin Sulfonic Acid Esters

Roehert et al. [76] described α-hydroxymethylbenzoin sulfonic acid esters as photoacid generators. Irradiation of the sulfonic acid ester to an excited triplet state leads to fragmentation via an α-cleavage (Norrish type I) into two radical intermediates, as shown in Scheme 7.14 [76, 77]. The benzoyl radical is stabilized via H-abstraction to yield a (substituted) benzylaldehyde almost quantitatively. The second radical intermediate is stabilized via cleavage of the carbon–oxygen bond, which is linked to the sulfonyl moiety ($-CH_2- OSO_2-$) and forms the respective acetophenone and sulfonic acid. The photoacid-generating efficiency of this type of sulfonate is compared with those of bis(arylsulfonyl) diazomethane (BAS-DM), 1,2,3-tris(methanesulfonyloxy)benzene (MeSB), and 2,1-diazonaphthoquinone-4-sulfonate (4-DNQ) using the tetrabromophenol blue indicator technique. The order of acid-generating efficiency is BAS-DM > α-hydroxymethylbenzoin sulfonic acid esters ~ MeSB > 4-DNQ.

7.3.1.7 α-Sulfonyloxyketones

α-Sulfonyloxyketones were reported as acid generators in chemical-amplification resist systems by Onishi et al. [78]. The photochemistry of α-sulfonyloxyketones is shown in Scheme 7.15 [77]. *p*-Toluenesulfonic acid is liberated after photoreduction.

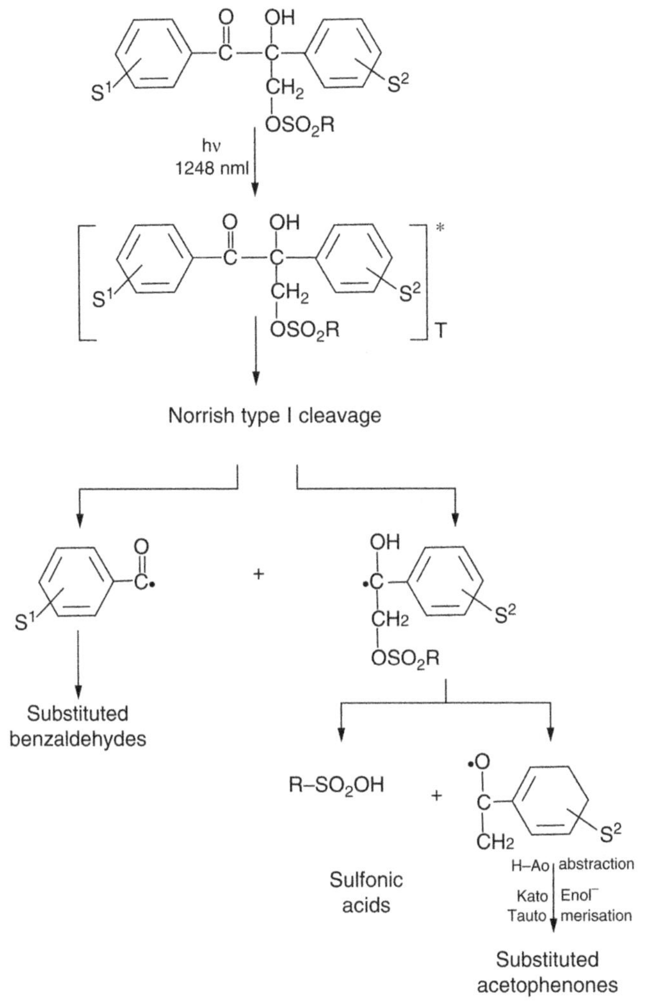

SCHEME 7.14 Photochemical reaction mechanism of α-hydroxymethylbenzoin sulfonic acid esters.

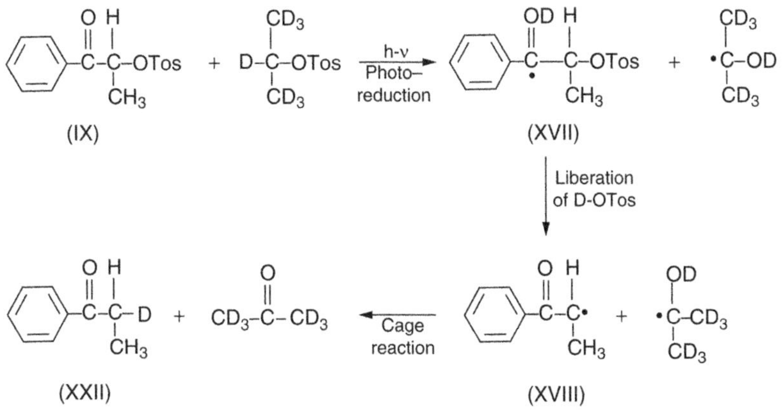

SCHEME 7.15 Photochemical reaction mechanism of α-sulfonyloxy ketones.

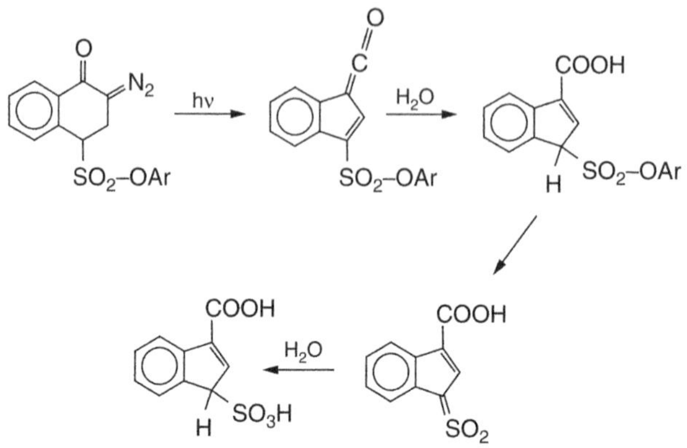

SCHEME 7.16 Sulfonic acid formation from direct photolysis of DNQ-4-sulfonates.

7.3.1.8 Diazonaphthoquinone-4-sulfonate (4-DNQ)

1,2-Diazonaphthoquinone-4-sulfonate can be used as an acid generator [79–81]. The photochemistry of 1,2-diazonaphthoquinone-4-sulfonate shown in Scheme 7.16 was proposed by Buhr et al. [79]. It is expected that the reaction follows its classical pathway from diazonaphthoquinone via the Wolff-rearranged ketene to the indenecarboxylic acid. In polar media, possibly with proton catalysis, the phenol ester moiety can be eliminated, leading to sulfene that adds water to generate the sulfonic acid. This reaction mechanism was also supported by Vollenbroek et al. [10]. 4-DNQ acid generators were used for acid-catalyzed crosslinking of image-reversal resists [79, 80] and for an acid-catalyzed deprotection reaction [81].

7.3.1.9 Iminosulfonates

Iminosulfonates, which produce sulfonic acids upon irradiation, are generally synthesized from a sulfonyl chloride and an oxime derived from ketones. Shirai and Tsunooka proposed the reaction mechanism (Scheme 7.17) [82, 83]. Upon irradiation with UV light, the cleavage of –O–N= bonds of

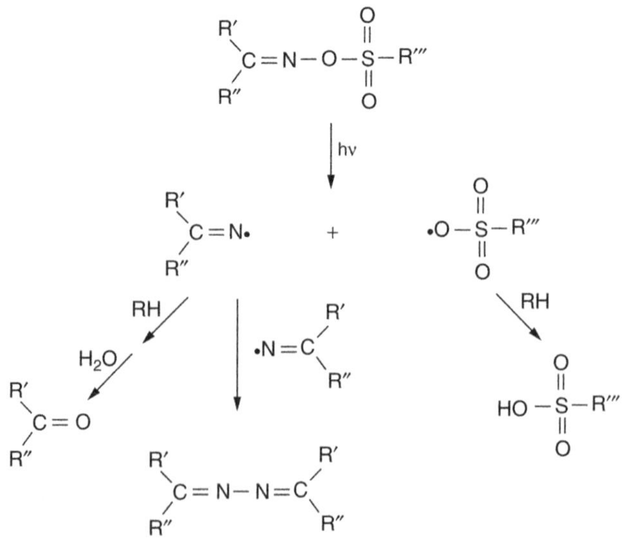

SCHEME 7.17 Photochemistry of iminosulfonates.

iminosulfonates and the subsequent abstraction of hydrogen atoms lead to the formation of sulfonic acids accompanying the formation of azines, ketones, and ammonia.

7.3.1.10 N-Hydroxyimidesulfonates

The photodecomposition of N-tosylphthalimide (PTS) [84] to give an acid is outlined in Scheme 7.18 [85, 86]. Irradiation with deep-UV light leads to homolytic cleavage of the N–O bond, giving a radical pair. The pair can either collapse to regenerate the starting PTS or undergo cage escape. Hydrogen abstraction from the polymeric matrix by the toluenesulfonyl radical gives toluenesulfonic acid. An alternative mechanism involves electron transfer from the excited-state polymer or other aromatic species to PTS to form a radical-cation–radical-anion pair. The PTS radical anion will be protonated, followed by homolytic decomposition to give toluenesulfonatyl radical. As in the mechanism for direct photolysis, the sulfonatyl radical abstracts a hydrogen atom from the medium to give a protic acid.

7.3.1.11 α,α′-Bisarylsulfonyl Diazomethanes

Pawlowski and coworkers [87] have reported the nonionic acid generators α,α′-bisarylsulfonyl diazomethanes, which generate sulfonic acids upon deep-UV irradiation. They analyzed photochemical products of α,α′-bis(4-t-butylphenylsulfonyl)diazomethane in acetonitrile/water solution using HPLC and proposed the photochemical reaction mechanism shown in Scheme 7.19 [87, 88]. The mechanism of acid generation is that the intermediate carbene formed during photolytic cleavage of nitrogen rearranges to a highly reactive sulfene, which then adds water present in the solvent mixture to give the sulfonic acid (8a). The main product is 4-t-butylphenylthiosulfonic acid 4-t-butylphenyl ester (4a), probably formed via elimination of nitrogen and carbon dioxide from the

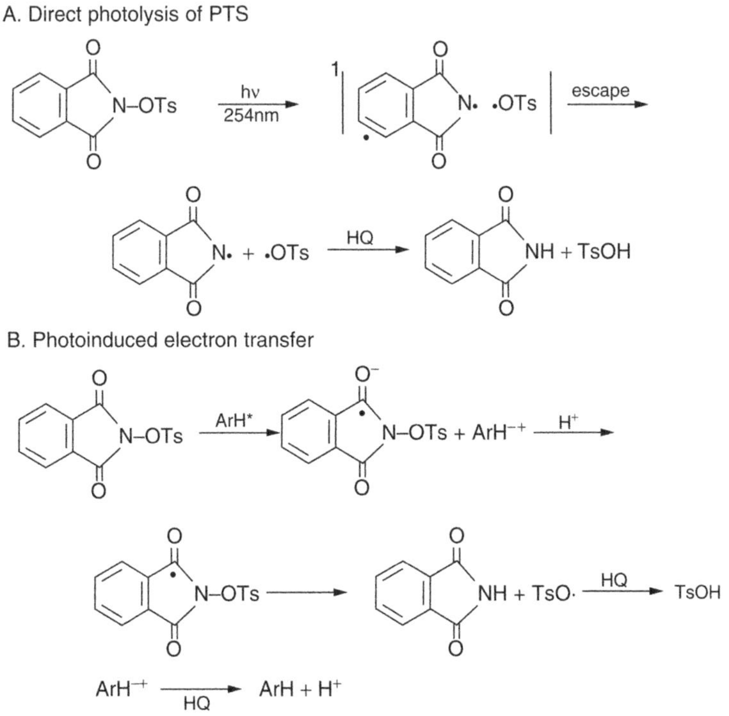

SCHEME 7.18 Mechanism of sulfonic acid formation from photolysis of N-hydroxyimidesulfonates.

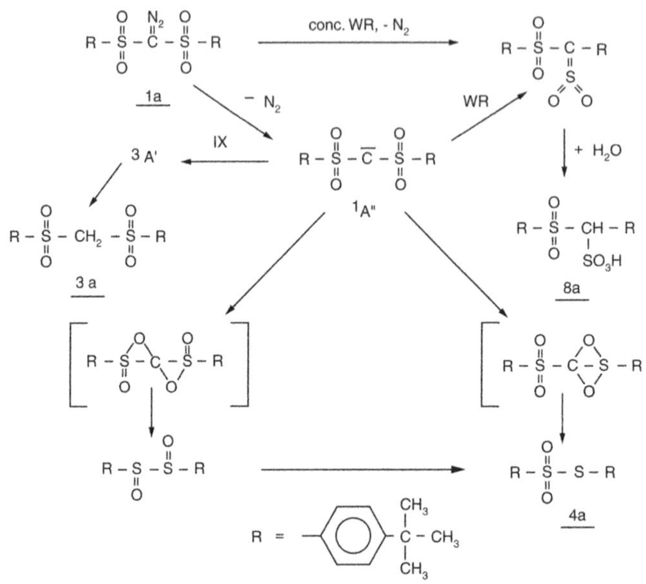

SCHEME 7.19 Photodecomposition mechanism of α,α′-bisarylsulfonyl diazomethanes.

parent compound. Although 4a does not contribute to the acid-catalyzed reaction, it is known that thiosulfonic acid esters undergo a thermally induced acid- or base-catalyzed decomposition reaction to yield the respective disulfones and sulfinic acids.

7.3.1.12 Disulfones

The reaction mechanism for disulfones is shown in Scheme 7.20 [89]. Homolytic cleavage of an S–S bond is followed by hydrogen abstraction, yielding two equivalents of sulfinic acid. Quantum yields of the photolysis of disulfone compounds in THF solution were determined to be in the range of 0.2–0.6, depending on chemical structures.

7.3.2 Acid-Catalyzed Reactions

7.3.2.1 Deprotection Reaction

The protection of reactive groups is a general method in synthesis [90]. There are several protecting groups for the hydroxy group of polyhydroxystyrene.

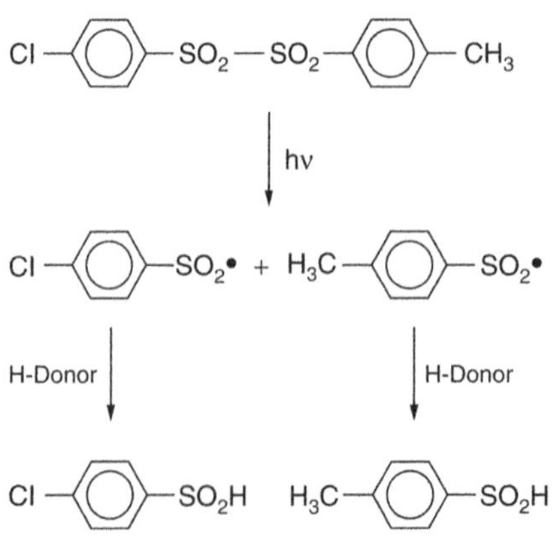

SCHEME 7.20 Photodecomposition mechanism of disulfones.

7.3.2.1.1 t-BOC Group

The IBM group has reported resist systems based on acid-catalyzed thermolysis of the side-chain *t*-BOC protecting group [44]. A resist formulated with poly(*p*-1-butoxycarbonyloxystyrene) (*t*-BOC-PHS) and an onium salt as a photoacid generator has been described [44, 45]. The acid produced by photolysis of an onium salt catalyzes acidolysis of the *t*-BOC group, which converts *t*-BOC-PHS to poly(hydroxystyrene) (PHS), as shown in Scheme 7.21. Because the photogenerated acid is not consumed in the deprotection reaction, it serves only as a catalyst. This system is named *chemical amplification*.

The exposed part is converted to a polar polymer that is soluble in polar solvents such as alcohols or aqueous bases, whereas the unexposed area remains nonpolar. This large difference in polarity between exposed and unexposed areas gives a large dissolution contrast in the developer, which also allows negative or positive images depending on the developer polarity. Development with a polar solvent selectively dissolves the exposed area to give a positive tone image. Development with a nonpolar solvent dissolves the unexposed area to give a negative tone image. Because this resist is based on the polarity change, there is no evidence of the swelling that is faced in gel formation–type resists such as cyclized rubber–based negative resists. The swelling phenomena cause pattern deformation and result in limited resolution.

Because the tertiary butyl ester group is sensitive to $A_{AL}1$-type hydrolysis (acid-catalyzed unimolecular alkyl cleavage) [91] in a reaction that does not require a stoichiometric amount of water, reactions related to the *t*-butyl group described in Scheme 7.22 have been reported to apply to resist systems [92, 115].

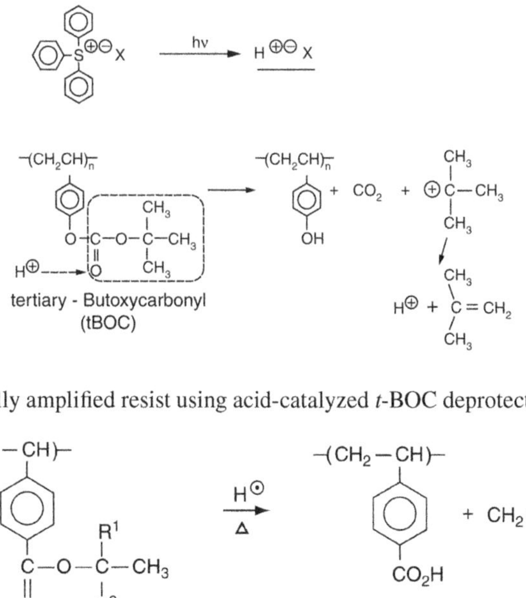

SCHEME 7.21 Chemically amplified resist using acid-catalyzed *t*-BOC deprotection reaction.

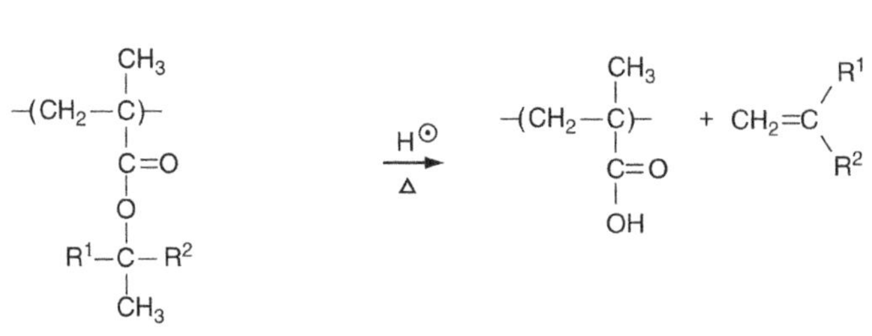

SCHEME 7.22 Acid-catalyzed deprotection (A_{al} — 1 acidolysis) for polarity change.

Workers of AT&T proposed a series of copolymers of tert-butoxycarbonyloxystyrene and sulfur dioxide prepared by radical polymerization [93]. The sulfone formulation exhibits both improved sensitivity and high contrast. It is expected that chain degradation of the matrix polymer will result in a chemically amplified resist with improved sensitivity. They also prepared a new terpolymer, poly[(*tert*-butoxycarbonyloxy)styrene-co-acetoxystyrene-co-sulfone], and found that the acetoxy group can be cleaved from acetoxystyrene monomer in aqueous base solution as shown in Scheme 7.23 [94]. This cleavage occurs only when sulfone is incorporated into the copolymer in the appropriate amounts. It is expected that this base-catalyzed cleavage during development will enhance the solubility of the exposed area where the *t*-BOC group is removed by acid. Incorporation of 50 wt.% acetoxystyrene into the polymer reduces the weight loss by approximately 18% in the solid film, whereas poly[(*tert*-butoxycarbonyloxy)styrene sulfone] shows 40% loss. The reduction of weight loss is advantageous for adhesion.

A disadvantage of onium salts such as diaryliodonium and triarylsulfonium salts is strong dissolution inhibition for alkali development. To improve the solubility of onium salts in the developer, Schwalm et al. [95] proposed a unique photoacid generator that completely converts to phenolic products upon irradiation, as shown in Scheme 7.24. They synthesized sulfonium salts containing an acid-labile *t*-BOC protecting group in the same molecule. After irradiation, unchanged hydrophobic initiator and hydrophobic photoproducts are formed in addition to the acid, but upon thermal treatment, the acid-catalyzed reaction converts all these compounds to phenolic products, resulting in enhanced solubility in aqueous bases.

Kawai et al. [96] described a chemically amplified resist composed of partially *t*-BOC-protected monodisperse PHS as a base polymer, *t*-BOC-protected bisphenol A as a dissolution inhibitor, and a photoacid generator. They reported that the use of monodisperse polymer and optimization of the degree of *t*-BOC protection improved resolution capability and surface inhibition.

Three-component resists using *t*-BOC-protected compounds as dissolution inhibitors in combination with a photoacid generator and a phenolic resin have been reported by several groups [97–100]. Aoai et al. [101] systematically investigated the effect of the chemical structures of the

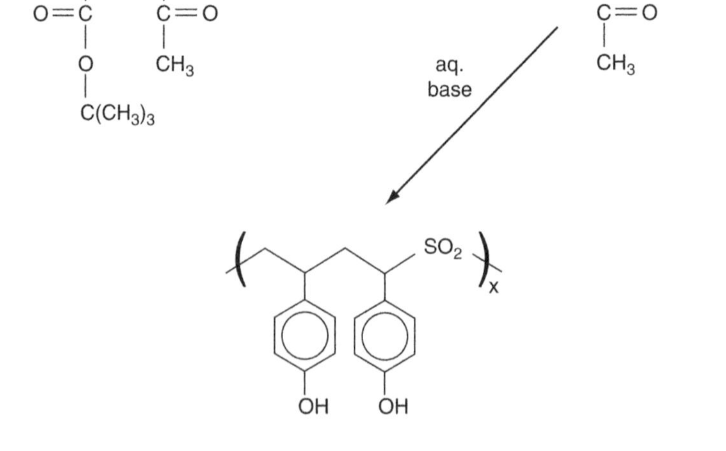

SCHEME 7.23   Acid-catalyzed deprotection reaction during postexposure baking and base-catalyzed cleavage during development.

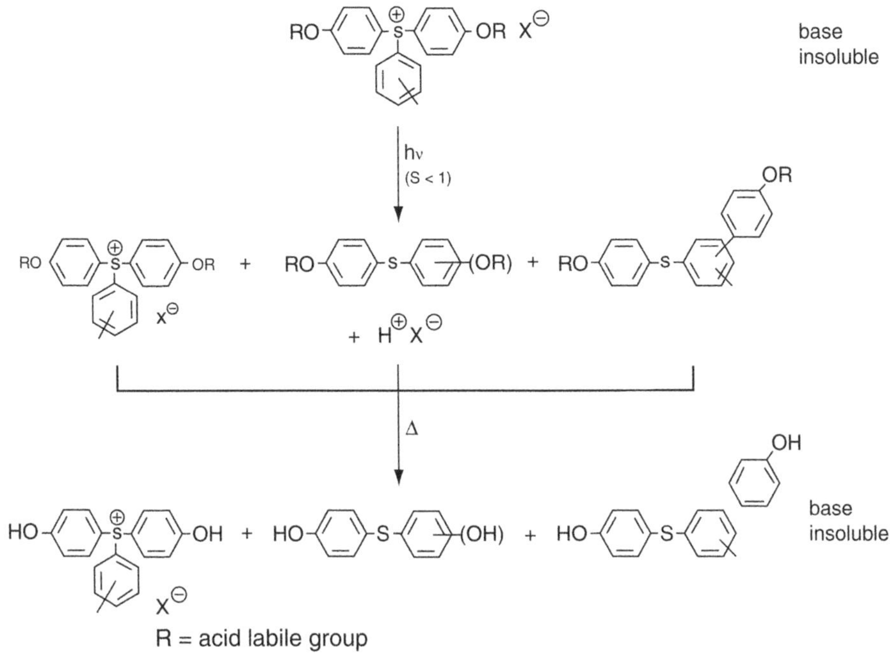

SCHEME 7.24 Reaction mechanism of *t*-BOC-protected acid generator.

backbone of *t*-BOC compounds on dissolution-inhibition capability. Similar chemical structural effects have been observed as reported in DNQ compounds. *t*-BOC compounds with large distances between *t*-BOC groups and high hydrophobicity show strong dissolution-inhibition effects.

Workers at Toshiba [102], Nihhon Kayaku [103], and NTT [104] reported 1-(3H)-isobenzofuranone derivatives protected with *t*-BOC as a new type of dissolution inhibitor. The concept of this resist system is shown in Figure 7.26. These inhibitors decomposed by an acid-catalyzed thermal reaction. In addition, the lactone ring of the decomposed products was cleaved by base-catalyzed reaction in the developer, which may enhance the dissolution rate of the exposed area.

7.3.2.1.2 THP Group

The tetrahydropyranyl (THP) group can be used as a protecting group of PHS [63, 105]. The deprotection reaction of the THP group has been investigated in detail by Sakamizu et al. [106]. The proposed mechanism is shown in Scheme 7.25. A proton first attacks the phenolic oxygen to produce PHS and a carbocation, 1. Carbocation 1 can react with water from the atmosphere or trapped in novolac to give 2-hydroxytetrahydropyran, or it loses a proton to give 3,4-hydropyran. 2-Hydroxytetrahydropyran is a hemiacetal and is in equilibrium with 5-hydroxypentanal.

A fully THP-protected PHS (THP-M) suffered from poor developability in aqueous base when THP-M was used as a base polymer. The effect of the deprotection degree on dissolution rate was investigated by Hattori et al. [107]. As shown in Figure 7.27, 30% protection degree is enough for negligible dissolution for alkali development. It should be noted that deprotection from 100% to 30% cannot induce a change in dissolution in 2.38% tetramethylammonium hydroxide solution. Optimization of the protection degree can provide alkali-developable two-component resists for KrF lithography (Figure 7.28). It is difficult, as shown in Figure 7.29, to deprotect THP groups completely for both high- (92%) and low- (20%) protected THP-M. It should be noted that 20% THP-protected THP-M gives a product of lower protected degree at fully exposed regions than 92% THP-protected THP-M. Therefore, it is expected that THP-M with low protection degree reduces the surface inhibition due to the high yield of alkali-soluble hydroxy groups.

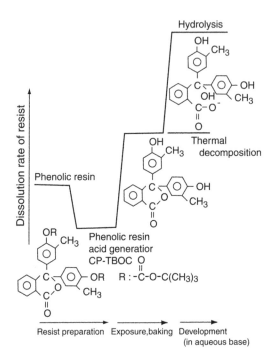

FIGURE 7.26 Concept of dissolution-rate enhancement during development. *t*-BOC compounds that show dissolution inhibition are deprotected by acid-catalyzed reaction. The deprotected compounds with lactone ring are cleaved by base-catalyzed reaction during development, leading to dissolution enhancement of the exposed region. (From Kihara, N., et al., *Proc. SPIE,* 1672, 194, 1992.)

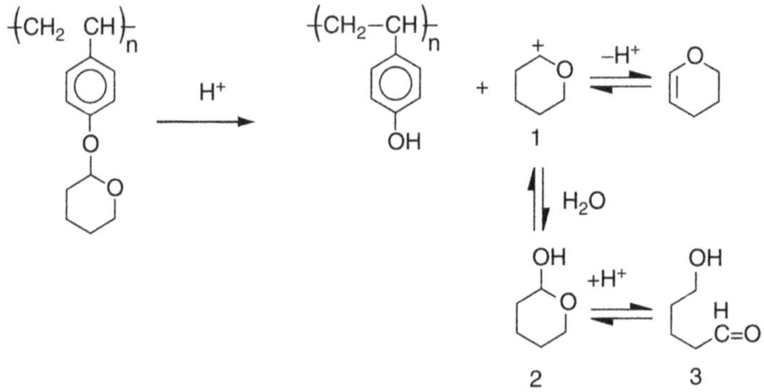

SCHEME 7.25 Acid-catalyzed deprotection reaction of tetrahydropyranyl-protected polyhydroxystyrene (THP-M).

Other polymers incorporating the THP protecting group have been reported. Taylor et al. [108] evaluated copolymers of benzylmethacrylate and tetrahydropyranylmethacrylate as deep-UV resists. Kikuchi and coworkers [109] described copolymers of styrene and tetrahydropyranylmeth-acrylate. Terpolymers of N-hydroxybenzylmethacrylamide, tetra-hydropyranyloxystyrene, and acrylic acid were applied to deep-UV resists, which showed good adhesion to silicon substrates and high glass transition temperatures [110].

7.3.2.1.3 Trimethylsilyl Group

Early work on the trimethylsilyl group as a protecting group was reported by Cunningham and Park [111]. Yamaoka et al. [69] made preliminary experiments on the rate of acid-catalyzed hydrolysis

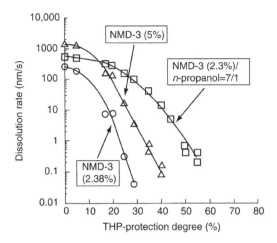

FIGURE 7.27 Dissolution rate of THP(tetrahydropyranyl)-protected polyhydroxystyrene, THP-M, as a function of THP protection degree for various developers. NMD is tetramethylammonium hydroxide aqueous solution. (From Hattori, T., et al., *Opt. Eng.,* 32, 2368, 1993.)

FIGURE 7.28 Line-and-space patterns of 0.3 μm using a resist composed of partially THP-protected poly-hydroxystyrene (THP-M) and an onium salt. (From Hattori, T., et al., *Opt. Eng.,* 32, 2368, 1993.)

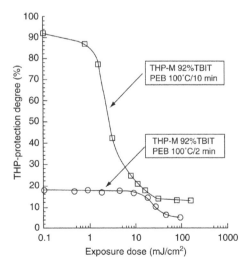

FIGURE 7.29 Change in THP protection degree with exposure dose for THP-Ms of low protection degree (20%) and high protection degree (92%). The deprotection degree was determined by IR after postexposure baking. (From Hattori, T., et al., *Opt. Eng.,* 32, 2368, 1993.)

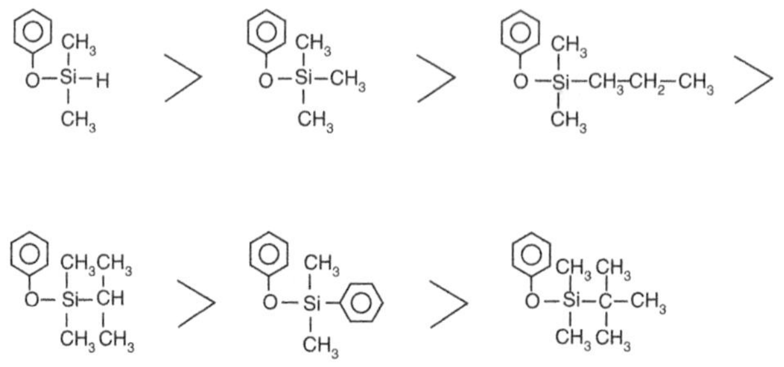

FIGURE 7.30 The order of acid-catalyzed hydrolysis rates for alkyl- and arylsilylated phenols.

for a series of alkylsilylated and arylsilylated phenols. The order of the rate is shown in Figure 7.30. It is likely that the rate of hydrolysis is governed by the steric hindrance of the trisubstituted silyl group rather by electronic induction effect, because no obvious correlation between the rate of the hydrolysis and Hammet's values of the silylating substituents is observed. Among the silylating substituents studied, the trimethylsilyl group was chosen as a protecting group for PHS because of its high rate of hydrolysis and good stability, whereas the highest rate of hydrolysis was obtained for the dimethylsilyl group. They reported a resist using trimethylsilyl-protected PHS combined with *p*-nitrobenzylsulfonate as an acid generator [112].

7.3.2.1.4   Phenoxyethyl Group

Phenoxyethyl was proposed as a protecting group of PHS by Jiang and Bassett [113]. As shown in Scheme 7.26, phenol is produced from the protection group via acid-catalyzed cleavage, as well as production of PHS. Because phenol is very soluble in aqueous bases, it acts as a dissolution promoter in exposed areas. In addition, phenol is not volatile under lithographic conditions, resulting in smaller film thickness loss than the *t*-BOC system.

7.3.2.1.5   Cyclohexenyl Group

Poly[4-(2-cyclohexenyloxy)-3,5-dimethylstyrene] has been prepared for a dual tone imaging system [114]. Poly[4-(2-cyclohexenyloxy)styrene] is less attractive due to the occurrence of some Claisen rearrangement and of other side reactions, as shown in Scheme 7.27. On the contrary, the polymer with ortho-position methylation is deprotected easily due to limitation of the side reaction by blocking of the reaction site.

7.3.2.1.6   t-Butoxycarbonylmethyl Group

It is generally accepted that a carboxylic acid shows higher dissolution promotion than the hydroxy group of a phenol. It is expected that the deprotected polymer containing carboxylic acid acts as a dissolution accelerator in aqueous base [115]. However, poly (vinylbenzoic acid) shows high absorbance at 248 nm. To avoid the strong absorption of the benzoyl group, Onishi et al. [116] reported partially *t*-butoxycarbonylmethyl-protected PHS, which involves a methylene group between phenyl

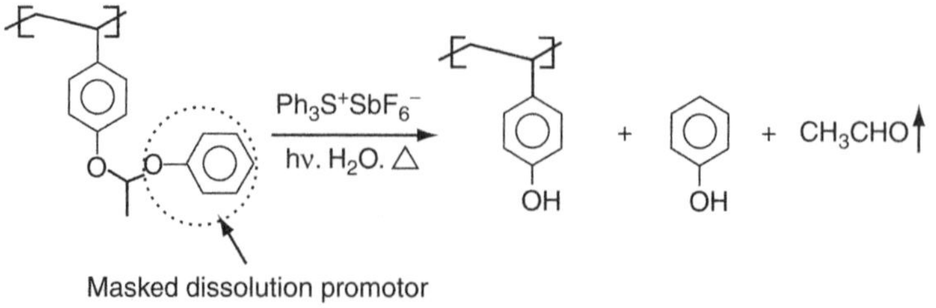

SCHEME 7.26 Acid-catalyzed reaction of poly(4-(1-phenoxyethoxy)styrene).

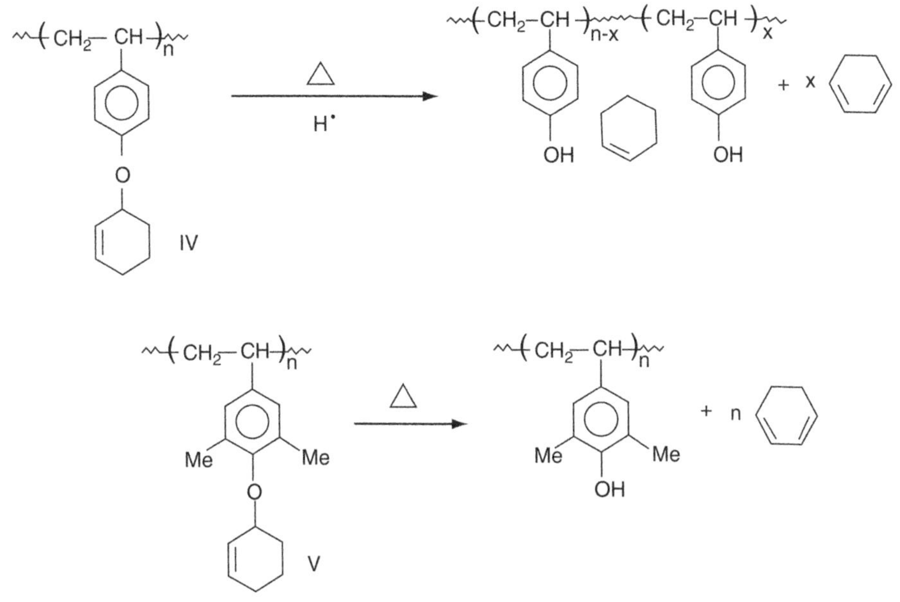

SCHEME 7.27 Acid-catalyzed reaction of cyclohexyl-protected polyhydroxystyrene and methyl-substituted polyhydroxystyrene.

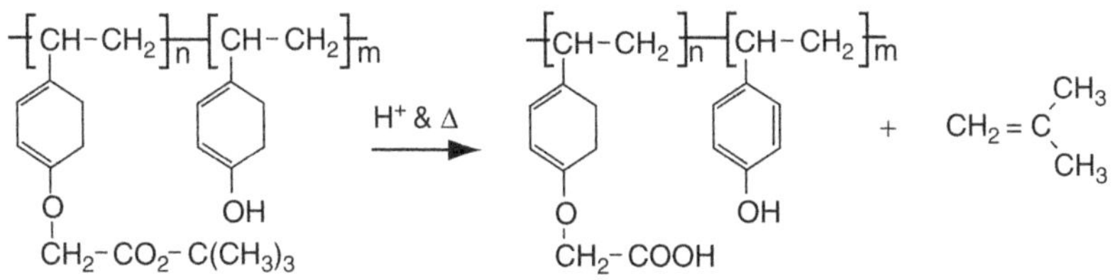

SCHEME 7.28 Chemical structure of BCM-PHS and its acid-catalyzed thermolysis.

and carboxylic acid. The reaction mechanism is shown in Scheme 7.28. Another advantage is that *t*-butoxycarbonylmethyl- protected PHS gave no phase separation after development in film that is faced in *t*-BOC protected PHS.

7.3.2.2 Depolymerization

7.3.2.2.1 Polyphthalaldehyde (PPA)

Ito and Willson [117] reported acid-catalyzed depolymerization of polyaldehydes as a first stage of chemical amplification resists. PPA is classified as O,O-acetal, which will be described below. The polymerization of polyaldehyde is known to be an equilibrium process of low ceiling temperature (T_c). Above T_c, the monomer is more thermodynamically stable than its polymer. During polymerization at low temperature, end-cap by alkylation or acylation terminates the equilibrium process and renders the polymer stable at approximately 200 °C. Although polymers of aliphatic aldehydes are highly crystalline substances that are not soluble in common organic solvents, PPA provides noncrystalline materials that are highly soluble and can be coated to provide clear isotropic films of high quality. The T_i of PPA is approximately −40°C. The resist composed of PPA and an onium salt can give positive tone images without subsequent processes. This is called *self-development* imaging. As shown in Scheme 7.29, the acid generated from onium salts catalyzes the cleavage of the main-chain acetal bond. After the bond is cleaved at above the ceiling temperature, the material spontaneously depolymerizes to monomers. PPA can be used as a dissolution inhibitor of novolac

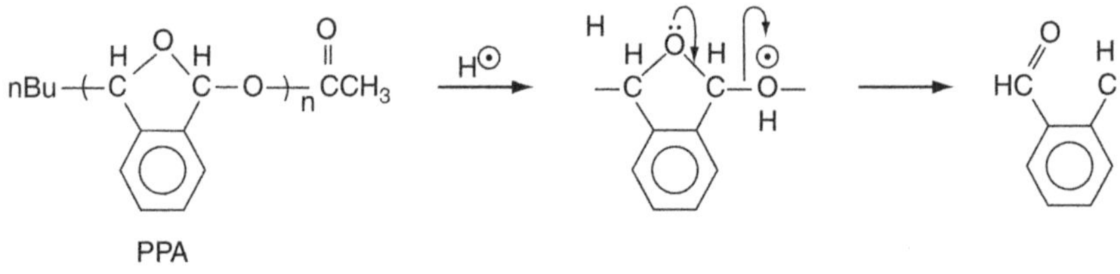

SCHEME 7.29 Acid-catalyzed depolymerization of polyphthalaldehyde.

resin in a three-component system. The photogenerated acid induces depolymerization of PPA, resulting in loss of dissolution-inhibition capability to give a positive image.

Although PPA materials are sensitive self-developing resists, they have drawbacks, such as liberation during exposure of volatile materials that could damage the optics of exposure tools and poor dry-etching resistance. Ito et al. [118, 119] found that poly(4-chlorophthalaldehyde) does not spontaneously depolymerize upon exposure to radiation; it requires a postexposure bake (PEB) step to obtain positive relief image, which can avoid the damage to optics. This system is called *thermal development*, as distinguished from *self-development*.

7.3.2.2.2 O,O- and N,O-Acetals

The workers of Hoechst AG reported three-component chemical-amplification resist systems using acid-catalyzed depolymerization of *O,O*- and *N,O*-acetals [120, 121]. The reaction mechanisms are shown in Scheme 7.30 and Scheme 7.31, respectively. Poly-*N,O*-acetal is protonated at the oxygen atom and liberates an alcohol, XOH. The intermediate formation of a carbocation is the rate-limiting step in hydrolysis, and its stability is influenced by mesomeric and inductive effects of the substituents $R1$ and $R2$. It is noteworthy that the liberation of XOH causes a decrease in the molecular weight of an inhibitor and that cleavage products such as alcohols and aldehydes show strong dissolution promotion. Because the novolac resin suffers from strong absorption at 248 nm, the authors developed methylated polyhydroxystyrene, poly(4-hydroxystyrene-*co*-3-methyl-4-hydroxystyrene), as a base resin that shows a reasonable dissolution-inhibition effect [122–124].

7.3.2.2.3 Polysilylether

The silicon polymer containing silylether groups in the main chain is hydrolyzed by acid and degraded to low–molecular weight compounds [125, 126]. This polymer can be used as a dissolution

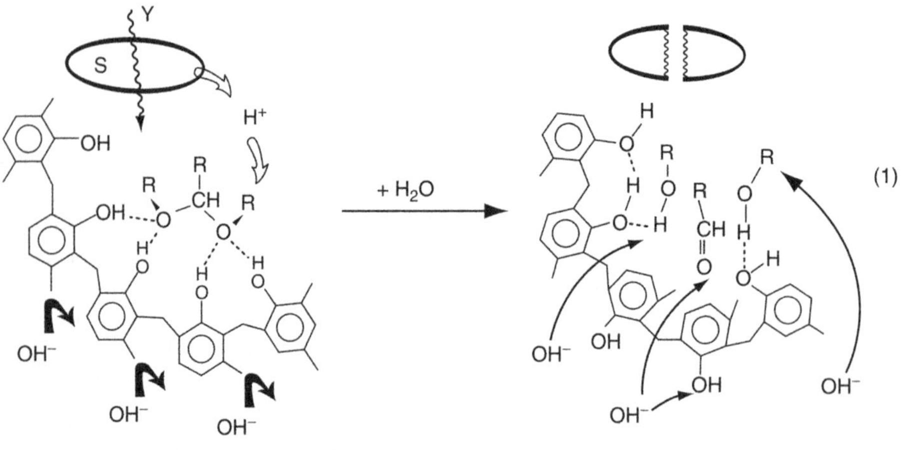

SCHEME 7.30 Acid-catalyzed depolymerization of acetals leading to their loss of dissolution-inhibition capability.

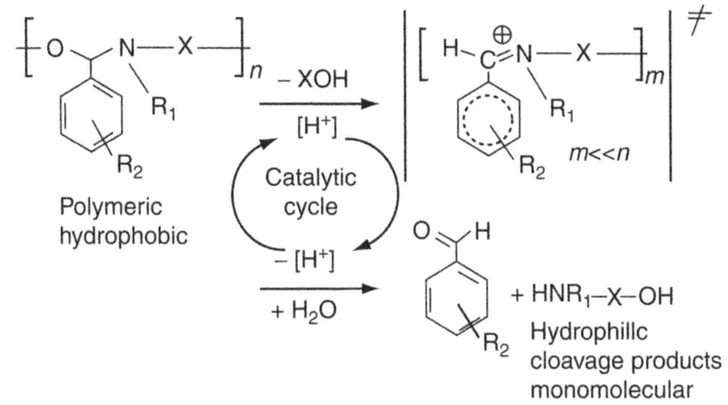

SCHEME 7.31 Reaction mechanism of acid-catalyzed N,O-acetal hydrolysis.

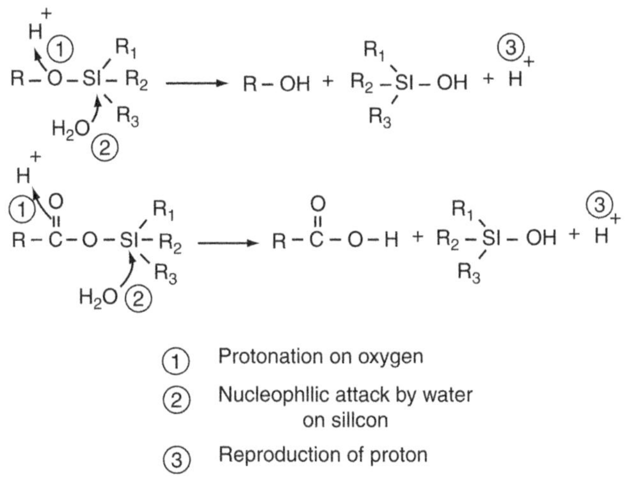

$\textcircled{1}$ Protonation on oxygen

$\textcircled{2}$ Nucleophilic attack by water on silicon

$\textcircled{3}$ Reproduction of proton

SCHEME 7.32 Acid-catalyzed depolymerization of polysilylethers.

inhibitor of a novolac resin in a three-component system. As shown in Scheme 7.32, the decomposition of the polymer includes protonation of oxygen, a nucleophilic attack of water to the Si atom, the cleavage of the Si–O bond, and reproduction of the proton, resulting in depolymerization and loss of inhibition capability. Investigation of the effect of chemical structure around silylether groups in the polymer on the hydrolysis rate indicated that the rate decreases with increasing bulkiness of alkyl and alkoxy groups.

7.3.2.2.4 Polycarbonate and Others

Frechet et al. [127–132] have designed, prepared, and tested dozens of new imaging materials based on polycarbonates, polyethers, and polyesters, which are all susceptible to acid-catalyzed depolymerization. The reaction mechanism of polycarbonates is E_1-like elimination, shown in Scheme 7.33. The protonation of the carbonyl group of a carbonate is followed by cleavage of the adjacent allylic carbon–oxygen bond to produce two fragments: a fragment containing a monoester of carboxylic acid, 2a, and another containing an allylic carbocation moiety, 2c. Elimination of a proton from this carbocationic moiety results in regeneration of the acid catalyst and formation of the terminal diene-containing fragment 2b. The unstable monoester of carbonic acid 2a decarboxylates, releasing a terminal alcohol fragment, 2d. The process continues at other carbonate sites with complete breakdown of the polymer chain, resulting in the eventual release of benzene, additional carbon dioxide, and a diol. The protons initially generated by irradiation are not consumed in the process.

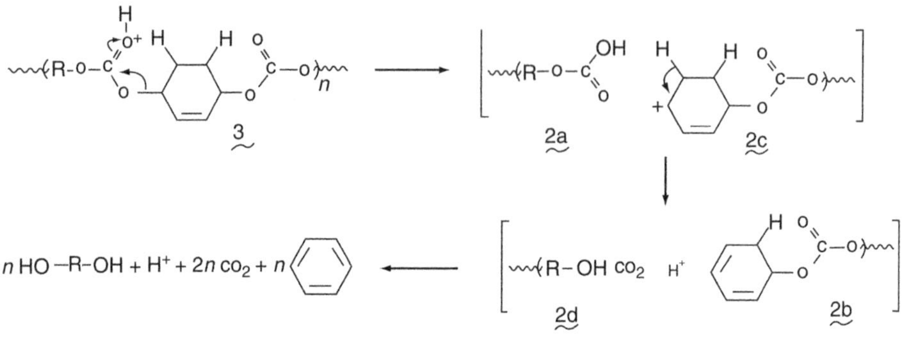

SCHEME 7.33 Acid-catalyzed depolymerization of polycarbonates.

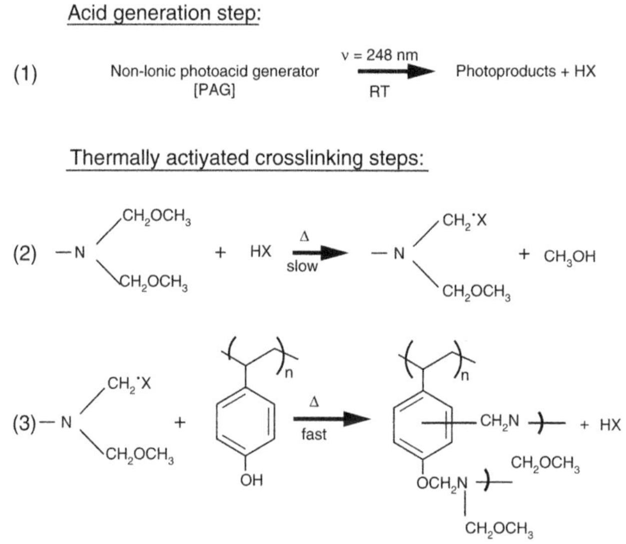

SCHEME 7.34 Acid -hardening mechanism in phenolic resins.

7.3.2.3 Crosslinking and Condensation

7.3.2.3.1 Acid Hardening of Melamine Derivatives

Negative tone resists based on acid-hardening resin (AHR) chemistry are three-component systems composed of a novolac resin, a melamine crosslinking agent, and a radiation-sensitive acid generator [133–136]. The reaction leading to a decrease in dissolution rate in alkali aqueous solution via crosslinking is shown in Scheme 7.34. The protonated melamine liberates a molecule of alcohol upon heating to leave a nitrogen-stabilized carbonium ion. Alkylation of novolac then occurs at either the phenolic oxygen (O-alkylation) or a carbon on the aromatic ring (C-alkylation), and a proton is regenerated. There are several reactive sites on the melamine, allowing it to react more than once per molecule to give a crosslinked polymer. The difference in dissolution rate between exposed and unexposed is quite high, as reported by Liu et al. [137]. The acid hardening–type resists have been widely evaluated as resists for deep-UV and electron-beam lithography. The effects of molecular weight and molecular-weight distribution of base resins, such as novolac or PHS, have been investigated [138–140].

7.3.2.3.2 Electrophilic Aromatic Substitution

Frehet et al. [141–146] applied another acid-catalyzed reaction—electrophilic aromatic substitution—to negative resists. The reaction mechanism is shown in Scheme 7.35. The photogenerated

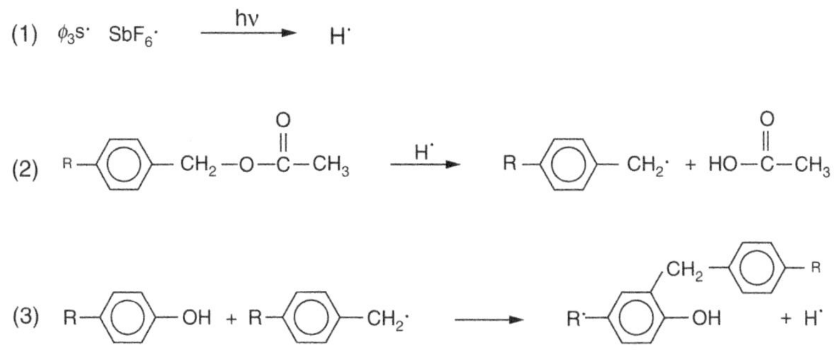

SCHEME 7.35   Acid-catalyzed electrophilic aromatic substitution in a negative chemically amplified resist.

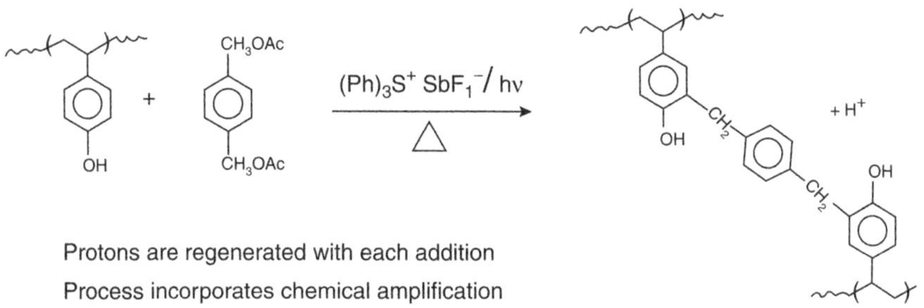

Protons are regenerated with each addition

Process incorporates chemical amplification

FIGURE 7.31   Crosslinking via nonpolymeric multifunctional latent electrophile.

acid reacts with a latent electrophile, such as a substituted benzylacetate, to produce a carbocationic intermediate, while acetic acid is liberated. The carbocationic intermediate then leads to electrophilic reaction with neighboring aromatic moieties, regenerating a proton. One application to chemical amplification is a resist composed of copolymers containing a latent electrophile and an electron-rich aromatic moiety and a photoacid generator. The copolymers of 4-vinylbenzylacetate and 4-vinylphenol are prepared. An alternate approach is to use a polyfunctional low–molecular weight latent electrophile in a three-component system including a photoacid generator and a phenolic polymer (Figure 7.31).

Another type of polyfunctional latent electrophile acting as a crosslinker, polyfunctional alcohol, has also been reported [144–147]. Its reaction mechanism is shown in Scheme 7.36.

7.3.2.3.3   Cationic Polymerization

The chain-reaction character of the epoxy group in acid-catalyzed ring-opening polymerization (crosslinking) can be used to obtain negative resists [148]. The use of epoxy-novolac resins is well known in negative resist systems but is limited by poor resolution due to image swelling when developed with organic solvent [45, 149]. Conley et al. [150] reported a resist using certain classes of polyepoxide and monomeric epoxide compounds as crosslinkers for alkaline-soluble phenolic resin to get a nonswelling resist. They showed that biscyclohexane epoxide used as a crosslinker shows high sensitivity, high resolution, and excellent resistance to thermal ring-opening polymerization, whereas some epoxy compounds show sensitivity to the acidity of phenolic resin even at room temperature.

7.3.2.3.4   Silanol Condensation

Silanol compounds undergo condensation to form siloxane in the presence of acid (Scheme 7.37). It has already been reported that silanols can act as dissolution promoters in a novolac resin in the design of a silicon-containing resist with oxygen plasma etching resistance for two-layer resist

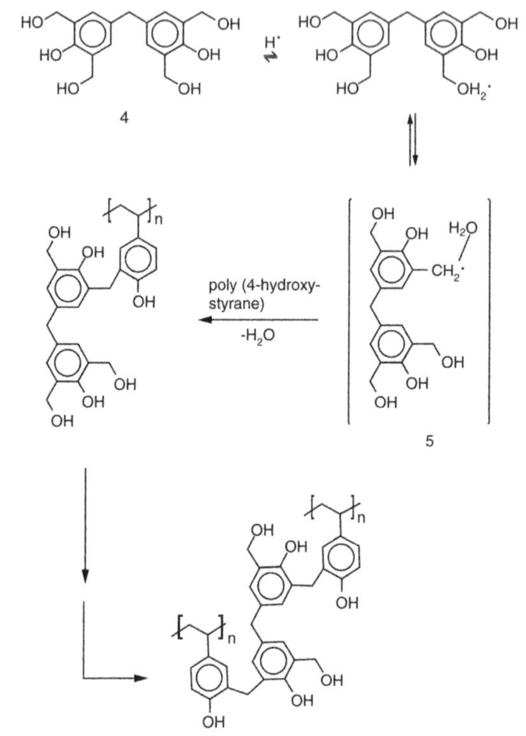

SCHEME 7.36   Crosslinking reaction using acid-catalyzed electrophilic aromatic substitution.

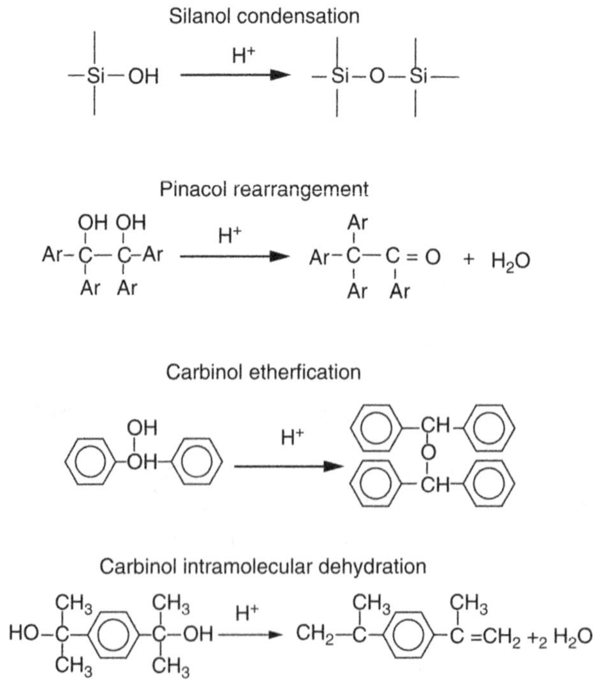

SCHEME 7.37   Acid-catalyzed reactions for polarity change in negative chemically amplified resists.

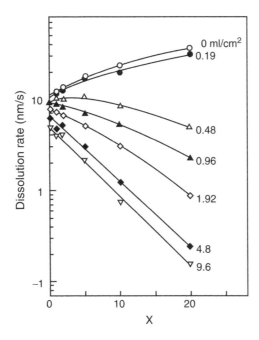

FIGURE 7.32 Effect of $Ph_2Si(OH)_2$ concentration and exposure dose on dissolution rates. The resist is composed of a novolac resin, $Ph_2Si(OH)_2$, and $Ph_3SCCF_3SOK_3$. The dissolution rates were measured after postexposure baking. (From Ueno, T., et al., *Proc. SPIE*, 1262, 26, 1990.)

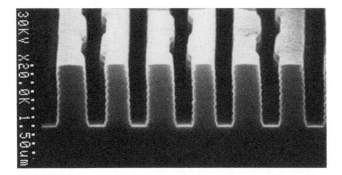

FIGURE 7.33 Space patterns of 0.3 μm of the resist using acid-catalyzed condensation of diphenylsilanediol. Exposure carried out with an i-line stepper in conjunction with a phase-shifting mask.

systems [151]. Therefore, it is expected that the acid-catalyzed reaction will produce a dissolution inhibitor, siloxane, that can be applied to a negative resist [152–154]. The resist based on acid-catalyzed silanol condensation consists of diphenylsilanediol, a photoacid generator, and a novolac resin. As shown in Figure 7.32, the dissolution rate of the film increases with increase in silanol content, whereas the exposed film shows a drastic decrease in dissolution rate after baking. The acid-catalyzed products in this system were expected to be siloxane oligomers, which was confirmed by IR spectroscopy and gel-permeation chromatography. This acid-catalyzed reaction can be applied to i-line phase-shifting lithography [153] when i-line sensitive photoacid generator is used (Figure 7.33). A similar system was also reported by McKean et al. [155].

Shiraishi et al. [154] investigated the dissolution-inhibition mechanism in a novolac matrix to explain the following interesting experimental results. The fully exposed and baked resist film showed a very small dissolution rate (less than 0.03 nm/s). However, the recoated film, after the exposed film was dissolved in the casting solvent, showed a large dissolution rate. As shown in

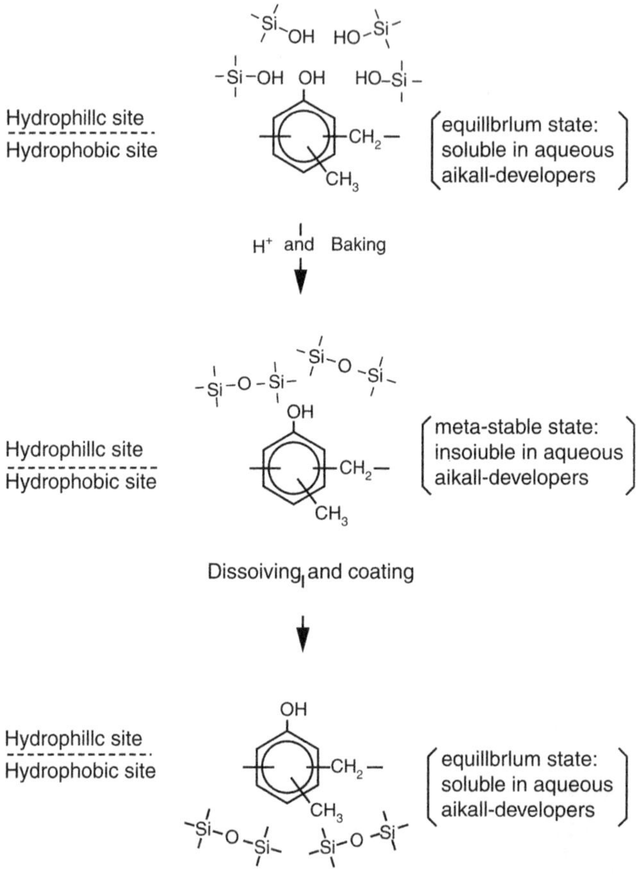

FIGURE 7.34   Model of dissolution promotion and inhibition mechanism in the resist composed of a novolac resin, $Ph_2Si(OH)_2$, and $Ph_3S^+CF_3SO_3^-$. Silanol compounds that surround OH groups of novolac resin promote dissolution. OH groups are blocked by siloxanes produced by acid-catalyzed condensation of silanols, leading to dissolution inhibition. (From Shiraishi, H., et al., *Chem. Mater.*, 3, 621, 1991.)

Figure 7.34, hydrophilic silanol groups of silanol compounds surround the hydrophilic hydroxy groups of the novolac resin matrix in the equilibrium state. When radiation-induced acid-catalyzed condensation takes place at a hydrophilic site, siloxanes are produced that surround the hydroxy group and shield or block the dissolution in alkali. When this metastable-state film is once dissolved and recoated from solution, hydrophobic siloxanes are surrounded by the hydrophilic sites of novolac resin, resulting in poor dissolution-inhibition effect. The insolubilization mechanism of silanol condensation is noncrosslinking but is affected by the site or orientation of silanol condensation products.

7.3.2.4   Polarity Change

Although the resist based on silanol condensation shows high resolution capability and high sensitivity, it causes the problem of SiO* residue formation by oxygen plasma removal processes. Uchino et al. [156–159] designed and tested several types of acid-catalyzed reactions that can be applied to negative resists, as shown in Scheme 7.37. The basic concept is the same as that of silanol condensation: a carbinol acts as a dissolution promoter in novolac resin, while acid-catalyzed reaction products act as a dissolution inhibitor. Ito et al. [160] called these systems *reverse polarity*.

Pinacol rearrangement using acid-catalyzed dehydration of pinacols was applied to an alkali-developable resist [156, 157]. The resist system was composed of a pinacol compound used as a dissolution-inhibition precursor, an onium salt, and a novolac resin. The insolubilization mechanism for the acid-catalyzed reaction of hydrobenzoin in a phenolic matrix is shown in Scheme 7.37 and Scheme 7.38. Hydrobenzoin converts to diphenylacetaldehyde (Scheme 7.37), which reacts with hydrobenzoin to form 2,2-diphenyl-4,5-diphenyl-1,3-dioxolane (Scheme 7.38), resulting in dissolution inhibition. Sooriyakumaran et al. [160] have independently reported a resist system using a polymeric pinacol and benzopinacol.

The use of acid-catalyzed etherification of carbinols gives alkali-developable, highly sensitive negative resists [158]. Certain carbinols can be dissolution promoters of novolac resin, whereas acid-catalyzed products inhibit novolac dissolution in aqueous base. One example of the system consists of diphenylcarbinol (DPC), *m,p*-cresol novolac resin, and diphenyliodonium triflate. The reaction mechanism for insolubilization to alkali developer is shown in Scheme 7.39. In lightly exposed regions, bimolecular etherification of DPC occurs, whereas in the heavily exposed region, DPC reacts with novolac resin to form *o*-diphenylmethyl novolac resin.

Intramolecular dehydration of the α-hydroxypropyl group was used for a negative chemical-amplification resist in deep-UV [161] and i-line phase-shifting lithography (Figure 7.35) [159]. Polymers or compounds with an α-hydroxypropyl group are soluble in aqueous base, and acid-catalyzed intramolecular dehydration reactions (Scheme 7.37) produce dissolution inhibition.

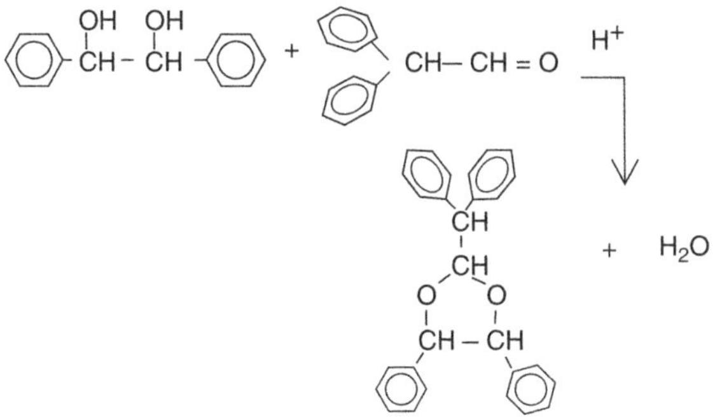

SCHEME 7.38 Reaction of hydrobenzoin with diphenylacetaldehyde. Hydrobenzoin converts to diphenylacetaldehyde in the presence of acid. Then, diphenylacetaldehyde reacts with hydrobenzoin to form 2,2-diphenyl-4,5-diphenyl-1,3-dioxolane, resulting in dissolution inhibition.

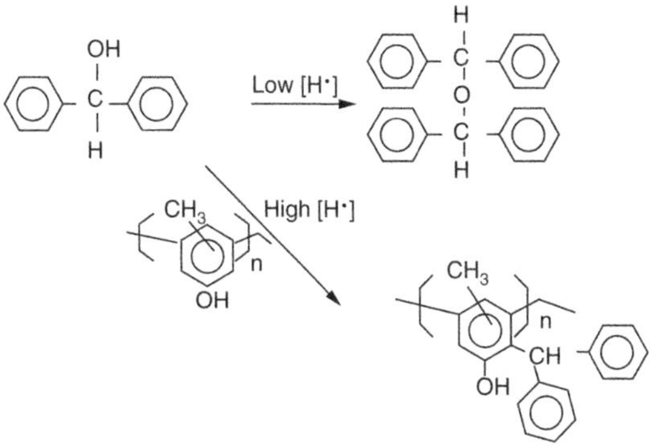

SCHEME 7.39 Acid-catalyzed reaction mechanism of diphenylcarbinol (DPC) in phenolic resin. In lightly exposed regions, bimolecular etherification of DPC occurs, while in the highly exposed region, DPC reacts with novolac resin to form *o*-diphenylmethyl novolac resin.

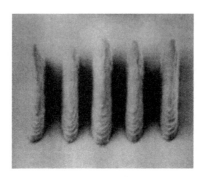

FIGURE 7.35   Line-and-space patterns of 0.275 μm of the resist using acid-catalyzed intramolecular dehydration of a carbinol exposed to i-line radiation using a phase-shifting mask.

7.3.3   ROUTE FOR ACTUAL USE OF CHEMICALLY AMPLIFIED RESISTS

Chemically amplified resists described above can be summarized as shown in Figure 7.36. Some of them have been extensively evaluated for actual use in deep-UV lithography, including KrF excimer laser lithography. It is very difficult to select from the commercially available resists, because many things have to be taken into consideration, such as sensitivity, resolution, depth of focus, heat resistance, dry-etch resistance, shelf life, impurities, etc. In addition, the compositions of commercially available resists are not disclosed. For positive resists, however, acid-catalyzed reaction is one of the candidates. There have been many reports on resists using t-BOC protecting groups. As for negative resists, resists using acid hardening of melamines from Shipley have been widely accepted.

Process people always compare the chemically amplified resists with DNQ-novolac resists. The sensitivity required for KrF lithography is not a critical issue as far as acid-catalyzed reactions are used. At present, it is considered that optimum sensitivity is in the range of 20–50 mJ/cm^2 for KrF excimer laser lithography. The resolution limitation of chemically amplified resists used in KrF lithography is usually better than that of DNQ-novolac resists used for i-line lithography, though more DOF latitude is required. The dry-etch resistance of chemically amplified resists is similar to that of DNQ-novolac resists because a phenolic resin is used as a base polymer. An acid hardening–type negative resist is expected to show high dry-etch resistance, as exposed areas are crosslinked.

The main issues for chemically amplified resists are some intrinsic problems associated with acid-catalyzed reactions. Chemically amplified resists are susceptible to process conditions such as airborne contamination, baking conditions, and delay time between exposure and postexposure baking (postexposure delay). Underlying substrates sometimes influence the resist profile. The shelf life is also a critical issue compared with DNQ-based resists. Most positive chemically amplified resists suffer from linewidth shift and/or formation of an insolubilization layer or "T-top" profiles (Figure 7.37), depending on the postexposure delay. Although it was reported that the delay effect of acid hardening–type negative resists is better than that of the positive ones [139], one should pay attention to process stability. Here, factors causing problems are classified into acid diffusion, airborne contamination, and solvent uptake of polymers, though these factors affect the resist performance together.

7.3.3.1   Acid Diffusion

Acid catalysis utilizes a reaction induced by acid diffusion in the resist film, but too much diffusion causes the resolution capability to deteriorate. Acid diffusion in the chemically amplified resists is a critical issue to understand in order to improve resolution and linewidth control and understand the reaction mechanism. Acid diffusion is similar to general diffusion phenomena in polymers [162]. The factors that determine acid diffusion length are baking temperature, residue content of casting solvent, acid size, and matrix polymer properties such as glass transition temperature, protection degree, etc.

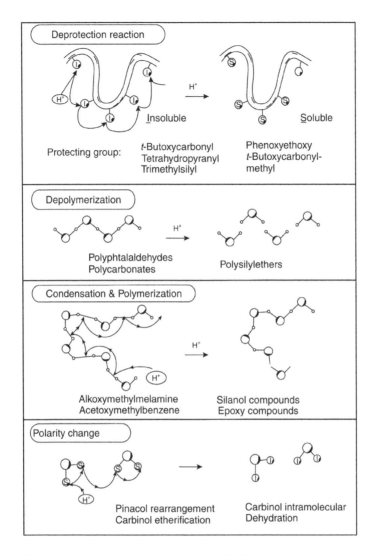

FIGURE 7.36 Acid-catalyzed reactions for chemically amplified resists.

FIGURE 7.37 Typical "T-top" profiles of a positive chemically amplified resist.

Acid diffusion has been systematically investigated by Schlegel and coworkers [163] using a simple technique described in Figure 7.38. The exposed film containing photogenerated acid is transferred onto a positive chemically amplified resist. The photogenerated acid diffuses into the underlying resist during the subsequent baking process, which causes an acid-catalyzed reaction that results in an increase in dissolution rate in the aqueous-base developer. After the development, the depth of the hole is measured to give the diffusion range of a photogenerated acid. The advantage of

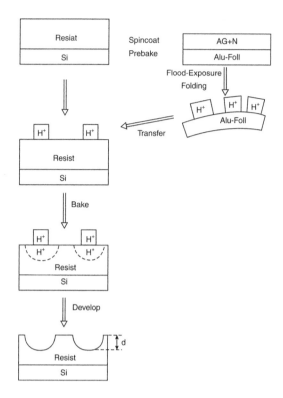

FIGURE 7.38 Process scheme for determination of the diffusion range of photogenerated acid in a positive resist. (From Schlegel, L., et al., *J. Vac. Sci. Technol. B*, 9, 278,1991; Schlegel, L., et al., *Jpn. J. Appl. Phys. B*, 30, 3132, 1991.)

this method is an exact boundary condition of acid concentration before diffusion; a constant higher acid concentration exists in the transferred exposed film, whereas no acid exists in the underlying film before baking. This systematic experiment shows that acid diffusion strongly depends on baking temperature, especially after spin coating. Acid diffusion is influenced by the amount of residual casting solvent and the glass transition temperature of the resist film. To minimize the diffusion range, higher prebaking temperatures and postexposure baking below the glass transition temperature are recommended.

Other studies on the determination of acid diffusion based on an electrochemical method, a contact-replication method using X-ray lithography [164], a method using a water-soluble layer [165], and measurement of linewidth change [166] have also been reported. The acid-diffusion range was diminished as the prebaking temperature was raised or the postexposure baking temperature was reduced.

7.3.3.2 Airborne Contamination

In a typical clean room, chemically amplified resists tend to suffer from instability of linewidth, which is manifested as a "T-top" (Figure 7.37). MacDonald et al. [167, 168] investigated the cause of T-top formation and linewidth change induced by holding after exposure. The experimental apparatus to see the effect of contaminated air on T-top formation is shown in Figure 7.39. Their conclusion is that a surface skin is formed by neutralization of photogenerated acid with airborne contamination such as hexamethyldisilazane, N-methylpyrrolidone (NMP), and bases from paint, adhesive, etc.

The airborne-contamination problems have been alleviated to some extent by purifying the enclosing atmosphere using activated charcoal filters, which allows the manufacture of 1 Mbit DRAM by deep-UV lithography [169]. The contamination was also protected by application of a protective overcoat. Some approaches for the overcoat were reported [170–172].

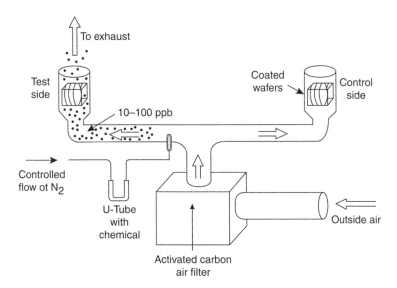

FIGURE 7.39 Dynamic flow system used to generate contaminated air. (From MacDonald, S. A., et al., *Proc. SPIE*, 1466, 2, 1991.)

7.3.3.3 N-Methylpyrrolidone (NMP) Uptake

To study the effect of residual casting solvent on the diffusion rate of the contamination into chemically amplified resist films, Hinsberg et al. [173–175] directly measured NMP uptake by a series of thin polymer films containing known amounts of residual casting solvent. The reasons for choosing NMP are: (1) it is widely used as a casting and stripping solvent in semiconductor manufacturing, (2) performance degradation of chemical amplified resists has been observed upon exposure to a low concentration of NMP vapor, and (3) it is readily available from a commercial source. The amount of residual solvent after bake was determined by a radiochemical method. NMP uptake was determined by mixing airborne methyl-C-N-methylpyrrolidone into the airstream. There is no simple relationship between the solvent content in the film and its ability to absorb NMP. Although the amount of casting solvent residue clearly increases with increasing polarity, the NMP uptake behavior does not follow a similar trend. However, NMP uptake seems to be influenced by polymer structure.

Therefore, the NMP vapor absorption properties of various polymer materials were studied under identical conditions [175]. NMP contents following storage in an NMP airstream were measured and plotted as a function of solubility parameter and glass transition temperature, as shown in Figure 7.40. It is seen that polymers with extremely high polarity or extremely low polarity show lower NMP uptake: polymers with a solubility parameter, <5, similar to that of NMP show higher NMP uptake. When the glass transition temperature of a polymer is lower than the baking temperature, NMP uptake of the polymer is negligible. These results suggest that design rules for chemically amplified resists are that the resist polymers should be baked at a higher temperature than their glass transition temperature and that the solubility parameters of the polymers should be different from NMP solubility parameters.

7.3.4 Improvement in Process Stability

7.3.4.1 Additives

Some additives to solve the delay problems were investigated by researchers at BASF [176]. The expected reactions of additives with airborne contamination, usually bases, are shown in Scheme 7.40. Among the additives, sulfonic acid esters and disulfones showed promising results.

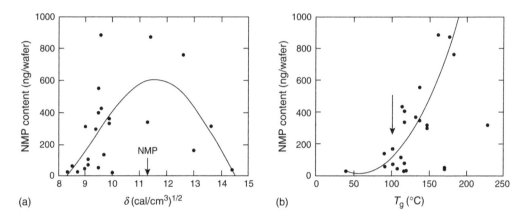

(a)   (b)

FIGURE 7.40 (a) Amount of NMP* under equivalent conditions by various polymers as a function of solubility parameter, <5, of the polymer. (b) Amount of NMP* absorbed under equivalent conditions by various polymers, as a function of the glass transition temperature (T_g) of the polymer. (From Hinsberg, W. D., et al., *Chem. Mater.*, 6, 481, 1994.)

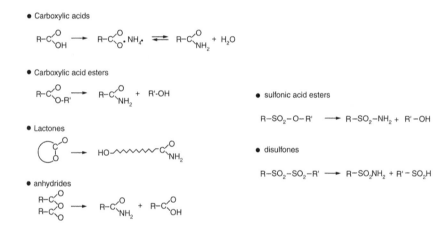

SCHEME 7.40 Expected reaction of additives with airborne contaminations.

Przybilla and coworkers [177] of Hoechst proposed a photosensitive base that loses basicity upon irradiation. The concept is shown in Figure 7.41. In the exposed region, acid is generated, and at the same time, a base is decomposed. When the acid diffuses into unexposed areas, it is neutralized by a photosensitive base. It was reported that this improves the linewidth stability for process conditions.

7.3.4.2 Polymer End Groups

The reaction of the *t*-BOC resists was based on acid-catalyzed deprotection chemistry. Although this reaction was expected to be independent of the molecular weight of a *t*-BOC-PHS, a difference in *t*-BOC deprotection yield was observed for different molecular weights of the polymer [178]. As shown in Figure 7.42, the deprotection yield (OH transmittance decrease at 3500 cm^{-1}) for $M_n = 63,000$ is higher than that for $M_n = 11,000$. This difference may be explained in terms of the different concentrations of poisoning CN end groups derived from 2,2-azobis(isobutyronitrile) (AIBN): as the molecular weight becomes lower, the end group concentration becomes higher. When *t*-BOC-PHS prepared by radical polymerization using benzoyl peroxide (BPO) or living anionic polymerization with sec-butyllithium was used, a higher extent of acid-catalyzed deprotection reaction was observed compared with the polymer obtained by radical polymerization with

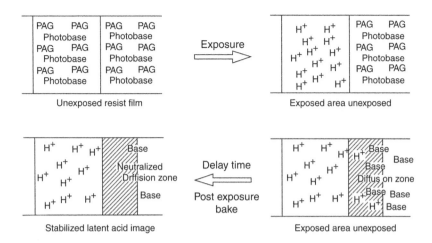

Unexposed resist film   Exposed area unexposed

Stabilized latent acid image   Exposed area unexposed

FIGURE 7.41 Concept for process stability using a photosensitive base. (From Przybilla, K.-J., et al., *Proc. SPIE*, 1925, 76,1993.)

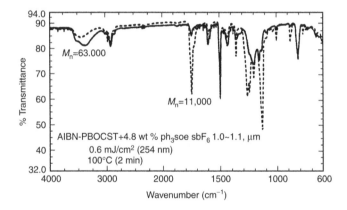

FIGURE 7.42 IR spectra of *t*-BOC resists consisting of *t*-BOC protected polyhydroxystyrene (PBOCST) of different molecular weight and Ph₃S+ SbF₆. PBOCST is obtained by radical polymerization with AIBN. The exposure dose is 0.6 mJ/cm², and it is postexposure baked at 100 °C for 2 min. (From Ito, H., et al., *Proc. SPIE*, 1672, 2, 1992.)

AIBN initiator. Polymer end groups are important to maximize the performance of chemical-amplification resists.

7.3.4.3 T_g of Polymers

As described earlier, the NMP uptake is primarily governed by glass transition temperature; lower-T_g polymer films absorb much less NMP due to better annealing. Ito and coworkers [179] proposed chemical-amplification resists formulated with lower-T_g *meta-* *t*-BOC-PHS for environmental stabilization. It was observed that *meta-t-BOC-PHS* resists are more resistant to postexposure delay than their *para-t-BOC-PHS* counterparts when prebake after spin-coating is performed at approximately 100 °C. This was ascribed to the reduced free volume when the polymer is prebaked at a higher temperature than its T_g. The reduction of the free volume has been confirmed by measuring the refractive indices of the two isomer films by a waveguide technique [179].

However, lowering T_g is not the best solution, because low-T_g resists suffer from serious thermal flow during high-temperature processes. It is necessary to design an environmentally robust resist system that can be processed at high temperature to obtain good annealing. It is not easy to design

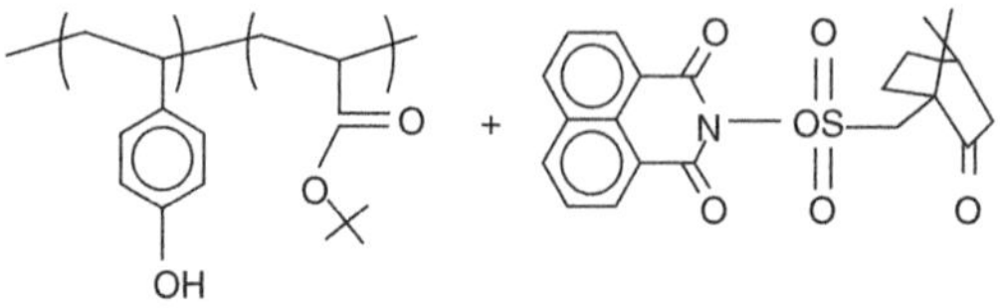

FIGURE 7.43 The compositions of ESCAP (environmentally stable chemical amplification positive resist).

such polymers using partially protected PHS, for example with *t*-BOC, because these polymers are thermally decomposed by acidic phenol hydroxy groups at a lower temperature than fully protected polymers. On the other hand, the T_g of partially protected polymers is higher than that of fully protected polymers due to the hydrogen-bonding interaction of the unprotected hydroxy group [180, 181]. Therefore, Ito and coworkers [180] showed the design concept of ESCAP (environmentally stable chemically amplification positive resist), composed of a thermally and hydrolytically stable resin and a thermally stable photoacid generator: a copolymer of 4-hydroxystyrene (HOST) with *t*-butylacrylate (TBA), and camphorsulfonyloxynaphthalimide (CSN) as an organic nonionic acid generator (Figure 7.43). ESCAP can employ a bake temperature of 150 °C or above. Using this resist, they showed 2 h stability for postexposure delay.

7.4 SURFACE IMAGING

The demand for improved resolution requires imaging systems with increasingly higher aspect ratios and better linewidth control over topography. Linewidth control and resist profile strongly depend on substrate steps and the reflectivity of substrates. Depth of focus (DOF) latitude is another critical issue. The use of shorter wavelengths in photolithography is one potential method for obtaining smaller feature sizes. Absorption of the basic polymer matrix increases in the deep-UV region, which affects the image profile in the resist film. In ArF excimer laser lithography, the absorption coefficient at 193 nm of aromatic polymers is so high that light absorption is confined to the top ~0.1 μm for polymers. On the other hand, the cost-effectiveness of manufacturing integrated circuits is pushing conventional single-layer resist processes. One solution to these demands is surface imaging.

Surface imaging utilizes the chemical change of a surface layer upon exposure to radiation. If a dry-etch-resistant material can be selectively incorporated into the surface layer of the exposed or unexposed region, the subsequent oxygen plasma etching affords pattern formation. In surface imaging, the thick layer can be used to cover the substrate topography, and reaction is induced only in the surface region. Therefore, the effects of the topography, reflection of substrates, and interference in the resist film on pattern profile and linewidth control are expected to be minimized. If the substrate topography is planarized with resist film, fine patterns can be obtained even in a limited DOF condition.

7.4.1 GAS-PHASE FUNCTIONALIZATION

Taylor and coworkers [182, 183] of AT&T Bell Laboratories first demonstrated the surface imaging called "gas-phase-functionalization." The concept is depicted in Figure 7.44. Reactive groups, A, convert to product groups, P, upon irradiation. Next, the exposed film is treated with a reactive gas, MR, containing reactive groups R and inorganic atoms M. The atoms can form nonvolatile compounds, MY, which are resistant to removal under oxygen-reactive ion-etching conditions using gas Y. When MR reacts with A, positive tone patterns can be obtained. Negative tone patterns can be obtained by selecting an MR that reacts with P.

The radiation-initiated reactions studied in the resist were the creation and destruction of unsaturated hydrocarbons (olefins, etc.) [182]. Diborane, which efficiently reacts with electron-rich C=C

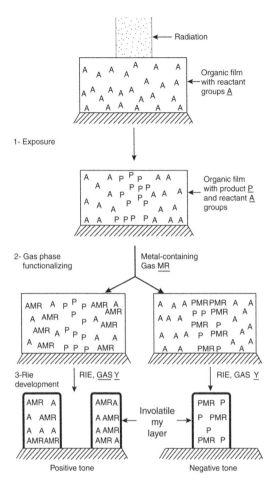

FIGURE 7.44 Schematic representation of gas-functionalized plasma developed resists showing (1) exposure a material with reactive groups A, (2) conversion of such groups to product group P, and (3) selective functionalization of either A or P with metal-containing reagent MR, where M is metal and R is reactive or inert groups or atoms. (4) Reactive-ion etch development using gas Y forming nonvolatile compounds MY and volatile AY, PY, and host Y compounds. (From Taylor, G. N., et al., *J. Electrochem. Soc.*, 131, 1658, 1984.)

double bonds to form organoboranes, was used as an inorganic functionalizing reagent. The formation of both positive and negative tone patterns was achieved using chloroacrylate homopolymers or copolymers with allylacrylate containing 12%–15% unsaturation.

The authors also described positive-acting resists based on a bisazide and a cyclized poly(isoprene) base polymer functionalized with gaseous inorganic halides [183]. Upon irradiation, a bisazide loses two molecules of nitrogen to give a dinitrene intermediate. The nitrene can crosslink a double bond of polyisoprene to give aziridine or insert into a carbon–hydrogen bond to give a secondary amine. Inorganic halides, such as silicon tetrachloride ($SiCl_4$) and tin tetrachloride ($SnCl_4$), are known to react with secondary amines. The transformation of the exposed area from an azide to either an aziridine or an amine is useful in gas-phase functionalization processes. Then, the subsequent oxygen reactive ion etching gave a positive relief image.

7.4.2 DESIRE

Coopmans and Roland [184, 185] proposed DESIRE (diffusion-enhanced silylating resist), whose process flow is depicted in Figure 7.45. After UV exposure, a selective silylation takes place in the

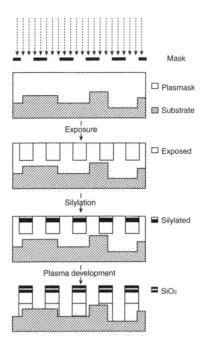

FIGURE 7.45 Process for diffusion-enhanced silylating resist (DESIRE). (From Coopman, F., and Roland, B., *Proc. SPIE,* 631, 34, 1986.)

exposed area. During dry development by oxygen plasma, silylated parts rapidly form a silicon dioxide–rich layer that retards further etching, while the unexposed region can be etched away by the anisotropic process, which results in a negative-type relief image. There have been increased reports including germylation [186, 187] since this work. The effect of chemical structure of silylating agents has also been evaluated in the DESIRE process [188, 189].

Visser et al. [190] studied the selective silylation mechanism of resists composed of DNQ compounds and a novolac resin using IR spectroscopy and Rutherford backscattering spectrometry. The model proposed is that some DNQs can act as physical crosslinkers between polymer chains via the formation of hydrogen bonds, whereas the corresponding indenecarboxylic acids cannot. It was found that during silylation, a swollen layer is formed with a sharp front separating it from the unreacted resin.

To obtain a positive tone image, the esterification reaction of photogenerated ketene from diazopiperidine with phenolic resin under a dry nitrogen atmosphere is used in image reversal processes (Figure 7.46) [191]. The number of hydroxy groups of the phenolic resin decreases by esterification in exposed areas, resulting in the reaction being blocked by silylating reagents. The following flood exposure renders the diffusivity for silylating reagents of unexposed areas in the first patternwise exposure, leading to a positive tone.

Resists containing azides were also utilized for positive tone [192]. The exposed area reduces the reactivity with silylating reagents by using crosslinking of polymers to decrease diffusivity, whereas unexposed areas can react with the silylating reagent.

7.4.3 Liquid-Phase Silylation

Sezi et al. [193, 194] of Siemens reported a resist that can be silylated in a standard puddle development track at room temperature. The resist is composed of an anhydride-containing copolymer and a DNQ compound. The silylation can be performed with an aqueous solution of bisaminosiloxane in water and a dissolution promoter. During the silylation, film thickness increases as swelling occurs.

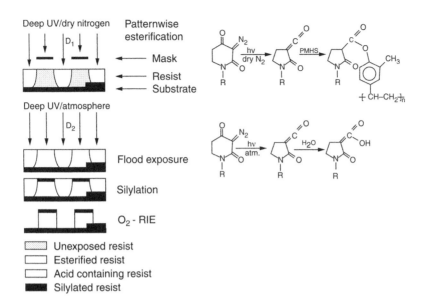

FIGURE 7.46 Imaging reversal of DESIRE process. (From Mutsaers, C. M. J., et al., *J. Vac. Sci. Technol. B*, 10, 729, 1992.)

It is difficult to explain the fast silylation (20 nm/s) by an increase in the free volume generated through molecular rearrangement and release of nitrogen molecules with the photolysis of DNQ. Salt formation between ICA and the primary amine moiety of aminosiloxane may induce the dissolution of the salt in the aqueous phase. After the dissolution of ICA in the exposed area into the aqueous phase, free space is created in the surface area, into which aminosiloxane molecules can diffuse and react with ICA or/and the anhydride groups of the base resin (Figure 7.47). The latter process leads to the formation of amide and carboxylic acid bound to the resin backbone. Although crosslinking and dissolution compete during silylation, crosslinking was confirmed in the exposed region, where the layer cannot be dissolved in any solvent. Liquid-phase silylation has also been reported by other groups [189, 195, 196].

7.4.4 Use of Chemical Amplification

MacDonald et al. [197] have reported a negative tone oxygen plasma–developable resist system that utilizes the change in chemical reactivity of the exposed area. The concept is outlined in Figure 7.48 using a resist composed of *t*-BOC-PHS and a photoacid generator. UV exposure and subsequent heating initiate acid-catalyzed deprotection of the *t*-BOC group to yield a phenolic hydroxy group. Because hexamethyldisilazane (HMDS) and (dimethylamino)trimethylsilane (DMATMS) are known to react with the phenolic hydroxy group and correspondingly, not to react with *t*-BOC-PHS, treating the exposed film with DMATMS vapor selectively incorporates an organosilicon species into the exposed region of the film. When the silylated film is exposed to oxygen plasma, the regions of the film that do not contain the organosilicon species are etched to the substrate, whereas those containing silicon are not etched, resulting in a negative relief image.

MacDonald et al. also have described a gas-phase image reversal process that generates a positive tone, plasma-developable image for a chemically amplified system [198]. The process is shown in Figure 7.49. The phenolic hydroxy groups that are produced by image exposure react with an organic reagent (which must not contain Si, Ge, Ti, etc.) to form a product that is thermally stable and unreactive toward silylation. Flood exposure and baking converts the remaining *t*-BOC-PHS to PHS. These hydroxy groups react with the silylating agent to selectively incorporate silicon. When

Resist polymer
Silylation
H2N Siloxane NH2
Crosslinked matrix
Siloxane
NH OH
OH NH
=Comonomer
Siloxane = – Si –(O – Si)n–
CH3

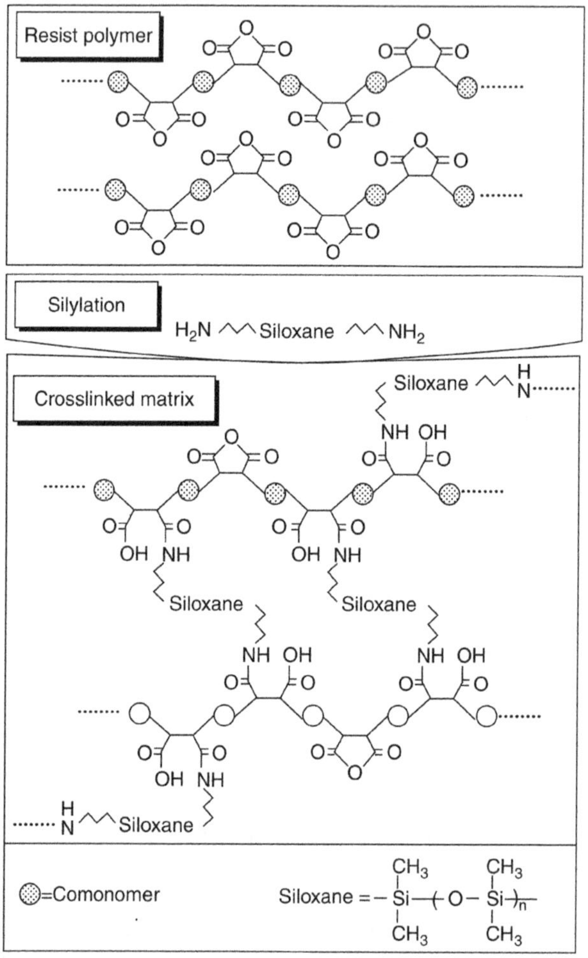

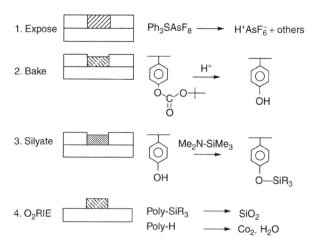

FIGURE 7.47 Silylation of an anhydride-containing polymer with a bisaminosiloxane in the aqueous phase. (From Sezi, R., et al., *Proc. SPIE*, 1262, 84, 1990.)

1. Expose
Ph3SAsF8
H+AsF6- + others
2. Bake
H+
OH
3. Silyate
Me2N-SiMe3
OH
O—SiR3
4. O2RIE
Poly-SiR3
Poly-H
SiO2
Co2- H2O

FIGURE 7.48 Negative tone dry development process using chemically amplified resist. (From MacDonald, S. A., et al., *Chem. Mater.*, 3, 435, 1991.)

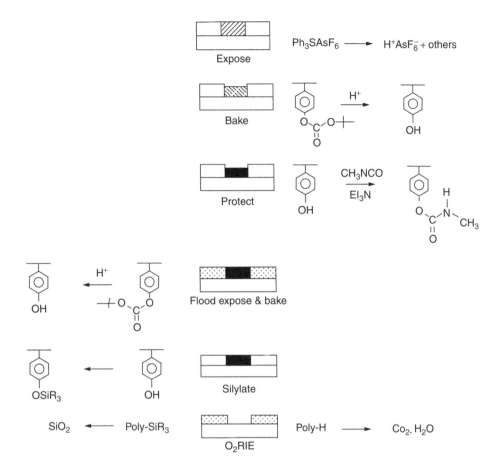

FIGURE 7.49 Positive tone dry development process using imaging reversal of chemically amplified resist. (From MacDonald, S. A., et al., *Chem. Mater.*, 4, 1346, 1992.)

the film is exposed to oxygen plasma, the regions that do not contain the organosilicon species are etched to the substrate, resulting in a positive tone relief image.

Workers of Philips [199] and Shipley [200] reported resists composed of phenolic resin, hexamethoxymethylmelamine as a crosslinker, and a photoacid generator. Photogenerated acid causes an acid-catalyzed condensation during postexposure baking, leading to crosslinking of phenolic resin as described in the previous section. A gaseous silylating reagent can diffuse selectively into unexposed areas to react with hydroxy groups of phenolic resin, whereas its diffusion is limited in exposed crosslinked areas. This gas-phase modification of exposed polymer film, followed by reactive oxygen-ion etching, leads to a positive tone relief image. The other negative resists using acid-catalyzed reaction, such as electrophilic aromatic substitution of 1,2,4,5-tetra(acetoxymethyl) benzene, can also be applied to gas-phase silylation plasma-developable resists [201].

7.4.5 Factors That Influence Pattern Formation

Although surface imaging has several advantages for future lithography, some factors should be discussed for actual use: silylation, plasma etching, and removal of silylated resists. In the silylation process, it is important to control the diffusion of silylating reagents, because swelling due to the diffused reagent may cause pattern deformation. There is also a possibility that silylation will occur in undesirable regions where the diffusion of silylating reagent is limited. Therefore, silylation selectivity is dependent on silylating reagents and process conditions such as reaction temperature and gas pressure.

During oxygen plasma treatment, silicon oxide–rich layer formation competes with etching, including sputtering, in the silylated region. SiCl formation depends on the silicon concentration and its distribution along the film thickness. Plasma etching conditions are key to determining the selectivity of the etching. Although the regions where silicon is not incorporated are to be etched away, "grass" of the sputtered silicon oxide on the substrate may cause residue problems.

7.5 RESISTS FOR ArF LITHOGRAPHY

For 193 nm lithography, deep-UV resists cannot be used as currently formulated. The aromatic polymers are not suitable due to their high absorption coefficients at 193 nm. Several approaches to obtain resists with improved transmittance at 193 nm have been reported. The use of acrylic polymers as base resins combined with acid-catalyzed reaction has been reported.

Allen et al. [202] demonstrated patterning with 193 nm exposure using acrylic polymers and a photoacid generator, which was initially developed for printed-circuit-board technology. The acrylic polymers are terpolymers of methylmethacrylate, *t*-butylmethacrylate, and methacrylic acid (Figure 7.50). The photogenerated acid induces acid-catalyzed deprotection of the *t*-butyl group to yield polymethacrylic acid. Carboxylic acid polymer is soluble in aqueous base, resulting in a positive image.

A disadvantage of acrylic polymers is poor dry-etch resistance. To improve the dry-etch resistance, researchers at Fujitsu [205, 204] proposed alicyclic polymers containing adamantane or norbornane groups (Figure 7.51). The alicyclic component without conjugated double bonds is desirable for transmittance at 193 nm and improved dry-etch resistance compared with aliphatic methacrylates

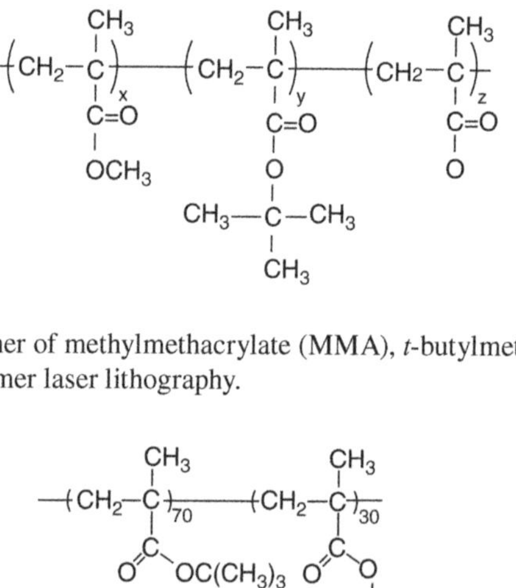

FIGURE 7.50 Acrylic terpolymer of methylmethacrylate (MMA), *t*-butylmethacrylate (TBMA), and methacrylic acid (MAA) for ArF excimer laser lithography.

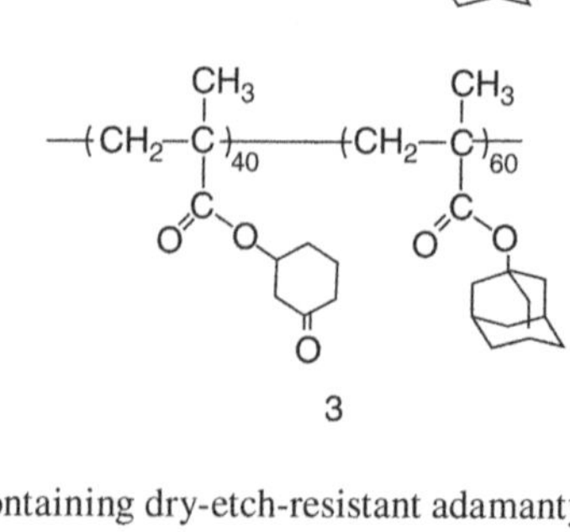

3

FIGURE 7.51 Acrylic polymers containing dry-etch-resistant adamantyl group for ArF resists.

such as poly(methylmethacrylate) (PMMA) or poly(*t*-butylmethacrylate). They prepared a copolymer of *t*-butyl methacrylate and adamantyl methacrylate as a base polymer for ArF excimer laser lithography. A similar approach for an ArF resist composed of poly(norbornylmethacrylate) and dicyclohexyldisulfone was reported by researchers at Matsushita [205]. Although it was reported that the dry-etch resistance of these polymers is compatible with that of DNQ-novolac resists, it is to be noted that the dry-etch resistance is always dependent on etching conditions, such as etching gas, pressure and flow rate of the etching gas, apparatus, power, shape and size of electrodes, pumping rate, etc.

Nakano et al. of NEC [206] reported an alkylsulfonium salt, cyclohexylmethyl-(2-oxocyclohexyl) sulfonium triflate (Figure 7.52), as an acid generator that shows high transmittance at 193 nm. A methacrylate terpolymer, poly(tricyclodecanylmethacrylate-co-tetrahydropyranylmethacrylate-co-methacrylic acid) (Figure 7.52), was synthesized and used as a base polymer. They obtained a negative image using a resist composed of the above and an organic solvent developer.

Researchers at Toshiba [207] found that compounds containing the naphthalene moiety afford a lower absorption coefficient at 193 nm than those with the phenyl group (Figure 7.53). This is ascribed to a shift of A_{max} (wavelength of absorption maximum) induced by conjugation extension from benzene to naphthalene. This finding leads to resists containing the naphthalene moiety rather than phenyl groups.

Using the high absorbance at 193 nm of phenolic polymers, Hartney et al. [208, 209] described surface imaging based on silylation. Irradiation at 193 nm produces direct crosslinking in phenolic polymers near the surface (Figure 7.54). The crosslinking prevents diffusion of organosilicon reagents to generate silylation selectivity. Subsequent oxygen plasma treatment forms an etch-resistant silicon oxide mask in the silylated area and etches the unsilylated areas to give a positive tone image.

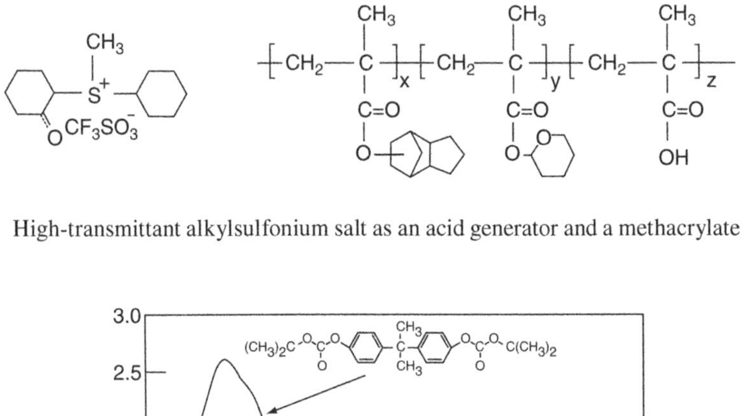

FIGURE 7.52 High-transmittant alkylsulfonium salt as an acid generator and a methacrylate terpolymer for ArF resists.

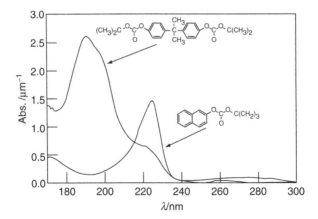

FIGURE 7.53 Visible-UV absorption spectra of *t*-BOC- protected naphthol and bisphenol A. (From Naito, T., et al., *Jpn. J. Appl. Phys,* 33, 7028, 1994.)

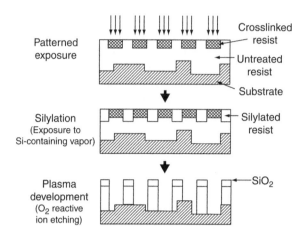

FIGURE 7.54 Schematic process flow for a 193 nm process. (From Hartney, M. A., et al., *Proc. SPIE,* 1466, 238, 1991.)

7.6 NEW APPROACHES OF CONTRAST ENHANCEMENT DURING DEVELOPMENT

Contrast enhancement during development using a reaction of base with base-labile compounds, such as DNQ sulfonic acid ester of phenolphthalein [42], *t*-BOC-protected phenolphthalein [102, 103] (Figure 7.26), and polymers containing acetoxystyrene [94] has been reported. Uchino et al. [210] proposed a new contrast enhancement method called *contrast-boosted resists* (CBRs), which consist of a phenolic resin, a photoactive compound, and a base-labile compound. The concept of a negative CBR is shown in Figure 7.55. The CBR offers high resolution because the photochemically induced solubility difference between exposed and unexposed areas is enhanced by the reaction of base-labile water-repellent compound with base. In exposed areas, the photochemically produced hydrophobic compounds and the water-repellent compound work together to retard the permeability of the base developer penetrating into the resist. Thus, the exposed area of the CBR is completely insolubilized in the developer. On the other hand, the base developer permeates into the unexposed area of the CBR and converts the water-repellent compound into a hydrophilic compound that gives no dissolution inhibition in a phenolic resin. Therefore, CBRs exhibit high contrast compared with conventional resists.

Uchino and coworkers [210] investigated halomethyl ketones listed in Table 7.2 as water-repellent compounds in a negative resist composed of 4,4′-diazido-3,3′dimethoxybiphenyl and *m,p*-cresol novolac resin. The expected products by reaction of halomethylketones in aqueous base are hydroxy-methylketones. It is interesting to note from Figure 7.56 that tris(bromoacetyl)benzene gives higher sensitivity and contrast than bis- and mono-bromoacetylbenzenes. This suggests that changing the three bromoacetyl (water-repellent) groups to hydroxyacetyl (hydrophilic) groups is necessary for high contrast enhancement. The concept can be also applied to positive resists.

7.7 UPDATE ON MODERN RESIST TECHNOLOGY

7.7.1 INTRODUCTION

The following sections will examine the "modern" resist materials used in lithography nodes beyond deep UV (DUV) (248 nm). Included in this section is an update of published research on advanced 248 nm materials. The revolutionary change in imaging mechanism called *chemical amplification* leads to high-speed, high-contrast resist materials that are the workhorse materials used today for photolithography. Although DUV resists have been covered in the preceding sections, a recent review from one of the pioneers of this field provides more detail [211].

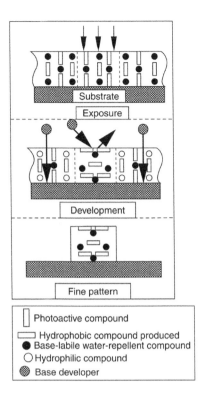

FIGURE 7.55 Schematic representation of contrast enhancement during development for negative CBR (contrast-boosted resist). (From Uchino, S., et al., in *Proceedings of the Regional Technical Conference on Photopolymers,* SPE, Ellenville, New York, 1994, 306.)

TABLE 7.2
Base-Labile Water-Repellent Compounds for CBR

Compound	Name	Chemical Structure
Bromoacetyl benzene	BAB	$COCH_2Br$
p-Bis(bromoacetyl)benzene	γ-BBAB	$COCH_2Br$ / $COCH_2Br$
m-Bis(bromoacetyl)benzene	m-BBAB	$COCH_2Br$ / $COCH_2Br$
1,3,5-Tris(bromoacetyl)benzene	TBAB	$COCH_2Br$ / BrH_2COC $COCH_2Br$
1,3,5-Tris(dibromoacetyl)benzene	TDBAB	$COCHBr_2$ / Br_2HCOC $COCHBr_2$
1,3,5-Tris(chloroacetyl)benzene	TCAB	$COCH_2Cl$ / ClH_2COC $COCH_2Cl$

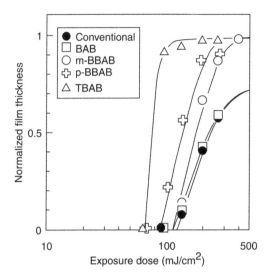

FIGURE 7.56 Sensitivity curves of negative CBRs using halomethylaryl ketones with a different number of bromoacetyl groups on a benzene ring. The exposure was carried out with i-line. The resist composition is 4,4′-diazido- 3,3′-dimethoxybiphenyl/halomethylaryl ketone/novolac = 10/20/100 (wt ratio). (From Uchino, S., et al., in *Proceedings of the Regional Technical Conference on Photopolymers*, SPE, Ellenville, New York, 1994, 306.)

The grand challenge of "post-DUV" (defined as resist technology for wavelengths below 248 nm [e.g., 193 nm, 157 nm, EUV]) is exploiting the benefits of the chemical amplification (CA) imaging mechanism (speed and contrast) with new polymer materials. Additionally, the desired feature sizes get smaller in successive generations. Thus, the new resist paradigm is "New materials, smaller features, but the constant is the CA imaging mechanism."

7.7.2 CHEMICAL-AMPLIFICATION RESIST UPDATE FOR 248 nm

7.7.2.1 PHS-Based Resists

First, recent developments in DUV resist materials as a jumping-off point for 193 nm resists are summarized. DUV resists are almost always based on homopolymers or copolymers of poly(4-hydroxystyrene), also known as PHS. Homopolymers of PHS are highly transparent at 248 nm, have high glass transition temperatures (T_g), and have rapid dissolution in standard resist developers. The properties of PHS materials are very different from those of their isomeric cousins—novolac resin. Although structurally similar, novolac (the workforce polymer in i-line and g-line photoresists) and PHS (workforce for DUV) materials are made via different polymerization chemistry. Novolac is derived from a condensation polymerization process (cresol and formaldehyde are the typical monomers) that leads to a complex mixture of polymers/oligomers. PHS is most typically prepared by the free-radical polymerization of hydroxystyrene or a protected hydroxyl styrene. The simple, straightforward polymerization produces a corresponding material that is quite simple, with a narrow molecular-weight distribution. Table 7.3 compares the properties of a typical novolac and a typical PHS homopolymer [212].

A wide variety of PHS-based DUV resists have achieved commercial use in the high-volume manufacture of integrated circuits. An excellent overview of the evolution of advanced 248 nm resists has recently been published by Lee and Pawlowski [213], who categorized the various PHS copolymer types. Figure 7.57 shows the various PHS material families in use today.

The PHS copolymer families differ in the type of acid-labile protecting groups used, a continuing theme of today's resist design strategy. High-activation protecting groups, typified by IBM's

TABLE 7.3

Novolac and PHS: Comparison of Properties

Property	Novolac	PHS
Molecular weight	Low	Medium
Polydispersity	Broad	Narrow
T_g	100 °C	160–180 °C
Dissolution rate (0.26 N TMAH)	Slow	Fast
Absorbance at 248 nm	ca. 1.0/μm	ca. 0.2/μm

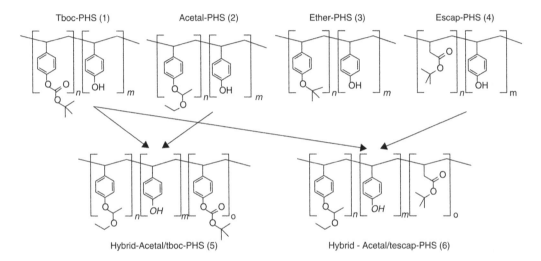

FIGURE 7.57 PHS material families in use today. The four major polymer types in DUV lithography (top) and two hybrid DUV polymer examples (bottom). (From Lee, D. and Pawloski, G., *J. Photopolym. Sci. Technol.*, 15(3), 427, 2002.)

ESCAP platform, achieved early commercial success; ESCAP has since become a workforce resist (commercialized by several resist vendors). The *t*-butyl ester functional group demands relatively high deprotection temperatures, so these resists are also referred to as *high-bake* resists. An advantage of the ESCAP platform is the high thermal stability that accompanies the choice of protecting groups, opening up the possibility for baking (postapply and postexposed) close to the T_g of the film. The "annealing" effect acts to densify the film and improve resolution, environmental stability, and etch resistance. Ito and coworkers [214], pioneers of the ESCAP platform, also observed another consequence of *t*-butyl ester protection: extreme dissolution contrast. Figure 7.58 shows the high dissolution contrast of the ESCAP resist. The production of carboxylic acid in the exposed film leads to extreme dissolution differentiation, with a dissolution rate ratio (R_{max}/R_{m}m) of 104. This design gave a dissolution rate unheard of in earlier DUV resists (or i-line resist technology, for that matter). The modern DUV resists rely on this conversion to carboxylic acid to boost contrast. A key point is that this extreme differentiation, first commercialized in the ESCAP platform, is commonplace in the world of 193 nm resists.

High-performance resists do not always require the enormous dissolution rate differentiation described above. For this reason, many modern DUV resists contain substantial amounts of other monomers. For example, nonactive monomers such as styrene are added (to improve etch resistance), or other, lower-activation protecting groups (e.g., *t*-BOC, acetals, etc.) are added to form terpolymers. These lower-activation groups apparently improve certain lithographic performance

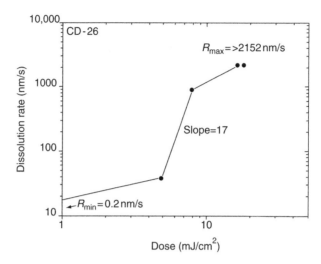

FIGURE 7.58   Dissolution contrast of the ESCAP resist. (From Ito, H., et al., *J. Photopolym. Sci. Technol.*, 9(4), 557, 1996.)

(e.g., isolated line performance). The wide availability of low-activation groups and third monomers allows resist designers to finely tune the performance of modern DUV resists, leading to a phenomenal level of performance of today's DUV lithography.

7.7.2.2   Limits of CA Resists

Before discussing 193 nm resists, fundamental research into the limits of the CA imaging mechanism, which uses 248 nm materials (note: the materials are also useful for EUV resists), will be described. The lessons learned through this fundamental research may also impact 193 nm materials, as these will presumably be used in the 45 nm node and possibly beyond.

Hinsberg and coworkers [215] addressed the extendibility of chemically amplified resists in a landmark paper in 2003. They noted that industrial roadmaps had assumed CA resists down to the 20 nm node and suggested that CA resist limits and the industry roadmaps are on a "collision course." Although PMMA (via high-energy radiation–induced chain scission) has achieved resolution to 20 nm, CA resists had not come close to this resolution. The very different imaging mechanism involving thermally driven acid-catalyzed processes is summarized in Figure 7.59. These limiting factors include the following:

1. Acid gradients at exposed/unexposed interface
2. Trace amount of background acid in "unexposed" regions much larger at lower k factors
3. Acid diffusion during post-exposed bake
4. Swelling and statistical dissolution effects at line edge
5. Small number of high-energy photons in small areas of high-sensitivity resists (shot noise)

In this work, the authors employed both simulation and experimentation. The simulation was based on parameters from t-BOC-protected PHS (a medium–activation energy protecting group) and a nonvolatile photoacid. The model has two effects dominating resolution: the chemical contribution (deprotection chemistry occurring outside the "exposed" areas) and the diffusion contribution of photoacid migrating outside the intended area. Interestingly, the contribution from deprotection chemistry dominates above 50 nm, whereas the diffusion contribution in this model system dominates below 50 nm. The authors also point out that medium- to high-activation groups show experimental resolution drop-off at linewidths of approximately 50 nm (ESCAP resists). Figure 7.60 summarizes this work.

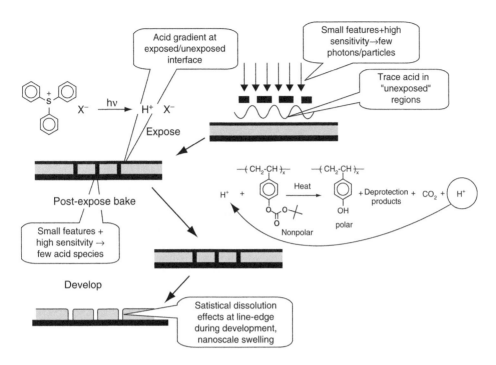

FIGURE 7.59 Origin of limiting factors in chemically amplified resists. (From Hinsberg, W., et al., *Proc. SPIE,* 5039, 1, 2003.)

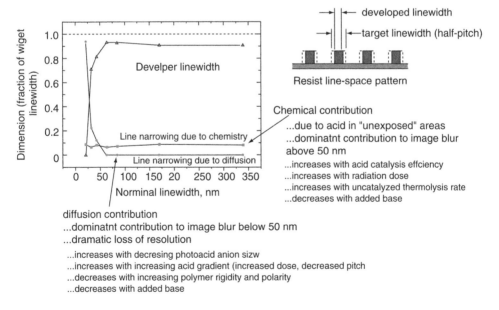

FIGURE 7.60 Image blur as a function of line/space dimension for the PFBS/*t*-BOC model resist system calculated using the PEB model. (From Hinsberg, W., et al., *Proc. SPIE,* 5039, 1, 2003.)

7.7.2.3 Resolution of CA Resists

Work at IBM published the following year (2004) further disclosed that they had indeed achieved nearly 20 nm resolution (using e-beam lithography for exposure) in a CA resist [216]. The resist used a low-activation protecting group bonded to PHS and is known as *KRS*. The resist undergoes deprotection at low bake temperatures. In fact, KRS can be operated with PEB temperatures of 20 °C;

FIGURE 7.61 30 nm line/space features printed in KRS using e-beam lithography at a dose range of 36–42 μC/cm². (From Wallraff, G. M., et al., *Proc. SPIE,* 5753, 309, 2005.)

that is, it does not need a postexpose bake. The room-temperature deprotection is carried out in the presence of stoichiometric water, so the acid diffusion is strongly suppressed. High-quality 30 nm features have been recently printed in KRS resists in both e-beam and EUV exposure experiments (see Figure 7.61).

Interestingly, in the absence of a base quencher, KRS shows substantial diffusion when baked at room temperature [217]. Even so, the low-bake KRS resist showed much less acid diffusion than the high-bake ESCAP processed above PEB, greater than 100 °C. However, when KRS was processed above 100 °C, it also showed high acid diffusion.

These IBM resist researchers also found, using e-beam imaging, that mid- to high-bake resists had resolution limits of 50 nm and above, whereas KRS easily printed features less than 40 nm. Subsequently, KRS has been shown to print to 30 nm in early EUV lithography studies [218]. Therefore, it is clear that resolution in CA resists can be extended beyond the 45 nm node.

The fantastic array of protecting groups, polymer types, formulation additives (base quenchers, etc.), and PAGs featuring low-diffusion photoacids may be assembled into high-performing resist formulations. Then, the resist process, coupled with the characteristics of the imaging tool, will allow lithographers to design manufacturable processes for the future. The future (at least the immediate future) takes advantage of the learning described above but uses completely different materials.

7.7.3   193 nm Resists

7.7.3.1   Backbone Polymers

Several excellent reviews cover the early development of 193 nm resists [212, 219]. As discussed above, the use of the chemical-amplification imaging mechanism and high-activation *t*-butyl ester protecting groups demonstrates a common link between DUV resists (e.g., ESCAP) and 193 nm resists. In fact, some of the earliest 193 nm resists (IBM Version 1) were used in experimental DUV formulations as polymeric dissolution inhibitors [220]. Methacrylate terpolymers (shown in Figure 7.62) were useful first-generation resists for testing the initial generation of 193 nm microsteppers. They also had utility in the design of 248 nm resists as they were miscible with a variety of phenolic resins (PHS, novolac, etc.).

These methacrylate polymers added significant transparency to 248 nm resists (allowing the use of novolac resin) and were similar in behavior to ESCAP-type copolymers. Interestingly, these methacrylate copolymers were not initially designed for 193 nm lithography but for PWB (printed

Structure 1

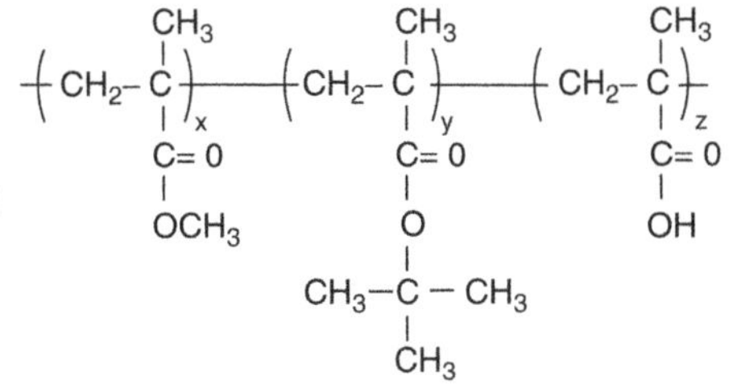

MMA-TBMA-MAA methacrylate terpolymer

FIGURE 7.62 IBM Version 1 terpolymer. (From Allen, R. D., et al., *J. Photopolym. Sci. Technol.*, 6(4), 575, 1993; Wallraff, G. et al., *J. Vac. Sci. Technol. B*, 11(6), 2783,1993.)

wiring boards) direct write lithography applications [221]. Although modern 193 nm resists are based on monomers of very different structures, many design elements of this material (IBM Version 1) are still in use today.

The material is based on simple, direct terpolymerization of methacrylate monomers, with no postpolymerization reactions as commonly found in 248 nm materials. Multiple monomers with separate functions are actively used today. Also, the use of moderate protecting-group concentrations (30–50 mol%) is in the same range as the 193 nm resists in use today. For example, the IBM Version 1 structure, the *t*-butyl ester, was used as the hydrophobic protecting group, and a combination of methylmethacrylate (MMA) and methacrylic acid (MAA) allowed the appropriate level of acidity and polarity. The "modern" replacements for these key functionalities are as follows:

1. Cyclic protecting groups for etch resistance
2. Cyclic polar groups (e.g., lactones) for polarity and etch resistance
3. Fluoroalcohols or carboxylic acids for acidity

The poor etch resistance of acrylic polymers (at least in comparison to phenolics) is predicted by the Ohnishi model [222] and more recently refined by Kunz and coworkers [223]. Essentially, the composition or molecular structure controls the etch resistance. Polymer etch rate is closely coupled to the fraction of resist mass contained in ring structures. Thus, aromatic or cycloaliphatic polymers have substantially lower etch rates than do acrylic polymers. The Fujitsu group perhaps first recognized the importance of cyclic polymers for etch-resistant 193 nm resists in the early 1990s [205]. In later work, the same group recognized the importance of lactone functional groups as excellent polar modifiers [224].

7.7.3.2 PAG Effects in 193 nm Resists

Initially, all 193 nm resists used triflic acid photoacid generators. Whereas DUV resists had significant, controllable unexposed film loss, most 193 nm materials had no dark erosion (no unexposed thinning). Additionally, early 193 nm resists had two characteristics: methacrylate polymers with no unexposed thinning and high-activation protecting groups. Workers at IBM [225] first established the importance of nonvolatile, low-diffusible PAGs in 193 nm resists. Higher–molecular weight perfluoroalkyl sulfonates (e.g., perfluorobutyl sulfonate) were shown to dramatically reduce volatility, improve resolution, and allow higher bake temperatures than triflic acid–containing PAGs. The volatility of triflic acid was shown to cause T-topping and webbing in nonthinning high-activation (high-bake) resists. This observation leads to breakthrough resolution, with high-quality 150 nm features being produced in etch-resistant methacrylate polymers (IBM Version 2 resists). Higher–molecular weight perfluoroalkyl photoacids of this type are in use today for high-resolution resist applications. Figure 7.63 shows the consequence of photoacid volatility on resist imaging, and Figure 7.64 shows clear image patterns without T-tops.

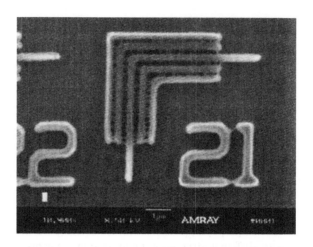

FIGURE 7.63   Consequence of photoacid volatility on resist imaging. (From Allen, R. D., et al., *J. Photopolym. Sci. Technol,* 10(3), 503, 1997.)

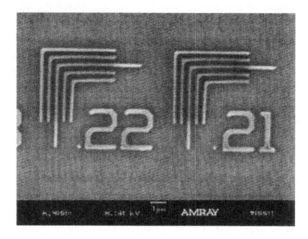

FIGURE 7.64   Clear image patterns without T-top. (From Allen, R. D., et al., *J. Photopolym. Sci. Technol.,* 10(3), 503,1997.)

To summarize the early exploratory work in 193 nm resists in terms of highlights and design principles that apply to twenty-first-century resists:

1. Observation that alicyclic methacrylate polymers have reasonably good etch resistance and transparency at 193 nm [203].
2. Observation that methacrylate ter- or tetrapolymers can form CA resists with high resolution. [220]
3. Combination of etch resistance and good imaging performance in an alicyclic modified methacrylate resist [226]; this work introduced the use of steroid dissolution inhibitors.
4. Strong polar groups (lactones) for compatibility of nonphenolic resists with industry standard developers [224].

7.7.3.3   Cyclic-Olefin-Backbone Polymers

Very rapidly after these initial findings, significant advances were made in cyclic olefin polymers that offered etch resistance superior to methacrylate-based resists. Researchers at AT&T [227] and the University of Texas [228] investigated resist materials based on maleic anhydride alternating copolymers. The AT&T group focused on tetrapolymers of norbornene, maleic anhydride, and a combination of *t*-butyl acrylate and acrylic acid. The Texas group pioneered the use of dinorbornene monomers for improved etch resistance. They used *t*-butyl esters of dinorbornene (also known as *tetracyclododecene)* in copolymers with maleic anhydride. Both these research groups had excellent success with 193 nm resist compositions that combined etch resistance, imaging performance, and process compatibility with standard developers (0.26 N TMAH). The AT&T group employed cholate ester dissolution inhibitors to produce developer compatibility, presumably required due to their use of acrylic acid. The Texas group used a carboxylic acid–free resist design and obtained good aqueous development properties, presumably due intrinsically to the anhydride group.

The copolymerization chemistry of maleic anhydride/cyclic olefin polymers was less well established than that of the methacrylate polymerization used to prepare methacrylate co- and terpolymers. Fundamental investigations of copolymerization of maleic anhydride with norbornene and norbornene esters were completed by researchers at Virginia Tech and IBM [229]. This work demonstrated that norbornene polymerizes very efficiently with maleic anhydride to afford high-yielding, medium-MW (8000–9000 Da) polymers with fairly narrow polydispersities (approximately 1.5). The polymerizations were very sensitive to total solids and behaved best under relatively concentrated conditions (60% solids). The use of substituted cyclic olefins (e.g., norbornene-*t*-butyl ester) had a pronounced negative impact on yield and molecular weight. Terpolymerization studies (norbornene/maleic anhydride/norbornene-1-butyl ester) using in situ Fourier transform infrared spectroscopy (FTIR) showed that the rate of polymerization sharply decreased with an increase in norbornene ester. The relative rate of norbornene incorporation was higher (1.7 times) than that of the norbornene ester. The use of protected acrylates and methacrylates was also investigated. These workers found that high conversions and uniform materials were produced when acrylate monomers (*t*-butyl acrylate) were used, whereas methacrylate monomers (*t*-butyl methacrylate) produced lower yields and nonuniform polymers, because the methacrylate monomer polymerized early in the polymerization. This data suggests that the AT&T workers selected the monomer types for optimal polymerization efficiency.

Because of the relatively straightforward synthesis by free-radical polymerization, tremendous attention was paid to the maleic anhydride polymers through the late 1990s. Commercial success did not follow, however. This was due, in part, to the success of low-oxygen methacrylate polymers (maleic anhydride carries three oxygen atoms) that gave an etch advantage to the methacrylates.

An even more complex polymerization forms linear polymers via cyclic olefin addition polymerization when catalyzed by appropriate metal compounds [230]. These materials have a significant etch advantage and as such, are still being actively pursued. This polymerization chemistry was pioneered by B.F. Goodrich (now Promerus) and uses nickel or palladium catalysts. The materials are highly transparent at 193 nm. An optimized resist based on this polymerization was recently described by IBM [231]. The central problem with these rigid materials is swelling during development. The use of fluorocarbinol functional groups to defeat this problem is an excellent example of the impact of 157 nm resist research on 193 nm materials.

7.7.3.4 Backbone Polymers with Hexafluoroalcohol

Further exploitation of 157 nm resist learning for high-performance 193 nm materials has been recently reported for negative and positive resists. Methacrylate polymers have been modified through the incorporation of hexafluoroalcohol (HFA) [232]. Researchers at IBM and JSR found that incorporation of fluorocarbonol into methacrylate polymers greatly improved the dissolution response and PEB sensitivity. The chemical structures of these new methacrylate monomers are shown in Figure 7.65.

The dissolution rate versus dose plots appear to be more similar to DUV (PHS-based) materials than the traditional methacrylate polymers used in 193 nm resists, which are prone to swelling during development. Another significant benefit from HFA incorporation was the remarkably reduced

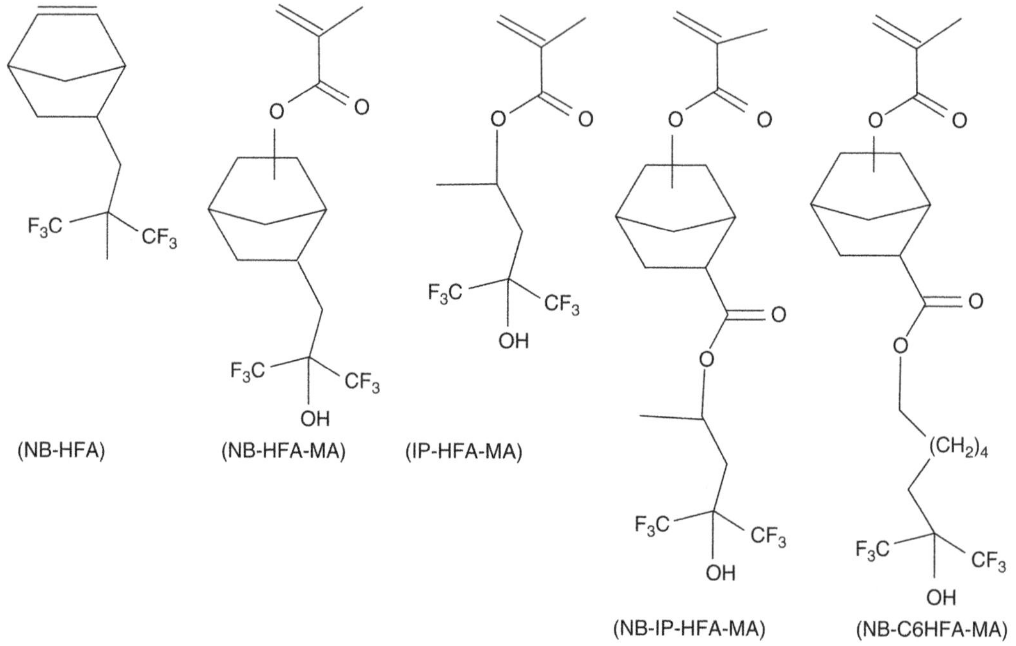

FIGURE 7.65 Chemical structures of HFA methacrylate monomers. (From Varanasi, P. R., et al., *Proc. SPIE*, 5753, 131, 2005.)

PEB sensitivity (linewidth change/°C). High PEB sensitivity was a notorious problem in early 193 nm resists. Figure 7.66 shows the reduction for two similar polymers; one is a lactone copolymer and the other is an HFA copolymer.

The HFA-containing methacrylate monomers used in this work have also been applied to the design of 193 nm negative resists. A necessary condition for negative resists is that the starting resist material is soluble in aqueous base. While carboxylate polymers are indeed soluble at a relatively high concentrations of COOH groups, swelling greatly limits resolution. Hattori and coworkers at Hitachi have tried to circumvent this problem by using an acid-catalyzed intra-molecular esterification reaction (ring closing to lactone) [233]. The best lithography was found using very dilute developer (0.05% TMAH). The fully reacted polymer has no acid groups (it is a lactone polymer). The starting material dissolves rapidly in developer. The difficulty arises in regions with high COOH content, but not so high that the material is soluble. This intermedi-ate composition can produce swelling. Nevertheless, high-quality imaging was produced in this polarity-switch resist.

Researchers at IBM recently published an HFA-based resist that also operates via a polarity-switch mechanism [234]. The imaging mechanism was the acid-catalyzed elimination of water (or other low–molecular weight polar species). The base polymer was comprised of an HFA acrylate (for dissolution) and tertiary alcohol containing cyclic olefin monomer for polarity switching. This material (Figure 7.67) was able to resolve features down to 120 nm without swelling or microbridg-ing (common problems with negative resists).

7.7.4 IMMERSION LITHOGRAPHY MATERIALS

Recently, the focus on advanced lithographic materials has been on 193 nm immersion lithography. Much attention has been given to the measurement, control, and elimination of component extrac-tion from the resist film into the immersion fluid (water). Several recent publications detail methods used to measure resist PAG extraction [235]. A recent publication shows that the use of protective

Arf methacrylate resist (peb sensitivity: –6 nm/C)

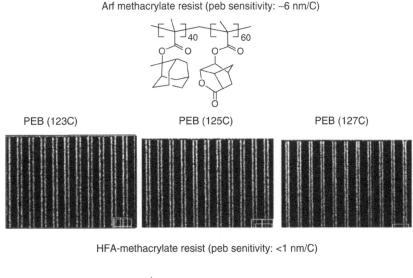

HFA-methacrylate resist (peb senitivity: <1 nm/C)

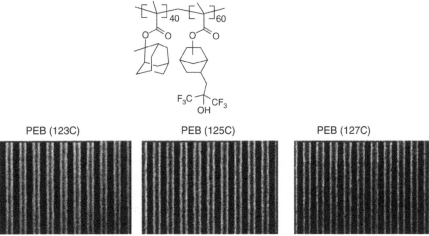

FIGURE 7.66 PEB sensitivity of fluoromethacrylate resist materials. (From Varanasi, P. R., et al., *Proc. SPIE*, 5753, 131, 2005.)

top coats reduces, and in some cases eliminates, extraction of PAG into water. The top-coat materials must meet the following requirements:

1. Materials are insoluble in water but are soluble in developer.
2. Materials can be spin-coated in solvents that are nonsolvents for resists.
3. Materials are transparent at 193 nm.

A class of top coats exists, based on HFA methacrylate polymers similar to those described earlier, that meets all these requirements [236]. Figure 7.68 shows the structures of some typical top-coat candidates.

7.7.5 Negative Tone Developable (NTD) ArF Resists

A new application using ArF resists in reverse tone imaging was developed recently [237]. Through the use of a solvent developer such as n-butyl acetate (NBA) and/or 2-heptanone, a negative tone

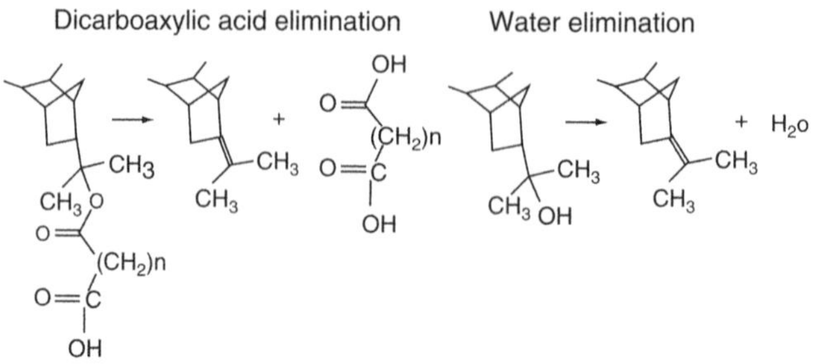

FIGURE 7.67 Polarity switch negative resist based on HFA-containing polymers. (From Sooriyakumaran, R., et al., *Proc. SPIE,* 5376, 71, 2004.)

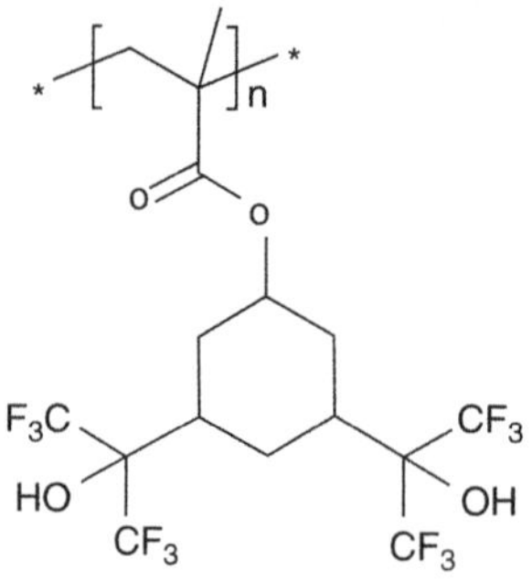

FIGURE 7.68 Chemical structure of immersion top-coat materials. (From Allen, R. D., et al., *J. Photopolym. Sci. Technol.,* 18(5), 615, 2005.)

pattern can be obtained with standard chemically amplified ArF resists. The development of negative tone developable (NTD) resists was driven by the improved image log slope for low-k_1 contact hole imaging using bright-field masks as opposed to dark-field masks for positive tone resists (Figure 7.69) [238]. The aerial image contrast extends to trench resolution as well. As evidenced by the extensive literature for NTD resists, carboxylic acid protecting groups, photoacid generators, and quenchers similar to those found in PTD resists can be used [239]. The development difference between TMAH and organic solvent is striking. Whereas TMAH must react with the carboxylic acids on the polymer before dissolution, the organic solvent only has to diffuse in the unexposed resist and solubilize the resist polymer. This simplified development process leads to lower surface roughness of the NTD resist, as evidenced by in situ high-speed atomic force microscopy (AFM) experiments (Figure 7.70) [240]. Figure 7.71 shows a PTD contrast curve versus an NTD contrast curve for the same ArF resist with only the developer type changed from TMAH for PTD to NBA for NTD development [239]. The extensive shrinkage of the NTD resist pattern is a result of the acid-catalyzed deprotection events in the exposed region as well as some solubility of the exposed resist in NBA. The extensive shrinkage may be contributing to the enhanced trench/contact resolution of NTD resists.

7.8 EUV RESISTS

With the introduction of extreme ultraviolet (EUV) lithography, a need exists for resists customized for ionizing radiation of 13.5 nm wavelength. Because of the high-energy nature of EUV radiation, a broad array of resist chemistries can be used [241]. However, the relatively weak source of the EUV radiation has led to the pursuit of resist chemistries that have very fast sensitivity, such as CARs [242]. CARs have been developed and utilized for longer wavelengths, most notably for exposure

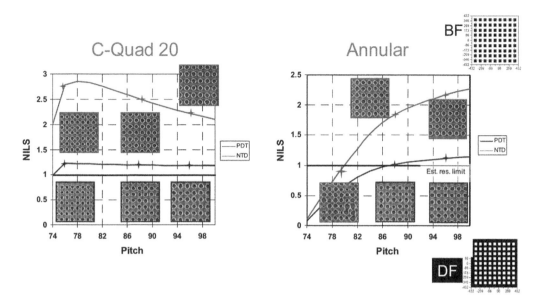

FIGURE 7.69 NILS @ 1.35 NA and resist images for two illuminations. Lithographic response is better for the bright-field mask. (From Reilly, M., et al., *Proc. SPIE*, 8325, 832507–1, 2012.)

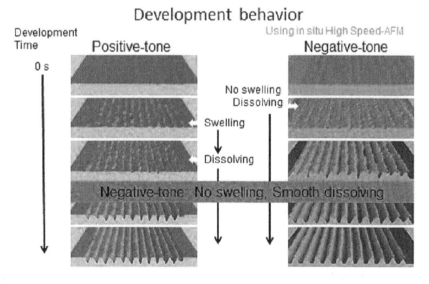

FIGURE 7.70 Development behavior of NTD compared with PTD, as observed using in situ high-speed atomic force microscopy. NTD shows no swelling and dissolves smoothly. (From Fujimori, T., et al., *Proc. SPIE*, 9425, 942505-1, 2015.)

with 248 and 193 nm radiation [243]. Due to the absorbing properties of the resins, specific resin materials had to be designed for 248 and 193 nm wavelengths, which are sufficiently discussed in the appropriate sections of this chapter. No such absorbing issue exists at 13.5 nm. Table 7.4 shows the linear absorption coefficients for a variety of lithographic resins, which leads one to conclude that at the thicknesses required for patterning, the resins are very transparent [244]. In fact, the opposite problem occurs at 13.5 nm: the lithographic resins are too transparent. Only perfluorinated polyethylene has sufficient absorption for a 40 nm film [244b]. However, good imaging can be obtained with CAR materials. We will discuss EUV CAR performance in the next section.

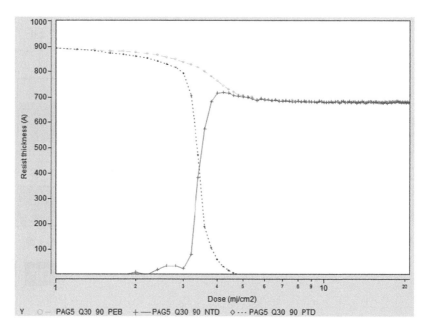

FIGURE 7.71 Contrast curves for a chemically amplified resist: showing deprotection shrinkage prior to development step, development in 0.26 N TMAH, and development in n-butyl acetate. (From Landie, G., et al., *Proc. SPIE*, 7972, 797206-3, 2011.)

7.8.1 EUV CAR Resists

EUV CAR resist development builds on an extensive learning curve from the well-understood lithographic patterning at longer wavelengths. The first widely available EUV resists were simply extensions of the previously developed ESCAP resists based on PHS/styrene/*t*-butyl acrylate polymers [245]. These resists are robust and meet the fast sensitivity and reasonable resolution of 30 nm half-pitch requirements [246]. These resists were used extensively in the early tool development stage. As more was learned during the early resist development stage, it became clear that specific resist materials optimized for the 13.5 nm wavelength were needed. Some of the best early work focused on cycloaliphatic polymers as used in 193 nm resists versus phenol-containing resists as developed for 248 nm. Fedynyshyn et al. elucidate a variety of matrix factors that impact overall resist sensitivity [247]. Yamamoto et al. found that proton donor sources such as vinyl phenol can enhance acid generation in EUV exposure [248]. One key learning was the necessity to control acid diffusion in order to enhance CAR resolution. Some of the reported approaches to lower acid diffusion included the use of a high-T_g polymer matrix [249], a hydrophilic polymer matrix [250], large, polar acid anion design [251], PAG anion incorporation in the polymer [252], and the photodestroyable quencher (PDQ) [253] concept. With these methods of diffusion control, 13 nm half-pitch resolution has been achieved with CAR resists! (Figures 7.72 and 7.73).

Other issues particular to EUV lithography have led to novel resist concepts. One aspect is the effect of out-of-band (OOB) radiation on resist performance. OOB is unwanted wavelengths exposing the EUV resist. OOB wavelengths of greater concern are 248 and 193 nm. It has been estimated that 4% of the radiation from the EUV exposure tool is unwanted OOB [254]. Roberts et al. found the OOB sensitivity of the resists they examined to be within an order of magnitude using the resist absorbance value [255]. Resist designers use extensively PAGs that comprise the triphenyl sulfonium [TPS] cation, which has high sensitivity to OOB radiation [256]. Hence, there was a need for the design of EUV-specific PAG cations that have high sensitivity to EUV but little to no sensitivity to the longer wavelengths. Indeed, some novel PAG cations with low sensitivity to OOB have been designed [257].

TABLE 7.4

Absorbance Properties of Common Lithographic Polymers

Name	Structure	Formula	Density (g/cm³)	Calculated linear absorption coefficient (μm⁻¹)
PMMA		$C_5H_8O_2$	1.18	5.19
PNB		C_7H_{10}	0.92	2.55
PSt		C_8H_8	1.05	2.95
PHOSt		C_8H_8O	1.16	4.05
PTMSSt		$C_{11}H_{16}Si$	1.14	2.78
PMPS		C_7H_8Si	1.12	2.60
PPSSQ		$C_{24}H_{20}Si_4O_4$	1.50	4.52
PAF		$C_{15}H_{14}O_3F_{12}$	1	6.97

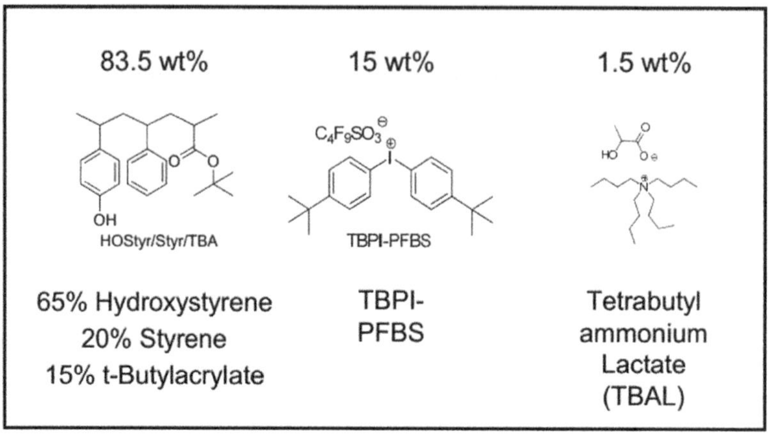

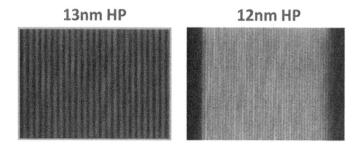

83.5 wt%   15 wt%   1.5 wt%

HOStyr/Styr/TBA   TBPI-PFBS

65% Hydroxystyrene   TBPI-   Tetrabutyl
20% Styrene   PFBS   ammonium
15% t-Butylacrylate   Lactate
                                             (TBAL)

FIGURE 7.72 Representative ESCAP resist formulated specifically for EUV lithography. (From Narasimhan, A., et al, *Proc. SPIE,* 9422, 942208, 2015.)

13nm HP   12nm HP

FIGURE 7.73 EUV CAR resist exposed on an ASM-L NXE3300B scanner. Other issues particular to EUV lithography have led to novel resist concepts. One aspect is the effect of out-of-band (OOB) radiation on resist performance. OOB is unwanted wavelengths exposing the EUV resist. OOB wavelengths of greater concern are 248 and 193 nm. It has been estimated that 4% of the radiation from the EUV exposure tool is unwanted OOB [254]. Roberts et al. found the OOB sensitivity of the resists they examined to be within an order of magnitude using the resist absorbance value [255]. Resist designers use extensively PAGs that comprise the triphenyl sulfonium [TPS] cation, which has high sensitivity to OOB radiation [256]. Hence, there was a need for the design of EUV-specific PAG cations that have high sensitivity to EUV but little to no sensitivity to the longer wavelengths. Indeed, some novel PAG cations with low sensitivity to OOB have been designed [257].

7.8.1.1 EUV CAR Resist Reaction Mechanism

It is important to distinguish between the CAR positive resist reaction mechanism for optical lithography and for EUV lithography. Tagawa and Kozawa have done significant work in the area of ionizing radiation (EUV and electron beam) versus optical lithography (193 and 248 nm) and its impact on resist patterning [258]. As Figure 7.74 depicts, ionizing radiation such as EUV is absorbed by the polymer matrix to generate secondary electrons of varying energies (5–80 eV), which can reduce the acid generator to produce acid [259]. This step is different from optical lithography, whereby the photoacid generator is directly photolyzed by the incoming exposure radiation. Once the acid is produced, the normal steps of acid-catalyzed deprotection and acid diffusion are carried out. Some workers have reported crosslinking reactions at higher dose leading to negative tone behavior [260].

Work to take advantage of this alternate mechanism has been reported. Incorporation of halides to increase the absorption cross section has been reported [261]. The use of more easily reduced acid generators has led to sensitivity enhancement as well [262].

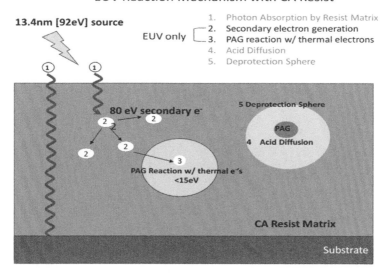

EUV Reaction Mechanism with CA Resist

13.4nm [92eV] source

EUV only
1. Photon Absorption by Resist Matrix
2. Secondary electron generation
3. PAG reaction w/ thermal electrons
4. Acid Diffusion
5. Deprotection Sphere

80 eV secondary e⁻

5 Deprotection Sphere

PAG

4 Acid Diffusion

3

PAG Reaction w/ thermal e⁻'s
<15eV

CA Resist Matrix

Substrate

FIGURE 7.74 EUV resist reaction mechanism.

7.8.1.2 Resolution–Linewidth Roughness–Sensitivity (RLS) Tradeoff

CAR resist performance has been driven by three main performance criteria: resolution, linewidth roughness (LWR), and sensitivity. These criteria are actively being pursued simultaneously for optimum performance. It has been recently noted that there may indeed be a difficult tradeoff in optimizing all three criteria simultaneously [263]. Some researchers are giving the so-called RLS tradeoff the term *triangle of death*. Since resolution is primarily improved by acid diffusion control and quencher concentration (higher is better), the sensitivity of a CAR is thereby limited. A similar tradeoff exists for LWR versus sensitivity. This issue is especially relevant for EUV lithography, where the lack of EUV photons is creating a need for very fast resists. Figure 7.75 shows the effect of the RLS tradeoff on LWR versus dose [264].

Many ideas have been attempted to improve the RLS tradeoff. One concept that has been explored is the use of acid amplifiers in EUV CARs [265]. Acid amplifiers are molecules that are triggered by photogenerated acids to yield a catalytic cascade of acid to effectively lower the dose necessary for acid production. The acid amplifier molecule must be thermally stable for adequate resist shelf life and must not diffuse too far during the PEB step. Kruger et al. reported on the first fluorinated sulfonate esters as acid amplifiers that can produce catalytic amounts of triflic acid [266]. They also designed a series of fluorosulfonic acids and successfully patterned them with a champion result 2.1 nm line edge roughness (LER) for a 50 nm dense pattern (Figure 7.76).

7.8.1.3 Positive Tone EUV CAR Resists

The first resists successfully patterned with EUV radiation were the so-called ESCAP-type resists developed by Ito for KrF lithography. The formulations were modified by Brainard and colleagues to thinner thicknesses and higher PAG loadings [245]. These resists served as early learning vehicles for patterning and measuring acid yields. Ultimately, ESCAP-type resists were capable of 30 nm hp patterning with exposure doses in the 15–20 mJ/cm² range. Meanwhile, EIDEC in Japan was doing early screening on novel resist concepts with its own EUV exposure unit [267]. It was an early proponent of the molecular glass–based resist. The molecular glass (MG) concept was built upon the hypothesis that polymers were too large for ultrafine patterning in the sub-30 nm regime. PHS polymer, for instance, with a molecular weight of 10,000, is estimated to have a radius of gyration of 3.0 nm, which is about 10% of the critical dimension target [268]. By making discrete MGs, such

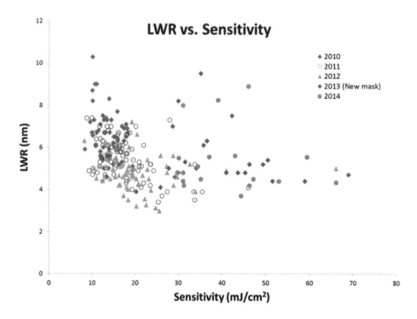

FIGURE 7.75 LWR and sensitivity trend chart through 2010–2014 from SEMATECH. Basically, the two parameters are a tradeoff. One of the challenges is to find a way to reduce both the LWR and dose sensitivity parameters simultaneously. (From Montgomery, C., et al., *Proc. SPIE,* 9425, 942526, 2015.)

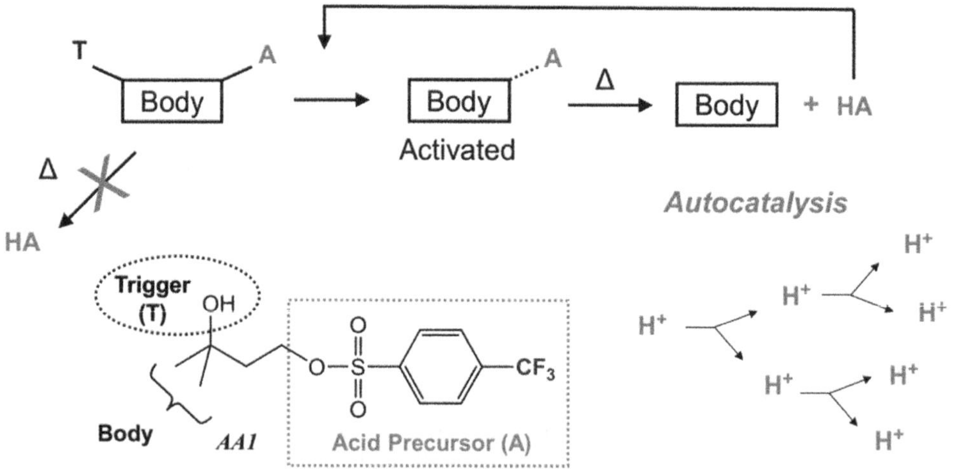

FIGURE 7.76 Generic representation of acid amplifiers (AAs). These compounds consist of three parts: a trigger (T), an acid precursor (A), and a body. (From Kruger, S., et al., *Proc. SPIE*, 8325, 832514, 2012.)

as calixarenes, one can reduce the average molecule size to less than 1 nm. Figure 7.77 shows some MG structural paradigms that have been used in EUV formulations [269]. Many workers have tried to design MGs, most notably Nishikubo, who developed the NORIA system [270]. One of the challenges of replacing polymer with MGs is that it is difficult to build thermal and mechanical properties into the MG. If the glass transition temperature is too low, for instance, the photoacid diffusion becomes too large to control image blur. Also, it is difficult to build dissolution contrast in an MG resist, which limits the overall lithographic performance. NORIA-based positive tone resists have been developed by Yamamoto et al. with an adamantyl ester protecting group [271]. This type of

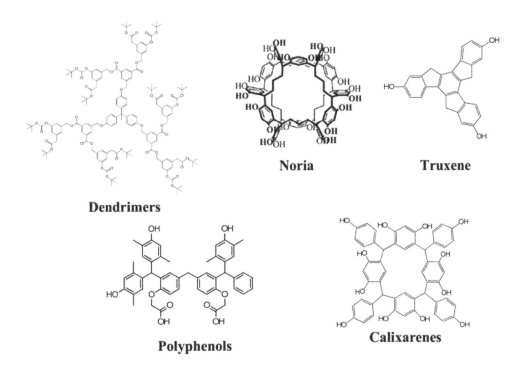

FIGURE 7.77 Examples of different cores for molecular glass resists for EUV lithography. (From D.P. Green, et al., *Proc. SPIE,* 8679, 867912, 2013.)

resist resolved 22 nm hp, as reported by Maruyama [272]. Frommhold et al. have recently reported on a fullerene-based resist whereby protecting groups were selectively grafted on the fullerene structure to yield high-resolution resists [273].

Another interesting class of EUV CAR resist has been developed: the so-called polymer-bound PAG systems. Initial work was done by Gonsalves, who reported the copolymerization of (phenyl methacrylate)dimethyl sulfonium triflate with ethyl adamantyl methacrylate (EADMA) and vinyl phenol to make PAG cation-bound resist polymers [274]. These polymers exhibited fast sensitivity and reasonable resolution down to 50 nm. The question became whether one wanted to attach the PAG through the cation and allow the acid anion to diffuse or to attach through the PAG anion to reduce acid diffusion. Thackeray reported on the improved resolution achieved when copolymerizing the acid anion of the PAG with methyl adamantyl methacrylate (MADMA), hydroxyvinyl naphthalene (HVN), and a hexafluoroalcohol unit, as shown in Figure 7.78 [275]. This polymer achieved 25 nm resolution easily and confirmed the advantage of tethering the acid anion to the resist polymer [276]. Further resolution improvements to the PBP system led to achievement of 15 nm hp with reasonable sensitivity of 20 mj/cm2 [277]. Tamaoki et al. reported a comparison of traditional EUV CAR with PAG blending versus anion-bound PBP. They reported that anion-bound PBP had less swelling during development as well as lower acid diffusion blur, leading to better LWR [278].

Hori et al. have reported on the use of a high-T_g polymer with a low–acid diffusion PAG to achieve 13 nm hp resolution at a dose of 34 mj [249]. Matsumaru et al. reported recently on a novel protecting group with various PAG cations optimized for acid yield [279]. They were able to resolve 17 nm hp with a dose of 41 mj.

7.8.1.4 Negative Tone Developable (NTD) EUV Resists

Recent progress has been made on NTD EUV resists. Fujimori et al. reported a 60% improvement in dose and LWR for NTD resist versus PTD resist [280]. Tsubaki et al. described the advantage

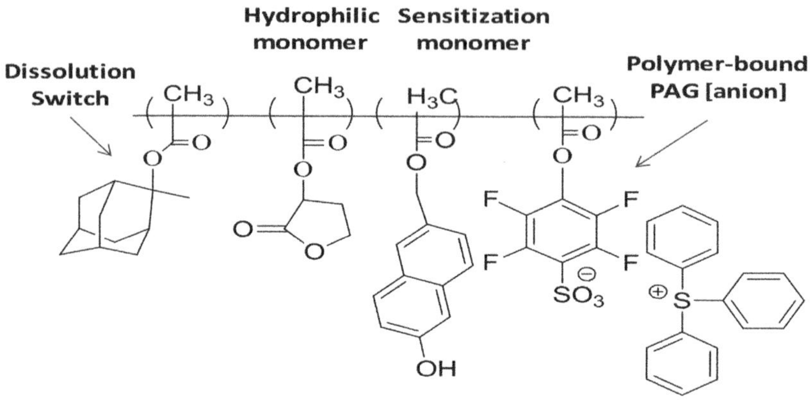

FIGURE 7.78   Exemplary polymer-bound PAG (PBP) system.

of higher photon flux for contacts and trenches using the NTD approach [281]. Of course, mask defects become a big issue for bright-field imaging. They did report 19 nm trench resolution at a dose of 17 mj.

7.8.2   EUV Metal-Containing Resists

Recently, we have seen a very active effort to develop metal-containing resists for EUV lithography. There are many advantages in the use of metal-containing resists for this application. The first advantage is the significant EUV absorption cross section of many metals, such as hafnium and tin [282]. With the high transparency of organic-based resists, a great portion of the EUV photon flux is wasted. With the incorporation of metals, the EUV photons are absorbed more easily. Also, the second advantage of metal-containing resists is that the etch resistance is superior to that of organic-based resists, thereby allowing thinner resist thickness. This advantage translates into lower–aspect ratio resists and therefore mitigates pattern collapse. There are also significant barriers to the use of metal-containing resists. First and foremost, concern about metal contamination of the device exists, and this will require careful use of these materials [283]. The removal of other residual metals is expected to be difficult as well [284]. A more subtle challenge is the lack of understanding of the resist reaction mechanism, unlike the well-developed understanding of the EUV CAR reaction mechanism.

Ober has spearheaded the development of HfO_2- and ZrO_2-based nanoparticles (NPs) for use in EUV resists [285]. The NPs are ~1–3 nm in size distribution and are combined with a PAG and MAA as one of the exemplary formulations. It is thought that the photoacid can displace the MAA ligand, thereby causing a solubility change in the developer, which leads to a negative tone pattern. There also exists the possibility of some NP aggregation. Nonetheless, this system has shown impressive EUV sensitivity of 4.2 mj and resolution down to 26 nm hp. Cardineau et al. extended the NP resists to other strongly binding ligands, particularly sulfonate salts, in order to enable TMAH development [286]. Ouyang et al. have studied NP development using organic solvents such as acetone and n-butyl acetate [287]. They were able to show 32 nm hp imaging using e-beam. Chakrabarty et al. reported on the use of dimethyl acrylic acid ligands with a ZrO_2 NP and PAG formulation [288]. With this chemistry, they reported a sensitivity of 1.2 mj with resolution of 30 nm hp. These sensitivities are very impressive, and it does beg the question of whether ligand displacement reactions are catalytic in nature. Chakrabarty et al. reported on the mechanism by which these unique resists work [289]. Through IR studies, they confirmed the ligand-exchange mechanism, whereby the MAA ligand attached to the ZrO_2 NP is displaced by

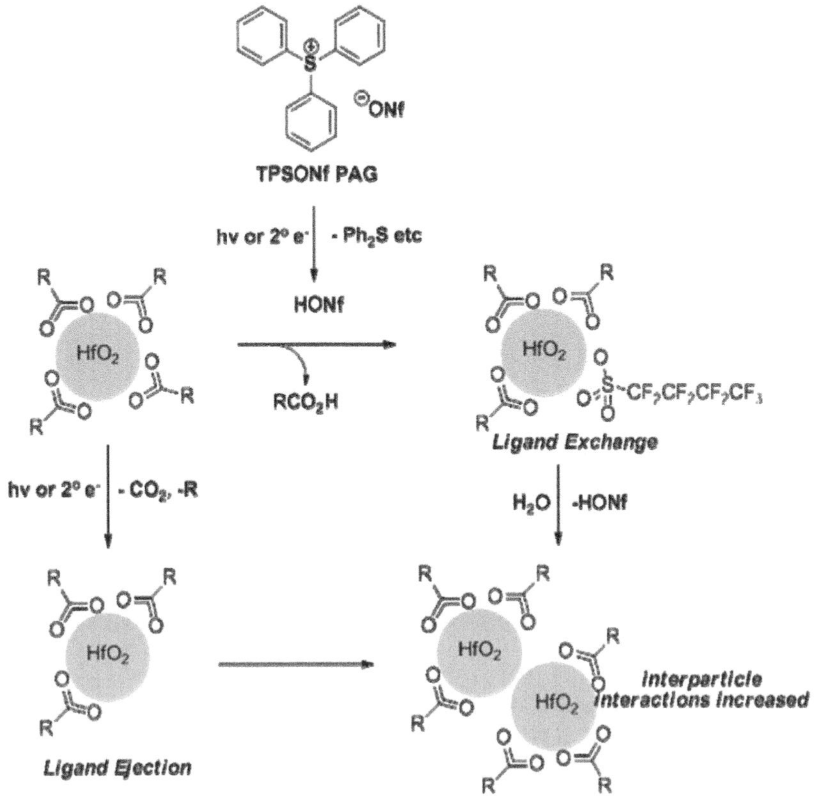

FIGURE 7.79 Schematic of proposed patterning mechanisms for NP-based resist. (From Krysak, M., et al., *Proc. SPIE,* 9048, 904805, 2014.)

the photochemically produced triflic acid. The lower solubility of the sulfonate-bound NP in the developer leads to negative tone patterning. Krysak et al. have also studied the patterning mechanism, and they report a complex set of ligand-exchange reactions as well as NP aggregation being operative at once (see Figure 7.79) [290]. They reported on a new NP formulation with the use of HfO_2 benzoic acid/MAA ligands and PAG. EUV patterning was reported down to 15 nm hp with a dose of 38 mj. Jiang et al. have reported on the use of aliphatic versus aromatic acid ligands [291]. They have reported that aromatic acid ligands give much worse sensitivity than aliphatic acid ligands.

Amador et al. have reported on the use of hafnium sulfonate peroxide complexes, which undergo dehydration leading to condensation to form insoluble hafnium sulfonate oxide complexes that allow negative tone patterning using 0.54 N TMAH [292]. The chemical mechanism was elucidated using IR techniques. Clark et al. showed impressive lithography with 13 nm hp resolution at a dose of 35 mj on an ASM-L NXE3300B EUV scanner [293, as shown in Figure 7.80].

Brainard's group at CNSE has extended the metal-containing resist work to a wide array of metals for screening. Cardineau et al. reported on tin-oxo clusters, $[(RSn)_{12}O_{14}(OH)_6]$ X_2, which upon exposure to EUV light can form negative tone patterns using aqueous isopropanol as developer. Resolution of 18 nm hp lines was possible, albeit at a large dose of 350 mj [294]. Passarelli et al. have studied bismuth complexes as potential EUV resist materials [295]. Initial materials are based on oligomers of diphenyl bismuth chloride co-condensed with phenyl bismuth dichloride (see

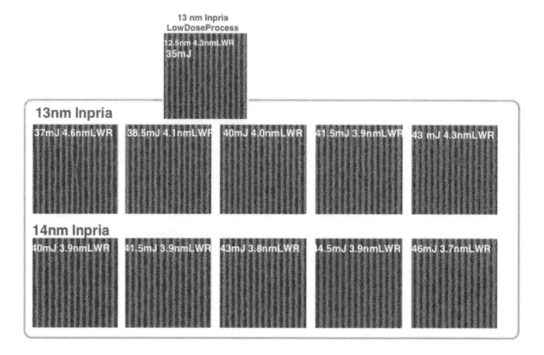

FIGURE 7.80 Exposures performed on an ASML NXE:3300B with YF-AA and Dipole45X illumination. Dose to size at 13 nm half-pitch was 39 mJ/cm², with 4.0 nm LWR. The process was extended to print 13 nm hp at 35 mJ/cm². (From J.T. Clark, et al., *Proc. SPIE,* 9425, 94251A, 2014.)

Figure 7.81). These oligomers are combined with a PAG to make working resist formulations. The photogenerated triflic acid can cleave phenyl bismuth bonds, thereby breaking up oligomers and yielding a positive tone image. EUV photons can cause bismuth clusters to condense, leading to negative tone patterns, as shown in Figure 7.82.

Sortland et al. have investigated Pt and Pd oxalate and carbonate complexes ($L_2M(CO_3)$ and $L_2M(C_2O_4)$; M = Pt or Pd) for use in EUV lithography [296]. In particular, (dppm)Pd(C_2O_4) (dppm = 1, 1-bis(diphenylphosphino)methane) prints 30 nm dense lines with E_{size} of 50 mJ/cm². Figure 7.82 illustrates the ability to achieve positive and negative tone behavior depending on the metal and the ligands chosen.

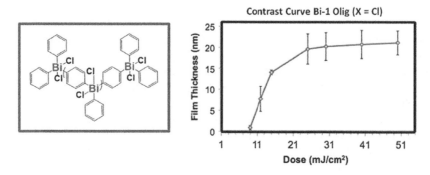

FIGURE 7.81 Bismuth complexes as potential EUV resist materials. (From Passarelli, J., et al., *Proc. SPIE,* 9051, 90512A, 2014.)

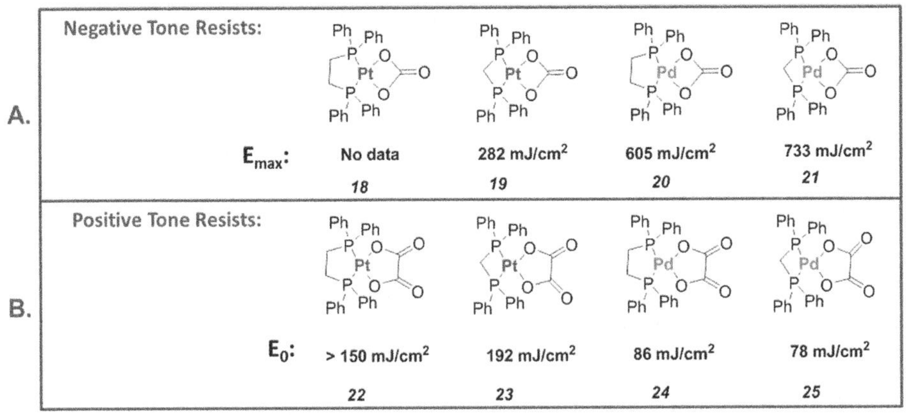

FIGURE 7.82 $L_2M(C_2O_4)$ and $L_2M(CO_3)$ evaluated for EUV sensitivity. A) Platinum and palladium carbonates showing negative tone behavior. B) Platinum and palladium oxalates showing positive tone. (From Sortland, M., et al., *Proc. SPIE*, 9422, 942207, 2015.)

REFERENCES

1. T. Ueno, H. Shiraishi, S. Uchino, T. Sakamizu, and T. Hattori. 1994. *Journal of Photopolymer Science and Technology*, **7**(3): 397.
2. S. Okazaki. 1993. *Applied Surface Science*, **70/71**: 603.
3. H. Fukuda, T. Terasaea, and S. Okazaki. 1991. *Journal of Vacuum Science and Technology, Part B*, **9**: 3113.
4. H. Fukuda, A. Imai, T. Terasawa, and S. Okazaki. 1991. *IEEE Transactions on Electronic Devices*, **38**(1): 67.
5. R. Dammel. 1991. *Diazonapthoquinone-Based Resists*, SPIE tutorial text TT11, SPIE Press, Bellingham, WA.
6. O. Suess. 1994. *Justus Liebigs Annalen der Chemie*, **556**: 65.
7. J. Packansky and J.R. Lyerla. 1979. *IBM Journal of Research and Development*, **23**(1): 42; J. Packansky. 1979. *Polymer Engineering and Science*, **20**: 1049.
8. N.J. Hacker and N.J. Turro. 1982. *Terahedron Letters*, **23**(17): 1771.
9. F.A. Vollenbrok, W.P.M. Nijssen, C.M.J. Mutsaers, M.J.H.J. Geomini, M.E. Reuhman, and R.J. Visser. 1989. *Polymer Engineering and Science*, **29**(14): 928.
10. F.A. Vollenbrok, C.M.J. Mutsaers, and W.P.M. Nijssen. 1989. *Polymeric Materials: Science and Engineering*, **61**: 283.
11. K. Nakamura, S. Udagawa, and K. Honda. 1972. *Chemistry Letters*, **20**: 763.
12. T. Shibata, K. Koseki, T. Yamaoka, M. Yoshizawa, H. Uchiki, and T. Kobayashi. 1988. *Journal of Physical Chemistry*, **92**(22): 6269.
13. M. Barra, T.A. Fisher, G.J. Cernigliaro, R. Sinta, and J.C. Scaiano. 1992. *Journal of the American Chemical Society*, **114**: 2680.
14. J. Andraos, Y. Chiang, C.-G. Huang, A.J. Kresge, and J.C. Scaiano. 1993. *Journal of the American Chemical Society*, **115**(23): 10605.
15. K. Tanigaki and T.W. Ebbsen. 1989. *Journal of the American Chemical Society*, **109**: 5883; K. Tanigaki and T.W. Ebbsen. 1989. *Journal of Physical Chemistry*, **93**: 4531.
16. K. Sheats. 1988. *IEEE Transactions on Electronic Devices*, **35**: 129.
17. A. Furuta, M. Hanabata, and Y. Uemura. 1986. *Journal of Vacuum Science and Technology, Part B*, **4**: 430; M. Hanabata, A. Furuta, and Y. Uemura. 1986. *Proceedings of SPIE*, **631**: 76; M. Hanabata, A. Furuta, and Y. Uemura. 1986. *Proceedings of SPIE*, **771**: 85.
18. M. Hanabata, Y. Uetani, and A. Furuta. 1988. *Proceedings of SPIE*, **920**: 349.
19. M. Hanabata, Y. Uetani, and A. Furuta. 1989. *Journal of Vacuum Science and Technology, Part B*, **7**: 640.
20. M. Hanabata and A. Furuta. 1990. *Proceedings of SPIE*, **1262**: 476.
21. M. Hanabata, F. Oi, and A. Furuta. 1991. *Proceedings of SPIE*, **1466**: 132.

22. M. Hanabata, F. Oi, and A. Furuta. 1992. *Polymer Engineering and Science*, **32**(20): 1494.
23. T. Kajita, T. Ota, H. Nemoto, Y. Yumoto, and T. Miura. 1991. *Proceedings of SPIE*, **1466**: 161.
24. K. Honda, B.T. Beauchemin, Jr., R.J. Hurditch, A.J. Blankeney, K. Kawabe, and T. Kokubo. 1990. *Proceedings of SPIE*, **1262**: 493.
25. K. Honda, B.T. Beauchemin, Jr., E.A. Fitzgerald, A.T. Jeffries, III, S.P. Tadros, A.J. Blankeney, R.J. Hurditch, S. Tan, and S. Sakaguchi. 1991. *Proceedings of SPIE*, **1466**: 141.
26. K. Honda, B.T. Beauchemin, Jr., R.J. Hurditch, A.J. Blankeney, and T. Kokubo. 1992. *Proceedings of SPIE*, **1672**: 305; K. Honda, A.J. Blankeney, R.J. Hurditch, S. Tan, and T. Kokubo. 1992. *Proceedings of SPIE*, **1925**: 197.
27. W.D. Hinsberg and M.L. Guitierrez. 1983. Kodak Microelectronics Interface Seminar, November.
28. J.-P. Huang, T.K. Kwei, and A. Resiser. 1989. *Proceedings of SPIE*, **1086**: 74.
29. T.-F. Yeh, H.-Y. Shih, A. Resier, M.A. Toukhy, and B.T. Beauchemin, Jr. 1992. *Journal of Vacuum Science and Technology, Part B*, **10**: 715.
30. T.-F. Yeh, H.-Y. Shih, and A. Resier. 1992. *Macromolecules*, **25**(20): 5345.
31. T.-F. Yeh, A. Resiser, R.R. Dammel, G. Pawlowski, and H. Roeschert. 1993. *Macromolecules*, **26**(4): 862.
32. H.-Y. Shih, T.-F. Yeh, A. Reiser, R.R. Dammel, H.J. Merrem, and G. Pawlowski. 1994. *Macromolecules*, **27**(12): 3330.
33. T. Hattori, T. Ueno, H. Shiraishi, N. Hayashi, and T. Iwayanagi. 1991. *Japanese Journal of Applied Physics*, **30**(1, No. 11B): 3125.
34. R.R. Dammel, M.D. Rahman, P.H. Lu, A. Canize, and V. Elango. 1994. *Proceedings of SPIE*, **2195**: 542.
35. S. Kishimura, A. Yamaguchi, Y. Yamada, and H. Nagata. 1992. *Polymer Engineering and Science*, **32**(20): 1550.
36. S. Tan, S. Sakaguchi, K. Uenishi, Y. Kawabe, T. Kokubo, and R.J. Hurditch. 1991. *Proceedings of SPIE*, **1262**: 513.
37. H. Nemoto, K. Inomata, T. Ota, Y. Yumoto, T. Miura, and H. Chaanya. 1992. *Proceedings of SPIE*, **1672**: 305.
38. K. Uenishi, Y. Kawabe, T. Kokubo, and A. Blakeney. 1991. *Proceedings of SPIE*, **1466**: 102.
39. K. Uenishi, S. Sakaguchi, Y. Kawabe, T. Kokubo, M.A. Toukhy, A.T. Jeffries, III, S.G. Slater, and R.J. Hurditch. 1992. *Proceedings of SPIE*, **1672**: 262.
40. R. Hanawa, Y. Uetani, and M. Hanabata. 1992. *Proceedings of SPIE*, **1672**: 231; R. Hanawa, Y. Uetani, and M. Hanabata. 1992. *Proceedings of SPIE*, **1925**: 227.
41. M. Hanabata. 1994. *Advanced Materials, Optics and Electronics*, **4**(2): 75.
42. W. Brunsvold, N. Eib, C. Lyons, S. Miura, M. Plat, R. Dammel, O. Evans, et al. 1992. *Proceedings of SPIE*, **1672**: 273.
43. P. Trefonas, III and B.K. Danielss. 1987. *Proceedings of SPIE*, **771**: 194.
44. H. Ito, C.G. Willson, and J.M.J. Frechet. 1982. Paper presented at the 1982 Symposium on VLSI Technology, Oiso, Japan; H. Ito, C.G. Willson, and J.M.J. Frechet. 1985. US Patent 4,491,628.
45. H. Ito and C.G. Willson. 1983. *ACS Symposium Series*, **242**: 11.
46. H. Steppan, G. Buhr, and H. Vollmann. 1982. *Angewandte Chemie, International Edition in English*, **21**(7): 455.
47. T. Iwayanagi, T. Ueno, S. Nonogaki, H. Ito, and C.G. Willson. 1988. *Advances in Chemistry Series*, 218, Chapter 3.
48. J. Lingnau, R. Dammel, and J. Theis. 1989. *Solid State Technology*, **32**: 9, 105; J. Lingnau, R. Dammel, and J. Theis. 1989. *Solid State Technology*, **32**: 10, 107.
49. J.R. Sheats. 1990. *Solid State Technology*, **33**: 6, 79.
50. A.A. Lamola, C.R. Szmanda, and J.W. Thackeray. 1990. *Solid State Technology*, **34**: 8, 53.
51. E. Rechmanis, F.M. Houlihan, O. Nalamasu, and T.X. Neenan. 1991. *Chemistry of Materials*, **3**(3): 394.
52. J.F. Cameron and J.M.J. Frechhet. 1990. *Journal of Organic Chemistry*, **55**: 5918.
53. C.G. Willson, J.F. Cammeron, S.A. MacDonald, C.-P. Niesert, J.M.J. Frechet, M.K. Leung, and A. Ackman. 1993. *Proceedings of SPIE*, **1925**: 354.
54. J.V. Crivello. 1983. *Polymer Engineering and Science*, **23**(17): 953; J.V. Crivello. 1983. *Advances in Polymer Science*, 62: 1.
55. N.P. Hacker and J.L. Dektar. 1990. *Journal of Organic Chemistry*, **55**: 639; N.P. Hacker and J.L. Dektar. 1990. *Polymeric Materials: Science and Engineering*, **61**: 76.
56. N.P. Hacker and J.L. Dektar. 1990. *Journal of the American Chemical Society*, **112**(16): 6004.
57. J.L. Dektar and N.P. Hacker. 1989. *Journal of Photochemistry and Photobiology A: Chemistry*, **46**(2): 233.

58. N.P. Hacker, D.V. Leff, and J.L. Dektar. 1990. *Molecular Crystals and Liquid Crystals*, **183**: 505.
59. N.P. Hacker and K.M. Welsh. 1991. *Macromolecules*, **24**(8): 2137.
60. D.R. Mckean, U. Schaedeli, and S.A. MacDonald. 1989. *ACS Symposium Series*, **412**: 27.
61. D.R. Mckean, U. Schaedeli, P.H. Kasai, and S.A. MacDonald. 1989. *Polymeric Materials Science and Engineering*, **61**: 81.
62. G. Buhr. 1985. European Patent Application EP 0 137 452.
63. G.H. Smith and J.A. Bonham. 1973. US Patent 3,779,778.
64. G. Buhr, R. Dammel, and C. Lindley. 1989. *Polymeric Materials Science and Engineering*, **61**: 269.
65. G. Calabrese, A. Lamola, R. Sinta, and J. Thackeray. 1990. *Polymers for Microelectronics*, Tokyo: Kodansha, 435.
66. F.M. Houlihan, A. Schugard, R. Gooden, and E. Reichmanis. 1988. *Macromolecules*, **21**(7): 2001.
67. T.X. Neenan, F.M. Houlihan, E. Reichmanis, J.M. Kometani, B.J. Bachman, and L.F. Thompson. 1989. *Proceedings of SPIE*, **1086**: 2.
68. F.M. Houlihan, T.X. Neenan, E. Reichmanis, J.M. Kometani, and T. Chin. 1991. *Chemistry of Materials*, **3**(3): 462.
69. T. Yamaoka, M. Nishiki, K. Koseki, and M. Koshiba. 1989. *Polymer Engineering and Science*, **29**(13): 856.
70. K. Naitoh, K. Yoneyama, and T. Yamaoka. 1992. *Journal of Physical Chemistry*, **96**(1): 238.
71. T. Ueno, H. Shiraishi, L. Schlegel, N. Hayashi, and T. Iwayanagi. 1990. In *Polymers for Microelectronics—Science and Technology*, Y. Tabata, I. Mita, S. Nonogaki, K. Horie, and S. Tagawa, Eds, Tokyo: Kodansha, p. 413.
72. H. Shiraishi, N. Hayashi, T. Ueno, T. Sakamizu, and F. Murai. 1991. *Journal of Vacuum Science and Technology, Part B*, **9**: 3343.
73. L. Schlegel, T. Ueno, H. Shiraishi, N. Hayashi, and T. Iwayanagi. 1990. *Chemistry of Materials*, **2**(3): 299.
74. T. Sakamizu, H. Yamaguchi, H. Shiraishi, F. Murai, and T. Ueno. 1993. *Journal of Vacuum Science and Technology, Part B*, **11**: 2812.
75. T. Ueno, L. Schlegel, N. Hayashi, H. Shiraishi, and T. Iwayanagi. 1992. *Polymer Engineering and Science*, **32**(20): 1511.
76. H. Roechert, C. Eckes, and G. Pawlowski. 1993. *Proceedings of SPIE*, **1925**: 342.
77. G. Berner, R. Kirchmayr, G. Rist, and W. Rutsch. 1986. *Journal of Radiation Curing*, **10**.
78. Y. Onishi, H. Niki, Y. Kobayashi, R.H. Hayase, N. Oyasato, and O. Sasaki. 1991. *Journal of Photopolymer Science and Technology*, **4**(3): 337.
79. G. Buhr, H. Lenz, and S. Scheler. 1989. *Proceedings of SPIE*, **1086**: 117.
80. J.J. Grunwald, C. Gal, and S. Eidelman. 1990. *Proceedings of SPIE*, **1262**: 444.
81. R. Hayase, Y. Onishi, H. Niki, N. Oyasato, and S. Hayase. 1994. *Journal of the Electrochemical Society*, **141**(11): 3141.
82. M. Shirai, T. Masuda, M. Tsunooka, and M. Tanaka. 1984. *Makromolecular Chemistry, Rapid Communications*, **5**(10): 689.
83. M. Shirai and M. Tsunooka. 1990. *Journal of Photopolymer Science and Technology*, **3**(3): 301.
84. C.A. Renner. 1980. US Patent 4,371,605.
85. W. Brunsvold, R. Kwong, W. Montgomery, W. Moreau, H. Sachdev, and K. Welsh. 1990. *Proceedings of SPIE*, **1262**: 162.
86. W. Brunsvold, W. Montgomery, and B. Hwang. 1990. *Proceedings of SPIE*, **1262**: 162.
87. G. Pawlowski, R. Dammel, C. Lindley, H.-J. Merrem, H. Roeschert, and J. Lingau. 1990. *Proceedings of SPIE*, **1262**: 16.
88. A. Poot, G. Delzenne, R. Pollet, and U. Laridon. 1971. *Journal of Photographic Science*, **19**(4): 88.
89. T. Aoai, A. Umehara, A. Kamiya, N. Matsuda, and Y. Aotani. 1989. *Polymer Engineering and Science*, **29**(13): 887.
90. T.W. Green. 1981. *Protective Groups in Organic Synthesis*, New York: Wiley.
91. J. March. 1985. *Advances in Organic Chemistry*, 3rd Ed., New York: Wiley; C.K. Ingold. 1985. *Structure and Mechanism in Organic Chemistry*, 3rd Ed., Ithaca, NY: Cornell University Press.
92. H. Ito. 1992. *Japanese Journal of Applied Physics*, **31**(1, No. 12B): 4273.
93. R.G. Tarascon, E. Reichmanis, F. Houlihan, A. Schugard, and L. Thompson. 1989. *Polymer Engineering and Science*, **29**(13): 850.
94. J.M. Kometani, M.E. Galvin, S.A. Heffner, F.M. Houlihan, O. Nalamasu, E. Chin, and E. Reichmanis. 1993. *Macromolecules*, **26**(9): 2165.
95. R. Schwalm. 1989. *Polymeric Materials: Science and Engineering*, **61**: 278.

96. Y. Kawai, A. Tanaka, and T. Matsuda. 1992. *Japanese Journal of Applied Physics*, **31**(1, No. 12B): 4316.
97. M.J. O'Brien. 1989. *Polymer Engineering and Science*, **29**(13): 846.
98. D.R. McKean, S.A. MacDonald, R.D. Johnson, N.J. Clecak, and C.G. Willson. 1990. *Chemistry of Materials*, **2**(5): 619.
99. T. Kumada, S. Kubota, H. Koezuka, T. Hanawa, S. Kishimura, and H. Nagata. 1991. *Journal of Polymer Science and Technology*, **4**: 469.
100. H. Ban, J. Nakamura, K. Deguchi, and A. Tanaka. 1991. *Journal of Vacuum Science and Technology, Part B*, **9**: 3387.
101. T. Aoai, T. Yamanaka, and T. Kokubo. 1994. *Proceedings of SPIE*, **2195**: 111.
102. N. Kihara, T. Ushirogouchi, T. Tada, T. Naitoh, S. Saitoh, and O. Sasaki. 1992. *Proceedings of SPIE*, **1672**: 194.
103. H. Koyanagi, S. Umeda, S. Fukunaga, T. Kitaori, and K. Nagasawa. 1992. *Proceedings of SPIE*, **1672**: 125.
104. H. Ban, J. Nakamura, K. Deguchi, and A. Tanaka. 1994. *Journal of Vacuum Science and Technology, Part B*, **12**: 3905.
105. S.A.M. Hesp, N. Hayashi, and T. Ueno. 1991. *Journal of Applied Polymer Science*, **42**(4): 877.
106. T. Sakamizu, H. Shiraishi, H. Yamaguchi, T. Ueno, and N. Hayashi. 1992. *Japanese Journal of Applied Physics*, **31**(1, No. 12B): 4288.
107. T. Hattori, L. Schlegel, A. Imai, N. Hayashi, and T. Ueno. 1993. *Optical Engineering*, **32**(10): 2368.
108. G.N. Taylor, L.E. Stillwagon, F.M. Houlihan, T.M. Wolf, D.Y. Sogah, and W.R. Hertler. 1991. *Journal of Vacuum Science and Technology, Part B*, **9**: 3348; G.N. Taylor, L.E. Stillwagon, F.M. Houlihan, T.M. Wolf, D.Y. Sogah, and W.R. Hertler. 1991. *Chemistry of Materials*, **3**(6): 1031.
109. H. Kikuchi, N. Kurata, and K. Hayashi. 1991. *Journal of Polymer Science and Technology*, **4**: 357.
110. C. Mertesdorf, B. Nathal, N. Munzel, H. Holzwarth, and H.T. Schacht. 1994. *Proceedings of SPIE*, **2195**: 246.
111. W.C. Cunningham, Jr. and C.E. Park. 1987. *Proceedings of SPIE*, **771**: 32.
112. M. Murata, T. Takahashi, M. Koshiba, S. Kawamura, and T. Yamaoka. 1990. *Proceedings of SPIE*, **1262**: 8.
113. Y. Jiang and D. Bassett. 1992. *Polymeric Materials Science and Engineering*, **66**: 41.
114. J.M.J. Frechet, E. Eichler, S. Gauthier, B. Kryczka, and C.G. Willson. 1989. *ACS Symposium Series*, **381**: 155.
115. H. Ito, L.A. Pederson, K.N. Chiong, S. Sonchik, and C. Tai. 1989. *Proceedings of SPIE*, **1086**: 11.
116. Y. Onishi, N. Oyasato, H. Niki, R.H. Hayase, Y. Kobayashi, K. Sato, and M. Miyamura. 1992. *Journal of Polymer Science and Technology*, **5**: 47.
117. H. Ito and C.G. Willson. 1983. *Polymer Engineering and Science*, **23**: 1013.
118. H. Ito, M. Ueda, and R. Schwalm. 1988. *Journal of Vacuum Science and Technology. Part B*, **6**: 2259.
119. H. Ito and R. Scwalm. 1989. *Journal of the Electrochemical Society*, **136**(1): 241.
120. J. Lingnau, R. Dammel, and J. Theis. 1989. *Polymer Engineering and Science*, **29**(13): 874; G. Pawlowski, K.-J. Przybilla, W. Spiess, H. Wengenroth, and H. Roeschert. 1989. *Journal of Photopolymer Science and Technology*, **5**: 55.
121. H. Roeschert, K.-J. Przybilla, W. Spiess, H. Wegenroth, and G. Pawlowski. 1992. *Proceedings of SPIE*, **1672**: 33.
122. D.R. Mckean, W.D. Hinsberg, T.P. Sauer, C.G. Willson, R. Vicari, and D.J. Gordon. 1990. *Journal of Vacuum Science and Technology, Part B*, **8**: 1466.
123. K.-J. Przybilla, H. Roeschert, W. Spiess, C. Eckes, G. Pawlowski, and R. Dammel. 1991. *Proceedings of SPIE*, **1466**: 174.
124. G. Pawlowski, T. Sauer, R. Dammel, D.J. Gordon, W. Hinsberg, D. McKean, C.R. Lindley, et al. 1990. *Proceedings of SPIE*, **1262**: 391.
125. T. Aoai, A. Umehara, A. Kamiya, N. Matsuda, and Y. Aoai. 1989. *Polymer Engineering and Science*, **29**(13): 887.
126. T. Aoai, Y. Aotani, and A. Umehara. 1992. *Journal of Photopolymer Science and Technology*, **3**: 389.
127. J.M.J. Frechet, F. Bouchard, F. Houlihan, B. Kryczka, and C.G. Willson. 1985. *Polymeric Materials: Science and Engineering*, **53**: 263.
128. J.M.J. Frechet, T. Iizawa, F. Bouchard, and M. Stanciulescu. 1986. *Polymeric Materials: Science and Engineering*, **55**: 299.
129. J.M.J. Frechet, E. Eichler, M. Stanciulescu, T. Iizawa, F. Bouchard, F.M. Houlihan, and C.G. Willson. 1987. *ACS Symposium Series*, **346**: 138.
130. J.M.J. Frechet, M. Stanciulescu, T. Iizawa, and C.G. Willson. 1989. *Polymeric Materials: Science and Engineering*, **60**: 170.

131. J.M.J. Frechet, C.G. Willson, T. Iizawa, T. Nishikubo, K. Igarashi, and J. Fahey. 1989. *ACS Symposium Series*, **412**: 100.
132. J.M.J. Frechet, J. Fahey, C.G. Willson, T. Iizawa, K. Igarashi, and T. Nishikubo. 1989. *Polymeric Materials: Science and Engineering*, **60**: 174.
133. W.E. Feely, J.C. Imhof, and C.M. Stein. 1986. *Polymer Engineering and Science*, **26**(16): 1101.
134. A. Burns, H. Luethje, F.A. Vollenbroek, and E.J. Spiertz. 1987. *Microelectronic Engineering*, **6**(1–4): 467.
135. H. Roechert, R. Dammel, Ch. Eckes, W. Meier, K.-J. Przybilla, W. Spiess, and G. Pawlowski. 1992. *Proceedings of SPIE*, **1672**: 157.
136. J.W. Thackeray, G.W. Orsula, E.K. Pavelchek, D. Canistro, L.E. Bogan, A.K. Berry, and K.A. Graziano. 1989. *Proceedings of SPIE*, **1086**: 34.
137. H.-Y. Liu, M.P. de Grandpre, and W.E. Feely. 1988. *Journal of Vacuum Science and Technology, Part B*, **6**: 379.
138. J.W. Thackeray, G.W. Orsula, M.M. Rajaratnam, R. Sinta, D. Herr, and E.K. Pavelchek. 1991. *Proceedings of SPIE*, **1466**: 39.
139. M.T. Allen, G.S. Calabrese, A.A. Lamola, G.W. Orsula, M.M. Rajaratnam, R. Sinta, and J.W. Thackeray. 1991. *Journal of Photopolymer Science and Technology*, **4**(3): 379.
140. A. Yamaguchi, S. Kishimura, K. Tsujita, H. Morimoto, K. Tsukamoto, and H. Nagata. 1993. *Journal of Vacuum Science and Technology, Part B*, **11**: 2867.
141. J.M.J. Frechet, S. Matuszczak, H.D.H. Stoever, C.G. Willson, and B. Reck. 1989. *ACS Symposium Series*, **412**: 74.
142. H.D.H. Stoever, S. Matuszczak, C.G. Willson, and J.M.J. Frechet. 1993. *Macromolecules*, **24**: 1741.
143. J.T. Fahey and J.M.J. Frechet. 1991. *Proceedings of SPIE*, **1466**: 67.
144. S.M. Lee, J.M.J. Frechet, and C.G. Willson. 1994. *Macromolecules*, **27**(18): 5154.
145. S.M. Lee and J.M.J. Frechet. 1994. *Macromolecules*, **27**(18): 5160.
146. S.M. Lee, S. Matuszczak, J.M.J. Frechet, C. Lee, and Y. Shacham-Diamand. 1994. *Chemistry of Materials*, **6**(10): 1796.
147. T. Kajita, E. Kobayashi, A. Tsuji, and Y. Kobayashi. 1993. *Proceedings of SPIE*, **1925**: 133.
148. J.C. Dubois, A. Eranian, and E. Datamanti. 1978. In *8th International Conference on Electron and Ion Science and Technology*, Seattle, 303.
149. M. Hatzakis, K.J. Stewart, J.M. Shaw, and S.A. Rishton. 1991. *Journal of the Electrochemical Society*, **138**(4): 1076.
150. W. Conley, W. Moreau, S. Perreault, G. Spinillo, R. Wood, J. Gelorme, and R. Martino. 1990. *Proceedings of SPIE*, **1262**: 49.
151. M. Toriumi, H. Shiraishi, T. Ueno, N. Hayashi, S. Nonogaki, F. Sato, and K. Kadota. 1987. *Journal of the Electrochemical Society*, **134**(4): 936.
152. T. Ueno, H. Shiraishi, N. Hayashi, K. Tadano, E. Fukuma, and T. Iwayanagi. 1990. *Proceedings of SPIE*, **1262**: 26.
153. N. Hayashi, K. Tadano, T. Tanaka, H. Shiraishi, T. Ueno, and T. Iwayanagi. 1990. *Japanese Journal of Applied Physics*, **29**(1, No. 11): 2632.
154. H. Shiraishi, E. Fukuma, N. Hayashi, K. Tadano, and T. Ueno. 1991. *Chemistry of Materials*, **3**(4): 621.
155. D.R. Mckean, N.J. Clecak, and L.A. Pederson. 1990. *Proceedings of SPIE*, **1262**: 110.
156. S. Uchino, T. Iwayanagi, T. Ueno, and N. Hayashi. 1991. *Proceedings of SPIE*, **1466**: 429.
157. S. Uchino and C.W. Frank. 1992. *Polymer Engineering and Science*, **32**(20): 1530.
158. S. Uchino, M. Katoh, T. Sakamizu, and M. Hashimoto. 1992. *Microelectronic Engineering*, **18**(4): 341.
159. T. Ueno, S. Uchino, K.T. Hattori, T. Onozuka, S. Shirai, N. Moriuchi, M. Hashimoto, and S. Koibuchi. 1994. *Proceedings of SPIE*, **2195**: 173.
160. R. Sooriyakumaran, H. Ito, and E.A. Mash. 1991. *Proceedings of SPIE*, **1466**: 419.
161. H. Ito, R. Sooriyakumaran, Y. Maekawa, and E.A. Mash. 1992. *Polymeric Materials: Science and Engineering*, **66**: 45.
162. J. Crank and G.S. Park, Eds. 1969. *Diffusion in Polymers*, New York: Academic Press.
163. L. Schlegel, T. Ueno, N. Hayashi, and T. Iwayanagi. 1991. *Journal of Vacuum Science and Technology, Part B*, **9**: 278; L. Schlegel, T. Ueno, N. Hayashi, and T. Iwayanagi. 1991. *Japanese Journal of Applied Physics B*, **30**(1, No. 11B): 3132.
164. J. Nakamura, H. Ban, K. Deguchi, and A. Tanaka. 1991. *Japanese Journal of Applied Physics*, **30**(1, No. 10): 2619.
165. K. Asakawa. 1993. *Journal of Photopolymer Science and Technology*, **6**(4): 505.
166. T.H. Fedynyshyn, J.W. Thackeray, J.H. Georger, and M.D. Denison. 1994. *Journal of Vacuum Science and Technology, Part B*, **12**: 3888.

167. S.A. MacDonald, N.J. Clecak, H.R. Wendt, C.G. Willson, C.D. Snyder, C.J. Knors, N.B. Deyoe, et al. 1991. *Proceedings of SPIE*, **1466**: 2.
168. S.A. MacDonald, W.D. Hinsberg, H.R. Wendt, N.J. Clecak, C.G. Willson, and C.D. Snyder. 1993. *Chemistry of Materials*, **5**(3): 348.
169. J.G. Maltabes, S.J. Homes, J.R. Morrow, R.L. Barr, M. Hakey, G. Reynolds, C.G. Willson, N.J. Clecak, S.A. MacDonald, and H. Ito. 1990. *Proceedings of SPIE*, **1262**: 2.
170. O. Nalamasu, E. Reichmanis, M. Cheng, V. Pol, J.M. Kometani, F.M. Houlihan, T.X. Nenan, et al. 1991. *Proceedings of SPIE*, **1466**: 13.
171. T. Kumada, Y. Tanaka, S. Kubota, H. Koezuka, A. Ueyama, T. Hanawa, and H. Morimoto. 1993. *Proceedings of SPIE*, **1925**: 31.
172. A. Oikawa, N. Santoh, S. Miyata, Y. Hatakenaka, H. Tanaka, and K. Nakagawa. 1993. *Proceedings of SPIE*, **1925**: 92.
173. W.D. Hinsberg, S.A. MacDonald, N.J. Clecak, and C.D. Snyder. 1992. *Proceedings of SPIE*, **1672**: 24.
174. W.D. Hinsberg, S.A. MacDonald, N.J. Clecak, and C.D. Snyder. 1993. *Proceedings of SPIE*, **1925**: 43.
175. W.D. Hinsberg, S.A. MacDonald, N.J. Clecak, and C.D. Snyder. 1994. *Chemistry of Materials*, **6**(4): 481.
176. D.J.H. Funhoff, H. Binder, and R. Schwalm. 1992. *Proceedings of SPIE*, **1672**: 46.
177. K.-J. Przybilla, Y. Kinoshita, S. Masuda, T. Kudo, N. Suehira, H. Okazaki, G. Pawlowski, M. Padmanabam, H. Roeschert, and W. Spiess. 1993. *Proceedings of SPIE*, **1925**: 76.
178. H. Ito, W.P. England, and S.B. Lundmark. 1992. *Proceedings of SPIE*, **1672**: 2.
179. H. Ito, W.P. England, R. Sooriyakumaran, N.J. Clecak, G. Breyta, W.D. Hinsberg, H. Lee, and D.Y. Yoon. 1993. *Journal of Photopolymer Science and Technology*, **6**(4): 547.
180. H. Ito, G. Breyta, D. Hofer, R. Sooriyakumaran, K. Petrillo, and D. Seeger. 1994. *Journal of Photopolymer Science and Technology*, **7**(3): 433.
181. P.J. Paniez, C. Rosilio, B. Mouanda, and F. Vinet. 1994. *Proceedings of SPIE*, **2195**: 14.
182. G.N. Taylor, L.E. Stillwagon, and T. Venkatesan. 1984. *Journal of the Electrochemical Society*, **131**(7): 1658.
183. T.M. Wolf, G.N. Taylor, T. Venkatesan, and R.T. Kraetsch. 1984. *Journal of the Electrochemical Society*, **131**(7): 1664.
184. F. Coopman and B. Roland. 1986. *Proceedings of SPIE*, **631**: 34.
185. F. Coopman and B. Roland. 1987. *Solid State Technology*, **30**: 6, 93.
186. H. Fujioka, H. Nakajima, S. Kishimura, and H. Nagata. 1990. *Proceedings of SPIE*, **1262**: 554.
187. Y. Yoshida, H. Fujioka, H. Nakajima, S. Kishimura, and H. Nagata. 1991. *Journal of Photopolymer Science and Technology*, **4**: 49.
188. K.-H. Baik, L. Van den hove, A.M. Goethals, M. Op de Beeck, and B. Roland. 1990. *Journal of Vacuum Science and Technology, Part B*, **8**: 1482.
189. J.M. Shaw, M. Hatzakis, E.D. Babich, J.R. Paraszczak, D.F. Witman, and K.J. Stewart. 1989. *Journal of Vacuum Science and Technology, Part B*, **7**: 1709.
190. R.-J. Visser, J.P.W. Schellekens, M.E. Reuhman-Huisken, and L.J. Van Ijzendoorn. 1987. *Proceedings of SPIE*, **771**: 111.
191. C.M.J. Mutsaers, W.P.M. Nijssen, F.A. Vollenbroek, and P.A. Kraakman. 1992. *Journal of Vacuum Science and Technology, Part B*, **10**: 729.
192. B.-J.L. Yang, J.-M. Yang, and K.N. Chiong. 1989. *Journal of Vacuum Science and Technology, Part B*, **7**: 1729.
193. R. Sezi, M. Sebald, R. Leuscher, H. Ahne, S. Birkle, and H. Borndoefer. 1990. *Proceedings of SPIE*, **1262**: 84.
194. M. Sebald, J. Berthold, M. Beyer, R. Leuscher, C. Noelschner, U. Scheler, R. Sezi, H. Ahne, and S. Birkle. 1991. *Proceedings of SPIE*, **1466**: 227.
195. K.-H. Baik, L. Van den hove, and B. Roland. 1990. *Journal of Vacuum Science and Technology, Part B*, **9**: 3399.
196. E. Babich, J. Paraszczak, J. Gelorme, R. McGouey, M. Brady, R. Nunes, and R. Smith. 1991. *Microelectronic Engineering*, **13**(1–4): 47.
197. S.A. MacDonald, H. Schlosser, H. Ito, N.J. Clecak, and C.G. Willson. 1991. *Chemistry of Materials*, **3**(3): 435.
198. S.A. MacDonald, H. Schlosser, H. Ito, N.J. Clecak, C.G. Willson, and J.M.J. Frechet. 1992. *Chemistry of Materials*, **4**: 1346.
199. J.P.W. Schellekens and R.-J. Visser. 1989. *Proceedings of SPIE*, **1086**: 220.
200. J.W. Thackeray, J.F. Bohland, G.W. Orsula, and J. Ferrari. 1989. *Journal of Vacuum Science and Technology, Part B*, **7**: 1620.
201. J. Fahey, J.M.J. Frechet, and Y. Schcham-Diamand. 1994. *Journal of Materials Chemistry*, **4**: 1533.

202. R.D. Allen, G.M. Wallraff, W.D. Hinsberg, and L.L. Simpson. 1991. *Journal of Vacuum Science and Technology, Part B*, **9**: 3357.
203. Y. Kaimoto, K. Nozaki, S. Takechi, and N. Abe. 1992. *Proceedings of SPIE*, **1672**: 66.
204. K. Nozaki, Y. Kaimoto, M. Takahashi, S. Takechi, and N. Abe. 1994. *Chemistry of Materials*, **6**(9): 1492.
205. K. Yamashita, M. Endo, M. Sasago, N. Nomura, H. Nagano, S. Mizuguchi, T. Ono, and T. Sato. 1993. *Journal of Vacuum Science and Technology, Part B*, **11**: 2692.
206. K. Nakano, K. Maeda, S. Iwasa, J. Yano, Y. Ogura, and E. Hasegawa. 1994. *Proceedings of SPIE*, **2195**: 194.
207. T. Naito, K. Asakawa, N. Shida, T. Ushirogouchi, and M. Nakase. 1994. *Japanese Journal of Applied Physics*, **33**(1, No. 12B): 7028.
208. M.A. Hartney, D.W. Johnson, and A.C. Spencer. 1991. *Proceedings of SPIE*, **1466**: 238.
209. M.A. Hartney, M.W. Horn, R.R. Kunz, M. Rothschild, and D.C. Shaver. 1992. *Microlithography World*, May/June: 16.
210. S. Uchino, T. Ueno, S. Migitaka, T. Tanaka, K. Kojima, T. Onozuka, N. Moriuchi, and M. Hashimoto. 1994. In *Proc. Reg. Tech. Conf. Photopolymers*, Ellenville, NY: SPE, 306.
211. H. Ito. 2005. *Advances in Polymer Science*, **172**: 37.
212. W. Hinsberg, G. Wallraff, and R. Allen. 1998. Lithographic resists. In *Kirk-Othmer Encyclopedia of Chemical Technology*, 4th Ed., pp. 233–280.
213. D. Lee and G. Pawloski. 2002. *Journal of Photopolymer Science and Technology*, **15**: 3, 427.
214. H. Ito, G. Breyta, W. Conley, P. Hagerty, J. Thackerey, S. Holmes, R. Nunes, D. Fenzel-Alex-ander, R. DiPietro, and D. Hofer. 1996. *Journal of Photopolymer Science and Technology*, **9**: 4, 557.
215. W. Hinsberg, F. Houle, M. Sanchez, J. Hoffnagle, G. Wallraff, D. Medeiros, G. Gallatin, and J. Cobb. 2003. *Proceedings of SPIE*, **5039**: 1.
216. G. Wallraff, D.R. Medeiros, M. Sanchez, K. Petrillo, W.-S. Huan, C. Rettner, B. Davis, et al. 2004. *Journal of Vacuum Science and Technology, Part B*, **22**: 3479.
217. G.M. Wallraff, D.R. Medeiros, C.E. Larson, M. Sanchez, K. Petrillo, W.-S. Huang, C. Rettner, et al. 2005. *Proceedings of SPIE*, **5753**: 309.
218. P. Naulleau, K.A. Goldberg, E. Anderson, J.P. Cain, P. Denham, B. Hoef, K. Jackson, A.-S. Morlens, S. Rekawa, and K. Dean. 2005. *Proceeding of SPIE*, **5751**: 56.
219. R. Allen, W. Conley, and R. Kunz. 1997. Deep UV resist technology. In: *Handbook of Microlithography*, Bellingham, WA: SPIE Press, pp. 321–376.
220. R.D. Allen, G.M. Wallraff, W.D. Hinsberg, W. Conley, and R.R. Kunz. 1993. *Journal of Photopolymer Science and Technology*, **6**: 4, 575; G. Wallraff, R.D. Allen, W.D. Hinsberg, C.F. Larson, R.D. Johnson, R. DiPietro, G. Breyta, N. Hacker, and R.R. Kunz. 1993. *Journal of Vacuum Science and Technology, Part B*, **11**: 6, 2783.
221. R.D. Allen, W.D. Hinsberg, C.F. Larson, R.D. Johnson, R. DiPietro, G. Breyta, and N. Hacker. 1991. *Journal of Vacuum Science and Technology, Part B*, **9**: 3357.
222. H. Gokan, S. Esho, and Y. Ohnishi. 1983. *Journal of the Electrochemical Society*, **130**(1): 143.
223. R.R. Kunz, S.C. Palmteer, A.R. Forte, R.D. Allen, G.H. Wallraff, R.A. DiPietro, and D.C. Hofer. 1996. *Proceedings of SPIE*, **2724**: 365.
224. K. Nozaki, K. Watanabe, E. Yao, A. Kotachi, S. Takechi, and I. Hanyu. 1996. *Journal of Photopolymer Science and Technology*, **9**: 3, 509.
225. R.D. Allen, J. Opitz, C.E. Larson, T.I. Wallow, R.A. DiPietro, G. Breyta, R. Sooriyakumaran, and D.C. Hofer. 1997. *Journal of Photopolymer Science and Technology*, **10**: 3, 503.
226. R.D. Allen, G.M. Wallraff, R.A. DiPietro, D.C. Hofer, and R.R. Kunz. 1995. *Proceedings of SPIE*, **2438**: 474.
227. T. Wallow, F. Houlihan, O. Nalamasu, E. Chandross, T. Neenan, and E. Reichmanis. 1996. *Proceedings of SPIE*, **2724**: 355.
228. K. Patterson, U. Okoroanyanwu, T. Shimokawa, S. Cho, J. Byers, and C.G. Willson. 1998. *Proceedings of SPIE*, **3333**: 425.
229. A. Pasquale, R. Allen, and T. Long. 2001. *Macromolecules*, **34**(23): 8064.
230. R. Allen, R. Sooriyakumaran, J. Optiz, G. Wallraff, R. DiPietro, G. Breyta, D. Hofer, et al. 1996. *Proceedings of SPIE*, **2724**: 334.
231. W. Li, R. Varanasi, M.C. Lawson, R.W. Kwong, K.-J. Chen, H. Ito, H. Trunong, et al. 2003. *Proceedings of SPIE*, **5039**: 61.
232. P.R. Varanasi, R.K. Kwong, M. Khojasteh, K. Patel, K.-J. Chen, W. Li, M.C. Lawson, et al. 2005. *Proceedings of SPIE*, **5753**: 131.
233. T. Hattori, Y. Yokoyama, K. Kimura, R. Yamanaka, T. Tanaka, and H. Fukuda. 2003. *Proceedings of SPIE*, **5039**: 175; Y. Yokoyama, T. Hattori, K. Kimura, T. Tanaka, and H. Shiraishi. 2003. *Journal of Photopolymer Science and Technology*, **13**: 579.

234. R. Sooriyakumaran, B. Davis, C.E. Larson, P.J. Brock, R.A. DiPietro, T.I. Wallow, E.F. Connor, et al. 2004. *Proceedings of SPIE*, **5376**: 71.

235. R. Dammel, G. Pawlowski, A. Romano, F.M. Houlihan, W.-K. Kim, R. Sakamuri, D. Abdallah, M. Padmanaban, M.D. Rahman, and D. McKenzie. 2005. *Journal of Photopolymer Science and Technology*, **18**: 5, 593; S. Kanna, H. Inabe, K. Yamamoto, S. Tarutani, H. Kanda, K. Mizutani, K. Kitada, S. Uno, and Y. Kawabe. 2005. *Journal of Photopolymer Science and Technology*, **18**: 5, 603.

236. R.D. Allen, P.J. Brock, L. Sundberg, C.E. Larson, G.M. Wallraff, W.D. Hinsberg, J. Meute, T. Shimokawa, T. Chiba, and M. Slezak. 2005. *Journal of Photopolymer Science and Technology*, **18**: 5, 615.

237. a) S. Tarutani, H. Tsubaki, and S. Kanna. 2008. Development of materials and processes for double patterning toward 32-nm ArF immersion lithography process. *Proceedings of the SPIE*, **6923**: 69230F; b) S. Kang, M. Reilly, T. Penniman, R. Bell, K. Spizuoco, L. Joesten, and Y. Bae. 2010. Sub-30nm contact hole patterning using immersion single exposure with negative tone development. *Journal of Photopolymer Science and Technology*, **23**(2): 211–215.

238. M. Reilly, C. Andes, T. Cardolaccia, Y.S. Kim, and J.K. Park. 2012. Evaluation of negative tone develop resists for ArF lithography. *Proceedings of the SPIE*, **8325**: 832507-1.

239. G. Landie, Y. Xu, S. Burns, K. Yoshimoto, M. Burkhardt, L. Zhuang, K. Petrillo, et al. 2011. Fundamental investigation of Negative Tone Development (NTD) for the 22nm node (and beyond). *Proceedings of the SPIE*, **7972**: 797206-3.

240. T. Fujimori, T. Tsuchihashi, and T. Itani. 2015. Recent progress of negative tone imaging with EUV exposure. *Proceedings of the SPIE*, **9425**: 942505-1.

241. H. Nakagawa, T. Naruoka, and T. Nagai. 2014. Recent EUV resists toward high volume manufacturing. *Journal of Photopolymer Science and Technology*, **27**(6): 739–746.

242. D.C. Brandt, et al. 2014. LPP EUV source readiness for NXE 3300B. *Proceedings of the SPIE*, **9048**: 90480C.

243. H. Ito. 1999. Chemically amplified resists: Past, present, and future. *Proceedings of the SPIE*, **3678**: 2–11.

244. a) Y.-J. Kwark, J.P. Bravo-Vasquez, M. Chandhok, H. Cao, H. Deng, E. Gullikson, and C.K. Ober. 2006. Absorbance measurement of polymers at extreme UV wavelength: Correlation between experimental and theoretical calculations. *Journal of Vacuum Science and Technology, Part B*, **24**(4), 1822–1826. b) J. Thackeray. 2011. Materials challenges for sub-20nm lithography. *Journal of Micro-Nanolithography MEMS and MOEMS*, **10**(3): 033009.

245. A. Narasimhan, et al. 2015. Studying secondary electron behavior in EUV resists using experimentation and modeling. *Proceedings of the SPIE*, **9422**: 942208.

246. T. Wallow, C. Higgins, R. Brainard, K. Petrillo, W. Montgomery, C.-S. Koay, G. Denbeaux, O. Wood, and Y. Wei. 2008. Evaluation of EUV resist materials for use at the 32 nm half-pitch node. *Proceedings of the SPIE*, **6921**: 69211F.

247. T. Fedynyshyn, R.B. Goodman, and J. Roberts. 2008. Polymer matrix effects on acid generation. *Proceedings of the SPIE*, **6923**: 692319.

248. H. Yamamoto, K. Kozawa, A. Nakano, K. Okamoto, S. Tagawa, T. Ando, M. Sato, and H. Komano. 2004. Modeling and simulation of chemically amplified electron beam, x-ray, and EUV resist process. *Journal of Vacuum Science and Technology, Part B*, **22**: 3522.

249. M. Hori, et al. 2015. Novel EUV resist development for sub-14 nm half pitch. *Proceedings of the SPIE*, **9422**: 942220.

250. H. Tsubaki, S. Tarutani, H. Takizawa, and T. Goto. 2012. EUV resist materials for 20 nm and below half pitch applications. *Proceedings of the SPIE*, **8325**: 832509.

251. H. Tsubaki, S. Tarutani, N. Inoue, H. Takizawa, and T. Goto. 2013. EUV resist materials design for 15 nm half pitch and below. *Proceedings of the SPIE*, **8679**: 867906.

252. a) H. Tamaoki, S. Tarutani, H. Tsubaki, T. Takahashi, N. Inoue, T. Tsuchihashi, H. Takizawa, and H. Takahashi. 2011. Characterizing polymer bound PAG type EUV resist. *Proceedings of the SPIE*, **7972**: 79720A. b) J.W. Thackeray, R.A. Nassar, R. Brainard, D. Goldfarb, T. Wallow, Y. Wei, J. Mackey, et al. 2007. Chemically amplified resists resolving 25 nm 1:1 line: Space features with EUV lithography. *Proceedings of the SPIE*, **6517**: 651719.

253. a) J.W. Thackeray, J. Cameron, V. Jain, P. LaBeaume, S. Coley, O. Ongayi, M. Wagner, A. Rachford, and J. Biafore. 2013. Progress in resolution, sensitivity and critical dimensional uniformity of EUV chemically amplified resists. *Proceedings of the SPIE*, **8682**: 868213; b) S. Chen, L.L. Chang, Y.H. Chang, C.C. Huang, C. Y. Chang; Y. Ku. 2012. Contrast improvement with balanced diffusion control of PAG and PDB. *Proceedings of the SPIE*, **8325**: 83250O-1.

254. C.-M. Park, et al. 2014. Prospects of DUV OoB suppression techniques in EUV lithography. *Proceedings of the SPIE*, **9048**: 90480S-1.

255. J.M. Roberts, et al. 2009. Sensitivity of EUV resists to out-of-band radiation. *Proceedings of the SPIE*, **7273**: 72731W-1.

256. J. Thackeray. 2011. Materials challenges for sub-20nm lithography. *Journal of Micro-Nanolithography MEMS and MOEMS*, **10**(3): 033009.

257. a) V. Jain, et al. 2011. Impact of polymerization process and OOB on lithographic performance of a EUV resist. *Proceedings of the SPIE*, **7969**: 796912; b) K. Inukai, et al. 2012. Proceedings of the SPIE, 8322: 83220X-1; c) J. Iwashita, et al. 2012. Out of band insensitive polymer bound PAG for EUV resist. Proceedings of the SPIE, 8322: 83220Y-1.

258. T. Kozawa and S. Tagawa. 2010. *Japanese Journal of Applied Physics*, 49: 030001.

259. J.W. Thackeray, et al. 2013. Progress in resolution, sensitivity and critical dimensional uniformity of EUV chemically amplified resists. *Proceedings of the SPIE*, **8682**: 868213.

260. Fedynyshyn, et al. 2010. Polymer photochemistry at the EUV wavelength. *Proceedings of the SPIE*, **7639**: 76390A-1.

261. a) T. Sasaki, et al. 2008. Development of partially fluorinated EUV-resist polymers for LER and sensitivity improvement. *Proceedings of the SPIE*, **6923**: 692347; b) O. Ongayi, et al. 2012. High sensitivity chemically amplified EUV resists through enhanced EUV absorption. *Proceedings of the SPIE*, **8322**: 83220T-1; c) M. Christianson, et al. 2013. High absorbing resists based on trifluoromethacrylate-vinyl ether copolymers for EUV lithography. *Proceedings of the SPIE*, **8682**: 868213.

262. Tarutani, et al. 2010. Study on approaches for improvement of EUV-resist sensitivity. *Proceedings of the SPIE*, **7639**: 763909.

263. C. Higgins, A. Antohe, G. Denbeaux, S. Kruger, J. Georger, and R. Brainard. 2009. RLS tradeoff vs. quantum yield of high PAG EUV resists. *Proceedings of the SPIE*, **7271**: 727147-1.

264. C. Montgomery, et al. 2015. Evaluation of novel processing approaches to improve extreme ultraviolet (EUV) photoresist pattern quality. *Proceedings of the SPIE*, **9425**: 942526.

265. K. Ichimura. 2002. *Chemical Record*, **2**(1): 46–55.

266. S. Kruger, et al. 2012. Stable, fluorinated acid amplifiers for use in EUV lithography. *Proceedings of the SPIE*, **8325**: 832514.

267. H. Oizumi, et al. 2009. Development of EUV resists at Selete. *Proceedings of the SPIE*, **7273**: 72731M.

268. R.L. Brainard, G.G. Barclay, E.H. Anderson, and L.E. Ocola. 2002. Resists for next generation lithography. *Microelectronic Engineering*, **61–62**, 707–715.

269. Green, et al. 2013. Development of molecular resist derivatives for EUV lithography. *Proceedings of the SPIE*, **8679**: 867912.

270. X. Andre, et al. 2007. Phenolic molecular glasses as resists for next generation lithography. *Proceedings of the SPIE*, **6519**: 65194B.

271. Yamamoto, et al. 2014. Study on resist performance of chemically amplified molecular resist based on noria derivative and calixarene derivative. *Proceedings of the SPIE*, **9051**: 90511Z.

272. Maruyama, et al. 2010. Development of EUV resist for 22 nm half pitch and beyond. *Proceedings of the SPIE*, **7636**: 76360T.

273. Frommhold, et al. 2013. EUV lithography performance of negative-tone chemically amplified fullerene resist. *Proceedings of the SPIE*, **8682**: 86820Q.

274. a) K. Dean, et al. 2006. Effects of material design on extreme ultraviolet (EUV) resist outgassing. *Proceedings of the SPIE*, **6153**: 61531E; b) M. Thiyagarajan, K. Dean, and K.E. Gonsalves. 2005. Improved lithographic performance of EUV resists based on polymers having a photoacid generator in the backbone. *Journal of Photopolymer Science and Technology*, **18**(6): 737.

275. J. Thackeray. 2011. Materials challenges for sub-20nm lithography. *Journal of Micro/Nanolithography, MEMS, and MOEMS*, **10**(3): 033009.

276. J.W. Thackeray, R.A. Nassar, R. Brainard, D. Goldfarb, T. Wallow, Y. Wei, J. Mackey, et al. 2007. Chemically amplified resists resolving 25 nm 1:1 line: Space features with EUV lithography. *Proceedings of the SPIE*, **6517**: 651719.

277. J.W. Thackeray, et al. 2014. Understanding EUV resist mottling leading to better resolution and line-width roughness. *Proceedings of the SPIE* **9048**: 904807-10.

278. H. Tamaoki, et al. 2011. Characterizing polymer bound PAG type EUV resist. *Proceedings of the SPIE* **7972**: 79720A.

279. S. Matsumaru, et al. 2015. Development of EUV chemically amplified resist which has novel protecting group. *Proceedings of the SPIE*, **9425**: 94250U.

280. Fujimori, et al. 2015. Recent progress of negative-tone imaging with EUV exposure. *Proceedings of the SPIE*, **9425**: 942505-1.

281. H. Tsubaki, et al. 2014. Novel EUV resist materials design for 14 nm half pitch and below. *Proceedings of the SPIE*, **9048**: 90481E.

282. Y.-J. Kwark, J.P. Bravo-Vasquez, M. Chandhok, H. Cao, H. Deng, E.Gullikson, and C.K. Ober. 2006. Absorbance measurement of polymers at extreme UV wavelength: Correlation between experimental and theoretical calculations. *Journal of Vacuum Science and Technology. Part B*, **24**(4): 1822–1826.

283. D. Ulieru. 1998. Trace contaniments from photoresist materials by modern spectrometry determination. *Proceedings of the SPIE*, **3332**: 721.

284. S. Yamada, et al. 2015. Development of spin-on metal Hardmask (SOMHM) for advanced node." *Proceedings of the SPIE*, **9425**: 94251X.

285. M. Trikeriotis, et al. 2010. Development of an inorganic photoresist for DUV, EUV, and electron beam imaging. *Proceedings of the SPIE*, **7639**: 76390E.

286. B. Cardineau, et al. 2012. Tightly-bound ligands for hafnium nanoparticle EUV resists. *Proceedings of the SPIE*, **8322**: 83220V.

287. C.Y. Ouyang, et al. 2013. Non-aqueous negative-tone development of inorganic metal oxide nanoparticle photoresists for next generation lithography. *Proceedings of the SPIE*, **8682**: 86820R.

288. S. Chakrabarty, et al. 2014. Oxide nanoparticle EUV resists: Toward understanding the mechanism of positive and negative tone patterning. *Proceedings of the SPIE*, **9048**: 90481C.

289. S. Chakrabarty, et al. 2013. Increasing sensitivity of oxide nanoparticle photoresists. *Proceedings of the SPIE*, **8679**: 867906.

290. M. Krysak, et al. 2014. Investigation of novel inorganic resists materials for EUV lithography. *Proceedings of the SPIE*, **9048**: 904805.

291. J. Jiang, et al. 2015. Systematic study of ligand structures of metal oxide EUV nanoparticle photoresists. *Proceedings of the SPIE*, **9422**: 942222.

292. J.M. Amador, et al. 2015. Patterning chemistry of HafSOx photoresist. *Proceedings of the SPIE*, **9051**: 90511A.

293. Clark, et al. 2014. Coater/developer process integration of metal-oxide based photoresist. *Proceedings of the SPIE*, **9425**: 94251A.

294. B. Cardineau, et al. 2014. EUV resists based on tin-oxo clusters. *Proceedings of the SPIE*, **9051**: 90511B.

295. J. Passarelli, et al. 2014. EUV resists comprised of main group organometallic oligomeric materials. *Proceedings of the SPIE*, **9051**: 90512A.

296. M. Sortland, et al. 2015. Positive-tone EUV resists: Complexes of platinum and palladium. *Proceedings of the SPIE*, **9422**: 942227.

8 Photoresist and Materials Processing

Bruce W. Smith

CONTENTS

8.1 INTRODUCTION

For the most part, conventional single-layer photoresists have been based on components with two primary functions. Whether considering diazonaphthoquinone (DNQ)/novolac g/i-line resists or chemically amplified (CA) 248 nm/193 nm deep-UV (DUV) resists, an approach has been utilized wherein a base resin material is modified for sensitivity to exposure by a photoactive compound or through photoinduced chemical amplification. The resist base resin is photopolymeric in nature and is responsible for etch resistance, adhesion, coatability, and bulk resolution performance. These resins generally do not exhibit photosensitivity on the order required for integrated circuit (IC) manufacturing. Single-component polymeric resists, including methacrylates, styrenes, and other polymers or copolymers, have been utilized for microlithography, but sensitization is generally low and limited to exposures at short ultraviolet (UV) wavelengths or with ionizing radiation. Inherent problems associated with low absorbance and poor radiation resistance (required, for example, during ion implantation or plasma etching steps) generally limit the application of these types of resists to low volumes or processes with unique requirements.

The sensitization of photoresist materials has been accomplished by several methods. In the case of conventional g/i-line resists, chemical modification of a base-insoluble photoactive compound (PAC), DNQ, to a base-insoluble photoproduct, indenecarboxylic acid (ICA), allows an increase in aqueous base solubility. For 248 and 193 nm chemically amplified resists, exposure of a photoacid generator (PAG) leads to the production of an acid, which subsequently allows polymer deprotection (positive behavior) or crosslinking (negative behavior). Other similar processes have been developed (as discussed in Chapter 7, Chemistry of Photoresist Materials) and may involve additional components or mechanisms.

For any resist system, the thermodynamic properties of polymeric resins play an important role in processability. During the coating, exposure, and development processes of a resist, an understanding of the thermodynamic properties is desirable, because the glass transition temperature (T_g) of a polymer influences planarizability, flow, and diffusion, Although reasonably high T_g values may be desirable, glassy materials with values above 200 °C are usually not suitable because of poor mechanical performance. After three-dimensional resist features are formed, however, a thermoset material may be desired, in which the polymer does not flow with temperature, and a T_g essentially does not exist. This ensures the retention of high–aspect ratio features through subsequent high-temperature and high-energy processes. By appropriate engineering of bake steps during single-layer resist processing, the control of polymer thermoplastic and thermoset properties can be made possible. For negative resists, the situation is somewhat simplified. Coated negative resists are thermoplastic in nature with a well-defined T_g range. Upon exposure and subsequent secondary reactions, crosslinking leads to a networked polymer that will not flow with temperature. At the temperature of decomposition (T_d), the polymer will break down and begin to lose significant volume. Imaging steps are therefore responsible for the production of thermally stable resist features. Operations are often included in the processing of positive resists that can accomplish similar thermal stability enhancements.

This chapter addresses the critical issues involved in the processing of single-layer resists materials. Process steps to be discussed include:

- Resist stability, contamination, and filtration
- Substrate priming
- Resist coat
- Soft bake
- Exposure
- Postexposure brake
- Development
- Swing effects
- Hard bake and postdevelopment treatment

The step-by-step process flow of photoresist processing has been covered elsewhere. More details regarding the chemistry involved with various photoresist materials are covered in Chapter 7.

8.2   RESIST STABILITY, CONTAMINATION, AND FILTRATION

8.2.1   RESIST STABILITY AND FILTRATION

DNQ/novolac resists have proved to be robust materials with respect to sensitivity to thermodynamic and aging effects while stored in uncast form. A resist shelf life of several months can be expected with no significant change in lithographic performance. Because resists are considered for application in production, the stability of materials at various points of the process also needs to be considered.

For DNQ/novolac resists, aging can lead to an increase in absorption at longer wavelengths. Resist materials are susceptible to several thermal and acid–base (hydrolytic) reactions when stored [1]. These include thermal degradation of the DNQ to ICA followed by acid-induced azo dye formation and azo coupling of the DNQ and novolac. A characteristic "red darkening" results from this coupling, induced by the presence of acids and bases in the resist. Although long-wavelength absorbance is altered by this red azo dye, the impact on UV absorbance and process performance is most often negligible. Degradation mechanisms can also result in crosslinking, leading to an increase in high–molecular weight components. Hydrolysis of DNQ may occur to form more soluble products, and hydrolysis of solvents is possible, which can lead to the formation of acids [2]. The practical limitation of shelf life for DNQ/novolac resists is generally on the order of 6 months to 1 year. Once coated, resist film can absorb water and exhibit a decrease in sensitivity, which can often be regained through the use of a second soft bake step. As will be described, process delays for chemically amplified polyhydroxylstyrene (PHOST) resists are much more critical than for DNQ/ novolac materials.

A larger problem encountered when storing DNQ/novolac resists is sensitizer precipitation. With time, DNQ PAC can fall out of solution, especially at high temperatures. These crystallized precipitates most readily form with high levels of DNQ. In addition, resist particulate levels can be increased by the formation of gel particles, a result of acid-induced novolac crosslinking via thermal decomposition of DNQ. Any of these routes to particulate formation can lead to levels exceeding that measured by the resist manufacturer. Because of this, point-of-use filtration has become common practice for most production applications to ensure photoresist consistency [3]. Resist materials are commonly filtered at a level of approximately 25% of the minimum geometry size. As the geometry size approaches sub-0.35 μm, filtration requirements approach 0.05 μm. Such ultrafiltering will have an impact on how resist can be manufactured and used. Filtration speed is dramatically reduced, and material preparation becomes more costly. Similar concerns can exist for pump throughput during resist dispensing. Fractionation of a resist material can also occur, resulting in the removal of long polymer chains and a change in process performance. To illustrate this, consider an i-line (DNQ/novolac) with a molecular weight on the order of 1020×10^3 g/mol (number average). The resulting average polymer chain size is nearly 5–6 nm with a maximum as large as 40 nm. In highly concentrated resist formulations (greater than 30 wt.%), intertwisting of polymers can result in chain sizes greater than 80 nm. If such a resist is filtered to 0.05 μm, the largest polymer chains can be removed. As technology has progressed toward smaller feature resolution, particulate and filtration issues have needed to be carefully considered.

8.2.2 Stability Issues for Chemically Amplified Resists (CARs)

Filtration concerns extend to DUV lithography, where the issue of ultrafiltering becomes an increasingly important problem. In addition, environmental stability issues are present for many chemically amplified resists (CARs), especially for resists based on acid-catalyzed reactions. Ion-exchange methods are conventionally used to reduce ion contamination levels in resists below 10 ppb. Ionic contamination reduction in both positive and negative CAR systems needs to be carefully considered. Deprotection of acid-labile components can result from reaction with cationic exchange resins. The catalytic acid produced upon exposure of these resists is also easily neutralized with base contamination at ppb levels. These contaminants can include such things as ammonia, amines, and NMP, which are often present in IC processing environments [4]. Any delay between exposure and postexposure bake (PEB) can result in a decrease in sensitivity and the formation of a less soluble resist top layer or "T-top."

To reduce the likelihood of base contamination of these resists, several improvements have been made in resist formulations. One method to reduce acid loss is the use of low–activation energy (E_a)

polymers with highly reactive protection groups. These resists are sufficiently active that deprotection can occur immediately upon exposure, significantly reducing the sensitivity to PEB delay effects [5, 6]. Additives have also been incorporated into DUV CARs to improve their robustness to contamination effects [7, 8], and resist top-coating approaches have been introduced [9]. By coating a thin water-soluble transparent polymeric film over a resist layer, protection from airborne contamination can be made possible. This "sealing" layer is removed prior to development with a water rinse. Although such a solution leads to minimal additional process complexity, it is still desirable to use resist techniques that do not require additional material layers. An alternative route for the reduction of contamination effect is the use of high–activation energy resist materials. By reducing the reactivity of a resist, higher bake-process temperatures are allowed. Increasing the bake temperatures above a polymer's T_g results in densification of the photoresist. This leads to a significant decrease in the diffusion rate of airborne base contamination prior to or after exposure [10, 11]. These high-E_a resists also require that a photoacid generator be chosen that can withstand high temperatures. Other methods used to reduce base contamination of acid-catalyzed resists include the use of activated charcoal air filtration during resist coating exposure and development operations [12]. This is now considered a requirement for the processing of PHOST CAR resists. Environmental base contamination can be neutralized and further reduced by adding weak acids to these filters.

The stability or shelf life of PHOST-based resists is also influenced by the structure of polymer protective groups. This is especially true for low–activation energy (high-reactivity) resists, for which the lability of protective groups may decrease usable resist life. Conversely, the more stable protective groups utilized with high–activation energy (low-reactivity) resists lead to a higher degree of stability.

8.3   RESIST ADHESION AND SUBSTRATE PRIMING

Adequate adhesion of photoresist to a wafer surface is critical for proper process performance. Resist adhesion failure can occur not only during photolithography operations but also in subsequent etch, implant, or other masking steps. Negative resists are less prone to adhesion failure, because crosslinking results in a networked polymer that is bound to the wafer surface. Positive resists (especially phenolic-based materials such as novolac or PHOST resists) are more likely to be single-polymer chains and rely on weaker physical and chemical forces for adhesion. Etch process undercutting can often result from inadequacies at the resist interface, resulting in loss of etch linewidth control. The causes of resist adhesion failure are generally related to dewetting of a photoresist film. This can result from a large discordance between the surface tension of the wafer and that of the resist material, especially when coating over silicon oxide. Silicon dioxide is an especially difficult layer to coat, because it provides a hydrophilic (water-attracting) surface to a hydrophobic (water-repelling) resist. The surface tension of thermal silicon dioxide may be on the order of 15 dynes/cm^2, whereas the surface tension of phenolic resists in casting solvent may be near 30 dynes/cm^2. Surface defects can also cause adhesion failure, as surface free energy can result in dewetting.

Methods of adhesion promotion can be used for most silicon oxide layers, whether they are thermally grown, deposited, native, or glasslike. Chemical passivation of these surfaces is generally carried out using silylating priming agents that act to modify the wafer surface. Some benefit can be realized with priming of layers other than oxides if techniques promote a closer matching of material surface tension. Alkylsilane compounds are generally used to prime oxide surfaces, leading to a lowering of surface hydrophilicity. The most commonly used silane-type adhesion promoter is hexamethyldisilazane (HMDS). Other similar promoters are available, including trimethylsilyldiethylamine (TMSDEA), which can be more effective but also less stable, resulting in lower shelf and coated lifetimes. Reduction of substrate surface tension is carried out in two stages, as shown in Figure 8.1. The figure depicts a silicon oxide surface with adsorbed water and OH⁻ groups. An initial reaction of water with an alkylsilane (HMDS) produces an inert hexamethyldisiloxane and

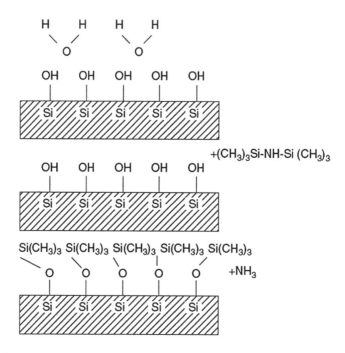

FIGURE 8.1 Adhesion promotion of a silicon oxide surface with HMDS surface priming. The substrate is first dehydrated upon reaction with silane promoter. Further reaction with heat leads to a hydrophobic surface.

ammonia, resulting in a dehydrated surface. Further reaction with HMDS produces a trimethylsilyl-substituted hydroxyl or oxide species and unstable trimethylsilylamine. With heat, this unstable compound reacts with other surface hydroxyl groups to produce further ammonia and a trimethylsiloxy species. The process continues until steric hindrance (via the large trimethylsilyl groups) inhibits further reaction.

Surface priming using HMDS, TMSDEA, or similar agents can be carried out in either liquid- or vapor-phase modes. In either case, elevated process temperatures (~100 °C) must be reached to complete the priming reaction. Substrates should be cleaned prior to application using UV ozone, HF dip, plasma, or other "oxidative" cleaning methods. Adhesion of a photoresist to silicon nitride or deposited oxide layers can be enhanced by using an oxygen/ozone plasma treatment. Priming agents are generally best applied using vapor prime methods, either inline or in batch vacuum ovens. Uniformity and reduced chemical usage make this more attractive than liquid methods.

Overpriming of a wafer surface can result in dewetting and lead to further adhesion problems. This can occur with repeated treatment or by using excessive vapor times. Problems are often noticed in isolated substance areas, depending on device topography or condition. A phenomenon known as resist "popping" can also occur as a result of overpriming; in this case, high-fluence exposure (such as that encountered with heavy UV overexposure in ion implantation steps) can cause failure of weakened resist adhesion. Deposition of resist debris onto adjacent substrate areas can result. Measurements of resist surface tension using water contact angle techniques can identify such overpriming problems. Remedies include the use of shorter priming times, resist solvents with lower surface tension, double resist coating steps, or a pretreatment of the wafer surface with the resist casting solvent. Oxygen or ozone plasma treatments can also correct an overprimed wafer surface and allow repriming under more appropriate conditions.

The strength of adhesion bonds between a photoresist and a substrate has also been shown to influence the T_g and thermal expansion coefficient of a thin film. The impact is greatest as resist films approach 1000 Å thicknesses [13].

8.4 RESIST COATING

8.4.1 RESIST SPIN-COATING TECHNIQUES AND CONTROL

A photoresist can be dispensed by several methods, including spin coating, spray coating, and dip coating. The most widely used methods for coating resists onto wafer substrates are spin coating methods. During spin coating, the resist is dispensed onto a wafer substrate (either statically or dynamically), accelerated to final spin speed, and cast to a desired film thickness. Variations on this process have been suggested, including the use of a short-term high-speed initial coating step followed by a slow drying stage [14]. Spin-coating processes use the dynamics of centrifugal force to disperse a polymeric resist material over the entire wafer surface. The flow properties (rheology) of the resist influence the coating process and need to be considered to achieve adequate results [15]. In addition, solvent transport through evaporation occurs, which can result in an increase in resist viscosity and shear thinning, both of which affect the final film properties. As a resist-solvent material is spin cast, the film thickness decreases uniformly at a rate dependent on the spin speed (ω), kinematic viscosity (v), solids concentration (c), solvent evaporation rate (e), and initial film thickness, expressed by the following rate equations:

$$\frac{dS}{dt} = \frac{-c2\omega^2 h^3}{3v} \tag{8.1}$$

$$\frac{dL}{dt} = (1-c)\frac{2\omega^2 h^3}{3v} - e \tag{8.2}$$

where dS/dt and dL/dt are the rate of change of solids (S) and solvents (L), respectively [16]. The results are shown in Figure 8.2 for a 1 μm film, where both solids and solvent volumes are plotted against spin time. Initially, the concentration changes little, as resist spread dominates. When the resist thickness drops to one-third of its original value, evaporation dominates, and the solvent content reaches its final value. The high viscosity of the resist eliminates further flow.

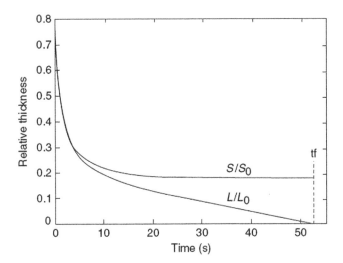

FIGURE 8.2 Calculated time dependence during spin coating on the volume of solids (S) and solvent (L) per unit area normalized to initial values. When the resist thickness drops to one-third of its original value, evaporation dominates, and the solvent content reaches its final value. (From Meyerhofer, D., *J. Appl. Phys.*, 49(7), 3993, 1978.)

The primary material factors that influence spin-coated film properties include the resist polymer molecular weight, solution viscosity, and solvent boiling point (or vapor pressure). Primary process factors include wafer spin speed, acceleration, temperature, and ambient atmosphere. The thickness of a resist film can be modified to some extent through control of the rotation speed of the substrate. Resist thickness is inversely proportional to the square root of spin speed (ω):

$$\text{Thickness} \propto \frac{1}{\sqrt{\omega}} \tag{8.3}$$

To achieve large thickness changes, modification of the resist solution viscosity is generally required, because coating at excessively low or high speeds results in poor coating uniformity. At excessively high speeds, mechanical vibration and air turbulence result in high levels of across-wafer nonuniformity. At low spin speeds, solvent loss of the resist front as it is cast over the substrate results in a situation of dynamic resist viscosity, also resulting in high levels of nonuniformity. The optimal spin range is dependent on wafer size. Wafers up to 150 mm can be coated at rotation speeds on the order of 4000–5000 rpm. Larger substrates require lower speeds.

The optimum coating thickness for a resist layer is determined by the position of coherent interference nodes within the resist layer. Standing waves (see Section 8.9) resulting from reflections at the resist–substrate interface result in a regular distribution of intensity from the top of the resist to the bottom. This distribution results in a "swing" in the required clearing dose (E_0) for a resist, as shown in Figure 8.3. Three curves are shown, for polysilicon, silicon nitride (1260 Å), and silicon dioxide (3700 Å) coated substrates. A general upward trend in E_0 is seen as resist thickness increases. This is due to the residual nonbleachable absorption of the resist, which can be significant. (For a resist with a residual absorption of 0.10 μm^{-1}, the intensity at the bottom of a 1 μm resist layer is 90% of that experienced at the top.) In addition to this E_0 trend, sensitivity oscillates from minimum to maximum values with thickness. Within one swing cycle, an exposure variation of 32% exists for the polysilicon substrate, 27% for silicon nitride, and 36% for silicon dioxide. When a resist is coated over a dielectric layer, such as silicon dioxide, silicon nitride, or an antireflective coating (ARC), there will be a shift in the phase of E_0 oscillations. Analysis of swing behavior may therefore be unique for various lithographic levels. Coated thickness optimization can be performed using these swing curves, determined experimentally through open frame exposure of a

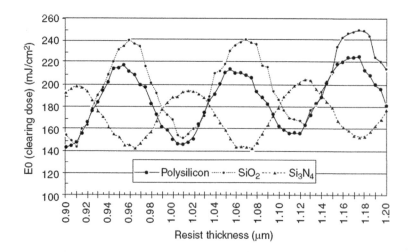

FIGURE 8.3 Clearing dose (E_0) swing curves for an i-line resist over polysilicon, silicon dioxide (3700 Å), and silicon nitride (1260 Å). The increasing trend in required dose is a function of residual absorption. Conditions of minimum interference lead to maximum E_0 values but minimal scumming.

resist coated within a small range of thicknesses. Lithographic modeling can aid in the generation of such relationships using knowledge of the resist refractive index, absorption properties (possibly dynamic), exposure wavelength, and resist–substrate reflectivity.

Inspection of the E_0 swing curve in Figure 8.3 suggests several possibilities for resist thickness, of which only a few are desirable. For polysilicon, there is a minimum dose requirement at a thickness of ~1.01 µm, where constructive interference occurs, and there is maximum intensity at the resist base. At a resist thickness over polysilicon of ~1.06 µm, destructive interference leads to a maximum E_0 requirement. Other alternatives might include positions on either side of these values (between nodes). Thicknesses corresponding to these midnodal positions allow the least amount of coating process latitude, as small deviations from the targeted film thickness lead to significant changes in dose requirements. Greater latitude exists at maximum interference positions, where there is a minimum requirement for exposure dose, which may be an attractive choice. Small changes in film thickness result only in small E_0 variations, but the direction of these changes is toward higher clearing dose values. The result may be scumming of resist features resulting from underexposure, a situation that is unacceptable. The best choice for targeted film thickness may be at a corresponding interference minimum, where small thickness changes result in a small decrease in the dose requirement. Slightly lower throughput may result (generally not a gating factor in today's exposure operations), but this will ensure no resist scumming related to underexposure. Image fidelity at the top surface of a resist film is also influenced by film thickness and positions on the interference curve. By coating at a midnodal thickness, top surface rounding, or T-topping, can result (see Section 8.9 for further discussion).

During spin coating, a large amount of resist "free volume" can be trapped within a resist layer. A simplified free-volume model of molecular transport can be quite useful for correlation and prediction of diffusion properties of resist materials [17, 18]. Volumetric expansion enhances polymer chain mobility and acts similarly to the addition of plasticizers. The resist's glass transition temperature (T_g) is lowered and the dissolution properties of novolac- and PHOST-based resist can be increased [19]. Coating-included free volume has been shown to affect acid diffusion as well and becomes a concern when considering reduction of airborne base contamination and postexposure delay.

8.4.2 SOLVENT CONTRIBUTION OF FILM PROPERTIES

Residual casting solvent can act as a plasticizer and can reduce the T_g of a resist. Resist solvent content has been shown to be dependent on film thickness. A 1000 Å resist film may, for example, exhibit 50% more solvent retention than a 10,000 Å film. Figure 8.4 shows residual solvent in PHOST polymer films coated at thicknesses of 12,000 and 1100 Å. Only near the resist T_g (135 °C) does the solvent content for the 1100 Å film approach that of the thicker film. Table 8.1 shows diffusion coefficients for propylene glycol methyl ether acetate (PGMEA) solvent in the same photoresist film thicknesses, determined by diffusion analysis during 2 h of baking. These results may be due to a smaller degree of inter- or intramolecular hydrogen bonding in thinner films [20–22], which can allow a stronger polymer interaction with the casting solvent and lower solvent evaporation rates. A higher solvent content leads to an increased dissolution rate and increased diffusivity levels. When considering various resist solvent systems, it might also be expected that lower–boiling point (T_b) solvents would lead to lower solvent retention than higher-T_b solvents. The opposite, however, has been demonstrated [23]. PGMEA, for instance, has a boiling point of 146 °C and an evaporation rate of 0.34 (relative to n-butyl acetate). Ethyl lactate has a higher T_b of 154 °C and an evaporation rate of 0.29. Despite its lower boiling point, PGMEA is more likely to be retained in a resist film. The reason for this is a skin formation that results from rapid solvent loss during the coating process [24]. The resist viscosity at the surface increases more rapidly for PGMEA as solvent is exhausted, leading to more residual solvent remaining throughout the resist film. If a resist film is then baked at temperatures below

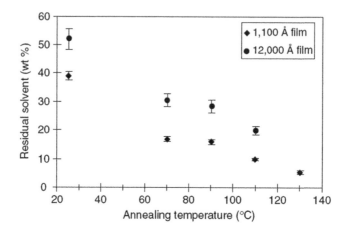

FIGURE 8.4   Back-temperature dependence of residual PGMEA solvent in 1100 and 12,000 Å spin-cast films annealed for 1300 min. (From Frank, C.W., Rao, V., Despotopoulou, M.M., Pease, R.F.W., Hinsberg, W.D., Miller, R.D., and Robolt, R.F., *Science,* 273, 972, 1996.)

TABLE 8.1
Diffusion Coefficient Values for PGMEA Solvent in CAR Resist Films with 2 Hours of Baking

Temperature (°C)	Diffusion Coefficient (cm²/s)	
	0.11 µm Film	1.2 µm Film
70	4.2×10^{-14}	1.2×10^{-12}
90	9.4×10^{-14}	4.4×10^{-12}
110	1.1×10^{-13}	1.4×10^{-11}

the bulk T_g, densification of surface free volume is allowed only at the top surface of the film and is prevented throughout the bulk. Entrapped solvent therefore leads to an apparent surface-induction effect, which can be reduced only if the resist is backed above its T_g. Because solvent content plays an important role in determining the ultimate glass transition temperature of the resist (and therefore, its dissolution properties), any postcoating incorporation of solvents can also have an adverse impact on performance. Such additional solvent may be encountered, for instance, when using an edge-bead removal process based on acetone, ethyl lactate, pentanone, or other organic solvents.

8.4.3   SUBSTRATE CONTRIBUTION TO RESIST CONTAMINATION

Continuous improvements have been made in the materials and processes used for DUV chemically amplified resists to reduce top-surface base contamination effects. An additional contamination problem occurs when processing resists over some substrates. Resists coated over Si_3N_4, boro-phospho-silicate glass (BPSG), spin-on-glass (SOG), Al, and TiN have seen shown to initiate a substrate contamination effect that can result in resist scumming or "footing." With TiN substrates, the problem has been attributed to surface N^{-3} and TiO_2, which can act to neutralize photogenerated acid, resulting in a lowering of the dissolution rate at the resist–substrate interface [25]. Sulfuric acid/hydrogen peroxide and oxygen plasma pretreatments have been shown to reduce contamination effects when coating over Si_3N_4 and other problematic substrates [26].

8.4.4 EDGE-BEAD REMOVAL

After a spin-coating process, a bead of hardened resist exists at the edge of a wafer substrate. The formation of this edge bead is caused in part by excessive resist drying and can result in resist accumulation up to 10 times the thickness of the coated film. Elimination of this edge bead is required to reduce contamination of process and exposure tools. Solvent edge-bead removal (EBR) techniques can be utilized to remove this unwanted resist by spraying a resist solvent on the backside and/or the frontside of the wafer substrate. Surface tension allows removal of a 2 to 3 mm resist edge from the front resist surface while removing any backside resist coating. Acetone, ethyl lactate, and pentanone are possible solvent choices for edge-bead removal processes.

8.5 RESIST BAKING—SOFT BAKE

8.5.1 GOALS OF RESIST BAKING

Baking processes are used to accomplish several functions and generally alter the chemical and/or physical nature of a resist material. The goals of resist baking operations may include the following, which are accomplished at various stages during resist processing:

- Solvent removal
- Stress reduction
- Planarization
- Reduction of resist voids
- Reductions of standing waves
- Polymer crosslinking and oxidation
- Polymer densification
- Volatilization of sensitizer, developer, and water
- Induction of (acid) catalytic reactions
- Sensitizer–polymer interactions

Polymeric resist resins are thermoplastic in nature, more amorphous than crystalline, and with glass transition temperatures in the 70–180 °C range. Thermal flow properties are taken advantage of during processing, in which baking steps at or near the T_g allow some degree of fluid-like resist behavior. In a fluid-like state, stress in a coated film can be reduced, and the diffusion of solvents, sensitizer, and photoproducts is enhanced. At high resist baking temperatures, sensitizer and polymer decomposition can occur. A proper choice of the baking temperature, time, and method must therefore take into account the evaporation and diffusion properties of solvents, the decomposition temperature of PAC or PAG, and the diffusion properties of the photoinduced acid and base contaminants for chemically amplified resist material.

8.5.2 RESIST SOLVENT AND T_G CONSIDERATIONS

There is a relationship between a solvent's evaporation rate or boiling point and the residual solvent content in a resist film. Obtaining such a relationship is difficult, however, because residual solvent is highly dependent on the T_g of the resist. If a baking process does not sufficiently reach the resist T_g, it is difficult to remove solvent to levels below a few percent [27]. The resist T_g therefore plays an important role in the evaporation mechanisms of resist solvent removal. Solvent removal has been shown to occur in two stages: the first stage is the diffusion of solvent molecules at temperatures near T_g and their accumulation at the film surface. The second stage is the evaporation of these adsorbed molecules at higher temperatures, dependent on hydrogen bonding properties.

No compaction would be expected if baking temperatures above the resist bulk T_g values were not reached. In fact, soft bake temperatures 10–20 °C below a bulk resist T_g are frequently employed with successful results. The reason for this is that the T_g of a resist film is not identical to that of bulk resist. It may actually be a great deal lower. This can be explained by the concept of resist free volume, which is minimized if soft bake temperatures above the bulk T_g values are used. If resists are baked more than 5–10 °C below the actual resist film T_g, compaction of this intrinsic free volume generally cannot occur [28].

Several resist material factors affect coated CAR film flow properties and may be modified for optimum performance. In general:

- Blocking of phenolic groups generally results in a lowering of T_g, a consequence of the decrease in hydrogen bonding between phenolic groups.
- A 40–50 °C decrease in polymer flow temperature (T_f) has been reported with 25% blocking [29].
- By modifying the molecular weight of the polymer, 20–25 °C of the loss in T_g or T_f due to blocking can be regained.
- As the number of phenolic groups is increased (relative to t-butoxycarbonyl [t-BOC] protective groups), the deprotection temperature is reduced.
- As the number of the phenolic group is increased, T_g is also increased.

These relationships can result in a difficult situation if the deprotection temperature is forced lower than the resist T_g. Further manipulation may be possible through the use of copolymerization or by introducing additional hydrogen bonding sites.

There is a tradeoff in determining the optimal soft bake temperature. A high soft bake temperature is desired so that the resist film T_g approaches that of the bulk, which leads to the best thermal performance. But at lower soft bake temperatures, an increase in postexposure acid diffusion is made possible, allowing a reduction in standing waves. To accommodate a lower soft bake temperature, a higher PEB may be needed to achieve adequate thermal and plasma etch performance properties (see Section 8.7 for an additional description).

8.5.3 SOFT BAKE

Resist films are coated from a polymer solution, making solvent reduction of a coated film a primary action of soft bake (or prebake) processes. Other consequences of soft baking include a reduction of free volume and polymer relaxation, which are important phenomena that affect resist process performance [31]. Prior to coating, photoresist contains between 65 and 85% solvents. Once cast, the solvent content is reduced to 10–20%, and the film can still be considered to be in a "liquid" state. If it were exposed and processed at this point, several adverse consequences would result. At this high solvent level, the film is tacky and highly susceptible to particulate contamination, which can be transferred through handling to subsequent steps. Also, inherent stress resulting from casting a thin film leads to adhesion problems. The most significant impact resulting from the elimination of a soft bake step is lack of dissolution discrimination between exposed and unexposed resist. With such high solvent levels, the expanded resist volume allows a high degree of dissolution regardless of the state of inhibition, protection, acceleration, and crosslinking. Ideally, a solvent content of a few percent would be desirable. Further densification could then be allowed to control the small molecular diffusion properties. To achieve this, the baking operation must approach the boiling point of the casting solvent (on the order of 140 °C for conventional solvent systems). At this elevated temperature, decomposition of the resist sensitizer is likely to occur, since DNQ decomposition temperatures are on the order of 100–120°C. Some residual solvent may also be desirable for DNQ/novolac to allow water diffusion and conversion of ketene to ICA. Complete solvent removal at prebake is therefore not attempted; instead,

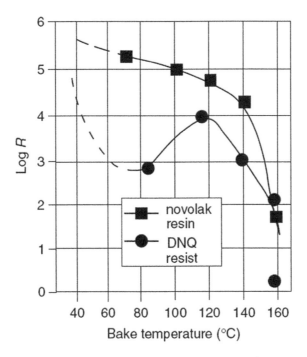

FIGURE 8.5 The influence of prebake temperature on dissolution rate (Å/min) for a novolac resin and a DNQ/novolac resist. Prebake time is 30 min; developer is 0.25 N KOH. (From Koshiba, M., Murata, M., Matsui, M., and Harita, Y., *Proc. SPIE*, 920, 364, 1998.)

adequate removal to allow the best exposure and dissolution properties is targeted. Figure 8.5 shows the development rate versus soft bake temperature for a DNQ/novolac resist. Four zones exist: (I) a no-bake zone, where residual solvent and dissolution rates are high; (II) a low-temperature zone (up to 80 °C), where the dissolution rate is reduced due to solvent removal; (III) a mid-temperature zone (80–110 °C), where DNQ is thermally converted to ICA, leading to an increase in development rate; and (IV) a high-temperature zone (greater than 120 °C), where film densification occurs, DNQ is further decomposed, and water is removed, leading to an increase in inhibition rather than acceleration. Also, at these temperatures, oxidation and crosslinking of the novolac begin [31].

Whereas a prebake for DNQ/novolac is required to bring the resist to a state suitable for subsequent exposure and development, control to ±1 °C is generally adequate to ensure consistent lithographic performance. This is not the case for chemically amplified resists based on poly(hydroxystyrene) (PHOST), however. Because acid diffusion is an integral part of both positive and negative acid-catalyzed resist, control of polymer composition and density is critical. Consideration for soft bake differs for high- and low-E_a CAR resists, depending on the reactivity of the resist to acid-induced deprotection. Resists with moderate activation energy (E_a) had been the conventional route to positive CAR resists. By making use of resists with lower reactivity (high E_a), elevated baking processes above the polymer T_g (~150 °C) allow maximum densification and reduction of the diffusion of acid-neutralizing contaminants. This resist design concept also reduces acid diffusion, which can result in an increase in resolution capability [32].

8.5.4 Resist Baking Methods

The preferred method of resist bake for IC applications utilizes conduction on a vacuum hotplate, usually in line with the entire resist process. Convection baking is another option, performed in an

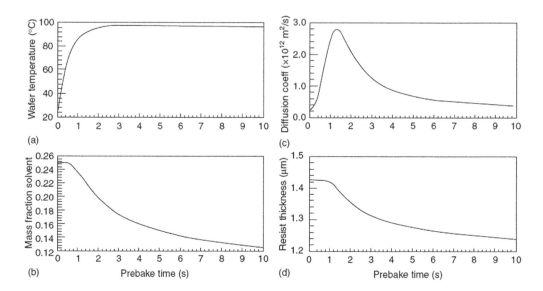

FIGURE 8.6 Mechanisms involved during the first 10 seconds of hotplate baking of a resist. (a) Rise in wafer temperature; (b) solvent loss; (c) diffusivity; and (d) typical thickness change.

oven and generally used in a batch mode. Loading effects, slow recovery, and poor air circulation uniformity present problems for convection baking with current resist process demands. Microwave and infrared (IR) baking methods have been utilized to a limited extent, but the practicality of these methods is limited because of substrate incompatibility and process control. Both temperature and uniformity control are important for baking processes, the latter becoming most critical as chemically amplified resist requirements are considered. To achieve uniformity, careful consideration of the vacuum plate baking mechanics is required, including temperature uniformity, airflow, and cycle time. To meet the high demands of wafer-to-plate contact across an entire wafer, strict requirements are also placed on contamination control and resist edge-bead removal.

Figure 8.6 shows the mechanisms involved during the first 10 seconds of baking on a vacuum hotplate. Figure 8.6a shows that the temperature of the wafer increases with a rapid rise during the first seconds of baking and levels off after 3 seconds. Figure 8.6b illustrates the solvent loss during this time, which continues to decrease well out to 10 seconds and beyond. Figure 8.6c shows the impact of baking time on diffusivity, which rises during the initial few seconds and then decreases. Figure 8.6d shows a typical change in resist thickness during baking, which decreases as the resist density increases.

8.6 PHOTORESIST EXPOSURE

8.6.1 RESIST EXPOSURE REQUIREMENTS

Exposure of a photoresist involves the absorption of radiation and subsequent photochemical change, generally resulting in a modification of dissolution properties. The absorption characteristics of a photoresist largely influence its resolution and process capabilities. Resists based only on exponential attenuation of radiation (i.e., with no mechanism for photobleaching or chemical amplification) can be limited by a maximum allowable contrast, sidewall angle, and ultimate resolution. This is because of the inherent absorption tradeoff required when imaging into a resist film. Both maximum transmission (to reach to the bottom of the resist) and maximum absorption (to achieve the highest sensitivity) are desired. There is therefore an optimum resist absorbance value for any resist thickness.

It a resist film has a thickness t, and dt is the thickness of the bottommost portion of the resist, the intensity transmitted through the film thickness to dt can be determined from Beer's law:

$$I = I_0 e^{-\varepsilon m t} \tag{8.4}$$

where

ε is the molar extinction coefficient of the sensitizer
m is the molar concentration

The energy density absorbed at the bottom of the resist is

$$E = I_0 e^{-\varepsilon m t} \frac{\left(1 - e^{-\varepsilon m \, dt}\right)}{dt} \tag{8.5}$$

Because dt is small, $e^{-\varepsilon m \, dt}$ can be approximated as $I - \varepsilon m \, dt$ and

$$E = I_0 \varepsilon m e^{-\varepsilon m t} \tag{8.6}$$

which is maximized when $\varepsilon m t = 1$. Converting to absorbance:

$$\text{Absorbance} = \log_{10} e^{\varepsilon m t} = 0.434 \tag{8.7}$$

This is the optimum absorbance for a resist film regardless of thickness. In other words, higher absorption is desired for thinner resists, and lower absorption is desired for thicker films. Resists for which chemical amplification or photobleaching is used introduce mechanisms that allow deviation from these constraints. The absorbance of a PAG for chemically amplified resist can be quite low because of the high quantum yield resulting from acatalytic reactions. Photobleaching resists (such as DNQ/novolac) exhibit dynamic absorption properties, which can allow increased transmission toward the base of a resist film. For these resists, other absorption considerations are required.

The dynamic absorption that exists for DNQ occurs as exposure leads to a more transparent ICA photoproduct. This bleaching phenomenon can be described in terms of Dill absorption parameters A, B, and C [33]. The A parameter describes the exposure-dependent absorption of the resist, the B parameter describes the exposure-independent absorption, and C describes the rate of absorption change or bleaching rate. For a DNQ/novolac resist, the C parameter is conveniently related directly to resist sensitivity, because photobleaching corresponds to the conversion of the PAC to the photoproduct (see Chapter 9 for a more detailed discussion of parameter characterization and modeling). The choice of a specific DNQ compound for mid-UV lithography needs to include evaluation of the unique A, B, and C parameters at the wavelength of exposure. It is generally desirable to have low exposure-independent absorption (B parameter) to achieve maximum exposure efficiency to the bottom of a resist layer. Several approaches exist to minimize residual B parameter absorption of DNQ/novolac resists [34]. These have included the use of highly transparent backbones for multifunctional PACs and binding of the sensitizer to the novolac polymer. A larger C parameter is also desirable for maximum sensitivity. Figure 8.7 shows absorbance spectra for two DNQ/novolac resists. Based only on the evaluation of the absorption properties of these two resists, it can be expected that the resist with the smaller B parameter and the larger A and C parameters may perform better (at a specific exposure wavelength) in terms of sensitivity, sidewall angle, and contrast.

Chemical amplification is another avenue that exists to improve the absorption characteristics of a resist. With quantum efficiencies several orders of magnitude higher than what can be achieved for direct photomodified resists, only a small amount of photon absorption is needed. This can be quantified in terms of A and B parameters, which may be on the order of 0.30 μm^{-1} compared with the 0.90 μm^{-1} levels encountered with i-line materials. The downside of such high transparency for

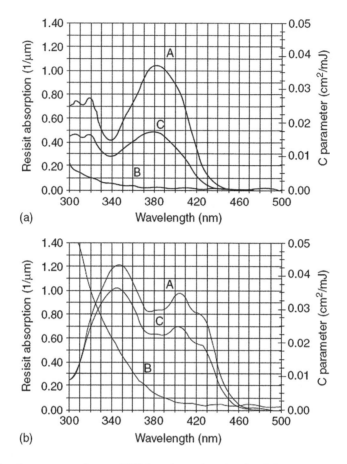

(a)

(b)

FIGURE 8.7 Absorbance curves for two DNQ/novolac resists showing Dill *A*, *B*, and *C* parameters. Resist (a) exhibits lower residual absorption (*B* parameter) at 365 nm, whereas resist (b) has higher bleachable absorbance and speed (*A* and *C* parameters) at 436 nm.

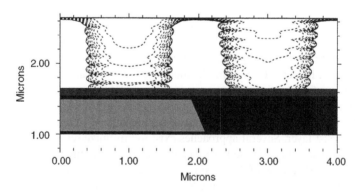

FIGURE 8.8 Resist standing waves resulting from coherent inference of incident and reflected radiation within a resist layer. Standing-wave phase and amplitude are dependent on the underlying substrate.

resist materials is the increased opportunity for substrate reflection to degrade performance. The effects can be manifested as a linewidth variation over reflective steps (nothing) and a sidewall standing wave. Figure 8.8 shows standing waves resulting from the coherent interference of incident and reflected radiation within a resist layer. Reduction of these standing waves is crucial to retain critical dimension (CD) control. This can be dealt with in either resist exposure or process stages and is ordinarily addressed in both. To reduce standing-wave effects during exposure, the reflected contribution to exposure must be controlled. This can be accomplished by incorporating a dye in the resist formulation. Dyes such as coumarin or curcumin compounds have been used as additives to DNQ/novolac resists and are very effective at reducing the reflected exposure contribution in a resist layer at g-line and i-line wavelengths. When a dye is added, the exposure-independent absorption (B parameter) increases. The result will be a decrease in reflection effects and standing waves and also a decrease in the amount of energy transferred toward the bottom of the resist. This will result in a decrease in sidewall angle, resist contrast, and sensitivity. Dyed resist for i-line use is therefore usually limited to highly reflective, noncritical layers.

Dyes that play a role in the chemistry of a resist system have also been used. By transferring the energy absorbed by a dye to the photosensitive component of a resist (for instance, through energy transfer or photoinduced electron transfer mechanisms), an active dye can allow reduction of reflection effects while maintaining sensitivity and resolution. This has been accomplished in chemically amplified resists [35]. The resultant increase in the resist B parameters in this case leads to an increase in photoinduced chemical activity, yielding minimal loss in resist performance. The loading of such photoactive dyes can be determined from the preceding analysis for static absorption resists. Resist absorbance on the order of 0.434 could be expected to be best suited for optimal throughput. An alternative to dye incorporation is the use of an antireflective layer on the resist. The requirements of current IC processes will soon demand that substrate reflectivity be reduced to levels below 15 for critical layers. The use of such ARCs has become a popular practice for i-line and DUV lithography and is discussed in more detail in Chapter 12.

8.6.2 Resist Exposure and Process Performance

The sensitivity (E_0) described earlier is a bulk resist characteristic that is useful for resist process comparisons or non-feature-specific process optimization. Resist E_0 swing curves allow, for instance, the determination of optimum coating thickness values. Monitoring of clearing dose requirements is also useful for characterization of exposure or development uniformity. Through the use of an exposure response curve, as shown in Figure 8.9, two additional bulk resist parameters can be obtained that allow further process characterization. Normalized thickness (normalized to

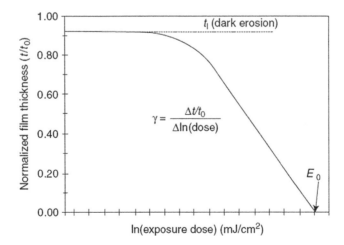

FIGURE 8.9 Normalized thickness as a function of $\log_e$ (exposure) for a positive resist of a given thickness, exposed and developed under specific process conditions. An area of linearity exists in the logarithmic relationship that can be characterized using a single contrast term (γ). Resist clearing dose (E_0) is the point on the exposure dose axis where normalized thickness becomes zero, and the normalized thickness loss (t_ℓ) is an indication of the amount of resist erosion that occurs in unexposed regions.

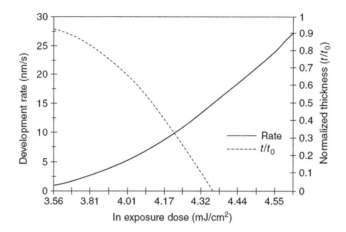

FIGURE 8.10 Resist dissolution rate (nm/s) plotted as a function of $\log_e$ (exposure dose) for a positive resist. Plotted also is an exposure response curve for this resist coated at 9000 Å and developed for 60 seconds.

initial predevelopment thickness) is plotted as a function of $\log_e$ exposure for a positive resist of a given thickness exposed and developed under process conditions. An area of linearity exists in the logarithmic relationship, which can be characterized using a single contrast term:

$$\gamma = \frac{\Delta t / t_0}{\Delta \ln(\text{dose})} \tag{8.8}$$

where
- t is thickness
- t_0 is the initial coating thickness

Resist sensitivity or clearing dose (E_0) is the point on the exposure dose axis at which the normalized thickness become zero. A third parameter that is useful for describing resist capability is a normalized thickness loss parameter (t_ℓ), which is an indication of the amount of resist erosion that occurs in unexposed regions (also called *dark erosion*). An exposure response curve can be generated uniquely for any resist/development process, can be obtained from a more general development process, or can be obtained from a more general development rate curve, described in greater detail in Section 8.8. Figure 8.10 shows a development rate curve, where resist dissolution rate is plotted as a function of $\log_e$ exposure dose for a positive resist. As exposure is increased, the dissolution rate increases. The linearity of this relationship can be realized on a log–log plot, but a lognormal plot is used here to demonstrate the relationship between a development rate and an exposure response curve. Plotted along with dissolution rate in Figure 8.10 is an exposure response curve for this resist coated at 9000 Å and developed for 60 seconds.

8.6.3 EXPOSURE AND PROCESS OPTIMIZATION

To determine the optimum exposure dose (or range of doses) for a resist process, bulk resist characterization needs to be augmented with feature-specific characterization. To accomplish this, process specifications must be defined uniquely for each feature type and size, mask level, substrate, resist, and process. This is often a difficult task, because definition on an optimum process requires operation at a near-optimum level. The task is therefore an iterative one that can often be made easier through the use of lithographic simulation tools.

Consider the focus-exposure matrix in Figure 8.11. Here, 500 nm dense lines are imaged into an i-line resist with a projection system at a partial coherence of 0.6. Resist CD is plotted against focal

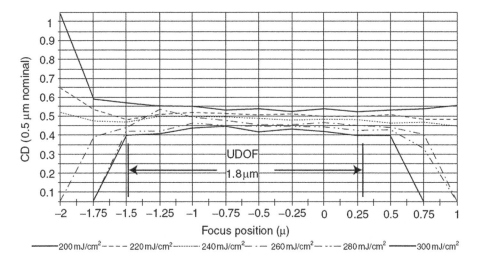

FIGURE 8.11 Focus-exposure matrix for 0.5 μm dense line features in a positive i-line resist over polysilicon using a numerical aperture of 0.45 and partial coherence of 0.60. Resulting usable depth of focus (UDOF) is 1.8 μm for a CD specification of ±1 and an exposure latitude requirement of 20%. The corresponding k_2 value for this single-layer resist process is near 0.5.

position for a series of exposure dose values. The nominal focus setting is labeled 0.0 and is typically at a point near 30% from the top surface, depending upon the resist refractive index and, to a lesser extent, upon NA and σ (for high NA values). At this position, there is an optimum exposure dose for printing a biased mask CD to its targeted size (see Chapter 2 for a description of mask biasing). As focal position is changed, the feature no longer prints at the target value; it is either too small or too large. The adverse influence of overexposure is also increased. This can be understood by considering an aerial image as it is varied through focus, as shown in Figure 8.12. At best focus, a relatively large range of exposure variation can be tolerated. As defocus increases, a small amount of overexposure can result in a large amount of energy in shadow areas. Underexposure can result in insufficient energy in clear areas. An exposure must be chosen, therefore, to allow the greatest range of usable focus that results in a CD that remains within the specified target range. More appropriately, a specification is placed on the required exposure latitude to account for within-field-to-field

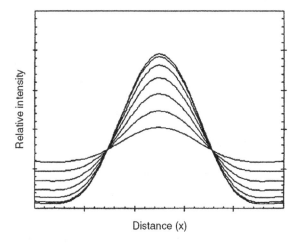

FIGURE 8.12 Through-focus aerial images for $\pm\lambda/NA^2$ of defocus. Images with no defocus can tolerate a large exposure variation. As defocus is increased, the choice of optimum exposure becomes more critical.

nonuniformities. Together with tolerance limits on CD, a resulting usable depth of focus (UDOF) can be determined. For instance, for a ±10% exposure latitude requirement and a ±10% specification on CD, the optimum exposure from Figure 8.11 is 240 mJ/cm^2, and the UDOF is near 1.8 μm. The resulting UDOF needs to be evaluated with respect to device topography. It should also be noted that any change in resist process will influence the focus. Any process enhancement methods, whether optical or chemical, have the potential to improve focal depth. These can include the use of reflection suppression (e.g., with an ARC), modified PEB, enhancements of contrast, and improvements in developer selectivity. For evaluation purposes, it is convenient to express a normalized UDOF through use of the Rayleigh DOF k_2 factor:

$$k_2 = \frac{(\text{UDOF})(\text{NA}^2)}{2\lambda} \tag{8.9}$$

The corresponding k_2 factor for the focus-exposure matrix of Figure 8.11 is near 0.5. Generally, single-layer resist processes can produce k_2 in the range of 0.4–0.6.

If a resulting k_2 does not correspond to UDOF values large enough for device topography, techniques of planarization may be required. Polymeric planarization is one alternative but is generally not considered robust enough to allow adequate process latitude for high-resolution IC applications. Chemical mechanical polishing (CMP) techniques, in which substrate surfaces are polished to reduce topography, are becoming commonplace in many IC process operations. From a lithographic standpoint, CMP can be considered as an operation that allows improvements in focal depth, CD tolerance, and exposure dose control.

A focus-exposure matrix can also be generated for resist sidewall angle, although this is a much more difficult task than for CD. Figure 8.13a shows a sidewall angle focus-exposure matrix, and Figure 8.13b shows the corresponding array or resist feature profiles. At an optimum exposure dose value, the base CD of the features remains tightly controlled. At negative values or focus (corresponding to the bottom of the resist), there is a widening at the base of features. At positive focus values (toward the lens), feature thinning occurs. To evaluate the impact of these changes through focus and exposure, subsequent etch process selectivity and isotropy need to be considered. If overexposure or underexposure is used to tune CD over topography (corresponding to a shift in focal position), it is likely that the resulting feature sidewall angle will not be adequate. A general requirement for sidewall angle specification is angles greater than 85°.

From these types of feature-specific process evaluation techniques, it can be understood why it is difficult to increase exposure throughput for a resist without adversely affecting overall performance. For features large enough that aerial image modulation is high (i.e., there is little energy in shadow regions and sufficient energy in clear areas), there will exist a good deal of exposure latitude and UDOF, so that some degree of exposure tuning is possible. As features approach $R = 0.5\lambda/\text{NA}$ in size, the situation becomes challenging for conventional binary masking and single-layer resists.

Linewidth linearity has already been discussed as it applies to optical imaging systems. Photoresist materials also behave with a nonlinear response to exposure, which can be seen in the exposure response and development rate curves in Figures 8.9 and 8.10. These nonlinearities can be used to take advantage of the lithographic process. Because the modulation need for small features is much lower than for large features, it would be expected that a resist that reacted linearly to exposure would do a poor job of faithfully reproducing mask features below λ/NA in size. By tailoring the exposure and dissolution response properties of a resist to operate in a specific desirable nonlinear fashion (nonlinearity itself does not imply added capability), the linearity of the entire lithographic process can be improved. Figure 8.14 shows a plot of resist CD versus mask CD for an i-line resist process. Only as features approach 0.45 μm in size does the linear relationship between the mask and the resist feature begin to break down. Features in this region require unique mask biasing to print to their targeted size. The situation becomes more complex as various feature types are considered, and resist linearity down to the minimum size of device geometry is desirable.

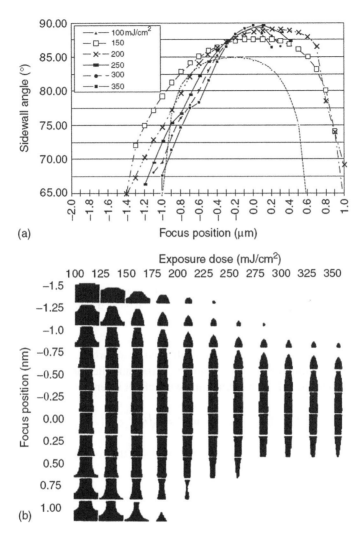

FIGURE 8.13 A sidewall angle focus-exposure matrix. UDOF can be determined from a sidewall angle specification and exposure latitude requirements. (b) Corresponding array of resist feature profiles.

8.7   POSTEXPOSURE BAKE

A PEB of a DNQ/novolac resist brings about chemical and physical actions similar to those for pre-bake. By subjecting resist films to a predevelopment bake step at a temperature higher than that used during prebake, some DNQ decomposition prior to exposure can be reduced. By baking exposed resists at temperatures on the order of 5–15 °C higher than prebake temperatures, solvent content can be reduced from 4–7% (prior to exposure) to 2–5%. Whereas prebake is generally performed to bring the resist into region II in Figure 8.5, the elevated temperatures used for PEB place the resist toward region III. The most beneficial consequence of a PEB step is, however, not an extended action of earlier bake steps but instead, a significant impact on standing-wave reduction via thermal flow [36]. During exposure over a reflective substrate, coherent interference produces a distribution of intensity within the resist film. The nodal spacing is a function of the resist refractive index (n_i) and wavelength:

$$\text{Distance between nodes} = \frac{\lambda}{2n_i} \tag{8.10}$$

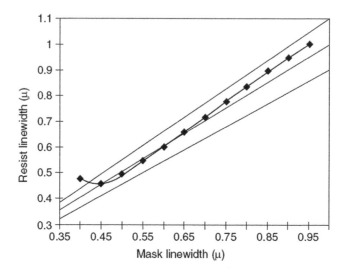

FIGURE 8.14   Plot of mask CD versus resist CD for an i-line resist process. As features approach 0.45 μm, the linear relationship between mask and resist features begins to break down. Features in this region require unique mask biasing to print to their targeted size.

The amplitude of the resist standing wave will be affected by absorbance of the resist (α, which may be dynamic if resist bleaching occurs), resist thickness (t), and reflectance at the resist–air (R_1) and resist–substrate (R_2) interfaces [37], resulting in a wing effect where

$$\text{Standing wave "swing"} = 4\sqrt{R_1 R_2}\exp\left(-\alpha t\right) \qquad (8.11)$$

Interface reflectance values (R_1 and R_2) are determined from the complex refractive index values of the media involved:

$$\text{Reflectance} = \left(\frac{n_a^* - n_b^*}{n_a^* + n_b^*}\right)^2 \qquad (8.12)$$

In addition to resist and substrate contributions to interference effects, exposure source characteristics need to be considered. The coherence length (l_c) for a radiation source is determined on the basis of the source wavelength (λ) and bandwidth ($\Delta\lambda$) as

$$l_c = \frac{\lambda^2}{\Delta\lambda} \qquad (8.13)$$

As the resist thickness approaches the source coherent length, interference effects (standing waves) become negligible. A typical coherence length for i-line stepper tools may be on the order of 10 μm. For a 248 nm lamp-based step-and-scan system, l_c may be on the same order, but with KrF and ArF excimer laser lithography, a 10^3–10^4 factor increase or more can exist. Although it is not expected that a resist would be coated to thicknesses approaching 10 μm or more, the analysis offers insight into the increasingly critical concern with standing-wave control. Considered together with the higher transparency (low α) of a chemically amplified resist, it becomes obvious that resist imaging over reflective substrates using an excimer laser source would be difficult without employing means of control or compensation.

To demonstrate the impact that a PEB step can have on resist standing waves, consider Figure 8.15. Unexposed regions of the photoresist contain a photoactive compound, exposed regions contain photoproduct, and the boundary between them is determined by the constructive and destructive

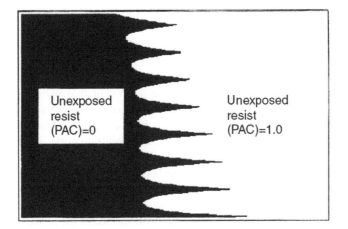

FIGURE 8.15 Distribution of photoactive compound (PAC) as a result of standing-wave resist exposure. Unexposed regions of the photoresist contain PAC; exposed regions contain photoproduct. Because the resist temperature is near or above its T_g, the PAC can effectively diffuse through the polymer matrix to reduce the effective standing wave in the developed resist sidewall.

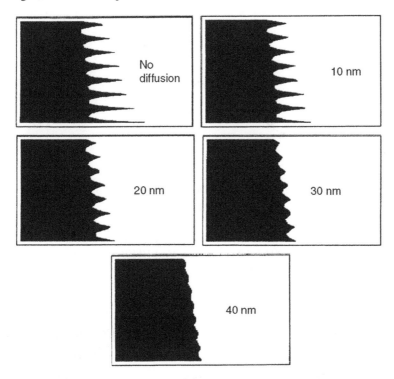

FIGURE 8.16 Simulated standing-wave pattern in developed resist as a function of increasing PAC diffusion length.

interference nodes of the exposure standing wave. If the resist temperature is near or above its T_g, the PCA can effectively diffuse through the polymer matrix and produce an averaging effect across the exposed/unexposed boundary. This PAC diffusion is a function of the resist T_g, the PAC size and functionality, bake time and temperature, remaining solvent concentration, resist free volume, and (for DNQ/novolac, for instance) any binding between the resin. Figure 8.16 shows the resist standing-wave pattern as a function of the PAC diffusion length.

For chemically amplified resists based on thermally induced acid reactions, PEB is used to accomplish critical chemical stage and must be considered nearly as critical as exposure in terms of control, uniformity, and latitude requirements. The concept of a bake "dose" control is a convenient comparative metric to exposure dose control for acid-catalyzed deprotection (positive resists) or crosslinking (negative resists). An Arrhenius relationship exists between PEB and exposure, such as that shown in Figure 8.17 [38]. Here, log (exposure dose) is plotted against 1/PEB, from which an effective E_a can be determined. Whereas exposure uniformity on the order of ~1% is a current requirement for resist exposure tools, it can be expected that PEB control should be at least as critical. The rate of these reactions is a function of the acid concentration, the activation energy of the protected polymer, the diffusivity of the acid, and the PEB conditions.

Acid diffusion in PHS chemically amplified resists is generally on the order of 50 to a few hundred angstroms, limited by resolution requirements. If neutralization of acid occurs, the exposed resist dissolution rate decreases. If environmental contamination occurs after exposure and prior to PEB, a thin, less soluble inhibition layer forms on the top surface of the resist, which results in the formation of a characteristic T-top upon development, as shown in Figure 8.18. Any delay between

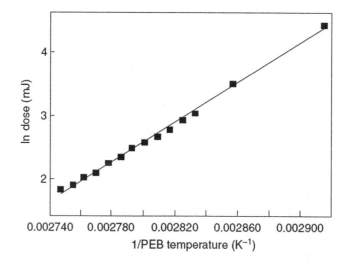

FIGURE 8.17 Arrhenius plot or resist sensitivity versus PEB temperature for a DUV CAR resist. Dose required to print 0.5 μm lines with a 90 s bake is plotted against 1/PEB temperature. An effective activation energy can be calculated from such a plot (130 kJ/mol in this case). (From Sturdevant, J., Holmes, S., and Rabidoux, P, *Proc. SPIE,* 1672, 114, 1992.)

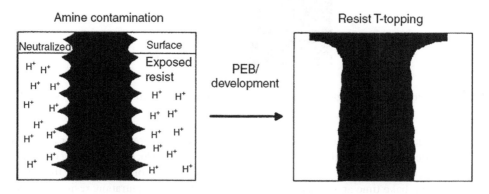

FIGURE 8.18 Environmental amine contamination of a chemically amplified (PHS) resist resulting in "T-top" formation after development.

exposure and PEB (known as *postexposure delay* [PED]) allows this phenomenon to occur. Early material PED time requirements were on the order of a few minutes. Current generation resists allow delays of up to several hours. It is important to realize, however, that acid neutralization can begin immediately upon exposure, and performance must be evaluated for the specific resist process conditions and process tolerances. Low-E_a resists do not require elevated PEB temperatures for reaction, which can minimize the control requirements for this process step. These resists control tolerate high baking temperatures, which may introduce additional stability problems. Choice of the PAG is less critical, allowing choices with smaller cross sections than those needed for high–activation energy materials. Diffusion properties are also based on size and should be evaluated uniquely for resists of either category.

When determining PEB requirements for deprotection, postexposure solvent content and polymer density also need to be taken into consideration. Temperature control on the order of 0.1 °C may be required, with uniformity needs on the same order over wafer diameters 300 mm and larger. Material improvements may relax this requirements somewhat, but control to a few tenths of a degree can be expected as an upper limit.

An additional improvement brought about through the use of a PEB step is a potential increase in resist development rate properties, specifically the dissolution rate log slope (DLS), as discussed later in the development section. This occurs as a result of the decrease in resist dissolution rate, especially at the resist surface, which is more pronounced in some resists than in other [39]. By reducing the surface dissolution rate, unexposed film erosion can be reduced, resulting in an increase in sidewall slope and an improvement in process latitude. The surface modification is a result of solvent loss and a possible surface "skin" effect, enhanced by resist modification at elevated temperatures.

For optimal process performance, PEB processes for chemically amplified PHOST resists cannot be considered separately from soft bake steps. There are tradeoffs between high and low levels of each. During soft bake of a chemically amplified PHOST-based resist, standing waves can be reduced through the use of lower temperatures, which allow an increase in acid diffusion across the exposed/unexposed resist boundary. To reduce T-top formation, a high PEB is also desirable. Through the use of high PEB temperatures, deprotection can occur even with some degree of reduced acid concentration at the resist top surface. This low–soft bake, high-PEB combination can result in additional undesirable phenomena. When the resist film is allowed to retain a relatively low T_g, pattern deformation can result from rapid PHOST deprotection and subsequent gas evolution at high PEB temperatures. A stepwise PEB has been suggested as a possible solution [40]. An initial low-temperature stage removes protective groups from the bulk of the resist with minimum deformation. A high-temperature stage follows to enhance top-surface deprotection and reduce T-top formation.

As discussed earlier, intrinsic or added free volume will affect the glass transition and dissolution properties and deprotection mechanisms of phenolic-based photoresists. The CAR deprotection reaction itself has also been shown to contribute additional free volume [19]. Exposure of PEB-induced variation in resist density can affect dissolution uniformity and CD control. This added influence of exposure and PEB can lead to complex relationships and increased control requirements. The deprotection and densification mechanism can best be separated in resist systems with low activation energies.

8.8 RESIST DEVELOPMENT

Resist systems based on solvent development, such as crosslinking bis(arylazide)-*cis*-polyisoprene negative resists, require some degree of swelling to allow the removal of soluble polymer chains or fragments [41]. To keep resist pattern deformation to a minimum, a series of solvents is generally needed for development, with careful consideration of kinetic and thermodynamic properties [42]. The aqueous base development of resists based on novolac, PHOST, or other acrylic resins involves similar dissolution stages but does not require such adverse swelling for dissolution [43–45].

8.8.1 Dissolution Kinetics of Phenolic Resin Resists

In novolac resins, a narrow penetration zone is formed as water and hydroxyl groups are incorporated in the novolac matrix. This zone is rate limited and does not encompass the entire resist layer, allowing dissolution of resist with minimal swelling. Following the formation of this intermediate layer, phenol is deprotonated, and the resulting phenolic ion reacts with water. The negative charge of the phenolate ion is balanced by the developer cations. Upon sufficient conversion of phenol groups, the polymer is soluble in aqueous alkaline developer. A three-zone model has also been suggested, as shown in Figure 8.19 [46]. These zones exist during novolac development:

1. A gel layer containing water, tetramethylammonium hydroxide (TMAH) (base), and partially ionized novolac. The thickness of this layer depends on agitation, novolac microstructure, and the developer cation.
2. A penetration zone with a low degree of ionization. The thickness of this zone also depends on the novolac structure as well as the developer cation size and hydrophobicity.
3. Unreacted bulk novolac resist.

The resulting dissolution rate of a resist material is determined by the formation of the gel and penetration layers and the subsequent dissolution of the gel layer into the developer.

In DNQ/novolac resists, exposure of the PAC leads to photoproduction of ICA, which acts as a dissolution accelerator. First-order kinetic models for development are based on this mechanism. DNQ in sufficiently high concentrations (5–25 wt%) also acts as a dissolution inhibitor for the phenolic novolac resin, decreasing its hydrophilic nature. This inhibition effect can also be accomplished with other compatible compounds and is not limited to DNQ. It might be reasoned that a similar mechanism would exist for other resins, but this is not the case. Although the dissolution of novolac can be decreased when combined with DNQ by as much as two orders of magnitude, the high dissolution rate of PHOST is not significantly reduced. This can be explained by considering the hydrophilic nature of both materials. Novolac is, for the most part, a hydrophobic resin with hydrophilic sites created when developer reacts with hydroxyl groups. These phenolate ion positions allow a diffusion path for development. The sites of hydroxyl groups are only potential hydrophilic sites that can be tied up or isolated by the polymer's large aromatic rings. PHOST is

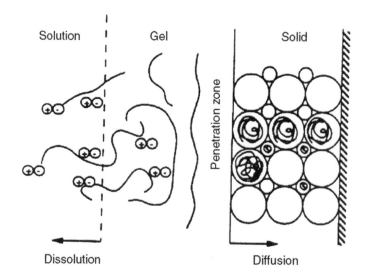

FIGURE 8.19 Diagram of a three-zone novolac dissolution model with a gel layer containing water, base, and hydrated, partially ionized novolac chains; a penetration zone.

a vinyl polymer with no aromatic rings in the backbone. Hydroxyl groups exist in a "corkscrew" configuration along the polymer chain, allowing very effective diffusion paths and extremely high dissolution rates.

The dissolution kinetics of chemically amplified resists based on PHOST are also quite different from those for DNQ/novolac. Whereas the dissolution of novolac resins is determined by two competitive reactions (inhibition by the DNQ and acceleration resulting from the developer-induced deprotonation of the novolac), there is one primary rate-determining stage for two-component PHOST chemically amplified resists. A similar developer-induced deprotonation of phenolic hydroxyl groups has been shown to dominate for negative resists, whereas developer penetration into hydrophobic t-BOC-protected PHS is rate determining for positive resists [47].

8.8.2 DEVELOPMENT AND DISSOLUTION RATE CHARACTERIZATION

The development of exposed photoresist is based on image-wise discrimination. To understand and optimize photoresist development, it is necessary to characterize exposure-dependent dissolution properties. As discussed earlier, the extraction of development rate and exposure relationships allows tremendous insight into the imaging capabilities of a resist process. Shown in Figure 8.20 is a family of normalized thickness versus development time curves for a resist material exposed at various dose levels. Such curves can be obtained using laser interferometry techniques and separate substrate exposure [48, 49] or a single substance and multiple exposures [50]. A development rate versus exposure curve such as that in Figure 8.21 can be produced from a family of these dissolution curves. To characterize fully the dissolution properties of a resist throughout its entire thickness, a rate curve may not suffice, because development rate is a function of resist thickness. Surface inhibition reduces the development rate significantly at the top surface of the resist. Figure 8.22 shows a development rate versus resist thickness curve demonstrating that development rate increases from the top surface of the resist toward the bottom.

When development rate and exposure are plotted in a log–log fashion, as shown in Figure 8.23, a linear region exists that can be described in terms of a DLS:

$$DLS = \frac{\partial \ln\left(\text{dev.rate}\right)}{\partial \ln\left(\text{dose}\right)} \tag{8.14}$$

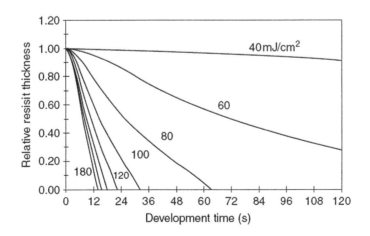

FIGURE 8.20 Normalized film thickness versus development time curves for resist exposed at 40 to 180 mJ/cm^2. Selection of single development time values can lead to development rate versus relative exposure curves.

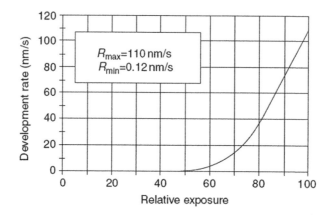

FIGURE 8.21 Development rate versus relative exposure curve from Figure 8.20.

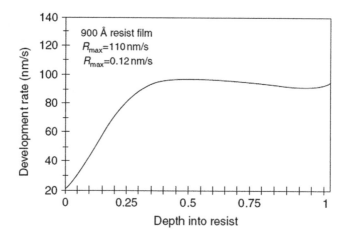

FIGURE 8.22 Development rate versus resist thickness curves, showing distribution of dissolution properties through a resist film.

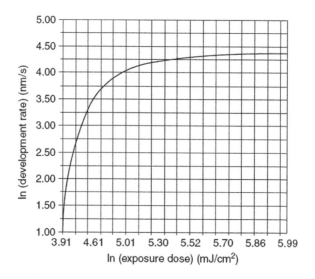

FIGURE 8.23 Log development rate versus log exposure dose. Dissolution rate performance can be evaluated in terms of the slope of the linear region of the curve and the R_{max}/R_{min} ratio.

Dissolution rate contrast (DRC) can be expressed as

$$\text{DRC} = \frac{R_{\max}}{R_{\min}} \tag{8.15}$$

where $R_{\max}$ and $R_{\min}$ are the maximum and the minimum development rate, respectively. Together, DLS and dissolution rate contrast are effective measures of the dissolution properties of a resist material. High levels of both metrics are desirable. Control of the resist and development properties can be used to some extent to influence performance. A moderate contrast value near 10,000 is generally sufficient. As development normality is changed, $R_{\max}$ and $R_{\min}$ are both affected (as long as base concentration does not fall below a critical level, pH ~ 12.5), but their ratio and the shape of the curve remain fairly constant [51]. The $R_{\max}$ is primarily a function of the polymer itself. Major resist, sensitizer, and solvent factors that influence development rate include

- Polymeric structure
- PAC/PAG/inhibitor structure
- PAC/PAG/inhibitor concentrations (DNQ/loading >20%)
- Protection ratio (PHS protection ration > 25%)
- Solvent concentration
- Polymeric molecular weight (typically greater than ~1×10^4 g/mol)
- Polydispersity (typically less than 2)
- Developer composition (metal ion vs. TMAH)
- Developer cation size and concentration
- Resist surfactant
- Resist dissolution inhibition/acceleration state
- Developer surfactant

Hydroxyl group positions on PHOST polymer also affect the dissolution rate, a function of hydrogen bonding and steric hindrance with the polymer backbone [52]. Copolymers of 2- and 4-hydroxystyrene have been shown to allow control of PHOST dissolution properties.

8.8.3 Developer Composition

Phenolic photoresists can be developed using buffered alkaline solutions such as sodium metal silicates. These metal silicate developers yield a maximum development rate and low dark erosion compared with most other developer choices. Possible metal ion contamination of devices has nearly eliminated the use of this class developer in favor of metal-ion-free TMAH. Furthermore, the standardization of TMAH developer formulations is becoming widespread, because having no requirements for coexistence of several developer types is more economical and allows better quality control. The larger cross section of TMH compared with NaOH leads to lower development rates, which may result in lower working sensitivities. TMAH developer concentrations in the range of 0.2–0.3 N allow sufficient sensitivity with high contrast and minimum erosion. A 0.26 N solution has become a standard for most photoresist processing.

In addition to TMAH, surfactants are added to a developer to reduce development time and scumming. Surfactants are used as additives to photoresists, chemical etchants, and developers to improve surface activity or wetting. TMAH developers often employ surfactants at the ppm level to reduce surface tension. Surfactants are especially useful for improving the dissolution of small–surface area features such as contacts or small space features. By increasing the effective wetting of the developer, scumming problems can be minimized, and overexposure- or overdevelopment-induced process biases can be reduced. Also, through the use of developer surfactants, initial development inhibition at the resist surface can be reduced. Additives used for resist developers are mainly of the nonionic variety with hydrophilic and hydrophobic structural characteristics, such as ethylene oxide/propylene oxide block polymer segments with molecular weights near 1000. Concentrations may be up to 800 ppm or 0.05 wt.%. Surface tension decreases as the concentration of surfactants increases until a critical micelle concentration (CMC) is reached. This CMC level represents an equilibrium state at which aggregation begins. Very small changes in the structure of surfactant molecules have been shown to result in large changes in resist development performance. The number and location of hydroxyl groups in the molecular structure of surfactant determine the activity of surface dissolution during development [53]. The behavior of developer surfactants also depends on resist material properties. Resists that have a larger degree of surface hydrophobicity benefit most from surfactant additives and have been shown to be less dependent on surfactant type. The benefits gained for a particular resist therefore depend on unique resist and surfactant properties and interactions. Figure 8.24 shows how surface agents acting at a resist/developer interface enhance the surface activity by decreasing surface free energy [54].

In addition to the chemical composition of the developer, the concentrations of hydroxyl ions, developer cations, and anions have been shown to strongly influence the dissolution rate of pure novolac films [55]. As the cation concentration is increased, a linear increase in dissolution rate occurs at a constant pH, apparently independently of anion concentration [55]. The structure of the developer cation and anion will determine the maximum allowable concentration before a decrease in dissolution rate begins. Along with the dependence of dissolution on the size of the developer cation, its hydrophilicity and the T_g of the partially ionized novolac also influence the development rate [46].

Development temperature is important and requires tight control. This is true not only for the bulk development reservoir and plumbing but also for the development bowl and development atmosphere. The rate of dissolution for TMAH developers follows an Arrhenius relationship:

$$k = A_0 e^{-E_a / RT} \tag{8.16}$$

with an apparent negative activation energy (E_a) [56]. The reason for this is a highly exothermic sequence of deprotonation steps. A decrease in development temperature results in increased

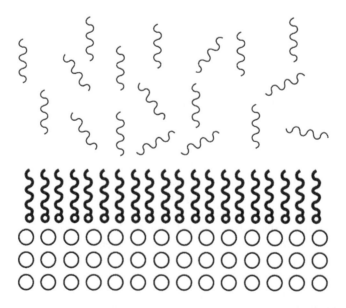

FIGURE 8.24 Orientation of surface molecules (center) at a hydrophobic (top)/hydrophilic (bottom) interface. The surfactant molecules are oriented so as to lower the interfacial free energy. (From Florus, G., and Loftus, J., *Proc. SPIE,* 1672, 328, 1992.)

activity, which may seem counterintuitive. This is not the case for development with NaOH or developers with similar chemistry, because no exothermic reactions exist.

8.8.4 DEVELOPMENT METHODS

Resist dissolution properties are highly dependent on the development method employed. Development methods commonly used for conventionally coated wafer substrates include static immersion, continuous spray with slow rotation (~500 rpm), and stationary or slow rotation puddle development.

Spray development involves one or more spray nozzles to dispense developer toward the wafer substrate. Processes using ultrasonic atomization of developer allow relatively low-velocity dispersion and minimal adiabatic cooling effects during the dispensing. For conventional spray development, the resist dissolution rate has been shown to be relatively insensitive to spray nozzle pressure but linearly dependent on spin speed [57]. With ultrasonic spray development, dissolution variation and resist erosion can result from poor control of the nozzle spray pattern. Dissolution rate and uniformity are also dependent on the uniformity of the spray pattern directed toward the wafer. To minimize nonuniformities, wafer chuck rotation must be carefully controlled at fairly low speeds (100–500 rpm) to ensure adequate developer replenishment. For these reasons, along with the excessive developer volume dispensed during processing, puddle processes are now most common for high-volume production.

Puddle development techniques can improve the tool-to-tool matching of development processes. In addition, the control of puddle development processes over time is significantly better. For optimal performance, flow is kept low to reduce variations in development rate at the edge of the wafer. The dispensed volume or developer should also be kept low to minimize chemical costs and backside wetting but not so low as to cause localized nonuniformities or scumming. The dispensed volume should be carefully controlled to ensure full and uniform coverage of the entire resist surface. During a puddle development process, slow rotation or single rotation may be used to disperse trapped air and provide agitation. If a surfactant-free developer is used, an increase in dispensing pressure may be required to enhance resist wetting.

There are several tradeoffs to consider when making a choice between spray, puddle, and immersion methods. Although immersion development can lead to minimum erosion by physical removal of a resist, it is not well suited to the inline processing that now dominates in production. During puddle development, chemical activity decreases with time as the small developer volume becomes exhausted, a situation that is not encountered during immersion processing. There is a more subtle difference between these two techniques as the depletion of developer additives is considered. At ppm levels, surfactants can be exhausted during static immersion development, significantly changing the chemical nature of the developer with time. Because puddle development introduces fresh chemistry for every wafer, consistent wafer-to-wafer development can be ensured. A multiple-puddle method of development [58] is now commonplace in wafer processing and usually involves two development stages. The process consists of dispensing developer onto a wafer and allowing it to remain for a short time (10–20 seconds). It is then spun off, and a second puddle is formed and allowed to complete the process (20–30 seconds). Although such multiple-puddle processes serve to replenish developer to the resist, the effect that makes this technique most attractive is an inhibition that occurs between development cycles. The dissolution of unexposed or partially exposed resist will be preferentially slowed at the location of the resist front after the end of the first development step. This phenomenon can be enhanced if rinse and spin-dry steps are added between development cycles [59]. The inhibition may be a result of base-induced oxidation or azo coupling of the novolac. It has also been suggested that a surface modification occurs with interrupted development, and an increase in surface energy can enhance resist dissolution discrimination [60]. Multiple–development step processing has been also reported, and variations on these techniques are now widespread [61]. Intermediate development baking processes have also been investigated [62]. Figure 8.25

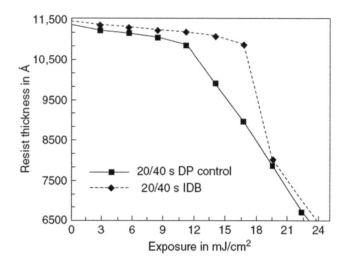

FIGURE 8.25   Resist film thickness versus exposure dose for interrupted double-puddle (DP) development and for 50 °C warm-water intermediate development bake (IDB) showing improvement in resist contrast with IDB. (From Damarakone, N., Jaenen, P., Van den Hove, L., and Hurditch, R., *J. Proc. SPIE,* 1262, 219, 1990.)

shows resist film thickness plotted as a function of exposure dose for double-puddle and warm-air intermediate development baking processes. Resist contrast is effectively increased with the use of the warm-air bake.

It is important to rinse remaining chemicals from the wafer surface after development, because inadequate removal can result in high defect levels. Both a topside rinse and a backside rinse are needed to remove developer and contaminants that could be transferred to subsequent process steps.

8.8.5   Development Rate Issues with Chemically Amplified Resists

Although similarities exist between the dissolution properties of DNQ/novolac and DUV chemically amplified resists, comparison of these materials demonstrates salient differences in their performance and capabilities. Development rate curves are the tool best suited for this type of comparison. Shown in Figure 8.26 are development rate curves for an i-line DNQ/novolac resist, a negative DUV chemically amplified resist, and a positive DUV chemically amplified resist [47]. The negative DUV resist is a three-component acid-hardening resist based on a phenolic resin (PHOST), a melamine crosslinker, and a photoacid generator [63]. The positive DUV resist is a t-BOC-protected PHOST and a photoacid generator [64]. The rate curve for the i-line resist (Figure 8.26a) shows a nonlinear dissolution rate that can be divided into three regions. The ability of this resist to achieve high contrast (γ) is evident from the steep DLS. A large DRC also exists, leading to high sensitivity and low erosion properties. Figure 8.27a shows an Arrhenius plot for this resist, where distinct regions also exist: (I) a high-dose, low-temperature region where E_a is small and positive, (II) an intermediate region where E_a is negative and decreasing, and (III) a low-dose, high-temperature region where E_a is positive and comparatively large. These results can be compared with the development rate curves for the negative DUV resist (Figures 8.26b and 8.27b). In this case, a moderately high DLS exists, and when compared with the i-line resist, it would be expected to coincide with worse lithographic performance. The Arrhenius plot shows a constant E_a and suggests that only one reaction mechanism governs development rate. Figures 8.26c and 8.27c show plots for the positive DUV resist. The DLS value is much greater than that for either the i-line or the negative resist. The Arrhenius plot also suggests a single dissolution mechanism, but a saturation of E_a appears to exist at higher doses. At these high doses, deprotonation of hydroxyl groups at the PHS becomes rate determining, as in the mechanism for the negative resist. Figure 8.28 shows the proposed dissolution model for the

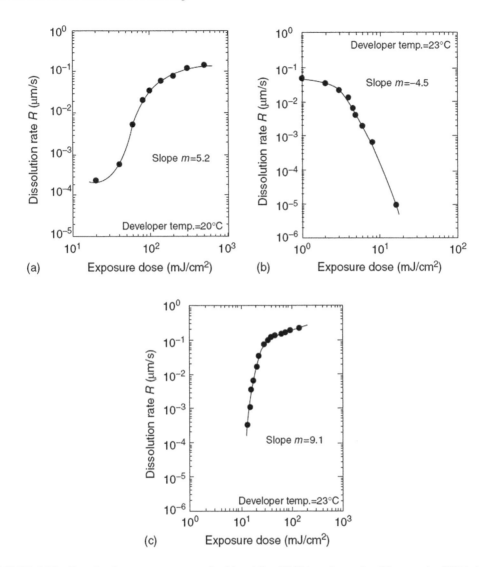

FIGURE 8.26 Log development rate curves for (a) an i-line DNQ/novolac resist; (b) a negative DUV chemically amplified resist; and (c) a positive DUV chemically amplified resist. Slope is the development rate log slope (DSL). (From Itani, T., Itoh, K., and Kasama, K., *Proc. SPIE,* 1925, 388, 1993.)

positive DUV chemically amplified resist. Figure 8.29a and b show development rate curves for a three-component positive DUV resist that includes an additional dissolution inhibitor and various t-BOC protection levels and molecular weights [65]. Shown in Figure 8.30 are plots of dissolution rate and log slope as a function of inhibitor concentration for a three-component PHOST chemically amplified resist [66]. Increasing both the protection ratio and the inhibitor concentration results in higher DRC and DLS values. Strong surface inhibition effects can limit the practical levels. The optimum protection ratio from Figure 8.30 is on the order of 30–35% with an inhibitor concentration of 3% PAG. Structure modification, molecular weight polydispersity, and polymer end groups can allow further improvements.

Preferential dissolution conditions for positive PHOST chemically amplified resists can be summarized. A high R_{max}/R_{min} contrast is desirable, but not so high as to result in severe resolution of sidewall standing waves (some low-contrast behavior is desired to reduce sensitivity to small oscillations in intensity at threshold levels). Because R_{max} is primarily a function of the polymeric resin,

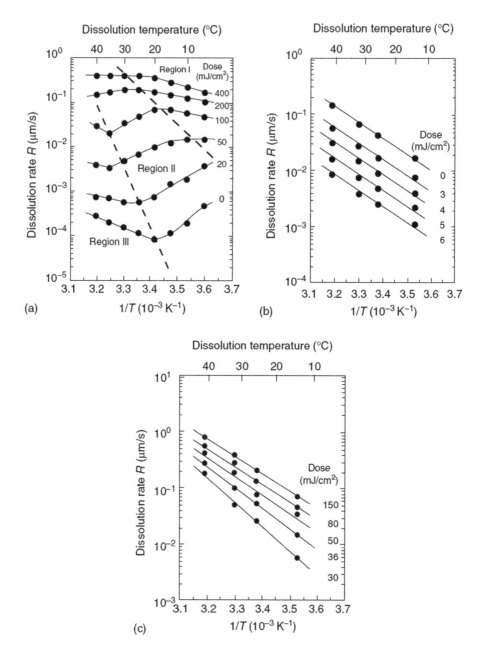

FIGURE 8.27 Arrhenius plots for the three resists in Figure 8.26: (a) an Arrhenius plot for the i-line resist where three distinct regions exist: (I) a high-dose/low-temperature region where E_a is small and positive, (II) an intermediate region where E_a is negative and decreasing. and (III) a low-dose/high-temperature region where E_a is positive and comparatively large; (b) a constant E_a for the negative CAR suggests that only one reaction mechanism governs development rate; and (c) suggests also a single dissolution mechanism for the positive CAR, but a saturation of E_a appears to exist at higher doses, where location of hydroxyl groups at the PHOST becomes rate determining, similar to the mechanism for the negative resist. (From Itani, T., Itoh, K., and Kasama, K., *Proc. SPIE,* 1925, 388, 1993.)

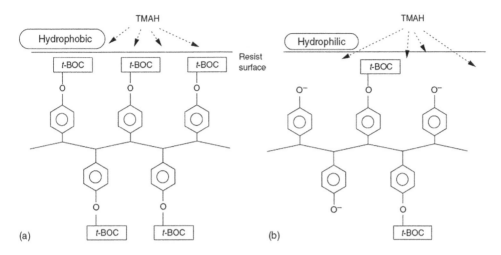

FIGURE 8.28 A proposed dissolution model for positive DUV chemically amplified resist derived from Arrhenius plot analysis. Two situations are considered: (a) before deblocking, when TMAH penetration is prevented by strong hydrophobicity, and (b) after deblocking. (From Itani, T., Itoh, K., and Kasama, K., *Proc. SPIE*, 1925, 388, 1993.)

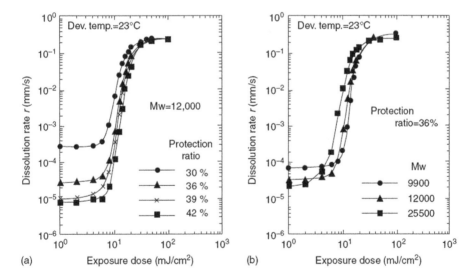

FIGURE 8.29 Development rate curves for a three-component positive DUV resist, which includes an additional dissolution inhibitor and various *t*-BOC protection levels and molecular weights. (From Itani, T., Iwasaki, H., Fujimoto, M., and Kasama, K., *Proc. SPIE*, 2195, 126, 1996.)

maximizing the R_{max}/R_{min} dissolution contrast ratio generally involves reduction of unexposed resist erosion. As shown earlier, a ratio above 10,000 is desirable, but probably no greater than 50,000 should be expected. A large DLS is more important and can be controlled to some extent by the resin molecular weight.

8.9 E_0 AND CD SWING CURVE

Coherent interference resulting from substrate reflectivity will result in a swing of intensity distributed within a resist film. This will correlate with variations in clearing dose (E_0) and CD throughout

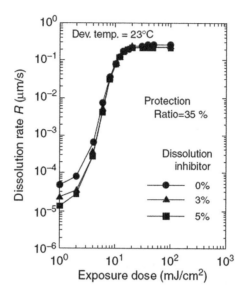

FIGURE 8.30 Plots of dissolution rate and log-slope plotted as a function of protection ratio for a three-component PHOST chemical resist. (From Itani, T., Iwasaki, H., Yoshin, H., Fujimoto, M., and Kasama, K., *Proc. SPIE,* 2438, 91, 1995.)

a resist layer. The swing in E_0 is dependent only on the coherent interference of radiation within the film and can be predicted on the basis of knowledge of the coupling efficiency between the resist and the exposure variation. The only requirement for predicting the E_0 swing is that the distribution and conversion efficiency of the sensitizer reaction be known, making it independent of resist dissolution characteristics. In other words, only the photospeed of the resist influences the E_0 swing. Figure 8.31 shows the relationship between the E_0 swing ratio and exposure clearing dose (E_0) for several resist systems [67]. The swing ratio has been calculated from Equation 8.11. Because higher

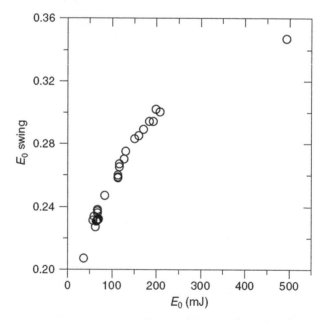

FIGURE 8.31 Plot of normalized E_0 vertical swing versus E_0 for several hypothetical resist systems. (From Hansen, S.G., Hurdich, R.J., and Brzozowy, D.J., *Proc. SPIE,* 1925, 626, 1993.)

bleaching efficiency leads to higher resist sensitivities, the relationship shown in Figure 8.31 makes sense from the standpoint of resist absorbance. Increased nonbleachable absorbance will lead to lower swing ratios, as can be demonstrated by adding a dye to a resist. Figure 8.32 is a plot of E_0 swing versus exposure-independent resist absorbance (B parameter). As resist absorbance increases, swing decreases.

The independence of E_0 swing on resist dissolution, along with the ease with which E_0 values can be measured, makes it an effective method for resist coat characterization, as seen in Section 8.4. The swing in CD is not as straightforward, because it is influenced by the resist process and dissolution properties. Resist process exposure latitude (or the amount of the allowable over- or underexposure) can have a significant influence on CD swing and can be predicted from the relationship

$$\text{CD swing} = \frac{E_0}{E_L^x} \qquad (8.17)$$

where
E_L is exposure latitude
x is an empirically determined exponent, on the order of 2.5 for a reasonably fast photoresist

Secondary swing effects are also present in a resist process. These include swings in resist contrast (γ) and surface contour effects. These secondary effects are a result of the spatial position of a standing-wave node at the top surface of the resist. If there is a destructive interference node near the top surface of the resist, a surface induction effect will result, leading to an increase in γ. Figure 8.33 shows how this can be manifested. At resist thicknesses slightly greater than those corresponding to an optimum maximum exposure (1.18 μm), PAC concentration is increased at the top surface of the resist, resulting in T-top formation. This results in a significant loss of linearity compared with the optimized condition. At thicknesses slightly lower than 1.18 μm, there is a decrease in PAC concentration, resulting in top rounding. These surface effects occur at regular swing periods. The extent to which these secondary surface effects influence a resist process depends on how

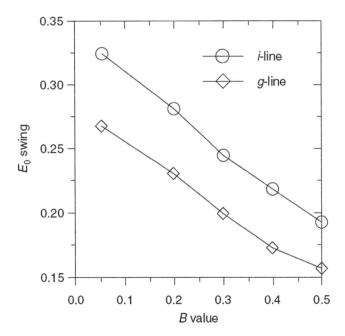

FIGURE 8.32 Effect of residual absorption (B parameter) on E_0 swing at constant sensitivity for g-line and i-line exposure. (From Hansen, S.G., Hurdich, R.J., and Brzozowy, D.J., *Proc. SPIE*, 1925, 626, 1993.)

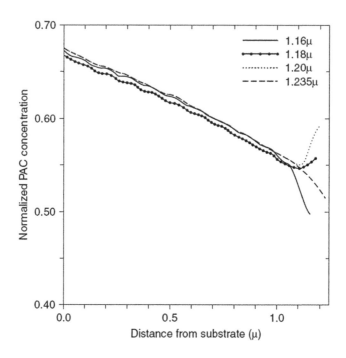

FIGURE 8.33 Calculated PAC distribution (at the mask edge) following PEB for four film thicknesses. (From Hansen, S.G., Hurdich, R.J., and Brzozowy, D.J., *Proc. SPIE,* 1925, 626, 1993.)

exposure requirements are affected. Surface effects influence clearing dose (E_0) more strongly than they influence dose to size ($E_{1:1}$). The exposure margin (*EM*), which is a ratio of these doses,

$$EM = \frac{E_{1:1}}{E_0} \tag{8.18}$$

is a measure of the overexposure requirements for a resist and is a good indicator of mask linearity [71]. Surface-induced swing in E_0 leads to a corresponding swing in EM. Resists that demonstrate large surface induction effects during development are most susceptible to these secondary swing effects.

8.10   POSTDEVELOPMENT BAKING AND RESIST TREATMENT

Postdevelopment baking is often utilized to remove remaining casting solvent, developer, and water within the resist. Baking above the resist's T_g also improves the adhesion of the resist to the substrate. Because photosensitivity is no longer required, the baking temperature can be elevated toward the solvent boiling point (T_b), effectively eliminating the solvent from the resist and allowing maximum densification. For a DNQ/novolac resist, any remaining DNQ can lead to problems in subsequent process steps. If the still-sensitized resist is subjected to high-energy exposure (such as ion implantation), rapid release of nitrogen can result from radiolysis of DNQ. In densified resist films, the nitrogen cannot easily diffuse outward and may result in localized explosive resist popping that disperses resist particles on the wafer surface. Baking above the DNQ T_d after development is therefore desired to volatilize the PAC.

Novolac resins generally suffer from thermal distortion (during subsequent high-temperature processes) more than PHOST polymers. PHOSTs used for DUV resisters have T_g values on the order of 140–180 °C; the T_g values for novolac resins are in the 80–120 °C range. Elevated baking temperatures also result in oxidation and crosslinking of the novolac, producing more "thermoset"

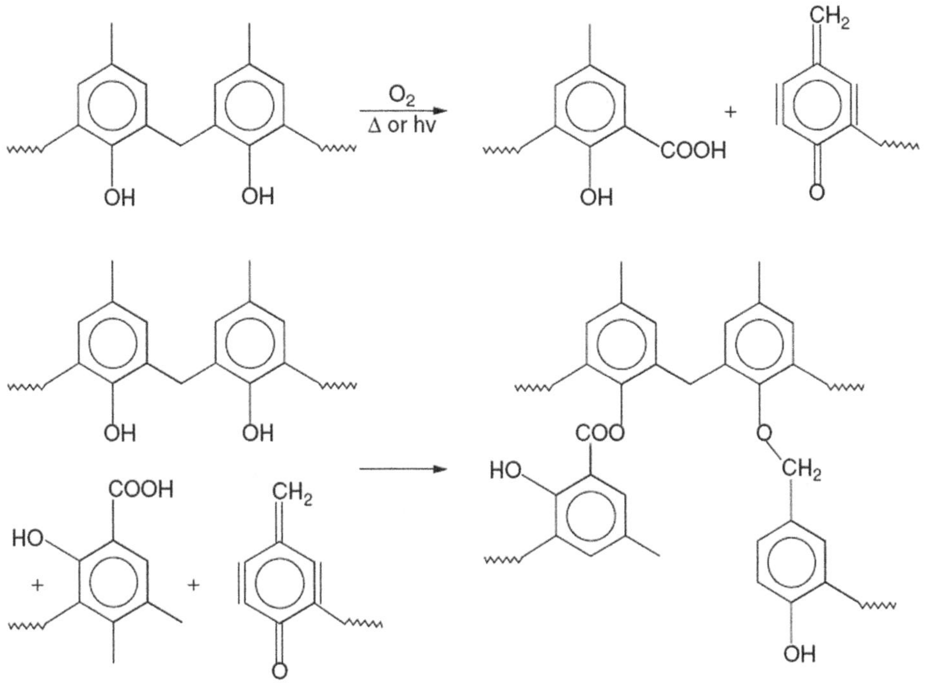

FIGURE 8.34 Radical reactions and oxidations induced in novolac from DUV curing. DUV hardening is effective in the presence of oxygen or with no oxygen (e.g., in nitrogen) through self-oxidation of novolac. Removal of DUV-cured novolac can be difficult, and particulate contamination can occur from rapid nitrogen outgassing at high exposure levels.

materials with flow temperatures higher than the original resist T_g (and less well defined). The temperatures required accomplish this, however, above the resist's T_g, thereby allowing the flow of patterned features in the process. Novolac resins with T_g values above 130 °C are now commonly used, and higher-T_g resists have been introduced [69].

To enhance the thermal properties of DNQ/novolac resists, the UV crosslinking properties of novolac can be utilized. Although the efficiency is quite low, novolac resin can be made to cross-link at DUV wavelengths. This is facilitated at high temperatures. The high optical absorbance of novolac at wavelengths below 300 nm (absorbance $>>1$ mm^{-1}) prevents crosslinking to substantial depths. If patterned resist features are subjected to DUV exposure at temperatures above 150 °C, a thermally stabilized surface crust can be formed. By elevating the temperature of the "DUV cure" process, oxidation of the bulk of the resist feature can be accomplished. The process is outlined in Figure 8.34. The now-networked resist feature can then withstand thermal processes up to 210 °C without significant resist flow.

8.11 PHOTORESIST FOR 193 NM LITHOGRAPHY

As photolithography has progressed below 130 nm design rules, the 193 nm ArF excimer laser has been employed as an enabling exposure source. The traditional phenolic polymers used for 248 nm and g/i-line application absorb heavily at 193 nm and thus are not usable. This is due to the structure of the phenolic ring, which possesses a carbon–carbon double bond that absorbs at wavelengths below 220 nm. Alternative platforms have been found that could retain suitable etch resistance and solubility but remain transparent at 193 nm. Resist systems based on polymethacrylates have consequently been developed. These systems can achieve adequate 193 nm transparency, high sensitivity, alkaline-develop-process compatibility, and good imaging properties for

TABLE 8.2

Summary of the Performance Comparison for Three Major Photoresist Platforms for 193 nm Applications

Platform	Advantages	Issues
COMA	Etch resistance, mechanical rigidity	Transparency (moderate), control of water content
CO addition polymer	Transparency, excellent etch resistance, mechanical rigidity	Polymerization control, resin purification
MA acrylic polymer	Transparency, design flexibility, excellent adhesion	Etch resistance, mechanical rigidity, line slimming

very low k_1 values [70]. The addition of alicyclic pendant groups has resulted in plasma etch resistance comparable to that of phenolic-based photoresists, leading to a generation of materials capable of supporting lithography well below 65 nm generations [71]. Most 193 nm photoresists are based on five main platforms: (1) methacrylate copolymers with alicyclic pendant groups (MA), (2) ring-opening metathesis polymer (ROMP) with alicyclics in the main backbone, (3) cyclo-olefin addition polymers (CO), (4) alternating copolymers of cyclo-olefins with maleic anhydride (COMA), and (5) vinyl ether/maleic anhydride (VEMA). Three of these platforms have gained the most attention: the MA, CO, and COMA approaches. The advantages and issues of these platforms are summarized in Table 8.2 [72].

The improvements in the etch resistance of these resist platforms have led to materials that are much more production worthy than the earliest formulations, but this remains an issue. The COMA resists exhibit the highest level of resistance but still do not compare to their 248 nm PHOST or i-line novolac counterparts. The integration of these materials may require the use of a hard mask pattern transfer layer, typically silicon oxinitride (SiON). Such a layer can be designed and deposited as an underlying bottom antireflective coating (BARC), as described later. The most recent advances in these resists have led to their reduced sensitivity to the postexposure bake impact on CD control. The issues associated with line slimming with scanning electron microscopy (SEM) inspection will always remain a problem with acrylate polymers, as the basis of these systems is a scissioning polymer that undergoes volumetric losses with exposure to such high-energy sources. COMA resists are less sensitive to this effect, but the problems can be minimized when they are used with low SEM acceleration voltages. The acrylates and methacrylates have received recent attention because of their flexibility in terms of both imaging performance and chemical synthesis. Because these polymers are of the free radical type, they are simpler to formulate and modify to tailor imaging and processing properties. Their maturity compared with the other systems has also led to their adoption for most applications.

The main challenges with 193 nm resist technology are those that are related to the continual shrinking of feature sizes as very-low-k_1 lithography is pursued. As linewidths move below current generations, the fundamental stochastic imaging limits of chemically amplified polymeric resists will eventually be reached.

8.12 CHALLENGES FOR CURRENT AND FUTURE PHOTORESISTS

In addition to the material attributes described earlier for the major 193 nm material platforms, there are at several fundamental challenges that will continue to require attention at every generational transition: resolution, outgassing, pattern collapse, and line-edge roughness (LER). Pattern collapse and LER may represent the largest near-term challenges with photoresist materials as the CD budget approaches the size scale of the polymers. The ultimate image resolution of chemically amplified resists will determine the longer-range limits of these materials.

8.12.1 IMAGE RESOLUTION AND RESIST BLUR

The application of chemical amplification has been the enabling factor in achieving highly sensitive photoresists for 248 and 193 nm applications. It was imperative in the early stages of DUV resist development that a route be identified by which the quantum efficiency of the photoresist materials could be increased to levels above 1.0. The acid-induced polymeric thermal deprotection and subsequent diffusion mechanisms that have allowed these resists to drive optical lithography for several decades will at some point lead to their detriment. This fundamental limit of these materials will be tied to the image blur or spread function that inherently results from this thermally activated deprotection and subsequent diffusion. As the targeted resolution for these materials approaches the physical blur function of diffusion, their usefulness will become limited.

The exposure of a CAR generates a catalytic photoacid species. This photogenerated catalyst goes on to initiate a deprotection reaction and, because it is generally not consumed, induce the deprotection of many sites until it reaches an average catalytic chain length. This catalytic chain length may be several hundred atoms long, leading to a quantum efficiency (the number of reactions per photon absorbed) much larger than one. The fundamental limit to resolution is tied to this chain length, because an image blur will result as the photoacid diffuses into unexposed resist regions. On a resolution scale where the physical diffusion is sufficiently smaller than the targeted geometry, an image bias results, where the printed line size is reduced. Photoacid size, exposure dose, mask bias, and bake-process conditions can be used to compensate for this effect to some extent, but often a base additive will be employed to the resist formulation to decrease the diffusion of the acid. At the point where no acid diffusion is tolerable, the quantum efficiency must be reduced to one or below, leading to a nonchemically amplified photoresist system.

The spatial spread of an image in a chemically amplified photoresist is commonly called the *resist blur*. Through chemical kinetics simulations, the form of this blur function has been fitted to a Lorentzian function, as shown in Figure 8.35 [73, 74]. In a general high–activation energy resist, the resulting diffused resist blur after PEB may be on the order of 50 nm (full width at half maximum [FWHM]) [75]. The blurring process can be approximated by convolution integrals, which have been modeled for some time. Fickian diffusion of an intensity profile in a photoresist has been

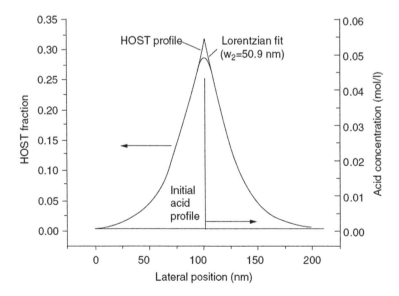

FIGURE 8.35 Resist blur function for a high–activation energy resist, modeled by a Lorentzian function. The full width at half maximum (FWHM) blur is on the order of 50 nm. (From Brunner, T.A., *J. Vac. Sci. Technol. B*, 21, 2632, 2003.)

used to account for acid diffusion, optical vibration, random aberration, and other effects [76]. A resist blur function such as the one in Figure 8.35 provides an additional level of understanding to the resolution limitations of a photoresist. A general rule of thumb can be used based on convolution requirements that image resolution is not possible when a blur function is larger than the image function itself. In other words, an image function can be made so small that it acts only as a probe for the blur function, constituting the resolution-limiting aspect of the process. A blur value of 50 nm could therefore limit resolution to values of 65 nm or greater.

There are improvements that can be made to the resist blur function. The obvious one is the decrease in the catalytic amplification factor by limiting PEB or acid size. Though high–activation energy resists exhibit blur values in the 40–60 nm range, values for low–activation energy materials can be lower than 20 nm. The ketal/acetal and carbonate/ester photoresists that fall into this category require little or no PEB to catalyze deprotection. Because the acid diffusion of these resists is reduced, blur can also be minimized [77]. The price paid for this can be detrimental, as the benefits of resist blur to line-edge roughness are not obtained. Additionally, these platforms can produce excessive outgassing, leading to contamination of optical elements during the exposure process.

8.12.2 Photoresist Outgassing

Resist outgassing, which has been an issue for vacuum lithography technologies (specifically electron beam and ion beam), has become an issue with optical lithography as well. When DNQ/novolac resists are exposed, nitrogen is released as a by-product. Although this would limit the application of these resists for an immersion lithography application, it has not posed a significant problem for conventional exposure in air. The chemically amplified resists used for 248 and 193 nm can be problematic, as by-products can lead to condensation on exposure optics, which dramatically reduces the lifetime of a projection lens. Careful control of the vapor pressure of solvent systems, photoacid generators, and additives can reduce these issues somewhat, as can the addition of radical scavengers to the resist formulation [78]. Although solvent outgassing can pose problems if a photoresist is not sufficiently baked near or above its glass transition temperature, careful bake-process design can nearly eliminate this as a practical problem [79]. The level of outgassing resulting from acid-catalyzed deprotection immediately after exposure is a function of the photoacid generator and the polymer protecting group. Outgassing rates of 193 nm photoresists have been measured at between 10^{11} and 10^{13} molecules/cm^{2}s. The bulk of the species outgassed are protecting-group fragments and PAG fragments, which can be reduced substantially when a high-activation resist platform is employed.

8.12.3 Line-Edge Roughness

LER is a primary concern for device geometries below 180 nm and becomes critical to understand and control for sub-90 nm generations. LER is different from linewidth roughness (LWR), which is related but is a measurement of the roughness effect on both sides of a feature. Although line thinning may not be indicated in a LER measurement, it would be measurable with LWR. LER and LWR of a patterned line are accumulations of variation from the photomask, the exposure source, the photoresist, and the resist-material stack. The metric used for LER is usually a 3σ measurement of variation, which generally should be no more than 5% of the CD in the photoresist. This reduces the CD budget and will also be transferred to underlying layers during subsequent etch process steps. The photoresist contribution to LER has not been fully quantified but has been found to be related to several factors. Photoresist process conditions have been identified as an important factor in LER control [80], but molecular weight and distribution have been found to play a less significant role [81]. The dissolution of partially protected polymers influences LER only to a minor extent, but the phase separation of protected and deprotected polymers can have a significant effect [82]. Polymer aggregates in the range of 23–30 nm in diameter can also lead to significant roughness [83]. Shot noise also becomes a contributor for small-pattern geometries. It would stand to reason that by increasing the acid diffusion length of a chemically amplified resist, LER could be decreased. Although this is true, increasing the diffusion length to 20% of the targeted feature pitch value saturates the improvement; beyond that, only losses in image resolution and process latitude will result [84].

8.12.4 RESIST PATTERN COLLAPSE

As the aspect ratio of photoresist features increases, so does their susceptibility to collapsing. This has especially become an issue for geometries below 130 nm, where the stability of 193 and 248 nm resists may not withstand the mechanical stresses of spinning or the capillary forces of drying during process stages, as seen in Figure 8.36 [85–87]. Mechanical forces play a large role in this behavior, where the tops and centers of line features (furthest from the substrate) may experience attraction, resulting in contact after rinse and dry steps. The mechanical properties of the photoresist material, including the Young's modulus, residual strain, residual stress, coefficient of thermal expansion, and glass transition temperature (T_g), will influence collapse. Several studies have determined the extent to which these factors influence patterning and allowed the calculation of the effect [91, 92]. The unbalanced capillary force surrounding a line (ΔF) can be determined as

$$\Delta F = 2\gamma \left(\frac{1}{s_1} - \frac{1}{s_2} \right) HA \cos\theta \qquad (8.19)$$

where
    s_1 and s_2 represent the space regions surrounding the line
    γ is the surface tension of the rinse fluid (where water is 72 mN/m)
    H is the film thickness
    A is the line edge area
    θ is the contact angle of the rinse liquid

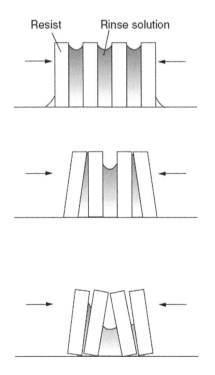

FIGURE 8.36 Resist pattern collapse for high–aspect ratio line-space features.

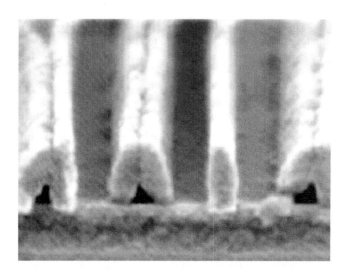

FIGURE 8.37 Pattern collapse for resist features with defocus, which increases the susceptibility for failure.

Uneven fluidic forces from rinse steps contribute most significantly to the collapse of resist lines. As a photoresist line becomes more compliant, lateral deflections can lead to the "shorting" of features from the top surface downward, which increases with aspect ratio. Collapse can also be influenced to a large extent by the adhesion of the photoresist to the substrate, feature profile (footing and undercut, as influenced by processing and reflection), pattern spacing, and pattern asymmetry. In general, symmetry on each side of a resist line feature will decrease the propensity for collapse, as attractive forces are equilibrated, as can be seen from Equation 8.18, where s_1 and s_2 are equal. Asymmetry, such as the condition for the last line of a grouping, will increase the likelihood of collapse. Imaging conditions will also play a large role in pattern collapse; the best focus and exposure conditions provide the maximum image modulation conditions and the lowest susceptibility to mechanical failure. As excursions from ideal focus and exposure are explored, degradation will result [90]. As an example, the SEM images of Figure 8.37 are under a condition of defocus.

To overcome the problems associated with resist pattern collapse, several approaches have been introduced for the reduction of the surface tension and attractive capillary forces during rinse and drying steps. Work with alcohol-based rinses followed by supercritical CO_2 (SCC) drying has shown promise, where surface tension vanishes in the supercritical phase [91]. Problems with the recondensation of water during the SCC drying process and the need for high-pressure chambers may present challenges. Alternative processes involving SCC as a solvent and development medium have also been introduced, but their practicality relies on the development of processing and tooling for their application [92]. A more adaptable solution to the problem may be the introduction of surfactants during the final rinse stage of processing [93]. Because capillary forces scale with the surface tension of the resist liquid and contact angle, rinse liquids with a low surface tension can minimize the unbalanced capillary forces. The challenge with rinse agents and surfactants is their compatibility with the variety of resist chemistries employed for sub-130 nm imaging. A thorough screening of materials for compatibility is required prior to implementation. An ideal strategy to reduce collapse problems is the design of resist materials so that the surface tension is optimized for use with water rinsing.

8.13 PHOTORESIST ISSUES FOR IMMERSION LITHOGRAPHY

As immersion lithography is employed for use at 193 nm, several issues arise that are unique to imaging in a fluidic environment. Additional process concerns include the leaching or extraction of resist components into the immersion fluid, the diffusion of the immersion fluid into the resist matrix, and the timescales of these processes related to imaging and subsequent mechanisms.

The leaching of PAG and base additives from a photoresist has been shown to be a surface phenomenon, where a small volume of molecules are extracted, on the order of 3–5% for a 200 nm film [94, 95]. The leaching occurs quickly, equilibrating after the first 10 s of contact between the photoresist and the fluid (where water has been most extensively studied) [96]. The simplest model assumes that the leaching of a PAG is a function of the number of molecules at the surface, which is quite adequate for understanding the mechanisms involved. Resist formulations optimized for use with water immersion imaging are needed to minimize this leaching; otherwise, top-coat materials will be needed as a barrier layer between the photoresist and the fluid [97]. As water is diffused into a photoresist film, it swells and experiences an increase in thickness. Using quartz crystal microbalance and optical reflectance techniques, the water uptake of a photoresist film during contact with water has been measured [98]. During the first 30 seconds of exposure to water, up to 4 parts per thousand can be carried into the photoresist film [99]. Despite the significant extraction of resist components into the immersion fluid and the relatively large amount of water driven into the photoresist film, the impact on imaging with the existing 193 nm resist platforms has been surprisingly minimal. Using interferometric immersion lithography for imaging various fluids and resists, a large number of materials have been tested and have achieved resolution well below 45 nm [100, 101]. Unlike the challenges that have been involved with the transitional requirements for photoresist materials with the changing of exposure wavelength, there has not been a need for a new class of materials for 193 nm water immersion lithography.

8.13.1 IMMERSION RESIST PROCESSING

The challenges with processing 193 nm immersion resists stem from the introduction of water into the space between the final lens element and the photoresist, normally filled with air. Because of this, there are concerns related to the influence of a fluid on both the optics and the photopolymer during the exposure operation and throughout scanning. If projection optics become contaminated (not only by the water but also by contaminants introduced into the water), the effects could be detrimental if not removed. If the fluid interacts with the photoresist, by diffusing into the resist or leaching things out of the resist, consequences will lead to the loss of process control.

8.13.1.1 Interactions with Optics

Optical contamination issues will arise when a lens element comes into contact with a fluid. The situation in a projection lithography system is worsened because of the high photon energy introduced to the solid–fluid interfaces during exposure. As the immersion fluid passes over a photoresist, any chemical species that can pass into the water will make contact with the lens surface. The energy in a 193 nm photon is sufficiently high to cause chemical decomposition of any such species, which may deposit onto the solid lens surface. As a lens becomes more contaminated, its throughput and uniformity can quickly degrade, leading to loss of process control. Without a cleaning method or an easily replaceable final lens element, the useful lifetime of projection optics will be impacted. In situ cleaning of the final element of immersion lithography optics has been introduced through several approaches. These have included the introduction of CO_2 gas into the immersion fluid, which can reverse lens contamination to a large extent [102].

8.13.2 CHEMICAL LEACHING

The small components in a resist formulation (such as the photo-acid generator, the PAG) can leach into an immersion fluid. The rate of leaching will depend on size and solubility kinetics and is generally nonlinear. Leaching rates can be measured and should be done so dynamically. Projection lithography system suppliers have set maximum leach rates at levels below about 2×10^{-12} mol/cm^{2}s [102].

8.13.3 Water Uptake in Photoresist

Just as chemical species from the photoresist can leach into an immersion fluid, so can the fluid diffuse into the photoresist. Studies show that the diffusion of water into a photoresist is fairly rapid and occurs within the first few seconds of contact. Although the water in a photoresist layer will be removed during subsequent bake steps, the influence on exposure through optical effects, thickness variation, and surface roughness can influence process control [102].

8.13.4 The Influence of a Scanned Meniscus

As described in Chapter 2, scanning immersion lithography involves the dynamic movement of a scanned meniscus. There is an advancing edge and a receding edge to the meniscus, which sets it apart from the scenario with simpler immersion microscopy. While the advancing edge has the ability to capture air during motion, the receding edge can break free and leave water stain defects on the resist as the wafer travels. Because of these potential (and likely) phenomena, the fluid dynamics of the solid–fluid–air interface need to be carefully controlled. Together with the consideration of the high rate of scanning over a 300–450 mm wafer, the surface tension effects influenced by the water and the photoresist composition need to be carefully controlled to prevent advancing and receding type defects. The design of the immersion fluid "head," where the introduction and removal of the fluid will have an influence, also comes into play. Surface modification approaches have been utilized through changes to the fluid, the lens, and the photoresist surface.

8.13.5 Top Coats for Immersion Photoresists

Top-coat layers have been used to modify the fluidic interaction effects at the solid–fluid–air boundaries in a scanning immersion meniscus. The optical effects of top-ARC resist layers described earlier can be utilized with immersion resists, combined with protective effects to reduce leaching, chemical diffusion, and adverse surface tension effects. Chemical compatibility between a top coat and a photoresist may be less challenging from a materials and design standpoint than the compatibility between the photoresist and the immersion fluid. Both top-coat and non-top-coat photoresists have been utilized in manufacturing.

8.14 RESISTS FOR PHOTOMASK PROCESSING

The patterning of a photomask requires a photomask blank that has been sensitized with a photoreactive coating. Often called a *photoresist*, such materials are likely to be exposed using nonoptical methods such as electron beam and will thus be referred to simply as *resists*. Photomask blanks have traditionally been precoated with resist by the blank supplier to be exposed, developed, and processed by the mask maker. This adds challenges that are not associated with wafer resist coating; that is, concerns about storing, handling, and contamination issues of thin photoresist films. This becomes more of a concern with CARs, where the acid generator compounds in the resist formulation can be sensitive to the environment. Careful control of coating and storage conditions is necessary in order to meet the dose and linewidth control requirements of modern lithographic generations. The alternative to precoated mask blanks is coating resist on demand. This places the responsibility for stringent control over coating thickness and uniformity, bake temperature and uniformity, and short-term storage with the mask maker.

8.14.1 Photomask Resists

Several categories of photomask resists have been used through the years, depending on requirements for such things as exposure method, tone, sensitivity, development chemistry, and etch selectivity. Resists used for electron beam exposure have included scissioning polybutene sulfone (PBS), with high sensitivity but little etch resistance; crosslinking epoxy copolymers, limited in resolution because of swelling; diazonaphthoquinone novolac resist in either positive or negative tone; and chemically amplified counterparts to modern optical photoresists. Each resist type has presented its own challenges with regard to a high-volume, manufacturable solution. Often, a compromise has been necessary between throughput and development method, where resolution and plasma etch resistance have been necessary to achieve the performance requirements of modern semiconductor device generations. Historically, the benchmark resist for resolution has been positive scissioning polymethyl methacrylate (PMMA). Its relatively low sensitivity and poor etch resistance have given rise to better alternatives. Up until the late 1990s, scissioning resists based on polyolefin sulfones, such as PBS, dominated for electron beam mask making. These are also scissioning resists but with sensitivity up to 100× that for PMMA. Being scissioning materials like PMMA, they offer no plasma etch resistance. Resolution has also limited their use for modern device generations, as sub-250 nm is difficult.

A resist based on a copolymer of chloromethyl acrylate and methyl styrene has for the most part replaced other polymeric (i.e., not chemically amplified) resists because of its etch resistance, sensitivity, and adequate resolution. ZEP 520A is a positive-acting resist that is developed in solvent developer (e.g., xylene or n-amyl acetate) with a sensitivity about 10× better than PMMA. The chemical structure for the copolymer is shown in Figure 8.38. It is used at a molecular weight around 5.5 kg/mol, cast in anisole or other suitable solvent. The chlorine attachment to the methacrylate provides high sensitivity, while the benzene (styrene) attachment provides the high etch resistance. The popularity of this copolymer for mask making has been driven by both etch resistance and sensitivity, which has been reduced to achieve dose values down to 5 $\mu C/cm^2$ [103].

Chemical amplification has been utilized for electron beam resists to enhance sensitivity in both positive- and negative-acting modes. Positive tone CARs for e-beam use follow similar formulations to those used for optical CARs, where protection groups on a polymer chain are cleaved through a PEB-induced catalytic reaction. The reaction is triggered by an acid produced by the exposure of a sensitizer (known as an *acid generator*). Since the acid is not consumed during the polymer deprotection, it can go on to create many additional reactions—hence the chemical amplification. There is an inherent tradeoff between the sensitivity of a CAR and its ultimate resolution. CARs used for mask making can achieve resolution below 100 nm with dose requirements on the order of 3–10 $\mu C/cm^2$. A variety of materials have been made available from various suppliers that offer high sensitivity and good plasma etch resistance by separating the exposure/reaction and deprotection/solubility stages of the polymer system. Several positive CARs fall into this category, including TOK EP-012M, Fujifilm FEP171, KRS-XE, and CAP209, using variations to the fundamental formulation of acid deprotection of a PHOST polymer or copolymer. The resulting polymer is soluble in an aqueous base, such as TMAH. The general chemistry involved is shown in Figure 8.39.

Negative-acting CARs generally utilize an exposure-produced acid to catalyze crosslinking, leading to insolubility in exposed regions. Negative CARs include TOK EN-012N and 024N, Sumitomo NEB-31and NEB-22, and Dow RHEM SAL 601/603. Negative resists such as these can exhibit high

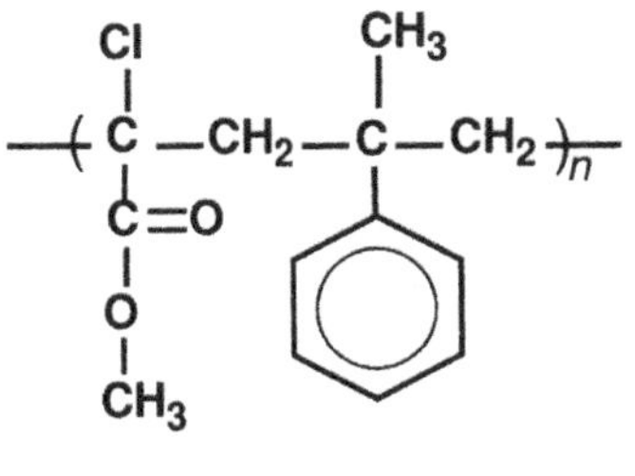

FIGURE 8.38 Chemical structure of the copolymer of chloromethyl acrylate and methyl styrene used in ZEP electron beam resists.

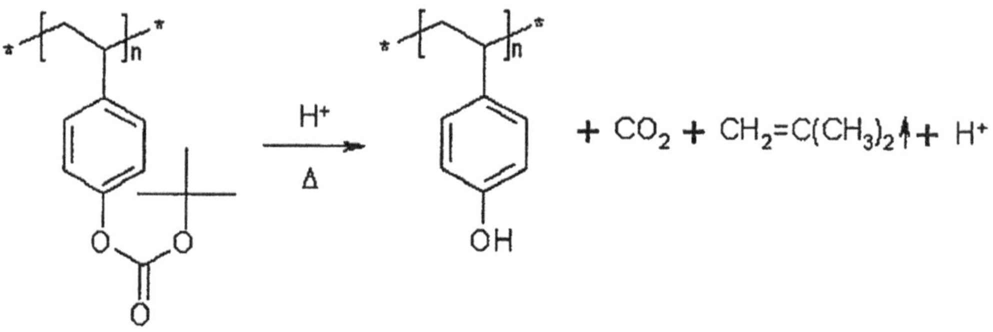

FIGURE 8.39 Chemical scheme for a positive-acting chemically amplified resist, where the tertiary group protecting the polymer is removed with acid (H+) and heat (PEB) to increase the solubility of the deprotected polymer in an aqueous base.

sensitivity, high resolution, and good plasma etch resistance and are developed in a metal ion–free aqueous base such as TMAH. The route toward the negative-acting properties of these materials is through acid-induced crosslinking of phenolic polymers such as novolacs and polyhydroxystyrenes. Examples of such acid-hardening resist systems are described in more detail in Chapter 7.

The photoresists used for laser mask writing are for the most part conventional optical resist formulations. Depending on the wavelength used, photoresists based on DNQ/novolac formulations described in Chapter 7 can be used with the DNQ sensitizer tuned to the laser wavelength range (405 nm, for example). For laser pattern generators operating in the 248–257 nm wavelength range, photoresists based on DUV CAR chemistry are suitable. Non-CARs are sometimes preferred to avoid the sensitivity that these resists can have to environmental contamination of the photoacid. An alternative is the use of a top-coat polymer layer over a DUV CAR resist.

8.14.2 Photomask Processing

Once the photoresist layer is exposed, the process sequence for mask fabrication involves the steps of development, etch, strip, and clean. Each of the operations may comprise several additional steps depending on the layers or complexity of the photomask layer stack. For a binary chromium-based mask, the etch step involves wet chemical or dry plasma etching of the CrON film left unmasked by the photoresist patterned above it. In the case of phase-shift or other multiple-level mask types, the sequence of steps would be replicated for each layer, as will be described later. Fundamental to most mask processes is the chromium etching processes, where the choice of methods depends on the tolerance to isotropy.

8.14.2.1 Wet-Etching of Chrome Masking Films

A wet-etch process is carried out by the chemical reaction of a film and a suitable etchant solution. When masked by a photoresist pattern, the transfer of that pattern can be made into the underlying film. Figure 8.40 shows how a wet chemical etch progresses through a photoresist window for thin CrON film. The traditional wet etchants used for chromium films are mixtures of ceric ammonium nitrite $(NH_4)_2[Ce(NO_3)_6]$ stabilized with perchloric acid $(HClO_4)$. The chemical reaction of this mixture with chromium produces chromium nitrate $(Cr(NO_3)_3)$, which is aqueous soluble and dissolved during the etch process. Polymeric photoresists are mostly unreactive in this chromium etch and therefore, protect underlying metal from being removed. But since the process is entirely chemical, dissolution proceeds both downward and laterally through the chromium layer. This isotropic removal of material results in the undercutting of the metal layer below the photoresist at a rate equivalent to the film removal rate. Shallow etched pattern sidewalls result, approaching 45° angles. Furthermore, the graded structure of

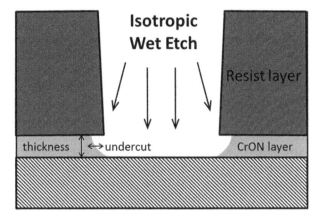

FIGURE 8.40 Depiction of isotropic pattern profiles with wet etching a thin film.

the CrON film creates a faceted edge because of the faster etch rate of the upper portions of the film. As overetch progresses to open a chrome window to a required dimension at the bottom of the layer, the dimension at the top of the film is made larger. For mask features that approach a few multiples of the chrome film thickness, the undercut becomes an appreciable amount of the feature size. At dimensions of 2× the film thickness, etch undercut results in complete loss of the feature. The limitations of isotropic wet etching eliminated this as an option for wafer processing for leading-edge device generations a few decades ago. But at the 4× magnification of dimensions at the photomask, wet etching could be used for several years beyond this. At a 1000 Å CrON layer thickness, wet etching is limited to mask dimensions greater than about 500 nm or more, or 125 nm dimensions at the wafer.

With wet chemical etching, the vertical and lateral etch rates are equivalent. In order to transfer smaller patterns through substantially thick films (i.e., more than 10–20% of the targeted feature dimension), an anisotropic etch must be used. Reactive ion–assisted plasma etching (or reactive ion etch [RIE]) allows such directional etching of thin films masked by a patterned photoresist later.

8.14.2.2 Plasma Etching of Chrome Masking Films

Plasma etching of metals began in the mid-1980s, primarily for the patterning of aluminum in chlorine and bromine etch chemistries. RIE allowed the anisotropy needed for direct pattern transfer with minimal lateral material removal. The halogen etch chemistries used in metal etch combined with the physical material removal via RIE led to aggressive photoresist loss compared with the processes used for dielectrics. Careful control of the chemistry, power, pressure, and bias is necessary to achieve optimized etch rates and selectivity to resist and underlying films. High–ion density plasma processing led to improved selectivity using electron coupled resonance (ECR), helicon radio-frequency (rf) plasma, or transformer coupled plasma (TCP). In the mid-1990s, inductively coupled plasma (ICP) etching was developed for high ion density and high selectivity while controlling damage to underlying materials. Magnetically enhanced RIE (ME-RIE) also offers improved etch rate and selectivity to standard rf RIE processes. As plasma etching was employed for chromium masking films, both ME-RIE and ICP RIE etching have been able to achieve pattern transfer for sub-250 nm lithography photomask generations. Currently, ICP dominates for many commercial applications, producing better results especially for the pattern process control for sub-180 nm generations.

A typical ICP etch process for a chromium mask film uses a mixture of chlorine (Cl_2) and oxygen (O_2) gases. Etch rates greater than 50 nm/min are typical, with a better than 5% uniformity, 0.5:1 selectivity to resist, and 50:1 selectivity to the mask (SiO_2) substrate.

8.15   BEYOND SINGLE-LAYER RESISTS

As higher-resolution approaches to microlithography are pursued, conventional single-layer resist schemes can fail to meet all process requirements. Multilayer resist techniques have been investigated for several years, but advances in single-layer technology had generally postponed their insertion into high-volume production operations. As long as single-layer resist materials could meet requirements for high–aspect ratio resolution, photosensitivity, plasma etch resistance, planarization, depth of focus, reflection control, and CD control, they would be preferred over most multiple-layer techniques. This becomes increasingly difficult, and at some point, the lithographer needs to consider the advantages of dividing the functions of a single-layer resist into separate layers. Fewer process steps are normally preferred, and the ultimate acceptance of any multitechnique will be determined by weighing the complexity of the overall process.

8.15.1   SINGLE-LAYER RESIST REQUIREMENTS

To understand the potential advantages of multiple-layer lithographic materials and processes, the general requirements of a photoresist should first be addressed. Although most resist requirements have existed for many generations of integrated circuit processing, the importance of a number of issues has recently increased dramatically.

8.15.1.1   Resist Sensitivity

Because resist sensitivity directly affects process throughput, it is a fundamental consideration for the evaluation of resist process capability. In general, resist sensitivity can be shown to be proportional to thickness. For a direct photochemical (not chemically amplified), nonbleaching resist material, this is an exponential relationship determined by resist absorption and chemical quantum efficiency. However, as resist bleaching mechanisms are considered (as with the photochemical conversion of diazonapthoquinone to indene carboxylic acid), dynamic absorption exists, which introduces some additional considerations to this exponential decay. With chemically amplified resists, quantum efficiency is sufficiently high that the dependence of sensitivity on resist thickness becomes less of an issue, and other considerations become of more concern.

8.15.1.2   Depth of Focus

As shown previously, the dependence of depth of focus on lens numerical aperture and wavelength can be expressed as

$$\text{DOF} = \pm k_2 \frac{\lambda}{n \sin^2 \theta} \tag{8.20}$$

where
  λ is wavelength
  $n \sin^2 \theta$ becomes NA^2 in air
  the parameter k_2 is the process-dependent UDOF factor, as described earlier

As optical lithographic technology has been pushed toward lower DUV wavelengths with higher NAs, the resulting DOF is reduced below a few hundred nanometers. This presents an interesting challenge for substrate topography and photoresist thickness issues. With such a small useful DOF, and without the use of some method of planarization, it is not easy to predict exactly how large a fraction of this range could be consumed by photoresist thickness.

8.15.1.3   Limitations of Resist Aspect Ratio

The physical and chemical nature of a polymeric resist material will determine its limitations for high–aspect ratio patterning. Additionally, the complex nature of development and process chemistry will influence limitations. As aspect ratio lower than 3:1 is common for conventional single-layer resists. The limit to how fine the resolution can be for a single-layer resist of a given thickness is influenced to a large extent by polymer flow properties, including glass transition temperature (T_g) and melting point (T_m). Because thermoplastic polymeric behavior is desired during processing, in which photoresist materials can go through cycles of heating, flowing, and cooling, they generally possess T_g values in the 70–180 °C range. Materials of lower T_g will inherently be capable of lower–aspect ratio imaging.

8.15.1.4 Reflection and Scattering Effects

Imaging over reflective substrates such as metal or polysilicon can allow significant intensity variation within a resist film. High levels of reflectivity may produce overexposure, manifested not only as a bulk effect over the entire imaged file but also at pattern-specific locations such as line boundaries and corners. This is often referred to as *reflective line notching* or *necking*, which is a result of the scattering of radiation to unwanted field regions. Substrate reflection will affect the overexposure latitude and ultimately lead to a reduction in focal depth, limiting the amount of tolerable image degradation. To understand the impact of exposure latitude on depth of focus, consider imaging a feature with poor modulation. If a resist process is capable of resolving such a feature, this is likely to be possible only within a limited range of exposure dose. For a positive resist, overexposure can result in complete feature loss, and underexposure can result in scumming. There is an intimate relationship, therefore, between depth of focus and exposure latitude. Decreasing the demands on focal depth increases exposure latitude. For a reflective substrate, if a large degree of overexposure latitude must be tolerated, the useful depth of focus will be reduced significantly. It is desirable to reduce any reflected contribution to exposure so as to eliminate feature distortion from scattering and to reduce detrimental effects on focal depth. This can be accomplished in a single-layer resist by several methods. First, because absorption is dependent on resist thickness, a thicker absorbing resist layer will decrease the impact of reflection. Other requirements drive resist toward thinner layers, however, reducing the practicality of this method. A second alternative is to increase the absorption of the resist, so that little radiation is allowed to penetrate to the resist–substrate interface and then be reflected back through the resist. The addition of dyes into a resist will accomplish this, but at the cost of resist-sidewall sensitivity and resolution. The beneficial dynamic bleaching mechanism of the DNQ/novolac materials is undermined by the addition of an absorbing dye that makes no direct contribution to the photochemical process. An alternative approach to reduction is the use of a multiplayer resist system incorporating a separate antireflective layer.

8.15.1.5 Reflective Standing-Wave Effects

An additional reflection phenomena that deserves consideration is the resist standing-wave effect. This is an exposure variation within a resist layer resulting from coherent interference between incident and reflected radiation. This has significant impact on exposure, CD control, DOF, and coating uniformity requirements. Minimization of standing wave is generally desired.

8.15.1.6 Plasma Etch Resistance

Postlithographic processing operations ultimately dictate the minimum acceptable resist thickness after development. For example, resist erosion during etch processing will increase any lithography-related thickness requirements. Furthermore, if new materials are considered for short-wavelength exposure application, their etch resistance in halogen-based plasma etch processes may be reduced. Postlithographic processes may place the most restrictive demands on resist performance and may preclude any consideration of thinner single-layer resists.

8.15.1.7 Planarization

Because the lithography operations involved in IC fabrication are rarely performed over a flat substrate, planarization of topography is a fundamental function of a resist material. The degree of planarization required for a specific level will be determined by step height, feature size and density, and substrate surface properties. Material properties of a resist, including polymer molecular weight, solids content, solvent type, coating-spin speed, acceleration, temperature, and exhaust, will contribute to the extent of substrate smoothing. Polymeric materials with low molecular weight and a high solids content are generally employed for maximum results [104]. A fluid-dynamics approach can be used to demonstrate the relationships between process factors and planarization:

$$P \propto \frac{t\gamma h_0^3}{\eta\omega^4} \tag{8.21}$$

where

    t is leveling time
    γ is surface tension
    h_0 is initial film thickness
    η is solution viscosity
    w is feature width

This relationship suggests that several factors can be modified to affect net results. Planarization of close-proximity features (local geometry) and of widely spaced features (global geometry) may be required, depending on substrate characteristics and process needs. Figure 8.41 illustrates that planarization by a polymeric material may be suitable for both situations. The extent of planarization can be quantified by considering the initial step height (z_0) and the final effective step height after smoothing (z_1) and determining the normalized ratio:

$$\text{Effective planarization} = \frac{z_0 - z_1}{z_0} \tag{8.22}$$

which can be calculated for local and global features [105].

Planarization can be accomplished by means of substrate overcoating (generally with an organic polymeric film), etch-back processing, or polishing of topographic substrate to reduce step height. Techniques of CMP are becoming widely accepted alternatives to additive planarization methods, reducing constraints on resist processing and requirement for focal depth [106]. Methods of CMP can allow global planarization of both insulator and conductor layers in multilevel-metallization-interconnect structure and of both deep and shallow trench isolation materials. These techniques have become a critical path for both logic and memory production, and a number of issues are receiving careful attention, including optimization of process techniques, cleaning considerations, and defects.

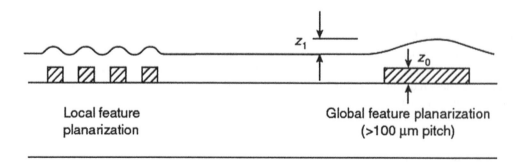

FIGURE 8.41 Polymeric planarization of local and global topography. The extent of planarization can be determined from measurement of initial and final step heights (z_0 and z_1).

8.15.1.8   Multilayer Resist Processes as Alternatives to Conventional Resist Patterning

The appeal of multilayer technology has increased because of the resolution and etch constraints of single-layer photoresists. By using surface or near-surface imaging, the photochemical modification necessary to affect the pattern transfer is restricted to a thin upper resist layer. Ideally, the imaging layer should be as thin and as planar as possible. In this way, maximum resolution can be obtained, and optimum use can be made of the entire available focus range of the exposure tool. The resist image is more free from the effects of device topography and substrate reflection, an advantage that is becoming increasingly important.

A large number of polymeric multilayer systems have been developed and utilized over several decades. Multilayer schemes can be divided into four basic categories, with some overlap in function. Specifically, approaches have allowed planarization, reduction of reflection, contrast enhancement, and surface imaging. These categories are not necessarily clearly divided, as a single multiplayer approach can accomplish several objectives. Details of these approaches will be explored in this chapter.

8.16   MULTILAYER PLANARIZING PROCESSES—EARLY APPROACHES

Various multilayer techniques have been introduced that employ polymeric planarization layers to reduce substrate topography and allow the use of a thin top-coated imaging resist layer, as fundamentally depicted in Figure 8.42 [107]. Methods have included a wet-processed thick planarization layer [108], a two-layer portable conformable mask (PCM) [109], and a three-layer plasma transfer process [110]. The wet-processed approach leads to isotropic dissolution of an underlying planarizing layer, limiting application generally to nonintegrated circuit use. The PCM process employs a DUV-sensitive planarizing layer, typically PMMA [111] or poly(dimethylgluterimide) (PMGI) [112], and a DNQ/novolac imaging layer. Because the DNQ/novolac is highly absorbing at wavelengths below 300 nm, once imaged, it acts as a surface contact mask over the bottom resist layer. Deep ultraviolet flood exposure and development of the bottom layer allow pattern transfer through the entire multilayer stack. This technique can be limited by interfacial mixing of the two resist layers, which is minimized when using PMGI materials. Poor contrast of the DUV planarizing layer

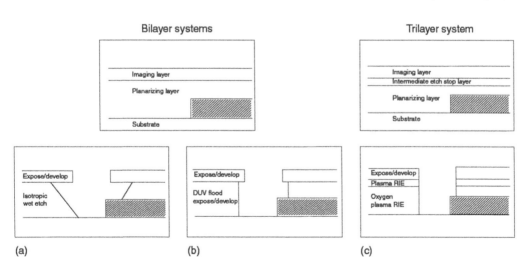

FIGURE 8.42   Multilayer approaches utilizing polymeric planarization layers. (a) A wet bilayer system using isotropic dissolution of the planarization layer; (b) a bilayer system using a blanket deep ultraviolet (DUV) exposure of a photosensitive planarization layer masked by a patterned DNQ/novolac layer; (c) a trilayer resist system utilizing a planarizing layer, an etch stop layer, and an imaging layer. (Particular film losses, degree of undercutting, and isotropy are not depicted to scale).

and reduced process control of this two-layer technique have limited resolution, making sub-0.5 μm imaging difficult.

Variations on the multiple-resist approach have also been used for electron beam T-gate fabrication [113, 114]. Three resist layers may be used to allow specific feature shaping through the depth of a resist stack. For example, a bottom layer of PMMA is overcoated by a layer of a methyl methacrylate–methacrylic acid copolymer (PMMA–MAA), followed by a top coat of PMMA. Exposure and wet development lead to larger pattern widths in the more sensitive PMMA–MAA layer, allowing the formation of T-shaped metal gate structures through a subsequent additive liftoff process.

8.17 DRY-PATTERN TRANSFER APPROACHES TO MULTILAYERS

Anisotropic pattern transfer can allow significant improvement over the isotropic processing of wet-etched multilayer approaches. Through the use of a plasma RIE pattern transfer process, near anisotropy can be approached, allowing high–aspect ratio, fine-feature resolution [115, 116]. The three-layer scheme, depicted in Figure 12.3, makes use of a polymeric planarizing layer (such as novolac resin or polyimide) and a thin intermediate etch-stop layer. This etch-stop layer can be a spin-on organosilicon compound (SOG), a low-temperature oxide, a silicon oxinitride, or a metallic layer, which provide oxygen etch resistance. A thin resist imaging layer is coated over this etch-stop, exposed, and wet developed. Pattern transfer into the intermediate etch-stop layer can be achieved with wet-etch or dry-plasma techniques with suitable chemistry. Anisotropic pattern transfer through the thick polymeric planarizing layer can be achieved via an oxygen RIE process. Variations on this technique have been used for both optical and electron beam applications [117].

The importance of polymeric planarization approaches historically has declined as single-layer resists and CMP techniques have steadily improved. As shorter-wavelength exposure technologies are pursued, however, it is likely that the application of multilayer approaches will become more viable. A bilayer resist technique that allows planarization, etch resistance, and reflection control with a thin imaging layer has many attractive properties. Such schemes will be addressed in detail as silicon-containing resists and top-surface imaging techniques are addressed.

8.18 RESIST REFLECTIVITY AND ANTIREFLECTIVE COATINGS

Reflection at resist–substrate interfaces has been a concern for many IC generations. The impact is most pronounced when using high-contrast resists, a result of increasing exposure thresholding effects. Figure 8.43 shows the effect that varying resist film thickness has on feature size (CD swing curves) for an i-line resist imaged over polysilicon and aluminum films. The reflectance from a resist–polysilicon interface at 365 nm can be above 30%, and at 245 nm, above 38%. Resist over aluminum can produce reflectivity values of above 86% at 365 nm and above 88% at 248 nm. Interface reflectance can be determined from a Fresnel relationship for two media at normal incidence as

$$R = \left| \frac{n_2^* - n_1^*}{n_2^* + n_1^*} \right|^2 . \tag{8.23}$$

where n^* is the complex refractive index or $n - ik$, where n is the real refractive index and k is the extinction coefficient. For nonabsorbing materials, $k=0$ and $n^*=n$, simplifying Equation 8.23. For nonnormal incidence, an additional $(\cos \theta)$ term is required, where θ is the angle of incidence. Inspection of the optical constants for materials in Table 8.3 gives an indication of the need to incorporate methods of reflectance control.

If the materials that make up the lithographic substrate possess low absorbance or are nonabsorbing, reflectance values at each interface must be uniquely considered to determine the net reflectance through the film stack. Figure 8.44 provides an example of a resist film over an SiC_2/Si

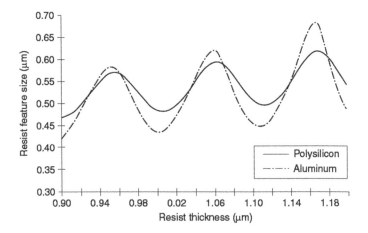

FIGURE 8.43 Critical dimension (CD) swing curve showing the effect of resist thickness variation on resist linewidth. Results for polysilicon and aluminum substrates are shown at 365 nm using a resist with a refractive index of 1.7.

TABLE 8.3
Optical Constants (*n* and *k*) for Several Materials at 436, 365, 248, and 193 nm

	193		248		365		436	
	n	*k*	*n*	*k*	*n*	*k*	*n*	*k*
Silicon	0.960	2.88	1.58	3.60	6.41	2.62	4.79	0.175
SiO$_2$	1.56	0.00	1.51	0.00	1.47	0.00	1.47	0.00
Si$_3$N$_4$	2.65	0.18	2.28	0.005	2.11	0.00	2.051	0.00
Aluminum	0.117	2.28	0.190	2.94	0.407	4.43	0.595	5.35
Polysilicon	0.970	2.10	1.69	2.76	3.90	2.66	4.46	1.60
DNQ/novolac					1.70	0.007	1.67	0.007
DUV CARs (PHOST and acrylates)	1.71	0.06	1.76	0.01				

substrate. Here, the contribution from the oxide–resist interface is low compared with the contribution from the silicon underlying material. In this case, the thickness of the SiO$_2$ film can be adjusted to minimize total reflectivity through destructive interference (the use of quarter-wave approaches for inorganic antireflection materials is discussed later).

Control of reflectivity at the resist–substrate interface to values near a few percent is generally required for critical lithography levels, leading to the need for some method of control. The situation becomes more critical as lithographic methods incorporate shorter-wavelength sources, a smaller spectral bandwidth, and more transparent resists. Reduction of reflection to values below 1% is needed for current-generation lithography. Dye incorporation into a resist can reduce the coherent interference effects but at the cost of exposure throughput and sidewall angle, leading ultimately to resolution loss. Dyed resists are, therefore, generally limited to noncritical reflective levels, at which the highest resolution is not necessary.

Instead of reducing reflection effects through modification of a resist material, methods that reduce reflectivity at the resist interface can provide control with minimal loss of resist performance. This can be accomplished through manipulation of thin-film optical properties and film thicknesses and by careful matching of the optical properties of each layer in an entire resist/substrate stack.

8.18.1 Control of Reflectivity at the Resist–Substrate Interface: Bottom Antireflective Coatings

To reduce the reflectivity at the interface between a resist layer and a substrate, an intermediate film can be coated beneath the resist. This is known as a *bottom antireflective coating* (BARC). Inspection of Equation 8.23 suggests one approach where the refractive index of this layer could be close to that of the resist at the exposing wavelength. To reduce reflectivity, the film could then absorb radiation incident from the resist film. Thin-film absorption (α) is related to the optical extinction coefficient (k) as

$$\alpha = \frac{4\pi k}{\lambda} \tag{8.24}$$

and transmission through an absorbing film is

$$T = \exp(-\alpha t) \tag{8.25}$$

where

T is transmission
t is film thickness

A high extinction coefficient may therefore be desirable, leading to high absorption and low transmission through the BARC layer. As k is increased, however, reflectivity at the resist–BARC interface is also increased, as seen in Equation 8.23. An extinction coefficient value in the range of 0.25–1.25 may be reasonable, based on these considerations and depending on resist material and film thickness demands. A series of plots showing substrate reflectance versus BARC thickness for real-index matched materials with extinction coefficient values from 0.1 to 1.2 is shown in Figure 8.45. Because the BARC layer must accommodate pattern transfer (using either a wet or a dry plasma etching approach), minimum thickness values are desirable. For these layer combinations, a BARC thickness between 500 and 800 produces a first reflectance minimum. These minima occur as reflectance from the BARC–substrate interface and interfere destructively with the reflection at the resist–BARC interface. This interference repeats at intervals of $\lambda/2n$, which can be explained by examining Figure 8.46. Here, radiation passes twice through the BARC layer with wavelength

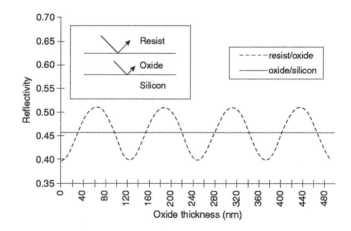

FIGURE 8.44 Reflection contribution at resist–oxide and oxide–silicon interfaces with increasing silicon dioxide thickness at a 365 nm wavelength using a resist with $n = 1.7$ and $k = 0.007$. Minimum reflectivity occurs at quarter-wave oxide thicknesses.

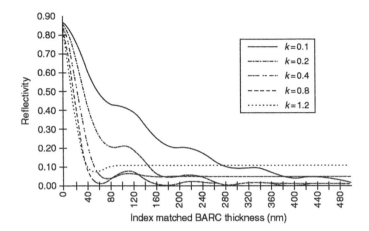

FIGURE 8.45 Substrate reflectivity versus bottom antireflective coating (BARC) thickness for real index matched materials ($n = 1.7$ at 365 nm) and extinction coefficient values from 0.1 to 1.2. The best performance for a thin BARC layer may be possible with k values near 0.8.

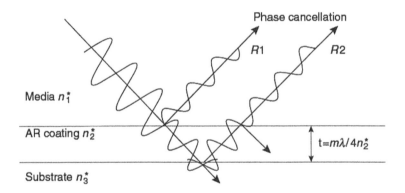

FIGURE 8.46 Diagram of the principle of a quarter-wave antireflective (AR) layer between two media. A thickness of the AR coating is chosen to produce destructive interference between reflected components, R_1 and R_2. The ideal refractive index is $\sqrt{n_1^* n_3^*}$.

compression corresponding to its refractive index (λ/n_i). Two passes through a quarter-wave thickness ($\lambda/4n$) result in a half-wave phase shift directed toward the resist–BARC interface. This phase-shifted wave will then interfere with the reflected wave at the resist. Complete destruction will occur only if the amplitude values of the waves are identical, which is possible only if the reflectance at the resist–BARC interface is exactly equal to the reflectance at the BARC–substrate interface or if

$$n_{arc} = \sqrt{n_{resist}^* \times n_{substrate^-}^*} \qquad (8.26)$$

This leads to a more complex route toward reflection reduction with a bottom ARC using both absorption and interference considerations.

Reflection control and aspect ratio requirements need to be considered to determine optimum ARC film thickness values. A first reflectance minimum corresponding to a relatively thin film may be chosen for fine feature pattern transfer. If thicker layers can be tolerated, a further reduction in reflectivity may be achieved by increasing the BARC film thickness. There is an exponential trend in reflection reduction with increasing thickness and BARC absorption. As extinction coefficient values increase toward 0.8, reflection begins to increase, and for values above 1.2, reflectivity below

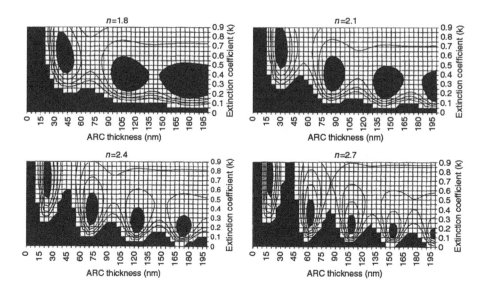

FIGURE 8.47 Contour plots of substrate reflectivity (over silicon) at 248 nm as a function of bottom anti-reflective coating (BARC) thickness for materials with refractive indices between 1.8 and 2.7 and extinction coefficient values to 0.9. Central contour areas correspond to 0–2% reflectivity, and 2% constant contours are shown. Results beyond 10% reflectivity are shown in black ($n_{resist}(248) = 1.75$).

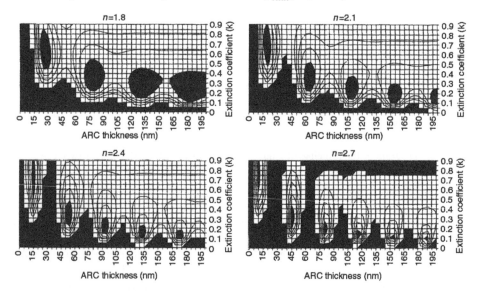

FIGURE 8.48 Contour plots of substrate reflectivity (over silicon) at 193 nm as a function of bottom anti-reflective coating (BARC) thickness for materials with refractive indices between 1.8 and 2.7 and extinction coefficient values to 0.9. Central contour areas correspond to 0–2% reflectivity, and 2% constant contours are shown. Results beyond 10% reflectivity are shown in black. Resist refractive index is 1.6.

a few percent becomes difficult. Figures 8.47 and 8.48 are a series of contour plots of substrate reflectivity for BARC films with extinction coefficient values from 0.0 to 0.9 and refractive index values from 1.8 to 2.7.

8.18.1.1 Organic BARCs

Organic BARC materials have been used for some time, typically in the form of spin-on polymeric materials [118]. These polymers contain highly absorbing dyes introduced at levels to deliver

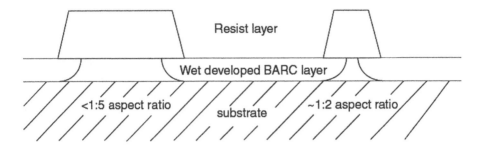

FIGURE 8.49 Process schematic for an aqueous base developed bottom antireflective coating (BARC). Exposure and development of a top resist layer allow development to continue through a suitably baked AR film. Isotropic pattern transfer limits this approach.

appropriate extinction coefficient values. Several classes of materials can be used, depending to a large extent on the pattern transfer requirements of a process [119–122]. Wet-developable organic BARCs based on partially cured polyamic acids (polyimide precursors) have been utilized for large feature geometry. These materials, with refractive index values near 1.7 and extinction coefficients near 0.3, are coated and partially cured prior to resist application. Partial curing of the dyed polyamic acid allows tailoring of the alkaline solubility of the layer to match that of exposed resist. Bottom antireflective coating materials made of dyed triazine derivatives have also been introduced [119].

As shown in Figure 8.49, exposure and development of an aqueous base-soluble resist layer exposes the underlying BARC material, which is also base soluble if cured appropriately. Materials have been formulated that provide a high degree of bake latitude (as high as +20 °C) [123] and exhibit a very low degree of interfacial mixing. The inherent problem with this approach for antireflection is the isotropy of wet pattern transfer. With no preferential direction for etching, undercutting results to the full extent of the BARC thickness. As shown in Figure 8.49, resist features are undercut by twice the BARC thickness, limiting application of wet-developed organic materials to a resolution above 0.5 µm.

Dry-etch-compatible organic BARC materials can allow control of the etch profile through use of plasma RIE methods of pattern transfer. The requirements then become resist to BARC etch selectivity to minimize resist erosion and the accompanying loss in process and CD control. Initial candidate materials for use as dry-etch BARCs may be polymers that undergo efficient scissioning with plasma exposure, leading to increased volatility. For instance, dyed polyolefin sulfone materials (frequently employed as electron beam resists) [124] could allow relatively high oxygen-based RIE etch selectivity to novolac or PHOST resist materials. Several dry-etch materials have been introduced [125, 126] for use in 248 nm and i-line application. Figure 8.50 shows lithographic results and reduction of reflective standing-wave effects with the use of a dry-etch BARC material.

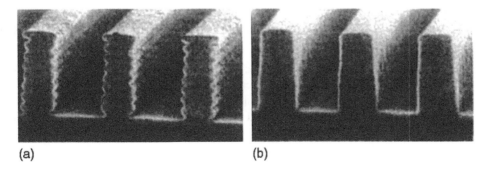

(a)   (b)

FIGURE 8.50 Comparison of reflective standing-wave reduction through use of a dry-etch bottom antireflective coating (BARC): (a) without a bottom ARC; (b) with a bottom ARC.

A potential problem with spin-on organic BARC materials is their planarizing nature. As seen in Figures 8.47 and 8.48, control of film thickness to a few tens of angstroms may be required for suppression of substrate reflectivity. If a polymeric material is spin coated over severe topography, film thickness can deviate substantially from a targeted value. The film is generally not conformal, which leads to significant variation in reflection reduction across a field. To increase the conformal properties of a BARC layer, alternative deposition methods can be explored. Also, through elimination of the polymeric nature of the BARC material, planarization can be further reduced. This leads to a class of inorganic antireflective materials that can be coated using chemical vapor or vacuum deposition methods.

8.18.1.2 Inorganic BARCs

Vapor-deposited ARC materials were first proposed for use over aluminum [127] and have since been applied over a variety of reflective substrate layers. The optical requirements for an inorganic layer are generally the same as for organic films. This is, however, a more difficult task with inorganic dielectric materials than it is with organic polymers, as practical material choices are generally limited to those that allow process compatibility. The challenges for organic and inorganic materials can therefore differ. For inorganic films, the flexibility of optical constants is made possible to some extent through material selection and variation in stoichiometry. Deposition thickness and uniformity can be controlled accurately to the nanometer level, and films generally conform to the underlying topography. The choice between inorganic and organic BARC materials, therefore, depends in part on the underlying substrate and processing that will be encountered. Titanium nitride [128], silicon nitride, silicon oxinitride [129], amorphous carbon [130], tantalum silicide [131], and titanium tungsten oxide [132] films have been used as inorganic antireflection layers at 365, 248, and 193 nm [133]. Substrate–resist interaction effects for 248 nm chemically amplified resists also need to be considered as candidates are evaluated, which may reduce the attractiveness of some materials for some ARC applications [25].

For nonstoichiometric materials such as silicon oxinitride, modifications in stoichiometry can be used to tailor optical properties. Traditionally, a chemical compound is thought of as having a fixed atomic ratio and composition. A wider range of properties is possible by relaxing this stoichiometric requirement. By controlling the ratios of material components during deposition, optical behavior can be modified. It is not immediately obvious that nonstoichiometric composite films of metal, insulator, or semiconductor combinations will exhibit predictable optical properties. Through analysis of the atomistic structure of materials, and by relating optical material properties to electrical properties, some conclusions can be drawn [134]. Optical constants can be related to electrical properties by neglecting material structure and considering macroscopic material quantities only:

$$n^2 = \frac{1}{2}\left(\sqrt{\varepsilon_1^2 + \varepsilon_2^2} + \varepsilon_1\right), \tag{8.27}$$

$$k^2 = \frac{1}{2}\left(\sqrt{\varepsilon_1^2 + \varepsilon_2^2} - \varepsilon_1\right) \tag{8.28}$$

where ε_1 and ε_2 are real and imaginary dielectric constants, respectively. To account for material structure, Drude analysis of optical and electrical constants describes free electron or metallic behavior quite well in the visible and IR region [135]. Equations 8.29 and 8.30 are Drude equations for optical and electrical constants, related to material plasma frequency (v_1) and damping frequency (v_2):

$$n^2 - k^2 = \varepsilon_1 = 1 - \frac{v_1^2}{v^2 + v_2^2} \tag{8.29}$$

$$2nk = \varepsilon_2 = \left(\frac{v_2}{v_1}\right)\frac{v_1^2}{v^2 + v_2^2} \tag{8.30}$$

To account for optical properties at shorter wavelengths, bound-electron theory needs to be utilized. For dielectric materials, no intraband transitions exist because of filled valence bands. Interband transitions are also limited in IR and visible regions because of large band gap energies. Bound-electron theory alone is sufficient to describe classical dielectric behavior. Characterization of metallic and noninsulating materials in UV and visible regions requires use of both free-electron and bound-electron theory. By assuming a given number of free electrons and a given number of harmonic oscillators, optical properties over a wide wavelength range can be described. Using bound-electron or harmonic-oscillator theory, relationships for optical and electrical constants can be determined from the following equations:

$$\varepsilon_1 = 1 + \frac{4\pi e^2 m N_a \left(v_0^2 - v^2\right)}{4\pi^2 m^2 \left(v_0^2 - v^2\right)^2 + \gamma' v^2} \tag{8.31}$$

$$\varepsilon_2 = \frac{2e^2 N_a \gamma' v}{4\pi^2 m^2 \left(v_0^2 - v^2\right)^2 + \gamma' v^2} \tag{8.32}$$

where
γ is the damping factor
N_a is the number of oscillators
m is electron mass
e is electron charge

From this analysis, it can be shown that the optical properties of a material can be described by metallic behavior combined with dielectric behavior. Figure 8.51 shows plots of optical constants for a metallic film using the Drude model for free-electron motion and metallic–dielectric composite films using combined free- and bound-electron models. These results suggest that the optical properties of a composite material can be modified by controlling the ratio of its components. It is expected, therefore, that the optical properties of nonstoichiometric materials would fall somewhere between those of their stoichiometric elemental or compound constituents.

Silicon nitride (Si_3N_4) possesses optical constants that may be a good starting point for use as an ARC at several wavelengths. Optical constant data for silicon nitride, silicon dioxide (SiO_2), and

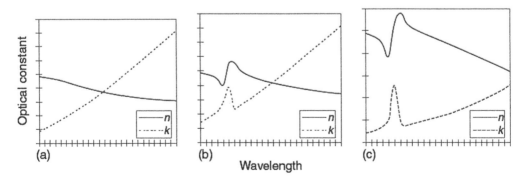

FIGURE 8.51 Plots of optical constants for a metallic film using the Drude model for free electron motion and metallic–dielectric composite films using combined free- and bound-electron models. (a) A metallic film; (b) a dielectric film; and (c) a dielectric composite film with increasing metallic content.

silicon at 436, 365, 248, and 193 nm are contained in Table 8.3 [136]. By adjusting the deposition parameters during film formation (gas flow ratios for chemical vapor deposition [CVD] and power, pressure, and gas flow for sputtering, for instance), thin-film materials can be produced with optical properties defined by these constituents. Figure 8.52 shows the reflectivity at the substrate interface for several 248 nm SiON ARC materials under a resist with a refractive index of 1.76. Shown also in Figure 8.53 are the optical constants (n and k) for a variety of materials at 193 nm [137]. From

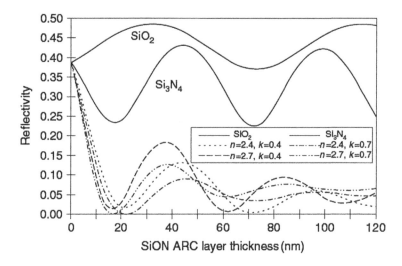

FIGURE 8.52 Reflectivity at the resist–substrate interface for several 248 nm understoichiometric SiON antireflective coating (ARC) materials under a resist with a refractive index of 1.76. The underlying substrate is polysilicon. Reflectivity for stoichiometric SiO_2 and Si_3N_4 is also shown.

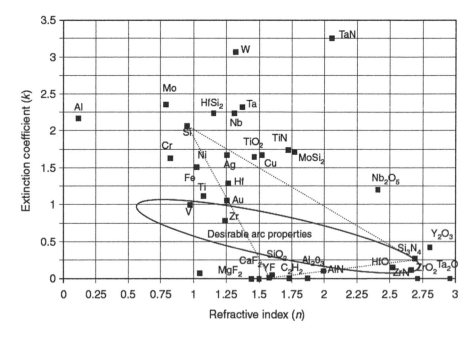

FIGURE 8.53 Plot of the optical constants (n and k) for various materials at 193 nm. A window of desirable antireflective coating (ARC) properties is shown, as is also shown in Figure 8.48. An understoichiometric SiON film would also suffice for use at 193 nm, with properties falling within the area defined by dashed lines.

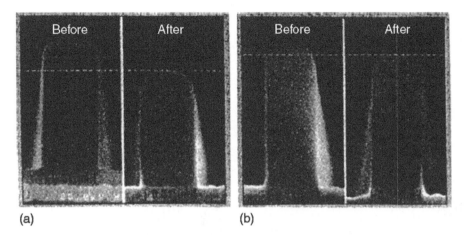

FIGURE 8.54 Comparison of resist loss during pattern transfer to bottom ARC layers. (a) Organic spin-on ARC; (b) dielectric inorganic ARC. (From Bencher, C., Ngai, C., Roman, B., Lian, S., and Vuong, T., *Solid State Technol.*, 111, 1997.)

these data, together with compatibility and process requirements, potential ARC films for 193 nm can be identified.

An additional advantage from the use of inorganic ARCs is the ability to grade the indices of materials to best match requirements of resist and substrate layers [138]. This is possible through control of process parameters during deposition. To achieve similar results with organic spin-on ARCs, multiple layers would be required. Pattern transfer for inorganic antireflective layers can also result in higher selectivity to resist, made possible, for instance, if fluorine-based etch chemistries are used. Shown in Figure 8.54 is a comparison of pattern transfer processes through an organic spin-on ARC and an inorganic ARC with high resist selectivity [139]. Minimum resist erosion during the etch process with the inorganic material can result in an increase in CD control. A major compromise when using inorganic materials is the increased complexity of deposition processes over spin coating. Process trends and requirements will probably lead to incorporation of both approaches for various lithographic operations.

8.18.2 MULTILAYER BARCs FOR IMMERSION LITHOGRAPHY AND POLARIZATION

As described in earlier chapters, polarization issues become increasingly important for high NA immersion lithography. The implications for immersion lithography are not limited to the imaging system optics and will extend to the resist and substrate stack at the image plane [140]. As radiation propagates through an imaging system, it is influenced by reflection, refraction, and interference. As the polarization state of radiation is considered, it is important to consider the components within the system that can be influenced. To reduce the reflectivity at an interface between a resist layer and a substrate, a bottom anti-reflective coating is coated beneath the resist, so that the light incident and the reflected light interfere destructively. As described earlier, this destructive interference thickness repeats at quarter-wave thickness. Figure 8.55 is a series of plots of substrate reflectivity for a 193 nm resist as a single-layer BARC's optical properties and thickness are varied. These plots are for normal incidence, which is not necessarily a good predictor for reflectivity effects at high NA. Figure 8.56 shows plots for unpolarized 45° incidence in the resist, which would be expected for a numerical aperture of 1.2 and a resist index of 1.70. As seen, the location and extent of the nodes are significantly impacted by the incident angle at the image plane.

Furthermore, the two linear states of polarization (transverse electric [TE] and transverse magnetic [TM]) behave differently. Figure 8.57 shows plots for TE polarization, and Figure 8.58 shows

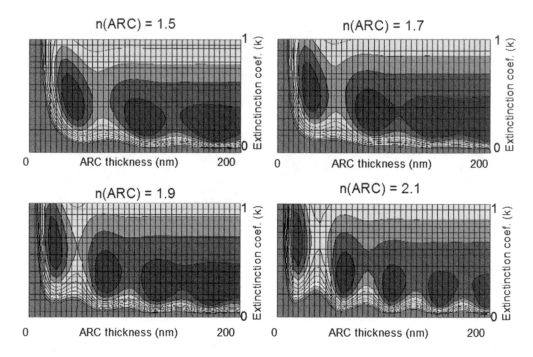

FIGURE 8.55   Reflectivity contour plots for 193 nm BARC films under resist at normal incidence.

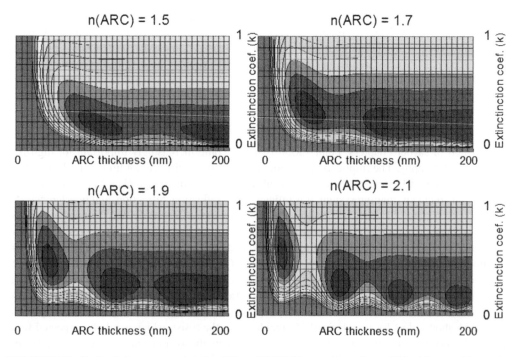

FIGURE 8.56   Reflectivity contour plots for 193 nm BARC films under resist at 45° incidence with unpolarized illumination.

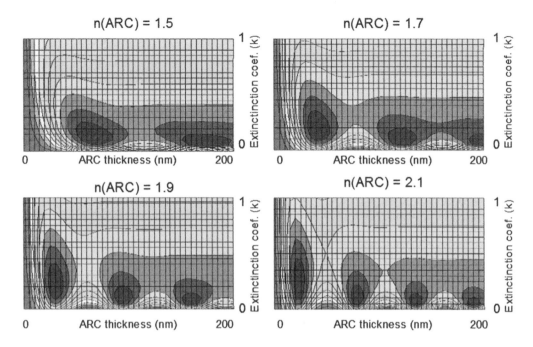

FIGURE 8.57   Reflectivity contour plots for 193 nm BARC films under resist at 45° incidence with TE polarized illumination.

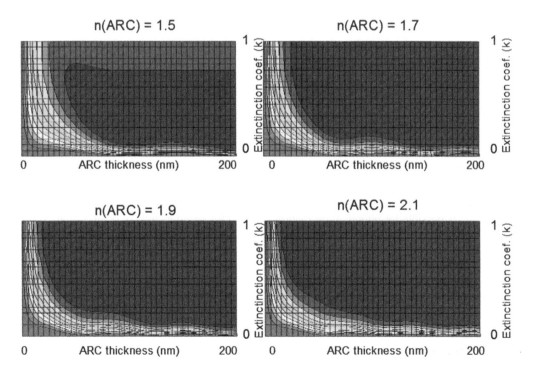

FIGURE 8.58   Reflectivity contour plots for 193 nm BARC films under resist at 45° incidence with TM polarized illumination.

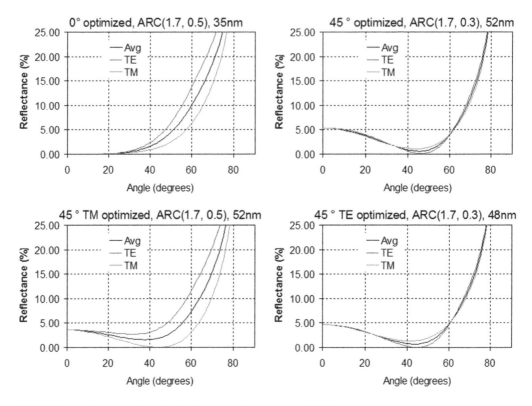

FIGURE 8.59   Optimization of reflectivity for single-layer BARC films.

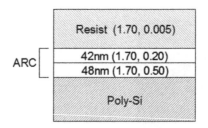

FIGURE 8.60   Optical stack configuration for a dual-layer BARC.

TM polarization (where the unpolarized case is the sum of TE and TM). Optimization of a single-layer BARC is possible for oblique illumination and also for specific cases of polarization, as seen in the plots of Figure 8.59. The issue with a single-layer AR film, however, is its inability to achieve low reflectivity across all angles and through both states of linear polarization. This can be achieved using a multilayer BARC design, as shown in Figure 8.60.

By combining two films in a stack and optimizing their optical and thickness properties, reflectivity below 0.6% can be made possible for angles to 45° (1.2 NA) for all polarization states, as shown in Figure 8.61.

8.19   TOP ANTIREFLECTIVE APPROACHES

The bottom antireflective approach leads to reflection reduction at the interface between a resist material and the substrate. Reflection also occurs at the top of the resist, at the resist–air interface, as shown in Figure 8.62. This leads to a situation similar to that of a Fabry–Perot etalon. An expression

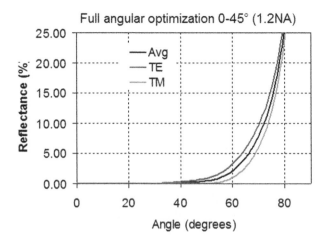

FIGURE 8.61 Optimization of reflectivity for a dual-layer BARC film.

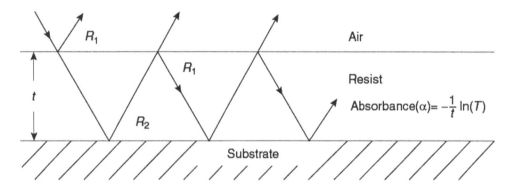

FIGURE 8.62 Diagram of reflection contribution from resist–substrate and resist–air interfaces.

for the reflective swing ratio can be utilized to address reflection effects within an entire resist film stack [141]:

$$\text{Swing} = 4\sqrt{R_1 R_2} \exp\left(-\alpha t\right) \tag{8.33}$$

where

R_1 is the reflectivity at the resist–air interface
R_2 is the reflection at the resist–substrate interface
α is resist absorbance
t is resist thickness

Here, the swing is the ratio of the peak-to-valley change in intensity to the average intensity, which is desired to be a minimum. A decrease in R_2 via a BARC layer can be used to accomplish this, as can an increase in absorption (through the use of a resist dye, for instance) or an increase in resist thickness. A reduction in R_1 is also desirable and can be addressed through the use of a top antireflective coating (TARC) [142]. Because there is a mismatch between refractive indices of the resist material and air, R_1 can be on the order of 7%. This can lead to resist exposure and CD control problems encountered as a result of internal reflectance via multiple interference effects, scattered light, and reflective standing wave. An exposure-tool alignment signal detection can also be degraded by top-surface reflection effects [143]. Like conventional

AR coatings for optical applications, TARC films are not absorbing materials but instead transparent thin-film interference layers that utilize destructive interference to eliminate reflectance. The ideal refractive index for such a film coated over a resist material is that which produces equivalent reflectance from the air and from the resist side of the interface. This leads to an optimum index of

$$n_{AR}^* = \sqrt{n_{air}^* \times n_{resist^-}^*}$$ (8.34)

If a film of this index is coated to a quarter-wave thickness ($\lambda/4n_{ARc}$), complete destructive interference can occur. For i-line resist materials with a refractive index of 1.70, an ideal TARC material would have an index of 1.30 and would be coated at 700 Å. Figure 8.63 shows the reduction in reflection for a quarter-wave top antireflective layer when the refractive index is varied from 1.1 to 1.5. For refractive index values below 1.3, there is a larger contribution to reflection from the resist–TARC interface. At values above 1.3, the contribution from the air–TARC interface is larger.

The refractive index and reflectance properties of several TARC materials are given in Table 8.4. As the refractive index of a material approaches the ideal value of $n_{air}^* \times n_{resist^-}^*$, the

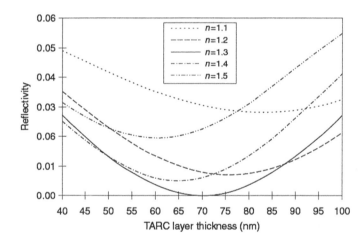

FIGURE 8.63 Reduction in surface reflection through use of a top antireflective coating (TARC) material at 365 nm. Reflection can be eliminated through use of a quarter-wave thickness (700 Å) of a material with a refractive index of 1.3.

TABLE 8.4

Refractive Index and Reflectance Properties of Several Top Antireflective Coating (TARC) Candidates

	Refractive Index, n (365 nm)	Departure from Ideal	Quarter-Wave Thickness at 365 nm (Å)	Reflectance at Air Interface (%)	Reflectance at Resist Interface (%)
Polyvinyl alcohol	1.52	0.32	600	4.26	0.3
Polyethylvinylether	1.46	0.16	625	3.5	0.6
Polyfluoroalkylpolyether	1.27	0.02	713	1.5	2.0
Water-based TAR [145]	1.41	0.11	647	2.9	0.9
Ideal[a]	1.30	—	702	1.8	1.8

[a] Assuming a resist refractive index of 1.70.

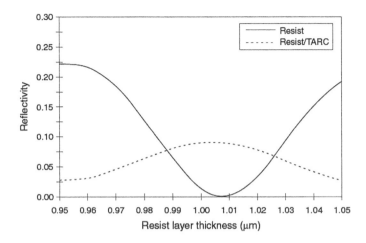

FIGURE 8.64 Reflectivity at the top surface of a resist material as a function of resist thickness with and without a top antireflective coating (TARC). Results are for 365 nm and a polysilicon substrate. The refractive index of the TARC is 1.41, and its thickness is 647 Å.

reflectances at the resist and air interfaces are equivalent, allowing destructive interference at a quarter-wave thickness. Residual reflectances result as TARC indices deviate from the ideal. To achieve refractive index values near the ideal 1.3, polyfluoroalkyl-polyethers and polytetrafluoroethylene-based materials cast in solvents that do not dissolve novolac resist materials have been utilized at TARC layers [141, 144]. These fluorinated polymers require removal with chlorofluorocarbons prior to development and have been replaced by water-based and water-soluble materials to improve process compatibility [145]. Although these materials do not possess a refractive index as close to the ideal values for use with DNQ/novolac resists at 436 and 365 nm (the refractive index is 1.41 at 365 nm), the reduced process complexity makes them a more attractive choice over solvent-based systems. Figure 8.64 shows the reflectivity of a resist film stack as resist film thickness is varied. Reflectivity varies by over 20% for resist over polysilicon over one swing period. This is reduced significantly with the addition of a water-soluble TAR coated to a quarter-wave thickness. Exposure and focus-latitude improvement has also been demonstrated with these materials.

8.20 THE IMPACT OF NUMERICAL APERTURE ON REFLECTANCE EFFECTS

In Equation 8.23 and Equation 8.33, the assumption of normal incidence is made. In projection imaging, the incident angle of illumination is a function of the NA of the optical system. Reflectance at an interface is a function of the angle of incidence (in air) I_1 and the angle of refraction I_2 as

$$R = \frac{1}{2}\left[\frac{\sin^2\left(I_1 - I_2\right)}{\sin^2\left(I_1 + I_2\right)} + \frac{\tan^2\left(I_1 - I_2\right)}{\tan^2\left(I_1 + I_2\right)}\right] \tag{8.35}$$

The first term in this equation corresponds to the reflection that is polarized in the plane of incidence, and the second term corresponds to the reflection in the perpendicular plane. The effective film thicknesses for resist, AR layers, and underlying dielectric films are also scaled by the angle of incidence as $(t \cos \theta)$, where t is the thickness of the film at normal incidence. Large angles of incidence (with high-NA optics) have been shown to contribute to reducing the reflective swing ratio [146].

8.21   PHOTORESIST HARD MASKS AND MATERIALS

Over the past several generations, photoresist films have been forced to become thinner to accommodate smaller geometry. Resist films below about 100 nm have had to make use of an underlying intermediate film to act as an etch transfer layer. As substrate topography has been substantially reduced though CMP, the need for additional planarization layers (as seen in Figure 8.42 for instance) has been for the most part eliminated. This has resulted in the predominant use of multilayer photoresist schemes employing a polymeric photoresist film over a thin etch stop layer, commonly also referred to as a *hard mask*. The "resistance" functionality of sub-100 nm photoresist films has thus been relegated mostly to an underlying hard mask. The materials used for such a layer need to allow efficient etch transfer both from the photoresist above and to the underlying films below. By only needing to survive pattern transfer to the hard mask, the etch resistance requirements of photoresists have been substantially reduced. The imaging of small features on the nanometer scale can then be carried out in photoresists coated to thicknesses corresponding to <3:1 aspect ratios, with the necessary etch resistance only to withstand erosion when used to pattern the thin hard-mask layers below. Several alternatives have been developed for use in lithography generations starting at about the sub-100 nm technology node.

8.21.1   Silicon-Based Hard Masks

By combining the etch stopping of a hard mask with the optical interference of a BARC antireflection layer, a single-layer material can serve two functions. The use of BARC coatings with "built-in" hard-mask properties can arise from the design of inorganic SiON films (as described in Section 8.18.1.2) or though potentially more challenging organic materials (as described in Section 8.18.1.1). Early work to optimize such layers included plasma-enhanced chemical vapor deposition (PECVD)-deposited titanium nitride and silicon oxinitride films with optical properties tuned for either 248 or 193 nm [147, 148]. Index graded ARCs have also been introduced to reduce the reflectance at the hard mask–resist interface [148].

8.21.2   Carbon-Based Hard Masks

Predominately three types of carbon-based materials and processes have been employed as hard masks: amorphous carbon layers (ACL), diamond-line carbon (DLC), and spin-on carbon (SOC) films. Each has been used and patterned with thin photoresist layers as well as in combination with other layers to enhance reflection suppression, pattern transfer, and selectivity.

The use of amorphous carbon to enhance resist stack etch selectivity was proposed several generations ago [149]. PECVD deposition of an ACL in a bilayer process can lead to etch rate selectivity better than 2× of those for photoresist materials. Early on, patterning of ACLs carried out by oxygen RIE had been shown to be capable of resolution to 40 nm. Being carbon based, these layers are also easily trimmed and stripped after patterning. Etch selectivity to silicon-based layers with fluorine chemistry (such as CF_4) has since been shown to be greater than 6:1 [150].

DLC films have been used for various IC and non-IC applications. Films are normally formed via CVD with gas mixtures of hydrocarbons (such as butadiene, benzene, or methane), hydrogen, and argon [151]. Etching is generally accomplished using CF_4 with O_2. In some instances, the optical properties of DLC films can be tailored better for ARC properties than ACLs, but the etch selectivity may be poorer, with etch characteristics similar to SiO_2 [152].

The use of SOCs for hard masks can in some instances offer advantages over other materials in terms of cost, defectivity, planarization, and process integration [153]. SOC materials are essentially organic polymer solutions, spin coated and cured to form carbon-based films with properties similar to deposited layers. As with any organic layer introduced into the substrate film stack, the temperature requirements for curing (which can be in the 300–500 °C range) need to be accounted for with regard to the compatibility with what lies below. Intermixing of organic layers should also be taken into account so as to not form interfacial layers that can pose problems during pattern transfer and removal, leading to susceptibility to defects.

REFERENCES

1. W. Moreau. 1988. *Semiconductor Lithography*, New York: Plenum, p. 49.
2. L. Katzisyma. 1966. *Russ. Chem. Rev.*, **35**: 388.
3. S. Asaumi, M. Futura, and A. Yokota. 1992. *Proc. SPIE*, **1672**: 616.
4. W.D. Hinsberg et al. 1993. *Proc. SPIE*, **1925**: 43.
5. W. Huang et al. 1994. *Proc. SPIE*, **2195**: 37.
6. O. Nalamasu et al. 1995. *ACS Symp. Ser.*, **614**: 4.
7. H. Roschert et al. 1992. *Proc. SPIE*, 1672: 33.
8. D.J.H. Funhoff, H. Binder, and R. Schwalm. 1992. *Proc. SPIE*, **1672**: 46.
9. O. Nalamasu et al. 1992. *J. Vac. Sci. Technol. B*, **10**(6): 2536.
10. H. Ito et al. 1994. *J. Photopoly. Sci. Technol.*, **7**: 433.
11. W. Conley et al. 1996. *Proc. SPIE*, **2724**: 34.
12. S.A. McDonald. 1991. *Proc. SPIE*, **1466**: 2.
13. J.H. van Zanten, W.E. Wallace, and W.L. Wu. 1996. *Phys. Rev. E*, **53**: R2053.
14. D. Lyons and B.T. Beauchemin. 1995. *Proc. SPIE*, **2438**: 726.
15. D.E. Bornside, C.W. Macosko, and L.E. Scriven. 1987. *J. Imag. Tech.*, **13**: 123.
16. D. Meyerhofer. 1978. *J. Appl. Phys.*, **49**(7): 3993.
17. H. Fujita. 1961. *Adv. Polym. Sci.*, **3**: 1.
18. J.S. Vrentas. 1977. *J. Polym. Sci. Polym. Phys. Ed.*, **15**: 403.
19. L. Pain, C. Le Cornec, C. Rosilio, and P.J. Paniez. 1996. *Proc. SPIE*, **2724**: 100.
20. V. Rao, W.D. Hinsberg, C.W. Frank, and R.F.W. Pease. 1994. *Proc. SPIE*, **2195**: 596.
21. C.W. Frank, V. Rao, M.M. Despotopoulou, R.F.W. Pease, W.D. Hinsberg, R.D. Miller, and J.F. Rabolt. 1996. *Science*, **273**: 912.
22. L. Schlegel, T. Ueno, N. Hayashi, and T. Iwayanagi. 1991. *J. Vac. Sci. Technol. B*, **9**: 278.
23. B. Beauchemin, C.E. Ebersol, and I. Daraktchiev. 1994. *Proc. SPIE*, **2195**: 613.
24. T.E. Salamy et al. 1990. *Proc. Electrochem. Soc.*, **90**(1): 36.
25. K.R. Dean, R.A. Carpio, and G.K. Rich. 1995. *Proc. SPIE*, **2438**: 514.
26. J. Chun, C. Bok, and K. Baik. 1996. *Proc. SPIE*, **2724**: 92.
27. E. Fields, C. Clarisse, and P.J. Paniez. 1996. *Proc. SPIE*, **2724**: 460.
28. P.J. Paniez, C. Rosilio, B. Monanda, and F. Vinet. 1994. *Proc. SPIE*, **2195**: 14.
29. R. Sinta, G. Barclay, T. Adams, and P. Medeiros. 1996. *Proc. SPIE*, **2724**: 238.
30. P.J. Paniez, G. Festes, and J.P. Cholett. 1992. *Proc. SPIE*, **1672**: 623.
31. M. Koshiba, M. Murata, M. Matsui, and Y. Harita. 1988. *Proc. SPIE*, **920**: 364.
32. H. Ito et al. 1995. *Proc. SPIE*, **2438**: 53.
33. F.H. Dill et al. 1975. *IEEE Trans. Electr. Dev.*, **ED-22**: 440.
34. R. Dammel. 1993. *Diazonapthoquinone-Based Resists*, Bellingham, WA: SPIE Press, p. 19.
35. J. Sturdevant, W. Conley, and S. Webber. 1996. *Proc. SPIE*, **2724**: 273.
36. J.M. Shaw and M. Hatzakis. 1978. *IEEE Trans. Electr. Dev.*, **ED-25**: 425.
37. T. Brunner. 1991. *Proc. SPIE*, **1463**: 297.
38. J. Sturdevant, S. Holmes, and P. Rabidoux. 1992. *Proc. SPIE*, **1672**: 114.
39. M.A. Toukhy and S.G. Hansen. 1994. *Proc. SPIE*, **2195**: 64.
40. T. Tanabe, Y. Kobayashi, and A. Tsuji. 1996. *Proc. SPIE*, **2724**: 61.
41. K. Überreiter and F. Asmussen. 1962. *J. Polym. Sci.*, **57**: 187.
42. A.E. Novembre and M.A. Hartney. 1985. *Proc. SPE Photopolym. Conf.*, New York: Ellenvile.
43. J.P. Huang, T.K. Kwei, and A. Reiser. 1989. *Proc. SPIE*, **1086**: 74.
44. A. Reiser. 1989. *Photoreactive Polymers*, New York: Wiley Interscience, pp. 211–223.
45. H. Shin, T. Yeu, and A. Reiser. 1994. *Proc. SPIE*, **2195**: 514.
46. K. Honda, A. Blakeney, and R. Hurditch. 1993. *Proc. SPIE*, **1925**: 197.
47. T. Itani, K. Itoh, and K. Kasama. 1993. *Proc. SPIE*, **1925**: 388.
48. K.L. Konnerth and F.H. Dil. 1975. *IEEE Trans. Electr. Dev.*, **ED-22**: 453.
49. S.D. Chowdhury, D. Alexander, M. Goldman, A. Kukas, N. Farrar, C. Takemoto, and B.W. Smith. 1995. *Proc. SPIE*, **2438**: 659.

50. For instance, the Perkin Elmer Development Rate Monitor (DRM).

51. R.A. Arcus. 1986. *Proc. SPIE*, **631:** 124.

52. R.R. Dammel et al. 1994. *Proc. SPIE*, **2195:** 542.

53. S. Shimomura, H. Shimada, R. Au, M. Miyawaki, and T. Ohmi. 1993. *Proc. SPIE*, **1925:** 602.

54. G. Flores and J. Loftus. 1992. *Proc. SPIE*, 1672: 328.

55. C.L. Henderson et al. 1996. *Proc. SPIE*, **2724:** 481.

56. C.M. Garza, C.R. Szmanda, and R.L. Fischer. 1988. *Proc. SPIE*, **920:** 321.

57. V. Marriott. 1983. *Proc. SPIE*, **394:** 144.

58. W. Moreau. 1988. *Semiconductor Lithography*, New York: Plenum Press, Ch. 10.

59. W.M. Moreau, A.D. Wilson, K.G. Chiong, K. Petrillo, and F. Hohn. 1988. *J. Vac Sci. Technol. B*, **6:** 2238.

60. E. Fadda, G.M. Amblard, A.P. Weill, and A. Prola. 1994. *Proc. SPIE*, **2195:** 576.

61. T. Yoshimura, E. Murai, H. Shiraishi, and S. Okazaki. 1990. *J. Vac. Sci. Technol. B*, **6:** 2249.

62. N. Damarakone, P. Jaenen, L. Van den Hove, and R. Hurditch. 1990. *Proc. SPIE*, **1262:** 219.

63. J.W. Thackeray et al. 1989. *Proc. SPIE*, **1086:** 34.

64. O. Nalamasu et al. 1990. *Proc. SPIE*, **1262:** 32.

65. T. Itani, H. Iwasaki, M. Fujimoto, and K. Kasama. 1994. *Proc. SPIE*, **2195:** 126.

66. T. Itani, H. Iwasaki, H. Yoshin, M. Fujimoto, and K. Kasama. 1995. *Proc. SPIE*, **2438:** 91.

67. S.G. Hansen, R.J. Hurdich, and D.J. Brzozowy. 1993. *Proc. SPIE*, **1925:** 626.

68. S.G. Hansen and R.H. Wang. 1993. *J. Electrochem. Soc.*, **140:** 166.

69. M.A. Toukhy, T.R. Sarubbi, and D.J. Brzozowy. 1991. *Proc. SPIE*, **1466:** 497.

70. R. Allen et al. 1993. *Solid State Technol.*, **36:** 11, 53–62.

71. S. Takechi et al. 1992. *J. Photopol. Sci. Technol.*, **5:** 439.

72. K. Ronse. 2001. *Future Fab Int.*, **10.**

73. F.A. Houle, W.D. Hinsberg, M.I. Sanchez, and J.A. Hoffnagle. 2002. *J. Vac. Sci. Technol. B*, **20:** 924.

74. J.A. Hoffnagle, W.D. Hinsberg, M.I. Sanchez, and F.A. Houle. 2002. *Opt. Lett.*, **27:** 1776.

75. T.A. Brunner. 2003. *J. Vac. Sci. Technol. B*, **21:** 2632.

76. T.A. Brunner and R.A. Fergusen. *Proc. SPIE*, **2726:** 198.

77. G.M. Walraff, D.R. Medeiros, M. Sanchez, K. Petrillo, and W.S. Huang. 2004. *J. Vac. Sci. Technol. B*, **22:** 3479.

78. O. Nalamasu and A.H. Gabor. 2000. *Future Fab Int.*, **8:** 159–163.

79. R.R. Kunz and D.K. Downs. 1999. *J. Vac. Sci. Technol. B*, **17:** 3330.

80. S. Palmateer et al. 1998. *Proc. SPIE*, **3333:** 634–642.

81. S. Masuda, X. Ma, G. Noya, and G. Pawlowski. 2000. *Proc. SPIE*, **3999:** 253.

82. Q. Lin, D. Goldfarb, M. Angelopoulos, S. Sriam, and J. Moore. 2001. *Proc. SPIE*, **4345:** 78.

83. T. Yamaguchi, H. Namatsu, M. Nagese, K. Yamazaki, and K. Kurihara. 1997. *J. Photopoly. Sci. Technol.*, **10:** 635.

84. U. Okoroanyanwu and J.H. Lammers. 2004. *Future Fab Int.*, **17.**

85. A.M. Goethals et al. 2001. *J. Photopoly. Sci. Technol.*, 3: 333.

86. W.D. Domke et al. 1999. *Proc. SPIE*, **3999:** 139.

87. S. Mori, T. Morisawa, N. Matsuzawa, Y. Kaimoto, M. Endo, T. Matsuo, M. Sasago. 1998. *J. Vac. Sci. Technol. B*, **16:** 3744.

88. L. Que and Y.B. Gianchandania. 2000. *J. Vac. Sci. Technol. B*, **18:** 3450.

89. G. Czech, E. Richter, and O. Wunnicke. 2002. *Future Fab Int.*, **12.**

90. O. Wunnickea, A. Henniga, K. Grundkeb, M. Stammb, and G. Czecha. 2002. *Proc. SPIE*, **4690:** 332.

91. J. Simons, D. Goldfarb, M. Angelopoulos, S. Messick, W. Moreau, C. Robinson, J. De Pablo, and P. Nealey. 2000. *Proc. SPIE*, **4345:** 19–29.

92. C.L. McAdams et al. 2000. *Proc. SPIE*, **4345:** 327–334.

93. S. Hien, G. Rich, G. Molina, H. Cao, and P. Nealy. 2002. *Proc. SPIE*, **4690:** 254.

94. W. Hinsberg et al. 2004. *Proc. SPIE*, **5376:** 21–33.

95. M. Yoshida, K. Endo, K. Ishizuka, and M. Sato. 2004. *J. Photopoly. Sci. Technol.*, **17:** 603.

96. R.R. Dammel, F.M. Houlihan, R. Sakamuri, D. Rentkiewicz, and A. Romano. 2004. *J. Photopoly. Sci. Technol.*, **17(4):** 587–602.

97. R. Dammel, G. Pawlowski, A. Romano, F. Houlihan, W.K. Kim, and R. Sakamuri. 2005. *Proc. SPIE*, **5753:** 95.

98. W. Hinsberg, F. Houle, H. Ito, K. Kanzawa, S-W. Lee. 2003. *Proceedings of the 13th International Conference on Photopolymers*, Society of Plastics Engineers.

99. W. Hinsberg et al. 2005. *Proc. SPIE*, **5376**: 21.
100. B.W. Smith, A. Bourov, Y. Fan, L. Zavyalova, N. Lafferty, and F. Cropanese. 2004. *Proc. SPIE*, **5377**: 273–282.
101. A. Raub and S. Brueck. 2003. *Proc. SPIE*, **5040**: 667.
102. Yayi Wei, *Advanced Processes for 193-nm Immersion Lithography*, Bellingham, WA: SPIE, 2009.
103. B. Shokouhi, J. Zhang, and B. Cui. 2011. "Very high sensitivity ZEP resist using MEK:MIBK developer." *Micro Nano Lett.*, **6(12)**: 992–994.
104. L.E. Stillwagon and G.N. Taylor. 1989. "Polymers," in *Microlithography: Materials and Processes, ACS Symposium Series 412*, Washington, DC: American Chemical Society, pp. 252–265.
105. L.E. Stillwagon and R.G. Larsen. 1988. *J. Appl. Phys. Lett.*, **63**: 5251.
106. J.J. Steigewald, S.P. Murarka, and R.J. Gutmann. 1997. *Chemical Mechanical Planarization of Microelectronic Materials*, New York: Wiley.
107. J.A. Bruce, B.J. Lin, D.L. Sundling, and T.N. Lee. 1987. *IEEE Trans. Electr. Dev.*, **34**: 2428.
108. B.J. Lin, E. Bassous, W. Chao, and K.E. Petrillo. 1981. *J. Vac. Sci. Technol. B*, **19**: 1313.
109. B.J. Lin. 1980. *J. Electrochem. Soc.*, **127**: 202.
110. E.D. Lin, M.M. O'Toole, and M.S. Change. 1981. *IEEE Trans. Electr. Dev.*, **28**: 1405.
111. C.H. Ting and K.L. Liauw. 1983. *J. Vac. Sci. Technol. B*, **1**: 1225.
112. M.P. de Grandpre, D.A. Vidusek, and M.W. Leganza. 1985. *Proc. SPIE*, **539**: 103.
113. Y. Todokoro. 1980. *IEEE Trans. Electr. Dev.*, **27**: 1443.
114. P.A. Lamarre. 1992. *IEEE Trans. Electr. Dev.*, **39**: 1844.
115. E.D. Liu. 1982. *Solid State Technol.*, **26**: 66.
116. C.H. Ting. 1983. *Proc. Kodak Interf.*, **83**: 40.
117. J.R. Havas. 1976. *Electrochem. Soc. Ext. Abstr.*, **2**: 743.
118. T. Brewer, R. Carlson, and J. Arnold. 1981. *J. Appl. Photogr. Eng.* **7**: 184.
119. R.D. Coyne and T. Brewer. 1983. *Proc. Kodak Interf.*, **83**: 40.
120. Y. Mimura and S. Aoyama. 1993. *Microelectron. Eng.*, **21**: 47.
121. W. Ishii, K. Hashimoto, N. Itoh, H. Yamazaki, A. Yokuta, and H. Nakene. 1986. *Proc. SPIE*, **631**: 295–301.
122. C. Nolscher, L. Mader, and M. Scheegans. 1989. *Proc. SPIE*, **1086**: 242.
123. Such as Brewer Science ARC-XLN, Brewer Science Inc.
124. A. Reiser. 1989. *Photoreactive Polymers*, New York: Wiley, p. 323.
125. T.S. Yang, T. Koot, J. Taylor, W. Josphson, M. Spak, and R. Dammel. 1996. *Proc. SPIE*, **2724**: 724.
126. E. Pavelchek, J. Meudor, and D. Guerrero. 1996. *Proc. SPIE*, **2724**: 692.
127. H. Van Den Berg and J. Van Staden. 1979. *J. Appl. Phys.*, **50**: 1212.
128. B. Martin and D. Gourley. 1993. *Microelectron. Eng.*, **21**: 61.
129. T. Ogawa, H. Nakano, T. Gocho, and T. Tsumori. 1994. *Proc. SPIE*, **2197**: 722–732.
130. Y. Tani, H. Mato, Y. Okuda, Y. Todokoro, T. Tatsuta, M. Sanai, and O. Tsuji. 1993. *Jpn. J. Appl. Phys.*, **322**: 5909.
131. B.L. Draper, A.R. Mahoney, and G.A. Bailey. 1987. *J. Appl. Phys.*, **62**: 4450.
132. H. Tompkins, J. Sellars, and C. Tracy. 1993. *J. Appl. Phys.*, **7**: 3932.
133. B. Smith, Z. Alam, and S. Butt. 1997. *Abstracts of the Second International Symposium on 193 nm Lithography*, Colorado Springs.
134. B.W. Smith, S. Butt, Z. Alam, S. Kurinec, and R. Lane. 1996. *J. Vac. Sci. Technol. B*, **14**: 3719.
135. R.E. Hummel. 1993. *Electronic Properties of Materials*, New York: Springer, pp. 186–230.
136. E.C. Palik. 1985. *Handbook of Optical Constants of Solids I*, New York: Academic Press.
137. B.W. Smith, S. Butt, and Z. Alam. 1996. *J. Vac. Sci. Technol. B*, **14(6)**: 3714.
138. R.A. Cirelli, G.R. Weber, A. Komblit, R.M. Baker, F.P. Klemens, and J. DeMacro. 1996. *J. Vac. Sci. Technol. B*, **14**: 4229.
139. C. Bencher, C. Ngai, B. Roma, S. Lian, and T. Vuong. 1997. *Solid State Technol.*, **20**: 109.
140. B.W. Smith, L.V. Zavyalova, and A. Estroff. 2004. *Proc. SPIE*, **5377**, Optical Microlithography XVII.
141. T.A. Brunner. 1991. *Proc. SPIE*, **1466**: 297.
142. T. Tanaka, N. Hasegawa, H. Shiraishi, and S. Okazaki. 1990. *J. Electrochem. Soc.*, **137**: 3900.
143. N. Bobroff and A. Rosenbluth. 1988. *J. Vac. Sci. Technol. B*, **6**: 403–408.
144. H. Shiraishi and S. Okazaki. 1991. *Proceedings of SPE Regional Technical Conference on Photopolymers*, Ellenville, NY, p. 195.
145. C.F. Lyons, R.K. Leidy, and G.B. Smith. 1992. *Proc. SPIE*, **1674**: 523.
146. D. Bernard and H. Arbach. 1991. *J. Opt. Soc. Am.*, **8**: 123.

147. A. Mahorowala et al. 2001. *Proc. SPIE*, **4343**, Emerging Lithographic Technologies V.
148. K. Babich et al. 2003. *Proc. SPIE*, **5039**, Advances in Resist Technology and Processing XX.
149. M. Kakuchi, M. Hikita, and T. Tamamura. 1986. *Appl. Phys. Lett.*, **48**(13): 835–837.
150. W. Liu et al. 2003. *Proc. SPIE*, **5040**: 841–848, Optical Microlithography XVI.
151. J. Seth, R. Padiyath, and S.V. Babu. 1991. *Mater. Sci. Monogr.*, **73**: 851–856.
152. K. Takahashi and R. Takahashi. 2018. *Phys. Sci. Technol.*, **5**(2): 4–9.
153. T. Kudo et al. 2014. *Proc. SPIE*, **9051**, Advances in Patterning Materials and Processes XXXI.

9 Optical Lithography Modeling

Chris A. Mack, John J. Biafore, and Mark D. Smith

CONTENTS

9.1   INTRODUCTION

Optical lithography modeling began in the early 1970s when Rick Dill started an effort at IBM Yorktown Heights Research Center to describe the basic steps of the lithography process with mathematical equations. At a time when lithography was considered a true art, such an approach was met with much skepticism. The results of this pioneering work were published in a landmark series of papers in 1975 [1–4], now referred to as the *Dill papers*. These papers not only gave birth to the field of lithography modeling; they represented the first serious attempt to describe lithography not as an art but as a science. These papers presented a simple model for image formation with incoherent illumination, the first-order kinetic "Dill model" of exposure, and an empirical model for development coupled with a cell algorithm for photoresist profile calculation. The Dill papers are still the most referenced works in the body of lithography literature.

While Dill's group worked on the beginnings of lithography simulation, a professor from the University of California at Berkeley, Andy Neureuther, spent a year on sabbatical working with Dill. Upon returning to Berkeley, Neureuther and another professor, Bill Oldham, started their own modeling effort. In 1979, they presented the first result of their effort, the lithography modeling program SAMPLE [5]. SAMPLE improved the state of the art in lithography modeling by adding partial coherence to the image calculations and by replacing the cell algorithm for dissolution calculations with a string algorithm. But more importantly, SAMPLE was made available to the lithography community. For the first time, researchers in the field could use modeling as a tool to help understand and improve their lithography processes.

The author began working in the area of lithographic simulation in 1983 and in 1985 introduced the model PROLITH (the Positive Resist Optical LITHography model) [6]. This model added an analytical expression for the standing-wave intensity in the resist, a prebake model, a kinetic model for resist development (now known as the Mack model), and the first model for contact and proximity printing. PROLITH was also the first lithography model to run on a personal computer (the IBM PC), making lithography modeling accessible to all lithographers, from advanced researchers to process development engineers to manufacturing engineers. Over the years, PROLITH advanced to include a model for contrast enhancement materials, the extended source method for partially coherent image calculations, and an advanced focus model for high–numerical aperture (NA) imaging.

Since the late 1980s, commercial lithography simulation software has been available to the semiconductor community, providing dramatic improvements in the usability and graphics capabilities of the models. Lithography modeling has now become an accepted tool for use in a wide variety of lithography applications.

9.2   STRUCTURE OF A LITHOGRAPHY MODEL

Any lithography model must simulate the basic lithographic steps of image formation, resist exposure, postexposure bake diffusion, and development to obtain a final resist profile. Figure 9.1 shows a basic schematic of the calculation steps required for lithography modeling. There follows a brief overview of the physical models found in a typical lithography simulator. More details on these models can be found in subsequent sections.

Aerial Image: The extended source method, or Hopkin's method, can be used to predict the aerial image of a partially coherent diffraction-limited or aberrated projection system based on scalar diffraction theory. Single-wavelength or broadband illumination is possible. The image model must account for the important effect of image defocus through the resist film at a minimum. Mask patterns can be one-dimensional lines and spaces or two-dimensional contacts and islands, as well as arbitrarily complex two-dimensional mask features. The masks often vary in the magnitude and phase of their transmission in what

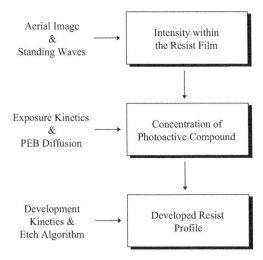

FIGURE 9.1 Flow diagram of a lithography model.

are called *phase-shifting masks*. The illumination source may be of a conventional disk shape or other more complicated shapes, as in off-axis illumination. For very high numerical apertures, vector calculations should be used.

Standing Waves: An analytical expression is used to calculate the standing-wave intensity as a function of depth into the resist, including the effects of resist bleaching, on planar substrates. Film stacks can be defined below the resist with up to many layers between the resist and substrate. Contrast enhancement layers or top-layer antireflection coatings can also be included. The high-NA models should include the effects of nonvertical light propagation.

Prebake: Thermal decomposition of the photoresist photoactive compound during prebake is modeled using first-order kinetics resulting in a change in the resist's optical properties (the Dill parameters A and B). Other important effects of baking have not yet been modeled.

Exposure: First-order kinetics are used to model the chemistry of exposure using the standard Dill ABC parameters. Both positive and negative resists can be used.

Postexposure Bake: A diffusion calculation allows the postexposure bake to reduce the effects of standing waves. For chemically amplified resists (CARs), this diffusion includes an amplification reaction, which accounts for crosslinking, blocking, or deblocking in an acid-catalyzed reaction. Acid loss mechanisms and nonconstant diffusivity could also be needed.

Development: A model relating the resist dissolution rate to the chemical composition of the film is used in conjunction with an etching algorithm to determine the resist profile. Surface inhibition or enhancement can also be present. Alternatively, a data file of development rate information could be used in lieu of a model.

CD Measurement: The measurement of the photoresist linewidth should give accuracy and flexibility to match the model to an actual critical dimension (CD) measurement tool.

The combination of the models described provides a complete mathematical description of the optical lithography process. Use of the models incorporated in a simulation software package allows the user to investigate many interesting and important aspects of optical lithography. The following sections describe each of the models in detail, including derivations of most of the mathematical models as well as physical descriptions of their basis.

Of course, more work has been done in the field of lithography simulation than it is possible to report in one chapter. Typically, there are several approaches, sometimes equivalent, sometimes not, that can be applied to each problem. Although the models presented here are representative of the possible solutions, they are not necessarily comprehensive reviews of all possible models.

9.3 AERIAL IMAGE FORMATION

9.3.1 BASIC IMAGING THEORY

Consider the generic projection system shown in Figure 9.2. It consists of a *light source*, a *condenser lens*, the *mask*, the *objective lens*, and finally, the resist-coated wafer. The combination of the light source and the condenser lens is called the *illumination system*. In optical design terms, a lens is a system of (possibly many) lens elements. Each lens element is an individual piece of glass (refractive element) or a mirror (reflective element). The purpose of the illumination system is to deliver light to the mask (and eventually into the objective lens) with sufficient intensity, the proper directionality and spectral characteristics, and adequate uniformity across the field. The light then passes through the clear areas of the mask and diffracts on its way to the objective lens. The purpose of the objective lens is to pick up a portion of the diffraction pattern and project an image onto the wafer, which, one hopes, will resemble the mask pattern.

The first and most basic phenomenon occurring here is the diffraction of light. Diffraction is usually thought of as the bending of light as it passes through an aperture, which is certainly an appropriate description for diffraction by a lithographic mask. More correctly, diffraction theory simply describes how light propagates. This propagation includes the effects of the surroundings (boundaries). Maxwell's equations describe how electromagnetic waves propagate, but with partial differential equations of vector quantities, which, for general boundary conditions, are extremely difficult to solve without the aid of a powerful computer. A simpler approach is to artificially decouple the electric and magnetic field *vectors* and describe light as a *scalar* quantity. Under most conditions, scalar diffraction theory is surprisingly accurate. Scalar diffraction theory was first rigorously used by Kirchoff in 1882 and involves performing one numerical integration (much simpler than solving partial differential equations!). Kirchoff diffraction was further simplified by Fresnel for the case when the distance away from the diffracting plane (that is, the distance from the mask to the objective lens) is much greater than the wavelength of light. Finally, if the mask is illuminated by a spherical wave that converges to a point at the entrance to the objective lens, Fresnel diffraction simplifies to *Fraunhofer diffraction*.

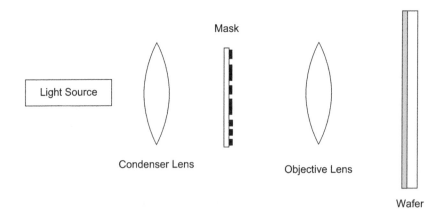

FIGURE 9.2 Block diagram of a generic projection system.

Let us describe the electric field transmittance of a mask pattern as $m(x,y)$, where the mask is in the x,y-plane, and $m(x,y)$ has in general both magnitude and phase. For a simple chrome-glass mask, the mask pattern becomes binary: $m(x,y)$ is 1 under the glass and 0 under the chrome. Let the x',y'-plane be the diffraction plane, that is, the entrance to the objective lens, and let z be the distance from the mask to the objective lens. Finally, we will assume monochromatic light of wavelength λ and that the entire system is in air (so that its index of refraction can be dropped). Then, the electric field of our diffraction pattern, $E(x',y')$, is given by the Fraunhofer diffraction integral:

$$E(x',y') = \int\limits_{-\infty}^{\infty} \int\limits_{-\infty}^{\infty} m(x,y)\,e^{-2\pi i(f_x x + f_y y)}\,dx\,dy \qquad (9.1)$$

where $f_x = x'/(z\lambda)$ and $f_y = y'/(z\lambda)$ are called the *spatial frequencies* of the diffraction pattern.

For many scientists and engineers (and especially electrical engineers), this equation should be quite familiar: it is simply a *Fourier transform*. Thus, the diffraction pattern (i.e., the electric field distribution as it enters the objective lens) is just the Fourier transform of the mask pattern. This is the principle behind an entire field of science called *Fourier optics* (for more information, consult Goodman's classic textbook [7]). Figure 9.3 shows two mask patterns, one an isolated space, the other a series of equal lines and spaces, both infinitely long in the y-direction. The resulting mask pattern functions, $m(x)$, look like a square pulse and a square wave, respectively. The Fourier transforms are easily found in tables or textbooks and are also shown in Figure 9.3. The isolated space gives rise to a *sinc* function diffraction pattern, and the equal lines and spaces yield discrete *diffraction orders*.

Let's take a closer look at the diffraction pattern for equal lines and spaces. Notice that the graphs of the diffraction patterns in Figure 9.3 use spatial frequency as the x-axis. Since z and λ are fixed for a given stepper, the spatial frequency is simply a scaled x'-coordinate. At the center of the objective lens entrance ($f_x = 0$), the diffraction pattern has a bright spot called the *zero order*. The zero order is the light that passes through the mask and is not diffracted. The zero order can be thought of as "D.C." light, providing power but no information as to the size of the features on the mask. To either side of the zero order are two peaks called the *first diffraction orders*. These peaks occur at spatial frequencies of $\pm 1/p$, where p is the pitch of the mask pattern (linewidth plus spacewidth). Since the position of these diffraction orders depends on the mask pitch, their position contains information about the pitch. It is this information that the objective lens will use to reproduce the image of the mask. In fact, in order for the objective lens to form a true image of the mask, it must have the zero order and at least one higher order. In addition to the first order, there can be many higher orders, with the nth order occurring at a spatial frequency of n/p.

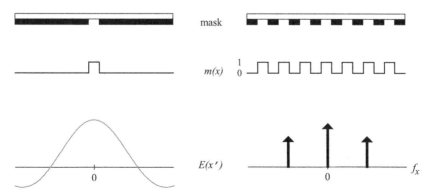

FIGURE 9.3 Two typical mask patterns, an isolated space and an array of equal lines and spaces, and the resulting Fraunhofer diffraction patterns.

Summarizing, given a mask in the x-y plane described by its electric field transmission $m(x,y)$, the electric field M as it enters the objective lens (the x'-y' plane) is given by

$$M\left(f_x, f_y\right) = \mathcal{F}\left\{m\left(x, y\right)\right\} \tag{9.2}$$

where
   the symbol $\mathcal{F}$ represents the Fourier transform
   f_x and f_y are the spatial frequencies and are simply scaled coordinates in the x'-y' plane

We are now ready to describe what happens next and follow the diffracted light as it enters the objective lens. In general, the diffraction pattern extends throughout the x'-y' plane. However, the objective lens, being only of finite size, cannot collect all the light in the diffraction pattern. Typically, lenses used in microlithography are circularly symmetric, and the entrance to the objective lens can be thought of as a circular aperture. Only those portions of the mask diffraction pattern that fall inside the aperture of the objective lens go on to form the image. Of course, we can describe the size of the lens aperture by its radius, but a more common and useful description is to define the maximum angle of diffracted light that can enter the lens. Consider the geometry shown in Figure 9.4. Light passing through the mask is diffracted at various angles. Given a lens of a certain size placed a certain distance from the mask, there is some maximum angle of diffraction, α, for which diffracted light just makes it into the lens. Light emerging from the mask at larger angles misses the lens and is not used in forming the image. The most convenient way to describe the size of the lens aperture is by its *numerical aperture*, defined as the sine of the maximum half-angle of diffracted light that can enter the lens times the index of refraction of the surrounding medium. In the case of lithography, all the lenses are in air, and the numerical aperture is given by $NA = \sin\alpha$. (Note that the spatial frequency is the sine of the diffracted angle divided by the wavelength of light. Thus, the maximum spatial frequency that can enter the objective lens is given by NA/λ.)

Obviously, the numerical aperture is going to be quite important. A large numerical aperture means that a larger portion of the diffraction pattern is captured by the objective lens. For a small numerical aperture, much more of the diffracted light is lost.

To proceed further, we must now describe how the lens affects the light entering it. Obviously, we would like the image to resemble the mask pattern. Since diffraction gives the Fourier transform of the mask, if the lens could give the inverse Fourier transform of the diffraction pattern, the resulting image would resemble the mask pattern. In fact, spherical lenses do behave in this way. We can define an ideal imaging lens as one that produces an image which is identically equal to the Fourier transform of the light distribution entering the lens. It is the goal of lens designers and manufacturers to create lenses as close as possible to this ideal. Does an ideal lens produce a perfect image? No.

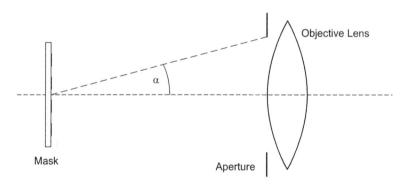

FIGURE 9.4   The numerical aperture is defined as $NA = \sin\alpha$, where α is the maximum half-angle of the diffracted light that can enter the objective lens.

Because of the finite size of the numerical aperture, only a portion of the diffraction pattern enters the lens. Thus, even an ideal lens cannot produce a perfect image unless the lens is infinitely big. Since in the case of an ideal lens the image is limited only by the diffracted light that does not make it through the lens, we call such an ideal system *diffraction limited*.

In order to write our final equation for the formation of an image, let us define the objective lens *pupil function P* (a pupil is just another name for an aperture). The pupil function of an ideal lens simply describes what portion of light enters the lens: it is one inside the aperture and zero outside:

$$P(f_x, f_y) = \begin{cases} 1, & \sqrt{f_x^2 + f_y^2} < NA/\lambda \\ 0, & \sqrt{f_x^2 + f_y^2} > NA/\lambda \end{cases} \tag{9.3}$$

Thus, the product of the pupil function and the diffraction pattern describes the light entering the objective lens. Combining this with our description of how a lens behaves gives us our final expression for the electric field at the image plane (that is, at the wafer):

$$E(x, y) = \mathcal{F}^{-1}\{M(f_x, f_y) P(f_x, f_y)\} \tag{9.4}$$

The *aerial image* is defined as the intensity distribution at the wafer and is simply the square of the magnitude of the electric field.

Consider the full imaging process. First, light passing through the mask is diffracted. The diffraction pattern can be described as the Fourier transform of the mask pattern. Since the objective lens is of finite size, only a portion of the diffraction pattern actually enters the lens. The numerical aperture describes the maximum angle of diffracted light that enters the lens, and the pupil function is used to mathematically describe this behavior. Finally, the effect of the lens is to take the inverse Fourier transform of the light entering the lens to give an image that resembles the mask pattern. If the lens is ideal, the quality of the resulting image is only limited by how much of the diffraction pattern is collected. This type of imaging system is called *diffraction limited*.

Although we have completely described the behavior of a simple ideal imaging system, we must add one more complication before we have described the operation of a projection system for lithography. So far, we have assumed that the mask is illuminated by *spatially coherent* light. Coherent illumination means simply that the light striking the mask arrives from only one direction. We have further assumed that the coherent illumination on the mask is normally incident. The result was a diffraction pattern that was centered in the entrance to the objective lens. What would happen if we changed the direction of the illumination so that the light struck the mask at some angle θ'? The effect is simply to shift the diffraction pattern with respect to the lens aperture (in terms of spatial frequency, the amount shifted is $\sin\theta'/\lambda$). Recalling that only the portion of the diffraction pattern passing through the lens aperture is used to form the image, it is quite apparent that this shift in the position of the diffraction pattern can have a profound effect on the resulting image. Letting f_x' and f_y' be the shift in the spatial frequency due to the tilted illumination, Equation 9.4 becomes

$$E(x, y, f_x', f_y') = \mathcal{F}^{-1}\{M(f_x - f_x', f_y - f_y') P(f_x, f_y)\} \tag{9.5}$$

If the illumination of the mask is composed of light coming from a range of angles rather than just one angle, the illumination is called *partially coherent*. If one angle of illumination causes a shift in the diffraction pattern, a range of angles will cause a range of shifts, resulting in broadened diffraction orders. One can characterize the range of angles used for the illumination in several ways, but the most common is the *partial coherence factor*, σ (also called the degree of partial coherence, or the pupil filling function, or just the partial coherence). The partial coherence is defined as the sine of the half-angle of the illumination cone divided by the numerical aperture of the objective lens. It is thus a measure of the angular range of the illumination relative to the angular acceptance of the lens. Finally, if the range of angles striking the mask extends from −90° to 90° (that is, all possible angles), the illumination is said to be *incoherent*.

The extended source method for partially coherent image calculations is based on dividing the full source into individual point sources. Each point source is coherent and results in an aerial image given by Equation 9.5. Two point sources from the extended source, however, do not interact coherently with each other. Thus, the contributions of these two sources must be added to each other incoherently (that is, the intensities are added together). The full aerial image is determined by calculating the coherent aerial image from each point on the source and then integrating the intensity over the source.

9.3.2 ABERRATIONS

Aberrations can be defined as the deviation of the real behavior of an imaging system from its ideal behavior (the ideal behavior was described earlier using Fourier optics as diffraction-limited imaging). Aberrations are inherent in the behavior of all lens systems and come from three basic sources: defects of construction, defects of use, and defects of design. Defects of construction include rough or inaccurate lens surfaces, inhomogeneous glass, incorrect lens thicknesses or spacings, and tilted or decentered lens elements. Defects of use include use of the wrong illumination or tilt of the lens system with respect to the optical axis of the imaging system. Also, changes in the environmental conditions during use, such as the temperature of the lens or the barometric pressure of the air, result in defects of use. Defects of design may be a bit of a misnomer, since the aberrations of a lens design are not mistakenly designed *into* the lens but rather, were not designed *out* of the lens. All lenses have aberrated behavior, since the Fourier optics behavior of a lens is only approximately true and is based on a linearized Snell's law for small angles. It is the job of a lens designer to combine elements of different shapes and properties so that the aberrations of each individual lens element tend to cancel in the sum of all of the elements, giving a lens system with only a small residual number of aberrations. It is impossible to design a lens system with absolutely no aberrations.

Mathematically, aberrations are described as a *wavefront deviation*, the difference in phase (or path difference) of the actual wavefront emerging from the lens compared with the ideal wavefront as predicted from Fourier optics. This phase difference is a function of the position within the lens pupil, most conveniently described in polar coordinates. This wavefront deviation is in general quite complicated, so the mathematical form used to describe it is also quite complicated. The most common model for describing the phase error across the pupil is the *Zernike polynomial*, a 36-term polynomial of powers of the radial position R and trigonometric functions of the polar angle θ. The Zernike polynomial can be arranged in many ways, but most lens design software and lens measuring equipment in use today employs the fringe or circle Zernike polynomial, defined as

$$W(R,\theta) = Z1 * R * \cos\theta$$

$$+Z2 * R * \sin\theta$$
$$+Z3 * \left(2 * R * R - 1\right)$$
$$+Z4 * R * R * \cos 2\theta$$
$$+Z5 * R * R * \sin 2\theta$$
$$+Z6 * \left(3 * R * R - 2\right) * R * \cos\theta$$
$$+Z7 * \left(3 * R * R - 2\right) * R * \sin\theta$$
$$+Z8 * \left(6 * R**4 - 6 * R * R + 1\right)$$
$$+Z9 * R**3 * \cos 3\theta$$
$$+Z10 * R**3 * \sin 3\theta$$

$$+Z11*(4*R*R-3)*R*R*\cos 2\theta$$
$$+Z12*(4*R*R-3)*R*R*\sin 2\theta$$
$$+Z13*(10*R**4-12*R*R+3)*R*\cos\theta$$
$$+Z14*(10*R**4-12*R*R+3)*R*\sin\theta$$
$$+Z15*(20*R**6-30*R**4+12*R*R-1)$$
$$+Z16*R**4*\cos 4\theta$$
$$+Z17*R**4*\sin 4\theta$$
$$+Z18*(5*R*R-4)*R**3*\cos 3\theta$$
$$+Z19*(5*R*R-4)*R**3*\sin 3\theta$$
$$+Z20*(15*R**4-20*R*R+6)*R*R*\cos 2\theta$$

$$+Z21*(15*R**4-20*R*R+6)*R*R*\sin 2\theta$$
$$+Z22*(35*R**6-60*R**4+30*R*R-4)*R*\cos\theta$$
$$+Z23*(35*R**6-60*R**4+30*R*R-4)*R*\cos\theta$$
$$+Z24*(70*R**8-140*R**6+90*R**4-20*R*R+1)$$
$$+Z25*R**5*\cos 5\theta$$
$$+Z26*R**5*\sin 5\theta$$
$$+Z27*(6*R*R-5)*R**4*\cos 4\theta$$
$$+Z28*(6*R*R-5)*R**4*\sin 4\theta$$
$$+Z29*(21*R**4-30*R*R+10)*R**3*\cos 3\theta$$

$$+Z30*(21*R**4-30*R*R+10)*R**3*\sin 3\theta$$
$$+Z31*(56*R**6-105*R**4+60*R**2-10)*R*R*\cos 2\theta$$
$$+Z32*(56*R**6-105*R**4+60*R**2-10)*R*R*\sin 2\theta$$
$$+Z33*(126*R**8-280*R**6+210*R**4-60*R*R+5)*R*\cos\theta$$
$$+Z34*(126*R**8-280*R**6+210*R**4-60*R*R+5)*R*\sin\theta \tag{9.6}$$
$$+Z35*(252*R**10-630*R**8+560*R**6-210*R**4+30*R*R-1)$$
$$+Z36*(924*R**12-2772*R**10+3150*R**8$$

$$-1680*R**6+420*R**4-42*R*R+1)$$

where
  $W(R,\theta)$ is the optical path difference relative to the wavelength
  Zi is called the ith Zernike coefficient

It is the magnitude of the Zernike coefficients that determines the aberration behavior of a lens. They have units of optical path length relative to the wavelength.

The impact of aberrations on the aerial image can be calculated by modifying the pupil function of the lens with the phase error due to aberrations given by Equation 9.6:

$$P(f_x, f_y) = P_{\text{ideal}}(f_x, f_y) e^{i2\pi W(R,\theta)} \tag{9.7}$$

9.3.3 ZERO-ORDER SCALAR MODEL

Calculation of an aerial image means, quite literally, to determine the image in air. Of course, in lithography, one projects this image into photoresist. The propagation of the image into resist can be quite complicated, so models usually make one or more approximations. This section and the sections that follow describe approximations made in determining the intensity of light within the photoresist.

The lithography simulator SAMPLE [5] and the 1985 version of PROLITH [6] used the simple imaging approximation first proposed by Dill [4] to calculate the propagation of an aerial image in photoresist. First, an aerial image $I_i(x)$ is calculated as if projected into air (x being along the surface of the wafer and perpendicular to the propagation direction of the image). Second, a standing-wave intensity $I_s(z)$ is calculated assuming that a plane wave of light is normally incident on the photoresist-coated substrate (where z is defined as zero at the top of the resist and is positive going into the resist). Then, it is assumed that the actual intensity within the resist film $I(x,z)$ can be approximated by

$$I(x,z) \approx I_i(x)I_s(z) \tag{9.8}$$

For very low numerical apertures and reasonably thin photoresists, these approximations are valid. They begin to fail when the aerial image changes as it propagates through the resist (i.e., it defocuses) or when the light entering the resist is appreciably nonnormal. Note that if the photoresist bleaches (changes its optical properties during exposure), only $I_s(z)$ changes in this approximation.

9.3.4 FIRST-ORDER SCALAR MODEL

The first attempt to correct one of the deficiencies of the zero-order model was made by the author [8] and, independently, by Bernard [9]. The aerial image, while propagating through the resist, is continuously changing focus. Thus, even in air, the aerial image is a function of both x and z. An aerial image simulator calculates images as a function of x and the distance from the plane of best focus, δ. Letting δ_o be the defocus distance of the image at the top of the photoresist, the defocus within the photoresist at any position z is given by

$$\delta(z) = \delta_o + \frac{z}{n} \tag{9.9}$$

where n is the real part of the index of refraction of the photoresist. The intensity within the resist is then given by

$$I(x,z) = I_i(x,\delta(z))I_s(z) \tag{9.10}$$

Here, the assumption of normally incident plane waves is still used when calculating the standing-wave intensity.

9.3.5 HIGH-NA SCALAR MODEL

The light propagating through the resist can be thought of as various plane waves traveling through the resist in different directions. Consider first the propagation of the light in the absence of diffraction by a mask pattern (that is, exposure of the resist by a large open area). The spatial dimensions of the light source determine the characteristics of the light entering the photoresist. For the simple case of a coherent point source of illumination centered on the optical axis, the light traveling into the photoresist would be the normally incident plane wave used in the calculations presented earlier. The standing-wave intensity within the resist can be determined analytically [10] as the square of the magnitude of the electric field given by

$$E(z) = \frac{\tau_{12}E_I\left(e^{-i2\pi n_2 z/\lambda} + \rho_{23}\tau_D^2 e^{i2\pi n_2 z/\lambda}\right)}{1 + \rho_{12}\rho_{23}\tau_D^2} \tag{9.11}$$

where

the subscripts 1, 2, and 3 refer to air, the photoresist, and the substrate, respectively
D is the resist thickness
E_I is the incident electrical field
λ is the wavelength
complex index of refraction of film j: $\mathbf{n}_j = n_j - i\kappa_j$
transmission coefficient from i to j: $\tau_{ij} = \dfrac{2\mathbf{n}_i}{\mathbf{n}_i + \mathbf{n}_j}$

reflection coefficient from i to j: $\rho_{ij} = \dfrac{\mathbf{n}_i - \mathbf{n}_j}{\mathbf{n}_i + \mathbf{n}_j}$

internal transmittance of the resist: $\tau_D = e^{-i2\pi \mathbf{n}_2 D/\lambda}$

A more complete description of the standing-wave equation (9.11) is given in Section 9.4.

This expression can be easily modified for the case of nonnormally incident plane waves. Suppose a plane wave is incident on the resist film at some angle θ_1. The angle of the plane wave inside the resist will be θ_2 as determined from Snell's law. An analysis of the propagation of this plane wave within the resist will give an expression similar to Equation 9.11 but with the position z replaced with $z\cos\theta_2$.

$$E(z, \theta_2) = \frac{\tau_{12}(\theta_2)E_I\left(e^{-i2\pi \mathbf{n}_2 z\cos\theta_2/\lambda} + \rho_{23}(\theta_2)\tau_D^2(\theta_2)e^{-i2\pi \mathbf{n}_2 z\cos\theta_2/\lambda}\right)}{1 + \rho_{12}(\theta_2)\rho_{23}(\theta_2)\tau_D^2(\theta_2)} \tag{9.12}$$

The transmission and reflection coefficients are now functions of the angle of incidence and are given by the Fresnel formulas (see Section 9.4). A similar approach was taken by Bernard and Urbach [11].

By calculating the standing-wave intensity at one incident angle θ_1 to give $I_s(z, \theta_1)$, the full standing-wave intensity can be determined by integrating over all angles. Each incident angle comes from a given point in the illumination source, so that integration over angles is the same as integration over the source. Thus, the effect of partial coherence on the standing waves is accounted for. Note that for the model described here, the effect of the nonnormal incidence is included only with respect to the zero-order light (the light that is not diffracted by the mask).

Besides the basic modeling approaches described, there are two issues that apply to any model. First, the effects of defocus are taken into account by describing defocus as a phase error at the pupil plane. Essentially, if the curvature of the wavefront exiting the objective lens pupil is such that it focuses in the wrong place (i.e., not where you want it), one can consider the wavefront curvature to be wrong. Simple geometry then relates the optical path difference (OPD) of the actual wavefront from the desired wavefront as a function of the angle of the light exiting the lens, θ.

$$\text{OPD}(\theta) = \delta(1 - \cos\theta) \tag{9.13}$$

Computation of the imaging usually involves a change in variables, where the main variable used is $\sin\theta$. Thus, the cosine adds some algebraic complexity to the calculations. For this reason, it is common in optics texts to simplify the OPD function for small angles (i.e., low numerical apertures).

$$\text{OPD}(\theta) = \delta(1 - \cos\theta) \approx \frac{\delta}{2}\sin^2\theta \tag{9.14}$$

Again, the approximation is not necessary and is only made to simplify the resulting equations. In this work, the approximate defocus expression is used in the standard image model. The high-NA model uses the exact defocus expression.

Reduction in the imaging system adds an interesting complication. Light entering the objective lens will leave the lens with no loss in energy (the lossless lens assumption). However, if there is reduction in the lens, the *intensity* distribution of the light entering will be different from that leaving, since the intensity is the energy spread over a changing area. The result is a radiometric correction well known in optics [12] and first applied to lithography by Cole and Barouch [13].

9.3.6 FULL SCALAR AND VECTOR MODELS

The method described for calculating the image intensity within the resist still makes the assumption of separability: that an aerial image and a standing-wave intensity can be calculated independently and then multiplied together to give the total intensity. This assumption is not required. Instead, one could calculate the full $I(x,z)$ at once, making only the standard scalar approximation. The formation of the image can be described as the summation of plane waves. For coherent illumination, each diffraction order gives one plane wave propagating into the resist. Interference between the zero order and the higher orders produces the desired image. Each point in the illumination source will produce another image, which will add incoherently (i.e., intensities will add) to give the total image. Equation 9.12 describes the propagation of a plane wave in a stratified media at any arbitrary angle. By applying this equation to each diffraction order (not just the zero order as in the high-NA scalar model), an exact scalar representation of the full intensity within the resist is obtained.

Light is an electromagnetic wave, which can be described by time-varying electric and magnetic field vectors. In lithography, the materials used are generally nonmagnetic, so that only the electric field is of interest. The electric field vector is described by its three vector components. Maxwell's equations, sometimes put into the form of the wave equation, govern the propagation of the electric field vector. The *scalar approximation* assumes that each of the three components of the electric field vector can be treated separately as scalar quantities, and each scalar electric field component must individually satisfy the wave equation. Further, when two fields of light (say, two plane waves) are added together, the scalar approximation means that the sum of the fields would simply be the sum of the scalar amplitudes of the two fields.

The scalar approximation is commonly used throughout optics and is known to be accurate under many conditions. There is one simple situation, however, in which the scalar approximation is not adequate. Consider the interference of two plane waves traveling past each other. If each plane wave is treated as a vector, they will interfere only if there is some overlap in their electric field vectors. If the vectors are parallel, there will be complete interference. If, however, their electric fields are at right angles to each other, there will be no interference. The scalar approximation essentially assumes that the electric field vectors are always parallel and will always give complete interference. These differences come into play in lithography when considering the propagation of plane waves traveling through the resist at large angles. For large angles, the scalar approximation may fail to account for these vector effects. Thus, a vector model would keep track of the vector direction of the electric field and use this information when adding two plane waves together [14, 15].

9.4 STANDING WAVES

When a thin dielectric film placed between two semi-infinite media (e.g., a thin coating on a reflecting substrate) is exposed to monochromatic light, standing waves are produced in the film. This effect has been well documented for such cases as antireflection coatings and photoresist exposure [1, 16–19]. In the former, the standing-wave effect is used to reduce reflections from the substrate. In the latter, standing waves are an undesirable side effect of the exposure process. Unlike the antireflection application, photolithography applications require a knowledge of the intensity of the light within the thin film itself. Previous work [1, 19] on determining the intensity within a thin photoresist film has been limited to numerical solutions based on Berning's matrix method [20]. This section presents an analytical expression for the standing-wave intensity within a thin film [10]. This

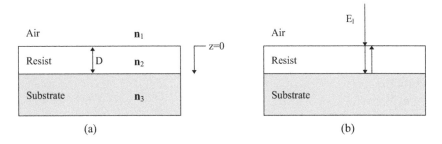

FIGURE 9.5 Film stack showing geometry for standing-wave derivation.

film may be homogeneous or of a known inhomogeneity. The film may be on a substrate or between one or more other thin films. The incident light can be normally incident or incident at some angle.

Consider a thin film of thickness D and complex index of refraction n_2 deposited on a thick substrate with complex index of refraction n_3 in an ambient environment of index n_1. An electromagnetic plane wave is normally incident on this film. Let E_1, E_2, and E_3 be the electric fields in the ambient, thin film, and substrate, respectively (see Figure 9.5). Assuming monochromatic illumination, the electric field in each region is a plane wave or the sum of two plane waves traveling in opposite directions (i.e., a standing wave). Maxwell's equations require certain boundary conditions to be met at each interface: specifically, E_j and the magnetic field H_j are continuous across the boundaries $z=0$ and $z=D$. Solving the resulting equations simultaneously, the electric field in region 2 can be shown to be [10]

$$E_2(x,y,z) = E_1(x,y)\frac{\tau_{12}\left(e^{-i2\pi n_2 z/\lambda} + \rho_{23}\tau_D^2 e^{i2\pi n_2 z/\lambda}\right)}{1 + \rho_{12}\rho_{23}\tau_D^2} \tag{9.15}$$

where

$E_1(x,y)$ = the incident wave at $z=0$, which is a plane wave
$\rho_{ij} = (n_i - n_j)/(n_i + n_j)$, the reflection coefficient
$\tau_{ij} = 2n_i/(n_i + n_j)$, the transmission coefficient
$\tau_D = \exp(-ik_2 D)$, the internal transmittance of the film
$k_j = 2\pi n_j/\lambda$, the propagation constant
$n_j = n_j - i\kappa_j$, the complex index of refraction
λ = vacuum wavelength of the incident light

Equation 9.15 is the basic standing-wave expression, where film 2 represents the photoresist. Squaring the magnitude of the electric field gives the standing-wave intensity. Note that absorption is taken into account in this expression through the imaginary part of the index of refraction. The common absorption coefficient α is related to the imaginary part of the index by

$$\alpha = \frac{4\pi\kappa}{\lambda} \tag{9.16}$$

It is very common to have more than one film coated on a substrate. The problem then becomes that of two or more absorbing thin films on a substrate. An analysis similar to that for one film yields the following result for the electric field in the top layer of an $m-1$ layer system:

$$E_2(x,y,z) = E_1(x,y)\frac{\tau_{12}\left(e^{-i2\pi n_2 z/\lambda} + \rho'_{23}\tau_D^2 e^{i2\pi n_2 z/\lambda}\right)}{1 + \rho_{12}\rho'_{23}\tau_D^2} \tag{9.17}$$

where

$$\rho'_{23} = \frac{n_2 - n_3 X_3}{n_2 + n_3 X_3}$$

$$X_3 = \frac{1 - \rho'_{34} \tau_{D3}^2}{1 + \rho'_{34} \tau_{D3}^2}$$

$$\rho'_{34} = \frac{n_3 - n_4 X_4}{n_3 + n_4 X_4}$$

$$X_m = \frac{1 - \rho_{m,m+1} \tau_{Dm}^2}{1 + \rho_{m,m+1} \tau_{Dm}^2}$$

$$\rho_{m,m+1} = \frac{n_m - n_{m+1}}{n_m + n_{m+1}}$$

$$\tau_{Dj} = e^{-ik_j D_j}$$

and all other parameters are defined previously. The parameter ρ'_{23} is the effective reflection coefficient between the thin film and what lies beneath it.

If the thin film in question is not the top film (layer 2), the intensity can be calculated in layer j from

$$E_j(x,y,z) = E_{Ieff}(x,y)\tau_{j-1,j}^* \frac{\left(e^{-ik_j z_j} + \rho'_{j,j+1}\tau_{Dj}^2 e^{ik_j z_j}\right)}{1 + \rho_{j-1,j}^* \rho'_{j,j+1}\tau_{Dj}^2} \tag{9.18}$$

where $\tau_{j-1,j}^* = 1 + \rho_{j-1,j}^*$. The effective reflection coefficient ρ^* is analogous to the coefficient ρ', looking in the opposite direction. E_{Ieff} is the effective intensity incident on layer j. Both E_{Ieff} and ρ^* are defined in detail by Mack [10].

If the film in question is not homogeneous, these equations are, in general, not valid. Let us, however, examine one special case in which the inhomogeneity takes the form of small variations in the imaginary part of the index of refraction of the film in the z-direction, leaving the real part constant. In this case, the absorbance Abs is no longer simply αz but becomes

$$Abs(z) = \int_0^z \alpha(z')dz' \tag{9.19}$$

It can be shown that Equations 9.15–9.18 are still valid if the anisotropic expression for absorbance (Equation 9.19) is used. Thus, $I(z)$ can be found if the absorption coefficient is known as a function of z. Figure 9.6 shows a typical result of the standing-wave intensity within a photoresist film coated on an oxide on silicon film stack.

Equation 9.15 can be easily modified for the case of nonnormally incident plane waves. Suppose a plane wave is incident on the resist film at some angle θ_1. The angle of the plane wave inside the resist will be θ_2 as determined from Snell's law. An analysis of the propagation of this plane wave within the resist will give an expression similar to Equation 9.15 but with the position z replaced with $z\cos\theta_2$:

$$E(z,\theta_2) = \frac{\tau_{12}(\theta_2)E_I\left(e^{-i2\pi n_2 z \cos\theta_2/\lambda} + \rho_{23}(\theta_2)\tau_D^2(\theta_2)e^{i2\pi n_2 z \cos\theta_2/\lambda}\right)}{1 + \rho_{12}(\theta_2)\rho_{23}(\theta_2)\tau_D^2(\theta_2)} \tag{9.20}$$

The transmission and reflection coefficients are now functions of the angle of incidence (as well as the polarization of the incident light) and are given by the Fresnel formulas

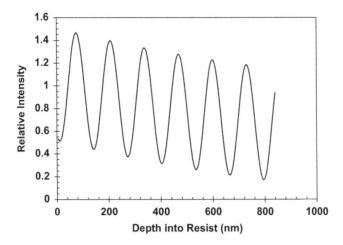

FIGURE 9.6 Standing-wave intensity within a photoresist film at the start of exposure (850 nm of resist on 100 nm SiO$_2$ on silicon, $\lambda = 436$ nm). The intensity shown is relative to the incident intensity.

$$\rho_{ij\perp}(\theta) = \frac{\mathbf{n}_i \cos(\theta_i) - \mathbf{n}_j \cos(\theta_j)}{\mathbf{n}_i \cos(\theta_i) + \mathbf{n}_j \cos(\theta_j)}$$

$$\tau_{ij\perp}(\theta) = \frac{2\mathbf{n}_i \cos(\theta_i)}{\mathbf{n}_i \cos(\theta_i) + \mathbf{n}_j \cos(\theta_j)}$$

$$\rho_{ij\parallel}(\theta) = \frac{\mathbf{n}_i \cos(\theta_j) - \mathbf{n}_j \cos(\theta_i)}{\mathbf{n}_i \cos(\theta_j) + \mathbf{n}_j \cos(\theta_i)}$$

$$\tau_{ij\parallel}(\theta) = \frac{2\mathbf{n}_i \cos(\theta_i)}{\mathbf{n}_i \cos(\theta_j) + \mathbf{n}_j \cos(\theta_i)} \tag{9.21}$$

For the typical unpolarized case, the light entering the resist will become polarized (but only slightly). Thus, a separate standing wave can be calculated for each polarization and the resulting intensities summed to give the total intensity.

9.5 PHOTORESIST EXPOSURE KINETICS

The kinetics of photoresist exposure is intimately tied to the phenomenon of absorption. The following discussion begins with a description of absorption, followed by the chemical kinetics of exposure. Finally, the chemistry of chemically amplified resists will be reviewed.

9.5.1 ABSORPTION

The phenomenon of absorption can be viewed on a macroscopic or a microscopic scale. On the macro level, absorption is described by the familiar Lambert and Beer laws, which give a linear relationship between absorbance and path length times the concentration of the absorbing species. On the micro level, a photon is absorbed by an atom or molecule, promoting an electron to a higher energy state. Both methods of analysis yield useful information needed in describing the effects of light on a photoresist.

The basic law of absorption is an empirical one with no known exceptions. It was first expressed by Lambert in differential form as

$$\frac{dI}{dz} = -\alpha I \tag{9.22}$$

where
  I is the intensity of light traveling in the z-direction through a medium
  α is the absorption coefficient of the medium and has units of inverse length

In a homogeneous medium (i.e., α is not a function of z), Equation 9.22 may be integrated to yield

$$I(z) = I_0 \exp(-\alpha z) \tag{9.23}$$

where
  z is the distance the light has traveled through the medium
  I_0 is the intensity at $z=0$

If the medium is inhomogeneous, Equation 9.23 becomes

$$I(z) = I_0 \exp(-Abs(z)) \tag{9.24}$$

where

$$Abs(z) = \int_0^z \alpha(z')dz' = \text{the absorbance}$$

When working with electromagnetic radiation, it is often convenient to describe the radiation by its complex electric field vector. The electric field can implicitly account for absorption by using a complex index of refraction $\boldsymbol{n}$ such that

$$\boldsymbol{n} = n - i\kappa \tag{9.25}$$

The imaginary part of the index of refraction, sometimes called the *extinction coefficient*, is related to the absorption coefficient by

$$\alpha = 4\pi\kappa/\lambda \tag{9.26}$$

In 1852, Beer showed that for dilute solutions, the absorption coefficient is proportional to the concentration of the absorbing species in the solution:

$$\alpha_{\text{solution}} = ac \tag{9.27}$$

where
  a is the molar absorption coefficient, given by $a = \alpha MW/\rho$
  MW is the molecular weight
  ρ is the density
  c is the concentration

The stipulation that the solution be dilute expresses a fundamental limitation of Beer's law. At high concentrations, where absorbing molecules are close together, the absorption of a photon by one molecule may be affected by a nearby molecule [21]. Since this interaction is concentration dependent, it causes deviation from the linear relation (Equation 9.27). Also, an apparent deviation from Beer's law occurs if the real part of the index of refraction changes appreciably with concentration. Thus, the validity of Beer's law should always be verified over the concentration range of interest.

For an N-component homogeneous solid, the overall absorption coefficient becomes

$$\alpha_T = \sum_{j=1}^{N} a_j c_j \tag{9.28}$$

Of the total amount of light absorbed, the fraction of light that is absorbed by component i is given by

$$\frac{I_{Ai}}{I_{AT}} = \left(\frac{a_i c_i}{\alpha_T} \right) \tag{9.29}$$

where

I_{AT} is the total light absorbed by the film
I_{Ai} is the light absorbed by component i

We will now apply the concepts of macroscopic absorption to a typical positive photoresist. For g-line (436 nm) and i-line (365 nm) lithography, the most common resists are of the diazonaphtho-quinone/novolac variety. These positive photoresists are made up of three major components; a base novolac resin that gives the resist its structural properties and etch resistance, a photoactive compound M (abbreviated PAC), exposure products P generated by the reaction of M with ultraviolet light, and a solvent S. Although photoresist drying during prebake is intended to drive off solvents, thermal studies have shown that a resist may contain 10% solvent after a 30 min 100 C prebake [22, 23]. The absorption coefficient α is then

$$\alpha = a_M M + a_P P + a_R R + a_S S \tag{9.30}$$

If M_o is the initial PAC concentration (i.e., with no UV exposure), the stoichiometry of the exposure reaction gives

$$P = M_o - M \tag{9.31}$$

Equation 9.30 may be rewritten as [2]

$$\alpha = Am + B \tag{9.32}$$

where

$A = (a_M - a_P) M_o$
$B = a_P M_o + a_R R + a_S S$
$m = M/M_o$

A and B are called the *bleachable* and *nonbleachable* absorption coefficients, respectively, and make up the first two Dill photoresist parameters [2].

The quantities A and B are experimentally measurable [2] and can be easily related to typical resist absorbance curves measured using a UV spectrophotometer. When the resist is fully exposed, $M=0$ and

$$\alpha_{\text{exposed}} = B \tag{9.33}$$

Similarly, when the resist is unexposed, $m=1$ ($M=M_o$) and

$$\alpha_{\text{unexposed}} = A + B \tag{9.34}$$

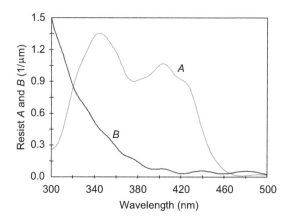

FIGURE 9.7   Resist parameters A and B as a function of wavelength measured using a UV spectrophotometer.

From this, A may be found by

$$A = \alpha_{unexposed} - \alpha_{exposed} \qquad (9.35)$$

Thus, $A(\lambda)$ and $B(\lambda)$ may be determined from the UV absorbance curves of unexposed and completely exposed resist (Figure 9.7).

As mentioned previously, Beer's law is empirical in nature and thus, should be verified experimentally. In the case of positive photoresists, this means formulating resist mixtures with differing photoactive compound to resin ratios and measuring the resulting A parameters. Previous work has shown that Beer's law is valid for conventional photoresists over the full practical range of PAC concentrations [24].

9.5.2   Exposure Kinetics

On a microscopic level, the absorption process can be thought of as photons being absorbed by an atom or molecule, causing an outer electron to be promoted to a higher energy state. This phenomenon is especially important for the photoactive compound, since it is the absorption of UV light that leads to the chemical conversion of M to P:

$$M \xrightarrow{\text{UV}} P \qquad (9.36)$$

This concept is stated in the first law of photochemistry: only the light that is absorbed by a molecule can be effective in producing photochemical change in the molecule. The actual chemistry of diazonaphthoquinone exposure is given below.

The chemical reaction (Equation 9.36) can be rewritten in general form as

$$M \underset{k_2}{\overset{k_1}{\rightleftarrows}} M^* \xrightarrow{k_3} P \qquad (9.37)$$

where

M = the photoactive compound (PAC)
M^* = molecule in an excited state
P = the carboxylic acid (product)
k_1, k_2, k_3 = the rate constants for each reaction

Simple kinetics can now be applied. The proposed mechanism (Equation 9.37) assumes that all reactions are first order. Thus, the rate equation for each species can be written:

$$\frac{dM}{dt} = k_2 M^* - k_1 M$$

$$\frac{dM^*}{dt} = k_1 M - (k_2 + k_3) M^*$$

$$\frac{dP}{dt} = k_3 M^* \tag{9.38}$$

A system of three coupled linear first-order differential equations can be solved exactly using Laplace transforms and the initial conditions

$$M(t = 0) = M_o$$

$$M^*(t = 0) = P(t = 0) = 0 \tag{9.39}$$

However, if one uses the steady-state approximation, the solution becomes much simpler. This approximation assumes that in a very short time, the excited molecule M^* comes to a steady state; that is, M^* is formed as quickly as it disappears. In mathematical form,

$$\frac{dM^*}{dt} = 0 \tag{9.40}$$

A previous study has shown that M^* does indeed come to a steady state quickly, on the order of 10^{-8} seconds or faster [25]. Thus,

$$\frac{dM}{dt} = -KM \tag{9.41}$$

where

$$K = \frac{k_1 k_3}{k_2 + k_3}$$

Assuming K remains constant with time,

$$M = M_o \exp(-Kt) \tag{9.42}$$

The overall rate constant K is a function of the intensity of the exposure radiation. An analysis of the microscopic absorption of a photon predicts that K is directly proportional to the intensity of the exposing radiation [24]. Thus, a more useful form of Equation 9.41 is

$$\frac{dm}{dt} = -CIm \tag{9.43}$$

where

the relative PAC concentration m $(=M/M_o)$ has been used

C is the standard exposure rate constant and the third Dill photoresist parameter

A solution to the exposure rate equation (Equation 9.43) is simple if the intensity within the resist is constant throughout the exposure. However, this is generally not the case for g- and i-line resists. In fact, many resists *bleach* upon exposure; that is, they become more transparent as the photoactive compound M is converted to product P. This corresponds to a positive value of A, as seen, for example, in Figure 9.7. Since the intensity varies as a function of exposure time, this variation must be known in order to solve the exposure rate equation. In the simplest possible case, a resist film coated on a substrate of the same index of refraction, only absorption affects the intensity within the resist. Thus, Lambert's law of absorption, coupled with Beer's law, could be applied.

$$\frac{dI}{dz} = -(Am + B)I \tag{9.44}$$

where Equation 9.32 was used to relate the absorption coefficient to the relative PAC concentration. Equations 9.43 and 9.44 are coupled and thus become first-order nonlinear partial differential equations, which must be solved simultaneously. The solution to Equations 9.43 and 9.44 was first carried out numerically for the case of lithography simulation [2] but in fact was solved analytically by Herrick [26] many years earlier. The same solution was also presented more recently by Diamond and Sheats [27] and by Babu and Barouch [28]. These solutions take the form of a single numerical integration.

Although an analytical solution exists for the simple problem of exposure with absorption only, in more realistic problems, the variation of intensity with depth in the film is more complicated than Equation 9.44. In fact, the general exposure situation results in the formation of standing waves, as discussed previously. In such a case, Equations 9.15–9.18 can give the intensity within the resist as a function of the PAC distribution $m(x,y,z,t)$. Initially, this distribution is simply $m(x,y,z,0) = 1$. Thus, Equation 9.15, for example, would give $I(x,y,z,0)$. The exposure equation (Equation 9.43) can then be integrated over a small increment of exposure time Δt to produce the PAC distribution $m(x,y,z,\Delta t)$. The assumption is that over this small increment in exposure time the intensity remains relatively constant, leading to the exponential solution. This new PAC distribution is then used to calculate the new intensity distribution $I(x,y,z,\Delta t)$, which in turn is used to generate the PAC distribution at the next increment of exposure time $m(x,y,z,2\Delta t)$. This process continues until the final exposure time is reached.

9.5.3 Chemically Amplified Resists

The formulation of state-of-the-art CARs is highly complex, with mixtures composed of polymer resins, usually "blocked" with functional groups to inhibit dissolution or affect other properties, photoacid generators (PAGs), base quenchers (which are possibly light sensitive, such as photodecomposable bases), crosslinking agents, dyes, surfactants, and other specialized additives. Modern CARs are customized for specific applications in IC manufacturing by photoresist designers.

As the name implies, the photoacid generator forms a strong acid when exposed to deep-UV light. Ito and Willson first proposed the use of an arylonium salt [29], and triphenylsulfonium salts have been studied extensively as PAGs. The reaction of a common PAG is shown below:

$$\begin{array}{c} Ph \\ | \\ Ph-S^+ \; CF_3COO^- \xrightarrow{\; h\nu \;} CF_3COOH + others \\ | \\ Ph \end{array}$$

The acid generated in this case (trifluoroacetic acid) is a derivative of acetic acid in which the electron-drawing properties of the fluorines are used to greatly increase the acidity of the molecule. The reaction is an example of the direct photolysis of PAG, the exposure mechanism most prevalent in CARs used in state-of-the-art 248 and 193 nm lithography. In direct photolysis, the absorption of one photon by the photosensitive molecule ideally results in the completion of the reaction and the release of the photoproduct. Upon absorption, the PAG enters an electronically excited state, possessing excess energy favorable for completion of the reaction. The excess energy can be dissipated either by radiative mechanisms such as fluorescence or phosphorescence or by radiationless mechanisms, in this case the completion of the reaction and the release of the acid. The probability that the reaction proceeds to completion is equal to the quantum efficiency ϕ with the probability of relaxation to the ground state equal to $1-\phi$.

$$\text{pag} \xrightarrow{hv} \text{pag}^*$$

$$\left\langle \text{acid} \,\middle|\, \text{pag}^* \right\rangle = \phi \tag{9.45}$$

$$\left\langle \text{pag} \,\middle|\, \text{pag}^* \right\rangle = 1 - \phi$$

The group of atoms that serve as a unit in the absorption of light are known as the *chromophore*. Considerable effort is expended in the optimization of the absorption properties of chromophores used in CARs. In the example used here, the triphenylsulfonium (TPS) cation serves as the chromophore. The direct photolytic reaction is dependent upon resonance between the oscillations of the light and the electrons of the PAG. Energy conservation requires that

$$\Delta E = hv \tag{9.46}$$

where
ΔE is an energy gap between two electronic states of the PAG
h is Planck's constant
v is the frequency of the light

The condition is imposed by the quantized nature of electronic states. The interaction of the light with the molecule depends on the energy gap or the frequency of the oscillations possible for the electrons of the molecule. The oscillating electric field produced by the passing light wave applies a vector force upon the electrons of the PAG, reshaping its electron distribution [30]. The total electromagnetic force on an electron produced by the passing light wave can be expressed by the Lorentz force on a moving charge:

$$F = e\left(\xi + v \times B\right) \tag{9.47}$$

where
e is the fundamental charge of the electron in coulombs
ξ represents the electric field in units of newtons per coulomb
$e\xi$ gives the electric force component in newtons and is independent of the motion of the electron
B represents the magnetic field vector with the magnetic force given by $ev \times B$, with the units of B one newton-second per coulomb-meter (or 1 tesla)

The work done on the electron is due to the electric force and not the magnetic force. The kinetics of the PAG exposure reaction are standard first order:

$$\frac{\partial G}{\partial t} = -CIG \tag{9.48}$$

where

    G is the concentration of PAG at time t

    I is the intensity of light

    C is the exposure rate constant

For constant intensity, the rate equation can be solved for G:

$$G = G_o e^{-CIt} \tag{9.49}$$

where G_o is the initial concentration of PAG. The acid concentration H is given by

$$H = G_o - G = G_o \left(1 - e^{-CIt}\right) \tag{9.50}$$

The quantum yield is given by the ratio of the number of acid molecules that are released to the number of photons that are absorbed:

$$\Phi = \frac{n_{\text{acids}}}{n_{\text{absorbed photons}}} \tag{9.51}$$

Since one photon absorption event may generate at most one acid, it is straightforward to see that the maximum possible quantum yield for the direct photolytic mechanism is 1.

The exposure rate constant C for direct photolysis can be calculated with knowledge of the molar absorptivity or molar extinction coefficient and the quantum efficiency of the PAG:

$$C = \phi \frac{a}{N_A} \left(hv\right)^{-1} = \phi \frac{2.303\,\varepsilon}{N_A} \left(hv\right)^{-1} \tag{9.52}$$

where

    C is expressed in units of area per joule (usually cm^2/mJ)

    a is the log base e molar absorptivity of the PAG in cm^2/mol

    N_A is Avogadro's number

Alternatively, we can use ε, the log base 10 molar extinction coefficient of the PAG, also in cm^2/mol. It is expressed as

$$\varepsilon = \log_{10}\left(I_0/I_T\right)/\left(dG\right) \tag{9.53}$$

The numerator on the right side of Equation 9.53 represents the unitless optical density of a thin film; the molar extinction coefficient is derived by dividing the optical density by the product of the known concentration G and the optical path length for absorption equivalent to d.

Exposure of the resist with an aerial image $I(x)$ results in an acid latent image $H(x)$. A postexposure bake (PEB) is then used to thermally induce a chemical reaction. This may be the activation of a crosslinking agent for a negative resist or the deblocking of the polymer resin for a positive resist. The reaction is catalyzed by the acid, so the acid is not consumed by the reaction and H remains constant. Ito and Willson first proposed the concept of deblocking a polymer to change its solubility [29]. A base polymer such as poly (p-hydroxystyrene) (PHS) is used, which is very soluble in an aqueous base developer. It is the hydroxyl groups that give the PHS its high solubility, so by "blocking" these sites (by reacting the hydroxyl group with some longer-chain molecule), the solubility can be reduced. Ito and Willson employed a t-butoxycarbonyl group (t-BOC), resulting in a very slowly dissolving polymer. In the presence of acid and heat, the t-BOC-blocked polymer will undergo acidolysis to generate the soluble hydroxyl group, as shown below.

One drawback of this scheme is that the cleaved t-BOC is volatile and will evaporate, causing film shrinkage in the exposed areas. Higher–molecular weight blocking groups can be used to reduce this film shrinkage to acceptable levels (below 10%). Also, the blocking group is such an effective inhibitor of dissolution that nearly every blocked site on the polymer must be deblocked in order to obtain significant dissolution. Thus, the photoresist can be made more "sensitive" by only partially blocking the PHS. Typical photoresists use 10–30% of the hydroxyl groups blocked, with 20% a typical value. Molecular weights for the PHS run in the range of 3000 to 5000, giving about 20 to 35 hydroxyl groups per molecule.

Using M as the concentration of some reactive site, these sites are consumed (i.e., are reacted) according to kinetics of some unknown order n in H and first order in M [31]:

$$\frac{\partial M}{\partial t'} = -K_{amp} M H^n \tag{9.54}$$

where
  K_{amp} is the rate constant of the amplification reaction (crosslinking, deblocking, etc.)
  t' is the bake time

Simple theory would indicate that $n = 1$, but the general form will be used here. Assuming H is constant, Equation 9.54 can be solved for the concentration of reacted sites X:

$$X = M_o - M = M_o \left(1 - e^{-K_{amp} H^n t'}\right) \tag{9.55}$$

(Note: Although H^+ is not consumed by the reaction, the value of H is not locally constant. Diffusion during the PEB and acid loss mechanisms cause local changes in the acid concentration, thus requiring the use of a reaction–diffusion system of equations. The approximation that H is constant is a useful one, however, which gives insight into the reaction as well as accurate results under some conditions.)

It is useful here to normalize the concentrations to some initial values. This results in a normalized acid concentration h and normalized reacted and unreacted sites x and m:

$$h = \frac{H}{G_o} \qquad x = \frac{X}{M_o} \qquad m = \frac{M}{M_o} \tag{9.56}$$

Equations 9.50 and 9.56 become

$$h = 1 - e^{-CIt}$$

$$m = 1 - x = e^{-\alpha h^n} \tag{9.57}$$

where α is a lumped "amplification" constant equal to $G_o{}^n K_{amp} t'$. The result of the PEB is an amplified latent image $m(x)$, corresponding to an exposed latent image $h(x)$, resulting from the aerial image $I(x)$.

The above analysis of the kinetics of the amplification reaction assumed a locally constant concentration of acid H. Although this could be exactly true in some circumstances, it is typically only an approximation and is often a poor approximation. In reality, the acid diffuses during the bake. In one dimension, the standard diffusion equation takes the form

$$\frac{\partial H}{\partial t'} = \frac{\partial}{\partial z}\left(D_H \frac{\partial H}{\partial z}\right) \tag{9.58}$$

where D_H is the diffusivity of acid in the photoresist. Solving this equation requires a number of things: two boundary conditions, one initial condition, and knowledge of the diffusivity as a function of position and time.

The initial condition is the initial acid distribution within the film, $H(x,0)$, resulting from the exposure of the PAG. The two boundary conditions are at the top and bottom surfaces of the photoresist film. The boundary at the wafer surface is assumed to be impermeable, giving a boundary condition of no diffusion into the wafer. The boundary condition at the top of the wafer will depend on the diffusion of acid into the atmosphere above the wafer. Although such acid loss is a distinct possibility, it will not be treated here. Instead, the top surface of the resist will also be assumed to be impermeable.

The solution of Equation 9.51 can now be performed if the diffusivity of the acid in the photoresist is known. Unfortunately, this solution is complicated by two very important factors: the diffusivity is a strong function of temperature and, most probably, the extent of amplification. Since the temperature is changing with time during the bake, the diffusivity will be time dependent. The concentration dependence of diffusivity results from an increase in free volume for typical positive resists: as the amplification reaction proceeds, the polymer blocking group evaporates, resulting in a decrease in film thickness but also an increase in free volume. Since the acid concentration is time and position dependent, the diffusivity in Equation 9.51 must be determined as a part of the solution of Equation 9.51 by an iterative method. The resulting simultaneous solution of Equations 9.47 and 9.51 is called a *reaction-diffusion system*.

The temperature dependence of the diffusivity can be expressed in a standard Arrhenius form:

$$D_o(T) = A_R \exp\left(-E_a/RT\right) \tag{9.59}$$

where
  D_o is a general diffusivity
  A_r is the Arrhenius coefficient
  E_a is the activation energy

A full treatment of the amplification reaction would include a thermal model of the hotplate in order to determine the actual time–temperature history of the wafer [32]. To simplify the problem, an ideal temperature distribution will be assumed: the temperature of the resist is zero (low enough for no diffusion or reaction) until the start of the bake, at which time it immediately rises to the final bake temperature, stays constant for the duration of the bake, and then instantly falls back to zero.

The concentration dependence of the diffusivity is less obvious. Several authors have proposed and verified the use of different models for the concentration dependence of diffusion within a polymer. Of course, the simplest form (besides a constant diffusivity) would be a linear model. Letting D_o be the diffusivity of acid in completely unreacted resist and D_f the diffusivity of acid in resist that has been completely reacted,

$$D_H = D_o + x\left(D_f - D_o\right) \tag{9.60}$$

Here, diffusivity is expressed as a function of the extent of the amplification reaction. Another common form is the Fujita–Doolittle equation [33], which can be predicted theoretically using free volume arguments. A form of that equation, which is convenient for calculations, is shown here:

$$D_H = D_o \exp\left(\frac{\alpha x}{1 + \beta x}\right) \tag{9.61}$$

where α and β are experimentally determined constants and are, in general, temperature dependent. Other concentration relations are also possible [34], but the Fujita–Doolittle expression will be used in this work.

Through a variety of mechanisms, acid formed by exposure of the resist film can be lost and thus not contribute to the catalyzed reaction to change the resist solubility. There are two basic types of acid loss: loss that occurs between exposure and post-exposure bake, and loss that occurs during the post-exposure bake.

The first type of loss leads to delay time effects—the resulting lithography is affected by the delay time between exposure and post-exposure bake. Delay time effects can be very severe and, of course, are very detrimental to the use of such a resist in a manufacturing environment [35, 36]. The typical mechanism for delay time acid loss is the diffusion of atmospheric base contaminates into the top surface of the resist. The result is a neutralization of the acid near the top of the resist and a corresponding reduced amplification. For a negative resist, the top portion of a line is not insolubilized, and resist is lost from the top of the line. For a positive resist, the effects are more devastating. Sufficient base contamination can make the top of the resist insoluble, blocking dissolution into the bulk of the resist. In extreme cases, no patterns can be observed after development. Another possible delay time acid loss mechanism is base contamination from the substrate, as has been observed on TiN substrates [36].

The effects of acid loss due to atmospheric base contaminants can be accounted for in a straightforward manner [37]. The base diffuses slowly from the top surface of the resist into the bulk. Assuming that the concentration of base contaminate in contact with the top of the resist remains constant, the diffusion equation can be solved for the concentration of base, B, as a function of depth into the resist film:

$$B = B_o \exp\left(-(z/\sigma)^2\right) \tag{9.62}$$

where
- B_o is the base concentration at the top of the resist film
- z is the depth into the resist ($z=0$ at the top of the film)
- σ is the diffusion length of the base in resist

The standard assumption of constant diffusivity has been made here so that diffusion length goes as the square root of the delay time.

Since the acid generated by exposure for most resist systems of interest is fairly strong, it is a good approximation to assume that all of the base contaminant will react with acid if there is sufficient acid present. Thus, the acid concentration at the beginning of the PEB, H^*, is related to the acid concentration after exposure, H, by

$$H^* = H - B \quad \text{or} \quad h^* = h - b \tag{9.63}$$

where the lower-case symbols again represent the concentration relative to G_o, the initial PAG concentration.

Acid loss during the PEB could occur by other mechanisms. For example, as the acid diffuses through the polymer, it may encounter sites that "trap" the acid, rendering it unusable for further amplification. If these traps were in much greater abundance than the acid itself (for example, sites on the polymer), the resulting acid loss rate would be first order:

$$\frac{\partial h}{\partial t'} = -K_{\text{loss}}\, h \tag{9.64}$$

where K_{loss} is the acid loss reaction rate constant. Of course, other more complicated acid loss mechanisms can be proposed, but in the absence of data supporting them, the simple first-order loss mechanism will be used here.

Acid can also be lost at the two interfaces of the resist. At the top of the resist, acid can evaporate. The amount of evaporation is a function of the size of the acid and the degree of its interaction with the resist polymer. A small acid (such as the trifluoroacetic acid discussed earlier) may have very significant evaporation. A separate rate equation can be written for the rate of evaporation of acid:

$$\left.\frac{\partial h}{\partial t'}\right|_{z=0} = -K_{\text{evap}}\left(h(0,t') - h_{\text{air}}(0,t')\right) \tag{9.65}$$

where

$z = 0$ is the top of the resist

h_{air} is the acid concentration in the atmosphere just above the photoresist surface

Typically, the PEB takes place in a reasonably open environment with enough air flow to eliminate any buildup of evaporated acid above the resist, making $h_{\text{air}} = 0$. If K_{evap} is very small, virtually no evaporation takes place, and we say that the top boundary of the resist is impenetrable. If K_{evap} is very large (resulting in evaporation that is much faster than the rate of diffusion), the effect is to bring the surface concentration of acid in the resist to zero.

At the substrate, there is also a possible mechanism for acid loss. Substrates containing nitrogen (such as titanium nitride and silicon nitride) often exhibit a foot at the bottom of the resist profile [36]. Most likely, the nitrogen acts as a site for trapping acid molecules, which gives a locally diminished acid concentration at the bottom of the resist. This, of course, leads to reduced amplification and a slower development rate for a positive resist, resulting in the resist foot. The kinetics of this substrate acid loss will depend on the concentration of acid trap sites at the substrate, S. It will be more useful to express this concentration relative to the initial concentration of PAG.

$$s = \frac{S}{G_o} \tag{9.66}$$

A simple trapping mechanism would have one substrate trap site react with one acid molecule.

$$\left.\frac{\partial h}{\partial t'}\right|_{z=D} = -K_{\text{trap}}\, h(D,t')\, s \tag{9.67}$$

Of course, the trap sites would be consumed at the same rate as the acid. Thus, knowing the rate constant K_{trap} and the initial relative concentration of substrate trapping sites s_o, one can include Equation 9.67 in the overall mechanism of acid loss.

The combination of a reacting system and a diffusing system is called a reaction–diffusion system. The solution of such a system is the simultaneous solution of Equations 9.54 and 9.59 using Equation 9.50 as an initial condition and Equation 9.61 or 9.62 to describe the reaction-dependent diffusivity. Of course, any or all of the acid loss mechanisms can also be included. A convenient and straightforward method to solve such equations is the finite difference method (see, for example,

Incropera and DeWitt [38]). The equations are solved by approximating the differential equations by difference equations. By marching through time and solving for all space at each time step, the final solution is the result after the final time step. A key part of an accurate solution is the choice of a sufficiently small time step. If the spatial dimension of interest is Δx (or Δy or Δz), the time step should be chosen such that the diffusion length is less than Δx (using a diffusion length of about one-third of Δx is common).

9.5.4 Stochastic Modeling of Exposure in Chemically Amplified Resists

Modeling strategies for optical lithography have historically (and successfully) relied on the *continuum approximation* to describe the physical world being simulated. Even though light and matter are quantized into photons and spatially distributed molecules, respectively, the calculations of aerial and latent images ignore the discrete nature of these fundamental units and instead use continuous mathematical functions and "classical" chemical rate equations. Continuum models allow the computational domain to be subdivided into ever-smaller cells, with each cell retaining the properties of the bulk. An alternative to continuum modeling and, in a real sense, a more fundamental approach is to attempt to build the quantization of light and matter directly into the models in what is called *stochastic resist modeling*. Stochastic models are advantageous when smaller features and shorter exposure wavelengths dominate the process of interest. Two examples of processes where stochastic modeling is routinely used are high-NA ArF and extreme ultraviolet (EUV) lithography. Stochastic modeling approaches involve the use of random variables and probability density functions to describe the expected statistical fluctuations. The lithographic metrics that we obtain through stochastic modeling are described by the first and second moments of their probability distributions—the average and the variance. Stochastic modeling can be used to investigate interesting and important phenomena that cannot be studied with continuum approaches. For example:

1. The counting statistics of absorbed photons, molecules of resist reactants, generated acids, or other photoproducts
2. The effects of photon shot noise or acid (chemical) shot noise
3. The effects of resist ionization and reactivity with scattering photoelectrons at EUV
4. The uncertainty in feature edge placement, CD, or uniformity

9.5.4.1 Photon Energy

Light is described by both wave and quantum theory. The concept that light travels as discrete packets of energy can be traced back to Sir Isaac Newton's corpuscular theory. Christian Huygens later discovered that light actually did behave like a wave, culminating in James Clerk Maxwell's groundbreaking discovery of the laws of electrodynamics. Yet in 1905, Albert Einstein, in his explanation of the photoelectric effect, showed that light does sometimes behave like a particle. According to quantum theory, light consists of individual photons, each small enough to be absorbed by a single electron and with kinetic energy equivalent to the product of the frequency of the light and Planck's constant:

$$E = hv = h\frac{c}{\lambda} \tag{9.68}$$

The wave–quantum duality is immediately evident in that the calculation of the kinetic energy of a photon depends on the frequency of the light wave. If we consider a monochromatic beam of light made up of photons all traveling with the same velocity c and energy E, the photons possess linear momentum p:

$$p = \frac{E}{c} = \frac{hv}{c} \tag{9.69}$$

with long-wavelength, low-frequency photons carrying much less energy and momentum than short-wavelength, high-frequency photons. Comparing the energy of 193 and 13.5 nm photons, used in ArF and EUV lithography, the photon energy for 193 nm is

$$E = h\nu = h\frac{c}{\lambda} = \frac{\left(6.626\times10^{-31}\,\text{mJ}\cdot\text{s}\right)\left(2.998\times10^{17}\,\text{nm/s}\right)}{193\,\text{nm}} \quad (9.70)$$

$$= 1.03\times10^{-15}\,\text{mJ} = 1.03\times10^{-18}\,\text{J}$$

Converting the energy to units of electron volts, eV, the photon energy is

$$\left(1.03\times10^{-18}\,\text{J}\right)\left(6.242\times10^{18}\,\frac{\text{eV}}{\text{J}}\right) = 6.4\,\text{eV}$$

The photon energy in units of joules for 13.5 nm light is

$$E = h\nu = h\frac{c}{\lambda} \cong \frac{\left(6.626\times10^{-31}\,\text{mJ}\cdot\text{s}\right)\left(2.998\times10^{17}\,\text{nm/s}\right)}{13.5\,\text{nm}}$$

$$= 1.47\times10^{-14}\,\text{mJ} = 1.47\times10^{-17}\,\text{J}$$

In units of electron volts, the energy of a 13.5 nm photon is

$$\left(1.47\times10^{-17}\,\text{J}\right)\left(6.242\times10^{18}\,\frac{\text{eV}}{\text{J}}\right) = 91.8\,\text{eV}$$

and we observe that the energy of one photon of 13.5 nm light is about 14 times greater than the energy of one photon of 193 nm light.

9.5.4.2 Photon Counting and Shot Noise

Quantum theory allows us to consider the photon as a reagent, which may collide, react, and initiate further events after absorption by the molecules of the photoresist. Therefore, it is the statistical fluctuations that occur during the absorption of light that will be of most interest.

Repeated independent trials are called *Bernoulli* trials if there are only two possible outcomes for each trial and their probabilities remain the same throughout the trials. The two probabilities are denoted by p and q and refer to the outcome with probability p as "success" and q as "failure." Clearly, p and q must be nonnegative, and $p+q=1$. We are mostly interested in the total number of successful absorption events produced in a succession of n Bernoulli trials and not necessarily in their order. There are two possible outcomes for each trial, in that either a photon is absorbed by the resist or it is not (photon transmission, reflection, refraction, and diffraction also represent repeated Bernoulli trials). The integer number of absorbed photons can be 0, 1, 2,…, n. Our problem is to determine the corresponding probabilities. The event "n trials result in k successes and $n - k$ failures" can happen in as many ways as k letters can be distributed among n places [39]. Our event therefore contains $\binom{n}{k}$ points, with each point having the probability $p^k q^{n-k}$.

Let $b(k;n,p)$ be the probability that n Bernoulli trials with probabilities p for success and $q=1-p$ for failure result in k successes and $n - k$ failures. The well-known *binomial* distribution is

$$b(k;n,p) = \binom{n}{k}p^k q^{n-k} \quad (9.71)$$

The binomial distribution becomes cumbersome to work with as the number of trials becomes large. It's more convenient to use Poisson's approximation to the binomial distribution, which is

$$p(k; L) = e^{-L} \frac{L^k}{k!} \tag{9.72}$$

where $L = np$ is equivalent to the average number of successful trials. One important property of the Poisson distribution is that the first and second moments, the average and the variance, are equal. The interested reader is referred to Feller [39] for a detailed description of the binomial and Poisson distributions.

$$\langle k \rangle = L, \quad \mathrm{var}(k) = L \tag{9.73}$$

Consider a volume of photoresist V of constant absorbance α irradiated with dose E_{inc} in a lithographic process. If the volume is small enough that we can treat the incident exposure dose inside V as constant, then the average number of photons that are absorbed by V is equivalent to the total energy absorbed in V divided by the energy of one photon:

$$\langle n \rangle = \frac{\alpha E_{inc} V}{h\nu} = \alpha E_{inc} V \left(\frac{\lambda}{hc} \right) \tag{9.74}$$

where
  $\langle n \rangle$ represents the average number of photons absorbed by V
  E_{inc} is the incident exposure dose in units of joules per area (mJ/cm^2)

It is easy to see that the average number of photons absorbed by the volume is proportional to the resist absorbance, the incident dose, the wavelength of light, and the size of the volume. As an example, consider the difference in the average number of photons absorbed in two cubes of resist 3 nm on a side, each with absorbance equal to 5/µm and irradiated at an exposure dose of 20 mJ/cm^2. One cube is exposed to 193 nm light, while the other cube is exposed to 13.5 nm light.

For the 193 nm case, the average number of photons absorbed in the cube is

$$\langle n \rangle = \alpha E_{inc} V (h\nu)^{-1}$$

$$= (0.005/\mathrm{nm})(2.0 \times 10^{-13}\,\mathrm{mJ/nm^2})(27\,\mathrm{nm^3})$$

$$\left(\frac{193\,\mathrm{nm}}{(6.626 \times 10^{-31}\,\mathrm{mJ \cdot s})(2.998 \times 10^{17}\,\mathrm{nm/s})} \right)$$

$$\langle n \rangle = 26.2\ \text{photons}$$

By the properties of the Poisson distribution noted earlier, the average and the variance are equal.

$$\mathrm{var}(n) = 26.2\ \text{photons}$$

For the 13.5 nm case, we have

$$\langle n \rangle = (0.005/\mathrm{nm})(2.0 \times 10^{-13}\,\mathrm{mJ/nm^2})(27\,\mathrm{nm^3})$$

$$\left(\frac{13.5\,\mathrm{nm}}{(6.626 \times 10^{-31}\,\mathrm{mJ \cdot s})(2.998 \times 10^{17}\,\mathrm{nm/s})} \right)$$

$$\langle n \rangle = 1.8 \text{ photons}$$

$$\text{var}(n) = 1.8 \text{ photons}$$

The average total amount of energy absorbed in the cube is $\alpha E_{\text{inc}} V = 2.7 \times 10^{-14}$ mJ and is identical for both the 193 and 13.5 nm cases (by construction), yet in the case of 13.5 nm light, we see that the energy is quantized into about 14× fewer photons.

We can calculate and plot the probability distribution function (pdf) of observing the integer numbers of absorbed photons using the averages calculated here and the equation for the Poisson distribution. The probability that k photons are absorbed in V is

$$p\left(k; \langle n \rangle \right) = e^{-\langle n \rangle} \frac{\langle n \rangle^k}{k!} \tag{9.75}$$

The average number of photons absorbed in V is $\langle n \rangle$, and the variance is also $\langle n \rangle$ (Figure 9.8).

Let us now compare the standard deviation of the absorbed energy for both cases. We can see from Figure 9.8 that at 13.5 nm, interestingly, there is a ~15% chance that no photons at all will be absorbed in V, while at 193 nm, the probability of absorbing no photons in V is about zero.

The standard deviation of the absorbed energy for each case is

$$\sigma_E = \sigma_n h v = \sqrt{\text{var}(n)} \, h v = \sqrt{\langle n \rangle} \, h v$$

$$\sigma_E, 193 \, \text{nm} = \sqrt{26.23} \times 6.42 \, \text{eV} = 32.9 \, \text{eV}$$

$$\sigma_E, 13.5 \, \text{nm} = \sqrt{1.83} \times 91.84 \, \text{eV} = 124.4 \, \text{eV}$$

Let us also calculate the relative uncertainty in observing the expected absorbed energy for each case. The relative uncertainty in observing the expected absorbed energy can be expressed as the

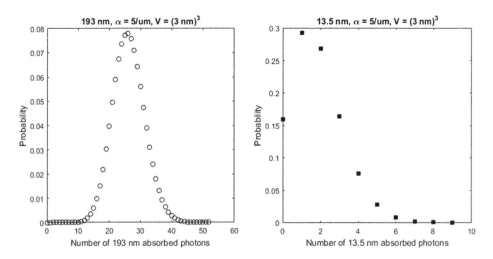

FIGURE 9.8 Comparative Poisson distributions of the probability of absorbing a given number of photons in a small cube of resist irradiated at ArF and EUV exposure wavelengths. Even though the average amount of absorbed energy is identical for both wavelengths, the plot on the right shows that there is about a 15% probability that no photons are absorbed in the cube for the 13.5 nm case.

standard deviation of the absorbed energy divided by the average absorbed energy (identical for each wavelength). At 193 nm, the relative uncertainty in absorbing the expected energy is

$$\frac{\sigma_E}{\langle E \rangle}, 193\,\text{nm} = 32.9\,\text{eV}/168.52\,\text{eV} = 0.195 \cong 20\%$$

and the relative uncertainty in absorbing the expected number of photons at 193 nm is also about 20%:

$$\frac{\sigma_n}{\langle n \rangle}, 193\,\text{nm} = \sqrt{26.2\,\text{photons}}/26.2\,\text{photons} = 0.195 \cong 20\%$$

At 13.5 nm, the relative uncertainty in absorbing the expected energy and the expected number of photons is

$$\frac{\sigma_E}{\langle E \rangle}, 13.5\,\text{nm} = 124.4\,\text{eV}/168.52\,\text{eV} = 0.738 \cong 74\%$$

$$\frac{\sigma_n}{\langle n \rangle}, 13.5\,\text{nm} = \sqrt{1.8\,\text{photons}}/1.8\,\text{photons} = 0.738 \cong 74\%$$

The relative uncertainty of absorbing the expected, average energy at 193 nm is observed to be about 20%, while at 13.5 nm, it's about 74%. This is a direct consequence of absorbing 14× fewer quanta at 13.5 nm. We see that the uncertainty in absorbing the expected number of photons is also ~20% at 193 nm versus ~74% at 13.5 nm. That the relative uncertainty in absorbing the expected number of photons grows as the wavelength is reduced highlights a naturally occurring phenomenon known as *photon shot noise* (PSN).

$$PSN = \frac{\sqrt{\text{var}\,n}}{\langle n \rangle} = \frac{\sigma_n}{\langle n \rangle} = \frac{1}{\sqrt{\langle n \rangle}} \tag{9.76}$$

In photoresists, shot noise and its consequences worsen as the resist absorbance, the wavelength, the incident exposure dose, and the volume of interest decrease. Figure 9.9 shows a comparative plot of the effects of photon shot noise upon the absorbed energy. Plotted is the relative uncertainty in absorbing the expected energy versus the incident dose for ArF and EUV, using the conditions described above.

As an aside, recall that as lithographic manufacturing processes advance, the dimensions of the functional elements within integrated circuits created by lithography have decreased, while the number of elements per device has continually increased, a trend commonly described as Moore's law [40]. But Moore's law is limited (at least) by the effects of shot noise; the consequences of shot noise are thought to be the ultimate limiter of optical lithographic technology, since modern processes seek to manipulate matter and energy at ever-smaller length scales.

9.5.4.3 Chemical Concentration

The idea of chemical concentration, the average number of molecules per unit volume, makes the assumption that the volume in which one is interested is large enough to contain many, many molecules, so the concentration in each subvolume or cell can be calculated using the bulk average number of molecules. However, chemical concentration loses its meaning as the cell becomes very small. In reality, we know that matter is composed of many discrete molecules with "empty space" between them. Photoresists are amorphous mixtures composed of resin polymers, photo-acid

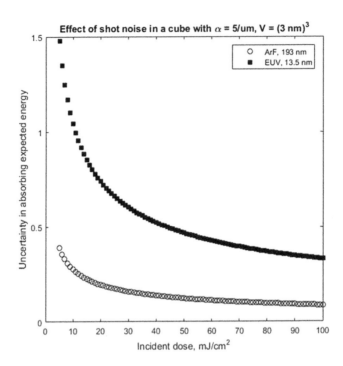

FIGURE 9.9 Comparison of the photon shot noise (PSN) effect in two small cubes of resist irradiated at 193 and 13.5 nm exposure wavelengths. The average amount of energy absorbed by each cube is identical for both wavelengths, but the uncertainty in observing the average absorbed energy is much worse at EUV than at ArF.

generators, base quenchers, and others, each a discrete molecule of finite size. Fluctuations in the local densities of these reactants may strongly modify the local properties of the resist; these fluctuations can produce effects that are easily observed, such as line-edge roughness (LER), but not easily mitigated and are quite problematic for lithography.

One way to partially account for these fluctuations in a stochastic modeling approach is to replace the bulk concentration of the reactants with local densities:

$$\rho_i = \frac{\langle n_i \rangle}{V_{\text{cell}}} \tag{9.77}$$

where $\langle n_i \rangle$ is the average number of particles of type i in a cell of volume V_{cell}. The density ρ represents the average number of particles in the cell averaged over many randomized trials but ignores any fluctuations from this average, or $\left(n_i - \langle n \rangle \right)^2$. Similarly to the shot noise effect described above, if $\langle n \rangle$ is large enough, then the fluctuations are negligible; if not, then the fluctuations may dominate the uncertainty of observing the expected (average) number of particles. Therefore, the cells in the computational domain ought to be small enough to justify the use of continuous functions to model the reactions in the cell yet large enough to minimize the effects of fluctuations—often a difficult compromise [41].

Molecules exhibit counting statistics that are identical to those of photons. Let C represent the average number of molecules per unit volume and dV a volume small enough that at most one molecule may be found in it. The probability of finding a molecule in dV is CdV. For some larger volume V, the probability of finding exactly n molecules in that volume will be given by a binomial distribution, which is well approximated by a Poisson distribution:

$$p(n; CV) = e^{-CV} \frac{(CV)^n}{n!} \tag{9.78}$$

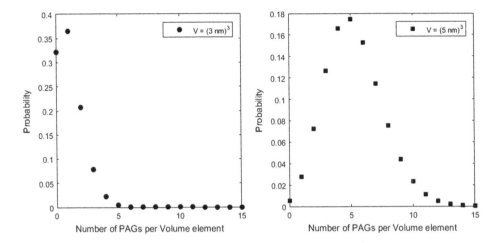

FIGURE 9.10 Comparison of the probabilities of the numbers of PAGs contained in two cubes of resist of different volumes. The average number density of PAGs is identical at 0.042 PAG molecules per cubic nanometer. The PAG counts follow Poisson statistics. Interestingly, for the smaller cube of volume $(3 \text{ nm})^3$, there is a ca. 32% probability that the cube will contain zero PAGs.

The average number of molecules in V is CV, and the variance is also CV. The relative uncertainty in the number of molecules found in V is

$$\frac{\sigma_n}{\langle n \rangle} = \frac{1}{\sqrt{\langle n \rangle}} = \frac{1}{\sqrt{CV}} \qquad (9.79)$$

As an example, consider a 193 nm photoresist with an initial PAG concentration of 3% by weight or a concentration of ca. 0.07 moles/liter, corresponding to a density of 1.2 g/ml and a molecular weight of 500 g/mole. Converting from moles to molecules using Avogadro's number, this corresponds to an average of 0.042 molecules of PAG per cubic nanometer. The probability that n molecules are found within a small cube of resist 3 nm on a side is

$$CV = \frac{0.042 \, \text{PAGs}}{\text{nm}^3} \times (3 \, \text{nm})^3 = 1.1 \, \text{PAGs}$$

$$p(n; 1.1 \, \text{PAGs}) = e^{-1.1 \text{PAGs}} \frac{(1.1 \, \text{PAGs})^n}{n!}, \quad n = 0, 1, \ldots N \qquad (9.80)$$

The probability that n molecules are found within a larger cube of resist 5 nm on a side is

$$CV = \frac{0.042 \, \text{PAGs}}{\text{nm}^3} \times (5 \, \text{nm})^3 = 5.3 \, \text{PAGs}$$

$$p(n; 5.3 \, \text{PAGs}) = e^{-5.3 \text{PAGs}} \frac{(5.3 \, \text{PAGs})^n}{n!}, \quad n = 0, 1, \ldots N \qquad (9.81)$$

The probability distributions of observing n molecules for each cube are plotted in Figure 9.10.

9.5.5 A Monte Carlo Scheme for the Stochastic Simulation of Direct Photolysis

The Monte Carlo (MC) method is a mathematical tool that uses random numbers to determine the outcome of probabilistic simulations. An MC method may be computationally expensive, and therefore we attempt to simulate only the most important effects that are present in the actual physical phenomenon to a greater or lesser degree. Let us introduce a simple, reliable Monte Carlo procedure to perform a stochastic simulation of photoresist exposure by the mechanism of direct PAG photolysis. Within statistical uncertainty, the method gives results in good agreement with experiment. The intent of the method is to reproduce the most important statistical fluctuations present in the direct photolytic mechanism. The precision of the results depends upon the number of randomized trials that are performed. In other words, extracting precise data from a stochastic simulator requires similar noise reduction techniques to those used in actual experiments in a lab. Evolution in the ability of computers to perform more calculations in less time allows us to perform many randomized trials reasonably fast and increase precision.

Since the Monte Carlo method depends upon the use of random numbers, it depends on securing reliable algorithms for random number generation. Specifically, we will need algorithms that produce streams of (virtually) nonrepeating uniformly distributed and Poisson-distributed random deviates. Uniform deviates are random numbers that lie uniformly within a specified range, usually 0.0 to 1.0. Within this range, one number is just as likely as any other, and the random number stream should converge to a sample mean of 1/2. The Poisson distribution is a discrete distribution, and a Poisson random deviate is therefore an integer drawn from a Poisson distribution with a known mean. Poor random number generators are more frequently encountered than one might expect and introduce artifacts, systematically biasing simulation data. The design of algorithms for random number generation is a vast topic that is beyond the scope of this text. The reader is advised to proceed with caution and due diligence and to consult some of the many useful references on this topic, among them Teukolsky et al. [42].

We adopt Cartesian coordinates to describe a rectangular, three-dimensional computational domain Ω representing the volume of photoresist to be (stochastically) exposed. Let us consider a simple model resist that is composed of a single species of PAG, a single species of conventional quencher (i.e., a quencher that is not light sensitive), and a blocked polymer resin. The discrete PAG and quencher molecules are dispersed randomly within Ω, and we ignore any volume-exclusion effects that in reality prevent excessive loadings of the reactants. To keep track of the positions of PAGs, quenchers, and any released acids, we record only the centers-of-mass of each molecule. The polymer resin and its protecting groups are assumed to be at a constant concentration throughout Ω and are not treated discretely. We take the following as given inputs to the MC scheme:

1. The normalized intensity of actinic light I in Ω
2. The incident exposure dose E_{inc} in mJ/cm^2
3. The quantum efficiency ϕ of the PAG
4. ε, the log base 10 molar extinction coefficient of the PAG in cm^2/mol
5. The imaginary part of the refractive index κ of the polymer resin

We first require calculations of the average loadings of each reactant as number-densities, for example in units of 1/nm^3. These can be calculated in many different ways, and we present two simple approaches here. The number density of PAG in a mM solution is calculated as

$$\left\langle \rho_{PAG} \right\rangle = \frac{\left\langle n\,PAGs \right\rangle}{nm^3} = mM_{PAG} \cdot 1.0 \times 10^{-30}\,\frac{mols}{nm^3} \cdot N_A \tag{9.82}$$

Alternatively, the number density of PAG in dry resist can be calculated from a weight percentage by

$$\left\langle \rho_{PAG} \right\rangle = \frac{\left\langle n\,PAGs \right\rangle}{nm^3} = wt\%_{PAG} \cdot \rho_{g\,resist/nm^3} \cdot \frac{1}{Mw_{g\,PAG/mol}} N_A \tag{9.83}$$

where

$wt\%_{PAG}$ is the weight-percent of PAG in dry resist

$\rho_{g\,resist/nm^3}$ is the density of dry resist in g/nm^3

$Mw_{g\,PAG/mol}$ is the molecular weight of the PAG in g/mol

Once the number densities of PAGs and quencher have been calculated, the centers-of-mass indicating the positions of the molecules must be randomly dispersed within Ω. The average total number of PAGs in Ω is the product of the average number density of PAG and the volume of Ω.

$$\langle nPAGs_\Omega \rangle = \langle \rho_{PAG} \rangle V_\Omega \tag{9.84}$$

A Poisson random deviate representing the total randomized, integer number of PAGs in Ω is drawn from a Poisson distribution with average $\langle nPAGs_\Omega \rangle$:

$$nPAGs_\Omega = \text{poissrand}\left(\langle nPAGs_\Omega \rangle\right) \tag{9.85}$$

where

$nPAGs_\Omega$ is the random deviate of the total number of PAGs in Ω

poissrand() indicates a function that returns the Poisson random deviate and takes as input the average number of PAGs in Ω

Each PAG has a distinct x, y, z-ordered triplet representing the position of its center-of-mass. To disperse the $nPAGs_\Omega$ randomly throughout Ω, we generate $nPAGs_\Omega$ ordered triplets within the x, y, z limits of Ω. For example, a randomized ordered triplet x_{random}, y_{random}, z_{random} within the inclusive limits $[x_{min}, x_{max}]$, $[y_{min}, y_{max}]$, $[z_{min}, z_{max}]$ is

$$x_{random} = \left(x_{max} - x_{min}\right)\mu_1 + x_{min}$$

$$y_{random} = \left(y_{max} - y_{min}\right)\mu_2 + y_{min} \tag{9.86}$$

$$z_{random} = \left(z_{max} - z_{min}\right)\mu_3 + z_{min}$$

where μ_1 μ_2 μ_3 are uniformly distributed random deviates drawn from the interval 0.0–1.0.

Once the domain Ω has been populated, and lists of the ordered triplets for both the PAGs and the quenchers have been generated, Ω may be discretized into cells of volume dV. For example, Ω may be discretized into cells 1 nm on a side. We then commence a loop over all cells in Ω to begin the exposure simulation.

The numbers of PAGs and quenchers in the cell are first counted, and the absorbance due to each species in the cell and the absorbance of the entire cell are calculated. The absorbance attributed to the PAGs within the cell is

$$\alpha_{PAG},\ 1/nm = \frac{\varepsilon_{PAG}\cdot\ln(10)\cdot 1.0\times 10^{14}}{N_A} \cdot \frac{n\,PAGs_{cell}}{V_{cell}} \tag{9.87}$$

The absorbance attributed to the quenchers within the cell is

$$\alpha_{quencher},\ 1/nm = \frac{\varepsilon_{quencher}\cdot\ln(10)\cdot 1.0\times 10^{14}}{N_A} \cdot \frac{n\,quenchers}{V_{cell}} \tag{9.88}$$

The absorbance due to polymer in the cell is calculated by knowledge of its imaginary refractive index κ:

$$\alpha_{\text{polymer}}, 1/\text{nm} = \frac{4\pi\kappa}{\lambda_{\text{nm}}} \tag{9.89}$$

By Beer's law, the total absorbance of the cell is

$$\alpha_{\text{cell}}, 1/\text{nm} = \alpha_{\text{PAG}} + \alpha_{\text{quencher}} + \alpha_{\text{polymer}} \tag{9.90}$$

The average number of photons absorbed by the cell is then calculated. We assume that the volume of the cell dV is small enough that we may treat the normalized intensity of light in the cell, I_{cell}, as constant and also that the total absorbance of the cell may be treated as constant within the cell. By Equation 9.74, the average number of photons absorbed in the cell is then

$$\langle n \rangle_{\text{cell}} = \frac{\alpha_{\text{cell}} E_{\text{inc}} I_{\text{cell}} V_{\text{cell}}}{h\nu} = \alpha_{\text{cell}} E_{\text{inc}} I_{\text{cell}} V_{\text{cell}} \left(\frac{\lambda}{hc} \right) \tag{9.91}$$

The average number of photons absorbed in the cell $\langle n \rangle_{\text{cell}}$ is used to generate a Poisson random deviate simulating the random integer number of photons absorbed in the cell:

$$n_{\text{cell}} = \text{poissrand}\left(\langle n \rangle_{\text{cell}} \right) \tag{9.92}$$

We next loop over each of the photons absorbed in the cell and calculate which of the photons, if any, were absorbed by PAGs in the cell. In this example, only the PAGs are photosensitive and are able to undergo electronic excitation and acid release.

For each absorbed photon, we calculate the probability that the photon was absorbed by one of the PAGs in the cell. This probability is the ratio of the absorbance attributed to the PAG in the cell to the total absorbance of the cell:

$$P_{\text{Abs}} = \frac{\alpha_{\text{PAG}}}{\alpha_{\text{cell}}} \tag{9.93}$$

We determine the outcome by comparing the probability P_{Abs} with μ_{abs}, a uniformly distributed random number in the interval 0.0–1.0. If $P_{\text{Abs}} \geq \mu_{\text{abs}}$, we conclude that one of the PAGs has absorbed the photon and that one of the PAGs has been placed into an electronically excited state.

One of the PAGs in the cell is then randomly selected. This PAG will be deemed to be electronically excited, and our simulation will determine whether or not it releases an acid. Since the PAGs are indistinguishable from each other, each PAG has an equal probability of absorbing the photon in question, and this probability is equal to $1/n\text{PAGs}_{\text{cell}}$. The random selection of a PAG does not depend on whether or not the PAG has previously released an acid, since the PAG's chromophore for absorption may still be intact even if the acid has been released. Also, CARs do not bleach appreciably, and experimentally determined values of the Dill's A parameter, the bleachable absorbance, are usually about zero (or even slightly negative for CARs, indicating a small amount of "darkening" upon exposure).

If the randomly selected PAG has already released its acid, it cannot release another, and we move to consider the outcome of the next photon absorbed in the cell, if there are any. If the randomly selected PAG has not yet released its acid, we test whether an acid is released by comparing a uniform random deviate μ_{phi} in the interval 0.0–1.0 with φ, the photolytic quantum efficiency of the PAG. If $\varphi \geq \mu_{\text{phi}}$, we determine that the PAG has released an acid at the position of its center-of-mass. Otherwise, we determine that the PAG relaxes to the ground state with no acid released. The

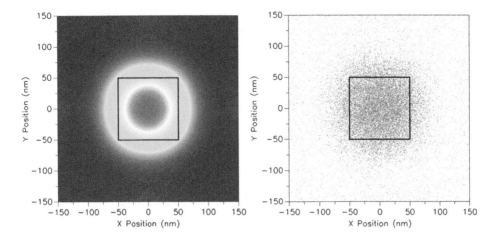

FIGURE 9.11   Results of the Monte Carlo for stochastic simulation of the direct photolysis of PAG. The left plot shows the top-down view of the normalized intensity of 193 nm light at the top of the resist. The black box indicates the mask edge of a 100 nm contact hole. The right plot shows the locations of the acid centers-of-mass after exposure. The PAG number density is 0.05 PAGs/nm^3, the PAG molar extinction coefficient is 4.4×10^7 cm^2/mol, the quantum efficiency is 0.2, and the incident dose is 50 mJ/cm^2. By Equation 9.52, the C parameter of the simulated resist is 0.034 cm^2/mJ.

outcome of the absorption of the next photon by the cell is determined, or if there are no others, we proceed to evaluate the next cell. In this way, the MC Carlo simulation of direct photolysis can be performed, cell by cell, photon by photon, until the entire domain Ω has been processed. Figure 9.11 shows an example result of this approach.

9.6   PHOTORESIST BAKE EFFECTS

9.6.1   PREBAKE

The purpose of a photoresist prebake (also called a postapply bake) is to dry the resist by removing solvent from the film. However, as with most thermal processing steps, the bake has other effects on the photoresist. When heated to temperatures above about 70C, the photoactive compound (PAC) of a diazo-type positive photoresist begins to decompose to a nonphotosensitive product. The reaction mechanism is thought to be identical to that of the PAC reaction during ultraviolet exposure [22, 23, 43, 44].

$$\text{(structure with } N_2) \xrightarrow{\Delta} \text{(structure with C=O)} + N_2 \uparrow \longrightarrow X \tag{9.94}$$

The identity of the product X will be discussed in a following section.

To determine the concentration of PAC as a function of prebake time and temperature, consider the first-order decomposition reaction

$$M \xrightarrow{\ \Delta\ } X \tag{9.95}$$

where M is the photoactive compound. If we let M'_o be the concentration of PAC before prebake and M_o the concentration of PAC after prebake, simple kinetics tells us that

$$\frac{dM_o}{dt} = -K_T M_o$$

$$M_o = M'_o \exp(-K_T t_b)$$

$$m' = \exp(-K_T t_b) \tag{9.96}$$

where

t_b = bake time
K_T = decomposition rate constant at temperature T
m' = M_o/M'_o

The dependence of K_T upon temperature may be described by the Arrhenius equation,

$$K_T = A_r \exp(-E_a/RT) \tag{9.97}$$

where

A_r = Arrhenius coefficient
E_a = activation energy
R = universal gas constant

Thus, the two parameters E_a and A_r allow us to know m' as a function of the prebake conditions, provided Arrhenius behavior is followed. In polymer systems, caution must be exercised, since bake temperatures near the glass transition temperature sometimes lead to non-Arrhenius behavior. For normal prebakes of typical photoresists, the Arrhenius model appears well founded.

The effect of this decomposition is a change in the chemical makeup of the photoresist. Thus, any parameters that are dependent upon the quantitative composition of the resist are also dependent upon prebake. The most important of these parameters fall into two categories: 1) optical (exposure) parameters, such as the resist absorption coefficient, and 2) development parameters, such as the development rates of unexposed and completely exposed resist. A technique will be described to measure E_a and A_r and thus quantify these effects of prebake.

In the model proposed by Dill et al. [2], the exposure of a positive photoresist can be characterized by the three parameters A, B, and C. A and B are related to the optical absorption coefficient of the photoresist, α, and C is the overall rate constant of the exposure reaction. More specifically,

$$\alpha = Am + B$$

$$A = (a_M - a_P) M_o \tag{9.98}$$

$$B = a_P M_o + a_R R + a_S S$$

where

a_M = molar absorption coefficient of the photoactive compound M
a_P = molar absorption coefficient of the exposure product P
a_S = molar absorption coefficient of the solvent S
a_R = molar absorption coefficient of the resin R
M_o = the PAC concentration at the start of the exposure (i.e., after prebake)
m = M/M_o, the relative PAC concentration as a result of exposure

These expressions do not explicitly take into account the effects of prebake on the resist composition. To do so, we can modify Equation 9.98 to include absorption by the component X.

$$B = a_P M_o + a_R R + a_X X \tag{9.99}$$

where a_X is the molar absorption coefficient of the decomposition product X, and the absorption term for the solvent has been neglected. The stoichiometry of the decomposition reaction gives

$$X = M'_o - M_o \tag{9.100}$$

Thus,

$$B = a_P M_o + a_R R + a_X \left(M'_o - M_o \right) \tag{9.101}$$

Let us consider two cases of interest, no bake (NB) and full bake (FB). When there is no prebake (meaning no decomposition), $M'_o = M_o$ and

$$\begin{aligned} A_{NB} &= \left(a_M - a_P \right) M'_o \\ B_{NB} &= a_P M'_o + a_R R \end{aligned} \tag{9.102}$$

We shall define full bake as a prebake that decomposes all PAC. Thus, $M_o = 0$ and

$$\begin{aligned} A_{FB} &= 0 \\ B_{FB} &= a_X M'_o + a_R R \end{aligned} \tag{9.103}$$

Using these special cases in our general expressions for A and B,

$$\begin{aligned} A &= A_{NB} m' \\ B &= B_{FB} - \left(B_{FB} - B_{NB} \right) m' \end{aligned} \tag{9.104}$$

The A parameter decreases linearly as decomposition occurs, and B typically increases slightly.

The development rate is, of course, dependent on the concentration of PAC in the photoresist. However, the product X can also have a large effect on the development rate. Several studies have been performed to determine the composition of the product X [22, 23, 44]. The results indicate that there are two possible products, and the most common outcome of a prebake decomposition is a mixture of the two. The first product is formed via Equation 9.105 and is identical to the product of UV exposure.

$$\tag{9.105}$$

As can be seen, this reaction requires the presence of water. A second reaction, which does not require water, is the esterification of the ketene with the resin:

$$\tag{9.106}$$

Both possible products have a dramatic effect on the dissolution rate. The carboxylic acid is very soluble in developer and enhances dissolution. The formation of carboxylic acid can be thought of as a blanket exposure of the resist. The dissolution rate of unexposed resist (r_{min}) will increase due to the presence of the carboxylic acid. The dissolution rate of fully exposed resist (r_{max}), however, will not be affected. Since the chemistry of the dissolution process is unchanged, the basic shape of the development rate function will also remain unchanged.

The ester, on the other hand, is very difficult to dissolve in aqueous solutions and thus retards the dissolution process. It will have the effect of decreasing r_{max}, although the effects of ester formation on the full dissolution behavior of a resist are not well known.

If the two mechanisms given in Equations 9.105 and 9.106 are taken into account, the rate Equation 9.96will become

$$\frac{dM_o}{dt} = -K_1 M_o - K_2 [H_2O] M_o \tag{9.107}$$

where K_1 and K_2 are the rate constants of Equations 9.105 and 9.106, respectively. For a given concentration of water in the resist film, this reverts to equation (3), where

$$K_T = K_1 + K_2 [H_2O] \tag{9.108}$$

Thus, the relative importance of the two reactions will depend not only on the ratio of the rate constants but also on the amount of water in the resist film. The concentration of water is a function of atmospheric conditions and the past history of the resist-coated wafer. Further experimental measurements of development rate as a function of prebake temperature are needed to quantify these effects.

Examining Equation 9.104, one can see that the parameter A can be used as a means of measuring m', the fraction of PAC remaining after prebake. Thus, by measuring A as a function of prebake time and temperature, one can determine the activation energy and the corresponding Arrhenius coefficient for the proposed decomposition reaction. Using the technique given by Dill et al. [2], A, B, and C can be easily determined by measuring the optical transmittance of a thin photoresist film on a glass substrate while the resist is being exposed.

Examples of measured transmittance curves are given in Figure 9.12, where transmittance is plotted versus exposure dose. The different curves represent different prebake temperatures. For every curve, A, B, and C can be calculated. Figure 9.13 shows the variation of the resist

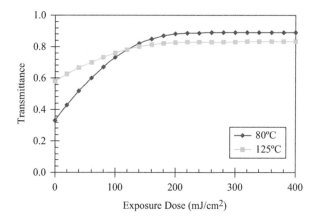

FIGURE 9.12 Two transmittance curves for Kodak 820 resist at 365 nm. The curves are for a convection oven prebake of 30 minutes at the temperatures shown [45].

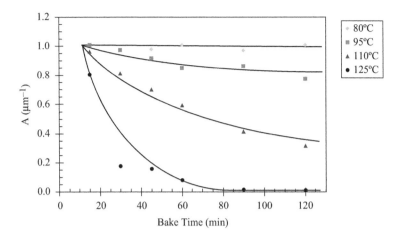

FIGURE 9.13 Variation of the resist absorption parameter A with prebake time and temperature for Kodak 820 resist at 365 nm [45].

parameter A with prebake conditions. According to Equations 9.96 and 9.104, this variation should take the form

$$\frac{A}{A_{NB}} = e^{-K_T t_b} \tag{9.109}$$

$$\ln\left(\frac{A}{A_{NB}}\right) = -K_T t_b \tag{9.110}$$

Thus, a plot of $\ln(A)$ versus bake time should give a straight line with a slope equal to $-K_T$. This plot is shown in Figure 9.14. Knowing K_T as a function of temperature, one can determine the activation energy and Arrhenius coefficient from Equation 9.97. One should note that the parameters A_{NB}, B_{NB}, and B_{FB} are wavelength dependent, but E_a and A_r are not.

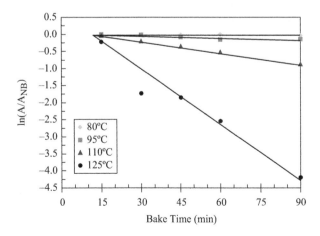

FIGURE 9.14 Log plot of the resist absorption parameter A with prebake time and temperature for Kodak 820 resist at 365 nm [45].

Figure 9.13 shows an anomaly in which there is a lag time before decomposition occurs. This lag time is the time it took the wafer and wafer carrier to reach the temperature of the convection oven. Equation 9.96 can be modified to accommodate this phenomenon:

$$m' = e^{-K_T(t_b - t_{wup})}$$

(9.111)

where t_{wup} is the warm-up time. A lag time of about 11 minutes was observed when convection oven baking a 1/4″ thick glass substrate in a wafer carrier. When a 60 mil glass wafer was used without a carrier, the warm-up time was under 5 minutes and could not be measured accurately in this experiment [45].

Although all the data presented thus far has been for convection oven prebake, this method of evaluating the effects of prebake can also be applied to hotplate prebaking.

9.6.2 Postexposure Bake

Many attempts have been made to reduce the standing-wave effect and thus increase linewidth control and resolution. One particularly useful method is the postexposure, predevelopment bake as described by Walker [46]. A 10°C oven bake for 10 minutes was found to reduce the standing-wave ridges significantly. This effect can be explained quite simply by the diffusion of PAC in the resist during a high-temperature bake. A mathematical model that predicts the results of such a PEB is described in the following.

In general, molecular diffusion is governed by Fick's Second Law of Diffusion, which states (in one dimension):

$$\frac{\partial C_A}{\partial t} = \mathcal{D} \frac{\partial^2 C_A}{\partial x^2}$$

(9.112)

where
  C_A = concentration of species A
  $\mathcal{D}$ = diffusion coefficient of A at some temperature T
  t = time that the system is at temperature T

Note that the diffusivity is assumed to be independent of concentration here. This differential equation can be solved given a set of boundary conditions, that is, an initial distribution of A. One possible boundary condition is known as the *impulse source*. At some point, x_o, there are N moles of substance A, and at all other points there is no A. Thus, the concentration at x_o is infinite. Given this initial distribution of A, the solution to Equation 9.112 is the Gaussian distribution function,

$$C_A(x) = \frac{N}{\sqrt{2\pi\sigma^2}} e^{-r^2/2\sigma^2}$$

(9.113)

where
  $\sigma = \sqrt{2\mathcal{D}t}$, the diffusion length
  $r = x - x_o$

In practice, there are no impulse sources. Instead, we can approximate an impulse source as having some concentration C_o over some small distance Δx centered at x_o, with zero concentration outside this range. An approximate form of Equation 9.113 is then

$$C_A(x) = \frac{C_o \Delta x}{\sqrt{2\pi\sigma^2}} e^{-r^2/2\sigma^2}$$

(9.114)

This solution is fairly accurate if $\Delta x < 3\sigma$. If there are two "impulse" sources located at x_1 and x_2, with initial concentrations C_1 and C_2 each over a range Δx, the concentration of A at x after diffusion is

$$C_A(x) = \left[\frac{C_1}{\sqrt{2\pi\sigma^2}} e^{-r_1^2/2\sigma^2} + \frac{C_2}{\sqrt{2\pi\sigma^2}} e^{-r_2^2/2\sigma^2} \right] \Delta x \tag{9.115}$$

where

$r_1 = x - x_1$

$r_2 = x - x_2$

If there are a number of sources, Equation 9.115 becomes

$$C_A(x) = \frac{\Delta x}{\sqrt{2\pi\sigma^2}} \sum C_n e^{-r_n^2/2\sigma^2} \tag{9.116}$$

Extending the analysis to a continuous initial distribution $C_o(x)$, Equation 9.116 becomes

$$C_A(x) = \frac{1}{\sqrt{2\pi\sigma^2}} \int_{-\infty}^{\infty} C_o(x - x') e^{-x'^2/2\sigma^2} dx' \tag{9.117}$$

where x' is now the distance from the point x. Equation 9.117 is simply the convolution of two functions:

$$C_A(x) = C_o(x) * f(x) \tag{9.118}$$

where

$$f(x) = \frac{1}{\sqrt{2\pi\sigma^2}} e^{-x^2/2\sigma^2}$$

This equation can now be made to accommodate two-dimensional diffusion:

$$C_A(x, y) = C_o(x, y) * f(x, y) \tag{9.119}$$

where

$f(x, y) = \dfrac{1}{2\pi\sigma^2} e^{-r^2/2\sigma^2}$

$r = \sqrt{x^2 + y^2}$

We are now ready to apply Equation 9.119 to the diffusion of PAC in a photoresist during a PEB. After exposure, the PAC distribution can be described by $m(x,z)$, where m is the relative PAC concentration. According to Equation 9.119, the relative PAC concentration after a PEB, $m^*(x,z)$, is given by

$$m^*(x, z) = \frac{1}{2\pi\sigma^2} \int_{-\infty}^{\infty}\!\!\int m(x - x', z - z') e^{-r^2/2\sigma^2} dx' dz' \tag{9.120}$$

In evaluating Equation 9.120, it is common to replace the integrals by summations over intervals Δx and Δz. In such a case, the restrictions that $\Delta x < 3\sigma$ and $\Delta z < 3\sigma$ will apply. An alternative solution is to solve the diffusion equation (Equation 9.112) directly, for example using a finite difference approach.

The diffusion model can now be used to simulate the effects of a PEB. Using the lithography simulator, a resist profile can be generated. By including the model for a PEB, the profile can be generated, showing how the standing-wave effect is reduced. The only parameter that needs to be specified in Equation 9.120 is the diffusion length σ, or equivalently, the diffusion coefficient $\mathcal{D}$ and the bake time t. In turn, $\mathcal{D}$ is a function of the bake temperature T and, of course, the resist system used. Thus, if the functionality of $\mathcal{D}$ with temperature is known for a given resist system, a PEB of time t and temperature T can be modeled. A general temperature dependence for the diffusivity $\mathcal{D}$ can be found using the Arrhenius equation (for temperature ranges that do not traverse the glass transition temperature):

$$\mathcal{D} = \mathcal{D}_o \, e^{-E_a/RT} \tag{9.121}$$

where

$\mathcal{D}_o$ = Arrhenius constant (units of nm^2/min)
E_a = activation energy
R = universal gas constant
T = temperature in kelvin

Unfortunately, very little work has been done in measuring the diffusivity of photoactive compounds in photoresist.

9.7 PHOTORESIST DEVELOPMENT

An overall positive resist processing model requires a mathematical representation of the development process. Previous attempts have taken the form of empirical fits to development rate data as a function of exposure [2, 47]. The model formulated below begins on a more fundamental level with a postulated reaction mechanism, which then leads to a development rate equation [48]. The rate constants involved can be determined by comparison with experimental data. An enhanced kinetic model with a second mechanism for dissolution inhibition is also presented [49]. Deviations from the expected development rates have been reported under certain conditions at the surface of the resist. This effect, called *surface induction* or *surface inhibition*, can be related empirically to the expected development rate, that is, to the bulk development rate as predicted by a kinetic model.

Unfortunately, fundamental experimental evidence of the exact mechanism of photoresist development is lacking. The model presented in the following is reasonable, and the resulting rate equation has been shown to describe actual development rates extremely well. However, faith in the exact details of the mechanism is limited by this dearth of fundamental studies.

9.7.1 KINETIC DEVELOPMENT MODEL

In order to derive an analytical development rate expression, a kinetic model of the development process will be used. This approach involves proposing a reasonable mechanism for the development reaction and then applying standard kinetics to this mechanism in order to derive a rate equation. We shall assume that the development of photoresist involves three processes: diffusion of developer from the bulk solution to the surface of the resist, reaction of the developer with the resist, and diffusion of the product back into the solution. For this analysis, we shall assume that the last step, diffusion of the dissolved resist into solution, occurs very quickly, so that this step may be ignored. Let us now look at the first two steps in the proposed mechanism. The diffusion of developer to the resist surface can be described with the simple diffusion rate equation, given approximately by

$$r_D = k_D \left(D - D_S \right) \tag{9.122}$$

where
  r_D is the rate of diffusion of the developer to the resist surface
  D is the bulk developer concentration
  D_S is the developer concentration at the resist surface
  k_D is the rate constant

We shall now propose a mechanism for the reaction of developer with the resist. The resist is composed of large macromolecules of resin R along with inhibitor M, which can be the PAC for g-line and i-line resists or the blocking group of a CAR. The resin is quite soluble in the developer solution, but the presence of the PAC or blocking group acts as an inhibitor to dissolution, making the development rate very slow. The reacted inhibitor P, the exposure product for a g/i-line resist or the deblocked site for a CAR, is very soluble in developer, enhancing the dissolution rate of the resin. Let us assume that n molecules of product P react with the developer to dissolve a resin molecule. The rate of the reaction is

$$r_R = k_R D_S P^n \tag{9.123}$$

where
  r_R is the rate of reaction of the developer with the resist
  k_R is the rate constant

(Note that the mechanism shown in Equation 9.123 is the same as the "polyphotolysis" model described by Trefonas and Daniels [50].) From the stoichiometry of the exposure reaction,

$$P = M_o - M \tag{9.124}$$

where M_o is the initial inhibitor concentration (i.e., before exposure and/or amplification).
The two steps outlined above are in series, i.e., one reaction follows the other. Thus, the two steps will come to a steady state such that

$$r_R = r_D = r \tag{9.125}$$

Equating the rate equations, one can solve for D_S and eliminate it from the overall rate equation, giving

$$r = \frac{k_D \, k_R \, D \, P^n}{k_D + k_R \, P^n} \tag{9.126}$$

Using Equation 9.124 and letting $m = M/M_o$, the relative inhibitor concentration, Equation 9.126 becomes

$$r = \frac{k_D \, D \, (1 - m)^n}{k_D / k_R M_o^n + (1 - m)^n} \tag{9.127}$$

When $m = 1$ (resist unexposed/unreacted), the rate is zero. When $m = 0$ (resist completely exposed/reacted), the rate is equal to r_{max}, where

$$r_{max} = \frac{k_D D}{k_D/k_R M_o^n + 1} \tag{9.128}$$

If we define a constant a such that

$$a = k_D/k_R M_o^n \tag{9.129}$$

the rate equation becomes

$$r = r_{max} \frac{(a+1)(1-m)^n}{a+(1-m)^n} \tag{9.130}$$

Note that the simplifying constant a describes the rate constant of diffusion relative to the surface reaction rate constant. A large value of a will mean that diffusion is very fast, and thus less important, compared with the fastest surface reaction (for completely exposed/reacted resist).

There are three constants that must be determined experimentally: a, n, and r_{max}. The constant a can be put in a more physically meaningful form as follows. A characteristic of some experimental rate data is an inflection point in the rate curve at about $m = 0.2$–0.7. The point of inflection can be calculated by letting

$$\frac{d^2 r}{dm^2} = 0$$

giving

$$a = \frac{(n+1)}{(n-1)}\left(1 - m_{TH}\right)^n \tag{9.131}$$

where m_{TH} is the value of m at the inflection point, called the *threshold inhibitor concentration*.

This model does not take into account the finite dissolution rate of unexposed resist (r_{min}). One approach is simply to add this term to Equation 9.130, giving

$$r = r_{max} \frac{(a+1)(1-m)^n}{a+(1-m)^n} + r_{min} \tag{9.132}$$

This approach assumes that the mechanism of development of the unexposed resist is independent of this proposed development mechanism. In other words, there is a finite dissolution of resin that occurs by a mechanism that is independent of the presence of exposed/reacted inhibitor. Also note that the true maximum development rate is now $r_{max} + r_{min}$, though typically, r_{min} is so small compared with r_{max} that it can be neglected.

Consider the case when the diffusion rate constant is large compared with the surface reaction rate constant. If $a \gg 1$, the development rate equation (Equation 9.132) will become

$$r = r_{max}(1-m)^n + r_{min} \tag{9.133}$$

The interpretation of a as a function of the threshold inhibitor concentration m_{TH} given by Equation 9.131 means that a very large a would correspond to a large negative value of m_{TH}. In other words, if the surface reaction is very slow compared with the mass transport of developer to the surface, there will be no inflection point in the development rate data, and Equation 9.133 will apply. It is quite apparent that Equation 9.133 could be derived directly from Equation 9.123 if the diffusion step were ignored.

9.7.2 Enhanced Kinetic Development Model

The previous kinetic model is based on the principle of dissolution enhancement. The carboxylic acid enhances the dissolution rate of the resin/PAC mixture for a g/i-line resist, or the phenolic group enhances the dissolution of the PHS for a chemically amplified resist. In reality, this is a simplification. There are really two mechanisms at work. The inhibitor acts to inhibit dissolution of the resin, while the reacted inhibitor acts to enhance dissolution. Thus, the rate expression should reflect both these mechanisms. A new model, called the *enhanced kinetic model*, was proposed to include both effects [49]:

$$R = R_{\text{resin}} \frac{1 + k_{\text{enh}}(1-m)^n}{1 + k_{\text{inh}}(m)^l} \tag{9.134}$$

where

k_{enh} is the rate constant for the enhancement mechanism
n is the enhancement reaction order
k_{inh} is the rate constant for the inhibition mechanism
l is the inhibition reaction order
R_{resin} is the development rate of the resin alone

For no exposure, $m = 1$, and the development rate is at its minimum. From Equation 9.134,

$$R_{\min} = \frac{R_{\text{resin}}}{1 + k_{\text{inh}}} \tag{9.135}$$

Similarly, when $m = 0$, corresponding to complete exposure/reaction, the development is at its maximum:

$$R_{\max} = R_{\text{resin}}(1 + k_{\text{enh}}) \tag{9.136}$$

Thus, the development rate expression can be characterized by five parameters: R_{max}, R_{min}, R_{resin}, n, and l.

Obviously, the enhanced kinetic model for resist dissolution is a superset of the original kinetic model. If the inhibition mechanism is not important, then $l = 0$. For this case, Equation 9.134 is identical to Equation 9.133 when

$$R_{\min} = R_{\text{resin}}, \quad R_{\max} = R_{\text{resin}} k_{\text{enh}} \tag{9.137}$$

The enhanced kinetic model of Equation 9.134 assumes that mass transport of developer to the resist surface is not significant. Of course, a simple diffusion of developer can be added to this mechanism, as was done above with the original kinetic model.

9.7.3 Surface Inhibition

The kinetic models given in the preceding sections predict the development rate of the resist as a function of inhibitor concentration remaining after the resist has been exposed to UV light and reacted during PEB. There are, however, other parameters that are known to affect the development rate but which were not included in this model. The most notable deviation from the kinetic theory is the surface inhibition effect. The inhibition, or surface induction, effect is a decrease in the expected development rate at the surface of the resist [43, 51, 52]. Thus, this effect is a function of the depth into the resist and requires a new description of development rate.

Several factors have been found to contribute to the surface inhibition effect. High-temperature baking of the photoresist has been found to produce surface inhibition and is thought to cause oxidation of the resist at the resist surface [43, 51, 52]. In particular, prebaking the photoresist may cause this reduced development rate phenomenon [43, 52]. Alternatively, the induction effect may be the result of reduced solvent content near the resist surface. Of course, the degree to which this effect is observed depends upon the prebake time and temperature. Finally, surface inhibition can be induced with the use of surfactants in the developer.

An empirical model can be used to describe the positional dependence of the development rate. If we assume that the development rate near the surface of the resist exponentially approaches the bulk development rate, the rate as a function of depth, $r(z)$, is

$$r(z) = r_B \left(1 - \left(1 - r_o \right) e^{-z/\delta} \right) \tag{9.138}$$

where

r_B is the bulk development rate
r_o is the development rate at the surface of the resist relative to r_B
δ is the depth of the surface inhibition layer

In several resists, the induction effect has been found to take place over a depth of about 100 nm [43, 52].

9.8 LINEWIDTH MEASUREMENT

A cross section of a photoresist profile has, in general, a very complicated two-dimensional shape (Figure 9.15). In order to compare the shapes of two different profiles, one must find a convenient description for the shapes of the profiles that somehow reflects their salient qualities. The most common description is to model the resist profile as a trapezoid. Thus, three numbers can be used to describe the profile: the width of the base of the trapezoid (linewidth, w), its height (resist thickness, D), and the angle that the side makes with the base (sidewall angle, θ). Obviously, to describe such a complicated shape as a resist profile with just three numbers is a great, though necessary, simplification. The key to success is to pick a method of fitting a trapezoid to the profile that preserves the important features of the profile, is numerically practical, and as a result is not overly sensitive to slight changes in the profile.

There are many possible algorithms for measuring the resist profile. One algorithm, called the *linear weight method*, is designed to mimic the behavior of a top-down linewidth measurement system. The first step is to convert the profile into a "weighted" profile as follows: at any given x-position (i.e., along the horizontal axis), determine the "weight" of the photoresist above it. The weight is defined as the total thickness of resist along a vertical line at x. Figure 9.16 shows a typical example. The weight at this x position would be the sum of the lengths of the line segments that are within the resist profile. As can be seen, the original profile is complicated and multivalued, whereas the weighted profile is smooth and single-valued.

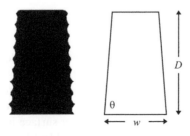

FIGURE 9.15 Typical photoresist profile and its corresponding trapezoid.

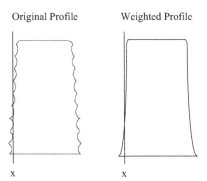

FIGURE 9.16   Determining the weighted resist profile.

A trapezoid can now be fitted accurately to the weighted profile. The simplest type of fit will be called the *standard* linewidth determination method: ignoring the top and bottom 10% of the weighted resist thickness, a straight line is fitted through the remaining 80% of the sidewall. The intersection of this line with the substrate gives the linewidth, and the slope of this line determines the sidewall angle. Thus, the standard method gives the best-fit trapezoid through the middle 80% of the weighted profile.

There are cases where one part of the profile may be more significant than another. For these situations, one could select the *threshold* method for determining linewidth. In this method, the sidewall angle is measured using the standard method, but the width of the trapezoid is adjusted to match the width of the weighted profile at a given threshold resist thickness. For example, with a threshold of 20%, the trapezoid will cross the weighted profile at a thickness of 20% up from the bottom. Thus, the threshold method can be used to emphasize the importance of one part of the profile.

The two linewidth determination methods deviate from one another when the shape of the resist profile begins to deviate from the general trapezoidal shape. Figure 9.17 shows two resist profiles at the extremes of focus. Using a 10% threshold, the linewidths of these two profiles are the same. Using a 50% threshold, however, shows profile (a) to be 20% wider than profile (b). The standard linewidth method, on the other hand, shows profile (a) to be 10% wider than profile (b). Finally, a 1% threshold gives the opposite result, with profile (a) 10% smaller than profile (b). The effect of changing profile shape on the measured linewidth is further illustrated in Figure 9.18, which shows CD versus focus for the standard and 5% threshold CD measurement methods. It is important to note that the sensitivity of the measured linewidth to profile shape is not particular to lithography simulation but is present in any CD measurement system. Fundamentally, this is the result of using the trapezoid model for resist profiles.

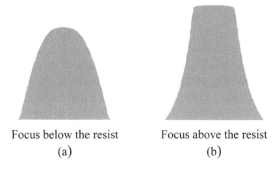

Focus below the resist   Focus above the resist
            (a)   (b)

FIGURE 9.17   Resist profiles at the extremes of focus.

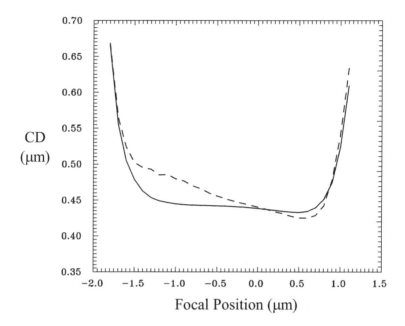

FIGURE 9.18 Effect of resist profile shape on linewidth measurement in a lithography simulator. CD measurement methods are standard (dashed line) and 5% threshold (solid line).

Obviously, it is difficult to compare resist profiles when the shapes of the profiles are changing. It is very important to use the linewidth method (and proper threshold value, if necessary), which is physically the most significant for the problems being studied. If the bottom of the resist profile is most important, the threshold method with a small (e.g., 5%) threshold is recommended. It is also possible to "calibrate" the simulator to a linewidth measurement system. By adjusting the threshold value used by the simulator, results comparable to actual measurements can be obtained.

9.9 LUMPED PARAMETER MODEL

Typically, lithography models make every attempt to describe physical phenomena as accurately as possible. However, in some circumstances, speed is more important than accuracy. If a model is reasonably close to correct and fast, many interesting applications are possible. With this tradeoff in mind, the lumped parameter model was developed [53–55].

9.9.1 DEVELOPMENT RATE MODEL

The mathematical description of the resist process incorporated in the lumped parameter model uses a simple photographic model relating development time to exposure, while the aerial image simulation is derived from the standard optical parameters of the lithographic tool. A very simple development rate model is used based on the assumption of a constant contrast. Before proceeding, however, let us define a few terms needed for the derivations that follow. Let E be the nominal exposure energy (i.e., the intensity in a large clear area times the exposure time), $I(x)$ the normalized image intensity, and $I(z)$ the relative intensity variation with depth into the resist. It is clear that the exposure energy as a function of position within the resist (E_{xz}) is just $E\,I(x)I(z)$, where $x=0$ is the center of the mask feature, and $z=0$ is the top of a resist of thickness D. Defining logarithmic versions of these quantities,

$$\varepsilon = \ln[E], \ i(x) = \ln\left[I(x)\right], \ i(z) = \ln\left[I(z)\right] \tag{9.139}$$

and the logarithm of the energy deposited in the resist is

$$\ln\left[E_{xz}\right] = \varepsilon + i(x) + i(z) \tag{9.140}$$

The photoresist contrast (γ) is defined theoretically as [56]

$$\gamma \equiv \frac{d \ln r}{d \ln E_{xz}} \tag{9.141}$$

where r is the resulting development rate from an exposure of E_{xz}. Note that the base e definition of contrast is used here. If the contrast is assumed constant over the range of energies of interest, Equation 9.141 can be integrated to give a very simple expression for development rate. In order to evaluate the constant of integration, let us pick a convenient point of evaluation. Let ε_o be the energy required to just clear the photoresist in the allotted development time, t_{dev}, and let r_o be the development rate that results from an exposure of this amount. Carrying out the integration gives

$$r(x,z) = r_o\, e^{\gamma(\varepsilon + i(x) + i(z) - \varepsilon_o)} = r_o\left[\frac{E_{xz}}{E_o}\right]^{\gamma} \tag{9.142}$$

As an example of the use of the above development rate expression and to further illustrate the relationship between r_o and the dose to clear, consider the standard dose to clear experiment, where a large clear area is exposed and the thickness of photoresist remaining is measured. The definition of development rate

$$r = \frac{dz}{dt} \tag{9.143}$$

can be integrated over the development time. If $\varepsilon = \varepsilon_o$, the thickness remaining is by definition zero, so that

$$t_{dev} = \int_0^D \frac{dz}{r} = \frac{1}{r_o}\int_0^D e^{-\gamma i(z)}\, dz \tag{9.144}$$

where $i(x)$ is zero for an open frame exposure. Based on this equation, one can now define an effective resist thickness, D_{eff}, which will be very useful in the derivation of the lumped parameter model that follows.

$$D_{eff} = r_o t_{dev}\, e^{\gamma i(D)} = e^{\gamma i(D)}\int_0^D e^{-\gamma i(z)}\, dz = \int_0^D \left[\frac{I(z)}{I(D)}\right]^{-\gamma} dz \tag{9.145}$$

As an example, the effective resist thickness can be calculated for the case of absorption only causing a variation in intensity with depth in the resist. For such a case, $I(z)$ will decay exponentially, and Equation 9.145 can be evaluated to give

$$D_{eff} = \frac{1}{\alpha\gamma}\left(e^{\alpha\gamma D} - 1\right) \tag{9.146}$$

If the resist is only slightly absorbing, so that $\alpha\gamma D \ll 1$, the exponential can be approximated by the first few terms in its Taylor series expansion.

$$D_{eff} \approx D\left(1 + \frac{\alpha\gamma D}{2}\right) \tag{9.147}$$

Thus, the effect of absorption is to make the resist seem thicker to the development process. The effective resist thickness can be thought of as the amount of resist of constant development rate that requires the same development time to clear as the actual resist with a varying development rate.

9.9.2 SEGMENTED DEVELOPMENT

Equation 9.142 is an extremely simple-minded model relating development rate to exposure energy based on the assumption of a constant resist contrast. In order to use this expression, we will develop a phenomenological explanation for the development process. This explanation will be based on the assumption that development occurs in two steps: a vertical development to a depth z, followed by a lateral development to position x (measured from the center of the mask feature) [57], as shown in Figure 9.19.

A development ray, which traces out the path of development, starts at the point $(x_o, 0)$ and proceeds vertically until a depth z is reached such that the resist to the side of the ray has been exposed more than the resist below the ray. At this point, the development will begin horizontally. The time needed to develop in both vertical and horizontal directions, t_z and t_x respectively, can be computed from Equation 9.142. The development time per unit thickness of resist is just the reciprocal of the development rate.

$$\frac{1}{r(x,z)} = \tau(x,z) = \tau_o \, e^{-\gamma(\varepsilon + i(x) + i(z))} \tag{9.148}$$

where

$$\tau_o = \frac{1}{r_o} e^{-\gamma \varepsilon_o} \tag{9.149}$$

The time needed to develop to a depth z is given by

$$t_z = \tau_o \, e^{-\gamma \varepsilon} \, e^{-\gamma i(x_o)} \int_0^z e^{-\gamma i(z')} dz' \tag{9.150}$$

Similarly, the horizontal development time is

$$t_x = \tau_o \, e^{-\gamma \varepsilon} \, e^{-\gamma i(z)} \int_{x_0}^x e^{-\gamma i(x')} dx' \tag{9.151}$$

The sum of these two times must equal the total development time.

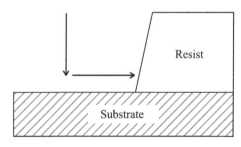

FIGURE 9.19 Illustration of segmented development: development proceeds first vertically, then horizontally, to the final resist sidewall.

$$t_{dev} = \tau_o\, e^{-\gamma\varepsilon} \left[e^{-\gamma i(x_o)} \int_0^z e^{-\gamma i(z')} dz' + e^{-\gamma i(z)} \int_{x_0}^x e^{-\gamma i(x')} dx' \right] \tag{9.152}$$

9.9.3 Derivation of the Lumped Parameter Model

Equation 9.152 can be used to derive some interesting properties of the resist profile. For example, how would a small change in exposure energy $\Delta\varepsilon$ affect the position of the resist profile x? A change in overall exposure energy will not change the point at which the development ray changes direction. Thus, the depth z is constant. Differentiating Equation 9.1452 with respect to log-exposure energy, the following equation can be derived:

$$\left.\frac{dx}{d\varepsilon}\right|_z = \frac{\gamma\, t_{dev}}{\tau(x,z)} = \gamma\, t_{dev}\, r(x,z) \tag{9.153}$$

Since the x position of the development ray endpoint is just one half of the linewidth, Equation 9.153 defines a change in CD with exposure energy. To put this expression in a more useful form, take the log of both sides and use the development rate expression 9.142) to give

$$\ln\left(\frac{dx}{d\varepsilon}\right) = \ln\left(\gamma\, t_{dev}\, r_o\right) + \gamma\left(\varepsilon + i(x) + i(z) - \varepsilon_o\right) \tag{9.154}$$

Rearranging,

$$\varepsilon = \varepsilon_o - i(x) - i(z) + \frac{1}{\gamma}\ln\left(\frac{dx}{d\varepsilon}\right) - \frac{1}{\gamma}\ln\left(\gamma\, t_{dev}\, r_o\right) \tag{9.155}$$

where ε is the (log) energy needed to expose a feature of width $2x$. Equation 9.155 is the differential form of the lumped parameter model and relates the CD versus log-exposure curve and its slope to the image intensity. A more useful form of this equation is given below; however, some valuable insight can be gained by examining Equation 9.155. In the limit of very large γ, one can see that the CD versus exposure curve becomes equal to the aerial image. Thus, exposure latitude becomes *image limited*. For small γ, the other terms become significant, and the exposure latitude is *process limited*. Obviously, an image-limited exposure latitude represents the best possible case.

A second form of the lumped parameter model can also be obtained in the following manner. Applying the definition of development rate to Equation 9.153 or, alternatively, solving for the slope in Equation 9.155 yields

$$\frac{d\varepsilon}{dx} = \frac{1}{\gamma\, t_{dev} r_o}\, e^{-\gamma(\varepsilon + i(x) + i(z) - \varepsilon_o)} \tag{9.156}$$

Before proceeding, let us introduce a slight change in notation that will make the role of the variable ε more clear. As originally defined, ε is just the nominal exposure energy. In Equations 9.154–9.156, it takes the added meaning of the nominal energy that gives a linewidth of $2x$. To emphasize this meaning, we will replace ε by $\varepsilon(x)$, where the interpretation is not a variation of energy with x but rather, a variation of x (linewidth) with energy. Using this notation, the energy to just clear the resist can be related to the energy that gives zero linewidth.

$$\varepsilon_o = \varepsilon(0) + i(x = 0) \tag{9.157}$$

Using this relation in Equation 9.156,

$$\frac{d\varepsilon}{dx} = \frac{1}{\gamma\, t_{\text{dev}} r_o}\, e^{-\gamma i(z)}\, e^{\gamma(\varepsilon(0)-\varepsilon(x))}\, e^{\gamma(i(0)-i(x))} \tag{9.158}$$

Invoking the definitions of the logarithmic quantities,

$$\frac{dE}{dx} = \frac{E(x)}{\gamma\, D_{\text{eff}}}\left[\frac{E(0)I(0)}{E(x)I(x)}\right]^{\gamma} \tag{9.159}$$

where Equation 9.145 has been used and the linewidth is assumed to be measured at the resist bottom (i.e., $z=D$). Equation 9.159 can now be integrated:

$$\int_{E(0)}^{E(x)} E^{\gamma-1}\,dE = \frac{1}{\gamma\, D_{\text{eff}}}\big[E(0)I(0)\big]^{\gamma}\int_{0}^{x} I(x')^{-\gamma}\,dx' \tag{9.160}$$

giving

$$\frac{E(x)}{E(0)} = \left[1 + \frac{1}{\gamma\, D_{\text{eff}}}\int_{0}^{x}\left(\frac{I(x')}{I(0)}\right)^{-\gamma}dx'\right]^{\frac{1}{\gamma}} \tag{9.161}$$

Equation 9.161 is the integral form of the lumped parameter model. Using this equation, one can generate a normalized CD vs. exposure curve by knowing the image intensity, $I(x)$, the effective resist thickness, D_{eff}, and the contrast, γ.

9.9.4 SIDEWALL ANGLE

The lumped parameter model allows the prediction of linewidth by developing down to a depth z and laterally to a position x, which is one-half of the final linewidth. Typically, the bottom linewidth is desired so that the depth chosen is the full resist thickness. By picking different values for z, different x positions will result, giving a complete resist profile. One important result that can be calculated is the resist sidewall slope and the resulting sidewall angle. To derive an expression for the sidewall slope, let us first rewrite Equation 9.152 in terms of the development rate.

$$t_{\text{dev}} = \int_{0}^{z}\frac{dz'}{r(0,z')} + \int_{x_o}^{x}\frac{dx'}{r(x',z)} \tag{9.162}$$

Taking the derivative of this expression with respect to z,

$$0 = \int_{0}^{z}\frac{d\tau}{dz}\,dz' + \frac{1}{r(0,z)} + \int_{x_o}^{x}\frac{d\tau}{dz}\,dx' + \frac{1}{r(x,z)}\frac{dx}{dz} \tag{9.163}$$

The derivative of the reciprocal development rate can be calculated from Equation 9.142 or 9.148:

$$\frac{d\tau}{dz} = -\gamma\,\tau(x,z)\frac{d\ln[E_{xz}]}{dz} \tag{9.164}$$

As one would expect, the variation of development rate with depth into the resist depends on the variation of the exposure dose with depth. Consider a simple example where bulk absorption is the only variation of exposure with z. For an absorption coefficient of α, the result is

$$\frac{d \ln[E_{xz}]}{dz} = -\alpha \tag{9.165}$$

Using Equations 9.164 and 9.165 in 9.163,

$$-\alpha \gamma \left(\int_0^z \tau \, dz' + \int_{x_0}^x \tau \, dx' \right) = \frac{1}{r(0,z)} + \frac{1}{r(x,z)} \frac{dx}{dz} \tag{9.166}$$

Recognizing the term in parentheses as simply the development time, the reciprocal of the resist slope can be given as

$$-\frac{dx}{dz} = \frac{r(x,z)}{r(0,z)} + \alpha \gamma t_{\text{dev}} \, r(x,z) = \frac{r(x,z)}{r(0,z)} + \alpha \frac{dx}{d\varepsilon} \tag{9.167}$$

Equation 9.167 shows two distinct contributors to sidewall angle. The first is the development effect. Because the top of the photoresist is exposed to developer longer than the bottom, the top linewidth is smaller, resulting in a sloped sidewall. This effect is captured in Equation 9.167 as the ratio of the development rate at the edge of the photoresist feature to the development rate at the center. Good sidewall slope is obtained by making this ratio small. The second term in Equation 9.167 describes the effect of optical absorption on the resist slope. High absorption or poor exposure latitude will result in a reduction of the resist sidewall angle.

9.9.5 Results

The lumped parameter model is based on a simple model for development rate and a phenomenological description of the development process. The result is an equation that predicts the change in linewidth with exposure for a given aerial image. The major advantage of the lumped parameter model is its extreme ease of application to a lithography process. The two parameters of the model, resist contrast and effective thickness, can be determined by the collection of linewidth data from a standard focus-exposure matrix. This data is routinely available in most production and development lithography processes; no extra or unusual data collection is required. The result is a simple and fast model that can be used as an initial predictor of results or as the engine of a lithographic control scheme.

Additionally, the lumped parameter model can be used to predict the sidewall angle of the resulting photoresist profile. The model shows the two main contributors to resist slope: development effects due to the time required for the developer to reach the bottom of the photoresist, and absorption effects resulting in a reduced exposure at the bottom of the resist.

Finally, the lumped parameter model presents a simple understanding of the optical lithography process. The potential of the model as a learning tool should not be underestimated. In particular, the model emphasizes the competing roles of the aerial image and the photoresist process in determining linewidth control. This fundamental knowledge lays the foundation for further investigations into the behavior of optical lithography systems.

9.10   USES OF LITHOGRAPHY MODELING

In the 20 years since optical lithography modeling was first introduced to the semiconductor industry, it has gone from a research curiosity to an indispensable tool for research, development, and manufacturing. There are numerous examples of how modeling has had a dramatic impact on the evolution of lithography technology, and many more ways in which it has subtly, but undeniably, influenced the daily routines of lithography professionals. There are four major uses for lithography simulation: 1) as a research tool, performing experiments that would be difficult or impossible to do any other way; 2) as a development tool, quickly evaluating options, optimizing processes, or saving time and money by reducing the number of experiments that have to be performed; 3) as a manufacturing tool, for troubleshooting process problems and determining optimum process settings; and 4) as a learning tool, to help provide a fundamental understanding of all aspects of the lithography process. These four applications of lithography simulation are not distinct—there is much overlap among these basic categories.

9.10.1 Research Tool

Since the initial introduction of lithography simulation in 1974, modeling has had a major impact on research efforts in lithography. Here are some examples of how modeling has been used in research.

Modeling was used to suggest the use of dyed photoresist in the reduction of standing waves [58]. Experimental investigation into dyed resists didn't begin until 10 years later [59, 60].

Since phase-shifting masks were first introduced [61], modeling has proved to be indispensable in their study. Levenson used modeling extensively to understand the effects of phase masks [62]. One of the earliest studies of phase-shifting masks used modeling to calculate images for Levenson's original alternating phase mask and then showed how phase masks increased defect printability [63]. The same study used modeling to introduce the concept of the outrigger (or assist slot) phase mask. Since these early studies, modeling results have been presented in nearly every paper published on phase-shifting masks.

Off-axis illumination was first introduced as a technique for improving resolution and depth of focus based on modeling studies [64]. Since then, this technique has received widespread attention and has been the focus of many more simulation and experimental efforts.

Using modeling, the advantages of having a variable numerical aperture, variable partial coherence stepper were discussed [64, 65]. Since then, all major stepper vendors have offered variable-NA, variable-coherence systems. Modeling remains a critical tool for optimizing the settings of these flexible new machines.

The use of pupil filters to enhance some aspects of lithographic performance has, to date, only been studied theoretically using lithographic models [66]. If such studies prove the usefulness of pupil filters, experimental investigations may also be conducted.

Modeling has been used in photoresist studies to understand the depth of focus loss when printing contacts in negative resists [67], the reason for artificially high values of resist contrast when surface inhibition is present [56], the potential for exposure optimization to maximize process latitude [68, 69], and the role of diffusion in chemically amplified resists [70]. Lithographic models are now standard tools for photoresist design and evaluation.

Modeling has always been used as a tool for quantifying optical proximity effects and for defining algorithms for geometry-dependent mask biasing [71, 72]. Most people would consider modeling to be a required element of any optical proximity correction scheme.

Defect printability has always been a difficult problem to understand. The printability of a defect depends considerably on the imaging system and resist used as well as the position of the defect relative to other patterns on the mask and the size and transmission properties of the defect. Modeling has proved itself to be a valuable and accurate tool for predicting the printability of defects [73, 74].

Modeling has also been used to understand the metrology of lithographic structures [75–78] and continues to find new application in virtually every aspect of lithographic research.

One of the primary reasons why lithography modeling has become such a standard tool for research activities is its ability to simulate such a wide range of lithographic conditions. While laboratory experiments are limited to the equipment and materials on hand (a particular wavelength and numerical aperture of the stepper, a given photoresist), simulation gives an almost infinite array of possible conditions. From high numerical apertures to low wavelengths, from hypothetical resists to arbitrary mask structures, simulation offers the ability to run "experiments" on steppers that you do not own with photoresists that have yet to be made. How else can one explore the shadowy boundary between the possible and the impossible?

9.10.2 Process Development Tool

Lithography modeling has also proved to be an invaluable tool for the development of new lithographic processes and equipment. Some of the more common uses include the optimization of dye loadings in photoresist [79, 80], the simulation of substrate reflectivity [81, 82], the applicability and optimization of top and bottom antireflection coatings [83, 84], and the simulation of the effect of bandwidth on swing curve amplitude [85, 86]. In addition, simulation has been used to help understand the use of thick resists for thin-film head manufacture [87] as well as other non-semiconductor applications.

Modeling is used extensively by makers of photoresist to evaluate new formulations [88, 89] and to determine adequate measures of photoresist performance for quality control purposes [90]. Resist users often employ modeling as an aid for new resist evaluations. On the exposure tool side, modeling has become an indispensable part of the optimization of the numerical aperture and partial coherence of a stepper [91–93] and in the understanding of the print bias between dense and isolated lines [94]. The use of optical proximity correction software requires rules on how to perform the corrections, which are often generated with the help of lithography simulation [95].

As a development tool, lithography simulation excels due to its speed and cost-effectiveness. Process development usually involves running numerous experiments to determine optimum process conditions, shake out possible problems, determine sensitivity to variables, and write specification limits on the inputs and outputs of the process. These activities tend to be both time consuming and costly. Modeling offers a way to supplement laboratory experiments with simulation experiments to speed up this process and reduce costs. Considering that a single experimental run in a wafer fabrication facility can take from hours to days, the speed advantage of simulation is considerable. This allows a greater number of simulations than would be practical (or even possible) in the fab.

9.10.3 Manufacturing Tool

Although you will find less published material on the use of lithography simulation in manufacturing environments [96–98], the reason is the limited publications by people in manufacturing rather than the limited use of lithography modeling. The use of simulation in a manufacturing environment has three primary goals: to reduce the number of test or experimental wafers that must be run through the production line, to troubleshoot problems in the fab, and to aid in decision-making by providing facts to support engineering judgment and intuition.

Running test wafers through a manufacturing line is costly, due not so much to the cost of the test as to the opportunity cost of not running product [99]. If simulation can reduce the time a manufacturing line is not running product even slightly, the return on investment can be significant. Simulation can also aid in the time required to bring a new process online.

9.10.4 Learning Tool

Although the research, development, and manufacturing applications of lithography simulation presented here give ample benefits of modeling based on time, cost, and capability, the underlying power of simulation is its ability to act as a learning tool. Proper application of modeling allows the user to learn efficiently and effectively. There are many reasons why this is true. First, the speed of simulation versus experimentation makes feedback much more timely. Since learning is a cycle (an idea, an experiment, a measurement, then comparison back to the original idea), faster feedback allows more cycles of learning. Since simulation is very inexpensive, there are fewer inhibitions and more opportunities to explore ideas. And, as the research application has shown us, there are fewer physical constraints on what "experiments" can be performed.

All these factors allow the use of modeling to gain an understanding of lithography. Whether learning fundamental concepts or exploring subtle nuances, the value of improved knowledge cannot be overstated.

REFERENCES

1. F. H. Dill, "Optical Lithography," *IEEE Trans. Electron Devices*, Vol. ED-22, No. 7 (July 1975) pp. 440–444.
2. F. H. Dill, W. P. Hornberger, P. S. Hauge, and J. M. Shaw, "Characterization of Positive Photoresist," *IEEE Trans. Electron Devices*, Vol. ED-22, No. 7 (July 1975) pp. 445–452.
3. K. L. Konnerth and F. H. Dill, "In-Situ Measurement of Dielectric Thickness during Etching or Developing Processes," *IEEE Trans. Electron Devices*, Vol. ED-22, No. 7 (1975) pp. 452–456.
4. F. H. Dill, A. R. Neureuther, J. A. Tuttle, and E. J. Walker, "Modeling Projection Printing of Positive Photoresists," *IEEE Trans. Electron Devices*, Vol. ED-22, No. 7 (1975) pp. 456–464.
5. W. G. Oldham, S. N. Nandgaonkar, A. R. Neureuther, and M. O'Toole, "A General Simulator for VLSI Lithography and Etching Processes: Part I – Application to Projection Lithography," *IEEE Trans. Electron Devices*, Vol. ED-26, No. 4 (April 1979) pp. 717–722.
6. C. A. Mack, "PROLITH: A Comprehensive Optical Lithography Model," *Proc. SPIE*, Vol. 538 (1985) pp. 207–220, Optical Microlithography IV.
7. J. W. Goodman, *Introduction to Fourier Optics*, McGraw-Hill (New York: 1968).
8. C. A. Mack, "Understanding Focus Effects in Submicron Optical Lithography," *Proc. SPIE*, Vol. 922 (March 1988) pp. 135–148, Optical/Laser Microlithography; and Vol. 27, No. 12 (December 1988) pp. 1093–1100, Optical Engineering.
9. D. A. Bernard, "Simulation of Focus Effects in Photolithography," *IEEE T. Semiconduct. M.*, Vol. 1, No. 3 (August 1988) pp. 85–97.
10. C. A. Mack, "Analytical Expression for the Standing Wave Intensity in Photoresist," *Appl. Opt.*, Vol. 25, No. 12 (15 June 1986) pp. 1958–1961.
11. D. A. Bernard and H. P. Urbach, "Thin-Film Interference Effects in Photolithography for Finite Numerical Apertures," *J. Opt. Soc. Am. A*, Vol. 8, No. 1 (January 1991) pp. 123–133.
12. M. Born and E. Wolf, *Principles of Optics*, 6th edition, Pergamon Press (Oxford, 1980) pp. 113–117.
13. D. C. Cole, E. Barouch, U. Hollerbach, and S. A. Orszag, "Extending Scalar Aerial Image Calculations to Higher Numerical Apertures," *J. Vac. Sci. Technol.*, Vol. B10, No. 6 (November/December 1992) pp. 3037–3041.
14. D. G. Flagello, A. E. Rosenbluth, C. Progler, and J. Armitage, "Understanding High Numerical Aperture Optical Lithography," *Microelectron. Eng.*, Vol. 17 (1992) pp. 105–108.
15. C. A. Mack and C-B. Juang, "Comparison of Scalar and Vector Modeling of Image Formation in Photoresist," *Proc. SPIE*, Vol. 2440 (1995) pp. 381–394, Optical/Laser Microlithography VIII.
16. S. Middlehoek, "Projection Masking, Thin Photoresist Layers and Interference Effects," *IBM J. Res. Dev.*, Vol. 14 (March 1970) pp. 117–124.
17. J. E. Korka, "Standing Waves in Photoresists," *Appl. Opt.*, Vol. 9, No. 4 (April 1970) pp. 969–970.
18. D. F. Ilten and K. V. Patel, "Standing Wave Effects in Photoresist Exposure," *Image Technol.* (February/March 1971) pp. 9–14.
19. D. W. Widmann, "Quantitative Evaluation of Photoresist Patterns in the 1 μm Range," *Appl. Opt.*, Vol. 14, No. 4 (April 1975) pp. 931–934.
20. P. H. Berning, "Theory and Calculations of Optical Thin Films," *Physics of Thin Films*, George Hass, ed., Academic Press (New York: 1963) pp. 69–121.
21. D. A. Skoog and D. M. West, *Fundamentals of Analytical Chemistry*, 3rd edition, Holt, Rinehart, and Winston (New York: 1976) pp. 509–510.
22. J. M. Koyler, F. Z. Custode, and R. L. Ruddell, "Thermal Properties of Positive Photoresist and Their Relationship to VLSI Processing," *Kodak Microelectronics Seminar Interface '79*, (1979) pp. 150–165.
23. J. M. Shaw, M. A. Frisch, and F. H. Dill, "Thermal Analysis of Positive Photoresist Films by Mass Spectrometry," *IBM J. Res. Dev.*, Vol. 21, No. 3, (May 1977) pp. 219–226.

24. C. A. Mack, "Absorption and Exposure in Positive Photoresist," *Appl. Opt.*, Vol. 27, No. 23 (1 December 1988) pp. 4913–4919.

25. J. Albers and D. B. Novotny, "Intensity Dependence of Photochemical Reaction Rates for Photoresists," *J. Electrochem. Soc.*, Vol. 127, No. 6 (June 1980) pp. 1400–1403.

26. C. E. Herrick, Jr., "Solution of the Partial Differential Equations Describing Photo-Decomposition in a Light-Absorbing Matrix Having Light-Absorbing Photoproducts," *IBM J. Res. Dev.*, Vol. 10 (January 1966) pp. 2–5.

27. J. J. Diamond and J. R. Sheats, "Simple Algebraic Description of Photoresist Exposure and Contrast Enhancement," *IEEE Electron Device Lett.*, Vol. EDL-7, No. 6 (June 1986) pp. 383–386.

28. S. V. Babu and E. Barouch, "Exact Solution of Dill's Model Equations for Positive Photoresist Kinetics," *IEEE Electron Device Lett.*, Vol. EDL-7, No. 4 (April 1986) pp. 252–253.

29. H. Ito and C. G. Willson, "Applications of Photoinitiators to the Design of Resists for Semiconductor Manufacturing," in *Polymers in Electronics, ACS Symp. Ser.*, 242 (1984) pp. 11–23.

30. N. J. Turro, *Modern Molecular Photochemistry*, University Science Books (California: 1991) pp. 79–81.

31. D. Seligson, S. Das, H. Gaw, and P. Pianetta, "Process Control with Chemical Amplification Resists Using Deep Ultraviolet and X-ray Radiation," *J. Vac. Sci. Technol.*, Vol. B6, No. 6 (November/December 1988) pp. 2303–2307.

32. C. A. Mack, D. P. DeWitt, B. K. Tsai, and G. Yetter, "Modeling of Solvent Evaporation Effects for Hot Plate Baking of Photoresist," *Proc. SPIE*, Vol. 2195 (1994) pp. 584–595, Advances in Resist Technology and Processing XI.

33. H. Fujita, A. Kishimoto, and K. Matsumoto, "Concentration and Temperature Dependence of Diffusion Coefficients for Systems Polymethyl Acrylate and n-Alkyl Acetates," *Trans. Faraday Soc.*, Vol. 56 (1960) pp. 424–437.

34. D. E. Bornside, C. W. Macosko and L. E. Scriven, "Spin Coating of a PMMA/Chlorobenzene Solution," *J. Electrochem. Soc.*, Vol. 138, No. 1 (January 1991) pp. 317–320.

35. S. A. MacDonald et al., "Airborne Chemical Contamination of a Chemically Amplified Resist," *Proc. SPIE*, Vol. 1466 (1991) pp. 2–12, Advances in Resist Technology and Processing VIII.

36. K. R. Dean and R. A. Carpio, "Contamination of Positive Deep-UV Photoresists," *OCG Microlithography Seminar Interface '94, Proc.* (1994) pp. 199–212.

37. T. Ohfuji, A. G. Timko, O. Nalamasu, and D. R. Stone, "Dissolution Rate Modeling of a Chemically Amplified Positive Resist," *Proc. SPIE*, Vol. 1925 (1993) pp. 213–226, Advances in Resist Technology and Processing X.

38. F. P. Incropera and D. P. DeWitt, *Fundamentals of Heat and Mass Transfer*, 3rd edition, John Wiley & Sons (New York: 1990).

39. W. Feller, *An Introduction to Probability Theory and Its Applications*, Vol. 1, Wiley (New York: 1968) pp. 147–159.

40. G. Moore, "Cramming More Components onto Integrated Circuits," *Electron. Mag.*, Vol. 38, No. 8 (19 April 1965).

41. D. ben-Avraham, S. Havlin, *Diffusion and Reactions in Fractals and Disordered Systems*, (Cambridge University Press: 2000).

42. W. Press, S. Teukolsky, W. Vetterling, B. Flannery, *Numerical Recipes*, 3rd edition, (Cambridge University Press: 2007).

43. F. H. Dill and J. M. Shaw, "Thermal Effects on the Photoresist AZ1350J," *IBM J. Res. Dev.*, Vol. 21, No. 3 (May 1977) pp. 210–218.

44. D. W. Johnson, "Thermolysis of Positive Photoresists," *Proc. SPIE*, Vol. 469 (1984) pp. 72–79, Advances in Resist Technology.

45. C. A. Mack and R. T. Carback, "Modeling the Effects of Prebake on Positive Resist Processing," *Kodak Microelectronics Seminar, Proc.* (1985) pp. 155–158.

46. E. J. Walker, "Reduction of Photoresist Standing-Wave Effects by Post-Exposure Bake," *IEEE Trans. Electron Devices*, Vol. ED-22, No. 7 (July 1975) pp. 464–466.

47. M. A. Narasimham and J. B. Lounsbury, "Dissolution Characterization of Some Positive Photoresist Systems," *Proc. SPIE*, Vol. 100 (1977) pp. 57–64, Semiconductor Microlithography II.

48. C. A. Mack, "Development of Positive Photoresist," *J. Electrochem. Soc.*, Vol. 134, No. 1 (January 1987) pp. 148–152.

49. C. A. Mack, "New Kinetic Model for Resist Dissolution," *J. Electrochem. Soc*, Vol. 139, No. 4 (April 1992) pp. L35–L37.

50. P. Trefonas and B. K. Daniels, "New Principle for Image Enhancement in Single Layer Positive Photoresists," *Proc. SPIE*, Vol. 771 (1987) pp. 194–210, Advances in Resist Technology and Processing IV.

51. T. R. Pampalone, "Novolac Resins Used in Positive Resist Systems," *Solid State Tech.*, Vol. 27, No. 6 (June 1984) pp. 115–120.

52. D. J. Kim, W. G. Oldham, and A. R. Neureuther, "Development of Positive Photoresist," *IEEE Trans. Electron Dev.*, Vol. ED-31, No. 12 (December 1984) pp. 1730–1735.

53. R. Hershel and C. A. Mack, "Lumped Parameter Model for Optical Lithography," Chapter 2, *Lithography for VLSI, VLSI Electronics – Microstructure Science*, R. K. Watts and N. G. Einspruch, eds., Academic Press (New York: 1987) pp. 19–55.

54. C. A. Mack, A. Stephanakis, R. Hershel, "Lumped Parameter Model of the Photolithographic Process," *Kodak Microelectronics Seminar, Proc.* (1986) pp. 228–238.

55. C. A. Mack, "Enhanced Lumped Parameter Model for Photolithography," *Proc. SPIE*, Vol. 2197 (1994) pp. 501–510, Optical/Laser Microlithography VII.

56. C. A. Mack, "Lithographic Optimization Using Photoresist Contrast," *KTI Microlithography Seminar, Proc.* (1990) pp. 1–12; and *Microelectron. Manuf. Technol.*, Vol. 14, No. 1 (January 1991) pp. 36–42.

57. M. P. C. Watts and M. R. Hannifan, "Optical Positive Resist Processing II, Experimental and Analytical Model Evaluation of Process Control," *Proc. SPIE*, Vol. 539 (1985) pp. 21–28, Advances in Resist Technology and Processing II.

58. A. R. Neureuther and F. H. Dill, "Photoresist Modeling and Device Fabrication Applications," *Optical And Acoustical Micro-Electronics*, Polytechnic Press (New York: 1974) pp. 233–249.

59. H. L. Stover, M. Nagler, I. Bol, and V. Miller, "Submicron Optical Lithography: I-line Lens and Photoresist Technology," *Proc. SPIE*, Vol. 470 (1984) pp. 22–33, Optical Microlithography III.

60. I. I. Bol, "High-Resolution Optical Lithography using Dyed Single-Layer Resist," *Kodak Microelectronics Seminar Interface '84* (1984) pp. 19–22.

61. M. D. Levenson, N. S. Viswanathan, R. A. Simpson, "Improving Resolution in Photolithography with a Phase-Shifting Mask," *IEEE Trans. Electron Devices*, Vol. ED-29, No. 12 (December 1982) pp. 1828–1836.

62. M. D. Levenson, D. S. Goodman, S. Lindsey, P. W. Bayer, and H. A. E. Santini, "The Phase-Shifting Mask II: Imaging Simulations and Submicrometer Resist Exposures," *IEEE Trans. Electron Devices*, Vol. ED-31, No. 6 (June 1984) pp. 753–763.

63. M. D. Prouty and A. R. Neureuther, "Optical Imaging with Phase Shift Masks," *Proc. SPIE*, Vol. 470 (1984) pp. 228–232, Optical Microlithography III.

64. C. A. Mack, "Optimum Stepper Performance Through Image Manipulation," *KTI Micro-electronics Seminar, Proc.* (1989) pp. 209–215.

65. C. A. Mack, "Algorithm for Optimizing Stepper Performance Through Image Manipulation," *Proc. SPIE*, Vol. 1264 (1990) pp. 71–82, Optical/Laser Microlithography III.

66. H. Fukuda, T. Terasawa, and S. Okazaki, "Spatial Filtering for Depth-of-focus and Resolution Enhancement in Optical Lithography," *J. Vac. Sci. Technol.*, Vol. B9, No. 6 (November/December 1991) pp. 3113–3116.

67. C. A. Mack and J. E. Connors, "Fundamental Differences Between Positive and Negative Tone Imaging," *Proc. SPIE*, Vol. 1674 (1992) pp. 328–338, Optical/Laser Microlithography V; and *Microlithogr. World*, Vol. 1, No. 3 (July/August 1992) pp. 17–22.

68. C. A. Mack, "Photoresist Process Optimization," *KTI Microelectronics Seminar, Proc.* (1987) pp. 153–167.

69. P. Trefonas and C. A. Mack, "Exposure Dose Optimization for a Positive Resist Containing Poly-functional Photoactive Compound," *Proc. SPIE*, Vol. 1466 (1991) pp. 117–131, Advances in Resist Technology and Processing VIII.

70. J. S. Petersen, C. A. Mack, J. Sturtevant, J. D. Byers, and D. A. Miller, "Non-constant Diffusion Coefficients: Short Description of Modeling and Comparison to Experimental Results," *Proc. SPIE*, Vol. 2438 (1995) pp. 167–180, Advances in Resist Technology and Processing XII.

71. C. A. Mack and P. M. Kaufman, "Mask Bias in Submicron Optical Lithography," *J. Vac. Sci. Technol.*, Vol. B6, No. 6 (November/December 1988) pp. 2213–2220.

72. N. Shamma, F. Sporon-Fielder, and E. Lin, "A Method for Correction of Proximity Effect in Optical Projection Lithography," *KTI Microelectronics Seminar, Proc.* (1991) pp. 145–156.

73. A. R. Neureuther, P. Flanner III, and S. Shen, "Coherence of Defect Interactions with Features in Optical Imaging," *J. Vac. Sci. Technol.*, Vol. B5, No. 1 (January/February 1987) pp. 308–312.

74. J. Wiley, "Effect of Stepper Resolution on the Printability of Submicron 5x Reticle Defects," *Proc. SPIE*, Vol. 1088 (1989) pp. 58–73, Optical/Laser Microlithography II.

75. L. M. Milner, K. C. Hickman, S. M. Gasper, K. P. Bishop, S. S. H. Naqvi, J. R. McNeil, M. Blain, and B. L. Draper, "Latent Image Exposure Monitor Using Scatterometry," *Proc. SPIE*, Vol. 1673 (1992) pp. 274–283.

76. K. P. Bishop, L. M. Milner, S. S. H. Naqvi, J. R. McNeil, and B. L. Draper, "Use of Scatterometry for Resist Process Control," *Proc. SPIE*, Vol. 1673 (1992) pp. 441–452.

77. L. M. Milner, K. P. Bishop, S. S. H. Naqvi, and J. R. McNeil, "Lithography Process Monitor Using Light Diffracted from a Latent Image," *Proc. SPIE*, Vol. 1926 (1993) pp. 94–105.

78. S. Zaidi, S. L. Prins, J. R. McNeil, and S. S. H. Naqvi, "Metrology Sensors for Advanced Resists," *Proc. SPIE*, Vol. 2196 (1994) pp. 341–351.

79. J. R. Johnson, G. J. Stagaman, J. C. Sardella, C. R. Spinner III, F. Liou, P. Tiefonas, and C. Meister, "The Effects of Absorptive Dye Loading and Substrate Reflectivity on a 0.5 μm I-line Photoresist Process," *Proc. SPIE*, Vol. 1925 (1993) pp. 552–563.

80. W. Conley, R. Akkapeddi, J. Fahey, G. Hefferon, S. Holmes, G. Spinillo, J. Sturtevant, and K. Welsh, "Improved Reflectivity Control of APEX-E Positive Tone Deep-UV Photoresist," *Proc. SPIE*, Vol. 2195 (1994) pp. 461–476.

81. N. Thane, C. Mack, and S. Sethi, "Lithographic Effects of Metal Reflectivity Variations," *Proc. SPIE*, Vol. 1926 (1993) pp. 483–494.

82. B. Singh, S. Ramaswami, W. Lin, and N. Avadhany, "IC Wafer Reflectivity Measurement in the UV and DUV and Its Application for ARC Characterization," *Proc. SPIE*, Vol. 1926 (1993) pp. 151–163.

83. S. S. Miura, C. F. Lyons, and T. A. Brunner, "Reduction of Linewidth Variation over Reflective Topography," *Proc. SPIE*, Vol. 1674 (1992) pp. 147–156.

84. H. Yoshino, T. Ohfuji, and N. Aizaki, "Process Window Analysis of the ARC and TAR Systems for Quarter Micron Optical Lithography," *Proc. SPIE*, Vol. 2195 (1994) pp. 236–245.

85. G. Flores, W. Flack, and L. Dwyer, "Lithographic Performance of a New Generation I-line Optical System: A Comparative Analysis," *Proc. SPIE*, Vol. 1927 (1993) pp. 899–913.

86. B. Kuyel, M. Barrick, A. Hong, and J. Vigil, "0.5 Micron Deep UV Lithography Using a Micrascan-90 Step-and-Scan Exposure Tool," *Proc. SPIE*, Vol. 1463 (1991) pp. 646–665.

87. G. E. Flores, W. W. Flack, and E. Tai, "An Investigation of the Properties of Thick Photoresist Films," *Proc. SPIE*, Vol. 2195 (1994) pp. 734–751.

88. H. Iwasaki, T. Itani, M. Fujimoto, and K. Kasama, "Acid Size Effect of Chemically Amplified Negative Resist on Lithographic Performance," *Proc. SPIE*, Vol. 2195 (1994) pp. 164–172.

89. U. Schaedeli, N. Münzel, H. Holzwarth, S. G. Slater, and O. Nalamasu, "Relationship Between Physical Properties and Lithographic Behavior in a High Resolution Positive Tone Deep-UV Resist," *Proc. SPIE*, Vol. 2195 (1994) pp. 98–110.

90. K. Schlicht, P. Scialdone, P. Spragg, S. G. Hansen, R. J. Hurditch, M. A. Toukhy, and D. J. Brzozowy, "Reliability of Photospeed and Related Measures of Resist Performances," *Proc. SPIE*, Vol. 2195 (1994) pp. 624–639.

91. R. A. Cirelli, E. L. Raab, R. L. Kostelak, and S. Vaidya, "Optimizing Numerical Aperture and Partial Coherence to Reduce Proximity Effect in Deep-UV Lithography," *Proc. SPIE*, Vol. 2197 (1994) pp. 429–439.

92. B. Katz, T. Rogoff, J. Foster, B. Rericha, B. Rolfson, R. Holscher, C. Sager, and P. Reynolds, "Lithographic Performance at Sub-300 nm Design Rules Using High NA I-line Stepper with Optimized NA and σ in Conjunction with Advanced PSM Technology," *Proc. SPIE*, Vol. 2197 (1994) pp. 421–428.

93. P. Luehrmann and S. Wittekoek, "Practical 0.35 μm I-line Lithography," *Proc. SPIE*, Vol. 2197 (1994) pp. 412–420.

94. V. A. Deshpande, K. L. Holland, and A. Hong, "Isolated-grouped Linewidth Bias on SVGL Micrascan," *Proc. SPIE*, Vol. 1927 (1993) pp. 333–352.

95. R. C. Henderson and O. W. Otto, "Correcting for Proximity Effect Widens Process Latitude," *Proc. SPIE*, Vol. 2197 (1994) pp. 361–370.

96. H. Engstrom and J. Beacham, "Online Photolithography Modeling Using Spectrophotometry and PROLITH/2," *Proc. SPIE*, Vol. 2196 (1994) pp. 479–485.

97. J. Kasahara, M. V. Dusa, and T. Perera, "Evaluation of a Photoresist Process for 0.75 Micron, G-line Lithography," *Proc. SPIE*, Vol. 1463 (1991) pp. 492–503.

98. E. A. Puttlitz, J. P. Collins, T. M. Glynn, L. L. Linehan, "Characterization of Profile Dependency on Nitride Substrate Thickness for a Chemically Amplified I-line Negative Resist," *Proc. SPIE*, Vol. 2438 (1995) pp. 571–582.

99. P. M. Mahoney and C. A. Mack, "Cost Analysis of Lithographic Characterization: An Overview," *Proc. SPIE*, Vol. 1927 (1993) pp. 827–832, Optical/Laser Microlithography VI.

10 Maskless Lithography

Kazuaki Suzuki

CONTENTS

10.1   INTRODUCTION

"Maskless" means "Direct Write." A maskless lithography tool has a pattern generator, which creates an exposure beam control converted from a designed pattern data format (such as GDS II) but does not use a mask. There are three kinds of exposure beam: electron beam (e-beam), ion beam, and optical laser beam.

Lithography using electron beams to expose the resist was one of the earliest processes used for integrated circuit fabrication, dating back to 1957 [1]. Today, essentially all high-volume production, even down to feature sizes smaller than 50 nm, is done with optical techniques due to the advances in optical stepper technology and process control described in this textbook. Nevertheless, electron beam systems continue to play two vital roles that will, in all probability, not diminish in importance for the foreseeable future. First, they are used to generate the masks that are used in all projection, proximity, and contact exposure systems, and second, they are used in the low-volume manufacture of ultrasmall features for very-high-performance devices. In addition, however, there is some activity in so-called "mix-and-match" lithography, in which the e-beam system is used to expose one or a few levels with especially small features, and optical systems are used for the rest. Thus, it is possible that as feature sizes move below about 40 nm (where optical techniques face substantial obstacles, especially for critical layers such as contacts and via-chain), electron beam systems might play a role in advanced manufacturing despite their throughput limitations as serial exposure systems. Recently, cutting lithography in a high-end device has been proposed and tried for a single-layer patterning in order to superpose line-and-space patterns by optical lithography and rectangular or square patterns by electron beam lithography. For these reasons, it is important for the lithographer to have some knowledge of the features of e-beam exposure systems, even though it is expected that optical lithography will continue to be the dominant manufacturing technique.

This chapter provides an introduction to such systems. It is intended to have sufficient depth for the reader to understand the basic principles of operation and design guidelines without attempting to be a principal source for a system designer or a researcher pushing the limits of the technique. For electron beam lithography, the treatment is based on a monograph by Owen [2], which the reader should consult for more detail, as well as background information and historical aspects. Originally, this chapter was written by Owen and Sheats for the first edition of *Microlithography* in 1998, and it was revised by Suzuki for the second edition of *Microlithography* in 2007. In this third edition, the technology developments in those 10 years are updated. In addition, useful equations for electron–electron interaction, ionization loss of electrons, and treatment of special relativity are introduced, and a description of ion beam lithography and laser direct write is added.

10.2   ELECTRON OPTICS OF ROUND-BEAM INSTRUMENTS

10.2.1   General Description

Figure 10.1 is a simplified ray diagram of a hypothetical scanned-beam electron lithography instrument, in which lenses have been idealized as thin optical elements. The electron optics of a scanning electron microscope (SEM) [3] would be similar in many respects.

Electrons are emitted from the source, whose crossover is focused onto the surface of the workplace by two magnetic lenses. The beam half-angle is governed by the beam shaping aperture. This intercepts current emitted by the gun that is not ultimately focused onto the spot. In order to minimize the excess current flowing down the column, the beam shaping aperture needs to be placed as near as possible to the gun and, in extreme cases, may form an integral part of the gun itself. This is beneficial, because it reduces electron–electron interactions, which have the effect of increasing the diameter of the focused spot at the workpiece. A second benefit is that the lower the current flowing through the column, the less opportunity there is for residual hydrocarbon or siloxane molecules to polymerize and form insulating contamination films on the optical elements. If present, these can acquire electric charge and cause beam drift and loss of resolution.

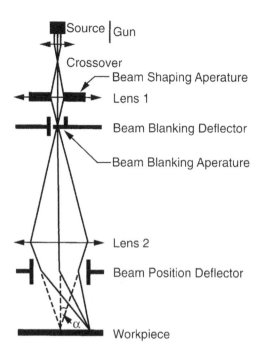

FIGURE 10.1 Simplified ray diagram of the electron optical system of a hypothetical round-beam electron lithography system.

A magnetic or electrostatic deflector is used to move the focused beam over the surface of the workpiece: this deflector is frequently placed after the final lens. The beam can be turned off by a beam blanker, which consists of a combination of an aperture and a deflector. When the deflector is not activated, the beam passes through the aperture and exposes the workpiece. However, when the deflector is activated, the beam is diverted, striking the body of the aperture.

A practical instrument would incorporate additional optical elements, such as alignment deflectors and stigmators. The arrangement shown in Figure 10.1 is only one possible configuration for a scanned-beam instrument. Many variations are possible; a number of instruments, for example, have three magnetic lenses.

If the beam current delivered to the workpiece is I, the area on the wafer to be exposed is A, and the charge density to be delivered to the exposed regions (often called the *dose*) is Q, it then follows that the total exposure time is

$$T = QA/I \tag{10.1}$$

Thus, for short exposure times, the resist should be as sensitive as possible, and the beam current should be as high as possible.

The beam current is related to the beam half-angle (α) and the diameter of the spot focused on the substrate (d_0) by the relationship

$$I = b\left(\frac{\pi d_0^2}{4}\right)(\pi \alpha^2) \tag{10.2}$$

where b is the brightness of the source. In general, the current density in the spot is not uniform but consists of a bell-shaped distribution; as a result, d_0 corresponds to an "effective" spot diameter. Note that the gun brightness and the beam half-angle need to be as high as possible to maximize current density.

Depending on the type of gun, the brightness can vary by several orders of magnitude (see Section 10.2.3): a value in the middle of the range is 10^5 A cm^{-2} sr^{-1}. The numerical aperture (NA) is typically about 5×10^{-3} rad. Using these values, and assuming a spot diameter of 0.5 μm, Equation 10.2 predicts a beam current of about 15 nA, a value that is typical for this type of lithography system.

For Equation 10.2 to be valid, the spot diameter must be limited only by the source diameter and the magnification of the optical system. This may not necessarily be the case in practice because of the effects of geometric and chromatic aberrations and electron–electron interactions.

The time taken to expose a chip can be calculated using Equation 10.1. As an example, a dose of 10 μC cm^{-2}, a beam current of 15 nA, and 50% coverage of a 5×5 mm^2 chip will result in a chip exposure time of 1.4 minutes. A 3 in.-diameter wafer could accommodate about 100 such chips, and the corresponding wafer exposure time would be 2.3 h. Thus, high speed is not an attribute of this type of system, particularly bearing in mind that many resists require doses well in excess of 10 μC cm^{-2}. For mask making, faster electron resists with sensitivities of up to 1 μC cm^{-2} are available: however, their poor resolution precludes their use for direct writing.

Equation 10.1 can also be used to estimate the maximum allowable response time of the beam blanker, the beam deflector, and the electronic circuits controlling them. In this case, corresponding to the area occupied by a pattern pixel, if the pixel spacing is 0.5 μm, then $A = 0.25\times10^{-12}$ m^2. Assuming a resist sensitivity of 10 μC cm^{-2} and a beam current of 15 nA implies that the response time must be less than 1.7 μs. Thus, the bandwidth of the deflection and blanking systems must be several MHz. Instruments that operate with higher beam currents and more sensitive resists require correspondingly greater bandwidths.

10.2.2   ELECTRON LENS ABERRATIONS

The following lens aberrations are important in electron beam instruments [3].

1. Geometric spot diameter
   Based on Equation 10.2, the geometric spot diameter (d_0) is derived as follows:

$$d_0 = \left(\frac{4I}{\pi^2 b}\right)^{1/2}\alpha^{-1} \tag{10.3}$$

2. Spherical aberration
   The electrons further from the electron optical axis are focused closer to the electron lens. The spherical aberration diameter d_s is expressed as follows:

$$d_s = 0.5C_s\alpha^3 \tag{10.4}$$

   where
       α is the beam half-angle (see Figure 10.1)
       C_s is the spherical aberration coefficient

3. Chromatic aberration
   The electrons with lower energy are focused closer to the electron lens. The chromatic aberration diameter d_c is expressed as follows:

$$d_c = C_c\frac{\Delta E}{E}\alpha \tag{10.5}$$

   where
       ΔE is the energy dispersion of the electron beam of energy E
       C_c is the chromatic aberration coefficient

Usually, the acceleration voltage is stable during exposure, and the chromatic aberration is not a dominant factor of the total spot diameter.

4. Fraunhofer diffraction disk
Fraunhofer diffraction at the aperture-limiting diaphragm in the final electron lens causes an Airy disk with a radius of d_d as follows:

$$d_d = 0.61 \frac{\lambda}{\alpha} \tag{10.6}$$

where λ is the de Broglie wavelength of the electron beam, which is given by,

$$\lambda = \frac{h}{mv} = \frac{(hc/qV)}{\sqrt{1 + 2(mc^2/qV)}} \tag{10.7}$$

λ becomes 5.4, 8.6, and 17 pm for the acceleration voltage of 50, 20, and 5 kV, respectively. If the beam half-angle is 5×10^{-3} rad as a typical value, d_d becomes 0.7, 1, and 2 nm for 50, 20, and 5 kV, respectively. Therefore, the diffraction is not a dominant factor of the total spot diameter.

5. Defocus
The diameter of the disk of confusion (d_z) caused by a defocus error Δz is given by the geometrical optical relationship

$$d_z = 2\alpha\Delta z \tag{10.8}$$

Astigmatism and focus controls of the beam (sometimes the height control of the sample as well) are causes of the defocus. If the beam half-angle of 5×10^{-3} rad and the defocus value of 0.2 μm are assumed, the induced spot blur d_z becomes 1 nm. Therefore, the defocus is not a dominant factor of the total spot diameter.

6. Total spot diameter
Total spot diameter (d_p) can be obtained from Equations 10.3–10.6 and 10.8 as follows:

$$d_p^2 = d_0^2 + d_s^2 + d_c^2 + d_d^2 + d_z^2 \tag{10.9}$$

As described above, d_c, d_d, and d_z are negligible in the case of electron lithography tool, so Equation 10.9 can be simplified as follows:

$$d_p^2 \approx d_0^2 + d_s^2 \tag{10.10}$$

When I is given, the beam half-angle is carefully chosen in order to minimize d_p.

10.2.3 ELECTRON EMISSION SOURCES

The electron guns used in scanning electron lithography systems are similar to those used in SEMs (for a general description, see, for example, Oatley [4]. There are four major types: thermionic guns using a tungsten hairpin as the source, thermionic guns using a lanthanum hexaboride source, tungsten field emission guns, and tungsten thermionic field emission (TF) guns.

Saturated emission current density (Js) has a relationship with the work function of the cathode surface (χ) and its temperature (T), as given by Equation 10.4, which is called Richardson–Dushman's formula [5]:

$$Js = CT^2 \exp(-\chi/kT) \tag{10.11}$$

where

    C is a material constant

    k is Boltzmann's constant

Then, the maximum brightness is given by Langmuir's formula [6] as follows:

$$b_{max} = \frac{Js}{\pi\alpha^2} \approx Js\left(\frac{eVa}{\pi kT}\right) \tag{10.12}$$

where Va represents the anode voltage.

Thermionic guns are commonly used, being simple and reliable. The source of electrons is a tungsten wire, bent into the shape of a hairpin, which is self-heated to a temperature of 2300–2700 °C by passing a direct current through it. The brightness of the gun and the lifetime of the wire depend strongly on temperature. At low heater currents, the brightness is of the order of 10^4 A cm^{-2} sr^{-1}, and the lifetime is of the order of 100 h. At higher heating currents, the brightness increases to about 10^5 Acm^{-2} sr^{-1}, but the lifetime decreases to a value of the order of 10 h (see, for example, Broers [7] and Wells [8]). Space charge saturation prevents higher brightness from being obtained. (The brightness values quoted here apply to beam energies of 10–20 keV.)

Lanthanum hexaboride is frequently used as a thermionic emitter by forming it into a pointed rod and heating its tip indirectly using a combination of thermal radiation and electron bombardment [7]. At a tip temperature of 1600 °C and a beam energy of 12 keV, Broers reported a brightness of over 10^5 Acm^{-2} sr^{-1} and a lifetime of the order of 1000 h. This represents an increase in longevity of a factor of two orders of magnitude over a tungsten filament working at the same brightness. This is accounted for by the comparatively low operating temperature, which helps to reduce evaporation. Two factors allow lanthanum hexaboride to be operated at a lower temperature than tungsten. The first is its comparatively low work function (approximately 2.7 eV as opposed to 4.5 eV). The second, and probably more important, factor is that the curvature of the tip of the lanthanum hexaboride rod is about 10 µm, whereas that of the emitting area of a bent tungsten wire is an order of magnitude greater. As a result, the electric field in the vicinity of the lanthanum hexaboride emitter is much greater, and the effects of space charge are much less pronounced.

Because of its long lifetime at a given brightness, a lanthanum hexaboride source needs to be changed only infrequently; this is a useful advantage for electron lithography, because it reduces the downtime of a very expensive machine. A disadvantage of lanthanum hexaboride guns is that they are more complex than tungsten guns, particularly as lanthanum hexaboride is extremely reactive at high temperatures, making its attachment to the gun assembly difficult. The high reactivity also means that the gun vacuum must be better than about 10^{-5} Pa (preferably 10^{-6} Pa) if corrosion by gas molecules is not to take place.

In the field emission gun, the source consists of a wire (generally of tungsten), one end of which is etched to a sharp tip with a radius of curvature of approximately 1 µm. This forms a cathode electrode, the anode being a coaxial flat disk, which is located in front of the tip. A hole on the axis of the anode allows the emitted electrons to pass out of the gun. To generate a 20 keV beam of electrons, the potential difference between the anode and the cathode is maintained at 20 kV, and the spacing is chosen so as to generate an electric field of about 10^9 V m^{-1} at the tip of the tungsten wire. At this field strength, electrons within the wire are able to tunnel through the potential barrier at the tungsten–vacuum interface, after which they are accelerated to an energy of 20 keV. An additional electrode is frequently included in the gun structure to control the emission current. (A general review of field emission has been written by Gomer [9].)

The brightness of a field emission source at 20 keV is generally more than 10^7 A cm^{-2} sr^{-1}. Despite this very high value, field emission guns have not been extensively used in electron beam lithography because of their unstable behavior and their high-vacuum requirements. In order to keep contamination of the tip and damage inflicted on it by ion bombardment to manageable proportions,

TABLE 10.1

Features of Electron Emission Source

Emission Type	Thermionic	Thermionic	Field Emission	Thermionic Field (TF)
Cathode material	Tungsten (W)	Lanthanum hexaboride (LaB$_6$)	Tungsten (W)	Zirconium oxide/W (ZrO/W)
Constant (A/cm^{2}K^2)	70	60	70	70
Work function (eV)	4.5	2.7	4.5	2.7
Emission source size	15–20 (μm)	10 (μm)	5–10 (nm)	15–20 (nm)
Energy dispersion (eV)	3	1.5	0.3	0.6
Brightness (A/cm^2 sr^2) (20 kV)	10^4	10^5	10^8	10^8
Cathode temperature (K)	2500	1800	300	1300
Lifetime	100 h	500 h	Several years	1–2 years
Current variation (/hour)	<1%	<2%	5%	<1%
Vacuum level (Pa)	10^{-5}	10^{-6}	10^{-8}	10^{-7}

the gun vacuum must be about 10^{-12} Torr. Even under these conditions, the beam is severely affected by low-frequency flicker noise, and the tip must be reformed to clean and repair it at approximately hourly intervals. Stille and Astrand [10] converted a commercial field emission scanning microscope into a lithography instrument. Despite the use of a servo-system to reduce flicker noise, dose variations of up to 5% were observed.

The structure of a TF gun is similar to that of a field emission gun, except that the electric field at the emitting tip is only about 10^8 V m^{-1} and the tip is heated to a temperature of 1000–1500 °C. Because of the Schottky effect, the apparent work function of the tungsten tip is lowered by the presence of the electric field. As a result, a copious supply of electrons is emitted thermionically at a comparatively low temperature. The brightness of a typical TF gun is similar to that of a field emission gun (at least 10^7 A cm^{-2} sr^{-1}), but the operation of a TF gun is far simpler. Because the tip is heated, it tends to be self-cleaning, and a vacuum of 10^{-9} torr is sufficient for stable operation. Flicker noise is not a serious problem, and lifetimes of many hundreds of hours are obtained, tip reforming being unnecessary. A description of this type of gun is given by Kuo and Siegel [11], and an electron lithography system using a TF gun is described below.

A thermionic gun produces a crossover whose diameter is about 50 μm, whereas field emission and TF guns produce crossovers whose diameters are of the order of 10 nm. For this reason, to produce a spot diameter of about 0.5 μm, the lens system associated with a thermionic source must be demagnifying, but that associated with a field emission or TF source needs to be magnifying.

The features of these emission sources are summarized in Table 10.1.

10.2.4 THE BEAM BLANKER

The function of the beam blanker is to switch on and off the current in the electron beam. To be useful, a beam blanker must satisfy three performance criteria:

1. When the beam is switched "off," its attenuation must be very great: typically, a value of 10^6 is specified.
2. Any spurious beam motion introduced by the beam blanker must be very much smaller than the size of a pattern pixel: typically, the requirement is for much less than 10 nm of motion.
3. The response time of the blanker must be very much less than the time required to expose a pattern pixel: typically, this implies a response time much less than 10 ns.

In practice, satisfying the first criterion is not difficult, but careful design is required to satisfy the other two, the configuration of Figure 10.1 being a possible scheme. An important aspect of its design is that the center of the beam blanking deflector is confocal with the workpiece.

Figure 10.2 is a diagram of the principal trajectory of electrons passing through an electrostatic deflector. The real trajectory is the curve ABC, which, if fringing fields are negligible, is parabolic. At the center of the deflector, the trajectory of the deflected beam is displaced by the distance $B'B$. The virtual trajectory consists of the straight lines AB' and $B'C$. Viewed from outside, the effect of the deflector is to turn the electron trajectory through the angle ϕ about the point B' (the center of the deflector). If, therefore, B' is confocal with the workpiece, the position of the spot on the workpiece will not change as the deflector is activated, and performance criterion 2 will be satisfied (the angle of incidence will change, but this does not matter).

The second important aspect of the design of Figure 10.1 is that the beam blanking aperture is also confocal with the workpiece. As a result, the cross section of the beam is smallest at the plane of the blanking aperture. Consequently, when the deflector is activated (shifting the real image of the crossover from B' to B), the transition from "on" to "off" occurs more rapidly than it would if the aperture were placed in any other position. This helps to satisfy criterion 3.

The blanking aperture is metallic, and so placing it within the deflector itself would, in practice, disturb the deflecting field. As a result, the scheme of Figure 10.1 is generally modified by placing the blanking aperture just outside the deflector; the loss in time resolution is usually insignificant.

An alternative solution has been implemented by Kuo et al. [12]. This is to approximate the blanking arrangement of Figure 10.2 by two blanking deflectors, one above and one below the blanking aperture. This particular blanker was intended for use at a data rate of 300 MHz, which is unusually fast for an electron lithography system, and so an additional factor, the transit time of the beam through the blanker structure, became important.

Considering the effect of special relativity, the velocity (v) of an electron traveling with a kinetic energy of qV electron volts is

$$v = c\beta = c\sqrt{1 - \frac{1}{\left(1 + qV/mc^2\right)^2}} \qquad (10.13)$$

where
  q is the charge of the electron
  m is the mass of the electron

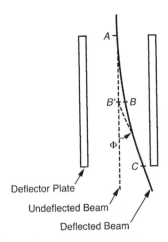

FIGURE 10.2   The principal trajectory of electrons passing through an electrostatic deflector.

Thus, the velocity of a 20 keV electron is approximately 8.2×10^7 m s^{-1}. The length of the blanker of Kuo et al. [12] in the direction of travel of the beam was approximately 40 mm, giving a transit time of about 0.5 ns. This time is significant compared with the pixel exposure time of 3 ns and if uncorrected, would have resulted in a loss of resolution caused by the partial deflection of the electrons already within the blanker structure when a blanking signal was applied.

To overcome the transit time effect, Kuo et al. [12] inserted a delay line between the upper and lower deflectors. This arrangement approximated a traveling wave structure in which the deflection field and the electron beam both moved down the column at the same velocity, eliminating the possibility of partial deflection.

10.2.5 DEFLECTION SYSTEMS

Figure 10.3a is a diagram of a type of deflection system widely used in SEMs, the "prelens double-deflection" system. The deflectors D1 and D2 are magnetic coils that are located behind the magnetic field of the final magnetic lens. In a frequently used configuration, L1 and L2 are equal, and the excitation of D2 is arranged to be twice that of D1 but acting in the opposite direction. This has the effect of deflecting the beam over the workpiece but not shifting the beam in the principal plane of the final lens, thus keeping its off-axis aberrations to a minimum. The size of this arrangement may be gauged from the fact that L1 typically lies between 50 and 100 mm.

The prelens double-deflection system is suitable for use in SEMs, because it allows a very small working distance (L) to be used (typically, it is less than 10 mm in SEM). This is essential in microscopy, because spherical aberration is one of the most important factors limiting the resolution, and the aberration coefficient increases rapidly with working distance. Thus, the prelens

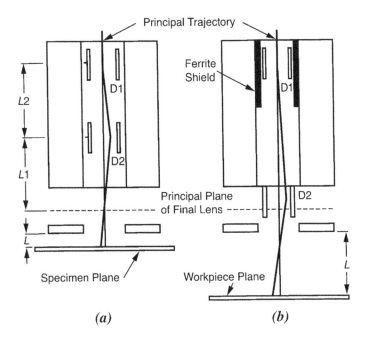

(a)   *(b)*

FIGURE 10.3 (a) A prelens double-deflection system, a type of deflector commonly used in scanning electron microscopes. (b) An in-lens double-deflection system, a type of deflector that is used in scanning electron lithography instruments. Note that the working distance (L) is made greater in (b) than it is in (a) because spherical aberration is not a limiting factor in scanning electron lithography; the longer working distance reduces the deflector excitation necessary to scan the electron beam over distances of several millimeters. The purpose of the ferrite shield in (b) is to reduce eddy current effects by screening D1 from the upper bore of the final lens.

double-deflection system allows a high ultimate resolution (10 nm or better) to be achieved; however, generally only over a limited field (about 10×10 µm). Outside this region, off-axis deflection aberrations enlarge the electron spot and distort the shape of the scanned area.

Although this limited field coverage is not a serious limitation for electron microscopy, it is for electron lithography. In this application, the resolution required is comparatively modest (about 50 nm), but it must be maintained over a field whose dimensions are greater than 1×1 mm². Furthermore, distortion of the scan field must be negligible.

Early work in scanning electron lithography was frequently carried out with a converted SEM using prelens double deflection. Chang and Stewart [13] used such an instrument and reported that deflection aberrations degraded its resolution to about 0.8 µm at the periphery of a 1×1 mm field at a beam half-angle of 5×10^{-3} rad. However, they also noted that the resolution could be maintained at better than 0.1 µm throughout this field if the focus and stigmator controls were manually readjusted after deflecting the beam. In many modem systems, field curvature and astigmatism are corrected in this way but under computer control, the technique being known as *dynamic correction* (see, for example, Owen [14]).

Chang and Stewart [13] also measured the deflection distortion of their instrument. They found that the nonlinear relationship between deflector current and spot deflection caused a positional error of 0.1 µm at a nominal deflection of 100 µm. The errors at larger deflections would be very much worse, because the relationship between distortion errors and nominal deflection consists of a homogeneous cubic polynomial under the conditions used in scanning electron lithography. Deflection errors are often corrected dynamically in modern scanning lithography systems by characterizing the errors before exposure, using the laser interferometer as the calibration standard. During exposure, appropriate corrections are made to the excitations of the deflectors.

Owen and Nixon [15] carried out a case study on a scanning electron lithography system with a prelens double-deflection system. On the basis of computer calculations, they showed that the source of off-axis aberrations was the deflection system, the effects of lens aberrations being considerably less serious. In particular, they noted that the effects of spherical aberration were quite negligible. This being the case, they went on to propose that for the purposes of electron lithography, postlens deflection was feasible. At working distances of several centimeters, spherical aberration would still be negligible for a well-designed final lens, and there would be sufficient room to incorporate a single deflector between it and the workpiece.

A possible configuration was proposed and built, the design philosophy adopted being to correct distortion and field curvature dynamically and optimize the geometry of the deflection coil so as to minimize the remaining aberrations. Amboss [16] constructed a similar deflection system, which maintained a resolution of better than 0.2 µm over a 2×2 mm² scan field at a beam half-angle of 3×10^{-3} rad. Calculations had indicated that the resolution of this system should have been 0.1 µm, and Amboss attributed the discrepancy to imperfections in the winding of the deflection coils.

A different approach to the design of low-aberration deflection systems was proposed by Ohiwa et al. [17]. This "in-lens" scheme, illustrated in Figure 10.3b, is an extension of the prelens double-deflection system, from which it differs in two respects:

1. The second deflector is placed within the polepiece of the final lens, with the result that the deflection field and the focusing field are superimposed.
2. In a prelens double-deflection system, the first and second deflectors are rotated by 180° about the optic axis with respect to each other. This rotation angle is generally not 180° in an in-lens deflection system.

Ohiwa et al. showed that the axial position and rotation of the second deflector can be optimized to reduce the aberrations of an in-lens deflection system to a level far lower than would be possible with a prelens system. The reasons for this are as follows:

1. Superimposing the deflection field of D2 and the lens field creates what Ohiwa et al. termed a "moving objective lens." The superimposed fields form a rotationally symmetrical distribution centered not on the optic axis but on a point whose distance from the axis is proportional to the magnitude of the deflection field. The resultant field distribution is equivalent to a magnetic lens, which, if the system is optimized, moves in synchronism with the electron beam deflected by D1 in such a way that the beam always passes through its center.
2. The rotation of DI with respect to D2 accounts for the helical trajectories of the electrons in the lens field.

A limitation of this work was that the calculations involved were based not on physically realistic lens and deflection fields but on convenient analytic approximations. Thus, although it was possible to give convincing evidence that the scheme would work, it was not possible to specify a practical design. This limitation was overcome by Munro [18], who developed a computer program that could be used in the design of this type of deflection system. Using this program, Munro designed a number of postlens, in-lens, and prelens deflection systems. A particularly promising in-lens configuration was one that had an aberration diameter of 0.15 μm after dynamic correction when covering a 5×5 mm^2 field at an angular aperture of 5×10^{-3} rad and a fractional beam voltage ripple of 10^{-4}.

Because the deflectors of an in-lens deflection system are located near metallic components of the column, measures have to be taken to counteract eddy current effects. The first deflector can be screened from the upper bore of the final lens by inserting a tubular ferrite shield as indicated in Figure 10.3b [19]. This solution would not work for the second deflector, because the shield would divert the flux lines constituting the focusing field. Chang et al. [20] successfully overcame this problem by constructing the lens pole pieces not of soft iron but of ferrite.

Although magnetic deflectors are in widespread use, electrostatic deflectors have several attributes for electron lithography. In the past, they have rarely been used because of the positional instability that has been associated with them, caused by the formation of insulating contamination layers on the surface of the deflection plates. However, recent improvements in vacuum technology now make the use of electrostatic deflection feasible.

The major advantage of electrostatic over magnetic deflection is the comparative ease with which fast response times can be achieved. A fundamental reason for this is that to exert a given force on an electron, the stored energy density associated with an electrostatic deflection field (U_E) is always less than that associated with a magnetic deflection field (U_M). If the velocity of the electrons within the beam is v and that of light in free space is c, the ratio of energy densities is

$$\frac{U_E}{U_M} = \left(\frac{v}{c}\right)^2 = \beta^2 \tag{10.14}$$

Thus, an electrostatic deflection system deflecting 20 keV electrons stores only 7.5% as much energy as a magnetic deflection system of the same strength occupying the same volume. It follows that the output power of an amplifier driving the electrostatic system at a given speed needs to be only 7.5% of that required to drive the magnetic system.

Electrostatic deflection systems have additional attractions in addition to their suitability for high-speed deflection:

1. They are not prone to the effects of eddy currents or magnetic hysteresis.
2. The accurate construction of electrostatic deflection systems is considerably easier than that of magnetic deflection systems. This is so because electrostatic deflectors consist of machined electrode plates, whereas magnetic deflectors consist of wires that are bent into shape. Machining is an operation that can be carried out to close tolerances comparatively simply, whereas bending is not.

An electron lithography system that uses electrostatic deflection is described below.

Computer-aided techniques for the design and optimization of electrostatic, magnetic, and combined electrostatic and magnetic lens and deflection systems have been described by Munro and Chu [21–24].

10.2.6 ELECTRON–ELECTRON INTERACTIONS (SPACE CHARGE EFFECT)

The mean axial separation between electrons traveling with a velocity v and constituting a beam current I is

$$\Delta = \frac{qv}{I} = \frac{qc}{I}\sqrt{1 - \frac{1}{\left(1 + qV/mc^2\right)^2}} \tag{10.15}$$

In a scanning electron microscope the beam current may be 10 pA, and for 20 keV electrons, this corresponds to a mean electron–electron spacing of 1.31 m. Because the length of an electron optical column is about 1 m, the most probable number of electrons in the column at any given time is less than one, and electron–electron interactions are effectively nonexistent.

In a scanning lithography instrument, however, the beam current has a value of between 10 nA and 1 μA, corresponding to mean electron spacings of between 1.34 mm and 13.4 μm. Under these circumstances, electron–electron interactions are noticeable. At these current levels, the major effect of the forces between electrons is to push them radially, thereby increasing the diameter of the focused spot. (In heavy-current electron devices, such as cathode-ray tubes or microwave amplifiers, the behavior of the beam is analogous to the laminar flow of a fluid, and electron–electron interaction effects can be explained on this basis. However, the resulting theory is not applicable to lithography instruments, in which the beam currents are considerably lower.)

Crewe [25] and Sasaki [26] used an analytic technique to estimate the magnitude of interaction effects in lithography instruments and showed that the increase in spot radius in nonrelativistic cases is given approximately by

$$\Delta r = \frac{1}{6\sqrt{2}\pi\varepsilon_0}\left(\frac{m}{q}\right)^{1/2}\frac{LI}{\alpha V^{3/2}} \tag{10.16}$$

where
ε_0 is permittivity of vacuum
the value of $\left(\dfrac{1}{4\pi\varepsilon_0}\right)$ is equal to 8.99×10^9 (Nm²/C² in SI units)

Using the electron mass of 9.1×10^{-31} kg and electron charge of 1.6×10^{-19} C, Equation 10.16 becomes

$$\Delta r = 1.0 \times 10^4 \frac{LI}{\alpha V^{3/2}} \tag{10.16'}$$

where
α represents the beam half-angle of the optical system
L represents the total length of the column
m and q are the mass and charge of an electron
ε_0 is the permittivity of free space

Note that neither the positions of the lenses nor their optical properties appear in Equation 10.16; it is only the total distance from source to workpiece that is important. For 20 keV electrons traveling down a column of length 1 m and beam half-angle 5×10^{-3} rad, the spot radius enlargement is 7 nm for a beam current of 10 nA, which is negligible for the purposes of electron lithography. However, at a beam current of 1 μA, the enlargement would be 0.7 μm, which is significant. Thus, great care must be taken in designing electron optical systems for fast lithography instruments that utilize comparatively large beam currents. The column must be kept as short as possible, and the beam half-angle must be made as large as possible.

Groves et al. [27] have calculated the effects of electron–electron interactions using not an analytic technique but a Monte Carlo approach. Their computations are in broad agreement with Equation 10.16. Groves et al. also compared their calculations with experimental data, obtaining reasonable agreement.

Another influence of electron–electron interaction is energy broadening, found by Boersch [28]. Theoretical study was done by Loeffler [29] and Crewe [30], and the analytical expression was given by Jansen et al. [31] as follows:

$$\frac{\Delta E_{\text{FWHM}}}{E} = \frac{\sqrt{2}}{\varepsilon_0} \left(\frac{m}{q} \right)^{1/2} \frac{I}{\alpha V^{3/2}} \tag{10.17}$$

Using the electron mass and electron charge, Equation 10.17 becomes

$$\frac{\Delta E_{\text{FWHM}}}{E} = 3.8 \times 10^5 \frac{I}{\alpha V^{3/2}} \tag{10.17'}$$

For 20 keV electrons, using the same parameter values, $(\Delta E_{FWHM}/E)$ becomes 0.27 ppm for 10 nA and 27 ppm for 1 μA, respectively. This Boersch effect induces chromatic aberration as expressed in Equation 10.5.

Defocusing is also induced. Jansen [32] reported the close agreement of space charge effects (spot radius, energy broadening, and defocus) between analytical expression from theory, Monte Carlo simulation, and the available experimental data.

10.2.7 ROUND-BEAM EXPOSURE SYSTEMS

The round-beam exposure system was designed and built primarily for mask making for optical lithography on a routine basis. It had a resolution goal of 2 μm linewidths and was designed in such a way as to achieve maximum reliability in operation rather than pushing the limits of capability.

The most unusual feature of this machine was that the pattern was written by moving the mask plate mechanically with respect to the beam. The plate was mounted on an X-Y table that executed a continuous raster motion, with a pitch (separation between rows) of 128 μm. If the mechanical raster were executed perfectly, each point on the mask could be accessed if the electron beam were scanned in a line, 128 μm long, perpendicular to the long direction of the mechanical scan. However, in practice, since mechanical motion of the necessary accuracy could not be guaranteed, the actual location of the stage was measured using laser interferometers, and the positional errors were compensated for by deflecting the beam appropriately. As a result, the scanned field was 140 × 140 μm, sufficient to allow for errors of ±70 μm in the x-direction and ±6 μm in the y-direction.

The advantage of this approach was that it capitalized on well-known technologies. The manufacture of the stage, although it required high precision, used conventional mechanical techniques. The use of laser interferometers was well established. The demands made on the electron optical system were sufficiently inexacting to allow the column of a conventional SEM to be used, although it had to be modified for high-speed operation [19].

Because the round-beam exposure system was not intended for high-resolution applications, it was possible to use resists of comparatively poor resolution but high sensitivity, typically 1 μC/cm². At a beam current of 20 nA, Equation 10.1 predicts that the time taken to write an area of 1 cm² would be 50 s (note that the exposure time is independent of pattern geometry in this type of machine). Therefore, the writing time for a 6 in. mask or reticle would be about 3.2 h, regardless of pattern geometry.

Because of its high speed, it is practicable to use the round-beam exposure system for making masks directly, without going through the intermediate step of making reticles [33]. However, with the advent of wafer steppers, a major use of these machines is now for the manufacture of reticles, and they are in widespread use. The writing speeds of later models have been somewhat increased, but the general principles remain identical to those originally developed.

10.3 SHAPED-BEAM INSTRUMENTS

10.3.1 FIXED SQUARE-SPOT INSTRUMENTS

Although it is possible to design a high-speed round-beam instrument with a data rate as high as 300 MHz, it is difficult, and its implementation is expensive. Pfeiffer [34] proposed an alternative scheme that allows patterns to be written at high speeds using high beam currents but without the need for such high data rates. In order to do this, he made use of the fact that the data supplied to a round-beam machine are highly redundant.

The spot produced by a round-beam machine is an image of the gun crossover, modified by the aberrations of the optical system. As a result, not only is it round but also the current density within it is nonuniform, conforming to a bell-shaped distribution. Because of this, the spot diameter is often defined as the diameter of the contour at which the current density falls to a particular fraction of its maximum value, this fraction typically being arbitrarily chosen as 1/2 or $1/e$. In order to maintain good pattern fidelity, the pixel spacing (the space between exposed spots) and the spot diameter must be relatively small compared with the minimum feature size to be written. A great deal of redundant information must then be used to specify a pattern feature (a simple square will be composed of many pixels).

Pfeiffer and Loeffler [35] pointed out that electron optical systems could be built that produced not round, nonuniform spots but square, uniformly illuminated ones. Thus, if a round-spot instrument and a square-spot instrument operate at the same beam current and expose the same pattern at the same dose, the data rate for the square-spot instrument will be smaller than that for the round-spot instrument by a factor of n^2, where n is the number of pixels that form the side of a square. Typically, $n = 5$ to get adequate uniformity, and the adoption of a square-spot scheme will reduce a data rate of 300 MHz to 12 MHz, a speed at which electronic circuits can operate with great ease.

To generate a square, uniformly illuminated spot, Pfeiffer and Loeffler [35] used Köhler's method of illumination, a technique well known in optical microscopy (see, for example, Born and Wolf [36]). The basic principle is illustrated in Figure 10.4a. A lens (L1) is interposed between the source (S) and the plane to be illuminated (P). One aperture (SA1) is placed on the source side of the lens, and another (BA) is placed on the other side. The system is arranged in such a way that the following optical relationships hold:

1. The planes of S and BA are confocal.
2. The planes of SA1 and P are confocal.

Under these circumstances, the shape of the illuminated spot at P (i-i) is similar to that of SA1 but demagnified by the factor d1/d2. (For this reason, SA1 is usually referred to as a *spot shaping aperture*.) The beam half-angle of the imaging system is determined by the diameter of the aperture BA. Thus, if SA1 is a square aperture, a square patch of illumination (i-i) will be formed at P, even though the aperture BA is round. The uniformity of the illumination stems from the fact that all trajectories emanating from a point such as on the source are spread out to cover the whole of (i-i).

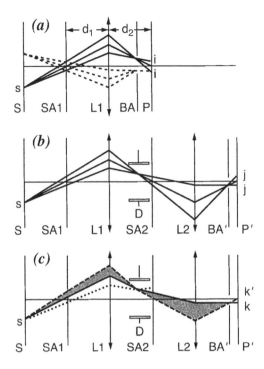

FIGURE 10.4 (a) The principle of the generation of a square, uniformly illuminated spot using Köhler's method of illumination. (b) The extension of the technique to the generation of a rectangular spot of variable dimensions, with the spot-shaping deflector D unactivated. (c) As for (b), but with D activated.

When Köhler's method of illumination is applied to optical microscopes, a second lens is used to ensure that trajectories from a given source point impinge on P as a parallel beam. However, this is unnecessary for electron lithography.

Pfeiffer [34] and Mauer et al. [37] have described a lithography system, the EL1, that used this type of illumination. It wrote with a square spot nominally measuring 2.5×2.5 μm containing a current of 3 μA. Because of the effects of electron–electron interactions, the edge acuity of the spot was 0.4 μm. The optical system was based on the principles illustrated in Figure 10.4a but was considerably more complex, consisting of four magnetic lenses. The lens nearest the gun was used as a condenser, and the spot shaping aperture was located within its magnetic field. This aperture was demagnified by a factor of 200 by the three remaining lenses, the last of which was incorporated in an in-lens deflection system.

The EL1 lithography instrument was used primarily for exposing interconnection patterns on gate array wafers. It was also used as a research tool for direct writing and for making photomasks [38].

The JEOL system adopts 100 kV acceleration voltage for wafer direct write systems.

10.3.2 VARIABLE SHAPED BEAM (VSB) INSTRUMENTS

A serious limitation of fixed square-spot instruments is that the linear dimensions of pattern features are limited to integral multiples of the minimum feature size. For example, using a 2.5×2.5 μm square spot, a 7.5×5.0 μm rectangular feature can be written, but a 8.0×6.0 μm feature cannot. An extension of the square-spot technique that removed this limitation was proposed by Fontijn [39] and first used for electron lithography by Pfeiffer [40].

The principle of the scheme is illustrated in Figure 10.4b. In its simplest form, it involves adding a deflector (D), a second shaping aperture (SA2), and a second lens (L2) to the configuration of

Figure 10.4a. The positions of these additional optical components are determined by the following optical constraints:

1. SA2 is placed in the original image plane, P.
2. The new image plane is P′, and L2 is positioned so as to make it confocal with the plane of SA2.
3. The beam shaping aperture BA is removed and replaced by the deflector D. The center of deflection of D lies in the plane previously occupied by BA.
4. A new beam shaping aperture BA′ is placed at a plane conjugate with the center of deflection of D (i.e., with the plane of the old beam shaping aperture BA).

SA1 and SA2 are both square apertures, and their sizes are such that a pencil from s that just fills SA1 will also just fill SA2 with the deflector unactivated. This is the situation depicted in Figure 10.4b, and under these circumstances, a square patch of illumination, j-j, is produced at the new image plane P′.

Figure 10.4c shows what happens when the deflector is activated. The unshaded portion of the pencil emitted from s does not reach P′, because it is intercepted by SA2. However, the shaded portion does reach the image plane, where it forms a uniformly illuminated rectangular patch of illumination k-k′. By altering the strength of the deflector, the position of k′ and hence, the shape of the illuminated patch can be controlled. Note that because the center of deflection of D is confocal with BA′, the beam shaping aperture does not cause vignetting of the spot as its shape is changed. Only one deflector is shown in the figure; in a practical system, there would be two such deflectors mounted perpendicular to each other so that both dimensions of the rectangular spot could be altered.

Weber and Moore [38] built a machine, the EL2, based on this principle, which was used as a research tool. Several versions were built, each with slightly different performance specifications. The one capable of the highest resolution used a spot whose linear dimensions could be varied from 1.0 to 2.0 μm in increments of 0.1 μm. A production version of the EL2, called the EL3, was built by Moore et al. [41]. The electron optics of this instrument were similar to those of the EL2, except that the spot shaping range was increased from 2:1 to 4:1. A version of the EL3 that was used for 0.5 μm lithography is described by Davis et al. [42]. In this instrument, the acceleration voltage was 50 kV [43], and the spot current density was reduced from 50 to 10 A cm^{-2} and the maximum spot size to 2×2 μm to reduce the effects of electron–electron interactions.

Equation 10.1 cannot be used to calculate the writing time for a shaped-beam instrument, because the beam current is not constant but varies in proportion to the area of the spot. A pattern is converted for exposure in a shaped-spot instrument by partitioning it into rectangular "shots." The instrument writes the pattern by exposing each shot in turn, having adjusted the spot size to match that of the shot. The time taken to expose a shot is independent of its area and is equal to Q/J (Q being the dose and J the current density within the shot), and so the time taken to expose a pattern consisting of N shots is

$$T = NQ/J \tag{10.18}$$

Thus, for high speed, the following requirements are necessary:

1. The current density in the spot must be as high as possible.
2. In order to minimize the number of pattern shots, the maximum spot size should be as large as possible. In practice, a limit is set by electron–electron interactions that degrade the edge acuity of the spot at high beam currents.
3. The pattern conversion program must be efficient at partitioning the pattern into as few shots as possible, given the constraint imposed by the maximum shot size.

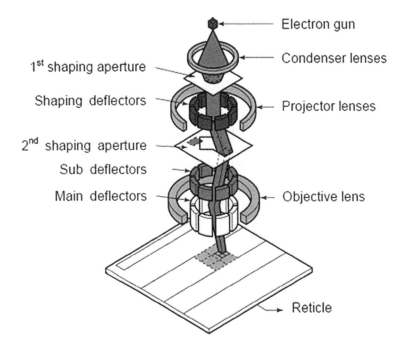

FIGURE 10.5 Example of VSB instrument (NuFlare's mask writer). (From Sunaoshi, H., et al., *Proc. SPIE*, 6283, 628306, 2006.)

In a modern ultrahigh-resolution commercial system such as that manufactured by JEOL, the area that can be scanned by the beam before moving the stage is of the order of 1×1 mm, and the minimum address size (pixel size) can be chosen to be 5 or 25 nm.

In addition, oblique patterns such as 45° and 135° orientations are prepared on SA2 (Hitachi, NuFlare, and Vistec), and those patterns can be written on wafer or mask as well as rectangular patterns as shown in Figure 10.5 [44].

Multiple-variable shaped beam (MSB) was proposed and its prototype was developed by Vistec [45]. The electron beam emitted from a single source is divided to multiple beams by the first shaping aperture array; then, those beams are deflected onto the second shaping aperture array independently by the deflector array. An arrangement of 4×4 beams was adopted in the proof of lithography system, and 8×8 is planned for the production system.

The acceleration voltage of the Hitachi system is 30 kV, whereas that of the NuFlare and the Vistec system is 50 kV.

10.3.3 Character Projection (CP) Instruments

Pfeiffer proposed to extend the shaped rectangular–spot method to a character projection method by replacing SA2 with a character plate with an array of complex aperture shapes that are frequently used in semiconductor device patterns [46]. The example of the character plate is introduced in Figure 10.6. A square-spot electron beam is deflected and projected onto the selected character on the character plate, and the beam through the aperture of this character is projected onto the wafer. This method can reduce the number of exposure shots and can make throughput higher. Hitachi and Fujitsu (succeeded by Advantest) successfully developed their instruments with this concept independently [47, 48].

As an extreme case, all magnified patterns of the semiconductor device can be prepared on the character plate. Because these magnified patterns have a one-to-one relationship with device patterns, the character plate can be called a *mask*. Periodic motion of electron beam deflection and

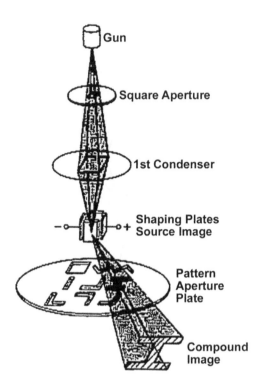

FIGURE 10.6 Example of character plate in a character projection instrument. (From Pfeiffer, H.C., *IEEE Trans. Electron. Dev.*, ED-26, 663, 1979.)

mask movement can realize efficient exposure. The difference of the scattered angle of incident electrons by Rutherford scattering between different materials can be used to form a contrast. This concept has been used in the area of transmission electron microscopy and was applied to image formation on an actinic film using a master stencil by Koops [49]. Berger [50] applied this concept to a lithography tool with a membrane-type mask that consists of heavy metal patterns on the thin membrane. A silicon stencil type can be also used as a mask for electron projection lithography (EPL) as well as a character plate [51–53]. Prototypes and related infrastructure have been developed in the past, but this research work has been discontinued.

10.3.4 Multiple-Beam Instruments

In order to make throughput higher, it is necessary to increase the beam current I in Equation 10.1 for a shorter exposure time. However, the spot radius becomes larger in the case of larger beam current (Equations 10.3 and 10.16). This is the reason why the multiple-beam concept has been considered for higher throughput and better resolution simultaneously. A multicolumn e-beam exposure system was developed by ASET (the Association of Super-Advanced Electronics Technologies), a Japanese consortium, as a Mask D2I (mask design, drawing, and inspection) project. This proof of concept system adopts 2×2 multicolumn cells [54, 55].

10.4 MASSIVELY PARALLEL ELECTRON BEAM INSTRUMENTS

10.4.1 Blanking Aperture Array (BAA)

In the recent technological development of microelectromechanical systems (MEMS), several kinds of programmable aperture array or blanking aperture array (BAA) have been proposed and

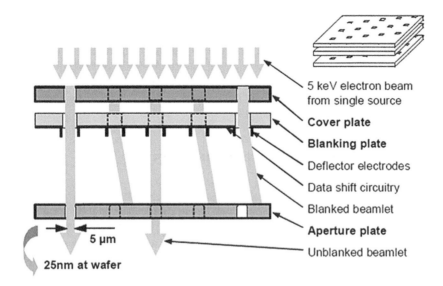

5 keV electron beam from single source

Cover plate

Blanking plate

Deflector electrodes

Data shift circuitry

Blanked beamlet

Aperture plate

Unblanked beamlet

5 μm

25nm at wafer

FIGURE 10.7 Principle of blanking aperture array (BAA). (From Brandstatter, C., et al., *Proc. SPIE*, 5374, 601, 2004.)

developed. Yasuda et al. reported two-dimensional BAA fabrication (for 1024 beams) in an early work [56]. Figure 10.7 shows the principle of BAA [57]. The electron beam emitted from a single source is divided into multiple beams by the first aperture array; then, these beams are independently deflected or undeflected by a programmable array of electrodes. The undeflected beam goes through the second aperture, but the deflected beam stops at the second aperture.

10.4.2 BAA PROJECTION INSTRUMENTS

IMS Nanofabrication has developed a projection maskless lithography (PML2) system with a magnification of 1/200 from the second blanking aperture onto an exposed sample plane. The acceleration voltage is 50 kV, and the number of beams is designed to be 512×512 in its eMET (electron Mask Exposure Tool) system. The total electrical current is 1 μA, and the electrical current of each beam is 3.8 pA [58]. Recently, the development of a new mask writer using a similar concept was reported by NuFlare [59].

KLA-Tencor developed the Reflective Electron Beam Lithography (REBL) system, which used a digital pattern generator (DPG) with a 248×4096 array. DPG works as a digital mirror array against an electron beam to reflect it to an exposed sample through electron optics or to cause it to be obstructed by the aperture [60]. The acceleration voltage of the prototype system is 75–100 kV, and the magnification was 1/85 from DPG onto the exposed sample [61]. Research work on REBL has now been discontinued.

10.4.3 MAPPER SYSTEM

Multiple Aperture Pixel by Pixel Enhancement of Resolution (MAPPER) was proposed and developed by MAPPER Lithography. At first, the beam from the electron source is divided into 13,260 multiple beams by the first aperture array. Then, the electron optics downstream for each divided beam works as one BAA system. Each divided beam is injected into BAA to form 7×7 sub-beams, and those 7×7 sub-beams are switched on and off individually by nondeflecting or deflecting a sub-beam onto the beam stop aperture. The total electrical current of 649,740 (=$13,260 \times 7 \times 7$) multiple beams becomes 170 μA, and the electrical current of each set of 7×7 beams and each sub-beam is 13 nA and 262 pA, respectively. A dispenser-type cathode can provide such a large electrical current. The required time for exposing one 300 mm wafer is 418 s for 100 µC/cm² resist sensitivity. In contrast to other systems, acceleration voltage is just 5 kV [62, 63]. The features of the low voltage are small energy deposition to wafer and small radius of proximity effect correction (see Section 10.8), but the space charge effect becomes large, and a thin resist process becomes necessary because of a shorter electron range in the resist material. In addition, this system might be affected by disturbance and contamination.

10.5 ELECTRON BEAM ALIGNMENT TECHNIQUES

10.5.1 PATTERN REGISTRATION

The first step in directly writing a wafer is to define registration marks on it. Commonly, these are grouped into sets of three, each set being associated with one particular chip site. The registration marks may be laid down on the wafer in a separate step before any of the chip levels are written, or they may be concurrently written with the first level of the chip.

Pattern registration is necessary because no lithography instrument can write with perfect reproducibility. Several factors, discussed in detail in the following, can lead to an offset of a given integrated circuit level from its intended position with respect to the previous one:

1. *Loading errors.* Wafers are generally loaded into special holders for exposure. Frequently, the location and orientation of a wafer are fixed by a kinematic arrangement of three pins or an electrostatic chuck. However, the positional errors associated with this scheme can be several tens of micrometers, and the angular errors can be as large as a few hundreds of microradians.

2. *Static and dynamic temperature errors.* Unless the temperature is carefully controlled, thermal expansion of the wafer can cause these errors to be significant. The thermal expansion coefficient of silicon is 2.4×10^{-6} °C^{-1}, and over a distance of 100 mm, this corresponds to a shift of 0.24 µm for a 1 °C change. When two pattern levels are written on a wafer at two different temperatures, but the wafers are in thermal equilibrium in each case, a static error results. However, if the wafer is not in thermal equilibrium while it is being written, a dynamic error results, whose magnitude varies as the temperature of the wafer changes with time.

3. *Substrate height variations.* These are important, because they give rise to changes in deflector sensitivity. For example, consider a postlens deflector, nominally 100 mm above the substrate plane and deflecting over a 5×5 mm scan field. A change of 10 µm in the height of the chip being written results in a maximum pattern error of 0.25 µm. Height variations can arise from two causes. The first is nonperpendicularity between the substrate plane and the undeflected beam, which makes the distance between the deflector and the portion of the substrate immediately below it a function of stage position. The second cause is curvature of the wafer. Even unprocessed wafers are bowed, and high-temperature processing steps can significantly change the bowing. The deviations from planarity can amount to several micrometers.

4. *Stage yaw.* The only type of motion that a perfect stage would execute would be linear translation. However, when any real stage is driven, it will rotate slightly about an axis perpendicular to its translational plane of motion. Typically, this motion, called yaw, amounts to several arcseconds. The positional error introduced by a yaw of 10″ for a 5×5 mm chip is 0.24 µm.

5. *Beam position drift.* The beam in a lithography instrument is susceptible to drift, typically amounting to a movement of less than 1 µm in 1 hour.

6. *Deflector sensitivity drift.* The sensitivities of beam deflectors tend to drift because of variations in beam energy and gain changes in the deflection amplifiers. This effect can amount to a few parts per million in 1 hour.

By aligning the pattern on registration marks, exact compensation is made for loading errors, static temperature errors, substrate height variations, and yaw, all of which are time independent. Usually, dynamic temperature errors, beam position drift, and deflector sensitivity drift are reduced to negligible levels by pattern registration, because although they are time dependent, the time scales associated with them are much greater than the time taken to write a chip.

A typical alignment scheme would consist of a coarse registration step followed by a fine registration step. The procedures are, in general, quite similar to those used in optical lithography, which are discussed at length in Chapter 5. Here, the aspects unique to electron lithography, primarily having to do with the nature of the alignment detection signals using electron beams, are the focus.

Using the wafer flat (or some other mechanical factor in the case of nonstandard substrates), a coarse positioning is carried out and the wafer scanned (not at high resolution) in the area of an alignment mark. Assuming it is detected, the coordinates of its center (with reference to an origin in the machine's system) are now known, and an offset is determined from the coordinates specified for it in the pattern data.

This level of accuracy is still inadequate for pattern writing, but it is sufficient to allow the fine registration step to be carried out. One purpose of fine registration is to improve the accuracy with which the coarse registration compensated for loading errors. In addition, it compensates for the remaining misregistration errors (temperature errors, substrate height variations, stage yaw, beam position drift, and deflector sensitivity drift). Because these will, in general, vary from chip to chip, fine registration is carried out on a chip-by-chip basis. Each chip is surrounded by three registration marks. Because the residual errors after coarse registration are only a few micrometers, the chip marks need to be only about 100 μm long. The steps involved in fine registration could be as follows:

1. The pattern data specify coordinates corresponding to the center of the chip. These are modified to account for the wafer offset and rotation measured during the coarse registration step. The stage is moved accordingly so as to position the center of the chip under the beam.
2. The electron beam is deflected to the positions of the three alignment marks in turn, and each mark is scanned. In this way, the position of each of the three marks is measured.
3. The pattern data are transformed linearly so as to conform to the measured positions of the marks, and the pattern is then written onto the chip.

This procedure is repeated for each chip on the wafer.

Numerous variations of the scheme described here exist. A serious drawback of this scheme is that it works on the assumption that each chip corresponds to a single scanned field. A registration scheme described by Wilson et al. [64] overcomes this limitation, allowing any number of scanned fields to be stitched together to write a chip pattern, making it possible to write chips of any size.

10.5.2 ALIGNMENT MARK STRUCTURES

Many types of alignment mark have been used in electron lithography, including pedestals of silicon or silicon dioxide, metals of high atomic number, and trenches etched into the substrate. The last type of mark is frequently used and will be considered here as an example to demonstrate how alignment signals are generated. A common method of forming trenches is by etching appropriately masked silicon wafers in aqueous potassium hydroxide. Wafers whose top surface corresponds to the (100) plane are normally used for this purpose. The etching process is anisotropic, causing the sides of the trenches to be sloped as illustrated in Figure 10.8a. Typically, the trench is 10 μm wide and 2 μm deep.

The way in which the resist film covers a trench depends on the dimensions of the trench, the material properties of the resist, and the conditions under which the resist is spun onto the wafer.

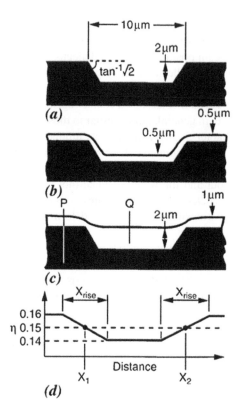

FIGURE 10.8   (a) Cross-sectional view of an alignment mark consisting of a trench etched into silicon. The wall angle of $\tan^{-1}(2)^{1/2}$ is the result of the anisotropic nature of the etch. (b) Coverage of the trench by a film of resist that has little tendency to "planarize." (c) Coverage by a resist that has a strong tendency to planarize. (d) Variation of backscatter coefficient η as a function of position for the situation depicted in (c).

Little has been published on this subject, but from practical experience, it is found that two extreme cases exist.

1. If the resist material shows little tendency to planarize the surface of the wafer and is applied as a thin film, the situation depicted in Figure 10.8b arises. The resist forms a uniform thin film whose top surface faithfully follows the shape of the trench.
2. If, on the other hand, a thick film of a resist that has a strong tendency to planarize is applied to the wafer, the situation shown in Figure 10.8c results. The top surface of the resist is nearly flat, but the thickness of the resist film increases significantly in the vicinity of the trench.

The mechanisms for the generation of alignment mark signals are different in these two cases.

10.5.3   ALIGNMENT MARK SIGNALS

The electrons emitted when an electron beam with an energy of several kiloelectron volts bombards a substrate can be divided into two categories:

1. The secondary electrons are those ejected from the substrate material itself. They are of low energy, and their energy distribution has a peak at an energy of a few electron volts. By convention, it is assumed that electrons with energies below 50 eV are secondaries.

2. The backscattered electrons are primaries that have been reflected from the substrate. For a substrate of silicon (atomic number $Z = 14$), their mean energy is approximately 60% of that of the primary beam [65].

The electron collectors used in SEMs are biased at potentials many hundreds of volts above that of the specimen in order to attract as many secondary electrons as possible. As a consequence, it is these electrons that dominate the formation of the resulting image. Everhart et al. [66] explain why this is done:

> the paths of (back-scattered) electrons from the object to the collector are substantially straight, whilst those of secondary electrons are usually sharply curved. It follows that (back-scattered) electrons cannot reveal detail of any part of the object from which there is not a straight-line path to the collector, while secondary electrons are not subject to this limitation. Thus, secondary electrons provide far more detail when a rough surface is under examination.

However, this argument does not apply to the problem of locating a registration mark, which is a comparatively large structure whose fine surface texture is of no interest. Consequently, no discrimination is made against backscattered electrons in alignment mark detection, and in fact, it is these electrons that most strongly contribute to the resulting signals. Backscattered electrons may be collected either by using a scintillator–photomultiplier arrangement or by using a solid-state diode as a detector. This is a popular collection scheme and is usually implemented by mounting an annular diode above the workpiece. Wolf et al. [67] used a solar cell diode 25 mm in diameter with a 4 mm diameter hole in it through which the primary electron beam passed, with the total solid angle subtended at the workpiece being 0.8 sr. Detectors of this type are insensitive to secondary electrons, because these electrons are not sufficiently energetic to penetrate down to the depletion region that is under the surface; the threshold energy for penetration is generally several hundred electron volts. The gain of the detector varies linearly with excess energy above the threshold, and the gradient of the relationship is approximately 1 hole–electron pair per 3.5 eV of beam energy.

A useful extension of this technique (see, for example, Reimer, 1984 [68]) is to split the detector into two halves. When the signals derived from the two halves are subtracted, the detector responds primarily to topographic variations on the substrate: this mode is well suited for detecting the type of mark depicted in Figure 10.8b. When the signals are added, the detector responds to changes in the backscattered electron coefficient (η). The type of mark shown in Figure 10.8c is best detected in this mode, because the values of η at locations such as P and Q are significantly different, whereas the topographic variations are small.

The backscattering coefficients of composite samples consisting of thin films supported on bulk substrates have been studied by electron microscopists (see, for example, Niedrig [69]. The composite backscattering coefficient varies approximately linearly from a value corresponding to the substrate material (η_S) for very thin films to a value corresponding to the film material (η_F) for very thick films. The value η_F is achieved when the film thickness is greater than about half the electron range in the film material.

A silicon substrate has a backscattering coefficient $\eta_S = 0.18$ (see, for example, Reed [70]). The widely used electron resist (polymethyl methacrylate) (PMMA) has the chemical formula $C_5O_2H_8$, and its mass concentration averaged atomic number is 6.2; as a rough approximation, it can be assumed that its backscattering coefficient is equal to that of carbon, that is, that $\eta_F = 0.07$. The density of PMMA is 1.2 g/cm³, and the data of Holliday and Sternglass [71] imply that the extrapolated range of 20 keV electrons in the material is about 8 μm. From these values. it follows that the backscattering coefficient of bulk silicon is reduced by roughly 0.02 for every 1 μm of PMMA covering it. A similar result has been calculated by Aizaki [72] using a Monte Carlo technique.

Figure 10.8d is a sketch of the variation of η along the fiducial mark depicted in Figure 10.8c. It has been assumed that at P and Q, well away from topographical changes, η has the values 0.16 and

0.14, respectively. The change in η caused by the sides of the trench is assumed to occur linearly over a distance of x_{rise} 5 μm. The exact form of the transition depends on the shape of the trench, the way the resist thickness changes in its vicinity, the range of elections in the resist, and most importantly, their range in silicon. (The extrapolated ranges of 20 and 50 keV electrons in silicon are about 2 μm and 10 μm, respectively.) Because a split backscattered electron detector connected in the adding mode responds to changes in η, Figure 10.8d also represents the signal collected as a well-focused electron beam is scanned over the registration mark. Despite the simplifying assumptions that have been made, this sketch is representative of the signals that are obtained in practice.

10.5.4 THE MEASUREMENT OF ALIGNMENT MARK POSITION

A threshold technique is frequently used to measure the position of an alignment mark. In Figure 10.8d, for example, the threshold has been set to correspond to $\eta=0.15$, and the position of the center of the trench is

$$x = (x_1 + x_2)/2 \tag{10.19}$$

The accuracy with which x_1 and x_2 can be measured is limited by electrical noise, of which there are three major sources:

1. The shot noise (see Section 10.6.3) associated with the primary electron beam
2. The noise associated with the generation of backscattered electrons
3. The noise associated with the detector itself

Wells et al. [73] analyzed the effects of shot noise and secondary emission noise on alignment accuracy. Their theory may be adapted to deal with backscatter noise by replacing the coefficient of secondary emission by the backscattering coefficient η. With this modification, the theory predicts that the spatial accuracy with which a threshold point such as x_1 or x_2 may be detected is

$$\Delta x = \left(\frac{128 q x_{rise}^2}{m^2 HQ} \frac{1+\eta}{\eta} \right)^{1/3} \tag{10.20}$$

The quantities appearing in this equation are defined as follows.

1. Δx is the measure of the detection accuracy. It is defined such that the probability of a given measurement being in error by more than Δx is 10^{-4}.
2. x_{rise} is the rise distance of the signal. A value of 5 μm is assumed in Figure 10.8d.
3. η is the mean backscattering coefficient. A value of 0.15 is assumed.
4. m is the fractional change in signal as the mark is scanned. In Figure 10.8d, $m=(0.16-0.14)/0.15=0.13$.
5. It is assumed that the beam oscillates rapidly in the direction perpendicular to the scan direction with an amplitude $(1/2)H$. In this way, the measurement is made, not along a line, but along a strip of width H. A typical value is $H=10$ μm.
6. Q is the charge density deposited in the scanned strip.

Equation 10.20 indicates that a compromise exists between registration accuracy (Δx) and charge density (Q) and that Δx may be made smaller than any set value provided that Q is large enough. For 0.5 μm lithography, an acceptable value of Δx could be 0.1 μm. To achieve this level of accuracy, Equation 10.20 predicts that a charge density $Q=2300$ μC/cm^2 has to be deposited. This calculation illustrates the point made earlier: because electron resists require doses in the range 1–100 μC/cm^2 for exposure, pattern features cannot be used as registration marks.

TABLE 10.2

Example of Test for a High-Resolution Electron Lithography System

Feature	Evaluation tool and purpose
25-step incremental dose pads in two versions: large and small shot sizes	Optical/Nanospec—at lowest resolving dose, beam current density distribution visible in pad section; track thickness vs. dose for resist shelf/film life
25-step incremental dose line/space features from 0.1 to 0.5 μm	SEM—determine shot-butting quality for fine-line exposure (0.2 μm); line width vs. dose
Single- and four-field grid pattern	System mark detection—measure field gain/distortion
10×10 array of single crosses	System mark detection—measure stage accuracy
Mask mode field butting	Optical—measure verniers to determine needed corrections
Aligned overlay test pattern	Optical—measure verniers to determine needed corrections
Custom	Add any feature needed for special exposures

Wells et al. pointed out that a threshold detection scheme is wasteful, because it uses only that part of the alignment signal that corresponds to the immediate vicinity of the threshold point. If the complete waveform is used, all the available information is utilized. The charge density necessary to achieve an accuracy of Δx is reduced by the factor $\Delta x/x_{\text{rise}}$. In the case of the example considered here, this would reduce Q from 2300 to 46 μC/cm². One way in which the complete waveform can be utilized is to use a correlation technique to locate the alignment mark; such schemes have been described by Cumming [74], Holburn et al. [75], and Hsu [76].

10.5.5 MACHINE AND PROCESS MONITORING

Although a modern commercial electron lithography exposure system comes with a great deal of computer control, close system monitoring is essential for obtaining optimum performance. Table 10.2 shows an example of a set of tests that have been found useful for a high-resolution system operating in an R&D mode. In addition, a weekly alignment monitor is run using an electrical test pattern. The first level consists of standard van der Pauw pads connected to resistors oriented in the X- and Y-directions. The second level cuts a slot in the resistor, dividing it in two equal parts. The difference between resistance values gives the amount of misalignment.

10.6 INTERACTION OF THE ELECTRON BEAM WITH THE SUBSTRATE

10.6.1 POWER BALANCE

An electron passing through matter (a primary electron) generates secondary electrons by an ionization process. The secondary electrons induce chemical reactions in resist material and thermal energy deposition. The ionization loss of the relativistic electron in the material $\left(-\dfrac{dE}{dx} \right)$ is given by Bethe's formula [77] as follows:

$$
-\frac{dE}{dx} = \left(\frac{1}{4\pi\varepsilon_0} \right)^2 \left(\frac{2\pi e^4}{mv^2} \right) \left(\frac{N_A Z \rho}{A} \right)
$$

$$
\left\{ \ln \frac{mv^2 E}{2J^2(1-\beta^2)} - \left(2\sqrt{1-\beta^2} - 1 + \beta^2 \right) \ln 2 + 1 - \beta^2 + \frac{1}{8}(1 - \sqrt{1-\beta^2})^2 \right\}
$$

(10.21)

where

ε_0 is the permittivity of vacuum

the value of $\left(\dfrac{1}{4\pi\varepsilon_0} \right)$ is equal to 8.99×10^9 (Nm^2/C^2 in SI units)

$\left(\dfrac{N_A Z \rho}{A} \right)$ represents the electron density of the matter, where ρ, Z, A, and N_A are the (resist)

material density, atomic number, atomic weight of the matter, and Avogadro constant (6.02×10^{23} mol^{-1}), respectively

J represents the ionization potential, the value of which is approximately assumed as 10 eV because the first ionization energy of hydrogen, carbon, oxygen, silicon, and so on is from 8 to 15 eV [78]

β means the ratio of the electron velocity against the light velocity in free space as shown in Equation 10.13

Assuming PMMA ($C_5O_2H_8$) as a resist material, values of 1.2 g/cm³, 6.2 and 10.5 can be adopted for ρ, Z, and A, respectively. Then, the electron density is approximately $3.9 \times 10^{23}/cm^3$. If the acceleration voltage of the electron is 20 kV, $v = 0.82 \times 10^8$ m/s ($\beta = 0.272$). The ionization loss per unit length $\left(-\dfrac{dE}{dx} \right)$ becomes 2.07 keV/µm from Equation 10.21. If the thickness of PMMA is 0.5 µm, the deposit energy becomes 1.04 keV, which generates approximately 100 secondary electrons.

In the nonrelativistic case, the energy loss as the beam passes in the forward direction through the resist can be calculated using the Thomson–Whiddington law, expressed by Wells [79] as

$$E_A^2 - E_B^2 = b'z \tag{10.22}$$

with $b' = 6.9 \times 10^9 \left(\rho E_A^{0.5} / Z^{0.2} \right)$

Then,

$$E_A - E_B = E_A - \sqrt{E_A^2 - b'z} \tag{10.23}$$

where

E_A is the energy (in eV) of the electrons as they enter the resist
E_B is the mean energy as they exit it
z is the thickness (in cm) of the resist film
ρ is the density of the resist (g/cm³),
Z is the effective atomic number of the resist.

Now, $\rho = 1.2$ g/cm³ and $Z = 6.2$ can be used for PMMA, approximately. Because the resist film is thin, it is assumed that the number of electrons absorbed or generated within it is negligible. Assuming PMMA thickness of 0.5 µm and an incident electron energy of 20 keV, the mean energy loss that occurs within the resist film in this "forward" direction becomes $\Delta E_f = E_A - E_B = 1.04$ keV from Equation 10.22. This is almost the same value as the one calculated by Equation 10.21. In the case of injected electron energy of 100 keV ($v = 1.64 \times 10^8$ m/s [$\beta = 0.548$]), ΔE_f becomes 0.306 keV from Equation 10.21 and 0.455 keV from Equation 10.22, respectively.

Consider a silicon substrate coated with 0.5 µm thick PMMA and an electrical current of 1 µA impinging on the top surface of the PMMA, with power flowing at 20 mW. Part of this power is dissipated chemically and thermally in the resist, part thermally in the substrate, and the remainder leaves the substrate. The power dissipated by the beam as it passes through the resist is given by

$$P_f = I_B \Delta E_f = I_B (E_A - E_B) \tag{10.24}$$

which for the conditions cited previously is 1.04 mW.

The beam, whose power is 19.0 mW (=20 mW − 1.04 mW), penetrates into the silicon, and a fraction $\eta = 0.18$ of the electron current is backscattered out of the substrate into the resist. The data of Bishop [65] indicate that the mean energy of the backscattered electrons is 60% of the incident energy (11.4 keV = 0.60 × 19.0 keV in this case). Then, the power of the backscattered electrons becomes 2.05 mW (=0.18 μA × 11.4 keV). Therefore, the power dissipated in the substrate is 16.9 mW, which is obtained from the difference between the incident power to the substrate and the power of backscatter electrons.

The backscattered electrons pass through the resist film toward the surface in the "backward" direction. It is assumed that the number of electrons absorbed or generated within the resist film is negligible, so the electrons constitute a current ηI_B. Although the mean backscattered electron energy into the resist is 11.4 keV as described in the previous paragraph, one cannot use the Thomson–Whiddington law to calculate the energy lost in the resist film, ΔE_B, for two reasons: first, the spread of electron energies is wide, and second, the directions of travel of the backscattered electrons are not, in general, normal to the resist–substrate interface.

The ΔE_B may be estimated from the results of Kanter [80], who investigated the secondary emission of electrons from aluminum (Z = 13). He proposed that because of the lower energies and oblique trajectories of electrons backscattered from the sample, they would be a factor γ more efficient at generating secondary electrons at the surface than was the primary beam. Kanter's estimated value for γ was 4.3, and his experimentally measured value was 4.9. Since secondary electron generation and resist exposure are both governed by the rate of energy dissipation at the surface of the sample, and since the atomic numbers of aluminum (Z = 13) and silicon (Z = 14) are nearly equal, it is reasonable to assume that these results apply to the exposure of resist on a silicon wafer. Thus,

$$\Delta E_B = \gamma \Delta E_f \tag{10.25}$$

where γ is expected to have a value between 4 and 5. The power dissipated in the resist film by the backscattered electrons is

$$P_b = \eta I_B \Delta E_B = \eta \gamma I_B \Delta E_f \tag{10.26}$$

Comparing Equations 10.24 and 10.26, P_b/P_f is given by the ratio

$$\eta_e = \frac{P_b}{P_f} = \eta \gamma \tag{10.27}$$

It is generally acknowledged that for silicon at a beam energy of 20 keV, η_e lies between 0.7 and 0.8, corresponding to $\gamma = 4$. (A number of experimental measurements of η_e have been collated by Hawryluk [81]: these encompass a range of values varying from 0.6 to1.0.) If it is assumed that $\gamma = 4.0$, then the mean energy lost by the backscattered electrons as they pass through the resist film is $\Delta E_B = 4.16$ keV.

Thus, in this example, of the incident power in the beam, approximately 5% is dissipated in the resist by forward-traveling electrons, approximately 4% is dissipated in the resist by backscattered electrons, 85% is dissipated as heat in the substrate, and the remaining 6% leaves the workpiece as the kinetic energy of the emergent backscattered electrons.

The quantity that controls the change in solubility of a resist is the total energy absorbed per unit volume (energy density), E (in J/m^3). Assuming that the energy is absorbed uniformly in a resist layer of thickness z, this is related to the exposure dose Q by the expression

$$E = \frac{\Delta E_f (1 + \eta_e)}{z} Q \tag{10.28}$$

For the example considered in this section, Equation 10.28 indicates that an exposure dose of 1 μC/cm^2 corresponds to an absorbed energy density of 3.6×10^7 J/m^3.

10.6.2 Spatial Distribution of Energy in the Resist Film

Monte Carlo techniques have been used to investigate the interactions of electrons with matter in the context of electron probe microanalysis and scanning electron microscopy by Bishop [65], Shimizu and Murata [82], and Murata et al. [83]. Kyser and Murata [84] investigated the interaction of electron beams with resist films on silicon using the same method, pointing out that its conceptual simplicity and the accuracy of the physical model were useful attributes.

Monte Carlo calculations indicate that the lateral distribution of energy dissipated by the forward-traveling electrons at the resist–silicon interface may be approximated closely as a Gaussian. Broers [85] has noted that the standard deviation computed in this way may be expressed as σ_f (measured in micrometers) as follows:

$$\sigma_f = \left(\frac{9.64z}{V} \right)^{1.75} \tag{10.29}$$

where:

V (measured in keV)   is the energy of the incident electron beam
z (measured in μm)   is the thickness of the resist film

For 20 keV electrons penetrating 0.5 μm of resist, Equation 10.29 predicts that $\sigma_f = 0.08$ μm. In order to make the forward-scattering radius small in recent electron beam exposure tools, a higher acceleration voltage and/or a thin resist process is necessary.

The electrons backscattered from the substrate expose the resist film over a region with a characteristic diameter σ_b, which is approximately twice their range in the substrate. Monte Carlo simulation techniques have been extensively used to simulate electron backscattering. Shimizu and Murata applied this technique to compute the backscatter coefficient and the lateral distribution of electrons backscattered from aluminum at a beam energy of 20 keV. Their calculations showed that $\eta = 0.18$ and that the electrons emerged from a circular region 4 μm in diameter. Because silicon and aluminum are adjacent elements in the periodic table, the results for silicon are almost identical.

In the context of electron lithography, it is important that the volume in the resist within which the forward-traveling electrons dissipate their energy is very much smaller than that within which the backscattered electrons dissipate theirs. The comparatively diffuse backscattered energy distribution determines the contrast of the latent image in the resist, and the more compact forward-scattered energy distribution determines the ultimate resolution.

Han et al. [86] proposed a detailed energy deposition model including the plasmon excitation as well as ionization.

10.6.3 Shot Noise

As research continues to make faster higher-resolution resists, a limit is reached due to the quantum nature of electrons, namely shot noise, because such local dosage fluctuation might induce line-edge roughness (LER). For an optimum dose Q, in other words the resist sensitivity, the number of electrons required to define the smallest feature in the area S is

$$N = \frac{QS}{e} \tag{10.30}$$

where e represents the elemental charge (1.6×10^{-19} C in SI units).

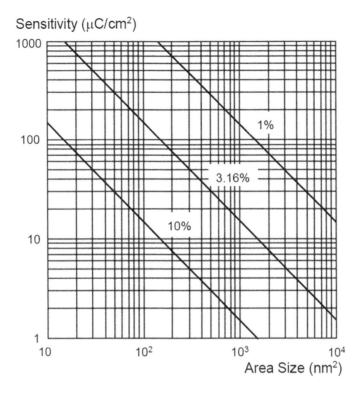

FIGURE 10.9 Shot noise diagram for 1.0, 3.16, and 10% local dose fluctuation. The upper right area of each threshold is the acceptable region.

The number of electrons N in the area S is described by a Poisson distribution, and its standard deviation is $\sqrt{N}$. The fluctuation of the dose becomes

$$\frac{\Delta Q}{Q} = \frac{\Delta N}{N} = \frac{\sqrt{N}}{N} = \frac{1}{\sqrt{N}} \tag{10.31}$$

A 10% fluctuation is the largest that can be tolerated for reliable exposure and determines a critical resist-sensitivity limit. From Equation 10.31, at least 100 electrons are required to reliably expose a pixel. If 1% fluctuation is required, at least 10,000 electrons become necessary.

It is interesting to compare the shot noise limit with the sensitivity of commercial resists. Figure 10.9 shows the critical sensitivity versus feature size calculated from Equations 10.30 and 10.31. The upper right area of each threshold is the acceptable region. In the case of a 10% fluctuation level, conventional high-resolution single-component resists (e.g., PMMA) of sensitivity 500–1000 μC/cm² would reach the shot noise limit at 1.3–1.8 nm feature sizes. Dimensions of 1 nm and below require very insensitive resists, at least 1600 μC/cm². Modern CAR resists have sensitivities of 5–20 μC/cm² on a Si substrate. From Figure 10.9, the critical sensitivity for 10 nm square (100 nm²) features is 16 μC/cm², and that for a 20 nm square (400 nm²) would be 4 μC/cm².

A lot of research including shot noise statistics, process, and development has been reported [87–90].

10.7 ELECTRON BEAM RESISTS AND PROCESSING TECHNIQUES

The main requirements for resist performance are resolution and sensitivity. However, there is a tradeoff between resolution and sensitivity, as shown in Figure 10.10 [91]. In addition, there is

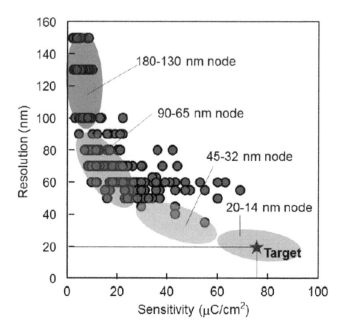

FIGURE 10.10 Resolution–sensitivity tradeoff of chemically amplified resist evaluated for several technology nodes. (From Kon, J., et al., *Proc. SPIE*, 8323, 832324, 2012.)

another tradeoff between sensitivity and LER, as described in Section 10.6.3. Therefore, the resist and its process must be optimized in the tradeoff between resolution, sensitivity, and LER.

PMMA, which was one of the first resists used with electron lithography, is still used because of its high resolution and because it is one of the best-known and best-understood positive e-beam resists. Its primary disadvantages are very low sensitivity and poor etch resistance under many important plasma conditions.

Some positive resists that eliminate these problems are now available. One of them is ZEP series, supplied by Zeon Corporation. ZEP series is a polymer-type resist (the same as PMMA). Its resolution is close to that of PMMA, but its sensitivity is much better.

Traditional negative resists possessed far higher sensitivities (because only one crosslink per molecule is sufficient to insolubilize the material) but at the expense of resolution. The resists were developed in a solvent appropriate for the noncrosslinked portion, and the crosslinked material would be swollen by this solvent, often touching neighboring patterns and resulting in extensive pattern distortion. As a result, they were limited to feature sizes greater than about 1 μm.

Hydrogen silsesquioxane (HSQ) (available from Dow Corning) has been used as a negative resist. It was reported that sub-30 nm resolution was achieved.

The introduction of Shipley's acid-catalyzed chemically amplified negative resist (SAL-601 or its subsequent versions) was a major advance, because it gave excellent resolution and good process latitude with etch resistance similar to that of conventional optical resists. Even though it relies on crosslinking for its action, it avoids the swelling problem, because it is developed in aqueous base in essentially the same manner as positive optical novolak resists. Water is not a good solvent for the polymer, so no swelling occurs.

10.8 THE PROXIMITY EFFECT

10.8.1 DESCRIPTION OF THE EFFECT

As briefly described in Section 10.6.1, electrons that penetrate the resist layer enter into the substrate.

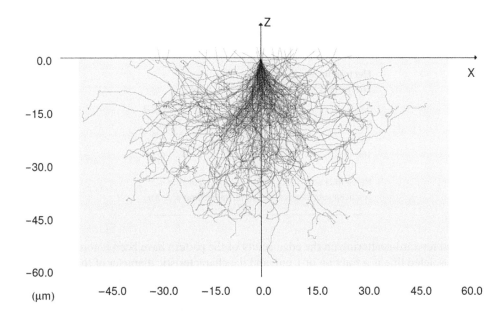

FIGURE 10.11   Monte Carlo simulation result of 100 keV electrons injected into bare silicon (100 event).

Then, the electrons are scattered by Coulomb force from the atomic nucleus (Rutherford scattering). The thickness of the substrate is several hundred microns in the case of silicon wafer or several millimeters in the case of mask blank, so the electrons are scattered many times within the substrate, losing their kinetic energy based on Bethe's formula as described in Equation 10.21. Therefore, electrons injected into the substrate stop in the substrate or are backscattered into the resist film. A Monte Carlo simulation result in the case of 100 keV electrons and silicon wafer is shown in Figure 10.11.

The proximity effect is the exposure of resist by electrons backscattered from the substrate, constituting a background on which the pattern is superimposed. If this background were constant, it would create no lithographic problem other than a degradation in contrast that, although undesirable, would be by no means catastrophic. However, the background is not constant, and the serious consequence of this was observed in positive resists by Chang and Stewart [13]: "Several problems have been encountered when working at dimensions of less than 1 μm. For example, it has been found that line width depends on the packing density. When line spacing is made less than 1 μm, there is a noticeable increase in the line width obtained when the spacing is large."

The simple model of the energy density $\varepsilon(r)$ deposited in the resist can be obtained from Equation 10.28 (a similar expression was shown by Owen [92]) as follows:

$$\varepsilon(r) = \frac{Q\Delta E_f}{z}\left[\frac{1}{2\pi\sigma_f^2}\exp\left(-\frac{r^2}{2\sigma_f^2}\right) + \frac{\eta_e}{2\pi\sigma_b^2}\exp\left(-\frac{r^2}{2\sigma_b^2}\right)\right] \tag{10.32}$$

where r and σ_b represent the distance from the injection point of the electrons and the standard deviation of the distribution of the backscattered electrons, respectively. Values of σ_f, σ_b, and η_e for 5, 20, 50, and 100 kV are shown in Table 10.3. The value of σ_f can be calculated by Equation 10.29, and the values of σ_b and η_e are obtained from Owen [92] and Jackel et al. [93].

Figure 10.12 illustrates the way in which the proximity effect affects pattern dimensions. In Figure 10.12a, it is assumed that an isolated narrow line (for example, of width 0.5 μm) is to be exposed. The corresponding energy density distribution is shown in Figure 10.12d. The energy density deposited in the resist by forward-traveling electrons is E, and the effects of lithographic

TABLE 10.3

Parameter Values Related to Scattering

Electron Beam Energy (keV)	σ_f (µm)	σ_b (µm)	η_e
5	0.94	[0.13]	[0.74]
20	0.083	1.4	0.74
50	0.017	6.7	0.74
100	0.005	22.1	0.74

Resist thickness of 0.5 µm is used for calculating σ_f.
A silicon substrate is assumed for calculating σ_b and η_e.

resolution and forward-scattering on the edge acuity of the pattern have been ignored. Because the width of the isolated line is a fraction of 1 µm, and the characteristic diameter of the backscattered electron distribution (σ_b) is several micrometers, the backscattered energy density deposited in the resist is negligibly small.

This, however, is not the case for the exposure of the isolated space depicted in Figure 10.12c. Since the width of the space is much smaller than σ_b, it is a good approximation to assume that the pattern is superimposed on a uniform background energy density $\eta_e E$, as is shown in Figure 10.12f. Figure 10.12b and e deal with an intermediate case—an infinite periodic array of equal lines and spaces. Here, the pattern is superimposed on a background energy density of $\eta_e E/2$.

Thus, the energy densities corresponding to exposed and unexposed parts of the pattern are strongly dependent on the nature of the pattern itself. This leads to difficulties in maintaining the

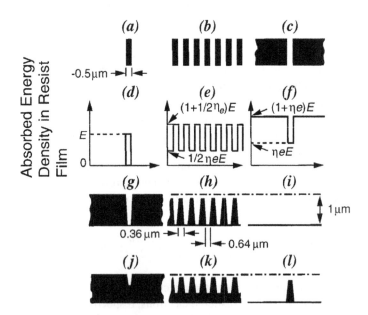

FIGURE 10.12 Influence of the proximity effect on the exposure of (a) an isolated line, (b) equally spaced lines and spaces, and (c) an isolated space (exposed regions are denoted by shading). The absorbed energy densities are shown in (d), (e), and (f). The profiles of the resist patterns are shown in (g), (h), and (i) after development time appropriate for the isolated line. The profiles shown in (j), (k), and (l) occur after a development time appropriate for the isolated space. The shapes of the developed resist patterns have been drawn to be roughly illustrative of what would be observed in practice.

fidelity of the pattern during development. In the drawings in Figure 10.12g through 10.12i, it has been assumed that a development time has been chosen that allows the isolated line (g) to develop out to the correct dimensions. However, because the energy densities associated with the isolated space are appreciably higher, this pattern may well be so overdeveloped at the end of this time that it has completely disappeared (i). The problem may not be so drastic for the equal lines and spaces, for which the energy densities have lower values. Nevertheless, the lines will be overdeveloped: in (h), they are depicted (arbitrarily) as having a width of 0.36 µm, not the required 0.5 µm.

The situation cannot be improved by reducing the development time. Figure 10.12j through 10.12l illustrate what happens if the patterns are developed for a time appropriate for the isolated space. Although this is now developed out to the correct dimension, the isolated line and equal line-and-space patterns are inadequately developed.

In general, because of the proximity effect, any one of these types of pattern may be made to develop out to the correct dimensions by choosing an appropriate development time. However, it is impossible to develop them all out to the correct dimensions simultaneously. Although the proximity effect poses serious problems, it does not constitute a fundamental resolution limit to electron lithography. A number of methods for compensating for the effect, or at least reducing the gravity of the problems, have been devised. These are described in the following.

10.8.2 Methods of Compensating for the Proximity Effect

The most popular form of proximity effect compensation is probably dose correction. To implement this method, the dose delivered by the lithography instrument is varied in such a way as to deposit the same energy density in all exposed regions of the pattern. With reference to Figure 10.12, this situation occurs if the dose delivered to the isolated line (a) is increased by the factor $(1 + \eta_e)$ and that to the lines of the line-and-space pattern (b) by a factor of $(1 + \eta_e)(1 + \eta_e/2)^{-1}$. If this is done, all three types of pattern will develop out to approximately the correct dimensions after the same development time (that is, appropriate for the isolated space in Figure 10.12c).

The dose correction scheme was first proposed by Chang et al. [94], who applied it to a vector scan round-beam lithography instrument, variations in dose being obtained by varying the speed with which the beam was scanned over the substrate. Submicrometer bubble devices were successfully made in this way, and because of the simplicity and highly repetitive nature of the pattern, the calculation of the necessary dose variations was not arduous. However, for complex, nonrepetitive patterns, the dose calculations become involved, making the use of a computer to carry them out essential. The calculations are time consuming and therefore expensive to carry out, several hours of CPU time sometimes being necessary even when using large fast computers. A set of programs designed for this purpose has been described by Parikh [95], and these have since become widely used.

Parikh's algorithm involves a convolution of the pattern data with the inverse of the response function associated with electron backscattering. In order to make the computation tractable, the approximations made are considerable. Kern [96] has pointed out that the dose compensation calculations can be carried out exactly and more conveniently in the spatial frequency domain using a Fourier transform technique. Further analysis of dose compensation algorithms has been published by Owen [91, 97].

Another popular form of compensation is shape correction. This would be applied to the patterns of Figure 10.12 by decreasing the widths of the exposed lines in (b), increasing the width of the isolated space in (c), and developing the pattern for a time appropriate for the isolated line pattern in (a). In practical applications, the magnitudes of the shape corrections to be applied are determined empirically from test exposures. For this reason, this technique is generally applied only to simple, repetitive patterns for which the empirical approach is not prohibitively time consuming.

A correction scheme that involves no computation other than reversing the tone of the pattern has been described by Owen and Rissman [98]. The scheme is implemented by making a correction exposure in addition to the pattern exposure. The correction exposure consists of the reversed field of the pattern and is exposed at a dose a factor $\eta_e (1+\eta_e)^{-1}$ less than the pattern dose using a beam defocused to a diameter a little less than σ_b. The attenuated, defocused beam deposits an energy density distribution in the resist that mimics the backscattered energy density associated with a pattern pixel. Since the correction exposure is the reverse field of the pattern exposure, the combination of the two produces a uniform background energy density, regardless of the local density of the pattern. If this correction technique were applied to the pattern of Figure 10.12, all pattern regions would absorb an energy density $(1+\eta_e)E$ and all field regions an energy density of $\eta_e E$. These are identical to the exposure levels of a narrow isolated space (to which a negligible correction dose would have been applied). As a result, after development, all the patterns (a), (b), and (c) would have the correct dimensions.

It has been reported that the use of beam energies much greater than 20 keV (e.g., 50 keV) reduces the proximity effect, and experimental and modeling data have been presented to support this claim (see, for example, Neill and Bull [99]). At first sight, it would seem that the severity of the proximity effect should not be affected by increasing the beam energy, because there is good evidence that this has a negligible effect on η_e [93].

A possible explanation for the claimed reduction of the proximity effect at high energies hinges on the small size of the test patterns on which the claims are based. At 20 keV, the standard deviation of backscattered electrons in silicon is about 1.4 μm, and at 50 keV, it is about 6.7 μm (see Table 10.3). A typical test structure used for modeling or experimental exposure may have linear dimensions of the order of a few micrometers. As a result, at 20 keV, nearly all the backscattered electrons generated during exposure are backscattered into the pattern region itself. However, at 50 keV, the electrons are backscattered into a considerably larger region, thereby giving rise to a lower concentration of backscattered electrons in the pattern region. Thus, the use of a high beam energy will reduce the proximity effect if the linear dimensions of the pattern region are considerably smaller than the electron range in the substrate.

A more general explanation, which applies to pattern regions of any size, is statistical. At 20 keV, a given point on a pattern receives backscattered energy from a region about 4 μm in diameter (1.4 μm as standard deviation), whereas at 50 keV, it would receive energy from a region about 20 μm in diameter (6.7 μm as standard deviation). It is therefore to be expected that the variation in backscattered energy from point to point should be less at 50 keV than at 20 keV, because a larger area of the pattern is being sampled at the higher beam energy. Consequently, because the proximity effect is caused by point-to-point variations in backscattered energy density, it should become less serious at higher beam energies.

Recently, the requirement for linewidth control has become tighter. Especially, the number of backscatter electrons from heavy metal parts is larger than that from silicon. Ogino, Hoshino, and Machida reported the trial of the proximity effect correction considering a multiple-layer structure [100].

10.9   ION BEAM NANOLITHOGRAPHY

10.9.1   FOCUSED ION BEAM SYSTEM

The motivation to develop a focused ion beam (FIB) system for lithography was no requirement for proximity effect correction due to the much lower level of backscattered ions from the substrate. One early work was reported in 1988 by Ochiai et al. using several kinds of ions from a liquid metal ion source (LMIS) (Au–Si–Be alloy) [101]. Figure 10.13 shows the 150 kV round-beam system used for this research; its resolution was lower than 0.1μm. The system configuration was similar to that using an electron beam.

Ar^+ ion and Ga^+ ion have more commonly been used for FIB. In 1998, Rau et al. reported their research work on shot noise and edge roughness at 10 nm features using 50 kV Ga^+ FIB with 8 nm spot size [102].

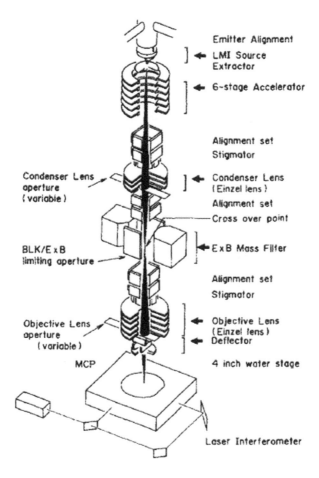

FIGURE 10.13 Example of focused ion beam system. (From Ochiai, Y., et al., *J. Vac. Sci. Technol.*, B6, 1055, 1988.)

IMS Nanofabrication developed a prototype multiple ion beam system with a blanking aperture array (43,000 apertures) [103], which was similar to their multiple electron beam system described in Section 10.4.2. H^+ ion or Ar^+ ion with 10 or 20 kV was used as an ion beam; 40 nm line features and 25 nm dots were successfully exposed.

10.9.2 Ion Beam Milling and Deposition

With a gas assist function, the ability to mill nanometer features by sputtering and depositing material to modify original shapes in FIB is more attractive, because those are unique as FIB features.

Yasaka et al. reported an FIB photomask repair system [104]. A Ga^+ ion beam from a Ga liquid metal source is focused and injected into the substrate after going through the ion beam optics. Secondary electrons from the substrate are detected to observe its surface. Figure 10.14 represents repair processes of mask defect patterns. Opaque defects are repaired with a gas assist FIB milling (etching) process, and clear defects are repaired with an FIB-induced carbon deposition method. There is a gas nozzle to supply assist gas for FIB milling or FIB deposition. When an assist gas such as a halogen is supplied, the milling rate increases. When a hydrocarbon assist gas is supplied, carbon is deposited to form a carbon film [105].

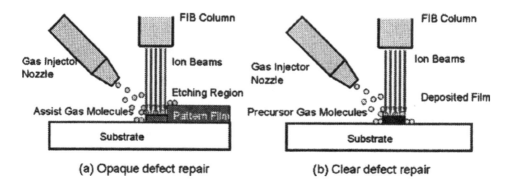

(a) Opaque defect repair   (b) Clear defect repair

FIGURE 10.14 Principle of mask repair by focused ion beam. (From Yasaka, A., et al., *J. Vac. Sci. Technol.*, B26, 2121, 2008.)

10.10   LASER DIRECT WRITE

10.10.1   System Configuration

Laser direct write (laser writer, laser pattern generator) for high-end patterns typically adopts DUV light sources with wavelength lower than 300 nm. There are two major systems, of which the platforms are quite different.

The first platform uses a conventional multiple-beam scanning system with a polygonal mirror and f-θ (or fsinθ) lens. A system with 257 nm continuous-wave laser supplied by ETEC Systems (an Applied Materials company) uses an acoustic-optic deflector (AOD), which introduces real-time corrections for both the intensity and the position by radiofrequency (RF) power change (the diffraction efficiency change) and RF frequency change (the angle of diffraction change), respectively [106].

The second platform is a small field stepper with a spatial light modulator (SLM) as a programmable mask supplied by Micronic Laser Systems [107] (Figure 10.15). A 248 or 193 nm excimer laser is adopted as a light source. The SLM has massive flat micromirrors, and the light reflected by SLM exposes the photoresist on the mask blank. An electrostatic voltage can be applied to micromirrors independently, and the micromirror to which the electrostatic voltage is supplied tilts in

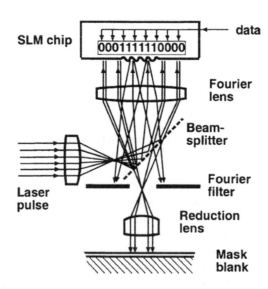

FIGURE 10.15 Schematic of pattern generation by spatial light modulator (SLM). (From Ljungblad, U., et al., *Proc. SPIE*, 4186, 16, 2001.)

order to scatter the light outside the Fourier aperture. Then, the light reflected by that tilted micro-mirror is blocked at the aperture stop in the projection optics.

10.10.2 Application

A typical direct laser write system has 0.7–1.0 μm resolution, which is worse than that of an electron beam direct write (EBDW) system. However, the writing speed of laser direct write is faster than that of EBDW. Therefore, a laser direct write system is used for writing a mask pattern of noncritical device layers or a mask pattern of a flat panel display in order to reduce the cost of mask writing.

With the shorter wavelength of the 193 nm ArF laser or higher-NA optics such as 0.9, the application of laser direct write is expanding due to its improved resolution, in the 0.4–0.5 μm range.

ACKNOWLEDGMENTS

The contributor would like to thank Geraint Owen and James R. Sheats for their work in the first edition, from which some of this chapter originates. The contributor also thanks Elizabeth A. Dobisz et al. for information about electron beam resist in their work in the second edition.

REFERENCES

1. D.A. Buck and K. Shoulders, *Proc. Eastern Joint Computer Conference*, ATEE, New York, 1957, p. 55.
2. G. Owen, "Electron lithography for the fabrication of microelectronic devices", *Rep. Prog. Phys.*, **48**, 795 (1985).
3. L. Reimer, *Scanning Electron Microscopy*, Springer-Verlag, 1985, pp. 23–31.
4. C.W. Oatley, *The Scanning Electron Microscope*, Cambridge University Press, Cambridge, UK, 1972, chapters 2 and 3.
5. R. Fowler, "The restored electron theory of metals and thermionic formula", *Proc. R. Soc. A*, **117**, 549 (1928).
6. D. B. Langmuir, "Theoretical limitations of cathode-ray tubes", *Proc. Inst. Radio Eng.*, **25**, 977 (1937).
7. A.N. Broers, "Some experimental and estimated characteristics of the lanthanum hexaboride rod cathode electron gun", *J. Phys. E: Sci. Instrum.*, **2**, 273 (1969).
8. O.C. Wells, *Scanning Electron Microscopy*, McGraw-Hill, New York, 1974, table 4.
9. R. Gomer, *Field Emission and Field Ionization*, Harvard University Press, Cambridge, MA, 1961, chapters 1 and 2.
10. G. Stille and B. Astrand, "A field emitter electron beam exposure system", *Phys. Scr.*, **18**, 367 (1978).
11. H.P. Kuo and B.M. Siegel, *Electron and Ion Beam Science and Technology, 8th International Conference* (R. Bakish, ed.), The Electrochemical Society, Princeton, NJ, 1978, pp. 3–10.
12. H.P. Kuo, J. Foster, W. Haase, J. Kelly and B.M. Oliver, *Electron and Ion Beam Science and Technology, 10th International Conference* (R. Bakish, ed.), The Electrochemical Society, Princeton, NJ, 1982, pp. 78–91.
13. T.H.P. Chang and A.D.G. Stewart, *Proceedings 10th Symposium on Electron, Ion and Laser Beam Technology* (L. Marton, ed.), IEEE, San Francisco, CA, 1969, p. 97.
14. G. Owen, "Automatic measurement and correction of deflection astigmatism and defocusing in the Hewlett-Packard 605 electron beam lithography system", *J. Vac. Sci. Technol.*, **19**, 1064 (1981).
15. G. Owen and W.C. Nixon, "Aberration correction for increased lines per field in scanning electron beam technology", *J. Vac. Sci. Technol.*, **10**, 983 (1973).
16. K. Amboss, "Design of fast deflection coils for an electron-beam microfabrication system", *J. Vac. Sci. Technol.*, **12**, 1152 (1975).
17. H. Ohiwa, E. Goto and A. Ono, "Elimination of third-order aberrations in electron-beam scanning systems", *Electron. Commun. Jpn.*, **54B**, 44 (1971).
18. E. Munro, "Design and optimization of magnetic lenses and deflection systems for electron beams", *J. Vac. Sci. Technol.*, **12**, 1146 (1975).
19. L.H. Lin and H.L. Beauchamp, "High speed beam deflection and blanking for electron lithography", *J. Vac. Sci. Technol.*, **10**, 987 (1973).

20. T.H.P. Chang, A.J. Speth, C.H. Ting, R. Viswanathan, M. Parikh and E. Munro, *Electron and Ion Beam Science and Technology, 7th International Conference* (R. Bakish, ed.), The Electrochemical Society, Princeton, NJ, 1982, pp. 376–391.

21. E. Munro and H.C. Chu, "Numerical analysis of electron beam lithography systems – 1. Computation of fields in magnetic deflectors", *Optik*, **60**, 371 (1982).

22. E. Munro and H.C. Chu, "Numerical analysis of electron beam lithography systems – 2. Computation of fields in electrostatic delfectors", *Optik*, **61**, 1 (1982).

23. H.C. Chu and E. Munro, "Numerical analysis of electron beam lithography systems – 3. Calculation of the optical properties of electron focusing systems and dual-channel deflection systems with combined magnetic and electrostatic fields", *Optik*, **61**, 121 (1982).

24. H.C. Chu and E. Munro, "Numerical analysis of electron beam lithography systems – 4. Computerized optimization of the electron optical performance of electron beam lithography systems using the damped least squares method", *Optik*, **61**, 213 (1982).

25. A.V. Crewe, "Some space charge effects in electron probe devices", *Optik*, **52**, 337 (1978).

26. T. Sasaki, "Study of space-charge effect by computer", *J. Vac. Sci. Technol.* **21**, 695 (1982).

27. T. Groves, D.L Hammon and H. Kuo, "Electron-beam broadening effects caused by discreteness of space charge", *J. Vac. Sci. Technol.*, **16**, 1680 (1979).

28. H. Boersch, *Zeitschrift Phys.*, **139**, 115 (1954).

29. K.H. Loeffler, *Z. Angew. Phys.*, **27**, 145 (1969).

30. A.V. Crewe, "Collision broadening in electron beams", *Optik*, **50**, 205 (1978).

31. G.H. Jansen, T.R. Groves and W. Stickel, "Energy broadening in electron beams: A comparison of existing theories and Monte Carlo simulation", *J. Vac. Sci. Technol., B*, **3**, 190 (1985).

32. G.H. Jansen, "Coulomb interactions in particle beams", *J. Vac. Sci. Technol., B*, **6**, 1977 (1988).

33. R.F.W. Pease, J.P. Ballantyne, R.C. Henderson, M. Voshchenkov and L.D. Yau, *IEEE Trans. Electron Dev.*, **ED-22**, 393 (1975).

34. H.C. Pfeiffer, "New imaging and deflection concept for probe-forming microfabrication systems", *J. Vac. Sci. Technol.* **12**, 1170 (1975).

35. H.C. Pfeiffer and K.H. Loeffler, *Proceedings 7th International Conference on Electron Microscopy*, Societé Française de Microscopie Electronique, 1970, pp. 63–64.

36. M. Born and E. Wolf, *Principles of Optics*, Pergamon, Oxford, 1975.

37. J.L. Mauer, H.C. Pfeiffer and W. Stickel, *IBM J. Res. Dev.*, **21**, 514 (1977).

38. E.V. Weber and R.D. Moore, "Variable spot-shaped e-beam lithographic tool", *J. Vac. Sci. Technol.* **16**, 1780 (1979).

39. L.A. Fontijn, *Ph.D. thesis*, Delft University Press, Delft (1972).

40. H.C. Pfeiffer, "Variable spot shaping for electron-beam lithography", *J. Vac. Sci. Technol.*, **15**, 887 (1978).

41. R.D. Moore, G.A. Caccoma, H.C. Pfeiffer, E.V. Weber, and O.C. Woodard, "EL-3: A high throughput, high resolution e-beam lithography tool", *J. Vac. Sci. Technol.*, **19**, 950 (1981).

42. D.E. Davis, S.J. Gillespie, S.L. Silverman, W. Stickel and A.D. Wilson, "EL-3 application to 0.5 µm semiconductor lithography", *J. Vac. Sci. Technol. B*, **1**, 1003 (1983).

43. H.C. Pfeiffer, "Advanced e-beam systems for manufacturing", *Proc. SPIE*, **1671**, 100 (1992).

44. H. Sunaoshi, et al., "EBM-5000: Electron beam mask writer for 45 nm node", *Proc. SPIE*, **6283**, 628306 (2006).

45. M. Slodowski, H.-J. Doering, T. Elster and I.A. Stolberg, "Coulomb blur advantage for a multi shaped beam lithography approach", *Proc. SPIE*, **7271**, 72710Q (2009).

46. H.C. Pfeiffer, "Recent advances in electron-beam lithography for the high-volume production of VLSI devices", *IEEE Trans. Electron Dev.*, **ED-26**, 663 (1979).

47. Y. Nakayama, Y. Sohda, N. Saitou, and H. Itoh, "Highly accurate calibration method of electron-beam cell projection lithography", *Proc. SPIE*, **1924**, 183 (1993).

48. H. Yasuda, K. Sakamoto, A. Yamada and K. Kawashima, "Electron beam block exposure", *Jpn. J. Appl. Phys.*, **30**, 3098 (1991).

49. W.H.P. Koops and J. Grob, *Springer Series in Optical Sciences Vol.43: X-ray Microscopy*, edited by G. Schmahl and D. Rudolph, Springer, Berlin, 1984, pp. 119–128.

50. S.D. Berger and J.M. Gibson, "New approach to projection-electron lithography with demonstrated 0.1 µm linewidth", *Appl. Phys. Lett.*, **57**, 153 (1990).

51. K. Suzuki, "EPL technology development", *Proc. SPIE*, **4754**, 775 (2002).

52. K. Suzuki and S. Shimizu, "Development status of EPL technology", *J. Photopolym. Sci. Tech.*, 15, 395 (2002).

53. K. Suzuki, et al., "Full-field exposure performance of electron projection lithography tool", *J. Vac. Sci. Technol. B*, **22**, 2885 (2004).
54. A. Yamada, Y. Oae, T. Okawa, M. Takizawa and M. Yamabe, "Evaluation of throughput improvement by MCC and CP in multicolumn e-beam exposure system", *Proc. SPIE*, **7637**, 76370C (2010).
55. A. Yamada, H. Yasuda and M. Yamabe, "Electron beams in individual column cells of multicolumn cell system", *J. Vac. Sci. Technol. B*, **26**, 2025 (2008).
56. H. Yasuda, et al., "Fast electron beam lithography system with 1024 beams individually controlled by blanking aperture array", *Jpn. J. Appl. Phys.*, **32**, 6012 (1993).
57. C. Brandstatter, et al., "Projection mask-less lithography", *Proc. SPIE*, **5374**, 601 (2004).
58. E. Platzgummer, C. Klein and H. Loeschner, "Electron multibeam technology for mask and wafer writing at 0.1 nm address grid", *J. Micro/Nanolith. MEMS MOEMS*, **12**(3), 031108 (2013).
59. H. Matsumoto, H. Inoue, H. Yamashita, H. Morita, S. Hirose, M. Ogasawara, H. Yamada and K. Hattori, "Multi-beam mask writer MBM-1000 and its application field", *Proc. SPIE*, **9984**, 998405 (2016).
60. M. McCord, S. Kojima, P. Petric, A. Brodie and J. Sun, "High-current electron optical design for reflective electron beam lithography direct write lithography", *J. Vac. Sci. Technol. B*, **28**, C6C1 (2010).
61. T. Gubiotti, et al., "Reflective electron beam lithography: lithography results using CMOS controlled digital pattern generator chip", *Proc. SPIE*, **8680**, 86800H (2013).
62. M.J. Wieland, H. Derks, H. Gupta, T. van de Peut, F. M. Postma, A. H. V. van Veen and Y. Zhang, "Throughput enhancement technique for MAPPER maskless lithography", *Proc. SPIE*, **7637**, 76371Z (2010).
63. G. de Boer, M. Dansberg, R. Jager, J.J.M. Peijster, E. Slot, S. Steenbrink and M.J.-J. Wieland, "MAPPER: Progress towards a high volume manufacturing system", *Proc. SPIE*, **8680**, 86800O (2013).
64. A.D. Wilson, T.W. Studwell, G. Folchi, A. Kern and H. Voelker, *Electron and Ion Beam Science and Technology, 8th International Conference* (R. Bakish, ed.), The Electrochemical Society, Princeton, NJ, 1978, pp. 198–205.
65. E. Bishop, "Electron scattering in thick targets", *Br. J. Appl. Phys.*, **18**, 703 (1967).
66. T.E. Everhart, O.C. Wells and C.W. Oatley, *J. Electron. Control*, **7**, 97 (1959).
67. E.D. Wolf, P.J. Coane and F.S. Ozdemir, "Composition and detection of alignment marks for electron-beam lithography", *J. Vac. Sci. Technol.*, **12**, 1266 (1975).
68. L. Reimer, *Electron Beam Interactions with Solids*, SEM Inc., Chicago, 1984, pp. 299–310.
69. H. Niedrig, "Film-thickness determination in electron microscopy: The electron backscattering method", *Opt. Acta*, **24**, 679 (1977).
70. S.J.B. Reed, *Electron Microprobe Analysis*, Cambridge University Press, 1975, table 13.1.
71. J.E. Holliday and E.J. Sternglass, "New method for range measurements of low-energy electrons in solids", *J. Appl. Phys.*, **30**, 1428 (1959).
72. N. Aizaki, "Monte Carlo simulation of alignment mark signal for direct electron-beam writing", *Jpn. J. Appl. Phys.*, **18**(suppl. 18-1), 319–325 (1979).
73. O.C. Wells, T.E. Everhart and R.K. Matta, "Automatic positioning of device electrodes using the scanning electron microscope", *IEEE Trans. Electron Dev.*, **ED-12**, 556 (1965).
74. D. Cumming, *Microcircuit Engineering 80, Proceedings International Conference on Microlithography*, Delft University Press, Delft, 1981, pp. 75–81.
75. D.M. Holburn, G.A.C. Jones and H. Ahmed, "A pattern recognition technique using sequences of marks for registration in electron beam lithography", *J. Vac. Sci. Technol.*, **19**, 1229 (1981).
76. T.J. Hsu, "Digital adaptive matched filter for fiducial mark registration", *Hewlett Packard J.*, **32**(5), 34–36 (1981).
77. R.D. Birkhoff, "The passage of fast electrons through matter", *Handbuch Phys.*, **19**, 53 (1958); Original: H.A. Bethe, *Ann. Physik*, **5**, 325 (1930).
78. C.W. Allen, *Astrophysical Quantities* (3rd ed), University of London, The Athlone Press, 1973, p. 36.
79. O.C. Wells, *Scanning Electron Microscopy*, McGraw-Hill, New York, 1974, section 3.2.1.
80. H. Kanter, "Contribution of backscattered electrons to secondary electron formation", *Phys. Rev.*, **121**, 681 (1961).
81. R.J. Hawryluk, "Exposure and development models used in electron beam lithography", *J. Vac. Sci. Technol.*, **19**, 1 (1981).
82. R. Shimizu and K. Murata, "Monte Carlo calculations of the electron-sample interactions in the scanning electron microscope", *J. Appl. Phys.*, **42**, 387 (1971).
83. K. Murata, T. Matsukawa and R. Shimizu, "Monte carlo calculations on electron scattering in a solid target", *Jpn. J. Appl. Phys.*, **10**, 678 (1971).
84. D.F. Kyser and K. Murata, *Electron and Ion Beam Science and Technology, 6th International Conference* (R. Bakish, ed.), The Electrochemical Society, Princeton, NJ, 1974, pp. 205–223.

85. A.N. Broers, "Resolution, overlay, and field size for lithography systems", *IEEE Electron. Dev.*, **ED-28**, 1268 (1981).

86. G. Han, M. Khan, Y. Fang and F. Cerrina, "Comprehensive model of electron energy deposition", *J. Vac. Sci. Technol. B*, **20**, 2666 (2002).

87. H.I. Smith, "A statistical analysis of ultraviolet, x-ray, and charged-particle lithographies", *J. Vac. Sci. Technol. B*, **4**, 148 (1986).

88. G.M. Gallatin, "Continuum model of shot noise and line edge roughness", *Proc. SPIE*, **4404**, 123 (2001).

89. M.L. Yu, A. Sagle and B. Buller, "Exploring the fundamental limit of CD control: A model for shot noise in lithography", *Proc. SPIE*, **5751**, 687 (2005).

90. L.H.A. Leunissen, M. Ercken and G.P. Patsis, "Determining the impact of statistical fluctuations on resist line edge roughness", *Microelectr. Eng.*, **78–79**, 2 (2005).

91. J. Kon, et al., "Optimization of chemically amplified resist for high-volume manufacturing by electron-beam direct writing toward 14 nm node and beyond", *Proc. SPIE*, **8323**, 832324 (2012).

92. G. Owen, "Methods for proximity effect correction in electron lithography", *J. Vac. Sci. Technol. B*, **8**, 1889 (1990).

93. L.D. Jackel, R.E. Howard, P.M. Mankiewich, H.G. Craighead and R.W. Epworth, "Beam energy effects in electron beam lithography: The range and intensity of backscattered exposure", *Appl. Phys. Lett.*, **45**, 698 (1984).

94. T.H.P. Chang, A.D. Wilson, A.J. Speth and A. Kern, *Electron and Ion Beam Science and Technology, 6th International Conference* (R. Bakish, ed.), The Electrochemical Society, Princeton, NJ, 1974, pp. 580–588.

95. M. Parikh, "Corrections to proximity effects in electron beam lithography. I. Theory", *J. Appl. Phys.*, **50**, 4371 (1979).

96. D. P. Kern, *Electron and Ion Beam Science and Technology, 9th International Conference* (R. Bakish, ed.), The Electrochemical Society, Princeton, NJ, 1980, pp. 326–329.

97. G. Owen, "Proximity effect correction in electron-beam lithography", *Opt. Eng.*, **32**, 2446 (1993).

98. G. Owen and P. Rissman, "Proximity effect correction for electron beam lithography by equalization of background dose", *J. Appl. Phys.*, **54**, 3573 (1983).

99. T.R. Neill and C.J. Bull, *Microcircuit Engineering 80, Proceedings International Conference on Microlithography*, Delft University Press, Delft, 1981, pp. 45–55.

100. K. Ogino, H. Hoshino, Y. Machida, M. Osawa, H. Arimoto, T. Maruyama and E. Kawamura, "Three-dimensional proximity effect correction for multilayer structures in electron beam lithography", *Jpn. J. Appl. Phys.*, **43**, 3762 (2004).

101. Y. Ochiai, Y. Kojima and S. Matsui, "Direct writing through resist exposure using a focused ion beam system", *J. Vac. Sci. Technol. B*, **6**, 1055 (1988).

102. N. Rau, F. Stratton, C. Fields, T. Ogawa, A. Neureuther, R.L. Kubena and G. Willson, "Shot-noise and edge roughness effects in resists patterned at 10 nm exposure", *J. Vac. Sci. Technol. B*, **16**, 3784 (1998).

103. E. Platzgummer and H. Loeschner, "Charged particle nanopatterning", *J. Vac. Sci. Technol. B*, **27**, 2707 (2009).

104. A. Yasaka, et al., "Image quality improvement in focused ion beam photomask repair system", *J. Vac. Sci. Technol. B*, **26**, 2121 (2008).

105. A. Yasaka, et al., "Application of vector scanning in focused ion beam photomask repair system", *J. Vac. Sci. Technol. B*, **26**, 2127 (2008).

106. R. Kiefer, et al., "Enhancement of the image fidelity and pattern accuracy of a DUV laser generated photomask through next-generation hardware", *Proc. SPIE*, **5754**, 1011 (2005).

107. U. Ljungblad, T. Sandstrom, H. Buhre, P. Duerr and H.K. Lakner, "New architecture for laser pattern generators for 130 nm and beyond", *Proc. SPIE*, **4186**, 16 (2001).

11 Imprint Lithography

Doug Resnick and Helmut Schift

CONTENTS

11.1   INTRODUCTION

Relative to the other lithographic techniques discussed in this book, nanoimprint lithography (NIL) is unusual as high-resolution lithography. Its underlying physical principle of molding a resist by mechanical contact instead of a masked exposure with light or electrons contradicts the credo of modern high-volume lithography of noncontact processing. In fact, its basics are very old, and imprinting (the generation of a topography in a thin layer of resist by molding) as well as printing (the transfer of ink from the surface protrusions of a stamp to a substrate) can be considered established technologies. While Johannes Gutenberg is generally credited with the invention of modern printing, imprinting had already been practiced for centuries in China [1]. As early as 500 BC, there is evidence of carved characters in stone and ceramic. Metal castings were commonly used for diplomatic purposes, and one of the more famous examples is the King of Na gold seal that was presented to Japan from China in the year AD 57. The seal has five Chinese characters, which were associated with the Han Dynasty. As an interesting sidelight, the seal was lost for centuries and accidently discovered by a Japanese farmer in 1784. The seal is now considered a Japanese Treasure and is on display in the Fukuoka prefecture of Japan.

With the invention of paper in approximately 200 BC, it became possible to reproduce works on large writing surfaces. Important work was etched into a stone slab and transferred onto materials such as hemp, silk rags, or bark. In the seventh century, stone gave way to woodblock, a technique that was also used in Europe during the Middle Ages. To form a printed image, a sheet of paper was placed over the block, and the image was transferred either by rubbing or by inking the block and abrading the paper.

Woodblock remained the primary means of imprinting well into the nineteenth century in China. However, in 1040, Bi Sheng began to experiment with movable type made from either clay or ceramics. The individual type was arranged on an iron plate and held in place with wax and resin. By melting the wax, the type could be removed and then rearranged as necessary. Eventually, clay gave way to wood and later, to more durable materials such as copper and brass. The technology never became popular in China because of the thousands of characters that were needed to convey information in the Chinese language.

Gutenberg began his work in the town of Strasbourg in 1436 and later moved to Mainz, both in the Rhine valley. As a goldsmith, he was a master of metallurgy, able to make and multiply hard letter types from carved metal originals within seconds. He also took advantage of the foundation of paper mills in Germany, enabling printing on paper instead of vellum. His invention was a combination of cast metal characters ("types") with a screw-type press, which was a takeoff of machines that were used to produce wine in the Rhine valley. By splitting individual components of language such as letters, numbers, and punctuation into individual units, it was now possible to quickly form different sentences and paragraphs. The press was completed in 1440, and in 1455, printed versions of a 42 Line Bible (42 lines per page) appeared. Two hundred copies of this Bible were printed, and 48 copies are known to exist today. A picture of the press, along with a page of the Bible, is shown in Figure 11.1. Gutenberg first wanted to create only high-quality (expensive) books with a quality

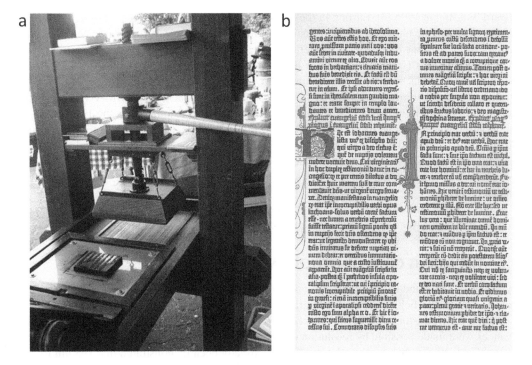

FIGURE 11.1   (a) The Gutenberg press. (b) A page from the 42 Line Bible.

similar to those copied by hand, that is, with colors and inlays. But printing was equally applicable for less expensive work, and finally, it became an all-purpose technique for the flyers of Christian reformation and newspapers.

Not unlike startups today, Gutenberg's work required venture capital, and a moneylender by the name of Johannes Fust realized the potential of Gutenberg's work. In 1449, Fust loaned Gutenberg 800 florins, which was used for the preparation of the imprinting tool [2]. Prior to the release of the Bible, several books and treatises were published, and the impact of the technology began to blossom. By the time of the printing of the Bible, however, the relationship between Gutenberg and Fust had become strained, and Fust took Gutenberg to court, claiming, among other things, embezzling of funds. The archbishop's court of worldly justice ruled primarily in favor of Fust, and in the same year that the 42 Line Bible was printed, Gutenberg lost his printing workshop and was effectively bankrupt.

Gutenberg continued his work, along with other printers, up until 1462, when the Archbishop of Nassau attacked the city of Mainz. Once the city was under control, the surviving citizens, including Gutenberg, were forced to leave. When printers and compositors settled in new towns, the knowledge of the printing process spread beyond the immediate vicinity of Mainz. By 1477, William Caxton published the first book in England. By the end of the century, printing technology was established in over 250 cities across Europe. Roman type was introduced in 1572, and Oxford University started a printing operation in 1587. In 1593, Shakespeare's "Venus and Adonis" appeared in print and began a new era in literature.

It should also be noted that although movable type is generally thought of as a European invention, a common use of movable type began in Korea at around the same time as Gutenberg's innovative work. An alphabetical script known as "Han'gul," originally consisting of 28 characters, was officially presented in 1444.

The modern version of imprint lithography is generally attributed to Stephen Chou, although there is evidence of imprinting at the nanoscale in the literature that predates the seminal work put forward by Chou in 1995 [3]. Earlier, in the 1970s, Susumi Fujimori from NTT labs had developed an imprint technique based on (thermoplastic) molding and the typical pattern transfer, while others were using molding and pattern transfer processes in the micro range [4]. Jan Haisma at Philips developed ultraviolet (UV)-based imprint lithography at about the same time as Chou [5]. It had long been known that materials can be molded with sub-20 nm resolution. In the 1960s, using transmission and scanning electron microscopy, polymethyl methacrylate (PMMA) replicas with a resolution of below 10 nm were cast from biological species and examined. In the 1970s, compact disc molding was attaining 1 μm wide pits in 120 mm polycarbonate discs within 3 s cycle time, and Blu-ray has reduced resolution to below 100 nm. It is safe to say that Chou's work spurred on an entire generation of scientists and engineers who explored the many aspects necessary to both understand and improve the process and search for applications and markets that could not be enabled otherwise.

In this chapter, we cover the cost of lithography, discuss variants of nanoimprint lithography in detail, and also cover some of the more promising markets and applications for the technology.

11.2   THE RISING COST OF LITHOGRAPHY

High-density semiconductor circuits are now being mass-produced with critical dimensions less than 20 nm. In the field of optical lithography (or photolithography [PL]), major advancements in resolution have historically been achieved through use of shorter wavelengths of light. Further advances were made possible by resolution enhancement techniques, immersion, and multiple patterning, Using phase-shift mask technology, it has already been demonstrated that 193 nm immersion photolithography can produce ~40 nm features. By applying self-aligned double patterning (SADP) processes to immersion-based 193 nm systems, devices such as NAND Flash memory are in production at half-pitches as small as 19 nm. By repeating the spacer process again (self-aligned quadruple patterning [SAQP]), device manufacturers are currently supplying devices with half pitches down to 15 and 16 nm.

Along this path toward ever-shrinking devices comes a potential increase in the cost of ownership (CoO), which for the semiconductor business, is typically expressed in terms of a cost per wafer level. Several models exist that can estimate the cost of ownership. In previous generations of devices, it was typically sufficient to include the two main ingredients of tool cost and mask cost.

It is interesting to note that the cost of photo tools has kept pace with data density, as depicted in Figure 11.2 [6]. The development of both light sources and optics to support the sources is one of

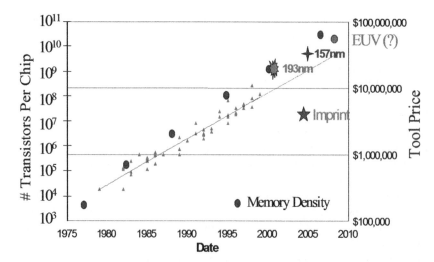

FIGURE 11.2   Transistor density and tool cost as a function of time.

the primary factors responsible for the rise in the cost of high-end photo tools. For example, 157 nm lithography (a technology that never made it to production required the use of CaF_2 as a lens material. In the case of extreme ultraviolet lithography (EUVL), source power is a constant challenge, and higher powers for future generations of EUV systems will always be required to meet the industry's throughput requirements. For this technology, a continuously running source of at least 250 W is required. To date, 80 W has been demonstrated, but with uptimes of only about 60%. While EUVL systems may eventually meet their targeted and power specification (along with a throughput of 125 wafers per hour), it is anticipated that the cost of these tools will likely exceed $100 million, a price tag considered prohibitive for many companies.

With the introduction of SADP and SAQP processes, cost of ownership considerations now need to take into account the added deposition, etch, and lithography steps associated with the process flow (for more detail on this subject, refer to Chapter 3).

To put this into perspective, Figure 11.3 is a plot of cost of ownership comparing a conventional ArF SAQP process with EUVL + SADP, EUVL, and NIL. Clearly, the combination of a simplified process flow and a reduced tool cost can have a significant impact on the cost per wafer level for a high-end semiconductor device.

Although the focus so far has been on semiconductor device fabrication (and in fact, this carries through as the main theme for many of the chapters in this book), it is important to recall that the unique physical and chemical phenomena at the nanoscale can lead to other types of novel devices that have significant practical value. Emerging nanoresolution applications include subwavelength optical components (antireflection gratings and wire-grid polarizers), biochemical analysis devices (nanofluidics), high-speed compound semiconductor chips, distributed feedback lasers (gratings), photonic crystals (on vertical emitting lasers and photovoltaic solar cells), patterned sapphire substrates for light-emitting diodes (LEDs), and high-density patterned magnetic media for storage [7]. To take advantage of these opportunities, it is necessary to be able to cost-effectively image features well below 100 nm. For some of these applications, the question of structural tolerance, overlay, and defectivity is more relaxed in comparison to semiconductor chip manufacturing: single-layer nanopatterning does not need machines enabling overlay; high-redundancy, read-out correction schemes (e.g., in patterned magnetic media) or even self-healing processes (e.g., with directed self-assembly using block copolymers) overcome the loss of local structural details. Even the demand for high speed may be relaxed if 80,000 vertical-cavity surface-emitting lasers (VCSEL) are patterned

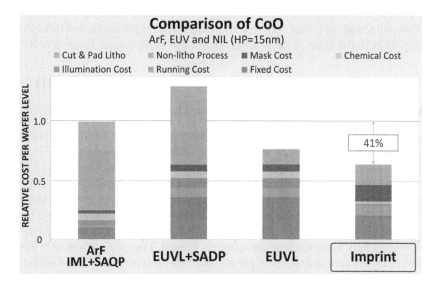

FIGURE 11.3 Cost of ownership comparison for four lithographic approaches.

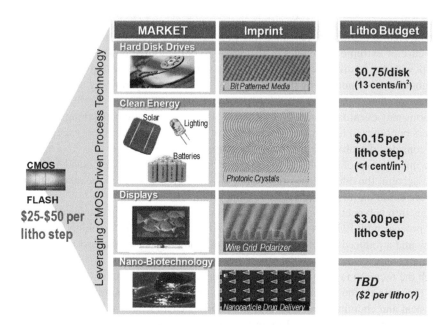

FIGURE 11.4 Cost of ownership for several markets requiring high-resolution patterning. Relative to semiconductors, the other markets have much smaller lithography budgets.

with polarizing gratings within one step. Other applications profit from the fact that NIL is capable of patterning multilevel resist structures, which can be used to reduce the number of levels in interconnect layers, and is even able to pattern functional materials (e.g., dielectrics) [8].

It is interesting to compare the cost of lithography for a high-density semiconductor device, such as NAND Flash memory, with some of the applications mentioned in the previous paragraph. As illustrated in Figure 11.4, the lithography cost per wafer level for a complementary metal–oxide–semiconductor (CMOS) NAND Flash critical layer is somewhere between $25 and $50. Other markets, including patterned media for hard drives, clean energy, displays, and biomedicine, have comparable needs for patterning (in terms of feature size and density) but require reductions in lithography costs by as much as three orders of magnitude [9]. This can only be accomplished with high-resolution patterning tools that deliver both lithographic performance and throughput as well as minimizing the material costs of the process. This is one of the unique benefits of NIL.

11.3 INTRODUCTION TO MODERN NANOIMPRINT LITHOGRAPHY

11.3.1 AN OVERVIEW OF IMPRINTING AND PRINTING LITHOGRAPHY

NIL is a high-throughput, high-resolution parallel patterning method in which a surface pattern of a stamp is replicated into a material by mechanical contact and three-dimensional (3-D) material displacement [10, 11]. This can be done by shaping a liquid followed by a curing process for hardening, by variation of the thermomechanical properties of a film by heating and cooling, or by any other kind of shaping process using the difference in hardness between a mold and a moldable material. The local thickness contrast of the resulting thin molded film can be used not only as a means to pattern an underlying substrate at wafer level by standard pattern transfer methods but also directly in applications where a bulk modified functional layer is needed. Therefore, it is mainly aimed toward fields in which electron beam and high-end photolithography are costly and do not provide sufficient resolution at reasonable throughput. This was demonstrated by Stephen Chou using stamps with 10 nm features (see Figure 11.5).

Nanoprinting (known as *microcontact printing* [μCP] [12]), in contrast to nanoimprinting, transfers an ink from stamp surface protrusions onto a substrate. Such an ink consists of (linear) chain molecules with the ability to adhere or bind to surface atoms, which often create monomolecular films that—due to their density and chemical properties—can act as resists. However, such films, in the case of self-assembled monolayers (SAMs) only 1 to 2 nm thick, present chemical contrast rather than topographical contrast against typical etchants (see Table 11.1). This process has demonstrated sub-100 nm resolution and in principle, can pattern structures with similar resolution to NIL, if lateral diffusion of densely printed molecules is inhibited. While NIL is clearly directed toward high-resolution applications, μCP is for research applications, mostly in the area of biotechnology. The main criteria for this are resolution, throughput, and the possibility to integrate the process into existing process chains.

NIL is essentially a nanomolding process in which the topography of a template defines the patterns created on a substrate. This is not only a 1:1 process in lateral dimensions, as in the case of contact or proximity photolithography, but an exact copy of the topography of the stamp—even defects will be transferred to the molded materials or inhibit the stamp from touching the surface at the location of contact. Investigations by several researchers indicate that resolution is only limited by the resolution of the template fabrication process, because most polymers can assume the shape of stamp structures in their liquid state and keep it after they are cured or frozen. The most famous example of this capability is the reproduction of a 2 nm carbon nanotube by the John Rogers Research group at the University of Illinois (Figure 11.6) [13]. NIL possesses other important advantages over conventional photolithography and other next-generation lithographic (NGL) techniques, since it does not require expensive projection optics, advanced illumination sources, or specialized resist materials that are central to the operation of these technologies.

There are three basic approaches to imprint lithography: thermal nanoimprint lithography (T-NIL), ultraviolet–assisted nanoimprint lithography (UV-NIL), and soft lithography (SL). SL has become a separate approach, because it uses a soft stamp for both printing and imprinting (molding). It will be treated separately in Section 11.4, as ink-based printing processes have found various applications outside semiconductor processing. All three topics can be performed on large areas with single stamp imprinting, step and repeat (S&R) approaches where a smaller stamp is printed sequentially onto the same substrate, and roll-to-roll (R2R) approaches, in which a stamp is bent around a roll and printed onto a foil-like substrate, which is continuously fed into a printing area between two rolls.

All three topics are discussed in detail. In addition, there is growing interest in a R2R-NIL process for large-area substrates such as flat panels. Although the imprint process is not unique in itself, the techniques used to create nanopatterns from a roller are different enough to warrant a separate discussion. Following this, we present, in more detail, specific applications using NIL.

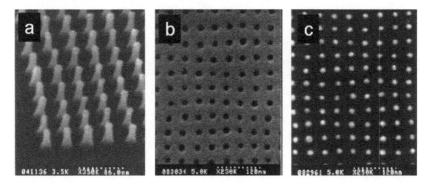

FIGURE 11.5 The initial demonstration of thermal NIL by Stephen Chou and coworkers showing the NIL process chain. SEM micrographs: (a) SiO_2 stamp with pillars of 10 nm diameter, 60 nm height, and 40 nm period; (b) imprinted pattern into a thermoplastic film (PMMA resist); (c) 10 nm metal dots after lift-off process. (Reprinted from Chou, S.Y. and Krauss, P.R., *Microelectron. Eng.*, 35, 237, 1997. With permission. © 1997 American Vacuum Society.)

TABLE 11.1

Comparison between Nanoimprint Lithography (NIL) and Nanoprinting Lithography via Microcontact Printing (μCP)

	Pattern definition	Pattern transfer	
	Resist patterning	Residual layer etch	Substrate patterning
Nanoimprint lithography (NIL)			
Nanoprint lithography (soft or micro-contact lithography [μCP])	Ink transfer	Gold etching	Substrate patterning
Issues to be considered	3-D rheology, inclusions, residual layer homogeneity, demolding, shrinkage, relaxation, distortion	Pattern fidelity and homogeneity, aspect ratio and contrast, surface contamination	Pattern fidelity, etch selectivity, topological and chemical contrast (hard mask), resist removal

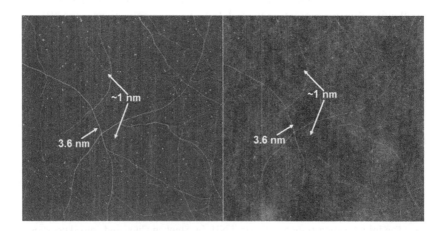

While NIL generates a topographical contrast in a resist, μCP generates a chemical contrast due to a dense, monomolecular ink layer.

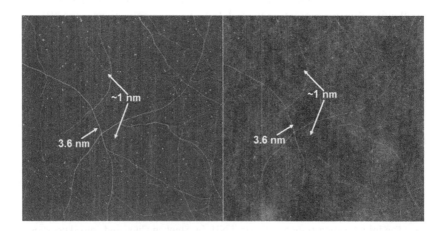

FIGURE 11.6 Replication of carbon nanotubes into a UV-curable resist material, demonstrating a resolution of less than 2 nm. (Reprinted with permission from Hua, F., et al., *Nano Lett.*, 4, 2004, 2487. © 2004 American Chemical Society.)

NIL relies on the same toolbox as that for typical replication processes but with two main differences: It uses cleanroom-based micromachining techniques for stamp fabrication and silicon or comparable materials (semiconductors, fused silica, glass, and sapphire) as substrates, and it relies on thin polymeric films where the sizes and heights of the structures become comparable to the films to be patterned. While it exhibits great potential as a manufacturing process for a range of nanoscale surface topographies, its definition as "lithography" is only valid for specific applications, in which the last step is the transformation of the surface topography in the thin polymer film into a different material; for example, by using it as a masking layer for etching into the substrate or for metallization. The focus of this chapter is on NIL as a lithographic process for integrated circuit (IC) manufacturing; however, care is taken that the engineer familiar with modern lithography is able to follow the main process routes established for chip manufacturing. However, NIL owes a part of its interest to other applications. Therefore, a more generalized concept of molding will be presented and the differences to photolithographic patterning explained.

11.3.2 IMPRINT PROCESS SEQUENCE

The NIL process chain comprises the main processes of origination, replication, and pattern transfer (equivalent to mask making and exposure in photolithography). It is often not complete if process cycles are iterated or combined with other processes (e.g., stamp copying) [14]. This sequence is valid for replication processes of polymeric components with different structure sizes and applications, such as the manufacturing of CDs, diffractive optical elements (DOEs), or microfluidic chips. In photolithography, the sequence would comprise mask making, lithography, and pattern transfer. Therefore, for high-volume chip manufacturing, different terms have been established than in the replication of micro- and nanocomponents. For example, a mask is also referred to as a *template*, *mold*, *stamp*, or *die*; the lithography step is named *replication*, *molding*, *imprint*, or *embossing*; and the pattern transfer often comprises an additional step before the substrate is ready for metallization, etching, etc. This step is called *descum*, but more common terms are *residual layer etch*, *breakthrough etch*, *window opening*, or simply *resist thinning*. These principal processes of the process chain involve many process steps: design, process simulation, stamp copying, transformation into a working tool with appropriate structural resolution, area enlargement, and even the integration of mixed micro- and nanostructures with 3-D features. In the last several years, the combination of imprint steps with bottom-up processes, that is, directed self-assembly (DSA), has become popular. Furthermore, R2R processes are integrated into the toolbox because of their use in flexible electronics and optics. They are not physically different from wafer-type manufacturing but are nearer to the fabrication of thin foils for holograms or packaging.

Figure 11.7 shows schematics with different processes for topography origination, replication, and pattern transfer.

As displayed in the schematic, the standard route for origination is electron beam lithography combined with pattern transfer processes suitable for generating topography in silicon and quartz substrates. These processes are complemented by so-called *tooling processes*, which can be the transfer of the original (often very costly) master template into a working stamp (copy, daughter, or replica) in a desired material (hard, flexible, bendable, and durable). Such daughter stamp fabrication is also needed to protect the master from damage and to ensure backup, parallel production, or simply transfer into a material (e.g. metal) that can be placed as an insert in metal frames, for example into CD injection molding tools. While copying by casting and electroplating is the fabrication of "monolithic" daughters with a negative (inverse polarity) surface topography, NIL can be used to modify the original topography or to combine it with topographies generated by other techniques. This is possible by postprocessing for enhancement of resolution by spacer-etching, block copolymer coating, or enhancement of aspect ratio. For UV-NIL processes using hard templates (e.g. J-FIL), the imprint can be done on prepatterned templates, which are difficult to host in electron beam tools (patterns on mesas). Today, not only flat wafer or die-like substrates are used but also bendable molds for R2R processes. The enlargement of area is no different from the step-and-repeat

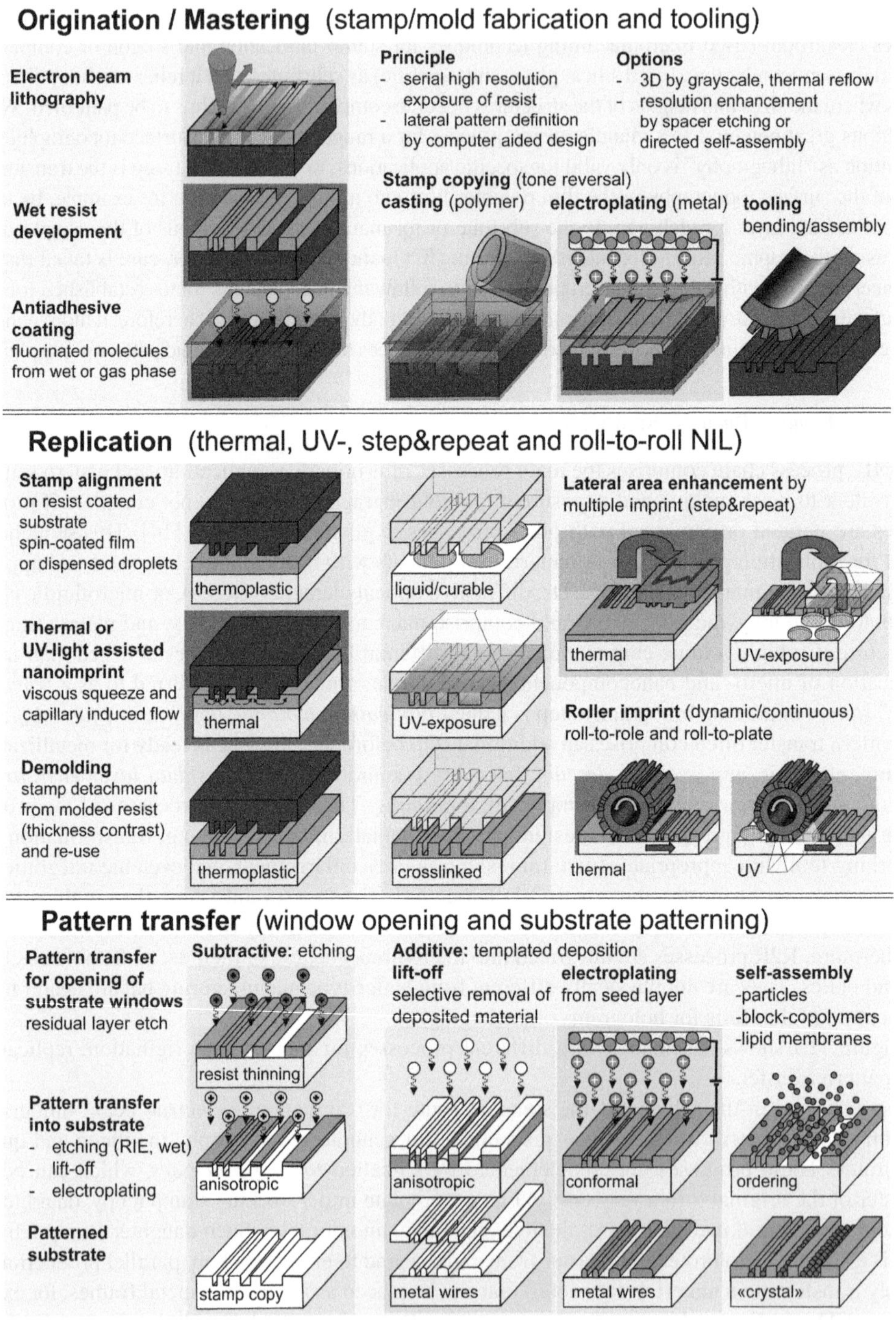

Origination / Mastering (stamp/mold fabrication and tooling)

Electron beam lithography

Principle
- serial high resolution exposure of resist
- lateral pattern definition by computer aided design

Options
- 3D by grayscale, thermal reflow
- resolution enhancement by spacer etching or directed self-assembly

Wet resist development

Stamp copying (tone reversal)
casting (polymer)   **electroplating** (metal)   **tooling**
bending/assembly

Antiadhesive coating
fluorinated molecules from wet or gas phase

Replication (thermal, UV-, step&repeat and roll-to-roll NIL)

Stamp alignment
on resist coated substrate
spin-coated film or dispensed droplets

thermoplastic   liquid/curable

Lateral area enhancement by multiple imprint (step&repeat)

thermal   UV-exposure

Thermal or UV-light assisted nanoimprint
viscous squeeze and capillary induced flow

thermal   UV-exposure

Roller imprint (dynamic/continuous)
roll-to-role and roll-to-plate

Demolding
stamp detachment from molded resist (thickness contrast) and re-use

thermoplastic   crosslinked   thermal   UV

Pattern transfer (window opening and substrate patterning)

Pattern transfer for opening of substrate windows
residual layer etch

Subtractive: etching

resist thinning

Additive: templated deposition
lift-off
selective removal of deposited material

electroplating
from seed layer

self-assembly
-particles
-block-copolymers
-lipid membranes

Pattern transfer into substrate
- etching (RIE, wet)
- lift-off
- electroplating

anisotropic   anisotropic   conformal   ordering

Patterned substrate

stamp copy   metal wires   metal wires   «crystal»

FIGURE 11.7 Main imprint processes: thermal NIL and UV-NIL, including step and repeat and roll-to-roll variants, and an overview on typical additive and subtractive pattern transfer processes. (Reproduced and adapted with permission from Schift, H., et al., *Proc. SPIE*, 9049, 2014. © 2014 SPIE.)

approaches used for high-volume manufacturing (HVM) and involves overlay and stitching issues. This process, called *recombination* because it was historically used for the assembly of different dies into a large (composite) stamp, is simply a combination of adjacent imprints on a larger surface covering the surface of an entire roll or a large-area stamp. The use of mix- and-match or top-down and bottom-up processes, along with 3-D topographies, enables combined functionality, for example stepped structures with different height levels or grayscale surfaces, which cannot be easily manufactured using one single process only.

11.3.3 THERMAL NIL

Thermoplastic materials can be shaped in their viscous state simply by pressing a stamp into a softened film [15]. By using pressure, the viscous polymer will flow into the cavities of the stamp until it has completely assumed the stamp surface. In NIL, when using a resist thickness similar to the lateral size of the stamp protrusions, this process is governed by so-called *squeeze flow* [16–20]. Due to the confinement between a stamp with its protrusions and a flat substrate, a lateral flow of material is induced below each protrusion, which leads to a thinning of the material below protrusions, while the material rises in the areas of cavities. The result is a thin polymer layer with a thickness contrast of exactly the height of the cavities, thus replicating the stamp topography. By cooling, the polymer hardens, and the stamp can be detached from the surface. Several issues have to be considered to achieve good results. By experiment and simulation, different generalized process parameters were found to be suitable [21–24]:

1) Imprint is performed at temperatures higher than the so-called *glass transition temperature* (T_g). This is a range rather than a defined temperature and defines the point above which the polymer is considered to be soft, while below T_g, it is considered to be hard. Therefore, thermocycles have to be employed, which raise the "mold–polymer–substrate sandwich" to a temperature $T_{imprint}$ 30–40 °C above T_g and after molding is completed, to a temperature T_{demold} 20–30 °C below T_g for detachment and handling. The imprint temperatures are still well below the typical temperatures used for injecting molding, in which a low-viscosity melt is injected into a colder mold cavity, where it freezes upon contact. For noncrosslinked materials, T_g of 40 to 120 °C is advisable, since in the following etch steps, any heating near T_g will cause distortions. During the imprint of the material at a typical pressure of 2 to 10 MPa (20 to 100 bar/atm or 1.4 to 6.9 psi), the viscous polymer film below the protrusions of the stamp undergoes a squeeze flow, which means it is pressed sideways into the recessed stamp areas (cavities). Due to relatively high viscosities in the range of 10^5–10^7 Pa·s, this process is slow for large protrusions. Therefore, auxiliary cavities are often needed in unpatterned areas of several 100 µm size to keep imprint times low enough.

2) A variothermal process scheme involves heating and cooling the entire "mold–polymer–substrate sandwich." Due to the intimate contact at high pressures, these cycles typically involve not only the sandwich but also a considerable amount of heat capacity given by the tool's steel plates. Therefore, it is advisable to keep this temperature range small, particularly the maximum temperature, because heat and cooling rates limit throughput. Typical cycle times in laboratory-type machines are 5 to 20 min, which are often much larger than the actual molding time of a few tens of seconds up to a few minutes, thereby adding a significant overhead to the overall process (see Figure 11.8). Newer developments with heatable (pulsed NIL) stamps allow short pulsed operation, in which temperatures of 400 °C are achieved in the area of direct contact of stamp to polymer (see Figure 11.8) [25–27]. This reduces the printing time to 100 µs and eliminates the tool's overhead, enabling a very fast cycle time determined mostly by the time of handling (see Figures 11.9 and 11.10). Other alternatives, such as laser-assisted direct imprint (LADI), employ short (250 µs) laser pulses that heat up substrate surfaces above 1000 °C [28]. This makes it possible to mold the upper

layer of silicon substrates through transparent quartz stamps (see Figure 11.11). While pulsed NIL has demonstrated 100 mm wafer imprint, LADI has patterned areas of a few square millimeters.

3) Thermal expansion leads to lateral shear and bending; therefore, a thermal expansion mismatch between stamp and substrate should be avoided or its effects reduced. This can be done by using the same material for stamp and substrate or by using small stamps. An expansion mismatch of a few microns or even nanometers can lead to lateral shearing and distortion. Here, also, pulsed NIL may have the advantage that only the thin resist layer is heated, while the substrate is—due to fast dissipation of heat into the bulk—kept at ambient.

4) Air bubbles can be avoided by printing in an evacuated tool or by pressing at high temperatures. At a pressure of 10 MPa (100 bar), air bubbles will be compressed to 100th of their initial volume at ambient (1 bar). For lower pressures, air bubbles will be present, but smaller bubbles of air will be taken up by the material.

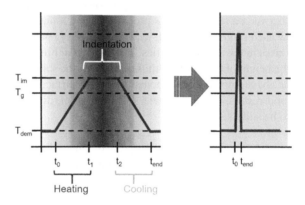

FIGURE 11.8 Imprint process scheme demonstrating the impact of the T-NIL "dilemma." Fast thermal imprint can be achieved if both heating and cooling times ("overhead") are minimized; at the same time, the actual indentation (imprint/embossing) time is strongly dependent on the temperature (low viscosity). (Reprinted from *Microelectron. Eng.*, 141, Tormen, M., et al., Sub-100 μs nanoimprint lithography at wafer scale, 21–26, © 2015, with permission from Elsevier.)

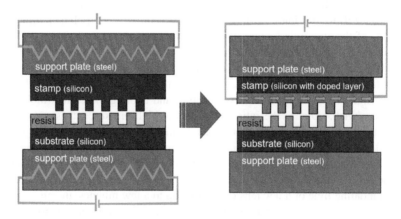

FIGURE 11.9 The high overhead typically associated with thermal NIL is due to the variothermal process scheme, in which not only stamp and substrate are kept in thermal equilibrium with the molded resist but also part of the (often bulky) support plates. This can be greatly reduced if localized ultrafast heating schemes are used, in which only the upper part of the stamp is heated (e.g., by pulsed resistive heating or by exposure to intense laser pulses). (Reprinted from *Microelectron. Eng.*, 155, Pianigiani, M., et al., Effect of nanoprint on the elastic modulus of PMMA: Comparison between standard and ultrafast thermal NIL, 85–91, © 2016, with permission from Elsevier.)

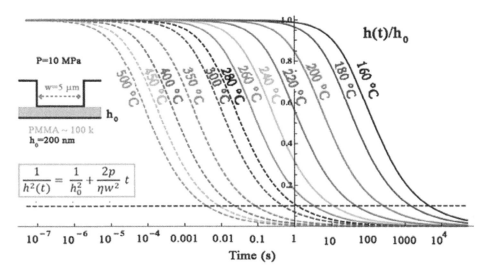

FIGURE 11.10 Typical imprint temperatures are about 50 to 80 °C higher than the glass transition temperature T_g, resulting in viscosities of about 10^5–10^7 Pa·s, but well below the degradation of resist or antisticking coatings. Fast temperatures of below 1 s can be achieved with ultrafast processes, which allow heating above the melting temperature without damaging materials. Also, only part of the resist is heated, allowing the use of high pressures for conformal contact. (Reprinted from *Microelectron. Eng.*, 155, Pianigiani, M., et al., Effect of nanoprint on the elastic modulus of PMMA: Comparison between standard and ultrafast thermal NIL, 85–91, © 2016, with permission from Elsevier.)

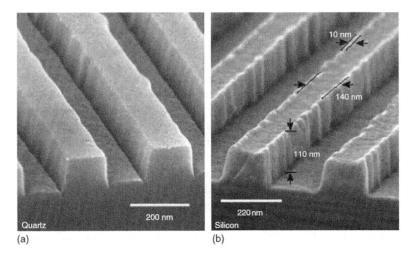

FIGURE 11.11 SEM micrographs of the template (a), and of the patterned silicon (b), using LADI. (Reprinted by permission from Macmillan Publishers Ltd. [*Nature*] Chou, S.Y, et al., Ultrafast and direct imprint of nanostructures in silicon, 417, 2002, 835, © 2002.)

In T-NIL, the high imprint pressure needed for the resist viscosities of 10^5 Pa·s and more enables compensation for the lack of flatness of substrate and stamp by conformal bending over large areas. Therefore, single polymer layers are often used. Thickness can be chosen from very thin layers for high pattern transfer fidelity (with a thickness lower than the effective stamp protrusions sizes used), where cavities are only partially filled, to thick layers, which can be used for the integration of lenses and microfluidic channels. Pressure equilibration is guaranteed by using a compliance layer at the backside of the stamp or in the case of air-pressurized NIL systems, by a flexible (soft) membrane. A good choice of process parameters by rules of thumb is often sufficient to get a very good insight into how a polymer imprints—often in the first test. These rules of thumb result from tradeoffs between structure height, resist thickness, pressure, and temperature to be used. Many polymers can be imprinted with high resolution, and even polymers without known thermoplastic behavior can sometimes be patterned using pressure and/or heat.

11.3.4 UV-Assisted NIL

UV-NIL processes are performed at room temperature, at which resist precursors are present as liquid films or droplets. The stamp either sinks down to the substrate or must be kept at a constant distance from the substrate during both filling and exposure due to the low resist viscosities. Patterning on non-flat substrates or over topography requires a planarization strategy and often is only possible by keeping extreme tolerances or using small stamps. Several types of UV-NIL systems have been realized for S&R and for large-area (single-wafer) NIL. The main difference from T-NIL is the integration of an exposure system into one side of the mechanical setup, which has to be able to compensate for wedge errors in a low–imprint pressure process. In terms of materials, two main process variants can be seen: (1) a moderate-viscosity process (with $\eta_0 = 50$ mPa·s and below) for providing liquid films by spin coating and (2) a low-viscosity process (with $\eta_0 \leq 10$ mPa·s), which uses local dispensing of defined quantities of a liquid resin by inkjetting prior to imprint. In both cases, patterning has to be performed immediately after coating. Several issues have to be considered to achieve good results [5, 29–33].

1) Due to the low pressures of 0.1–0.5 MPa (0.1–0.5 bar/atm) involved, the filling of cavities is entirely driven by capillary forces, which involves the instant wetting of the surfaces. The good wetting behavior of the stamp is in contradiction to the need to achieve high hydrophobicity for good demolding; therefore, compromises are needed. During the filling process, the dynamic properties of displacement and merging of both resin droplets and air bubbles have to be taken into account.

2) Air bubbles are not compressed during imprint and either need to be taken up by the liquid or have to be displaced during the contact phase. This depends on the ability of the ambient to occupy the free volume in a polymer in its viscous state along with materials adjacent to the polymer. Porous stamp materials made from polydimethylsiloxane (PDMS) facilitate the uptake. If ink-drops need to merge, care has to be taken that enclosed air bubbles are small enough to be absorbed by the circumventing material [34]. Materials with no solvent and higher viscosity enable the use of vacuum prior to molding. A particular development was achieved by AIST, where bubbles were eliminated by using a condensable gas that liquefies at pressures of 1.5 bar (see Figures 11.12 and 11.13) [35–38].

3) Various materials and mechanisms are available for the curing and crosslinking of oligomers into a dense polymeric matrix (epoxy, acrylic). They are different in terms of curing speed (from a few seconds down to a fraction of a second), volumetric shrinkage (typically 1–10%), reaction with the silane-based antisticking layer (ASL, reducing its lifetime), and inhibition of crosslinking in the presence of oxygen. The crosslinking is finished when the mechanical stability allows mechanical separation, which can be done—depending on the dose—in less than a fraction of a second. In general, non-oxygen-inhibiting materials need to be used with PDMS. In contrast to this, oxygen-inhibiting materials cannot completely cure in ambient, which frequently results in tacky, uncured surfaces. This is may be advantageous if superfluous material is extruded at the borders of a hard quartz stamp and therefore can be removed by solvents. Similarly, during some step & repeat processes, it is useful if adjacent imprints

do not print on hardened excess material and inhibit the hard stamp from touching the substrate surface.

4) Crosslinked, silicon-containing resists will, when left behind on quartz stamps, be difficult to remove with existing wet cleaning (RCA) protocols. Even when ASL-coated stamps are cleaned with oxygen plasma (silane coupling group), silicon traces will remain on the surface.

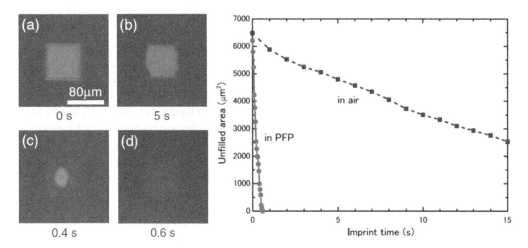

FIGURE 11.12 Time evolution of (a) 45 nm L/S pattern block filled with UV-curable resin for UV-NIL, (b) in air, and (c) in PFP. Schematic: Complete filling of cavities (d) is achieved for PFP after 0.6 s, while in air, only the onset of dissolution can be seen. Dashed line, UV-NIL in air; solid line, UV-NIL in PFP. (Reproduced and adapted with permission from Hiroshima, H. and Komuro, M., *Jpn. J. Appl. Phys.*, 46, 2007, 6391, © 2007 IOP *Jpn. J. Appl. Phys.* and Hiroshima, H. and Suzuki, K., *Jpn. J. Appl. Phys.*, 51, 2012, 06FJ10, © 2012 The Japan Society of Applied Physics.)

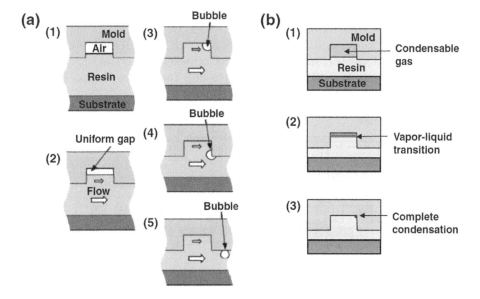

FIGURE 11.13 Control of bubble defects in UV nanoimprint for (a) air and for (b) condensable gas. (Reproduced with permission from Hiroshima, H. and Komuro, M., *Jpn. J. Appl. Phys.*, 46, 2007, 6391, © 2007 The Japan Society of Applied Physics.)

11.3.5 Combined Processes

In the case of thermoplastic UV-curable materials, which are solid at room temperature (due to solvent evaporation during spin coating and postbaking) and imprinted in an isothermal process, several of the abovementioned problems are elegantly solved. The resist with ability to crosslink can be imprinted by T-NIL and cured either by heat or by exposure. Such combined thermal and UV-NIL processes enable isothermal processing, that is, imprint at constant temperature with demolding at elevated temperature, because the initial T_g rises or disappears during the process [39, 40]. However, even more than with dispensable resist, the additional requirements imposed by the curing process will impose new restrictions on the choice of materials. This process is employed in Obducat's STU (simultaneous thermal and UV) NIL process and combined with the fabrication of a daughter stamp, called IPS (intermediate polymer stamp) (see Figure 11.14) [18, 41]. The latter is produced for each imprint cycle, which has inherent advantages for defectivity control but needs cost models whereby the high waste of resources and material is justified. Overall, this process is nearer to the hot embossing approach. It allows not only combined nanoimprint and photolithography (CNP) but also simultaneous exposure of the imprinted structures through the stamp. As resist, a multifunctional epoxidized novolac resin and a photoacid generator can be used.

11.3.6 Materials

NIL profits from the availability of a vast range of polymers with a range of properties [42–48]. This is particularly the case for T-NIL, because most linear polymers are thermoplastic in nature. If such polymers are dissolved in suitable solvents, they can be spin coated and form a hard film simply by evaporation of the solvent. They can be easily prepared for a range of thicknesses in a spin-coating process. For UV-NIL resists, which need to be liquid at room temperature, monomers or oligomers with the ability to crosslink upon exposure have to be used. The ability to modify polymers by using different base materials for polymerization and by adding functional groups and chemical components such as surfactants, photoinitiators, and crosslinking agents enables tailoring of their physical and chemical properties. Apart from standard materials, for example for T-NIL, PMMA and polystyrene (PS), which are available in a wide range of molecular weights from different chemical suppliers, commercial material formulations have found their way into research and manufacturing. While NIL tool manufacturers such as Obducat, Nanonex, EVG, and MII have developed their own resists, specific material suppliers such as micro resist technology GmbH and Toyo Gosei provide different material solutions for T-NIL and UV-NIL [49–56]. The three main tasks for materials in NIL can be subdivided into resists, stamps, and functional materials.

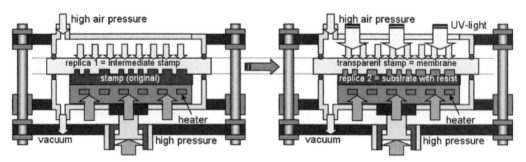

FIGURE 11.14 The advantage of spin coating of resists and UV curing has been exploited by using combined processes in which the thermoplastic polymer can be made viscous by heating and crosslinked by UV light. The isothermal process scheme allows low overhead and short imprint times. This is illustrated by a schematic of a tandem tool setup with IPS production by thermal NIL (left side) and STU process for resist patterning (right side), both imprinted with a polymer foil acting as a soft pressurized membrane. (Reprinted with permission from Schift, H., *J. Vac. Sci. Technol. B*, 26, 2008, 458, © 2008 American Vacuum Society.)

- Resists are the basis for masking substrate surfaces and enabling pattern transfer processes. They are functional films only used as intermediate masking layers and are typically removed after the pattern transfer is done. The main issues are the ability to spin coat homogeneous layers with desired thickness and dispense or inkjet desired amounts correctly distributed over the surface. After patterning, the polymer needs to be stable enough to support the pattern transfer step.
- Polymers are more and more used as working stamps; for example, for Obducat's IPS or EVG's SmartNIL stamps. Apart from pure polymeric solutions, hybrid organic–inorganic materials such as Ormocers, sol-gel materials, and polydimethylsiloxane (PDMS) are most common in use.
- Functional NIL materials are defined as materials that exhibit a function after patterning and remain on the substrate. This is different from normal resists. Apart from polymers, metals, dielectrics, ceramics, plastics, and nanomaterials can be patterned.

For thermal imprinting of thermoplastic resists, the major characteristic is the glass transition temperature T_g, at which the thermomechanical properties of a polymer change significantly from glassy (below T_g) to rubbery (above T_g). Resists with different T_g are now available with etch selectivity higher than that of PMMA and PS. This can be achieved by the choice of the right polymer chemistry or by the addition of silicon compounds into a thermoplastic matrix.

For UV-assisted NIL, different curing mechanisms are available. Depending on the monomer used, an imprint resin can have different properties. As resists for NIL, polymers obtained via acrylate, vinyl ether, and epoxide formulations have been used. The most widely used UV-curing formulations are based on free-radical polymerization of acrylic and methacrylic monomers because of their high reactivity and also their availability. They are known to have a low viscosity and a fast curing rate. However, their shrinkage during curing is rather large, about 10%, and their radical-based photopolymerization is inhibited by the presence of oxygen. Dissolved oxygen scavenges free-radical species and thus inhibits the polymerization process at the resist surface at the onset of exposure. This prolongs the required exposure time or even prevents curing. Jay Guo reports that his group developed a UV-curable epoxysilicone material based on the cationic crosslinking of cycloaliphatic epoxies [49]. This resist combines a number of desired features for nanoimprinting. Because cationic polymerization is not prone to oxygen inhibition, as compared with the free-radical polymerization of acrylate monomers, fewer defects due to noncured material are expected. The resist exhibits very good dry-etching resistance because of the high silicon content. In addition, with a suitable undercoating polymer, a uniform liquid precursor can be formed simply by spin coating. Currently, different UV-curable resists are commercialized. In contrast to the widely used acrylic-based resists, AMO [50] is offering AMONIL® with a base material viscosity of 50 mPas, which is a low viscosity NIL resist. It can therefore be spin coated and imprinted in vacuum. Similarly, micro resist technology GmbH [51] offers mr-NIL210 as a purely organic, photocurable NIL resist with outstanding dry-etch characteristics. Both resists are particularly suitable for soft NIL using soft stamp materials such as PDMS.

For both thermal and UV-NIL, release properties have been improved by specific (fluorinated) additives in the resist. Since ASL coatings on stamps degrade over time, it is advantageous if the quality of the demolding is mainly dependent on the release properties of the resist, which will be constant for every substrate. This will enable a long lifetime of the stamps and—in the case of an additional ASL on the stamp—will slow down its degradation.

11.3.7 THREE-DIMENSIONAL PATTERNING AND HYBRID PROCESSING

One of the specific advantages of NIL over photolithography is the ability to generate a 3-D surface relief with one imprint step. Even more, the entire molding step is governed by a volume displacement of material and therefore, is 3-D by definition [57]. After imprint, if the molded material fills

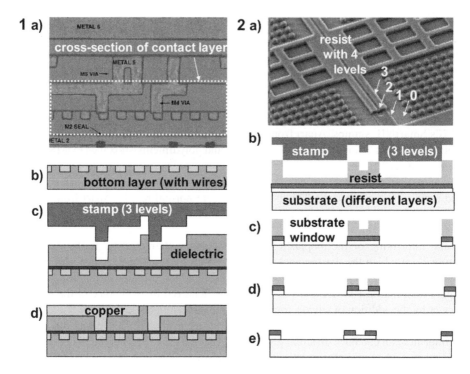

FIGURE 11.15 Two process schemes for the patterning of multilevel resist structures. In the case of the dual damascene process (left side), the dielectric is directly patterned and filled with metal. In the case of the self-aligned imprint lithography process (right side), it enables the patterning of two or more levels, which are used in a multistep etching step to define the different level patterns. (Reproduced and adapted with permission from Stewart, M.D., et al., *Proc. SPIE*, 5751, 2005, 210, [58, 59] © 2005 SPIE, and Kim, H.-J. et al., Roll-to-roll manufacturing of electronics on flexible substrates using self-aligned imprint lithography (SAIL). *J. Soc. Inf. Disp.* 2009. 17. 963–970. © Wiley-VCH Verlag GmbH & Co. KGaA. Reproduced with permission. Assembled in, with kind permission from Springer Science+Business Media: *Appl. Phys. A*, Nanoprint lithography: 2D or not 2D? A review, 121, 2015, 415, Schift, H.)

all surface cavities of the stamp, the resist pattern presents an exact replica of the stamp surface. Such 3-D surface topographies can be used for different purposes, as was outlined for materials: as resist, as stamps, or as functional materials.

- Resist multiple-layer patterning is possible by using only one imprint step. This enables the reduction of process steps and the use of self-alignment schemes, needed, for example, for R2R processing of flexible electronics, where alignment is most critical (see Figure 11.15). In the case of the dual damascene process, presented by Grant Willson's group, both the lateral wire layer and the vertical through hole of a semiconductor device are patterned with one stamp. The dielectric is directly patterned and can be filled by metal CMP (chemical mechanical polishing) processes or by electroplating. In this way, the through holes are exactly aligned with the wire layer. The process, proposed with the aim of reducing the number of lithographic steps for the interconnection layer of a chip, is currently being actively considered but holds promise for future developments [58]. Similarly, the self-aligned imprint lithography (SAIL) process, developed by Hewlett Packard, enables patterning of two or more levels, which are then followed by a multistep etching process to define the different levels in the material stack [59]. Such processes are the basis for new masking schemes but require selective pattern transfer processes.

- Multilevel and hybrid stamps are commonly needed if micro- and nanostructures have to be combined, for example, for the generation of mesas for S&R applications, or for micro-optical or fluidic devices in which the channels need to be decorated with nanostructures. Apart from stamps with specific 3-D topographies, new processes are enabled by hybrid stamps and processes that combine UV-NIL and photolithography or bottom-up and top-down approaches (i.e., lithography and self-ordering processes) [60].
- 3-D patterning of functional materials can range from surface patterning of prism-like structures in lightguides for backlight illumination of displays or for biomimetic structures, e.g. for superhydrophobic surfaces. Apart from NIL, other molding processes such as injection molding, thermoforming, and casting are used to create such structures.

11.3.8 UPSCALING

Upscaling means size, throughput, and parallel manufacturing—often combined. As an example, for semiconductor manufacturing, 300 mm wafer-sized substrates with throughputs of 60 wafers per hour, in S&R fashion with 33 mm × 26 mm fields, means 185 imprints within 2 minutes, including alignment. For the required 60 wafers per hour, this can partly be achieved by using clusters of machines (two or four) with 15–30 wafers per hour. The footprint of such clusters is still smaller than that of current EUVL tools. In contrast to this, 80,000 VCSELs on 75 mm (3 inch) wafers with some $100 \times 100 \ \mu m^2$ area each can be patterned simultaneously with one 75 mm wide large stamp, for which process time constraints are of minor importance. A compact disc with 120 mm diameter is molded within a cycle time of 3 seconds, replicating 10^9 bits with lateral sizes smaller than 1 μm [61, 62]. Nearly 10,000 copies of one music album can be replicated within an 8 hour working shift. Roll embossing can print up to 5 m/s with web sizes of 1.2 m. Inkjetting for printing newspapers has now reached 2 m/s, which means that within 8 hours, more than 50 km of paper are printed. Screens for high-definition television (HDTV) are printed in sizes of a few square meters, and float glass for windows is fabricated in sizes of $3 \times 6 \ m^2$. Single windows or screens can be imprinted using flat stamp or roll imprinting using up to $1.2 \times 1.2 \ m^2$ stamps (see Figure 11.16) [63–65]. Areas of this size have been imprinted by Scivax in Japan, but with diffraction gratings of unknown resolution,

FIGURE 11.16 Stamp with subwavelength lines generated by interference lithography at Fraunhofer ISE with substrate size of $1.2 \times 1.2 \ m^2$. This is achieved with optical path lengths up to 20 m and exposure times up to 5 h. (Reproduced from Fraunhofer Institute for Solar Energy Systems ISE, Freiburg, Germany. With permission. © Fraunhofer ISE/Holger Vonderlind.)

FIGURE 11.17 Large-area replication. Substrates with areas up to 1.3×1.1 m² have been manufactured by UV-NIL with soft stamps on glass substrates. Reproduced from Scivax Nanoimprint Solutions, Japan. With permission.

most probably for window or large display applications (see Figure 11.17) [66]. The vacuum chambers for physical vapor deposition (PVD) processes such as sputtering are adapted to these sizes. For hard disks using patterned magnetic media, double-sided imprint is needed to achieve high storage capacity in a compact housing. Some of these applications will be discussed in detail later in this chapter.

11.4   SOFT LITHOGRAPHY

11.4.1   Process Background

SL uses a soft stamp, which is able to conform to nonideal substrates, that is, to surface undulations and deviations in structure height, due to the relative softness of its micro- and nanosized stamp protrusions as well as its backbone (typically a plate several 100 μm to some millimeters thick). It is used in NIL for molding low-viscosity resists and in μCP for ink transfer. The stamp is elastomeric, which means that it can deform upon (moderate) pressure and restore its original shape when the pressure is released. As μCP, it generally refers to the process of transferring long-chained thiol molecules from the stamp protrusions to a gold film (see Figure 11.18), thus forming a densely packed SAM. The invention of the technology dates back to 1994 and is the result of work from the laboratory of George Whitesides at Harvard [12, 67]. In parallel, a similar technique was developed by Bruno Michel at IBM Rüschlikon, based on IBM's flexoprinting approaches, and termed *soft lithography*, with resolutions down to 50 nm [68]. The MIMIC process (micromolding in capillaries), also presented by Xia and Whitesides, is a molding process that resembles NIL [69, 70]. It uses capillary force to fill gaps between a stamp and a substrate with a liquid, which then can be cured either by UV light or by solvent evaporation. Apart from this, micro- and nanofluidic components, such as channels or cover plates, as well as backing cushions for pressure equilibration are manufactured using elastomeric materials.

11.4.2 LITHOGRAPHIC PATTERNING OF SELF-ASSEMBLED MONOLAYERS (SAMs)

The μCP technology surfaced mainly as a quick and easy way for students to print small geometries in a laboratory environment. Whitesides et al. formed a template by applying a liquid precursor to PDMS over a master mask produced using either electron beam or optical lithography. Sylgard® 184 from Dow Corning [71, 72] enables the fabrication of a number of transparent templates with sub-μm details. The liquid is cured by heat, and the PDMS solid is peeled away from the stamp original. The PDMS is essentially an elastomeric material consisting of a polymer chain of silicon containing oils (silicone). Typically, its mechanical properties include a tensile strength of 7.1 MPa, an elongation at break of 140%, and tear strength of 2.6 kN/m. As a result, relative to either silicon or fused silica, it is quite pliant. Since Sylgard is quite soft for high-resolution applications, harder variants of PDMS (h-PDMS and X-PDMS) were developed. The hardness ranges from 3.5–4 MPa for Sylgard 184 to 13 MPa for h-PDMS, and even higher, depending on curing conditions. Furthermore, other materials can be fabricated using UV curing and can be applied to hard backbones [73–75]. Apart from the stiffness, which helps to prevent collapse and achieve higher resolution, PDMS materials and curing can be optimized for higher resolution and reduced shrinkage [76, 77]. Due to this shrinkage in size, a difference occurs in the desired dimensions of the patterned substrate. UV-PDMS KER-4690 from Shin-etsu, with viscosity of 2.7 Pa s and a tensile strength of 7.7 MPa, similar to Sylgard, can be molded with only 0.02% volume shrinkage instead of the 10% observed with Sylgard [51, 78, 79]. Since the resolution of the process depends largely on a good balance between softness and rigidity of both surface protrusions and backbone, current applications use hybrid (composite) stamps with soft and rigid elements as well as inflatable and bendable stamps [61]. More details about the formation of the master are covered in Section 11.5.

Once prepared, the PDMS template can then be coated with a thiol ink solution, such as an alkanethiol. The printing process is depicted in Figure 11.18. The thiol molecules are subsequently transferred to a substrate coated with a thin layer of gold, thereby forming an SAM on the gold surface, meaning a dense parallel alignment of all alkane chains with headgroups oriented toward the substrate. Similarly to amphiphilic lipids on water, they are mobile and therefore have a high ability to reduce defects. The nature of the gold–sulfur bond is still not completely understood. Kane et al. postulate that the species present at the surface of the gold is a gold thiolate [80]:

$$R\text{-}SH + Au(0)_n \rightarrow RS\text{-}Au(I)Au(0)_{n-1} + \rightarrow \uparrow\tfrac{1}{2}H_2 \rightarrow \uparrow \tag{11.1}$$

To prevent adhesion between the master stamp (made from silicon, quartz, or replicated polymer templates) and a PDMS daughter, the master surface is passivated by the gas-phase deposition of a long-chain, fluorinated alkylchlorosilane ($CF_3(CF_2)_6(CH_2)_2SiCl_3$). The fluorosilane reacts with the free silanol groups on the surface of the master to form a Teflon-like surface with a low interfacial free energy. The passivated surface acts as a release layer that facilitates the removal of the PDMS stamp from the master.

The pattern transfer process starts with a wet-etch of the thin gold film. A wet-etch is typically used, since gold is not readily reactively ion etched. Although it is possible to sputter etch gold, the thin thiol layer would not hold up to such a process. The gold film then acts as an etch mask for any underlying materials. Because gold is a soft metal, it is often necessary to include a second hard mask beneath the gold.

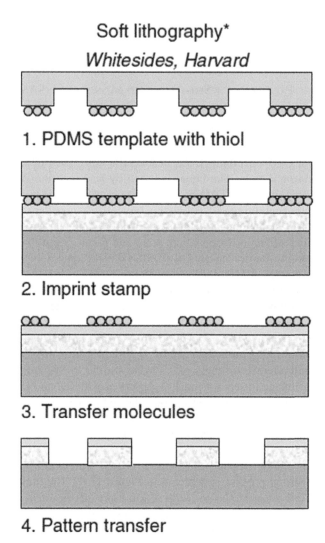

Soft lithography*

Whitesides, Harvard

1. PDMS template with thiol

2. Imprint stamp

3. Transfer molecules

4. Pattern transfer

FIGURE 11.18   Fabrication sequence for microcontact (or soft lithography) printing.

The range of feature sizes that can be printed with this technology is quite broad [81]. While squares and lines with geometries of several microns are easily achieved, smaller circular features with dimensions as small as 30 nm have also been demonstrated (Figure 11.19) [74].

11.4.3   INK MOBILITY AND SPEED OF INK TRANSFER

Ink diffusion from the PDMS bulk to the surface occurs during the formation of the patterned SAM on the substrate. The mobility of the ink can cause lateral spreading to unwanted regions. Upon the transfer, this spreading can influence the desired pattern.

By decreasing the stamp–substrate contact times to the range of milliseconds, Wolf and coworkers improved the uniformity and reproducibility of the printed monolayer [82]. This millisecond printing time is three orders of magnitude shorter than the usual contact time, and it appeared to be sufficient to transfer uniform and etch-protective hexadecane thiol SAMs onto a gold surface. At these very low contact times, the surface spreading of the thiol and the diffusion via ambient vapor phase did not occur. Positioning, printing, and retraction of the 100 µm-thick PDMS stamp with 1 µm wide features were realized with an automated piezoelectric actuator mounted on a motorized

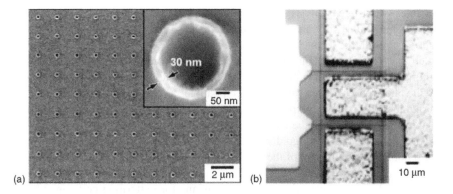

FIGURE 11.19 (a) A 30 nm ring created with μCP. (b) A working field effect transistor. (Reprinted with permission from Odom, T.W., et al., *J. Am. Chem. Soc.*, 124, 2002, 12112. © 2002 American Chemical Society.)

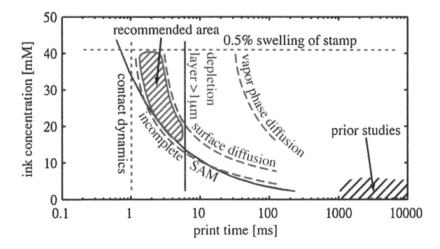

FIGURE 11.20 High-speed μCP process window as a function of ink concentration and contact time. (Reprinted with permission from Helmuth, J.A., et al., *J. Am. Chem. Soc.*, 128, 2006, 9296. © 2006 American Chemical Society.)

two-axis stage. A process window of high-speed μCP was defined (Figure 11.20), in which the recommended printing conditions are mapped by limits of the contact dynamics, the distortion of the stamp due to swelling, and the conditions for complete SAM formation and surface spreading.

11.4.4 Applications of μCP

The μCP technology has been used to make working field effect transistors, but complex electronic devices are not likely to be the strength of this particular printing technology. Although the SAMs made from thiols are easily transferred and tend to self-heal during deposition (because their binding is noncovalent), the thickness of the molecule (~1 nm) causes problems for routine pattern transfer of the thicker films typically required for semiconductor devices. Thinner films are also subject to defects, and yield is critical in the semiconductor field. In addition, the elastic nature of the PDMS template makes it impractical for the very precise layer-to-layer overlay tolerances required in the semiconductor industry. Interestingly enough, however, the same feature that renders it difficult to align one level to another also allows printing on surfaces that are not planar. This attribute is relatively unique to μCP.

For templates with dimensions as small as a few 100 nm, the master can be made at relatively low cost. PDMS material is very inexpensive, and template replication can cost as little as $0.25 per square inch. As a result, the technology may be very attractive for biological applications. It should also be noted that PDMS is biocompatible and permeable to many gases and can also be utilized for cell cultures. Conversely, many photosensitive materials used in the semiconductor industry, in addition to the photo process itself, are not biocompatible.

A large number of biological experiments have been made possible with soft lithography. As an example, patterned SAMs can be used to control the absorption of protein on certain surfaces. Using µCP, Lopez et al. patterned gold surfaces into regions terminated in oligo(ethylene glycol) and methyl groups [83]. Immersion of the patterned SAMs in proteins resulted in adsorption of the proteins on the methyl terminated regions.

It is also possible to use soft lithography as a means for patterning cells on substrates [84]. Furthermore, the shape of a cell can also be affected by the patterns created with soft lithography. Singhvi et al. used this ability to explore the effect of cell shape on cell function [85]. It was noted that cells attached preferentially to treated islands and in many cases, conformed to the shape of the island. In addition, the size and shape of the island played a role in the size and shape of the cells.

There has much recent attention in the biological field to microfluidic testing, and several groups have done groundbreaking work in this field. In the past, channels have been fabricated in fused silica using photolithographic processing and wet chemical etching of the glass. The challenge has always been to find a cost-effective method for producing channels and developing a robust process for sealing the channels. Kim et al. developed a three-dimensional micromolding process by bringing a PDMS mold in contact with a conformal substrate, thereby forming microfluidic channels [86]. The technique is extremely useful for directing specific types of cells through different channels.

A limitation for micron-size channels is the amount of turbulent flow encountered in the device. Flow is typified by a Reynolds number; a lower number being characteristic of a more laminar flow. By operation under laminar flow conditions, the opportunities for intermixing are reduced. Whitesides et al. introduced the concept of a network of capillaries to better direct cells in micron-sized microfluidic channels [87]. A simple network is shown in Figure 11.21. The technique is very useful for cell capture and subsequent delivery of chemicals for culturing. Meanwhile, different companies offer stamp replication and µCP in an automated way, although still at laboratory scale. For example, the µContactPrinter from GeSiM replicates stamp patterns in soft polymers. It is a versatile benchtop instrument for doing microfluidics by NIL and surface patterning on the nanoscale [88, 89]. In addition, some NIL tools enable soft contact imprint on large surfaces. More details about the formation of the master are covered in Section 11.5.

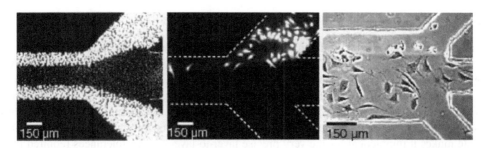

FIGURE 11.21 Laminar flow patterns in microfluidic channels created using soft lithography. (Reprinted with permission from Takayama, S., et al., *Proc. Natl. Acad. Sci. U.S.A.*, 96, 1999, 5545. © 2002 American Chemical Society.)

11.4.5 NANOIMPRINT LITHOGRAPHY

The ideal case of NIL for high-resolution patterning, that is, by using rigid stamps on rigid substrates, while the only significantly deformable material is the resist between stamp and substrate, is often only valid if the polymer layer is thick enough. In this case, the polymer flows fast enough both vertically and laterally that a constant contact between stamp and moldable material is ensured. This is basically the case for high–aspect ratio microstructures on thick polymer plates, where flexibility of stamp structures would lead to immediate failure during molding or demolding. For thin polymer layers used as resists, conformal contact is compromised if the nonideal substrates exhibit thickness variations (10 to 50 μm from center to border in typical silicon wafers), substrates are prepatterned, or particles lead to local thickness variations, which induce wedge effects that inhibit contact of the stamp with the substrate (see Figure 11.22). Molding speed is impacted by pattern size and density variations (and thus, the sinking of the stamp). To overcome these effects, the stamp has to be locally (single protrusions) or globally (backbone) flexible. With some restrictions, soft lithography can also be used for T-NIL, particularly if dense, low–aspect ratio structures have to be transferred, in which collapse can be avoided. The ability of the elastomeric stamp to restore during long imprint processes is helpful to imprint difficult surface structures even with sidewall roughness. A new application of soft lithography is the use of hybrid PDMS metal–film photomasks for large-area photolithography [90].

The elastomeric property of the stamp is helpful for the demolding of "difficult" structures, that is, structures with considerable sidewall roughness or even undercuts [91]. Due to a small elongation of the individual features during the detachment process, the structures release from the sidewalls more easily, and the stamp structures are restored after the demolding process. However, high–aspect ratio structures may rip if the demolding force exceeds the cohesion strength of PDMS.

FIGURE 11.22 Particle or defect issues on semiconductor substrates can be minimized with a flexible stamp. The two examples show a defect sticking out of the pattern and a defect entirely covered by nanostructures (SCIL composite stamp). (Reprinted from *Microelectron. Eng.*, 87, Ji, R., et al., UV enhanced substrate conformal imprint lithography (UV-SCIL) technique for photonic crystals patterning in LED manufacturing, 963–967, © 2010, with permission from Elsevier.)

11.4.6 RELATIVE DEFORMATION AND STAMP ISSUES

The main problems of resolution are as follows: a) large cavities will bow if not supported by auxiliary structures; b) high–aspect ratio pillars (>1) will deform upon pressure or collapse due to shear [92, 93]. Apart from this, there are difficulties in replicating the viscous PDMS from small stamp structures, such as small cavities, and shrinkage of a few percent is detrimental to precise patterning. Harder silicone rubbers with mechanical strength of about 20 MPa are becoming popular (h-PDMS) but are more brittle and therefore prone to damage during demolding. Therefore, particularly for NIL processes, hybrid stamps, that is, a hard surface topography on a soft backbone, are now used. Often, these soft layers are coated on hard or flexible backbones, such as thin glass sheets. For specific NIL processes such as substrate-conformal imprint lithography (SCIL), inflatable molds have been developed, which enable the subsequent patterning from one side of an extended substrate to the other while squeezing out the air to one side. In general, PDMS is considered porous, which means that it is permeable to different gases (including oxygen, which may inhibit polymerization for specific polymers), and also, small molecules (thiols and resist monomers) can be absorbed (soaked in). This makes it possible to transfer ink not only from the surface but also from the bulk, which is needed as a reservoir for repeated printing before recoating is done. In the case of NIL, the porosity of the PDMS has two consequences: a) it can absorb small resist molecules that may alter the stamp after some imprints, and b) the oxygen penetration makes it necessary to use specific resists that are not oxygen-inhibiting. Otherwise, only partial polymerization will occur. Nowadays, UV-curable PDMS variants are available with a range of mechanical properties, and resists are available with adapted molecular sizes, reducing uptake.

11.4.7 STAMP DEFORMATION

During direct contact between the stamp and the surface, one must be careful, because the stamp can easily be physically deformed, causing printed features that are different from the original stamp features. Horizontally stretching or compressing the stamp will cause deformations in the raised and recessed features. Also, applying too much vertical pressure on the stamp during printing can cause the raised relief features to flatten against the substrate. These deformations can yield submicron features even though the original stamp has a lower resolution.

Deformation of the stamp can occur during removal from the master and during the substrate-contacting process. When the aspect ratio of the stamp is high, buckling of the stamp can occur. When the aspect ratio is low, roof collapse can occur. A correlation between the minimum feature size for linear gratings and the system energy responsible for stability and collapse is drawn in Figure 11.23 and demonstrates the need for h-PDMS with 11 MPa or even x-PDMS (coined by Philips for extreme PDMS) with up to 40 MPa. According to Ji [61], structures collapse if $E_{surface} < E_{mechanical\ deformation}$, which in the case of h-PDMS occurs for grating pitches below 200 nm and for x-PDMS below 40 nm. While silicone rubbers with Young's modulus 60–80 MPa can be manufactured, they are still able to conform to surfaces but are very brittle.

For unwanted surface defects and for imprinting over corrugations, different deformation schemes have been developed; see Figure 11.24.

11.4.8 CONTAMINATION AND SWELLING

During the curing process, some fragments can potentially be left uncured and contaminate the process. When this occurs, the quality of the printed SAM decreases. When the ink molecules contain certain polar groups, the transfer of these impurities is increased. Swelling of the stamp may also occur. Most organic solvents induce swelling of the PDMS stamp. Ethanol in particular has a very

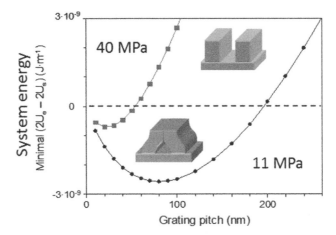

FIGURE 11.23 System energy of h-PDMS with 11 MPa and x-PDMS with 40 MPa. Collapse is likely to occur in the case of h-PDMS for grating pitches below 200 nm and for x-PDMS below 40 nm. (Reprinted from *Microelectron. Eng.*, 87, Ji, R., et al., UV enhanced substrate conformal imprint lithography (UV-SCIL) technique for photonic crystals patterning in LED manufacturing, 963–967, © 2010, with permission from Elsevier.)

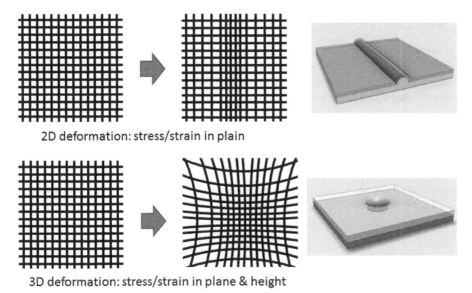

FIGURE 11.24 Schematic of a full-wafer soft UV-NIL process for nonideal substrates, for example, 2-D deformation on cylindrical surfaces and 3-D deformation on lens-like spherical surfaces. (Reprinted from *Microelectron. Eng.*, 87, Ji, R., et al., UV enhanced substrate conformal imprint lithography (UV-SCIL) technique for photonic crystals patterning in LED manufacturing, 963–967, © 2010, with permission from Elsevier.)

small swelling effect, but many other solvents cannot be used for wet inking because of high swelling. Because of this, the process is limited to polar inks that are soluble in ethanol. When PDMS is used in NIL, the patterning of viscous resists with solvents or monomers and oligomers that can be crosslinked can lead to soaking of the stamp due to its porosity. Specific resist chemistry can help to reduce the risk of penetration into the stamp and thus, an accumulation of residues that will impair the surface properties and the dimensional stability of the stamp.

11.5 NANOIMPRINT LITHOGRAPHY—BASIC CONCEPTS AND TOOLS

In the past, various tool concepts have been developed to fulfill the needs of customers (research and industry) and end users (for a focused or wide range of applications). Although often different in terms of capabilities and processes, the basic concepts have much in common, and many efforts have been undertaken to adapt different processes to the requirements of various applications (see Table 11.2). The J-FIL tool, initially developed by Molecular Imprints (MII), now Canon Nanotechnologies Inc. (CNT) [56, 94], is not different in this respect. It is currently the only process directed toward the HVM of semiconductor chips, and because of the direct comparison with other (NGL) processes covered in this book, it is described separately in Section 11.6. In a similar way, R2R NIL is in many respects different from wafer-based lithography and is therefore described in Section 11.7. The following aspects will be covered:

- Tools for thermal and UV-NIL, large area, S&R
- Stamps, masks, and molds, including antisticking layers (ASL)

After a general introduction to basic concepts, examples are given of tools currently advertised in the suppliers' web pages as well as some older tools still in operation (see Table 11.3). This includes upscaling, large-area imprint, high volumes manufacturing and automation.

11.5.1 BASIC CONCEPTS

A mold is considered as the component of the molding tool (a tool insert) that is in touch with the moldable material during the entire process, that is, initial contact and/or pressing resulting in material displacement and separation (demolding). In HVM, it survives the molding process unaltered and uncontaminated and thus, can be reused many times after each molding step. In this way, many

---

TABLE 11.2
Hard and Soft Tool Concepts and Setups for NIL

	Hard stamper + compliant layer	Soft pressurized membrane	Soft (flexible) stamp	Hard stamper step and stamp
Mode	Pressure equilibration due to elastic compliance and bendable stamp	Elastic and inelastic pressure equilibration and bendable stamp	Conformal contact of elastic protrusions on bendable (hard) backbone	Hard stamp active + passive gap equilibration + planarization strategy
	Full-wafer (large-area) single imprint		**Segmented multiple imprint**	
Schematic				
Specific Advantage	Air-induced automated demolding	Isotropic pressure by compliant membrane	Conformal contact by elastic stamp	Inkjet dispensing with density modulation

Source: adapted from Schift, H., *J. Vac. Sci.*, 26, 458, 2008. © 2008 with permission from *Journal of Vacuum Science and Technology.*

TABLE 11.3
Tools Available from Different Suppliers

Name (Tool)	Basic Specifications	Concept	Remarks
Aurotek Corp.			
Nanosis 610	Ø 150–200 mm wafers		For PSS (high-volume manufacturing)
EVGroup			
EVG520HE, semi-automated hot embossing system	Up to Ø 200 mm, T-NIL	Hard stamper; converted anodic bonder	Pneumatic de-embossing options for high–aspect ratio structures; prealignment possible in combination with mask aligner
EVG620, UV NIL system	Ø 200 mm wafers, SmartNIL process	Converted mask aligner, SmartNIL	Normal use as mask aligner
EVG720, semi-automated UV NIL system	Ø 200 mm wafers, 7200LA for Ø 370×470 mm^2 (Gen2)	SmartNIL with a hydrophobic polymer foil	Integrated into HERCULES®NIL within INSPIRE cluster for PSS (high-volume manufacturing)
EVG770, automated UV-NIL stepper	Ø 100 mm up to Ø 300 mm wafers	Vacuum imprinting on a spun-on resist	Alignment system within ±500 nm, contact-free wedge compensation
GDnano			
Soft NIL Tool GD-N-03	Ø 100 mm wafers, UV-NIL	Membrane for pressure equilibration	
Hitachi			
Photonanoimprint system for nanofabrication	Ø 300 mm wafers, UV-NIL	HiDAF (high-pressure direct air flow) cushion	C2C cassette-to-cassette wafer handling
Nanonex			
NX-2608BA	Ø 200 mm, T-/UV-NIL	Air Cushion Press™ (ACP)	With front- and backside alignment
NX-3000	Ø 200 mm, S&R, UV-NIL	Enhanced ACP	Sub-70 s per imprint field, upgradable to resist-drop dispensing
NILT			
CNI v2.0	Ø 100 mm and Ø 150 mm wafer NIL tool, T-/UV-NIL vacuum	Desktop, manual stamp loading and unloading	T-NIL (up to 200 °C), laboratory, new version CNI v3.0 with Ø 200 mm
Obducat			
EITRE® LA NIL	T-/UV-NIL	IPS® material is both stamp and membrane	Roll application and demolding semi-automatic
SINDRE®, 400/600/800	Ø 100-200 mm	IPS®/STU® process	For PSS (high-volume manufacturing), 30 w/hour, can be upscaled to Ø 150 or 200 mm
Philips SCIL Nanoimprint Solutions			
AutoSCIL, C2C cassette-to-cassette tool	Ø 150 mm, UV NanoGlass (sol-gel resist)	Designed for 40 wafers/h, composite stamps	AutoSCIL 150 mm high-volume machine, C2C cassette-to-cassette, application patterned sapphire, tool, upscalable to Ø 200 mm wafers

(Continued)

TABLE 11.3 (CONTINUED)
Tools Available from Different Suppliers

Name (Tool)	Basic Specifications	Concept	Remarks
SUSS			
SUSS Smile, based on MA/BA8 mask aligner	Ø 200 mm wafers	Hybrid PDMS stamps (x-/ soft PDMS on glass)	For microlenses, wafer-level cameras and image sensors, normal use as mask aligner
SUSS SCIL, based on MA/BA8 mask aligner	Ø 200 mm wafers	Converted mask aligner	For high-brightness LEDs, photovoltaics, MEMS, normal use as mask aligner, cascaded pressure system
SUSS/SET NPS300	Ø 100 mm wafers, T-/UV-NIL	Converted flip-chip bonder with rotation head	Small stamp areas (a few square millimeters) for T-NIL due to limited press force
ThunderNIL			
Ulysses	Ø 200 mm wafers, T-NIL	Pulsed NIL with heatable stamps	150 µs fast imprint with T > 300 °C, different applications, now also available as S&R
Toshiba Machine			
ST50S-LED, Gigalane, Korea	Ø 150 mm wafers, UV-NIL	Imprint with polymer mold	Polymer stamps made by R2R for PSS; high-temperature T-NIL machine for glass molding, micropatterns
PTMTEC (R2R equipment)			
R2R100T and R2R100UV	R2R (T- and UV-NIL)	Reverse gravure coating unit	Imprinting on plastic films, laboratory

identical replicas can be drawn (copied) from one mold. Due to the conformal molding, i.e., the intimate (conformal) contact of the stamp with the surface, the surface of these copies is the negative structure of the original (topography with inverted polarity). Therefore, a true replica of the mold is generated when a negative is again molded into a positive structure. Here, we use the terms *replica* and *copy* in the more general sense by which negatives also are considered as true copies of an original. While the stamp is mechanically "passive," the imprint tool is the "actuator," which ensures the contact of both stamp and substrate and enables a small movement in the vertical (Z) direction. This transfer is parallel, i.e. synchronized over the entire stamp surface (X and Y), ensuring contact of all stamp protrusions and the initiation of the molding by which the cavities of the stamp topography are homogeneously filled with the viscous material. As said before, this requires either totally flat and parallel surfaces or a certain ability of the stamp, substrate, or moldable material to equilibrate any undulations and deviations from the ideal form. It further involves handling issues such as insertion and exchange of the stamp and/or substrate, alignment, and demolding, and often also control of defects and molding quality. In the best case, this is all included into a single tool. However, tool providers have also found solutions in which the alignment and the demolding can be done in separate tools. For example, they have converted their anodic bonding tools into thermal imprint machines and used the corresponding mask aligners for alignment of stamp and substrate. This is then—clamped as a sandwich in a holder—transferred to the bonding tool. Often, tools are placed within an automated process chain, which includes robotic manipulation of wafers and prealigned stacks. Furthermore, a proper macro- and microenvironment has to be ensured, which ranges from reduction of contamination and stray light to the inclusion of vacuum and specific gases and the addition of stamp cleaning facilities. The term *embossing* is often used if polymer sheets or plates are structured with microstructures by thermal molding processes, such as fluidic devices or refractive/diffractive optical elements.

11.5.2 IMPRINT TOOLS

NIL is a highly dynamic process whereby the vertical sinking movement of a stamp is transformed into a 3-D flow with large lateral flow components. In T-NIL, the speed of the pressure buildup at the backside of the substrate and stamp and equilibration of local homogeneities over a large surface can influence the mode of cavity filling. This determines how the stamp will bend during sinking. To cope with this inhomogeneity, different machine and tool concepts were developed, which involve hard (such as silicon and metal) and soft elements (such as silicone rubber and bendable membranes) [18]. Mechanical nanofabrication techniques based on molding therefore need tools and materials with matched mechanical properties. The mold has to be made from a material that is sufficiently hard to maintain the structural resolution and fidelity during the mechanical contact and to sustain a required number of processing cycles without altering both structure topography and surface chemistry. This is around 100 for research but several thousands or tens of thousands in production. At the same time, the imprint tool has to ensure an intimate contact between stamp surface and the material to be deformed or displaced, that is, the thin resist layer. This contact has to be maintained while the resist surface conforms to the topography of the stamp, thus replicating it into the exact negative of the topography. It means that the imprint tool has to ensure this conformal contact during the imprint time, under hard constraints of lateral alignment, sometimes with a precision much lower than the smallest feature size, and also of short imprint cycles, which can be down to a fraction of a second. The tool has to keep the stamp either leveled (i.e., a hard die with constant wedge) or adapted (i.e., a bendable stamp in pressure-controlled, equilibrated distance).

Any kind of conformal contact can be ensured with a high enough pressure: silicon stamps can bend 100 nm down or up within a lateral distance of few 100 μm, which can be visualized in the interference color variations present in imprints of 0.5 mm silicon wafers [19]. The bending can be ensured by using a pressure mechanism, which includes soft elements that translate the hardness of a solid plate into a non-flat surface, or by bending membranes or the entire stamp using air pressure. Table 11.2 shows schematics of different concepts in which hard and soft elements are combined. Typically, for T-NIL and resists with moderate viscosity at elevated temperatures, pressure equilibration is done using compliance layers and pressurized cushions with up to 10 MPa (100 bar or 1450 psi). For low-pressure imprint (e.g., using liquid or low-viscosity resists), active and passive gap equilibration is needed. All setups allow for the integration of UV-exposure setups for resist curing, which is done by shining collimated light through the transparent stamp. Therefore, combined processes are possible. In addition to components for heating, exposure, and pressure, alignment (either prealignment in a mask aligner or within the press), demolding, and handling subsystems are often included in the system.

11.5.3 THERMAL IMPRINT TOOLS

Thermoplastic molding processes are widely used in industry, and tools range from simple "pill-presses" to sophisticated R2R imprint tools [18]. T-NIL profits from the huge toolbox of industrial thermoplastic processing; it is therefore often nearer to well-established injection molding of CDs (120 mm diameter, 1.2 mm thick transparent disks in polycarbonate with micron-sized pits on the surface) than to semiconductor wafer processing. The tools and the process can be inherently cheaper due to the lack of expensive optics and sources, particularly if they are used in a research-like environment for small-scale production. T-NIL has found its application mostly in areas where large areas need to be patterned at reasonable cost; flexibility is needed, such as when patterning functional materials with thermoplastic behavior; or micro- and nanostructures need to be combined (e.g., microfluidics). Several tools have been commercialized, and some are now used for mass fabrication. Some came from micromachining or microembossing, for example, of microfluidic or micro-optical devices (such as the Jenoptik [95] HEX line; see Figure 11.25), but some were also derived from anodic bonders (such as EVG [55] and SUSS MicroTec [96]). S&R tools were derived from flip-chip bonders, such as SUSS MicroTec, which transferred the technology to SET [97]. The first Obducat [16, 17, 54] tool was derived from a CD molding machine. The requirements for optical elements, photonic devices, filters, and other high-end products, along with the development of emerging markets, are likely to sustain the imprint tool business regardless of whether it can address the needs of CMOS chips. Several vendors are supplying thermal imprint systems to customers. Most of them are for full-wafer single-imprint use; that is, the stamp has the same size as the substrate, and the entire surface area is patterned in a single step, where the substrate and the stamp are sandwiched between two plates (see Figure 11.26). Since typical thermoplastic materials have glass transition temperatures T_g above the temperature of use (i.e., T_g of 60 to 140 °C), an imprint temperature of up to 200 °C at a pressure up to 10 MPa is often sufficient to achieve a patterning by squeeze flow, that is, with significant ability to shape the material's surface without excessive lateral flow. Below 200 °C, the degradation of standard thermoplastic polymers, fittings, antiadhesive coatings, etc. can be avoided or greatly reduced. Furthermore, oil or even water cooling is still applicable. Temperatures up to 400 °C and even higher can be employed, for example, for glass embossing. A variothermal process scheme involves heating and cooling the entire "mold–polymer–substrate sandwich". Due to the intimate contact at high pressure, these cycles typically involve not only the heating and cooling of the "sandwich" but also the tool's steel plates with their considerable heat capacity. Therefore, it is advisable to keep this temperature range small, particularly the maximum temperature, because heat and cooling rates are machine limited. Typical cycle times are 5 to 20 min in laboratory-type machines, which includes the time of a few tens of seconds up to a few minutes for imprint and a large "overhead." Newer developments with heatable stamps from ThunderNIL [25, 26, 99] allow short pulsed operation. Here, temperatures of above 400 °C are achieved within microseconds only in the area of direct contact of stamp to polymer. This is an

FIGURE 11.25 Jenoptik HEX03 thermal imprint (hot embossing) tool at PSI, Switzerland, representing the hard stamper + compliant layer principle in Table 11.2.

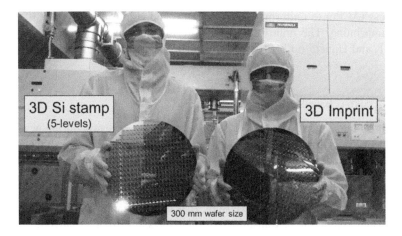

FIGURE 11.26   300 mm stamp and imprinted substrate processed with an EVG 620HE thermal NIL tool. (Courtesy of S. Landis, CEA-LETI, Grenoble, France.)

advantage, as the heat is confined not only to the stamp surface but also to the polymer in contact with the stamp. Therefore, the amount of material to be displaced is limited. This reduces the printing time to 100 μs and eliminates the tool's overhead, thus enabling very fast cycle time, determined mostly by the time of handling. This would be particularly attractive for S&R applications, where interferences with neighboring imprints can be excluded if heat dissipation stays local.

A non–room temperature process is particularly critical for all processes where lateral alignment needs to be taken into account. At the same time, heating and cooling cycles can be easily controlled by simple setups consisting of heatable plates that can be scaled up (using hot oil or electrically heated cartridges). Large areas can be controlled, because the high pressure ensures conformal contact over entire wafer areas (which are not perfectly flat due to warping). Particularly difficult is the achievement of a homogeneous residual layer thickness in the imprinted resist. This is dependent on the ability of the stamp to displace the resist below the stamp protrusions and thus, the stamp design (area fill factor of protrusions and their lateral size, depth, and distribution), mostly in the micro range (i.e., protrusion sizes of a few tens of μm). Designs need to be adapted to ensure sinking of all stamp protrusions with similar speed and thus, a moderate distribution of residual layer thicknesses. For this, rules of thumb are often sufficient, along with imprint strategies from ensuring a high enough height budget or by imprinting into very thin layers of resist with high etch selectivity. Since this is sometimes not intuitive, in the future, computer-based design optimization will play an important role.

As many as four different companies in Europe are currently (2020) selling imprint tools: Obducat [54], EV Group [55], SUSS MicroTec [96], and NILT [100], with others, such as Jenoptik [95], having left this field. Entry into the imprint business was an obvious extension for all four European companies, with many institutes adding NIL tools as a complement to their mask aligner cleanroom equipment. In Northern America, tools from companies such as Nanonex [53] have found wide distribution. From Asia, there are several tools available (Aurotek [101], AimCore [102] [Taiwan], Gdnano [103] [China], SaeHan [104] and Hunet Plus [105] [South Korea], Toshiba [106], DNK [107], Scivax [66], Kyodo [108], and Hitachi [109] [Japan]), and probably many more (because some of them only sell tools to their local market).

One concern during the imprint process is pressure nonuniformity. Pressure nonuniformity can lead to serious errors, such as unfilled imprint areas and nonuniform residual layers [53]. This can easily occur in a system that relies on alignments of two parallel plates to make contact between the template and the wafer. Other issues that may occur during the process include surface roughness and bow in either the template or the wafer. To minimize these effects, the Nanonex tools apply an air-cushion press (ACP*) to uniformly apply pressure and conformally contact the template and wafer during imprinting. The improvement in uniformity is substantial. It should be noted that Obducat, NILT, and Hitachi (to be discussed) use similar technologies to achieve good pressure uniformity.

11.5.4 UV-Assisted Imprint Tools

The integration of lamps and exposure systems for the curing of resists during or after imprint is an integral part of UV-NIL. The requirements are similar to those of mask aligners, and some of the machines have been developed on the basis of aligners, with μm-alignment, moderate pressure (0.1 MPa), and highly collimated i-line (365 nm) and g-line exposure ($\lambda = 435$ nm) from highly collimated light from Hg-gas discharge lamps. However, since no mask-related proximity effects have to be considered, the requirements for exposure, that is, collimation and uniformity, are slightly relaxed. Collimated light has to be transmitted through the stamp (or alternatively, through the substrate). Fresnel-type diffraction at borders and structural protrusions and interference effects within the resist have to be excluded, for example, near vertical sidewalls or due to thicker residual layers in which standing waves can form. Monochromatic, multichromatic, and wideband light sources can be used, with wideband sources advantageous because of elimination of diffraction and interference effects. High-brightness UV-LEDs with 365 nm wavelength have become options for compact designs with limited collimation requirements.

In contrast to T-NIL, where thermal expansion has to be compensated for sub-50 nm alignment, alignment for UV-NIL has several advantages. Because of the transparent molds and the liquid resists, lateral alignment is done through the stamp, in proximity or in "lubricated" liquid contact while the stamp and the wafer are held at a fixed vertical distance. A self-leveling flexure aids the parallel alignment. The forces on the stamp protrusion are induced by capillary action and are therefore low.

Different groups have performed alignment studies. Sub-150 nm lateral alignment was achieved over an area of 25×25 mm² and sub-250 nm over the entire area of a 100 mm wafer using simple low-resolution stages without temperature control or wafer–stamp mismatch compensation [110]. With advanced Moiré fringe techniques, a sub-20 nm overlay in NIL is possible [111].

11.5.5 Stamps for Thermal and UV-Assisted NIL

As has been done with imprint tools, different companies are offering know-how and services for the fabrication of stamps. Often, they are offering tooling capabilities and resources from public or clean rooms shared with universities. Apart from the imprint tool providers NILT and Nanonex, these include, among others, Eulitha [112], Soken [113], temicon [114], and IMS Chips [115].

Stamp material: Micro- and nanomachining using standard silicon wafers is the preferred way of manufacturing wafer-sized stamps. They can be coated and patterned with standardized holders and do not need any precautions due to impurities. Vertical sidewalls are needed for high resolution, but some applications require draft angles of 2–5°. While the etching of silicon is well established in many research labs, the processing of quartz and fused silica is often difficult. This is mostly because silicon can be manufactured more purely than fused silica, in which inhomogeneity (impurities and crystalline areas) is often present. Meanwhile, specialized companies offer fused silica templates with high mechanical and optical quality, including mesas needed for step and repeat processes. An additional advantage of silicon for stamps in thermal imprint is the identical thermal expansion of both stamp and substrate in the variothermal process (i.e., long thermal cycles in which stamp, resist, and substrate are kept in thermal equilibrium). In contrast to this, the use of nickel stamps on silicon substrates would lead to significant lateral stress, particularly if large-area stamps

---

* ACP is a trademark of Nanonex.

are used. Composite stamps, that is, with a backbone made from silicon or borosilicate glass (Pyrex/Borofloat) with thermal expansion adapted to silicon, are advantageous. The lateral thermal expansion is defined by the (thick) backbone rather than by the thin coating made of sol-gel, hydrogen silsesquioxane (HSQ) [71], Ormocers (hybrid organic–inorganic material) [51], polymer, PDMS, diamond, diamond-like carbon (DLC), and glassy carbon [116]. Composite and layered stamps are therefore a good choice if cost issues are prerequisite, because they enable both good thermomechanical properties and ease of replication, that is, for use as auxiliary stamps. One particular approach is that of a "lost mold," which means that the stamp is only used once, or—in the case of a material that can be easily dissolved—demolding with undercuts can be done while a new stamp is needed in every cycle.

While silicon has a Young's modulus similar to that of steel, its brittleness and susceptibility to breaking due to notching are often prohibitive for use under high mechanical and thermal stress. Stamp breaking can often be avoided by careful handling (no cracks and scratches), insertion without excessive mechanical stress (clamping), and contamination control (large particles exerting point stress). During demolding, the delamination is enabled by the ability of the stamp to bend, which is different if an entirely stiff setup is used. Soft stamps, that is, stamps made from elastomeric materials, have become highly attractive for UV-NIL. However, the popular Sylgard 184 from Dow Corning [71], a UV-transparent rubber, has severe limitations in resolution and dimensional stability, which is limited to about 1 μm. By diluting the PDMS precursor (with toluene) or pressurized filling of stamps, resolutions down to 100 nm can be achieved. Further improvements down to 10 nm can be achieved in a hard material (Ormostamp, h-PDMS) on a soft backplane.

Antisticking layers (ASL): While a spin-coated polymer needs to adhere well to the substrate, it needs to be released without damage from the stamp after imprint. Typically, due to the surface topography, the surface of the stamp is larger, and the sidewalls of each cavity will lead to enhanced friction. Therefore, mold release strategies have to be employed, such as an antiadhesion coating on the stamp or internal mold release agents in the resist [18, 51]. For stamps, monomolecular films based on perfluorinated hydrocarbon chains serve as an ultrathin antiadhesive coating; with a typical thickness around 1 nm, they do not modify the surface topography of the stamp. The main goal of surface energy engineering of a stamp is to achieve good wetting with the viscous polymer during molding, low adhesion to polymer during release, and low friction during sliding, for example, during the separation of stamps with vertical sidewalls. A good antiadhesive property will also reduce contaminants from adhering to the stamp and enable cleaning with dry air or solvents. In many cases, a hydrophobic property (associated with high water contact angle) can be associated with a good antiadhesive property against standard polymers. In this way, a 120° water contact angle, typically achieved with fluorinated carbon chains, results in a 90° contact angle against poly(methyl methacrylate) (a standard electron beam and thermal NIL resist). Silane-based coupling groups are particularly suitable for silicon and quartz, while phosphate- or phosphonate-based coupling groups are suitable for metals such as nickel [118]. Degradation of this binding or of the carbon chain leads to reduction of the antiadhesive properties and eventually to failure. This is due to the high thermal and mechanical load in T-NIL and also in UV-NIL due to the interaction of the coupling group with the photochemistry of the molding material. This was investigated by the groups of Helmut Schift, Frances Houle, and Marc Zelsmann [117–119]. The chemicals can be purchased from chemical vendors (ABCR, Aldrich, etc.) or from specialized companies (Optool DSX from Daikin [120, 121] and BGL-GZ-83 from Profactor [122, 123]). Today, apart from good ASL coatings and stamps with inherent antisticking properties (glassy carbon, fluorinated diamond-like carbon, and fluorinated polymers), antiadhesive agents are added to the resist materials to ensure a constant demolding property irrespective of the quality of the stamp [124, 125]. Apart from this, flexibility (softness) of the stamp helps to ensure release and avoid breaking due to localized excessive forces during demolding.

Fluorinated ASLs on stamps by silane chemistry need to be renewed after a number of imprints (typically 100 to 1000). With the current resolution of 10 nm or more, the added thickness of the silanes (1 nm) is still negligible. The same accounts for the "etching" during repeated cleaning and recoating cycles that modify the stamp geometries. In contrast to this, resists with internal release properties enable better control of release over time, with no additional ASL needed on the stamp surface.

11.5.6 NANOIMPRINT LITHOGRAPHY SYSTEMS

In the following, three systems are presented in more detail. The current development is going toward automated handling or even wafer track systems. Most tools can be upgraded with a Class 1 (ISO 3)-capable mini-environment to guarantee the lowest defect rates. For achieving high throughput, often clusters are arranged that allow parallel manufacturing.

Obducat was one of the first companies to address the pressure uniformity issue using pressurized membrane-based technology, called SoftPress®. NanoNex (Air Cushion Press™ [ACP]), Hitachi, and NILT have adapted similar schemes, either with thin membranes behind the hard stamp (e.g., aluminum, polymer foil, or a thick rubber-like membrane) or by patterning an intermediate polymer stamp (IPS®*) from an original by T-NIL into a thin thermoplastic foil. Such a bendable membrane-based stamp can be easily demolded from the imprinted resist by delamination. The disadvantages are that only one side of the substrate is heated and that sticking of the foil at the wafer boundaries has to be avoided. The technique was first used for T-NIL, but for higher throughput and stability of the polymer matrix, it was converted into a combined thermal and photolithographic step, called simultaneous thermal and UV (STU®), in which a curable thermoplastic resist (e.g., mr-NIL 6000 from micro resist technology GmbH) is patterned by thermal NIL and crosslinked by simultaneous or consecutive UV exposure. The process has the additional advantage that spin-coated films can be used; however, the use and disposal of single-use stamps presents a severe drop-back in terms of cost and waste.

Obducat set up an automated wafer handling system. The SINDRE® cluster uses Obducat's proprietary STU and IPS process on 100 mm sapphire substrates up to 60 wafers/hour. Here, the IPS is produced in a continuous foil and used in the second process as a single-use stamp. Such a track system is used in Taiwan for LED manufacturing for micropatterned sapphire substrates (μPSS). The additional cost for patterning a LED using μPSS was US$0.5 in in 2012 and is projected to be US$0.21 in 2018, which means that the severe cost issues call for inexpensive solutions that are upscalable [126]. The industry, which currently (2020) uses 100 mm wafers in Taiwan, has already adopted 150 mm wafers (mostly in Europe). For this purpose, the currently used "retired" photolithography steppers with 50 mm wafers and a throughput of 25 wafers/h are too costly. Obducat has also confirmed the use of its SINDRE system by Kimberly Clark for the fabrication of surface-patterned microneedle arrays.

SUSS MicroTec's substrate conformal imprint lithography (SCIL)† circumvents problems associated with the controlled imprinting of viscous UV-curable resists [61, 62]. It combines the advantages of both soft and rigid stamps, allowing large-area patterning and sub-50 nm resolution to be achieved at the same time. Cushion techniques are typically used for ensuring a homogeneous pressure on stamps by equilibrating pressure drops (e.g., by fast sinking of specific stamp areas) using constant pressure on the back of the stamp. In contrast to this, the division of the total area of a rigid backbone into subareas that can be locally addressed by using valves combines movement and buildup of pressure while the advantage of constant local pressure can be maintained. Such a system was built at Philips by Marc Verschuuren and termed *SCIL*. It is based on a locally inflatable mold in which different compartments of the mold can be moved toward the substrate and retracted (for a process scheme, see Figure 11.27). If this is done in a coordinated way (e.g., by addressing consecutive segments that move toward or away from the substrate), the imprint

---

* IPS and STU are trademarks of Obducat.
† SCIL is trademark of Philips and SUSS.

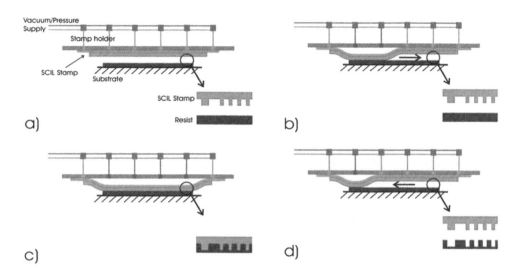

FIGURE 11.27 Schematic of the SCIL process developed by Philips. In the initial state (a), the stamp is separated from the resist. In step (b), the stamp is pushed toward the resist. In (c), after complete molding is achieved, the resist is cured either by UV exposure (through the stamp) or by drying (water take-up by stamp). In step 4, the stamp is delaminated from the hardened resist. (Reprinted from *Microelectron. Eng.*, 87, Ji, R., et al., UV enhanced substrate conformal imprint lithography (UV-SCIL) technique for photonic crystals patterning in LED manufacturing, 963–967, © 2010, with permission from Elsevier.)

can be directed from one side of the substrate to the other. This approach needs a hybrid stamp fabrication with a hard, 200 µm-thick glass backbone and a 200 µm-thick PDMS layer as the membrane (see Figure 11.28). For high resolution, this soft cushion is coated with a hard PDMS (h-PDMS or x-PDMS). As a resist, either UV-curable materials are used or sol-gel materials, which can be directly patterned without the use of light. Resolutions of below 10 nm were obtained (see Figure 11.29). The main advantages are:

a) During molding, air can be pushed out to the sides of the substrate. Deformation of single stamp protrusions can be controlled because shear is minimized.
b) Demolding is done by peeling; that is, the local retraction of the stamp makes demolding in a line-like fashion possible, enabling the filling of cavities with air.

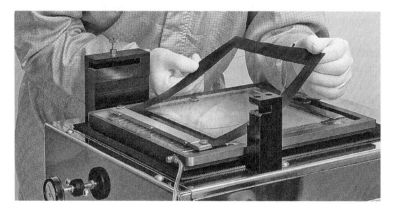

FIGURE 11.28 A SCIL composite stamp inserted into a SUSS mask aligner with SCIL insert for UV-NIL. (Courtesy of E. Storace, SUSS MicroTec Lithography GmbH.)

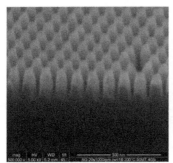 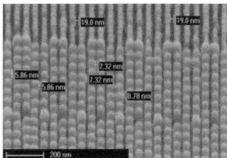

Etched silicon: 10 nm gaps   Etched quartz: 6 nm gaps

FIGURE 11.29 Results of the SCIL process after imprint and etching. (Reprinted from *Microelectron. Eng.*, 87, Ji, R., et al., UV enhanced substrate conformal imprint lithography (UV-SCIL) technique for photonic crystals patterning in LED manufacturing, 963–967, © 2010, with permission from Elsevier.)

The entire setup can be integrated into typical mask aligners. This was done by Philips in conjunction with SUSS MicroTec. Similar techniques, often only using a central valve behind a PDMS stamp, enable the printing from the center of a stamp to the border. The advantage of SCIL is that the movement of the stamp can be ensured by retracting only the area that is being addressed.

The SCIL process was implemented as an option in a standard SUSS mask aligner in 2009 and is available on MA/BA6, MA/BA6, and MA/BA8 aligners. With this technology, substrates up to 200 mm can be patterned with features down to a few nanometers' resolution, delivering unique uniformity of the imprint and the residual layer. SCIL is applied in diverse fields, including HB LEDs (patterned sapphire substrates and photonic crystals), VCSELs (polarization gratings), photovoltaics (antireflective surfaces), microelectromechanical systems (MEMS), nanoelectromechanical systems (NEMS), and mass production of optical gratings for gas sensing and telecommunications. While for VCSEL production, 600 patterned wafers within 4 years with up to 80,000 VCSELs each does not yet represent mass fabrication, SCIL is currently adapted for other applications, too. For patterning of PSS, sol-gel resist has proved to have an etch selectivity of 1. Any kind of solvent can be absorbed by the porous PDMS. The process has the potential to be used for large-area NIL, and the enterprise Philips SCIL Nanoimprint Solutions is now working on a tool with 150 mm that can be upscaled up to 200 mm [127, 128]. In comparison to other techniques, the stamp manufacturing is more complicated than with standard molding schemes. The advantage is that once the alignment between the backplane of the stamp and the substrate has been done, the lateral displacement of the stamp during the process can be largely controlled. Additionally to SCIL, SUSS MicroTec has also implemented simpler process scheme, called SMILE (SUSS MicroTec imprint lithography equipment).

EVG's SmartNIL process: The UV-NIL system called HERCULES®NIL is based on a modular platform for proximity lithography with an integrated UV-NIL solution for wafers up to 200 mm (see Figure 11.30). EVG employs a proprietary SmartNIL™* imprinting technology, which uses multiple-use soft stamps (made from imprinted thermoplastic polymer foil with hydrophobic properties) for HVM and a wafer track with integrated cleaning, resist coating, and baking preprocessing modules. The tool can be upgraded with a Class 1 (ISO 3)-capable mini-environment to guarantee the lowest defect rates. According to the company's data, it can handle substrate sizes up to 200 mm with a throughput of up to 40 wafers/hour and alignment < 3 µm. The polymer stamps can be used for >100 imprints, and automated separation is supported.

---

* SmartNIL is a trademark of EVG.

FIGURE 11.30 HERCULES® Lithography Track System from EVG for pressure sensor manufacturing. The tool for 200 mm NIL looks similar from outside. (Courtesy of T. Glinsner, EV Group (EVG), St. Florian, Austria.)

In 2015, CEA-LETI (France) and EV Group (Austria) launched a new program in NIL, called INSPIRE, to demonstrate the benefits of NIL and spread its use for applications beyond semiconductors [98, 129]. In addition to creating an industrial partnership to develop NIL process solutions, the INSPIRE program is designed to demonstrate the technology's CoO benefits for a wide range of application domains, such as photonics, plasmonics, lighting, photovoltaics, wafer-level optics, and biotechnology. CEA-LETI and EVG jointly support the development of new applications from the feasibility-study stage to supporting the first manufacturing steps on EVG platforms and transferring integrated process solutions to their industrial partners, thus significantly lowering the entry barrier to the adoption of NIL for manufacturing novel products.

Apart from these new NIL equipment platforms, replication-based imprint-like processes have already been used in manufacturing for years.

Jenoptik's hot embossing process: As part of the large community using the so-called *LiGA* process (stands for lithography, electroplating, and molding), Jenoptik [95] was one of the first suppliers to address the lithography and replication of high–aspect ratio microstructures. Its HEX hot embossing tools are based on material testing machines and due to their screw-type actuator scheme, enable distance-controlled imprint and demolding, particularly suited for molds with deep, vertical trenches. Particularly important in molding high–aspect ratio microstructures is the distance and force–controlled embossing (imprint) and the equally gentle retraction of the stamp, often supported by pressing air between the stamp and the molded polymer for the generation of a homogeneously distributed demolding force. For a couple of years, it has been used in the mass manufacturing of fluidic devices (lactate sensors for blood testing) with channels of below 50 μm width and 200 μm depth. Such dimensions are difficult to replicate with thermal injection molding. With 80 chips on a Ø 150 mm substrate area of a HEX02, within 8 min cycle times, 4800 chips/8 h are manufactured in polymer foils. With its last development, the HEX04, using a Ø 300 mm substrate area and 4 min faster cycle times, a sixfold increase of throughput would be possible. The production still continues, although Jenoptik gave up imprint activity in 2014, leaving the field for others. For example, EV Group is now entering the market, with a different machine concept but an ability to pattern microfluidic devices. It also employs a pneumatic de-embossing step that helps to demold stamps from polymer substrates without distortion.

Step and Repeat NIL: S&R approaches are directed either toward mass fabrication, as done, for example, in Canon's J-FIL tools, or simply for the fabrication of stamp copies, multiplying small-area structures by repeating them over large areas. This process, which is similar to the recombination of individual stamps, is useful if identical small areas are needed over the required large substrate area. The enlargement of stamps is needed by many manufacturing, because while large-area stamps range from 100 to 300 mm wafer sizes (and beyond), roll embossing needs stamps with sheets of 370×470 mm^2 (Gen2 panel) and larger. This requires either high-throughput serial writing or parallel patterning, for example, with interference lithography. Fraunhofer ISE [65] has achieved 1.2×1.2 m^2 (see Figure 11.17). For the use of multiplying structures from a few square millimeters up to several square centimeters, few commercial stepper NIL machines are available. The two models from SUSS MicroTec and EV Group are derived (as was the case with their anodic bonders converted into T-NIL tools) from flip-chip bonders. Other machines, such as those from Soken Chemicals & Engineering Co., Ltd. [113] and temicon GmbH [114], are custom made. Such companies offer replication services that allow surface enlargement with specific designs provided by the customer. Their systems are based on UV-NIL and enable recombination with minimal stitching errors. Such errors would include visible gaps between adjacent fields of imprints, damage to already imprinted fields by the next imprint, mismatch (X,Y,θ) of microstructures (far field), mismatch between lines (near field), visible defects due to interference, thickness variation, and phase mismatch.

Other S&R machines have been optimized for specific tasks [130–135]. The SET NPS300 with its rotating head has been used to manufacture lightguides with outcoupling prisms or gratings with custom-designed placement (X,Y,θ) (see Figure 11.31) [136]. For this, small stamplets were individually rotated and placed over a large area, thus enabling the generation of outcoupling characteristics over a lightguide with side-illumination. Stitching is often critical because of crosstalk due to thermal heating or imprint over hardened resist (due to excess material at the borders and stray light). For zero or even negative stitching, strategies have been developed, such as remolding of already molded structures in T-NIL, and partial curing and overprinting of the still soft material [132].

11.6 UV-IMPRINT LITHOGRAPHY AND JET AND FLASH IMPRINT LITHOGRAPHY

11.6.1 UV-Assisted Imprint Lithography with Hard Templates

Devices that require several lithography steps and precise overlay will need an imprinting process capable of addressing registration issues. A derivative of NIL, UV-NIL, addresses the issue of alignment by using a transparent template, thereby facilitating conventional overlay techniques. In addition, the imprint process is performed at low pressures and at room temperature, which minimizes magnification and distortion errors. Two types of approaches are being considered for UV-NIL. The first method uses conventional spin-on techniques to coat a wafer with a UV-curable

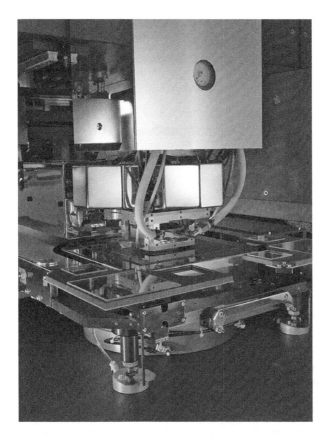

FIGURE 11.31 NPS Thermal S&R NIL tool by SET. It is equipped with a rotating head and trays for multiple stamp use and can be modified for UV-NIL. Courtesy of Smart Equipment Technology, Saint Jeoire, France.

resist [31]. Although it is possible to uniformly coat the wafer, there are concerns that the viscosity of the resist will be too high to facilitate the formation of very thin residual layers. If the residual layer is too thick, the critical dimension uniformity may suffer as a result of the subsequent pattern transfer process. In addition, a uniform coating of resist cannot account for variations in pattern densities on the template or mask, thereby leading to nonuniform residual layers. This problem is addressed by locally dispensing a low viscosity. This second approach was first disclosed by Willson et al. in 1999 and is generally referred to as step and flash imprint lithography, or S-FIL (Figure 11.32) [29]. Advances in S-FIL have led to inkjet based dispensers with drop volumes as small as 1.0 pl. This inkjetting approach has been successfully employed for both semiconductor applications, where an step and repeat approach is required, and large-area imprinting, where the throughput demands a process using a single imprint step to pattern the entire substrate. Because both applications use inkjetting to deposit the resist, the technology has been rebranded as Jet and Flash* Imprint Lithography (J-FIL[5]).

The J-FIL process relies on photopolymerization of a low-viscosity, acrylate-based monomer solution. Acrylate polymerization is known to be accompanied by volumetric shrinkage as a result of chemical bond formation. Consequently, the size and shape of the replicated features may be affected. Volumetric shrinkage is found to be less than 10% (v/v) in most cases [136]. Most acrylate-based imprint fluids that have been used to date have a viscosity in the range of 4–12 cps (mPa·s) [44].

---

* Jet and Flash Imprint Lithography and J-FIL are trademarks of Molecular Imprints, Inc.

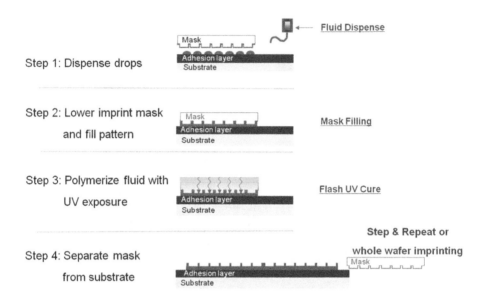

FIGURE 11.32   Jet and Flash Imprint Lithography process.

Relative to the spin-on UV-NIL approach, the J-FIL stepper has unique cost advantages. In spin-on UV or thermal imprint processes, a separate spin-coating tool is needed for resist deposition. Capital cost in a J-FIL tool is controlled by the inclusion of a self-contained material dispense module. With respect to the cost of consumables, the drop-dispense approach has virtually no waste. It is estimated that the drop-dispense approach will consume 1% to 0.1% of the volume that a spin-coating process will consume. Because semiconductor fabrication requires highly purified materials with parts per billion contamination levels, the cost of these materials is high and is generally proportional to the volume of the material used.

11.6.2   JET AND FLASH IMPRINT LITHOGRAPHY

Imprint tools based on drop-on-demand UV nanoimprinting are broadly divided into wafer steppers and whole substrate tools. Wafer steppers are used for applications requiring nanoresolution overlay and mix-and-match with photolithography (such as silicon ICs and thin film heads for magnetic storage), wherein the field size printing in one patterning step is the same as the industry standard advanced photolithography field size (26×33 mm^2). Whole substrate tools are used for applications that do not require nanoresolution overlay (such as patterned media and photonic crystals for LEDs). Since the material is dispensed only where needed just prior to patterning, the basic drop-on-demand process shown in Figure 11.32 can be integrated into either tooling platform.

Because so much of this book is devoted to processes associated with the fabrication of semiconductor devices, the J-FIL approach specifically designed for semiconductor fabrication is discussed in detail in this section. Other applications for the J-FIL process can be found in Section 11.8. The three primary building blocks that contribute to a UV imprint lithography process are the imprint tool, imprint materials, and an imprint mask. Because the tool and materials are so intimately connected during the imprint process, they are discussed as one subject in the next section. Following this discussion, imprint mask attributes and performance results are described.

11.6.3   IMPRINT TOOL

Imprint lithography relies on the parallel orientation of the imprint template and the substrate. Inaccurate orientation may yield a residual layer that is nonuniform across the imprint field. Thus, it is necessary to develop a mechanical system whereby template and substrate are brought into co-parallelism during etch barrier exposure. In contrast with the ACP process discussed in the previous section, this was originally achieved in S-FIL by way of a two-step orientation scheme. In Step 1, the template stage and wafer chuck are brought into coarse parallelism via micrometer actuation. The second step uses a passive flexure-based mechanism that takes over during the actual imprint [137, 138].

The first step-and-repeat system was built at the University of Texas in Austin by modifying a 248 nm Ultratech stepper that had been donated by IBM (see Figure 11.33). Key system attributes include a microresolution Z-stage that controls the imprint force, an automated X-Y stage for step-and-repeat positioning, a precalibration stage that enables parallel alignment between the template and the substrate, a fine-orientation flexure stage that provides a highly accurate, automatic parallel alignment of the template and wafer, an exposure source that is used to cure the etch barrier, and an automated fluid delivery system that accurately dispenses known amounts of the liquid etch barrier.

A commercialized version of this tool, the Imprio 100, was made available by Molecular Imprints Inc. (MII), a startup company that received its initial funding back in 2001 [56]. The Imprio 100 was designed as an S&R patterning tool and could accommodate wafer sizes up to 200 mm in diameter [139]. The standard die size was originally 25×25 mm^2, although both smaller and larger die sizes were possible. In order to minimize defect issues during the imprint process, the tool was equipped with a class 0.1 mini-environment. Although the Imprio 100 from MII is a substantial improvement relative to the first university tool, it had neither the throughput nor the overlay specifications (~250 nm 3σ) necessary for silicon IC fabrication. Instead, the system was primarily designed and manufactured to address the needs of the compound semiconductor, mechanical microstructures, advanced packaging, thin film head, and photonics markets. These markets require high-resolution

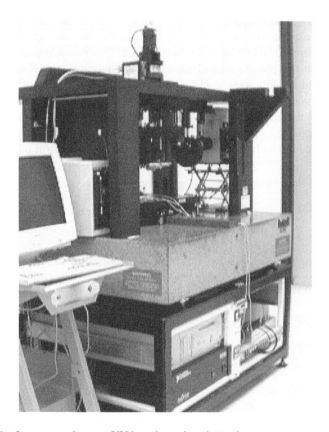

FIGURE 11.33 The first step-and-repeat UV-based nanoimprint tool.

features but are typically less sensitive to defects. They also operate at low volumes of wafers and are hence more sensitive to costs, particularly tool costs. The tool had a throughput capacity of approximately two 200 mm wafers per hour.

The Imprio 100 was developed in partnership with several key original equipment manufacturer (OEM) suppliers for the stage technology, the UV source, and the control architecture. The extremely complicated and costly imaging optics, source, and step and scan mechanical systems associated with other NGL techniques were not required for this technology. It was essentially a precise mechanical system with specialized fluid mechanics subsystems and a mercury arc lamp as its source. Therefore, it was a much simpler system, with a significantly smaller footprint and a cost structure that was appealing for companies looking to understand the advantages of an imprint approach.

Of particular interest is the resist delivery system, which incorporated a microsolenoid nozzle capable of dispensing drops less than 5 nl in volume. This type of control was essential for the control of the residual layer formed during the imprint process. When integrated with a well-designed flexure stage and wafer chuck, it was possible to print resist with residual layers well under 100 nm. Figure 11.34 depicts the data for residual layer uniformity in a single die. In this case, a mean thickness of less than 60 nm was achieved, with a 10 nm 3σ variation.

For patterning at dimensions below 32 nm, significant improvements to throughput, overlay, and residual layer thickness were required, and the next Molecular Imprints S&R system was an automated Imprio 250 operating at ~5 wafers per hour [140]. The system used an inkjet head to deposit resist, leading to residual layers as small as 10–15 nm. An optical photograph of the imprinted die with the inkjet dispense head is shown in Figure 11.35.

Figure 11.36 shows the schematic of a photolithography stepper and a UV imprint stepper based on drop dispense of the imprint material. The imprint tools include precision self-leveling flexure systems to passively align the imprint mask and substrate to be parallel during the imprint process [141]. In addition, by using a drop-dispense approach that can be tailored based on mask pattern variation, a highly uniform residual layer can be achieved. This film needs to be thin and uniform to achieve a subsequent etch process with a high degree of long-range critical dimension uniformity.

The Imprio 250 and the following model, the Imprio 300, used precision mechanical systems to achieve nanoresolution alignment and overlay. The alignment subsystem that aligns the imprint mask in the X, Y, and theta directions with respect to the wafer is based on a field-to-field Moiré detection alignment scheme originally developed at MIT for X-ray lithography [142]. In addition to alignment, magnification and shape corrections are required to perform nanoresolution overlay,

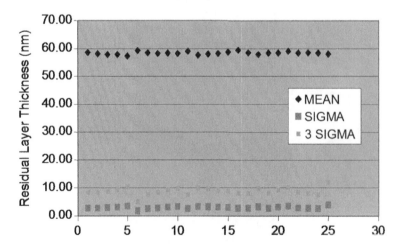

FIGURE 11.34 Residual layer thickness uniformity obtained across a 25 mm×25 mm field using a microsolenoid valve dispense system.

FIGURE 11.35 Residual layer uniformity for several dies using an inkjet dispense system. Note the color uniformity.

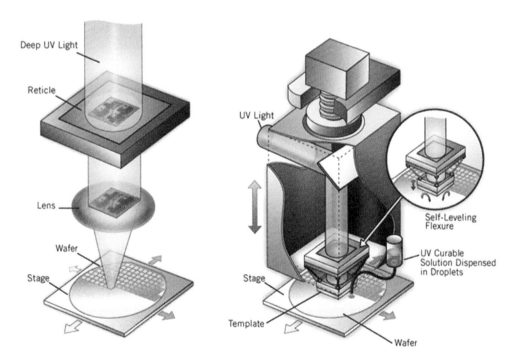

FIGURE 11.36 Comparison between an optical projection stepper and a J-FIL stepper.

particularly when mix-and-matching to optical lithography projection tools. A precision mechanical deformation system that deforms the imprint mask has been developed and implemented as part of the stepper system [137, 138]. This is the key step that allows the stepper to achieve the overlay required for CMOS fabrication.

In 2014, the semiconductor business of MII was acquired by Canon Inc., and the company was renamed Canon Nanotechnologies (CNT) [94]. Together with Canon Inc., new S&R tools were designed to address both pilot and production semiconductor applications. The FPA-1100 NZ2 has a footprint smaller than an i-line tool and has throughput and overlay specifications of 10 wafers

FIGURE 11.37   Conceptual layout of a four station cluster tool.

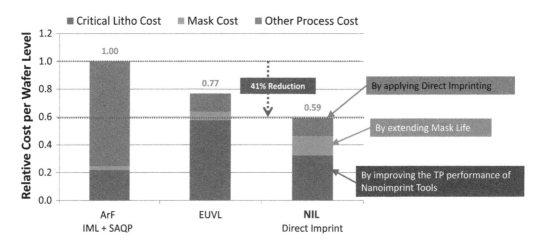

FIGURE 11.38   Cost of ownership comparison for various lithographic processes.

per hour and 8 nm, respectively. The system is suitable for pilot-scale operation, device fabrication demonstration, and process module development.

To address HVM, a cluster approach will be used in order to meet throughput and CoO requirements. A conceptual layout of a system containing four imprint stations is shown in Figure 11.37.

Clustering of modules is an approach that is widely adopted by the semiconductor industry and is extensively used for both deposition and etch systems. By clustering imprint stations, it is straightforward to meet throughput requirements that result in superior CoO. As an example, a four-station nanoimprint system with a throughput of 15 wafers per imprint station has a more attractive CoO than either ArF immersion multiple patterning approaches or an EUVL system running at 125 wafers per hour. A basic model depicting CoO for each lithographic approach is shown in Figure 11.38 [143].

In the next section, the mask or template is described in detail. Following this section, specific performance criteria for the technology are discussed, along with recent results.

11.6.4   THE IMPRINT MASK (TEMPLATE)

Early semiconductor mask fabrication schemes started with a $6'' \times 6'' \times 0.25''$ (6025) conventional photomask plate and used established Cr and phase-shift etch processes to define features in the glass substrate [139]. Although sub-100 nm geometries were demonstrated, critical dimension losses during the etching of the thick Cr layer etch make the fabrication scheme impractical for 1× templates. It was not unusual, for example, to see etch biases as high as 100 nm [144].

More recently, a much thinner (<15 nm) layer of Cr has been used as a hard mask. Thinner layers still suppress charging during the e-beam exposure of the template and have the advantage that CD losses encountered during the pattern transfer through the Cr are minimized. Because the etch selectivity of glass to Cr is better than 18:1 in a fluorine-based process, the Cr layer is also sufficient as a hard mask during the etching of the glass substrate. Other mask fabrication schemes, including the incorporation of a conductive and transparent layer of indium tin oxide (ITO) on the glass substrate, have also been tested, but the process flow is not easily compatible with the infrastructure available in commercial mask shops. The experimental details of this alternative approach have been covered in previous publications [145, 146].

An example of a patterned J-FIL imprint mask is shown in Figure 11.39. A mesa defines the field size and is set to have maximum dimensions of 26×33 mm^2, thereby making it compatible with existing optical projection lithography tools. Typical mesa heights are on the order of 15–30 μm. The mesa height is critical, as it also provides an impediment to any liquid resist extruding at the edge of the field.

As previously discussed, imprint resolution is defined by the relief image in the mask. Early work at Motorola took advantage of the high resolution of a Gaussian beam writer and ZEP520A resist to produce imprinted features as small as 30 nm [147]. Resist patterning is followed by a descum step, a dry Cr etch (using chlorine and oxygen) and a reactive ion etch of the fused silica (with chemistries such as CHF$_3$ or CF$_4$). More recently, Dai Nippon Printing (DNP) has demonstrated 14 nm half-pitch patterns on a mask. Although impressive, Gaussian beam writers suffer from long write times. Commercial mask suppliers typically prefer to limit write times to less than 24 hours and therefore use shaped beam systems in which an aperture is employed to define minimum resolution. In these systems, in addition to aperture size, a combination of current density and beam blur through the aperture also impacts minimum feature size. Dense 20 nm line patterns have been resolved using advanced NuFlare tools [148]. Further refinements in the electron optics combined with a reduction in the Coulomb interaction should result in sub-20 nm feature resolution. Recently, tool architectures using multibeam approaches have the potential not only to reduce writing times but also to allow resolutions of about 15 nm. An example of DNP's mask fabrication capability is shown in Figure 11.40.

FIGURE 11.39   Example of a 6025 imprint mask.

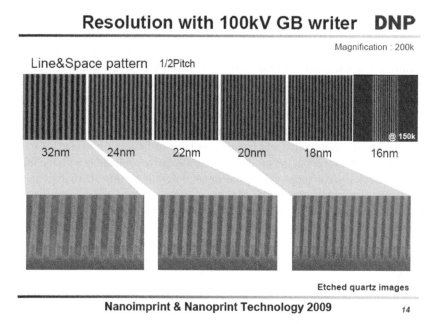

FIGURE 11.40 Dense line/space patterns after imaging and after pattern transfer into the fused silica.

Image placement on a mask also affects device yield, as noncorrectable errors result in layer-to-layer overlay misregistration on a device. Because a typical optical photomask is written at a magnification of 4×, image placement errors on the mask of 8 nm are reduced to 2 nm on the wafer. Fortunately, an imprint mask has a writing area 1/16th that of a photomask, and beam drift can be better controlled across the smaller area. Sub-2 nm feature placement has been demonstrated on fully patterned imprint masks [149].

In addition to resist patterning and dry-etch pattern transfer, the final mask must go through inspection and repair steps in order to identify and eliminate defects. These steps are typically accomplished using electron beam inspection and repair equipment. Both KLA-Tencor and Hermes Microvision have demonstrated sub-10 nm resolution on 32 nm patterns, but at the cost of throughput [150]. Optical inspection is possible, and NuFlare mask inspection tools have been used to capture breaks or shorts in 22 nm line-and-space patterns; however, it is likely that smaller and more subtle defects are not seen with this method, meaning that possible critical defects may not be detected and subsequently repaired. Pritschow et al. studied repair capability on 40 and 32 nm patterns and were successful at repairing a variety of different defect types [151]. Figure 11.41 shows an example of repaired features at a 32 nm half-pitch. Additional work is still needed to understand repair at sub-20 nm dimensions. Pritschow et al. [151], Myron et al. [152], and Selinidis et al. [153] cover some of the work that has been done in this field.

In the actual imprint process, mask lifetime is dictated by mainly by the interaction of particles with the mask. It is anticipated that lifetimes will fall somewhere between 10,000 and 100,000 imprints, thereby creating a need for additional masks (or templates). It is not feasible to electron beam write multiple numbers of masks because of the high cost associated with this process. Replica masks, however, can be fabricated using the same type of imprinting tools used for wafer imprinting [154, 155]. Throughputs for this type of tooling can be quite high, since only a single field is needed on the replica mask. Critical dimensions are generally well controlled because of the "copy exactly" nature of the imprint process. Finally, image placement errors are minimized, since both the master mask and the replica mask blank are made of fused silica, thereby minimizing any thermal mismatches during the process that would normally be induced by two materials with dissimilar thermal coefficients of expansion. An example of a replicated mask pattern is shown in Figure 11.42,

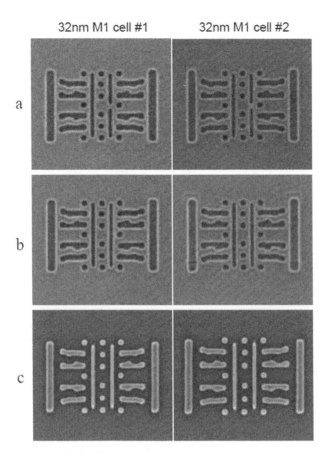

FIGURE 11.41 SEM images of two different 32 nm SRAM Metal 1 cells with a 32 × 32 nm mousebite defect before repair (a), after repair (b), and the corresponding wafer imprint (c).

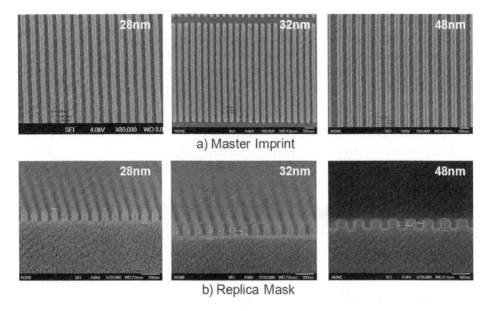

FIGURE 11.42 Pattern transfer of a semiconductor replica mask. (a) Imprints of the master onto the mesa of the replica substrate. (b) The same features etched into the replica mask.

11.6.5 J-FIL Performance Results

In addition to defectivity, the primary criteria that must be addressed are critical dimension control, residual layer thickness (RLT) control, overlay, and throughput.

11.6.5.1 Critical Dimension Control and Residual Layer Thickness

While critical dimension uniformity control is required for any lithographic system, the critical dimension is typically a function of mask resolution, and there is very little contribution to the critical dimension error budget as it relates to the actual printed feature. Examples of high-resolution patterning are shown in Figure 11.43. Residual layer thickness control is critical, however, for a couple of reasons; first, it is important to minimize the RLT in order to minimize critical dimension bias during pattern transfer, and second, it is critical to control the residual layer thickness uniformity in order to avoid any significant CD uniformity resulting from the descum of the residual layer itself. Residual layer control is affected by a variety of factors, including

- Alignment errors between the imprint mask and the substrate:
- Field flatness of wafers, imprint masks, and chucks:
- Drop-on-demand resist placement relative to the pattern on the mask

In order to minimize critical dimension biases and variations during the removal of the residual layer, a sub-32 nm pattern typically requires a process with a mean residual layer of less than 15 nm. Therefore, in the presence of pattern density variations within a field, resist dispense that is uncorrelated with the mask pattern will require the liquid to travel distances over millimeters in nanoscale channels. In practice, this situation leads to very high localized pressures in the fluid, causing wafer or mask deformations, highly varying residual layers, and longer filling times [156]. In one demonstration, 6 pl drops were dispensed in correlation with the mask design, leading to several thousand drops per 26 mm × 33 mm^2 field. This process was automated and was based on offline volume computations using mask design information available in GDS-II format. By incorporating the three factors listed into the tool, a residual layer variation of <5 nm, 3σ, was achieved over a 200 mm wafer, as shown in Figure 11.44 [157].

11.6.5.2 Overlay

Alignment is defined as the accuracy with which an imprinted field can be registered relative to a previously lithographed field at the four corners of the patterned fields. Overlay refers to the accuracy with which an imprinted field can be registered relative to the previous field at the four corners and many points (typically ~100) over a field. The imprint stepper has three key subsystems that contribute to alignment and overlay:

- *Interferometric Moiré Alignment Technique (i-MAT):* This is the approach used to obtain real-time relative overlay errors between points on the imprint mask and the corresponding points on the wafer. The system is capable of measuring alignment errors at a single point well below 1 nm.

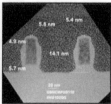

FIGURE 11.43 J-FIL SEMs. From left to right: 38 nm NAND Flash gate layer, 30 nm storage class memory cross section, 22 nm half-pitch resist lines, and 11 nm half pitch resist lines.

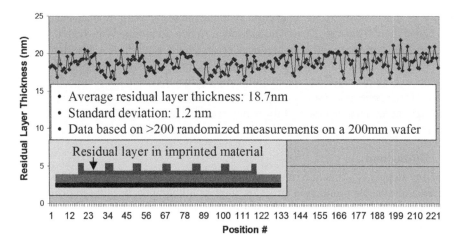

FIGURE 11.44 Residual layer thickness (RLT) uniformity across a 200 mm wafer.

- *Magnification actuator system:* The magnification actuator works on the basis of imparting elastic deformations to fused silica over a small range of motion (typically on the order of a few parts per million). The system incorporates an array of force feedback–controlled actuators that are mounted around the imprint mask. This strategy allows in-plane corrections in X and Y, orthogonality, and to some extent, higher-order distortion corrections, as discussed by Melliar-Smith [9].
- *Wafer stage motion for rigid body corrections (X, Y, and Θ):* An air-bearing wafer stepper stage is used to align the mask and the wafer. The stage needs to hold position with very low noise to maximize overlay performance. The low-viscosity liquid resist, in turn, acts as a damping agent for any stage vibrations.

Overlay requirements for a NAND Flash memory device are typically on the order of one-third of the half-pitch. As an example, for a 16 nm NAND Flash device, a mix-and-match overlay of approximately 5 nm is required. Previous results have demonstrated mix-and-match overlay to a 193 nm immersion scanner of approximately 10 nm [158]. A newer "through the mask" (TTM) approach to alignment has been implemented to improve overlay. Figure 11.45 shows the measurement

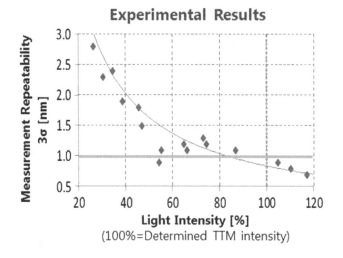

FIGURE 11.45 Through the Mask (TTM) system with <1.0 nm measurement repeatability.

repeatability of a TTM system as a function of light intensity. With a sufficient signal, repeatability of better than 1 nm can clearly be achieved. When combined with better wafer flatness, enhanced wafer temperature control and reduced errors on the wafer have allowed overlay errors to be reduced to as little as 4.2 nm in X and 4.8 nm in Y (mean + 3σ), as shown in Figure 11.46. Further improvements to both the mask and the magnification actuator system are planned in order to meet the demands of the semiconductor roadmap. Better overlay may be achieved in future generations of tooling through thermal management of the wafer, reducing noise from the magnification actuators, and reducing the image placement errors on the imprint mask itself.

11.6.5.3 Throughput

The cost of any stepper is a function of the process throughput. A throughput of 20 wafers per hour is considered a reasonable production target for imprint lithography, because imprint tools are significantly lower in capital cost as compared with 193 nm immersion and EUV tools. Based on this target, one shot has to be completed in about 1.4 seconds. The major contributors to this target are shown below:

Stage move, fluid dispense time	~0.15 seconds
Alignment, mask fill time	~1.00 seconds
UV cure time	~0.15 seconds
Separation time	~0.10 seconds
TOTAL	~1.40 seconds

Of all the steps outlined, only fluid filling is believed to be a fundamental technical risk. The other targets can be met using extensions of known engineering approaches. Fluid filling in a drop-on-demand UV imprint process is affected by several factors, such as fluid viscosity, dispense-drop resolution, control of fluid front dynamics, and targeting of drops relative to the mask design. Also, changes in pattern density in the mask design can cause regions where drop tailoring is suboptimal due to limitations in drop resolution and drop placement.

By applying the filling guidelines described earlier, a test pattern consisting of 24 nm device patterns, dummy fill, and metrology marks was printed and inspected for non-fill defects [159]. In this work, further optimization of the drop pattern was achieved by using gridded drop patterns in areas where the features were essentially parallel lines. In addition, the drop volume was set as low as 1.5 picoliters. The results are shown in Figure 11.47. Figure 11.47a depicts the location of each defect within the printed mask field for a fill time of 1.5 seconds. Although the defectivity ($1.2/cm^2$) was higher than the targeted value of $1.0/cm^2$, it should be noted that the defects are systematic. Two types of non-fill defects were observed. The first were the non-fill defects within the printed Moiré align mark (see Figure 11.47b). The second defect always occurred in transition areas between a repeating structure in a die and another pattern type. Both defects can likely be addressed with specific imprint patterns designed to enhance filling in these areas and by further reductions in drop volumes.

More recent data has been presented demonstrating the feasibility of filling fields in less than 1 second with a non-fill defectivity of zero [160]. Currently, throughput specifications are typically limited by the ability to fill partial fields without the presence of non-fill defects. The issue is primarily a function of controlling resist spreading in smaller areas, where the process cannot begin at the center of the field. Partial fields with areas equivalent to at least 35% of the total area of a full field are generally quite manageable. An example of a fill sequence for a partial field is shown in Figure 11.48.

As the relative area drops, non-fill defectivity for shorter fill times begins to rise, and an example of a field still containing detected non-fill defects is shown in Figure 11.50. In this example, non-fill

	X	Y
Average	-0.3 nm	0.1 nm
3σ	3.9 nm	4.7 nm
Ave.+3σ	4.2 nm	4.8 nm

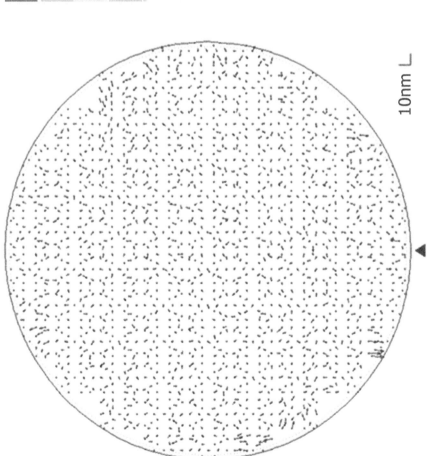

10nm └

FIGURE 11.46 Mix and match overlay to an ArF immersion scanner demonstrating <5 nm overlay.

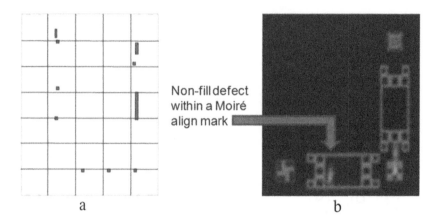

FIGURE 11.47 (a) Non-fill defects in a 26 mm × 33 mm field for a resist spread time of only 1.5 seconds. (b) The non-fill defects were confined to transition regions and to the area near the Moiré align mark.

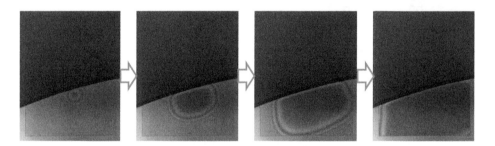

FIGURE 11.48 Filling characteristics of a partial field with a relative area of 52%.

FIGURE 11.49 A fully imprinted wafer (left) and a close-up of several imprinted edge fields (right).

defectivity in a field with a 22% area was 5.2/cm^2 after a fill time of 2 seconds [161]. In general, a fill time of 2 seconds is sufficient for fields with areas greater than 25%. It should be noted that the non-fill defectivity will continue to decrease as fill time is extended. An example of a fully patterned wafer including edge fields is shown in Figure 11.49.

11.6.5.4 Defectivity
Because the mask comes into contact with the liquid resist, there are defect mechanisms for imprint lithography that are somewhat unique to the process. Defect mechanisms that are important to address (besides non-fill defectivity, which was discussed in the preceding section) include contamination

defects, separation defects, and most importantly, particle-induced defects. Figure 11.50 shows examples of each defect type.

The contamination voids referenced in Figure 11.51 are believed to result from environmental factors that locally degrade the adhesive properties of the adhesion deposited prior to the imprint process. It is believed that moisture may be adsorbed on the surface of the adhesion layer, thereby causing local adhesion failure or disrupting the filling of the liquid resist. By taking precautions such as storing wafers in a nitrogen environment prior to printing and by adding carbon filtration systems, these defects are virtually eliminated.

In addition to on-tool resist filtration, the separation algorithms of newer J-FIL platforms have been upgraded to produce a uniform separation front. Inconsistent separation caused by either variations in pattern density or poorly controlled separation velocity is known to cause shearing defects. By modification of the tool algorithms, the separation velocity front is now well controlled (Figure 11.52). Pictured is the release of the template field from silicon wafer over a period of 0.30 seconds.

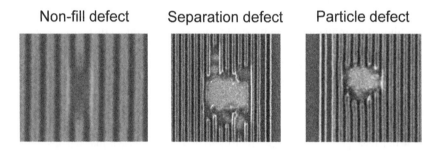

FIGURE 11.50 Three defect types common to imprint processes.

FIGURE 11.51 Left: Defects caused during the separation process. Right: By optimizing the control systems, shearing defects can be eliminated.

FIGURE 11.52 Separation front between mask and wafer over 0.30 seconds.

Particles represent a much more serious problem for imprint lithography, and it is important to address all sources of particle-related defects. Most of these defects can be traced back to some portion of the imprint tool itself, and two major sources include particles generated within the inkjetting system and particles generated within the chamber of the imprint tool.

In a series of experiments done by Ye et al., it was noted that when 50 nm dense lines were printed, there was almost no occurrence of line break defects [162]. At 28 nm, the number of line breaks increased significantly. To understand whether line breaks could be caused by very small particles, the number of particles residing on the silicon substrate was substantially increased by deliberately introducing ~30 nm diameter polystyrene spheres so that almost any high-resolution SEM image, with a field of view of only a few microns, would contain polystyrene spheres. The wafers were then imprinted, and hundreds of SEM images were captured and inspected for line breaks. Every image contained line break defects. The majority of these defects were confined to a single line.

To remedy this, the first step taken was to filter the imprint resist multiple times using a 10 nm filter. The combination of a 30% improvement in material strength and reduced particulates in the resist immediately reduced the number of line break defects, dropping the defect density by more than two orders of magnitude. To further address defects generated in the dispensing process, an on-tool 10 nm filtered recirculation system located adjacent to the inkjet dispenser was incorporated into the resist dispense system.

In 2015, Iwamoto et al. examined particulates generated from within the imprint tool itself. Previous studies had demonstrated that the movement within the imprint tool itself creates 0.1 to 1.0 defects per wafer pass, a specification that is considered too high for device manufacturing. By implementing several particle reduction methods, such as material surface processing, air flow control, and ionization techniques, it is possible to drive the particle adders down to fewer than 0.05 particles per wafer pass [163]. More recently, Emoto et al. have shown that particle adders can be significantly reduced by incorporating an air curtain that steers particles away from the wafer plane. Initial accelerated studies indicate that particle adders can be reduced to levels lower than 0.001 particles per wafer pass [164]. The data is shown in Figure 11.53.

11.6.6 Production Readiness

Recent reports from various groups summarize the progress made up through the end of 2016 and can be summarized as follows [165]:

- Throughput: 60 wafers per hour for a four-station cluster tool
- Overlay: <5 nm, 3σ
- Defectivity: 5 defects/cm^2 up to ~100 wafers (or four wafer lots)

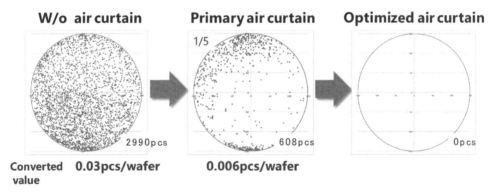

FIGURE 11.53 Particle adders per wafer based on three different operating conditions. An optimized air curtain is very effective in reducing particles at the wafer plane.

Acceptance of the technology for device manufacturing will require small improvements in throughput; however, defectivity will likely need to come down another order of magnitude and mask lifetime. Throughput can be addressed by improving resist fill speed. The defectivity targets are more challenging; however, Emoto et al. recently described an air curtain system that addresses particle adders with a path forward to achieving levels better than 0.001 adders per wafer pass [164]. The implementation of such a system will be critical for readying the technology for production applications.

11.7   ROLL-TO-ROLL NANOIMPRINT LITHOGRAPHY

As previously discussed, the ability to pattern materials at the nanoscale can enable a variety of applications, ranging from high-density data storage, displays, photonic devices- and CMOS integrated circuits to emerging applications in the biomedical and energy sectors. These applications require varying levels of pattern control and short- and long-range order, and have varying cost tolerances. Continuous processing methods, such as newspaper printing, sheet extrusion, cloth weaving, etc., are considered to have cost and processing advantages and are typically employed if speed and throughput have to be increased, and also if a constant quality over a large quantity of devices has to be achieved. However, they are physically identical to noncontinuous (sequential and stationary) processing and require similar processing times for pattern formation and transfer. Since such processes have to be applied to a moving substrate, this results in a complex interplay of process steps that needs a careful balance of process parameters and material properties. This is the reason why dynamic processes often have restrictions in terms of resolution and overlay, and lack freedom to vary process and material relative to stationary processes. Furthermore, inline control is much more difficult than offline control with single substrates, and in a sequence of processes, all processes have to be done at the same speed (synchronization is needed). Vacuum processes are more difficult to employ because of the constant feed from ambient into areas with defined atmosphere and vice versa. In contrast to this, the advantages of continuous processing are dynamic stability (e.g., variothermal processing with a fixed tool temperature but rapid heating upon contact and cooling after demolding) and inline processing of a continuous substrate, in which handling of individual substrates is avoided.

Roll-to-roll or reel-to-reel (R2R) processing involves the patterning of flexible materials such as plastics or metal foils. If surface topographies are transferred, the term *roll embossing* is often used, which here—in accordance with the scope of the chapter—is named *roller imprint* or *roll-to-roll nanoimprint lithography* (R2R-NIL) [166–168]. The flexible material, or web, is unwound from a core, processed, and then returned to a second core at the end of the sequence. R2R processing is in use today by industry (packaging, flexible solar cells, and counterfeit holograms). Many R2R processes already exist for etching and deposition (e.g., metallization and printing) and have been optimized for high speed and throughput. The key step is the feeding of the web into the gap between two rotating cylinders, which are either pressed together or kept at a constant distance. Since the contact of both rolls with the web is line-like, the feeding into the gap results in a compression of material during feeding and the separation of the web from the rolls during widening of the gap. This is similar to printing with bendable stamps, in which the pressure is applied from one side of the substrate to the other, and demolding happens by delamination. The area of highest pressure in which dislocation of material is possible is—due to the flexibility of either cylinders or web—in a "widened line," or length of the web, which is called the *nip*. This nip, from a few millimeters to centimeters, defines, combined with the speed of feeding, the process time available for patterning.

Lithographic processes are also established for micron-scale manufacturing and for applications that only require polymer embossing without any subsequent processing [169, 170]. One of the most prominent examples is the manufacturing of large-area brightness enhancement films (BEFs) for LCD displays by the 3M Display and Graphics Laboratory [171,172]. In contrast to this, in the VTT Technical Research Centre [173], self-aligned metal electrodes in organic transistors were fabricated in a continuous R2R pilot line. The complete process is a combination of several techniques, including evaporation, reverse gravure, flexography, lift-off, UV-exposure, and development methods [174]. In conjunction with μCP, devices requiring metal etching were investigated, but again at a micron scale [175]. However, R2R patterning of arbitrary patterns with thin residual layer control (needed for subsequent pattern transfer) at the nanoscale is far more challenging, particularly at a cost structure suited for commodity applications. The challenge is to create a process that is scalable and meets defectivity, throughput, and CoO requirements. If successful, the potential applications include thin film transistors, flexible displays, wire grid polarizers (WGPs), color filters, and solar devices. The challenge, as always, is to create a process that meets both defectivity (yield) and throughput (cost of ownership) requirements. Apart from R2R, hybrid processes such as roll-to-plate (R2P) and plate-to-roll (P2R) sometimes have specific advantages over R2R. For example, R2P allows the use of a bent stamp to print over rigid, wafer-like substrates, thus employing the advantages of high line pressure, sequential (directive) continuous processes, and delamination by bending. P2R allows the use of flat stamps on a continuous web, which enables high precision similarly to S&R processing (see Figure 11.54) [176].

In general, the advantage of roll processing is that once the process is stabilized, it can be run as long as contamination does not accumulate, ASLs are preserved, and processes are kept within a narrow process window. The disadvantage is that sometimes long optimization runs are needed before stability is achieved, causing waste of material, and the difficulty of interrupting processes for process control.

As is the case with NIL, R2R processes for both thermal and UV-assisted processing can be employed. T-NIL imprints either directly into the thermoplastic web or into a coating (with lower T_g than the web). The heating is applied either to the foil before it is fed into the nip or by using heated cylinders and thus upon contact to the rolls. This means that the heating of the moldable material above T_g has to be ensured before the peak pressure is reached, thus enabling molding of the material, and the cooling below T_g has to be effectuated before demolding takes place. In this mode, the overhead (heating and cooling) typically associated with noncontinuous processes is reduced at the expense of highly interlinked process steps. UV-assisted molding, in contrast, requires the curing of the material before the stamp is demolded from the viscous resin. Therefore, either the UV lamp has to be integrated into the cylinder and focused in a widened line onto the molded resin or the

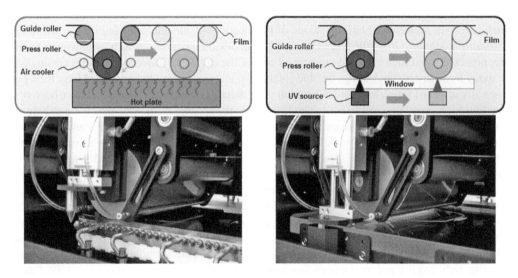

FIGURE 11.54 Roll-to-plate and plate-to-roll processes in a tool with an exchangeable base-plate. (Reprinted from *Microelectron. Eng.*, 88, Lim, H.J., et al., Roller nanoimprint lithography for flexible electronic devices of a sub-micron scale, 2017–2020, © 2011, with permission from Elsevier.)

lamp is illuminated from outside onto the web before it is released from the web. Instead of using patterned cylinders, belts connecting neighboring cylinders have been employed, which enables enlargement of the process length in which the belt-like stamp is in contact with the resist before complete curing is achieved.

The cost of manufacturing is typically driven by speed (or throughput), tool complexity, cost of consumables (materials used, mold or master cost, etc.), substrate cost, and the downstream processing required (annealing, deposition, etching, etc.). In order to achieve low-cost nanopatterning, it is imperative to move toward high-speed imprinting, less complex tools, near-zero waste of consumables, and low-cost substrates. Several research groups are currently investigating roll-based NIL using either thermal or UV-NIL processes established for planar applications [177–180]. These approaches are limited in their ability to simultaneously address the challenges noted above.

There are additional requirements for the adoption of R2R-NIL. One requirement is achieving the lithographic performance required for pattern transfer of nanoscale structures (as opposed to a strictly functional pattern such as an embossed film). Typical lithography metrics such as aspect ratio, minimum critical dimension, pattern complexity dependence, RLT, and consumables costs are especially demanding at the scale required for realizing patterned nanostructures over large areas and at high throughput. In additional, a manufacturing infrastructure must be established to support production processes. Large-area nanostructured devices will require master patterns written using high-end lithography (e-beam, 193 immersion) and replication to create large-area daughter imprint templates. Also, processing steps such as descum etching must be implemented over a large area and at a throughput similar to that of the lithographic process.

A roll-based NIL process should also allow precise and easy control over the resist thickness. NIL (both thermal and UV based) has been used by others to pattern very small features, but like traditional lithographic techniques, it requires the use of spin coaters, necessitating significant resist material waste as well as spinning of the substrate, which can limit the size and format of the substrate. Additionally, because the nanoimprint process is essentially a molding technique, the spin-coated resist must fill voids in the nanostructured imprint template to accomplish patterning. Since the thickness of the spin-cast resist film is uniform over the area of the substrate, changes in pattern density/depth will result in nonuniform residual resist thickness at the base of the patterned features. Such nonuniformities can lead to pattern transfer difficulties when using etch process techniques [181]. In addition, spin coating is not optimal for R2R coating, and knife coating or spray coating is generally employed [168]. Newer developments, such as by Inmold A/S, use R2R extrusion coating, in which a molten polymer is cast onto a polymeric web and immediately printed (via slot die coating into the gap of the patterned cylinder) [182, 183]. However, volume control of the imprint resist to a level suitable for thin RLT is extremely challenging. In the following, we review some of the earlier efforts in establishing R2R-NIL solutions for a variety of applications.

The Hewlett-Packard Company (HP) recently focused its efforts on flexible displays and created an active matrix backplane for driving a display using a R2R process and an imprint method referred to as *self-aligned imprint lithography* (SAIL). The idea of the SAIL process is to create a three-dimensional multilevel roller stamp and print the 3-D relief image into a resist coated over a film stack consisting of aluminum for the bottom gate metal, a dielectric, an undoped amorphous silicon layer, a microcrystalline silicon layer, and a final layer of chromium. This is patterned to form both the data lines and pixel electrodes [184, 185]. A schematic of the SAIL process is shown in Figure 11.55. It is based on the fact that multilevel resist patterns can be used in a multistep mode that allows patterns to be etched with different masking designs by sequentially etching the layers of the 3-D topography down to the substrate (window opening), followed by a substrate-selective etch. Because the 3-D pattern is given by the stamp, the different etching levels are aligned with respect to each other. Such processes allow the use of a one-step imprint followed by multiple etching steps, which is particularly suited for R2R processing of polymer webs. The advantage of this process is that the imprinted web can be collected and stored, and the multiple pattern transfer steps can be done offline on a separate reel at a different speed.

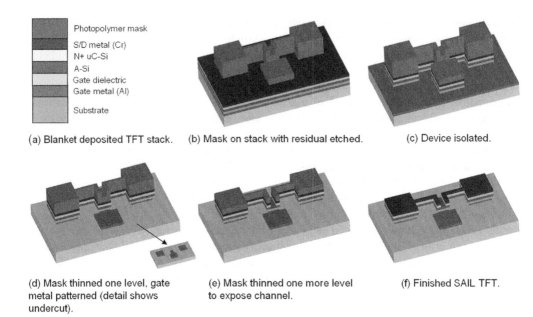

Photopolymer mask
S/D metal (Cr)
N+ uC-Si
A-Si
Gate dielectric
Gate metal (Al)

Substrate

(a) Blanket deposited TFT stack.   (b) Mask on stack with residual etched.   (c) Device isolated.

(d) Mask thinned one level, gate metal patterned (detail shows undercut).   (e) Mask thinned one more level to expose channel.   (f) Finished SAIL TFT.

FIGURE 11.55 Process sequence for HP's self-aligned imprint lithography (SAIL) process. (Reproduced from Jeans, A., et al., *Proc. SPIE*, 7637, 2010, 763710 and Holland, E.R., et al., *Proc. SPIE*, 7970, 2011, 797016. © SPIE. With permission from Hewlett Packard.)

The roller mask is often fabricated by first creating a master mask on a planar silicon or glass substrate and then, transferring the pattern to a bendable shim that can be wrapped around a cylinder. Most common in industry is the creation of a thin metal shim (100–300 µm thick nickel) by electroplating. Although the process is only used for the replication of surface topographies with moderate aspect ratio, often the original is destroyed during separation. Other forms of replication are either casting PDMS or imprint into TeflonAF flexible film. The film is then peeled away from the silicon, attached to the roller, and used to print the 3-D structures. A photo of the prototype R2R imprint tool is shown in Figure 11.56. Working color displays consisting of 160×120 pixels with a pixel size of 480×480 µm^2 have been successfully fabricated using this process.

Nanoscale applications include both color filters and WGPs, and Se Hyun Ahn and Jay Guo have concentrated their efforts on high-throughput/resolution R2R systems. Their initial tool focused on simple gratings, and they were able to demonstrate a 100 nm pitch grating using an epoxysilicone resist across a 10 mm-width roll [177]. More recently, they have scaled their process and created a large-area (150 mm) tool capable of printing a 4 ″ wide film. The system is capable of printing in both R2R and R2P mode. A schematic and photo of the system are shown in Figure 11.57. Web speeds of 1 m/min were obtained and were limited by the motor used to drive the web.

The epoxysilicone resist was chosen for this study, since the material exhibits a cationic curing mechanism and is not subject to the oxygen poisoning typical of acrylate-based resists. In addition, the shrinkage of the resist after curing is substantially smaller than for acrylate resists. The typical sensitivity of the epoxysilicone resist is about 100 mW/cm^2. RLT was studied, and the key parameters affecting RLT were web speed and the force applied between the rollers. Not surprisingly, RLT decreases with decreasing web speed and increasing force.

It should be noted that the viscosity of the epoxysilicone resist prohibits the printing of very thin RLTs. Typical values of RLT ranged anywhere from 200 to 300 nm. Clearly, this is a limiting factor for device applications requiring pattern transfer through the underlying substrate, and

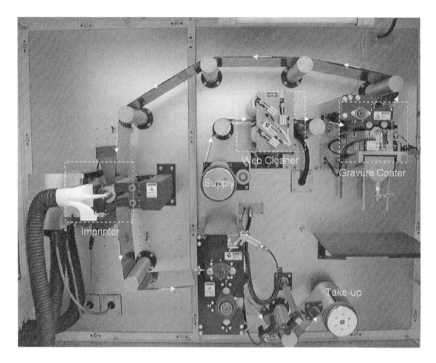

FIGURE 11.56 Web-based imprint tool used to create flexible displays. (Reproduced from Jeans, A., et al., *Proc. SPIE*, 7637, 2010, 763710. © 2010 SPIE. With permission.)

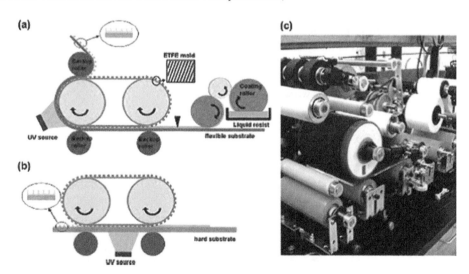

FIGURE 11.57 (a) Schematic of a roll-to-roll and roll-to-plate imprinter. (b) Photograph of the tool. Maximum throughput is 1 m/s. (Reprinted with permission from Ahn, S.H. and Guo, L.J., *ACS Nano*, 3, 2009, 2304. © 2009 American Chemical Society. Photo courtesy of J. Guo.)

low-viscosity resist materials and processes will need to be implemented in order to reduce RLTs to less than 20 nm.

At the University of Massachusetts Amherst [186, 187], the James Watkins group's interests include the synthesis, characterization, and utilization of nanoscale and hybrid materials and the fabrication of devices that exploit the unique properties of the materials that they create. There is a large focus on self-assembly materials and hybrid materials to create functional patterns. Their fundamental studies are then applied to more practical manufacturing schemes, including the R2R printing of materials specifically for flexible electronic applications. Their work is done in the Center for Hierarchical Manufacturing, an NSF nanoscale science and engineering center.

Imprinting is done with a UMass NANOemBOSS R2R-NIL tool fabricated together with Carpe Diem Technologies [188]. The tool is capable of printing sub-50 nm features at rates on the order of a few feet per minute. A photograph of their R2R UV-NIL tool is shown in Figure 11.58. Two different applications using R2R-NIL are discussed below.

One example of their work centered on the R2R deposition of transparent mesoporous hybrid titanium dioxide by applying a UV process to films that combined polymers and TiO_2 nanoparticles [186]. This approach utilized a UV-curable polymer in conjunction with the photocatalytic activity of TiO_2 to form composite films in one step and to produce films with well-controlled porosity and refractive index. By adjustment of the loading of TiO_2 nanoparticles in the polymer, the refractive index could be tuned between 1.53 and 1.73. These hybrid films were then applied toward the fabrication of robust one-dimensional photonic crystals on both silicon and flexible poly(ethylene terephthalate) (PET) substrates. The resultant Bragg mirrors were then shown to be effective chemical vapor sensors.

A second application focused on the fabrication of microfluidic devices that could serve as biosensors. As pointed out by Chen et al., very few examples of soft material microfluidic devices have been fabricated using high-throughput processes, with even fewer demonstrating any meaningful commercial success [179]. In their study, UV-curable thioene polymers and PET films were applied to create a flexible microfluidic chip via nanoimprint lithography. The final devices, complete with operational electrovalves, were then used to detect Salmonella in a liquid sample [184]. The data from these tests is shown in Figure 11.59.

Several other research centers that are currently developing processes and applications using R2R NIL include the Nascent Center [189], Joanneum Research [190], VTT [173], PTMTEC [191], and FHNW [192]. Apart from many research groups developing their own tools and molds, different companies are now offering services and tools. Interested readers are referred to EV Group [55], temicon GmbH [114], Tsai et al. [193], Maruyama et al. [194], and Asahi Kasei E-materials Corp. [195] for additional details on their work.

FIGURE 11.58 R2R UV-NIL tool located at the University of Massachusetts Amherst. (Courtesy of J.J. Watkins and K. Carter, The UMass Roll to Roll Fabrication and Processing Facility, Amhurst, MA.)

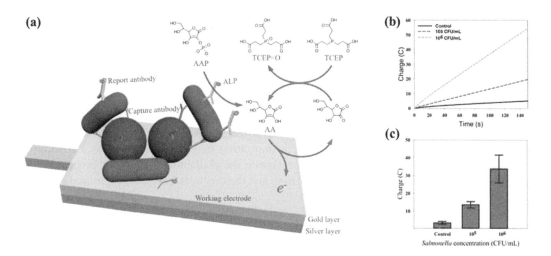

FIGURE 11.59 (a) Schematic representation of electrochemical detection of Salmonella. (b) Chronocoulograms obtained at 0.35 V on a working electrode with varying concentrations. (c) Concentration dependence of charge at 100 seconds. (From Chen, J., et al., *Lab. Chip*, 15, 2015, 3086. Reproduced by permission of The Royal Society of Chemistry.)

11.8 MARKETS AND APPLICATIONS

Although much of the focus so far has been on semiconductor device fabrication (and in fact, this carries through as the main theme for many of the chapters in this book), it is important to recall that the unique physical and chemical phenomena at the nanoscale can lead to other types of novel devices that have significant practical value. Emerging nanoresolution applications include subwavelength optical components (antireflection gratings, wire-grid polarizers and metalenses), nanofluidics (biochemical analysis devices), patterned flowcells (DNA analytics and sequencing), high-speed compound semiconductor chips, photonic crystals and polarization gratings (VCSELs, distributed feedback lasers and photovoltaic solar cells), diffractive optical elements (augmented/virtual reality applications and mobile displays), PSSs for high-brightness LEDs, and high-density bit patterned magnetic media (BPM) for storage. To take advantage of these opportunities, it is necessary to be able to cost-effectively image features that range from a few hundred nanometers to well below 100 nm. For some of these applications, the question of structural tolerance, overlay, and defectivity is more relaxed in comparison to semiconductor chip manufacturing: Single-layer nanopatterning does not always need machines enabling overlay, high redundancy, read-out correction schemes (e.g. in patterned magnetic media), or even self-healing processes (e.g., with directed self-assembly using block copolymers) that allow for the loss of local structural details. Even the demands on high speed may be relaxed if 80,000 VCSELs are patterned within one step.

Applications can generally be distinguished in terms of markets, volume, complexity of process, and requirements with respect to resolution, alignment, defectivity, throughput, and area. Then, there are "flagship" applications such as semiconductor devices, where NIL tries to replace existing patterning methods or to become one of the NGL methods for high-volume fabrication (the next "node"). Or, there are new applications where patterning has not yet been used, when this becomes the additional step that has to be included into an existing process chain. Examples include BPM and PSS, where a single layer has to be patterned with relaxed requirements on alignment and defectivity. Finally, there are applications where new materials need to be patterned, either by a resist-based process or by patterning functional materials, for example, in printed electronics or for bioanalytical sensors.

In general, NIL has to become a general-purpose patterning method that is able to compete with existing techniques, where its main advantages are cost, scalability (in terms of both substrate size and pattern size), versatility (flexible, prepatterned, and wavy substrates), and ability for upscaling (no showstoppers in terms of resolution, overlay, or large area). It is important to succeed in more traditional high-volume markets, since this will significantly lower the threshold for other applications. This success may allow penetration into diversified markets such as photonics or bioanalytics and may even lead to the development of combined processes that also use conventional photolithography, 3-D patterning, and other hybrid micro- and nanostructures. Functional patterning including inks and resists, along with reverse tone imprinting, may also come into play. Particularly important during the research and development stages is the availability of tools and processes that can generate enough volume to accelerate learning and thereby intersect a particular market in the required time.

While integrated chip manufacturing will be a market where only a specialized provider such as Canon will succeed in delivering an S&R solution in accordance with the International Technology Roadmap for Semiconductors, large-area single-imprint patterning is likely to be satisfied by several providers using soft stamp patterning with UV-NIL or combined thermal and UV-NIL. Imprint is also widely used for the surface patterning of thermoplastic materials and glass, such as for counterfeit tags, microfluidics, micro-optics, and biomimetics. However, these techniques do not often involve pattern transfer; they are nearer to classical replication methods such as injection molding, roll embossing, and thermoforming and are therefore not presented in more detail here.

In the end, the availability of a technique does not automatically mean that it will be employed. Equally important to any company attempting to satisfy the printing requirements of a particular technology is the size of the market and the opportunity to sell multiple numbers of tools over an extended period of time. Without this incentive, it is unlikely that a company would commit the very high nonrefundable engineering costs associated with a new imprint platform.

A number of markets have been mentioned already. Table 11.4 summarizes several markets and requirements. This chapter does not allow a thorough investigation of all the possible applications, however, but included in the following are some brief discussions on BPM (or PM) for hard disk drives (HDDs), flat panel display (FPD) applications, solar devices, and enhanced-performance LEDs and PSS.

11.8.1 PATTERNED MEDIA

The history of magnetic recording dates back over 100 years and is driven by the interaction of an external field on a magnetic material. The technology has been practically applied to the fabrication of hard drives, and the increase in area density has spanned eight orders of magnitude in only 50 years. In the last 10 years alone, the areal density has increased nearly two orders of magnitude, and disk drives with capacities greater than 1 Tb (10^{12} bytes, i.e., 8×10^{12} data bits) are readily available. As a result, the placement of hard drives is ubiquitous: routine applications include cloud storage, servers, desktop computers, laptops, digital video recorders, game consoles, video cameras, and so on. By the end of 2016, global internet traffic exceeded the zettabyte (1000 exabytes = 10^9 terabytes = 10^{21} bytes) threshold, so this certainly qualifies as an attractive market [196]!

This remarkable progress has been driven by the precise deposition of thin magnetic films. As shown on the left side of Figure 11.60, data is stored in circular tracks on a disk [197]. Within each track, a stream of data bits is recorded as regions of opposite magnetization. Each track consists of equally spaced bit cells, with a digital "1" being indicated by a boundary (called a *magnetic transition*) between regions of opposite magnetization within a bit cell, and a "0" being indicated by a continuous region without such a boundary.

At high magnification, it becomes apparent that within each bit cell, there are many tiny magnetic grains. These grains are randomly created during the deposition of the magnetic film. Each grain behaves like an independent magnet whose magnetization can be flipped by the write head

TABLE 11.4
Requirements and Achievements for Different Applications Using NIL

	Resolution	Throughput	Area	Overlay	Defectivity	Requirement	Substrate
Integrated circuits (IC)	11 nm	60 per hour	33×25 mm² on 300 mm wafers	3 nm	0.1/cm²	50 layers	Patterned silicon wafer
Bit patterned media (BPM)	7 nm	350 per hour	Ø 2.5″ disc	–	Relaxed	Double sided	Glass or aluminum
Photovoltaic solar cells (PV)	200 nm	60 per hour	Ø 156×156 mm²	–	Relaxed	Sol-gel resist, 3-D	180 μm silicon
Flat panel displays (FPD)	50–500 nm	60 per hour	Up to Gen8 panels	Micrometers	Relaxed, for smaller defects	Integrate on glass panel	Glass or flexible films
Patterned sapphire substrates (PSS)	5 μm → 200 nm	60 per hour	Ø 100 mm Ø 150 mm	No stitching	Relaxed	Sol-gel resist	Sapphire
PC on light-emitting diodes (LED)	200 nm	Not relevant	Ø 75 mm wafers with 200×200 μm² each	Micrometers	High redundancy, 80,000 on one wafer	Waviness, non-flat substrate	GaN, GaAs
Grating on lasers (VCSEL)		1000/year	VCSEL				
Window light redirection (LDIR)	200 nm	Huge	Up to 6×3 m²	Stitching of large fields, subwavelength	No visible defects (seam lines, color)	Sol-gel materials compatible with glass	Float glass
Bio substrates for DNA analysis	200 nm	>60 per hour	Rectangular 150×80 mm²	–			Glass/quartz

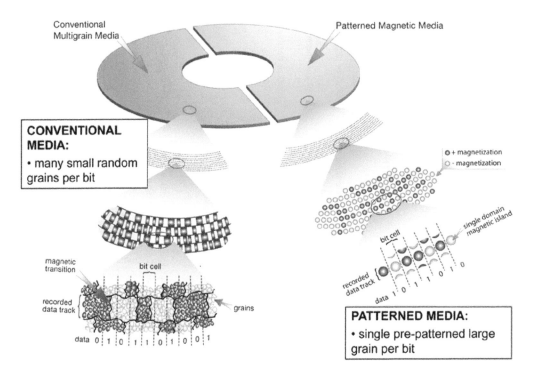

CONVENTIONAL MEDIA:
• many small random grains per bit

PATTERNED MEDIA:
• single pre-patterned large grain per bit

FIGURE 11.60 Comparison between conventional hard drive media and BMP.

during the data writing process. For decades, films with smaller grains have been deposited to support high-density recording. Today, however, grain sizes have become so small that further shrinkage would cause the magnetization of the individual grains to be unstable. If a significant fraction of the grains on the disk flip spontaneously, the data stored on the disk erases itself!

The International Disk Drive Equipment and Materials Association (IDEMA) has published a roadmap (see Figure 11.61) for the advancement of device density, and the two most promising paths include heat assisted magnetic recording (HAMR), which utilizes more robust magnetic materials, and BPM. With BPM (see the right-hand side of Figure 11.60), each bit is stored in a single deliberately formed magnetic switching volume rather than a collection of grains. Since each island is a single magnetic domain, patterned media are thermally stable even at densities far higher than can be achieved with conventional media.

The introduction of a lithographic process to the fabrication of the media represents a paradigm for the hard drive industry. Previous process flows required no lithographic step. The deposition of the magnetic thin film is followed by the application of a hard mask. A dense array of islands is then lithographically formed in a resist layer. Magnetic islands are formed by sequentially etching through the hard mask and magnetic material. Although the fabrication process is relatively standard when compared with the steps used to pattern semiconductor devices, the size and pitch of the magnetic array of dots, along with the cost requirements of the patterning sequence, are substantially different for the two industries. For the semiconductor industry, the lithographic cost per wafer level of a critical patterning layer can vary between $25 and $75, that is, $0.10 to $0.30 per die on a 300 mm wafer (total number of 250 dies assumed). For the hard drive media, the added cost per disk is limited to ~$1, a 25×–75× reduction in cost.

Pattern resolution has historically been driven by optical projection systems that use complex lens stacks to reduce the pattern generated on a mask by a factor of 4× or 5×. These systems are limited in resolution to a half-pitch of about 40 nm. By comparison, the crossover point between conventional media and patterned media is expected to occur at about 1.5 Tb/in². At this density,

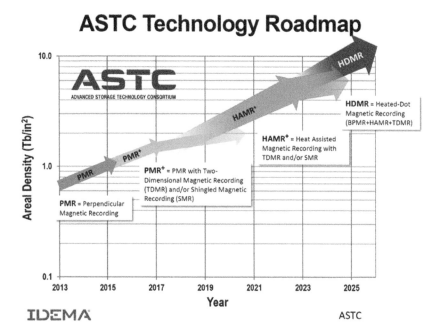

FIGURE 11.61 IDEMA roadmap, in which heat assisted magnetic recording (HAMR) is first introduced and later combined with BMP.

the half-pitch requirements will be <12 nm! Other lithographic solutions being considered for future generations of semiconductor ICs include EUVL, electron beam lithography (EBL), and NIL. Throughput and cost prohibit the use of both EUVL and EBL. There is total consensus in the hard drive industry that the only patterning process that can address both resolution and throughput requirements for BPM is NIL. However, the only way to meet the cost targets for a PM disk is to minimize material usage by only depositing the imaging resist on the disk surface.

J-FIL, discussed in Section 11.6, is a derivative UV-NIL process that uses inkjet-based dispensers with drop volumes as small as 1.0 pl. Inkjetting allows resist to be deposited at densities that match the relief image density of the template, thereby ensuring a uniform residual layer of resist. This is critical for disk drive media that contain both dense data zones and relatively sparse servo zones. An inkjet approach also avoids the generation of resist waste by conventional spin-coated resist processes. For a typical 65 mm disk, the J-FIL system applies only 100 nl to either side of the disk. For a conventional spin-coat process, approximately 1 ml is required per disk side, and all but 100 nl of the resist is collected in a waste receptacle, where it is either disposed or put through a costly recycling process. Spin-on resists typically cost about $1000 per liter, meaning that the entire budget for creating the patterned media would be exceeded by the spin-on resist material itself.

The primary processing steps for fabricating conventional continuous media consist of a disk clean, magnetic material sputter deposition, carbon overcoat, lubrication, burnish, and flight test. To create the patterned media disk, a second set of steps is integrated into the current media fabrication process flow (Figure 11.62) [198]. The discussion below focuses on the fabrication of a template for this application and some examples of both the imprint process and the device performance.

The fabrication of a master template for patterned media is interesting, because it requires a combination of lithographic technologies discussed in this chapter and in more detail elsewhere in this book. Because magnetic bits are read by the thin film head as the disk spins, a rotary stage electron beam system such as the Pioneer Corp. EBR-401 is needed to expose the pattern [199].

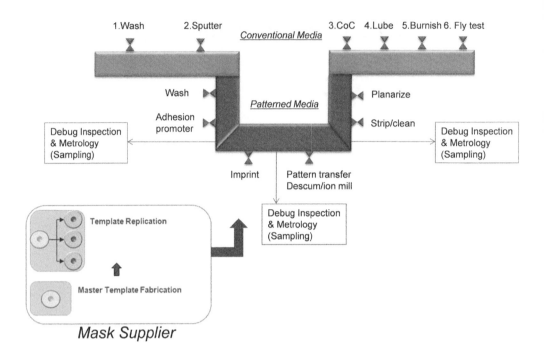

FIGURE 11.62 Process flow to fabricate BPM. The new processing steps needed in the process flow are indicated in in the diagram.

These systems typically employ a Gaussian beam writing strategy, and therefore, the time needed to expose a master template is measured in days rather than hours. In addition, the uniform writing of lines or dots at half-pitches below 12 nm across a 65 mm diameter is a daunting task even for a high-resolution electron beam writer.

To mitigate both write time and patterning issues, it is possible to increase pattern density by using either spacer multiple patterning processes (Chapter 3) or directed self-assembly (DSA) processes (Chapter 13). Using a spacer process, for example, 24 nm half-pitch tracks can be reduced to 12 nm. The tracks can then be converted to individual bits as a second masking pattern to cut the lines into isolated islands [200]. Alternatively, a relaxed dot pattern can be densified using a DSA approach [201]. One final point of interest is that servo zone patterns must also be defined during the creation of the dense array areas. Xiao et al. [202] discuss this process in detail.

Just as in the semiconductor case, electron beam templates cannot be used for the final patterning of the media because of the high cost associated with the mastering process. Instead, replica templates must be created, and Molecular Imprints Inc. also offered template replication tools to the hard drive industry. To complete the process, a high-throughput imprint tool that can print on both sides of a disk is required. MII offered a 350 disk per hour system under the name NuTera HD7000. At this throughput, the imprint CoO per disk is about 30 cents. Figure 11.63a shows a 1.5 Tb/in^2 pattern after imprint and pattern transfer into the CoCrPt medium. Figure 11.63b depicts the typical hysteresis curve for a single bit [203].

Realization of the technology for production will be dictated by the maturity of the template fabrication process and the demonstration of a bit pattern with a low enough data error rate. As the technology continues to be developed, media companies (i.e., Toshiba, Western Digital, and Seagate) have first pursued the insertion of HAMR, primarily because of the lower capital outlay relative to a lithographic approach. It is anticipated that both approaches will be integrated for media in order to further higher densities. A conclusive review of the state of the art in 2015 by Western Digital is given by Albrecht et al. [204].

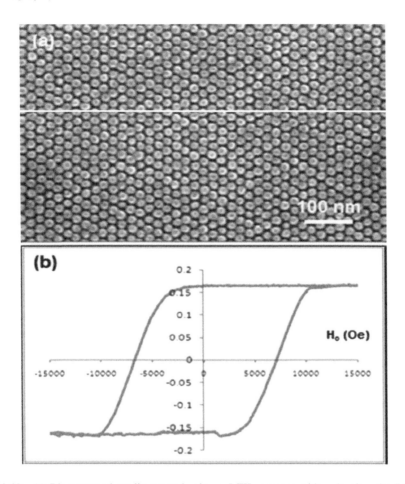

FIGURE 11.63 (a) Bit patterned media created using a J-FIL process with a density of 1.5 Tbit/in². (b) Corresponding magnetic behavior demonstrating bit switching characteristics.

11.8.2 FLAT PANEL DISPLAYS

The flat panel market now has an impact on everyday life. While the technology displaced conventional cathode-ray tubes and was immediately picked up by the television industry, flat panel plays have found their way into monitors, smartphones, tablets, watches, and other wearable devices. Display applications, including liquid crystal displays (LCD), organic light-emitting diode (OLED), and flexible displays, are particularly interesting because of the ability to impact multiple levels in the basic display. Figure 11.64 depicts the key components in the LCD panel. Of particular interest are the polarizer, dual-brightness enhancement film (DBEF), thin film transistor, and color filter. Roll-based imprinting, as introduced in the previous section, has the opportunity to create high-performance components within the display while improving the CoO of the panel. As an example, an integrated wire grid polarizer can potentially replace both the existing organic film polarizer and the DBEF, which is used for light recycling [205, 206]. Recent publications have also reported on the application of plasmonic color filters as a replacement for low-transmission dye-based color filters [207, 208]. In addition, thin film transistor arrays have been fabricated with a single–lithography step process (SAIL) by Hewlett Packard [184, 185]. Outside LCDs, equally attractive applications exist for the OLED market (such as a circular polarizers), as well as the patterning required for flexible display devices. In this section, we limit the discussion to polarizers and color filters, since nanoscale patterning with NIL technology provides a path forward to scaling the technology as well as meeting the lithography demands for both applications.

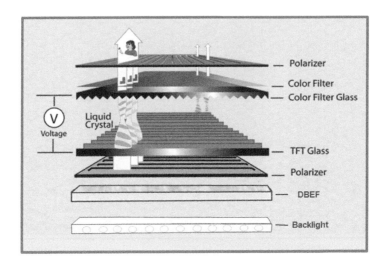

FIGURE 11.64 Schematic illustration of the component within a liquid crystal flat panel display.

There are two major criteria for introducing new technology into the flat panel process flow. The first is cost; the second is performance. Performance has a profound impact on battery life for mobile applications, as each component in the display reduces the amount of light, thereby requiring greater power from the initial backlight source.

11.8.3 POLARIZERS

The LCD polarizer market revenue is well over $10 billion. Films are supplied by vendors such as Nitto-Denko and LG Chemical, and polarization effects are achieved by preferentially embedding iodine within the films to achieve high contrast and transmission. To achieve higher transmission, panels are typically fitted with both a polarizer and a dual-brightness enhancement film (DBEF). The DBEF film processes the rejected polarized light off the first polarizer and returns it back to the polarizer. While this is effective, the cost is impacted by the need to provide a polarizer film that must be attached to the panel and by the cost of a second DBEF film to enhance the light output.

WGPs are already used in digital projectors [209]. The combination of performance and temperature durability makes their use an attractive choice for this market. Their application to larger displays, including mobile phones, tablets, monitors and TVs has been limited by an inability to scale the WGP to the required areas for these markets. A roll based printing process enables printing over substantially larger areas and therefore addresses the requirements of both performance and CoO. In addition, it may be possible to integrate the WGP directly onto a flat panel, thereby eliminating the requirement of a separate film and an extra lamination step.

WGPs typically consist of a grating pattern of aluminum lines on a transparent substrate. Pitches below about 150 nm are necessary in order to achieve both good transmission and polarization. Aluminum features can be formed either through a standard subtractive etch process or by first patterning a grating on a resist film and subsequently depositing the aluminum. Several companies have fabricated large-area WGPs. Asahi-Kasei has employed a deposition process to create WGPs on 200 mm wide flexible films [210]. In 2014, MII introduced a LithoFlexTM 350 imprint system capable of patterning on either webs or flat panels up to 370×470 mm^2 (Gen2 panel) [211]. An example of a web-based WGP made by Asahi-Kasei is shown in Figure 11.65a. The extinction ratio (up to 50,000) and transmission (up to 44%) of a high-end WGP made on a glass substrate by MII are shown in Figure 11.65b and c [212]. The two curves in each graph demonstrate the advantages of decreasing the pitch of the aluminum wires from 130 to 100 nm. Similar large-area processes have also been implemented by other companies.

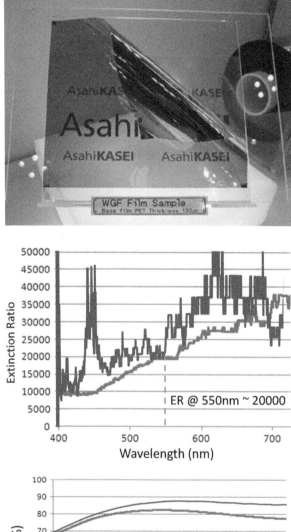

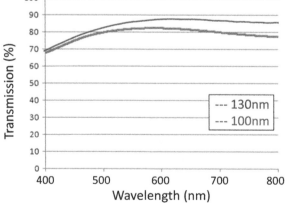

FIGURE 11.65   (a) Web-based WGP from Asahi-Kasei. (b) and (c) Extinction ratio and transmission of a high-end WGP made on a glass substrate.

11.8.4 Color Filters

Color filters are used in LCD panels as a means of defining the color of the final display image. Wells are typically imaged on a panel and filled with dye-based materials to define the standard red, green, and blue color of the display. Transmission loss is again an issue, and alternative methods for defining colors, such as plasmonic color filters, are being explored by several companies and researchers.

A plasmonic device takes advantage of the interaction between electromagnetic fields and free electrons in a metal. Free electrons in the metal can be excited by the electric component of light to have collective oscillations, thereby enhancing light output at particular wavelengths. The wavelength of the device is determined by both the pitch and the duty cycle of the metal used. Because of the relatively short wavelengths (~400–800 nm) that are relevant for displays, Al-based structures are again a good choice, as the short plasmon resonance wavelength of Al enables nanostructures embedded in a dielectric material to have surface plasmon–polariton resonances in the appropriate wavelength range. Al also offers a cost benefit in comparison to materials such as Ag and Au, which are more typically employed for plasmonic structures.

Arrays of either subwavelength holes or lines can be used to create an effective plasmonic filter. Both transmissive and reflective filters are possible. Kaplan et al. fabricated high-performance reflective filters using an NIL process [213]. Transmissive devices have been published by both Yokogawa et al. and Inoue et al. [207, 208] Patterns were defined using either electron beam or focused ion beam processes, which while effective for small-area demonstration, cannot be scaled to address any practical manufacturing. Figure 11.66 shows the effect of varying pitch and duty cycle on an aluminum-based device to create a large array of colors [208].

11.8.5 Photovoltaic Solar Cells

Photovoltaic (PV) cells have similarities with CMOS wafers, but in comparison to semiconductor chip and LED patterning, the entire substrate (typically squares covering an area of 156×156 mm^2) is used as a device, often combined with other wafers in a square meter–sized panel. Therefore, the entire surface of a cell has to be patterned in a single step. The wafers are 180 μm thick, are very fragile, and are sometimes made of material that is not entirely flat. Due to its high resolution and applicability for large-area patterning, NIL is a promising technology for PV applications, too. Possible applications in this field concern the optical properties (light in-coupling [213], light trapping [214, 215], and plasmonics [216, 217]) or the patterning of device features (e.g., metallization or diffusion barriers). However, the requirements set by the PV industry, especially for the fabrication of wafer-based silicon solar cells, are not easy to meet. Although wafer thicknesses today are 180 μm, the International Technology Roadmap for Photovoltaic [218] calls for thicknesses to be decreased further to 150 μm over the next few years. Although the printing on such thin and brittle substrates will be challenging for NIL, it also represents an opportunity to introduce novel processes into fabrication through the introduction of sophisticated photon management structures. Such structures will be required to maintain high absorption and thus, high quantum efficiencies (see Figure 11.67). Other processing challenges include patterning on wire saw–cut substrates that introduce a total thickness variation as high as 10% and have a surface roughness substantially greater than a silicon wafer used for IC manufacturing.

At Fraunhofer ISE, the development of NIL processes on the novel SmartNIL technology from EVG with a focus on PV applications is described. In this tool, a soft stamp is used to imprint onto a resist-coated substrate. The demolding of the stamp in particular is crucial when handling very thin substrates. Therefore, this step has to be realized as a very controlled sequential process, whereby the stamp is demolded smoothly, starting on one side, and peeled off to the other side. Aspects such as adaptability to uneven surfaces are tackled by using soft stamps made of PDMS or other materials. Homogeneous residual layers can be achieved by applying a uniform pressure using soft polymeric stamps. Other institutes, such as Holtz together with Philips, use SCIL for PV patterning.

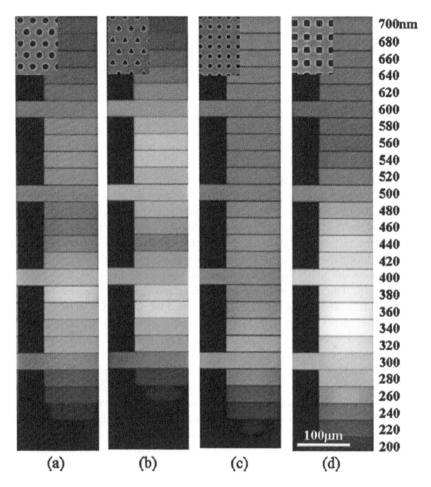

FIGURE 11.66 Plasmonic-based transmissive color filter results. (a) Circular holes arranged on a hexagonal lattice, (b) triangular holes, (c) circular holes arranged on a square lattice, and (d) square holes.

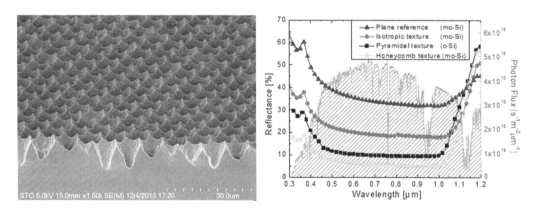

FIGURE 11.67 SEM micrograph of honeycomb texture in multicrystalline silicon. The corresponding reflectance is depicted in the graph below, demonstrating a huge reduction in reflectance in comparison with no or different surface patterns. (Copyright Fraunhofer. Reproduced and adapted from *Energy Procedia*, 6, Bläsi, B., et al., Photon management structures originated by interference lithography, 712–718, © 2011, with permission from Elsevier and *Microelectron. Eng.*, 98, Wolf, A.J., et al., Origination of nano- and microstructures on large areas by interference lithography, 293–296, © 2012, with permission from Elsevier.)

Applications that that are being pursued include the realization of honeycomb textures on multi-crystalline silicon solar cells, the integration of features for fine line metallization (width ~10 μm) into the templates for such textures, and the fabrication of very fine metal contacts with prismatic profiles (width ~2 μm) for use in highly efficient multi-junction solar cells [219–221]. In the context of the honeycomb texturing, interference lithography (IL) as a mastering technology will be introduced, and the efficiency enhancement of more than 0.5% absolute on large-area multicrystalline silicon solar cells (156 × 156 mm²) processed using an industrial-scale pilot line will be presented. For the integration of features for metal contacts already in the template used for the stamp fabrication for NIL processes, different lithographic processes are combined (IL, photolithography, and NIL).

Similarly to LED surface patterning, for a textured surface for silicon solar cells, two effects for absorption enhancement are used: reduced front surface reflectivity and internal path-length enhancement. The state of the art in industry is stochastic textures by etching, realizing an anti-reflective moth eye effect, which turns a reflective silicon surface into a dark, absorptive surface. Higher efficiencies are achieved by defined textures, for which good results were achieved on the lab scale. The benefit of defined textures is particularly pronounced for multicrystalline silicon. A photolithography-based solution with micrometer-sized honeycomb frontside texture was achieved using photolithography. The high cost of the photolithography process was considered prohibitive for a PV production line, and it was subsequently replaced by using a UV-NIL process and a soft stamp, The soft method has excellent adaptability to rough surfaces, yields homogeneous and low residual layer thickness (<100 nm) on large substrates, and was successfully tested on very thin wafer substrates (50 μm). A 3-D structure created on a 100 μm thick solar cell featured diffractive grating and exhibited a higher short circuit current (JSC) than a planar cell with 250 μm thickness. Apart from frontside texturing, other possibilities for use of NIL in PV manufacturing could be envisaged. These include the reduction of back reflection and deflection into the solar cell and patterning of wire metallization, for example, by using transparent fishnet electrodes or 3-D metal wires [222, 223]. Apart from this, polymer solar cells will profit from the patterning capabilities of interdigitated heterojunctions to enable both efficient charge separation and transport [224, 225].

11.8.6 Patterned Sapphire Substrates

GaN-based white LEDs have many advantages over incandescent light sources, including lower energy consumption, longer lifetime, improved physical robustness, smaller size, and faster switching. One of the major drivers for cost reduction is the increase of efficiency. However, the large difference in refractive index through the smooth interface between air (n = 1) and GaN (n = 2.5) leads to considerable Fresnel loss and total internal reflection, with a subsequent decrease in light extraction efficiency. In order to produce more efficient LEDs, manufacturers have developed PSS technology. If the polished sapphire wafer is patterned before GaN growth, more light is extracted, therefore improving the LED substrate efficiency (see Figure 11.68) [226–228]. In 2014, 87% of fabricated LEDs were processed on PSS [229].

The use of micropatterned PSS (μPSS) has been widely adopted in the industry, because epitaxial quality and light extraction efficiency can be improved simultaneously. This is done by patterning a resist using (1×) projection photolithography steppers and etching a regular microtopography (e.g., an orthogonal pattern with about 3 μm period) into a sapphire substrate. This pattern can be transferred into sapphire by (wet or dry) etching. After etching, the rectangular shape of the resist results in square pyramids, cones, or domes with sizes of about 3 μm and 1.0–2.0 μm height due to the crystalline orientation of sapphire. Using this, both micro- and nanopatterned PSS can be fabricated, see Figure 11.69 [230]. The controlled texture increases the efficiency by as much as 30% if compared with nonpatterned sapphire LEDs [226], and it leads to a lower density of defects within the film due to growth on a three-dimensional landscape that spurs earlier coalescence of GaN epitaxial islands during metal–organic chemical vapor deposition (MOCVD) growth. Apart from PSS,

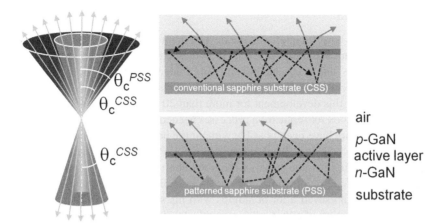

air
p-GaN
active layer
n-GaN
substrate

FIGURE 11.68 Schematics of the light paths in an unpatterned and a patterned sapphire substrate (left side). The emission cone can be greatly enhanced by patterned surfaces.

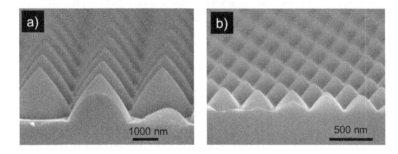

FIGURE 11.69 Surface patterns on sapphire substrates. (a) Micropatterned PSS and (b) nanopatterned PSS. (Reproduced from Ulvac Inc., Japan. © *Ulvac Vacuum Magazine*.)

the light-emitting surface of the high-brightness LEDs can be patterned with a quasi-crystalline (photonic crystal) array. For both cases, PSS and PC, the ability to create wafer-scale, stitching error–free patterns of nanostructures is essential. For this, NIL is ideally suited, and it may be the most economic choice if production has to be scaled up to 150 mm substrates. In addition to this, NIL is able to place patterns on bowed and warped surfaces and even on topography. Since the light extraction efficiency is increased further by the use of smaller patterns on PSS, the micrometer-sized domes are too large for patterning based on projection lithography steppers. Therefore, several NIL companies have delivered production machines to PSS manufacturing, and thus, NIL may be able replace conventional photolithography.

To keep the LED industry competitive, the epitaxy CoO must fall 50% every 5 years, according to the U.S. Department of Energy [231]. Different NIL tool providers have entered the PSS manufacturing business. Obducat has confirmed the installation of a SINDRE wafer track system for μPSS manufacturing in 100 mm sapphire substrates in a production environment. Other NIL companies such as EVG, SUSS/Philips, and AimCore are also active in this field. While Ø 50 mm sapphire substrates are less and less used, Ø 100 mm has become standard in Asia, with prospects of Ø 150 mm, which is already used in Europe. For this, a scalable technology is needed. NIL has the advantage that it is not only capable of scaling up to larger substrate sizes but also able to enter into the market once nanopatterned PSS is needed. Besides the availability of old refurbished optical steppers, the non-flat characteristic of sapphire wafers is another reason for manufacturers to switch to NIL; the bow and warp of wafers of 20 μm (Ø100 mm wafers) calls for a conformal lithographic technique with no restrictions on depth of focus.

ACKNOWLEDGMENTS

The basis for writing this book chapter was the development of nanoimprint lithography, starting with Steve Chou's first demonstration of NIL in 1995 down to 10 nm resolution. Many thanks are due to all those researchers, engineers, technicians, and students who contributed to the continuous development of the NIL technology in industry and research labs. Within Paul Scherrer Institut, we were able to participate in this development for more than 20 years via technology, tools, and processes, that is, the full toolbox for replication processes needed for academic research and industrial applications. This would not have been possible without the help of our excellent collaborators. We want to thank Jens Gobrecht, Laura Heyderman, Sunggook Park, Arne Schleunitz, Mirco Altana, Christian Spreu, Robert Kirchner, the entire staff of the Laboratory for Micro- and Nanotechnology, and particularly Konrad Vogelsang, who as a dedicated technician has continuously helped to keep our technology running and saved a lot of stamps with his skills and good humor. The development of NIL for the fabrication of advanced semiconductor devices has been a long and satisfying journey. Our thanks go to Grant Willson for his pioneering work on Step and Flash Nanoimprint Lithography. Molecular Imprints was founded to develop the concept. Canon recognized the potential of the technology and has taken it to the point where the first production tools are now out in the field. As a result, we would also like to acknowledge the efforts of all the folks from Molecular Imprints Inc., Canon Nanotechnologies Inc., and Canon Inc.

REFERENCES

1. A. Kapr and Johann Gutenberg, *The Man and His Invention*, Scolar Press, Aldershot, and Ashgate Publishing Company, Brookfield, VT, 3rd edition (1996).
2. J.H. Lienhard, Engines of our ingenuity, No. 735, *Johann Gutenberg 1988–1997*, http://www.uh.edu/engines/epi753.htm.
3. S.Y. Chou, P.R. Krauss, and P.J. Renstrom, Imprint of sub-25nm vias and trenches in polymers, *Appl. Phys. Lett.* **67**(21) (1995) 3114–3116.
4. S. Fujimori, Fine pattern fabrication by the molded mask method (nanoimprint lithography) in the 1970s, *Jap. J. Appl. Phys.* **48**(6) (2009) 06FH01 (7 pp).
5. J. Haisma, M. Verheijen, K. van den Heuvel, and J. van den Berg, Mold-assisted lithography: A process for reliable pattern transfer, *J. Vac. Sci. Technol. B* **14**(6) (1996) 4124–4128.
6. D.J. Resnick, et al., Imprint lithography: Lab curiosity or the real NGL?, *Proc. SPIE* **5037** (2003) 12.
7. J.A. Liddle and G.M. Gallatin, Nanomanufacturing: A perspective, *ACS Nano* (online) (Feb. 2016) doi: 10.1021/acsnano.5b03299.
8. W. Trybula, Cost of ownership – Projecting the future, *Microelectron. Eng.* **83**(4–9) (2006) 614–618.
9. M. Melliar-Smith, presented at the panel discussion on Economics of Lithography for Alternative Applications, SPIE Advanced Lithography Symposium, March 2, 2011.
10. S.Y. Chou and P.R. Krauss, Imprint lithography with sub-10 nm feature size and high throughput, *Microelectron. Eng.* **35**(1–4) (1997) 237–240.
11. H. Schift and A. Kristensen, *Nanoimprint Lithography – Patterning Resists Using Molding. Chapter (Part A/9) in Handbook of Nanotechnology*, ed. B. Bhushan, Springer Verlag, Berlin Heidelberg, Germany. ISBN: 978-3-642-02524-2, XLVIII, with DVD, 3rd edition, 271–312 (2010).
12. Y. Xia, E. Kim, and G. Whitesides, Microcontact printing of alkanethiols on silver and its application in microfabrication, *J. Electrochem. Soc.* **143**(3) (1996) 1070–1079.
13. F. Hua, Y. Sun, A. Gaur et al., Polymer imprint lithography with molecular-scale resolution, *Nano Lett.* **4**(12) (2004) 2487–2471.
14. H. Schift, P. Urwyler, P.M. Kristiansen, and J. Gobrecht, Nanoimprint lithography process chains for the fabrication of micro- and nanodevices, *Proc. SPIE* **9049** (2014).
15. M. Heckele and W.K. Schomburg, Review on micro molding of thermoplastic polymers, *J. Micromech. Microeng.* **14**(3) (2004) R1.
16. B. Heidari, I. Maximov, E.-L. Sarwe, and L. Montelius, Large scale nanolithography using imprint lithography, *J. Vac. Sci. Technol. B* **17**(6) (1999) 2961–2964.
17. B. Heidari, I. Maximov, and L. Montelius, Nanoimprint lithography at the 6 in. wafer scale, *J. Vac. Sci. Technol. B* **18**(6) (2000) 3557–3560.

18. H. Schift, Nanoimprint lithography: An old story in modern times? A review, *J. Vac. Sci. Technol. B* **26**(2) (2008) 458–480.

19. L.J. Heyderman, H. Schift, C. David, J. Gobrecht, and T. Schweizer, Flow behaviour of thin polymer films used for hot embossing lithography, *Microelectron. Eng.* **54**(3–4) (2000) 229–245.

20. H.-C. Scheer and H. Schulz, A contribution to the flow behaviour of thin polymer films during hot embossing lithography, *Microelectron. Eng.* **56**(3–4) (2001) 311–332.

21. Y. Hirai, Y. Onishi, T. Tanabe, et al., Pressure and resist thickness dependency of resist time evolutions profiles in nanoimprint lithography, *Microelectron. Eng.* **85**(5–6) (2008) 842–845.

22. H.D. Rowland, A.C. Sun, P.R. Schunk, and W.P. King, Impact of polymer film thickness and cavity size on polymer flow during embossing: Toward process design rules for nanoimprint lithography, *J. Micromech. Microeng.* **15**(12) (2005) 2414–2425.

23. T. Leveder, S. Landis, L. Davoust, and N. Chaix, Flow property measurements for nanoimprint simulation, *Microelectron. Eng.* **84**(5–8) (2007) 928–931.

24. T. Leveder, S. Landis, L. Davoust, and N. Chaix, Optimization of demolding temperature for throughput improvement of nanoimprint lithography, *Microelectron. Eng.* **84**(5–8) (2007) 953–957.

25. M. Tormen, R. Malureanu, R.H. Pedersen, et al., Fast thermal nanoimprint lithography by a stamp with integrated heater, *Microelectron. Eng.* **85**(5–6) (2008) 1229–1232.

26. M. Tormen, E. Sovernigo, A. Pozzato, M. Pianigiani, M. Tormen, Sub-100 μs nanoimprint lithography at wafer scale, *Microelectron. Eng.* **141** (2015) 21–26.

27. M. Pianigiani, R. Kirchner, E. Sovernigo, et al., Effect of nanoimprint on the elastic modulus of PMMA: Comparison between standard and ultrafast thermal NIL, *Microelectron. Eng.* **155** (2016) 85–91.

28. S.Y. Chou, C. Keimel, and J. Gu, Ultrafast and direct imprint of nanostructures in silicon, *Nature* **417**(6891) (2002) 835–837.

29. M. Colburn, et al., Step and flash imprint lithography: A new approach to high resolution patterning, *Proc. SPIE* **3676** (1999) 379–385.

30. M. Komuro, J. Taniguchi, S. Inoue, et al., Imprint characteristics by photo-induced solidification of liquid polymer, *Jpn. J. Appl. Phys.* **39**(1, No. 12B) (2000) 7075–7079.

31. M. Otto, M. Bender, B. Hadam, B. Spangenberg, H. Kurz, Characterization and application of a UV-based imprint technique, *Microelectron. Eng.* **57–58** (2001) 361–366.

32. B. Vratzov, A. Fuchs, M. Lemme, W. Henschel, H. Kurz, Large scale ultraviolet-based nanoimprint lithography, *J. Vac. Sci. Technol. B* **21**(6) (2003) 2760–2764.

33. Y. Hirai, H. Kikuta, and T. Sanou, Study on optical intensity distribution in photocuring nanoimprint lithography, *J. Vac. Sci. Technol. B* **21**(6) (2003) 2777–2782.

34. X. Liang, H. Tan, Z. Fu, and S.Y. Chou, Air bubble formation and dissolution in dispensing nanoimprint lithography, *Nanotechnology* **18** (2007) 025303 (7 pp).

35. H. Hiroshima, M. Komuro, N. Kasahara, Y. Kurashima, J. Taniguchi, Elimination of pattern defects of nanoimprint under atmospheric conditions, *Jpn. J. Appl. Phys.* **42**(1, No. 6B) (2003) 3849.

36. H. Hiroshima and M. Komuro, Control of bubble defects in UV nanoimprint, *Jpn. J. Appl. Phys.* **46**(9B) (2007) 6391–6394.

37. H. Hiroshima and K. Suzuki, Throughput of ultraviolet nanoimprint in pentafluoropropane using spin coat films under thin residual layer conditions, *Jpn. J. Appl. Phys.* **51** (2012) 06FJ10.

38. S. Matsui, H. Hiroshima, Y. Hirai, and M. Nakagawa, Innovative UV nanoimprint lithography using a condensable alternative chlorofluorocarbon atmosphere, *Microelectron. Eng.* **133** (2015) 134–155.

39. K. Pfeiffer, F. Reuther, M. Fink, et al., A comparison of thermally and photochemically cross-linked polymers for nanoimprinting, *Microelectron. Eng.* **67–68** (2003) 266–273.

40. X. Cheng and L.J. Guo, A combined-nanoimprint-and-photolithography patterning technique, *Microelectron. Eng.* **3–4** (2004) 277–282.

41. M. Beck and B. Heidari, *OnBoard Technology* (2006) 52–55, http://www.Onboard-Technology.com/.

42. C.G. Willson, R.A. Dammel, and A. Reiser, Photoresist materials: A historical perspective, *Proc. SPIE* **3049** (1997) 28–41.

43. M.D. Stewart and C.G. Willson, Photoresists, *Encyclopedia of Materials: Science and Technology*, 6973–6978 (2001).

44. F. Xu, N. Stacey, M. Watts, et al., Development of imprint materials for the step and flash imprint lithography process, *Proc. SPIE* **5374** (2004) 232–241.

45. M. Vogler, S. Wiedenberg, M. Mühlberger, et al., Development of a novel, low-viscosity UV-curable polymer system for UV-nanoimprint lithography, *Microelectron. Eng.* **84**(5–8) (2007) 984–988.

46. P. Voisin, M. Zelsmann, R. Cluzel, et al., Characterisation of ultraviolet nanoimprint dedicated resists, *Microelectron. Eng.* **84**(5–8) (2007) 967–972.

47. H. Schmitt, L. Frey, H. Ryssel, M. Rommel, C. Lehrer, UV nanoimprint materials: Surface energies, residual layers, and imprint quality, *J. Vac. Sci. Technol. B* **25**(3) (2007) 785–790.
48. C.-C. Yu and H.-L. Chen, Nanoimprint technology for patterning functional materials and its applications, *Microelectron. Eng.* **132** (2015) 98–119.
49. L.J. Guo, Nanoimprint lithography: Methods and material requirements, *Adv. Mat.* **19**(4) (2007) 495–513.
50. AMO GmbH, *Gesellschaft für Angewandte Mikro- und Optoelektronik*, Aachen, Germany, http://www.amo.de/de/.
51. Micro resist technology GmbH, Berlin, Germany, www.microresist.com.
52. Toyo Gosei Co., Ltd, Tokyo, Japan, www.toyogosei.co.jp/eng/.
53. Nanonex, Princeton, NJ, USA, http://www.nanonex.com/.
54. Obducat AB, Lund, Sweden, http://www.obducat.com/.
55. EV Group (EVG), St. Florian, Austria, http://www.evgroup.com/.
56. Molecular Imprints Inc., Austin, Texas, http://www.molecularimprints.com/.
57. H. Schift, Nanoimprint lithography: 2D or not 2D? A review, *Appl. Phys. A* **121**(2) (2015) 415–435.
58. M.D. Stewart, J.T. Wetzel, G.M. Schmid, et al., Direct imprinting of dielectric materials for dual dama-scene processing, *Proc. SPIE* **5751** (2005) 210–218.
59. H.-J. Kim, M. Almanza-Workman, B. Garcia, et al., Roll-to-roll manufacturing of electronics on flexible substrates using self-aligned imprint lithography (SAIL), *J. Soc. Inf. Disp.* **17**(11) (2009) 963–970.
60. X. Cheng and L.J. Guo, One-step lithography for various size patterns with a hybrid mask-mold, *Microelectron. Eng.* **3–4** (2004) 288–293.
61. R. Ji, M. Hornung, M.A. Verschuuren, et al., UV enhanced substrate conformal imprint lithography (UV-SCIL) technique for photonic crystals patterning in LED manufacturing, *Microelectron. Eng.* **87**(5–8) (2010) 963–967.
62. M.A. Verschuuren, P. Gerlach, H.A. van Sprang, and A. Polman, Improved performance of polariza-tion-stable VCSELs by monolithic sub-wavelength gratings produced by soft nano-imprint lithography, *Nanotechnol.* **22** (2011) 505201 (9pp).
63. B. Bläsi, H. Hauser, O. Höhn, et al., Photon management structures originated by interference lithogra-phy, *Energy Procedia* **6** (2011) 712–718.
64. A.J. Wolf, H. Hauser, V. Kübler, et al., Origination of nano- and microstructures on large areas by inter-ference lithography, *Microelectron. Eng.* **98** (2012) 293–296.
65. Fraunhofer Institute for Solar Energy Systems ISE, Freiburg, Germany, http://www.ise.fraunhofer.de/.
66. SCIVAX Nanoimprint Solutions, Japan, http://www.scivax.com/nano_print.html.
67. J.L. Wilbur, A. Kumar, E. Kim, and G. Whitesides, Microcontact printing microfabrication by micro-contact printing of self-assembled monolayers, *Adv. Mater.* **6**(7–8) (1994) 600–604.
68. B. Michel, A. Bernard, A. Bietsch et al., Printing meets lithography: Soft approaches to high-resolution patterning, *IBM J. Res. Dev.* **45**(5) (2001) 697–719.
69. Y. Xia and G.M. Whitesides, Soft lithography, Angew, *Chem. Int.* **37**(5) (1998) 550–575.
70. D. Qin, Y. Xia, and G.M. Whitesides, Soft lithography for micro- and nanoscale patterning, *Nat. Protoc.* **5**(3) (2010) 491–502.
71. Dow Corning Corp., Auburn MI, USA, http://www.dowcorning.com/.
72. J.J. Kennan, Siloxane copolymers. In: *Siloxane Polymers*, eds. S.J. Clarson and J.A. Semlyen, Prentice Hall, Englewood Cliffs, NJ, 121–124 (1993).
73. H. Schmid and B. Michel, Siloxane polymers for high-resolution, high-accuracy soft lithography, *Macromolecules* **33** (2000) 3042–3049.
74. T.W. Odom, Venkat R. Thalladi, J. Christopher Love, and George M. Whitesides, Generation of 30–50 nm structures using easily fabricated, composite PDMS masks, *J. Am. Chem. Soc.* **124**(41) (2002) 12112–12113.
75. U. Plachetka, M. Bender, A. Fuchs, et al., Comparison of multilayer stamp concepts in UV-NIL, *Microelectron. Eng.* **83**(4-9) (2006) 944–947.
76. M. Bender, U. Plachetka, J. Ran, et al., High resolution lithography with PDMS molds, *J. Vac. Sci. Technol. B* **22**(6) (2004) 3229–3232.
77. N. Koo, U. Plachetka, M. Otto, et al., The fabrication of a flexible mold for high resolution soft ultravio-let nanoimprint lithography, *Nanotechnology* **19** (2008) 225304 (4pp).
78. K.M. Choi and J.A. Rogers, A photocurable poly(dimethylsiloxane) chemistry designed for soft litho-graphic molding and printing in the nanometer regime, *J. Am. Chem. Soc.* **125**(14) (2003) 4060–4061.
79. Shin-etsu Chemicals, Tokyo, Japan, http://www.shinetsu.co.jp, and Shin-etsu silicones, http://www.shi-netsusilicones.com/shindex.html.

80. R.S. Kane, S. Shuichi, T. Takayama, D.E. Ingber, G.M. Whitesides, Patterning proteins and cells using soft lithography, *Biomaterials* **20**(23–24) (1999) 2363–2376.
81. A. Perl, D.N. Reinhoudt, and J. Huskens, Microcontact printing: Limitations and achievements, *Adv. Mater.* **21**(22) (2009) 2257–2268.
82. J.A. Helmuth, H. Schmid, R. Stutz, A. Stemmer, and H. Wolf, High-speed microcontact printing, *J. Am. Chem. Soc.* **128**(29) (2006) 9296–9297.
83. G.P. Lopez, H.A. Biebuyck, R. Harter, A. Kumar, G.M. Whitesides, Fabrication and imaging of two-dimensional patterns of proteins adsorbed on self-assembled monolayers by scanning electron microscopy, *J. Am. Chem. Soc.* **115**(23) (1993) 10774–10781.
84. P.M. St. John, R. Davis, N. Cady, et al., Diffraction-based cell detection using a microcontact printed antibody grating, *Anal. Chem.* **70**(6) (1998) 1108–1111.
85. R. Singhvi, A. Kumar, G. Lopez, et al., Engineering cell shape and function, *Science* **264**(5159) (1994) 696–698.
86. E. Kim, Y. Xia, and G.M. Whitesides, Micromolding in capillaries: Applications in materials science, *J. Am. Chem. Soc.* **118**(24) (1996) 5722–5731.
87. S. Takayama, J.C. McDonald, E. Ostuni, et al., Patterning cells and their environments using multiple laminar fluid flows in capillary networks, *Proc. Natl. Acad. Sci. U.S.A.* **96**(10) (1999) 5545–5548.
88. GeSiM Gesellschaft für Silizium-Mikrosysteme mbH, Radeberg, Germany, http://gesim-bioinstruments-microfluidics.com/.
89. P. Kim, K.W. Kwon, M.C. Park, et al., Soft lithography for microfluidics: A review, *BioChip J.* **2** (2008) 1–11.
90. Y.-T. Hsieh and Y.-C. Lee, A soft PDMS/metal-film photo-mask for large-area contact photolithography at sub-micrometer scale with application on patterned sapphire substrates, *IEEE/ASME JMEMS* **23** (2014) 719–726.
91. S. Möllenbeck, N. Bogdanski, M. Wissen, et al., Investigation of the separation of 3D-structures with undercuts, *Microelectron. Eng.* **84**(5–8) (2007) 1007–1010.
92. N. Koo, M. Otto, J.W. Kim, J. Jeong, H. Kurz, Press and release imprint: Control of the flexible mold deformation and the local variation of residual layer thickness in soft UV-NIL, *Microelectron. Eng.* **88**(6) (2011) 1033–1036.
93. J.A. Rogers, K.E. Paul, and G.M. Whitesides, Quantifying distortions in soft lithography, *J. Vac. Sci. Technol. B* **16**(1) (1998) 88–97.
94. Canon Nanotechnologies, Austin, Texas, http://cnt.canon.com/.
95. Jenoptik Mikrotechnik, Jena, Germany, http://www.jenoptik.com/.
96. SUSS MicroTec Lithography GmbH, http://www.suss.com/.
97. SET – Smart Equipment Technology, Saint Jeoire, France, http://www.set-sas.fr/.
98. CEA-Leti, Grenoble, France, http:// www.leti-cea.com/.
99. ThunderNIL srl, Padova, Italy, http://www.thundernil.com/.
100. NIL Technology ApS, Kongens Lyngby, Denmark, http://www.nilt.com/.
101. Aurotek Corp., Taipei, Taiwan, http://www.aurotek.com/.
102. AimCore Technology, Hsinchu, Taiwan, http://www.aimcore.com.tw/.
103. Gdnano Ltd., ChangShu Economic Development Zone, China, http://www.gdnano.com/, European Distributor 5 Microns GmbH, of Gdnano Ltd., Ilmenau, Germany, http://5microns.de/.
104. SeaHan Nanotech, Gyeonggi-do, Korea, http://www.21saehan.com/ and KIMM, Korea Institute of Machinery and Materials, Daejeon, Korea, http://www.kimm.re.kr/.
105. Hunet Plus from (Seoul), Korea, http://www.naver.com/.
106. Toshiba Machine Co., Ltd., http://www.toshiba-machine.co.jp/en/product/nano/lineup/st/st50s.html.
107. DNK Co. Ltd. Japan, Subcontract of DNP (Dai Nippon Printing) Co. Ltd. (Only Japanese Pages), http://www.dnp.co.jp/dnk/index.html (Japanese only).
108. Kyodo International Co. Inc., Kanagawa, Japan, http://www.kyodo-inc.co.jp/english/electronics/nanoimprint/pss.html/.
109. Hitachi Co. Ltd., Tokyo, Japan, http://www.hitachi.com, and http://www.hitachi.com/rev/pdf/2008/r2008_technology_tp.pdf.
110. A. Fuchs, B. Vratzov, T. Wahlbrink, Y. Georgiev, H. Kurz, Interferometric in situ alignment for UV-based nanoimprint, *J. Vac. Sci. Technol. B* **22**(6) (2004) 3242–3245.
111. N. Li, W. Wu, and S.Y. Chou, Sub-20-nm alignment in nanoimprint lithography using moiré fringe, *Nano Lett.* **6**(11) (2006) 2626–2629.
112. Eulitha, Würenlingen, Switzerland, http://www.eulitha.com/.
113. Soken Chemical & Engineering Co., Ltd.,Tokyo, Japan, http://www.soken-ce.co.jp.

114. temicon GmbH, Dortmund, Germany, http://www.temicon.com/.

115. IMS, Stuttgart, Germany, http://www.ims-chips.dev.

116. Y. Hirai, S. Yoshida, N. Takagi, et al., High aspect pattern fabrication by nano imprint lithography using fine diamond mold, *Jpn. J. Appl. Phys.* **42**(1, No. 6B) (2003) 3863–3866.

117. H. Schift, S. Saxer, S. Park, et al., Controlled co-evaporation of silanes for nanoimprint stamps, *Nanotechnology* **16** (2005) S171–S175.

118. S. Bellini, C. Padeste, D. Siewert, and H. Schift, Anti-sticking layers for nickel-based nanoreplicati on tools, *Microelectron. Eng.* **123** (2014) 23–27.

119. F.A. Houle, Eric Guyer, D.C. Miller, and R. Dauskardt, Adhesion between template materials and UV-cured nanoimprint resists, *J. Vac. Sci. Technol. B* **25**(4) (2007) 1179–1185.

120. D. Truffier-Boutry, R. Galand, A. Beaurain, et al., Mold cleaning and fluorinated anti-sticking treatments in nanoimprint lithography, *Microelectron. Eng.* **86**(4-6) (2009) 689–672.

121. Daikin Industries, Ltd., Osaka, Japan, http://www.daikin.com/.

122. M. Mühlberger, I. Bergmair, A. Klukowska, et al., UV-NIL with working stamps made from Ormostamp, *Microelectron. Eng.* **86**(4–6) (2009) 691–693.

123. Profactor, Steyr-Gleink, Austria, http://www.profactor.at/.

124. F.A. Houle, C.T. Rettner, D.C. Miller, and R. Sooriyakumaran, Antiadhesion considerations for UV nanoimprint lithography, *Appl. Phys. Lett.* **90**(21) (2007) 213103.

125. M. Bossard, J. Boussey, B. Le Drogoff, and M. Chaker, Alternative nano-structured thin-film materials used as durable thermal nanoimprint lithography templates, *Nanotechnology* **27** (2016) 075302 (14pp).

126. Y.C. Lee, Talk at NIL Industrial Day 2016, Vienna, Austria.

127. M.A. Verschuuren, M. Megens, J. Ni, H. van Sprang, A. Polman, Large area nanoimprint by substrate conformal imprint lithography (SCIL), *Adv. Opt. Technol.* **6**(3–4) (2017) 243–264.

128. SCIL Nanoimprint Solutions, http://www.ip.philips.com/data/static/scil/.

129. H. Teyssedre, S. Landis, C. Thanner, et al., A full process chain assessment for nanoimprint technology on 200 mm industrial platform, *Adv. Opt. Technol.* **6**(3–4) (2017) 277–292.

130. T. Haatainen, J. Ahopelto, G. Gruetzner, et al., Step & stamp imprint lithography using commercial flip chip bonder, *Proc. SPIE* **3997** (2000) 874–880.

131. R. Kirchner, L. Nüske, A. Finn, B. Lu, W.-J. Fischer, Stamp-and-repeat UV-imprinting of spin-coated films: Pre-exposure and imprint defects, *Microelectron. Eng.* **97** (2012) 117–121.

132. M. Otto, M. Bender, J. Zhang, et al., Dimensional stability in step & repeat UV-nanoimprint lithography, *Microelectron. Eng.* **84**(5–8) (2007) 980–983.

133. C. Peroz, S. Dhuey, M. Vogler, et al., Step and repeat UV nanoimprint lithography on pre-spin coated resist film: A promising route for fabricating nanodevices, *nanotechnology* **21** (2010) 445301 (5pp).

134. K. Ishibashi, H. Goto, J. Mizuno, and S. Shoji, Large-scale atmospheric step-and-repeat UV nanoimprinting, *J. Nanotechnol.* (2012) 103439 (9 pp).

135. T. Haatainen, T. Mäkelä, A. Schleunitz, G. Grenci, M. Tormen, Integration of rotated 3-D structures into pre-patterned PMMA substrate using step & stamp nanoimprint lithography, *Microelectron. Eng.* **98** (2012) 180–183.

136. M. Colburn, I. Suez, B. J. Choi, et al., Characterization and modeling of volumetric and mechanical properties for step and flash imprint lithography photopolymers, *J. Vac. Sci. Technol. B* **19**(6) (2001) 2685–2689.

137. J. Choi, K. Nordquist, A. Cherala, et al., Distortion and overlay performance of UV step and repeat imprint lithography, *Microelectron. Eng.* **78–79** (2005) 633–640.

138. A. Cherala et al., An apparatus for varying the dimensions of a substrate during nano-scale manufacturing, US Patent No. 7,170,589.

139. M. Colburn, T. Bailey, B.J. Choi, et al., Development and advantages of step and flash imprint lithography, *Solid State Technol.* **46**(7) (2001) 67.

140. M. Melliar-Smith, Lithography beyond 32nm – A role for imprint?, *Proc. SPIE* **6519** (2007) U9–U22.

141. B.J. Choi, S. Johnson, M. Colburn, et al., Design of orientation stages for step and flash imprint lithography, *J. Int. Soc. Precis. Eng. Nanotechnol.* **25**(3) (2001) 192–199.

142. E.E. Moon, J. Lee, P. Everett, and H.I. Smith, Application of interferometric broadband imaging alignment on an experimental x-ray stepper, *J. Vac. Sci. Technol. B* **16**(6) (1998) 3631–3636.

143. T. Takashima, Y. Takabayashi, N. Nishimura, et al., Nanoimprint system development and status for high-volume semiconductor manufacturing, *Proc. SPIE* **9777** (2016) 977706.

144. K.H. Smith, J.R. Wasson, P.J.S. Mangat, W.J. Dauksher, D.J. Resnick, *J. Vac. Sci. Technol. B* **19**(6) (2001) 2906.

145. T.C. Bailey, D.J. Resnick, D. Mancini, et al., Template fabrication schemes for step and flash imprint lithography, *Microelectron. Eng.* **61–62** (2002) 461–467.

146. W.J. Dauksher, K.J. Nordquist, D. Mancini, et al., Characterization of and imprint results using ITO-based step and flash imprint lithography templates, *J. Vac. Sci. Technol. B* **20**(6) (2002) 2857–2861.

147. D.J. Resnick, W.J. Dauksher, D. Mancini et al., High-resolution templates for step and flash imprint lithography, *Proc. SPIE* **4688** (2002) 205–213.

148. N. Hayashi, T. Abe, T. Shimomura, et al., NGL masks: Development status and issue, *Proc. SPIE* **7985** (2011) 798505.

149. K. Ichimura, K. Yoshida, S. Harada, et al., Development of nanoimprint lithography templates toward high-volume manufacturing, *J. Micro./Nanolith. MEMS MOEMS* **15**(2) (2016) 8021006.

150. K. Selinidis, E. Thompson, I. McMackin, et al., Defect inspection of imprinted 32 nm half pitch patterns, *Proc. SPIE* **7122** (2008) 71222K.

151. M. Pritschow, H. Dobberstein, K. Edinger, et al., High-resolution e-beam repair for nanoimprint templates, *Proc. SPIE* **7488** (2009) 74880V.

152. L.J. Myron, E. Thompson, I. McMackin, et al., Defect inspection for imprint lithography using a die to database electron beam verification system, *Proc. SPIE* **6151** (2006) 61510M.

153. K. Selinidis, E. Thompson, S.V. Sreenivasan, and D.J. Resnick, Inspection and repair of imprint masks at 32 nm and below, *Proc. SPIE* **7379** (2009) 73790N.

154. C. Brooks, K. Selinidis, G. Doyle, et al., Development of template and mask replication using jet and flash imprint lithography, *Proc. SPIE* **7823** (2010) 78230O.

155. K.S. Selinidis, C.B. Brooks, G.F. Doyle, Mask replication using jet and flash imprint lithography, *Proc. SPIE* **7970** (2011) 797009.

156. V. Sirotkin, A. Svintsov, S. Zaitsev, and H. Schift, Coarse-grain method for modeling of stamp and substrate deformation in nanoimprint, *Microelectron. Eng.* **84**(5–8) (2007) 868.

157. S.V. Sreenivasan and P.D. Schumaker, Critical dimension control, overlay, and throughput budgets in UV nanoimprint stepper technology, B.J. Choi, ASPE, *Spring Proceedings*, 2008.

158. T. Higashiki, T. Nakasugi, and I. Yoneda, Nanoimprint lithography and future patterning for semiconductor devices, *J. Micro/Nanolith. MEMS MOEMS* **10**(4) (2011) 043008.

159. L. Singh, K. Luo, Z. Ye, et al., Defect reduction of high-density full-field patterns in jet and flash imprint lithography, *Proc. SPIE* **7970** (2010) 797007.

160. N. Khusnatdinov, Z. Ye, K. Luo, et al., High throughput jet and Flash Imprint Lithography for advanced semiconductor memory, *Proc. SPIE* **9049** (2014) 904910.

161. H. Takeishi and S.V. Sreenivasan, Nanoimprint system development and status for high volume semiconductor manufacturing, *Proc. SPIE* **9423** (2015) 94230C.

162. Z. Ye, K. Luo, X. Lu, et al., Defect reduction for semiconductor memory applications using jet and flash imprint lithography, *J. Micro./Nanolith. MEMS MOEMS* **11**(3) (2012) 031404.

163. K. Iwamoto, T. Iwanaga, S.V. Sreenivasan, and J. Iwasa, Nanoimprint system development and status for high-volume semiconductor manufacturing, *Proc. SPIE* **9635** (2015) 96350P.

164. K. Emoto, F. Sakai, C. Sato, et al., Defectivity and particle reduction for mask life extension, and imprint mask replication for high volume semiconductor manufacturing, *SPIE Proc.* **9777** (2016) 97770C.

165. D.J. Resnick and J. Choi, A review of nanoimprint lithography for high-volume semiconductor device manufacturing, *Adv. Opt. Technol.* **6**(3–4) (2017) 229–241.

166. H. Schift, Roll embossing and roller imprint, *Science and New Technology in Nanoimprint. Advanced Technology and Application of Nanoimprint*, ed. Y. Hirai. Frontier Publishing Co., Ltd., Japan. ISBN: 4-902410-09-5, 74–89 (June 2006).

167. H. Tan, A. Gilbertson, and S.Y. Chou, Roller nanoimprint lithography, *J. Vac. Sci. Technol. B* **16**(6) (1998) 3926–3928.

168. N. Kooy, K.L. Tze Pin, O. Su Guan, A review of roll-to-roll nanoimprint lithography, *Nanoscale Res. Lett.* **9**(1) (2014) 320 (13 pp).

169. T. Mäkelä, T. Haatainen, and J. Ahopelto, Roll-to-roll printed gratings in cellulose acetate web using novel nanoimprinting device, *Microelectron. Eng.* **88**(8) (2011) 2045–2047.

170. T. Mäkelä and T. Haatainen, Roll-to-roll pilot nanoimprinting process for backlight devices, *Microelectron. Eng.* **97** (2012) 89–91.

171. V.W. Jones, S. Theiss, M. Gardiner, et al., Roll to roll manufacturing of subwavelength optics, *Proc. SPIE* **7205** (2009) 72050T-1.

172. 3M Display and Graphics Laboratory, Maplewood, MN, USA, http://www.3m.com/.

173. VTT Technical Research Centre, Espoo, Finland, http://www.vttresearch.com/services/smart-industry/printed-and-hybrid-manufacturing-services and http://www.vtt.fi/.

174. M. Vilkman, T. Ruotsalainen, K. Solehmainen, E. Jansson, J. Hiitola-Keinänen, Self-aligned metal electrodes in fully roll-to-roll processed organic transistors, *Electronics* **5**(2) (2016).

175. Y. Xia, D. Qin, and G.M. Whitesides, Microcontact printing with a cylindrical rolling stamp: A practical step toward automatic manufacturing of patterns with submicrometer-sized features, *Adv. Mater.* **8**(12) (1996) 1015–1017.

176. H.J. Lim, K.-B. Choi, G.H. Kim et al., Roller nanoimprint lithography for flexible electronic devices of a sub-micron scale, *Microelectron. Eng.* **88**(8) (2011) 2017–2020.

177. S.H. Ahn and L.J. Guo, Large-area roll-to-roll and roll-to-plate nanoimprint lithography: A step toward high-throughput application of continuous nanoimprinting, *ACS Nano* **3**(8) (2009) 2304–2310.

178. J.J. Dumond, K.A. Mahabadi, Y.S. Yee, et al., High resolution UV roll-to-roll nanoimprinting of resin moulds and subsequent replication via thermal nanoimprint lithography, *Nanotechnology* **23** (2012) 485310 (9pp).

179. J. Chen, Y. Zhou, D. Wang, et al., UV-nanoimprint lithography as a tool to develop flexible microfluidic devices for electrochemical detection, *Lab Chip* **15**(14) (2015) 3086–3094.

180. M.W. Thesen, S. Ruttloff, R.P.F. Limberg, et al., Photo-curable resists for inkjet dispensing applied in large area and high throughput roll-to-roll nanoimprint processes, *Microelectron. Eng.* **123** (2014) 121–125.

181. C.-L. Wu, C.-K. Sung, P.-H. Yao, and C.-H. Chen, Sub-15 nm linewidth gratings using roll-to-roll nanoimprinting and plasma trimming to fabricate flexible wire-grid polarizers with low colour shift, *Nanotechnology* **24** (2013) 265301 (7pp).

182. S. Murthy, M. Matschuk, Q. Huang, et al., Fabrication of nanostructures by roll-to-roll extrusion coating, *Adv. Eng. Mater.* **18**(4) (2015) 1–6.

183. Inmold Biosystems A/S, Taastrup, Denmark, http://inmoldbiosystems.com/services-2/r2r-processing/.

184. A. Jeans, M. Almanza-Workman, R. Cobene, et al., Advances in roll-to-roll imprint lithography for display applications, *Proc. SPIE* **7637** (2010) 763719 (12pp).

185. E.R. Holland, A. Jeans, P. Mei, et al., Adaptation of roll-to-roll imprint lithography: From flexible electronics to structural templates, *Proc. SPIE* **7970** (2011) 797016.

186. J. John, Y.Y. Tang, J.P. Rothstein, et al., Large-area, continuous roll-to-roll nanoimprinting with PFPE composite molds, *Nanotechnology* **24** (2013) 505307 (9pp).

187. The UMass Roll to Roll Fabrication and Processing Facility, Amhurst, MA, USA, http://www.pse.umass.edu/.

188. Carpe Diem Technologies, Franklin, MA, USA, https://www.carpediemtech.com/.

189. Nascent Center, Austin, TX, USA, http://nascent-erc.org/.

190. Joanneum Research Materials, Weiz, Austria, http://www.joanneum.at/.

191. PTMTEC Oy, Helsinki, Finland, http://www.ptmtec.com/.

192. University of Applied Sciences and Arts Northwestern Switzerland, Institute for Polymer Nanotechnology, Windisch, Switzerland, http://www.fhnw.ch/technik/inka.

193. S.-W. Tsai, P.-Y. Chen, S.-R. Huang, and Y.-C. Lee, Fabrication of seamless roller mold with 3D micropatterns using inner curved surface photolithography, *Microelectron. Eng.* **150** (2016) 19–25.

194. H. Maruyama, N. Unno, and J. Taniguchi, Fabrication of roll mold using electron-beam direct writing and metal lift-off process, *Microelectron. Eng.* **97** (2012) 113–116.

195. Asahi Kasei E-materials Corp., Tokyo, Japan, http://www.asahi-kasei.co.jp/ake-mate/wgf/en/product/.

196. http://www.c4isrnet.com/story/military-tech/cyber/2016/01/27/data-growth-internet-things-require-security-cyberattacks-espionage/79351308/.

197. E.A. Dobisz, Z.Z. Bandic, T.W. Wu, and T. Albrecht, Patterned media: Nanofabrication challenges of future disk drives, *Proc. IEEE* **96**(11) (2008) 1836–1846.

198. Z. Ye, R. Ramos, C. Brooks, et al., High volume jet and flash imprint lithography for discrete track patterned media, *Proc. SPIE* **7970** (2010) 79700L (10 pp).

199. M.T. Moneck, T. Okada, J. Fujimori et al., Fabrication and recording of bit patterned media prepared by rotary stage electron beam lithography, *IEEE Trans. Magn.* **47**(10) (2011) 2656–2659.

200. J. Lille, K. Patel, R. Ruiz, et al., Imprint lithography template technology for bit patterned media (BPM), *Proc. SPIE* **8166** (2011) 816626 (6pp).

201. K.C. Patel, R. Ruiz, J. Lille, et al., Line-frequency doubling of directed self-assembly patterns for single-digit bit pattern media lithography, *Proc. SPIE* **8323** (2012) 83230U.

202. S. Xiao, X. Yang, P. Steiner, et al., Servo-integrated patterned media by hybrid directed self-assembly, *ACS Nano* **8**(11) (2014) 11854–11859.

203. X. Yang, S. Xiao, Y. Hsu, et al., Fabrication of servo-integrated template for 1.5 Teradot/inch^2 bit patterned media with block copolymer directed assembly, *J. Micro/Nanolith. MEMS MOEMS* **13**(3) (2014) 031307.

204. T.R. Albrecht, H. Arora, V. Ayanoor-Vitikkate, et al., Bit patterned magnetic recording: Theory, media fabrication, and recording performance, *Proc. IEEE Trans. Magn.* **51**(5) (2015).

205. Y. Ekinci, H.H. Solak, C. David, and H. Sigg, Bilayer Al wire-grids as broadband and high-performance polarizers, *Opt. Express* **14**(6) (2006) 2323.

206. L. Chen, J.J. Wang, F. Walters, et al., Large flexible nanowire grid visible polarizer made by nanoimprint lithography, *Appl. Phys. Lett.* **90**(6) (2007) 063111.

207. S. Yokogawa, S.P. Burgos, and H.A. Atwater, Plasmonic color filters for CMOS image sensor applications, *Nanoletters* **12**(8) (2012) 4349–4354.

208. D. Inoue, A. Miura, T. Nomura, et al., Polarization independent visible color filter comprising an aluminum film with surface-plasmon enhanced transmission through a subwavelength array of holes, *Appl. Phys. Lett.* **98**(9) (2011) 093113.

209. D. Hansen, E. Gardner, R. Perkins, et al., The display applications and physics of the ProFlux wire grid polarizer, *SID 02 DIGEST* (2002) 730–733.

210. H. Yamaki, Y. Sato, T. Namatame, and Y. Kawazu, European Patent, EP 2 090 909 B1 (August 29, 2012).

211. S.H. Ahn, M. Miller, S. Yang, et al., High volume nanoscale roll-based imprinting using jet and flash imprint lithography, *Proc. SPIE* **9049** (2014) 90490G.

212. A.F. Kaplan, T. Xu, Y.-K. Yu, and L.J. Guo, Multi-layer pattern transfer for plasmonic color filter application, *J. Vac. Sci. Technol. B* **28**(6) (2010) C6060–C6063.

213. A. Mellor, H. Hauser, C. Wellens, et al., Nanoimprinted diffraction gratings for crystalline silicon solar cells: Implementation, characterization and simulation, *Opt. Express* **21**(S2) (2013) A295.

214. C. Heine and R.H. Morf, Submicrometer gratings for solar energy applications, *Appl. Opt.* **34**(14) (1995) 2476–2482.

215. N. Tucher, J. Eisenlohr, H. Hauser, et al., Crystalline silicon solar cells with enhanced light trapping via rear side diffraction grating, *Energy Procedia* **77** (2015) 253–262.

216. A. Polman and H.A. Atwater, Plasmonics for improved photovoltaic devices, *Nat. Mater.* **9**(3) (2010) 205–213.

217. S.-K. Meisenheimer, S. Jüchter, O. Höhn, et al., Large area plasmonic nanoparticle arrays with well-defined size and shape, *Opt. Mater. Express* **4**(5) (2014) 944–952.

218. Semi PV, International Technology Roadmap for Photovoltaic (ITRSV) (2014), http://www.itrpv.net/.

219. A. Hauser, I. Melnyk, P. Fath, et al., A simplified process for isotropic texturing of mc-Si, *J. Photovolt.* **2** (2003) 21447–21450.

220. H. Hauser, B. Michl, S. Schwarzkopf, et al., Honeycomb texturing of silicon via nanoimprint lithography for solar cell applications, *IEEE J. Photovolt.* **2**(2) (2012) 114–122.

221. H. Hauser, A. Mellor, A. Guttowski, et al., Diffractive backside structures via nanoimprint lithography, *Energy Procedia* **27**(0) (2012) 337–342.

222. A. Polman and H.A. Atwater, Photonic design principles for ultrahigh-efficiency photovoltaics, *Nat. Mater.* **11**(3) (2012) 174–177.

223. M.-G. Kang, M.-S. Kim, J. Kim, and L.J. Guo, Organic solar cells Using nanoimprinted transparent metal electrodes, *Adv. Mater.* **20**(23) (2008) 4408–4413.

224. D. Cheyns, K. Vasseur, C. Rolin, et al., Nanoimprinted semiconducting polymer films with 50 nm features and their application to organic heterojunction solar cells, *Nanotechnology* **19** (2008) 424016.

225. Y. Yang, K. Mielczarek, M. Aryal, A. Zakhidov, W. Hu, Nanoimprinted polymer solar cell, *ACS Nano* **6**(4) (2012) 2877–2892.

226. Y.-C. Lee and S.-H. Tu, Improving the light-emitting efficiency of GaN LEDs using nanoimprint lithography, *Recent Advances in Nanofabrication Techniques and Applications*, ed. B. Cui, Intech, 173–196 (2011).

227. Y.-C. Lee, S.-C. Yeh, Y.-Y. Chou, et al., High-efficiency InGaN-based LEDs grown on patterned sapphire substrates using nanoimprinting technology, *Microelectron. Eng.* **105** (2013) 86–90.

228. LEDs magazine, online, http://www.ledsmagazine.com/articles/print/volume-11/issue-6/features/patterned-wafers/patterned-substrates-enhance-led-light-extraction.html.

229. Yole Developement LED Front-end Manufacturing Trends (2014), http://de.slideshare.net/Yole_Developpement/led-frontend-manufacturing-trends-report.

230. Ulvac Inc., Japan, https://www.ulvac.co.jp/wiki/en/led_pss_process/.

231. M. LaPedus, How to lower LED costs, semi engineering (July 17th, 2014), http://semiengineering.com/how-to-lower-led-costs/.

12 Metrology for Nanolithography

Kazuaki Suzuki and Eran Amit

CONTENTS

12.1 INTRODUCTION

Metrology is the science of measurement, which is a *system of measures*. It is necessary for the development of new processes, the control and monitoring of existing processes, and as a product qualification yardstick. The importance of measurement fidelity was eloquently summarized long ago by William Thomson, also known as Lord Kelvin (Figure 12.1):

> When you can measure what you are speaking about, you know something about it. But when you cannot measure it, your knowledge is of a meager and unsatisfactory kind. It may be the beginning of knowledge, but you have scarcely advanced to the stage of science.

Microscopes are primarily used for metrology dealing with dimensional measurements. They are especially important in the semiconductor industry where an integrated circuit's (IC) minimum feature size, referred to as the critical dimension (CD), has continued to shrink to the nanometer scale. The conventional requirement in integrated circuit metrology that the fabrication tolerance be 10% of the critical dimension, also known as the "Gauge Maker's rule," now translates to a demanded measurement precision (between 10 and 30% of the fabrication tolerance), which is on the single nanometer level for technologies having CDs of less than 50 nm. For example, the fabrication tolerance for a 20 nm nominal gate length would be 2 nm with a required metrology precision of about 0.2 nm. Similar arguments can also be made for the relative (overlay) and absolute (placement) position of features. Of course, modification of the Gauge Maker's rule will be necessary as the dimensions to be measured become smaller. There has also been an increased need for more structure profile information during the development of IC device fabrication processes, which has made the ability to perform accurate three-dimensional measurements even more important. The increased number of material layers and structure aspect ratios in patterned masks or wafers used in the fabrication of current ICs has also made the theory and understanding of metrology tool behavior more complicated. In addition to measuring feature dimensions with sub-nanometer precision, high throughput and full automation are also desirable for an inline IC manufacturing environment. As the theoretical limits of measurement capability are approaching, it is reasonable to believe that the current paradigm of stand-alone metrology tool platforms used for post-process sampling will shift toward more integrated in-situ measurement architectures in order to achieve the required level of process control and yield in future IC manufacturing (advanced equipment control: AEC).

FIGURE 12.1   Photograph of William Thomson, Lord Kelvin.

However, this will require metrology tools as robust and user-friendly as the processing equipment used for fabrication.

12.2 TERMS AND DEFINITIONS

12.2.1 DIMENSIONAL METRICS

The concept of *critical dimension* metrology becomes obscured when the quantity to be measured is not well defined. For example, the definition of "line width" is unclear for the structure shown in Figure 12.2. The ideal structures for line width measurement would consist of flat features with uniform height and vertical sidewalls having smooth edges. In reality, the line edges are usually not well defined. Structures typically encountered in IC technology may have ragged edges and sidewalls that are asymmetric or even re-entrant. Line "profile" or "contour" would be more appropriate terms to describe the physical surface height in the X direction at a Y–Y' location along the line. Alternatively, the profile could be given by the width X–X' at each height position Z of the structure (see Figure 12.3). These are important considerations due to the effect of line shape variation on the final device's electrical performance and are also major components in the uncertainty of cross section–based reference measurements, as well as for any proposed line width reference standards.

Sidewall shape (see Figure 12.3) is the predominant factor to complicate the concept of "line width." Furthermore, the choice of Z height to acquire the width might also depend on the context of the current process. From the front view (cross-sectional) of the line structure in Figure 12.3, the top width might be chosen if a highly anisotropic plasma etch process (with low resist selectivity) of an underlying layer of material were to follow. The line width might be defined closer to the bottom, if a more isotropic wet etch process were to ensue. Defining the term "line width" becomes even more complicated if the resist profile changes over time, as is the case when there is not infinite selectivity between the resist and the material to be patterned. As feature sizes continue to shrink and possess more complicated profile shapes, a completely clear and exact definition of accuracy will become more elusive.

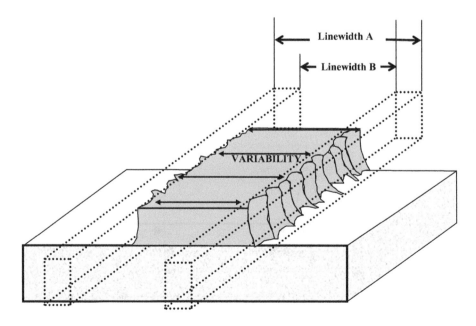

FIGURE 12.2 Three-dimensional "line" structure.

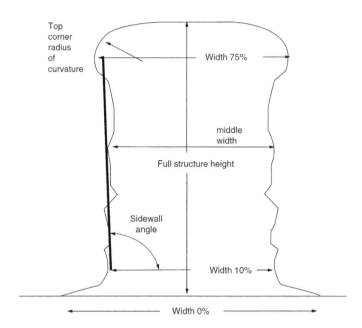

FIGURE 12.3   Parameterization of the full structure contour to set dimensional metrics.

12.2.2   PRECISION, ACCURACY, AND UNCERTAINTY

A random measurement distribution (Figure 12.4) is obtained when making repeated measurements on a given sample while holding all factors constant. The mean $(\overline{X})$ is obtained when an average of the measurement values is taken such that

$$\overline{X} = \sum_{n=1}^{N} \frac{x_n}{N}, \tag{12.1}$$

where x_n is the n-th measurement and N is the total number of measurements. Repeatability (σ_{rpt}) is the short-term variation in the measurements of the same feature under identical conditions. These are also referred to as *static* measurements. If the variation in these measurements is random with

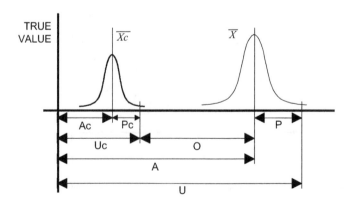

FIGURE 12.4   Illustration of accuracy, precision, offset, and uncertainty for a given sample type.

normal distribution and no systematic components, an estimate of the repeatability can be obtained from the standard deviation:

$$\sigma_{rpt} = \left[\frac{\sum_{n=1}^{N} (x_n - \bar{x})^2}{N-1} \right]^{1/2}. \tag{12.2}$$

The level of variation is considered to represent the best case of an instrument's performance. Reproducibility (σ_{rpd}) is defined as the fluctuation in static mean values if the measurements are made while certain conditions are changed. The factors usually chosen are those affecting an instrument's ability to reproduce results, such as sample loading. Reproducibility tests allow one to determine the dependence of systematic, or non-random, error components on the factors chosen to vary. The choice of factors to vary depends on the underlying physics of the instrument's operation. Precision is the total variation of a metrology system's measurements. This is usually obtained by summing the repeatability and reproducibility in quadrature:

$$\sigma = \sqrt{\sigma_{rpt}^2 + \sigma_{rpd}^2}. \tag{12.3}$$

Precision values are usually quoted in terms of an integral multiple of standard deviations, such as 3σ. These expressions are only valid if the sample population (number of measurements) is sufficiently large and the distribution is Gaussian, or normal. Normal distributions typically result when the errors are of a random nature, which can therefore be reduced by averaging. Systematic errors, on the other hand, either skew the data or produce multiple modes in a measurement distribution. In addition, obtaining a large number of measurements per site may be difficult when there are many sites to test or the sample material is sensitive to the metrology technique being used (electron beam–induced shrinkage of photoresist for argon fluoride [ArF] 193 nm, for example). Therefore, particular attention must be paid to the experimental design and instrumentation setup so that valid results can be obtained without requiring a prohibitive number of measurements to be taken or changing the feature under observation. Accuracy refers to how closely measurements conform to a defined standard, or absolute reference, assuming there is a fundamental basis of comparison (see Figure 12.4).

The mean value $(X \equiv \bar{X})$ of a measurement set with spread $2P$ $(P \equiv 3\sigma)$ has an offset O relative to an accepted standard with a mean value of X_c' and a spread of $2P_c'$. Calibration of an instrument involves minimizing the offset O between X and X_c'. However, calibration never completely yields absolute accuracy since a reference measurement system (RMS) has its own uncertainty U_c'. The net result can be simplified by taking the sum of A' and P'. The accuracy and precision of a measurement set can then be combined to give the total uncertainty U as

$$U = Ac + A + \sqrt{P_c^2 + P^2}. \tag{12.4}$$

Systematic errors are typically added in a linear fashion and random errors are summed in quadrature. When a quantity is measured as $X \pm U$, the true value may be anywhere in the interval $\pm U$. For some measurement quantities, the systematic errors have opposite signs and cancel each other so that the offset in accuracy between X and X' can actually be nullified to some extent, as with the measurement of the pitch (spatial period) of a line grating.

12.2.3 Feature Shape Variation and Roughness

The variation of a feature's "width" along its length is often referred to as the "line width roughness" (LWR). The single edge equivalent, the meandering of a single edge along its length, is called the "line edge roughness" (LER). Roughness is a statistical phenomenon and it can cause variations in the channel length from transistor to transistor that may subsequently result in integrated circuit timing issues. Off-state leakage currents and device drive currents are thought to be affected as well [1]. Since the polycrystalline silicon gate acts as the mask for dopant implantation, a rough edge will also affect the final distributions of a dopant. Roughness at wavelengths large compared to diffusion lengths affects the shape of the doped volume, while shorter wavelength roughness affects the dopant concentration gradient. The International Technology Roadmap for Semiconductors (ITRS) [2] specifies LWR over a window of spatial frequencies and that the amount of LWR tolerance be 8% of the etched gate length (Figure 12.5). The lithography roadmap definition for LWR is 3σ of the total variation, evaluated with a distance that allows the assessment of spatial wavelengths up to two times the technology node, while sampling the low-end spatial wavelengths down to a limit x_j. The range of spatial frequencies is $1/x_j$ to $1/$pitch, because the smallest spatial frequency (longest wavelength) is chosen to distinguish LWR from CD variation. Changes in width from transistor to transistor are considered CD variation, while changes in width for a single transistor are considered in LWR. The "E" and "W" subscripts refer to the edge position and width measures.

The quadratic, or mean squared, standard deviation measures of roughness for LER are

$$R_{Eq} = \left[\frac{1}{N-1} \sum_{i=0}^{N-1} (x_i - \bar{x})^2 \right]^{1/2} \tag{12.5}$$

and for LWR:

$$R_{Wq} = \left[\frac{1}{N-1} \sum_{i=0}^{N-1} (w_i - \bar{w})^2 \right]^{1/2}. \tag{12.6}$$

The ITRS LWR definition is stated as $3R_{Wq}$ and the proposed total LER is

$$R_T^2 = R_{Eq\_Left}^2 + R_{Eq\_Right}^2, \tag{12.7}$$

where R_{Wq} is a function of R_{Eq_right} and R_{Eq_left} and their correlation c.

The correlated edges shown on the left-hand side of Figure 12.6 ($c=1$, in phase) produce $R_{Wq}=0$ because the CD (distance between edges) is constant. The non-correlated edges in the center of Figure 12.6 ($c=0$, 90° out of phase) result in $R_{Wq}=R_{Eq}\sqrt{2}$. Finally, anti-correlated edges on the right-hand side of Figure 12.6 ($c = -1$, 180° out of phase) correspond to $R_{Wq}=2R_{Eq}$, as when two periodic functions superpose additively [3]. Anti-correlated edges are more of an issue because they result in a larger R_{Wq}. The correlation c might be significantly non-zero if non-random factors are present, such as optical proximity, topography, granularity of resist or poly-Si, or proximity to random defects. It should be noted that c is just an estimate and is itself a random variable. A set of edges

Year of Production	2013	2014	2015	2016	2017	2018	2019	2020	2022	2024
DRAM 1/2 Pitch (nm) (= Technology node)	28	26	24	22	20	18	17	15	13	11
DRAM CD Control 3σ (nm)	2.8	2.6	2.4	2.2	2.0	1.8	1.7	1.5	1.3	1.1
DRAM Overlay Control 3σ (nm)	5.7	5.2	4.8	4.4	4.0	3.7	3.4	3.1	2.6	2.2
MPU printed gate length (nm)	28	25	22	19.8	17.7	15.7	14.0	12.5	9.9	7.9
MPU etchedl gate length (nm)	20	18	17	15.3	14.0	12.8	11.7	10.6	8.9	7.4
3σ LWR control, <8% of etched gate (nm)	1.61	1.47	1.34	1.23	1.12	1.02	0.93	0.85	0.71	0.59
Uncertainty of LWR measurement (nm)	0.32	0.29	0.27	0.25	0.22	0.20	0.19	0.17	0.14	0.12

FIGURE 12.5 Selected sections of the 2013 International Roadmap for Semiconductors (ITRS) for CD metrology that are applicable to roughness. In addition, a selected section of ITRS 2013 for lithography that is applicable to CD control and overlay control is shown.

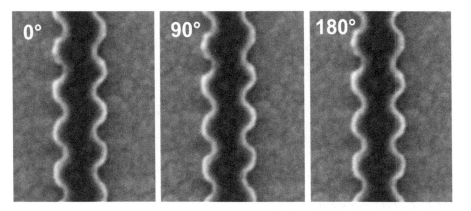

FIGURE 12.6 Left: Correlated (in phase), center: non-correlated (90° out of phase), and right: anti-correlated (180° out of phase) line edges.

that appear to have an average correlation coefficient close to 0 may nevertheless have individual members with rather large correlation coefficients. This implies that R_{Wq} is a better measure than R_{Eq} since correlation is included [3]. Standard deviation is one of the most commonly used estimators of LWR. However, a substantial amount of information about the LWR is ignored if that is the only measurement. Recently, it has been pointed out that a measurement of line width roughness by itself does not guarantee the complete characterization of the physical phenomenon under investigation, and that a full spectral analysis is required. It is important to have measures that include such information as roughness wavelengths or characteristic sizes of roughness asperities in the direction parallel to the line, because the effect on a device may depend on its wavelength as well as its amplitude. LWR is generally involved with a variety of apparently unrelated problems in microlithography, such as resist composition [4], aerial image contrast [5, 6], and development [7] and process conditions [8]. In the case of self-affine edges, it has been demonstrated that a set of three parameters is needed to define the physical system: the LWR standard deviation (σ), the correlation length (ξ) containing spectral information on the edges, and the roughness exponent (α) related to the relative contribution of high-frequency fluctuations [9–11]. In addition to standard deviation (σ), the spatial frequency dependence is also described by two additional parameters that quantify the spatial aspects of LWR: the roughness exponent (α) and the correlation length (ξ). The roughness exponent is associated with the fractal dimension D by the relationship $\alpha = 2 - D$ [12], and its physical meaning is the relative contribution of high-frequency fluctuations to LWR and large α signify less high-frequency fluctuations. The correlation length denotes the distance after which the edge points can be considered uncorrelated. The power spectral density (PSD) [13–15] is related to the discrete Fourier transform of a series, whose coefficients w_j are given by [16]

$$C_k = \sum_{j=0}^{N-1} w_j e^{\frac{2\pi I j k}{N}}, \tag{12.8}$$

which can be calculated using a Fast Fourier transform (FFT) algorithm. The positive and negative frequencies present in the Fourier transform are summed to produce a one-sided (positive frequencies only) PSD (Figure 12.7). The PSD is not a single number but a curve. The factor for sampling interval, Δ, is also required on dimensional grounds when dealing with real profiles. The discrete form of Parseval's theorem is

$$\sum_{k=0}^{N-1} w_k^2 = \frac{1}{N\Delta} \sum_{k=0}^{N/2} P_k, \tag{12.9}$$

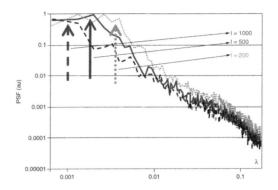

FIGURE 12.7 Fourier power spectral density.

so that the PSD is related to LWR:

$$R_{Wq}^2 = \frac{1}{(N-1)N\Delta} \sum_{k=0}^{N/2} P_k. \tag{12.10}$$

The expression for R_{Wq}^2 is similar, except that the factor of $N-1$ in the denominator becomes $N-2$ and P_k must be defined in terms of the edge residuals instead of the width residuals. This means that the area under the PSD curve is related to the RMS measure of roughness. We can generalize this by summing only those P_k between specified limits corresponding to $f_{min} < f < f_{max}$ to determine the RMS roughness contributed only by relevant frequencies. Note the $1/f^{2.3}$ slope on the plot.

The autocorrelation function can be calculated from the inverse Fourier transform of the PSD. Alternatively, it can be computed directly from the measured widths by

$$c(i) = \frac{1}{(N-1)R_{Wq}^2} \sum_k w_{l+k} w_k. \tag{12.11}$$

A similar definition applies for the edge correlation function except for the use of edge instead of width residuals and a factor of $N-2$ in the denominator instead of $N-1$. The amount i, by which one copy of the curve is shifted with respect to itself before multiplying, is referred to as the lag.

The lag may be positive or negative. The autocorrelation has its maximum value of 1 at a lag of 0. For randomly rough (non-periodic) edges produced by a stationary process, the autocorrelation is expected to tend toward zero for increasing lag. (There are, however, practical issues in the estimation of correlation functions from a finite length series. For instance, background subtraction can produce artifacts in the curve [17, 18].) The decrease in (c) occurs over a characteristic distance, called the correlation length. This length may be characteristic of a grain size or other physical phenomenon that sets a lateral distance scale for the roughness. As with line edge roughness, there are different metrics for the correlation length. It may be defined as the point at which the correlation decreases below a threshold such as $1/e$, or it may be determined by fitting an exponential around zero lag (see Figure 12.8). The $1/e$ (~0.36) crossing is a measure of the "correlation length."

Errors in the width determination result from a set of conditions to be, ε_i, and are distributed with a standard deviation σ_ε. The sum of "measured width" squares, as opposed to the actual widths, is used to obtain a measured roughness given by

$$(n-1)R_{Wq\_meas}^2 = \sum_{l=0}^{N-1} \left[W_l + \varepsilon_l - \overline{(W+\varepsilon)} \right]^2 = \sum_{l=0}^{N-1} \left[w_l + \varepsilon_l - \overline{\varepsilon} \right]^2. \tag{12.12}$$

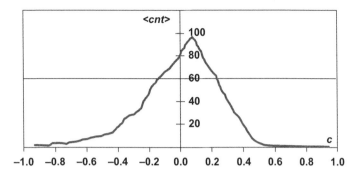

FIGURE 12.8 A smoothed histogram of correlation values. (From Figure 3 in B. Bunday, et al., "Determination of optical parameters for CDSEM line edge roughness," *Proc. SPIE*, 5375, 515 [2004].)

"W" in the subscript has been omitted because the result can be shown to apply for the case of edge roughness, provided σ_ε is understood to be the standard deviation of the distribution of edge errors instead of width errors. This result means random errors in the edge roughness bias the measurement and add in root sum of squares fashion with the true roughness to bias the measured result (Figure 12.9a). The thick meandering line represents the real edge. The thinner meandering line represents one measured edge out of infinite possibilities. The Gaussian curves overlaid on each line scan represent the distribution of measured edge locations [3]. If the σ_ε are known, σ_ε^2 can be subtracted from the measured square of the roughness to obtain a corrected estimate of the actual squared roughness. Such correction is reasonable when the edge assignment repeatability is smaller than or comparable to the roughness. However, when it becomes much larger than the roughness, discerning the roughness above the noise "background" is likely to become increasingly difficult. Noise also affects the repeatability of the roughness measurement. Since R_{Wq} is essentially a standard deviation, its repeatability is the standard deviation of a standard deviation. Assuming σ_ε is small compared to the roughness yields:

$$\sigma_{R_{Wq\_meas}} = \frac{\sigma_\varepsilon}{\sqrt{(N-1)}} = \frac{\sigma_\varepsilon}{\sqrt{2(N-1)}} \sqrt{1 + \frac{R_{Wq}^2}{\sigma_\varepsilon^2 + R_{Wq}^2}}, \tag{12.13}$$

where σ_ε is comparable to R_q.

The behavior of $\langle R_{q\_meas}^2 \rangle$ through $\sigma_{R_{Wq\_meas}}$ is illustrated graphically in Figure 12.9b. The central line is the expected value for infinitely many repeated measurements. Note that it is biased with respect to the true value (10 in this example). The spread between the outer lines represents the scatter (± 1 standard deviation) for a single LWR measurement [3]. Some tools allow edge positions with N "bins" with the results from n line scans averaged. This reduces σ_ε as $\sqrt{n}$, but at less spatial frequency sensitivity.

12.2.4 PRECISION GAUGE STUDY DESIGNS

Carefully designed experimental plans provide an efficient way to study the effects of tool operating conditions on measurement performance. Results from each experiment also help to focus on more significant factors affecting metrologic capability. There are several models to choose from when designing an experiment to gauge the precision of a measurement system. A traditional model has been the "crossed-effect" model, where one factor (A) is common to each cycle of the other factor (B), as in Figure 12.10a. For example, the impact of process operators (A) and parts (B) on instrument performance could yield a model to estimate the operator-by-part interaction.

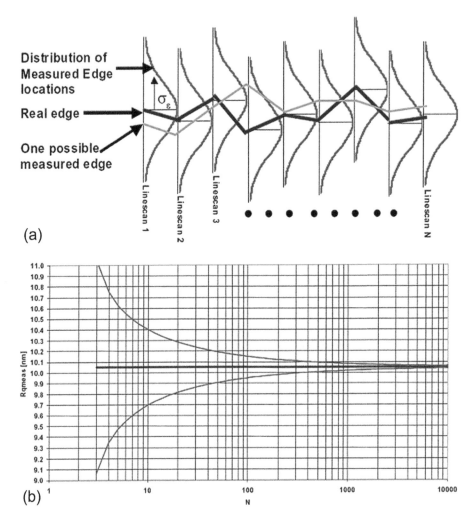

FIGURE 12.9 (a) Effect of noise on the apparent measured position of a real line edge, and (b) bias and uncertainty versus number of lines for an LWR measurement with noise $0.1R_{Wq}$ and table listing of minimum line scans necessary to meet the ITRS LWR measurement precision specifications, for different values of σ_e. (From Figures 6 and 7 in B. Bunday et al., "Determination of optical parameters for CDSEM line edge roughness," *Proc. SPIE*, 5375, 515 [2004].)

Another design, known as the nested effects model (see Figure 12.10b), uses the analysis of variations (ANOVA) approach to separate effects due to each factor and estimate each variance component in the total precision [19]. Static repeats can be nested within each load cycle that is iterated while the effects of a particular factor of interest are studied. This model allows multiple levels of nested factors. The total number of factors must be limited, due to the factorial growth in the number of required measurements with each level added to the model.

12.2.5 MORE ON THE NATURE OF ACCURACY

A constant measurement offset (O) only affects the true accuracy of a metrology tool, but does not affect the stability of IC device dimensional control. Unfortunately, variation of the offset (O) with respect to sample type (i.e. material or pattern) does occur and is presently a serious issue encountered after the precision has been gauged on a given reference sample. The variable offsets should be

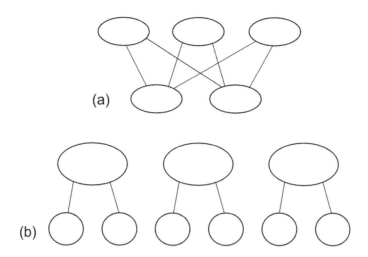

FIGURE 12.10 (a) The crossed effects model and (b) the nested effects model.

regarded as systematic error sources affecting accuracy and uncertainty. If a precise reference measurement system with constant offset across different sample types provides calibration values (X_c'), the previous illustration can be plotted for multiple sample types and combined to clarify the systematic error component resulting from sample-dependent fluctuations of the offset (Figure 12.11). Sample-dependent offset variations are now comparable to the total estimated uncertainty (U), as devices have continued to scale downward in size (Figure 12.11).

To avoid missing important process shifts, it follows from the previous discussion that precision is indeed a necessary, but not entirely sufficient metric for the evaluation of a CD metrology tool. Subtle variations in the manufacturing process over time can actually be masked by the periodic procedures of tuning a metrology tool solely for the optimization of precision. A balance must be achieved between precision and accuracy to effectively monitor the fabrication processes of more advanced technologies involving devices whose critical dimensions fall below the 90 nm node. The classical definition of accuracy has usually been quoted as the offset between a measurement distribution mean and the physical value that corresponds to the absolute truth. Attempts to improve the metrology tool's level of accuracy are then made through the process of calibration, where measurements from the tool under test (TUT) are correlated to more accurate values obtained using an accepted reference measurement system (RMS) at the same sites. In order to perform an effective calibration, a valid regression analysis technique must be applied for correlation between the TUT and RMS measurement values. A simple linear relation (with slope and offset) between the two variables (X' and X_c') is usually provided by performing a linear regression, or least squares

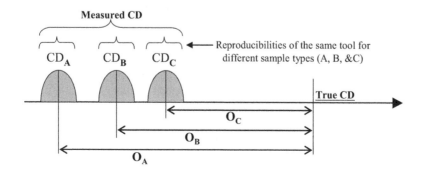

FIGURE 12.11 Offset variation with sample type and its influence on estimated total uncertainty.

fit. However, the first problem to occur with this type of analysis involving ordinary least squares (OLS) regression is the requirement to make an assumption that the independent reference value (X_c') has little or no variation. Increasing the sample size of (X_c') to maximize the confidence limits is typically not practical due to the more manually intensive nature of (X_c') measurements, as these are typically acquired using cross-sectional transmission electron microscopy (TEM) and scanning electron microscopy (SEM) RMS techniques. To make matters worse, a large amount of uncertainty is intrinsic to cross-sectional techniques due to their inability to sample multiple profiles for the spatial averaging of variation along the structure. The recent introduction and proposed application of critical dimension atomic force microscopy (CD-AFM) and dual-beam focused ion beam (FIB)/SEM techniques to serve as RMS sources (described in Section 12.3.4 and 12.3.5) promises to offer the possibility of reference measurements (X_c') at much improved throughput rates and spatial averaging of multiple profiles acquired in small increments along the structure (i.e. higher sampling frequency). Although these recently proposed RMS techniques will hopefully improve the precision and accuracy of the set of reference values (X_c'), variation and finite levels of uncertainty within the feature size range of interest remain. Fortunately, in this regard, significant improvements have also recently been made to the methodologies employed for analysis of measurement data during the evaluation of metrology tool performance [20]. Most of these recent advances have originated from the initial work done by John Mandel [21] in 1964, who introduced a least squares technique that allowed treatment of the case when both regression variables $(X'$ and $X_c')$ are subject to error. Input of the ratio of the variances of the regression variables and their correlation to each other is required for this technique. The estimated slope and intercept of the best-fit line resulting from the Mandel analysis, as well as the uncertainty of the predicted values, are strongly dependent on these initial inputs. In order to regress one variable onto another, an assumption must be made regarding the relationship between these two variables. This section will rely on the assumption that measurements (X') from the TUT respond in a linear fashion (at least to first order) with respect to the corresponding values (X_c') provided by the RMS at the exact same site locations. Representation of such a model by a slope (β) and intercept (α) yields the following equation:

$$y_i = \alpha + \beta x_i + \varepsilon_i, \tag{12.14}$$

where (y_i) and (x_i) represent the i-th dependent (X') and independent (X_c') variables, respectively, and (ε_i) is the residual at each point. The notion of a "corrected precision" can then be defined as a combination of the precision (σ_x) and the estimated slope of the correlation:

$$\text{Corrected Precision} \equiv \hat{\beta}\sigma_x, \tag{12.15}$$

where (σ_x) is the precision of the TUT and $(\hat{\beta})$ is the slope estimated from regression of the RMS values onto the TUT measured values. Thus, a smaller $(\hat{\beta})$ implies a greater change in the TUT measurement for a given increment in RMS values. A change in the TUT offset (O) across some trend in features would then be indicated by an estimated precision not equal to unity. The notion of a corrected precision makes sense because a TUT could exhibit better precision values than other tools and yet have a larger slope (i.e. more nonlinear). Nonlinearity (in the case of slope <1) could imply a less sensitive measurement tool, while at the same time, better precision would seem to indicate a more resolute measurement and the ability to sense smaller changes in the process being monitored. The product of these two metrics then represents a balance for the raw uncorrected precision. The linearity of the relationship between x and y should be checked by considering the residual error (d_i) at each ordered pair of data, defined as

$$d_i = y_i - \hat{a} - \hat{\beta}x_i, \tag{12.16}$$

where $(\hat{a})$ and $(\hat{\beta})$ are the estimated intercept and slope. The net residual error squared (D^2) is then the mean squared error of these residuals, as expressed by

$$D^2 = \frac{\sum_{i=1}^{N} d_i^2}{N-2}. \tag{12.17}$$

This net residual error is comprised of both systematic and random components of error. From the precision estimates (σ_x) and (σ_y) of the x and y variables obtained by using the sampling procedures described in Section 12.2.4, an estimate of the input variance of the data set can be made:

$$\text{Var(input)} = \sigma_y^2 + \hat{\beta}^2 \sigma_x^2. \tag{12.18}$$

The ratio of the square of the residual error to the input variance is the metric that distinguishes systematic from random errors in the complete data set. This is essentially the F statistic, but for clarity, will be referred to as the nonlinearity metric:

$$\text{Nonlinearity} \equiv \frac{D^2}{\text{Var(input)}}. \tag{12.19}$$

It should be emphasized that the validity of the nonlinearity test requires the precision of the TUT to be within specification. The estimated intercept $(\hat{a})$ is dependent on the estimated slope, so that the two parameters of the first-order regression analysis are not statistically independent of each other. It is difficult to form an intuitive understanding of the estimated intercept parameter, since ordinarily intercepts are defined by the value of $(y$ at $x=0)$. To avoid this, the parameter known as offset will be expressed more specifically as

$$\text{Offset} \equiv \Delta = \bar{y} - \bar{x}, \tag{12.18}$$

where $(\bar{x})$ and $(\bar{y})$ are the measurement averages from a calibration effort. Since the TUT measurements will be regressed against the RMS values during calibration, the offset is indeed a reflection of the accuracy. Ordinary least squares (OLS) assumes that no error exists in the independent variable, but this is not the case for real RMSs. Fortunately, there are indicators as to when the conditions permit use of OLS. One is based on the precision of the independent variable (σ_x) being much smaller than the standard deviation of all the x values:

$$\frac{\sigma_{\text{allRMSvalues}}}{\sigma_x} \gg 1. \tag{12.21}$$

Another criterion for acceptable use of the OLS fit is

$$|\beta| \times \frac{\sigma_x}{\sigma_y} \ll 1, \tag{12.22}$$

and is similar to the corrected precision criterion. An important assumption in the estimation of the accuracy of an unknown TUT is the effect of uncertainty in the RMS values. To address this, a method of linear regression that allows for significant measurement error in both the x and y variables is needed. A means for handling errors in both the x and y variables was introduced in 1964 and further developed in 1984 by John Mandel [21]. Although the details will not be covered here, Mandel developed a methodology for performing a least squares fit when both variables are subject to error. This more generalized regression analysis can also be used for all permutations in the degree of error in x and y variables. The key parameter affecting the degree of importance in using the Mandel regression method is a parameter that is usually denoted by

$$\lambda \equiv \frac{\sigma_y^2}{\sigma_x^2}, \tag{12.23}$$

where (σ_x) and (σ_y) are the precisions of the TUT and RMS measurements. Mandel's method assumes that both of the regression variables x (TUT) and y (RMS) are subject to errors. The following pair of equations can be used to represent these errors:

$$x_i = E(x_i) + \delta_i \text{ and } y_i = E(y_i) + \varepsilon_i, \tag{12.24}$$

where (E) is the expectation value which represents a point lying on the estimated calibration curve with (δ_i) and (ε_i) defined as the deviations for the i-th data point in the x and y directions, respectively, and are assumed to be statistically independent. This assumption is valid during the stages of calibration of each variable. The variance (Var) of each variable can also be assumed constant across the range of ordered data pairs. Bridging to the earlier notations for variances now yields

$$\mathrm{Var}(\delta_i) = \hat{\sigma}_\delta^2 = \hat{\sigma}_x^2 \text{ and } \mathrm{Var}(\varepsilon_i) = \hat{\sigma}_\varepsilon^2 = \hat{\sigma}_y^2, \tag{12.25}$$

therefore,

$$\hat{\beta} = \frac{S_{yy} - \lambda S_{xx} + \sqrt{\left(S_{yy} - \lambda S_{xx}\right)^2 + 4\lambda S_{xy}^2}}{2S_{xy}}, \tag{12.26}$$

where S_x, S_y, and S_{xy} are the sum of squares from the raw data as defined by

$$S_{xx} = \sum_{i=1}^{N}(x_i - \bar{x})^2, \quad S_{yy} = \sum_{i=1}^{N}(y_i - \bar{y})^2, \text{ and } S_{xy} = \sum_{i=1}^{N}(x_i - \bar{x})(y_i - \bar{y}). \tag{12.27}$$

The number of ordered data pairs is (N). In the "classical" case, where OLS applies, the uncertainty of the independent (RMS) variable goes to zero and $\lambda \rightarrow \infty$. The estimated slope in this case is S_{xy}/S_{xx}, but when most of the error is contained within the x variable, $\lambda \rightarrow 0$ and the slope becomes S_{yy}/S_{xy}. This would essentially consist of x regression onto y, which indicates that the Mandel regression is symmetrical with respect to both variables. The utility of this feature has recently been demonstrated for the case when two sets of calibration data can be coupled [20]. The primary importance of the Mandel-based analysis is that uncertainty in both of the TUT-X' and RMS-X_c' variables is allowed and the assumption of absolutely true reference values (X_c') is no longer required. Once an RMS with a fixed offset variation that does not depend on sample type and that also has adequate precision can be established, the goodness of fit between a tool under test (TUT) and the RMS can be obtained using the previously described Mandel regression analysis to gauge the metrology performance of the TUT, using the concept of total measurement uncertainty (TMU) [22]. A proposed flow for evaluating metrology tool performance based on our prior discussions is shown in Figure 12.12. Due to variation in imaging and measurement performance for different samples, a tool should be gauged across a variety of sample types. The robustness of the calibration can be further improved by matching across a series of features having a trend with fine increments of a metric of interest (i.e. sidewall angle, bottom width, height, etc.), by design or change of fabrication conditions to yield a worst-case process variation range.

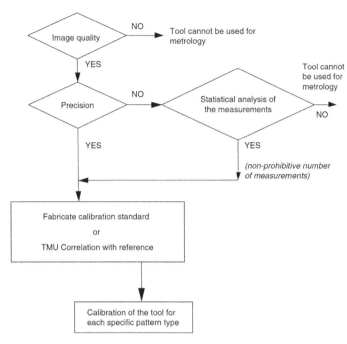

FIGURE 12.12 Measurement performance evaluation flow chart.

12.3 TOOLS AND TECHNIQUES

12.3.1 THE NOTION OF IMAGING

The optical system was never invented: it is one of the five senses and for a very long time man has tried to improve his ability to see. The first chronicle describing the use of an optical component for improving human eyesight was due to Pliny the Elder who wrote in 23–79 AD that *"emeralds are usually concave so that they may concentrate the visual rays. The emperor Nero used to watch through an emerald the gladiatorial combats."* The physical mechanisms describing the capacity of glass to bend light rays have been known since the time of Ptolemy, who described a method to determine the refractive index of water by measuring the apparent bending of a partially immersed stick. In the 10th century AD, the Arabian scholar Alhazen published the first major optical work—*Opticae Thesaurus*—in which the basic principles of optics were discussed. R. Bacon wrote in his *Perspectiva* (1267): *"Great things can be performed by refracted vision. If the letters of a book, or any minute object, be viewed through a lesser segment of a sphere of glass or crystal, whose plane is laid down upon them, they will appear far better and larger."* For this pertinent approach, R. Bacon is generally considered as the inventor of the glass lens. Later in the 17th century, A. Van Leeuwenhoeck created a new invention, known as the "microscope," and reported the first magnified images of biological samples. Although the first microscopes were composed of a very simple single lens that was held up to the eye, their overall concept was not far from our modern tools. The first attempt to increase resolution in a "non-classical" way was made by G.B. Amici, who introduced the concept of immersion in the 19th century. Later that century, the scientists E. Abbe and Lord Rayleigh proposed the existence of a limit to resolution, thereby dashing hopes to observe the ultimate structure of matter by light-optical-based microscopy. Finally, in the 1950s, F. Zernike invented the phase microscope and thus demonstrated the first case of image improvement by encoding (no longer a mere magnified copy of the sample observed by the eye).

The notion of image is as old as the history of science itself. Basically, an image can be defined as the transform of a physical object under observation. However, the relation between an object and its image is not simple, as most images are two-dimensional intensity representations resulting from the transform of a three-dimensional material object. One of the first applications of the wave theory of light to imaging systems was demonstrated by E. Abbe, who asserted that every object acts as a grating that diffracts incident light into angles that are inversely related to the spatial period. Following Abbe's theory, a beam of light impinging on a flat (bi-dimensional) object will be diffracted at higher orders for smaller grating periods (i.e. finer object details). The acceptance angle of an imaging system determines how many of the diffracted orders can be collected in order to resolve the finest of object details. Abbe further established that a fundamental limitation then exists for any microscope that relies on lenses to focus light because diffraction obscures the ability to resolve those object details that are smaller in size than approximately one-half of the wavelength of the light used for illumination. This limitation is referred to as "resolution" and is one of the most, if not the most, important notions in imagery and essentially refers to the resolving capability of an imaging system. Resolution is undoubtedly the most widely used parameter chosen for assessing image quality. However, other metrics affect the overall image quality, such as noise and contrast, with which resolution should not be confused. The empirical treatment of Abbe's principle involves finding the circumstances under which a microscope resolves the lines of a grating object with certain spatial frequency (p). Abbe examined the limiting case for the smallest resolvable spacing (p_{min}). It was found to be a function of the light wavelength and the angle (α) subtended by the collection aperture with respect to the object plane normal, namely,

$$p_{min} = \frac{\lambda}{2\sin(\alpha)}.$$ (12.28)

This limit has come to be known as the Abbe barrier, or the far-field diffraction limit. A concise form of Abbe's principle, known as the *Rayleigh Criterion*, will be derived in Section 12.3.2 on total image quality and has historically been the most widely used approach for the quantitative estimation of resolution. A recent plethora of criteria based on analysis of the final image have also been proposed, due to the advent of scanning electron microscopy.

In addition to conventional full-field imaging, where all of the intensity is collected simultaneously in parallel, the advent of scanned image microscopy has further extended the ability to gain more information about the sample with even finer detail. Namely, by sequential sampling of the energy scattered from the sample discretely on a point-by-point basis at each position within the scan trajectory of the primary illumination spot (or probe). Although some of the first working scanning optical and electron microscopes were described in the 1930s, a number of writers put forward proposals that, if technology had been sufficient at the time, anticipated many of these developments well in advance. However, there have also been many instances where much of the groundwork had been completed and the technology did exist, but no one made the imaginative leap until much later. A primary example is that of electron optics, which had to await Busch's paper in 1926 for the real beginning and application of the components that were already available (i.e. electron guns, phosphor screens, vacuum systems, etc.). But in general, *it is indeed worthwhile to study the early literature for ideas that were truly before their time or whose significance was not appreciated.* The idea of scanning an image to enable its transmission electrically appears to have been first proposed by Alexander Bain in a patent published back in 1843. Bain, a Scot in London, was not formally educated and worked as an apprentice for a clockmaker. After hearing a lecture on "Light, heat, and the electric fluid," his attention was drawn to see if any of the recently popularized electromagnetic apparatus could be applied to more useful purposes rather than just a novelty. He soon became convinced that a telegraph that could print the messages, instead of merely presenting evanescent signals to the eye, would be far superior in every way. He first proposed a scanning printing telegraph in a patent taken out in 1843, before pursuing his primary interest of working on the printing telegraph published in his 1866 pamphlet. This apparatus appears to have been the first to demonstrate dissection of an image for electrical transmission through the use of object scanning. Mention should also be made of P. Nipkow, who invented the scanning disk in 1884. Although it turned out to be a dead-end, as far as television is concerned, it has recently found use in the scanning confocal optical microscope.

The first proposal for a scanned image light-optical microscope appeared in a paper written by E.H. Synge in 1928, in which the goal was to actually overcome Abbe's wavelength limitation on resolution by using a proposed method that is now referred to as "near-field" optical microscopy. He put forward the notion of producing a very small probe of light by collimation through an aperture smaller than the illumination wavelength. The specimen of interest would be placed very close to the aperture and raster scanned relative to it. Also realizing the importance of providing a visible image, he wrote in a paper published in 1932: "Practically speaking, the method might be termed a visual one if a picture of the field could be formed on a phosphorescent screen... in the course of a few seconds." He then proposed the use of piezoelectric quartz crystals for rapid scanning of the specimen. Unfortunately, the technology required for reduction to practice of an imaging tool using his flat aperture plate did not exist in 1928. Synge also approached Einstein in 1928 for feedback on the feasibility of his concept, referred to by then as a "hypermicroscope." After a year of written correspondence with Einstein, Synge then proposed "constructing a little cone or pyramid of glass having its point (P) brought to a sharpness of 10^{-6} cm. One could then coat the sides and point with some suitable metal (eg in a vacuum tube) and then remove the metal from the point until P was just exposed." The imaging probes on modern-day near-field scanning optical microscopes are now manufactured in this way. Almost 26 years later, in 1956, J.A. O'Keefe at NASA independently proposed the same concept but decided that the realization would be too difficult because of the requirement to bring an object so close to the pinhole and then maintain their separation. In 1972, Ash and Nichols were finally able to demonstrated $\lambda/60$ resolution using 3 cm microwave radiation. Twelve years later in 1984, Lewis *et al.* published the first widely acknowledged paper on the practical demonstration of a visible light–based scanning near-field optical microscope.

12.3.2 IMAGE QUALITY: RESOLUTION, CONTRAST, AND NOISE

The explosive growth of computer graphics and internet technologies has also resulted in the proliferation of digital images almost everywhere. This has demanded the availability of high resolution and image quality. However, real images are usually not perfect, so that sufficient details about the object may be lacking. In the area of research known as image analysis and processing, the notion of resolution is further divided into three sub-classifications:

(1) *Spatial* refers to the image pixel, or sampling, density as per our earlier discussion. An image is said to suffer from *aliasing* artifacts when the finer details of an object are not represented spatially with sufficient pixel sampling density.
(2) *Brightness* is also referred to as gray-level resolution and is related to the number of brightness levels that can be used to represent a pixel. For instance, a monochrome image is usually quantized with 256 levels for an 8-bit digital imaging system.
(3) *Temporal* represents the maximum rate of frame acquisition, or the number of frames that can be captured per second. The typical frame rate for comfortable "real-time" viewing by the human eye is about 25 frames per second.

In this section, the term resolution will unequivocally refer to spatial resolution. We will also avoid quantification of the inter-relationship between brightness and spatial resolutions in this chapter. Images acquired from real imaging systems generally suffer from aliasing, blurring, and the effects of noise. A model matrix (M) can be used to provide a general description of the real image formation process. Specifically, the three main terms that will be used to capture an image formation process are image motion, optical blur, and the sampling process. A dynamic scene with continuous intensity distribution $X(x,y)$ is seen to be warped at the detector input due to environmental factors such as mechanical noise (e.g. relative motion between the scene and image-recording device), electromagnetic fluctuations (e.g. interference from stray fields), or thermal instabilities. The resultant captured images are blurred both by environmental variations and the imaging system performance characterized by a continuous point spread function (PSF). They are subsequently sampled and discretized at the imaging detector (e.g. charge-coupled device [CCD] camera) resulting in a digitized noisy frame $Y[m,n]$. A forward model [23], illustrated in Figure 12.13, is represented by

$$Y\big[m,n\big]=\Big[H_{\mathrm{cam}}\big(x,y\big)**F\big(H_{\mathrm{atm}}\big(x,y\big)**X\big(x,y\big)\big)\Big]\downarrow+V\big[m,n\big], \qquad (12.29)$$

in which ** is the two-dimensional convolution operator, (F) is the warping operator, $\downarrow$ is the downsampling or discretization operator, $V[m,n]$ is the system noise, and $Y[m,n]$ is the resulting discrete noisy and blurred image. In ideal situations, these modeling terms would capture the actual effects of the image formation process. In practice, however, the models used reflect a combination of computational and statistical limitations. For example, it is common to assume simple parametric space-invariant blurring functions for the imaging system. This allows one to utilize efficient and stable algorithms for estimating an unknown blurring function. Although this approach is reasonable, it must be understood that incorrect approximations can lead to significant reductions in the overall image analysis algorithm performance. The concept of quantitative analysis of image quality was proposed by Shannon [24] in terms of the information content contained. The information content of an image is primarily characterized by two factors: resolution and the signal-to-noise ratio (SNR). These are also the two primary characteristics used in determining overall image quality. Therefore, it is possible to represent the quality of an image in terms of its information content. According to this concept, an imaging system can be regarded as a function that transfers information from the sample to a final image output. Image quality can thus be represented by its own information-passing capacity (IPC). IPC only depends on resolution if the signal-to-noise ratio

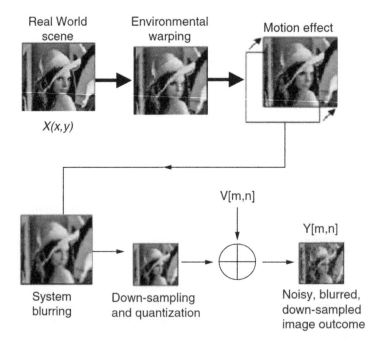

FIGURE 12.13 Block diagram representation of observed image formation model using the famous lady's image. (From Figure 1 in S. Farsiu et al., "Advances and challenges in super-resolution," *JIST*, March 2004.)

remains fixed. Conversely, the imaging resolution of a system can be determined from the IPC at each signal-to-noise ratio.

The topic of validating and establishing quantitative methodologies for assessing resolution and, more generally, the relationship of total image quality on metrologic precision and accuracy is more complex than would initially appear. Any proposed methodology for achieving this goal must begin with an appropriate standard, or in our case, a set of well-characterized images of known resolution of objects commensurate with the operational range of interest. It is not practical to produce such a set of images experimentally, due to the effects of sample morphology on resolution estimates. This can be accomplished by creating multiple groups of synthetic images (Figure 12.13). The morphology of each set is then perfectly known and can be modeled to emulate real-life resolution standards consisting of nano-scale metallic spheres (usually tin or gold) on a carbon substrate. The images are produced by generating binary arrangements of spheres in agreement with the size distributions of interest. The binary images are then convolved with Gaussian point spread functions having standard deviations also within the range of interest in order to create identical sets of images having different, but exactly known resolutions [25]. The field of view (FOV), or conversely magnification, should be chosen such that the pixel sampling density is at least twice the expected spatial frequency associated with the resolution, as per the Nyquist Criterion. The binary images can then be convolved with different Gaussian point spread functions having standard deviations that vary across the range of expected resolution estimate extremes. An adequate size for the convolution kernel should be at least 100 pixels in order to be large enough to avoid truncation errors for any of the PSF. As mentioned earlier, the nominal resolution of each image is related to the PSF width (a la Rayleigh Criterion), so that in the case of a Gaussian PSF with standard deviation σ, the resolution associated with a 25–75% threshold transition is approximately 0.96σ, while the 50% density diameter is 1.66σ [26]. In order to be more specific to our interests of dimensional metrology below the 90 nm node (i.e. estimation of spatial resolution in the range between 1 and 6 nm), the resolution of each synthetic image is reported in nanometer units instead of pixels with the relationship of 1 pixel equal to 0.782 nm. It should be noted that this procedure does not perfectly model the conditions obtained in a metrology SEM. For the case of low-voltage SEM (landing energy <1 KV), field-emission source, and immersion optics with effective lens aberration coefficients <200 μm and a primary spot diameter of only a few nanometers, the beam probe has a point spread function that is highly sensitive to focus and is only approximately Gaussian within the usual region of best focus [27]. Magnification of the image sets should not be varied because the resolution estimate should always be performed at the smallest possible pixel size (i.e. maximum practical magnification or smallest field of view), such that the spatial sampling density and the resultant information-passing capacity (IPC) of the image are maximized for a given system under test. The addition of specific levels of noise (Gaussian or Poisson) to the various images allows for adjustment of the signal-to-noise ratio. Once a series of calibration images are available, it becomes possible to evaluate the different resolution algorithms by application of each to the control set of synthetic images for final comparison results. The remainder of this section will primarily focus on comparing the efficacies of commonly implemented resolution algorithms including the gap, spectral FFT, derivative, contrast-to-gradient, and correlation methods of analysis for quantitative estimation of image resolution.

One of the most venerable and still commonly used ways of assessing resolution is the so-called "gap" method. The operator records an image from the resolution standard, such as metallic spheres on a carbon background, and then manually finds the smallest gap discernable between the high-contrast features. In this way, the resolution value is taken to be equal to this gap distance. The subsequent results are only meaningful if the magnification is high enough to ensure a pixel size that is much smaller than the anticipated resolution, as per the Nyquist Criterion. The result is very dependent on the operator who performs this procedure and is usually of poor precision. This is compounded by the fact that it only uses about a half dozen or so pixels out of an image typically composed of one quarter million or more pixels. A more recent implementation of this approach utilizes the selection of a small region of interest (ROI) in which subsequent interpolation is performed to apparently provide higher pixel density. The operator then manually extracts a line profile from the ROI to measure the width of an intensity transition where the beam crossed a feature edge. The resolution is then interpreted as being equal to the horizontal distance traveled by the scanned beam probe to raise the intensity profile from 10 to 90% of its maximum value [28, 29]. These results are displayed in Figure 12.14, where the gap method is seen to be relatively insensitive to the actual nominal resolution of the reference image. This is attributed to an artifact that arises from the intensity interpolation process, which essentially makes image noise indistinguishable from the real surface detail. Poor precision exhibited by the gap process and interpolation is further exacerbated when the operator is allowed to adjust the image contrast and brightness. The gap method is therefore an unreliable approach to resolution estimation. It is possible that the accuracy of this method could be improved somewhat if a suitable resolution standard having a well-known and controlled feature edge gap structure could be fabricated and characterized.

The derivative method also estimates resolution by estimating the signal rise across an edge, but has the advantage of being a fully automated technique. This algorithm normally involves initial computation of horizontal and vertical derivative images. Direct computation of the derivative from an original digital image is avoided, due to the high levels of noise that result from such an operation. The issue of derivative-induced noise susceptibility is mitigated to some extent by treating the original image as a three-dimensional surface (X, Y, Intensity) where analytic functions, such as cubic splines, are fitted to ROIs within the overall image to smooth random noise and facilitate the computation of the intensity signal derivative along any chosen direction. The algorithm subsequently locates the highest slope values in the derivative images, which correspond to the sharpest intensity transitions. The line profile is extracted from these locations and fit to a Gaussian integral. The local resolution is then equated with the scanned distance required to raise the integral from 25 to 75% or 10–90% of its maximum value. The procedure is iterated many times in both the horizontal and vertical directions to produce a final result. A typical set of results from the derivative approach to resolution estimation after operation on the synthetic reference image standards is displayed in the second plot of Figure 12.15. This derivative method does show slightly more sensitivity to resolution and maintains estimate accuracy for resolutions >3 nm (~4 pixels), although the precision is still relatively poor at 1.2 nm (~2 pixels). The lack of sensitivity below 3 nm is due to the use of image processing steps, such as smoothing prior to the derivative operation, which also removes

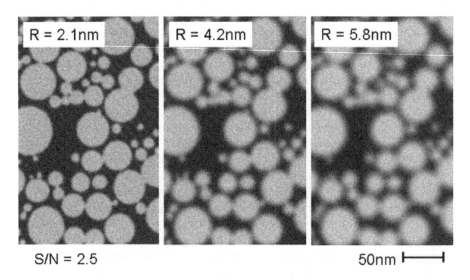

FIGURE 12.14   Resolution standards of 2.1, 4.2, and 5.8 nm. All images have an SNR of 2.5. (From Figure 1 in G.F. Lorusso and D.C. Joy, "Experimental resolution measurement in critical dimension scanning electron microscope metrology," *Scanning*, 25, 175 [2003].)

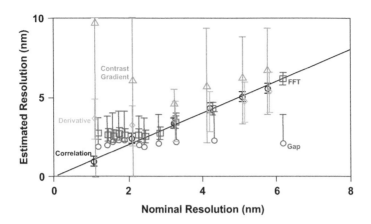

FIGURE 12.15 Accuracy plots for the various methods used to estimate resolution. (From Figure 2 in G.F. Lorusso and D.C. Joy, "Experimental resolution measurement in critical dimension scanning electron microscope metrology," *Scanning*, 25, 175 [2003].)

portions of the high-frequency information about the sample contained within the original image. The accuracy might be improved by less aggressive filtering techniques, such as median filtering instead of kernel smoothing, but would further deteriorate the already poor precision. The derivative method might still provide some value for metrology systems having resolutions larger than 3 nm, but would not be relevant to the current specifications for lithography metrology (ITRS 2013).

The Fourier transform is a mathematical operation that is performed in order to decompose a spatial image into its frequency domain components, otherwise known as the power spectrum of the image [30–32]. The highest spatial frequency that exists within the image is then assumed to correspond with the smallest detectable distance between features, which is another way of interpreting the Rayleigh Criterion. On a pixel-by-pixel basis, noise and signal are indistinguishable. The SNR also begins to degrade at the higher spatial frequencies, due to a roll-off equal to at least $1/\omega$ of the microscope optical transfer function [33]. The resultant discrimination between signal and noise becomes more and more imprecise as the spatial frequency increases. A recent implementation [34] of this technique introduced an automated procedure to infer the threshold between signal and noise in the frequency domain, thereby removing any operator dependency effects on the precision. A superposition diffractogram technique [31, 32, 35] has also been applied for automatic threshold selection, although it requires two sequentially recorded copies of the same image. An image is then converted to binary form after the threshold has been set. The signal region of the power spectrum is compacted by means of opening and closing cycles (i.e. repeated pairs of dilation and erosion operations) to measure the dimensions of an area occupied by the signal in Fourier space by fitting to an ellipse. The average lengths of the major and minor axes of the ellipse are then calculated and an average radius R_{FFT} (pixels) that corresponds to the cutoff in real space units is used to define:

$$\text{Spatial Resolution}\left(\text{nm}\right) = \left(\text{ROIW}/R_{FFT}\right) * \text{pixel size}\left(\text{nm}\right) \tag{12.30}$$

where ROIW is the width in pixels of the region of interest analyzed for this measurement. The maximum value of R_{FFT} is ROIW/2, which implies that the best spatial resolution that can be determined with this method is 2 pixel units. The results obtained by the operation of the spectral FFT method on the reference images are displayed in the third plot of Figure 12.15. Once again, good agreement between the estimated and nominal resolution values is only found above 3 nm (~4 pixels). In the case of the spectral FFT method, this loss of accuracy below 3 nm can be attributed to the undesired removal of high-frequency information that occurs during the dilation and erosion cycles of image processing. However, the precision is better (0.3 nm or 0.4 pixels) because all of the image pixels are utilized in this type of analysis.

A recently proposed method, known as contrast-to-gradient, is a more sophisticated variation of the standard derivative method [36]. This approach is also fully automated and defines the average resolution of an image as a weighted harmonic mean of the local resolutions. An image map of gradients and curvatures is obtained analytically through interpolation of the image with a bi-quadratic function typically originated from a 5×5 pixel subset which is moved throughout the ROI. This method contains two main assumptions: (1) The change in contrast required to estimate local resolution is proportional to the dynamic range of the image (defined as the range of intensity $I_{max}-I_{min}$). This allows one to calculate the 25–75% resolution from a slope that extends over the same range of $I_{max}-I_{min}$. (2) Local resolution estimates larger than the local radius of curvature are meaningless and should be discarded, thereby reducing the dependence of resolution values on the SNR. Results obtained by using the contrast-to-gradient method for operation on the reference images are shown in the fourth plot of Figure 12.15. This method appears to possess appropriate accuracy for resolution values greater than 3 nm (~4 pixels). The precision is poor, with an average of about 7.9 nm (~10 pixels) variation, possibly due to the effect of noise on the interpolation procedure.

The correlation method [37–39] provides a viable alternative to the previously discussed spectral FFT approach. The finite width of an autocorrelation peak is utilized to reflect correlations introduced by the image-forming system by effectively approximating the autocorrelation of the PSF. The cross-correlation function of two regions within the same total image is computed first. A profile of this cross-correlation function is then assessed using the Rayleigh Criterion to deduce the effective spatial limit of the correlated image detail to produce an effective image resolution estimate. This approach is also fully automated and requires no operator intervention. The primary algorithm employed calculates the autocorrelation by convolution of the image of interest with itself in the frequency domain before an inverse Fourier transform is subsequently applied. Results from the application of this correlation method to the reference images are plotted in the fifth and final graph of Figure 12.15. This method is capable of estimating resolution values down to at least 1 nm (~1.2 pixels) while demonstrating a precision of 0.3 nm (~0.4 pixels). It has also been shown that these estimated resolution values appear to give no indication of any systematic trend or degradation in precision with SNR [25]. The full potential of this method was also tested by reconstructing the PSF from a set of images produced with different PSFs.

A relatively new area of research, known as "super-resolution" (SR) image processing and reconstruction, has recently been explored for improving the final image detail and information content. Super-resolution is the term generally applied to the problem of transcending limitations of the optical imaging systems through the use of image processing algorithms, which presumably are relatively less inexpensive to implement. Super-resolution image processing requires the availability and subsequent fusion of several low-resolution (LR) acquisitions of a scene to obtain a final resultant description having increased information content over its individual constituents. *Super-resolution* essentially refers to the process of producing final images of high resolution from multiple real-life LR acquisitions, through the removal of system-related imaging degradations and the maximization of spatial sampling frequency. SR processing essentially extrapolates the high-frequency image components, while seeking to concurrently minimize aliasing and blurring. This approach avoids the noise amplification encountered with interpolation-based processes, where the image is simply convolved with a filter function designed to boost higher-frequency components. Super-resolution algorithms attempt to extract the high-resolution image corrupted by limitations of the physical optical imaging system. This is primarily an *inverse* problem, wherein the source of information (high-resolution image) is estimated from the observed data (low-resolution image or images). Solving an inverse problem, in general, first requires constructing a forward model, which was done at the beginning of this section (Figure 12.14).

12.3.3 LIGHT-OPTICAL MICROSCOPY

12.3.3.1 Conventional Full-Field, Confocal, and Interferometric

Even as the dimensions to be measured extend below the 90 nm range, optical measurement techniques still play an important role in dimensional metrology. They offer the advantages of being non-destructive and fast, with little or no requirement for modification of the specimen. Optical tools can even possess surprising repeatability ($P < 5$ nm) for features approaching the quarter-micron regime. In the 19th century, the precise natural sciences experienced an enormous upswing. Even in the twenties and thirties, the science of light and the theory of optical imaging were placed on a sound foundation. One of the most successful researchers in this field—and not only in theory—was Joseph von Fraunhofer (1787–1826). We have him to thank for the creation of what is now the most common optical lens system with chromatic error correction—known as the Achromat system—and for basic knowledge on the diffraction of light. Abbe proposed the notion of an ultimate resolution limit for conventional lens-based light-optical imaging systems governed by a phenomenon known as diffraction. However, even when a fine structure of interest can be resolved for making repeatable measurements, good measurement precision may not necessarily be equivalent to accuracy. Linearity between the measured and actual values is frequently lost to interference, resonance, and shadowing effects well before the diffraction limit and ultimate resolution are reached. A more detailed derivation of the diffraction limit will be useful for understanding and comparing the different types of optical microscopes presented in this section. The observed distribution of image intensity changes with the distance propagated from an aperture object (Figure 12.16a) at which it originally passed through or was reflected. The direction of light propagation is unaltered as it initially passes through the aperture. When the image plane is very near to the aperture, there has been insufficient opportunity for interference between the light modes to have occurred. The result is a sharp image representation of the aperture by the wave front. As the image plane is propagated further away from the aperture, each point on a wave front acts as an independent light source from which other wave fronts originate. This is known as *Huygen's Principle*. Diffraction is then said to occur when the wave fronts begin to interfere with each other to create a more complex series of waves. The set of nodes corresponding to constructive interference that define a trajectory with angular direction $\hat{\theta}$ is referred to as a diffraction *order*. The relative difference in angular direction or spreading of the diffraction orders (Figure 12.16b) increases as the initial aperture width becomes more narrow (i.e. they diverge more rapidly). Each order carries information about the aperture. The higher diffracted orders (i.e. larger angular deviation, or "spread") carry information about the finer spatial details of the object aperture. High resolution and the ability to image fine details of the object are only possible if the imaging lens is able to capture these higher diffraction orders (Figure 12.17).

Conventional optical microscopy systems are usually configured such that the image collection lens is located at a distance from the aperture object that is much larger than that of the wavelength. In this case, the lens plane is said to lie in the "Fraunhoffer" region of space, otherwise known as far-field imaging. A second-order differential equation, known as the wave equation, can be derived from Maxwell's laws to describe the behavior of electromagnetic waves in space and time. To solve this equation, we assume that all information necessary for the calculation of the intensity is contained in the scalar wave function:

$$\Psi(r,t) = \Psi(r)e^{i\omega t}. \tag{12.31}$$

In a linear, homogeneous, isotropic media, the wave equation results in

$$\left(\nabla^2 - \frac{1}{c}\frac{\partial^2}{\partial t^2}\right)\Psi(r,t) = 0. \tag{12.32}$$

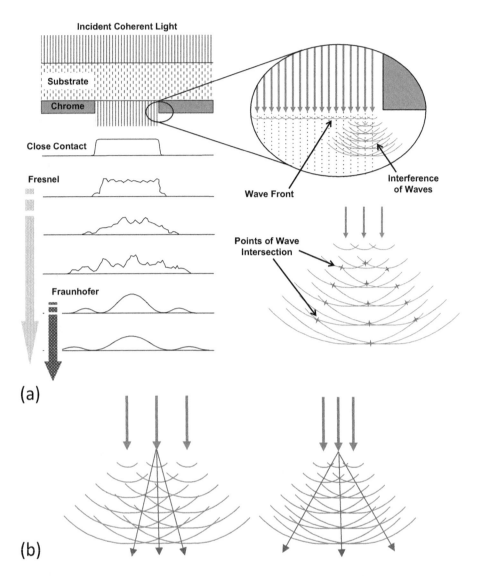

FIGURE 12.16 (a) Diffraction, interference, and the intensity profile through propagation. (b) The effect of aperture size on the diffraction peak spreading.

Substituting the wave function into this relationship yields the Scalar Helmholtz Equation (SHE):

$$\left[\nabla^2 + k^2\right]\psi\left(r\right),$$

(12.33)

where $k = \omega/c$. A Green's Function solution of the Scalar Helmholtz Equation can be performed if we assume that $\psi(r)$ is zero on the scattering object surface and all outgoing waves have only radial variation (isotropic radiation). The far-field image at a distance r from the circular aperture seen in (Figure 12.18) can then be found from the resulting Fraunhoffer integral [40]:

$$\Psi(r) = \frac{iA_o e^{-ikr}}{\lambda r} \int_{S'} e^{ik(\alpha x' + \beta y')} dS',$$

(12.34)

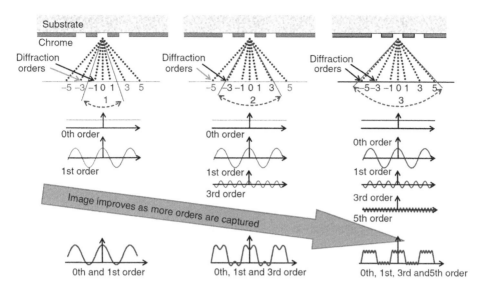

FIGURE 12.17   Concept of diffraction orders and resolution.

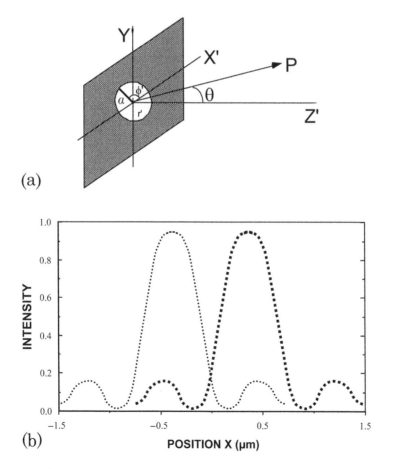

FIGURE 12.18   (a) Circular aperture in an opaque screen and (b) overlapping Airy intensity disks and the Rayleigh Criterion.

where A_o is the initial field amplitude, λ is the wavelength, k is the spatial frequency, $x' = r'\cos\phi'$, $y' = r'\sin\phi'$, and s' is the surface of integration $(ds' = r'dr'd\phi')$. Evaluation of the integral and multiplication of Ψ with the complex conjugate Ψ^* yields the image intensity:

$$I(r,\theta) = I_o(r)\left[\frac{2J_1(ka\sin\theta)}{ka\sin\theta}\right]^2, \tag{12.35}$$

where J_1 is a first-order Bessel function. The intensity of a diffraction-limited spot is described as $I(r,\theta)$, often referred to as the "Airy disk." The Airy disk radius is defined by the first zero of $I(r,\theta)$, which occurs when $ka \sin\theta = 1.22\pi$. Therefore, the Airy disk radius is given by

$$r_a = 1.22\lambda\left(z/2a\right). \tag{12.36}$$

For monochromatic, collimated, and incoherent source illumination, the minimum resolvable separation between the center of one disk and the first minimum of an adjacent light spot is then given by the familiar Rayleigh Criterion:

$$D = \frac{1.22\lambda}{2NA}, \tag{12.37}$$

where NA denotes the numerical aperture of the lens. The Fraunhoffer integral changes for different apertures (i.e. sample features) and will not yield the same resolution for each type of feature. Accuracy and precision, therefore, depend on the resolution and thus will vary with the feature type observed.

The ability of a microscope to produce sharp image intensity profiles determines the certainty at which the edge location can be specified within the intensity transition edge range, hence the accuracy. The linerity of the measured CDs versus the actual structure size X_c' (provided by RMS) is also strongly dependent on the intensity transition shape produced by diffraction at the object feature edge. Even if an image can be resolved, CD measurements cannot be performed if the correspondence between the image size and the actual feature dimensions becomes too nonlinear. Line width uncertainty is heavily affected by the choice of edge position in an intensity profile. Actually, this holds true for any microscope whether it be an optical, electron, or scanning probe. The graph in Figure 12.19 shows the measured intensity distribution as a function of position across the edges of a transparent space in an anti-reflective chromium background (i.e. clear aperture in an opaque

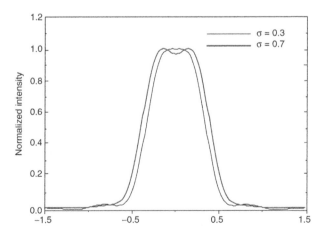

FIGURE 12.19 Intensity versus position of a 1 μm-wide space for two source coherences.

screen) on a quartz substrate observed in transmitted light with a scanning-slit optical microscope. It is not clear where the line edge is actually located on the intensity profile. For the microscope in this example, the illumination was provided by a broadband light source that was partially coherent. The difference in line edge position between the 25 and 50% intensity points is about 0.1 µm, a considerable fraction of the feature width. Other factors that affect the intensity distribution of the image are the numerical aperture of the objective lens, coherence, lens aberrations, flare, and the wavelength spectrum of the source [41]. Coherence is often used to describe the interference aspects of a microscope and is defined as the ratio of the condenser to the objective lens NA for a conventional optical microscope. The measured line width depends on the coherence of the source illumination. If the actual coherence of the light is unknown, the choice of intensity threshold for locating the line edge position could typically be anywhere between 25 and 75% of maximum for each profile.

The focus level is another factor affecting the determination of the line edge position in the intensity profile as well as the measurement repeatability. Usually, the focus is adjusted for optimum image quality. Attempts to accurately determine the line edge position from intensity profiles are being made through the use of theoretical calculations [42]. One starts with calculating the image intensity as a function of the lateral position across the feature. The calculated intensity profile is then fit to the experimental distribution with the desired critical dimension as an adjustable parameter. Once agreement between the theoretical and measured intensity profiles has been found, the position of the edges can be determined. The desired critical dimension is then obtained as the distance between the edges.

Unfortunately, the intensity distribution in diffraction patterns depends on the optical properties and physical dimensions of all the materials on the sample. For instance, phase-shifting masks (PSMs) contain features whose thickness exceeds 1/4 of the wavelength of the illuminating light and therefore cannot be considered thin in the mathematical analysis. In this case, the image intensity profile becomes strongly dependent on the thickness and shape of the feature (considerably increasing the difficulty of computing the intensity profiles). Scattering theory must be used for thick-layer features, such as phase-shifting lines. The mathematical model used to predict scattering profiles is currently under investigation at various labs, such as the National Institute of Standards and Technology. Even if the modeling results are known to agree with experiment, the model could be used to determine the edge locations only if all other parameters used by the model were known and measured simultaneously. Therefore, the basic principle of instrument design for realistic submicron dimensional metrology of thick-layer features, such as those on phase-shifting masks, should be to design the measurement system for improved resolution, thereby providing precision and accuracy for simplicity in modeling and design.

A diagram of a conventional optical microscope is shown in Figure 12.20. Intensity profiles are obtained from the gray-level brightness along a line in the video image. These systems typically

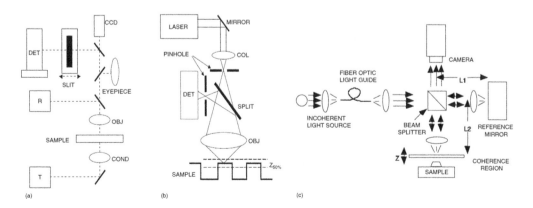

FIGURE 12.20 (a) A combined reflective and transmission mode scanning-slit optical microscope. Optical ray diagram of a sample, (b) scanning confocal, and (c) Linnik interferometric microscopes.

have autofocusing capability to improve the measurement precision. The ultimate resolution limit for this conventional optical microscope was given previously. A common objective lens for measurements may have a magnification of 150× and NA of 0.9. If the illumination wavelength for this lens is centered at 550 nm, D is approximately 0.35 μm. A partially coherent white light is typically used. Although the nonlinearities due to sample properties affect the linearity before the diffraction limit is reached, it is still instructive to compare the fundamental limit of each optical configuration to gain insight into its performance. In the scanning-slit system of Figure 12.20a, the intensity in the image plane is scanned by a slit whose length and width can be varied. The intensity transmitted through the slit is then detected and recorded versus the slit scan position. A commercially available scanning-slit microscope [43] has a laser autofocusing system, which locates the focal position interferometrically. The interferometric laser autofocusing technique is capable of finding the focal position more often and more repeatable than the video analytic technique of the system described previously. The scanning-slit width determines the resolution at which the image intensity can be sampled, but the optical resolution and hence performance are still limited by diffraction and non-linearity in the same way as that for other conventional optical microscopes.

In a scanning laser confocal microscope, a diffraction-limited spot is focused onto the sample and then re-imaged by a pinhole before detection. Figure 12.20b illustrates the arrangement of a scanning confocal microscope. A point source of laser light (defined by a pinhole) is focused onto the surface by an objective lens, which creates a diffraction-limited intensity spot of the same form as in the conventional microscope. This spot is focused and spatially filtered (using the same pinhole) twice, once before and once after scattering from the sample. This double focusing/filtering is referred to as a confocal operation and produces a narrower image point spread function than in the conventional microscope. A convolution of the source and detector apertures results from the confocal operation so that the image intensity is now given by

$$I(r,\theta) = I_o(r) \left[\frac{2J_1(ka\sin\theta)}{ka\sin\theta} \right]^4. \tag{12.38}$$

The fourth power results from this convolution of source and detector aperture intensity. The spatial sensitivity of the detection is equal to that of the source and the minimum resolvable linear distance is now given by

$$D = \frac{1.22\lambda}{2\text{NA}} \left(1 - \frac{1}{\pi} \right). \tag{12.39}$$

Therefore, the point resolution of a confocal microscope is $1/\pi$ (33%) better than that of conventional optical microscopes. If the illumination is coherent, the minimum resolvable distance is then less (i.e. also a smaller PSF). If the same NA and wavelength are used as the conventional microscope example earlier, then D becomes approximately 0.24 μm. For many years, a scanning ultraviolet (UV)-laser confocal microscope [44] with λ equal to 325 nm has usefully served in imaging metallic photomask lines 0.25 μm wide. The minimum resolvable distance D for the UV system is approximately 0.14 μm. Interestingly, interference and resonance effects produce a loss of linearity at feature sizes <0.5 μm. The confocal microscope can also measure vertical dimensions to some extent, due to its extremely small depth of focus (~100 nm). The intensity transmitted through the detector aperture changes as a function of focal distance z. The detected intensity is maximum at the optimum focal position and decays rapidly on either side. The full width at half the maximum of $I(z)$ is then

$$2Z_{50\%} = \frac{0.89\lambda}{\text{NA}^2}. \tag{12.40}$$

The relative step height between two points can be found by measuring the z difference in the $I(z)$ maximum (i.e. focus) for each. Additional benefits to contrast are also gained because of the tight depth of focus inherent to the confocal microscope. The majority of the intensity through the detector pinhole is from the reflection off the surface at the focal plane z. If the focal level is set at the top of the feature (top focus), variations in the CD measurements due to the substrate can be reduced. One can also obtain an image at each z height (within the depth of focus). Bottom focus imaging could be used for improved discrimination of a feature's bottom width. However, this is affected by shadowing and substrate effects more than with top focus. The line appears bright in the top focus image when the substrate is out of the focal range. Conversely, the substrate appears bright in the bottom focus image and the line appears dark. Scattering and shadowing by the feature edges, however, will also inject an offset between top- and bottom-based width measurements [45]. Another type of confocal microscope exists whereby the aperture or light spot is scanned instead of the sample for real-time imaging using a Nipkow disk [46]. The Linnik interferometer microscope [47] (Figure 12.20c) uses white light interference when imaging samples and is capable of performing measurements in both the vertical and lateral directions. A Linnik interferometer is formed by placing an objective lens into each arm of a Michelson interferometer. When broadband illumination of partial coherence is used, interference at the camera only occurs when the optical path lengths in each leg of the interferometer are approximately equal. An "optical coherence region" can then be defined as the planar region in space located exactly by the focal point of the sample objective. If the sample is scanned vertically (z-axis) through the coherence region, an interference signal (phase signal) is detected at the camera as the surface passes through the coherence region. As the sample is scanned vertically, each pixel in the XY image creates a "phase signal." An entire three-dimensional volume of data is produced as an entire XY image is recorded by the camera at each vertical scan position. A cross-sectional representation of the sample is then provided using digital signal processing techniques. Conversely, the sample could be stationary, while the reference optical path is varied. Image contrast is generated by the degree of coherence between corresponding XY pixels in the object and reference image planes. A similar improvement in the lateral resolution to that of the confocal is also realized, because the field from each point in the object plane is convolved with a corresponding point in the reference plane. Therefore, the PSF is of the same form as that of the scanning confocal. However, broadband illumination helps to reduce undesired interference effects in thin films, such as photoresist. Patterns consisting of chromium features on an un-etched quartz background (i.e. binary) were used to estimate the relative spreading from diffraction [48]. In this case, an edge transition is defined as the change from 90 to 10% (or 10–90%) of the maximum intensity. The width of this transition is defined as a percentage of the pitch (Figure 12.21). Combinations of chromium, quartz, and etched quartz were used to create permutations of a line. The first type is shown in Figure 12.22a. This consists of a transparent inner line with an etched quartz phase shifter on either side in an opaque chromium background. The intensity line scans (Figure 12.22b) for the phase-shifted isolated line are quite different from the binary isolated line, even though they have the same amplitude functions.

For an etched quartz center line and un-etched quartz rims on a chromium background (Figure 12.23a), the instruments perform differently than with the previous type of line. The reflected intensity from the etched inner line is now less than that due to the un-etched quartz shoulders (Figure 12.23b). This produces a subtle inflection in the intensity profile that complicates the determination of the shifter edge position. For transmission, the choices of edge location are further obscured by diffraction effects between the inner line and outer quartz rim. Hence, only a reversal of the shifter tone produces marked changes in the image intensity profiles (image quality) and precisions.

12.3.3.2 Optical Scatterometric Encoding for Dimensional Metrology

Critical dimension and profile information are critical parameters that need to be measured accurately for use in semiconductor manufacturing process control. As device geometry shrinks,

Instrument	Transition Width (% of Pitch) Flat Lines	Transition Width (% of Pitch) Etched Lines
Scanning slit (trans)	33	30
Scanning slit (refl)	28	27
Conventional	30	28
Linnik	25	26
325nm (UV) confocal	20	9

FIGURE 12.21   Image intensity broadening (f) as a percentage of the 1 µm pitch.

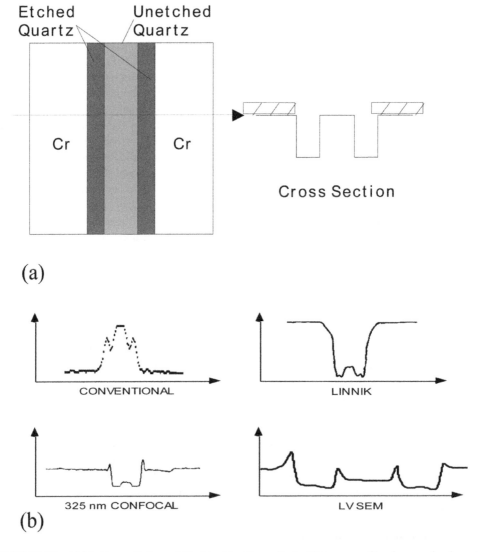

FIGURE 12.22   (a) First type of phase-shifted isolated line—IL A. (b) Image profiles from each microscope of IL A.

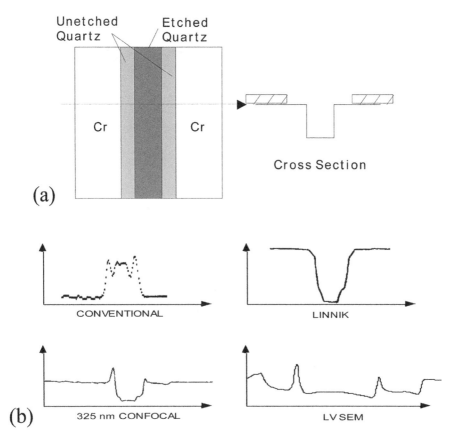

FIGURE 12.23 (a) Second type of phase-shifted isolated line—IL B. (b) Image profiles from each microscope of IL B.

traditional metrology methods such as CD-SEM are fast reaching their limits in providing accurate and repeatable CD measurements. While charging due to the electron beam is a major concern for CD-SEMs, profile measurement normally uses destructive methods such as cross-section SEM or TEM, or the very slow atomic force microscopy (AFM). Additionally, these off-line approaches limit the cycle time to results and sampling rate, thereby leading to high wafer cost. As technology nodes move from 90 nm to 65 nm and beyond, lithography process windows continue to shrink, especially so for contact holes for which the process windows are already smaller than those for line-and-space structures of comparable size. This problem is made even more serious by 300 mm wafer production due to the higher wafer cost, increased sampling rate requirements, and the need for integrated metrology on process tools to provide on-tool process control. Scatterometry provides a solution that addresses these issues to meet the challenges of the 90 nm node and beyond.

Optical digital profilometry (ODP) is one of several scatterometry-based commercial solutions that are available in the market today. It is an optical, non-destructive, inline measurement technique that generates digital cross-sectional representations of IC features. ODP employs a light source to extract CD and profile information from a periodic grating structure. A beam of broadband (including both visible and UV) light is incident on the wafer, striking a test grating that is designed into the scribe line between adjacent dies on the wafer, as shown in Figure 12.24. When the light strikes the periodic structure, the light undergoes diffraction due to the periodic grating, and this diffracted light is reflected into an optical detector. Some of the light is absorbed into the films, and some of the light is transmitted through the thin film layers, part of which ends up at

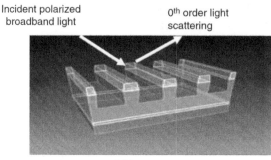

Incident polarized
broadband light

0th order light
scattering

Grid test pattern

FIGURE 12.24   Typical grating pattern for scatterometry measurements.

the optical detector, a result of being reflected at film interfaces. The optical detector collects the reflected light (typically intensity and/or polarization), generating spectral information as a function of wavelength or energy. The behavior of light through these patterned layers changes with the grating structure. The spectral "signature" from the structure depends on many factors, including the grating pitch, structure profile, thicknesses of the constituent films, film material properties, and angle of incidence. As the CD changes from 60 nm to 65 nm, the spectral signature changes, so that a unique signature exists for each combination of variables. For a typical structure, the angle of incidence and the pitch may be fixed, with the CD and film thicknesses undergoing variation due to processing conditions.

Scatterometry utilizes Maxwell's equations of light to simulate what the light signature might look like, based on input such as grating pitch, film properties, angle of incidence, CD, and film thickness. The practical application is to first simulate a large set of possible parameter combinations, generating a large set of spectral signatures, and when the measurements are actually taken on the wafer's grating structures, find the simulated signature with the closest match to the measured signature. This "match" corresponds to a unique set of CD and film thickness parameters, which then becomes the measured result. The quality of the results depends not only on the technology used to measure the grating, but also on the algorithms used by the analysis software. This section discusses the various hardware technology choices available, as well as the analysis algorithms.

Scatterometry was first developed using a single-wavelength light beam (i.e. a laser), with a number of measurements being made on a single grating, using various angles of incidence [49, 50]. This technology is still in use today, although it is far less common than the single-angle, variable (broadband) wavelength systems that are most prevalent in the major semiconductor manufacturing fabs. The early research of Timbre Technologies, Inc. showed how thin film measurement tools, common in every fab, which already used a single-angle, broadband wavelength design, could be used successfully to provide spectra for scatterometry measurements [51, 52]. The value of existing equipment, connected to a computer to generate measurements, led to the predominance of spectral broadband (single-angle) scatterometry.

In general, spectral scatterometry uses either spectroscopic ellipsometry or reflectometry. Spectroscopic ellipsometry involves the use of broadband light incident on the sample surface at an angle, typically around 65°. The light reflected off the sample is collected by a detector that is spatially fixed at an angle that collects the zeroth order of the reflected light. The light typically has a wavelength range from 190 nm to around 900 nm. Figure 12.25 shows a schematic of spectroscopic ellipsometry and unpolarized reflectometry technology. Reflectometry itself can be either polarized or unpolarized. Each of these flavors of technologies has its advantages and disadvantages.

For spectroscopic ellipsometry the light collected from the sample contains information from the grating sample and is represented by two quantities: $\tan\psi$ and $\cos\Delta$. The $\tan\psi$ signal represents

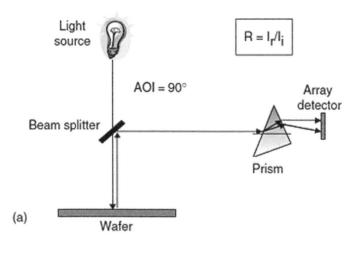

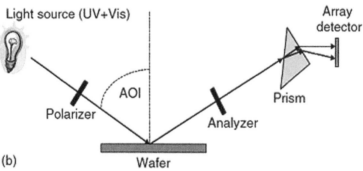

FIGURE 12.25   (a) Optical paths for reflectometry and (b) spectroscopic ellipsometry.

the ratio of the reflected complex magnetic to complex electric fields. This ratio is also known as ρ, where

$$\tan\psi = |\rho| = |R_p/R_s|,$$
(12.41)

R_p (p stands for parallel, the German word for parallel) is the reflectance in the plane of the incident and reflected beams, and R_s (s stands for senkrechte, the German word for perpendicular) is the reflectance normal to the plane of the incident and reflected beams. Simply stated, $\tan\psi$ is a measure of the change in light intensity or magnitude as a result of being reflected off a grating structure. The formal equation is represented as

$$\rho = R_p/R_s = \tan\psi \cdot e^{i\Delta}$$
(12.42)

The $\cos\Delta$ signal represents the phase difference between the reflected and incident beams (the phase shift due to reflection), where $\Delta = \delta_1 - \delta_2$. The phase difference between the p and s waves of the incident beam is δ_1, and the phase difference between the p and s waves of the reflected beam is δ_2. The p wave is the part of the wave that is in the plane of incidence, and the s wave is the part of the wave that is perpendicular to the plane of incidence. The plane of incidence is formed by the incident beam and the normal to the surface.

In unpolarized reflectometry, the beam coming from the source is incident on the sample normal to the surface. The reflected light is then collected at a diode array detector. In polarized reflectometry, a polarizer and compensator are introduced into the light path, effectively providing the transverse electric (TE) and transverse magnetic (TM) component of the reflected light. Reflectance itself is simply the ratio of reflected to incident intensities of light, where $R = I_R / I_I$. Since SE measures the change in intensity *and* the change in phase, the additional information provides the capability to determine more complex profiles. Polarized reflectometry can also provide significant sensitivity to complex profiles, and both of these techniques are now routinely used in lithography and etch applications. The unpolarized reflectometer has a significant role to play in integrated metrology. Its relatively simple optics makes it an ideal tool to integrate onto fab process tools such as the track. This technology is now used in leading-edge fabrication facilities. Lensing et al. [53] provide a comprehensive review of these different technologies and their practical applicability in a semiconductor fab environment. The information obtained from these tools is adequate for a detailed reconstruction of the profiles of periodic gratings.

The theory behind typical scatterometry analysis engines is known as rigorous coupled-wave analysis (RCWA). RCWA is an analytical technique that uses a state-variable method for determining the numerical solution. The details of RCWA theory and several of its variations can be found in many reference works [52]. The conventional approach of RCWA involves solving hundreds of high-order differential equations simultaneously for large amounts of variables and boundary conditions. It requires a large amount of computing power to provide the results in a reasonable time; high-speed computers have dealt with this issue effectively. Optical CD metrology is, in part, possible today because of the availability of powerful and inexpensive computing technology that allows libraries to be stored and compared with incoming data. These libraries are fast, and can be integrated into the etch or track tools to get immediate feedback, thereby facilitating feedback or feed-forward process control.

Numerous applications of scatterometry have been demonstrated in the semiconductor manufacturing process. Major processes include mask processing, lithography, etch, and chemical mechanical polishing (CMP). Two examples are provided here: the first involves use of scatterometry for detecting open and closed contacts; the second describes an application in mask process metrology. ODP scatterometry has demonstrated this technology's ability to measure three-dimensional contact holes as well as lines and spaces [54]. The response correctly responds to profile changes caused by variations in focus and dose. Combined with excellent precision, scatterometry can detect process excursions and closed contacts [55]. Figure 12.26 demonstrates its ability to provide a die-based contour map showing deformed/closed contacts (shown in images at +0.25 μm and +0.35 μm defocus). Scatterometry is now seeing applications in the photomask industry as well. Binary and phase shift masks have been measured successfully [50, 56, 57].

The ability of scatterometry to provide line width and trench depth measurements together, in a single measurement, provides huge time savings compared to conventional CD-SEMs. Figure 12.27 shows CD uniformity (CDU) measurements from a mask from both a CD-SEM and an ODP scatterometer. Scatterometer-based CDs provide a more accurate representation of the uniformity across the mask by averaging out local variations, thereby showing fewer hot spots. A 20–30% improvement in CDU values is reported. Embedded phase shift masks (EPSM) final check CD data were obtained from both a polarized reflectometer and a CD-SEM. A mask error enhancement factor (MEEF) calculation based on a polarized reflectometer CD shows about a 40% improvement over that from a CD-SEM. Scatterometry applications are now being used in the manufacturing environment not only for CD metrology, but also for enabling feedback and feed-forward control in etchers and coaters developers [58].

12.3.3.3 Near-Field and Evanescent Mode Optical Imaging

As suggested previously in the discussion on "super-resolution" image processing techniques and as E.H. Singhe indeed suggested over 75 years ago, the diffraction limit in optical microscopy is not *fundamental*; rather, it arises from the assumption that a lens is used to focus a spot of light onto an object surface. If one introduces a spot of light onto an object's surface by using a sub-wavelength

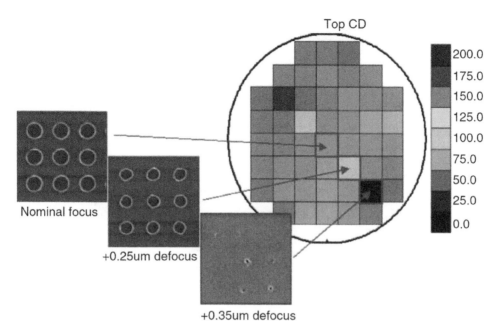

FIGURE 12.26 Optical digital profilometry (ODP) scatterometry can detect deformed or closed contacts.

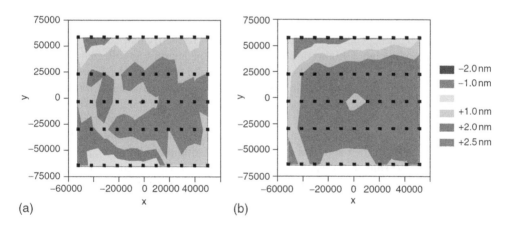

FIGURE 12.27 Critical dimension (CD) uniformity comparison with 56 targets from an EPSM. (a) CD-SEM and (b) optical digital profilometry (ODP) scatterometry.

aperture instead of a lens, the classical far-field diffraction limit does not apply and imaging beyond the Rayleigh limit is possible if the spacing between the aperture and the surface is much less than the illumination wavelength. For many years, it has been assumed that imaging could only occur with classical propagating mode solutions to Maxwell's equations. However, imaging with the so-called "non-propagating" exponential (evanescent) modes is also possible. This rather non-conventional technique is referred to as "near-field" optical microscopy. In near-field optics, the evanescent decay of intensity with distance from a highly localized source or detector (i.e. tiny aperture) is utilized for the transduction of super-resolution information to far-field propagating modes that are detected.

There are several techniques for imaging with highly localized evanescent light sources. In an illumination-mode type of near-field scanning optical microscope (NSOM), a light-emitting aperture near the apex of a sharp probe is brought to within nanometers of a sample surface and then scanned laterally to produce images (Figure 12.28). The amount of light that tunnels through the

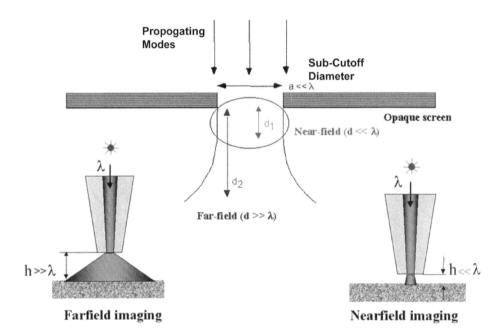

FIGURE 12.28 Injection of super-resolution information into collected far-field carriers via near-field scanning optical microscope (NSOM).

aperture is exponentially dependent on changes in the surface refractive index as well as tip-to-sample separation (i.e. topography). Light transmitted through the aperture is collected in the far field with a conventional optical microscope. The far-field collection optics can also be used for viewing and positioning the probe at the sample area of interest before fine scanning. If force detection is used to maintain constant aperture-to-sample distance during fine scanning, the effect of topography is eliminated and the resulting image is primarily due to the optical properties of refraction. Both the 3D surface topography and optical index properties of surface features can be mapped in parallel. The aperture is formed and light is confined by coating the sides of a tapered fiber probe with metal (typically 100 nm of Al), as illustrated in Figure 12.29. The lateral resolution is primarily limited by the smallest aperture diameter that can be fabricated to confine the intensity, which can approach 10 nm; however, SNR usually becomes unacceptable below diameters of 20 nm.

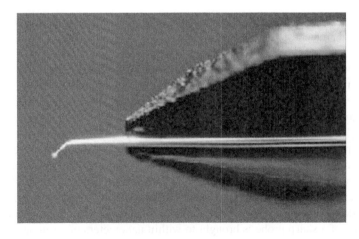

FIGURE 12.29 Near-field scanning optical microscope "fiber tip" from Nanonics, Inc.

For the case of an aperture in close proximity ($d \ll \lambda$) to the sample surface, the assumption of a far-field solution to the wave equation is no longer valid and

$$\left(\nabla^2 - \frac{1}{c}\frac{\partial^2}{\partial t^2}\right)\Psi(r,t) = 0 \;\rightarrow\; \left[\nabla^2 + k^2\right]\psi(r), \tag{12.43}$$

now yields a different solution of the form:

$$\Psi(x,z) = \int_{S'} \frac{dk_x}{\partial x} e^{ik_x} A(k_x,z), \tag{12.44}$$

where

$$A(k_x,z) = \int dx\, e^{-ik_x} \Psi(x,z), \tag{12.45}$$

with A as the near-field amplitude, λ the wavelength, k the spatial frequency, $x' = r'\cos\phi'$, $y' = r'\sin\phi'$, and s' the surface of integration ($ds' = r'dr'd\phi'$). In the near field regime:

$$\Psi(x,z) \rightarrow 0 \quad \text{for} \quad x > \frac{\lambda}{N}, \quad \text{where} \quad N \approx 20, \tag{12.46}$$

so that

$$A(k_x,z) = A(k_x,z=0)e^{-\sqrt{k_x^2 - \left(\frac{2\pi}{\lambda}\right)^2}\,z} \tag{12.47}$$

and

$$\Psi(x,z) = \int \frac{dk_x}{2\pi} e^{ik_x x} A(k_x,z=0)e^{-\sqrt{k_x^2 - \left(\frac{2\pi}{\lambda}\right)^2}\,z} \approx e^{-k_x z}. \tag{12.48}$$

The main idea contained within the above derivation can be summarized conceptually by the diagrams in Figure 12.30. The first diagram (Figure 12.30a) shows the far-field spatial representation for an Airy intensity pattern due to diffraction and the second diagram (Figure 12.30b) has the corresponding frequency domain spectrum. The maximum spatial frequency at which power can be collected is denoted by k_{max}:

$$k_{\max} \equiv 2\pi \sin(\theta)/\lambda \approx \frac{2\pi}{\lambda}. \tag{12.49}$$

The spatial and frequency domain representations shown in Figure 12.30a and b are related by what is referred to as a transform pair. The transform pair for the case of an infinitely localized spatial image point (i.e. the Kronecker Delta function) is illustrated by Figure 12.30c and d, where the corresponding frequency domain power spectrum has infinite extent along the k dimension. This inverse relationship between spatial and frequency domain members of a transform pair is a consequence of the uncertainty principle and basically means that more spatial localization of an intensity distribution will correspond to a more spread-out frequency spectrum. The main "question" posed in near-field optics is: *What is the maximum spectral frequency that can be collected for final imaging, beyond that classically attainable for diffraction-limited resolution?*

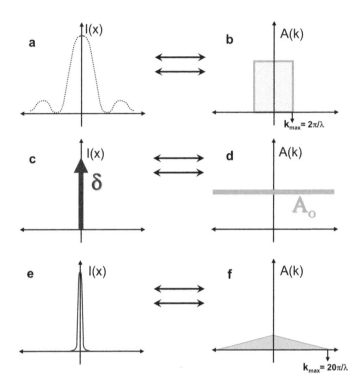

FIGURE 12.30 Fourier transform pairs for (a, b) FF diffraction, (c, d) perfect delta, and (e, f) near-field scanning optical microscope (NSOM) aperture case.

The answer for real near-field scanning microscopy exists between transform pair extremes (Figure 12.30e and f), due to practical constraints imposed by the physical aperture condition and coating material.

A key to unraveling the unknowns between initial exposure and the definition of the final device patterns lies with the ability to measure dimensional changes after each step in the fabrication process sequence in order to allow the possible control. Near-field optical techniques can provide a new way to obtain highly spatially resolved images of both topographic and chemical changes for improved characterization and control of the entire lithography process [59]. In addition to scanned apertures, differences in frustrated total internal reflection across a membrane in close proximity $(z < \lambda)$ to the surface can be detected using immersion objectives with annular illumination for real-time full-field "projection" near-field optical imaging. This technique is referred to as "photon tunneling microscopy," or PTM (Figure 12.31) [60]. The primary contrast mechanisms in PTM images are due to changes in the surface topography (i.e. changes in separation) and optical properties that affect photon tunneling probabilities. A basic diagram of a PTM system is shown in Figure 12.32, as well as the resultant images. PTM seems very similar to conventional full-field imaging, except that the immersion lens is actually focused above the sample in order to image the membrane distal face and not the sample surface itself [59]. Lateral spatial resolution of PTM is more difficult to quantify than the other techniques, due to its convolution of many contrast mechanisms (i.e. topographic, refractive index, immersion PSF, membrane morphology, etc.) and is therefore still under assessment.

12.3.3.4 Auto Macro Inspection (Polarized Diffraction Light Detection)

A unique technique to detect exposed pattern conditions macroscopically is called automatic macro inspection (AMI) [61–63]. The advantage of this system is quick measurement of the entire wafer. Figure 12.33 shows its system configuration. A wafer is placed on a stage and the illumination

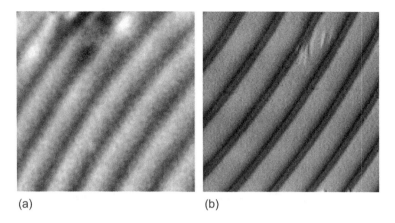

(a)   (b)

FIGURE 12.31 (a) Refractive index and (b) topographic volume changes in chemically amplified deep ultra-violet (DUV) photoresist after post-exposure bake, before development.

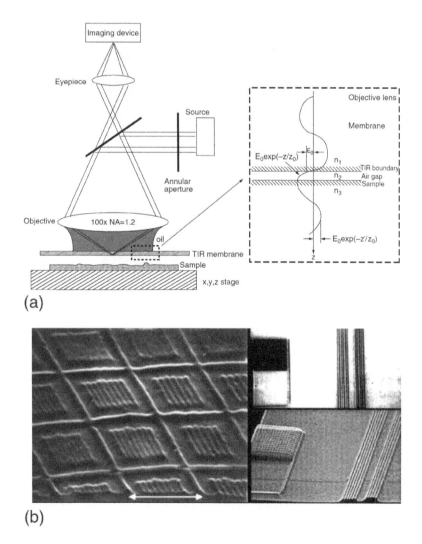

FIGURE 12.32 (a) Schematic of PTM and (b) latent DUV resist images.

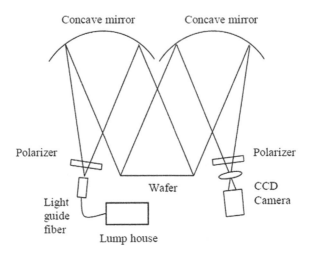

FIGURE 12.33   Optical system of auto macro inspection. (From Figure 7 in T. Omori et al., "Novel inspection technology for half pitch 55 nm and below," *Proc. SPIE*, 5752, 174 [2005].)

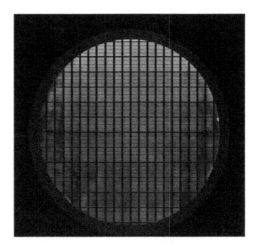

FIGURE 12.34   CCD image of processed wafer. (From Figure 13 in T. Omori et al., "Novel inspection technology for half pitch 55 nm and below," *Proc. SPIE*, 5752, 174 [2005].)

optical system of which components are a lamp house, a light guide fiber, a polarizer, and a concave mirror. A wavelength of 546 nm (e-line) is picked out and is led to the exit of the light guide fiber. Because the exit of the light fiber is located at the focal point of the concave mirror, the illumination light on the entire wafer becomes the parallel light. A polarizer at the exit side of the fiber converts the illumination light from random to linearly polarized light. The stage can rotate around the vertical axis which goes through the wafer center. The components of the optical system that observes an entire image of a wafer are a concave mirror, a polarizer, a camera lens, and a CCD device. The polarizer picks out an ingredient which crosses at right angle to the polarization made by the polarizer in the illumination optical system. This arrangement is called "crossed-nicols."

A process wafer with a 90 nm line and space patterns with a focus shift from −0.5 μm to +0.4 μm (0.1 μm step) along the horizontal axis and a dose shift from −45 mJ/cm² to +30 mJ/cm² (15 mJ/cm² step) along the vertical axis is produced. When the focus condition or the dosage condition changes, the birefringence of the exposed pattern becomes different and the average gray level of the pattern area changes, as shown in Figure 12.34. Therefore, there are correlations between the focus shift or

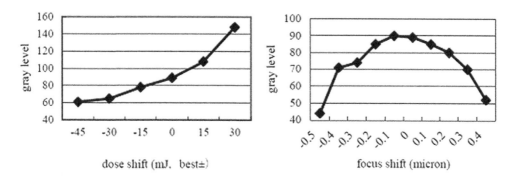

FIGURE 12.35 Correlation between dose or focus shift and gray level. (a) Correlation between dose and gray level at a focus of 0 μm. (b) Correlation between focus and gray level at the dose shift of 0 mJ/cm². (From Figures 15 and 16 in T. Omori et al., "Novel inspection technology for half pitch 55 nm and below," *Proc. SPIE*, 5752, 174 [2005].)

dose shift and the average gray level, as shown in Figure 12.35. These correlation graphs should be obtained in advance and can be used to determine defocus and/or dosage error.

12.3.4 SCANNING ELECTRON MICROSCOPY (SEM)

12.3.4.1 Principles of Operation

The electron was the first particle with a finite rest mass experimentally shown to have wave properties. The electron was initially classified as a particle following its discovery by J.J. Thomson in 1897. This was primarily because electrons exhibited well-defined trajectories, charge-to-mass ratios (for $v \ll c$), and spatial localization of their momentum and energy. Louis de Broglie later conjectured that particles may also have wave properties. It was further assumed that the physical laws and equations that describe the particle characteristics of electromagnetic waves also give the wave characteristics of material particles, such as electrons. The wave nature of electrons was verified in 1927 by the electron diffraction experiments of Davisson and Germer. For non-relativistic, free particles, the wavelength λ of a material particle having a momentum p is

$$\lambda = \frac{h}{p} = \frac{h}{mv}. \tag{12.50}$$

The de Broglie wavelength of the electron is given by Eq. 10.7 and can be simplified as follows in the non-relativistic case:

$$\lambda \, (\text{nm}) = \frac{h}{\sqrt{2mqV}} = \frac{1.226}{\sqrt{V}}. \tag{12.51}$$

Here, the electron rest mass (m), the charge of electron (q), and Plank's constant (h) are 9.109×10^{-31} kg, 1.602×10^{-19} C, and 6.626×10^{-34} J/sec, respectively.

For an accelerating voltage of 1000 V, the electron wavelength λ is about 0.4 Angstroms. The wavelength of electrons can easily be made much shorter than that of visible light. Hence, the limit of the resolution of a microscope may be extended to a value several hundred times smaller than that obtainable with conventional far-field light-optical instruments by using electrons, rather than light waves, to form an image of the object being examined. A beam of electrons can be focused by either a magnetic or an electric field, and both are used in electron microscope designs. Later in this section, it will become evident that by the proper design of such electron lenses, the elements of an optical microscope such as its condenser, objective, and eyepiece can all be electrically duplicated. The scanning electron microscope currently serves as the primary instrument of choice for measuring submicrometer features because of its nanometer-scale spatial resolution. However, despite its high spatial resolution, several issues still exist which currently limit the efficacy of the SEM in dimensional metrology. In fact, the practical resolution of an SEM is affected by such factors as source extension, chromatic dispersion and spherical aberration of the electron lens, beam-to-sample interactions, and the efficiency of the image signal detection system. A cross-sectional diagram of an SEM column is shown in Figure 12.36.

Electrons originate from thermionic or field-emission-type sources. An aperture is inserted to spatially filter the original electron source flux, so that a well-defined source "spot" can be provided for subsequent demagnification by electrostatic or electromagnetic lenses before striking the sample. A "column" is basically composed of the electron source with focusing optics to achieve beam formation and delivery of the primary electron (PE) probe, or spot, onto the sample surface at some known lateral position. Electrostatic deflection plates are used to deflect the beam so that the primary electron spot can be rastered laterally across a set range, or field of view. The minimum size of a primary beam spot at the sample surface is directly related to the point spread function of the microscope, and is therefore a primary factor determining the spatial imaging resolution. Secondary electrons emitted from the beam-to-sample interaction scattering volume also contribute to the detected image signal intensity, so that image resolutions equal to the primary spot size are rarely obtained. The intensity of the electrons emitted by the interaction volume is represented as a gray level at each pixel, in synchronism with the beam spot position.

Several types of electron emitters are represented in the energy diagram of Figure 12.37. Thermionic emitters are heated to temperatures above 2500°K for directly heated tungsten hairpin cathodes and 1400°K for indirectly heated LaB_6 rods. Heating must occur for electrons in the tail of the Fermi distribution to overcome the surface work functions $\phi = 4.5$ eV for tungsten and 2.7 eV

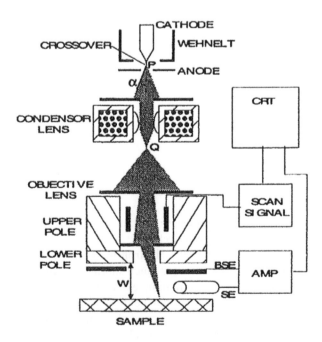

FIGURE 12.36   Schematic cross-sectional diagram of a scanning electron microscope.

for LaB_6 (see Table 10.1 in Chapter 10). The current density of emission at the cathode is given by Richardson's law [64]:

$$j_c = AT_c^2 e^{-\phi/kT_c}, \tag{12.52}$$

where A is a material constant and k is Boltzmann's constant. In addition to the emission current, cathodes are also often characterized by the axial gun brightness:

$$\beta = \frac{\Delta I}{\Delta A \Delta \Omega} = \frac{j_c}{\pi \alpha^2} \cong \frac{j_c E}{\pi k T_c}. \tag{12.53}$$

Axial gun brightness is defined as the current passing through an area ΔA into a solid angle $\Delta \Omega = \pi \alpha^2$ ($\alpha \ll 1$). The beam aperture α is the semi-apex angle of the incident electron cone. The axial gun brightness remains constant for all points along the optic axis through lenses and apertures. A thermionic electron gun (see Figure 12.37) is composed of a negatively biased cathode, a Wehnelt cup that is biased a few hundred more volts negatively than the cathode, and an anode at ground potential. The electron trajectories form a cross-over point between the cathode and anode that produces an electron source having an approximately Gaussian intensity profile. The relatively low energy at the cross-over point results in stochastic Coulomb interactions that increase the electron beam energy spread (Boersch effect). A large energy spread will result in a more chromatic aberration and degraded resolution of the electron optical lens. Schottky emission cathodes use a crystalline tip with a special coating in order to lower the work function from 4.5 eV (W) to 2.7 (ZrO/W(100)). This allows electrons to overcome the work function ϕ at a lower

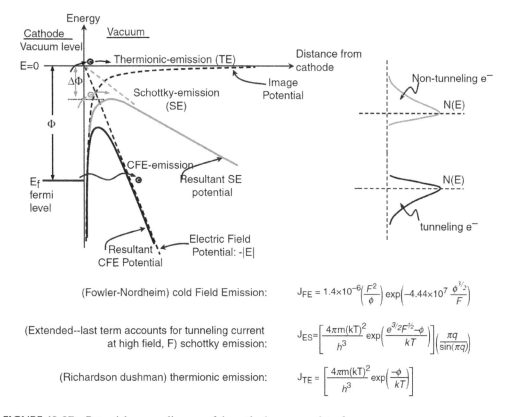

FIGURE 12.37 Potential energy diagram of the cathode–vacuum interface.

cathode temperature. The electrons are extracted by a much higher electric field strength E at the cathode than with thermionic emitters but not enough to produce tunneling. The emitters are tapered to a sharp point in order to increase the field gradient and enhance the emission current from the metal. The potential barrier is lowered by $\Delta\phi$ (Figure 12.37), known as the Schottky effect, and the electrons must overcome the remaining barrier using their thermal energy. The energy spread of this type of gun is not increased by the Boersch effect and has several orders of magnitude larger emission current density than thermal emitters [65]. A field-emission gun (FEG) uses an electric field E to decrease the potential barrier width to less than approximately 10 nm (Figure 12.37). A field-emission gun needs two anodes: one to regulate the field strength at the tip (extraction voltage) and the other to accelerate the electrons to their final kinetic energy (acceleration voltage). The emission current can be varied by the voltage of the first extraction anode. Electrons are accelerated to the final energy $E = eU$ by the second acceleration anode. The dependence of the field-emission current density on the electric field strength is given by the Fowler–Nordheim equation [66]:

$$j_c = \frac{k_1 |E|^2}{\phi} e^{\left(-k_2 \phi^{3/2}/|E|\right)},$$ (12.54)

where k_1 and k_2 are weakly dependent on E. FEGs use crystal-oriented tips and have a gun brightness that can be up to three orders of magnitude larger than thermionic emitters. Since FEGs operate with the tip at room temperature, adsorbed gas layers must be cleaned from them by heating (flashing) the tip approximately every 8 hours of operation. Waiting periods between 1 and 3 hours must follow after each tip flash for the emission current to stabilize. An advantage to cold cathode FEGs is their low electron energy spread and hence less susceptibility to lens chromatic aberrations.

Fluctuation of the emission current typically occurs with cold cathode FEGs, which can be compensated for by dividing the recorded signal with adsorbed currents at the diaphragms [67]. Due to the lack of heating, cold cathode FEG tip lifetimes are much longer than thermionic or Schottky emitters.

Electrons from the gun pass through the aperture diaphragm and are focused by the electron lens. An electron lens (see Figure 12.36) basically consists of an axial magnetic field having rotational symmetry. The flux of a coil is concentrated by iron pole pieces that form a magnetic field B. The z-component of a magnetic field has a Gaussian distribution, so that electrons travel along spiral trajectories, due to the Lorentz force. The electron beam diverging from the virtual source cross-over point is focused at the intermediate cross-over such that the source diameter is demagnified by the factor $M = p/q$, where p is the virtual source and q the intermediate cross-over axial positions. The top lens in Figure 12.36 is known as a condenser lens. After the condenser, electrostatic parallel plates are used to deflect the beam to create the scan motion on the sample. Before it reaches the sample, the beam is focused again by a final lens, known as the objective. The working distance is the physical distance between the lower objective pole piece and the specimen.

As in light optics, electron lenses also have aberrations as explained in Section 10.2.2. Assuming a resulting Gaussian beam diameter in the plane of least confusion [68]:

$$d_s = 0.5 C_s \alpha^3.$$ (12.55)

The weak lenses used in the SEM have a large spherical aberration coefficient $C_s = 10$–20 mm and the strong objective lenses in TEM can be operated with $C_s = 1$–2 mm.

An accelerating voltage determines the focal length. A disc of least confusion results from the energy spread ΔE of the electron gun. The diameter of the disc is

$$d_c = C_c \left(\frac{\Delta E}{E} \right) \alpha, \tag{12.56}$$

where C_c is the chromatic aberration coefficient.

Fraunhofer diffraction at the aperture-limiting diaphragm in the final lens causes an Airy disc of half-width given by

$$d_d = \frac{0.6\lambda}{\alpha}, \tag{12.57}$$

where λ is the de Broglie wavelength of the electron as described in Eq. 10.7 or Eq. 12.50.

The final electron probe is formed on the sample by successive demagnifications of the cross-over or virtual point source of diameter d_0 by several lenses. An intermediate image is formed at a large distance L in front of each lens. The lens magnification is found from the ratio of the object focal length f to the intermediate image distance L on the primary side of the lens. The overall system magnification is simply the product of geometric magnifications of each lens. The geometric probe diameter d_g is found from the product of the magnification (actually demagnification) and of the original cross-over diameter d_0 at the virtual source:

$$d_g = \frac{f_1 f_2 f_3}{L_1 L_2 L_3} d_0 = M d_0. \tag{12.58}$$

The geometric beam diameter can also be expressed in terms of the electron probe aperture [69]:

$$d_g = C_0 \alpha_p^{-1}. \tag{12.59}$$

The geometric probe diameter is broadened by aberrations of the lenses. A quadratic superposition of the different beam components results in an effective probe diameter:

$$d_p^2 = d_g^2 + d_d^2 + d_s^2 + d_c^2 \tag{12.60}$$

or

$$d_p^2 = \left[C_o^2 + (0.6\lambda)^2 \right] \alpha_p^{-2} + \frac{1}{4} C_s \alpha_p^6 + \left(C_c \frac{\Delta E}{E} \right)^2 \alpha_p^2. \tag{12.61}$$

A double-logarithmic plot of the final electron probe diameter d_p versus the probe aperture α_p for a thermionic emission SEM is shown in Figure 12.38a. The accelerating potential $V = 20$ kV, $C_s = 45$ mm, $C_c = 25$ mm, $\Delta E = 2$ eV, and $I_p = 10^{-11}$ A. The brightness β is calculated to be 7.38×10^8 A cm^2/sr. Thermionic emitters have a relatively large energy spread ΔE due to the Boersch effect. Decreased brightness with smaller energy, and hence larger d_g, implies that thermionic cathodes are not as effective for low-voltage (<3 kV) SEMs as either Schottky or field-emission sources. A logarithmic plot of the final electron probe diameter d_p versus the electron probe aperture α_p for a cold cathode field-emission SEM (the accelerating potential, $V = 1$ kV) is shown in Figure 12.38b. The field-emission current density was calculated by Eq. 12.54 and then used to determine the brightness in Eq. 12.53. An energy spread of $\Delta E = 0.3$ eV was assumed for the cold cathode field-emission case. The second and fourth terms in d_p^2 (see Eq. 12.60) dominate for Schottky and field-emission SEMs and the probe diameter d_p is minimized by optimizing diffraction and chromatic aberration terms. The electron optics can be further enhanced by immersion of the sample in the field between the objective lens pole pieces (in-lens operation) [70], so as to decrease C_s and C_c. Retardation fields can

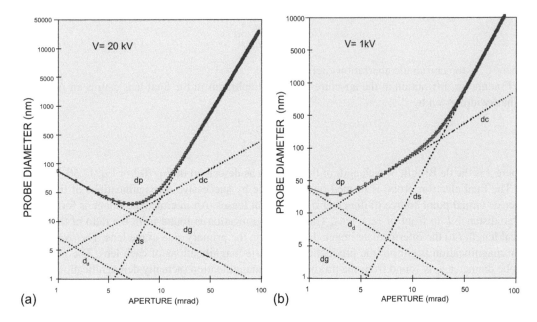

FIGURE 12.38 Logarithmic plot of electron probe diameter d_p versus aperture α_p in the case of (a) thermionic emission (20 kV potential) and (b) field emission (1 kV potential).

also be applied to permit higher electron energies in the column to reduce the effects of stray fields while allowing less sample damage and charging inherent with the resultant lower landing energies at the sample.

In order to obtain a smaller diffraction, a higher acceleration voltage is effective. Historically, the extremely high-resolution SEM (around 0.1 nm) had to adopt an acceleration voltage over 1 MV, but the system size became huge because of the necessity for strong insulation against discharge. Simultaneously, the achievable resolution was reaching its practical limit because its aberration could not be compensated for by a conventional magnetic lens that could behave as a convex lens only. In the 1990s, the electron beam optics, which can compensate for spherical aberration, was realized by using multiple pole lenses such as hexapole [71] or quadrupole-octupole [72] lenses. Now, a resolution of less than 0.1 nm is commercially available at the acceleration voltage of around 300 kV.

Once the final electron probe has been formed, it is raster scanned across the sample surface to render a spatial map of the detected electron scatter intensity (i.e. an image). Mechanisms for SEM image contrast are quite complex. However, a quick overview should hopefully aid in providing a basic understanding of these scattering intensity signals from which dimensional metrology "measurements" are derived. Electrons from the incident beam are first scattered by atoms within the sample, as shown in Figure 12.39. Electrons can undergo a wide variety of sample interactions over distances ranging from tens to hundreds of angstroms following entry to the target surface. Elastic scattering occurs when electrons are deflected through some angle relative to their initial trajectory without energetic loss, usually due to attraction or repulsion with the atom's negatively charged electrons or positively charged nucleus. On the other hand, inelastic scattering events occur when incident electrons lose some of their energy to the sample material. Energy lost during inelastic scattering occurs in discrete events. These occur through ionization of an atom by removal of inner shell electrons, thus generating characteristic x-rays and Auger electrons. Other inelastic scattering events include valence electron collisions to produce secondary electrons (SE) or the generation of phonons by interaction with the lattice. Electrons will undergo successive scattering events (elastic or inelastic) as they continue their travel within the interaction volume

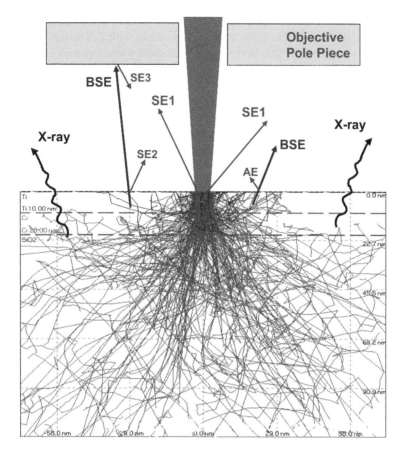

FIGURE 12.39 Diagrams of electron beam scattering with sample material.

until all of their energy is given up to the material and thermal equilibrium is reached or an escape from the sample is made in some way. For CD metrology, where maximum throughput and minimal electron exposure dose (i.e. less damage and charging) are desired, the image signal mainly depends on the detection of secondaries from inelastic scattering events. Energy losses from inelastic scattering processes can be described by the Bethe continuous energy loss relation (see Eq. 10.21). In non-relativistic cases and ignoring the last term of Eq. 10.21, it can be simplified as follows:

$$-\frac{dE}{dx} = \left(\frac{1}{4\pi\varepsilon_0}\right)^2 2\pi e^2 N_A \left(\frac{Z\rho}{A}\right)\left(\frac{1}{E}\ln\frac{1.166E}{J}\right)$$

$$= 7.8\times10^4 \frac{Z\rho}{AE}\ln\frac{1.166E}{J} \quad (\text{keV/cm}),$$

(12.62)

where dE is the change in electron energy as the electron traverses a path of length dx, N_A is Avogadro's number, ε_0 is the vacuum permittivity, Z is the atomic number, A is the atomic weight, ρ is the material density, and J is the mean ionization potential. The mean ionization potential increases with atomic number and can be approximated as

$$J = 1.15\times10^{-2}Z \quad (\text{keV}).$$

(12.63)

The rate of energy loss with path length traveled depends on the atomic number, due to the relation between Z/A and density, so that dE/dx generally increases with Z. The rate of energy loss also increases with decreasing energy. The Bethe continuous energy loss equation (Eq. 12.62) consists of three primary terms. The first term is a constant, the second is material dependent, and the third relies on both material and electron landing energy. The second term primarily varies with material density, which correlates to the position of the element within the periodic table. However, the change with density can be accounted for if the distance is measured in terms of mass thickness ρx (g/cm^2). The electron range can be calculated by integrating the rate of change in electron kinetic energy along the path x (i.e. Bethe continuous energy loss relation) from $E = E_0$ at $x=0$ at the electron point of entrance into the material:

$$R_B = \int_{E_0}^{0} \frac{1}{\left(\dfrac{dE}{d(\rho x)}\right)} \, dE. \tag{12.64}$$

Eq. 12.64 is known as the Bethe range R_B, which can be larger than the practical range R due to the elastic scattering of electrons in the material.

The practical electron range R_E (g/cm^2) is found by measuring the transmission of electrons through thin films and can be approximated by

$$R_E = 10E_0^{1.43} \; (\mu g/cm^2), \tag{12.65}$$

where E_0 is in kV. The practical and Bethe ranges are on the same order of magnitude for low-Z materials, such as photoresists and dielectrics, due to the small probability of elastic scattering from the factor of Z^2 in the Rutherford scattering formula [73].

The electron–sample interactions are used for both imaging and surface analysis with the SEM. As illustrated in Figure 12.39, the primary electrons (PE) from the incident electron probe produce secondary electrons (SE) through inelastic scattering, which emit from a shallow depth between about 1 and 10 nm. The energy spectrum of SEs has a peak at 2–5 eV and a long tail. Generally, electrons that are emitted with energies less than 50 eV are considered to be SE. The SE1 group of electrons excited by the PE contribute to high-resolution imaging on the order of the probe diameter. The so-called backscattered electrons (BSE) have energies ranging from 50 eV to the energy of the PE and can also produce SE2 electrons excited in a larger surface layer with a diameter that is approximately 100–1000 nm at beam energies above 10 keV, depending on the surface material. This range decreases to around 5–50 nm at low beam energies. The high energy BSE can also produce SE3 electrons through collisions with the specimen chamber and the lower pole piece of the objective lens. Another group of electrons, known as SE4, can actually diffuse out from the electron beam forming a column itself. After all is said, it is usually preferable to collect SE1 and SE2 electrons for the best imaging resolution. Although this is usually the case, it should be noted that the BSE interaction range actually approaches those of the SE1 and SE2 interaction diameters at low beam voltages (<2 keV). An advantage is that the exit energies of BSEs remain sufficiently high so that the effects of sample charging on the collected electrons can be reduced. Unfortunately, the BSE image signal-to-noise ratio is less than with SE.

Finally, a detector (see Figure 12.40) must be used to collect the electrons that are emitted from the interaction volume and convert them into signals for imaging. The low-energy SE can be collected using a positively biased grid on the front of a scintillator. The fraction of SE actually collected depends on the working distance. Light generated by the scintillator is guided to a photomultiplier where it is converted to current and then digitized to an image's gray level. The merit of

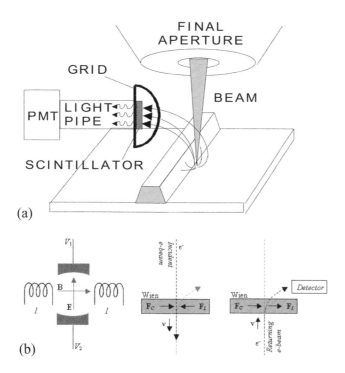

FIGURE 12.40 (a) Oblique mounted Everhart–Thornley and (b) normal angle "in-lens" detectors used for scattered electron collection during SEM imaging.

a detector system is determined mainly by the ratio of output-to-input root-mean-square (rms) noise amplitudes [74]:

$$r = \frac{N_{out}}{N_{in}} = \left(\frac{S}{N}\right)_{in}\left(\frac{N}{S}\right)_{out} \leq 1, \tag{12.66}$$

where

$$\left(\frac{S}{N}\right)_{in}^2 = \frac{I_p}{2e\Delta f} \tag{12.67}$$

is the mean shot noise of the electrons and Δf is the bandwidth.

Two main detector configurations have emerged throughout successive generations of SEM models. The first and most widely known is the Everhardt–Thornley (E-T) detector mounted inside the specimen chamber at an oblique angle of collection with respect to the sample plane (Figure 12.40a). Although this was one of the first methods for collecting scattered electrons, lack of symmetry and shadowing of the image intensity are serious drawbacks encountered with this technique. The second method of detection allows a top-down perspective (i.e. normal angle) collection of scattered electrons and is illustrated schematically in Figure 12.40b. An apparatus known as the Wein filter is used to route the scattered electrons (streaming back through in the opposite direction) toward a scintillator mounted inside the column. However, the primary beam electrons traveling toward the sample are not deflected. The Wein filter uses orthogonal electric and magnetic fields to deflect electrons returning from the sample to the detector. The first inset diagram of Figure 12.40b illustrates the basic geometry of the Wien filter. An analysis of the forces seen by electrons traveling through the Wein filter will clarify the balance between orthogonal electric (**E**) and magnetic (**B**) fields. A particle with a given charge, q, in an electric field experiences an electrostatic force, often called the Coulomb force:

$$\mathbf{F}_c = q\mathbf{E} \tag{12.68}$$

Since an electron always has the same charge, we can say that the force felt by the electron is proportional to the electric field. The electric field in a Wien filter is generated by applying opposite voltages to two plates, as shown by V_1 and V_2 in Figure 12.40b. The Wien filter's magnetic field is generated by the coils shown in Figure 12.40b and will exert a force on charged particles that are moving with a velocity, **v**, relative to the magnetic field. This force is called the Lorentz or magnetic force:

$$\mathbf{F}_L = q(\mathbf{v} \times \mathbf{B}). \tag{12.69}$$

This equation uses the vector cross product, and basically states that the magnetic force on a charged particle will be exerted perpendicular to the plane formed by the velocity and magnetic field vectors. Simply stated, in the geometry of the Wien filter, the magnetic field exerts a force parallel to the Coulomb force. Whether the magnetic force adds to or subtracts from the Coulomb force depends on the direction the electrons are traveling. Side views of the Wein filter show how the Coulomb and Lorentz forces cancel each other for incident (primary) electrons, but add to each other for returning electrons.

There are two general rules that should be followed when determining the strengths of the electric and magnetic fields in the Wien filter:

(1) The Coulomb and Lorentz forces must cancel each other completely for the primary beam.
(2) They must add such that returning electrons are deflected to the detector.

Practically speaking, Rule 1 states that the Wien filter voltages and current must be balanced such that the primary beam remains aligned. The strength of the electric field is adjusted with respect to the magnetic field to ensure that the primary beam remains aligned to either the magnetic lens or the landing energy retardation bias. Rule 2 states that the voltages and current in the Wien filter must be strong enough to fully deflect returning electrons onto the detector. During primary electron beam alignment, the ratio of voltage to current suitable for satisfaction of Rule 1 is determined and then set such that the SEM image is uniformly bright. Theoretically, the above electric and magnetic force equations add together and form what is called the Lorentz force law:

$$\mathbf{F} = \mathbf{F}_C + \mathbf{F}_L = q(\mathbf{E} + \mathbf{v} \times \mathbf{B}). \tag{12.70}$$

In addition to the E-T and in-lens detectors, which rely on scintillation, semiconductor and micro-channel-plate (MCP) detectors have also been used for the conversion of scattered electron intensities. Semiconductor detectors provide internal amplification of the incident electron current by means of the collection and separation of electron–hole pairs in the depletion layers. However, bandwidth and signal gain are not that of the E-T detector at low electron energies. The micro-channel-plate detector has become of increasing interest for IC metrology where low beam currents and low acceleration voltages ($0.5 < V < 3$ kV) are desired. In this operating range, semiconductor detectors decrease in sensitivity and the E-T detector shows even more asymmetry in its intensity profiles. The MCP is composed of a slice from a tightly packed group of fused tubes of lead-doped glass. The inner diameters of the tubes are typically 10–20 μm and the thickness of the slice is about 3 mm. Incident electrons are amplified by the MCP in a photomultiplier-like action where SE are produced at the inner wall of each tube and accelerated by a continuous voltage drop along the tube with a bias of a several kilovolts. The front plate can be biased such that both BSE and SE are collected or just BSE. In addition to the type of detector, there are various strategies for the placement of each detector to increase collection efficiency and maximize intensity scan symmetry.

As described previously in this section, electrons are re-emitted from the surface and detected. The recorded intensity signals from the electron detector can be displayed on a cathode ray tube (CRT) rastered in synchronism with the primary beam deflection to form essentially real-time images. Each gray level of the image corresponds to a detector intensity level, as affected by the SE yield. The SE yield is defined as the ratio of SEs leaving the surface per incident primary electron. The SE yield of the interaction volume is primarily a function of the change in surface height and/or atomic number (i.e. material transitions) [75]. Peaks in the detected electron intensity typically occur at abrupt changes in the surface height, such as the feature edges. The dependence of the electron yield on the surface height rate of change is illustrated in Figure 12.41 and is known as the *Secant Effect* [75]. Critical dimension measurements are obtained from these intensity profiles. Actually, an average is taken across several line profiles to provide spatial averaging in order to reduce the effects of sample variation, edge roughness. Intensity profile shapes are also affected by the SEM operating conditions, such as detector grid bias (i.e. electron energy filtering) in Figure 12.42. The actual topography (black trace) was measured using a 3D metrology AFM, which is described in Section 12.3.5.

Unfortunately, the SEM image intensity does not correspond directly to the surface height. This presents major problem in CD-SEM metrology – where in the intensity transition is the physical feature edge actually located? This presents ambiguity in the location of the feature edges relative to the SEM image intensity profile. Techniques for determining the actual edge position from an intensity profile are still somewhat arbitrary. One generally attempts to approximate the intensity

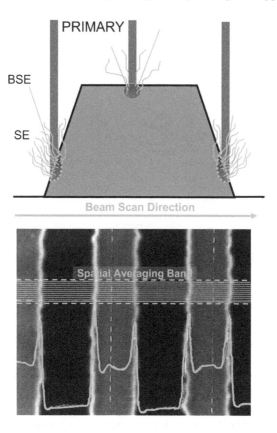

FIGURE 12.41   Topographic image contrast at feature edges.

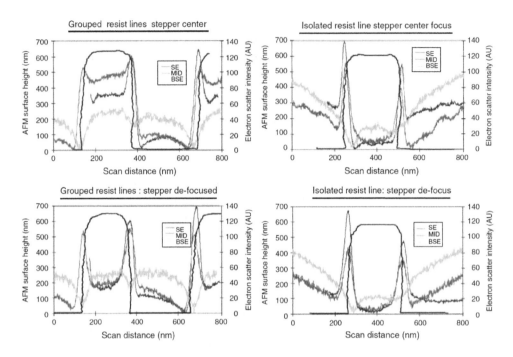

FIGURE 12.42 SE, mixed SE/BSE, BSE intensity scans for dense and isolated lines at different detector energy filter biases.

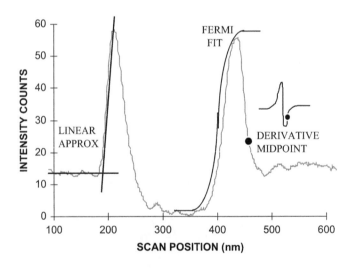

FIGURE 12.43 SEM image intensity line scan with representative edge detection algorithms.

transition with some mathematical function, such as a tangent line (linear approximation), a nonlinear step (Fermi) function, or a threshold in the first derivative of the detected intensity signal (see Figure 12.43). The line width, or CD, measurement can vary by as much as 100% depending on the edge detection algorithm used [76]. Offset differences in the edge detection algorithms are one of the primary factors in SEM-to-SEM matching. Once the microscope resolution, linearity, and precision are shown to be adequate for the particular feature size of interest, the most appropriate algorithm and threshold must be determined by comparison with a suitable reference measurement system, such as the 3D AFM or inline sectioning dual-beam FIB/SEM [77].

12.3.4.2 Beam–Sample Interactions

Beam-to-sample interactions can also significantly affect the measurement performance of an SEM. The detected low-energy SE1 and SE2 electrons can be deflected or totally recaptured by the sample if enough surface charge is present. The amount and polarity of charge accumulation on the surface is primarily dependent on the secondary electron yield δ. The electron yield can be brought close to unity at certain electron landing energies in order to minimize charge buildup to some extent. These energies are often referred to as E-points and are shown in Figure 12.44 (see E1 and E2). In low-voltage scanning electron microscopy, the landing energy at the sample is between the two cross-over points of the yield curve. This is the regime where the substrate emits an electron current higher than the impinging beam current or the regime in which the substrate must source electrons. When the beam kinetic energy at the sample is greater than the second cross-over point of the yield curve, the substrate sinks electrons. As the line scan moves over the FOV to form the image, the secondary electron emission is affected by the transverse field created by the surface charge density deposited by the scan. This transverse local field (i.e. orthogonal to the applied field) is directly proportional to the instantaneous dosage which is defined by the magnification, beam current, and scan rate. The effects of charging on image scale are illustrated in Figure 12.45 and width is illustrated in Figure 12.46. In addition to the apparent change in scale, the line also appears to have a dark center region surrounded by a bright frame. The apparent border is not just an image artifact, but it is also physically real and is due to actual physical deposition of adsorbed contaminants. Prior to ionization and subsequent polymerization (mostly by the low-energy SE), the concentration of this adsorbed surface contamination layer is modulated by electro-migration. The migration of adsorbed contaminants is induced by the electric field produced by the spatial distribution of the accumulated negative surface charge around the FOV border. This effect manifests itself as a systematic growth in the width of a line during successive measurement exposures, and is known as carry-over. A plot of the line width versus measurement number is shown in Figure 12.47. The first measurement was obtained prior to acquisition of the image frame shown in Figure 12.46a. The last measurement in our series was obtained after the image of Figure 12.46b.

12.3.4.3 CD-SEM Metrology

One of the most serious issues related to scanned electron beam–based measurements of patterned lithographic materials to arise recently is shrinkage of the ArF 193 nm resist. When the polymer is bombarded with electrons with energy greater than several volts, it can induce ionization of the polymer backbone and side groups. This ionization, or bond breakage, can result in the creation of shorter chain products (some of which are volatile), cross-linking of the polymer chains, or formation of double bonds and reformation of the initial bond. The first two mechanisms can provoke

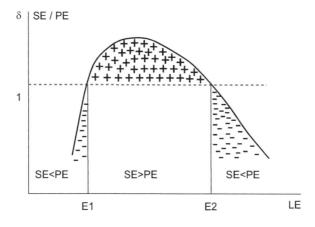

FIGURE 12.44 Secondary electron (SE) yield δ versus primary electron landing energy (PE).

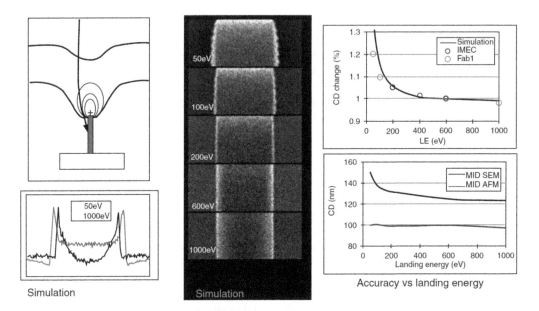

FIGURE 12.45 Effect of local fields due to charging on image scaling error.

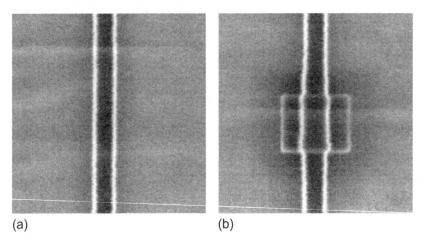

FIGURE 12.46 Combined image scale distortion and physical deposition due to charge accumulation.

shrinkage (one by physical loss of material, the other by tightening of the polymer matrix causing a net reduction in volume). In addition to the ionization process, there is the possibility that low-energy electrons will excite vibrational states in the polymers, thereby transferring energy as heat. In this case, the shrinkage will be caused by annealing of the resist. The first step in developing a solution to mitigate the issue of CD-SEM-induced shrinkage of resist involves quantification of the different components involved within the observed phenomenon. Numerous studies have been reported previously in the literature, where characterization of the effects of electron beam exposure on lithographic materials has been attempted with electron beam–based imaging. The results reported from such tests are difficult to interpret, since the imaging technique affects the feature to be measured. In addition, there is no direct correlation between the detected electron scattering intensity image and the actual physical shape changes. As discussed previously, charging effects can cause the image to appear to change in ways that are not physically occurring. The algorithms used for determining CD width are also sensitive to the intensity waveform shape, which can change

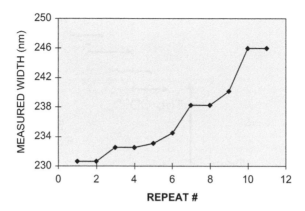

FIGURE 12.47 Systematic trend in measured width.

dramatically with imaging beam conditions. A new reference measurement methodology has been developed to allow separation and quantification of changes in the feature size after electron beam exposure directly from measurements of the three-dimensional physical surface shape [78].

12.3.5 Scanned Probe Microscopy (SPM)

12.3.5.1 Instrumentation

Since their introduction almost three decades ago, scanned probe microscopes have deeply impacted broad areas of basic science and have become an important new analytical tool for advanced technologies such as those used in the semiconductor industry. In the latter case, the metrology and characterization of integrated circuit and photomask features have been greatly facilitated over the last several years by the family of methods associated with proximal probes. As IC design rules continue to scale downward, the technologies associated with SPM will have to keep pace if their utility to this industry is to continue. Increased packing densities and aspect ratios and use of lithographic resolution enhancement techniques have created the need for topographic measurements in all three dimensions on both wafers and masks. The concept of "line profile," which describes the surface height along the scan direction at a particular line location (Figure 12.48), has become more appropriate than the two-dimensional notion of "line width." Alternatively, the profile could be given as the width ΔX at a particular height Z.

Structural information is essential when the choice of how to define a width (i.e. Z threshold) depends on the context of the fabrication process. Edge roughness and line width variation are also significant factors affecting the uncertainty of dimensional measurements. For example, profiles obtained with cross-sectional SEM imaging are highly variable due to their dependence on the particular location of sampling along the feature. Unlike light and electron optical microscopes, atomic force microscopes do not use wave optics to obtain images. A needle-like probe is brought very close (<2 nm) to the sample surface and traversed in a raster fashion. The probe rides up and down at a constant height above the sample, so that a topographic image of the surface is obtained. High resolution in all three dimensions is achieved simultaneously in a non-destructive manner. Regulation of the tip-to-sample distance is achieved by sensing the small forces between the atoms on the surface and those in the tip (Figure 12.49). A strong repulsive force is encountered by the tip at distances very near to the surface atoms, due to the Exclusion Principle. Contact mode scanning is said to occur when these repulsive forces are used for regulating the tip-to-sample distance. An attractive force is encountered as the tip is retracted.

The characteristics of different force sensing modes have significant implications on their application to dimensional metrology. The increased tip-to-sample separation of an attractive mode helps

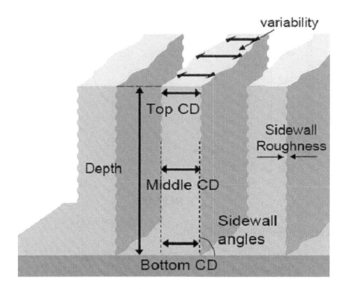

FIGURE 12.48 Three-dimensional representation of a line profile.

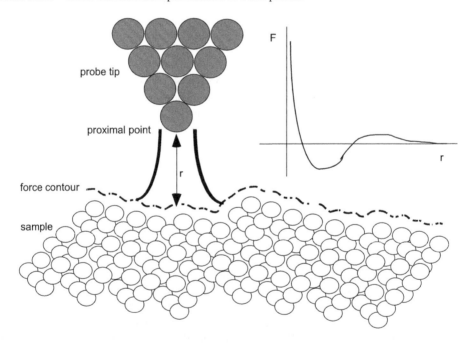

FIGURE 12.49 Atomic force microscope (AFM) tip-to-sample junction and force curve.

to minimize unwanted tip-to-sample contact, which damages and wears the probe. However, the repulsive force contact mode will track the physical surface more accurately in the presence of a static charge or fluid layer build up on the surface (remember that AFM really maps force contours—not the actual surface). AFMs use an amazing variety of sensors for detecting the presence of surface forces in either mode. The large number of surface proximity detectors invented in recent years attests to the importance, and difficulty, of surface detection in scanned probe microscopy [79]. We will briefly discuss two different force sensors commonly used for measurements on masks and wafers: (1) resonant microcantilever and (2) electrostatic force balance beam. In the resonant microcantilever approach [80, 81], an imaging tip resides on the face of a micro-mechanical beam

(Figure 12.50) that is excited at its natural frequency. Interactions between the probe and sample introduce variation in the amplitude, frequency, and phase of the cantilever oscillation. Any of these signals can be fed back into the microscope control electronics and used to maintain constant tip-to-sample separation. The force of interaction is not directly available, but Spatz *et al.* have obtained an estimate through modeling [82] of typically a few tenths of a micronewton.

Martin and Wickramasinghe have developed a more sophisticated two-dimensional force sensor whose design is also based on a microcantilever [83]. This system is supplied with a flared probe tip called a boot tip (Figure 12.51a), which allows improved access to vertical sidewalls and undercut regions [84, 88]. The force sensor operates in a non-contact fashion that exploits resonance enhancement of the microcantilever in two directions. A microcantilever with a spring constant of ≈ 10 N/m and $Q \approx 300$ is vibrated both vertically and laterally at different frequencies with an amplitude of ≈ 1 nm. Cantilever vibration is detected using an optical interferometer. The force sensitivity is

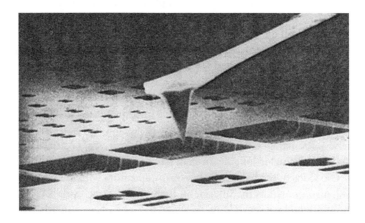

FIGURE 12.50 Atomic force microscope (AFM) imaging tip mounted on a microcantilever.

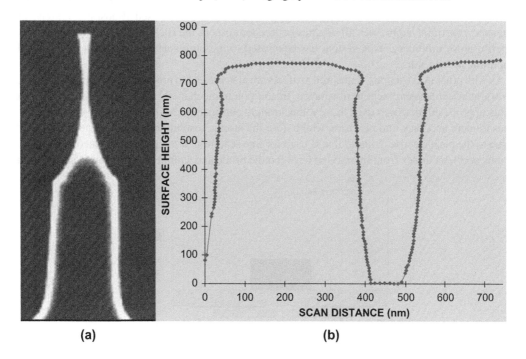

(a)   **(b)**

FIGURE 12.51 Undercut tip and profile scan for 2D resonant microcantilever atomic force microscope (AFM).

approximately 3×10^{-12} N, which allows it to detect the presence of the much weaker attractive forces (not necessarily van Der Walls only). The separation of the vertical and horizontal force components is accomplished through independent detection of each vibration frequency. Servo control of both the horizontal and vertical tip-to-sample distances now occurs via digitally controlled feedback loops with piezo actuators for each direction. When the tip encounters a sidewall, their separation is controlled by the horizontal feedback servo system. Conversely, tip height (z direction) is adjusted by the vertical force regulation loop. The scan algorithm uses force component information to determine the surface normal at each point, and subsequently deduces the local scan direction [83]. This allows the scan direction to be continually modified as a function of the actual topography in order for the motion to stay parallel to the surface at each point. In this way, data are no longer acquired at regular intervals along x, but at controlled intervals along the surface contour itself (Figure 12.51b). This is essential for scanning undercut regions, because in addition to having a flared tip, it is necessary to have a two-dimensional scanning algorithm that can provide servoed tip motion in both the lateral and vertical scan axis directions. One could imagine what would happen if a flared tip were to move vertically straight up while still under a feature overhang. The data set collected is stored as a three-dimensional mesh that maintains a relatively constant material density of data points irrespective of the sample slope. It is especially well suited for measuring the angle of a sidewall or its roughness. Another unique capability of this system is its ability to automatically adjust scan speed to surface slope, so that it slows when scanning a nearly vertical topography and speeds up when scanning a nearly flat surface. Data point density (in the x direction) also increases at sudden changes in surface height, such as feature edges (or most interest), because the sampling rate is fixed.

In the electrostatic force balance beam approach, high sensitivity is still achieved with a larger sensor. An example is a centimeter-long beam held balanced on a weak pivot by electrostatic force (Figure 12.52). The two sides of the balance beam form capacitors with the base. This balance-beam force sensor [85, 88], also known as the interfacial force microscope [86], is an inherently unstable mechanical system stabilized by a servo loop. By using force balance rather than a weak spring, this method of force sensing uncouples the sensitivity from the stiffness. The pivot, developed by Miller and Griffith [87], is a pair of steel ball bearings, one on each side of the beam, held to the substrate with a small magnet. The ball bearings are rugged, so a beam can be used almost indefinitely. The magnetic constraint suppresses all degrees of freedom except the rocking motion, which provides superior noise immunity. This system has been used to measure surface roughness with an RMS amplitude of <0.1 nm.

As with other microscopes, undesired artifacts are also present in probe microscope images which adversely affect measurement performance. In addition to mechanical noise susceptibility, two elements of probe microscopes exist that exhibit strongly nonlinear behavior which can seriously affect measurement accuracy and precision well before the atomic resolution has been reached. The first is due to the piezoelectric actuator that is used for fine scanning of the probe. The second, and more serious problem, arises from interaction between the probe and sample. Piezoceramic actuators are

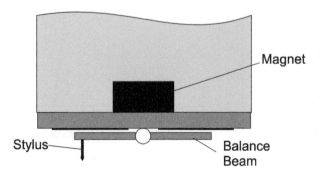

FIGURE 12.52 Scanned nanoprobe (SNP) balanced beam head assembly.

used to generate the probe motion because of their stiffness and ability to move in arbitrarily small steps. Being ferroelectrics, they suffer from hysteresis and creep so their motion is not linear with the applied voltage [88]. Therefore, any attempt to plot the surface height data versus the piezo scan signal results in a curved or warped image and does not reflect the true lateral position of the tip. A variety of techniques have been employed to compensate for the nonlinear behavior of piezos. In many instruments, the driving voltage is altered to follow a low-order polynomial in an attempt to linearize the motion. This technique is only good to several percent and does not really address the problem of creep. Attempting to address nonlinearities with a predetermined driving algorithm will not be adequate for dimensional metrology because of the complicated and non-reproducible behavior of piezoelectric materials. Another approach is to independently monitor the motion of piezoactuators with a reliable sensor. Several types of systems using this approach have been reported. One monitors the motion of a flexure stage with an interferometer [89]. Another measures the position of a piezotube actuator with capacitance-based sensors [90]. A third group employs an optical slit to monitor the piezotube scanner motion [91]. Electrical strain gauges have also been used for position monitoring as well. These techniques monitor the position of the piezoactuator and not the actual point of the probe that is in closest proximity to the surface (known as the proximal point). The error associated with the sensing position in a different plane from that of the proximal point, referred to as Abbe's offset, is illustrated in Figure 12.53. Abbe's offset error is given by

$$\text{Abbe's Error} = D\tan(\alpha), \tag{12.71}$$

which increases with the tilt of the scan head as the probe moves. To minimize this error, sensors are designed to be as close as possible to the proximal point. Ideally, the position of the probe apex should be directly measured; a tool developed by Marchman achieved this ideal with an optical position sensor and a light-emitting optical fiber probe [92].

A universal problem that is common to most one-dimensional scanning techniques, even optical and SEM, arises from the imaging algorithm itself. As seen in Figure 12.54, data points in a single

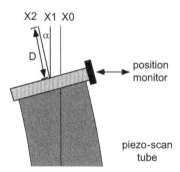

FIGURE 12.53   Abbe offset with a tube-based scanner.

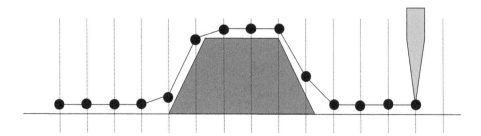

FIGURE 12.54   Surface height data points resulting from a one-dimensional raster scan.

line scan are equally spaced in increments of ΔX in the lateral scan direction. The size of each ΔX increment is constant, regardless of the topography, so that there is usually insufficient pixel density when abrupt changes in surface height (e.g. feature sidewall) are encountered. Scanning re-entrant profiles with even a flared-shaped probe is also forbidden, because the probe always moves forward or upward and is unable to reverse its direction without a 2D algorithm.

The scanning methodology employed with a two-dimensional resonant microcantilever, described in Section 12.3.5.1 on force sensors, does address the issue of data point density at abrupt surface height transitions. One can see from the line scan shown in Figure 12.54, that the point density is maintained along the surface contour in the XZ plane (instead of the X scan direction only), resulting in an abundance of topographic information at the feature edges. Regardless of the specific force sensing mode or probe scanning method, all AFMs share a common working element: a solid tip that is intimately involved in the measurement process. In fact, the probe tip size and shape primarily determine whether it is possible to image a desired feature on the mask or wafer. While there have been, and continue to be, important advances in other aspects, such as detection schemes and position monitoring, improvements in the tip offer the greatest potential for improving metrology performance and the extendibility of SPM techniques. As described earlier, current-generation probe tips are made of etched or milled silicon/silicon nitride (these make up by far the bulk of commercial tips), etched quartz fibers, or else they are built up from electron beam–deposited materials such as carbon. While all these tips have enabled significant metrological and analytical advances, they suffer serious deficiencies, the greatest being wear, fragility, and an uncertain and inconsistent structure. In particular, the most serious problem facing SPM dimensional metrology is the effect of probe shape on accuracy and precision. Very sharp probe tips are necessary to scan areas having high aspect ratios (>1:1) and abrupt surface changes [76]. When features having high aspect ratios are scanned (see Figure 12.55), they appear to have sloped walls or curtains [93]. The apparent surface is generated by the conical probe riding along the upper edge of the feature.

Even if the exact tip shape were known, there is no way to recover the true shape of the feature sidewalls. The fraction of the surface that is unrecoverable depends on the topography of the surface and the sharpness or aspect ratio of the tip. Furthermore, no universal probe shape exists that is appropriate for all surfaces. In most cases, a cylindrical or flared tip will be the preferred shape for scanning high aspect features [94]. Dilation of the image may still occur even with these probes. The cross section in Figure 12.56 demonstrates the effect of the cylindrical probe width on the pitch and line width measurements. The pitch measurement is unaffected by the probe width, but the line and trench width values are offset by the width of the probe.

Simple probe shapes provide greater ease of image correction and unrecoverable regions are vastly reduced. It is essential to make the probe as long and slender as possible in order to increase the maximum aspect ratio of a feature that can be scanned. However, a real cylindrical probe bottom has corners with a finite radius resulting in the inability to image the cusp region at the intersection of the feature bottom and substrate. High aspect ratios aside, conical probes having a sharp

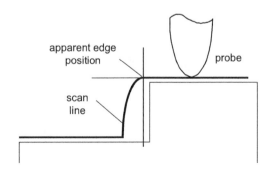

FIGURE 12.55   Apparent surface generated by probe tip shape mixing.

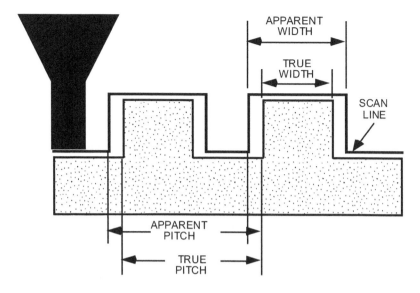

FIGURE 12.56 Scanned probe microscopy (SPM) image dilation resulting from cylindrical probe width.

apex are still useful for high lateral resolution imaging of surfaces with a gentle topography. A cone with a radius of curvature at the apex of the structure, shown in Figure 12.57, can perform surface roughness measurements that cover regions of wavelength-amplitude space unavailable to other tools [95–97]. Although a conical probe tip may not be able to access all parts of a feature, its interaction with an edge has some advantages over cylindrical probes [88, 98]. At the upper corners of a feature with a rectangular cross section, size measurement becomes essentially equivalent to the pitch measurement. The uncertainty in the position of the upper edge becomes comparable to the uncertainty in the radius of curvature of the probe apex, which can be very small. If the sidewalls are known to be nearly vertical, then the positions of the upper edges give a good estimate of the size of the feature. Sharper tips can also be obtained through ion beam milling techniques [99].

In semiconductor manufacturing, features having aspect ratios as high as 4:1 are being encountered. The probe must be narrow enough to fit into the feature without being so slender that it becomes mechanically unstable. To achieve a precise measurement, the apex of the probe must remain fixed relative to the probe shank. In other words, the probe must not flex. Flexing introduces measurement errors and causes instability of the feedback control loop, which leads to tip damage and wear. The analysis of probe stiffness is identical to that of any cantilevered beam. Imagine a

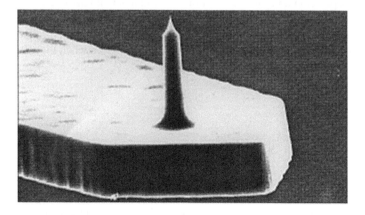

FIGURE 12.57 Conical probe tip on the underside of a resonant microcantilever.

probe tip with a uniform circular cross section having radius R and length L. Let the elastic modulus of the probe material be Y. If a lateral force F is impressed on the probe apex, the probe will deflect a distance Δx, which obeys the expression [100]:

$$\Delta x = \frac{4FL^3}{3\pi YR^4}. \tag{12.72}$$

Note that the geometrical factors, L and R, have the strongest influence on the deflection. This seems to indicate that efforts to find materials with a higher modulus would only produce marginal gains compared with the effects of size and shape. The need to precisely control the probe shape and achieve maximum possible stiffness has motivated recent efforts to employ carbon nanotubes as probes. Carbon nanotubes are recently discovered materials made purely of carbon arranged in one or more concentric graphene cylinders, having hollow interiors. This arrangement makes them essentially giant, elongated fullerenes, a relationship made clear by the strategy they share with spheroidal fullerenes (e.g. C_{60}—buckminsterfullerene) for closing by incorporating a total of 12 pentagons into their hexagonal network. Single-wall nanotubes (SWNTs) come closest of all carbon fibers to the fullerene ideal due to their remarkably high degree of perfection; their poverty of defects; and the special nature of carbon–carbon bonding that confers on them material properties, such as strength, stiffness, toughness, and electrical and thermal conductivities, that are far superior to those found in any other type of fiber [101]. There also exists larger-diameter cousins of single-wall (fullerene) nanotubes, typically having 4–20 concentric layers, known as multi-wall nanotubes (MWNTs). These may offer extra advantages as SPM probes due to their increased radial diameter, typically between 20 and 100 nm. In addition to having a perfectly cylindrical shape, these tubes exhibit an unusually high elastic modulus—in the terapascal range [102]. Dai *et al.* were the first to mount a nanotube on a microcantilever and scan with it [103]. The most important characteristic of carbon nanotube tips is their immunity to wear and erosion. Measurement precision would be greatly enhanced if their shape remained unchanged after accidental contact with the surface during scanning. The ability to measure their exact shape and size will provide a new level of accuracy and precision in SPM metrology. The TEM micrograph in Figure 12.58 shows an MWNT that has been mounted onto an AFM microcantilever tip base. Currently, the actual imaging performance is being gauged through studies using manually mounted nanotube tips in different AFM force imaging modes on sample features of interest [103].

Currently, probe microscopes image too slowly to compete with optical or electron microscopes in terms of speed. Since the protection of the probe tip is of paramount importance, the system should not be driven faster than its ability to respond to sudden changes in surface height. This is a source of complaint from those used to faster microscopes. In many instances, the probe microscope is, however, providing information unavailable from any other tool, so the choice is between slow and never. Preparation and modification of the sample should also be considered with cross-sectional SEM throughput comparisons.

The probe–sample interaction is a source of error for all measuring microscopes, but the interaction of a solid body (stylus) with a sample offers several advantages when high accuracy is needed.

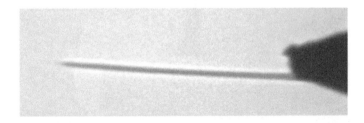

FIGURE 12.58 TEM image of a multi-wall nanotube (MWNT) mounted on a commercially available atomic force microscopes (AFM) tip.

The most important advantage arises from the relative insensitivity of a force microscope to sample characteristics such as the index of refraction, conductivity, composition, and pattern proximity (assuming that an adequate probe aspect ratio exists). Optical and electron beam tools are sensitive to these characteristics [104, 105], so errors can arise if a reference artifact has a composition different from the sample to be measured. For instance, calibrating an SEM-CD measurement with a metal line may give inaccurate results if the line to be measured consists of photoresist. However, in the calibration of a probe microscope there are two fundamental problems: measuring the behavior of the position sensors and finding the shape of the probe tip. Calibrating position sensors is the easier of the two chores and we will discuss it first. Lateral position sensor calibration is equivalent to a pitch measurement, and an excellent reference material is a grating, a two-dimensional one if possible. Periodic reference materials are available commercially, or they can be reliably fabricated through, for instance, holographic techniques [106]. Some samples are self-calibrating in that they contain periodic structures with a well-known period. Vertical calibration is more troublesome because most step height standards contain only one step. This makes it impossible to generate a calibration curve. In addition, the step height sometimes differs by orders of magnitude from the actual feature height to be measured. Surface roughness measurements often involve height differences of less than a nanometer, though they are sometimes calibrated with step heights of 100 nm or more. The vertical gain of a piezoscanner may be substantially different in the two height ranges.

The image produced by a scanning probe microscope is simply a record of the motion (obtained from the position sensors) of the probe apex as it is scanned across the sample. The image will faithfully represent the sample as long as the apex and the point of interaction coincide. If the interaction point wanders away from the apex, the relationship between the image and sample becomes more complicated. An example of this is shown in Figure 12.59a. As the probe moves across the step, the apex follows the dashed line rather than the true step profile. Note that the interaction point stays fixed at the upper edge of the step until the apex reaches the lower plane. The dashed line is, in fact, an inverted image of the probe.

Among several researchers, Villarrubia has conducted the most thorough study, making use of results from mathematical morphology. The behavior represented by the dashed line is often called "convolution," which is a misnomer. Convolution is a linear process, while the interaction of the probe with the sample is strongly nonlinear. The correct term for this process is *dilation*. The mathematical description of dilation uses concepts from set theory. Imagine two objects **A** and **B**, represented by sets of vectors. The individual vectors in the objects will be denoted **a** and **b**. The dilation of **A** by **B** is given by

$$\mathbf{A} \oplus \mathbf{B} = \bigcup_{\mathbf{b} \in \mathbf{B}} (\mathbf{A} + \underline{\mathbf{b}}), \tag{12.73}$$

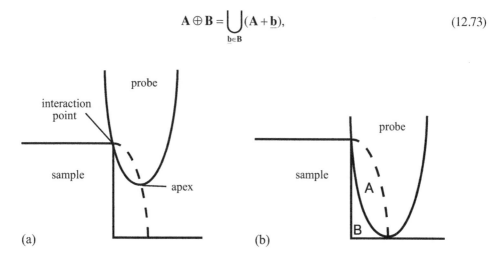

FIGURE 12.59 Trajectory of tip proximal point during probe shape mixing.

where $\mathbf{A} + \underline{\mathbf{b}}$ is the set $\mathbf{A}$ translated by the vector $\underline{\mathbf{b}}$. This operation replaces each point of $\mathbf{A}$ with an image of $\mathbf{B}$ and then combines all of the images to produce an expanded, or dilated, set. In scanning probe microscopy, we work with the following sets: $\mathbf{S}$, the sample; $\mathbf{I}$, the image; and $\mathbf{T}$, the probe tip. In the analysis, one frequently encounters $-\mathbf{T}$, the reflection of the probe tip about all three-dimensional axes. Villarrubia denotes the reflected tip as $\mathbf{P}$. It can be shown that

$$\mathbf{I} = \mathbf{S} \oplus (-\mathbf{T}) = \mathbf{S} \oplus \mathbf{P}. \tag{12.74}$$

In other words, the image is the sample dilated by the reflected probe tip. We can see this in Figure 12.59a, where the image of the step edge includes a reflected copy of the probe shape. One must convert this abstract notation into the following:

$$i(x) = \max_{x'}\left[s(x') - t(x' - x) \right], \tag{12.75}$$

where i, s, and t represent the surfaces of the sets. If the probe shape is known, then it is often possible to arrive at an estimate of the sample shape substantially better than that represented by the raw image. An example of this analysis is shown in Figure 12.59b. We know that the probe reached the position shown, because the dashed line represents the collection of points visited by the probe apex. The sample cannot extend into Region A because the sample and the probe are not allowed to overlap. We cannot make the same claim about Region B, however, because the probe was not capable of occupying that space. The subtraction of Region A from the image is an example of *erosion*. The erosion of set $\mathbf{A}$ by set $\mathbf{B}$ is defined as

$$\mathbf{A} - \mathbf{B} = \bigcap_{\mathbf{b} \in \mathbf{B}} (\mathbf{A} - \underline{\mathbf{b}}). \tag{12.76}$$

In our context, it can be shown that the best estimate of the true surface we can obtain from a probe tip $\mathbf{T}$ is $\mathbf{I} - \mathbf{P}$. Clearly, it is important to know the precise shape of $\mathbf{P}$. For many years, it has been known that scanning probe microscope scans can be used to measure the probe shape. In fact, it is the best way to measure the probe shape. Villarrubia has thoroughly analyzed this process, showing the manner in which any image limits the shape of the probe. His argument is founded on the deceptively simple identity:

$$(\mathbf{I} - \mathbf{P}) \oplus \mathbf{P} = \mathbf{I}. \tag{12.77}$$

If one erodes the image with the probe and then dilates the result with the same probe, one gets the image back. Villarrubia shows how this expression can be used to find an upper bound for the probe. It is based on the observation that the inverted probe tip must be able to touch every point on the surface of the image without protruding beyond that image. The algorithm implementing this, however, is subtle and complex, so we will not give it here. One of Villarrubia's publications provides the full C source code for implementing the algorithm. Some sample shapes reveal more about the probe than others [107, 108]. Those shapes specially designed to reveal the probe shape are called probe tip characterizers. The characterizer shape depends on the part of the probe that is to be measured. To measure the radius of curvature of the apex, a sample with small spheres of known size might be used. If the sides of the probe are to be imaged, then a tall structure with re-entrant sidewalls should be used. A re-entrant test structure for calibrating cylindrical and flared probes is shown in Figure 12.60a, and the resultant tip mirror image produced after scanning is shown in Figure 12.60b.

The AFM scans in Figure 12.61 show the same rectangular periodic line structure before (dashed) and after (diamonds) tip width removal from the image data, respectively. As a sanity check, one can overlay the corrected AFM line scan data on top of an SEM cross-section image of the feature to verify tip shape removal and calibration.

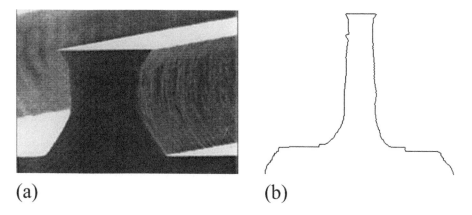

(a) (b)

FIGURE 12.60 Tip characterization structure and mirror image of undercut probe shape.

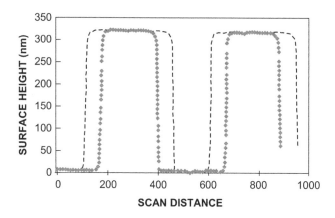

FIGURE 12.61 Atomic force microscope (AFM) scan of line profile structures before and after tip shape removal.

12.3.5.2 Measurement Precision

A gauge study experimental design for assessing the metrologic precision of AFM tools will be described in this section [109]. As described previously, AFMs can be operated primarily in two modes—standard (1D) and critical dimension modes. The standard mode operates with one-dimensional resonance force sensing and simple one-dimensional raster scanning, but the CD uses two-dimensional resonant force sensing along with the 2D surface contour scanning algorithm and undercut tip, as described earlier. Initially, screening experiments were performed in order to determine the amount of averaging necessary during each measurement and the relative weighting of different factors in the precision. AFM images are composed of a discrete number of line scans (in the X–Z plane)—one at each value of Y. For better precision estimates, each measurement was performed as close to the same location as possible (to within the stage precision) in order to minimize the effects of sample non-uniformity. Another important screening task was to determine how many line scans per image are necessary to provide adequate spatial averaging of the line edge roughness in order to reduce the effects of sample variation on the instrument precision estimate. Gauge precision was obtained by making measurements under different conditions. Determining the most significant factors to vary depends on one's understanding of the physical principles of the instrument's operation. Based on the principles of scanning probe microscopy outlined earlier, the main factors that should be included for this study are the scanning mode, tip shape, and sample loading effects. A diagram illustrating the nested model used in this study

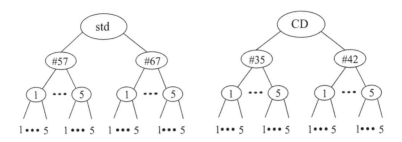

FIGURE 12.62 Atomic force microscope (AFM) precision gauge experimental designs. (From Figure 5 in H.M. Marchman, "Nanometer-scale dimensional metrology with noncontact atomic force microscopy," *Proc. SPIE*, 2725, 527 [1996].)

is shown in Figure 12.62. The highest level contains the scanning mode, with the tip next. There were two scanning modes (standard and CD), two tips to each mode, and three to five loading cycles. A loading cycle value consisted of a mean of three or five static measurements. We found it advantageous to use a separate model for each scanning mode. Due to the three-dimensional nature of the AFM data, it was possible to obtain width (at each Z), height, and wall angles. The reproducibility measurements were obtained on etched oxide lines and spaces. Tip width calibration was also performed before each cycle. Precision estimates for the standard and CD modes are shown in Figures 12.63 and 12.64, respectively. The nested experimental design allows the use of an analysis of variation (ANOVA) for calculating these estimates and actually separating the various components of precision:

x_{ijk} = k-th measurement of the j-th cycle on the i-th tip.
$\bar{x}_{ij}$ = average of n measurements in the k-th cycle on the i-th tip.
$\bar{x}_i$ = average of $b \times n$ measurements on the i-th tip.
$\bar{x}$ = total average of $a \times b \times n$ measurements taken in the experiment.

HEIGHT

Tip57 (i=1) Rep k = 1	2	3
Cycle j = 1	318	318	316
2	316	317	315
3	317	316	314
Tip67 (i=2) Rep k = 1	2	3
Cycle j = 1	317	317	317
2	315	315	316
3	315	316	316

σ_e = 0.9
σ_c = 0.9
σ_t = ~ 0
σ = 1.2

LEFT WALL ANGLE (DEG)

Tip57 (i=1) Rep k = 1	2	3
Cycle j = 1	79.1	79.2	78.9
2	78.7	78.9	79.3
3	79.3	79.3	78.9
Tip67 (i=2) Rep k = 1	2	3
Cycle j = 1	76.2	76	76.3
2	75.4	75.9	75.9
3	75.3	75.6	75.9

σ_e = 0.2
σ_c = 0.2
σ_t = 2.3
σ = 2.3

RIGHT WALL ANGLE (DEG)

Tip57 (i=1) Rep k = 1	2	3
Cycle j = 1	77.7	77.7	78.1
2	77.3	77.3	77.2
3	77.3	77.3	77.3
Tip67 (i=2) Rep k = 1	2	3
Cycle j = 1	79.4	79.7	79.9
2	80.4	79.8	79.6
3	80.5	79.8	80.1

σ_e = 0.3
σ_c = 0.2
σ_t = 1.7
σ = 1.8

TOP WIDTH

Tip57 (i=1) Rep k = 1	2	3
Cycle j = 1	170.2	171.4	174.1
2	174.7	178.3	176.0
3	176.3	170.2	174.0
Tip67 (i=2) Rep k = 1	2	3
Cycle j = 1	181.5	185.3	189.2
2	169.2	181.1	174.6
3	181.2	179.6	187.2

σ_e = 3.7
σ_c = 3.5
σ_t = 4.4
σ = 6.8

MIDDLE WIDTH

Tip57 (i=1) Rep k = 1	2	3
Cycle j = 1	211.2	210.7	216.6
2	224.2	227.4	222.8
3	220.0	222.5	225.3
Tip67 (i=2) Rep k = 1	2	3
Cycle j = 1	257.6	256.4	258.1
2	257.1	252.5	254.5
3	253.2	253.8	254

σ_e = 2.2
σ_c = 4.5
σ_t = 24.7
σ = 25.2

BOTTOM WIDTH

Tip57 (i=1) Rep k = 1	2	3
Cycle j = 1	281.7	273.6	270.3
2	288.4	290.3	281.4
3	275.0	285.1	293.3
Tip67 (i=2) Rep k = 1	2	3
Cycle j = 1	297.1	293.7	301.3
2	297.4	292.7	299.9
3	296.4	298.2	293.5

σ_e = 5.4
σ_c = 3
σ_t = 10
σ = 11.7

FIGURE 12.63 Standard 1D mode atomic force microscope (AFM) precision results. (From Figure 6 in H.M. Marchman, "Nanometer-scale dimensional metrology with noncontact atomic force microscopy," *Proc. SPIE*, 2725, 527 [1996].)

HEIGHT

Tip35 (i=1) Rep k = 1	2	3
Cycle j = 1 315.1	315.1	315.2
2 315.2	315.1	315.1
3 315.6	315.5	315
4 315.9	315.9	315.7
5 316	315.2	315.7
Tip42 (i=2) Rep k = 1	2	3
Cycle j = 1 315.7	315.5	315.3
2 315.4	315.8	315.5
3 315.3	316.2	316
4 316.7	316.5	316.1
5 315.8	315.8	315.1

σ_e = 0.3
σ_c = 0.3
σ_t = 0.2
σ = 0.5

LEFT WALL ANGLE (DEG)

Tip35 (i=1) Rep k = 1	2	3
Cycle j = 1 82.9	83	83.3
2 82.9	83.2	82.9
3 83.1	82.8	83.1
4 83.1	82.8	82.7
5 83.4	83	83.1
Tip42 (i=2) Rep k = 1	2	3
Cycle j = 1 82.7	83	82.9
2 82.8	82.8	82.8
3 82.7	82.9	82.8
4 82.7	82.8	82.9
5 82.8	83	83.4

σ_e = 0.18
σ_c = 0.04
σ_t = 0.1
σ = 0.21

RIGHT WALL ANGLE (DEG)

Tip35 (i=1) Rep k = 1	2	3
Cycle j = 1 84.9	84.9	85
2 85.1	84.9	85.1
3 84.8	84.9	84.9
4 84.9	84.8	84.9
5 85	84.9	85
Tip42 (i=2) Rep k = 1	2	3
Cycle j = 1 84.7	85	85
2 85	85.1	85.1
3 84.8	85	85
4 85.1	85.1	84.9
5 85	84.9	84.8

σ_e = 0.09
σ_c = 0.06
σ_t = ~ 0
σ = 0.11

TOP WIDTH

Tip35 (i=1) Rep k = 1	2	3
Cycle j = 1 180.1	181.3	179.8
2 181	181.7	181
3 182.1	180.4	181
4 181.9	178.6	181.1
5 178.9	181.8	179.3
Tip42 (i=2) Rep k = 1	2	3
Cycle j = 1 178.9	178.1	180.3
2 179.9	179.6	178.1
3 180.6	179	177.3
4 178.7	178.4	175.9
5 179	175.5	177.5

σ_e = 1.3
σ_c = ~ 0
σ_t = 1.5
σ = 2

MIDDLE WIDTH

Tip35 (i=1) Rep k = 1	2	3
Cycle j = 1 205.3	203.9	203.4
2 204.2	204.9	203.1
3 205.8	204.4	204.5
4 205.4	203.5	205.2
5 203.2	206.1	202.1
Tip42 (i=2) Rep k = 1	2	3
Cycle j = 1 204.1	204	205.6
2 205.6	204.4	204
3 208	205.1	204
4 204.3	205.2	204.5
5 205.4	200.5	202.5

σ_e = 1.4
σ_c = 0.4
σ_t = 0.2
σ = 1.5

BOTTOM WIDTH

Tip35 (i=1) Rep k = 1	2	3
Cycle j = 1 233.5	233.9	231.8
2 233.3	232.7	232.4
3 233.9	233.4	233.3
4 234.5	232.5	233.6
5 230.9	234.8	232.1
Tip42 (i=2) Rep k = 1	2	3
Cycle j = 1 233.9	232.1	235.5
2 234.7	233.4	231.9
3 235.9	235.1	234.6
4 232.5	232.8	229.3
5 234.1	229.4	229.9

σ_e = 1.5
σ_c = 0.9
σ_t = ~ 0
σ = 1.7

FIGURE 12.64 CD mode atomic force microscope (AFM) precision results. (From Figure 6 in H.M. Marchman, "Nanometer-scale dimensional metrology with noncontact atomic force microscopy," *Proc. SPIE*, 2725, 527 [1996].)

The precision of the AFM in this report is defined as

$$\sigma = \sqrt{\sigma_e^2 + \sigma_c^2 + \sigma_t^2}, \tag{12.78}$$

where σ_e^2 is the component for error (repeatability), σ_c^2 is the cycle variance component (reproducibility), and σ_t^2 is the tip-to-tip variance component. Estimates of these variance components are calculated using the sum of mean squares:

$$MS_e = \frac{\sum_{i=1}^{2}\sum_{j=1}^{b}\sum_{k=1}^{n}\left(x_{ijk}-\bar{x}_{ij}\right)^2}{ab(n-1)},$$

$$MS_c = \frac{n\sum_{i=1}^{2}\sum_{j=1}^{b}\left(\bar{x}_{ij}-\bar{x}_{i}\right)^2}{a(b-1)}, \tag{12.79}$$

$$MS_t = \frac{bn\sum_{i=1}^{2}\left(\bar{x}_{i}-\bar{x}\right)^2}{a-1}$$

and

$$\breve{\sigma}_e = \sqrt{MS_e}, \quad \breve{\sigma}_c = \sqrt{\frac{MS_c - MS_e}{n}}, \quad \breve{\sigma}_t = \sqrt{\frac{MS_t - MS_c}{bn}}. \tag{12.80}$$

The carets over each sigma indicate that they are only estimates of the variance components. A negative variance component estimate usually indicates a non-significant variance component and is set to zero. Variance component estimates can be biased, so the results should be interpreted with caution. The standard mode height precision (see Figure 12.63) is on the single nanometer level, as one would expect. Height data can be gathered very well with the standard mode tip. It can also be seen that the error and cycle components of the wall angle variance are quite good. As one would expect, tip variance is the main contributor to the overall imprecision. As described earlier, the angle of the feature sidewall was larger than that of the probe, so that imaging of the probe shape occurred instead of the feature. The CD mode height measurements (Figure 12.64) were more precise than in the standard mode and had a variance of $\sigma=0.5$ nm. The total measurement precision for the wall angle was also much better in the CD mode and were a few tenths of a degree. The top width precision was 2 nm and had no systematics. A dramatic improvement in the middle and bottom width measurements was realized by scanning in the 2D mode with the flared boot-shaped probes. Top and middle width measurement precision values were on the single nanometer level. The improvement in the CD mode precision values was mainly due to the ability to image the feature walls with the bottom corners of the flared-shaped probes. This eliminated the effect of probe shape on overall precision. The increased number of data points at the feature edges due to the 2D scanning algorithm also helped to improve the CD measurement precision. Automatic algorithm threshold levels for top and bottom width locations can also be set more reliably when there is a greater density of data points at the edges.

12.3.5.3   Application to Evaluation of SEM Measurement Linearity

Linearity describes how well an image corresponds to actual changes in the real feature size. One must determine if the measurements obtained from an image correspond in a direct fashion to reality. CD-SEM linearity is typically determined by comparing measurements of test features to those obtained by a more accurate reference, the AFM in this case (the cross-sectional SEM in another case). Typically, SEM width measurements are first plotted versus the corresponding reference values. A regression analysis yields the equation for a straight line that best fits the distribution of measured points:

$$Y_l = \alpha X_l + \beta + \varepsilon_l, \tag{12.81}$$

where α is the slope defect, β is the offset, and ε is the error component. Therefore, first-order regression analysis can be used to determine the accuracy (offset), magnification calibration (slope defect), and variation (error term). This provides a means of quantifying the linearity and accuracy.

A series of etched oxide lines ranging in width from 50 to 1000 nm will be used in this discussion. CD-SEM measurements of the same structures were then plotted against the AFM reference values, as shown in Figure 12.65. The degree of linearity and accuracy is given in terms of the slope

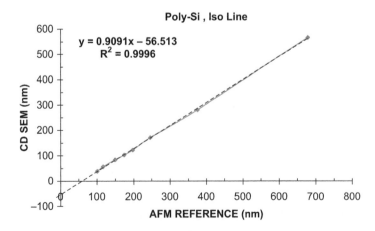

FIGURE 12.65   CD-SEM measurements plotted versus atomic force microscope (AFM) reference values.

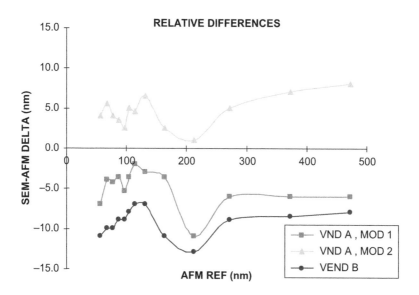

FIGURE 12.66 SEM-to-AFM offsets for various vendors.

defect (α), goodness of fit (R^2), and offset (β) in the figure. The extremely small sizes of the etched lines present a challenge for any SEM to resolve. In addition to having a measurement offset of –56 nm, the SEM in this example was not able to resolve the 60 nm line. It is important to note that the assumption of process linearity is not necessary, because we are comparing the SEM measurements to values obtained from an accurate reference tool. By plotting the SEM widths versus the actual reference values, we should obtain a linear trend in the data—even if the actual distribution of feature sizes is not linear. The relative matching, or tool-induced shift (TIS), between CD-SEMs can be studied by repeating this process for each system. Once the most linear algorithm has been determined, the additive offset needed to make the measurement curves (hopefully of the same shape) overlap with the reference curve must be found. The SEM-to-AFM measurement offsets are illustrated in Figure 12.66. All three SEMs continued to track changes in line width down to 50 nm in a linear fashion, but offsets existed between the three tools. The curves are fairly constant with respect to each other for feature sizes larger than 250 nm. The tool from Vendor B actually matched Model 1 from Vendor A better than Model 2 from Vendor A. Even if an SEM is from the same vendor, it may not be easier to match. The dip in all three curves at the 200 nm width suggests an error in the AFM tip width calibration at that site.

12.3.6 OVERLAY AND REGISTRATION

12.3.6.1 Historical Overview

Overlay is the accepted term to denote the overlay error or misregistration in a semiconductor integrated circuit structure. It measures the lateral deviation of the layer-to-layer alignment from the intended, perfect, alignment. The metrology of overlay has been driven by a simple hierarchy of rules of thumb: If we denote the size of the smallest feature in the integrated circuit as A, the maximum allowable overlay is 1/3–1/5 of this smallest dimension, or A/3–A/5 as shown in Figure 12.5. The metrology instrument is expected to measure the overlay to an accuracy better than 1/5–1/10 of the permitted overlay [110], i.e. to an accuracy of A/25–A/50. Over the past three decades, both the definition for the smallest feature size and the definition of overlay metrology accuracy have evolved. However, the basic rule of thumb has continued to peg the requirements for overlay metrology capability to the ever-decreasing feature size in integrated circuits. Until the mid-1980s, overlay was measured using Vernier-type test structures, viewed by a skilled operator through an optical microscope for reading the magnitude of the overlay. The Vernier-based solution became unviable when the overlay metrology uncertainty requirements shrank to the same magnitude as the subjective reading errors (about 50 nm). This opened up the market for automated overlay metrology, which for several years has relied exclusively on symmetric box-in-box targets [111] located in the scribe lines. The capabilities of automated overlay metrology tools have made steady progress, improving their accuracy from the original 50 nm to a single nanometer level, while simultaneously vastly enhancing their capabilities in the equally important areas of throughput, factory automation, and ease of use.

In the 20 nm process nodes, fabs started using multiple lithography steps on the same physical layer ("multiple patterning") to improve the resolution and maintain the demand for shrinkage. Since alignment is now to structures in the same layer as well as to structures in other physical layers, the overlay specs were tightened by a factor of two [112, 113].

Overlay control is done using optical metrology of the resist pattern. The major advantage of the optical overlay technique is its fast measurement time. The latest tools measure a target in less than half a second. This allows the fab to use more complex overlay models which are based on dense sampling. These models are then fed to the scanner which uses them to reduce the misregistration error.

In order to meet the metrology requirements, periodic targets were introduced in the imaging domain of metrology (advanced imaging metrology [AIM] targets) [114]. Furthermore, a new optical technique was introduced: scatterometry-based overlay (SCOL) [115, 116]. In this technique, overlay is calculated based on how the interaction between the gratings breaks the symmetry rather than the difference between each grating location as done in imaging.

Both optical overlay techniques are now being measured based on "accuracy" rather than "reproducibility," which reflects the transition from using overlay metrology results for process monitoring to more advanced control schemes.

12.3.6.2 Overlay Metrology Use Cases

Semiconductor and related manufacturers measure overlay to drive the decision-making necessary to run efficient and profitable wafer processing. Through analysis software systems, the overlay data are transformed into the information and knowledge necessary to drive a decision. Over time, the degree of automation for these decisions has increased remarkably, allowing sophisticated real-time decision-making. The many types of decisions that are driven by overlay metrology data can generally be categorized into several major types: line monitoring, tool monitoring, and off-line engineering.

12.3.6.2.1 Line Monitoring

A line monitor addresses disposition and control questions, typically using data from production material on a highly frequent basis with a fixed sampling. Questions such as

- Is the process (tools and materials) good?
- Is the wafer (lot) good?
- Should the lot be reworked?
- If so, what should be done differently, and how much improvement can we expect?
- Can we feed-forward any information so that problems can be fixed at a later processing step?
- Can we feedback any information for optimal equipment control on subsequent lots?
- How much of the error is correctable and how much is not?
- Is a correction justified?
- Which litho-cell should I use?
- What are my sources of variation?
- Is the measurement (metrology tool, target) good?
- Am I sampling sufficiently to minimize material at risk?

12.3.6.2.2 Tool Monitoring

A tool monitor addresses similar questions specifically geared toward equipment disposition and control typically using (non-production) test wafers with moderate frequency:

- Is the scanner qualified (i.e. within specifications)?
- Is the scanner matched to the other scanners?
- Is the overlay tool good, and is it matched to the other overlay tools and calibrated to a standard?

12.3.6.2.3 Off-Line Engineering

Off-line engineering and troubleshooting is typically exception based and is performed when, for example, either the line monitor or the tool monitor activities indicate some type of problem, the end of line or other sensor data indicate a problem, or the like. For this use case, specific wafers are often created and measured to augment line and tool monitor data as part of a drill-down process. It addresses questions such as

- Does the issue correlate to specific processes, tools, times, layers, etc.?
- Can I find the root cause of the issue?

12.3.6.3 Overlay Metrology Tool

There are two major optical overlay techniques: imaging and scatterometry (or diffraction-based) overlay [115, 116]. In imaging, the center of each layer is measured independently and the overlay is the difference between them. In scatterometry, the interaction between the gratings is measured and the overlay is extracted from the way it affects the signal.

12.3.6.3.1 System Architecture

In imaging, the image obtained on the camera is the enlarged image of the target since both the camera and the target are on the same optical plane (the field plane). An example for tool architecture of this type is shown in Figure 12.67. In scatterometry, the light is manipulated to enable independent measurements of diffraction orders. This order separation can be done by placing the image sensor in the pupil conjugate plane and using image processing algorithms. An example for tool architecture of this type is shown in Figure 12.68. Another option is to use a flexible barrier in the collection path to block undesired diffraction orders.

Modern architectures include modules for better control of the light wavelength, shape, and polarization. Additional improvements related to the light intensity allow faster measurement time of targets with worse reflectivity.

12.3.6.3.2 Illumination Subsystem

Overlay metrology systems may utilize visible broadband illumination, although spectral band-pass reduction is frequently used for contrast enhancement. In imaging, Köhler illumination is generally the architecture of choice, in which the light source is conjugated to the imaging system pupil. In scatterometry overlay, coherent lasers can be used in order to gain an intensity boost. The illuminator design typically sacrifices optical throughput in favor of pupil illumination uniformity in the battle against tool-induced shifts. As the metrology tool advances, it enables enhanced wavelength flexibility.

12.3.6.3.3 Optical Head

12.3.6.3.3.1 Imaging Optics   The imaging column is the heart of the optical system, generally comprising an objective lens, a tube lens, and two beam splitters, one for coupling the illumination and one for the focus system. The key optimization metrics in the optical design of the imaging column are minimization of (i) asymmetric aberrations and (ii) chromatic aberrations. In order to minimize sensitivity to aberrations, the numerical aperture is a critical trade-off parameter. In

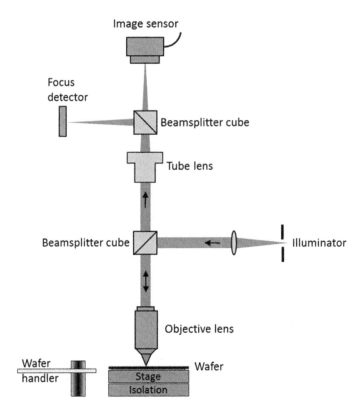

FIGURE 12.67 Schematic representation of imaging overlay metrology tool.

general, aberrations are controlled more easily for lower NA optics at the price of a reduced image resolution, with the additional benefit of depth of focus. In principle, it is possible to ascertain the center of symmetry of a feature to an exceedingly tight tolerance well below the resolution limit of the system. In practice, however, there are benefits to a higher resolution in terms of cross-talk reduction between adjacent features, especially when the mark size is reduced. For periodic overlay targets, higher resolution enables metrology on smaller pitches.

12.3.6.3.3.2 Scatterometry Optics In this technique, the scattered signal is usually collected as a function of wavelength or scattering angle. The optical head may provide multiple wavelengths simultaneously to support the first option. In order to measure in multiple angles, an apodizer shapes the spatial light distribution in the pupil conjugate plane. This provides control in which diffraction orders will be collected into the sensor. Some apodization may also be applied in the collection column (after interacting with the target).

12.3.6.3.4 Focus System

A number of different focus systems have been implemented in commercially available overlay metrology tools, including interferometric, astigmatic, and contrast-based methods. The key metrics are focus repeatability, robustness, and speed.

12.3.6.3.5 Image Sensor

The CCD is the sensor of choice in overlay metrology tools. A number of system parameters are of importance in the selection of the sensor, including dark noise, shot noise, frame rate, and pixel size. These parameters combine in a non-trivial and target-dependent manner, making sensor selection an exercise in system engineering.

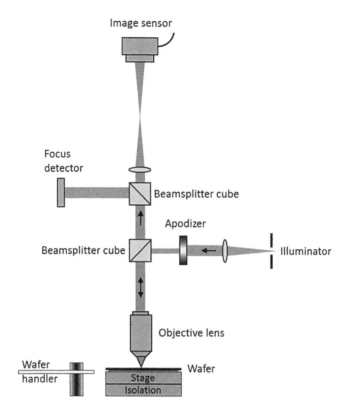

FIGURE 12.68   Schematic representation of scatterometry overlay metrology tool.

12.3.6.3.6   *Stage, Isolation, and Wafer Handling*

In addition to the foregoing list of sensor subsystem hardware, the metrology tool also includes significant platform components. Wafer handler, motion control, and vibration isolation systems must be seamlessly integrated to provide fast, contamination free, and highly reliable material handling. The cost of ownership of the metrology tool is typically dominated by the wafer throughput and reliability, which in turn are often governed by these platform components.

12.3.6.3.7   *Overlay Metrology Mark*

The overlay metrology mark is a critical and intrinsic component of the overlay metrology system architecture. It serves the dual functions of recording the local overlay misregistration and transferring that information with high fidelity to the metrology tool. The mark generally comprises information from the underlying process layer in the form of topographic or structural features which are generated by a broad range of semiconductor manufacturing processes. The position of these features is determined during one of the previous lithographic steps. Typically, the mark incorporates additional spatial information in the form of an after-develop photoresist pattern from the current lithographic step. One frequent exception to this rule is when overlay metrology is performed after subsequent etching and resist strip in order to characterize final feature placement and ascertain the nature of discrepancies from the after-develop results.

Today, the standard imaging workhorse is the AIM mark [117, 118]. The design of this target is shown in Figure 12.69. In advanced designs, all features are segmented, i.e. constructed of much smaller design rules features (e.g. lines with device-like pitch).

A scatterometry overlay mark consists of two interlaced gratings generally of the same pitch. The scattering intensity is sensitive to the displacement between the two gratings, and thus contains overlay information; however, the sensitivity is generally weak compared with that of other

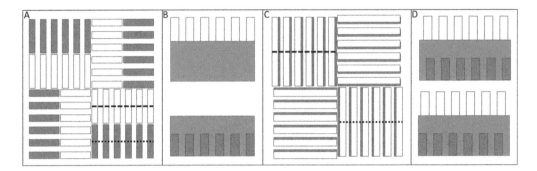

FIGURE 12.69 Examples of overlay target designs. Figures A and C show a schematic plan view of AIM (imaging) and scatterometry targets, respectively. Resist lines are white, process layer lines are dark gray. Figure B (D) shows cross plots of Figure A (C); the top and bottom figures correspond to the dashed and dotted lines, respectively.

structural parameters such as the pitch or sidewall angle. In the most straightforward approach, the determination of the overlay error from the scattered light intensity involves rigorous modeling. However, such accurate modeling requires substantial computing resources, due to the complex structure of the overlay mark, rendering the method a major technical challenge using existing commercially available electromagnetic simulation tools.

This difficulty has been overcome by the differential signal method. Several overlay marks, with different intentional offsets between the two gratings of each mark, are formed in close proximity. The difference between the intensities of light scattered from these overlay marks allows a model-free determination of the overlay error. This approach creates several inherent advantages. First, the pitch of the overlay mark can be much smaller than the imaging mark pitch. Such marks are more compatible with IC manufacturing processes, and are therefore more robust than large feature marks [119]. Another advantage of scatterometry overlay technology is the high density of information encoding associated with the small pitch grating of grating target architecture.

12.3.6.3.8 Overlay Algorithm

Enabling both precise and accurate overlay metrology requires automated and robust solutions to a wide range of algorithmic challenges. The list of tasks includes, but is not limited to, wafer alignment, target centering, illumination control, and the selection of focus and number of frames, not to mention the overlay measurement algorithm itself. These parameters, once the realm of the application or metrology engineer, are transitioning to the domain of intelligent optimization procedures, reducing the recipe setup load on engineers who are spread over a larger number of metrology and inspection applications in the fab. For imaging overlay targets, the actual overlay measurement itself boils down to the determination of the offset between the centers of symmetry of the features from the two layers in question. Probably the most widely used algorithms are correlation based, whereas edge detection methods can still be found in older metrology systems. Phase-based methods are also making an appearance as periodic structure overlay marks enter the industry. In scatterometry, the intensity corresponding to each diffraction order and target cell is identified on the image sensor. The overlay is calculated from the difference between these intensity values (if there is no difference in the intensities, the overlay is zero). In principle, the overlay can be calculated based on a single pixel in the image sensor; since there are many pixels (on the order of hundreds or thousands), more sophisticated algorithms are used to do "smart averaging" of the overlay results. Another important type of algorithm provides additional information regarding the measurement quality and can be used to monitor the stability of the overlay value errors between sites, wafers, and lots.

12.3.6.3.9   Overlay Analysis and Sample Plan

The overlay metrology system provides the basic input to the data analysis software necessary to drive the use cases and decisions previously listed in an on-line, real-time fashion. The basic input is typically the overlay error (relative displacement of one layer to a previous layer) in the x- and y-directions of the wafer plane at various target locations on the wafer, as well as the corresponding x- and y-location of those targets within the scanner field, and the x- and y-location of the scanner field on the wafer. Advanced data quality analysis and flyer (outlier) rejection is then automatically performed prior to and during linear regression analysis using a data model. The models applied are typically based on real physical spatial quantities relevant to the equipment involved, such as translation, rotation, magnification, etc., on both the field and wafer scale. Generally, the so-called inter-field and intra-field model terms can be obtained separately from sequential regressions or simultaneously from a single regression. In addition to the model-term values (e.g. rotation), indications of quality of fit and residual data, including process noise, metrology noise, and unmodeled systematics, can be generated in the modeling process. The output of the models, as well as statistical quantities, drives the decisions. Significant engineering expertise is required to choose the proper sampling, modeling, and statistical methods involved. Over the years, the analysis has become more sophisticated to include more systematic effects such as the contribution of lens distortions, lens aberrations, reticle errors, scanner stage errors, and asymmetries of disposition and etch. The additional robustness information reported by the overlay metrology tool is used as additional process monitoring parameters.

12.3.6.4   Metrics for Metrology Uncertainty

12.3.6.4.1   Tool-Induced Shift

The capabilities of overlay metrology depend on the two main systems participating in the metrology process: the overlay metrology tool and the overlay metrology target. One of the fundamental difficulties in assessing the capability of overlay metrology tools has been the lack of appropriate reference standards. One may analyze the various factors contributing to this state of affairs, such as the absence of calibration standards with the same stack structure as the specific process step under measurement [120], the inherent difficulty of taking process variations into account, and the fact that overlay metrology has progressed in lock-step with the leading-edge technology. The fact remains that in assessing the overlay metrology capability, both the tool suppliers and the tool users have had to resort to self-calibrating methods, without the *primus arbiter* of an absolute standard.

   The description of the overlay metrology tool metrics will begin with the tool-induced shift. TIS is defined, for x- and y-directions separately, by measuring the overlay of a given target in two rotational orientations 180° apart, the so-called 0° measurement and 180° measurement:

$$\mathrm{TIS}_x = \left(\mathrm{Overlay}_x\left(0°\right) - \mathrm{Overlay}_x\left(180°\right)\right)\!\big/2, \tag{12.82}$$

$$\mathrm{TIS}_y = \left(\mathrm{Overlay}_y\left(0°\right) - \mathrm{Overlay}_y\left(180°\right)\right)\!\big/2, \tag{12.83}$$

where the subscripts refer to the Cartesian xy-coordinate system of the wafer. The expectation for a perfect metrology system is that rotating the target by 180° simply changes the sign of the overlay, thereby giving a TIS of zero. Conversely, any deviation from zero indicates a violation of the basic symmetry assumptions inherent in the overlay metrology task. It is exactly this sensitivity to symmetry that gives TIS its special place among the performance metrics of the overlay metrology tool. As TIS can be interpreted as the error in the overlay measurement due to tool asymmetry, it is common practice to correct the measured overlay by subtracting TIS from it.

   A metric derived from TIS is the so-called TIS variability, or TIS-3σ. TIS-3σ is commonly defined as three times the standard deviation of TIS across a number of targets on a given wafer. A variant of this definition relates to the variation of TIS across separate wafers from a given lot. It is the authors' experience that the more challenging process variations take place across a wafer rather than across a lot, and thus the variant definition may yield overly optimistic results. As a non-zero TIS-3σ is caused by minute process-related variations from target to target, its magnitude indicates the sensitivity of the overlay metrology tool to these variations.

In overlay metrology, the term precision is used for the variation of repeated measurements of a given overlay target. A differentiation is made between so-called static and dynamic precision: static precision refers to repeat measurements of an overlay target without removing it from the field of view of the tool, whereas dynamic precision refers to a repeated cycling of unloading from and reloading the wafer to the tool between each measurement. There are variants to this terminology, such as repeatability and reproducibility for static and dynamic precision, respectively [121].

Because a typical lithography area in a fab is equipped with several overlay metrology tools, tool-to-tool matching is a significant metric, attesting to the (perceived) accuracy of the tools as well as to the ability to use the tools interchangeably. The two metrics used to describe tool-to-tool matching are average matching and matching variability (also called matching-3σ or site-by-site matching). The raw data for determining these metrics are generated by measuring a given wafer on two tools, using the same sampling plan. Assuming that N sites are measured within the sampling plan, and that $\Delta_{x,i}$ is the difference for the i-th site between the two tools in the x-direction, the matching metrics can be defined as follows (definitions in the y-direction are identical except for the subscript):

$$\text{Average matching in } x\text{-direction} \quad \overline{\Delta}_x = \frac{1}{N} \sum_{i=1}^{N} \Delta_{x,i}, \tag{12.84}$$

$$\text{Matching variability in } x\text{-direction} \; 3\sigma_{\text{Match}} = 3 \times \sqrt{\frac{\sum_{i=1}^{N} \left(\Delta_{x,i} - \overline{\Delta}_x\right)^2}{N-1}}. \tag{12.85}$$

When assessing the capabilities of a given tool, whether for comparison to other tools of similar make, for comparison to competing tools, or for trying to estimate a tool's basic capabilities, one would like to have a single metric describing the "goodness" of the tool. Such a metric, called the total measurement uncertainty (TMU), has been evolving over the years [122]. Several variants of TMU exist, but they are usually of the following type:

$$\text{TMU} = 3 \times \sqrt{\sigma_{\text{DP}}^2 + \sigma_{\text{TIS}}^2 + \sigma_{\text{Match}}^2}, \tag{12.86}$$

where the standard deviations refer to dynamic precision (DP), TIS, and site-to-site matching, respectively. This definition assumes that each contributor is measured in a way that guarantees its independence from the other contributors, e.g. that TIS is measured by averaging over several dynamic measurements. The omission of average terms, such as TIS or average matching, is based on the assumption that these averages can be calibrated out. This may not always be either possible or acceptable in a specific environment. An example is the implementation of advanced process control (APC), with a high stress on data integrity [123]. In this situation, offsets between individual overlay metrology tools, treated as fully interchangeable, will affect the data integrity, and by feeding into the control loop, will affect the operating point of the process. Another example is the use of the sum of mean matching and 3σ of site-to-site matching compared to a fixed reference tool as a fast qualifier for a new tool. A more rigorous approach is offered by the gauge repeatability and reproducibility (GR&R) method, where—in the common GR&R terminology—each overlay tool is the Operator, and each specific measurement site is the Part.

12.3.6.4.2 Measurement Accuracy

Up to now, we have been discussing the effects of the metrology tool. However, as mentioned above, the overlay targets also have an effect on the metrology results. Wafer-induced shift (WIS) is a distortion of the topography of the overlay target, which in turn causes the apparent overlay to be different from the true overlay [124–126]. In 20 nm nodes and below this became the dominant error source in overlay metrology. As such, it is desired that the metrology tool setup (or recipe) is optimized to minimize the inaccuracy error. This can be done either by information coming from the metrology tool or by external knowledge such as reference metrology (similar to Section 12.3.6.3) or designated experiments [127–130]. This error handling is even more complicated since there are no universally accepted standards for the quantitative evaluation of overlay measurement inaccuracy.

12.3.6.5   Recent Trends in Overlay Metrology

12.3.6.5.1   In-Die Overlay

As explained above, the tool and mark–related contributors to overlay metrology uncertainty are readily quantified. These contributors have been relentlessly reduced by aggressive product development to keep pace with the ITRS roadmap. Despite this trend, the overlay model residuals, i.e. the distribution of the discrepancies between the model and the measured overlay, site by site, continue to be significantly larger than those directly explainable from the tool and mark–related sources. In many cases, the unmodeled contributors dominate the residuals [118]. Some studies indicate that the major constituents of the unmodeled errors originate at the individual lithographic field level as opposed to the wafer level [131]. The high residual problem stems from two effects. The first is mismatch between the linear terms in the standard overlay models in use today and the higher-order sources of systematic variation, across the exposure tool field. The second is field-to-field variability of the intra-field signature.

High-order intra-field overlay behavior has its roots primarily in two areas. Firstly, reticle write processes require absolute position control of the exposing beam at the nanometer level over the full reticle field area without drift over a time period long enough to enable the full reticle to be written. Despite the impressive progress of technology, higher-order spatial displacements are inevitable, and can contribute variations of the order of several nanometers over the full lithographic field [132]. Secondly, the lithographic process is itself limited by high-order field distortions, which can be partially removed from the lithography process by the practice of scanner dedication in which subsequent lithographic steps are performed on the same exposure tool. However, manufacturing productivity in the lithography cell can be adversely affected by this practice, creating the need to enable so called scanner mix and match methodologies [133, 134]. Under these conditions, the distortion characteristics of different scanners need to be matched to minimize the generation of high-order intra-field overlay which cannot be corrected for, or even characterized, by standard production sample plans and models. This situation is further exacerbated when matching is required between scanners from different vendors and/or imaging wavelengths. A further complication arises from the fact that in previous scanner generations overlay models allowed for only a single set of intra-field correctibles assumed to be constant for all fields. In step-and-scan systems, the intra-field correctibles can vary from field to field and this effect can only result in larger residuals.

The inevitable outcome of all of the above is the need to perform in-die overlay metrology, which includes multiple measurements within the field. On the scanner side, high-order field models are supported.

The requirements for metrology targets that are to be inserted into the die are considerably more stringent than for those in the scribe line. Since space is at a premium in the die, size reduction becomes of increased importance, but even more crucial is the requirement of process compatibility. The standard targets with their relatively large open areas (compared to the device) become a flagrant design rule violation and will literally be out of their depth in this new environment. Work toward process-compatible targets is continuous and must include adaptation as the process advances and changes [135, 136]. As part of this target design effort, metrology and process simulation tools were developed [137, 138]. These tools are now commonly used in order to provide optimal targets for a given layer and process. The next level in target design optimization is described in the following section.

12.3.6.5.2   Device-Correlated Metrology

Traditionally, as stated at the beginning of Section 12.3.6.1, the overlay is related to the lateral deviation of the layer-to-layer alignment from the intended, perfect, alignment. This statement has become challenging in two senses:

(1) Can a single control parameter represent multiple local shifts in a complex structure?
(2) Can the metrology tool provide a good measure of this control parameter?

If these two conditions are not met, the metrology does not provide reliable parameters required to control the process and provide good yield.

In past nodes, the overlay budget was a measure of the misalignment of two layers, as illustrated in Figure 12.70, A1 and A2 (perfectly aligned and somewhat misaligned layers, respectively). In the more complex multi-patterning processes, the new layer may be aligned to multiple previous patterns simultaneously. These previous patterns may already be misaligned with respect to each other. As a result, the new layer positioning may be a compensation for both alignments; in an even worse scenario, such a common location does not exist. The different alignments to different structures within a single layer are described using the term "edge placement error" (EPE). EPE is the local distance between two close edges, a parameter which is related to the electrical properties of the feature.

Part of the first challenge is described in Chapter 3: Multiple Patterning Lithography, and is demonstrated in Figure 12.70. It is illustrated in row B, where the dark gray and middle gray features were produced by different lithography steps. The EPE of the current white bar depends on the previous layer's lithography.

Another example is alignment to a pattern produced in self-aligned quadrupole patterning. This example is illustrated in row C of Figure 12.70. Due to the nature of this process, bars marked in different grayscale may have different locations, or different EPEs with respect to the top layer. The EPEs are direct functions of parameters related to the CD of deposited spacers and different etch step qualities in addition to the lithography imperfections.

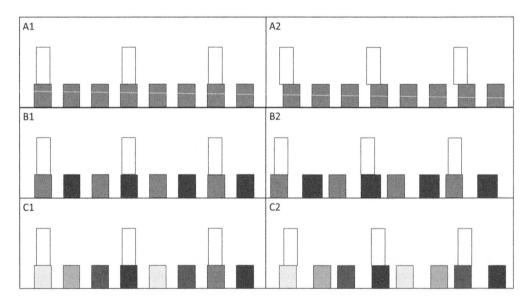

FIGURE 12.70   Edge placement error control challenge in advance process nodes. Cross sections of different alignment schemes are illustrated: The features in the top layers (which originated in a latter single lithography step) are to be aligned to the bottom features. The left column (A1, B1, C1) illustrates ideal cases (no OVL or CD errors); in the right column (A2, B2, C2) there are some OVL and CD errors in the process. Top line: ideal litho-litho alignment; intermediate line: litho-LELE alignment; bottom line: litho-SAQP alignment.

Traditionally, EPE was considered a lithography error and was corrected using optical proximity correction (OPC) and sub-resolution assist features (SRAF), which corrected the lithography step errors (described in Chapter 1: Lithography, Etch, and Silicon Process Technology). Newer OPC models may include etch effects [139, 140]. The self-aligned quadruple patterning (SAQP) example in Figure 12.70 demonstrates why EPE can no longer be solely solved by modification of the reticles (OPC, SRAF). Since it depends on additional process steps, it can no longer be solved based on lithography correction [125, 141, 142]. As a consequence, a single "overlay" can no longer effectively describe the full behavior of the patterns.

The second challenge mentioned earlier can be rephrased as: "Is the reported overlay identical to a specific EPE of interest?". In optical metrology, special targets are measured in order to predict the device behavior. Since EPE depends on both the feature geometry and the process environment of its production, manufacturing and measuring device-representative targets become a major challenge. Measuring may require new targets, algorithms, and hardware as well as new control schemes and simulators.

12.4 ADVANCED EQUIPMENT CONTROL FOR CRITICAL DIMENSION AND OVERLAY

12.4.1 CORRECTION SCHEME OF DOSE, FOCUS, AND OVERLAY (INTER- AND INTRA-FIELD)

In the semiconductor manufacturing process, stabilization controls against various factors of variations have been performed by adopting feedback and/or feed-forward, etc. Such control is called advanced process control (APC) in a wide sense and is called advanced equipment control (AEC) from the viewpoint of controlling equipment.

Metrology tools had been used in acquiring quality management data and determining the optimum dose, focus position, and overlay offset in order to increase the manufacturing yield as much as possible. Recently, ArF immersion lithography has been widely used under the 45 nm half-pitch node, but its process window is very narrow. The table in Figure 12.5 shows the required tolerance of the CD control and overlay control, which are approximately one-tenth and one-fifth of dynamic random-access memory (DRAM) half pitch, respectively. In parallel, inter- and intra-field controls of the exposure tool have been improved.

In order to obtain sufficient CD uniformity, the optimization of dose and focus controls is indispensable in determining the proper inter- and intra-field control parameters. In order to obtain sufficient overlay accuracy, the adjustments of the inter-field map on the wafer (grid) and intra-field distortion are very effective. Figure 12.71 shows the data flow from a metrology

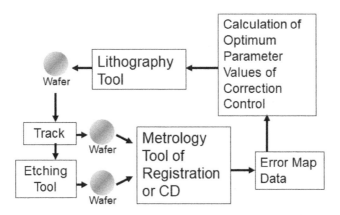

FIGURE 12.71 Data flow from metrology tool to lithography tool.

tool to an exposure tool. A wafer coated with resist is exposed by the lithography tool and is then developed in the track. Next, the developed wafer or the etched wafer is measured by the metrology tool. Then, the measured data are sent to a computer to calculate the optimum parameter values of the correction control. Subsequently, those parameter values are sent to the lithography tool.

12.4.2 DOSE-FOCUS INTRA-FIELD HIGHER ORDER CORRECTION FOR CD UNIFORMITY

C.P. Ausschnitt and T.A. Brunner reported a technique for the simultaneous measurement of dose, focus, and blur from scatterometry (optical critical dimension [OCD]) measurement data [143]. T. Toki et al. applied a similar technique for dose and focus in order to obtain better CD uniformity. Blur analysis is not included here because a flexible blur control is not realistic. The procedure for the improvement of CD uniformity is as follows [144]:

(1) Wafer exposure with dose/focus variation and optical critical dimension measurement.
(2) Modeling of (Dose, Focus) = NN(MCD, Height), (NN: neural network).
(3) Wafer exposure with nominal dose/focus and OCD measurement.
(4) Making error maps of dose and focus using modeled dose and focus functions.
(5) Determination of inter- and intra-field control parameters for dose and focus.
(6) Wafer exposure with optimized control parameter values.

Figure 12.72 shows neural network modeling using optical CD measurement data (MCD, Height) for the Dose/Focus matrix (parameter of curves in graphs are dose value [9.5–12.0 mJ/cm²,

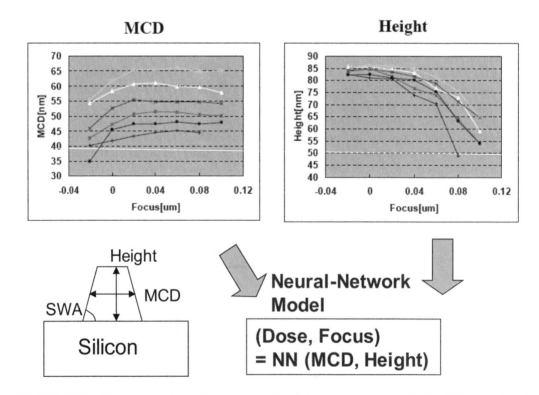

FIGURE 12.72 Neural network modeling using optical CD measurement data (MCD, Height) for Dose/Focus matrix (parameter of curves in graphs are dose value [9.5–12.0 mJ/cm², 0.5 mJ/cm² pitch]).

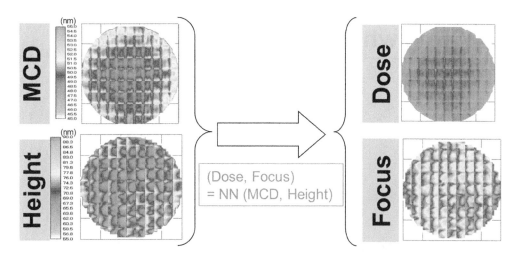

FIGURE 12.73 MCD and Height maps from optical CD measurements and Dose and Focus error maps obtained through neural network modeling. (From Figure 8 in T. Toki et al., "Simultaneous optimization of dose and focus controls in advanced ArF immersion scanners," *Proc. SPIE*, 7640, 764016 [2010].)

0.5 mJ/cm^2 pitch]). When a wafer is exposed with nominal dose and focus, MCD and Height maps are obtained by OCD measurements and then converted to Dose and Focus error maps through neural network modeling as shown in Figure 12.73. Consequently, MCD error maps in wafer and in the exposure field can be corrected by using Dose and Focus error maps as shown in Figure 12.74. A-L. Charley et al. reported a similar technique to obtain Dose and Focus error maps [145]. Focus and dosage control functions within field (intra-field) with a higher-order polynomial have been reported by T. Fujiwara et al. [146] and P. Vanoppen et al. [147], independently.

Similar attempts to determine Focus and Dose error maps using the SEM measurement results of top-CD and bottom-CD have been introduced by S. Hotta et al. [148]. Higher-order focus control is obtained by height and leveling controls at the wafer stage during scanning exposure. Dosage correction can be achieved by independent controls along scanning and cross-scanning directions. J. van Schoot et al. reported the effectiveness of pulse-by-pulse energy control of excimer laser light during scanning exposure and a gray filter with a varying profile of the slit width (usually several mm) of the static image field [149]. The former is related to the scanning direction and the latter is related to the cross-scanning direction.

12.4.3 WAFER GRID CORRECTION, INTRA-FIELD HIGHER-ORDER CORRECTION FOR OVERLAY

Step-and-scan exposure tools have a correction function of exposure positions on a wafer and intra-field distortion of each exposure field. Usually, those functions are expressed by a higher-order polynomial. Overlay data of the entire wafer obtained by a registration measurement tool are fitted to higher-order polynomials which express wafer grid and intra-field distortion of each exposure field. Then, the fitted parameter values are sent to the exposure tool for higher-order control. S. Wakamoto et al. reported those functions were effective in overlay, as shown in Figure 12.75 [150]. A. Gabor et al. also reported the effectiveness of wafer grid correction [151].

Recently, overlay accuracy has affected CD uniformity in the case of the pitch-split double-patterning process. Therefore, these correction functions have become increasingly important [152].

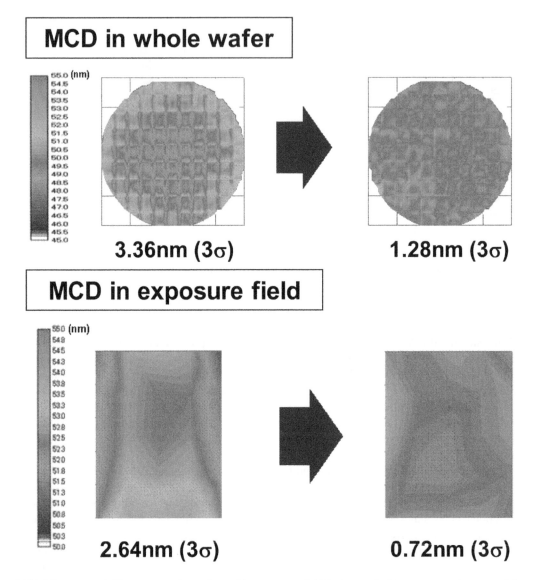

FIGURE 12.74 MCD error map in wafer and in exposure field before and after compensation control by Dose and Focus error maps. (From Figure 10 and Figure 11 in T. Toki et al., "Simultaneous optimization of dose and focus controls in advanced ArF immersion scanners," *Proc. SPIE*, 7640, 764016 [2010].)

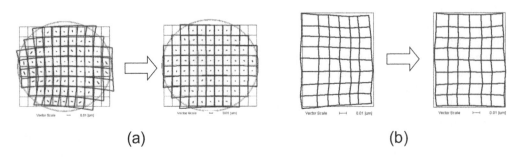

FIGURE 12.75   (a) Inter-field (wafer grid) correction. (b) Intra-field distortion correction.

ACKNOWLEDGMENTS

The basis of this chapter is "Critical-Dimensional Metrology for Integrated-Circuit Technology" in *Microlithography* (2nd edition). The authors have added new explanations and corrections to the original chapter.

The authors would like to thank H. Marchman, G. Lorusso, M. Adel and S. Yedur for their efforts as contributors in the previous edition and J. Seligson, Pavel Izikson, Daniel Kandel, J.C. Robinson, Kelly Barry, and S. DeMoor for cooperating with these contributors.

REFERENCES

1. C. Diaz, H. Tao, Y. Ku, A. Yen, and K. Yound, "An experimentally validated analytical model for gate line-edge roughness (LER) effects on technology scaling," *IEEE Electron Dev. Lett.*, 22, 287 (2001).
2. International Technology Roadmap for Semiconductors (ITRS), 2003 Edition, http://member.itrs.net.
3. B. Bunday, M. Bishop, D. McCormack, J. Villarrubia, A. Vladar, R. Dixson, T. Vorburger, and N. Orji, "Determination of optical parameters for CDSEM line edge roughness," *Proc. SPIE*, 5375, 515 (2004).
4. W.G. Lawrence, "Spatial frequency analysis of line edge roughness in nine chemically related photoresists," *Proc. SPIE*, 5039, 713 (2003).
5. H.P. Koh, Q.Y. Lin, X. Hu, and L. Chan, "Effect of process parameters on edge roughness of chemically amplified resists," *Proc. SPIE*, 3999, 240 (2000).
6. S. Masuda, X. Ma, G. Noya, and G. Pawlowski, "Lithography and line edge roughness of high activation energy resists," *Proc. SPIE*, 3999, 252 (2000).
7. M. Ercken, G. Storms, C. Delvaux, N. Vandenbroeck, P. Leunissen, and I. Pollentier, "Line edge roughness and its increasing importance," *Proc. Interface* (2002).
8. G.P. Patsis, V. Constantoudis, A. Tserepi, E. Gogolides, Grozdan Grozev, and T. Hoffmann, "Roughness analysis of lithographically produced nanostructures: Off-line measurement and scaling analysis," *Microelectron. Eng.*, 67–68, 319 (2003).
9. G.P. Patsis, V. Constantoudis, A. Tserepi, E. Gogolides, and G. Grozev, "Quantification of line edge roughness of photoresists," *J. Vac. Sci. Technol. B*, 21(3), 1008 (2003).
10. V. Constantoudis, G.P. Patsis, L.H.A. Leunissen, and E. Gogolides, "Line edge roughness and critical dimension variation: Fractal characterization and comparison using model functions," *J. Vac. Sci. Technol. B*, 22(4), 1974 (2004).
11. V. Constantoudis, G.P. Patsis, L.H.A. Leunissen, and E. Gogolides, "Toward a complete description of line width roughness: A comparison of different methods for vertical and spatial LER and LWR analysis and CD variation," *Proc. SPIE*, 5375, 967 (2004).
12. L.H.A. Leunissen, G.F. Lorusso, and T. Dibiase, "Full spectral analysis of line-edge roughness," *Proc. SPIE*, 5752, 499 (2005).
13. ASTM F1811-97, *Standard Practice for Estimating the Power Spectral Density Function and Related Finish Parameters from Surface Profile Data*. American Society for Testing and Materials, West Conshohocken, PA, 1997.
14. E. Marx, I.J. Malik, Y.E. Strausser, T. Bristow, N. Poduje, and J.C. Stover, "Power spectral densities: A study of different surfaces," *J. Vac. Sci. Technol. B*, 20(1), 31 (2002).
15. N.G. Orji, M.I. Sanchez, J. Raja, and T.V. Vorburger, "AFM characterization of semiconductor line edge roughness," In: *Applied Scanning Probe Methods*, B. Bhushan, H. Fuchs, and S. Hosaka, eds., Springer Verlag, Berlin, chap. 9, 2004.
16. W.H. Press, B.P. Flannery, S.A. Teukolsky, and W.T. Vetterling, *Numerical Recipes in C*, Cambridge University Press, Cambridge, 1988.
17. E.L. O'Neill and A. Walther, "A problem in the determination of correlation functions," *J. Opt. Soc. Am.*, 67(8), 1125 (1977).
18. E.R. Freniere, E.L. O'Neill, and A. Walther, "Problem in the determination of correlation functions. II," *J. Opt. Soc. Am.*, 69, 634 (1979).
19. G. Box, W. Hunter, and J. Hunter, *Statistics for Experimenters*, John Wiley and Sons, New York, 1978.
20. W. Banke and C. Archie, "Characteristics of accuracy for CD metrology," *Proc. SPIE*, 3677, 291 (1999).
21. J. Mandel, *The Statistical Analysis of Experimental Data*, Interscience Publishers, John Wiley & Sons, New York, 1964.
22. M. Sendelbach and C. Archie, "Scatterometry measurement precision and accuracy below 70nm," *Proc. SPIE*, 5038, 224 (2003).

23. S. Farsiu, D. Robinson, M. Elad, and P. Milanfar, "Advances and challenges in super-resolution," *JIST*, March 2004.

24. C.E. Shannon, "A mathematical theory of communication," *Bell Syst. Tech. J.*, 27(3), 379 (1948).

25. G.F. Lorusso and D.C. Joy, "Experimental resolution measurement in critical dimension scanning electron microscope metrology," *Scanning*, 25(4), 175 (2003).

26. S.A. Rishton, S.P. Beaumont, and C.W.D. Wilkinson, "Measurement of the profile of finely focused electron beams in a scanning electron microscope," *J. Phys. E: Sci. Instrum.*, 17(4), 296 (1984).

27. D.C. Joy, "The future of e-beam metrology : Obstacles and opportunities," *Proc. SPIE*, 4689, 1 (2002).

28. D.C. Joy, "Measurements of SEM parameters," *Proc. 7th SEM Symposium*, O. Johari, ed., IITRI, Chicago, pp. 327–34, 1974.

29. E. Kratschmer, S.A. Rishton, D.P. Kern, and T.H.P. Chang, "Quantitative analysis of resolution and stability in nanometer e-beam lithography," *J. Vac. Sci. Technol. B*, 6(6), 2074 (1988).

30. T.A. Dodson and D.C. Joy, "Fast Fourier transform techniques for measuring SEM resolution," In: *Proc. XIIth Int. Cong. EM*, Seattle, WA, G.W. Bailey, ed., San Francisco Press, San Francisco, CA, vol. 1, pp. 406–407, 1990.

31. S.J. Erasmus, D.M. Holburn, and K.C.A. Smith, "On-line computation of diffractograms for the analysis of SEM images," *Inst. Phys. Conf. Ser.*, 52, 73–77 (1980).

32. D.C. Joy, "SMART – A program to measure SEM resolution and performance," *J. Microsc.*, 208(1), 24 (2002).

33. C.B. Johnson, "Circular aperture diffraction limited modulation transfer functions," *App. Opt.*, 11, 1875–1876 (1975).

34. I.J. Rosenberg, "Evaluating the resolution of a CD-SEM," *Proc. SPIE*, 4689, 336 (2002).

35. E. Oho and K. Toyomura, "Strategies for optimum use of superposition diffractograms," *Scanning*, 23(5), 351 (2001).

36. T. Ishitani and M. Sato, "A method for personal expertise-independent evaluation of image resolution in scanning electron microscopy," *Scanning*, 24(4), 191 (2002).

37. J. Frank, *Computer Processing of Electron Microscope Images*, J. Frank and P. Hawkes, eds., Springer-Verlag, Berlin, pp. 197–233, 1980.

38. D.C. Joy, Y.-U. Ko, and J.J. Hwu, "Metrics of resolution and performance for CD-SEMs," *Proc. SPIE*, 3998, 108 (2000).

39. H.K. Youn and R.F. Egerton, "Resolution measurement of SEM using cross correlation function," *Proc. Microsc. Soc. Canada*, 24, 27–28 (1997).

40. M. Born and E. Wolf, *Principles of Optics*, 6th Ed., Pergamon Press, New York, p. 556, 1991.

41. D. Nyyssonen and R. Larrabee, "Submicrometer linewidth metrology in the optical microscope," *NIST J. Res.*, 92(3), 187 (1987).

42. D. Nyyssonen, "Theory of optical edge detection and imaging of thick layers," *J. Opt. Soc. Am.*, 72(10), 1425 (1982).

43. Leica CD 200, Wild Leitz USA, Inc., Rockleigh, NJ 07647.

44. Siscan 7325 Confocal Microscope, Siscan Systems, Campbell, CA 95008.

45. H. Marchman, S. Vaidya, C. Pierrat, and J. Griffith, "Metrology for phase-shifting masks", *J. Vac. Sci. Technol. B*, 11(6), 2482 (1993).

46. G. Kino, T. Corle, and G. Xiao, "The scanning optical microscope: An overview," *Proc. SPIE*, 897, 32 (1988).

47. KLA 5000 Coherence Probe Microscope, KLA Instruments, San Jose, CA 95161.

48. H.M. Marchman, Metrology section of the 1993 Sematech PSM Development Report.

49. C. Raymond, "Multiparameter grating metrology using optical scatterometry," *J. Vac. Sci. Technol. B*, 15(2), 361 (1997).

50. S. Naqvi, R. McNeil, S. Wilson, H. Marchman, and M. Blain, "Phase-shifting mask metrology using scatterometry", *Diffractive Optics*, 11, 342 (1994).

51. X. Niu, N. Jakatdar, J. Bao, and C.J. Spanos, "Specular spectroscopic scatterometry," *IEEE Trans. Semicond. Manuf.*, 14, 97 (2001).

52. X. Niu, N. Jakatdar, J. Bao, C.J. Spanos, and S.K. Yedur, "Specular spectroscopic scatterometry in DUV lithography," *Proc. SPIE*, 3677, 159 (1999).

53. K. Lensing, B. Stirton, B. Starnes, J. Synoradzki, B. Swain, and L. Lane, "A comprehensive comparison of spectral scatterometry hardware," *Proc. SPIE*, 5752, 337 (2005).

54. K. Barry, A. Viswanathan, X. Niu, and J. Bischoff, "Scatterometry for contact hole lithography," *Proc. SPIE*, 5375, 1081 (2004).

55. K. Barry, S. Cheng et al., "Integrated scatterometry for contact hole metrology and process monitoring," *Proceedings of the AEC/APC Symposium XVIII*, 2004.

56. K. Lee, S. Yedur, S. Henrichs, and M. Tavassoli, "CD-etch depth measurement from advanced phase-shift masks and wafers using optical scatterometry," *Proc. SPIE*, 6152, 61521P (2006).

57. S.M.G. Wilson, H.M. Marchman, S. Sohail, S.S. Naqvi, and J.R. McNeil, "Phase-shift mask metrology using scatterometry," *Proc. SPIE*, 2322, 305 (1994).

58. M. Sendelbach, A. Munoz, K.A. Bandy, D. Prager, and M. Funk, "Integrated scatterometry in high volume manufacturing for polysilicon gate etch control," *Proc. SPIE*, 6152, 61520F (2006).

59. H. Marchman and A. Novembre, "Near field optical imaging with the photon tunneling microscope," *Appl. Phys. Lett.*, 66(24), 3269 (1995).

60. J. Guerra, "Photon tunneling microscopy," *Appl. Opt.*, 29(26), 3741 (1990).

61. K. Komatsu, T. Omori, T. Kitamura, Y. Nakajima, A. Aiyer, and K. Suwa, "Automatic macro inspection system," *Proc. SPIE*, 3677, 764 (1999).

62. T. Kitamura, Y. Nakajima, H. Matsumoto, T. Omori, and K. Komatsu, "Automatic macro inspection system," *Proc. SPIE*, 3998, 615 (2000).

63. T. Omori, K. Fukazawa, T. Mikami, K. Yoshino, and Y. Yamazaki, "Novel inspection technology for half pitch 55 nm and below," *Proc. SPIE*, 5752, 174 (2005).

64. R. Fowler, "The restored electron theory of metals and thermionic formulae," *Proc. R. Soc. A*, 117(778), 549 (1928).

65. L. Reimer, *Image Formation in Low Voltage Scanning Electron Microscopy*, SPIE Optical Eng. Press v.TT12, Bellingham, WA, p. 15, 1993.

66. R. Fowler and L. Nordheim, "Electron emission in intense electric fields," *Proc. R. Soc. A*, 117, 173 (1928).

67. S. Saito, Y. Nakaizumi, H. Mori, and T. Nagatani, "A field emission SEM controlled by microcomputer", *J. Electron Microsc.*, 31, 378 (1982).

68. V. Cosslett, "Probe size and probe current in the scanning transmission electron microscope," *Optik*, 36, 85 (1972).

69. L. Reimer, *Image Formation in Low Voltage Scanning Electron Microscopy*, SPIE Optical Eng. Press v.TT12, Bellingham, WA, p. 22, 1993.

70. L. Reimer, *Image Formation in Low Voltage Scanning Electron Microscopy*, SPIE Optical Eng. Press v.TT12, Bellingham, WA, p. 27, 1993.

71. H. Rose, "Outline of a spherically corrected semiaplanatic medium-voltage transmission electron microscope," *Optik*, 85, 19 (1990).

72. O.L. Krivanek, N. Dellbya, and A.R. Lupinic, "Towards sub-A electron beams," *Ultramicroscopy*, 78, 1 (1999).

73. L. Reimer, *Scanning Electron Microscopy*, Springer-Verlag, Berlin, pp. 59–62, 1985.

74. W. Baumann and L. Reimer, "Comparison of the noise of different electron detection systems using a scintillator-photomultiplier combination," *Scanning*, 4, 141 (1981).

75. L. Reimer, *Image Formation in Low-Voltage Electron Microscopy*, SPIE Tutorials, Bellingham, WA, p. 64, 1993.

76. H.M. Marchman, J.E. Griffith, J.Z.Y. Guo, J. Frackoviak, and C.K. Celler, "Nanometer-scale dimensional metrology for advanced lithography," *J. Vac. Sci. Technol. B*, 12(6), 3585 (1994).

77. H.M. Marchman and J.E. Griffith, "Scanned probe microscope dimensional metrology," *Handbook of Silicon Semiconductor Metrology*, Alain C. Diebold, ed., Marcell Dekker, New York, 2001.

78. H. Marchman, "Electron beam based modification of lithographic materials and the impact on critical dimensional metrology," *Proc. SPIE*, 6152, 615227 (2006).

79. D. Sarid, *Scanning Force Microscopy*, 2nd Ed., Oxford University Press, New York, 1994.

80. T.R. Albrecht, S. Akamine, T.E. Carver, C.F. Quate, "Microfabrication of cantilever styli for the atomic force microscope," *J. Vac. Sci. Technol. A*, 8(4), 3386 (1990).

81. O. Wolter, Th. Bayer, and J. Greschner, "Micromachined silicon sensors for scanning force microscopy," *J. Vac. Sci. Technol. B*, 9(2), 1353 (1991).

82. J.P. Spatz, S. Sheiko, M. Moller, R.G. Winkler, P. Reineker, and O. Marti, "Process affecting the substrate in resonant tapping force microscopy", *Nanotechnology*, 6(2), 40–44 (1995).

83. Y. Martin and H.K. Wickramasinghe, "Method for imaging sidewalls by atomic force microscopy," *Appl. Phys. Lett.*, 64(19), 2498 (1994).

84. H.K. Wickramasinghe, "Scanned-probe microscopes," *Sci. Am.*, 261(4), 98 (1989).

85. G.L. Miller, J.E. Griffith, E.R. Wagner, and D.A. Grigg, *Rev. Sci. Instrum.*, 62(3), 705–709 (1991).

86. S.A. Joyce and J.E. Houston, "A new force sensor incorporating force-feedback control for interfacial force microscopy," *Rev. Sci. Instrum.*, 62(3), 710 (1991).

87. J.E. Griffith and G.L. Miller, U. S. Patent, No. 5,307,693 (1994).

88. J.E. Griffith, H.M. Marchman, G.L. Miller, and L.C. Hopkins, "Dimensional metrology with scanning probe microscopes," *J. Vac. Sci. Technol. B*, 13(3), 1100 (1995).

89. H. Yamada, T. Fujii, and K. Nakayama, "Linewidth measurement by a new scanning tunneling microscope," *Jpn. J. Appl. Phys.*, 28(1, No. 11), 2402 (1989).

90. J.E. Griffith, G.L. Miller, C.A. Green, D.A. Grigg, and P.E. Russell, "A scanning tunneling microscope with a capacitance-based position monitor," *J. Vac. Sci. Technol. B*, 8(6), 2023 (1990).

91. R.C. Barrett and C.F. Quate, "Optical scan-correction system applied to atomic force microscopy," *Rev. Sci. Instrum.*, 62(6), 1393 (1991).

92. H.M. Marchman, J.E. Griffith, and J.K. Trautman, "Optical probe microscope for nondestructive metrology of large sample surfaces," *J. Vac. Sci. Technol. B*, 13(3), 1106 (1995).

93. J.E. Griffith, H.M. Marchman, and L.C. Hopkins, "Edge position measurement with a scanning probe microscope," *J. Vac. Sci. Technol. B*, 12(6), 3567 (1994).

94. H.M. Marchman, J.E. Griffith, and R.W. Filas, "Fabrication of optical fiber probes for nanometer-scale dimensional metrology," *Rev. Sci. Instrum.*, 65(8), 2538 (1994).

95. M. Stedman, "Limits of topographic measurement by the scanning tunneling and atomic force microscopes," *J. Microsc.*, 152(3), 611 (1988).

96. M. Stedman and K. Lindsey, "Limits of surface measurement by stylus instruments," *Proc. SPIE*, 1009, 56 (1989).

97. K.L. Westra, A.W. Mitchell, and D.J. Thomson, "Tip artifacts in atomic force microscope imaging of thin film surfaces," *J. Appl. Phys.*, 74(5), 3608 (1993).

98. J.E. Griffith, H.M. Marchman, G.L. Miller, L.C. Hopkins, M.J. Vasile, and S.A. Schwalm, "Line profile measurement with a scanning probe microscope," *J. Vac. Sci. Technol. B*, 11(6), 2473 (1993).

99. M.J. Vasile, D. Grigg, J.E. Griffith, E. Fitzgerald, and P.E. Russell, "Scanning probe tip geometry optimized for metrology by focused ion beam ion milling," *J. Vac. Sci. Technol. B*, 9(6), 3569 (1991).

100. R.P. Feynman, R.B. Leighton, M. Sands, and S.B. Treiman, *The Feynman Lectures on Physics*, vol. 2, Addison-Wesley, Reading, MA, 1964.

101. B.I. Yakobson and R.E. Smalley, *Am. Sci.*, 85, 324 (1997).

102. M.M.J. Treacy, T.W. Ebbesen, and J.M. Gibson, "Exceptionally high Young's modulus observed for individual carbon nanotubes," *Nature*, 381(6584), 678 (1996).

103. H. Dai, J.H. Hafner, A.G. Rinzler, D.T. Colbert, and R.E. Smalley, "Nanotubes as nanoprobes in scanning probe microscopy," *Nature*, 384(6605), 147 (1996).

104. D. Nyysonen and R.D. Larrabee, "Submicrometer linewidth metrology in the optical microscope", *J. Res. Natl. Bur. Stand.*, 92(3), 187–204 (1987).

105. M.T. Postek and D.C. Joy, "Submicrometer microelectronics dimensional metrology: scanning electron microscopy", *J. Res. Natl Bur. Stand.*, 92(3), 205–228 (1987).

106. E.H. Anderson, V. Boegli, M.L. Schattenburg, D. Kern, and H.I. Smith, "Metrology of electron-beam lithography systems using holographically produced reference samples," *J. Vac. Sci. Technol. B*, 9(6), 3606 (1991).

107. J.E. Griffith, D.A. Grigg, M.J. Vasile, P.E. Russell, and E.A. Fitzgerald, "Characterization of scanning probe microscope tips for linewidth measurement," *J. Vac. Sci. Technol. B*, 9(6), 3586 (1991).

108. D.A. Grigg, P.E. Russell, J.E. Griffith, M.J. Vasile, and E.A. Fitzgerald, "Probe characterization for scanning probe metrology," *Ultramicroscopy*, 42–44, 1616 (1992).

109. H.M. Marchman, "Nanometer-scale dimensional metrology with noncontact atomic force microscopy," *Proc. SPIE*, 2725, 527 (1996).

110. International Technology Roadmap for Semiconductors (ITRS), *Metrology_2013Tables, MET3 Lithography Metrology (Wafer)Technology Requirements*, 2013.

111. *Specification for Overlay-Metrology Test Patterns for Integrated Circuit Manufacture*, Semiconductor Equipment and Materials International, Mountain View, CA, 1996 (SEMI P28-96).

112. N.M. Felix, A.H. Gabor, V.C. Menon, P.P. Longo, S.D. Halle, C.-S. Koay, and M.E. Colburn, "Overlay improvement roadmap: Strategies for scanner control and product disposition for 5-nm overlay," *Proc. SPIE*, 7971, 79711D (2011).

113. W. Arnold, "Toward 3nm overlay and critical dimension uniformity: An integrated error budget for double patterning lithography," *Proc. SPIE*, 6924, 692404 (2008).

114. M. Adel, J.A. Allgair, D.C. Benoit, M. Ghinovker, E. Kassel, C. Nelson, J.C. Robinson, and G.S. Seligman, "Performance study of new segmented overlay marks for advanced wafer processing," *Proc. SPIE*, 5038, 453 (2003).

115. W. Yang, R. Lowe-Webb, S. Rabello, J. Hu, J. Lin, J.D. Heaton, M.V. Dusa, A.J. den Boef, M. van der Schaar, and A. Hunter, "Novel diffraction-based spectroscopic method for overlay metrology," *Proc. SPIE*, 5038, 200 (2003).

116. H. Huang, G. Raghavendra, A. Sezginer, K. Johnson, F.E. Stanke, M.L. Zimmerman, C. Cheung, M. Miyagi, and B. Singh, "Scatterometry-based overlay metrology," *Proc. SPIE*, 5038, 126 (2003).

117. M. Adel, M. Ghinovker, B. Golovanevsky, P. Izikson, E. Kassel, D. Yaffe, A.M. Bruckstein, R. Goldenberg, Y. Rubner, and M. Rudzsky, "Optimized overlay metrology marks: Theory and experiment," *IEEE Trans. Semicond. Manuf.*, 17(2) (May 2004).

118. S. Gruss, A. Teipel, C. Fülber, E. Kassel, M. Adel, M. Ghinovker, and P. Izikson, "Test of a new sub 90 nm DR overlay mark for DRAM production," *Proc. SPIE*, 5375, 881 (2004).

119. A. Ueno, K. Tsujita, H. Kurita, Y. Iwata, M. Ghinovker, J.M. Poplawski, E. Kassel, and M.E. Adel, "Improved overlay metrology device correlation on 90-nm logic processes," *Proc. SPIE*, 5375, 222 (2004).

120. R.M. Silver, M. Stocker, R. Attota, M. Bishop, J. Jun, E. Marx, M. Davidson, and R. Larrabee, "Calibration strategies for overlay and registration metrology," *Proc. SPIE*, 5038, 103 (2003).

121. A.C. Diebold, "Silicon semiconductor metrology," In: *Handbook of Silicon Semiconductor Metrology*, Alain C. Diebold, ed., Marcel Dekker, New York, p. 4, 2001.

122. A.F. Plambeck, "Overlay metrology as it approaches the gigabit era," *Microlith. World*, Winter, pp. 17–22 (1996).

123. C. Gould, "Advanced process control: Basic functionality requirements for lithography," *IEEE/SEMI Advanced Semiconductor Manufacturing Conference*, pp. 49–53, 2001.

124. S. Bae, Y. Kim, K. Park, J. Kim, W. Lee, S. Lee, and D. Lee, "The reduction of wafer scale error between DI and FI in multi-level metallization by adjusting edge detection method," *Proc. SPIE*, 3998, 460 (2000).

125. M. Ruhm, B. Schulz, E. Cotte, R. Seltmann, and T. Hertzsch, "Overlay leaves litho: Impact of non-litho processes on overlay and compensation," *Proc. SPIE*, 9231, 92310O (2014).

126. D. Kandel, V. Levinski, N. Sapiens, G. Cohen, E. Amit, D. Klein, and I. Vakshtein, "Overlay accuracy fundamentals," *Proc. SPIE*, 8324, 832417 (2012).

127. B.-H. Ham, S. Yun, M.-C. Kwak, S.M. Ha, C.-H. Kim, and S.-W. Nam, "New analytical algorithm for overlay accuracy," *Proc. SPIE*, 8324, 83240A (2012).

128. S.C.C. Hsu, et al., "Innovative fast technique for overlay accuracy estimation using archer self-calibration (ASC)," *Proc. SPIE*, 9050, 90501N (2014).

129. W.J. Tzai, et al., "Techniques for improving overlay accuracy by using device correlated metrology targets as reference," *J. Micro/Nanolith. MEMS MOEMS*, 13(4), 041412 (2014).

130. T.-B. Chan, A.A. Kagalwalla, and P. Gupta, "Measurement and optimization of electrical process window," *J. Micro/Nanolith. MEMS MOEMS*, 10, 013014 (2011).

131. A. Frommer, E. Kassel, P. Izikson, M. Adel, P. Leray, and B. Schulz, "In field overlay uncertainty contributors," *Proc. SPIE*, 5752, 51 (2005).

132. M. Tanaka, H. Ito, H. Takahashi, K. Oonuki, H. Kadowaki, H. Sato, H. Kawano, Z. Wang, K. Mizuno, and G. Matsuoka, "Improved image placement performance of HL-7000M," *Proc. SPIE*, 5256, 646 (2003).

133. S.J. DeMoor, J.M. Brown, J.C. Robinson, S. Chang, and C. Tan, "Scanner overlay mix and match matrix generation: Capturing all sources of variation," *Proc. SPIE*, 5375, 66 (2004).

134. S. Baek, A. Wei, D.C. Cole, G. Nellis, M.S. Yeung, A.Y. Abdo, R.L. Engelstad, M. Rothschild, and M. Switkes, "Simulation of the coupled thermal/optical effects for liquid immersion nanolithography," *J. Microlith. Microfab., Microsyst.*, 4, 013002 (2005).

135. R. Attota, R. Silver, M. Bishop, E. Marx, J.J. Jun, M. Stocker, M.P. Davidson, and R.D. Larrabee, "Evaluation of new in-chip and arrayed line overlay target designs," *Proc. SPIE*, 5375, 395 (2004).

136. Yi-Sha Ku, Chi-Hong Tung, and Nigel P. Smith, "In-chip overlay measurement by existing bright-field imaging optical tools," *Proc. SPIE*, 5752, 438 (2005).

137. G. Ben-Dov et al., "Metrology target design simulations for accurate and robust scatterometry overlay measurements," *Proc. SPIE*, 9778, 97783B (2016).

138. Y.-S. Kim et al., "Improving full-wafer on-product overlay using computationally designed process-robust and device-like metrology targets," *Proc. SPIE*, 9424, 942414 (2015).

139. C. Tabery, H. Morokuma, A. Sugiyama, and L. Page, "Evaluation of OPC quality using automated edge placement error measurement with CD-SEM," *Proc. SPIE*, 6152, 61521F (2006).

140. J. Sturtevant, R. Gupta, S. Shang, V. Liubich, and J. Word, "Characterization and mitigation of relative edge placement errors (rEPE) in full-chip computational lithography," *Proc. SPIE*, 9635, 963505 (2015).

141. Y. Borodovsky, "EUV lithography at insertion and beyond," *International Workshop on EUV Lithography*, Maui, HI, 2012.

142. P. Zhang, C. Hong, and Y. Chen, "A generalized edge-placement yield model for the cut-hole patterning process," *Proc. SPIE*, 9052, 90521Q (2014).

143. C.P. Ausschnitt and T.A. Brunner, "Distinguishing dose, focus and blur for lithography characterization and control," *Proc. SPIE*, 6520, 65200M (2007).

144. T. Toki, P. Izikson, J. Kosugi, N. Sakasai, K. Saotome, K. Suzuki, D. Kandel, J.C. Robinson, and Y. Koyanagi, "Simultaneous optimization of dose and focus controls in advanced ArF immersion scanners," *Proc. SPIE*, 7640, 764016 (2010).

145. A.-L. Charley, K. D'have, P. Leray, D. Laidler, S. Cheng, M. Dusa, P. Hinnen, and P. Vanoppen, "Focus and dose de-convolution technique for improved CD control of immersion clusters," *Proc. SPIE*, 7638, 763808 (2010).

146. T. Fujiwara, T. Toki, D. Tanaka, J. Kosugi, T. Susa, N. Sakasai, and A. Tokui, "Advanced CDU control for 22 nm node and below," *Proc. SPIE*, 7973, 797310 (2011).

147. P. Vanoppen, T. Theeuwes, H. Megens, H. Cramer, T. Fliervoet, M. Ebert, and D. Satriasaputra, "Lithographic scanner stability improvements through advanced metrology and control," *Proc. SPIE*, 7640, 764010 (2010).

148. S. Hotta, T. Brunner, S. Halle, K. Hitomi, T. Kato, and A. Yamaguchi, "Dose-focus monitor technique using a critical-dimension scanning electron microscope and its application to local variation analysis," *J. Micro/Nanolith. MEMS MOEMS*, 11(4), 043011 (2012).

149. J. van Schoot, O. Noordman, P. Vanoppen, F. Blok, D. Yim, C.-H. Park, B.-H. Cho, T. Theeuwes, and Y.-H. Min, "CD uniformity improvement by active scanner corrections," *Proc. SPIE*, 4691, 304 (2002).

150. S. Wakamoto, Y. Ishii, K. Yasukawa, A. Sukegawa, S. Maejima, A. Kato, J.C. Robinson, B.J. Eichelberger, P. Izikson, and M. Adel, "Improved overlay control through automated high order compensation," *Proc. SPIE*, 6518, 65180J (2007).

151. A. Gabor, B. Liegl, M. Pike, E. Hwang, and T. Wiltshire, "The GridMapper challenge: How to integrate into manufacturing for reduced overlay error," *Proc. SPIE*, 7640, 764015 (2010).

152. A.J. Hazelton, S. Wakamoto, S. Hirukawa, M. McCallum, N. Magome, J. Ishikawa, C. Lapeyre, I. Guilmeau, S. Barnola, and S. Gaugiran, "Double-patterning requirements for optical lithography and prospects for optical extension without double pattering," *J. Micro/Nanolith. MEMS MOEMS*, 8, 011003 (2009).

13 Directed Self-Assembly of Block Copolymers

Chi-chun Liu, Kenji Yoshimoto, Juan de Pablo, and Paul Nealey

CONTENTS

13.1 INTRODUCTION

Directed self-assembly (DSA) of block copolymers (BCPs) is an alternative patterning technique to complement and extend optical lithography. DSA utilizes a topographical or chemical guiding pattern (GP) to direct the BCPs into a desired morphology at a pre-determined location while the material properties of the BCPs control the feature size and uniformity of the resulting structures. Such a technique has drawn great attention to semiconductors, hard disk drive (HDD), and memory devices such as dynamic random-access memory (DRAM) and non-volatile memory due to its capability for pattern density multiplication and pattern quality rectification. Recent studies on 193i/high-volume manufacturing (HVM) compatibility, defectivity, and device demonstration of DSA further reinforce its role as a potential candidate for lithography extension rather than merely a lab-scale nanofabrication method [1–5]. In its 2015 reports, the International Technology Roadmap of

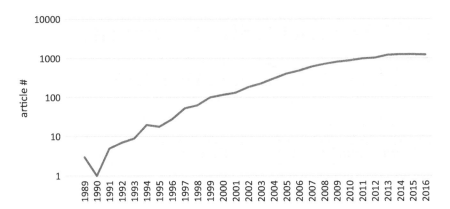

FIGURE 13.1 Number of "self-assembly" and "block copolymer"-related journal articles published from 1989 to 2016. (From Web of Science.)

Semiconductors (ITRS) also includes DSA as one of the potential lithography solutions, in addition to maskless, nanoimprint, and extreme ultraviolet lithography (EUV) [6].

The self-assembly of block copolymers in bulk and in thin films has been studied for decades [7–12]. The use of self-assembled block copolymers as a template for pattern transfer was first introduced as "block copolymer lithography" in the literature in 1997 [8]. Block copolymer lithography can achieve periodic patterns at the length scale of 5–50 nm over a large area at a relatively low cost without using any advanced photolithography patterning tools [13] Since the early 1990s, the study of the self-assembly of BCPs has grown exponentially, as shown in Figure 13.1. The number of publications started to plateau around 2010–2011, implying that the DSA technology had gradually transitioned from the "research" phase into the "development" phase. Nevertheless, as some of the DSA applications have matured and are being perfected in the field, there are yet many other exciting new areas not fully explored.

The aim of this chapter is to build the foundations for readers to understand, implement, and potentially further evolve the DSA work in which they are interested. The first half of the chapter focuses on fundamental knowledge such as the polymer physics of self-assembly, materials and processes commonly used in directed self-assembly, and the basics of numerical models for DSA. The second half of the chapter then focuses on the practical perspectives of DSA, such as pattern transfer, applications, and potential challenges.

13.1.1 SELF-ASSEMBLY OF BLOCK COPOLYMERS

For those who are not familiar with polymer chemistry, here is a short review of the terms that will be used throughout the chapter. Polymers are composed of one or more repeating units, i.e. monomers. A homopolymer is a polymer made of only one type of monomer, for example, a polystyrene homopolymer is made of styrene monomers, as shown in Figure 13.2. A copolymer refers to a polymer composed of more than one monomer. Given a simple linear copolymer made of A and B monomers, based on the sequence of monomers in the chain, which is determined by the polymer synthesis method [14], one could imagine that there exist several different types of copolymers. Two commonly seen examples are the random copolymer and the block copolymer: the former has no specific order for the monomer position in the chain while the latter has all the same type of monomers grouped together. Furthermore, diblock is the simplest form of block copolymer structure while triblock, such as ABA or ABC, multi-block, or non-linear block copolymers can be synthesized as well. In this chapter, we will only discuss random and diblock copolymers, thus the term "block copolymer" (or BCP) will simply refer to the diblock copolymer.

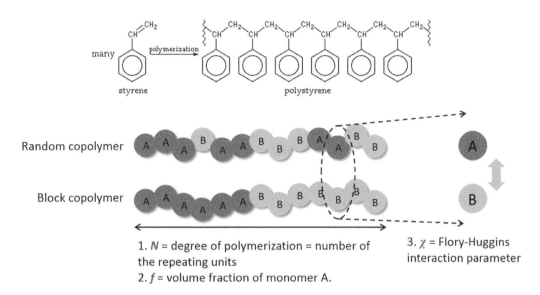

1. N = degree of polymerization = number of the repeating units
2. f = volume fraction of monomer A.

3. χ = Flory-Huggins interaction parameter

FIGURE 13.2 Commonly used polymer terminology.

BCPs with two chemically distinct blocks can phase segregate into a variety of micro-structures under certain conditions. Phase segregation or self-assembly of a BCP melt can be understood as a physical phenomenon of chain configuration rearrangement. The driving force behind this phenomenon is the free energy penalty of mixing two distinct polymers. Because in most cases the interfacial energy between two chemically distinct blocks is larger than the interfacial energy between the same kind, it is favorable to reduce the interfacial area between different kinds of blocks, hence, micro-domains of A and B will form. Because the two blocks on a BCP chain are covalently bound together, the natural size and period of the micro-domains are limited and will have a certain correlation with the molecular weight of the blocks. Furthermore, the morphology of the resulting micro-domains is determined by three key parameters: degree of polymerization N, volume fraction of one of the blocks f, and Flory–Huggins monomer–monomer interaction parameter χ (Greek letter *chi*) [15]. N and f are controlled by polymer synthesis, while χ is determined by the choice of monomer pair. Typically, χ has a weak temperature dependency, such as $\chi = \frac{\alpha}{T} + \beta$, where α is usually >0 meaning χ decreases with temperature.

There are a few criteria for phase segregation to occur: (1) polymer chains need to have sufficient mobility, for example, heated above glass transition temperature (T_g); and (2) the reduction in interfacial energy outweighs the increase in entropy-related free energy, which means the BCP melt could stay in disordered state if the driving force is not sufficiently strong. The product of χ and N can be seen as an approximation of the interfacial energy per unit area times the interfacial area of a BCP chain, which represents the interfacial energy penalty when placing a BCP chain at the interface of micro-domains, but the two blocks are surrounded by different kind of polymers, such as illustrated in the first drawing in Figure 13.3. Thus, χN is commonly used as an indicator of the driving force for phase segregation. It is found that the order–disorder transition (ODT) occurs at $\chi N \sim 10.5$ for symmetric ($f=0.5$) diblock copolymers. When $\chi N \gg 10.5$, it is known as the strong segregation limit (SSL) regime, where the AB interface is sharp and the concentration profile of A or B across the interface is like a square wave. The width of the interface scales with $\chi^{-1/2}$ and the period of the micro-domain $L_0 \propto N^{2/3}\chi^{1/6}$. When $\chi N > 10.5$ but still in the vicinity of ODT, it is referred to as the weak segregation limit (WSL) regime where the A–B interface becomes more diffuse compared to SSL and the period of micro-domain is $L_0 \propto N^{1/2}$, i.e. no longer χ dependent. The concentration profile in the WSL is closer to a sine wave than to a square wave compared to the

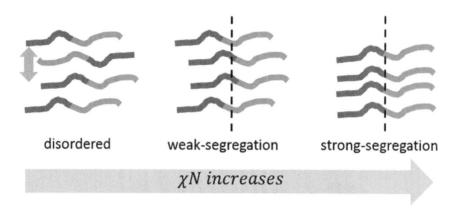

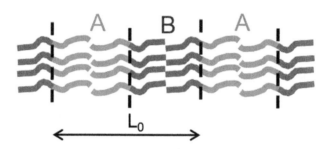

FIGURE 13.3 An illustration of the phase segregation behavior of BCPs.

profile in SSL. One should note that in either SSL or WSL, the physical dimension of the micro-domains, L_0, is mainly, if not solely, determined by N.

Furthermore, the equilibrium morphology of self-assembled BCPs will be determined by χN and f of the BCP chain. Phase diagrams based on these parameters have been constructed using simulations (Figure 13.4b) and experiments (e.g. Figure 13.4c). For a given χN, four distinct morphologies can usually be observed in a bulk A–B BCP (composed of monomers A and B) as the volume fraction of monomer A, f_A, increases: body-centered cubic spheres (S), hexagonally packed cylinders (C), bicontinuous gyroids (G), and lamellae (L). As f_A keeps increasing and block B becomes the minority, reversed gyroids (G′), reversed cylinders (C′), and reversed spheres (S′) will form. Sometimes, metastable structures, such as the perforated lamellae (PL) in Figure 13.4c, can be observed experimentally. One should note that these morphologies are observed in bulk polymer without boundary effect. In the case of the morphology of thin films of BCP, the interfacial energy between each block and the underlying substrate as well as the free surface at the top of the film cannot be ignored. As a result, additional BCP phases may appear [16].

Here, we use a simple case to demonstrate the boundary effect on thin film BCP self-assembly. Given a thin film of lamella-forming BCP coated on a substrate that is preferential to block B while the polymer–air surface is A-preferential, this system will self-assemble upon annealing and form lamellae structures parallel to the substrate, as shown in Figure 13.5. Depending on the initial film thickness, h, one may observe rounded, micron-size features with a quantized film height. These features result from the rearrangement of the film thickness in order to minimize the total free energy of the system by minimizing the area of unfavorable interfaces, such as A–substrate and B–air. As illustrated in Figure 13.5, apparently the ideal film thickness for such an "asymmetrical wetting" condition, i.e. when the free surface and substrate have different preferences for the two blocks in the BCP, is $h = \left(n + \dfrac{1}{2}\right)L_0$ where n is an integer. If the initial h does not satisfy this criterion, e.g.,

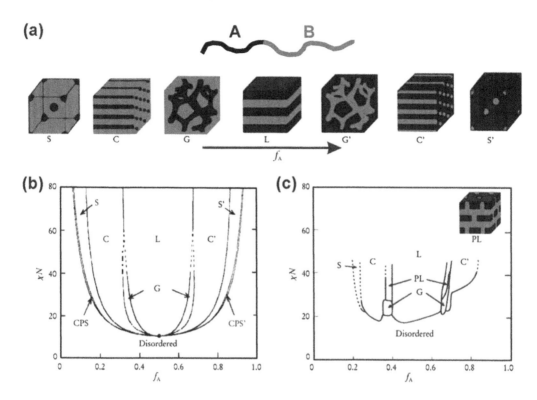

FIGURE 13.4 Morphology as a function of volume fraction of block A (f_A) of self-assembled *AB* diblock copolymers in bulk. (a) Three-dimensional schematics of the different morphologies. (b) Theoretical phase diagram. (c) Experimental phase diagram. (Reproduced from Ref. [17] with the permission of the American Institute of Physics.)

$h = \left(n + \dfrac{1}{2} + r\right)L_0$ where r is a real number between 0 and 1, the excess material, $r*L_0*a2$, would be rearranged into a structure with a smaller area of $a1$ and a height of L_0, so that $h = \left(n + \dfrac{1}{2}\right)L_0$ will hold locally at both thinner and thicker portions of the film. One can further derive that $a1/a2 = r$, which means the initial coating thickness would determine the surface topography after the self-assembly. When $r < 0.5$, the continuous phase observed with atomic force microscopy (AFM) could be either lower than the protruding structures, like islands, or the opposite "hole"-like structures will be observed when $r > 0.5$, as illustrated in Figure 13.5. This type of self-assembled structure is commonly referred to as the "islands/holes" structure [18, 19]. One can further consider the case of "symmetrical wetting," i.e. the substrate–polymer and air–polymer interface have the same preference, the quantized film thickness condition now becomes $h = n*L_0$. If the air–polymer preference is known, the substrate preference can be determined from an "islands/holes" test using a few samples with various known BCP film thickness. Such experiments can also estimate L_0, although it may not provide as much precision as other techniques. We will further discuss the case in which the substrate is not preferential to either block in Section 13.2.2.

13.1.2 BLOCK COPOLYMER LITHOGRAPHY AND ORIENTATION CONTROL

The term "block copolymer lithography" was first used by Park et al. and refers to the use of self-assembled BCPs as an etch template for pattern transfer [20]. A self-assembled BCP in thin film was noticed for its capability to achieve nanometer-scale, highly uniform arrays without using any

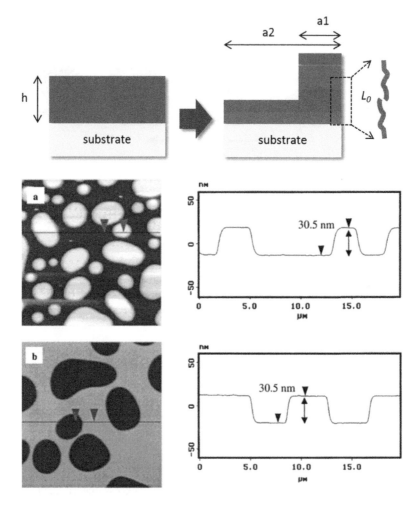

FIGURE 13.5 Example of islands/holes structures resulted from a thin film of self-assembled BCP over a preferential surface. (Reprinted with permission from Ref. [19]. Copyright (2000) American Chemical Society.)

advanced lithography tool. The first step to convert a BCP thin film into an etch template is the selective removal of one of the blocks. Park et al. demonstrated two approaches to utilize assembled spheres and parallel cylinders of polystyrene-b-polybutadiene (PS-b-PB) on silicon nitride substrate: (1) selectively remove the PB block with ozone and then use the remaining PS as an etch template, or (2) decorate the PB block with osmium tetroxide (OsO4) to enhance the etch resistance and then selectively remove the PS block. Although metal infusion could be a promising approach to enhance etch contrast, one should note that OsO4 is highly toxic and not suitable for large-scale use. Inorganic material can also be added to one of the blocks during the BCP synthesis process, so that the etch contrast can be inherently achieved in O2-based plasma. Cheng et al. demonstrated the assembly and etching of an iron/Si-containing, sphere-forming BCP, PS-b-polyferrocenyldimethylsilane (PS-b-PFS), and the subsequent pattern transfer, as shown in Figure 13.6 [21, 22]. The use of an inorganic block in the BCP can make the post-assembly metal incorporation process unnecessary. However, organic/organic BCPs are, in general, easier to assemble because the difference in surface energy between the blocks is usually larger when inorganic atoms are involved. Furthermore, when metal atoms/ions are involved, one would need to consider the compatibility with the intended application. For example, non-Si metal, e.g. Os and Fe, are mostly prohibited in front-end-of-the-line processes

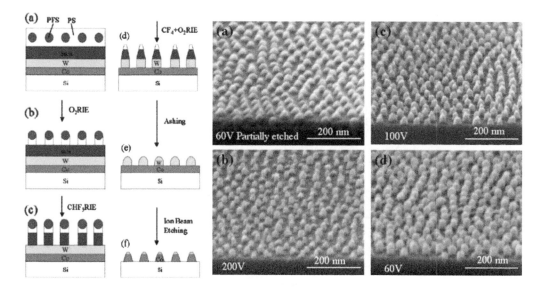

FIGURE 13.6 Example of block copolymer lithography using an inorganic block copolymer. (Reprinted with permission from Ref. [21]. Copyright (2001) John Wiley and Sons.)

in semiconductor manufacturing. Consequently, Si-containing BCP systems are investigated more than other metal-containing BCP systems.

Park et al. and Cheng et al. utilized sphere- and cylinder-forming BCPs to create an array of dots and lines, respectively, simply because those are the most natural morphologies in thin film when the substrate does not have any additional treatment, i.e. is preferential to one of the blocks. The thin film morphologies for different BCP systems on a preferential substrate are illustrated in the first column of Figure 13.7. However, from the pattern transfer point of view, it is preferred to have an etch template with a cross-sectional profile vertical to the substrate instead of a rounded profile as in the previous cases. In addition, the aspect ratio of the template, i.e. height divided by critical dimension (CD), in the case of parallel cylinders is only 0.5, which will impose challenges in the subsequent etch process. Thus, the preferred etch template for a hexagonal dot array would be perpendicular cylinders instead of spheres. Similarly, an ideal template for an array of lines would be perpendicular lamellae instead of parallel cylinders, as illustrated in the column "Neutral Wetting" of Figure 13.7. In the following paragraphs, we will discuss how to achieve perpendicular morphologies with a non-preferential "neutral" surface and what "neutral" means from a free energy point of view.

Surface modification for hydrophobicity control has been studied for several decades. A self-assembled monolayer (SAM) is one of the commonly used techniques [19, 23, 24]. SAM is typically a linear polymer with 10–20 carbons in the backbone and a functional group on one end of the chain. The end-functional group is designed to anchor the polymer onto the surface and is thus surface chemistry specific. For example, thiol and disulfide can be used for Au-coated surfaces and trichlorosilane can be used for Si surfaces. Because there is only one anchoring group per chain, only one monolayer can be bound onto the surface while excess materials can be removed by solvent rinsing. Furthermore, the surface property of a SAM-modified substrate, e.g. water contact angle, can be modulated by the chemistry of the polymer chain and the grafting density. For example, Whitesides et al. demonstrated that the incorporation of hydroxyl groups in an alkyl SAM polymer chain will lower the water contact angle. The same group of researchers also demonstrated the use of a blend of two types of SAMs to fine-tune the surface property. Nealey et al. utilized the kinetics of SAM formation to employ process time as a method of controlling the grafting density of SAMs.

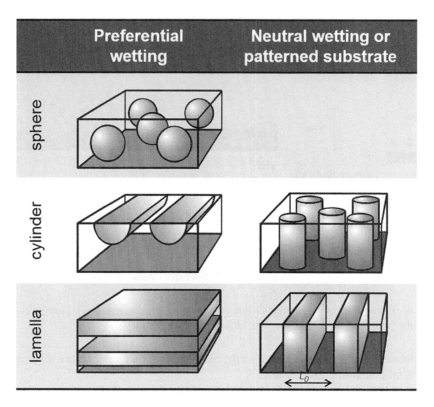

FIGURE 13.7 Different morphologies of sphere-, cylinder-, and lamella-forming BCPs in thin film when the substrate is preferential (left column) or non-preferential (right column).

A generalized surface modification method was introduced by Mansky et al. [25]. Their system was composed of a random copolymer of styrene and methyl methacrylate (MMA) with a hydroxyl group at the end, sometimes denoted as P(S-r-MMA)-OH and referred to as a "random copolymer brush." The brush was "grafted" onto an Si substrate surface through a condensation reaction between the hydroxyl group on the polymer chains and the native oxide on the substrate. By adjusting the styrene fraction in the random copolymer, a non-preferential surface for the lamella-forming PS-b-PMMA system was achieved. This system has several advantages over a blend of SAMs. First, because the random copolymer uses the same components as the BCP, the existence of a non-preferential composition can almost be ensured. Second, a brush molecule has ~10–20 times higher molecular weight compared to a SAM molecule, hence, the brush system usually has better surface coverage, especially when the substrate has defects and a lower density of grafting sites. Third, if not carefully controlled, a blend system could have a micellular structure, causing local compositional variation. In contrast, random copolymer chains will mix well and have no such concerns. On the other hand, the composition variation and molecular weight distribution of the random copolymer rely on the polymer synthesis method employed. Mansky et al. utilized "living free radical" polymerization which can provide good polydispersity control compared to conventional polymerization approaches. In addition, the brush system inherits a few good properties from SAM, such as conformality and being self-limiting.

Evolved from the end-functionalized brush system, a variety of surface modification systems were reported in order to address issues such as grafting kinetics, surface selectivity, and ease of synthesis [26, 27]. One of the variants was to incorporate a cross-linker in the random copolymer in addition to styrene and MMA, so that the polymer chains can react with each other and form a network. In this case, the anchoring group at the chain end could be optional. This type of self-cross-linking polymer material is usually referred to as a "mat," whose final thickness after rinsing is determined by the initial coating thickness and the cross-linking density.

Because the random copolymer has only two components (styrene and MMA in the case of P(S-r-MMA)-OH), the surface energy of the brush-grafted substrate can simply be described by the styrene fraction, f_{st}. One should note that the "surface energy" represents the interfacial energy between the brush surface and air. When studying a non-preferential or "neutral" surface for BCP, it is more accurate and important to consider the brush–PS and brush–PMMA interfacial energy instead. Furthermore, because the interfacial energy between the same units is zero, when $f_{st}=0$ (pure PMMA) or $f_{st}=1$ (pure PS), the brush would be PMMA- or PS-preferential, respectively. One can further infer that there exists a certain f_{st} that will result in identical brush–PS and brush–PMMA interfacial energy, resulting in a non-preferential condition of the brush substrate to the two blocks of the BCP.

While the explanation offered in the previous paragraph is helpful, it is oversimplified for understanding the self-assembly of a BCP. A more comprehensive way to understand the BCP behavior is the thermodynamics equilibration viewpoint, which describes the self-assembly of a thin film of BCP as a thermodynamic equilibration process affected by the given boundary conditions, e.g. the interfacial and surface energies at the annealing temperature, the interfacial area, and the intrinsic properties of the block copolymer itself. As a consequence of the equilibration process, the morphology that has the minimum total free energy will be the equilibrium structure. However, if several structures have comparable total free energies, a mixture of those morphologies may coexist and be observed randomly at different locations. Based on this viewpoint, an effective way to control the self-assembly is to create a large enough difference in the total free energy between the desired morphology and other possible morphologies so that only the desired morphology will form. According to thermodynamic principles, the larger the free energy difference, the lower the chance that undesired structures will occur, and also the faster the kinetics will be to achieve the most stable morphology. In the following paragraphs, we use the equilibration viewpoint to explain the optimum "non-preferential" condition together with experimental observations.

A phenomenological model of surface/interfacial free energies of the system has been frequently used when describing the self-assembly behavior of BCPs. The total free energy F of an arbitrary, small control volume of block copolymers, as illustrated in Figure 13.8, can be approximated, considering only interfacial and elastic energies, as

$$F = F_{A,B} + (F_{A,\text{air}} + F_{B,\text{air}}) + (F_{A,\text{sub}} + F_{B,\text{sub}}) + F_{\text{elastic}} + F_{\text{conformation}}. \tag{13.1}$$

The first five terms on the right-hand side of Eq. 13.1 are associated with the interfacial energy between (1) different domains in the BCP, (2) the polymer–free surface interface, and (3) the

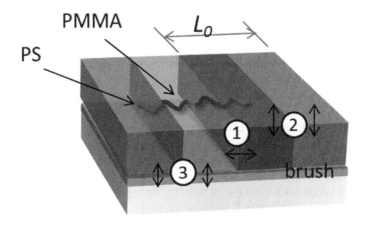

FIGURE 13.8   An example of a BCP thin film on a homogeneous brush.

polymer–brush/substrate interface, as illustrated in Figure 13.8. The sixth term is associated with stretching or compression of the polymer chain, e.g. due to the incommensurability between BCP and guiding patterns, which will be discussed in later sections. The last term is associated with the conformational entropy of the polymer chain. One common approach to manipulate the morphology of the assembled block copolymer film is to tune the polymer–substrate interfacial energy, $F_{A,\text{sub}} + F_{B,\text{sub}}$, which can be controlled by changing the composition of the random brush, f_{st} [25], or introducing a chemical pattern, so that the desired morphology will have the lowest total free energy among all the possible conformations [13, 28]. If the processing conditions and parameters of the system, besides f_{st}, are kept constant, and given that within a range of f_{st} the BCP will maintain perpendicular morphologies [25, 26, 28], $F_{A,\text{sub}} + F_{B,\text{sub}}$ will be the only terms that vary with f_{st} in Eq. 13.1. Hence, to design a brush to facilitate the desired perpendicular structure, one will need to find the optimum f_{st} that minimizes $F_{A,\text{sub}} + F_{B,\text{sub}}$.

For a PS-b-PMMA system on top of a homogeneous, unpatterned P(S-r-MMA) brush layer with PS content of f_{st}, polymer–brush/substrate interfacial energy can be expressed as

$$F_{\text{polymer,sub}} = F_{A,\text{sub}} + F_{B,\text{sub}} = \gamma_{\text{PS},f} * dA_{\text{PS}} + \gamma_{\text{PMMA},f} * dA_{\text{PMMA}}, \tag{13.2}$$

where $\gamma_{X,f}$ is the interfacial energy between domain X, which is either PS or PMMA, and the brush with a composition of f_{st}. The interfacial area between domain X ($X = $ PS or PMMA) and the brush in the control volume of interest is dA_X. The interfacial energy can be further approximated by the polar and the dispersive component of the surface energy of each material through the harmonic-mean equation [29]:

$$\gamma_{X,f} = \gamma_X + \gamma_f - 4 * \frac{\gamma_X^d * \gamma_f^d}{\gamma_X^d + \gamma_f^d} - 4 * \frac{\gamma_X^p * \gamma_f^p}{\gamma_X^p + \gamma_f^p}, \tag{13.3}$$

where $\gamma_X = \gamma_X^d + \gamma_X^p, \gamma_f = \gamma_f^d + \gamma_f^p$. Note that a surface energy, for example, γ_X or γ_f, is the interfacial energy between the material and the free surface (air or vacuum). For a P(S-r-MMA) random copolymer, the dispersive and polar component of a surface energy can be calculated through a linear combination [29] of the corresponding components of its constituents by Eqs. 13.4 and 13.5:

$$\gamma_f^d = f_{st} * \gamma_{\text{PS}}^d + \left(1 - f_{st}\right) * \gamma_{\text{PMMA}}^d, \tag{13.4}$$

$$\gamma_f^p = f_{st} * \gamma_{\text{PS}}^p + \left(1 - f_{st}\right) * \gamma_{\text{PMMA}}^p. \tag{13.5}$$

The polar and the dispersive components in Eqs. 13.4 and 13.5 can be found in the literature and extrapolated to the processing temperature of interest [29].

Figure 13.9 shows that the behavior of $\gamma_{\text{PS},f}$ and $\gamma_{\text{PMMA},f}$ as a function of f_{st} estimated by Eqs. 13.3–13.5 agrees well with the experimental measurements in the literature [25], except that the experimental results suggested that the two interfacial energy curves are slightly asymmetric. Therefore, one should be able to use the calculated interfacial energy from Eq. 13.3 to estimate $F_{\text{PS,sub}} + F_{\text{PMMA,sub}}$ by Eq. 13.2 and determine the optimum brush composition, f_{opt}, that results in the lowest total free energy. Although the exact f_{opt} may not be extracted from these calculations due to the difference between the experimental data and the harmonic-mean equations, the approximated values of f_{opt} will still be meaningful and can help us to understand the assembly process under different boundary conditions. In the following section, we will examine several experimental results from the literature to demonstrate the use of this free energy minimization analysis and reinvestigate the meaning of a "non-preferential" substrate.

Mansky et al. [25] experimentally showed that $\gamma_{\text{PS},f} \cong \gamma_{\text{PMMA},f}$ when $f_{st} \sim 0.58$, and a perpendicular lamellar structure can be obtained with this "neutral" brush. One might expect that this

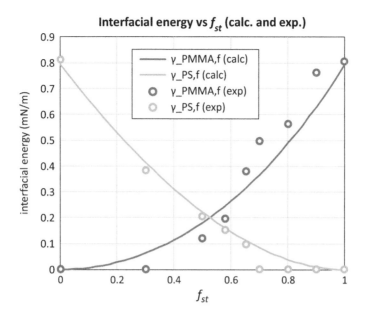

FIGURE 13.9 The calculated (line) and experimental (circle) interfacial energy of PS–brush and PMMA–brush interfaces as a function of styrene content in the brush. The experimental data (circles) was extracted from Manksy's work [25] (which was obtained at 170°C) and the lines were calculated from the harmonic-mean equation.

equalization in interfacial energies would be a universal "neutral" condition for all BCP systems and is independent of the length of the polymer chains, and thus the f_{opt} should be independent of the morphology of the BCP system. However, Han et al. [26, 30] carefully compared the behavior of cylinder-forming and lamella-forming PS-b-PMMA on a P(S-r-MMA) brush-grafted substrate and reported that a PMMA cylinder-forming system requires a brush with higher f_{st} than a lamella-forming system would need, i.e. f_{opt} ~0.7 for a cylinder-forming PS-b-PMMA and ~0.6 for a lamella-forming PS-b-PMMA. This dependency of f_{opt} with copolymer morphologies can be understood through the concept of free energy minimization.

As discussed earlier, the f_{opt} can be determined from the minimization of Eq. 13.2. A thin film having a perpendicular morphology of symmetric, lamella-forming PS-b-PMMA has $dA_{PS} \cong dA_{PMMA} = 0.5 * dA$, where dA is the total interfacial area between the brush and the copolymers in the control volume. Eq. 13.2 for a lamellar system becomes $F_{polymer,sub}^{lam} = (0.5 * \gamma_{PS,f} + 0.5 * \gamma_{PMMA,f}) * dA$. Similarly, a typical cylinder-forming PS-b-PMMA, whose minority block is PMMA, has a volume ratio of ~0.7 for the PS block. Thus, one can write $F_{polymer,sub}^{PMMA\ cyl} = (0.7 * \gamma_{PS,f} + 0.3 * \gamma_{PMMA,f}) * dA$. Using the calculated values of $\gamma_{PS,f}$ and $\gamma_{PMMA,f}$ as shown in Figure 13.9, we can plot the values of $F_{polymer,sub}^{lam}$ and $F_{polymer,sub}^{PMMA\ cyl}$ (per unit area) as a function of f_{st}, as shown in Figure 13.10. One can clearly see that f_{opt}, the f_{st} resulting in the lowest free energy, for a PMMA cylinder system is indeed greater than the lamellar system, as observed in the experiments. From a purely mathematical point of view, because the weighting factor of $\gamma_{PS,f}$ in $F_{polymer,sub}^{PMMA\ cyl}$ is larger than $F_{polymer,sub}^{lam}$, the minimum point of $F_{polymer,sub}^{PMMA\ cyl}$ will shift toward the direction that is more $\gamma_{PS,f}$ favorable, which means higher f_{st}. From this free energy minimization viewpoint, one can understand why the optimum brush composition is not a constant that results in $\gamma_{PS,f} \cong \gamma_{PMMA,f}$ but varies with the volume fraction in the BCP. One should note that $F_{polymer,sub}$ is not the only term in Eq. 13.1 that varies with the BCP volume fraction, $F_{polymer,air}$ also varies with the BCP volume fraction. However, because $\gamma_{PS,air}$ and $\gamma_{PMMA,air}$ are almost identical (difference ~0.1%), the contribution from $F_{polymer,air}$ to the total free energy is negligible.

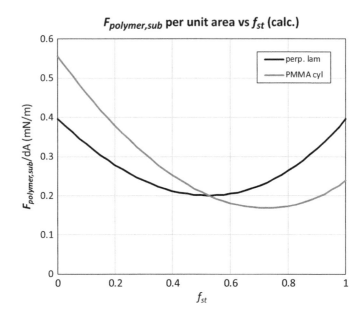

FIGURE 13.10   The calculated interfacial energy per unit area associated with the polymer–substrate interface for different BCP morphologies.

Even though this phenomenological model is simplified and includes some approximations, it provides a succinct and powerful way to understand the self-assembly behavior of BCP systems. It can also be applied to compare parallel versus perpendicular morphology as well as DSA on chemical patterns with and without density multiplication [31]. The main limitation of this simplified model is that the detailed 3-D profile of the resulting structures, such as discontinuous domains within the film, which can exist in some cases, would be difficult to predict and analyze by the phenomenological model. Other methods that provide insight into the 3-D domain structure, such as Monte Carlo (MC) and self-consistent field theory (SCFT) numerical models, will be discussed in later sections.

13.1.3   High-χ Block Copolymer Systems

As mentioned in Section 13.1.2, the natural period, L_0, of the BCPs (equivalent to "pitch," as commonly used in lithography terms) is mainly determined by the degree of polymerization, N. At the same time, χN needs to be greater than 10.5 in order for phase segregation to occur. Consequently, when the polymer pair that constitutes the BCP is determined, so is the χ and the smallest possible L_0. Using PS-b-PMMA as an example, Cheng et al. experimentally determined that self-assembly of PS-b-PMMA with $M_n = 44$ kg/mol resulted in $L_0 = 24.8$ nm [32]. With this data, and using $\chi = \sim 0.037$ at an annealing temperature of 180°C [33] and the weak segregation limit correlation of $L_0 \propto N^{1/2}$, the smallest achievable L_0 of PS-b-PMMA BCP can be estimated to be ~20 nm. Wan et al. experimentally verified that the phase segregation limit for PS-b-PMMA on a chemical pattern is indeed between 18.5 nm and 22 nm [34].

If a smaller L_0 is desired, a polymer pair with higher χ than PS/PMMA will be needed. A variety of material combinations have been studied and reported. The polymer systems of interest can be categorized into two major groups: (1) Si-containing systems, such as PS-b-polydimethylsiloxane (PDMS) and PS-b-polytrimethylsilylstyrene (PTMSS) [35–37]; and (2) organic-organic systems, such as PS-PEO, PS-P4VP, and other commercially available systems [38, 39]. Table 13.1 summarizes several common BCP systems and their reported characteristic dimensions, molecular weight, and estimated χ value determined by different methods [40].

TABLE 13.1

Several High-χ BCP Systems and Their Reported χ Value. PS-b-PMMA Is Included as a Reference

Block copolymer	Morphology	Characteristic dimension (nm)	Molecular weight (kg/mol)	χ_{eff}	Reference
PS-PDMS	sphere	15.2	14.8	0.26	[43]
PTMSS-PLA	cylinder	12.1	9.2	0.41	[44]
PS-PEO	cylinder	6.9	7	0.047	[45, 46]
PS-PMMA	lamella	15	27.6	0.037–0.04	[47]
PtBS-PMMA	lamella	14.4	17.6	0.053	[48]
PMOST-PTMSS	lamella	14.4	16.8	0.046	[36]
PS-P4VP	lamella	10.3	6.4	0.4	[49]

The PS-b-PMMA system has a unique property that makes the self-assembly of PS-b-PMMA much easier compared to most of the BCP systems: the interfacial energy of PS–air and PMMA–air is almost identical over a wide range of temperatures. The similar interfacial energies mean that the free surface automatically acts as a non-preferential ("neutral") layer. As discussed in Section 13.1.2, this is the main reason that only the polymer–substrate interfacial energy needs to be controlled with a brush or chemical patterns for PS-b-PMMA. Most other BCP systems do not have this unique property, leading to one block preferentially coating the top surface, thereby blocking the formation of a through-film, perpendicular morphology. Even though under certain conditions, parallel morphologies of high-χ systems might suffice as a template for pattern transfer, if perpendicular morphologies are desired, the interfacial energy at the free surface also needs to be carefully tuned. The free surface modification, similar to the substrate modification, can be done through the introduction of a polymer layer over the BCP film [41]. This technique will be discussed in a later section.

Another consideration when implementing high-χ systems is the compatibility with thermal annealing. Self-assembly can occur between the glass transition temperature (T_g) and the order–disorder transition temperature, but usually a higher temperature is preferred because the diffusivity of the BCP molecules will be higher and the assembly kinetics will be faster. However, some polymer systems may inherit a relatively low degradation temperature from the components, and hence, may not be practical or effective enough for thermal annealing. Solvent vapor annealing provides an alternative approach to mobilize the BCP molecules under room temperature [42]. This technique will also be discussed in a later section.

13.2 DIRECTED SELF-ASSEMBLY

Block copolymer lithography is very attractive due to its simplicity in process and the resulting highly uniform, material-controlled CD and pitch. Thus, this method has been utilized in a variety of research fields, such as air gap low-k dielectric [11], thermoelectric materials [50], quantum dot lasers [51, 52], and graphene band-gap opening [53–55]. However, the lack of long-range order in BCP lithography is the most significant limiting factor to the applications in semiconductor and electronics-related areas. Great endeavors have been made to search for a reliable approach to direct the self-assembly into a desired morphology at pre-determined locations. Several methods have been employed to achieve long-range order, for example, by applying an external field such as an electrical field demonstrated by Jaeger et al. [56], a temperature gradient by Hashimoto et al. [57], and shear flow by Kornfield et al. [58]. Due to the limitations of the physical size of the apparatus, these approaches would be suitable for controlling the BCP orientation of an area on the length scale of millimeters or larger. In order to further precisely control the local behavior at the micron, or even smaller, length scale, guiding patterns created by lithography processes were introduced. The most commonly studied methods consist of either topographically or chemically patterned substrates, known as graphoepitaxy (Grapho) and chemoepitaxy (Chemo) DSA, respectively. Because of the promising results that have been demonstrated using these two DSA approaches, and the potential to be integrated with the state-of-the-art tools of the semiconductor industry, in the following sections we will review the evolution of Grapho and Chemo DSA and the achievable morphologies and structures.

13.2.1 Graphoepitaxy DSA

Segalman et al. [59] first utilized a lithographically patterned substrate to direct a thin film of a sphere-forming PS-PVP block copolymer into a well-ordered hexagonal array of micro-domains. The substrate had a spatial period much larger than the natural period of a block copolymer, L_0, and a step height similar or greater than L_0, as shown in Figure 13.11. In order to determine the order of the copolymer domains achieved by this method, besides using a Fourier transform to qualitatively visualize the order of the structures, they also quantitatively studied the translational and orientational order parameter of the assembled structures under different annealing conditions [60]. Segalman et al. showed that the assembled copolymers under certain annealing conditions will retain a high orientational order, but the translational order will decay in most cases, which may imply the lack of a long-range translational order. Note that a later study by Liu et al. [61] showed that this decay in translational order could be an inaccuracy derived from the lattice vector calculation. Segalman et al. described their work using the term "graphoepitaxy" based on the similarity to a strategy used in the crystal growth field [62], where the researchers induced the growth of single

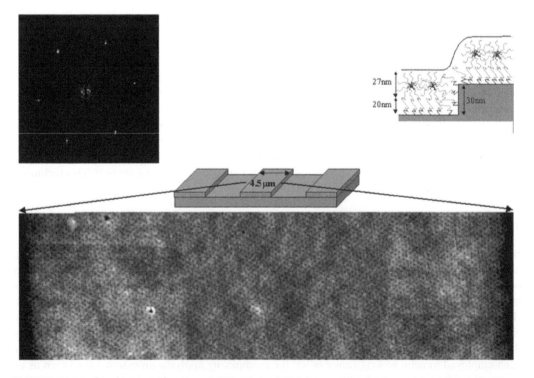

FIGURE 13.11 Fourier transform (upper left) of the AFM image (bottom) by graphoepitaxy of sphere-forming PS-b-PVP block copolymers and an illustration of copolymer molecules (upper right). (Reproduced from Ref. [59] with the permission of John Wiley and Sons.)

crystals on an amorphous substrate with sub-micron grating patterns. Graphoepitaxy of block copolymers provides a simple and attractive route for fabricating sublithographic, nanoscale structures.

Cheng et al. [22] successfully demonstrated the assembly of another sphere-forming block copolymer, polystyrene-b-polyferrocenyldimethylsilane, using graphoepitaxy and the subsequent pattern transfer. Although the geometry of a sphere is not ideal for pattern transfer, i.e. lacking a high aspect ratio and a vertical sidewall profile, due to the high etching selectivity between PFS and Si, Si pillars with an aspect ratio ≥3 were demonstrated. Similarly, Jung et al. assembled sphere-forming PS-b-PDMS copolymers on a PDMS brush–treated graphoepitaxy substrate and demonstrated the pattern transfer with a tone-inversion process [63]. It is worth mentioning that patterns with both tones, i.e. metal nano-mesh and nano-dot, were illustrated using the same copolymer template in Jung's work.

The Grapho-guiding effect is derived from the sidewall preference to one of the blocks in BCP. However, when envisioning this DSA process progressing along time, it is important to remember that self-assembly will also occur locally at the same time, especially away from the edges of the guiding patterns. At the beginning of the Grapho DSA, BCP molecules are close to amorphous when spin-coated onto the substrate. Upon annealing, molecules away from the boundaries will undergo the phase segregation process similar to what would happen without any guiding patterns. For example, a thin film of cylinder-forming BCPs on a substrate without any surface modification would result in fingerprint-like structures, as shown in Figure 13.12e. At the same time, the PMMA blocks near the boundaries of the guiding pattern will rearrange themselves and preferentially wet the sidewall, because it is more energetically favorable than the PS blocks wetting the sidewall. Once the first layer of the PMMA blocks has been rearranged to follow the guiding pattern, the PS blocks on the same BCP chains are forced to align accordingly. Similarly, the PS blocks on the next BCP chains will prefer to be in contact with the previous PS block in order to reduce the PS/PMMA interfacial area. As a result, the fingerprints from local phase segregation will be gradually directed into ordered structures. Black et al. experimentally demonstrated this evolution with cylinder-forming BCPs in guiding patterns with different shapes and sizes [64]. Figure 13.12e–g shows the BCP rearrangement and propagation from the edge to the center of a guiding pattern. Figure 13.12b–d has a similar behavior, except that the film thickness at the center was too thin to generate self-assembled structures.

In contrast to sphere-forming BCPs, Black et al. [64] and Xiao et al. [65] applied cylinder-forming PS-b-PMMA copolymers to a topographical substrate modified with a non-preferential random copolymer brush and obtained a perpendicular cylinder array in the trench. The wetting preference of the surface is crucial for the resulting morphology when using a cylinder-forming BCP with the graphoepitaxy method, as demonstrated by Black et al. [64]. After the assembly of the BCP, a template of nano-mesh composed of PS can be obtained by selectively removing the PMMA block. Xiao demonstrated the fabrication of metal nano-dots in the trench using this nano-mesh template together with a lift-off process. The resulting structure was similar to that of Cheng's work, but it was shown to have better pattern uniformity. Xiao further studied the relationship between the groove width and the number of rows, as shown in Figure 13.13, and found that the copolymers tend to form a specific number of rows instead of having a mixture of different numbers of rows, i.e. defects.

In most of the graphoepitaxy methods, a certain portion of the guiding patterns usually cannot be utilized to create dense array structures, such as the raised area of the trench-type guiding patterns. To address this issue, Bita et al. successfully miniaturized the guiding patterns to the size of the micro-domains of the block copolymer and incorporated the guiding patterns into the hexagonal spot array [35]. Consequently, a continuous hexagonal array, shown in Figure 13.14, was obtained over the entire patterned area without any disruptive structures, which would have been present if large guiding patterns/trenches were used as in previous graphoepitaxy methods. It is worth mentioning that the substrate used in Bita's work was treated with PDMS brushes, and therefore was chemically homogeneous.

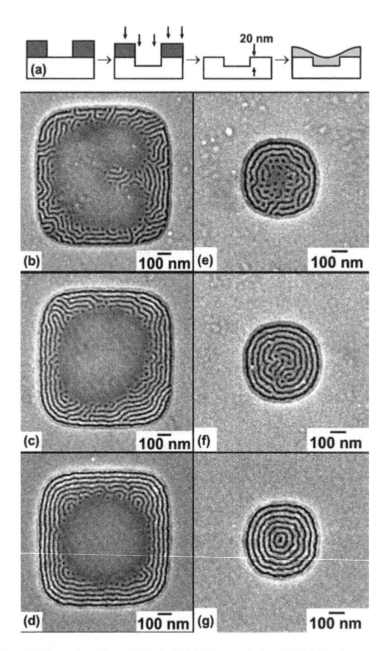

FIGURE 13.12   (a) Schematic of flow of Grapho DSA. Time evolution of DSA in a 1 μm square (b–d), and in a 0.5 μm circle (e and f). (b and e) After 10 min. (c and f) After 1 h. (d and g) After 17 h. (Reproduced from Ref. [64] with the permission of IEEE.)

In addition to hexagonal arrays, line/space arrays were also demonstrated using graphoepitaxy DSA of parallel-oriented cylinders, which were employed as a straightforward choice based on studies with hexagonal arrays. However, Black et al. [64] found that the copolymer structure formed in the trench area is very sensitive to the film thickness, which could be affected by the width of the trench as well as the mesa area next to the trench, as shown in Figure 13.15. Consequently, this thickness effect compromises the control of Black's approach and a more reliable DSA approach would be needed.

The same line/space patterns could also be achieved using perpendicular-oriented lamellae-forming block copolymers, similar to using perpendicular cylinders for a hexagonal array instead

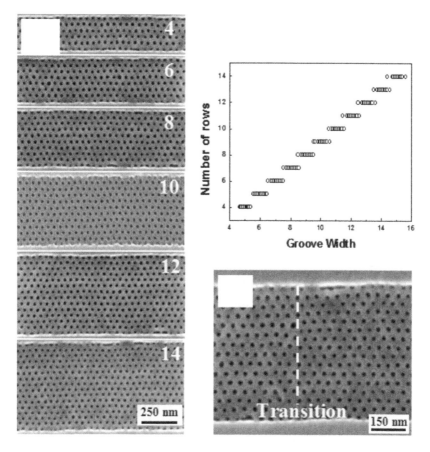

FIGURE 13.13 The number of row changes with the width of the groove in a step-like behavior. (Reproduced from Ref. [65] with the permission of IOP Publishing.)

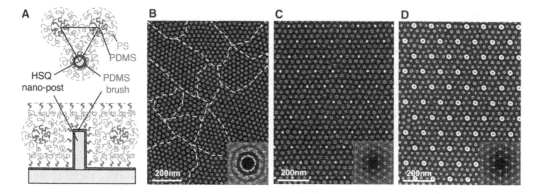

FIGURE 13.14 A graphoepitaxy method using guiding posts at about the same size as the micro-domain to achieve pattern density multiplication. (Reproduced from Ref. [35] with the permission of *Science*, AAAS.)

of spheres. From the discussion in the previous section, one could expect a better pattern transfer property from lamellae as opposed to parallel cylinders. Indeed, Park et al. [66] proved that perpendicular lamellae have a cleaner profile and lower line-edge roughness than parallel cylinders. However, even though perpendicular lamellae can be achieved using the same surface neutralization technique as used by Xiao et al., the orientation of the lines will not align with the trench, as

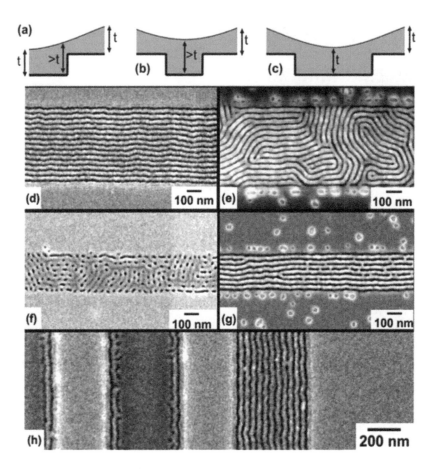

FIGURE 13.15 The orientation and morphology of the cylinder-forming copolymers in the graphoepitaxy method vary with the film thickness. (Copyright 2004 IEEE. Reprinted, with permission, from Ref. [64].)

shown in Figure 13.16b and c. Apparently, the neutral wetting condition at the sidewall of the trench does not create any energetic preference/penalty for either block of the BCP; therefore, both blocks can wet the sidewall of the trench freely. If one looks again closely at Xiao's work, i.e. Figure 13.13, half cylindrical domains were indeed created on the sidewall due to the neutral condition.

In order to obtain well-aligned perpendicular lamellae in the trench, ideally one would need to introduce a selective sidewall to either block of BCP, while maintaining the non-preferential trench bottom. Park et al. validated this idea by fabricating such substrates with a lift-off process and successfully guiding the copolymers to follow the direction of the trench, as shown in Figure 13.17a [66].

Considering the control and pattern quality of the lift-off process, Han et al. revisited the homogeneous surface modification approach and found that a weakly preferential surface together with a proper BCP film thickness can also lead to well-aligned lamellae [67]. On the other hand, a Grapho substrate with a selective sidewall and a non-preferential trench bottom can be achieved by simply patterning a photoresist lithographically on a neutral layer. However, there are two major challenges: (1) the typical photoresist is soluble in the solvent used for BCPs, so the Grapho template will be damaged during the spin-coating of BCP; and (2) neutrality may be altered during the lithography, e.g. due to the resist residue or the base developer. Kim et al. employed a negative tone resist to address both issues at once [68]. Because the exposed area of this negative tone resist will be cross-linked and therefore insoluble in the solvent developer, the subsequent spin-coating of the BCP will not cause any damage to the resist guiding patterns. Additionally, not only is the solvent

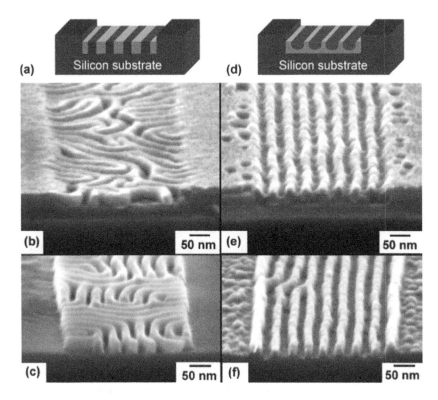

FIGURE 13.16 Perpendicular lamellae and parallel cylinders achieved by the graphoepitaxy method. (Reproduced from Ref. [66] with the permission of John Wiley and Sons.)

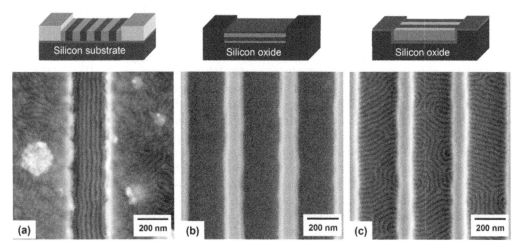

FIGURE 13.17 Lamellae-forming copolymers assembled on three types of graphoepitaxy substrate: (a) preferential sidewall and neutral bottom, (b) preferential, and (c) neutral. (Reproduced from Ref. [66] with the permission of John Wiley and Sons.)

developer considered to be more gentle than the base developer used in the positive tone process, but also the trench area is not directly exposed to UV irradiation during the negative tone lithography, hence the neutral material is considered less affected, if at all. However, the specific resist that Kim used is only sensitive to a wavelength of 300–400 nm, and thus is not suitable for the most commonly used 193 nm immersion system. Cheng et al. demonstrated DSA with a 193 nm negative tone development (NTD) process on a cross-linked neutral material [32]. The 193 nm NTD resist is normally non-cross-linking but changes polarity/solubility upon exposure. In addition, the NTD process also uses solvent developer instead of base developer, i.e. exposed resist is not soluble in the solvent of BCPs. Furthermore, the use of a cross-linked, non-preferential (neutral) mat instead of a grafted, neutral brush enhances the stability of the neutral material through the lithography processes. For example, Tsai et al. demonstrated a Grapho DSA made with the same cross-linking neutral mat that Cheng used together with an e-beam resist, hydrogen silsesquioxane (HSQ), which shares the base developer with conventional positive tone resist [69].

Here, we only discussed the evolution of the basic Grapho DSA process for hexagonal array and line/space array morphologies. A few examples of applications using Grapho DSA will be discussed in Section 13.5.

13.2.2 Chemoepitaxy

In contrast to the concept of graphoepitaxy, which was initially intended to drive the assembly solely by topography, chemoepitaxy directs the self-assembly of a block copolymer by making a pattern of alternating surface chemistries on the substrate. The first example of a type of chemoepitaxy was provided by Rockford et al. who directed the assembly of lamella-forming BCPs on a chemically heterogeneous surface with a topography of only several nm. The substrate was composed of alternating Au and SiO_2 stripes fabricated with a mis-cut Si wafer [70]. As one might expect, the line width and pitch of the mis-cut structures are not easily controllable at nm accuracy over a large area. Kim et al. first utilized lithography and etch techniques to pattern a SAM and created a well-controlled, chemically patterned substrate with minimum topography. Chemoepitaxy (Chemo) DSA of lamella-forming BCP was performed and resulted in dense line/space arrays [13]. Kim et al.'s work built the foundation of chemical pattern fabrication methods which deeply influenced the Chemo DSA field. The initial studies with chemoepitaxy focused on non-density-multiplied DSA, i.e. the assembled structure was at the same pitch as the guiding pattern. As a result, while those DSA methods offered improved process control of the lithographic process, they could not create sub-lithographic patterns. Shortly after Kim et al.'s study, Black et al. successfully generated sub-lithographic line arrays with parallel cylindrical BCPs with the graphoepitaxy method, followed by several demonstrations of prototype applications [64, 71]. At that time, graphoepitaxy DSA was more attractive than chemoepitaxy DSA for the lithography/patterning community.

Further studies on Chemo DSA in the following years consolidated the understanding of the physics behind DSA on lithographically generated chemical patterns. Edwards et al. utilized polymer brushes instead of self-assembled monolayers for chemical pattern fabrication [28]. Polymer brushes typically have >10× higher molecular weight than SAMs and therefore provide better control in chemistry via polymer synthesis. With such fine control in brush composition, Edwards was able to investigate the DSA behavior on chemical patterns composed of a variety of brush compositions. This work demonstrated how to maximize the chemical contrast of guiding patterns both experimentally and theoretically. Stoykovich et al. utilized the same chemical pattern approach but further explored structures beyond dense, unidirectional line/space arrays using ternary blends [72, 73]. Bent structures with angles of 45°, 90°, and 135°, arrays of jogs and isolated jogs, T-junctions, and isolated lines were all demonstrated using Chemo DSA of homopolymer/BCP blends, as shown in Figure 13.18. Homopolymers were introduced to the system to balance and compensate the local volume change at non-grating areas. Welander et al. studied the kinetics of DSA on chemical patterns and found that the diffusivity of BCP molecules exhibits an Arrhenius dependency, and that the annealing time can be greatly reduced at elevated temperatures, as illustrated by Figure 13.19 [74]. Ruiz et al. also reported similar behavior of cylinder- and lamella-forming BCP on homogeneous substrates without chemical patterns [75].

Chemo DSA with density multiplication was first published concurrently with graphoepitaxy work by Bita et al., i.e. Figure 13.14, which utilized mini pillars instead of trenches as the guiding

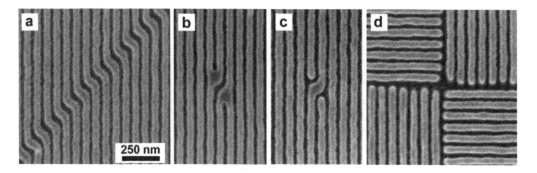

FIGURE 13.18   Non-gratings structures achieved by Chemo DSA of ternary blends. (Reprinted with permission from Ref. [72]. Copyright (2007) American Chemical Society.)

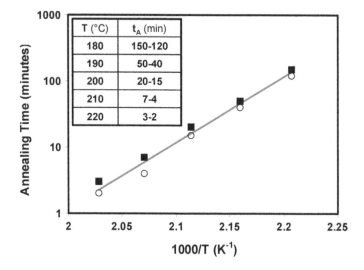

FIGURE 13.19   Temperature dependency of required annealing time for Chemo DSA of PS-*b*-PMMA. The solid square is the minimum time observed for defect-free assembly and the open circle is the maximum time for which abundant defects persisted. (Reprinted with permission from Ref. [74]. Copyright (2008) American Chemical Society.)

pattern. Ruiz et al. and Cheng et al. demonstrated hexagonal array and line/space array, respectively [76, 77]. More importantly, both Chemo DSA studies pointed out an important difference between Grapho and Chemo DSA: the rectification effect. Ruiz et al. showed that Chemo DSA could neglect some of the imperfections of the guiding pattern created by lithography, and instead result in an improved pattern quality. By comparing the pattern quality on the photoresist (Figure 13.20A and C) and the resulting patterns after DSA (Figure 13.20E and G), one can clearly see the effect of pattern rectification during DSA. Liu et al. later quantitatively showed that not only did the improvement occur in the BCP layer but it was also retained after pattern transfer [61]. One may also notice that higher pattern densities, such as those without density multiplication (Figure 13.20A and C), will enhance the imperfection in the resist; thus, the rectification effect is more discernible in Figure 13.20A/E and C/G than in Figure 13.20B/F and D/H.

The main difference between Ruiz's work and Bita's work is that the guiding pattern in Bita's work is part of the resulting pattern and Ruiz's guiding pattern is thin enough compared to the BCP layer such that the pattern transfer behavior is dominated by the BCP layer. For example, if any imperfect guiding post exists, the copolymers can only arrange themselves according to the existing boundary conditions instead of rectifying it. Another good example of pattern rectification effect is

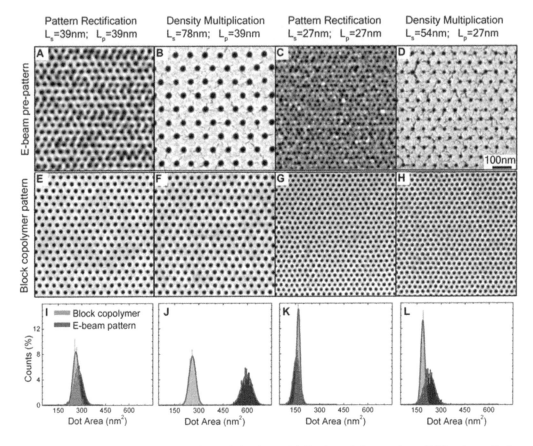

FIGURE 13.20 The pattern rectification and density multiplication of chemoepitaxy DSA using cylinder-forming BCPs. (Reproduced from Ref. [76] with the permission of *Science*, AAAS.)

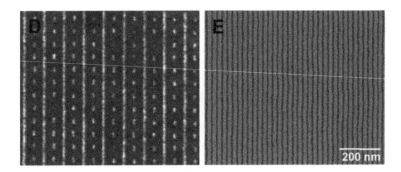

FIGURE 13.21 The pattern rectification and density multiplication of chemoepitaxy DSA using lamella-forming BCPs. (Reproduced from Ref. [77] with the permission of John Wiley and Sons.)

shown in Figure 13.21, in which Cheng created a guiding pattern with solid and dashed lines, but still obtained a uniform line/space array after DSA. Therefore, it is believed that graphoepitaxy is less efficient in pattern rectification than chemoepitaxy under certain boundary conditions, such as those discussed above.

Moreover, in addition to rectification, chemoepitaxy with density multiplication also provides an economic incentive for incorporating DSA with both e-beam and optical lithography. Assuming one could achieve an acceptable pattern quality at sub-30 nm pitch with e-beam direct write, i.e.

better than Figure 13.20C, the time to write a full 2.5" wafer at such density would take more than a month [76]. Indeed, pattern writing time is one of the most challenging issues for e-beam lithography. When combined with chemoepitaxy DSA with density multiplication, the writing time could be reduced to 1/4 or 1/9 and apparently it would be more amenable for manufacturing than writing the entire pattern without subsequent DSA. In the case of patterning the substrate with projection lithography, the exposure time is not a concern. The primary benefits of DSA with optical lithography are resolution enhancement, i.e. density multiplication, and improved process control, i.e. pattern rectification. In addition, the chemoepitaxy process usually has fewer steps compared to conventional multiple patterning techniques required to achieve the same resolution, such that cost reduction could be a possible benefit as well. Many researchers, including Park et al. [78], Cheng et al. [32], and Liu et al. [79], have proposed integration schemes of DSA on chemical patterns based on 193 nm immersion (193i) lithography tools. Here, we use Liu et al.'s method (aka LiNe flow, as shown in Figure 13.22) as an example to illustrate the fundamentals of Chemo DSA with density multiplication [31].

A chemical pattern is typically composed of two chemically distinct regions. For example, the two chemistries generated by the methods used in the works of Edwards, Stoykovich, and Ruiz are the brush and regions where the brush has been etched with a plasma, leaving either etched brush or exposed substrate. The main benefit of these methods is that only one brush material is needed to create the chemical contrast. However, as a result of using only one brush material, the chemistry of the second area cannot be freely chosen. The LiNe flow process introduces a second material in order to achieve independent control in chemistry. In order to avoid the two materials intermixing with each other, the LiNe flow process begins with a cross-linkable polymer (referred to as a mat) that is spin-coated and then cross-linked, followed by lithography, etch, resist stripping, and the backfill of a brush. It has been shown that the brush will not affect the chemistry of the mat if the mat is sufficiently cross-linked, which cannot be achieved simply with a two-brush system [80]. Many mat/brush combinations have been demonstrated since Liu et al.'s initial work, such as PS mat/P(S-r-MMA) brush, PMMA mat/P(S-r-MMA) brush, P(S-r-MMA) mat/PS brush, and PS mat/PVP brush, and several others based on the needs of different BCP systems, including some examples for applications other than DSA [31, 39, 81, 82].

The LiNe flow process enables insight into the optimum brush composition for density multiplication. It showed experimentally and analytically that not only is the previously known "neutral" composition not optimum for DSA with density multiplication, but also the optimum composition varies with the chemistry of the preferential stripe and the multiplication factor. The reason for the optimum composition being "off-neutral" is actually similar to what has been discussed in Section 13.1.2. According to the total free energy minimization viewpoint, higher PS content in

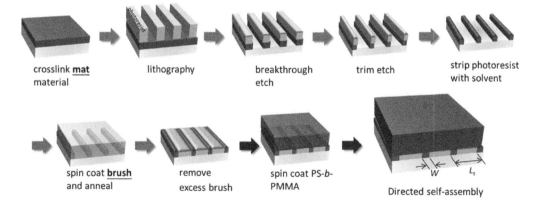

FIGURE 13.22 LiNe flow of chemical pattern fabrication for DSA with density multiplication. (Reprinted with permission from Ref. [31]. Copyright (2013) American Chemical Society.)

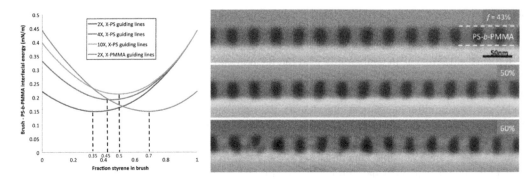

FIGURE 13.23 (Left) Estimated interfacial free energy using different guiding stripe chemistries and different multiplication factors. (Right) Cross-sectional SEM images of 2× density multiplied DSA using different brushes near the "neutral" composition (~60%). (Reprinted with permission from Ref. [31]. Copyright (2013) American Chemical Society.)

the BCP, such as a cylinder-forming system, would require higher styrene content in the brush. As illustrated by the final drawing in Figure 13.22, in the case of DSA over a chemical pattern, part of the BCPs is assembled over the guiding stripes. The equivalent BCP volume ratio over the brush region is actually different from the bulk BCP, e.g. in this case, two PMMA domains and one PS domain were assembled over the brush region, which implies a lower styrene content brush would be needed. One would also expect that as the density multiplication factor increases, the equivalent volume ratio over the brush becomes closer to the bulk ratio, e.g. a 4× case will have four PMMA domains and three PS domains over the brush. Hence, the optimum composition will be closer to the previously known "neutral" composition. In addition, if the guiding stripe is PMMA-preferential instead of PS-preferential, the equivalent BCP volume ratio over the brush region will become PS-rich instead of PMMA-rich, and therefore a PS-rich brush would be needed. These cases were demonstrated by interfacial free energy estimations and Monte Carlo simulations, and verified experimentally, i.e. Figure 13.23 [31]. Lastly, one should note the difference between Grapho and Chemo guiding patterns for line/space DSA in terms of brush composition. In the case of Grapho DSA, an equivalent BCP volume ratio over the brush region is always the same as the bulk, because the BCPs are assembled between the guiding stripes instead of partially over the guiding patterns as in Chemo cases.

13.3 DSA SIMULATIONS

13.3.1 INTRODUCTION

Simulation can be a very powerful tool to predict complicated, three-dimensional morphologies of the self-assembled block copolymers on a pre-patterned surface. Note, however, that the length scale of the self-assembled morphology is typically on the order of 10–100 nm; it is computationally demanding to simulate the self-assembly of dense block copolymers with fully atomistic models. A practical approach is coarse-graining the atomistic description of block copolymers on the basis of "particle" or "field." In the particle-based models, groups of atoms are represented by a single coarse-grained particle. It is relatively straightforward to conduct simulations with the particle-based models because they are compatible with traditional simulation techniques such as molecular dynamics (MD), Monte Carlo, and dissipative particle dynamics (DPD) [83, 84]. In addition, the particle-based approaches often enable modeling of complex systems, and more importantly, they include the effects of thermal fluctuations on the morphology of self-assembled block copolymers. On the other hand, particle-based simulations still tend to be highly computationally demanding due to the calculations of the interaction energy for each pair of coarse-grained particles.

The other type of coarse-graining descriptions relies on a field-based model. The most notable field-based model is the self-consistent field theory which has been the standard for numerical and analytical treatments of block copolymers [85–89]. In the field-based approach, the fundamental degrees of freedom are the local densities of a block copolymer. The parameters of the field-based model, e.g. the Flory–Huggins interaction parameter and inverse compressibility, are the macroscopic physical properties of block copolymers, which are not used in the conventional particle-based models. Another difference from the particle-based model is that in the field-based models, the polymeric chains are not explicitly described and with few exceptions, fluctuation effects are not taken into account.

To fill the gap between the field-based and particle-based models while keeping the strength of both models, a combinatorial approach has been proposed by de Pablo and his coworkers that is referred to as the "theoretically informed coarse-grained (TICG)" model [90–92]. In the TICG model, each polymer chain is described as a collection of coarse-grained particles, whereas the interaction energy between the chains is calculated from the local densities of the coarse-grained particles, as is done in the traditional field-based approaches. The TICG model has already been applied to a wide range of block copolymer systems that include nanoparticles [93], homopolymers, and solvents [94]. Importantly, the TICG model includes fluctuation effects that play an important role in the order–disorder transition of diblock copolymers [95], and the morphology and thermodynamic properties of self-assembled block copolymers [96]. In what follows, we review some basics and applications of the TICG model for simulations of DSA.

13.3.2 THEORETICALLY INFORMED COARSE-GRAINED MODEL

Consider a system that contains n AB diblock copolymer chains in volume V and at temperature T. Each diblock copolymer chain is assumed Gaussian and is composed of N coarse-grained segments (N_A for block A, and N_B for block B). All segments are assumed to have the same diameter b. The distance between the two bonded segments is constrained by a harmonic spring [97, 98] (Figure 13.24):

$$\frac{E_b\{r_i(s)\}}{k_BT} = \frac{3}{2b^2}\sum_{i=1}^{n}\sum_{s=1}^{N-1}\left|r_i(s+1)-r_i(s)\right|^2, \tag{13.6}$$

where k_B is the Boltzmann constant, and $r_i(s)$ is the position vector of the sth segment ($s=1,\ldots, N$) in the ith segment ($i=1,\ldots, N$). The bonded energy, E_b, accounts for the fluctuations concerning the

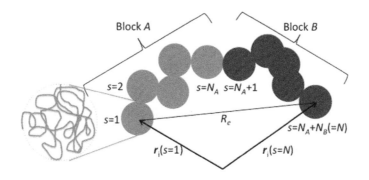

FIGURE 13.24 Schematic illustration of a Gaussian block copolymer chain.

internal degrees of freedom of the coarse-grained polymer chains. For the non-bonded interactions, a "field-based" energy function, E_{nb}, is employed that has essentially the same formula as the one proposed for the free energy of the AB binary mixture [99]:

$$\frac{E_{nb}\{\phi_A(\boldsymbol{r}),\phi_B(\boldsymbol{r})\}}{k_BT} = \rho_0 \iiint_V dxdydz\left[\chi\phi_A(\boldsymbol{r})\phi_B(\boldsymbol{r}) + \frac{\kappa}{2}(1-\phi_A(\boldsymbol{r})-\phi_B(\boldsymbol{r}))^2\right], \quad (13.7)$$

where $\phi_K(\boldsymbol{r})$ represents the normalized number density of type K (A or B) segments at a position $\boldsymbol{r}$:

$$\phi_A(\boldsymbol{r}) = \frac{1}{\rho_0}\sum_{i=1}^{n}\sum_{s=1}^{N_A}\delta(\boldsymbol{r}-\boldsymbol{r}_i(s)), \ \phi_B(\boldsymbol{r}) = \frac{1}{\rho_0}\sum_{i=1}^{n}\sum_{s=N_A+1}^{N}\delta(\boldsymbol{r}-\boldsymbol{r}_i(s)), \quad (13.8)$$

with ρ_0 as the bulk number density ($=nN/V$). The first term on the right-hand side of Eq. 13.7 represents the energy associated with the incompatibility between A and B segments, whose strength is characterized by the Flory–Huggins parameter, χ [100]. The second term constrains the local densities to around the bulk average. The parameter κ denotes the inverse compressibility (i.e. modulus). By increasing κ, the system becomes stiffer with the density oscillation more pronounced [90].

The Hamiltonian of the bulk system, H, is represented as a sum of the bonded and non-bonded energy functions:

$$\frac{H}{k_BT} = \frac{3(N-1)}{2R_e^2}\sum_{i=1}^{n}\sum_{s=1}^{N-1}|\boldsymbol{r}_i(s+1)-\boldsymbol{r}_i(s)|^2$$

$$+\sqrt{\bar{N}}\int_V\frac{d\boldsymbol{r}}{R_e^3}\left[\chi N\phi_A(\boldsymbol{r})\phi_B(\boldsymbol{r}) + \frac{\kappa N}{2}(1-\phi_A(\boldsymbol{r})-\phi_B(\boldsymbol{r}))^2\right] \quad (13.9)$$

where R_e is the root mean square of the end-to-end distance of the Gaussian chain $\left(=b\sqrt{N-1}\right)$ [97]. The so-called invariant degree of polymerization is denoted as $\bar{N}$ and $\sqrt{\bar{N}}$ $\left(=\rho_0R_e^3/N\right)$ represents the average number of Gaussian chains contained in the unit cube of volume R_e^3. Since the Gaussian chains are invariant to the choice of monomer units, all the model parameters in Eq. 13.9, i.e. R_e, $\sqrt{\bar{N}}$, χN, and κN, are represented with the unit of "per-chain" [99].

The Hamiltonian in Eq. 13.9 requires the calculations of the local number densities defined in Eq. 13.8. One approach is to represent each segment as a probability distribution function, $w(|\boldsymbol{r}-\boldsymbol{r}_i(s)|)$, that rapidly decays with the distance from the center of the segment [93, 101]. Then, $\phi_K(\boldsymbol{r})$ can be expressed as a summation of the probability weights over all type K segments.

$$\phi_A(\boldsymbol{r}) = \frac{1}{\rho_0}\sum_{i=1}^{n}\sum_{s=1}^{N_A}w(|\boldsymbol{r}-\boldsymbol{r}_i(s)|), \ \ \phi_B(\boldsymbol{r}) = \frac{1}{\rho_0}\sum_{i=1}^{n}\sum_{s=N_A+1}^{N}w(|\boldsymbol{r}-\boldsymbol{r}_i(s)|). \quad (13.10)$$

This weighting-function method makes the density field continuous and compatible with the local stress calculations [92]. However, the summation over all segments tends to be computationally demanding [102]. Another approach is a grid-based (or particle-mesh) scheme, where the system is subdivided into cubic cells with the edge of ΔL [93, 94]. It is suggested that the discretization parameter ΔL be comparable to the segment size ($=b$) and be smaller than the width of the interface between the self-assembled domains [102]. In the grid-based approach, the local density is defined as

$$\phi_K(i,j,k) = \frac{n_K(i,j,k)}{n_0}, \qquad (13.11)$$

where n_0 is the average number of segments per cubic cell ($=nN/(V/\Delta L^3)$), and $n_K(i,j,k)$ is the number of type K segments in the cubic space located from $x=(i-1)\Delta L$ to $x=i\Delta L$, from $y=(j-1)\Delta L$ to $y=j\Delta L$, and from $z=(k-1)\Delta L$ to $z=k\Delta L$. Since the grid-based approach requires only a conversion of the particle positions to the cell indices, it can significantly reduce the calculation time of the local densities.

Here, we show the procedures for determining the parameters of the TICG model, whose details can be found in [103]. Consider a PS-b-PMMA melt with a mass density ρ of 1.1 g/cm^3 and a molecular weight of 100.0 kg/mol (50.0 kg/mol for PS and 50.0 kg/mol for PMMA). At first, it is assumed that the averaged size of the coarse-grained segments, b, is 0.87 nm, and that the total number of coarse-grained segments per chain, N, is the same as the number of chemical repeating units (see [104]). Then, N is estimated from

$$N = \frac{M_{PS}}{m_{PS}} + \frac{M_{PMMA}}{m_{PMMA}} = \frac{5.0\times10^4 \text{ g/mol}}{104.0 \text{ g/mol}} + \frac{5.0\times10^4 \text{ g/mol}}{100.0 \text{ g/mol}} = 981,$$

where M_K and m_K (K=PS or PMMA) are the molecular weights of the block and monomer unit, respectively. Next, the end-to-end distance, R_e, is calculated from

$$R_e = b\sqrt{N-1} = (0.87 \text{ nm})\sqrt{981-1} = 27.2 \text{ nm}.$$

Furthermore, $\sqrt{\bar{N}}$ is obtained from

$$\sqrt{\bar{N}} = \frac{\rho N_{Av} R_e^3}{M_{PS} + M_{PMMA}}$$

$$= \frac{\left(1.05\times10^6 \text{ g/m}^3\right)\left(6.02\times10^{23} \text{ chains/mol}\right)\left(27.2\times10^{-9} \text{ m}\right)^3}{5.0\times10^4 \text{ g/mol} + 5.0\times10^4 \text{ g/mol}}$$

$$= 128.$$

The inverse compressibility, κN, should be calculated from the isothermal compressibility of polymer melts. However, it has been shown that the theoretical κN causes an unrealistic structure with some prominent peaks in the radial distribution function of the coarse-grained segments [92]. To avoid such artificial effects, a considerably lower κN (e.g. 35–50) has been employed in previous TICG simulations.

The last parameter is the interactive parameter, χN. The Flory–Huggins parameter, χ, for a pair of common polymers, e.g. PS and PMMA, can be found in the literature and handbooks [98, 105]. It is suggested that the value of χN used in the TICG model, $(\chi N)_{\text{TICG}}$, be smaller than that in the SCFT, $(\chi N)_{\text{SCFT}}$, to match their morphological results [92]. This is mainly due to the fluctuation effects. The ratio of $(\chi N)_{\text{TICG}}$ to $(\chi N)_{\text{SCFT}}$ approaches unity by increasing the average number of coarse-grained segments per cubic cell, n_0; however, the MC simulations become more computationally demanding with a larger n_0. As a practical choice, the n_0 was set at ~14–16 in the previous TICG models, where the ratio $(\chi N)_{\text{TICG}}/(\chi N)_{\text{SCFT}}$ was found to be ~0.82–0.85 [92].

13.3.3 MC SIMULATION FLOW

Unlike atomistic models, the Gaussian chain model allows the coarse-grained segments to overlap with one another. Therefore, even for a dense system, an initial configuration can be readily generated in a random manner. Once an initial configuration is generated, each segment is moved in a way that reduces the system's energy. Several different types of MC trial moves are proposed for effectively equilibrating a dense polymer melt, e.g. random displacement, cut-and-grow, chain translation, and block inversion [84]. A standard trial move is "random displacement," whose basic scheme is described as follows: (1) randomly select a segment, (2) propose a new position for the segment, (3) calculate the energy difference ΔE before and after the trial movement, and (4) accept or reject the new position based on the Metropolis criteria, $\min\left(1, \exp\left[-\Delta E/k_B T\right]\right)$ [83, 84]. For the calculation of ΔE, the old and new bonded energies associated with the selected segment are obtained from

$$\frac{\Delta E_b}{k_B T} = \frac{3(N-1)}{2R_e^2}$$

$$\left\{\left[1-\delta_{sN}\right]\left[\left|r_i(s+1)-r_i^{\text{new}}(s)\right|^2 - \left|r_i(s+1)-r_i^{\text{old}}(s)\right|^2\right]\right. \tag{13.12}$$

$$\left.+\left[1-\delta_{s1}\right]\left[\left|r_i^{\text{new}}(s)-r_i(s-1)\right|^2 - \left|r_i^{\text{old}}(s)-r_i(s-1)\right|^2\right]\right\},$$

where the Kronecker delta, δ_{sN}, becomes unity at $s=N$, and zero at $s \neq N$. The calculation of the non-bonded energies is required only if the segment is moved to a different cell. For example, when type A segment is displaced from the cell (i,j,k) to another cell (I,J,K), the local number densities in the two cells are changed from

$$\phi_A^{\text{old}}(i,j,k) = \frac{n_A(i,j,k)}{n_0}, \quad \phi_A^{\text{old}}(I,J,K) = \frac{n_A(I,J,K)}{n_0} \tag{13.13}$$

to

$$\phi_A^{\text{new}}(i,j,k) = \frac{n_A(i,j,k)-1}{n_0}, \quad \phi_A^{\text{new}}(I,J,K) = \frac{n_A(I,J,K)+1}{n_0}. \tag{13.14}$$

Note that the movement of the type A segment does not change the local densities of type B segments. Accordingly, the change in non-bonded energy is expressed as

$$\frac{\Delta E_{nb}}{k_B T} = f\left\{\phi_A^{\text{new}}(i,j,k), \phi_B^{\text{new}}(i,j,k)\right\} - f\left\{\phi_A^{\text{old}}(i,j,k), \phi_B^{\text{old}}(i,j,k)\right\}$$

$$+ f\left\{\phi_A^{\text{new}}(I,J,K), \phi_B^{\text{new}}(I,J,K)\right\} - f\left\{\phi_A^{\text{old}}(I,J,K), \phi_B^{\text{old}}(I,J,K)\right\} \tag{13.15}$$

with

$$f\left\{\phi_A(i,j,k), \phi_B(i,j,k)\right\} = \chi N \phi_A(i,j,k)\phi_B(i,j,k)$$

$$+ \frac{\kappa N}{2}\left\{1-\phi_A(i,j,k)-\phi_B(i,j,k)\right\}^2. \tag{13.16}$$

The acceptance of the movement is judged from the total energy change, $\Delta E = \Delta E_b + \Delta E_{nb}$. At $\Delta E \leq 0$, the new position is accepted. Even at $\Delta E > 0$, the movement is accepted if the Boltzmann factor, $\exp\left[-\Delta E_b/k_B T\right]$, is greater than a uniform random number generated between 0 and 1. Once

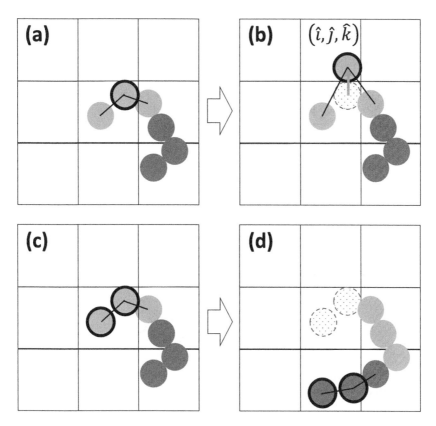

FIGURE 13.25 Schematic diagrams of MC trial moves: (a and b) random displacement and (c and d) reptation. The left and right figures are before and after the trial move, respectively.

the trial movement is accepted, the segment position and the local densities at the old and new cells are updated.

The random displacements are useful to rearrange the local packing behavior of the segments, but not efficient to drastically change a chain conformation. Alternatively, reptation moves can accelerate the formation of the self-assembled morphology of diblock copolymers. In the reptation scheme, the n_{cut} (≥ 1) segments are truncated from one end (head or tail), and then the same number of segments is attached to the other end (tail or head). The bond lengths associated with the newly added segments are kept the same as those of the removed segments. For the bulk of symmetric diblock copolymers with $N=32$, $n_{cut}=3-4$ was found very efficient to generate a lamella morphology from a random initial configuration [91]. Since there is no change in the bond lengths, energy change before and after the reptation move results only from the non-bonded interactions (Figure 13.25).

13.3.4 DSA Applications of MC Simulation

In this section, two typical examples of the TICG MC simulations are introduced: (1) density multiplications on a chemically patterned surface [106] and (2) the DSA hole shrink process using graphoepitaxy [96].

A thin film of symmetric AB diblock copolymers is coated on a chemically pre-patterned surface, whose thickness is L_z (Figure 13.26). The top surface of the film at $z=L_z$ is exposed to an air that is assumed to be neutral to the AB diblock copolymers. On the bottom surface at $z=0$, the film is contacted with the chemically pre-patterned surface composed of the guiding patterns (stripes)

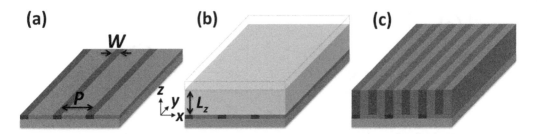

FIGURE 13.26 Schematic flow of chemoepitaxy DSA with lamella-forming symmetric diblock copolymers. The natural period of the lamellae is 1.0 L_0. (a) Chemically pre-patterned surface. Each "stripe" (with the width W of 0.5 L_0 and the pitch P of 2.0 L_0) represents the chemical guide pattern. The area between the stripes, which is referred to as "background," is filled with random copolymer brushes. (b) Casting of the symmetric diblock copolymers. The film is assumed to be flat with the thickness L_z of 1.0 L_0 and the top surface is covered with the neutral hard wall (dashed line). (c) Self-assembled morphology after annealing.

and the non-guiding area (background). The interaction energy between the pre-patterned surface and the segments, E_s, is described as a sum of the wall potential, U_s, over all segments [16, 106]:

$$\frac{E_s}{k_B T} = \sum_{i=1}^{nN} \frac{U_s(r_i, K_i)}{k_B T} = \sum_{i=1}^{nN} \frac{\Lambda(x_i, y_i, K_i)}{d_s/R_e} \exp\left[-\frac{z_i^2}{2d_s^2}\right], \qquad (13.17)$$

where d_s is the characteristic length of the wall potential. The other parameter, Λ, represents the interactive strength of the chemically pre-patterned surface, which varies with the location of the surface. It is assumed that the stripes are attractive to type A segments but repulsive to type B segments, and that the backgrounds are attractive to type B segments but repulsive to type A segments. For simplicity, the magnitudes of the attractive and repulsive interactions are set to be the same:

$$\Lambda = \begin{cases} -\Lambda_s/N \text{ for } A \\ \Lambda_s/N \text{ for } B \end{cases} \text{ on the stripes,}$$

$$\Lambda = \begin{cases} \Lambda_b/N \text{ for } A \\ -\Lambda_b/N \text{ for } B \end{cases} \text{ on the backgrounds,} \qquad (13.18)$$

where $\Lambda_s \geq 0$ and $\Lambda_b \geq 0$.

The model parameters used in the TICG simulations are summarized in Table 13.2. The values of χN, κN, and $\sqrt{\bar{N}}$ were chosen to match symmetric PS-b-PMMA with M_n of 74.0 kg/mol ($L_0 \cong 45.0$ nm) [106]. The cubic cell size ΔL was set at 0.166 R_e, which resulted in a reasonable number of coarse-grained segments per cell ($n_0 \cong 16$). With these model parameters, the L_0 in the bulk was estimated at ~1.66 R_e (i.e. ~10 cubic cells) [92].

In order to investigate the effects of chemical patterns on the self-assembled morphology of diblock copolymers, MC simulations were performed in a box of $L_x(4.0\,L_0) \times L_y(4.0\,L_0$ or $8.0\,L_0) \times L_z(1.0\,L_0)$ under periodic boundary conditions in the x and y directions [106]. The chemical pattern pitch, P,

TABLE 13.2
List of the Model Parameters

Parameter	χN	κN	$\sqrt{\bar{N}}$	$N(N_A/N_B)$	d_s/R_e	L_z/L_0	P/L_0
Value	25	35	128	32 (16/16)	0.15	1.0	2.0

was fixed at $2.0\,L_0$, whereas the width of the chemical stripes, W, was varied from $0.0\,L_0$ to $2.0\,L_0$. At each W, each of the two chemical affinities, Λ_s and Λ_b, was varied from 0.0 to 2.0. The system was equilibrated using several different types of MC moves, e.g. random displacement, reptation, whole-chain translation, and swapping of the two blocks [84].

Figure 13.27 illustrates the five distinct morphologies of symmetric AB diblock copolymers obtained from the TICG MC simulations. The color indicates the value of the local order parameter, $\psi = (\phi_A - \phi_B)/(\phi_A + \phi_B)$, which was averaged over the y direction for the xz cross-section image (left), and over the z direction for the top-down xy image (right). The three different depths, d, were selected for averaging ψ in the z direction from the top surface; $0.125\,L_0$, $0.500\,L_0$, and $1.000\,L_0$. The first distinct morphology is the horizontal lamellae that alternate the A-rich and B-rich layers in the z direction. The vertical lamellae are categorized into two types (VL or VL′). If the B-rich domains are vertically oriented to the substrate and not connected with each other, the morphology is classified as VL; otherwise, it is classified as VL′. The mixed lamellae (ML) are composed of horizontal

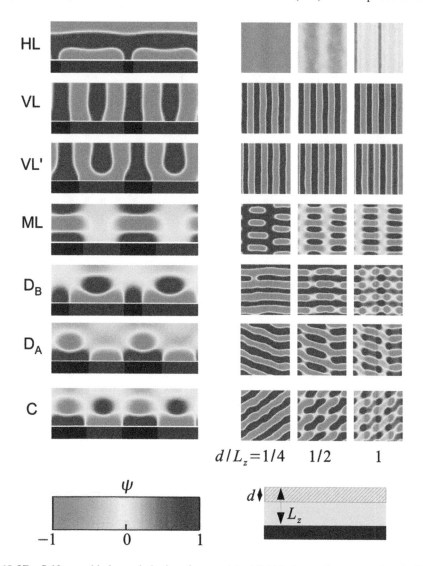

FIGURE 13.27 Self-assembled morphologies of symmetric AB diblock copolymers on chemically pre-patterned surfaces with various guide widths and guide/background affinities. (Reprinted with permission from Ref. [16]. Copyright (2010) American Chemical Society.)

lamellae above the stripes and vertical lamellae above the background. The checkerboard (*C*) and dot (*D*) morphologies change with the *z* position. For example, in the checkerboard morphology, the bottom layer replicates the guide patterns, but the layer on top of the bottom has opposite domains. In the "*B*-dots" morphology (D_B), the *B*-rich domains straddle over the *A*-rich domain which covers each stripe. In both morphologies, the lamella pattern is observed on top of the film whose orientation is not matched with that of the guide pattern.

Figure 13.28 summarizes the representative morphology at a given set of three parameters: *W*, Λ_s, and Λ_b. For the narrowest stripes ($W=0.25\ L_0$), the morphology at $\Lambda_b \geq 0.5$ is horizontal lamellae, regardless of Λ_s (Figure 13.28a). This is because the horizontal lamellae can maximize the contact area between the *B*-rich domains and the *B*-attractive backgrounds which occupy 87.5% of the pre-patterned surface. As the stripe width is increased from $0.25\ L_0$, some *B*-dots morphologies appear at $W=0.50\ L_0$ (Figure 13.28b), and they become predominant at $W=0.75\ L_0$ (Figure 13.28c). At $W=1.00\ L_0$, several different types of morphologies are observed, including the *A*-dots and the checkerboards (Figure 13.28d). It is clearly seen in Figure 13.28d that vertical lamellae are formed in a small portion of the parameter spaces explored here.

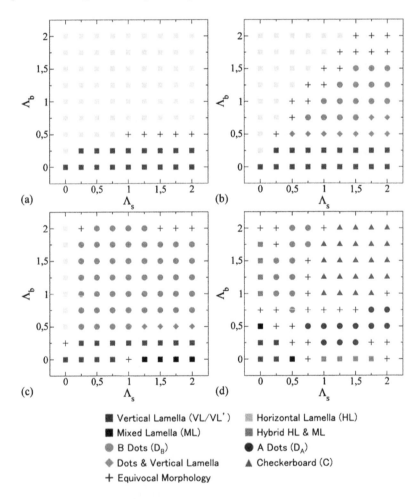

FIGURE 13.28 Basic morphologies arising from chemically pre-patterned surfaces of various Λ_s and Λ_b. Four different stripe widths were employed: $W/L_0 =$ (a) 0.25, (b) 0.50, (c) 0.75, and (d) 1.00. The stripe pitch and the film thickness were set at $2.00\ L_0$ and $1.00\ L_0$, respectively. (Reprinted with permission from Ref. [16]. Copyright (2010) American Chemical Society.)

The second example of the TICG MC simulations is the DSA hole shrink process (see Section 13.5.3 for details). The DSA hole shrink has been extensively investigated in the past few years, along with density multiplication. The basic process flow of the DSA hole shrink is illustrated in Figure 13.29. First, the guide holes (e.g. contact or via holes) are etched into the spin-on-glass (SOG) and spin-on-carbon (SOC) hard masks (HMs), which are commonly used in conventional lithographic processes. Next, an asymmetric PS-*b*-PMMA is spin-coated and annealed on the etched wafer. The volume fraction of the PMMA block is often selected to be ~0.30, which is suitable for generating a PMMA-rich cylindrical domain in the middle of a PS-rich matrix. The diameter of the self-assembled PMMA-rich domain ranges from ~15 nm to ~30 nm, depending on the molecular weight of the PS-*b*-PMMA. The sidewall of the guide hole is usually wetted by a thin PMMA layer, due to the hydrophilic SOG/SOC surface. Finally, after the selective removal of the PMMA, the remaining PS structure can be used as an etch template for fabricating even smaller holes.

One of the most crucial problems in the DSA hole shrink process is the morphological defect (Figure 13.29d) [107]. The PMMA-rich cylindrical domain does not reach the substrate due to a relatively thick PS-rich layer. Since the PS residual layer is difficult to observe in top-down scanning electron microscope (SEM) images, computer simulations have been performed to predict the three-dimensional morphology of PS-*b*-PMMA in the guide hole, including the DPD [108, 109], SCFT [110], Ohta–Kawasaki (OK) [107], and TICG MC [96]. In the following paragraphs, we will review an example of the application of the TICG MC simulations to the DSA hole shrink process.

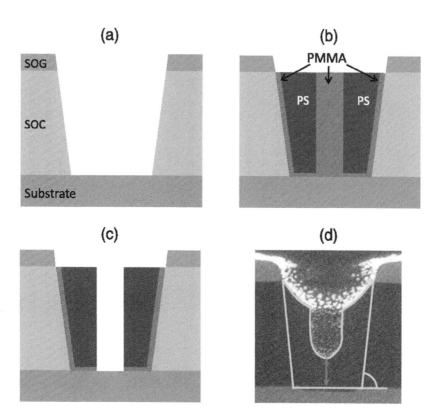

FIGURE 13.29 DSA hole shrink process. (a–c) Schematic images and (d) cross-sectional TEM image taken after the PMMA removal. In (d) the outline of the remaining PS domain is highlighted while the thickness of the PS residual layer is marked with an arrow. (Reprinted from Ref. [107] with permission.)

The model parameters considered here correspond to the asymmetric PS-b-PMMA (24 kg/mol for PS block and 47 kg/mol for PMMA block): $f=0.33$, $\sqrt{\bar{N}}=116$, $\chi N=26$, and $\kappa N=35$. To reflect the cylindrical geometry to the interactive energy, Eq. 13.17 was modified to

$$\frac{E_s}{k_B T} = \sum_{i=1}^{nN}\left[\frac{(2\delta_{KA}-1)\Lambda_s}{d_s/R_e}\exp\left(-\frac{r_i^2}{2d_s^2}\right) + \frac{(2\delta_{KA}-1)\Lambda_b}{d_s/R_e}\exp\left(-\frac{z_i^2}{2d_s^2}\right)\right], \quad (13.18)$$

where the first and second terms represent the interactive energy of a statistical segment with the sidewall and with the bottom surface, respectively. The d_s was set at 0.15 R_e. Kronecker's delta, δ_{KA}, becomes unity if the segment is type A, otherwise it is zero. The distance from the sidewall, r_i, is calculated from $r_i = D/2 - \sqrt{x_i^2 + y_i^2}$ with D as the diameter of the guide hole. In Eq. 13.18, there are two key parameters, i.e. Λ_s and Λ_b, for the interactive strength between the segment and the sidewall, and that between the segment and the substrate, respectively.

Figure 13.30 shows the results of particle-based MC simulations for the self-assembled morphology of asymmetric PS-b-PMMA in the cylindrical guide hole. The height of the guide hole was fixed at 1.5 L_0, whereas the diameter of the guide hole was varied from $D/L_0=2.00$ to $D/L_0=2.44$. Within the parameter range considered here, the PMMA-rich domain became either a "full" cylinder that extended through the hole (Figure 13.30a), or a "partial" cylinder that was disconnected from the substrate (Figure 13.30b). As the diameter of the guide hole was increased from $D=2.00\ L_0$ to $D=2.44\ L_0$, the PMMA chains in the full cylinder were gradually stretched. As a result, the full cylinders were observed only at considerably lower Λ_b (i.e. almost neutral surface) (Figure 13.30c–f).

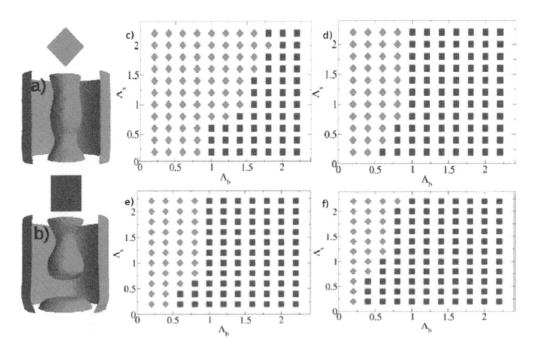

FIGURE 13.30 Self-assembled morphology of asymmetric PS-b-PMMA in the cylindrical guide hole with a height of 1.5 L_0. (a and b) Images of the PMMA-rich domains obtained from the simulations: (a) full cylinder that is connected to the substrate and (b) partial cylinder that is disconnected from the substrate. (c–f) Morphologies in the guide hole with four different diameters: $D/L_0 =$ (c) 2.00, (d) 2.12, (e) 2.28, and (f) 2.44. The values of the interactive strengths, Λ_s (sidewall) and Λ_b (substrate), are varied from 0.2 to 2.2. The full and partial cylinders are represented by diamonds and squares, respectively. (Reproduced from Ref. [96] with the permission of John Wiley and Sons.)

13.3.5 OUTLOOK

The results from the TICG MC simulations (e.g. Figures 13.28 and 13.30) have been useful in optimizing the chemoepitaxy DSA processes in experiments. Recently, the TICG MC simulations have been extended to optimize the design and interactive parameters of chemical guiding patterns using the covariance matrix adaptation evolution approach [111]. It was demonstrated that the evolution approach is much more efficient in finding the optimized parameters than conventional algorithms (e.g. random sampling of the parameter space). Such an evolution approach may be useful to identify optimal DSA materials and processes to achieve defect-free patterns. In addition, the TICG simulations have been used to reconstruct a three-dimensional lamella structure on a chemically pre-patterned surface, with the data obtained from small-angle X-ray scattering (SAXS) and grazing-incident small-angle X-ray scattering (GI-SAXS) measurements [112, 113]. The SAXS and GI-SAXS data reflects the three-dimensional structure of the self-assembled lamellae and of the guide patterns, whose details are very hard to see in conventional SEM or TEM images. Some of the resulting morphologies reveal the presence of relatively large edge roughness or irregularity in the vertical orientation [113]. Such buried defects in the film may significantly affect the following pattern transfer step (see Section 13.4).

TICG MC simulations are not limited to predicting the self-assembled morphology of BCPs on a pre-patterned surface at equilibrium. They can also estimate the free energy of the defect formation in DSA [114, 115]. One of the challenges for the TICG MC approach is to simulate a dynamic change (i.e. kinetics) in the morphology of BCPs over time. Such simulations require large time scales and length scales. The recent progress on the multi-scaling technique shows that the TICG MC simulations can be coupled with continuum dynamics models and thus the computational speed can be accelerated [102, 116].

13.4 PATTERN TRANSFER

Pattern transfer using self-assembled block copolymers typically begins with selective removal of one of the blocks, which is analogous to the development step in the photolithography process. The main considerations for this "BCP development" step is (1) the etch selectivity between the two blocks and (2) the etch selectivity between the remaining block and the immediate underlying layer. In other words, this process needs to leave sufficient material of the remaining block, after removal of the other block, so that the remaining material can serve as an etch mask for the subsequent pattern transfer step. In addition to selectivity, uniformity and process margin are also essential for all patterning techniques. Uniformity refers to feature size distribution and roughness within wafer, within batch, and between batches, which usually depend on the nature of the etch process and tool stability. Process margin is sometimes referred to as process window, which means how much process error or variation can be tolerated while still achieving the desired results. For example, a successful BCP development etch would have a thicker remaining block A than the minimum requirement after block B is completely removed. The amount of excess A compared to the minimum requirement determines how much additional etch time (process error) can be tolerated. In this case, a high selectivity BCP development process that consumes less material A while etching the same amount of B would have a larger process margin or "process window" than a low selectivity process.

In this section, we will focus on the options for BCP development of the most commonly used PS-*b*-PMMA system.

13.4.1 WET AND DRY PMMA REMOVAL

Selective PMMA removal after self-assembly without altering the assembled structure can be achieved in several different ways. Thurn-Albrecht et al. utilized the actinic property of PMMA to deep UV irradiation and demonstrated the degradation of the PMMA block in perpendicularly aligned cylinder-forming PS-*b*-PMMA followed by dissolving the degraded PMMA segments in acetic acid while leaving the PS structures intact [117]. This approach utilizes very simple tools, materials, and processes, and results in nano-structures with a feature size and uniformity that even the most advanced lithographic tools cannot easily achieve, and hence, it has been widely adopted [76]. Since then, many applications have been demonstrated using this method, such as (1) a high aspect ratio structure in Si achieved by etching the PS structure into an inorganic hard mask followed by a high aspect ratio Si etch [9]; (2) semiconductor nano-dots fabricated by deposition of a semiconductor material into the hard mask template as in (1) followed by a blanket "etch back" to remove excess material on top of the hard mask template [118]; (3) metal nano-dots achieved by the evaporation of metal into a PS template followed by a lift-off process, also using a further Si etch, as shown in Figure 13.31 [76, 119]; (4) fabrication of III-V semiconductor quantum dots by metalorganic chemical vapor deposition (MOCVD) from a III-V substrate with an inorganic template patterned by BCP [52]; and (5) isolated magnetic material used in bit-patterned media (BPM) [120].

This "wet PMMA development" process has almost infinite selectivity for PMMA over PS and excellent process control due to the large process window for DUV dose and wet chemical process time. However, acetic acid has a National Fire Protection Association (NFPA) health hazard rating of 3 and a flash point of 40°C, which means that when it is used in a high-volume manufacturing environment, ventilation and chemical waste disposal need to be carefully handled. To address this issue, several commonly used chemicals were identified as a replacement for acetic acid [121, 122]. Further defectivity studies of the wet development process using one of the new chemicals have been successfully demonstrated on 300 mm wafers in a HVM environment [123–125]. It is also worth noting that these studies show a very low percentage of DSA-induced defects under optimized conditions.

Not only can this wet development process be used for cylinder-forming PS-*b*-PMMA perpendicular to the substrate, but it is also applicable to parallel cylinders and lamellar-forming PS-*b*-PMMA [66]. Even though both parallel cylinders and perpendicular lamellae can provide grating-like structures, as previously shown in Figure 13.16, the cross-sectional profile is quite different. As discussed previously, perpendicular lamellae are usually better etch templates; however,

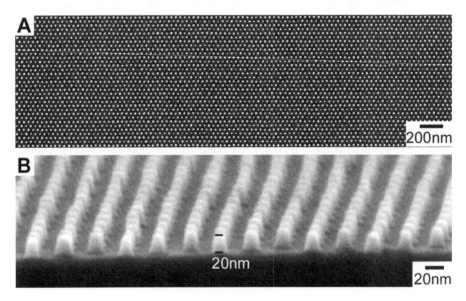

FIGURE 13.31 Hexagonal array of (A) Cr and (B) Si pillars derived from the DSA of cylindrical PS-*b*-PMMA. (Reproduced from Ref. [76] with the permission of *Science*, AAAS.)

an important concern when preparing a high aspect ratio (i.e. feature height over width) structure using a wet process is pattern collapse induced by capillary forces. A collapse-free perpendicular lamellar structure with an aspect ratio of ~1–1.5 was demonstrated when using acetic acid as the developer and H_2O for the rinse. Lower surface tension and contact angle developer and rinsing solvent may result in a higher aspect ratio [122, 126]. One should note that the mechanical properties of polymers are known to degrade as M_w. When smaller L_0 BCP is used, the achievable aspect ratio may decrease.

To address the pattern collapse issue, plasma etching, which is a gaseous phase reaction, could be a better option because capillary forces will be less of a concern. The general mechanism of plasma etching is to use reactive species or physical ion bombardment to break polymers into small volatile segments. Based on this viewpoint, the etch behavior of most linear organic polymers is similar to a certain extent because the backbone of organic polymers is composed of C–C bonds. As a result, the PMMA/PS selectivity of a plasma etching process (referred to as the "dry development process" below) is usually much lower than the wet development. Ohnishi et al. studied the etch rate of several organic polymers under a high ion energy condition and found strong correlation between the etch rate and the "Ohnishi parameter," i.e. the total number of atoms divided by the number of carbon atoms minus the number of oxygen atoms in the repeating units [127]. They also pointed out that this empirical correlation does not hold under a low ion energy, radical-only etch condition. The two extreme cases in Ohnishi's study represent pure physical and pure chemical etching, respectively, which implies the lower and upper limit of PMMA/PS selectivity is achievable. Since neither high ion energy nor radical-only conditions are commonly used in commercially available etchers, many researchers further studied the etch behavior of PS-b-PMMA under different types of plasma systems and chemistries.

Ting et al. studied the effect of etch selectivity and surface roughness of different etch chemistry, i.e. Ar, O_2/Ar, and fluorine-based gases, using a high density plasma system [128]. They showed that pure-Ar plasma may provide higher selectivity on a blanket film of homopolymers than other chemistries, but it exhibits deformation and high roughness on self-assembled structures. O_2/Ar plasma was recommended due to its roughness performance, slightly better PMMA/PS selectivity, and outstanding BCP/inorganic selectivity compared to other fluorine-based chemistries. The high selectivity to inorganic materials is crucial because any spatial non-uniformity in the etch rate of the dry development process can be mitigated by using an inorganic etch stop layer directly underneath the BCP layer. Other etch chemistries have also been studied, such as CO/H_2 [129, 130]. Higher selectivity was achieved through the deposition of plasma-generated polymers on PS domains. However, if the balance between etch rate and deposition rate is not well tuned, this type of etch process usually tends to induce additional roughness.

Furthermore, as pointed out by Ohnishi and other researchers that PMMA/PS selectivity depends on the ion energy and radical/ion ratio, one can infer that the distribution of the ion energy is also a key factor in the resulting selectivity. It is well known that the ion energy distribution (IED) in a dual-frequency capacitively coupled plasma (CCP) tool is much wider than high density plasma tools such as inductively coupled plasma (ICP) or helicon, and therefore it is more difficult to optimize the selectivity in CCP tools. Fortunately, Yamashita et al. found that the ion energy distribution in a CCP tool can be tuned by the frequency of the source power, such that the IED from a 100 MHz source is narrower than a 40 MHz source and similar to an ICP tool [131]. In addition, the IED will not deteriorate when modulating the ion energy by applying a different source power. Chan et al. further utilized this low ion energy, tight IED with an O_2/Ar plasma system and proposed a cyclic process using Ar/O_2 and Ar-only plasma to maximize the PS height, demonstrating a few techniques for Si etching, as shown in Figure 13.32 [132].

In short, if the desired pattern geometry does not have a potential pattern collapse concern, e.g. holes or low aspect ratio structure, wet PMMA development could be a better option because of its outstanding selectivity. Otherwise, dry development with proper process optimization could provide sufficient selectivity to support the subsequent pattern transfer.

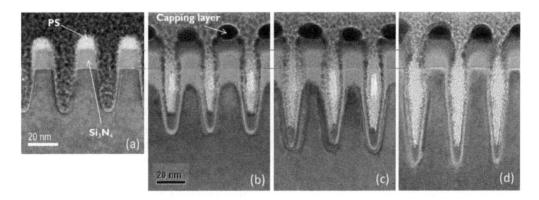

FIGURE 13.32   An example of deep Si etch, where the PMMA removal was done by using low ion energy, tight ion energy distribution O_2/Ar plasma and cyclic process using Ar/O_2 and Ar-only plasma. (Reprinted from Ref. [132]. Copyright (2014), with permission from Elsevier.)

13.4.2   SEQUENTIAL INFILTRATION SYNTHESIS

An alternative approach to achieve selectivity between PS and PMMA is to introduce an etch-resistant atom into one of the blocks. Peng et al. first demonstrated the use of an atomic layer deposition (ALD)-like process to selectively incorporate metal precursors into a PMMA domain, a process termed "sequential infiltration synthesis" (SIS). The selectivity in the SIS process is achieved by the chemical absorption of the ALD precursor with the carbonyl groups in PMMA [133]. Trimethyl aluminum (TMA) or titanium tetrachloride (TiCl4) were employed to demonstrate the fabrication of Al_2O_3 and TiO_2, respectively. The major difference between an ALD process and a SIS process is illustrated in Figure 13.33, where the ALD process is orders of magnitude lower in process pressure and shorter in exposure time. Consequently, the ALD process leads to surface reaction only because under such conditions the precursors cannot penetrate into the BCP film for uptake by the PMMA domain [134].

Further studies of the SIS mechanism, especially the reaction between the TMA precursor and PMMA, were conducted by Biswas et al. using *in situ* Fourier transform infrared (FTIR) spectroscopy [135]. A two-step reaction was suggested, where the first step allows TMA and PMMA to form an unstable intermediate complex in a relatively short amount of time and the second step converts

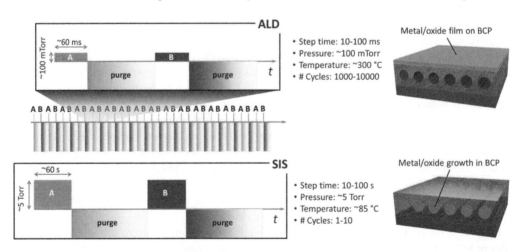

FIGURE 13.33   A comparison of ALD and SIS for BCP metallization. (Reproduced from Ref. [134] with the permission of The Royal Society of Chemistry.)

the intermediate product into covalently bonded structures. It was shown that the second step was the rate-limiting step and the forward/reverse reaction of the first step was kinetically balanced. Therefore, higher pressure, longer TMA exposure time, and shorter TMA purge time will promote the forward reaction and hence the Al_2O_3 growth within PMMA.

Ruiz et al. demonstrated the compatibility of SIS and DSA with sub-14 nm features and further pattern transfer into an Si substrate, as shown in Figure 13.34a–d [136]. In addition to etch selectivity improvement, Singh et al. observed that the resulting line edge roughness (LER) using the SIS process could be better than the typical selective dry etch process [137]. Even though the SIS structure is derived from the PMMA domain while the dry etch leaves the PS domain behind, it is believed that the intrinsic LER of the two domains should be the same. The mechanism of the LER smoothing effect by the SIS process is still unclear and requires further investigation. Additionally, the SIS process can also be applied to several other BCPs besides PS-b-PMMA, such as PS-b-P2VP, where TMA is selectively adsorbed to the P2VP domain due to the pyridine group [138].

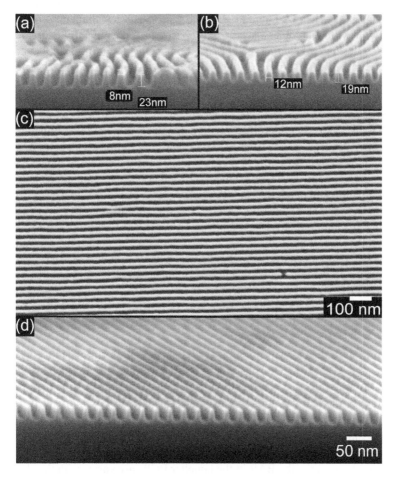

FIGURE 13.34 BCP films of a lamellae-forming PS-b-PMMA with L_0 of 27 nm were treated with three TMA/water cycles for sequential infiltration synthesis, etched by O_2 plasma for polymer removal, and etched by fluorine plasma for pattern transfer to the underlying Si substrate. (a and b) Cross-sectional SEM images of the resulting Si lines from BCP fingerprint patterns before (a) and after (b) removal of the remaining Al_2O_3 mask. (c) Top-down and (d) cross-sectional SEM images of the resulting Si lines from BCP patterns self-assembled on chemical patterns. (Reprinted with permission from Ref. [136]. Copyright 2012, American Vacuum Society.)

13.5 POTENTIAL LITHOGRAPHIC APPLICATIONS

13.5.1 BIT-PATTERNED MEDIA

For conventional hard disk drive or magnetic recording media, the disk or media contains a layer of polycrystalline, granular, magnetic material where a data "bit" is stored in an ensemble of grains. The higher storage capacity of a fixed size disk, often referred to as the areal density, requires a smaller number of grains per bit, resulting in larger bit size variation and lower signal-to-noise ratio (SNR). Consequently, the grain size needs to be scaled together with the areal density. However, smaller grain size magnetic material is less thermally stable due to superparamagnetism, i.e. the higher probability for a bit to flip on its own, and data reliability over a reasonable amount of time, e.g. 10 years, can no longer be guaranteed [139]. Using a material with high magnetic anisotropy could improve the thermal stability; however, the difficulty of writing data increases. The conflicting objectives from media signal-to-noise ratio, thermal stability, and media writeability are sometimes referred to as the media design "trilemma" in the magnetic recording field.

In recent years, the areal density scaling trend has been decelerating significantly and it seems that achieving an areal density >1 terabit per inch square is very challenging based on current technologies. Two new technologies have been proposed to fundamentally address the trilemma. Energy-assisted recording, including heat-assisted magnetic recording [140], and microwave-assisted magnetic recording [141] enable the use of high anisotropic magnetic material for small grains because the writing resistance, i.e. coercivity, can be reduced momentarily when energy is applied to the media. Miniaturization of the heating device, integration with the write head, and cooling after writing are the current challenges of this technology. The second approach is bit-patterned media [142, 143], in which data bits are stored in lithographically patterned magnetic islands instead of continuous granular media. The islands are uniformly patterned, better isolated, and significantly larger than conventional media grains, thereby exhibiting higher SNR and better thermal stability compared to conventional media. However, the extra manufacturing cost incurred by the additional patterning processes, e.g. lithography and etch tools, materials, and potentially reduced throughput, is a major concern for BPM. Several cost-effective BPM disk fabrication processes, which involve multiple novel nanofabrication elements, were proposed by Albrecht et al., Xiao et al., and other research groups [144, 145]. To address the patterning cost and support the required throughput, nanoimprint lithography (NIL) is a natural choice for creating the patterns on the final BPM disks. On the other hand, the master template for NIL still has to be prepared with a specialized e-beam tool. As previously shown in Figure 13.20A and C, the e-beam pattern quality at the target area density is insufficient for BPM use, but DSA rectification and density multiplication could work. This section will focus on the DSA process involved in BPM master template fabrication.

Sphere-forming [146], cylinder-forming [76, 147], and lamella-forming [148] BCP systems have all been considered for BPM use. One major design consideration is the bit-cell aspect ratio (BAR), which is defined as crosstrack pitch to downtrack pitch, i.e. pitch in the r-direction in a polar coordinate system to pitch in the θ-direction. For example, spherical and cylindrical BCP systems form circular bits in a hexagonal array with a BAR of $\sqrt{3}/2 = 0.87$. On the other hand, forming bits with a lamellar system requires two line/space DSA processes: a zoned radial line pattern and a circumferential track pattern. Rectangular bits in a rectangular array can be obtained from the intersection of these two patterns. Apparently, the BAR in the lamellar BCP case varies with the choice of BCP(s), e.g. BAR = 1, if the same BCP is used to create the line/space patterns in both directions, or BAR $\neq$ 1, if different BCPs are used. Due to the design of conventional write heads, the magnetic field gradient in the downtrack direction is typically at least 50% larger than in the crosstrack direction [144]. Therefore, the optimum BAR in this case would be greater than 1. One should note that the magnetic field gradient may vary with write head design and writing mechanisms, which may further affect the optimum BAR as well. Figure 13.35 illustrates BPM magnetic dots in a hexagonal or rectangular bit-cell array with different areal density and BAR.

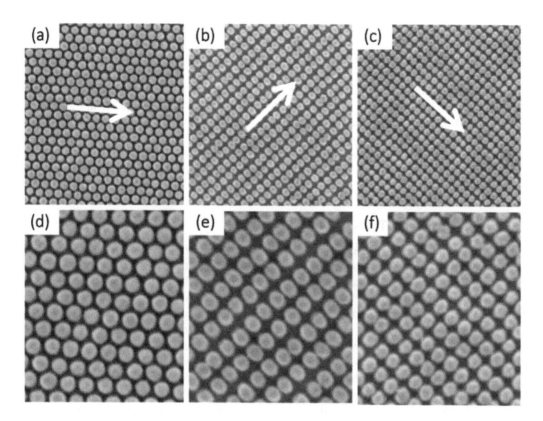

FIGURE 13.35 SEM images of BPM magnetic dots with an areal density of (a) 1.0 Td/in^2 in a 27.3 nm hexagonal array, (b) 1.2 Td/in^2 in a 27 nm circumferential $\times$ 20.5 nm radial rectangular array, and (c) 1.6 Td/in^2 in a 22 nm circumferential $\times$ 18.5 nm radial rectangular array. Arrows indicate the downtrack direction. (d–f) Higher magnification images of the same samples as (a–c). (Copyright 2015 IEEE. Reprinted with permission from [144].)

A few additional considerations for BPM using DSA need to be taken into account. First, the tracks are usually grouped in "zones" and bits are placed at a constant angular pitch within a zone for ease of recording. In other words, the angular velocity is constant within a zone but the linear velocity still varies at different r. Therefore, the bit spacing changes proportionally to the radius. Unlike granular media, the bit-cell array formed by BCP systems is mainly determined by the material property L_0. To accommodate the continuous change in pitch, compression or stretching of the BCP from its L_0 is needed. In addition, due to the mechanical design of the write head actuator, which is centered outside of the disk, the trajectory of the write head movement is an off-centered arc relative to the disk, which requires the lattice to be further skewed by up to $\pm 15°$. The lattice distortion is an important consideration for BPM because polymers can only tolerate a certain amount of stretching/compression, or assembly defects will readily appear. If needed, one could reduce the width of the zone, which will further reduce the maximum distortion of the lattice. Second, the majority of a BPM disk is for data storage and thus in the form of a high density array. However, for positioning purposes, a non-regular structure called a servo pattern is needed every few degrees along the track, which provides the necessary information to locate the data. For granular media, servo patterns are recorded onto the media in the factory with write heads, which is time-consuming and costly. For BPM, it is possible and beneficial to create servo patterns as a separate region/shape from the bit-cell during the fabrication process. Nevertheless, because BCP has inherent, regular geometries, the integration of conventional servo patterns in the DSA process will be challenging and defective. If the bit-cell and servo patterns are formed in separate lithography/etch steps, the alignment between the two patterns needs to be carefully handled. A more ideal option would be to write both pattern regions in one lithography step. In that case, a slightly more complicated process would be needed, i.e. first protect the data region and transfer the servo patterns, and then protect the servo region and finish the DSA process, as demonstrated by Albrecht et al. [144]. A third option would be to redesign the servo patterns and make them DSA-friendly. Several DSA-friendly servo patterns have been proposed and studied, such as "offset burst" and chevron shapes [144].

Great progress has been made in the BPM field in the past few years. Since the introduction of DSA for BPM master template fabrication, the density multiplication and pattern rectification of the DSA process have greatly improved the pattern quality and have made this technology more amenable for high-volume manufacturing. In addition, HDD is especially suitable for DSA because of the high defect tolerance of the recording media. BPM is considered one of the most promising applications of DSA.

13.5.2   Fin Field Effect Transistor (FinFET)

One potential application of DSA for semiconductor manufacturing is to create a dense array of fins for FinFET devices. The conventional approach for fin formation relies on the sidewall image transfer (SIT, aka self-aligned double/quadruple patterning, SADP/SAQP) process to create a "sea-of-fins" followed by two or more lithographic customization steps that remove or preserve part of the array. Figure 13.36 illustrates one of the possible approaches to form fins. In this process flow, a DSA process is used to create a uniform fin array, which will then be transferred to a hard mask layer. The customized lithography and etch are implemented after the hard mask formation, followed by Si fin etch. Many variations of this process could be envisioned, for example, based on the needs and considerations of the process, different DSA processes could be used, the "fin cut" could be done at earlier or later stages, and even the silicon-on-insulator (SOI) substrate could be replaced by bulk Si. Figure 13.37 is an integrated example of using this flow for 42 nm pitch fin formation, customization, and gate formation on an SOI substrate. A fully integrated FinFET device using a HVM-compatible toolset was demonstrated by Liu et al. [5]. One should note that those customization patterns in parallel with the fins are especially critical because the edges of the shapes need to be accurately placed in between two adjacent fins. The tolerance for placement error, including

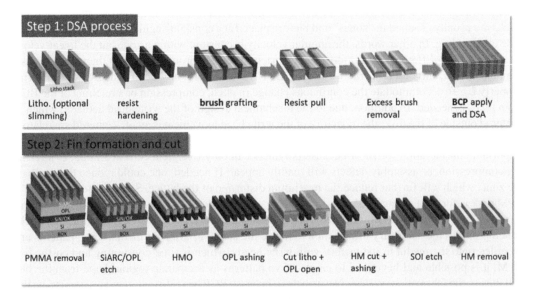

FIGURE 13.36   An example process flow for fin formation using DSA. (Reproduced from Ref. [5] with permission.)

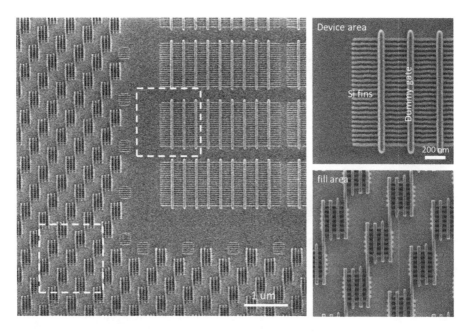

FIGURE 13.37 An integrated example of using the flow in Figure 13.36 for fin formation, customization, and gate formation on an SOI substrate. (Reproduced from Ref. [5] with permission.)

overlay, critical dimension uniformity (CDU), and LER/line width roughness (LWR) of both fin and "cut" pattern, is approaching the capability of current manufacturing toolsets for the 7 nm node, i.e. sub-30 nm fin pitch. Therefore, advanced process control (APC), e.g. feedforward/feedback between lithography and etch tools as well as novel within-wafer uniformity control tools, needs to be implemented in order to ensure manufacturability and high yield. Innovation and improvement of the limiting factors, e.g. CD/pitch uniformity, placement error, and LER, are highly valuable.

In order to address the edge placement error (EPE) issue for sub-30 nm pitch fins, several different DSA processes have been considered, in which a BCP with $L_0 = 27$ nm was used as an example. As mentioned in previous sections, many DSA process flows have been proposed and investigated. The three DSA options chosen and compared here are simply those that have been reported with fin and cut studies, all are 193i based with an integrated film stack, improved materials, and 300 mm HVM-compatible toolsets. Figure 13.38 illustrates the basic process flows and further details can be found in the references. The three options are (1) Grapho DSA with a tone-inversion process [69, 149], (2) a "sea-of-fins" approach by Chemo DSA [150, 151], and (3) a self-aligned cut by a "Hybrid" DSA process [152]. One should note that these DSA approaches aim to improve different terms in EPE. For example, in Grapho DSA, the guiding patterns become the fin cut after tone inversion, so that the fins are "self-aligned" to the cut shapes and the overlay issue is resolved. Tsai et al. employed an e-beam version of this Grapho process to demonstrate scaled FinFET at 28 nm pitch and further characterized the resulting device performance electrically [69]. Chemo DSA is similar to the SAQP process, which still relies on a later lithographic cut process for customization. Apparently, this approach does not relax the requirement for alignment accuracy. However, Chemo DSA provides a much more uniform pitch compared to SAQP and eliminates the commonly seen "pitch-walking" problem, which is also a dominant factor in EPE [153]. It is worth mentioning that Sayan et al. studied the effect of the patterning sequence of the fin and the cut. Instead of overlaying the cut shape on the fin array, as shown in Figure 14–36 Step 2, they demonstrated patterning the cut shape before the fin array is formed. Because the surface is planarized and the cut is buried underneath during the DSA process, the alignment marks will not create islands/holes and thus could be beneficial for later lithography processes. Hybrid DSA has the critical cut information embedded

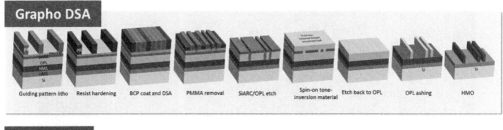

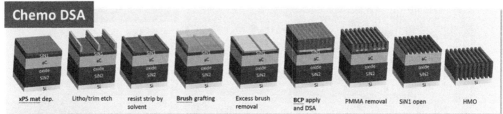

FIGURE 13.38 Process flow for fin formation and customization using different DSA approaches. (Reproduced from Ref. [151] with permission.)

in the guiding patterns, so that some of the fins will be removed during the etch transfer process without an additional lithography step. Therefore, fins are also self-aligned to the cuts as in Grapho DSA. The main differences between Grapho and Hybrid DSA is that Hybrid DSA still requires a gridded design, i.e. sea-of-fins-like array as in Chemo DSA, while the Grapho DSA could use a non-gridded design, as long as the trench width is commensurate with L_0. However, the etch resistance, reflow temperature, and profile of the resist guiding pattern plus the tone-inversion process would aggravate the pattern transfer challenges for Grapho DSA.

Figure 13.39 demonstrates the Si fins created by Chemo DSA and Hybrid DSA processes. Both are at similar fin CD and depth; however, in the case of Hybrid DSA, one can see the difference in etch depth and profile between the fins with a narrow opening to the next fin and the fins next to a large opening. This etch loading effect could be further optimized, but would require a certain amount of process fine-tuning, and new etch tool capability, such as gas/power pulsing, would be needed. This structural demonstration nicely points out the pros and cons of fin formation and customization using these two processes. Further fin cut demonstrations of this Chemo DSA process can be found in the literature [151].

13.5.3 Hole Shrink

Directed self-assembly of block copolymers has become a promising patterning technique for the advanced node hole shrink process due to its material-controlled CD uniformity and process simplicity [154]. For such an application, the cylinder-forming BCP system has been extensively investigated and compared to its counterpart, the lamella-forming system, mainly because cylindrical BCPs will form multiple vias in non-circular elongated guiding patterns, which is considered to be closer to technological needs [124, 125, 155, 156]. The technological need to generate

FIGURE 13.39 Examples of pattern transfer using Chemo DSA (upper) and Hybrid DSA (lower) into Si substrates. (Reproduced from Ref. [151] with permission.)

multiple DSA domains in a bar-shaped GP originated from the resolution limit of lithography, i.e. the bar-shape or sometimes "peanut-shape" GP was derived from merged vias placed too close to each other. In practice, in order to avoid unintentional merged vias shorting the circuit, multiple patterning and/or self-aligned via (SAV) processes have been implemented in semiconductor manufacturing to address this resolution issue [157]. The former approach separates one pattern layer with unresolvable dense features into several layers with lower density, resolvable features, while the latter approach simply utilizes selective etching and the superposition of via bars and the pre-defined metal trench patterns in a thin hard mask layer to resolve individual vias. For example, looking from a top-down view, a horizontal line/space array is defined in a layer of hard mask where the space can be later (but has not yet) etched into the underlying dielectric layer and become the metal lines after the Cu dual-damascene process. Here, we define the metal layer in the hard mask as Metal_x, the metal layer underneath as Metal_$(x − 1)$, and the thickness of the dielectric layer as h_Mx. If a vertical via bar is patterned in a lithography stack over this hard mask pattern and undergoes a dielectric etch, apparently only the intersections of the via bar and the space in the hard mask will be transferred into the dielectric layer. One should note that this etch step only etches $1/2\ h\_Mx$ deep into the dielectric layer. The lithography stack with the via patterns will then be stripped away, followed by another dielectric etch process with a similar target etch depth. Now the metal trenches are defined at $1/2\ h\_Mx$ deep in the dielectric layer and those intersections previously etched will be fully etched to h_Mx deep and will become the vias connecting Metal_x and Metal_$(x − 1)$ after the Cu process.

The conventional use of DSA for via applications is to recover merged vias back to separated ones so that fewer multiple patterning steps would be needed. Great efforts have been made in metrology, modeling, materials, process development, and designs [32, 154, 155, 158, 159]. One of the challenges in this type of DSA process is that the placement error of the resulting DSA vias is highly sensitive to the GP shape and the GP CD variation [160]. Placement error affects the contact area between the vias and the metal lines and the corresponding contact resistance, which has a significant impact on circuit performance and reliability. The SAV process greatly improves the overlay tolerance and is thus widely adopted. With a proper design, using a lamella-forming BCP

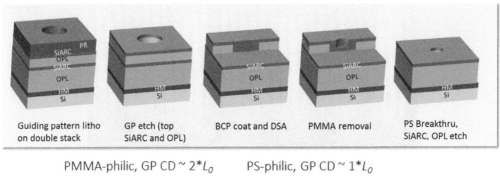

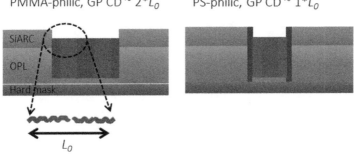

FIGURE 13.40 (Upper) Process flow of graphoepitaxy DSA via shrink and (lower) examples of sidewall preference. (Reproduced from Ref. [161] with permission.)

system to generate via bars compatible with the SAV process could provide an alternative solution to address the overlay and resolution issue.

Furthermore, the advanced technology node requires the via pitch to be less than 50 nm with a critical dimension of ~25 nm. Via patterns with a tight pitch can be achieved by either 193i multi-patterning or EUV single exposure. The 193i case would use DSA to shrink the CD from GP CD to the target (say ~20–25 nm) in each pass of 193i multi-patterning where GP is printed at a relaxed pitch and CD. With EUV, tight pitch GP and small GP CD can be resolved in one exposure; however, a constraint on the largest achievable GP CD exists, which would be ~10–15 nm smaller than the pitch or else most GPs will merge. This GP CD constraint together with the sidewall preference further determines the L_0 of the BCP to use in the DSA process. An example of a different sidewall preference is shown in Figure 13.40 (lower), where a neutral layer (thin layer between hardmask and BCPs) is assumed at the bottom of the GP in both cases to simplify the structural illustration. With all of these criteria in mind, to implement a GP design at 35 nm, the L_0 of the BCP needs to be ~17.5 nm if the sidewall is PMMA-philic, or ~35 nm if it is PS-philic. Because typical PS-PMMA cannot phase-separate at 17.5 nm, PS-philic sidewalls or high-χ BCP systems will be necessary. One should note that sidewall modification is typically achieved by a brush layer, so that the effective GP CD will be smaller than the post-etch GP CD by two times the brush thickness, which will consequently affect the optimum L_0 to use.

Two major differences were observed when using lamellar BCP and cylindrical BCP for via shrinking: (1) resulting structure in non-circular GPs and (2) 3-D morphology. As shown in Figure 13.41, the lamellar BCP system will form bar structures when directed to assemble in elongated GPs, while the cylindrical BCP system tends to form multiple vias, in which case the number and location of the resulting vias depend on the aspect ratio and the shape of the GP. In the case of circular GPs, even though the DSA structures derived from the two BCP systems look alike from a top-down view, both simulations and experimental results show that the cylindrical system tends to form "neckings" in the upper portion of the PMMA domain. This necking problem often occurs when the BCP film is slightly over-coated, i.e. the meniscus is higher than the top surface of the

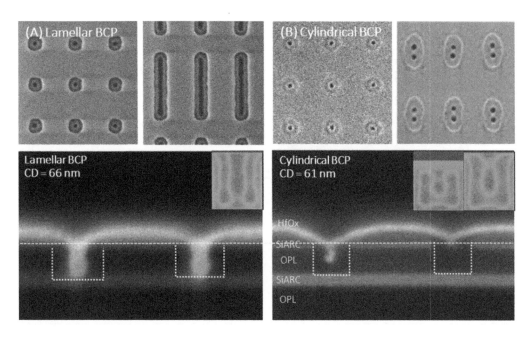

FIGURE 13.41 Comparison of graphoepitaxy DSA via shrink using lamellar BCP (group A on the left) and cylindrical BCP (group B on the right). In each group of images are: top-down SEM image of DSA in circular GP (upper left) and in higher aspect ratio GP (upper right), and cross-sectional SEM image of DSA in a circular GP (lower). In the cross-sectional SEM, PMMA has been removed and a metal oxide layer was deposited with ALD for contrast enhancement. (Reproduced from Ref. [161] with permission.)

GP, as shown in Figure 13.41B—lower image. As a result, the lower portion of the PMMA domain will be blocked by the PS necking and cannot be removed when a high selectivity process, such as the wet development process, is employed. On the other hand, the lamellar system has a tendency to form a PMMA domain reaching the bottom of the GP instead of forming a PS residual layer. Based on our simulations, the existence of the PS residual layer has a correlation with the GP CD. This tendency is preferable from a pattern transfer point of view, but one should not interpret it as a behavior controllable enough to eliminate the need for a neutral layer at the bottom of GPs.

As mentioned earlier, when GP CD is below 40 nm, due to the minimum L_0 achievable by the PS-b-PMMA system, a PS-philic sidewall instead of a PMMA-philic one is needed for DSA to succeed. Sidewall preference not only changes the optimum GP CD corresponding to a given BCP, but the DSA rectification effect also changes correspondingly. Figure 13.42 illustrates the DSA behavior in circular and elongated PS-philic GPs. Here, we define the rectification factor as the slope on a PMMA CD versus a GP CD plot. The rectification factor is an important indicator of DSA hole shrink because one of the main objectives of using DSA is to rectify the CD variation. A smaller rectification factor implies that the resulting CD derived from DSA is less sensitive to the variation in GP, and thus is preferred. The rectification factor is known to be affected by the GP aspect ratio (demonstrated in Figure 13.42), sidewall preference, and BCP composition. For example, it has been shown that the CD rectification effect of circular GP with a PS-philic sidewall is less effective than the PMMA-philic sidewall [162].

In addition to 193i lithography, DSA can also be used with EUV lithography. Chi et al. employed an electrical testable "via-chain" structure to illustrate the integration of the DSA and EUV [163]. The combination of a PS-philic sidewall and a GP with a tight pitch, e.g. <60 nm pitch circular "via" pattern, was printed by EUV with a post-etch GP CD of ~30 nm. A 25 nm L_0 BCP was used for DSA and further pattern transfer was demonstrated into the underlying hard mask layer and optical planarization layer (OPL). The local CD uniformity (LCDU) of the GP and the bottom HM/OPL

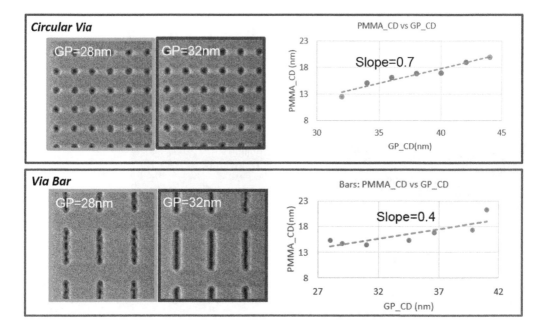

FIGURE 13.42 CD rectification effect of DSA with PS-philic sidewall. (Reproduced from Ref. [161] with permission.)

are 3.7 nm and 3.0 nm, respectively. The improvement in LCDU is ~30%, which is consistent with what was previously shown in the circular case in Figure 13.42. The DSA experiment was found to have a comparable electrical yield with the baseline (non-DSA) process in this study, which is a good validation of the use of DSA. However, not being able to fully utilize the improved LCDU by DSA may also imply other yield-limiting factors hiding in levels other than via formation, which will require further investigation of the baseline process.

The DSA hole shrink process can be used not only for forming via patterns, but also for cutting line patterns, such as gate or metal patterns [164]. Using the metal "cut/block" as an example, the minimum tip-to-tip CD determines the minimum design pitch for the previous metal layer, which is a critical parameter for back-end-of-the-line (BEOL) device scaling. The limit for small tip-to-tip derives from the lithography resolution. After the direct print resolution limit is reached, a "line plus cut" kind of process is the prevailing solution to address the need for smaller tip-to-tip. However, the smallest tip-to-tip is still determined by the minimum CD/CDU and overlay of the cut shape. The process details and material selections depend on the specific needs, but this type of application shares the same core concept as DSA via shrink, i.e. to utilize the DSA rectification effect while shrinking the CD, because frequently near- or sub-resolution CD and good LCDU cannot be achieved concurrently using lithography alone.

13.5.4 DRAM

As demonstrated in previous sections, DSA is very suitable for creating regular arrays, including a hexagonal spot array and line/space arrays. Interestingly, in the advanced 14 nm DRAM process, both types of DSA structures are considered as potential patterning process candidates [165]. The first potential insertion point for DSA is the active area, which defines the FET device that controls the storage cell. The active area is a line/space array typically made by SAQP. It is straightforward that a chemoepitaxy DSA process for line/space patterns, such as the LiNe process, could be used in place of SAQP for the active area. In contrast, the storage cell array, i.e. the capacitors, is typically an angled rectangular spot array instead of a hexagonal spot array; therefore, a slight rearrangement of the cell array layout is needed in order to implement the DSA process. The active area is at the bottom and the storage cell is at the top [165]. The major benefits to introducing DSA into the DRAM process are (1) process variations reduction, such as the pitch-walking derived from SAQP; and (2) process simplification. For example, the formation of a storage array may require two SADP, or even SAQP, processes, which would be much less expensive if they could be replaced by one DSA process. The line/space DSA process has been discussed in previous sections. Here, we will focus on the 193i compatible chemoepitaxy DSA process for hexagonal array formation.

Well-aligned hexagonal arrays formed by chemoepitaxy DSA of cylindrical BCPs was demonstrated by Ruiz et al. with a PS-brush system and e-beam lithography [76]. A similar process with EUV lithography was later demonstrated by Singh et al., including density multiplication [166]. In both demonstrations, the holes were printed on the resist and then transferred to the underlying brush with oxygen plasma. This process in theory is compatible with 193i. However, the main challenge is that the hole size after oxygen plasma usually needs to be roughly the same as the BCP domain size, which is around $1/2\ L_0$ of the BCP. For example, if one is targeting a 40 nm pitch at the final DSA stage, the guiding pattern CD on the brush would be roughly ~20 nm, which is too small for 193i to resolve, especially given that the oxygen plasma typically enlarges the resist CD. Therefore, it would be preferable to print pillar patterns instead of holes, followed by a trim etch to shrink the CD, i.e. a similar concept as in the LiNe process starting with cross-linkable PMMA. This chemoepitaxy-induced by pillar structures (CHIPS) flow was proposed and studied by Singh et al., as illustrated in Figure 13.43. Different multiplication factors and guiding pattern arrangements were also demonstrated.

Instead of a uniform CD distribution, the histogram of the resulting DSA CD from a CHIPS flow shows an interesting bimodal distribution, as shown in Singh's work. Given that this experiment employed a guiding pattern with a pitch of 90 nm and a BCP with $L_0 = 45$ nm pitch, i.e. a 4× in feature density, one can understand that the minority group in the histogram is derived from the guiding pattern, while the majority group is derived from self-assembled BCPs. One may notice that as the guiding pillar CD decreases, the mean CD of the minority group increases while the CD of the majority group remains unchanged. As illustrated in the schematic in Figure 13.43 (upper), the CD at the bottom of the PMMA domain, i.e. on top of the guiding pattern, is dictated by the pillar CD. However, the total volume of the PMMA domain is determined by the material and needs to be conserved. As a result, if the bottom CD is larger, a tapered PMMA domain is expected, hence the top CD, which is observed under CD-SEM, will be smaller. To be more precise, a Monte Carlo simulation of such cases reported by Detcheverry et al. shows that only the lower portion of the PMMA domain will be tapered and the rest of the domain will be relatively uniform [76]. This is based on the assumption that the non-pillar region of the guiding pattern has no strong preference to either block and the influence of the guiding pattern is only effective within a certain distance. Under such an assumption, only a fixed amount of volume needs to be compensated for by reducing the CD near the upper portion of the PMMA domain. Therefore, if a thicker BCP film was used, the CD loss observed from the top would be smaller compared to a thinner BCP film.

13.6 CHALLENGES AND OUTLOOK

Before we begin a discussion of the values and challenges of DSA, one should keep in mind that the foundation of DSA, i.e. the self-assembly of block copolymers, is also a very attractive patterning technology that is available for researchers from university labs to large-scale manufacturers. Several interesting applications were illustrated in previous sections utilizing the nanometer-scale structures with high uniformity and short-range order derived from the self-assembly of BCPs. When considering new applications for lithography using DSA, one could sometimes overlook the possibility of simply using self-assembly to achieve the goal.

In the past decade, tremendous progress has been made in the field of DSA, including understanding the underlying polymer physics, numerical modeling methods, and experimental

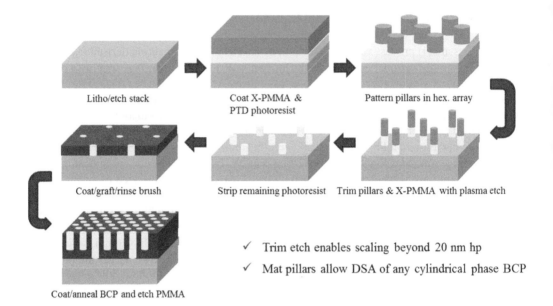

✓ Trim etch enables scaling beyond 20 nm hp

✓ Mat pillars allow DSA of any cylindrical phase BCP

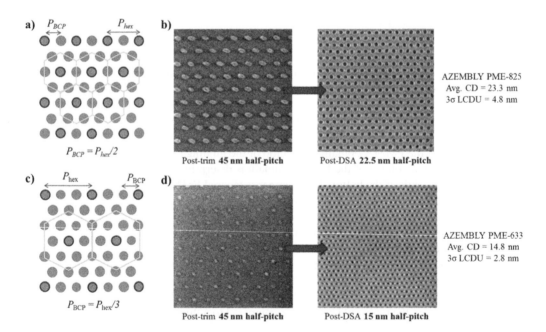

FIGURE 13.43 (Upper) The CHIPS process flow and (lower) the guiding pattern design, CD, and BCP L_0 effect on the resulting structure. (Reproduced from [167] with permission.)

verification. Many innovative processes have been developed to utilize the attractive benefits, i.e. density multiplication and rectification, of this material-based patterning technique. Highly promising demonstrations have been presented in multiple potential areas. However, the barrier to inserting a new process element into semiconductor manufacturing is exceedingly high. Here, we borrow a simple concept from the field of business strategy to illustrate the considerations and challenges for DSA insertion, i.e. the value of the system can be determined from the benefits it offers based on the costs that needed to be expended to bring the system into production.

First, the most ideal DSA application is a special case of this formula where the benefits dominate and outweigh the costs. This applies to the cases that are almost impossible to achieve without DSA so that the benefits term is almost infinite as there is no baseline process with which to compare. However, this kind of application is rarely found because lithographers and patterning researchers are very innovative and can almost always find a solution to a given problem. One DSA application that is close to this category is the creation of a very small and uniform metal line cut/block. As explained in Section 13.5.3, the tip-to-tip distance of metal lines is critical for BEOL scaling, and a line and cut process scheme would be more advantageous compared to direct print in terms of resolution. Cuts/blocks smaller than the lithography resolution limit can be made by lithography combined with trim etch, but maintaining CD uniformity especially over a range of designs is quite challenging. DSA shrink may provide an attractive solution; however, there are design restrictions and material/defectivity-related issues yet to be verified. One should also note that the tip-to-tip of the current BEOL design rule is still within the manageable range of existing lithography techniques; therefore, a new solution is not needed immediately in current technology node. (But it would be good timing to start considering and working on it.)

Second, we can use sub-30 nm pitch fin formation with SAQP and DSA as an example, noting that a similar analysis can be done for other applications. Fin pitch-walking has a significant impact on yield and device performance. The conventional solution is to implement an advanced process control system with feedback/feedforward within and between lithography and etch tools. SAQP has the following factors that increase its cost: the complexity of the process, purchase, implementation, and maintenance of the APC system. In contrast, DSA provides a simpler process scheme and eliminates the pitch-walking issue; however, development cost, defectivity concerns, and intrinsic pattern quality such as LER and its impact on yield could be less favorable than SAQP. (Even though yield-limiting LER and defectivity issues require further research and development, promising results have recently been published [168, 169].) Additionally, should a uniform gridded circuit design be implemented to accommodate certain DSA processes, a penalty in device density would result in a larger cell/chip area which means higher cost per chip. The cost to redesign a common component library and to enforce these strict design rules should also be considered. One should note that the density and scaling impact from a gridded design depends on applications. For example, memory devices could have a layout that is already very close to a gridded design. Also, a non-gridded DSA could be possible using graphoepitaxy-based processes [69].

Even though the formula is simple, it is quite difficult to estimate quantitatively the value term, unless serious development work is performed and the real gain in wafer yield can be approximated. Moreover, the net value has to be greater than a certain "activation/adoption barrier" for industry to be willing to make the change, and DSA would then be closer to viable implementation.

One should remember that it took roughly a decade to realize the "simple" idea of including a film of water between the resist and the lenses to make an immersion lithography tool. It took roughly three decades for EUV technology to start to see the light at the end of the tunnel. It is well known that time and resources and patience are required for any new technologies to mature, and DSA is no exception. DSA researchers will continue to improve existing processes and explore novel applications that could provide a higher net value. Although it may still take years, with the profound understanding of the nature of DSA that has accumulated over the past decade, the DSA research community is now ready to facilitate the technology.

REFERENCES

1. M. Somervell et al., "Comparison of directed self-assembly integrations," *Proceedings of SPIE*, vol. 8325, no. 1, p. 83250G, 2012.
2. P. Rincon Delgadillo et al., "Defect source analysis of directed self-assembly process (DSA of DSA)," *Proceedings of SPIE*, vol. 8680, pp. 86800L-9, 2013.

3. P. A. R. Delgadillo et al., "Implementation of a chemo-epitaxy flow for directed self-assembly on 300-mm wafer processing equipment," *Journal of Micro-Nanolithography MEMS and MOEMS*, vol. 11, no. 3, pp. 031302, July–Sept 2012.
4. C. Bencher et al., "Self-assembly patterning for sub-15nm half-pitch: A transition from lab to fab," in *Proceedings of SPIE – The International Society for Optical Engineering*, Conf., vol. 7970, The Society of Photo-Optical Instrumentation Engineers (SPIE), 2011.
5. C.-C. Liu et al., "Towards electrical testable SOI devices using directed self-assembly for fin formation," *Proceedings of SPIE*, vol. 9049, pp. 904909-12, 2014.
6. ITRS. (2015). International Technology Roadmap for Semiconductors. Available: http://www.itrs.net/
7. C. J. Hawker and T. P. Russell, "Block copolymer lithography: Merging "bottom-up" with "top-down" processes," *MRS Bulletin*, vol. 30, no. 12, pp. 952–966, Dec 2005.
8. M. Park, C. Harrison, P. M. Chaikin, R. A. Register, and D. H. Adamson, "Block copolymer lithography: Periodic arrays of similar to 10(11) holes in 1 square centimeter," *Science*, vol. 276, no. 5317, pp. 1401–1404, May 1997.
9. C. T. Black et al., "Polymer self assembly in semiconductor microelectronics," *IBM Journal of Research and Development*, vol. 51, no. 5, pp. 605–633, Sept 2007.
10. R. A. Segalman, "Patterning with block copolymer thin films," *Materials Science & Engineering R-Reports*, vol. 48, no. 6, pp. 191–226, Feb 2005.
11. H. C. Kim, S. M. Park, and W. D. Hinsberg, "Block copolymer based nanostructures: Materials, processes, and applications to electronics," *Chemical Reviews*, vol. 110, no. 1, pp. 146–177, Jan 2010.
12. M. W. Matsen and F. S. Bates, "Unifying weak- and strong-segregation block copolymer theories," *Macromolecules*, vol. 29, no. 4, pp. 1091–1098, Feb 1996.
13. S. O. Kim, H. H. Solak, M. P. Stoykovich, N. J. Ferrier, J. J. de Pablo, and P. F. Nealey, "Epitaxial self-assembly of block copolymers on lithographically defined nanopatterned substrates," *Nature*, vol. 424, no. 6947, pp. 411–414, July 2003.
14. M. Szwarc, M. Levy, and R. Milkovich, "Polymerization initiated by electron transfer to monomer. A new method of formation of block polymers," *Journal of the American Chemical Society*, vol. 78, no. 11, pp. 2656–2657, 1956.
15. F. S. Bates and G. H. Fredrickson, "Block copolymer thermodynamics—Theory and experiment," *Annual Review of Physical Chemistry*, vol. 41, pp. 525–557, 1990.
16. F. A. Detcheverry, G. L. Liu, P. F. Nealey, and J. J. de Pablo, "Interpolation in the directed assembly of block copolymers on nanopatterned substrates: Simulation and experiments," *Macromolecules*, vol. 43, no. 7, pp. 3446–3454, Apr 2010.
17. F. S. Bates and G. H. Fredrickson, "Block copolymers—Designer soft materials," *Physics Today*, vol. 52, no. 2, pp. 32–38, 1999.
18. D. Walton, G. Kellogg, A. Mayes, P. Lambooy, and T. Russell, "A free energy model for confined diblock copolymers," *Macromolecules*, vol. 27, no. 21, pp. 6225–6228, 1994.
19. R. D. Peters, X. M. Yang, T. K. Kim, B. H. Sohn, and P. F. Nealey, "Using self-assembled monolayers exposed to X-rays to control the wetting behavior of thin films of diblock copolymers," *Langmuir*, vol. 16, no. 10, pp. 4625–4631, May 2000.
20. M. Park, C. Harrison, P. M. Chaikin, R. A. Register, and D. H. Adamson, "Block copolymer lithography: Periodic arrays of~ 1011 holes in 1 square centimeter," *Science*, vol. 276, no. 5317, pp. 1401–1404, 1997.
21. J. Y. Cheng, C. Ross, V. H. Chan, E. L. Thomas, R. G. Lammertink, and G. J. Vancso, "Formation of a cobalt magnetic dot array via block copolymer lithography," *Advanced Materials*, vol. 13, no. 15, pp. 1174–1178, 2001.
22. J. Y. Cheng, C. Ross, E. Thomas, H. I. Smith, and G. Vancso, "Fabrication of nanostructures with long-range order using block copolymer lithography," *Applied Physics Letters*, vol. 81, no. 19, pp. 3657–3659, 2002.
23. C. D. Bain, H. A. Biebuyck, and G. M. Whitesides, "Comparison of self-assembled monolayers on gold: Coadsorption of thiols and disulfides," *Langmuir*, vol. 5, no. 3, pp. 723–727, 1989.
24. A. Kumar, H. A. Biebuyck, and G. M. Whitesides, "Patterning self-assembled monolayers – Applications in materials science," *Langmuir*, vol. 10, no. 5, pp. 1498–1511, May 1994.
25. P. Mansky, Y. Liu, E. Huang, T. P. Russell, and C. J. Hawker, "Controlling polymer–surface interactions with random copolymer brushes," *Science*, vol. 275, no. 5305, pp. 1458–1460, Mar 1997.
26. E. Han, K. O. Stuen, Y.-H. La, P. F. Nealey, and P. Gopalan, "Effect of composition of substrate-modifying random copolymers on the orientation of symmetric and asymmetric diblock copolymer domains," *Macromolecules*, vol. 41, no. 23, pp. 9090–9097, 2008.

27. D. Y. Ryu, K. Shin, E. Drockenmuller, C. J. Hawker, and T. P. Russell, "A generalized approach to the modification of solid surfaces," *Science*, vol. 308, no. 5719, pp. 236–239, 2005.

28. E. W. Edwards, M. F. Montague, H. H. Solak, C. J. Hawker, and P. F. Nealey, "Precise control over molecular dimensions of block-copolymer domains using the interfacial energy of chemically nanopatterned substrates," *Advanced Materials*, vol. 16, no. 15, pp. 1315–1319, Aug 2004.

29. S. Wu, *Polymer interface and adhesion*. New York: M. Dekker, 1982.

30. E. Han, K. O. Stuen, M. Leolukman, C. C. Liu, P. F. Nealey, and P. Gopalan, "Perpendicular orientation of domains in cylinder-forming block copolymer thick films by controlled interfacial interactions," *Macromolecules*, vol. 42, no. 13, pp. 4896–4901, Jul 2009.

31. C.-C. Liu et al., "Chemical patterns for directed self-assembly of lamellae-forming block copolymers with density multiplication of features," *Macromolecules*, vol. 46, no. 4, pp. 1415–1424, 2013.

32. J. Y. Cheng et al., "Simple and versatile methods to integrate directed self-assembly with optical lithography using a polarity-switched photoresist," *ACS Nano*, vol. 4, no. 8, pp. 4815–4823, Aug 2010.

33. T. P. Russell, R. P. Hjelm, and P. A. Seeger, "Temperature dependence of the interaction parameter of polystyrene and poly(methyl methacrylate)," *Macromolecules*, vol. 23, no. 3, pp. 890–893, 1990.

34. L. Wan et al., "The limits of lamellae-forming PS-*b*-PMMA block copolymers for lithography," *ACS Nano*, vol. 9, no. 7, pp. 7506–7514, 2015.

35. I. Bita, J. Yang, Y. Jung, C. Ross, E. Thomas, and K. Berggren, "Graphoepitaxy of self-assembled block copolymers on two-dimensional periodic patterned templates," (in English), *Science*, vol. 321, no. 5891, pp. 939–943, 2008.

36. W. J. Durand et al., "Design of high-χ block copolymers for lithography," *Journal of Polymer Science Part A: Polymer Chemistry*, vol. 53, no. 2, pp. 344–352, 2015.

37. D. Quach et al., "Impact of materials selection on graphoepitaxial directed self-assembly for line-space patterning," *Proceedings of SPIE*, vol. 9423, pp. 94230N-9, 2015.

38. S. Park, B. Kim, O. Yavuzcetin, M. T. Tuominen, and T. P. Russell, "Ordering of PS-b-P4VP on patterned silicon surfaces," *ACS Nano*, vol. 2, no. 7, pp. 1363–1370, 2008.

39. E. Hirahara et al., "Directed self-assembly of topcoat-free, integration-friendly high-x block copolymers," *Proceedings of SPIE*, vol. 9425, pp. 94250P-11, 2015.

40. C. Sinturel, F. S. Bates, and M. A. Hillmyer, "High χ-low N block polymers: How far can we go?," *ACS Macro Letters*, vol. 4, no. 9, pp. 1044–1050, 2015.

41. C. M. Bates et al., "Polarity-switching top coats enable orientation of sub-10-nm block copolymer domains," *Science*, vol. 338, no. 6108, pp. 775–779, Nov 2012.

42. K. W. Gotrik and C. A. Ross, "Solvothermal annealing of block copolymer thin films," *Nano Letters*, vol. 13, no. 11, pp. 5117–5122, Nov 2013.

43. X. Yang et al., "Fabrication of servo-integrated template for 1.5-teradot/inch2 bit patterned media with block copolymer directed assembly," *Journal of Micro/Nanolithography, MEMS, and MOEMS*, vol. 13, no. 3, pp. 031307, 2014.

44. J. D. Cushen et al., "Thin film self-assembly of poly(trimethylsilylstyrene-b-D,L-lactide) with sub-10 nm domains," (in English), *k*, vol. 45, no. 21, pp. 8722–8728, Nov 2012.

45. S. Park et al., "Macroscopic 10-terabit-per-square- inch arrays from block copolymers with lateral order," (in English), *Science*, vol. 323, no. 5917, pp. 1030–1033, Feb 2009.

46. E. W. Cochran, D. C. Morse, and F. S. Bates, "Design of ABC triblock copolymers near the ODT with the random phase approximation," *Macromolecules*, vol. 36, no. 3, pp. 782–792, 2003.

47. S. H. Anastasiadis, T. P. Russell, S. K. Satija, and C. F. Majkrzak, "Neutron reflectivity studies of the surface-induced ordering of diblock copolymer films," (in English), *Physical Review Letters*, vol. 62, no. 16, pp. 1852–1855, Apr 1989.

48. J. G. Kennemur, M. A. Hillmyer, and F. S. Bates, "Synthesis, thermodynamics, and dynamics of poly(4-tert-butylstyrene-b-methyl methacrylate)," (in English), *Macromolecules*, vol. 45, no. 17, pp. 7228–7236, Sept 2012.

49. A. Chaudhari et al., "Formation of sub-7 nm feature size PS-b-P4VP block copolymer structures by solvent vapour process," *Proceedings of SPIE*, vol. 9051, pp. 905110-10, 2014.

50. J. Y. Tang et al., "Holey silicon as an efficient thermoelectric material," (in English), *Nano Letters*, vol. 10, no. 10, pp. 4279–4283, Oct 2010.

51. J. H. Park, J. Kirch, L. J. Mawst, C. C. Liu, P. F. Nealey, and T. F. Kuech, "Controlled growth of InGaAs/InGaAsP quantum dots on InP substrates employing diblock copolymer lithography," *Applied Physics Letters*, vol. 95, no. 11, Sep 2009.

52. T. F. Kuech and L. J. Mawst, "Nanofabrication of III-V semiconductors employing diblock copolymer lithography," (in English), *Journal of Physics D: Applied Physics*, vol. 43, no. 18, p. 18, May 2010.

53. J. W. Bai, X. Zhong, S. Jiang, Y. Huang, and X. F. Duan, "Graphene nanomesh," *Nature Nanotechnology*, vol. 5, no. 3, pp. 190–194, Mar 2010.

54. B. H. Kim et al., "Surface energy modification by spin-cast, large-area graphene film for block copolymer lithography," *ACS Nano*, vol. 4, no. 9, pp. 5464–5470, Sept 2010.

55. M. Kim, N. S. Safron, E. Han, M. S. Arnold, and P. Gopalan, "Fabrication and characterization of large-area, semiconducting nanoperforated graphene materials," *Nano Letters*, vol. 10, no. 4, pp. 1125–1131, Apr 2010.

56. T. L. Morkved et al., "Local control of microdomain orientation in diblock copolymer thin films with electric fields," *Science*, vol. 273, no. 5277, pp. 931–933, Aug 1996.

57. T. Hashimoto, J. Bodycomb, Y. Funaki, and K. Kimishima, "The effect of temperature gradient on the microdomain orientation of diblock copolymers undergoing an order–disorder transition," *Macromolecules*, vol. 32, no. 3, pp. 952–954, 1999.

58. R. M. Kannan and J. A. Kornfield, "Evolution of microstructure and viscoelasticity during flow alignment of a lamellar diblock copolymer," *Macromolecules*, vol. 27, no. 5, pp. 1177–1186, 1994.

59. R. A. Segalman, H. Yokoyama, and E. J. Kramer, "Graphoepitaxy of spherical domain block copolymer films," *Advanced Materials*, vol. 13, no. 15, pp. 1152–1155, 2001.

60. R. A. Segalman, A. Hexemer, R. C. Hayward, and E. J. Kramer, "Ordering and melting of block copolymer spherical domains in 2 and 3 dimensions," *Macromolecules*, vol. 36, no. 9, pp. 3272–3288, May 2003.

61. C. C. Liu, G. S. W. Craig, H. M. Kang, R. Ruiz, P. F. Nealey, and N. J. Ferrier, "Practical implementation of order parameter calculation for directed assembly of block copolymer thin films," *Journal of Polymer Science Part B: Polymer Physics*, vol. 48, no. 24, pp. 2589–2603, Dec 2010.

62. H. I. Smith and D. C. Flanders, "Oriented crystal-growth on amorphous substrates using artificial surface-relief gratings," *Applied Physics Letters*, vol. 32, no. 6, pp. 349–350, 1978.

63. Y. S. Jung and C. A. Ross, "Well-ordered thin-film nanopore arrays formed using a block-copolymer template," *Small*, vol. 5, no. 14, pp. 1654–1659, July 2009.

64. C. T. Black and O. Bezencenet, "Nanometer-scale pattern registration and alignment by directed diblock copolymer self-assembly," *IEEE Transactions on Nanotechnology*, vol. 3, no. 3, pp. 412–415, 2004.

65. S. G. Xiao, X. M. Yang, E. W. Edwards, Y. H. La, and P. F. Nealey, "Graphoepitaxy of cylinder-forming block copolymers for use as templates to pattern magnetic metal dot arrays," *Nanotechnology*, vol. 16, no. 7, pp. S324–S329, July 2005.

66. S. M. Park, M. P. Stoykovich, R. Ruiz, Y. Zhang, C. T. Black, and P. E. Nealey, "Directed assembly of lamellae-forming block copolymers by using chemically and topographically patterned substrates," *Advanced Materials*, vol. 19, no. 4, pp. 607–611, Feb 2007.

67. E. Han, H. Kang, C.-C. Liu, P. F. Nealey, and P. Gopalan, "Graphoepitaxial assembly of symmetric block copolymers on weakly preferential substrates," *Advanced Materials*, vol. 22, no. 38, pp. 4325–4329, 2010.

68. S. J. Jeong et al., "Soft graphoepitaxy of block copolymer assembly with disposable photoresist confinement," (in English), *Nano Letters*, vol. 9, no. 6, pp. 2300–2305, June 2009.

69. H. Tsai et al., "Two-dimensional pattern formation using graphoepitaxy of PS-*b*-PMMA block copolymers for advanced FinFET device and circuit fabrication," *ACS Nano*, vol. 8, no. 5, pp. 5227–5232, 2014.

70. L. Rockford, Y. Liu, P. Mansky, T. P. Russell, M. Yoon, and S. G. J. Mochrie, "Polymers on nanoperiodic, heterogeneous surfaces," *Physical Review Letters*, vol. 82, no. 12, pp. 2602–2605, Mar 1999.

71. C. T. Black, "Self-aligned self assembly of multi-nanowire silicon field effect transistors," *Applied Physics Letters*, vol. 87, no. 16, Oct 2005.

72. M. P. Stoykovich et al., "Directed self-assembly of block copolymers for nanolithography: Fabrication of isolated features and essential integrated circuit geometries," *ACS Nano*, vol. 1, no. 3, pp. 168–175, Oct 2007.

73. M. P. Stoykovich et al., "Directed assembly of block copolymer blends into nonregular device-oriented structures," *Science*, vol. 308, no. 5727, pp. 1442–1446, June 2005.

74. A. M. Welander et al., "Rapid directed assembly of block copolymer films at elevated temperatures," *Macromolecules*, vol. 41, no. 8, pp. 2759–2761, Apr 2008.

75. R. Ruiz, J. K. Bosworth, and C. T. Black, "Effect of structural anisotropy on the coarsening kinetics of diblock copolymer striped patterns," *Physical Review B*, vol. 77, no. 5, Feb 2008.

76. R. Ruiz et al., "Density multiplication and improved lithography by directed block copolymer assembly," *Science*, vol. 321, no. 5891, pp. 936–939, Aug 2008.

77. J. Y. Cheng, C. T. Rettner, D. P. Sanders, H. C. Kim, and W. D. Hinsberg, "Dense self-assembly on sparse chemical patterns: Rectifying and multiplying lithographic patterns using block copolymers," *Advanced Materials*, vol. 20, no. 16, pp. 3155–3158, Aug 2008.

78. S. H. Park et al., "Block copolymer multiple patterning integrated with conventional ArF lithography," *Soft Matter*, vol. 6, no. 1, pp. 120–125, 2010.

79. C. C. Liu et al., "Integration of block copolymer directed assembly with 193 immersion lithography," *Journal of Vacuum Science & Technology B*, vol. 28, no. 6, pp. C6B30–C6B34, Nov 2010.

80. C. C. Liu et al., "Fabrication of lithographically defined chemically patterned polymer brushes and mats," *Macromolecules*, vol. 44, no. 7, pp. 1876–1885, Apr 2011.

81. J. Kim et al., "The SMART TM process for directed block co-polymer self-assembly," *Journal of Photopolymer Science and Technology*, vol. 26, no. 5, pp. 573–579, 2013.

82. M. S. Onses, P. Pathak, C. C. Liu, F. Cerrina, and P. F. Nealey, "Localization of multiple DNA sequences on nanopatterns," *ACS Nano*, vol. 5, no. 10, pp. 7899–7909, Oct 2011.

83. M. P. Allen and D. J. Tildesley, *Computer simulation of liquids*. Oxford University Press, 1989.

84. D. Frenkel and B. Smit, *Understanding molecular simulation: From algorithms to applications*. Academic Press, 2002.

85. E. Helfand, "Theory of inhomogeneous polymers: Fundamentals of the Gaussian random-walk model," *The Journal of Chemical Physics*, vol. 62, no. 3, pp. 999–1005, 1975.

86. M. W. Matsen, "The standard Gaussian model for block copolymer melts," *Journal of Physics: Condensed Matter*, vol. 14, no. 2, p. R21, 2001.

87. G. Fredrickson, *The equilibrium theory of inhomogeneous polymers*. Clarendon Press, 2006.

88. G. Gompper and M. Schick, *Soft matter: Volume 1 – Polymer melts and mixtures*. Wiley-VCH, 2006.

89. M. Müller and F. Schmid, "Advanced computer simulation approaches for soft matter science II," *Advances in Polymer Science*, 2005.

90. F. A. Detcheverry, D. Q. Pike, U. Nagpal, P. F. Nealey, and J. J. de Pablo, "Theoretically informed coarse grain simulations of block copolymer melts: Method and applications," *Soft Matter*, vol. 5, no. 24, pp. 4858–4865, 2009.

91. F. A. Detcheverry, D. Q. Pike, P. F. Nealey, M. Muller, and J. J. de Pablo, "Monte Carlo simulation of coarse grain polymeric systems," *Physical Review Letters*, vol. 102, no. 19, May 2009.

92. D. Q. Pike, F. A. Detcheverry, M. Müller, and J. J. de Pablo, "Theoretically informed coarse grain simulations of polymeric systems," *The Journal of Chemical Physics*, vol. 131, no. 8, p. 084903, 2009.

93. F. A. Detcheverry, H. M. Kang, K. C. Daoulas, M. Muller, P. F. Nealey, and J. J. de Pablo, "Monte Carlo simulations of a coarse grain model for block copolymers and nanocomposites," *Macromolecules*, vol. 41, no. 13, pp. 4989–5001, July 2008.

94. S.-M. Hur, G. S. Khaira, A. Ramírez-Hernández, M. Müller, P. F. Nealey, and J. J. de Pablo, "Simulation of defect reduction in block copolymer thin films by solvent annealing," *ACS Macro Letters*, vol. 4, no. 1, pp. 11–15, 2014.

95. M. Müller and K. C. Daoulas, "Calculating the free energy of self-assembled structures by thermodynamic integration," *The Journal of Chemical Physics*, vol. 128, no. 2, p. 024903, 2008.

96. B. L. Peters, B. Rathsack, M. Somervell, T. Nakano, G. Schmid, and J. J. de Pablo, "Graphoepitaxial assembly of cylinder forming block copolymers in cylindrical holes," *Journal of Polymer Science Part B: Polymer Physics*, vol. 53, no. 6, pp. 430–441, 2015.

97. M. Doi and S. F. Edwards, *The theory of polymer dynamics*. Oxford University Press, 1988.

98. M. Rubinstein and R. H. Colby, *Polymer physics*. New York: Oxford University, 2003.

99. E. Helfand and Z. Wasserman, "Block copolymer theory. 4. Narrow interphase approximation," *Macromolecules*, vol. 9, no. 6, pp. 879–888, 1976.

100. P. J. Flory, "Thermodynamics of high polymer solutions," *The Journal of Chemical Physics*, vol. 10, no. 1, pp. 51–61, 1942.

101. K. C. Daoulas, M. Muller, J. J. de Pablo, P. F. Nealey, and G. D. Smith, "Morphology of multi-component polymer systems: Single chain in mean field simulation studies," *Soft Matter*, vol. 2, no. 7, pp. 573–583, 2006.

102. M. Müller and J. J. de Pablo, "Computational approaches for the dynamics of structure formation in self-assembling polymeric materials," *Annual Review of Materials Research*, vol. 43, pp. 1–34, 2013.

103. M. Müller, "Speeding-up particle simulations of multicomponent polymer systems by coupling to continuum descriptions," *Hybrid Particle-Continuum Methods in Computational Materials Physics*, vol. 46, p. 127, 2013.

104. M. Sferrazza, C. Xiao, R. A. L. Jones, D. G. Bucknall, J. Webster, and J. Penfold, "Evidence for capillary waves at immiscible polymer/polymer interfaces," *Physical Review Letters*, vol. 78, no. 19, p. 3693, 1997.

105. T. A. Callaghan and D. R. Paul, "Interaction energies for blends of poly(methyl methacrylate), polystyrene, and poly(.alpha.-methylstyrene) by the critical molecular weight method," *Macromolecules*, vol. 26, no. 10, pp. 2439–2450, 1993.

106. F. A. Detcheverry, D. Q. Pike, P. F. Nealey, M. Muller, and J. J. de Pablo, "Simulations of theoretically informed coarse grain models of polymeric systems," *Faraday Discussions*, vol. 144, pp. 111–125, 2010.

107. K. Yoshimoto et al., "Optimization of directed self-assembly hole shrink process with simplified model," *Journal of Micro/Nanolithography, MEMS, and MOEMS*, vol. 13, no. 3, pp. 031305, 2014.

108. K. Kodera et al., "Novel error mode analysis method for graphoepitaxial directed self-assembly lithography based on the dissipative particle dynamics method," in *SPIE Advanced Lithography*. International Society for Optics and Photonics, pp. 868015-7, 2013.

109. T. Nakano, M. Matsukuma, K. Matsuzaki, M. Muramatsu, T. Tomita, and T. Kitano, "Dissipative particle dynamics study on directed self-assembly in holes," in *SPIE Advanced Lithography*. International Society for Optics and Photonics, pp. 86801J-6, 2013.

110. N. Laachi et al., "The hole shrink problem: Theoretical studies of directed self-assembly in cylindrical confinement," in *SPIE Advanced Lithography*. International Society for Optics and Photonics, pp. 868014-9, 2013.

111. G. S. Khaira et al., "Evolutionary optimization of directed self-assembly of triblock copolymers on chemically patterned substrates," *ACS Macro Letters*, vol. 3, no. 8, pp. 747–752, 2014.

112. H. S. Suh et al., "Characterization of the shape and line-edge roughness of polymer gratings with grazing incidence small-angle X-ray scattering and atomic force microscopy," *Journal of Applied Crystallography*, vol. 49, no. 3, pp. 823–834, 2016.

113. D. F. Sunday et al., "Determination of the internal morphology of nanostructures patterned by directed self assembly," *ACS Nano*, vol. 8, no. 8, pp. 8426–8437, 2014.

114. U. Nagpal, M. Müller, P. F. Nealey, and J. J. De Pablo, "Free energy of defects in ordered assemblies of block copolymer domains," *ACS Macro Letters*, vol. 1, no. 3, pp. 418–422, 2012.

115. S.-M. Hur et al., "Molecular pathways for defect annihilation in directed self-assembly," *Proceedings of the National Academy of Sciences*, vol. 112, no. 46, pp. 14144–14149, 2015.

116. M. Müller and K. C. Daoulas, "Speeding up intrinsically slow collective processes in particle simulations by concurrent coupling to a continuum description," *Physical Review Letters*, vol. 107, no. 22, p. 227801, 2011.

117. T. Thurn-Albrecht et al., "Ultrahigh-density nanowire arrays grown in self-assembled diblock copolymer templates," *Science*, vol. 290, no. 5499, pp. 2126–2129, 2000.

118. K. W. Guarini, C. T. Black, Y. Zhang, I. V. Babich, E. M. Sikorski, and L. M. Gignac, "Low voltage, scalable nanocrystal FLASH memory fabricated by templated self assembly," *Technical Digest – International Electron Devices Meeting*, Conf., pp. 541–544, 2003.

119. A. J. Hong et al., "Metal nanodot memory by self-assembled block copolymer lift-off," *Nano Letters*, vol. 10, no. 1, pp. 224–229, Jan 2010.

120. O. Hellwig et al., "Bit patterned media based on block copolymer directed assembly with narrow magnetic switching field distribution," *Applied Physics Letters*, vol. 96, no. 5, Feb 2010.

121. Y. Seino et al., "Contact hole shrink process using graphoepitaxial directed self-assembly lithography," *Journal of Micro/Nanolithography, MEMS, and MOEMS*, vol. 12, no. 3, pp. 033011, 2013.

122. M. Muramatsu et al., "Nanopatterning of diblock copolymer directed self-assembly lithography with wet development," *Journal of Micro/Nanolithography, MEMS, and MOEMS*, vol. 11, no. 3, pp. 031305-1–031305-6, 2012.

123. R. Harukawa et al., "DSA hole defectivity analysis using advanced optical inspection tool," in *SPIE Advanced Lithography*. International Society for Optics and Photonics, pp. 86811A-7, 2013.

124. R. Gronheid et al., "Process optimization of templated DSA flows," *Proceeding of SPIE*, vol. 9051, pp. 90510I-7, 2014.

125. M. Somervell et al., "High-volume manufacturing equipment and processing for directed self-assembly applications," in *SPIE Advanced Lithography*. International Society for Optics and Photonics, pp. 90510N-11, 2014.

126. C. C. Liu, P. F. Nealey, Y. H. Ting, and A. E. Wendt, "Pattern transfer using poly(styrene-block-methyl methacrylate) copolymer films and reactive ion etching," *Journal of Vacuum Science & Technology B*, vol. 25, no. 6, pp. 1963–1968, Nov 2007.

127. H. Gokan, S. Esho, and Y. Ohnishi, "Dry etch resistance of organic materials," *Journal of the Electrochemical Society*, vol. 130, no. 1, pp. 143–146, 1983.

128. Y. H. Ting et al., "Plasma etch removal of poly(methyl methacrylate) in block copolymer lithography," *Journal of Vacuum Science & Technology B*, vol. 26, no. 5, pp. 1684–1689, Sept–Oct 2008.

129. M. Omura, T. Imamura, H. Yamamoto, I. Sakai, and H. Hayashi, "Highly selective etch gas chemistry design for precise DSAL dry development process," in *SPIE Advanced Lithography*. International Society for Optics and Photonics, pp. 905409-7, 2014.

130. P. Pimenta Barros et al., "Etch challenges for DSA implementation in CMOS via patterning," *Proceedings of SPIE*, vol. 9054, pp. 90540G-10, 2014.

131. F. Yamashita, E. Nishimura, K. Yatsuda, H. Mochiki, and J. Bannister, "Exploration of suitable dry etch technologies for directed self-assembly," *Proceedings of SPIE*, vol. 8328, pp. 83280T-9, 2012.

132. B. T. Chan et al., "28 nm pitch of line/space pattern transfer into silicon substrates with chemo-epitaxy directed self-assembly (DSA) process flow," *Microelectronic Engineering*, vol. 123, no. 0, pp. 180–186, 2014.

133. Q. Peng, Y.-C. Tseng, S. B. Darling, and J. W. Elam, "Nanoscopic patterned materials with tunable dimensions via atomic layer deposition on block copolymers," *Advanced Materials*, vol. 22, no. 45, pp. 5129–5133, 2010.

134. M. Ramanathan, Y.-C. Tseng, K. Ariga, and S. B. Darling, "Emerging trends in metal-containing block copolymers: Synthesis, self-assembly, and nanomanufacturing applications," *Journal of Materials Chemistry C*, vol. 1, no. 11, pp. 2080–2091, 2013.

135. M. Biswas, J. A. Libera, S. B. Darling, and J. W. Elam, "New insight into the mechanism of sequential infiltration synthesis from infrared spectroscopy," *Chemistry of Materials*, 2014.

136. R. Ruiz et al., "Image quality and pattern transfer in directed self assembly with block-selective atomic layer deposition," (in English), *Journal of Vacuum Science & Technology B*, vol. 30, no. 6, Nov 2012.

137. A. Singh et al., "Impact of sequential infiltration synthesis on pattern fidelity of DSA lines," *Proceedings of SPIE*, vol. 9425, pp. 94250N-7, 2015.

138. S. Xiong et al., "Directed self-assembly of triblock copolymer on chemical patterns for sub-10-nm nano-fabrication via solvent annealing," *ACS Nano*, vol. 10, no. 8, pp. 7855–7865, 2016.

139. D. Weller and A. Moser, "Thermal effect limits in ultrahigh-density magnetic recording," *IEEE Transactions on Magnetics*, vol. 35, no. 6, pp. 4423–4439, Nov 1999.

140. M. H. Kryder et al., "Heat assisted magnetic recording," *Proceedings of the IEEE*, vol. 96, no. 11, pp. 1810–1835, Nov 2008.

141. J.-G. Zhu, Z. Xiaochun, and T. Yuhui, "Microwave assisted magnetic recording," *IEEE Transactions on Magnetics*, vol. 44, no. 1, pp. 125–131, 2008.

142. S. Y. Chou, M. S. Wei, P. R. Krauss, and P. B. Fischer, "Single-domain magnetic pillar array of 35 nm diameter and 65 Gbits/in.2 density for ultrahigh density quantum magnetic storage," *Journal of Applied Physics*, vol. 76, no. 10, pp. 6673–6675, 1994.

143. R. M. H. New, R. F. W. Pease, and R. L. White, "Submicron patterning of thin cobalt films for magnetic storage," *Journal of Vacuum Science & Technology B*, vol. 12, no. 6, pp. 3196–3201, 1994.

144. T. R. Albrecht et al., "Bit-patterned magnetic recording: Theory, media fabrication, and recording performance," *IEEE Transactions on Magnetics*, vol. 51, no. 5, pp. 1–42, 2015.

145. S. Xiao et al., "Servo-integrated patterned media by hybrid directed self-assembly," *ACS Nano*, vol. 8, no. 11, pp. 11854–11859, 2014.

146. S. G. Xiao, X. M. Yang, S. J. Park, D. Weller, and T. P. Russell, "A novel approach to addressable 4 teradot/in.(2) patterned media," *Advanced Materials*, vol. 21, no. 24, pp. 2516–2519, June 2009.

147. X. Yang, L. Wan, S. Xiao, Y. Xu, and D. Weller, "Directed block copolymer assembly versus electron beam lithography for bit patterned media with areal density of 1 terabit/inch2 and beyond," *ACS Nano*, vol. 3, pp. 1844–1858, 2009.

148. R. Ruiz, E. Dobisz, and T. R. Albrecht, "Rectangular patterns using block copolymer directed assembly for high bit aspect ratio patterned media," (in English), *ACS Nano*, vol. 5, no. 1, pp. 79–84, Jan 2011.

149. C.-C. Liu et al., "Fin formation using graphoepitaxy DSA for FinFET device fabrication," *Proceedings of SPIE*, vol. 9423, pp. 94230S-10, 2015.

150. S. Sayan et al., "Directed self-assembly process integration: Fin patterning approaches and challenges," in *SPIE Advanced Lithography*. International Society for Optics and Photonics, pp. 94250R-12, 2015.

151. C.-C. C. Liu et al., "DSA patterning options for FinFET formation at 7nm node," in *SPIE Advanced Lithography*. International Society for Optics and Photonics, pp. 97770R-15, 2016.

152. G. S. Doerk et al., "Enabling complex nanoscale pattern customization using directed self-assembly," *Nature Communications*, vol. 5, 1–8, 2014.

153. T. Kato et al., "Advanced CD-SEM metrology for pattern roughness and local placement of lamellar DSA," in *Conference on Metrology, Inspection, and Process Control for Microlithography XXVIII*, San Jose, CA, vol. 9050, 2014.

154. R. Gronheid et al., "Implementation of templated DSA for via layer patterning at the 7nm node," *Proceedings of SPIE*, vol. 9423, pp. 942305-10, 2015.

155. K. Fukawatase et al., "DFM for defect-free DSA hole shrink process," *Proceedings of SPIE*, vol. 9049, pp. 90491K-8, 2014.

156. C.-C. Liu et al., "Progress towards the integration of optical proximity correction and directed self-assembly of block copolymers with graphoepitaxy," in *SPIE Advanced Lithography*. International Society for Optics and Photonics, pp. 83230X-7, 2012.

157. S. Narasimha et al., "22nm high-performance SOI technology featuring dual-embedded stressors, epi-plate high-K deep-trench embedded DRAM and self-aligned via 15LM BEOL," in *2012 IEEE International Electron Devices Meeting (IEDM)*. IEEE, pp. 3.3.1–3.3.4, 2012.

158. Y. Ma et al., "Directed self assembly (DSA) compliant flow with immersion lithography: From material to design and patterning," *Proceedings of SPIE*, vol. 9777, pp. 97770N-11, 2016.

159. S. Yamaguchi et al., "New robust edge detection methodology for qualifying DSA characteristics by using CD SEM," in *Conference on Metrology, Inspection, and Process Control for Microlithography XXVIII*, San Jose, CA, vol. 9050, 2014.

160. J. Bekaert et al., "N7 logic via patterning using templated DSA: Implementation aspects," *Proceedings of SPIE*, vol. 9658, pp. 965804-11, 2015.

161. C. Chi et al., "DSA via hole shrink for advanced node applications," *Proceedings of SPIE*, vol. 9777, pp. 97770L-9, 2016.

162. K. Schmidt et al., "Strategies to enable directed self-assembly contact hole shrink for tight pitches," *Proceedings of SPIE*, vol. 9777, pp. 97771U-10, 2016.

163. C. Chi et al., "Electrical study of DSA shrink process and CD rectification effect at sub-60nm using EUV test vehicle," in *SPIE Advanced Lithography*, 2017.

164. C.-C. Liu et al., "DSA patterning options for logics and memory applications," *Proceedings of SPIE*, vol. 10146, pp. 1014603-13, 2017.

165. M. Kamon et al., "Virtual fabrication using directed self-assembly for process optimization in a 14nm DRAM," *Proceedings of SPIE*, vol. 9777, pp. 977710-15, 2016.

166. A. Singh, B. T. Chan, Y. Cao, G. Lin, and R. Gronheid, "Using chemo-epitaxial directed self-assembly for repair and frequency multiplication of EUVL contact-hole patterns," *Proceedings of SPIE*, vol. 9049, pp. 90492F-8, 2014.

167. A. Singh et al., "Patterning sub-25 nm half-pitch hexagonal arrays of contact holes with chemo-epitaxial DSA guided by ArFi pre-patterns," *Proceedings of SPIE*, vol. 9425, pp. 94250X-7, 2015.

168. H. Pathangi et al., "Improved cost-effectiveness of the block co-polymer anneal process for DSA," *Proceedings of SPIE*, vol. 9777, pp. 97771Z-5, 2016.

169. M. Muramatsu et al., "Pattern fidelity improvement of chemo-epitaxy DSA process for high-volume manufacturing," *Proceedings of SPIE*, vol. 9777, pp. 97770F-9, 2016.

Index

Printed and bound by CPI Group (UK) Ltd, Croydon, CR0 4YY

18/10/2024

01776204-0018